HUMAN ANATOMY & PHYSIOLOGY

HUMAN ANATOMY & PHYSIOLOGY

sixth edition

John W. Hole, Jr.

Wm. C. Brown Publishers

Dubuque, Iowa · Melbourne, Australia · Oxford, England

Book Team

Editor *Colin H. Wheatley*
Developmental Editor *Jane DeShaw*
Production Editor *Connie Balius-Haakinson*
Designer *Jeff Storm*
Photo Editor *Carol Judge*
Permissions Editor *Gail I. Wheatley*
Art Processor *Brenda A. Ernzen*
Visuals/Design Developmental Consultant *Donna Slade*

Wm. C. Brown Publishers
A Division of Wm. C. Brown Communications, Inc.

Vice President and General Manager *Beverly Kolz*
National Sales Manager *Vincent R. Di Blasi*
Assistant Vice President, Editor-in-Chief *Edward G. Jaffe*
Director of Marketing *John W. Calhoun*
Marketing Manager *Christopher T. Johnson*
Advertising Manager *Amy Schmitz*
Director of Production *Colleen A. Yonda*
Manager of Visuals and Design *Faye M. Schilling*

Design Manager *Jac Tilton*
Art Manager *Janice Roerig*
Publishing Services Manager *Karen J. Slaght*
Permissions/Records Manager *Connie Allendorf*

Wm. C. Brown Communications, Inc.

Chairman Emeritus *Wm. C. Brown*
Chairman and Chief Executive Officer *Mark C. Falb*
President and Chief Operating Officer *G. Franklin Lewis*
Corporate Vice President, Operations *Beverly Kolz*
Corporate Vice President, President of WCB Manufacturing *Roger Meyer*

Cover photo © Scott Chaney

Copyedited by *Sarah Aldridge*

The credits section for this book begins on page C-1 and is considered
an extension of the copyright page.

For Shirley,
Karen,
Michael,
Larry, Andrea,
Lilia, and
Natalie with
love.

Brief Contents

Contents

UNIT 1

Levels of Organization 2

1 Introduction to Human Anatomy and Physiology 4

2 Chemical Basis of Life 38

3 Cells 62

UNIT 2
Support and Movement 168

8 Joints of the Skeletal System 244

9 Muscular System 268

UNIT 3
Integration and Coordination 328

U N I T 4
Processing and Transporting 500

14 Digestive System 502

15 Nutrition and Metabolism 548

16 Respiratory System 580

17 Blood 618

18 Cardiovascular System 650

19 Lymphatic System and Immunity 714

20 Urinary System 746

21 Water, Electrolyte, and Acid-Base Balance 778

UNIT 5
Human Life Cycle 802

22 Reproductive Systems 804

23 Human Growth and Development 852

24 Human Genetics 882

Clinical, Practical, and Laboratory Applications

Preface

The sixth edition of **Human Anatomy and Physiology** is designed to provide accurate, current information about the structure and function of the human body in an interesting and readable manner. It is especially planned for students pursuing careers in allied health fields, who have minimal backgrounds in physical and biological sciences.

Organization

The text is organized in units, each containing several chapters. These chapters are arranged traditionally, beginning with a discussion of the physical basis of life and proceeding through levels of increasing complexity.

Unit 1 introduces the human body and its major parts. It is concerned with the structure and function of cells and tissues; it introduces membranes as organs and the integumentary system as an organ system.

Unit 2 deals with the skeletal and muscular systems that support and protect body parts, and make movements possible.

Unit 3 concerns the nervous and endocrine systems that integrate and coordinate body functions.

Unit 4 discusses the digestive, respiratory, circulatory, lymphatic, and urinary systems. These systems obtain nutrients and oxygen from the external environment; transport these substances internally; use them as energy sources, structural materials, and essential components in metabolic reactions; and excrete the resulting wastes.

Unit 5 describes the male and female reproductive systems, their functions in producing offspring, and the growth and development of the offspring. The final chapter concerns genetics and explains the determination of individual traits and their passage from parents to offspring.

Special Features of the Sixth Edition

The chapters of the sixth edition of **Human Anatomy and Physiology** have been carefully reviewed by several instructors who have used the textbook with their classes, and numerous changes have been made in the narrative in response to their very helpful suggestions. The illustrations were also examined by these reviewers, and as a result of their critical comments, many of the figures have been improved or replaced by new artwork, and a variety of new micrographs have been included.

In addition, several new summary charts have been prepared. These concern such topics as a comparison of DNA and RNA molecules, the characteristics of neuroglial cells, the classification of neurons, the components of the brain, the nerve tracts of the spinal cord, the plasma lipoproteins, the steps in antibody production, and the role of ADH in urine production.

Fifteen new boxed asides have been written. They deal with such subjects as DNA "fingerprinting," heat exhaustion, bone cancer, intraosseous infusion, total knee arthroplasty, H_2 blockers, inhibition of platelet aggregation, ACE inhibitors, anabolic steroids, and fragile chromosomes.

The clinical applications and clinical case studies have been checked and updated, and references to the cadaver plates that appear following chapter 24 have been added to the appropriate chapters.

Readability

Readability is an important asset of this text. The writing style is intentionally informal and easy to read. Technical vocabulary has been minimized, and

summary paragraphs and review questions occur frequently within the narrative. Numerous illustrations, summary charts, and flow diagrams are carefully positioned near the discussions they complement.

Pedagogical Devices

The text includes an unusually large number of pedagogical devices intended to increase readability, to involve students in the learning process, to stimulate their interests in the subject matter, and to help them relate their classroom knowledge to their future clinical experiences. For an annotated listing of these devices, see To the Reader, which follows this preface.

Supplementary Materials

Supplementary materials designed to help the instructor plan class work and presentations, and to aid students in their learning activities are also available:

1. *Instructor's Resource Manual and Test Item File* by John W. Hole, Jr., and Karen A. Koos contains chapter overviews, instructional techniques, suggested schedules, discussions of chapter elements, lists of related films, and directories of suppliers of audiovisual and laboratory materials. It also contains test items for each chapter of the text, which are designed to evaluate student understanding of the subject matter presented.

2. *WCB TestPak* is a computerized testing service offered free upon request to adopters of this textbook. It provides a call-in/mail-in test preparation service. A complete test item file is also available on computer diskette for use with IBM compatible, Apple IIe or IIc, or Macintosh computers.

3. *A Student Study Guide to Accompany Human Anatomy and Physiology* by Nancy A. Sickles Corbett, Rutgers, The State University, Camden, NJ, contains chapter overviews, chapter objectives, focus questions, mastery tests, study activities, and answer keys corresponding to the chapters of the text.

4. *Transparencies* include a set of 150 acetate transparencies designed to complement classroom lectures or to be used for short quizzes.

5. A *supplemental transparency* set of 100 acetates is also available.

6. *Color slides* include a set of light micrographs of tissues, organs, and other body features described in the textbook to complement classroom instruction.

7. *Laboratory Manual to Accompany Human Anatomy and Physiology* by John W. Hole, Jr., is designed specifically to accompany the sixth edition of *Human Anatomy and Physiology*.

8. *Dissection Video* demonstrates for students the procurement, maintenance, and utilization of cadavers, as well as prosection of abdominal and thoracic cavities, and arm and leg extremities. The video was prepared by Terry Martin and Hassan Rastegar of Kishwaukee College.

9. *Visuals Testbank* is a set of 50 transparency masters that are available for use by instructors. These feature line art from the text with labels deleted for student quizzing or for student practice.

10. *Extended Lecture Outline Software* consists of detailed outlines of each chapter on disk. Instructors can add their own lecture notes for convenience in lecture preparation. Available for use with IBM, Apple, or Macintosh.

Also available from WCB . . .

• Study Cards for Human Anatomy and Physiology by Van De Graff/Rhees/Creek

• *The Coloring Review Guide to Human Anatomy* by McMurtrie/Rikel

• *Atlas of the Skeletal Muscles* by Robert and Judith Stone

• The WCB Anatomy and Physiology Video Series

• *Anatomy and Physiology of the Heart* videodisc

• *Computer Review of Human Anatomy and Physiology* software by Davis/Zimmerman/Van De Graaff

• *Knowledge Map of Human Anatomy Systems* software (Macintosh) by Craig Gundy of Weber State College

• *Slice of Life, Vol. V.* videodisc

To the Reader

This textbook includes a variety of aids to the reader that should make your study of human anatomy and physiology more effective and enjoyable. These aids are included to help you master the basic concepts of human anatomy and physiology that are needed before progressing to more difficult material.

Unit Introductions

Each unit opens with a brief description of the general content of the unit and a list of chapters included within the unit (see page 2 for an example). This introduction provides an overview of the chapters and tells how the unit relates to the other aspects of human anatomy and physiology.

Chapter Introductions

Each chapter introduction previews the chapter's contents and relates that chapter to the others within the unit (see page 4).

After reading an introduction, browse through the chapter, paying particular attention to topic headings and illustrations so that you get a feeling for the kinds of ideas that are included in the chapter.

Chapter Objectives

Before you begin to study a chapter, carefully read the chapter objectives (see page 5). These indicate what you should be able to do after mastering the information within the narrative. The review activities at the end of each chapter (see page 29) are phrased as detailed objectives, and it is helpful to read them also before beginning your study. Both sets of objectives are guides that indicate important sections of the narrative.

Key Terms

The list of terms and their phonetic pronunciations given at the beginning of each chapter help build your science vocabulary. The words included in these lists are used within the chapter and are likely to be found in subsequent chapters as well (see page 5). An explanation of phonetic pronunciation is provided on page 944 of the glossary.

Aids to Understanding Words

Aids to understanding words at the beginning of each chapter also helps build your vocabulary. This section includes a list of word roots, stems, prefixes, and suffixes that help you discover word meanings. Each root and an example word that uses the root are defined (see page 5).

Knowing the roots from these lists will help you discover and remember scientific word meanings.

Review Questions
within the Narrative

Review questions occur at the ends of major sections within each chapter (see page 6). When you reach such questions, try to answer them. If you succeed, then you probably understand the previous discussion and are ready to proceed. If you have difficulty, reread that section before proceeding.

Illustrations and Charts

Numerous illustrations and charts occur in each chapter and are placed near their related textual discussion. They are designed to help you visualize structures and processes, to clarify complex ideas, to summarize sections of the narrative, or to present pertinent data.

As mentioned previously, it is a good idea to skim through the chapter before you begin reading, paying particular attention to these figures. Then, as you read for detail, carefully look at each figure and use it to gain a better understanding of the material presented.

Sometimes the figure legends contain questions that will help you apply your knowledge to the object or process the figure illustrates. The ability to apply information to new situations is of prime importance. These questions concerning the illustrations will provide practice in this skill.

There are also sets of special reference plates that you may want to refer to from time to time. One set (see page 30) is designed to illustrate the structure and location of the major internal organs of the body. Another set (see page 228) depicts the structural detail of the human skull, and still others (see page 323 and page 908) will help you locate major features on the body surface and visualize organs exposed by the dissection of a cadaver.

Boxed Information

Short paragraphs set off in boxes of colored type occur throughout each chapter (see page 8). These boxed asides often contain information that will help you apply the ideas presented in the narrative to clinical situations. Others contain information about changes that occur in the body's structure and function as a person passes through the various phases of the human life cycle. These will help you understand how certain body conditions change as a person grows older.

Clinical, Practical, and Laboratory Applications

Other longer asides are entitled Clinical, Practical, or Laboratory Applications. They discuss pathological disorders, laboratory techniques, and pertinent information of more general interest (see page xv for a list of these topics).

Clinical Terms

At the ends of most chapters are lists of related terms that are sometimes used in clinical situations, along with phonetic pronunciations for those terms (see page 26). Although these lists and the word definitions are often brief, they will be a useful addition to your understanding of medical terminology.

Chapter Summaries

A summary in outline form at the end of each chapter will help you review the major ideas presented in the narrative (see page 26). Scan this section a few days after you have read the chapter. If you find portions that seem unfamiliar, reread the related sections of the narrative.

Clinical Application of Knowledge

The questions at the end of each chapter (see page 28) will help you gain experience in applying information to a few clinical situations. Discuss your answers with other students or with an instructor. (Suggested answers to these questions are included in Appendix D.)

Review Activities

The review activities at the end of each chapter (see page 29) will check your understanding of the major ideas presented in the narrative. After studying the chapter, read the review activities; if you can perform the tasks suggested, you have accomplished the goals of the chapter. If not, reread the sections of the narrative that need clarification.

Clinical Case Studies

Fictitious clinical case studies occur at the ends of chapters 6, 12, 14, 16, 18, and 20. Each study describes a medical problem and explains how the problem might be treated. The purpose of the case study is to help you relate some of the ideas presented in the previous chapter to a clinical situation and, at the same time, to add to your understanding of medical terminology. Questions are included at the end of each study to stimulate thought about the problems and procedures described.

Appendixes, Glossary, and Index

The appendixes contain a variety of useful information. They include the following:

1. Periodic Table of Elements (see page 930).
2. Lists of various units of measurements and their equivalents together with a description of how to convert one unit into another (see page 931).
3. Lists of clinical laboratory tests commonly performed on human blood and urine. These lists include test names, normal adult values, and brief descriptions of their clinical significance (see page 933).
4. Answers to clinical application of knowledge questions (see page 937). These answers were prepared by Professor Anne Lesak, RN, MS, MSN, Moraine Valley Community College.

The glossary defines the more important textual terms and provides their phonetic pronunciations (see page 944). It also contains an explanation of phonetic pronunciation on page 944.

The index is complete and comprehensive.

Acknowledgments

Once again, I want to express my deep gratitude to the many users of earlier editions of **Human Anatomy and Physiology**, who supplied thoughtful suggestions for improving the textbook. I also want to acknowledge the valuable contributions of the reviewers for the sixth edition, who read portions of or the entire manuscript as it was being prepared, and who provided detailed criticisms and ideas for improving the narrative and the illustrations. They include the following:

David Logan
York University

Terry R. Martin
Kishwaukee College

Aaron E. James
Gateway Community College

Dr. Louis A. Giacinti
Milwaukee Area Technical College

Clarence C. Wolfe
Northern Virginia Community College

Dale A. DesLauriers
Chaffey College

Jean S. Helgeson
Collin County Community College

Nancy Ann S. Corbett
Camden College of Arts & Sciences

Edwin J. Bessler
Franciscan University of Steubenville

Ed Krol
Henry Ford Community College

Dwight Kamback
Northhampton Community College

Robert Smith
Forest Park Community College

John H. Dustman
Indiana University Northwest

The changes that have been made in the sixth edition of **Human Anatomy and Physiology** were in large part responses to the input of these users and reviewers. Without their kind assistance the task of preparing this revision would have been much more difficult.

UNIT

1

Levels of
Organization

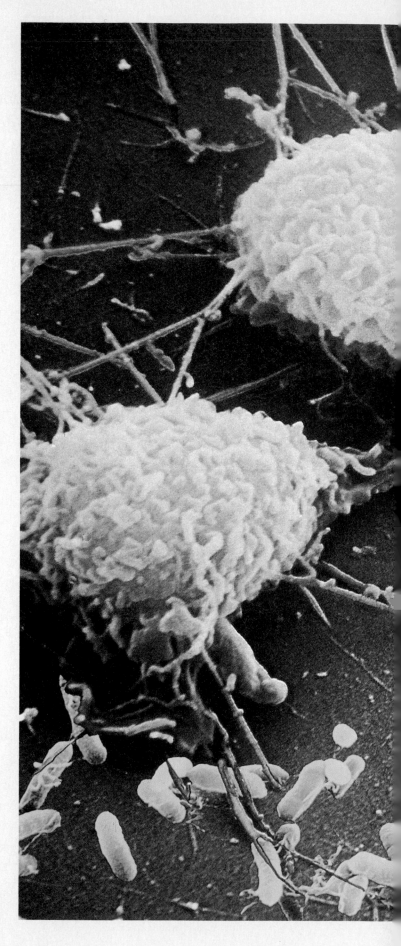

The chapters of unit 1 introduce the study of human anatomy and physiology. They present the similarities between humans and other living things, the interdependent relationship between structure and function of body parts, and the ways these parts interact with each other and with environmental factors to help ensure survival of the whole structure.

Unit 1 describes each of the levels of organization within the body—chemical substances, cellular organelles, cells, tissues, organs, organ systems, and the human organism—and prepares the reader for more detailed study of the organ systems in units 2–5.

A falsely colored scanning electron micrograph showing two large body cells (macrophages) that are capable of engulfing and destroying various bacterial cells (blue rods).

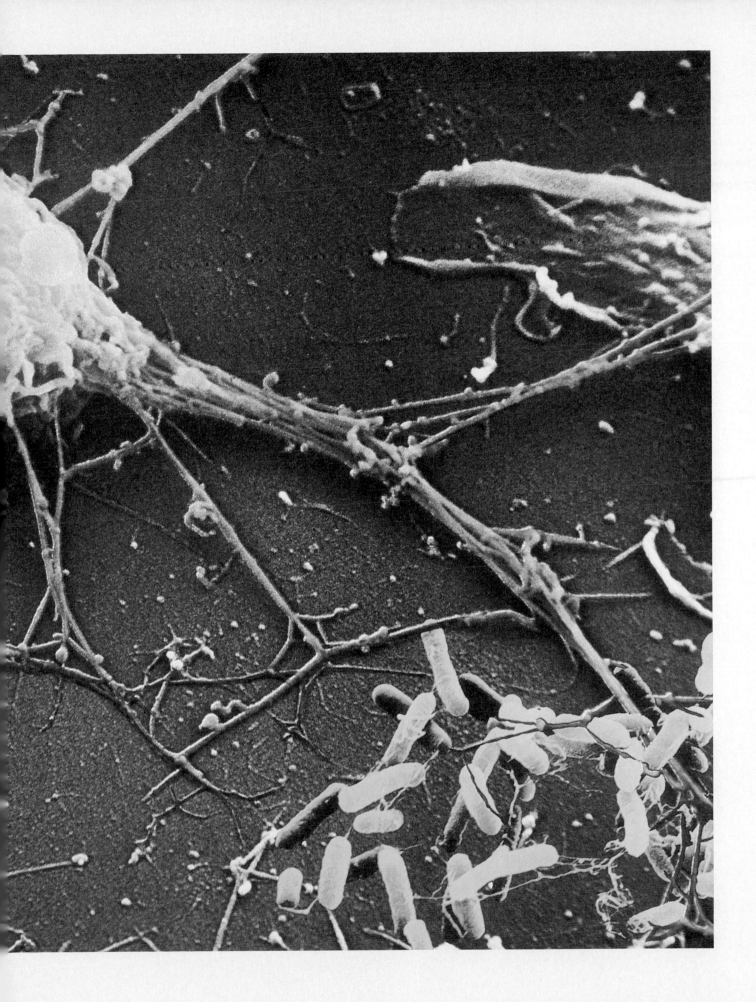

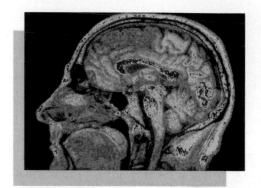

CHAPTER

1

Introduction to Human Anatomy and Physiology

*A*lthough a human represents a particular kind of living organism, it shares certain traits with other organisms. For example, it carries on life processes and demonstrates the characteristics of life. It also has needs that must be met if it is to survive, and its life depends upon maintaining a relatively stable internal environment.

A discussion of the traits that humans have in common with other organisms and of the way the complex human body is organized provides a beginning for the study of human anatomy and physiology in chapter 1.

This chapter also introduces the organization of the human body and a special set of terms used to describe its major parts ∎

Chapter Objectives	Key Terms	Aids to Understanding Words

After you have studied this chapter, you should be able to:

1. Define *anatomy* and *physiology*, and explain how they are related.

2. List and describe the major characteristics of life.

3. List and describe the major needs of organisms.

4. Define *homeostasis* and explain its importance to survival.

5. Describe a homeostatic mechanism.

6. Explain what is meant by levels of organization.

7. Describe the locations of the major body cavities.

8. List the organs located in each major body cavity.

9. Name the membranes associated with the thoracic and abdominopelvic cavities.

10. Name the major organ systems and list the organs associated with each.

11. Describe the general functions of each organ system.

12. Properly use the terms that describe relative positions, body sections, and body regions.

13. Complete the review activities at the end of this chapter. Note that the items are worded in the form of specific learning objectives. You may want to refer to them before reading the chapter.

---

absorption (ab-sorp′shun)

anatomy (ah-nat′o-me)

appendicular (ap″en-dik′u-lar)

assimilation (ah-sim″ĭ-la′shun)

axial (ak′se-al)

circulation (ser-ku-la′shun)

digestion (di-jest′yun)

excretion (ek-skre′shun)

homeostasis (ho″me-ō-sta′sis)

metabolism (mĕ-tab′o-lizm)

negative feedback (neg′ah-tiv fēd′bak)

organelle (or″gan-el′)

organism (or′gah-nizm)

parietal (pah-ri′ĕ-tal)

pericardial (per″ĭ-kar′de-al)

peritoneal (per″ĭ-to-ne′al)

physiology (fiz″e-ol′o-je)

pleural (ploo′ral)

reproduction (re″pro-duk′shun)

respiration (res″pĭ-ra′shun)

thoracic (tho-ras′ik)

visceral (vis′er-al)

The accent marks used in the pronunciation guides are derived from a simplified system of phonetics standard in medical usage. The single accent (′) denotes the major stress, meaning that emphasis is placed on the most heavily pronounced syllable in the word. The double accent (″) indicates secondary stress. A syllable marked with a double accent receives less emphasis than the syllable that carries the main stress, but more emphasis than neighboring unstressed syllables.

---

append-, to hang something: *append*icular—pertaining to the arms and legs.

cardi-, heart: peri*cardi*um—a membrane that surrounds the heart.

cran-, helmet: *cran*ial—pertaining to the portion of the skull that surrounds the brain.

dors-, back: *dors*al—a position toward the back of the body.

homeo-, same: *homeo*stasis—the maintenance of a stable internal environment.

-logy, the study of: physio*logy*—the study of body functions.

meta-, change: *meta*bolism—the chemical changes that occur within the body.

nas-, nose: *nas*al—pertaining to the nose.

orb-, circle: *orb*ital—pertaining to the portion of skull that encircles an eye.

pariet-, wall: *pariet*al membrane—a membrane that lines the wall of a cavity.

pelv-, basin: *pelv*ic cavity—a basin-shaped cavity enclosed by the pelvic bones.

peri-, around: *peri*cardial membrane—a membrane that surrounds the heart.

pleur-, rib: *pleur*al membrane—a membrane that encloses the lungs within the rib cage.

-stasis, standing still: homeo*stasis*—the maintenance of a relatively stable internal environment.

-tomy, cutting: ana*tomy*—the study of structure, which often involves cutting or removing body parts.

The study of the human body has a long and interesting history. It began with our earliest ancestors, who must have been as curious about their body parts and functions as we are today. At first their interests most likely concerned injuries and illnesses, because healthy bodies demand little attention from their owners. Certainly primitive people suffered from occasional aches and pains, injured themselves, bled, broke bones, and developed diseases. When they were sick or felt pain, these people probably sought relief by visiting shamans. However, the treatment they received may have been less than satisfactory because primitive doctors relied heavily on superstitions and notions about magic. As they tried to help the sick, these early medical workers began to discover useful ways of examining and treating the human body. They observed the effects of injuries, noticed how wounds healed, and attempted to determine the causes of deaths by examining dead bodies. They also began to learn how certain herbs and potions affected body functions and could sometimes be used to treat coughs, headaches, and other common problems.

In ancient times, it was generally believed that natural processes were controlled by spirits and supernatural forces that humans could not understand. Then, about 2,500 years ago, attitudes began to change, and the belief that humans could understand natural processes grew in popularity.

This new idea stimulated people to look more closely at the world around them. They began asking more questions and seeking answers. In this way, the stage was set for the development of modern science. As techniques for making accurate observations and performing careful experiments evolved, knowledge of the human body expanded rapidly.

At the same time, many new terms were devised to name body parts, describe their locations, and explain their functions. These terms, most of which originated from Greek and Latin, formed the basis for the language of anatomy and physiology. (See figure 1.1.) A list of some of the modern medical and applied sciences appears on page 26.

1. What factors probably stimulated an early interest in the human body?
2. What idea encouraged people to begin studying the natural world?
3. What kinds of activities helped promote the development of modern science?

Anatomy and Physiology

Anatomy is the branch of science that deals with the structure (morphology) of body parts—their forms and arrangements. Anatomists observe body parts grossly and microscopically, and describe them as accurately

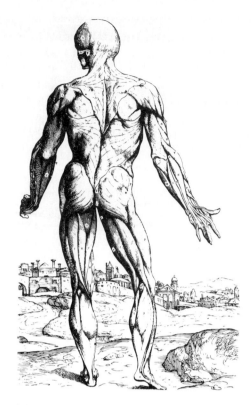

Figure 1.1   The study of the human body has a long history, as indicated by this illustration from the second book of *De Humani Corporis Fabrica* by Andreas Vesalius, issued in 1543.

and in as much detail as possible. **Physiology,** on the other hand, is concerned with the functions of body parts—what they do and how they do it. Physiologists are interested in finding out how such parts carry on life processes. In addition to using the same observational techniques as anatomists, physiologists are likely to conduct experiments and use complex laboratory equipment.

Actually, it is difficult to separate the topics of anatomy and physiology because the structures of body parts are so closely associated with their functions. These parts are arranged to form a well-organized unit—the **human organism**—and each part plays a role in the operation of the unit as a whole. This role, which is the part's function, depends upon the way the part is constructed—that is, the way its subparts are organized. For example, the arrangement of parts in the human hand with its long, jointed fingers is related to the function of grasping objects. The tubular blood vessels are designed to transport blood to and from the hollow chambers of the heart; the heart's powerful muscular walls are structured to contract and cause the blood to move out of the chambers and into blood vessels; and the valves associated with the vessels and chambers ensure that the blood will move in the proper direction. The shape of the mouth is related to the function of receiving food; the teeth are designed to

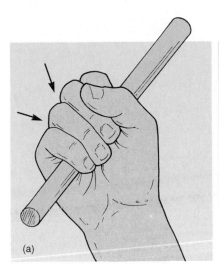 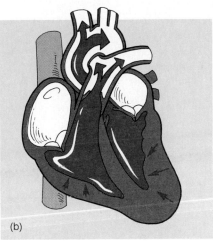 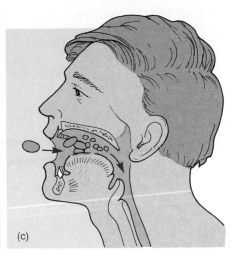

Figure 1.2 The structures of body parts are closely related to their functions. (a) The hand is adapted to grasping; (b) the heart for pumping blood; and (c) the mouth for receiving food.

break solid foods into smaller pieces; and the muscular tongue and cheeks are constructed to help mix food particles with saliva and prepare it for swallowing (figure 1.2).

1. What are the differences between an anatomist and a physiologist?
2. Why is it difficult to separate the topics of anatomy and physiology?
3. List several examples to illustrate the idea that the structure of a body part is closely related to its function.

Characteristics of Life

Before beginning a more detailed study of anatomy and physiology, it is helpful to consider some of the traits humans share with other organisms. These traits, which are called *characteristics of life,* include the following:

1. **Movement** usually refers to a self-initiated change in an organism's position or to its traveling from one place to another. However, the term also applies to the motion of internal parts, such as a beating heart.
2. **Responsiveness** (or irritability) is the ability of an organism to sense changes taking place inside or outside its body and to react to these changes. Seeking water to quench thirst is a response to a loss of water in body tissues. Moving away from a hot fire is another example of responsiveness.
3. **Growth** refers to an increase in body size, usually without any important change in shape. It occurs whenever an organism produces new

body materials faster than the old ones are worn out or depleted.

4. **Reproduction** is the process of making a new individual, as when parents produce an offspring. It also indicates the process by which microscopic cells produce others like themselves, as they do when body parts are repaired or replaced following an injury.
5. **Respiration** is the process of obtaining oxygen, using oxygen in the release of energy from food substances, and removing the resulting gaseous wastes (carbon dioxide).
6. **Digestion** is the process by which various food substances are chemically changed into simpler forms that can be absorbed and used by body parts.
7. **Absorption** refers to the passage of substances through membranes, as when digestive products pass through the membrane that lines the digestive organs and enter the body fluids.
8. **Circulation** is the movement of substances from place to place within the body by means of the body fluids.
9. **Assimilation** is the changing of absorbed substances into forms that are chemically different from those that entered the body fluids.
10. **Excretion** is the removal of wastes that are produced by body parts as a result of their activities.

Each of these characteristics of life—in fact, everything an organism does—depends upon physical and chemical changes that occur within body parts.

CHART 1.1 Characteristics of life

Process	Examples	Process	Examples
Movement	Change in position of the body or of a body part; motion of an internal organ	Digestion	Breakdown of food substances into simpler forms
Responsiveness	Reaction to a change taking place inside or outside the body	Absorption	Passage of substances through membranes and into body fluids
Growth	Increase in body size without change in shape	Circulation	Movement of substances from place to place in body fluids
Reproduction	Production of new organisms and new cells	Assimilation	Changing of absorbed substances into chemically different forms
Respiration	Obtaining oxygen, using oxygen in releasing energy from foods, and removing carbon dioxide	Excretion	Removal of wastes produced by metabolic reactions

Taken together, these physical and chemical changes or reactions are referred to as **metabolism.**

The characteristics shared by all living organisms are summarized in chart 1.1.

Vital signs are among the more common observations made by physicians and nurses working with patients. Assessment of vital signs includes measuring body temperature and blood pressure, and monitoring rates and types of pulse and breathing movements. There is a close relationship between these signs and the characteristics of life, since vital signs are the results of *metabolic activities.* In fact, death is recognized by the absence of such signs. A person who has died displays no spontaneous muscular movements (including those of the breathing muscles and beating heart), does not respond to stimuli (even the most painful that can be ethically applied), exhibits no reflexes (such as the knee-jerk reflex and pupillary reflexes of the eye), and generates no brain waves (demonstrated by a flat encephalogram, which reflects a lack of metabolic activity in the brain).

1. What are the characteristics of life?
2. How are the characteristics of life related to metabolism?

Maintenance of Life

With the exception of an organism's reproductive structures, which function to ensure that its particular form of life will continue into the future, the structures and functions of all body parts are directed toward achieving one goal—maintaining the life of the organism.

Needs of Organisms

Life is fragile, and it depends upon the presence of certain environmental factors for its existence. These factors include the following:

1. **Water** is the most abundant substance in the body. It is required for a variety of metabolic processes, and it provides the environment in which most of them take place. Water also transports substances within organisms and is important in regulating body temperature.
2. **Food** refers to substances that provide organisms with necessary chemicals in addition to water. Some of these chemicals are used as energy sources, others supply raw materials for building new living matter, and still others enter into vital chemical reactions.
3. **Oxygen** is a gas that makes up about one-fifth of the air. It is used in the process of releasing energy from food substances. The energy, in turn, is needed to drive metabolic processes.
4. **Heat** is a form of energy. It is a product of metabolic reactions, and the rate at which these reactions occur is partly governed by the amount of heat present. Generally, the greater the amount of heat, the more rapidly chemical reactions take place. *Temperature* is a measurement of the amount of heat present.
5. **Pressure** is an application of force on something. For example, the force acting on the outside of a land organism due to the weight of air above it is called *atmospheric pressure.* In humans, this pressure plays an important role in breathing. Similarly, organisms living under water are subjected to *hydrostatic pressure*—a

CHART 1.2 Needs of organisms

Factor	Characteristic	Use	Factor	Characteristic	Use
Water	A chemical substance	Needed for metabolic processes, as a medium for metabolic reactions, to transport substances, and to regulate temperature	Oxygen	A chemical substance	Needed to release energy from food substances
			Heat	A form of energy	Needed to help regulate the rates of metabolic reactions
Food	Various chemical substances	Needed to supply energy and raw materials for the production of living matter and for the regulation of vital reactions	Pressure	A force	Atmospheric pressure needed for breathing; hydrostatic pressure needed to help move blood

pressure exerted by a liquid—due to the weight of water above them. In complex organisms, such as humans, the heart action produces blood pressure (another form of hydrostatic pressure), which forces the blood through certain blood vessels.

Although organisms need water, food, oxygen, heat, and pressure, the presence of these factors alone is not enough to ensure their survival. Both the quantities and the qualities of such factors are also important. For example, the amount of water entering and leaving an organism must be regulated, as must the concentration of oxygen in the body fluids. Similarly, survival depends on the quality as well as the quantity of food available—that is, food must supply the correct chemicals in adequate amounts. Chart 1.2 summarizes the major needs of organisms.

Homeostasis

As an organism moves from place to place, or as the climate around it changes, factors in its external environment change. However, if the organism is to survive, conditions within the fluids surrounding its body cells must remain relatively stable. In other words, body parts function efficiently only when the concentrations of water, food substances, and oxygen and the conditions of heat and pressure remain within certain narrow limits. This tendency to maintain a stable internal environment is called **homeostasis.**

To better understand this idea of maintaining a stable internal environment, imagine a room equipped with a furnace and an air conditioner. Suppose the room temperature is to remain near 20°C (68°F), so the thermostat is adjusted to a *set point* of 20°C. Because

a thermostat is sensitive to temperature changes, it will signal the furnace to start and the air conditioner to stop whenever the room temperature drops below the set point. If the temperature rises above the set point, the thermostat will cause the furnace to stop and the air conditioner to start. As a result, a relatively constant temperature will be maintained in the room (figure 1.3).

A similar *homeostatic mechanism* regulates body temperature in humans. The "thermostat" is a temperature-sensitive region in a temperature control center of the brain. In healthy persons, the set point of the brain's thermostat is at or near 37°C (98.6°F).

If a person is exposed to a cold environment and the body temperature begins to drop, the temperature control center of the brain senses this change and triggers heat-generating and heat-conserving activities. For example, small groups of muscles may be stimulated to contract involuntarily, an action called *shivering.* Such muscular contractions produce heat, which helps warm the body. At the same time, blood vessels in the skin may be signaled to constrict so that less warm blood flows through them, and heat that might otherwise be lost is held in deeper tissues.

If a person is becoming overheated, the temperature control center of the brain may trigger a series of changes that promote the loss of body heat. For example, it may stimulate the sweat glands in the skin to secrete watery perspiration. As this water evaporates from the surface, some heat is carried away and the skin is cooled. At the same time, the brain center causes blood vessels in the skin to dilate. This allows the blood that carries heat from deeper tissues to reach the surface where some heat is lost to the outside. Also, the brain center stimulates an increase in heart rate,

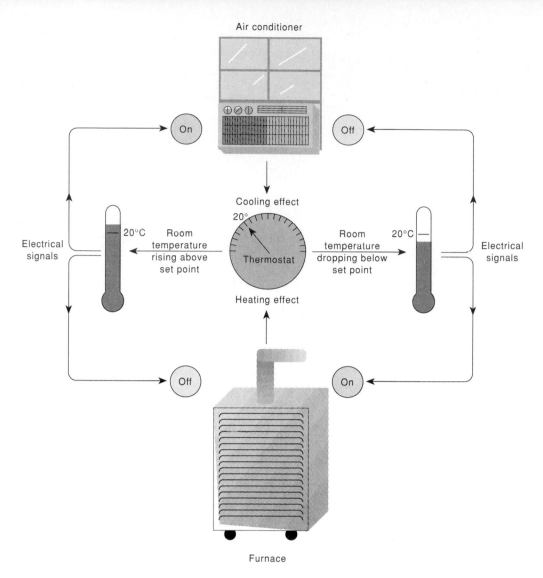

Air conditioner

On

Off

Cooling effect

20°

Electrical
signals

20°C

Room
temperature
rising above
set point

Thermostat

Room
temperature
dropping below
set point

20°C

Electrical
signals

Heating effect

Off

On

Furnace

Figure 1.3 A thermostat that can signal an air conditioner and a furnace to turn on or off maintains a relatively stable room temperature. This system provides an example of a homeostatic mechanism.

causing a greater volume of blood to move into the surface vessels, and it stimulates an increase in breathing rate, allowing more heat-carrying air to be expelled from the lungs. Body temperature regulation is discussed in more detail in chapter 6.

Another homeostatic mechanism regulates the blood pressure in the blood vessels (arteries) leading away from the heart. In this instance, pressure-sensitive parts (sensors) in the walls of these vessels are stimulated if the blood pressure increases above normal. When this happens, the sensors signal a pressure control center in the brain, which in turn signals the heart, causing its chambers to contract more slowly and with less force. Because of decreased heart action, less blood enters the blood vessels, and the pressure inside the vessels decreases. If the blood pressure is dropping below normal, the brain center signals the heart to contract more rapidly and with greater force so that the pres-

sure in the vessels increases. The regulation of blood pressure is described in more detail in chapter 18.

Similarly, a homeostatic mechanism regulates the concentration of sugar (glucose) in the blood. If, for example, the amount of blood sugar increases following a meal, an organ called the pancreas detects this change and releases a chemical (insulin) into the blood. This substance causes sugar to move from the blood into various body cells and to be stored in the liver and muscles. As this occurs, the concentration of blood sugar decreases, and when it reaches the normal set point, the pancreas ceases its release of insulin. (See figure 1.4.) If, on the other hand, the blood sugar concentration becomes abnormally low, the pancreas detects this change and releases a different chemical (glucagon) that causes sugar to be released from storage into the blood. The regulation of the blood sugar concentration is discussed in more detail in chapter 13.

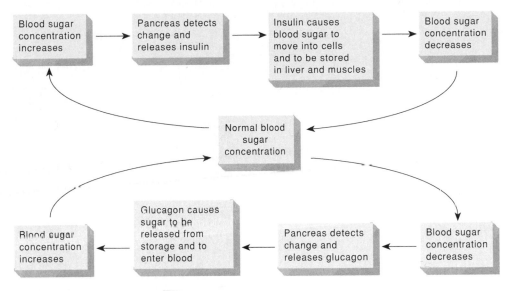

Figure 1.4 Homeostatic mechanism that regulates blood sugar concentration.

In these examples, homeostasis is maintained by a self-regulating control mechanism that can sense changes away from the normal set point and can cause reactions that tend to return conditions to normal. Since the changes away from the normal state stimulate changes to occur in the opposite direction, the responses are called *negative,* and the homeostatic control mechanism is said to act by a process of **negative feedback,** which is discussed in chapter 4.

Details about these and other homeostatic mechanisms are presented in later chapters. The maintenance of homeostasis is of primary importance to survival, and it is not surprising that metabolic activities are largely directed toward maintaining stable internal conditions.

> Sometimes changes occur that stimulate still other similar changes. Such a process that causes movement away from the normal state is called a *positive feedback mechanism.*
>
> Although most feedback mechanisms in the body are negative, a positive system operates for a short time when a blood clot forms, because the chemicals present in a clot promote still more clotting. (See chapter 17.)
>
> Because positive feedback mechanisms usually produce unstable conditions, they are most commonly associated with diseases and may lead to death.

1. What needs of organisms are provided from the external environment?
2. What is the relationship between the use of oxygen and the production of heat?
3. Why is homeostasis so important to survival?
4. Describe three homeostatic mechanisms.

Levels of Organization

Early investigators focused their attention on the larger body structures, for they were limited in their ability to observe small parts. Studies of small parts had to wait for the invention of magnifying lenses and microscopes, which came into use about 400 years ago. Once these tools were available, it was discovered that larger body structures were made up of smaller parts, which, in turn, were composed of even smaller ones.

Today, scientists recognize that all materials, including those that comprise the human body, are composed of chemicals. These substances are made up of tiny, invisible particles called **atoms,** which are commonly bound together to form larger particles called **molecules;** small molecules may be combined in complex ways to form larger molecules called **macromolecules.**

Within the human organism, the basic unit of structure and function is a microscopic part called a **cell.** Although individual cells vary in size, shape, and specialized functions, all have certain traits in common. For instance, all cells contain tiny parts called **organelles** that carry on specific activities. These organelles are composed of aggregates of large molecules, including those of such substances as proteins, carbohydrates, lipids, and nucleic acids.

Cells are organized into layers or masses that have common functions. Such a group of cells forms a **tissue.** Groups of different tissues form **organs**—complex structures with specialized functions—and groups of organs that function closely together comprise **organ systems.** Organ systems make up an **organism.** Thus, various body parts occupy different levels of organization, such as the *atomic level, molecular level,* or

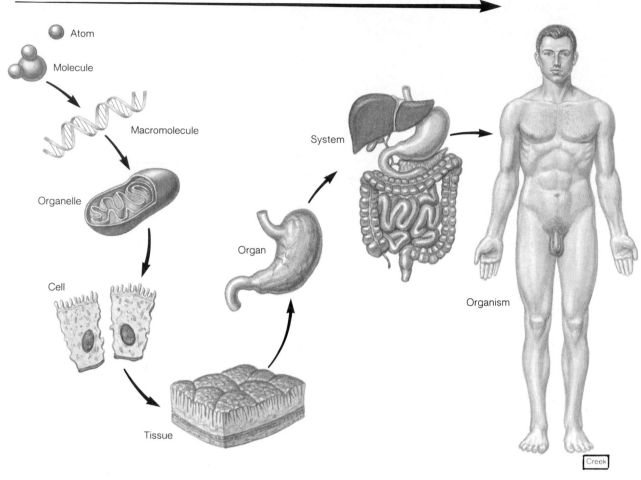

Increasing complexity

Atom

Molecule

Macromolecule

Organelle

Cell

Tissue

Organ

System

Organism

Creek

Figure 1.5 A human body is composed of parts within parts, which vary in complexity.

cellular level. Furthermore, body parts vary in complexity from one level to the next. That is, atoms are less complex than molecules, molecules are less complex than organelles, organelles are less complex than cells, cells are less complex than tissues, tissues are less complex than organs, organs are less complex than organ systems, and organ systems are less complex than the whole organism (figure 1.5).

Chapters 2–6 discuss these levels of organization in more detail. Chapter 2, for example, describes the atomic and molecular levels; chapter 3 deals with organelles and cellular structures and functions; chapter 4 explores cellular metabolism; chapter 5 describes tissues; and chapter 6 presents membranes as examples of organs, and the skin and its accessory organs as an example of an organ system. Beginning with chapter 7, the structure and functions of each of the organ systems are described in detail. (See chart 1.3.)

1. How can the human body be used to illustrate the idea of levels of organization?
2. What is an organism?
3. How do body parts that occupy different levels of organization vary in complexity?

Organization of the Human Body

The human organism is a complex structure composed of many parts. The major features of the human body include various cavities, a set of membranes, and a group of organ systems.

Body Cavities

The human organism can be divided into an **axial portion,** which includes the head, neck, and trunk, and an **appendicular portion,** which includes the arms and legs. Within the axial portion are two major cavities—a

CHART 1.3   Levels of organization

Level	Example	Illustration
Atom	Hydrogen atom, oxygen atom	Figure 2.1
Molecule	Water molecule, carbon dioxide molecule	Figure 2.9
Macromolecule	Protein molecule, DNA molecule	Figure 2.27
Organelle	Mitochondrion, Golgi apparatus, nucleus	Figure 3.12
Cell	Muscle cell, nerve cell	Figure 3.2
Tissue	Simple squamous epithelium, loose connective tissue	Figure 5.1
Organ	Skin, femur, heart, kidney	Figure 7.2
Organ system	Integumentary system, skeletal system, digestive system	Figure 7.18
Organism	Human	Figure 23.31

dorsal cavity and a larger ventral cavity. The organs contained within such a cavity are called *visceral organs*. The dorsal cavity can be subdivided into two parts—the cranial cavity, which houses the brain, and the vertebral canal (spinal cavity), which contains the spinal cord and is surrounded by sections of the backbone (vertebrae). The ventral cavity consists of a thoracic cavity and an abdominopelvic cavity. These major body cavities are shown in figure 1.6.

The thoracic cavity is separated from the lower abdominopelvic cavity by a broad, thin muscle called the *diaphragm*. When it is at rest, this muscle curves upward into the thorax like a dome. When it contracts during inhalation, it presses down upon the abdominal visceral organs. The wall of the thoracic cavity is composed of skin, skeletal muscles, and various bones. The viscera within it include the lungs and a region between the lungs, called the *mediastinum*. The mediastinum separates the thorax into two compartments that contain the right and left lungs. The remaining thoracic viscera—heart, esophagus, trachea, and thymus gland—are located within the mediastinum.

The abdominopelvic cavity, which includes an upper abdominal portion and a lower pelvic portion, extends from the diaphragm to the floor of the pelvis. Its wall consists primarily of skin, skeletal muscles, and bones. The visceral organs within the *abdominal cavity* include the stomach, liver, spleen, gallbladder, and most of the small and large intestines.

The *pelvic cavity* is the portion of the abdominopelvic cavity enclosed by the pelvic bones. It contains the terminal end of the large intestine, the urinary bladder, and the internal reproductive organs.

Smaller cavities within the head include the following:

1. *Oral cavity,* containing the teeth and tongue.
2. *Nasal cavity,* located within the nose and divided into right and left portions by a nasal septum. Several air-filled *sinuses* are connected to the nasal cavity. These include the sphenoidal and frontal sinuses shown in figure 1.7. (See chapter 7.)
3. *Orbital cavities,* containing the eyes and associated skeletal muscles and nerves.
4. *Middle ear cavities,* containing the middle ear bones (figure 1.7).

Thoracic and Abdominopelvic Membranes

The walls of the right and left thoracic compartments, which contain the lungs, are lined with a membrane called the *parietal pleura*. The lungs themselves are covered by a similar membrane called the *visceral pleura*. (Note: *Parietal* refers to the membrane attached to the wall of a cavity, and *visceral* refers to one that is deeper and associated with an internal organ, such as a lung.)

The parietal and visceral pleural membranes are separated by a thin film of watery fluid (serous fluid) that they secrete. Although there is normally no actual space between these membranes, the potential space between them is called the *pleural cavity*.

The heart, which is located in the broadest portion of the mediastinum, is surrounded by pericardial membranes. A thin *visceral pericardium* (epicardium) covers the heart's surface and is separated from a much thicker, fibrous *parietal pericardium* by a small amount of serous fluid. The potential space between these membranes is called the *pericardial cavity*. Figure 1.8 shows the membranes associated with the heart and lungs.

In the abdominopelvic cavity, the lining membranes are called peritoneal membranes. A *parietal peritoneum* lines the wall, and a *visceral peritoneum* covers each organ in the abdominal cavity. The potential space between these membranes is called the *peritoneal cavity* (figure 1.9).

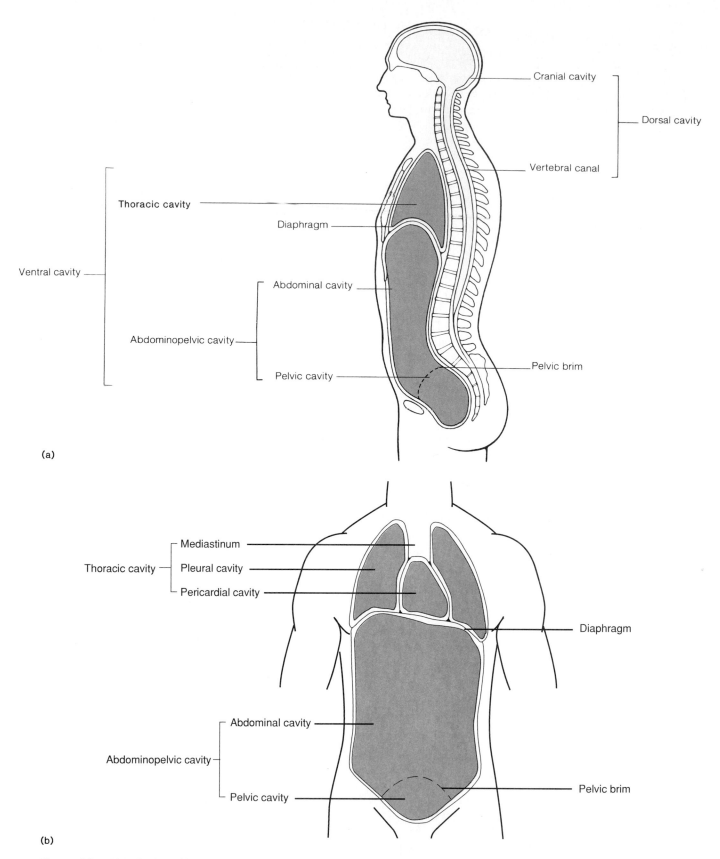

Figure 1.6   Major body cavities, as viewed from (a) the side and from (b) the front.

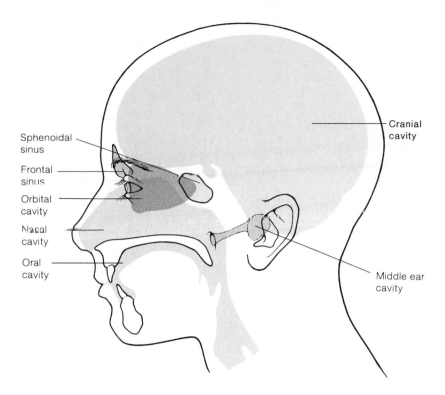

Figure 1.7 The cavities within the head include the oral, nasal, orbital, and middle ear cavities, as well as several sinuses.

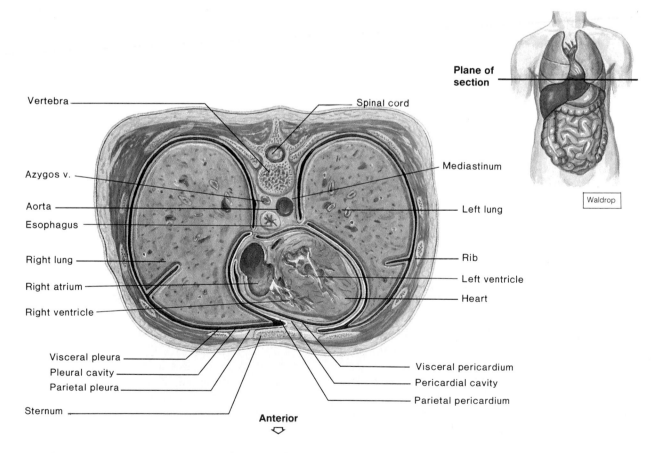

Figure 1.8 A transverse section through the thorax reveals the serous membranes associated with the heart and lungs (superior view).

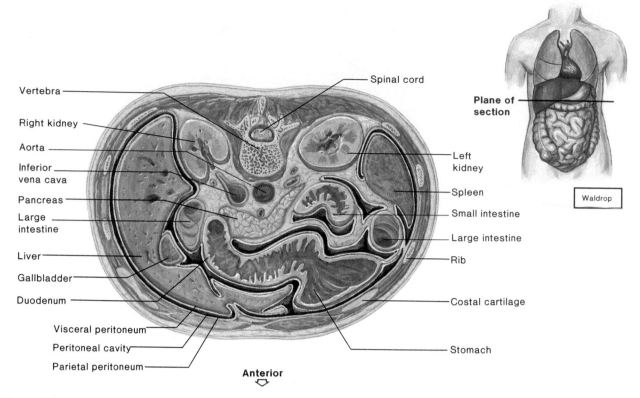

Figure 1.9 A transverse section through the abdomen (superior view). How would you distinguish between the parietal and the visceral peritoneums?

1. What is meant by the term *visceral organ*?
2. What organs occupy the dorsal cavity? The ventral cavity?
3. Name the cavities of the head.
4. Describe the membranes associated with the thoracic and abdominopelvic cavities.

Organ Systems

The human organism consists of several organ systems. Each system includes a set of interrelated organs that work together to provide specialized functions. As you read about each organ system, you may want to consult the illustrations of the human torso provided in reference plates 1–7 and locate some of the features listed in the descriptions.

Body Covering

The organs of the **integumentary system** include the skin and various accessory organs such as the hair, nails, sweat glands, and sebaceous glands. These parts protect underlying tissues, help regulate the body temperature, house a variety of sensory receptors, and synthesize certain products (figure 1.10). The integumentary system is discussed in chapter 6.

Support and Movement

The organs of the skeletal and muscular systems function to support and move body parts (figure 1.11).

The **skeletal system** consists of the bones as well as the ligaments and cartilages that bind the bones together at joints. These parts provide frameworks and protective shields for softer tissues, serve as attachments for muscles, and act together with muscles when body parts move. Tissues within bones also function to produce blood cells and store inorganic salts.

The muscles are the organs of the **muscular system.** By contracting and pulling their ends closer together, they provide the forces that cause body movements. They also function in maintaining posture and are the main source of body heat.

The skeletal and muscular systems are discussed in chapters 7, 8, and 9.

Integration and Coordination

For the body to act as a unit, its parts must be integrated and coordinated. That is, their activities must be controlled and adjusted from time to time so that homeostasis is maintained. This is the general function of the nervous and endocrine systems (figure 1.12).

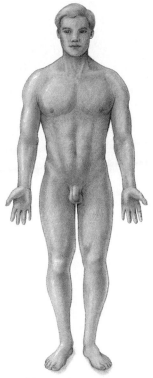

Integumentary system

Figure 1.10 The integumentary system forms the body covering.

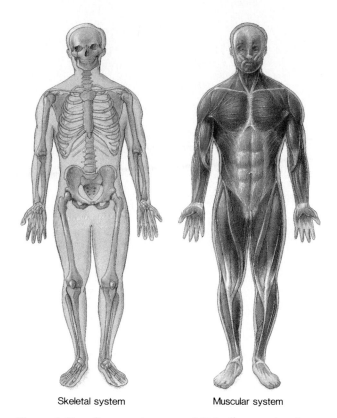

Skeletal system   Muscular system

Figure 1.11 Organ systems associated with support and movement.

The **nervous system** consists of the brain, spinal cord, nerves, and sense organs. Nerve cells within these organs use electrochemical signals called *nerve impulses* (action potentials) to communicate with one another and with muscles and glands. Each impulse produces a relatively short-term effect on the part it influences. Some nerve cells act as specialized sensory receptors that can detect changes occurring inside and outside the body. Others receive the impulses transmitted from these sensory units and interpret and act on the information received. Still others carry impulses from the brain or spinal cord to muscles or glands and stimulate these parts to contract or to secrete various products. The nervous system is discussed in chapters 10 and 11, and the organs associated with sensory reception are discussed in chapter 12.

The **endocrine system** includes all the glands that secrete chemical messengers called *hormones*. The hormones, in turn, travel away from the glands in body fluids such as blood or tissue fluid. Usually a particular hormone only affects a particular group of cells, which is called its *target tissue*. The effect of a hormone is to alter the metabolism of the target tissue. Compared to nerve impulses, hormonal effects occur over a relatively long time period.

The organs of the endocrine system include the pituitary, thyroid, parathyroid, and adrenal glands, as

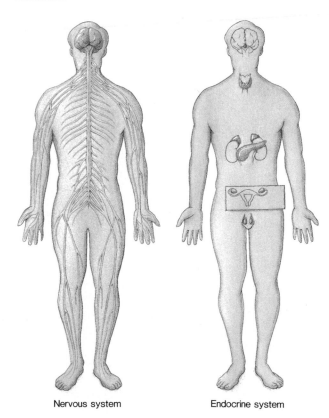

Nervous system   Endocrine system

Figure 1.12 Organ systems associated with integration and coordination.

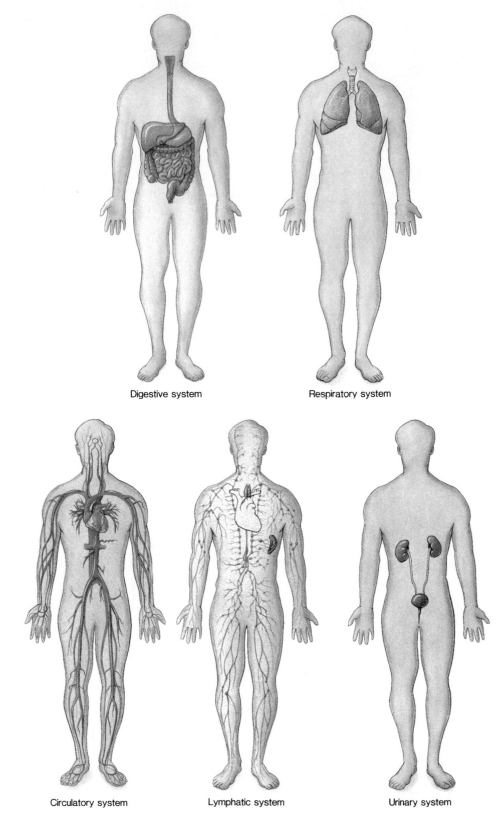

Digestive system

Respiratory system

Circulatory system

Lymphatic system

Urinary system

Figure 1.13   Organ systems associated with processing and transporting.

well as the pancreas, ovaries, testes, pineal gland, and thymus gland. They are discussed further in chapter 13.

Processing and Transporting

The organs of several systems are involved with processing and transporting nutrients, oxygen, and various wastes. (See figure 1.13.) The organs of the **digestive system,** for example, receive foods from the outside. Then they convert various food molecules into simpler forms that can pass through cell membranes and thus be absorbed. Materials that are not absorbed are eliminated by being transported back to the outside. Certain digestive organs also produce hormones and thus function as parts of the endocrine system (described in chapter 13).

The digestive system includes the mouth, tongue, teeth, salivary glands, pharynx, esophagus, stomach, liver, gallbladder, pancreas, small intestine, and large intestine. This system is discussed in chapter 14. Food molecules and other nutrients are discussed in chapter 15.

The organs of the **respiratory system** provide for the intake and output of air, and for the exchange of gases between the blood and the air. More specifically, oxygen passes from air within the lungs into the blood, and carbon dioxide leaves the blood and enters the air. The nasal cavity, pharynx, larynx, trachea, bronchi, and lungs are parts of this system, which is discussed in chapter 16.

The **circulatory system** includes the heart, arteries, veins, capillaries, and blood. The heart functions as a muscular pump that helps force blood through the blood vessels. The blood serves as a fluid for transporting gases, nutrients, hormones, and wastes. It carries oxygen from the lungs and nutrients from the digestive organs to all body cells, where these substances are used in metabolic processes. The blood also transports hormones from various endocrine glands to their target tissues, and carries wastes from body cells to the excretory organs, where the wastes are removed from the blood and released to the outside. The circulatory system is discussed in chapters 17 and 18.

The **lymphatic system** is sometimes considered part of the circulatory system. It is composed of the lymphatic vessels, lymph fluid, lymph nodes, thymus gland, and spleen. This system transports some of the fluid from the spaces within tissues (tissue fluid) back to the bloodstream and carries certain fatty substances away from the digestive organs. Lymphatic organs also aid in defending the body against infections by re-moving particles, such as microorganisms, from the tissue fluid and by supporting the activities of certain cells (lymphocytes) that produce immunity by reacting against specific disease-causing agents. The lymphatic system is discussed in chapter 19.

The **urinary system** consists of the kidneys, ureters, urinary bladder, and urethra. The kidneys remove various wastes from the blood, and assist in maintaining the body's water and electrolyte balance. The product of these activities is urine. Other portions of the urinary system function in storing urine and transporting it to the outside of the body. The urinary system is discussed in chapter 20.

Sometimes the urinary system is called the *excretory system.* However, excretion, or waste removal, is also a function of the respiratory, digestive, and integumentary systems.

Reproduction

Reproduction is the process of producing offspring (progeny). Cells reproduce when they divide and give rise to new cells. The **reproductive system** of an organism, however, is involved with the production of whole new organisms like itself. (See chapter 22.)

The male reproductive system includes the scrotum, testes, epididymides, vasa deferentia, seminal vesicles, prostate gland, bulbourethral glands, urethra, and penis. These parts are concerned with producing and maintaining the male sex cells, or sperm cells (spermatozoa). They also function to transfer these cells from their site of origin into the female reproductive tract.

The female reproductive system consists of the ovaries, uterine tubes, uterus, vagina, clitoris, and vulva. These organs produce and maintain the female sex cells, or egg cells (ova), receive the male cells, and transport the male and female cells within the female system. The female reproductive system also provides for the support and development of embryos and functions in the birth process.

Figure 1.14 illustrates the male and female reproductive systems.

The organ systems, the major organs that comprise them, and their major functions are summarized in chart 1.4.

---

1. Name the major organ systems, and list the organs of each system.
2. Describe the general functions of each organ system.

Anatomical Terminology

To communicate effectively with one another, investigators over the ages have developed a set of terms with precise meanings. Some of these terms concern the relative positions of body parts, others refer to imaginary planes along which cuts may be made, and still others describe various body regions.

When such terms are used, it is assumed that the body is in the **anatomical position;** that is, it is standing erect, the face is forward, and the arms are at the sides, with the palms forward.

Relative Position

Terms of relative position are used to describe the location of one body part with respect to another. They include the following:

1. **Superior** means a part is above another part, or closer to the head. (The thoracic cavity is superior to the abdominopelvic cavity.)
2. **Inferior** means situated below another part, or toward the feet. (The neck is inferior to the head.)

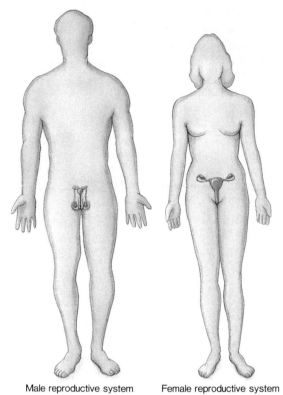

Male reproductive system   Female reproductive system

Figure 1.14   Organ systems associated with reproduction.

CHART 1.4	Organ systems	
Organ system	Major organs	Major functions
Integumentary	Skin, hair, nails, sweat glands, sebaceous glands	Protect tissues, regulate body temperature, support sensory receptors
Skeletal	Bones, ligaments, cartilages	Provide framework, protect soft tissues, provide attachments for muscles, produce blood cells, store inorganic salts
Muscular	Muscles	Cause movements, maintain posture, produce body heat
Nervous	Brain, spinal cord, nerves, sense organs	Detect changes, receive and interpret sensory information, stimulate muscles and glands
Endocrine	Glands that secrete hormones (pituitary gland, thyroid gland, parathyroid glands, adrenal glands, pancreas, ovaries, testes, pineal gland, and thymus gland)	Control metabolic activities of body structures
Digestive	Mouth, tongue, teeth, salivary glands, pharynx, esophagus, stomach, liver, gallbladder, pancreas, small and large intestines	Receive, break down, and absorb food; eliminate unabsorbed material
Respiratory	Nasal cavity, pharynx, larynx, trachea, bronchi, lungs	Intake and output of air, exchange of gases between air and blood
Circulatory	Heart, arteries, veins, capillaries	Move blood through blood vessels and transport substances throughout body
Lymphatic	Lymphatic vessels, lymph nodes, thymus, spleen	Return tissue fluid to the blood, carry certain absorbed food molecules, defend the body against infection
Urinary	Kidneys, ureters, urinary bladder, urethra	Remove wastes from blood, maintain water and electrolyte balance, store and transport urine
Reproductive	Male: scrotum, testes, epididymides, vasa deferentia, seminal vesicles, prostate gland, bulbourethral glands, urethra, penis	Produce and maintain sperm cells, transfer sperm cells into female reproductive tract
	Female: ovaries, uterine tubes, uterus, vagina, clitoris, vulva	Produce and maintain egg cells; receive and transport sperm cells; support development of an embryo and function in birth process

3. **Anterior** (or *ventral*) means toward the front. (The eyes are anterior to the brain.)

4. **Posterior** (or *dorsal*) is the opposite of anterior; it means toward the back. (The pharynx is posterior to the oral cavity.)

5. **Medial** relates to an imaginary midline dividing the body into equal right and left halves. A part is medial if it is closer to this line than another part. (The nose is medial to the eyes.)

6. **Lateral** means toward the side with respect to the imaginary midline. (The ears are lateral to the eyes.) **Ipsilateral** pertains to the same side (the spleen and the descending colon are ipsilateral), while **contralateral** refers to the opposite side (the spleen and the gallbladder are contralateral).

7. **Proximal** is used to describe a part that is closer to a point of attachment or closer to the trunk of the body than another part. (The elbow is proximal to the wrist.)

8. **Distal** is the opposite of proximal. It means a particular body part is farther from the point of attachment or farther from the trunk than another part. (The fingers are distal to the wrist.)

9. **Superficial** means situated near the surface. (The epidermis is the superficial layer of the skin.) **Peripheral** also means outward or near the surface. It is used to describe the location of certain blood vessels and nerves. (The nerves that branch from the brain and spinal cord are peripheral nerves.)

10. **Deep** is used to describe parts that are more internal. (The dermis is the deep layer of the skin.)

Body Sections

To observe the relative locations and arrangements of internal parts, it is necessary to cut or section the body along various planes (figures 1.15 and 1.16). The following terms are used to describe such planes and sections.

1. **Sagittal** refers to a lengthwise cut that divides the body into right and left portions. If a sagittal section passes along the midline and divides the body into equal parts, it is called *median* (midsagittal).

2. **Transverse** (or horizontal section) refers to a cut that divides the body into superior and inferior portions.

3. **Frontal** (or *coronal*) refers to a section that divides the body into anterior and posterior portions.

Sometimes a cylindrical organ such as a blood vessel is sectioned. In this case, a cut across the structure is called a *cross section,* an angular cut is called an *oblique section,* and a lengthwise cut is called a *longitudinal section* (figure 1.17).

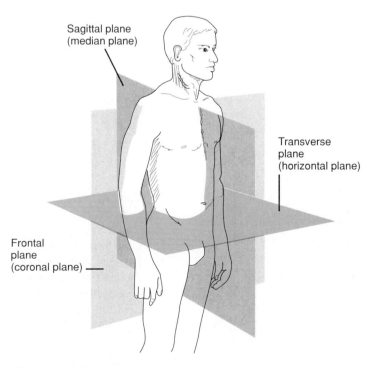

Figure 1.15 To observe internal parts, the body may be sectioned along various planes.

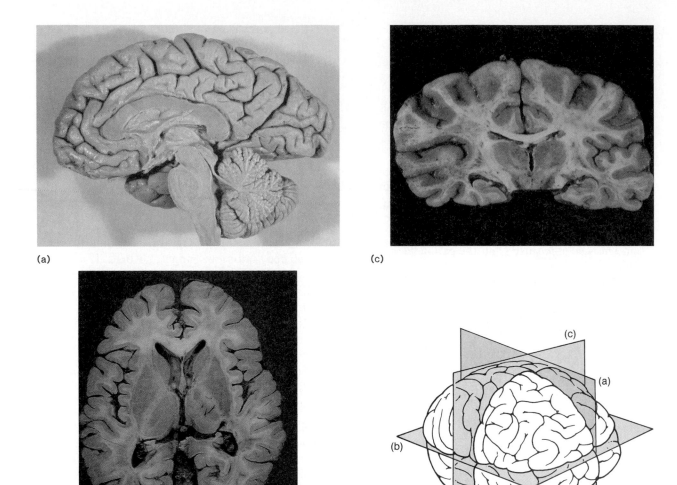

Figure 1.16 A human brain sectioned along (*a*) the sagittal plane, (*b*) the transverse plane, and (*c*) the frontal plane.

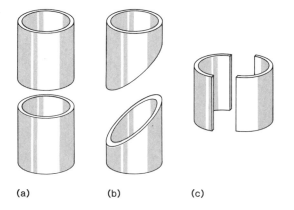

Figure 1.17 Cylindrical parts may be cut in (*a*) cross section, (*b*) oblique section, or (*c*) longitudinal section.

Body Regions

A number of terms are used to designate various body regions. The abdominal area, for example, is subdivided into the following nine regions, as shown in figure 1.18:

1. **Epigastric region** The upper middle portion.
2. **Left** and **right hypochondriac regions** On each side of the epigastric region.
3. **Umbilical region** The central portion.
4. **Left** and **right lumbar regions** On each side of the umbilical region.
5. **Hypogastric region** The lower middle portion.
6. **Left** and **right iliac (or inguinal) regions** On each side of the hypogastric region.

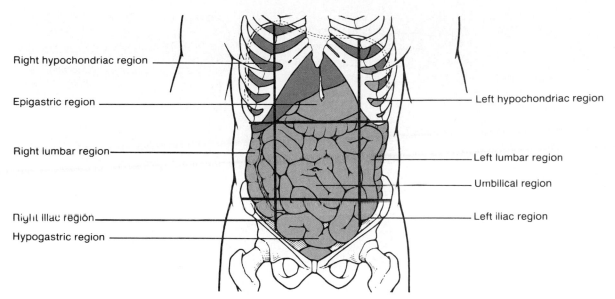

Right hypochondriac region

Epigastric region

Right lumbar region

Right iliac region

Hypogastric region

Left hypochondriac region

Left lumbar region

Umbilical region

Left iliac region

Figure 1.18 The abdominal area is subdivided into nine regions. How do the names of these regions describe their locations?

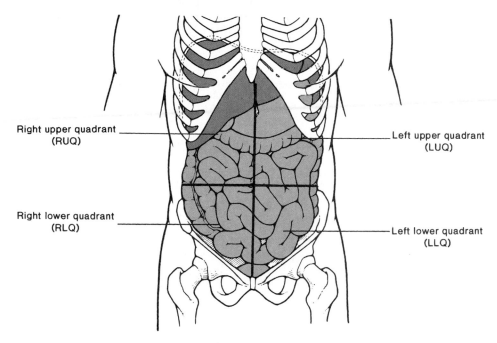

Right upper quadrant (RUQ)

Right lower quadrant (RLQ)

Left upper quadrant (LUQ)

Left lower quadrant (LLQ)

Figure 1.19 The abdominal area may be subdivided into four quadrants.

The abdominal area also may be subdivided into the following four *quadrants,* as illustrated in figure 1.19:

1. **Right upper quadrant** (RUQ).
2. **Right lower quadrant** (RLQ).
3. **Left upper quadrant** (LUQ).
4. **Left lower quadrant** (LLQ).

The following terms are commonly used when referring to various body regions: (Figure 1.20 illustrates some of these regions.)

abdominal (ab-dom′ĭ-nal)   the region between the thorax and pelvis.
acromial (ah-kro′me-al)   the point of the shoulder.

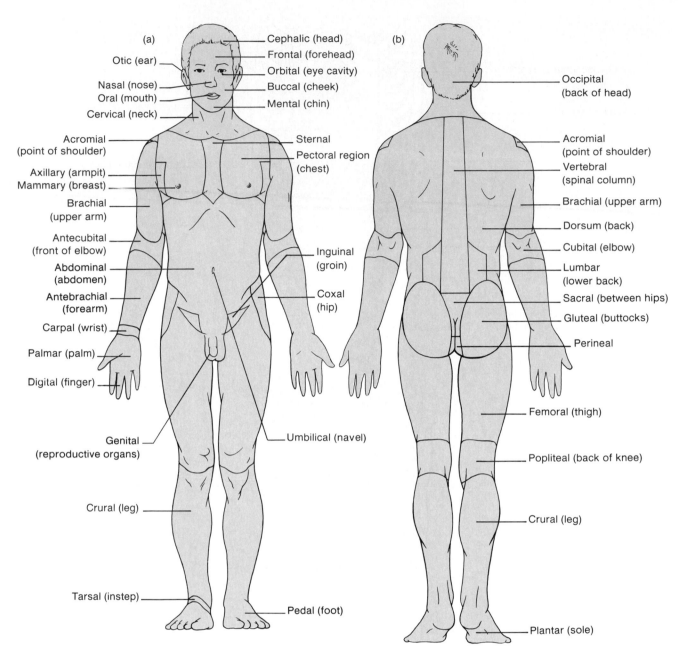

Figure 1.20 Some terms used to describe body regions. (a) Anterior regions; (b) posterior regions.

antebrachial (an″te-bra′ke-al) the forearm.

antecubital (an″te-ku′bĭ-tal) the space in front of the elbow.

axillary (ak′sĭ-ler″e) the armpit.

brachial (bra′ke-al) the upper arm.

buccal (buk′al) the cheek.

carpal (kar′pal) the wrist.

celiac (se′le-ak) the abdomen.

cephalic (sĕ-fal′ik) the head.

cervical (ser′vĭ-kal) the neck.

costal (kos′tal) the ribs.

coxal (kok′sal) the hip.

crural (krōōr′al) the leg.

cubital (ku′bĭ-tal) the elbow.

digital (dij′ĭ-tal) the finger.

dorsum (dor′sum) the back.

femoral (fem′or-al) the thigh.

frontal (frun′tal) the forehead.

genital (jen′i-tal) the reproductive organs.

gluteal (gloo′te-al) the buttocks.

inguinal (ing′gwĭ-nal) the depressed area of the abdominal wall near the thigh (groin).

lumbar (lum′bar) the region of the lower back between the ribs and the pelvis (loin).

mammary (mam′er-e) the breast.

mental (men′tal) the chin.

Ultrasonography and Magnetic Resonance Imaging

Often noninvasive procedures that allow internal organs to be visualized are employed to help diagnose abnormal body conditions. Two such procedures are ultrasonography and magnetic resonance imaging.

Ultrasonography makes use of high frequency sound waves—those that are beyond the range of human hearing. In this procedure, a transducer that emits sound waves is pressed firmly against the skin and moved over the surface of the region being examined. (See figure 1.21.)

The sound waves travel into the body, and when they reach a border (interface) between structures of slightly different densities, some of the waves are reflected back to the transducer. Other sound waves continue into deeper tissues, and some of them are reflected back by still other interfaces.

As the reflected sound waves reach the transducer, they are converted into electrical impulses that are amplified and used to create a sectional image of the body's internal structure on a viewing screen.

Ultrasonography is usually not used to examine very compact organs (such as bones) or those containing air-filled spaces (such as lungs), because the resulting images are of poor quality. However, useful images of medium density organs often can be obtained, making it possible to visualize structures of the heart, abnormal masses of tissues, accumulations of fluids, or various types of stones. It also may be used to study a fetus in the uterus. (See figure 1.22.)

In the procedure called **magnetic resonance imaging** (MRI), the body or part being examined is placed in a chamber surrounded by a powerful magnet and a special radio antenna. When the device is operating, the magnetic field created by the magnet affects the alignment and spin of certain types of atoms within the living materials. At the same time, a second rotating magnetic field is adjusted to cause particular kinds of atoms (such as the hydrogen atoms in body fluids and organic compounds) to release weak radio waves with characteristic frequencies. The radio waves are received by the nearby antenna and amplified. The amplified signals are then processed by a computer, and within a few minutes, the computer generates a sectional image that is based upon the locations and concentrations of the particular atoms being investigated. (See figure 1.23.)

In addition to visualizing anatomical structures, MRI can provide information concerning the physiology of body parts by revealing something about their chemical composition. Also, MRI can be used to obtain sectional views in any plane, so that the head, for example, can be viewed in transverse, coronal, or sagittal section.

MRI is particularly useful for studying soft tissues, such as those of the brain and spinal cord. It is also used for distinguishing between normal and cancerous tissues, depicting blood vessels, and assessing damage sustained by the heart muscle as a result of a heart attack.

Figure 1.21 Ultrasonography uses reflected sound waves to visualize internal body structures.

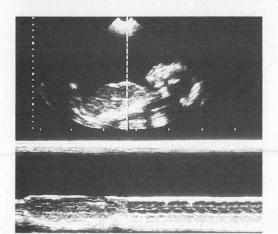

Figure 1.22 This image resulting from an ultrasonographic procedure reveals the presence of a fetus in the uterus.

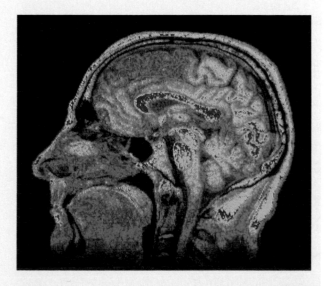

Figure 1.23 Falsely colored MRI image of a human head and brain (sagittal section).

nasal (na'zal)   the nose.

occipital (ok-sip'ĭ-tal)   the lower posterior region of the head.

oral (o'ral)   the mouth.

orbital (or'bi-tal)   the eye cavity.

otic (o'tik)   the ear.

palmar (pahl'mar)   the palm of the hand.

pectoral (pek'tor-al)   the chest.

pedal (ped'al)   the foot.

pelvic (pel'vik)   the pelvis.

perineal (per''ĭ-ne'al)   the region between the anus and the external reproductive organs (perineum).

plantar (plan'tar)   the sole of the foot.

popliteal (pop''lĭ-te'al)   the area behind the knee.

sacral (sa'kral)   the posterior region between the hipbones.

sternal (ster'nal)   the middle of the thorax, anteriorly.

tarsal (tahr'sal)   the instep of the foot.

umbilical (um-bil'ĭ-kal)   the navel.

vertebral (ver'te-bral)   the spinal column.

1. Describe the anatomical position.
2. Using the appropriate terms, describe the relative positions of several body parts.
3. Describe three types of sections.
4. Describe the nine regions of the abdomen.

Some Medical and Applied Sciences

cardiology (kar''de-ol'o-je)   Branch of medical science dealing with the heart and heart diseases.

cytology (si-tol'o-je)   Study of the structure, function, and diseases of cells.

dermatology (der''mah-tol'o-je)   Study of skin and its diseases.

endocrinology (en''do-kri-nol'o-je)   Study of hormones, hormone-secreting glands, and the diseases involving them.

epidemiology (ep''ĭ-de''me-ol'o-je)   Study of the factors determining the distribution and frequency of the occurrence of various health-related conditions within a defined human population.

gastroenterology (gas''tro-en''ter-ol'o-je)   Study of the stomach and intestines, as well as the diseases involving them.

geriatrics (jer''e-at'riks)   Branch of medicine dealing with elderly persons and their medical problems.

gerontology (jer''on-tol'o-je)   Study of the process of aging and the various problems of elderly persons.

gynecology (gi''nĕ-kol'o-je)   Study of the female reproductive system and its diseases.

hematology (hem''ah-tol'o-je)   Study of blood and blood diseases.

histology (his-tol'o-je)   Study of the structure and function of tissues.

immunology (im''u-nol'o-je)   Study of the body's resistance to disease.

neonatology (ne''o-na-tol'o-je)   Study of newborn infants and the treatment of their disorders.

nephrology (nĕ-frol'o-je)   Study of the structure, function, and diseases of the kidneys.

neurology (nu-rol'o-je)   Study of the nervous system in health and disease.

obstetrics (ob-stet'riks)   Branch of medicine dealing with pregnancy and childbirth.

oncology (ong-kol'o-je)   Study of tumors.

ophthalmology (of''thal-mol'o-je)   Study of the eye and eye diseases.

orthopedics (or''tho-pe'diks)   Branch of medicine dealing with the muscular and skeletal systems, and their problems.

otolaryngology (o''to-lar''in-gol'o-je)   Study of the ear, throat, larynx, and diseases of these parts.

pathology (pah-thol'o-je)   Study of structural and functional changes within the body produced by diseases.

pediatrics (pe''de-at'riks)   Branch of medicine dealing with children and their diseases.

pharmacology (fahr''mah-kol'o-je)   Study of drugs and their uses in the treatment of diseases.

podiatry (po-di'ah-tre)   Study of the care and treatment of the feet.

psychiatry (si-ki'ah-tre)   Branch of medicine dealing with the mind and its disorders.

radiology (ra''de-ol'o-je)   Study of X rays and radioactive substances, as well as their uses in diagnosing and treating diseases.

toxicology (tok''si-kol'o-je)   Study of poisonous substances and their effects upon body parts.

urology (u-rol'o-je)   Branch of medicine dealing with the urinary system, male reproductive system, and diseases of these systems.

Chapter Summary

Introduction (page 6)

1. Early interest in the human body probably developed as people became concerned about injuries and illnesses.
2. Primitive doctors began to learn how certain herbs and potions affected body functions.
3. In ancient times, it was believed that natural processes were caused by spirits and supernatural forces.
4. The belief that humans could understand forces that caused natural events led to the development of modern science.
5. A set of terms originating from Greek and Latin formed the basis for the language of anatomy and physiology.

Anatomy and Physiology (page 6)

1. Anatomy deals with the form and arrangement of body parts.
2. Physiology deals with the functions of these parts.
3. The function of a part depends upon the way it is constructed.

Characteristics of Life (page 7)

Characteristics of life are traits shared by all organisms.

1. These characteristics include:
    a. Movement—changing body position or moving internal parts.
    b. Responsiveness—sensing and reacting to internal or external changes.
    c. Growth—increasing in size without changing in shape.
    d. Reproduction—producing offspring.
    e. Respiration—obtaining oxygen, using oxygen to release energy from foods, and removing gaseous wastes.
    f. Digestion—changing food substances into forms that can be absorbed.
    g. Absorption—moving substances through membranes and into body fluids.
    h. Circulation—moving substances through the body in body fluids.
    i. Assimilation—changing substances into chemically different forms.
    j. Excretion—removing body wastes.
2. Together these activities constitute metabolism.

Maintenance of Life (page 8)

The structures and functions of body parts are directed toward maintaining the life of the organism.

1. Needs of organisms
    a. Water is used in a variety of metabolic processes, provides the environment for metabolic reactions, and transports substances.
    b. Food is needed to supply energy, to provide raw materials for building new living matter, and to supply chemicals necessary in vital reactions.
    c. Oxygen is used in releasing energy from food materials; this energy drives metabolic reactions.
    d. Heat is a product of metabolic reactions and helps govern the rates of these reactions.
    e. Pressure is an application of force to something; in humans, atmospheric and hydrostatic pressures help breathing and blood movements, respectively.
    f. Survival of an organism depends upon the quantities and qualities of these factors.
2. Homeostasis
    a. If an organism is to survive, the conditions within its body fluids must remain relatively stable.
    b. The tendency to maintain a stable internal environment is called homeostasis.
    c. Homeostatic mechanisms regulate body temperature, blood pressure, and blood sugar concentration.
    d. Homeostatic mechanisms employ negative feedback.

Levels of Organization (page 11)

The body is composed of parts that occupy different levels of organization.

1. Material substances are composed of atoms.
2. Atoms join together to form molecules.
3. Organelles contain aggregates of large molecules.
4. Cells, which are composed of organelles, are the basic units of structure and function within the body.
5. Cells are organized into layers or masses called tissues.
6. Tissues are organized into organs.
7. Organs are arranged into organ systems.
8. Organ systems constitute the organism.
9. These parts vary in complexity progressively from one level to the next.

Organization of the Human Body (page 12)

1. Body cavities
    a. The axial portion of the body contains the dorsal and ventral cavities.
        (1) The dorsal cavity includes the cranial cavity and vertebral canal.
        (2) The ventral cavity includes the thoracic and abdominopelvic cavities, which are separated by the diaphragm.
    b. The organs within a body cavity are called visceral organs.
    c. Other body cavities include the oral, nasal, orbital, and middle ear cavities.
2. Thoracic and abdominopelvic membranes
    a. Thoracic membranes
        (1) Pleural membranes line the thoracic cavity and cover the lungs.
        (2) Pericardial membranes surround the heart and cover its surface.
        (3) The pleural and pericardial cavities are potential spaces between these membranes.
    b. Abdominopelvic membranes
        (1) Peritoneal membranes line the abdominopelvic cavity and cover the organs inside.
        (2) The peritoneal cavity is a potential space between these membranes.
3. Organ systems
    The human organism consists of several organ systems. Each system includes a set of interrelated organs.
    a. Integumentary system
        (1) The integumentary system provides the body covering.
        (2) It includes the skin, hair, nails, sweat glands, and sebaceous glands.
        (3) It functions to protect underlying tissues, regulate body temperature, house sensory receptors, and synthesize various substances.

b. Skeletal system
   (1) The skeletal system is composed of bones and the ligaments and cartilages that bind bones together.
   (2) It provides framework, protective shields, and attachments for muscles; it also produces blood cells and stores inorganic salts.
c. Muscular system
   (1) The muscular system includes the muscles of the body.
   (2) It is responsible for body movements, maintenance of posture, and production of body heat.
d. Nervous system
   (1) The nervous system consists of the brain, spinal cord, nerves, and sense organs.
   (2) It functions to receive impulses from sensory parts, interpret these impulses, and act on them by causing muscles or glands to respond.
e. Endocrine system
   (1) The endocrine system consists of glands that secrete hormones.
   (2) Hormones help regulate metabolism by stimulating target tissues.
   (3) It includes the pituitary, thyroid, parathyroid, and adrenal glands; the pancreas, ovaries, testes, pineal gland, and thymus gland.
f. Digestive system
   (1) The digestive system receives foods, converts food molecules into forms that can pass through cell membranes, and eliminates the materials that are not absorbed.
   (2) Some digestive organs produce hormones.
   (3) It includes the mouth, tongue, teeth, salivary glands, pharynx, esophagus, stomach, liver, gallbladder, pancreas, small intestine, and large intestine.
g. Respiratory system
   (1) The respiratory system provides for the intake and output of air, and for the exchange of gases between the blood and the air.
   (2) It includes the nasal cavity, pharynx, larynx, trachea, bronchi, and lungs.
h. Circulatory system
   (1) The circulatory system includes the heart, which pumps blood, and the blood vessels, which carry blood to and from body parts.
   (2) Blood transports oxygen, nutrients, hormones, and wastes.
i. Lymphatic system
   (1) The lymphatic system is composed of lymphatic vessels, lymph nodes, thymus, and spleen.
   (2) It transports lymph from tissue spaces to the bloodstream, carries certain fatty substances away from the digestive organs, and aids in defending the body against disease-causing agents.
j. Urinary system
   (1) The urinary system includes the kidneys, ureters, urinary bladder, and urethra.
   (2) It filters wastes from the blood and helps maintain fluid and electrolyte balance.
k. Reproductive systems
   (1) The reproductive systems are concerned with the production of new organisms.
   (2) The male reproductive system includes the scrotum, testes, epididymides, vasa deferentia, seminal vesicles, prostate gland, bulbourethral glands, urethra, and penis, which produce, maintain, and transport male sex cells.
   (3) The female reproductive system includes the ovaries, uterine tubes, uterus, vagina, clitoris, and vulva, which produce, maintain, and transport female sex cells.

Anatomical Terminology (page 20)

Terms with precise meanings are used to help investigators communicate effectively with one another.

1. Relative position
   These terms are used to describe the location of one part with respect to another part.
2. Body sections
   Body sections are planes along which the body may be cut to observe the relative locations and arrangements of internal parts.
3. Body regions
   Various body regions are designated by special terms.

Clinical Application of Knowledge

1. In many states, death is defined as "irreversible cessation of total brain function." How is death defined in your state? How is this definition related to the characteristics of life?
2. In health, the body parts function together effectively to maintain homeostasis. In illness, the maintenance of homeostasis may be threatened, and various treatments may be needed to reduce this threat. What treatments might be used to help control a patient's (a) body temperature, (b) blood oxygen concentration, and (c) water content?
3. Suppose two individuals are afflicted with benign (noncancerous) tumors that produce symptoms because they occupy space and crowd adjacent organs. If one of these persons has the tumor in the ventral cavity and the other has a tumor in the dorsal cavity, which would be likely to develop symptoms first? Why?

4. If a patient complained of a "stomachache" and pointed to the umbilical region as the site of the discomfort, what organs located in this region might be the source of the pain?
5. How could the basic needs of a human be provided for a patient who is unconscious?
6. Assuming that the same information could be obtained by either method, what would be the advantage of using ultrasonography rather than X rays to visualize a fetus in the uterus?

Review Activities

Part A

1. Briefly describe the early development of knowledge about the human body.
2. Distinguish between the activities of anatomists and physiologists.
3. Explain the relationship between the form and function of human body parts.
4. List and describe ten characteristics of life.
5. Define *metabolism*.
6. List and describe five needs of organisms.
7. Explain how the idea of homeostasis is related to the five needs you listed in item 6.
8. Distinguish between heat and temperature.
9. Define two types of pressures that may act upon the outsides of organisms.
10. Explain how body temperature, blood pressure, and blood sugar concentration are controlled.
11. Explain why homeostatic mechanisms are said to act by negative feedback.
12. Explain what is meant by *levels of organization*.
13. List the levels of organization within the human.
14. Distinguish between the axial and appendicular portions of the body.
15. Distinguish between the dorsal and ventral body cavities, and name the smaller cavities that occur within each.
16. Explain what is meant by a *visceral organ*.
17. Describe the mediastinum.
18. Describe the locations of the oral, nasal, orbital, and middle ear cavities.
19. Distinguish between a parietal membrane and a visceral membrane.
20. Name the major organ systems, and describe the general functions of each.
21. List the major organs that comprise each organ system.

Part B

1. Name the body cavity in which each of the following organs is located:
   a. Stomach   f. Rectum
   b. Heart   g. Spinal cord
   c. Brain   h. Esophagus
   d. Liver   i. Spleen
   e. Trachea   j. Urinary bladder

2. Write complete sentences using each of the following terms correctly:
   a. Superior   h. Contralateral
   b. Inferior   i. Proximal
   c. Anterior   j. Distal
   d. Posterior   k. Superficial
   e. Medial   l. Peripheral
   f. Lateral   m. Deep
   g. Ipsilateral

3. Prepare a sketch of a human body, and use lines to indicate each of the following sections:
   a. Sagittal   c. Frontal
   b. Transverse

4. Prepare a sketch of the abdominal area and indicate the location of each of the following regions:
   a. Epigastric   d. Hypochondriac
   b. Umbilical   e. Lumbar
   c. Hypogastric   f. Iliac

5. Prepare a sketch of the abdominal area and indicate the location of each of the following regions:
   a. Right upper quadrant
   b. Right lower quadrant
   c. Left upper quadrant
   d. Left lower quadrant

6. Provide the common name for the region to which each of the following terms refers:
   a. Acromial   n. Orbital
   b. Antebrachial   o. Otic
   c. Axillary   p. Palmar
   d. Buccal   q. Pectoral
   e. Celiac   r. Pedal
   f. Coxal   s. Perineal
   g. Crural   t. Plantar
   h. Femoral   u. Popliteal
   i. Genital   v. Sacral
   j. Gluteal   w. Sternal
   k. Inguinal   x. Tarsal
   l. Mental   y. Umbilical
   m. Occipital   z. Vertebral

Reference Plates

The Human Organism

*T*he following series of illustrations shows the major organs of the human torso. The first plate illustrates the anterior surface and reveals the superficial muscles on one side. Each subsequent plate exposes deeper organs, including those in the thoracic, abdominal, and pelvic cavities.

The purpose of chapters 6–22 of this textbook is to describe the organ systems of the human organisms in detail. As you read them, you may want to refer to these plates to help yourself visualize the locations of various organs and the three-dimensional relationships that exist among them.

You may also want to study the photographs of human cadavers in the reference plates that follow chapter 24. These photographs illustrate many of the larger organs of the human body.

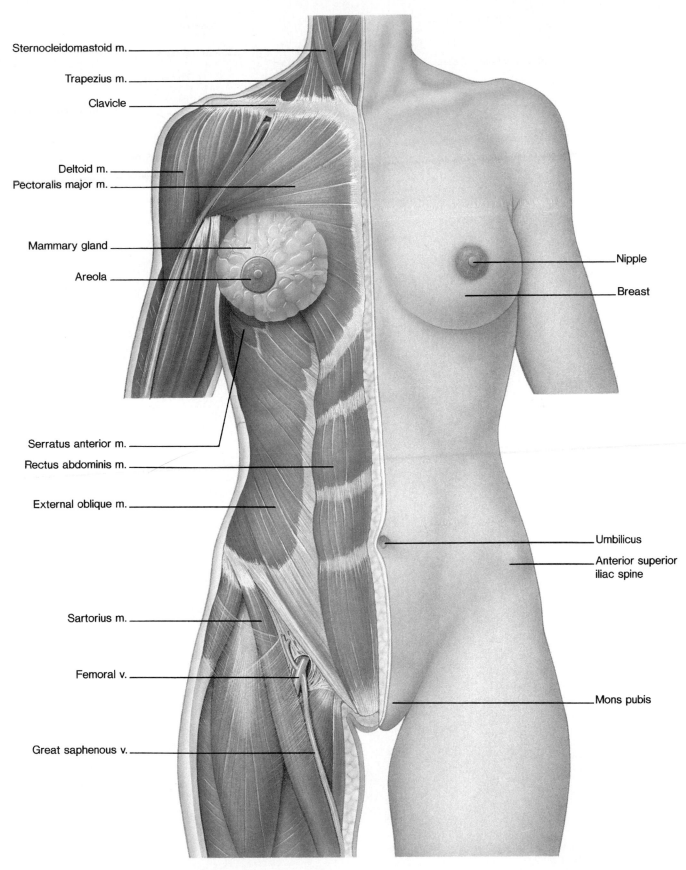

Sternocleidomastoid m.

Trapezius m.

Clavicle

Deltoid m.

Pectoralis major m.

Mammary gland

Areola

Serratus anterior m.

Rectus abdominis m.

External oblique m.

Sartorius m.

Femoral v.

Great saphenous v.

Nipple

Breast

Umbilicus

Anterior superior
iliac spine

Mons pubis

Plate 1 Human female torso, showing the anterior surface on
one side and the superficial muscles exposed on the other side.
(*m.* stands for *muscle*, and *v.* stands for *vein*.)

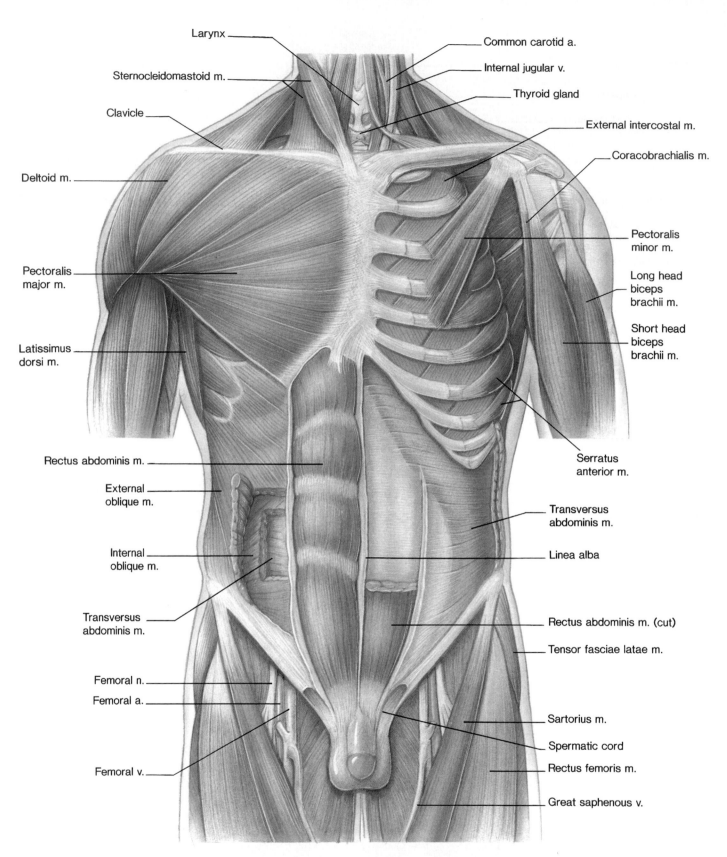

Larynx

Common carotid a.

Internal jugular v.

Sternocleidomastoid m.

Thyroid gland

Clavicle

External intercostal m.

Coracobrachialis m.

Deltoid m.

Pectoralis minor m.

Long head biceps brachii m.

Pectoralis major m.

Short head biceps brachii m.

Latissimus dorsi m.

Serratus anterior m.

Rectus abdominis m.

External oblique m.

Transversus abdominis m.

Internal oblique m.

Linea alba

Transversus abdominis m.

Rectus abdominis m. (cut)

Tensor fasciae latae m.

Femoral n.

Femoral a.

Sartorius m.

Spermatic cord

Rectus femoris m.

Femoral v.

Great saphenous v.

Plate 2   Human male torso, with the deeper muscle layers exposed. (*n.* stands for *nerve.*)

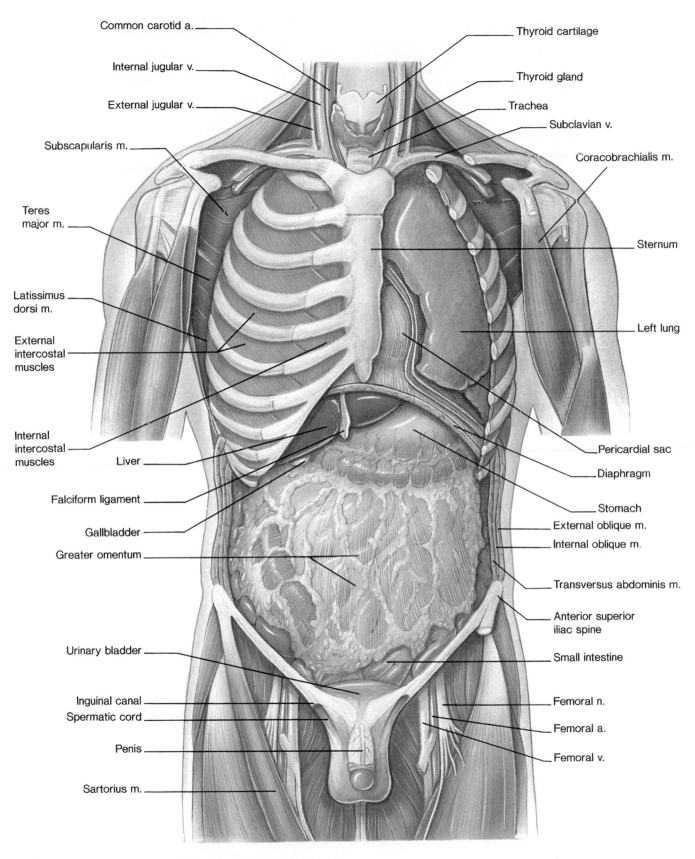

Common carotid a.

Internal jugular v.

External jugular v.

Subscapularis m.

Teres major m.

Latissimus dorsi m.

External intercostal muscles

Internal intercostal muscles

Liver

Falciform ligament

Gallbladder

Greater omentum

Urinary bladder

Inguinal canal

Spermatic cord

Penis

Sartorius m.

Thyroid cartilage

Thyroid gland

Trachea

Subclavian v.

Coracobrachialis m.

Sternum

Left lung

Pericardial sac

Diaphragm

Stomach

External oblique m.

Internal oblique m.

Transversus abdominis m.

Anterior superior iliac spine

Small intestine

Femoral n.

Femoral a.

Femoral v.

Plate 3   Human male torso, with the deep muscles removed and the abdominal viscera exposed. (*a.* stands for *artery.*)

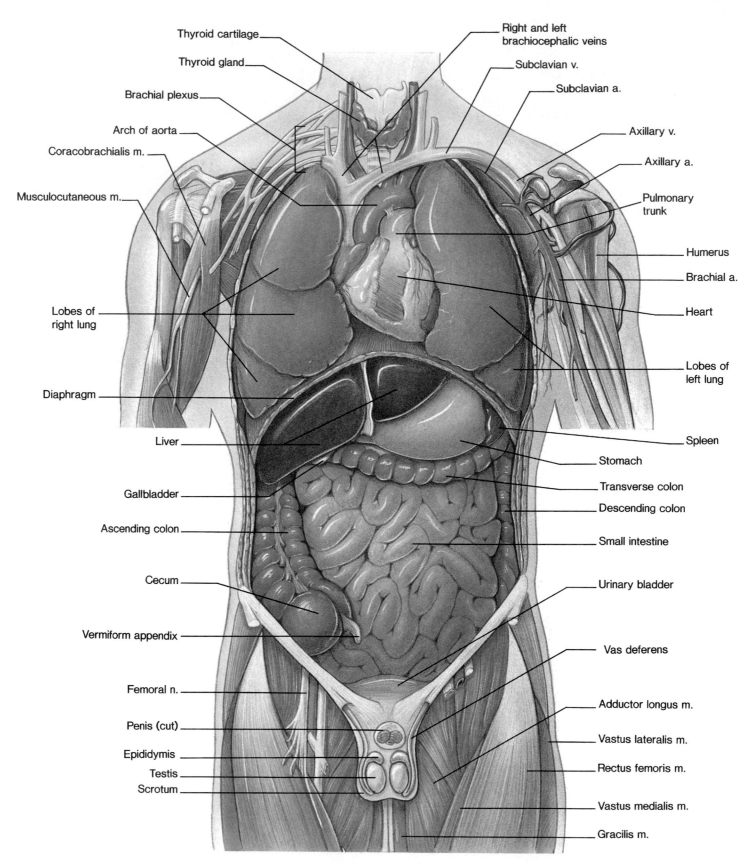

Thyroid cartilage

Thyroid gland

Brachial plexus

Arch of aorta

Coracobrachialis m.

Musculocutaneous m.

Lobes of right lung

Diaphragm

Liver

Gallbladder

Ascending colon

Cecum

Vermiform appendix

Femoral n.

Penis (cut)

Epididymis

Testis

Scrotum

Right and left brachiocephalic veins

Subclavian v.

Subclavian a.

Axillary v.

Axillary a.

Pulmonary trunk

Humerus

Brachial a.

Heart

Lobes of left lung

Spleen

Stomach

Transverse colon

Descending colon

Small intestine

Urinary bladder

Vas deferens

Adductor longus m.

Vastus lateralis m.

Rectus femoris m.

Vastus medialis m.

Gracilis m.

Plate 4 Human male torso, with the thoracic and abdominal viscera exposed.

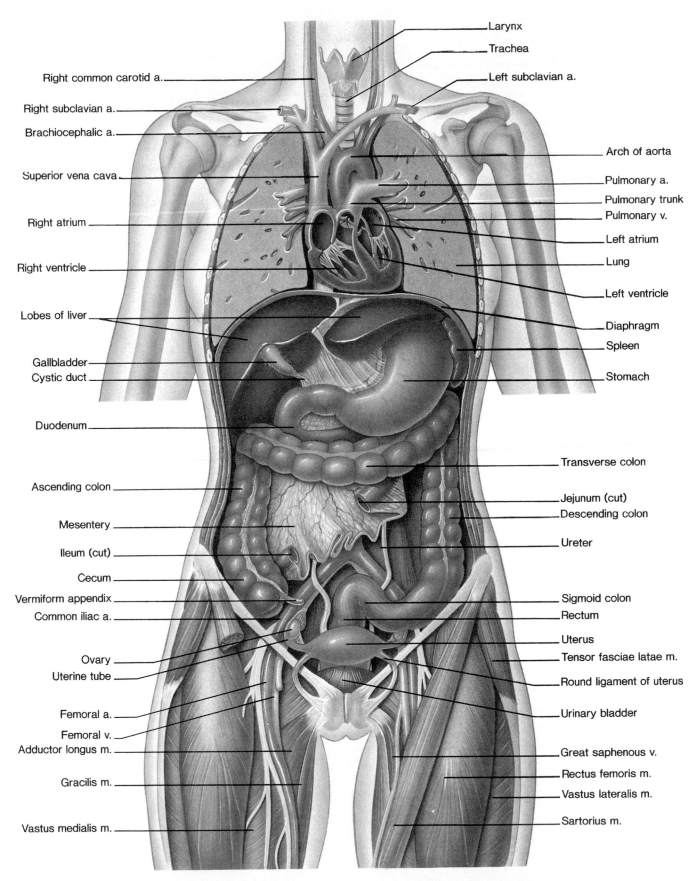

Larynx

Trachea

Right common carotid a.

Left subclavian a.

Right subclavian a.

Brachiocephalic a.

Arch of aorta

Superior vena cava

Pulmonary a.

Pulmonary trunk

Pulmonary v.

Right atrium

Left atrium

Lung

Right ventricle

Left ventricle

Lobes of liver

Diaphragm

Spleen

Gallbladder

Cystic duct

Stomach

Duodenum

Transverse colon

Ascending colon

Jejunum (cut)

Descending colon

Mesentery

Ureter

Ileum (cut)

Cecum

Vermiform appendix

Sigmoid colon

Common iliac a.

Rectum

Uterus

Ovary

Tensor fasciae latae m.

Uterine tube

Round ligament of uterus

Femoral a.

Urinary bladder

Femoral v.

Adductor longus m.

Great saphenous v.

Gracilis m.

Rectus femoris m.

Vastus lateralis m.

Vastus medialis m.

Sartorius m.

Plate 5   Human female torso, with the lungs, heart, and small
intestine sectioned.

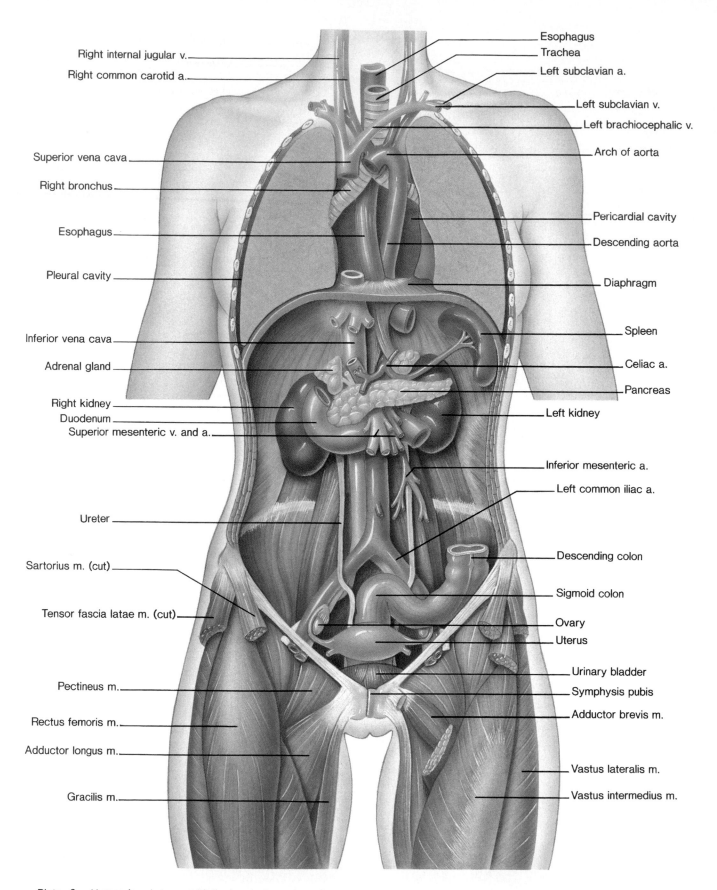

Right internal jugular v.

Right common carotid a.

Superior vena cava

Right bronchus

Esophagus

Pleural cavity

Inferior vena cava

Adrenal gland

Right kidney

Duodenum

Superior mesenteric v. and a.

Ureter

Sartorius m. (cut)

Tensor fascia latae m. (cut)

Pectineus m.

Rectus femoris m.

Adductor longus m.

Gracilis m.

Esophagus

Trachea

Left subclavian a.

Left subclavian v.

Left brachiocephalic v.

Arch of aorta

Pericardial cavity

Descending aorta

Diaphragm

Spleen

Celiac a.

Pancreas

Left kidney

Inferior mesenteric a.

Left common iliac a.

Descending colon

Sigmoid colon

Ovary

Uterus

Urinary bladder

Symphysis pubis

Adductor brevis m.

Vastus lateralis m.

Vastus intermedius m.

Plate 6 Human female torso, with the heart, stomach, and parts of the intestine and lungs removed.

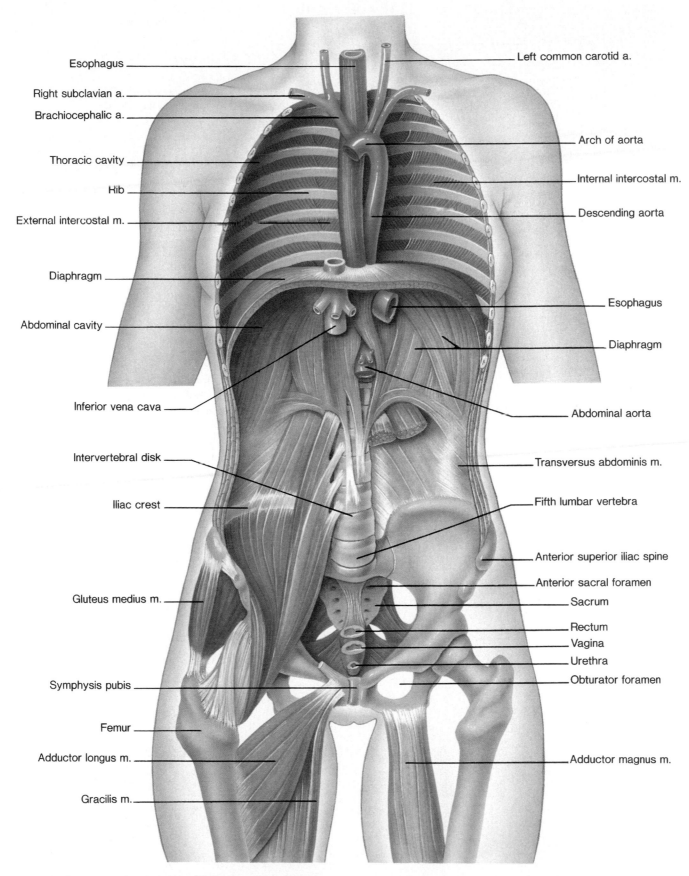

Esophagus

Right subclavian a.

Brachiocephalic a.

Thoracic cavity

Rib

External intercostal m.

Diaphragm

Abdominal cavity

Inferior vena cava

Intervertebral disk

Iliac crest

Gluteus medius m.

Symphysis pubis

Femur

Adductor longus m.

Gracilis m.

Left common carotid a.

Arch of aorta

Internal intercostal m.

Descending aorta

Esophagus

Diaphragm

Abdominal aorta

Transversus abdominis m.

Fifth lumbar vertebra

Anterior superior iliac spine

Anterior sacral foramen

Sacrum

Rectum

Vagina

Urethra

Obturator foramen

Adductor magnus m.

Plate 7 Human female torso, with the thoracic, abdominal, and pelvic visceral organs removed.

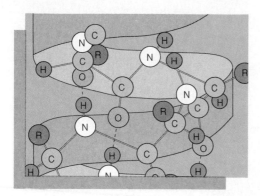

Chemical Basis of Life

*A*s chapter 1 explained, an organism is composed of parts that vary in level of organization. These parts include organ systems, organs, tissues, cells, and cellular organelles.

Organelles are, in turn, composed of chemical substances made up of atoms and groups of atoms called molecules. Similarly, atoms are composed of even smaller units.

A study of living material at the more fundamental levels of organization must be concerned with the chemical substances that form the structural basis of all matter and interact in the metabolic processes of organisms. This is the theme of chapter 2 ■

Chapter Objectives

After you have studied this chapter, you should be able to:

1. Explain how the study of living material is dependent on the study of chemistry.

2. Describe the relationships between matter, atoms, and molecules.

3. Discuss how atomic structure is related to the ways in which atoms interact.

4. Explain how molecular and structural formulas are used to symbolize the composition of compounds.

5. Describe three types of chemical reactions.

6. Discuss the concept of pH.

7. List the major groups of inorganic substances that are common in cells.

8. Describe the general roles played in cells by various types of organic substances.

9. Complete the review activities at the end of this chapter. Note that the items are worded in the form of specific learning objectives. You may want to refer to them before reading the chapter.

Key Terms

atom (at'om)

carbohydrate (kar''bo-hi'drāt)

decomposition (de''kom-po-zish'un)

electrolyte (e-lek'tro-līt)

formula (fōr'mu-lah)

inorganic (in''or-gan'ik)

ion (i'on)

isotope (i'so-tōp)

lipid (lip'id)

molecule (mol'ĕ-kūl)

nucleic acid (nu-kle'ik as'id)

organic (or-gan'ik)

protein (pro'te-in)

radiation (ra''de-a'shun)

synthesis (sin'thĕ-sis)

Aids to Understanding Words

bio-, life: *bio*chemistry—branch of science dealing with the chemistry of life forms.

di-, two: *di*saccharide—a compound whose molecules are composed of two saccharide units bound together.

glyc-, sweet: *glyc*ogen—a complex carbohydrate composed of sugar molecules bound together.

iso-, equal: *iso*tope—atom that has the same atomic number as another atom but a different atomic weight.

lip-, fat: *lip*ids—group of organic compounds that includes fats.

-lyt, dissolvable: electro*lyte*—substance that dissolves in water and releases ions.

mono-, one: *mono*saccharide—a compound whose molecule consists of a single saccharide unit.

nucle-, kernel: *nucle*us—central core of an atom.

poly-, many: *poly*unsaturated—molecule that has many double bonds between its carbon atoms.

sacchar-, sugar: mono*sacchar*ide—sugar molecule composed of a single saccharide unit.

syn-, together: *syn*thesis—process by which substances are united to form a new type of substance.

-valent, having power: electro*valent* bond—chemical bond produced by the attraction between two ions with opposite electrical charges.

Chemistry is the branch of science dealing with the composition of substances and the changes that take place in their composition. Although it is possible to study anatomy without much reference to this subject, knowledge of chemistry is essential for understanding physiology, because body functions involve chemical changes that occur within cells.

As interest in the chemistry of living organisms grew, and knowledge of the subject expanded, a new subdivision of science called biological chemistry or *biochemistry* emerged.

Biochemical studies have been important not only in helping explain physiological processes, but also in making possible the development of many new drugs and methods for treating diseases.

1. Why is a knowledge of chemistry essential to an understanding of physiology?
2. What is biochemistry?

Structure of Matter

Matter is anything that has weight and takes up space. This includes all the solids, liquids, and gases in our surroundings as well as in our bodies.

Elements and Atoms

Studies of matter have revealed that all things are composed of basic substances called **elements.** At present, 108 such elements are known, although naturally occurring matter on earth includes only ninety-one of them. Among these elements are such common materials as iron, copper, silver, gold, aluminum, carbon, hydrogen, and oxygen. Although some elements exist in a pure form, they occur more frequently in mixtures or united in chemical combinations.

About twenty elements are needed by organisms, and of these, oxygen, carbon, hydrogen, and nitrogen make up more than 95% (by weight) of a human body. A list of the major elements in the body is found in chart 2.1. Notice that each element is represented by a one- or two-letter symbol.

Elements are composed of tiny particles called **atoms,** which are the smallest complete units of the elements. Although the atoms that make up each element are similar to each other, they differ from the atoms that make up other elements. Atoms vary in size, weight, and the way they interact with each other. Some, for instance, are capable of combining with atoms like themselves or of combining with other kinds of atoms; other types of atoms lack these abilities.

Atomic Structure

An atom consists of a central portion called the **nucleus** and one or more **electrons** that are in constant motion

CHART 2.1	Major elements in the human body (by weight)	
Major elements	Symbol	Approximate percentage of the human body
Oxygen	O	65.0
Carbon	C	18.5
Hydrogen	H	9.5
Nitrogen	N	3.2
Calcium	Ca	1.5
Phosphorus	P	1.0
Potassium	K	0.4
Sulfur	S	0.3
Chlorine	Cl	0.2
Sodium	Na	0.2
Magnesium	Mg	0.1
	Total	99.9%

Trace elements		
Cobalt	Co	
Copper	Cu	
Fluorine	F	Together
Iodine	I	less than 0.1%
Iron	Fe	
Manganese	Mn	
Zinc	Zn	

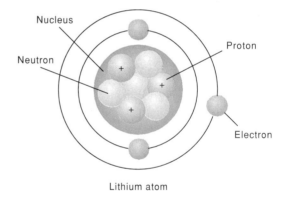

Lithium atom

Figure 2.1 An atom of lithium includes three electrons in motion around a nucleus that contains three protons and four neutrons.

around the nucleus. The nucleus contains some relatively large particles called **protons** and **neutrons,** whose weights are about equal (figure 2.1).

Electrons, which are extremely small and have almost no weight, carry a single, negative electrical charge (e^-), but protons each carry a single, positive electrical charge (p^+). Neutrons are uncharged and thus are electrically neutral (n^0).

Because the nucleus contains the protons, this part of an atom is always positively charged. However, the number of electrons outside the nucleus is equal to the number of protons, so a complete atom is electrically uncharged or *neutral.*

CHART 2.2 Atomic structure of elements 1 through 12

Element	Symbol	Atomic number	Atomic weight	Protons	Neutrons	Electrons in shells		
						First	Second	Third
Hydrogen	H	1	1	1	0	1		
Helium	He	2	4	2	2	2	(inert)	
Lithium	Li	3	7	3	4	2	1	
Beryllium	Be	4	9	4	5	2	2	
Boron	B	5	11	5	6	2	3	
Carbon	C	6	12	6	6	2	4	
Nitrogen	N	7	14	7	7	2	5	
Oxygen	O	8	16	8	8	2	6	
Fluorine	F	9	19	9	10	2	7	
Neon	Ne	10	20	10	10	2	8	(inert)
Sodium	Na	11	23	11	12	2	8	1
Magnesium	Mg	12	24	12	12	2	8	2

The atoms of different elements contain different numbers of protons. The number of protons in the atoms of a particular element is called its *atomic number.* Hydrogen, for example, whose atoms contain one proton, has the atomic number 1; and carbon, whose atoms have six protons, has the atomic number 6.

The weight of an atom of an element is due primarily to the protons and neutrons in its nucleus, because the electrons have so little weight. For this reason, an atom of carbon with six protons and six neutrons weighs about twelve times as much as an atom of hydrogen, which has only one proton and no neutrons.

The number of protons plus the number of neutrons in each of its atoms is approximately equal to the *atomic weight* of an element. Thus, the atomic weight of hydrogen is 1, and the atomic weight of carbon is 12 (chart 2.2).

Isotopes

All the atoms of a particular element have the same atomic number because they have the same number of protons and electrons. However, the atoms of an element vary in the number of neutrons in their nuclei; thus, they vary in atomic weight. For example, all oxygen atoms have eight protons in their nuclei. Some, however, have eight neutrons (atomic weight 16), others have nine neutrons (atomic weight 17), and still others have ten neutrons (atomic weight 18). Atoms that have the same atomic numbers but different atomic weights are called **isotopes** of an element. Because a sample of an element is likely to include more than one isotope, the atomic weight of the element is the average weight of the isotopes present.

The ways atoms interact with one another are due largely to the number of electrons they possess. Because the number of electrons in an atom is equal to its number of protons, all the isotopes of a particular element have the same number of electrons and react chemically in the same manner. Therefore, any of the isotopes of oxygen can play the same role in the metabolic reactions of an organism.

Although some of the isotopes of an element may be stable, others may have unstable atomic nuclei that decompose, releasing energy or pieces of themselves. Such unstable isotopes are called *radioactive,* and the energy or atomic fragments they give off are called *atomic radiations.* Examples of elements that have radioactive isotopes include oxygen, iodine, iron, phosphorus, and cobalt.

1. What is the relationship between matter and elements?
2. What elements are most common in the human body?
3. How are electrons, protons, and neutrons positioned within an atom?
4. What is an isotope?
5. What is meant by atomic radiation?

Bonding of Atoms

When atoms combine with other atoms, they gain or lose electrons, or share electrons with other atoms.

The electrons of an atom are arranged in one or more *shells* around the nucleus. The maximum number of electrons that each of the first three inner shells can hold for elements up to atomic number 20 is as follows:

First shell (closest to the nucleus)   2 electrons
Second shell   8 electrons
Third shell   8 electrons

(More complex atoms may have as many as eighteen electrons in the third shell.)

Uses of Radioactive Isotopes

Atomic radiation includes three common forms called alpha (α), beta (β), and gamma (γ). Alpha radiation consists of particles from atomic nuclei, each of which includes two protons and two neutrons, that travel relatively slowly and have weak ability to penetrate matter. Beta radiation consists of much smaller particles (electrons) that travel more rapidly and penetrate matter more deeply. Gamma radiation is similar to X-ray radiation and is the most penetrating of these forms.

Each kind of radioactive isotope produces one or more of these forms of radiation with particular energy levels. Furthermore, each of these isotopes tends to lose its radioactivity at a particular rate. The time required for an isotope to lose one-half of its activity is called its *half-life*. Thus, iodine-131, which emits one-half of its radiation in 8.1 days, has a half-life of 8.1 days. Similarly, the half-life of iron-59 is 45.1 days; that of phosphorus-32 is 14.3 days; that of cobalt-60 is 5.26 years; and that of radium-226 is 1,620 years.

Because it is possible to detect the presence of atomic radiation by using special equipment, such as a scintillation counter, radioactive substances are useful in studying life processes. A radioactive isotope, for example, can be introduced into an organism and then traced as it enters into metabolic activities. Since the human thyroid gland makes special use of iodine in its metabolism, radioactive iodine-131 has been used to study its functions and to evaluate patients with thyroid disease. Likewise, a form of thallium-201 with a half-life of 73.5 hours is commonly used to assess heart conditions, and gallium-67 with a half-life of 78 hours is used to detect and monitor the progress of certain cancers and inflammatory diseases. For example, thallium-201 may be used to detect disorders in the blood vessels supplying the heart muscle or to locate regions of damaged heart tissue following a heart attack. Such procedures involve injecting the isotope into the blood and following its path using detectors that record images on paper or film. (See figures 2.2 and 2.3.)

Other clinical uses of radioactive isotopes include measuring kidney functions, estimating the concentration of hormones in body fluids, measuring blood volume and red blood cell mass, and studying changes in the density of bone.

Atomic radiations also can cause changes in the structures of various chemical substances, and in this way they can alter vital cellular processes. This is the reason radioactive isotopes, such as cobalt-60, are sometimes used to treat cancers. The radiation coming from the cobalt can affect the chemicals within cancerous cells and cause their deaths. Because cancer cells are more susceptible to such damage than normal cells, the cancer cells are destroyed more rapidly than the normal ones.

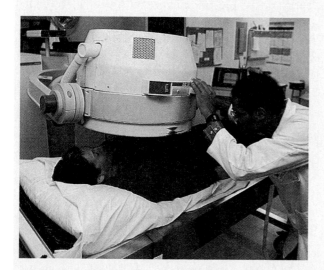

Figure 2.2   Scintillation counters such as this are used to detect the presence of radioactive isotopes.

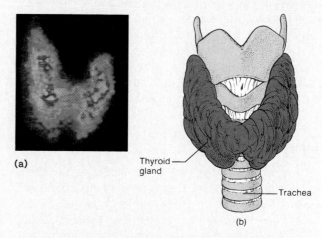

(a)

Thyroid gland

Trachea

(b)

Figure 2.3   (*a*) A scan of the thyroid gland twenty-four hours after the patient was administered radioactive iodine; (*b*) shape of the thyroid gland.

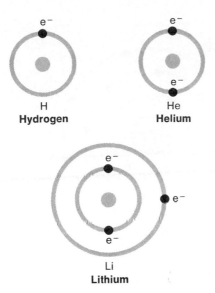

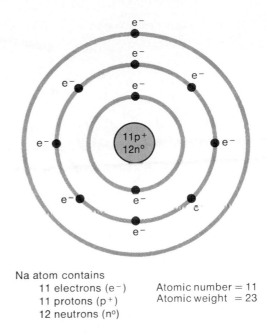

Figure 2.4 The single electron of a hydrogen atom is located in its first shell. The two electrons of a helium atom fill its first shell. The three electrons of a lithium atom are arranged with two in the first shell and one in the second shell.

Na atom contains
    11 electrons (e^-)   Atomic number = 11
    11 protons (p^+)   Atomic weight = 23
    12 neutrons (n^o)

Figure 2.5 A diagram of a sodium atom.

The arrangements of electrons within the shells of atoms can be represented by simplified diagrams such as those in figure 2.4. Notice that the single electron of a hydrogen atom is located in the first shell, the two electrons of a helium atom fill its first shell, and the three electrons of a lithium atom are arranged with two in the first shell and one in the second shell.

Atoms such as helium, whose outermost electron shells are filled, have stable structures and are chemically inactive (inert). Atoms with incompletely filled outer shells, such as those of hydrogen or lithium, tend to gain, lose, or share electrons in ways that empty or fill their outer shells. In this way they achieve stable structures.

An atom of sodium, for example, has eleven electrons arranged as shown in figure 2.5: two in the first shell, eight in the second shell, and one in the third shell. This atom tends to lose the single electron from its outer shell, which leaves the second shell filled and the form stable.

A chlorine atom has seventeen electrons arranged with two in the first shell, eight in the second shell, and seven in the third shell. An atom of this type will tend to accept a single electron, thus filling its outer shell and achieving a stable form.

Because each sodium atom tends to lose a single electron and each chlorine atom tends to accept a single electron, sodium and chlorine atoms will react together. During this reaction, a sodium atom loses an electron and is left with eleven protons (11^+) in its nucleus and only ten electrons (10^-). As a result, the atom develops a net electrical charge of 1^+ and is symbolized Na^+. At the same time, a chlorine atom gains an electron, which leaves it with seventeen protons (17^+) in its nucleus and eighteen electrons (18^-). Thus, it develops a net electrical charge of 1^- and is symbolized Cl^-.

Atoms that have become electrically charged by gaining or losing electrons are called **ions,** and ions with opposite electrical charges are attracted to one another. When this happens, a chemical bond called an *electrovalent bond* (ionic bond) is formed between them, as shown in figure 2.6. When sodium ions (Na^+) and chloride ions (Cl^-) unite in this manner, the compound sodium chloride (NaCl, table salt) is formed.

Similarly, a hydrogen atom may lose its single electron and become a hydrogen ion (H^+). Such an ion can bond with a chloride ion (Cl^-) to form hydrogen chloride (HCl, hydrochloric acid).

Atoms may also bond together by sharing electrons rather than by gaining or losing them. A hydrogen atom, for example, has one electron in its first shell, but needs two to achieve a stable structure. It may fill this shell by combining with another hydrogen atom in such a way that the two atoms share a pair of electrons. As figure 2.7 shows, the two electrons then encircle the nuclei of both atoms, and each achieves a stable form. In this case the chemical bond between the atoms is called a *covalent bond.*

Carbon atoms, with two electrons in their first shells and four electrons in their second shells, form covalent bonds when they unite with other atoms. In fact, carbon atoms may bond to other carbon atoms in

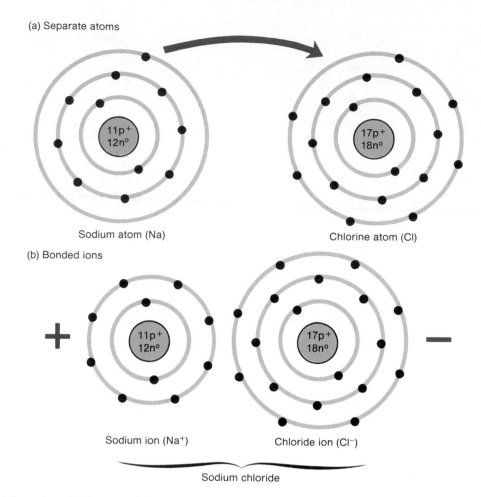

(a) Separate atoms

Sodium atom (Na)   Chlorine atom (Cl)

(b) Bonded ions

Sodium ion (Na$^+$)   Chloride ion (Cl$^-$)

Sodium chloride

Figure 2.6 (*a*) If a sodium atom loses an electron to a chlorine atom, the sodium atom becomes a sodium ion and the chlorine atom becomes a chloride ion. (*b*) These oppositely charged particles are attracted electrically to one another and become united by an electrovalent bond.

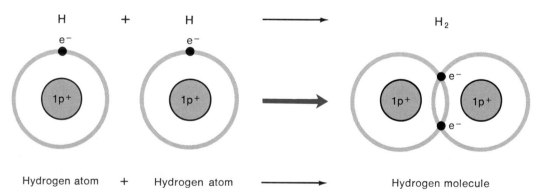

H   +   H   ⟶   H$_2$

Hydrogen atom   +   Hydrogen atom   ⟶   Hydrogen molecule

Figure 2.7 A hydrogen molecule is formed when two hydrogen atoms share a pair of electrons and are united by a covalent bond.

Ionizing Radiation

The various forms of atomic radiation described previously (alpha, beta, and gamma radiation) are said to be **ionizing** because they are capable of causing the formation of ions. Such ions are produced when radiation is absorbed by matter. At the same time, the radiation imparts enough energy to add or remove electrons from electrically neutral atoms or molecules. (See figure 2.8.) Other forms of *ionizing radiation* include X rays from X-ray tubes, cosmic rays from outer space, and neutrons that sometimes escape from atomic nuclei as a result of nuclear reactions.

The ions produced or the electrons dislodged from atoms by ionizing radiation may, in turn, cause changes in other nearby substances. Such changes may lead to a variety of effects in living things, including clouding of the eye lens, formation of cancers, and interference with normal growth and development. Consequently, ionizing radiation can constitute a health hazard, and unnecessary exposure to it should be limited whenever possible.

The quantity of ionizing radiation to which individuals are exposed varies greatly; however, in the United States, members of the general population are exposed to very low levels. Most of this exposure comes from *background radiation,* which originates from natural environmental sources. These sources include cosmic rays that arrive continually from outer space; radioactive elements that occur in the earth's surface, such as radium that gives rise to the radioactive gas called *radon;* rocks and clays used in building materials that contain small amounts of radioactive elements; and radioactive substances such as phosphorus-40 and carbon-14 that are found naturally in the body. Because background radiation is part of the natural environment, little if anything can be done to reduce exposure to it.

The next larger source of exposure to ionizing radiation comes from X rays and radioactive substances used in medical and dental procedures to diagnose and treat various

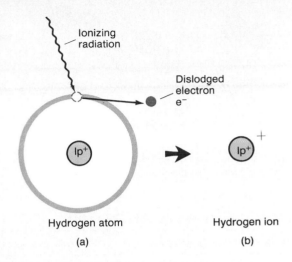

Figure 2.8 (*a*) Ionizing radiation may dislodge an electron from an electrically neutral hydrogen atom; (*b*) without its electron, the hydrogen atom becomes a positively charged hydrogen ion (H^+).

disorders and diseases. Although this source of exposure can be controlled, the benefits derived from the medical and dental uses of ionizing radiation are thought to outweigh any accompanying health risks.

The remaining exposure to ionizing radiation comes from a variety of sources. They include: (a) the development, production, and testing of nuclear weapons; (b) activities involved with mining and processing radioactive minerals; (c) the use of radioactive fuels in nuclear power plants; and (d) the inclusion of radioactive substances in consumer products, such as luminescent dials, certain smoke detectors, and some components of color television sets.

---

such a way that two atoms share one or more pairs of electrons. If one pair of electrons is shared, the resulting bond is called a *single covalent bond;* if two pairs of electrons are shared, the bond is called a *double covalent bond.*

Another type of chemical bond, called a *hydrogen bond,* is described in a subsequent section of this chapter.

Molecules and Compounds

When two or more atoms bond together, they form a new kind of particle called a **molecule.** If atoms of the same element combine, they produce molecules of that element. The gases of hydrogen (H_2), oxygen (O_2), and nitrogen (N_2) contain such molecules.

If atoms of different elements combine, molecules of substances called **compounds** are formed. Two atoms of hydrogen, for example, can combine with one atom of oxygen to produce a molecule of the compound water (H_2O), as shown in figure 2.9. Table sugar, baking soda, natural gas, beverage alcohol, and most medical drugs are examples of compounds.

A molecule of a compound always contains definite kinds and numbers of atoms. A molecule of water (H_2O), for instance, always contains two hydrogen atoms and one oxygen atom. If two hydrogen atoms combine with two oxygen atoms, the compound formed is not water, but hydrogen peroxide (H_2O_2).

Chart 2.3 lists some particles of matter and their characteristics.

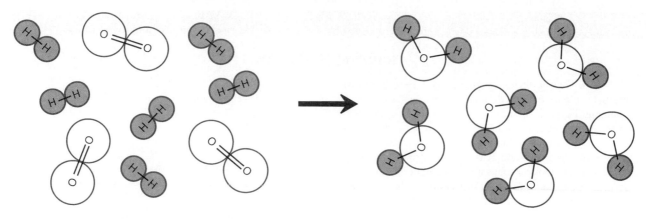

Figure 2.9 Hydrogen molecules combine with oxygen molecules to form water molecules.

CHART 2.3 Some particles of matter

Name	Characteristic	Name	Characteristic
Atom	Smallest particle of an element that has the properties of that element	Neutron (n^0)	Particle with about the same weight as a proton; uncharged and thus electrically neutral; found within a nucleus
Electron (e^-)	Extremely small particle with almost no weight; carries a negative electrical charge and is in constant motion around an atomic nucleus	Ion	Atom that is electrically charged because it has gained or lost one or more electrons
Proton (p^+)	Relatively large atomic particle; carries a positive electrical charge and is found within a nucleus	Molecule	Particle formed by the chemical union of two or more atoms

1. What is an ion?
2. Describe two ways that atoms may combine with other atoms.
3. Distinguish between a molecule and a compound.

Formulas

The numbers and kinds of atoms in a molecule can be represented by a *molecular formula*. Such a formula consists of the symbols of the elements in the molecule together with numbers to indicate how many atoms of each element are present. For example, the molecular formula for water is H_2O, which means there are two atoms of hydrogen and one atom of oxygen in each molecule. The molecular formula for the sugar called glucose is $C_6H_{12}O_6$, which means there are six atoms of carbon, twelve atoms of hydrogen, and six atoms of oxygen in a molecule.

Usually the atoms of each element will form a specific number of chemical bonds. Hydrogen atoms form single bonds, oxygen atoms form two bonds, nitrogen atoms form three bonds, and carbon atoms form four bonds. The bonding capacities of these atoms can be represented by using symbols and lines as follows:

$$ -\text{H} \quad -\text{O}- \quad -\overset{|}{\underset{}{\text{N}}}- \quad -\overset{|}{\underset{|}{\text{C}}}- $$

$H-H$	$O=O$	$H_{\diagdown O \diagup}H$	$O=C=O$
H_2	O_2	H_2O	CO_2

Figure 2.10 Structural formulas of molecules of hydrogen, oxygen, water, and carbon dioxide.

These representations can be used to show how atoms are bound and arranged in various molecules. Illustrations of this type are called *structural formulas*. (See figure 2.10.) Three-dimensional models that reflect these formulas can be built of colored parts representing different kinds of atoms. (See figure 2.11.)

Chemical Reactions

When chemical reactions occur, bonds between atoms, ions, or molecules are formed or broken. As a result, new combinations are produced. For example, when two or more atoms (reactants) bond together to form a more complex structure (end product), as when hydrogen and oxygen atoms bond to form molecules of water, the reaction is called **synthesis.** Such a reaction can be symbolized this way:

$$ A + B \rightarrow AB $$

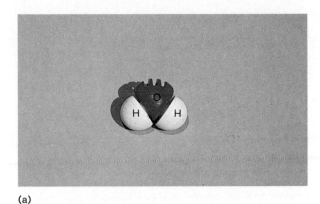

(a)

(b)

Figure 2.11   A water molecule (H_2O) can be represented by a three-dimensional model. (*a*) In this model, the white parts represent hydrogen atoms, and the red part represents oxygen. (*b*) In this model of a carbon dioxide molecule (CO_2), the black part represents carbon.

If the bonds within a reactant molecule break so that simpler molecules, atoms, or ions form, the reaction is called **decomposition.** Thus, molecules of water can decompose to yield the products hydrogen and oxygen. Decomposition is symbolized as follows:

$$AB \rightarrow A + B$$

Synthetic reactions are particularly important in the growth of body parts and the repair of worn or damaged tissues, which involve the build-up of larger molecules from smaller ones. On the other hand, when food substances are digested or energy is released from food molecules, the processes involve decomposition reactions.

A third type of chemical reaction is an **exchange reaction.** In this reaction, parts of two different kinds of molecules trade positions. The reaction is symbolized as follows:

$$AB + CD \rightarrow AD + CB$$

As is explained in the following section, when an acid reacts with a base, water and a salt are produced. This is an example of an exchange reaction.

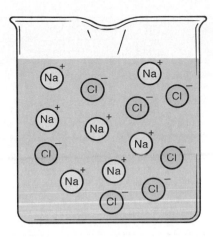

Figure 2.12   When crystals of table salt (NaCl) dissolve in water, sodium ions (Na^+) and chloride ions (Cl^-) are released.

Many chemical reactions are reversible. This means the end product (or products) of the reaction can change back to the reactant (or reactants) that originally underwent the reaction. A **reversible reaction** can be symbolized using a double arrow, as follows:

$$A + B \rightleftharpoons AB$$

Whether a reversible reaction proceeds in one direction or another depends on such factors as the relative proportions of reactant (or reactants) and end product (or end products), the amount of energy available for the reaction, and the presence or absence of *catalysts*. A catalyst is a particular atom or molecule that can change the rate of a reaction without itself being consumed by the reaction.

Acids, Bases, and Salts

Some compounds release ions (ionize) when they dissolve in water or react with water molecules. Sodium chloride (NaCl), for example, releases sodium ions (Na^+) and chloride (Cl^-) when it dissolves (figure 2.12). This reaction is represented by the following:

$$NaCl \rightarrow Na^+ + Cl^-$$

Because the resulting solution contains electrically charged particles (ions), it will conduct an electric current. Substances that release ions in water are, therefore, known as **electrolytes.** Electrolytes that release hydrogen ions (H^+) in water are called **acids.** For example, in water the compound hydrochloric acid (HCl) releases hydrogen ions (H^+) and chloride ions (Cl^-):

$$HCl \rightarrow H^+ + Cl^-$$

Electrolytes that release ions that combine with hydrogen ions are called **bases.** The compound sodium

CHART 2.4 Types of electrolytes

	Characteristic	Examples
Acid	Ionizes to release hydrogen ions (H⁺).	Carbonic acid, hydrochloric acid, acetic acid, phosphoric acid
Base	Ionizes to release ions that can combine with hydrogen ions.	Sodium hydroxide, potassium hydroxide, magnesium hydroxide, aluminum hydroxide
Salt	Substance formed by the reaction between an acid and a base.	Sodium chloride, aluminum chloride, magnesium sulfate

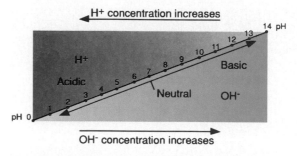

Figure 2.13 As the concentration of hydrogen ions (H^+) increases, a solution becomes more acidic, and the pH value decreases. As the concentration of hydrogen ion acceptors (such as hydroxyl or bicarbonate ions) increases, a solution becomes more basic, and the pH value increases.

hydroxide (NaOH) in water releases hydroxyl ions (OH^-):

$$NaOH \rightarrow Na^+ + OH^-$$

(Note: Some ions, such as OH^-, contain two or more atoms. However, such a group behaves like a single atom and usually remains unchanged during a chemical reaction.)

The hydroxyl ions, in turn, can combine with hydrogen ions to form water. Thus, sodium hydroxide is a base.

Acids that release hydrogen ions and bases that release hydroxyl ions can react to form water and electrolytes called **salts.** For example, hydrochloric acid and sodium hydroxide react to form water and sodium chloride:

$$HCl + NaOH \rightarrow H_2O + NaCl$$

Chart 2.4 summarizes the three types of electrolytes.

Acid and Base Concentrations

The chemical reactions involved with many life processes, such as those functioning in the regulation of blood pressure and breathing rate, are affected by the concentrations of acids and bases. Thus, the concentrations of these substances in body fluids are of special importance.

Acid and base concentrations are measured in units called **pH,** in which the pH value is equal to the negative logarithm of the hydrogen ion concentration. Therefore, these units indicate the concentration of *hydrogen ions* expressed in grams of ions per liter of solution. For example, a solution with a hydrogen ion concentration of 0.1 grams per liter has a pH value of 1.0; a concentration of 0.01 g H^+/l has pH 2.0; 0.001 g H^+/l has pH 3.0, and so forth. Note that between each whole number on the pH scale, which extends from pH 0 to pH 14.0, there is a tenfold difference in hydrogen ion concentration. Also note that as the hydrogen ion concentration increases, the pH value decreases.

In pure water, which ionizes only slightly, the hydrogen ion concentration is 0.0000001 g/l, and the pH is 7.0. Because water ionizes to release equal numbers of acidic hydrogen ions and basic hydroxyl ions, it is said to be *neutral.*

$$H_2O \rightarrow H^+ + OH^-$$

Therefore, solutions with more hydrogen ions than hydroxyl ions are said to be *acidic;* that is, they have pH values of less than 7.0. (See figure 2.13.) Solutions with fewer hydrogen ions than hydroxyl ions are said to be *basic* (alkaline); that is, they have pH values of more than 7.0.

Chart 2.5 illustrates the relationship between hydrogen ion concentration and pH. Figure 2.14 lists the pH values of some common substances. The regulation of hydrogen ion concentrations is discussed in chapter 21.

Note in figure 2.14 that the pH of human blood is normally about 7.4. If this pH value drops below 7.4, the person is said to have *acidosis;* and if it rises above 7.4, the condition is called *alkalosis.* Without medical intervention, a person usually cannot survive if the pH drops to 6.8 or rises to 8.0 for more than a few hours.

---

1. What is meant by a molecular formula? A structural formula?
2. Describe three kinds of chemical reactions.
3. Compare the characteristics of an acid with those of a base.
4. What is meant by pH?

CHART 2.5	Hydrogen ion concentrations and pH	
Grams of H$^+$ per liter	pH	
1.0	0	
0.1	1	
0.01	2	
0.001	3	Increasingly acidic
0.0001	4	
0.00001	5	
0.000001	6	
0.0000001	7	Neutral—neither acidic nor basic
0.00000001	8	
0.000000001	9	
0.0000000001	10	
0.00000000001	11	Increasingly basic
0.000000000001	12	
0.0000000000001	13	
0.00000000000001	14	

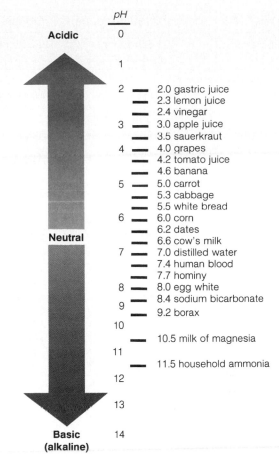

Figure 2.14   Approximate pH values of some common substances.

Chemical Constituents of Cells

The chemicals that enter into metabolic reactions or are produced by them can be divided into two large groups. Those that contain both carbon and hydrogen atoms are called **organic;** the rest are called **inorganic.**

Generally, inorganic substances will dissolve in water or react with water to release ions; thus they are *electrolytes*. Some organic compounds will dissolve in water also, but as a group they are more likely to dissolve in organic liquids like ether or alcohol. Those that will dissolve in water usually do not release ions, and are therefore called *nonelectrolytes*.

Inorganic Substances

Among the inorganic substances common in cells are water, oxygen, carbon dioxide, and a group called inorganic salts.

Water

Water (H_2O) is the most abundant compound in living material and is responsible for about two-thirds of the weight of an adult human. It is the major component of blood and other body fluids, including those within cells.

When substances dissolve in water, relatively large pieces of the substances break into smaller ones, and eventually, molecular-sized particles or ions result. These tiny particles are much more likely to react with one another than were the original large pieces. Consequently, most metabolic reactions occur in water.

Water also plays an important role in the transportation of chemicals within the body. The aqueous portion of blood, for example, carries many vital substances, such as oxygen, sugars, salts, and vitamins, from the organs of the digestive and respiratory systems to the body cells. Water also carries waste materials, such as carbon dioxide and urea, from these cells to the lungs and kidneys, respectively, which remove them from the blood and release them to the outside of the body.

In addition, water can absorb and transport heat. Thus, the heat released from muscle cells during exercise can be carried by the blood (which is 91.5% water) from deeper parts to the surface. At the same time, water released by skin cells in the form of perspiration can carry heat away by evaporation.

Oxygen

Molecules of oxygen gas (O_2) enter the body through the respiratory organs and are transported throughout the body by the blood and specialized blood cells (red blood cells). The oxygen is used by cellular organelles

CHART 2.6   Inorganic substances common in cells

Substance	Symbol or formula	Functions
I. Inorganic molecules		
Water	H_2O	Major component of body fluids (chapter 17); medium in which most biochemical reactions occur; transports various chemical substances (chapter 17); helps regulate body temperature (chapter 6)
Oxygen	O_2	Used in the release of energy from glucose molecules (chapter 4)
Carbon dioxide	CO_2	Waste product that results from metabolism (chapter 4); reacts with water to form carbonic acid (chapter 16)
II. Inorganic ions		
Bicarbonate ions	HCO_3^-	Helps maintain acid-base balance (chapter 21)
Calcium ions	Ca^{+2}	Necessary for bone development (chapter 7); needed for muscle contraction (chapter 9) and blood clotting (chapter 17)
Carbonate ions	CO_3^{-2}	Component of bone tissue (chapter 7)
Chloride ions	Cl^-	Helps maintain water balance (chapter 21)
Magnesium ions	Mg^{+2}	Component of bone tissue (chapter 7); needed for certain metabolic processes (chapter 15)
Phosphate ions	PO_4^{-3}	Needed for synthesis of ATP, nucleic acids, and other vital substances (chapter 4); component of bone tissue (chapter 7); helps maintain polarization of cell membranes (chapter 10)
Potassium ions	K^+	Needed for polarization of cell membranes (chapter 10)
Sodium ions	Na^+	Needed for polarization of cell membranes (chapter 10); helps maintain water balance (chapter 21)
Sulfate ions	SO_4^{-2}	Helps maintain polarization of cell membranes (chapter 10); helps maintain acid-base balance (chapter 21)

in the process of releasing energy from a sugar (glucose) and certain other molecules. The released energy is used to drive the cell's metabolic activities. A continuing supply of oxygen is necessary for cell survival and, ultimately, for the survival of the organism.

Carbon Dioxide

Carbon dioxide (CO_2) is a simple, carbon-containing compound of the inorganic group. It is produced as a waste product when energy is released during certain metabolic processes. As it moves from the cells into the surrounding body fluids and blood, most of the carbon dioxide reacts with water to form a weak acid (carbonic acid, H_2CO_3). This acid ionizes, releasing hydrogen ions (H^+) and bicarbonate ions (HCO_3^-), which the blood carries to the respiratory organs. There the chemical reactions reverse, and carbon dioxide gas is given off into the air within the lungs.

Inorganic Salts

Inorganic salts are abundant in body parts and fluids. They are the sources of many necessary ions, including ions of sodium (Na^+), chloride (Cl^-), potassium (K^+), calcium (Ca^{+2}), magnesium (Mg^{+2}), phosphate (PO_4^{-3}), carbonate (CO_3^{-2}), bicarbonate (HCO_3^-), and sulfate (SO_4^{-2}). These ions play important roles in metabolic processes, including those involved in maintenance of proper water concentrations in body fluids,

blood clotting, bone development, energy transfer within cells, and muscle and nerve functions.

These electrolytes must not only be present, but they must be present in the proper concentrations, both inside and outside cells, if homeostasis is to be maintained. Such a condition is called **electrolyte balance.** In certain diseases, this balance tends to be lost, and modern medical treatment places considerable emphasis on restoring it.

Chart 2.6 summarizes the functions of some of the inorganic substances that commonly occur in cells.

1. What is the difference between an organic molecule and an inorganic molecule?
2. What is the difference between an electrolyte and a nonelectrolyte?
3. Define electrolyte balance.

Organic Substances

Important groups of organic substances found in cells include carbohydrates, lipids, proteins, and nucleic acids.

Carbohydrates

Carbohydrates provide much of the energy that cells require. They also supply materials that are needed to

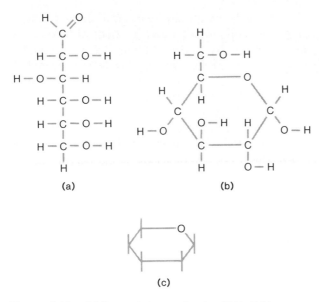

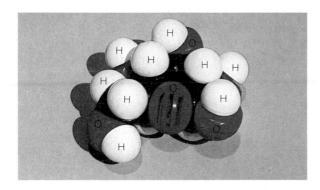

Figure 2.15 (*a*) Some glucose molecules ($C_6H_{12}O_6$) have a straight chain of carbon atoms. (*b*) More commonly, glucose molecules form a ring structure. (*c*) The ring structure of a glucose molecule can be symbolized by this shape.

Figure 2.16 Model of a glucose molecule ($C_6H_{12}O_6$).

build certain cell structures, and they often are stored as reserve energy supplies.

Carbohydrate molecules contain atoms of carbon, hydrogen, and oxygen. These molecules usually have twice as many hydrogen as oxygen atoms, the same ratio of hydrogen to oxygen that occurs in water molecules (H_2O). This ratio is easy to see in the molecular formulas of the carbohydrates glucose ($C_6H_{12}O_6$) and sucrose ($C_{12}H_{22}O_{11}$).

The carbon atoms of carbohydrate molecules are joined in chains whose lengths vary with the kind of carbohydrate. For example, those with shorter chains are called **sugars.** Sugars with six carbon atoms (hexoses) are known as *simple sugars* or **monosaccharides,** and they are the building blocks of more complex carbohydrate molecules. The simple sugars include glucose (dextrose), fructose, and galactose. Figure 2.15 illustrates the structural formulas of glucose; figure 2.16 shows a three-dimensional model of a glucose molecule.

In complex carbohydrates, a number of simple sugar molecules are bound together to form molecules of varying sizes, as shown in figure 2.17. Some, like sucrose (table sugar) and lactose (milk sugar), are *double sugars* or *disaccharides,* whose molecules each contain two simple sugar building blocks. Others are made up of many simple sugar units joined together to form *polysaccharides,* such as plant starch. Starch molecules consist of highly branched chains, each containing from twenty-four to thirty glucose units.

Animals, including humans, synthesize a polysaccharide similar to starch called *glycogen.* Its molecules also consist of branched chains of sugar units, but each chain contains only about a dozen glucose units.

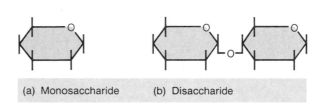

(a) Monosaccharide   (b) Disaccharide

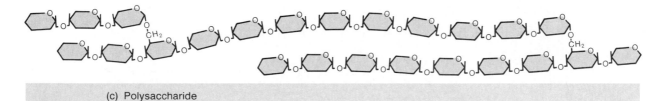

(c) Polysaccharide

Figure 2.17 (*a*) A monosaccharide molecule consists of one 6–carbon atom building block; (*b*) a disaccharide molecule consists of two of these building blocks; (*c*) a polysaccharide molecule consists of many building blocks.

Lipids

Lipids represent a group of organic substances that are insoluble in water, but are soluble in organic solvents, such as ether and chloroform. Lipids include a number of compounds, such as fats, phospholipids, and steroids, that have vital functions in cells and are important constituents of cell membranes. (See chapter 3.) The most common lipids, however, are the *fats*. They are used to build cell parts and to supply energy for cellular activities. In fact, fat molecules can supply more energy gram for gram than can carbohydrate molecules.

Like carbohydrates, fat molecules are composed of carbon, hydrogen, and oxygen atoms. However, they contain a much smaller proportion of oxygen than do carbohydrates. This is illustrated by the formula for the fat, tristearin, $C_{57}H_{110}O_6$.

The building blocks of fat molecules are **fatty acids** and **glycerol**. These smaller molecules are united so that each glycerol molecule is combined with three fatty acid molecules. The result is a single fat molecule or *triglyceride* (figures 2.18 and 2.19).

Figure 2.18   Model of a fatty acid molecule.

Although the glycerol portion of every fat molecule is the same, there are many kinds of fatty acids and, therefore, many kinds of fats. Fatty acid molecules contain a carboxyl group (COOH) at the end of a chain of carbon atoms. However, these molecules differ in the lengths of their carbon atom chains, although such chains usually contain an even number of carbon atoms. The chains also may vary in the ways the carbon atoms are combined. In some cases, the carbon atoms are all joined by single carbon-carbon bonds. This type of fatty acid is said to be *saturated;* that is, each carbon atom is bound to as many hydrogen atoms as possible and is thus saturated with hydrogen atoms. Other fatty acid chains have one or more double bonds between carbon atoms. Those with one double bond are called *monounsaturated fatty acids,* and those with two or more double bonds are said to be *polyunsaturated.* Similarly, fat molecules that contain only saturated fatty acids are called **saturated fats,** and those that include unsaturated fatty acids are called **unsaturated fats.** (See figure 2.20.)

Thus, each kind of fat molecule has particular kinds of fatty acids joined to its glycerol portion, and each kind has special properties.

A *phospholipid* molecule is similar to a fat molecule in that it contains a glycerol portion and fatty acid chains. The phospholipid, however, has only two fatty acid chains, and in place of the third one there is a portion containing a phosphate group. This phosphate portion is soluble in water (hydrophilic) and forms the "head" of the molecule, while the fatty acid portion is insoluble in water (hydrophobic) and forms a "tail." Figure 2.21 illustrates the molecular structure of cephalin, a phospholipid that occurs in blood.

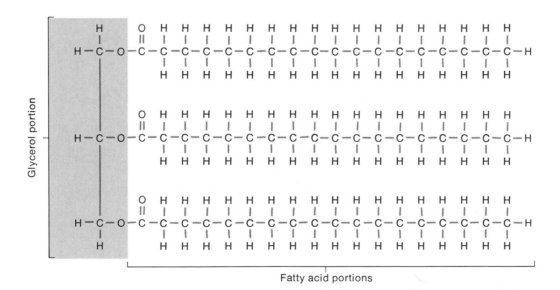

Figure 2.19   A triglyceride molecule consists of a glycerol portion and three fatty acid portions.

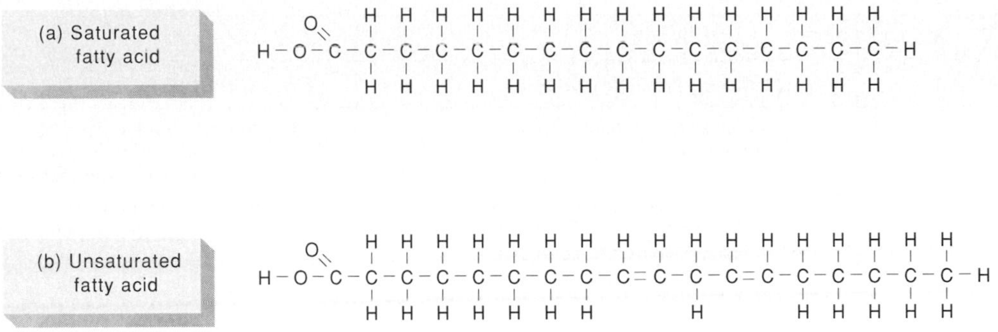

(a) Saturated fatty acid

(b) Unsaturated fatty acid

Figure 2.20 What are the major differences between (a) a molecule of saturated fatty acid and (b) a molecule of unsaturated fatty acid?

Glycerol portion

(a) A fat molecule (triglyceride)

Phosphoric acid portion

(b) A phospholipid molecule (cephalin)

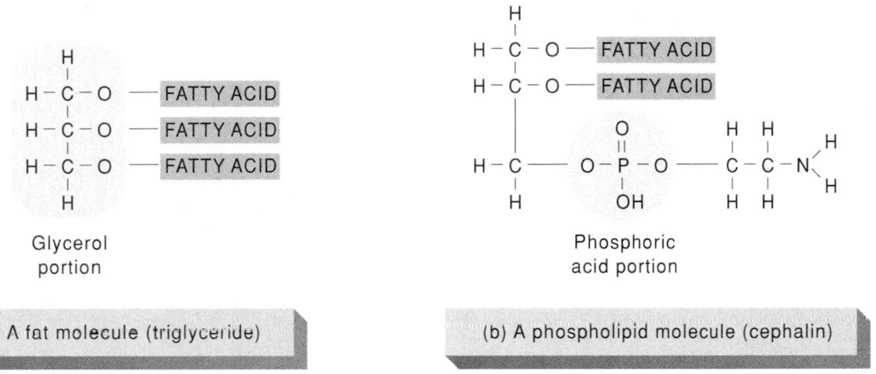

Figure 2.21 (a) A fat molecule (triglyceride) contains a glycerol portion and three fatty acids. (b) In a phospholipid molecule, one fatty acid is replaced by a phosphoric acid-containing group.

A diet that contains a high proportion of saturated fat seems to increase a person's chance of developing atherosclerosis, a serious disease in which certain blood vessels called arteries become obstructed. For this reason, many nutritionists recommend that polyunsaturated fats be substituted for some dietary saturated fats.

As a rule, saturated fats are more abundant in fatty foods that are solids at room temperature, such as butter, lard, and most other animal fats. Unsaturated fats, on the other hand, are likely to be plentiful in fatty foods that are liquids at room temperature, such as soft margarine and various seed oils, including corn oil and soybean oil. There are exceptions, however, since coconut and palm oils are relatively high in saturated fat.

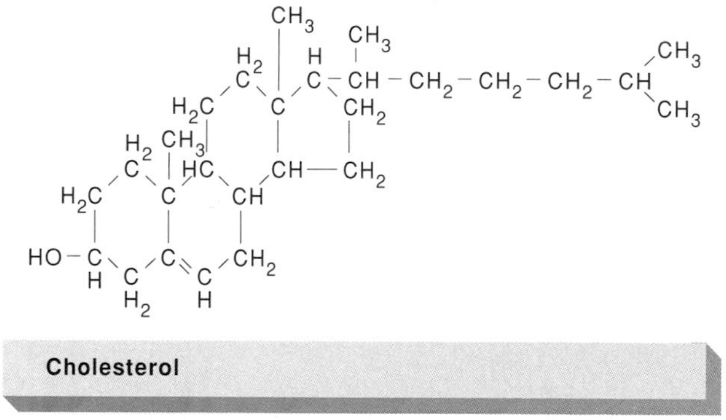

Cholesterol

Figure 2.22 Structural formula for cholesterol, a steroid that is widely distributed within the body.

Steroid molecules are complex structures that include interconnected rings of carbon atoms. (See figure 2.22.) Among the more important steroids are cholesterol, which occurs in all body cells and is used in the synthesis of other steroids; sex hormones, such as estrogen, progesterone, and testosterone; and several hormones from the adrenal glands. These steroids are discussed in chapters 13, 17, and 22.

Chart 2.7 summarizes the molecular structures and characteristics of lipids.

Proteins

Proteins serve as structural materials, energy sources, and chemical messengers (hormones). Others function as receptors on various surfaces that bond to particular kinds of molecules, or act as weapons (antibodies)

CHART 2.7 Important groups of lipids

Group	Basic molecular structure	Characteristics
Triglycerides	Three fatty acid molecules bound to a glycerol molecule	Most common lipid in the body; stored in fat tissue as an energy supply. Fat tissue also provides insulation beneath the skin.
Phospholipids	Two fatty acid molecules and a phosphate group bound to a glycerol molecule (may also include a nitrogen-containing molecule attached to the phosphate group)	Used as structural components in cell membranes; large amounts occur in the liver and parts of the nervous system.
Steroids	Four interconnected rings of carbon atoms	Widely distributed in the body with a variety of functions; includes cholesterol, sex hormones, and certain hormones of the adrenal glands.

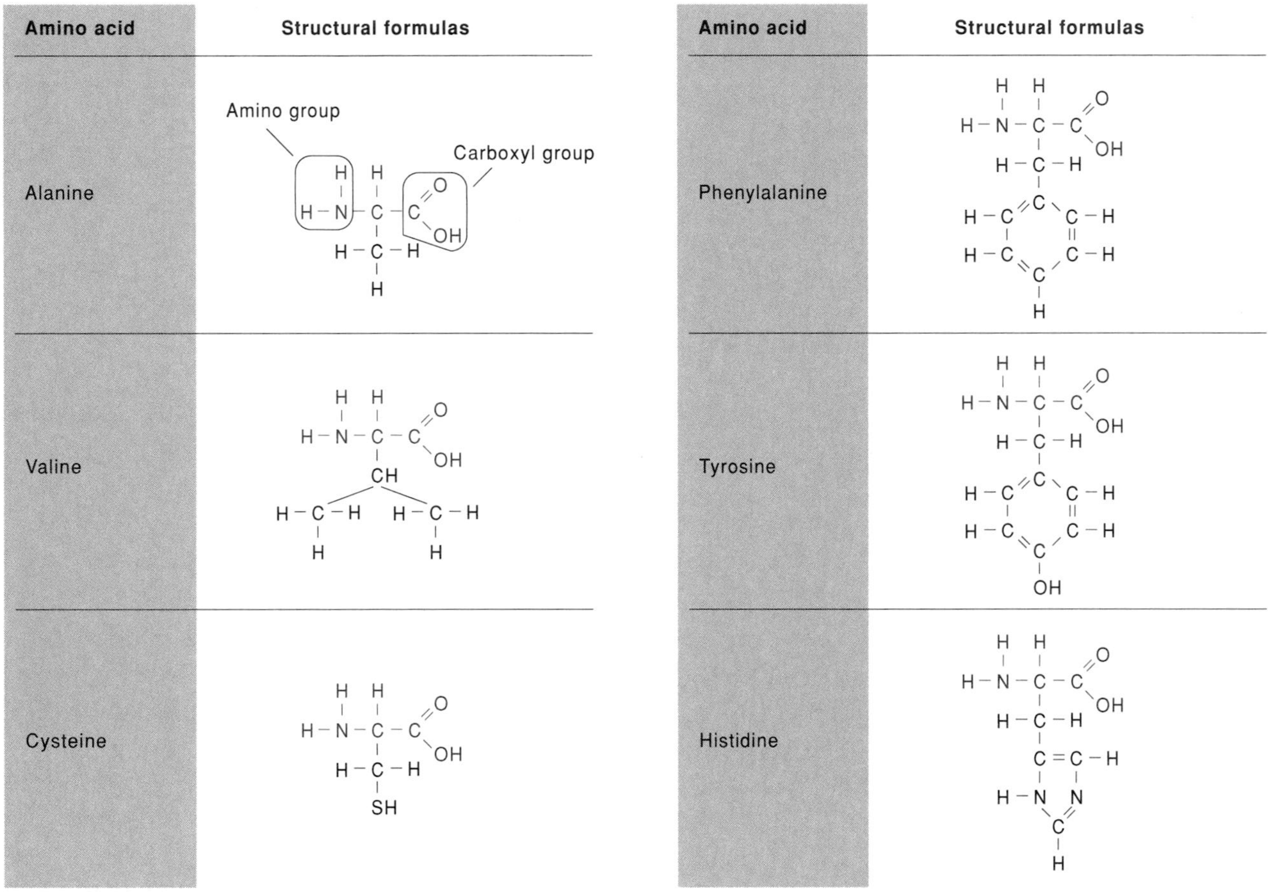

Amino acid	Structural formulas	Amino acid	Structural formulas
Alanine		Phenylalanine	
Valine		Tyrosine	
Cysteine		Histidine	

Figure 2.23 Some representative amino acids and their structural formulas. Each amino acid molecule has a particular shape due to the arrangement of its parts. Note the portion that is common to all amino acid molecules.

against substances that are foreign to the body. Still other proteins play vital roles in metabolic processes as **enzymes.** Enzymes are proteins that act as catalysts in living systems. That is, they have special properties and can speed up specific chemical reactions without being consumed by these reactions. (Enzymes are discussed in chapter 4.)

Like carbohydrates and lipids, proteins contain atoms of carbon, hydrogen, and oxygen. In addition, proteins always contain nitrogen atoms, and sometimes contain sulfur atoms as well. The building blocks of proteins are smaller molecules called **amino acids.**

Each amino acid molecule has an amino group (NH_2) at one end and a carboxyl group (COOH) at the other end. Between these end groups is a single carbon atom (alpha carbon), and it, in turn, is bound to a hydrogen atom and to another group of atoms called a *side chain.* It is the composition of the side chain that distinguishes one kind of amino acid from another kind. (See figures 2.23 and 2.24.)

CHART 2.8 Amino acids of proteins

Name	Abbreviation
Alanine	Ala
Arginine	Arg
Asparagine	Asn
Aspartic acid	Asp
Cysteine	Cys
Glutamic acid	Glu
Glutamine	Gln
Glycine	Gly
Histidine	His
Hydroxylysine	Hyl
Hydroxyproline	Hyp
Isoleucine	Ile
Leucine	Leu
Lysine	Lys
Methionine	Met
Phenylalanine	Phe
Proline	Pro
Serine	Ser
Threonine	Thr
Tryptophan	Trp
Tyrosine	Tyr
Valine	Val

Figure 2.24 Model of the amino acid, glycine. The blue part represents a nitrogen atom.

About twenty different kinds of amino acids occur commonly in proteins. These amino acid molecules are joined together in chains, varying in length from less than 100 to more than 50,000 amino acids. Biochemists commonly use three-letter abbreviations for amino acids. (See chart 2.8.)

An example of such a complex protein structure occurs in hemoglobin molecules, which are responsible for the red color of blood. Each hemoglobin molecule contains a protein portion that includes 574 amino acid units arranged in four chains. The amino acid sequence of another protein (parathyroid hormone) is shown in figure 13.3.

Each type of protein molecule contains specific numbers and kinds of amino acids arranged in a particular linear sequence (primary structure), as shown in figure 2.25. The amino acid chain is usually twisted to form a coil (secondary structure) (figure 2.26), and it, in turn, may be folded up into a unique three-dimensional form (tertiary structure) (figure 2.27). Consequently, different kinds of protein molecules have different shapes that are related to their particular functions. For example, the molecules of some proteins (fibrous proteins) are long and threadlike. They have special functions, such as binding body parts together or forming the fine threads of a blood clot. Other protein molecules (globular proteins) are coiled or folded into compact masses. Enzyme molecules are usually of this type.

The special shape of a protein molecule is maintained largely by *hydrogen bonds,* by which parts of the molecule are held to other parts of itself. A hydrogen bond is a weak chemical bond that occurs between a hydrogen atom that is covalently bound to a nitrogen or oxygen atom and at the same time is shared with another nitrogen or oxygen atom nearby. (See figure 2.26.)

The complex shapes of protein molecules can be altered by having hydrogen bonds broken as a result of exposure to excessive heat, radiation, electricity, or various chemicals. When this occurs, the molecules become disorganized and are said to be *denatured.* At the same time, the proteins lose their special properties. For example, the protein in egg white (albumin)

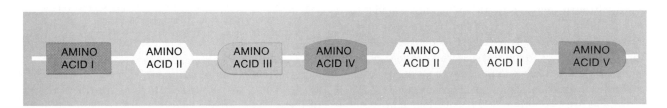

Figure 2.25 Each different shape in this chain represents a different amino acid molecule. The whole chain represents a portion of a protein molecule.

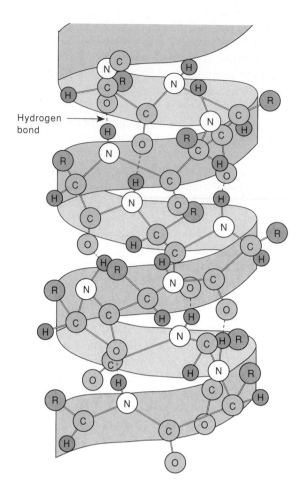

Hydrogen bond →

Portion of a protein molecule

Figure 2.26 The amino acid chain of a protein molecule is sometimes twisted to form a coil.

(a)

(b)

Figure 2.27 (a) The coiled amino acid chain of a protein molecule is folded into a unique three-dimensional structure; (b) a model of a portion of the protein called collagen.

is denatured when it is heated. Heating the protein causes it to change from a liquid to a solid—a change that cannot be reversed. Similarly, if cellular proteins are denatured, they are permanently changed and become nonfunctional.

Nucleic Acids

The most fundamental compounds in cells are the **nucleic acids.** These substances control cellular activities in very special ways.

Nucleic acid molecules are generally very large and complex. They contain atoms of carbon, hydrogen, oxygen, nitrogen, and phosphorus, which are bound into building blocks called **nucleotides.** Each nucleotide consists of a 5-carbon *sugar* (ribose or deoxyribose), a *phosphate group,* and one of several *organic bases* (figure 2.28). A nucleic acid molecule consists of many nucleotides united in a chain, as shown in figure 2.29.

There are two major types of nucleic acids. One type is composed of molecules whose nucleotides contain ribose sugar; it is called **RNA** (ribonucleic acid). The nucleotides of the second type contain deoxyribose

sugar; nucleic acid of this type is called **DNA** (deoxyribonucleic acid). Figure 2.30 compares the structure of these two 5-carbon sugars.

DNA molecules store information in a type of molecular code. This information is used by cells in the process of constructing specific protein molecules, and

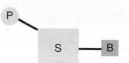

Figure 2.28 A nucleotide consists of a 5–carbon sugar (S), a phosphate group (P), and an organic base (B).

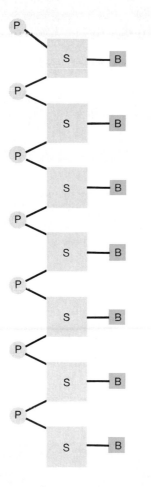

Figure 2.29 A nucleic acid molecule consists of nucleotides joined in a chain.

these proteins, many acting as enzymes, control the metabolic reactions that occur in cells. RNA molecules help to synthesize proteins.

DNA molecules also have a unique ability to make copies of, or replicate, themselves. They replicate prior to cell reproduction, and each newly formed cell receives an exact copy of its parent's DNA molecules. These molecules contain the instructions for synthesizing the proteins needed for carrying on vital life processes.

Chapter 4 discusses the storage of information in nucleic acid molecules, its use in the manufacture of protein molecules, and the function of these proteins in controlling metabolic reactions.

Chart 2.9 summarizes the four groups of organic compounds.

1. Compare the chemical composition of carbohydrates, lipids, proteins, and nucleic acids.
2. How does an enzyme affect a chemical reaction?
3. What is likely to happen to a protein molecule that is exposed to excessive heat or radiation?
4. What are the functions of DNA and RNA?

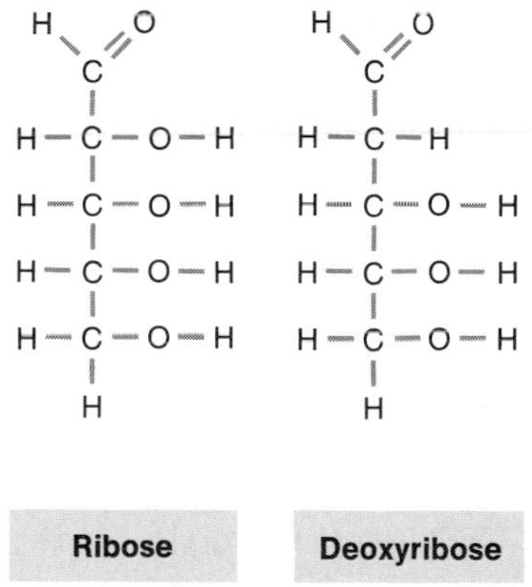

Ribose   **Deoxyribose**

Figure 2.30 How are the molecules of ribose and deoxyribose different? How are they similar to a molecule of glucose?

CHART 2.9	Organic compounds in cells			
Compound	Elements present	Building blocks	Functions	Examples
Carbohydrates	C,H,O	Simple sugar	Provide energy, cell structure	Glucose, starch
Lipids	C,H,O (often P)	Glycerol, fatty acids, phosphate groups	Provide energy, cell structure	Triglycerides, phospholipids, steroids
Proteins	C,H,O,N (often S)	Amino acids	Provide cell structure, enzymes, energy	Albumins, hemoglobin
Nucleic acids	C,H,O,N,P	Nucleotides	Store information for the synthesis of proteins, control cell activities	RNA, DNA

CT Scanning and PET Imaging

Two techniques for producing images of internal structures without invading the body are CT scanning and PET imaging.

CT, or **computerized tomography,** scanning, involves the use of X rays. In this procedure, which takes but a few seconds, an X-ray-emitting part is moved around the region of the body being examined. At the same time, an X-ray detector is moved in the opposite direction on the other side of the body. As the parts move, an X-ray beam is passed through the body from hundreds of different angles. Because tissues and organs of varying composition within the body absorb X rays differently, the amount of X ray reaching the detector varies from position to position.

The measurements made by the X-ray detector are recorded in the memory of a computer, which later combines them mathematically and creates a sectional image of the internal body parts on a viewing screen.

While ordinary X-ray techniques produce two-dimensional images, a CT scan provides three-dimensional information. The CT scan also makes it possible to differentiate clearly between soft tissues of slightly different densities, such as liver and kidneys, which cannot be seen in a conventional X ray. Thus, abnormal tissue growths, such as tumors, often can be located in a CT scan (figure 2.31).

PET, or **positron emission tomography,** makes use of a special group of radioactive isotopes that naturally give rise to *positrons*—positively charged electrons. These isotopes include carbon-11, nitrogen-13, oxygen-15, and fluorine-18. Whenever a positron is released by one of these substances, it interacts with a nearby negatively charged electron, and the two particles are destroyed (annihilated). At the moment of destruction, two gamma rays appear and travel away from each other in opposite directions. As was mentioned in the previous discussion of radioactive isotopes, such gamma radiation can be detected using special equipment.

To produce an image of a body part, a person is given a metabolically active compound to which a positron-emitting isotope has been bound chemically. For example, if the brain is being studied, a form of glucose containing fluorine-18 (flurodeoxyglucose) might be used. After the isotope-tagged compound is taken up by the brain tissues, the head is placed within a circular array of radiation detectors.

Whenever the radiation detectors register two gamma rays being emitted simultaneously and traveling in exactly opposite directions (the result of positron-electron annihilation), the event is recorded. Such data are collected and combined by a computer, which in turn generates a cross-sectional image. This image indicates the location and relative concentration of the radioactive isotope in various regions of the brain, and can be used to study the particular parts metabolizing glucose. Such information, of course, is useful for investigating the physiology as well as identifying functional regions of the brain. (See figure 2.32.)

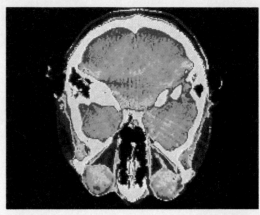

(a)

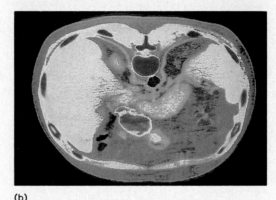

(b)

Figure 2.31 CT scans of (a) the head and (b) the abdomen. What features can you identify?

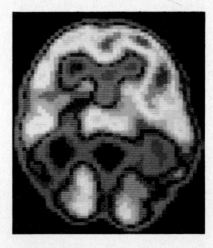

Figure 2.32 PET image of a human brain.

PET imaging is also used to evaluate various brain abnormalities and diseases, and to study blood flow in vessels supplying the brain and heart.

Clinical Terms Related to the Use of Radioactive Isotopes

alpha particle (al'fah par'ti-kl) Positively charged particle, consisting of two protons and two neutrons, given off by certain radioactive isotopes.

beta particle (ba'tah par'ti-kl) Negatively charged particle (electron) given off by certain radioactive isotopes; sometimes used to treat diseased tissues on or near the surface of organs.

cobalt-60 (ko'bawlt) Radioactive isotope of the element cobalt; used as a source of gamma radiation in the treatment of cancer.

cobalt unit (ko'bawlt u'nit) Therapeutic apparatus that contains cobalt-60.

curie (ku're) A unit of radioactivity used to measure the amount of radiation emitted from a radioactive isotope.

dosimetry (do-sim'e-tre) The process of measuring radiation exposure or dose.

gamma ray (gam'ah ra) A form of radiation, similar to X ray, that is given off by certain radioactive isotopes; the most penetrating form of radiation produced by isotopes.

half-life (haf ' lif) The time required for a radioactive isotope to lose one-half of its radioactivity.

iodine-131 (i'o-din) A radioactive isotope of the element iodine; sometimes used in testing or treating the thyroid gland.

ionizing radiation (i'on-i-zing ra-de-a'shun) Any form of radiation capable of causing atoms to become electrically charged ions.

iron-59 (i'ern) A radioactive isotope of the element iron; sometimes used to test bone marrow.

irradiation (i-ra''de-a'shun) The process of exposing something to radiation.

nuclear medicine (nu'kle-ar med'i-sin) The branch of medicine concerned with the use of radiation.

rad (rad) A unit used to measure the amount of radiation exposure; stands for *r*adiation *a*bsorbed *d*ose.

radiation sickness (ra''de-a'shun sik'nes) A disease usually resulting from the effects of radiation on the digestive organs.

rem (rem) A unit used to measure ionizing radiation; quantity that produces the same biological effect as one rad of X ray; stands for *r*oentgen-*e*quivalent *m*an.

scan (skan) A recording of the radiation emitted by radioactive isotopes within certain tissues.

tracer (trās'er) A substance, such as a radioactive isotope, that can be administered to a person and followed by using a detector.

Chapter Summary

Introduction (page 40)

Chemistry deals with the composition of substances and changes in their composition. Biochemistry is the chemistry of living things.

Structure of Matter (page 40)

Matter is anything that has weight and takes up space.

1. Elements and atoms
   a. Naturally occurring matter on earth is composed of ninety-one elements.
   b. Elements occur most frequently in mixtures or united in chemical combinations.
   c. Elements are composed of atoms.
   d. Atoms of different elements vary in size, weight, and ways of interacting.
2. Atomic structure
   a. An atom consists of electrons surrounding a nucleus, which contains protons and neutrons.
   b. Electrons are negatively charged, protons positively charged, and neutrons uncharged.
   c. A complete atom is electrically neutral.
   d. The atomic number of an element is equal to the number of protons in each atom; the atomic weight is equal to the number of protons plus the number of neutrons in each atom.
3. Isotopes
   a. Isotopes are atoms with the same atomic number but different atomic weights (due to differing numbers of neutrons).
   b. All the isotopes of an element react chemically in the same manner.
   c. Some isotopes are radioactive and release atomic radiation.
4. Bonding of atoms
   a. When atoms combine, they gain, lose, or share electrons.
   b. Electrons are arranged in shells around a nucleus.
   c. Atoms with completely filled outer shells are inactive, while atoms with incompletely filled outer shells tend to gain, lose, or share electrons and thus achieve stable structures.
   d. Atoms that lose electrons become positively charged; atoms that gain electrons become negatively charged.
   e. Ions with opposite charges are attracted to one another and become bound by electrovalent bonds; atoms that share electrons become bound by covalent bonds.

5. Molecules and compounds
   a. When two or more atoms of the same element are united, a molecule of that element is formed; when atoms of different elements are united, a molecule of a compound is formed.
   b. Molecules contain definite kinds and numbers of atoms.
6. Formulas
   a. The numbers and kinds of atoms in a molecule can be represented by a molecular formula.
   b. The arrangement of atoms within a molecule can be represented by a structural formula.
7. Chemical reactions
   a. When a chemical reaction occurs, bonds between atoms, ions, or molecules are broken or formed.
   b. Three kinds of chemical reactions are: synthesis, in which larger molecules are formed from smaller particles; decomposition, in which smaller particles are formed from larger molecules; and exchange reactions, in which the parts of two different molecules trade positions.
   c. Many reactions are reversible, and the direction of a reaction depends upon the proportion of reactants and end products present, the energy available, and the presence or absence of catalysts.
8. Acids, bases, and salts
   a. Compounds that ionize when they dissolve in water are electrolytes.
   b. Electrolytes that release hydrogen ions are acids, and those that release hydroxyl or other ions that react with hydrogen ions are bases.
   c. Acids and bases that release hydroxyl ions react together to form water and electrolytes called salts.
9. Acid and base concentrations
   a. The concentration of hydrogen ions (H^+) and hydroxyl ions (OH^-) in a solution can be represented by pH.
   b. A solution with equal numbers of H^+ and OH^- is neutral and has a pH of 7.0; a solution with more H^+ than OH^- is acidic (pH less than 7.0); a solution with fewer H^+ than OH^- is basic (pH more than 7.0).
   c. There is a tenfold difference in hydrogen ion concentration between each whole number in the pH scale.

Chemical Constituents of Cells (page 49)

Molecules containing carbon and hydrogen atoms are organic and are usually nonelectrolytes; other molecules are inorganic and are usually electrolytes.

1. Inorganic substances
   a. Water is the most abundant compound in cells and serves as a substance in which chemical reactions occur; it also transports chemicals and heat, and helps release excess body heat.
   b. Oxygen is used in releasing energy needed for metabolic activities from glucose and other molecules.
   c. Carbon dioxide is produced when energy is released during metabolic processes.
   d. Inorganic salts provide ions needed in a variety of metabolic processes.
   e. Electrolytes must be present in proper concentrations inside and outside of cells.
2. Organic substances
   a. Carbohydrates provide much of the energy needed by cells; their basic building blocks are simple sugar molecules.
   b. Lipids, such as fats, phospholipids, and steroids, supply energy and are used to build cell parts; their basic building blocks are molecules of glycerol and fatty acids.
   c. Proteins serve as structural materials, energy sources, hormones, surface receptors, antibodies, and enzymes.
      (1) Enzymes initiate or speed chemical reactions without being consumed themselves.
      (2) The building blocks of proteins are amino acids.
      (3) Proteins vary in the numbers and kinds of amino acids they contain, the sequences in which these amino acids are arranged, and their three-dimensional structures.
      (4) Protein molecules can be denatured by exposure to excessive heat, radiation, electricity, or various chemicals.
   d. Nucleic acids are the most fundamental compounds in cells because they control cell activities.
      (1) The two major kinds are RNA and DNA.
      (2) Nucleic acid molecules are composed of building blocks called nucleotides.
      (3) DNA molecules store information that is used by cell parts to construct specific kinds of protein molecules, including enzymes.
      (4) RNA molecules help synthesize proteins.
      (5) DNA molecules are replicated and passed from parent to offspring cells during cell reproduction.

Clinical Application of Knowledge

1. What acidic and alkaline substances do you encounter in your everyday life? What foods do you eat regularly that are acidic? What alkaline foods do you eat?

2. Using the information on page 52 to distinguish between saturated from unsaturated fats, try to list all of the sources of saturated and unsaturated fats you have eaten during the past twenty-four hours.

3. How would you reassure a patient who is about to undergo CT scanning for evaluation of a tumor, and who fears becoming a radiation hazard to family members?

4. Various forms of ionizing radiation, such as that released from X-ray tubes and radioactive substances, are commonly used in the treatment of cancer. Since such exposure can cause adverse effects, including the development of cancers, how would you explain the value of radiation therapy to a cancer patient in light of this seemingly contradictory situation?

5. How would you explain the importance of amino acids and proteins in a diet to a person who is following a diet composed primarily of carbohydrates?

6. What clinical laboratory tests with which you are acquainted involve a knowledge of chemistry?

Review Activities

1. Distinguish between chemistry and biochemistry.
2. Define *matter*.
3. Explain the relationship between elements and atoms.
4. Define *compound*.
5. List the four most abundant elements in the human body.
6. Describe the major parts of an atom.
7. Distinguish between protons and neutrons.
8. Explain why a complete atom is electrically neutral.
9. Distinguish between atomic number and atomic weight.
10. Define *isotope*.
11. Explain what is meant by *atomic radiation*.
12. Describe how electrons are arranged within atoms.
13. Explain why some atoms are chemically inert.
14. Distinguish between an electrovalent bond and a covalent bond.
15. Distinguish between a single covalent bond and a double covalent bond.
16. Explain the relationship between molecules and compounds.
17. Distinguish between a molecular formula and a structural formula.
18. Describe three major types of chemical reactions.
19. Explain what is meant by *reversible reaction*.
20. Define *catalyst*.
21. Define *acid, base, salt,* and *electrolyte*.
22. Explain what is meant by *pH,* and describe the pH scale.
23. Distinguish between organic and inorganic substances.
24. Describe the roles played by water and by oxygen in the human body.
25. Explain how carbon dioxide is transported within the body.
26. List several ions needed by cells, and describe their general functions.
27. Define *electrolyte balance*.
28. Describe the general characteristics of carbohydrates.
29. Distinguish between simple and complex carbohydrates.
30. Describe the general characteristics of lipids.
31. Distinguish between saturated and unsaturated fats.
32. Describe the general characteristics of proteins.
33. Define *enzyme*.
34. Explain how protein molecules may become denatured.
35. Describe the general characteristics of nucleic acids.
36. Explain the general functions of nucleic acids.

3

Cells

*T*he human body is composed entirely of cells, the products of cells, and various fluids. These cells represent the basic structural units of the body; they are the building blocks from which all larger parts are formed. They are also the functional units of the body, because whatever a body part can do is the result of activities within its cells.

Cells account for the shape, organization, and construction of the body and carry on its life processes. In addition, they can reproduce and thus provide the new cells needed for growth, development, and the replacement of worn and injured tissues ■

Chapter Objectives	Key Terms	Aids to Understanding Words
After you have studied this chapter, you should be able to:	active transport (ak′tiv trans′port)	cyt-, cell: *cyt*oplasm—the fluid that occupies the space between the cell membrane and nuclear envelope.
1. Explain how cells vary from one another.	centrosome (sen′tro-sōm)	endo-, within: *endo*plasmic reticulum—complex of membranous structures within the cytoplasm.
2. Describe the general characteristics of a composite cell.	chromosome (kro′mo-sōm)	
	cytoplasm (si′to-plazm)	hyper-, above: *hyper*tonic—a solution that has a greater concentration of dissolved particles than another solution.
3. Explain how the structure of a cell membrane is related to its function.	differentiation (dif″er-en″she-a′shun)	
	diffusion (dĭ-fu′zhun)	
4. Describe each kind of cytoplasmic organelle and explain its function.	endocytosis (en″do-si-to′sis)	hypo-, below: *hypo*tonic—a solution that has a lesser concentration of dissolved particles than another solution.
	endoplasmic reticulum (en′do-plaz′mik rĕ-tik′u-lum)	
5. Describe the cell nucleus and its parts.	equilibrium (e″kwĭ-lib′re-um)	inter-, between: *inter*phase—the stage that occurs between mitotic divisions of a cell.
6. Explain how substances move through cell membranes.	facilitated diffusion (fah-sil″ĭ-tat′-ed dĭ-fu′zhun)	
	filtration (fil-tra′shun)	iso-, equal: *iso*tonic—a solution that has a concentration of dissolved particles equal to that of another solution.
7. Describe the life cycle of a cell.	Golgi apparatus (gol′je ap″ah-ra′tus)	
8. Explain how a cell reproduces.	lysosome (li′so-sōm)	lys-, to break up: *lys*osome—organelle that contains enzymes capable of breaking up molecules of protein, carbohydrate, and nucleic acid.
9. Complete the review activities at the end of this chapter. Note that the items are worded in the form of specific learning objectives. You may want to refer to them before reading the chapter.	micrometer (mi′kro-me″ter)	
	mitochondrion (mi″to-kon′dre-on); plural, mitochondria (mi″to-kon′dre-ah)	
	mitosis (mi-to′sis)	mit-, thread: *mit*osis—the process of cell division during which threadlike chromosomes appear within a cell.
	nucleolus (nu-kle′o-lus)	
	nucleus (nu′kle-us)	
	osmosis (oz-mo′sis)	phag-, to eat: *phag*ocytosis—the process by which a cell takes in solid particles.
	permeable (per′me-ah-bl)	
	phagocytosis (fag″o-si-to′sis)	pino-, to drink: *pino*cytosis—the process by which a cell takes in tiny droplets of liquid.
	pinocytosis (pi″no-si-to′sis)	
	ribosome (ri′bo-sōm)	-som, body: ribo*som*e—a tiny, spherical organelle that contains protein and RNA.
	vesicle (ves′ĭ-k′l)	

It is estimated that an adult human body consists of about seventy-five trillion cells. These **cells** have much in common, yet those in different tissues vary in a number of ways. For example, cells vary considerably in size.

Cell sizes are measured in units called *micrometers*. A micrometer equals one thousandth of a millimeter and is symbolized μm. Measured in micrometers, a human egg cell is about 140 μm in diameter and is just barely visible to an unaided eye. This is large when compared to a red blood cell, which is about 7.5 μm in diameter, or the most common white blood cells, which vary from 10 to 12 μm in diameter. On the other hand, smooth muscle cells can be between 20 and 500 μm long. (See figure 3.1.)

Cells also vary in shape, and typically their shapes are closely related to their functions (figure 3.2). For instance, nerve cells often have long, threadlike extensions that transmit nerve impulses from one part of the body to another. Epithelial cells that line the inside of the mouth are thin, flattened, and tightly packed, somewhat like floor tiles. They are protective cells that shield underlying cells. Muscle cells, which function to pull parts closer together, are slender and rodlike, with their ends attached to the parts they move.

A Composite Cell

It is not possible to describe a typical cell, because cells vary so greatly in size, shape, content, and function. For purposes of discussion, however, it is convenient to imagine that one exists. Such a cell would contain parts observed in many kinds of cells, even though some of these cells lack parts included in the imagined structure. Thus, the cell shown in figure 3.3 and described in the following sections is not real. Instead, it is a *composite cell*—one that includes many known cell structures.

Commonly, a cell consists of two major parts—the nucleus and the cytoplasm. The *nucleus* is the innermost part and is enclosed by a thin, double-layered membrane called a *nuclear envelope*. The *cytoplasm* is

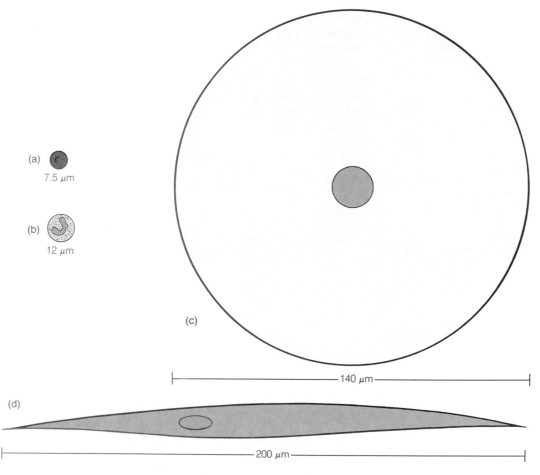

Figure 3.1   Cells vary considerably in size, as indicated in this illustration that shows the relative sizes of four types of cells. (*a*) Red blood cell, 7.5 μm in diameter; (*b*) white blood cell, 10–12 μm in diameter; (*c*) human egg cell, 140 μm in diameter; (*d*) smooth muscle cell, 20–500 μm in length.

a mass of fluid that surrounds the nucleus and is itself encircled by a *cell membrane* (also called plasma membrane or cytoplasmic membrane).

Within the cytoplasm are specialized cell parts called *cytoplasmic organelles*. These organelles perform specific metabolic functions. The nucleus, on the other hand, directs the overall activities of the cell.

1. Give two examples to illustrate that the shape of a cell is related to its function.
2. Name the two major parts of a cell.
3. What are the general functions of these two parts?

Cell Membrane

The **cell membrane** is the outermost limit of a cell, but it is more than a simple envelope surrounding the cellular contents. It is an actively functioning part of the living material, and many important metabolic reactions take place on its surfaces.

General Characteristics

The cell membrane is extremely thin—visible only with the aid of an electron microscope—but it is flexible and somewhat elastic. It typically has complex surface features with many outpouchings and infoldings that provide extra surface area. The membrane quickly seals minute breaks, but if it is extensively damaged, the cell contents escape, and the cell dies.

In addition to maintaining the wholeness of the cell, the membrane controls the entrance and exit of substances, allowing some substances to pass through it while excluding others. A membrane that functions in this manner is called *selectively permeable*. A *permeable membrane,* on the other hand, allows the substances involved to pass through.

Membrane Structure

Chemically, the cell membrane is composed mainly of lipids and proteins, although it also contains a small quantity of carbohydrate. Its basic framework consists

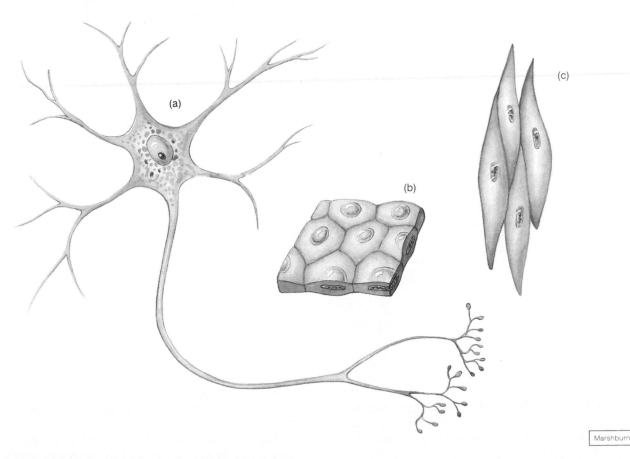

Marshburn

Figure 3.2 Cells vary in structure and function. (*a*) A nerve cell transmits impulses from one body part to another. (*b*) Epithelial cells protect underlying cells. (*c*) Muscle cells pull parts closer together.

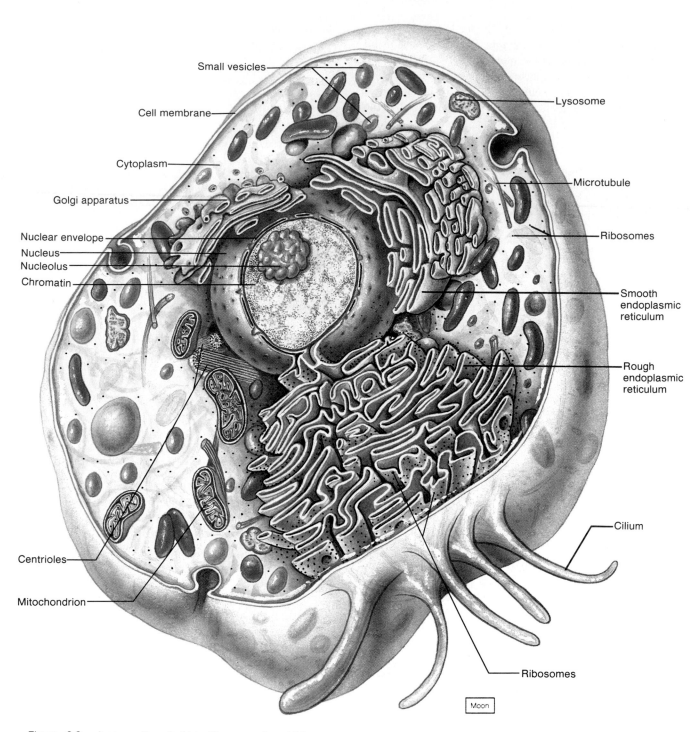

Small vesicles
Cell membrane
Cytoplasm
Golgi apparatus
Nuclear envelope
Nucleus
Nucleolus
Chromatin
Lysosome
Microtubule
Ribosomes
Smooth endoplasmic reticulum
Rough endoplasmic reticulum
Cilium
Centrioles
Mitochondrion
Ribosomes
Moon

Figure 3.3 A composite cell. (Note: The organelles of this composite cell are not drawn to scale.)

of a double layer (bilayer) of phospholipid molecules. (See chapter 2 and figure 2.21*b*.) These molecules are arranged so that their water soluble (hydrophilic) "heads," containing phosphate groups, form the surfaces of the membrane, and their water insoluble (hydrophobic) "tails," consisting of fatty acid chains, make up the interior of the membrane. (See figure 3.4.)

The lipid molecules are relatively free to move sideways within the plane of the membrane, and to-

gether they form a thin, but stable liquid film. The film is also soft and flexible.

Because the interior of the membrane consists largely of the fatty acid portions of the phospholipid molecules, it has an oily characteristic. Molecules that are soluble in lipids, such as oxygen and carbon dioxide molecules, can pass through this layer easily; however, the layer is impermeable to water soluble molecules, such as amino acids, sugars, proteins, nucleic acid mol-

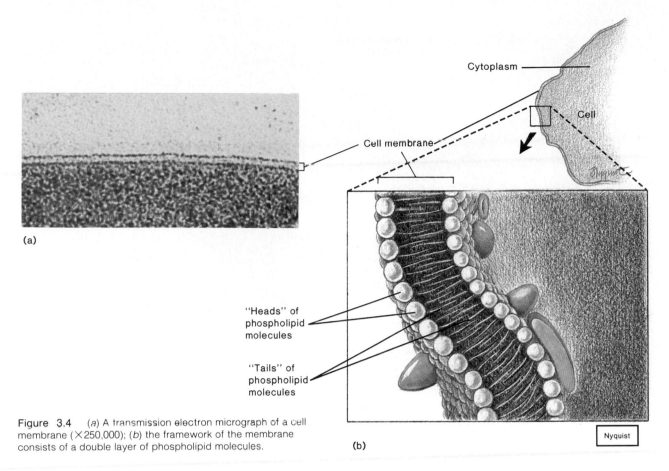

Figure 3.4   (a) A transmission electron micrograph of a cell membrane (×250,000); (b) the framework of the membrane consists of a double layer of phospholipid molecules.

(a)

Cytoplasm

Cell

Cell membrane

"Heads" of phospholipid molecules

"Tails" of phospholipid molecules

Nyquist

(b)

ecules, and various ions. Many cholesterol molecules embedded in the interior of the membrane also help make the membrane less permeable to water soluble substances. In addition, the relatively rigid structure of the cholesterol molecules helps make the membrane more stable than it would be otherwise.

Although the cell membrane includes only a few types of lipid molecules, it contains many kinds of proteins. These proteins are responsible for the special functions of the membrane, and they can be classified according to their shapes. One group of proteins, for example, consists of tightly coiled, rodlike molecules embedded in the bilayer of phospholipids. These rodlike proteins may completely span the membrane; that is, they may extend outward from its surface on one side, while their opposite ends communicate with the cell's interior. Such proteins often function as *receptors* that are specialized to combine with specific kinds of molecules, such as hormones (chapter 13).

Another group of proteins have molecules with more compact, globular shapes. Some of these proteins, called *integral proteins,* are embedded in the interior of the phospholipid bilayer. Typically, they span the membrane and form narrow passageways, or *channels,* through which various molecules and ions can cross the otherwise impermeable phospholipid bilayer. For example, some of these integral proteins form "pores" in the membrane that allow water molecules to pass

through. Others serve as selective channels that allow only particular substances to enter. In muscle and nerve cells, for example, such selective channels control the movements of sodium and potassium ions. These movements play important roles in the functions of the cells, including the establishment of electrical charges on their cell membranes, which are necessary for muscle contraction and nerve impulse conduction. These functions are described in chapters 9 and 10.

Other globular proteins, called *peripheral proteins,* are associated with the inside surface of the cell membrane. These proteins function as enzymes (see chapter 4) that promote specific chemical reactions. (See figure 3.5.)

Intercellular Junctions

Many cells, such as blood cells, are not in direct contact with one another because fluid-filled spaces (intercellular spaces) separate them. In other cases, however, the cells are tightly packed, and their cell membranes commonly are connected by **intercellular junctions.**

In one type of intercellular junction, called a *tight junction,* the membranes of adjacent cells converge and become fused together. The area of fusion surrounds the cell like a belt, and the junction closes the intercellular space between the cells. Cells that form sheet-like layers, such as those that line the inside of the digestive tube, often are joined by tight junctions.

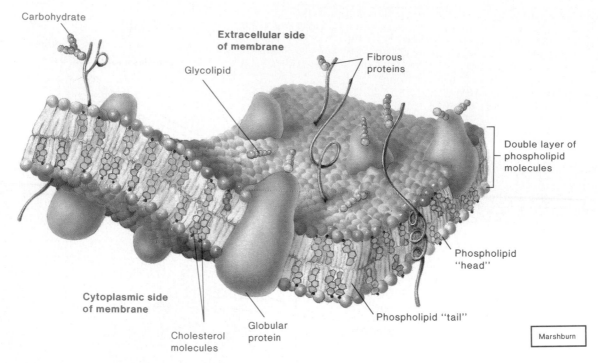

Carbohydrate

**Extracellular side
of membrane**

Glycolipid

**Fibrous
proteins**

Double layer of
phospholipid
molecules

Phospholipid
"head"

Phospholipid "tail"

**Cytoplasmic side
of membrane**

Cholesterol
molecules

Globular
protein

Marshburn

Figure 3.5 The cell membrane is composed primarily of
phospholipids, with proteins scattered throughout the lipid layer
and associated with its surfaces.

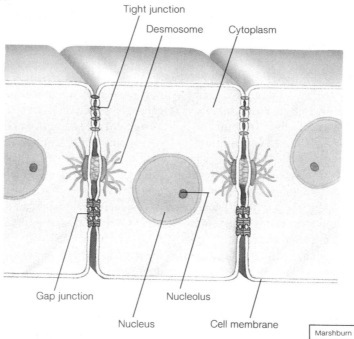

Tight junction

Desmosome

Cytoplasm

Gap junction

Nucleolus

Nucleus

Cell membrane

Marshburn

Figure 3.6 Some cells are joined by intercellular junctions,
such as tight junctions that fuse neighboring membranes
together, desmosomes that serve as "spot welds," or gap
junctions that allow movement of small molecules between the
cytoplasm of adjacent cells.

Another type of intercellular junction, called a
desmosome, serves to rivet or "spot weld" adjacent skin
cells, so they form a reinforced structural unit. The
membranes of certain other cells, such as those in heart
muscle and muscle of the digestive tube, are interconnected by tubular channels called *gap junctions.* These
channels link the cytoplasm of adjacent cells and allow
ions and nutrients (such as sugars, amino acids, and
nucleotides) and certain other relatively small molecules to be exchanged between them (figure 3.6). Chart
3.1 summarizes these intercellular junctions.

CHART 3.1 Types of intercellular junctions

Type	Function	Location
Tight junctions	Close intercellular space between cells by fusing cell membranes together	Cells that line inside of the small intestine
Desmosomes	Bind cells together by forming "spot welds" between adjacent membranes	Cells of the outer skin
Gap junctions	Form tubular channels between cells that allow substances to be exchanged	Muscle cells of the heart and digestive tube

Cytoplasm

When viewed through a light microscope, **cytoplasm** usually appears as a structureless substance with specks scattered throughout. However, a transmission electron microscope (figure 3.7), which produces much greater magnification and ability to distinguish fine detail (resolution), reveals that cytoplasm contains networks of membranes and other organelles suspended in a clear liquid (cytosol).

The maximum effective magnification that can be achieved using a light microscope is about 1,200× (diameters). A transmission electron microscope (TEM) can achieve an effective magnification of nearly 1,000,000×, while another type of electron microscope, the scanning electron microscope (SEM), can provide about 50,000×.

Although photographs of microscopic objects (micrographs) produced using the light microscope and the transmission electron microscope are typically two-dimensional, those obtained with the scanning electron microscope have a three-dimensional quality (figure 3.8).

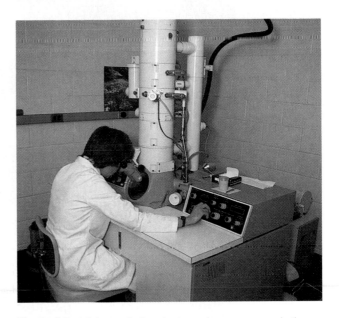

Figure 3.7 A transmission electron microscope reveals the cytoplasm of cells to be highly structured.

The activities of a cell occur largely in its cytoplasm, where food molecules are received, processed, and used. In other words, cytoplasm is a site of metabolic reactions, in which the following **cytoplasmic organelles** play specific roles:

1. **Endoplasmic reticulum.** The endoplasmic reticulum (ER) is a complex organelle composed of membrane-bound flattened sacs, elongated canals, and fluid-filled vesicles. These membranous parts are interconnected, and they communicate with the cell membrane, the nuclear envelope, and certain cytoplasmic organelles. Thus, endoplasmic reticulum is widely distributed through the cytoplasm. It functions as a tubular communication system through which molecules can be transported from one cell part to another.

   The endoplasmic reticulum also plays a role in the synthesis of protein and lipid molecules, some of which may be assembled into new membranes. Commonly, the outer membranous surface of the endoplasmic reticulum has numerous tiny, spherical organelles, called *ribosomes,* attached to it. These particles cause the endoplasmic reticulum to have a textured appearance when visualized using an electron microscope. Such endoplasmic reticulum is termed *rough* ER. Endoplasmic reticulum that lacks ribosomes is called *smooth* ER (figure 3.9).

   The ribosomes of rough endoplasmic reticulum synthesize proteins. The resulting molecules may be transported through tubules of the endoplasmic reticulum to the Golgi apparatus for further processing. Smooth endoplasmic reticulum, on the other hand, contains enzymes involved in manufacturing various lipid molecules, including steroid hormones.

2. **Ribosomes.** Although many ribosomes are attached to membranes of the endoplasmic reticulum, others occur as free particles scattered throughout the cytoplasm. In both cases, these particles are composed of protein

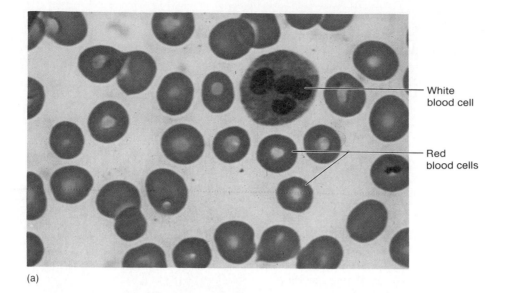

(a)

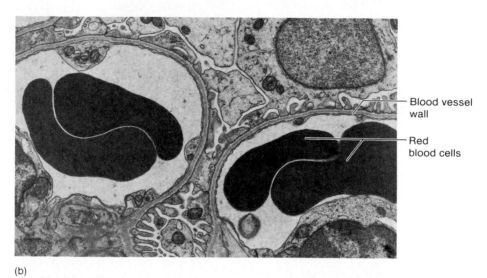

(b)

(c)

Figure 3.8   A human red blood cell as viewed using (*a*) a stained light microscope (×1,000), (*b*) a transmission electron microscope (×5,000), and (*c*) a falsely colored scanning electron microscope (×3,500).

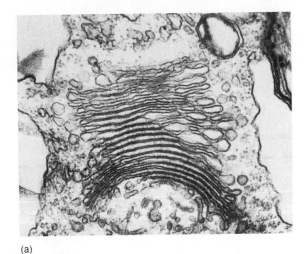

(a)

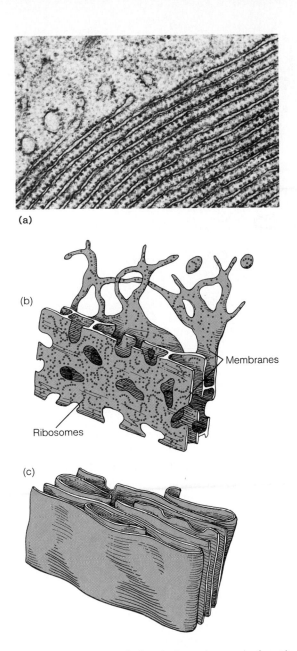

(b)

Membranes

Ribosomes

(c)

Figure 3.9   (a) A transmission electron micrograph of rough endoplasmic reticulum (×100,000). (b) Rough endoplasmic reticulum has ribosomes attached to its surface, while (c) smooth endoplasmic reticulum lacks ribosomes.

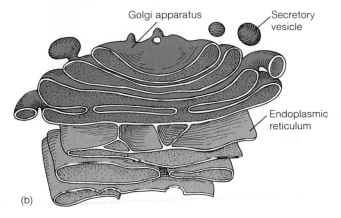

(a)

Golgi apparatus

Secretory vesicle

Endoplasmic reticulum

(b)

Figure 3.10   (a) A transmission electron micrograph of a Golgi apparatus (×36,000). (b) The Golgi apparatus consists of membranous sacs that are continuous with the endoplasmic reticulum.

and RNA molecules, and they function in the synthesis of proteins. This function is described in chapter 4.

3. **Golgi apparatus.** The Golgi apparatus is usually located near the nucleus. It is composed of a stack of half a dozen or so flattened, membranous sacs called *cisternae*. This apparatus is involved in refining, "packaging," and delivering the proteins that were synthesized by the ribosomes associated with the endoplasmic reticulum. (See figure 3.10.)

Proteins arrive at the Golgi apparatus enclosed in tiny sacs composed of membrane from the endoplasmic reticulum. These sacs fuse to the membrane at the end of the Golgi apparatus, which is specialized to receive proteins. Previously, these protein molecules were combined with sugar molecules, and thus they are *glycoproteins*.

The glycoproteins pass from layer to layer through the Golgi stacks and are modified chemically. For example, some sugar molecules may be added or removed from them. When they reach the outermost layer, the altered glycoproteins are packaged in bits of Golgi apparatus membrane that bud off and form transport vesicles. Such a vesicle may then move to the cell membrane, where it fuses with the membrane and releases its contents to the outside of the cell as a secretion. Other vesicles may transport glycoproteins to various cytoplasmic organelles within the cell. (See figure 3.11.) Note that the new membrane is first assembled in the endoplasmic reticulum. Portions of this membrane may then be

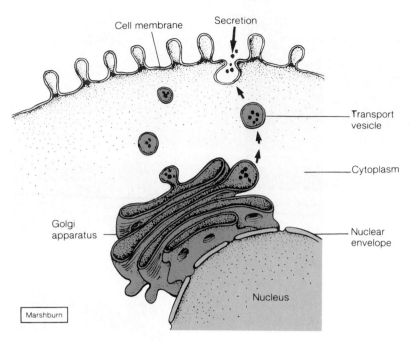

Cell membrane

Secretion

Transport vesicle

Cytoplasm

Nuclear envelope

Golgi apparatus

Nucleus

Marshburn

Figure 3.11 The Golgi apparatus functions in packaging glycoproteins, which then may be moved to the cell membrane and released as a secretion.

transferred to the Golgi apparatus. From there, some of the membrane may move outward as part of a vesicle and be added to the cell membrane.

In some cells, including certain liver cells and white blood cells (lymphocytes), glycoprotein molecules are secreted as rapidly as they are synthesized. However, in other cells, such as those that manufacture protein hormones, the vesicles containing the newly synthesized molecules remain in the cytoplasm and release the stored substances only when the cells receive special stimulation from the outside. (Hormone secretion is discussed in chapter 13.)

1. What is meant by a selectively permeable membrane?
2. Describe the chemical structure of a cell membrane.
3. What are the functions of the endoplasmic reticulum?
4. Describe how the Golgi apparatus functions.

4. **Mitochondria.** Mitochondria are elongated, fluid-filled sacs 2–5μm long. They often move about slowly in the cytoplasm and can reproduce by dividing. A mitochondrion also contains a supply of DNA that encodes information for making a few kinds of RNA and protein molecules. However, most of the structural and functional proteins of

mitochondria are encoded in the DNA of the nucleus. Thus, most mitochondrial proteins are synthesized elsewhere in the cell and are later taken up by the mitochondria.

The membrane of a mitochondrion has two layers—an outer membrane and an inner membrane. The inner membrane is folded extensively to form shelflike partitions called *cristae*. Small, stalked particles that contain enzymes are connected to the cristae. These enzymes and others dissolved in the fluid within the mitochondrion control some of the chemical reactions by which energy is released from glucose and other organic molecules. The mitochondria also function in transforming this newly released energy into a chemical form that is usable by various cell parts. For this reason mitochondria are sometimes called the "powerhouses" of cells. This energy-releasing function, involving molecules of the substance called adenosine triphosphate (ATP), is described in chapter 4 (figure 3.12).

5. **Lysosomes.** Lysosomes are sometimes difficult to identify because their shapes vary so greatly. However, they commonly appear as tiny, membranous sacs. (See figure 3.13.) These sacs contain powerful enzymes that are capable of breaking down protein, carbohydrate, and nucleic acid molecules, as well as foreign particles that sometimes enter cells. Certain white blood cells, for example, can engulf

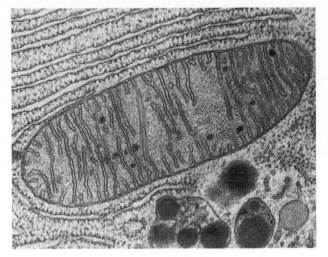

(a)

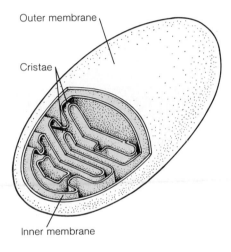

Outer membrane

Cristae

Inner membrane

(b)

Figure 3.12   (*a*) A transmission electron micrograph of a mitochondrion (×79,000). (*b*) What is the function of the cristae that form partitions within this saclike organelle?

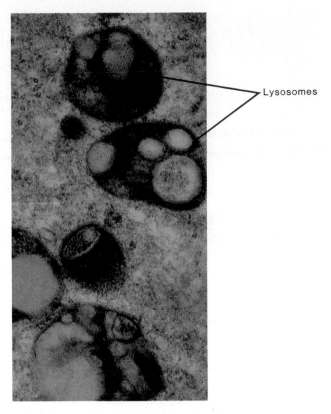

Lysosomes

Figure 3.13   In this falsely colored transmission electron micrograph, lysosomes appear as membranous sacs (×14,137).

bacteria that are then digested by the lysosomal enzymes. Consequently, white blood cells help prevent bacterial infections.

Lysosomes also function in the destruction of worn cellular parts. In fact, lysosomes of some scavenger cells may digest entire body cells that have been injured and engulfed. How the lysosomal membrane is able to withstand being digested itself is not well understood.

Lysosomal digestive activity seems to be responsible for decreasing the size of body tissues at certain times. Such regression in size occurs in the maternal uterus following the birth of an infant, in the maternal breasts after the weaning of an infant, and in skeletal muscles during periods of prolonged inactivity.

6.   **Peroxisomes.** Peroxisomes are membranous sacs that resemble lysosomes in size and shape. They occur most commonly in cells of the liver and kidneys, and they contain enzymes called *peroxidases.* These enzymes promote certain metabolic reactions that give rise to hydrogen peroxide (H_2O_2) as a by-product. Peroxisomes also contain an enzyme called *catalase.* This enzyme acts to decompose hydrogen peroxide, which is toxic to cells. Enzymes are discussed in chapter 4.

Although the specific functions of peroxisomes are not well understood, they play a role in the breakdown of fatty acid molecules and in the oxidation of, and thus the detoxification of, alcohol.

7.   **Centrosome.** A centrosome (central body) is located in the cytoplasm near the Golgi apparatus and nucleus. It is nonmembranous and consists of two hollow cylinders called *centrioles,* which in turn contain tiny, tubelike parts (microtubules). The centrioles usually lie at right angles to each other and function in cell reproduction. During this process, the centrioles move away from one another and take positions on either side of the nucleus.

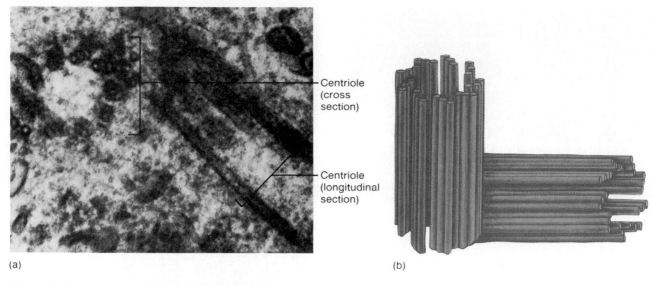

(a)

(b)

Figure 3.14   (a) A transmission electron micrograph of the two centrioles in a centrosome (×142,000). (b) Note that the centrioles lie at right angles to one another.

There they aid in distributing structures called *chromosomes,* which carry DNA information, to the newly forming cells (figure 3.14).

Centrioles also function in initiating the formation of hairlike cellular projections called *cilia* and *flagella.*

8.   **Cilia** and **flagella.** Cilia and flagella are motile processes that extend outward from the surfaces of certain cells. They are structurally similar and differ mainly in their length and the number present. Both contain a constant number of tubelike components (microtubules) arranged in a distinct pattern.

Cilia occur in large numbers on the free surfaces of some epithelial cells. Each cilium is a tiny, hairlike structure about 10 μm long, which is attached just beneath the cell membrane to a modified centriole called a *basal body.*

Cilia are arranged in precise patterns, and they have a "to-and-fro" type of movement. This movement is coordinated so that rows of cilia beat one after the other, producing a wave of motion that sweeps across the ciliated surface. This action propels fluids, such as mucus, over the surface of certain tissues, including those that form the lining of the respiratory tubes. (See figure 3.15.)

Flagella are considerably longer than cilia, and usually there is only a single flagellum on a cell. This projection has an undulating, wavelike motion that begins at its base. The tail of a sperm cell, for example, is a flagellum that causes the sperm's swimming

(a)

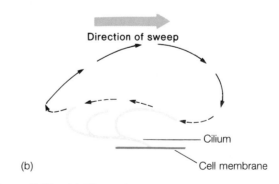

Direction of sweep

Cilium

(b)   Cell membrane

Figure 3.15   (a) Cilia, such as these (arrow), are common on the surface of certain cells that form the inner lining of respiratory tubes (×10,000). (b) Cilia have a "to-and-fro" movement.

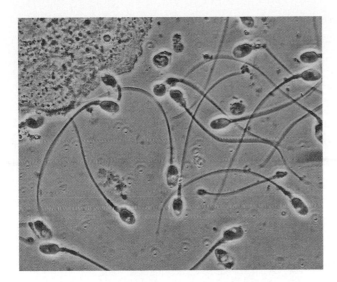

Figure 3.16 Light micrograph of human sperm cells (×1,200). Flagella form the tails of these cells.

Figure 3.17 A transmission electron micrograph of microfilaments and microtubules within the cytoplasm of a cell.

movements. (See figure 3.16 and chapter 22.) Flagella also occur in certain cells of the kidneys, testes, and various glands.

9. **Vesicles.** Vesicles (vacuoles) are membranous sacs that vary in size. They may be formed by an action of the cell membrane in which a portion of the membrane folds inward and pinches off. As a result, a tiny, bubblelike vesicle, containing some liquid or solid material that was formerly outside the cell, appears in the cytoplasm.

   The process of endocytosis, which involves vesicle formation, is discussed in a subsequent section of this chapter.

10. **Microfilaments** and **microtubules.** Two types of thin, threadlike structures found within the cytoplasm are microfilaments and microtubules.

    Microfilaments are tiny rods of protein (actin), arranged in meshworks or bundles. They function to cause various kinds of cellular movements. In muscle cells, for example, they are highly developed as *myofibrils,* which help these cells shorten or contract. In other cells, microfilaments are usually associated with the inner surface of the cell membrane and seem to aid in cell motility. (See figure 3.17.)

    Microtubules are long, slender tubes with diameters two or three times greater than microfilaments, and are composed of globular protein (tubulin). They are usually stiff, forming an "internal skeleton" within a cell, which helps maintain the shape of the cell or its parts. For example, microtubules provide strength to the structure of cilia and flagella. (See figures 3.17 and 3.18.)

Microtubules also aid in moving organelles from one place to another. For instance, they appear in the cytoplasm during cellular reproduction and are involved in the distribution of DNA-containing parts (chromosomes) to the newly forming cells. This process is described in more detail in a subsequent section of this chapter.

In addition to these functional organelles, cytoplasm may contain masses of lifeless chemical substances called *inclusions.* Most commonly, inclusions consist of cellular products that remain in a cell only temporarily. Examples of inclusions are stored nutrients (such as glycogen and various lipids) and pigments (such as melanin in the skin).

1. Why are mitochondria sometimes called the "powerhouses" of cells?
2. How do lysosomes within certain white blood cells function?
3. Describe the functions of microfilaments and microtubules.
4. Distinguish between organelles and inclusions.

Cell Nucleus

A **nucleus** is a cellular organelle that is usually located near the center of a cell. It is a relatively large, spherical structure that directs the activities of the cell.

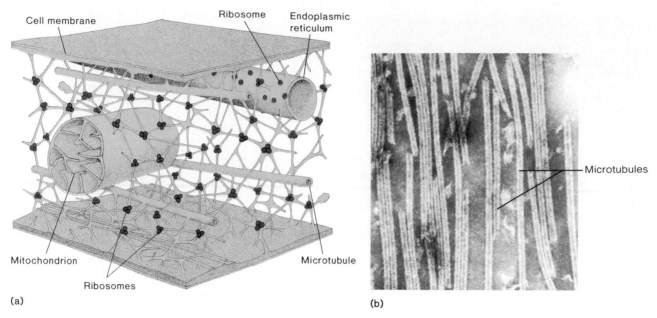

Cell membrane   Ribosome   Endoplasmic reticulum

Mitochondrion

Ribosomes

Microtubule

(a)

Microtubules

(b)

Figure 3.18 (a) Microtubules help maintain the shape of a cell by forming an "internal skeleton" within the cytoplasm; (b) a transmission electron micrograph of microtubules. (a) Adapted

from "The Ground Substance of the Living Cell," by Keith Porter and Jonathan Tucker. Copyright © 1981 by Scientific American, Inc. All rights reserved.

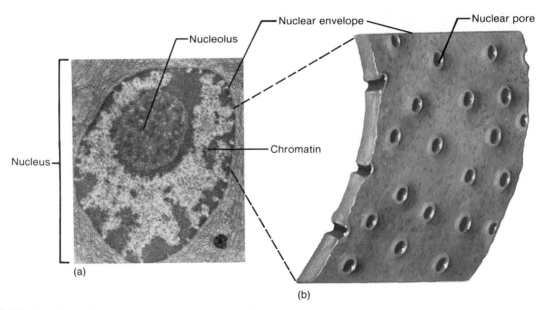

Nuclear envelope   Nuclear pore

Nucleolus

Nucleus

Chromatin

(a)

(b)

Figure 3.19 (a) A transmission electron micrograph of a cell nucleus (×8,000). It contains a nucleolus and masses of chromatin. (b) The nuclear envelope is porous and allows substances to pass between the nucleus and the cytoplasm.

The nucleus is enclosed in a double-layered **nuclear envelope,** which consists of an inner and an outer membrane. These two membranes have a narrow space between them, but are joined at various places that surround relatively large openings or "pores." The pores in the nuclear envelope allow certain dissolved substances to move between the nucleus and the cytoplasm (figure 3.19).

The nucleus contains a fluid (nucleoplasm) in which other structures float. These structures include the following:

1. **Nucleolus.** A nucleolus ("little nucleus") is a small, dense body composed largely of RNA and protein. It has no surrounding membrane and is formed in specialized regions of certain

CHART 3.2 Structures and functions of cellular organelles

Organelle	Structure	Function
Cell membrane	Membrane composed mainly of protein and lipid molecules	Maintains integrity of the cell and controls the passage of materials into and out of the cell
Endoplasmic reticulum	Complex of interconnected membrane-bound sacs, canals, and vesicles	Transports materials within the cell, provides attachment for ribosomes, and synthesizes lipids
Ribosomes	Particles composed of protein and RNA molecules	Synthesize proteins
Golgi apparatus	Group of flattened, membranous sacs	Packages protein molecules for transport and secretion
Mitochondria	Membranous sacs with inner partitions	Release energy from food molecules and transform energy into usable form
Lysosomes	Membranous sacs	Contain enzymes capable of digesting worn cellular parts or substances that enter cells
Peroxisomes	Membranous vesicles	Contain enzymes called peroxidases
Centrosome	Nonmembranous structure composed of two rodlike centrioles	Helps distribute chromosomes to new cells during cell reproduction and initiates formation of cilia
Cilia and flagella	Motile projections attached to basal bodies beneath the cell membrane	Propel fluids over cellular surface and enable sperm cells to move
Vesicles	Membranous sacs	Contain various substances that recently entered the cell and store and transport newly synthesized molecules
Microfilaments and microtubules	Thin rods and tubules	Provide support to cytoplasm and help move substances and organelles within the cytoplasm
Nuclear envelope	Porous double membrane that separates the nuclear contents from the cytoplasm	Maintains the integrity of the nucleus and controls the passage of materials between the nucleus and cytoplasm
Nucleolus	Dense, nonmembranous body composed of protein and RNA molecules	Forms ribosomes
Chromatin	Fibers composed of protein and DNA molecules	Contains cellular information for synthesizing proteins needed in carrying on life processes

chromosomes. It functions in the production of ribosomes, and once the ribosomes are formed, they migrate through the pores in the nuclear envelope and enter the cytoplasm. The nuclei of cells that synthesize large amounts of protein, such as those of glands, may contain especially large nucleoli.

2. **Chromatin.** Chromatin consists of loosely coiled fibers present in the nuclear fluid. When the cell begins to undergo the reproductive process, these fibers become more tightly coiled and are transformed into tiny, rodlike *chromosomes*. Chromatin fibers are composed of protein (histones) and DNA molecules, which in turn are organized into tiny, beadlike particles (nucleosomes). The DNA molecules contain the information for synthesis of proteins that are needed to promote cellular life processes.

Chart 3.2 summarizes the structures and functions of the cellular organelles.

1. How are the nuclear contents separated from the cytoplasm of a cell?
2. What is the function of the nucleolus?
3. What is chromatin?

Although the effects of aging on cells are poorly understood, studies of aged tissues indicate that certain organelles may be altered with time. For example, within the nucleus, the chromatin and chromosomes may show changes such as clumping, shrinking, or fragmenting, and the number and size of the nucleoli may increase. In the cytoplasm, the mitochondria may undergo changes in shape and number, and the Golgi apparatus may become fragmented. Lipid inclusions tend to accumulate in the cytoplasm, while glycogen-containing inclusions tend to disappear.

Movements through Cell Membranes

The cell membrane provides a surface through which various substances enter and leave the cell. More specifically, oxygen and food molecules enter a cell through this membrane, while carbon dioxide and other wastes leave through it. These movements involve *physical* (or nonliving) processes such as diffusion, facilitated diffusion, osmosis, and filtration, and *physiological* (or living) mechanisms such as active transport and

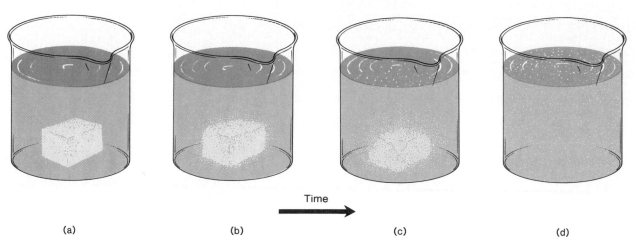

Time

(a)　　　　　　　(b)　　　　　　　(c)　　　　　　　(d)

Figure 3.20 An example of diffusion. (a, b, and c) If a sugar cube is placed in water, it slowly disappears. As this happens, the sugar molecules diffuse from regions where they are more concentrated toward regions where they are less concentrated. (d) Eventually, the sugar molecules are distributed evenly throughout the water.

endocytosis. These processes are important for understanding many of the topics presented in chapters throughout this textbook.

Diffusion

Diffusion is the process by which molecules or ions become scattered or are spread spontaneously from regions where they are in higher concentrations toward regions where they are in lower concentrations.

Under natural conditions, molecules and ions are constantly moving at high speeds. Each particle travels in a separate path along a straight line until it collides and bounces off some other particle. Then it moves in another direction, only to collide again and change direction once more. Such random motion accounts for the mixing of molecules that commonly occurs when different kinds of substances are put together.

For example, when sugar (a solute) is put into a glass of water (a solvent), as illustrated in figure 3.20, the sugar seems to remain in high concentration at the bottom of the glass. Then it slowly disappears into solution. As this is happening, the moving water and sugar molecules are colliding with one another, and in time the sugar and water molecules will be evenly mixed. This mixing process, by which the sugar molecules spread from the region where they are in higher concentration toward the regions where they are less concentrated, is diffusion. As a result of the process, the sugar eventually becomes uniformly distributed in the water. This state of uniform distribution of molecules is called *equilibrium,* and although the molecules continue to move after equilibrium is achieved, their concentrations no longer change.

To better understand how diffusion accounts for the movement of various molecules through a cell membrane, imagine a container of water that is separated into two compartments by a permeable membrane (figure 3.21). This membrane has numerous pores that are large enough for water and sugar molecules to pass through. The sugar molecules are placed in one compartment (A) but not in the other (B). Although the sugar molecules move randomly in all directions, more diffuse from compartment A (where they are in greater concentration) through the pores in the membrane and into compartment B (where they are in lesser concentration) than move in the other direction. At the same time, water molecules tend to diffuse from compartment B (where they are in greater concentration) through the pores into compartment A (where they are in lesser concentration). Eventually, equilibrium will be achieved, with equal numbers of water and of sugar molecules in each compartment.

Similarly, oxygen molecules diffuse through cell membranes and enter cells if the molecules are more highly concentrated on the outside than on the inside of the membranes. Carbon dioxide molecules, too, diffuse through cell membranes and leave cells if they are more concentrated on the inside than on the outside. It is by diffusion that oxygen and carbon dioxide molecules are exchanged between the air and the blood in the lungs, and between the blood and the cells of various tissues.

Several factors influence the rate at which diffusion occurs. These include the distance over which the diffusion will occur, the concentrations of the diffusing substances, the weight of the diffusing molecules, and the temperature of the diffusing molecules. Generally, diffusion occurs more rapidly over a shorter distance, when the concentration of the diffusing substance is greater, when the molecular weight is lower, and when the temperature is higher.

Dialysis and Its Use

Dialysis is the process of separating smaller molecules from larger ones in a liquid by means of diffusion. This process is employed in the clinical procedure called *hemodialysis,* when an artificial kidney is used to treat patients with kidney dysfunctions. When this device is operating, blood from a patient passes through a long, coiled tubing composed of porous cellophane. The size of the pores allows smaller molecules carried in the blood, such as urea, to pass out through the cellophane, while larger molecules, such as those of blood proteins, remain inside the tubing. The tubing is submerged in a tank of dialyzing fluid (wash solution) that contains varying concentrations of chemicals. For instance, the solution is low in concentrations of substances that should leave the blood, and higher in concentrations of those that should remain in the blood.

Because it is desirable for an artificial kidney to remove blood urea, the dialyzing fluid must have a lower concentration of urea than does the blood; it is also desirable to maintain the blood glucose concentration, so the concentration of glucose in the dialyzing fluid must be kept at least

equal to that of the blood. Thus, by altering the concentrations of molecules in the dialyzing fluid, it is possible to control those molecules that will diffuse out of the blood and those that will remain in it.

In order to prevent the blood that is entering the artificial kidney from coagulating, an anticoagulative substance such as heparin is added to it. Then, as the blood returns to the patient, a drug that will counteract the effect of the anticoagulant is administered to prevent subsequent excessive bleeding.

Another procedure sometimes used to help patients with kidney problems is called *peritoneal dialysis* (PD). In this case, dialyzing fluid is infused into the patient's abdominal cavity through an artificial opening. This cavity is naturally lined with a thin membrane called the peritoneum (see chapter 1), which serves as a dialyzing membrane. As before, the composition of the dialyzing fluid is controlled, so that unwanted substances will diffuse from the blood of blood vessels in the peritoneum into the dialyzing fluid. After some time, the fluid is drained from the abdominal cavity and replaced with fresh fluid.

- • Water molecule
- ○ Sugar molecule

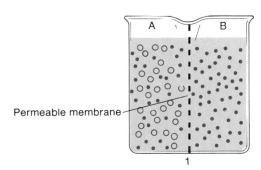
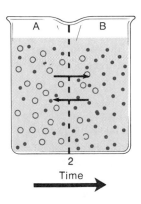
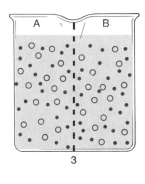

Time →

Figure 3.21 (*1*) The container is separated into two compartments by a permeable membrane. Compartment A contains water molecules and sugar molecules, while compartment B contains only water molecules. (*2*) As a result of molecular motions, sugar molecules will tend to diffuse from compartment A into compartment B. Water molecules will tend to diffuse from compartment B into compartment A. (*3*) Eventually, equilibrium is achieved.

Permeable membrane

Facilitated Diffusion

Most sugars are insoluble in lipids, and they have molecular sizes that prevent them from passing through membrane pores. However, the sugar glucose may still enter through the lipid portion of the membrane of some cells by a process called *facilitated diffusion.* In this process, the glucose molecule combines with a special protein carrier molecule at the surface of the membrane. This union of glucose and carrier molecule forms a compound that is soluble in lipids and that can diffuse to the other side. There the glucose portion is released, and the carrier molecule can return to the opposite side of the membrane and pick up another glucose molecule. The hormone called *insulin,* which is discussed in chapter 13, promotes facilitated diffusion of glucose through the membranes of certain cells.

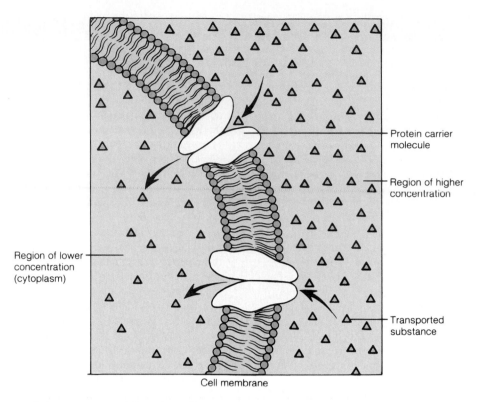

Region of lower
concentration
(cytoplasm)

Protein carrier
molecule

Region of higher
concentration

Transported
substance

Cell membrane

Figure 3.22 Some substances are moved through a cell membrane by facilitated diffusion, a process by which carrier molecules transport a substance from a region of higher concentration to one of lower concentration.

Facilitated diffusion is similar to simple diffusion in that it can only cause movement of molecules from regions of higher concentration toward regions of lower concentration. The rate at which facilitated diffusion can occur, however, is limited by the number of carrier molecules in the cell membrane (figure 3.22).

Osmosis

Osmosis is a special case of diffusion. It occurs whenever *water* molecules diffuse from a region of higher water concentration (where the solute concentration is lower) to a region of lower water concentration (where the solute concentration is higher) through a selectively permeable membrane, such as a cell membrane. In the following examples, it is assumed that the membranes involved are permeable to water molecules but impermeable to glucose molecules.

Ordinarily, the concentrations of water molecules are equal on either side of a cell membrane. Sometimes, however, the water on one side has more solute dissolved in it than does the water on the other side. For example, if there is a greater concentration of glucose (solute) in the water outside a cell, there must be a lesser concentration of water there because the glucose molecules occupy space that would otherwise contain water molecules. Under such conditions, water molecules diffuse from the inside of the cell (where they are in higher concentration) toward the outside (where they are in lesser concentration) of the cell.

This process, shown in figure 3.23, is similar to diffusion (figure 3.21). In osmosis, however, the membrane involved is selectively permeable—that is, water molecules pass through readily, but glucose molecules do not. In diffusion, the membrane is *permeable* and allows both water and glucose molecules to pass through.

Note in figure 3.23 that as osmosis occurs, the volume of water on side A increases. This increase in volume would be resisted if pressure were applied to the surface of the liquid on side A. The amount of pressure needed to stop osmosis in such a case is called *osmotic pressure*. Thus, the osmotic pressure of a solution is a potential pressure and is due to the presence of non-diffusible solute particles in that solution. Furthermore, the greater the number of solute particles in the solution, the greater the osmotic pressure of that solution.

When osmosis occurs, water tends to move toward the region of greater osmotic pressure. Because the amount of osmotic pressure depends upon the difference in concentration of solute particles on opposite sides of the membrane, the greater the difference, the greater the tendency for water to move toward the region of higher solute concentration.

If some cells were put into a water solution that had a greater concentration of solute particles (higher osmotic pressure) than did the cells, there would be a net movement of water out of the cells. Consequently,

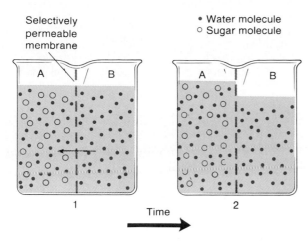

Selectively permeable membrane

• Water molecule
○ Sugar molecule

Time

Figure 3.23   An example of osmosis. (1) The container is separated into two compartments by a selectively permeable membrane. At first, compartment A contains water and sugar molecules, while compartment B contains only water molecules. As a result of molecular motions, water molecules will tend to diffuse by osmosis from compartment B into compartment A. The sugar molecules will remain in compartment A because they are too large to pass through the pores of the membrane. (2) Also, because more water is entering compartment A than is leaving it, water will accumulate in this compartment, and the level of the liquid will rise on this side.

the cells would begin to shrink. In time, equilibrium might be achieved, and the shrinking would cease. A solution of this type, in which more water leaves a cell than enters it because the concentration of solute particles is greater outside the cell, is called **hypertonic.**

Conversely, if there is a greater concentration of solute particles inside a cell than in the water around it, water will diffuse into the cell. As this happens, water accumulates within the cell, and it begins to swell. Although cell membranes are somewhat elastic, they may swell so much that they burst. A solution in which more water enters a cell than leaves it (because of a lesser concentration of solute particles outside the cell) is called **hypotonic.**

It is important to control the concentration of solute in solutions that are infused into body tissues or blood. Otherwise, osmosis may cause cells to swell or shrink, and they may be damaged. For instance, if red blood cells are placed in distilled water (which is hypotonic to them), water will diffuse into the cells and they will burst (hemolyze). On the other hand, if red blood cells are exposed to 0.9% NaCl solution (normal saline), the cells will remain unchanged because this solution is isotonic to human cells. Similarly, a 5% solution of glucose is isotonic to human cells. (The lower percentage is needed with NaCl to produce an isotonic solution, in part because NaCl ionizes in solution more completely and gives rise to a greater number of solute particles than does glucose.)

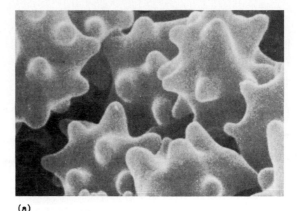

(a)

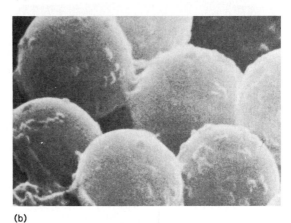

(b)

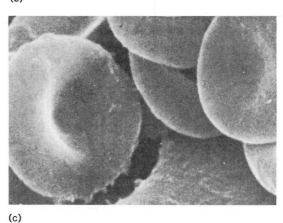

(c)

Figure 3.24   (a) If red blood cells are placed in a hypertonic solution, more water leaves the cells than enters them, and they shrink. (b) In a hypotonic solution, more water enters than leaves the cells, and they swell and may burst. (c) In an isotonic solution, water enters and leaves the cells in equal amounts, and their sizes remain unchanged. (Note: The magnification of the cells in (b) is lower than in the other two scanning electron micrographs.)

A solution that contains the same concentration of solute particles as a particular cell is said to be **isotonic** to that cell. In such a solution, water enters and leaves the cell in equal amounts, and the cell's size remains unchanged. Figure 3.24 illustrates the three types of solutions just discussed.

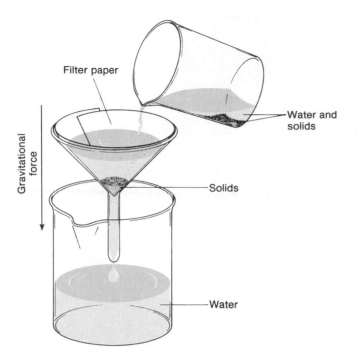

Figure 3.25 In this example of filtration, gravity provides the force that pulls water through filter paper, while the tiny openings in the paper hold the solids behind.

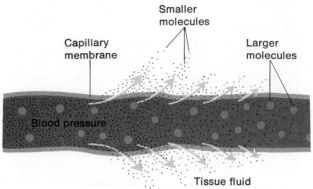

Figure 3.26 In this example of filtration, smaller molecules are forced through tiny openings in the wall of a capillary by blood pressure. The larger molecules remain inside.

1. What kinds of substances move most readily through a cell membrane by diffusion?
2. Explain the differences between diffusion, facilitated diffusion, and osmosis.
3. Distinguish among hypertonic, hypotonic, and isotonic solutions.
4. Explain how filtration occurs within the body.

Filtration

The passage of substances through membranes by diffusion or osmosis is due to movements of the molecules of those substances. In other instances, molecules are forced through membranes by the hydrostatic pressure, called blood pressure, that is greater on one side of the membrane than on the other. This process by which molecules are forced through membranes is called **filtration.**

Filtration is commonly used to separate solids from water. One method is to pour a mixture of solids and water onto filter paper in a funnel (figure 3.25). The paper serves as a porous membrane through which the small water molecules can pass, leaving the larger solid particles behind. Hydrostatic pressure, which is created by the weight of water due to gravity, forces the water molecules through to the other side.

Similarly, in the body, tissue fluid is formed when water and dissolved substances are forced out through the thin, porous walls of blood capillaries, but larger particles such as blood protein molecules are left inside (figure 3.26). The force for this movement comes from blood pressure, created largely by heart action, which is greater within the vessel than outside it. Filtration also takes place in the kidneys when water and various dissolved substances are forced out of blood vessels and into kidney tubules by blood pressure. This is the first step in the formation of urine, which is discussed in chapter 20.

Active Transport

When molecules or ions pass through cell membranes by diffusion, facilitated diffusion, or osmosis, their net movement is from regions of higher concentration to regions of lower concentration. Sometimes, however, the net movement of particles passing through membranes is in the opposite direction; that is, the particles move from a region of lower concentration to one of higher concentration.

It is known, for example, that sodium ions can diffuse through cell membranes. Yet, the concentration of these ions typically remains many times greater on the outside of cells (in the extracellular fluid) than on the inside (in the intracellular fluid) of cells. Furthermore, sodium ions are continually moved through the cell membrane from the regions of lower concentration (inside) to the regions of higher concentration (outside). Movement of this type is called **active transport.** It depends on life processes within cells and requires energy that is released by cellular metabolism. In fact, it is estimated that up to 40% of a cell's energy supply may be used for active transport of particles through its membranes.

The mechanism of active transport is similar to facilitated diffusion in that it involves specific carrier molecules found within cell membranes. As figure 3.27 shows, these carrier molecules are proteins that have binding sites that combine with the particles being transported. Such a union triggers the release of cellular energy, and this energy causes the shape of the

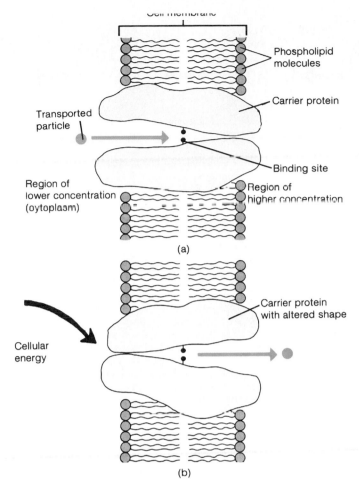

Cell membrane

Phospholipid
molecules

Carrier protein

Transported
particle

Binding site

Region of
lower concentration
(cytoplasm)

Region of
higher concentration

(a)

Carrier protein
with altered shape

Cellular
energy

(b)

Figure 3.27 (*a*) During active transport, a molecule or ion combines with a carrier protein, whose shape is altered as a result. (*b*) This action helps move the transported particle through the cell membrane.

carrier protein to be altered. As a result, the "passenger" molecules are moved through the membrane. Once on the other side, the transported particles are released as a result of an enzyme action, and the carrier molecules can accept other passenger molecules at their binding sites.

Particles that are carried across cell membranes by active transport include various sugars and amino acids, as well as a variety of ions such as those of sodium, potassium, calcium, and hydrogen. Movements of this type are important to cell survival, and are involved in the maintenance of homeostasis. Some of these movements are described in subsequent chapters.

Endocytosis

Another physiological process by which substances may move through cell membranes is called **endocytosis.** In this case, molecules or other particles that are too large to enter a cell by diffusion or active transport may be conveyed within a vesicle formed from a section of the cell membrane.

There are three forms of endocytosis—pinocytosis, phagocytosis, and receptor-mediated endocytosis.

Pinocytosis

Pinocytosis refers to the process by which cells take in tiny droplets of liquid from their surroundings (figure 3.28). When this happens, a small portion of cell membrane becomes indented (invaginated). The open end of the tubelike part thus formed seals off and produces a small vesicle about 0.1 μm in diameter. This tiny sac becomes detached from the surface and moves into the cytoplasm.

For a time, the vesicular membrane, which was part of the cell membrane, separates its contents from the rest of the cell; but eventually, the membrane breaks down and the liquid inside becomes part of the cytoplasm. In this way, a cell is able to take in water and

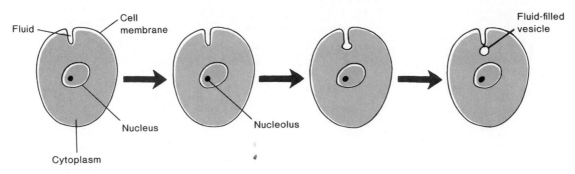

Figure 3.28 A cell may take in a tiny droplet of fluid from its surroundings by pinocytosis.

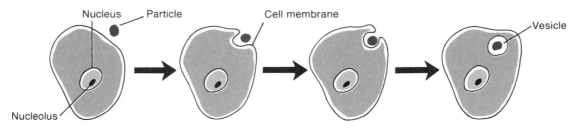

Figure 3.29 A cell may take in a solid particle from its surroundings by phagocytosis.

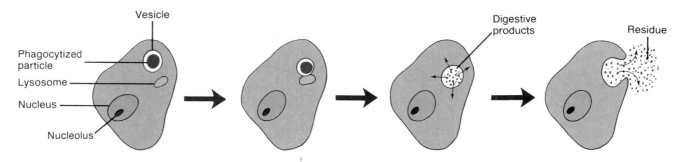

Figure 3.30 When a lysosome combines with a vesicle that contains a phagocytized particle, its digestive enzymes may destroy the particle. The products of digestion diffuse into the cytoplasm. Any residue may be expelled from the cell.

the particles dissolved in it, such as proteins, that otherwise might be unable to enter because of their relatively large molecular size.

Phagocytosis

Phagocytosis is a process that is essentially the same as pinocytosis. In phagocytosis, however, the material taken into the cell is solid rather than liquid.

Certain kinds of cells, including some white blood cells, are called *phagocytes* because they can take in tiny solid particles such as bacterial cells and cellular debris. When a phagocyte first encounters such a particle, the particle becomes attached to the phagocyte's cell membrane. This stimulates a portion of the membrane to project outward, surround the particle, and slowly draw it inside. The part of the membrane surrounding the solid detaches from the cell's surface, and

a vesicle containing the particle is formed (figure 3.29). Such a vesicle may be several micrometers in diameter.

Commonly, a lysosome soon combines with such a newly formed vesicle, and lysosomal digestive enzymes cause the vesicular contents to be decomposed (figure 3.30). The products of this decomposition may then diffuse out of the lysosome and into the cell's cytoplasm, where they may be used as raw materials in metabolic processes. Any remaining residue may be expelled from the cell (exocytosis). In this way, phagocytic cells can dispose of foreign objects like dust particles, remove damaged cells or cell parts that are no longer functional, or destroy bacteria that might otherwise cause infections.

Although pinocytosis is thought to provide only a minor route for substances entering cells, phagocytosis is an important line of defense against invasion by disease-causing microorganisms.

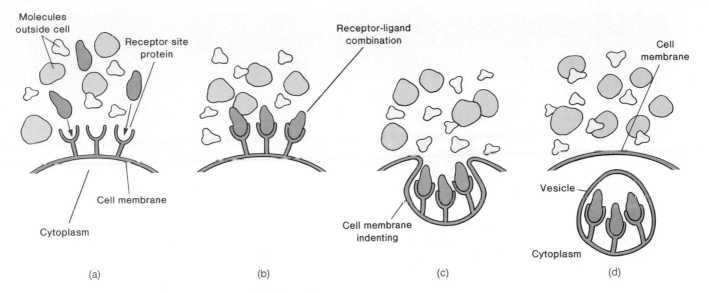

Figure 3.31 Receptor-mediated endocytosis. (a) A specific substance binds to a receptor site protein; (b and c) the combination of the substance with the receptor site protein stimulates the cell membrane to indent; (d) the resulting vesicle moves the transported substance into the cytoplasm.

As a consequence of an inherited trait, the lysosomes of some individuals fail to produce various digestive enzymes. This leads to one of several "storage diseases," which are characterized by an accumulation of undigested materials in certain types of cells. Tay-Sachs disease is an example of such a condition. In this disease, lipids accumulate abnormally in certain nerve cells, producing serious neurological disorders.

Receptor-Mediated Endocytosis

Although a variety of substances may enter cells by pinocytosis or phagocytosis, **receptor-mediated endocytosis** moves very specific kinds of particles through membranes. This process involves the presence of particular protein molecules that extend through the cell membrane and are exposed on its outer surface. These proteins serve as *receptor sites* to which specific substances (ligands) from the fluid surroundings of the cell can bind. Thus, when molecules that are capable of binding to the receptor sites are present, they are the ones that are selected to enter the cell, and other kinds of molecules are left outside.

For example, cholesterol molecules that have been synthesized in liver cells are packaged in relatively large spherical particles called *low-density lipoproteins* (LDL). An LDL particle is surrounded by a membrane that contains a binding protein called *apoprotein-B*. The membranes of various body cells possess receptor sites for apoprotein-B. When LDL particles are released by the liver into the blood, the body cells with apoprotein-B receptor sites can recognize the LDL particles and bind to them.

The formation of such a receptor-ligand combination somehow stimulates the body cell membrane to indent and form a vesicle that contains the LDL particle—an action similar to that described previously in the discussion of pinocytosis. The vesicle transports the LDL particle to a lysosome, where the lysosomal enzymes digest it and release the cholesterol molecules for cellular uses.

Receptor-mediated endocytosis is of particular importance because it allows a cell to remove specific kinds of substances, such as hormones and certain blood proteins, from its surroundings, even when these substances are present in very low concentrations. (See figure 3.31.)

Chart 3.3 summarizes the types of movement through cell membranes.

1. What type of mechanism is responsible for maintaining unequal concentrations of ions on opposite sides of a cell membrane?
2. How are the processes of facilitated diffusion and active transport similar? How are they different?
3. What is the difference between pinocytosis and phagocytosis?
4. Describe receptor-mediated endocytosis.

Life Cycle of a Cell

The series of changes that a cell undergoes from the time it is formed until it reproduces is called its *life cycle* (figure 3.32). Superficially, this cycle seems rather simple—a newly formed cell grows for a time and then divides in half to form two new cells, which in turn may

CHART 3.3   Movements through cell membranes

Process	Characteristics	Source of energy	Example
I. Physical processes			
A. Diffusion	Molecules or ions move from regions of higher concentration toward regions of lower concentration.	Molecular motion	Exchange of oxygen and carbon dioxide in the lungs
B. Facilitated diffusion	Molecules are moved through a membrane by carrier molecules from a region of higher concentration to one of lower concentration.	Molecular motion	Movement of glucose through a cell membrane
C. Osmosis	Water molecules move from regions of higher concentration toward regions of lower concentration through a selectively permeable membrane.	Molecular motion	Distilled water entering a cell
D. Filtration	Smaller molecules are forced through porous membranes from regions of higher pressure to regions of lower pressure.	Hydrostatic pressure	Molecules leaving blood capillaries
II. Physiological processes			
A. Active transport	Molecules or ions are carried through membranes by other molecules from regions of lower concentration toward regions of higher concentration.	Cellular energy	Movement of various ions and amino acids through membranes
B. Endocytosis			
1. Pinocytosis	Membrane acts to engulf minute droplets of liquid from surroundings.	Cellular energy	Membrane-forming vesicles containing particles of relatively large molecular size dissolved in water
2. Phagocytosis	Membrane acts to engulf solid particles from surroundings.	Cellular energy	White blood cell membrane engulfing bacterial cell
3. Receptor-mediated endocytosis	Membrane acts to engulf selected molecules combined with receptor proteins.	Cellular energy	Cell removing cholesterol-containing LDL particles from its surroundings

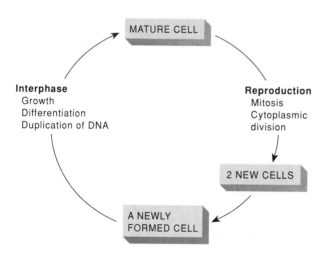

Figure 3.32   Life cycle of a cell.

grow and divide. Yet the details of the cycle are quite complex, involving mitosis, cytoplasmic division, interphase, and differentiation.

Mitosis

Cell reproduction involves the dividing of a cell into two portions and includes two separate processes: (1) di-

vision of the nuclear parts, which is called **mitosis** (karyokinesis), and (2) division of the cytoplasm (cytokinesis). Another type of cell division called *meiosis* occurs during the maturation of sex cells and is described in chapter 22.

Division of the nuclear parts by mitosis is, of necessity, very precise, because the nucleus contains information (in the form of DNA molecules) that "tells" cell parts how to carry on life processes. Each new cell resulting from mitosis must have a copy of this information in order to survive.

Although mitosis is often described in stages, the process is actually continuous without marked changes between one step and the next (figure 3.33). The idea of stages is useful, however, to indicate the sequence in which major events occur. The stages of mitosis include the following:

1. **Prophase.** One of the first indications that a cell is going to reproduce is the appearance of *chromosomes*. These structures are formed from chromatin in the nucleus, as fibers of chromatin condense into tightly coiled, rodlike parts. The resulting chromosomes contain DNA and protein molecules. Sometime earlier

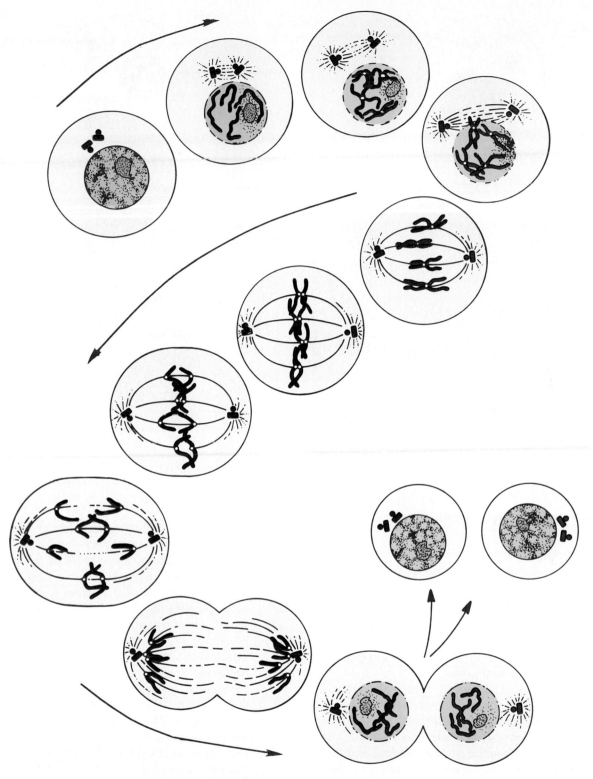

Figure 3.33 Mitosis is a continuous process during which the nuclear parts of a cell are divided into two equal portions. After reading about mitosis, identify the phases of the process and the cell parts shown in this diagram.

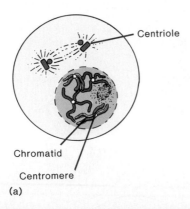

Centriole

Chromatid

Centromere

(a)

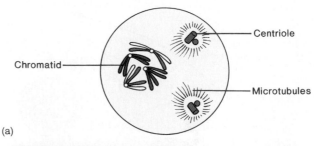

Chromatid

Centriole

Microtubules

(a)

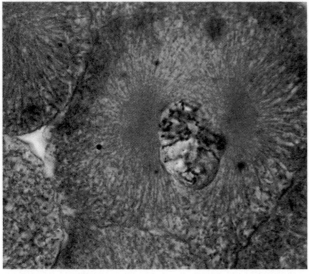

(b)

Figure 3.34 (a) In prophase, chromosomes form from chromatin in the nucleus, and the centrioles move to opposite sides of the cell. (b) What features can you identify in this micrograph of a cell in prophase (×250)?

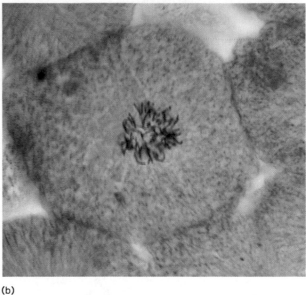

(b)

Figure 3.35 (a) Later in prophase, the nuclear envelope and nucleolus disappear. (b) A micrograph of a cell in late prophase.

(during interphase), the DNA molecules became replicated (duplicated), and consequently each chromosome is composed of two identical portions (chromatids). These parts are temporarily attached by a region on each called a *centromere*.

The centrioles of the centrosome replicate just before the onset of mitosis, and during prophase, the two newly formed pairs of centrioles move to opposite sides of the cytoplasm. Soon the nuclear envelope and the nucleolus become dispersed and are no longer visible. Microtubules are assembled from proteins in the cytoplasm, and these structures become associated with the centrioles and chromosomes (figures 3.34 and 3.35). A spindle-shaped group of microtubules (spindle fibers) forms between the centrioles as they move apart.

2. **Metaphase.** The chromosomes line up in an orderly fashion about midway between the centrioles, apparently as a result of microtubule activity. Spindle fibers become attached to the centromeres of the chromosomes so that a fiber accompanying one pair of centrioles is attached to one side of a centromere, and a fiber accompanying the other pair of centrioles is attached to the other side (figure 3.36).

3. **Anaphase.** Soon the centromeres of the chromosome parts (chromatids) separate, and these identical chromatids become individual chromosomes. The separated chromosomes now move in opposite directions, and once again the movement results from microtubule activity. In this case, the spindle fibers appear to shorten and pull their attached chromosomes toward the centrioles at opposite sides of the cell (figure 3.37).

4. **Telophase.** The final stage of mitosis is said to begin when the chromosomes complete their migration toward the centrioles. It is much like prophase, but in reverse. As the chromosomes

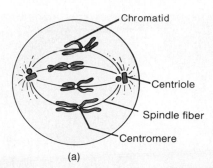

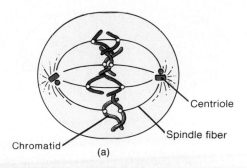

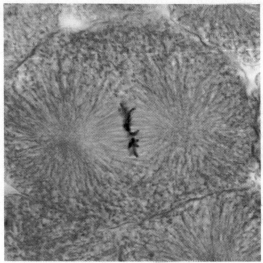

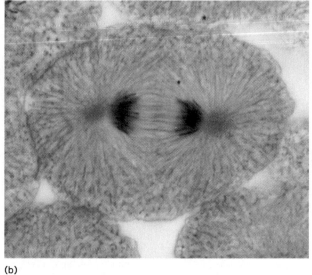

Figure 3.36 (a) In metaphase, the chromosomes become lined up midway between the centrioles. (b) A micrograph of a cell in metaphase (×250).

Figure 3.37 (a) In anaphase, the centromeres divide, and the spindle fibers that have become attached to them pull the chromatids toward the centrioles. (b) A micrograph of a cell in anaphase (×250).

approach the centrioles, the chromosomes begin to elongate and change from rodlike structures to threadlike structures. A nuclear envelope forms around each chromosome set, and nucleoli appear within the newly formed nuclei. Finally, the microtubules disappear (figure 3.38).

Chart 3.4 summarizes the phases of mitosis.

Cytoplasmic Division

Cytoplasmic division (cytokinesis) begins during anaphase when the cell membrane starts to constrict. The membrane continues to constrict through telophase. This process involves the musclelike contraction of a ring of microfilaments, which are composed of protein (actin). These filaments are assembled from the cytoplasm and are attached to the inner surface of the cell membrane. This contractile ring is positioned at right angles to the microtubules that pulled the chromosomes to opposite ends of the cell during mitosis. As the ring pinches inward, it separates the two newly

formed nuclei and divides about half of the cytoplasmic organelles into each of the new cells.

Although the newly formed cells may differ slightly in size and number of cytoplasmic parts, they have identical chromosomes and thus contain identical DNA information. Except for size, they are copies of the parent cell (figure 3.39).

Interphase

Once formed, the new cells usually begin growing. This requires that the young cells obtain nutrients and use them to manufacture new living materials and to synthesize many vital compounds. At the same time, various cell parts become duplicated. In the nucleus, the chromosomes may be doubling (at least in cells that will soon reproduce); and in the cytoplasm, new ribosomes, lysosomes, mitochondria, and various membranes are appearing. This stage in the life cycle is called **interphase,** during which the cell continues to carry on its usual metabolic functions. Interphase lasts until the cell begins to undergo mitosis. (See figure 3.40.)

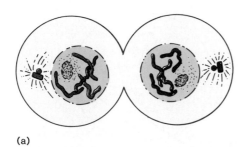

(a)

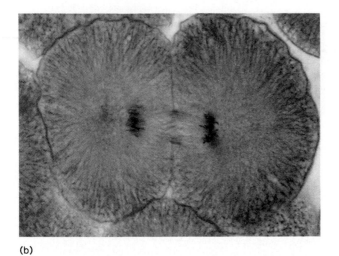

(b)

Figure 3.38 (a) In telophase, the chromosomes elongate to become chromatin threads, and the cytoplasm begins to divide. (b) A micrograph of a cell in telophase (×250).

CHART 3.4	Major events in mitosis
Stage	Major events
Prophase	Chromatin differentiates into chromosomes; centrioles move to opposite sides of cytoplasm; nuclear membrane and nucleolus become dispersed; microtubules appear and become associated with centrioles and duplicate chromatids of the chromosomes.
Metaphase	Chromosomes become arranged midway between the centrioles; spindle fibers from the centrioles become attached to the centromeres of each chromosome.
Anaphase	Centromeres separate, and duplicate chromatids of the chromosomes become separated; spindle fibers shorten and pull the individual chromosomes toward centrioles.
Telophase	Chromosomes elongate and form chromatin threads; nuclear membranes appear around each chromosome set; nucleoli appear; microtubules disappear.

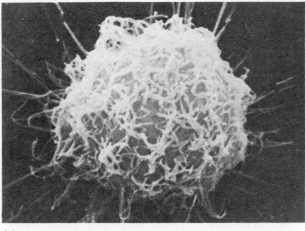

(a)

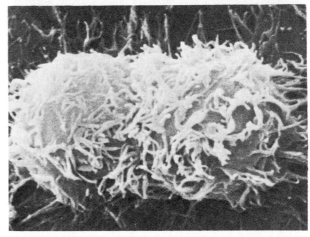

(b)

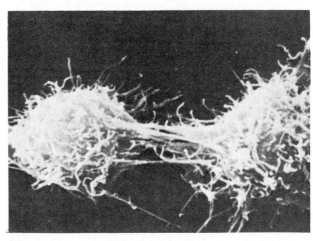

(c)

Figure 3.39 Following mitosis, the cytoplasm of the parent cell is divided into two portions as seen in these scanning electron micrographs (figure a ×3,750; figure b ×3,750; figure c ×3,190). (From *Scanning Electron Microscopy in Biology*, by R. G. Kessel and C. Y. Shih. © 1976 Springer-Verlag.)

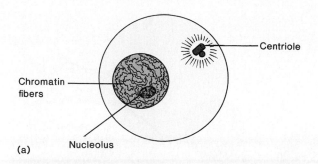

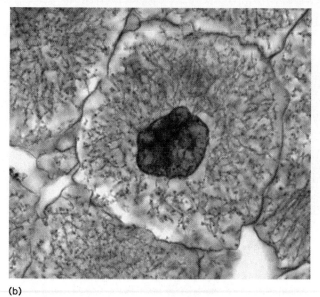

(b)

Figure 3.40 (a) Interphase lasts until a cell begins to undergo mitosis. (b) A micrograph of a cell in interphase (×250).

Many kinds of body cells are constantly growing and reproducing, thus increasing the number of cells that are present. Such activity is responsible for the growth and development of an embryo into a child, and of a child into an adult. It also is necessary for replacing cells that have relatively short life spans as well as cells that are worn out or lost due to injury or disease.

Cell Differentiation

Because all body cells are formed by mitosis and contain the same DNA information, they might be expected to look and act alike; obviously, they do not.

A human begins life as a single cell—a fertilized egg cell. This cell reproduces by mitosis to form two new cells; they, in turn, divide into four cells; the four become eight; and so forth. Then, sometime during development, the cells begin to *specialize*. That is, they develop special structures or begin to function in dif-

ferent ways. Some become skin cells, others become bone cells, and still others become nerve cells (figure 3.41).

The process by which cells develop different characteristics in structure and function is called **differentiation.** The mechanism responsible for this phenomenon is not well understood, but it involves the presence of specialized protein molecules that bind to particular portions of DNA molecules, causing the activation of some of the DNA information and the repression of other DNA information. Thus, the DNA information needed for general cell activities may be *activated* or "switched on" in both nerve and bone cells. The information related to specific bone cell functions, however, may be *repressed* or "switched off " in the nerve cells. Similarly, the information related to specific nerve cell functions may be repressed in bone cells.

Although the mechanism of differentiation is obscure, the results are obvious. Cells of many kinds are produced, and each kind carries on specialized functions; for example, skin cells protect underlying tissues, red blood cells carry oxygen, and nerve cells transmit impulses. Each type of cell somehow helps the others and aids in the survival of the organism.

1. Why is it important that the division of nuclear materials during mitosis be so precise?
2. Describe the events that occur during mitosis.
3. Name the process by which some cells become muscle cells and others become nerve cells.
4. How does DNA seem to control this process?

Control of Cell Reproduction

The existence of a control mechanism for cell reproduction is evident in the fact that some cells reproduce continually, others occasionally, and still others not at all.

Skin cells, blood-forming cells, and cells that line the intestine, for example, reproduce continually throughout life. Cells that compose some organs, such as the liver, seem to reproduce until a particular number of cells is present, and then they cease reproducing. Interestingly, if the number of liver cells is reduced by injury or surgery, the remaining cells are stimulated to reproduce again. Still other cells, such as nerve cells, apparently lose their ability to reproduce as they become differentiated; therefore, damage to nerve cells is likely to result in permanent loss of nerve function.

How the reproductive capacities of cells are controlled is not well understood, but the mechanism may involve the release of a *growth-inhibiting substance.*

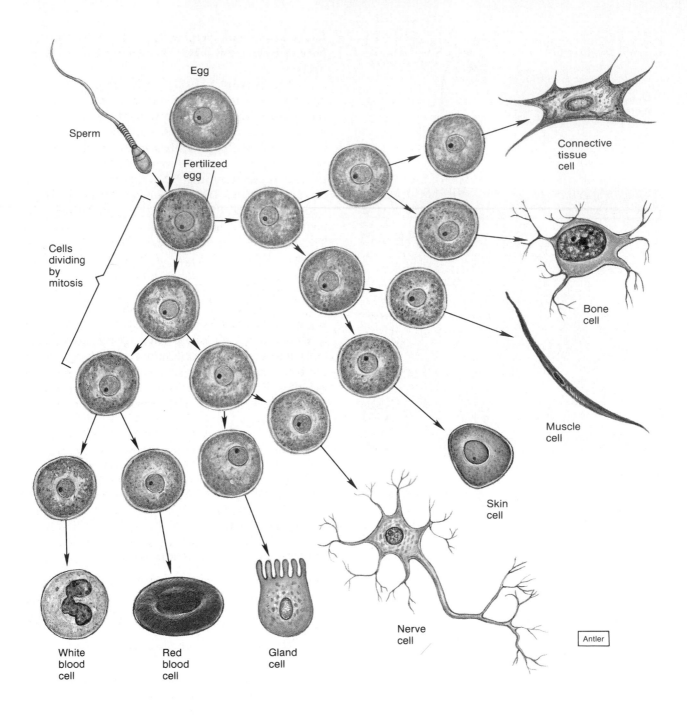

Figure 3.41 During development, the numerous body cells are produced from a single fertilized egg cell by mitosis. As these cells undergo differentiation, they become different kinds of cells that carry on specialized functions.

Such a substance might slow or stop the growth and reproduction of particular cells when their numbers reach a certain level.

Surface Area-Volume Relationships

Another factor involved in the control of cell reproduction is the relationship between a cell's membrane surface area and its volume. The quantity of nutrients needed to maintain a cell is directly related to the volume of its living material. At the same time, the quantity of nutrients that can enter the cell is related directly to the surface area of its membrane. As a cell grows, however, its membrane surface area increases proportionately less than its volume. Consequently, the surface area will eventually become inadequate for the needs of the living material inside.

Cancer

Cancer is a group of closely related diseases. It can occur in many different tissues, and it results from changes in cells that inactivate some of the normal control mechanisms. Cancers have certain common characteristics, including the following:

1. **Hyperplasia.** Hyperplasia is the uncontrolled reproduction of cells. Although the rate of reproduction among cancer cells is usually unchanged, they are not responsive to normal controls on cell numbers. As a result, cancer cells eventually give rise to very large cell populations.

2. **Anaplasia.** The word anaplasia is used to describe the appearance of abnormalities in cellular structure. Typically, cancer cells resemble undifferentiated or primitive cells. That is, they fail to develop the specialized structure of the kind of cell they represent. Also, they fail to function in expected ways. Certain white blood cells, for example, normally function in resisting bacterial infections. When cells of this type become cancerous, they do not function effectively, and the cancer patient becomes more subject to infectious diseases. Cancer cells are also likely to form disorganized masses rather than to become arranged in orderly groups like normal cells.

3. **Metastasis.** Metastasis is a tendency to spread into other tissues. Normal cells are usually cohesive; that is, they stick together in groups of similar kinds. Cancer cells often become detached from their cellular mass and may then be carried away from their place of origin to establish new cancerous

growths in other parts of the body. Metastasis is the characteristic most closely associated with the word *malignant,* which suggests a power to threaten life. This characteristic is often used to distinguish a cancerous growth from a noncancerous (benign) one.

The cause or causes of cancers are poorly understood, although increasing evidence indicates that the cause or causes are complex. Among the factors that may be involved are exposure to various chemicals or to harmful radiation, virus infections, changes in DNA structure, the presence of altered genes (oncogenes) that encode abnormal proteins (see chapter 4), or deficiencies in the body's ability to resist disease. Two or more of these factors may be needed to cause a cancer, or perhaps different factors or combinations of factors cause different kinds of cancers.

When untreated, cancer cells eventually accumulate in large numbers. They may damage normal cells by effectively competing with them for nutrients. At the same time, cancer cells may invade vital organs and interfere with their functions, obstruct important passageways, or penetrate blood vessels and cause internal bleeding.

If detected in its early stages, a cancerous growth may be removed surgically. At other times, cancers are treated with radiation or drugs (chemotherapy), or with some combination of surgery and treatments. The drugs most commonly used act on the structure of DNA molecules, affecting the DNA of rapidly reproducing normal cells as well as the DNA of cancer cells. However, the normal cells seem able to repair DNA damage more effectively than cancer cells can. Thus, the cancer cells are more likely to be destroyed by the drug's action.

A cell can solve this growth problem by dividing. The resulting cells are smaller than the parent cell and thus have a more favorable surface area-volume relationship.

1. How do cells vary in their rates of reproduction?
2. What factors seem to control the rate at which cells reproduce?

Chapter Summary

Introduction (page 64)

Cells vary considerably in size, shape, and function. The shapes of cells are closely related to their functions.

A Composite Cell (page 64)

1. A cell includes a nucleus, cytoplasm, and a cell membrane.
2. Cytoplasmic organelles perform specific vital functions, but the nucleus controls the overall activities of the cell.

3. Cell membrane
   a. The cell membrane forms the outermost limit of the living material.
   b. It acts as a selectively permeable passageway that controls the movements of substances between the cell and its surroundings.
   c. It includes protein, lipid, and carbohydrate molecules.
   d. The cell membrane framework consists mainly of a double layer of phospholipid molecules.
   e. Molecules that are soluble in lipids pass through the membrane easily, but the membrane is a barrier to the passage of water-soluble molecules.
   f. Cholesterol molecules help make the membrane stable.
   g. Proteins are responsible for the special functions of the membrane.
      (1) Rodlike proteins function as receptors on cell surfaces.
      (2) Globular proteins form channels for the passage of various ions and molecules.

      h. Some cells are connected by means of specialized intercellular junctions called tight junctions, desmosomes, and gap junctions.

    4. Cytoplasm
      a. Cytoplasm contains networks of membranes and organelles suspended in fluid.
      b. Endoplasmic reticulum is composed of interconnected membranous sacs, canals, and vesicles that provide a tubular communication system and an attachment for ribosomes; it also functions in the synthesis of proteins, lipids, and hormones.
      c. Ribosomes are particles of protein and RNA that function in protein synthesis.
      d. The Golgi apparatus is composed of a stack of flattened, membranous sacs that package glycoproteins for secretion.
      e. Mitochondria are membranous sacs that contain enzymes involved with releasing energy from food molecules and transforming energy into a usable form.
      f. Lysosomes are membranous sacs containing digestive enzymes that can destroy substances that enter cells.
      g. Peroxisomes are membranous, enzyme-containing vesicles.
      h. Cilia and flagella are motile processes that extend outward from some cell surfaces.
        (1) Cilia are numerous tiny, hairlike parts that serve to move fluids across cell surfaces.
        (2) Flagella are longer processes; the tail of a sperm cell is a flagellum that enables the cell to swim.
      i. The centrosome is a nonmembranous structure consisting of two centrioles that aids in the distribution of chromosomes during cell reproduction.
      j. Vesicles are membranous sacs containing substances that recently entered the cell.
      k. Microfilaments and microtubules are threadlike processes that aid cellular movements, and provide support and stability to cytoplasm.
      l. Cytoplasm may contain nonliving cellular products, such as nutrients and pigments, called *inclusions*.

    5. Cell nucleus
      a. The nucleus is enclosed in a double-layered nuclear envelope that controls the movement of substances between the nucleus and cytoplasm.
      b. A nucleolus is a dense body of protein and RNA that functions in the production of ribosomes.
      c. Chromatin is composed of loosely coiled fibers of protein and DNA that become chromosomes during cell reproduction.

Movements through Cell Membranes (page 77)

Movement of substances through membranes may involve physical or physiological processes.

    1. Diffusion
      a. Diffusion is a scattering of molecules or ions from regions of higher concentration toward regions of lower concentration.
      b. It is responsible for exchanges of oxygen and carbon dioxide within the body.
      c. Factors that increase the rate of diffusion include short distance, high concentration of diffusing molecules, low molecular weight, and high temperature.

    2. Facilitated diffusion
      a. Facilitated diffusion involves the use of carrier molecules in the cell membrane.
      b. This process only moves substances from regions of higher concentration to regions of lower concentration.

    3. Osmosis
      a. Osmosis is a special case of diffusion in which water molecules diffuse from regions of higher water concentration to lower water concentration through a selectively permeable membrane.
      b. Osmotic pressure increases as the number of particles dissolved in a solution increases.
      c. Cells lose water when placed in hypertonic solutions and gain water when placed in hypotonic solutions.
      d. A solution is isotonic when it contains the same concentration of dissolved particles as the cell contents.

    4. Filtration
      a. Filtration involves movement of molecules from regions of higher hydrostatic pressure toward regions of lower hydrostatic pressure.
      b. Blood pressure causes filtration of water and dissolved substances through porous capillary walls.

    5. Active transport
      a. Active transport is responsible for movement of molecules or ions from regions of lower concentration to regions of higher concentration.
      b. It requires cellular energy and involves action of carrier molecules in the cell membrane.

    6. Endocytosis
      a. Pinocytosis is a process by which a cell membrane engulfs tiny droplets of liquid.
      b. Phagocytosis is a process by which a cell membrane engulfs solid particles.
      c. Receptor-mediated endocytosis is a process by which receptor proteins combine with specific kinds of molecules in the cell surroundings, and the combinations are engulfed.

Life Cycle of a Cell (page 85)

    1. The life cycle of a cell includes mitosis, cytoplasmic division, interphase, and differentiation.
    2. Mitosis
      a. Mitosis is the division and distribution of nuclear parts to new cells during cell reproduction.
      b. The stages of mitosis include prophase, metaphase, anaphase, and telophase.
    3. Cytoplasmic division is a process by which cytoplasm is divided into two portions following mitosis.

4. Interphase
   a. Interphase is the stage in the life cycle when a cell grows and forms new organelles.
   b. It terminates when the cell begins to undergo mitosis.
5. Cell differentiation involves the development of specialized structures and functions.

Control of Cell Reproduction (page 91)

1. Cellular reproductive capacities vary greatly.
2. Reproductive capacities of cells may involve the release of growth-inhibiting substances.
3. Surface area-volume relationships
   a. As a cell grows, its surface area increases to a lesser degree than its volume, and eventually the area becomes inadequate for the needs of the living material within the cell.
   b. When a cell divides, the new cells have more favorable surface area-volume relationships.

Clinical Application of Knowledge

1. Which process—diffusion, osmosis, or filtration—is most closely related to each of the following situations?
   a. The injection of a drug that is hypertonic to the tissues stimulates pain.
   b. A person with extremely low blood pressure stops producing urine.
   c. The concentration of urea in the dialyzing fluid of an artificial kidney is kept low.
2. What characteristic of cell membranes may account for the observation that fat soluble substances like chloroform and ether cause rapid effects upon cells?
3. A person who has been exposed to excessive amounts of X ray may develop a decrease in white blood cell number and an increase in susceptibility to infections. In what way are these effects related?
4. Exposure to tobacco smoke causes cilia to become immobile and perhaps to disappear. How would you relate this to the fact that tobacco smokers have an increased incidence of respiratory infections?
5. How would you explain the function of phagocytic cells to a patient with a bacterial infection?
6. How is knowledge of cellular reproduction important to an understanding of:
   a. growth
   b. wound healing
   c. cancer

Review Activities

1. Use specific examples to illustrate how cells vary in size.
2. Describe how the shapes of nerve, epithelial, and muscle cells are related to their functions.
3. Name the major portions of a cell, and describe their relationships to one another.
4. Discuss the structure and functions of a cell membrane.

5. Distinguish between organelles and inclusions.
6. Define *selectively permeable*.
7. Describe the chemical structure of a membrane.
8. Explain how the structure of a cell membrane is related to its permeability.
9. Explain the function of membrane proteins.
10. Describe three kinds of intercellular junctions.
11. Describe the structures and functions of each of the following:
    a. endoplasmic reticulum
    b. ribosome
    c. Golgi apparatus
    d. mitochondrion
    e. lysosome
    f. peroxisome
    g. cilium
    h. flagellum
    i. centrosome
    j. vesicle
    k. microfilament
    l. microtubule
12. Define *inclusion*.
13. Describe the structure of the nucleus and the functions of its parts.
14. Distinguish between diffusion and facilitated diffusion.
15. Name four factors that increase the rate of diffusion.
16. Explain how diffusion aids in the exchange of gases within the body.
17. Define *osmosis*.
18. Explain what is meant by *osmotic pressure*.
19. Explain how the number of solute particles in a solution affects its osmotic pressure.
20. Distinguish between solutions that are hypertonic, hypotonic, and isotonic.
21. Define *filtration*.
22. Explain how filtration is involved with the movement of substances through capillary walls.
23. Explain why active transport is called a physiological process, while diffusion is called a physical process.
24. Explain the function of carrier molecules in active transport.
25. Distinguish between pinocytosis and phagocytosis.
26. Describe the process called *receptor-mediated endocytosis*.
27. List the phases in the life cycle of a cell.
28. Name the two processes included in cell reproduction.
29. Describe the major events of mitosis.
30. Explain how the cytoplasm is divided during cellular reproduction.
31. Explain what happens during *interphase*.
32. Define *differentiation*.
33. Explain how differentiation may involve the repression of DNA information.

4

Cellular Metabolism

*C*ellular metabolism includes all the chemical reactions that occur within cells. These reactions are of two major types, and for the most part, they involve the use of food substances. In one type of reaction, molecules of nutrients are converted into simpler forms through changes that are accompanied by the release of energy. In the other type, the nutrient molecules are used in constructive processes that produce the cell's structural and functional materials or the molecules that store energy.

In either case, metabolic reactions are controlled by proteins called enzymes that are synthesized in cells. The production of enzymes is, in turn, controlled by genetic information held within molecules of DNA, which instructs a cell how to synthesize specific kinds of enzyme molecules ■

Chapter Objectives	Key Terms	Aids to Understanding Words
After you have studied this chapter, you should be able to:	**aerobic respiration** (a-er-ō′bik res″pĭ-ra′shun)	**aer-**, air: *aer*obic respiration—a respiratory process that requires oxygen.
1. Distinguish between anabolic and catabolic metabolism.	**anabolic metabolism** (an″ah-bol′ik mĕ-tab′o-lizm)	**an-**, without: *an*aerobic respiration—a respiratory process that proceeds without oxygen.
2. Explain how enzymes control metabolic processes.	**anaerobic respiration** (an″a-er-ō′bik res″pĭ-ra′shun)	**ana-**, up: *ana*bolic metabolism—cellular processes in which smaller molecules are used to build up larger ones.
3. Explain how chemical energy is released by cellular respiration.	**catabolic metabolism** (kat″ah-bol′ik mĕ-tab′o-lizm)	**cata-**, down: *cata*bolic metabolism—cellular processes in which larger molecules are broken down into smaller ones.
4. Describe how energy is made available for cellular activities.	**coenzyme** (ko-en′zīm)	**co-**, with: *co*enzyme—a substance that unites with a protein and completes the structure of an enzyme molecule.
5. Describe the general metabolic pathways of carbohydrates, lipids, and proteins.	**deamination** (de-am″ĭ-na′shun)	
6. Explain how metabolic pathways are regulated.	**dehydration synthesis** (de″hi-dra′shun sin′thĕ-sis)	**de-**, undoing: *de*amination—a process by which nitrogen-containing portions of amino acid molecules are removed.
7. Describe how genetic information is stored within DNA molecules.	**DNA**	
8. Explain how genetic information is used in the synthesis of proteins.	**energy** (en′er-je)	**mut-**, change: *mut*ation—a change in the genetic information of a cell.
	enzyme (en′zīm)	
	gene (jēn)	**-strat**, spread out: sub*strat*e—a substance upon which an enzyme acts.
9. Describe how DNA molecules are replicated.	**genetic code** (jĕ-net′ik kōd)	
10. Explain how genetic information can be altered and how such a change may affect an organism.	**glycolysis** (gli-kol′ĭ-sis)	**sub-**, under: *sub*strate—a substance upon which an enzyme acts.
	hydrolysis (hi-drol′ĭ-sis)	
	mutation (mu-ta′shun)	**-zym**, causing to ferment: en*zym*e—a protein that initiates or speeds up a chemical reaction without itself being consumed.
	oxidation (ok″sĭ-da′shun)	
11. Complete the review activities at the end of this chapter. Note that the items are worded in the form of specific learning objectives. You may want to refer to them before reading the chapter.	**replication** (re″pli-ka′shun)	
	RNA	
	substrate (sub′strāt)	

A lthough a living cell may appear to be idle, it is actually very active. Numerous metabolic processes needed to maintain its life are occurring all the time. These processes involve numerous chemical reactions, each of which is controlled by a specific protein and depends upon the release of cellular energy.

Metabolic Processes

Metabolic processes, which include all of the chemical reactions that take place within a cell, can be divided into two major types. One type, called *anabolic metabolism,* involves the build-up of larger molecules from smaller ones and utilizes energy. The other type, called *catabolic metabolism,* involves the breakdown of larger molecules into smaller ones and releases energy.

Anabolic Metabolism

Anabolic metabolism includes all the constructive processes used to manufacture substances needed for cellular growth and repair. For example, cells often join many simple sugar molecules (monosaccharides) to form larger molecules of glycogen by an anabolic process called *dehydration synthesis.* In this process, the larger carbohydrate molecule is built by bonding monosaccharide molecules together into a complex chain. When adjacent monosaccharide units are joined, an $-OH$ (hydroxyl group) from one monosaccharide molecule and an $-H$ (hydrogen atom) from an $-OH$ group of another are removed. These particles react to produce a water molecule, and the monosaccharides are united by a shared oxygen atom, as shown in figure 4.1. As the process is repeated, the molecular chain becomes longer.

Similarly, glycerol and fatty acid molecules are joined by dehydration synthesis in fat (adipose) tissue cells to form fat molecules. In this case, three hydrogen atoms are removed from a glycerol molecule, and an $-OH$ group is removed from each of three fatty acid molecules, as shown in figure 4.2. The result is three water molecules and a single fat molecule, whose glycerol and fatty acid portions are bound by shared oxygen atoms.

Cells also build protein molecules by joining amino acid molecules by dehydration synthesis. When two amino acid molecules are united, an $-OH$ from one and an $-H$ from the $-NH_2$ group of another are removed. A water molecule is formed, and the amino acid molecules are joined by a bond between a carbon atom and a nitrogen atom (figure 4.3). This type of bond, which is called a *peptide bond,* holds the amino acids together. Two amino acids bound together form a *dipeptide molecule,* and many joined in a chain form a *polypeptide molecule.* Generally, a polypeptide consisting of 100 or more amino acid molecules that has

a specific function is called a *protein,* although the boundary distinguishing between polypeptides and proteins is not defined precisely.

Catabolic Metabolism

Physiological processes in which larger molecules are broken down into smaller ones constitute **catabolic metabolism.** An example of such a process is *hydrolysis,* which can bring about the decomposition of carbohydrates, lipids, and proteins.

When a molecule of one of these substances is hydrolyzed, a water molecule is used, and the original molecule is split into two simpler parts. The hydrolysis of a disaccharide such as sucrose, for instance, results in molecules of two monosaccharides—glucose and fructose:

$$C_{12}H_{22}O_{11} + H_2O \rightarrow C_6H_{12}O_6 + C_6H_{12}O_6$$
(sucrose)   (water)   (glucose)   (fructose)

In this case, the bond between the simple sugars within the sucrose molecule is broken, and the water molecule supplies a hydrogen atom to one sugar molecule and a hydroxyl group to the other. Thus, hydrolysis is the reverse of dehydration synthesis:

Hydrolysis
Disaccharide + Water ⇌ Monosaccharide + Monosaccharide
Dehydration
synthesis

Hydrolysis, which occurs during the process of *digestion,* is responsible for the breakdown of carbohydrates into monosaccharides; fats into glycerol and fatty acids; proteins into amino acids; and nucleic acids into nucleotides (figure 4.4). Digestion is discussed in chapter 14.

Both catabolic and anabolic metabolism are carried on continually within cells. However, these activities must be carefully controlled so that the breakdown or energy-releasing reactions occur at rates that are adjusted to the needs of the building up or energy-utilizing reactions. Any disturbance in this balance is likely to cause abnormalities or death of cells.

1. What general functions do anabolic metabolism and catabolic metabolism serve?
2. What substance is formed by the anabolic metabolism of monosaccharides? Of amino acids? Of glycerol and fatty acids?
3. Distinguish between dehydration synthesis and hydrolysis.

Control of Metabolic Reactions

Although different kinds of cells may conduct specialized metabolic processes, all cells perform certain basic reactions, such as the build-up and breakdown of carbohydrates, lipids, proteins, and nucleic acids. These

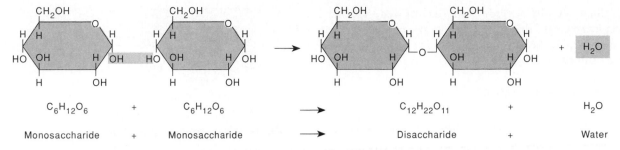

$$C_6H_{12}O_6 \quad + \quad C_6H_{12}O_6 \quad \longrightarrow \quad C_{12}H_{22}O_{11} \quad + \quad H_2O$$

Monosaccharide + Monosaccharide ⟶ Disaccharide + Water

Figure 4.1 What type of reaction is represented in this diagram?

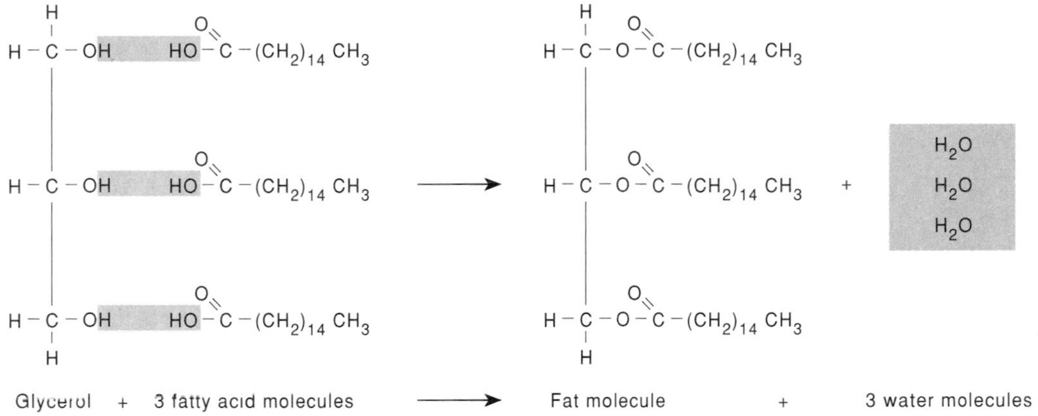

Glycerol + 3 fatty acid molecules ⟶ Fat molecule + 3 water molecules

Figure 4.2 A glycerol molecule and three fatty acid molecules may join by dehydration synthesis to form a fat molecule (triglyceride).

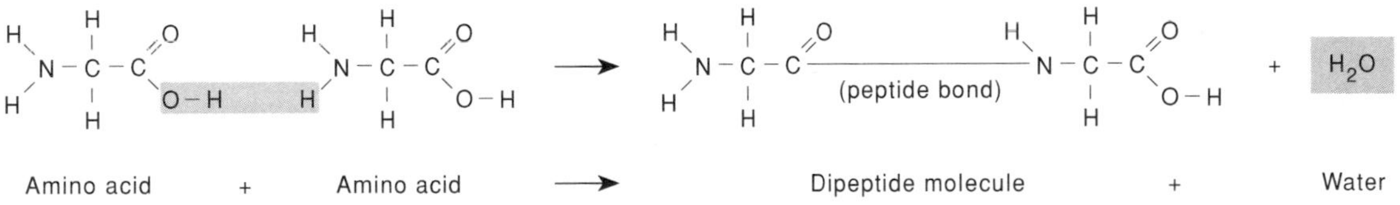

Amino acid + Amino acid ⟶ Dipeptide molecule + Water

Figure 4.3 When two amino acid molecules are united by dehydration synthesis, a peptide bond is formed between a carbon atom and a nitrogen atom.

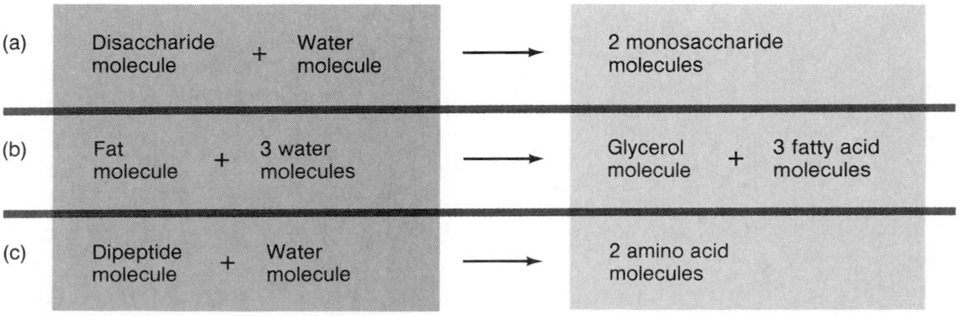

(a)	Disaccharide molecule + Water molecule	⟶	2 monosaccharide molecules
(b)	Fat molecule + 3 water molecules	⟶	Glycerol molecule + 3 fatty acid molecules
(c)	Dipeptide molecule + Water molecule	⟶	2 amino acid molecules

Figure 4.4 Hydrolysis causes the decomposition of (a) carbohydrates into monosaccharides; (b) fats into glycerol and fatty acids; and (c) proteins into amino acids.

reactions actually include hundreds of very specific chemical changes that must occur in an orderly fashion. They are controlled by mechanisms involving the presence of *enzymes*.

Enzymes and Their Actions

Like other chemical reactions, metabolic reactions generally require a certain amount of energy (activation energy) before they will occur. This is why heat is commonly used to increase the rates of chemical reactions in laboratories. Heat energy increases the rate at which molecules move and the frequency of molecular collisions. These collisions increase the likelihood of interactions among the electrons of the molecules that can form new chemical bonds. The temperature conditions that exist naturally in cells are usually too mild to adequately promote the reactions needed to support life, but this is not a problem because cells contain enzymes.

Enzymes are globular proteins (see chapter 2) that promote specific chemical reactions within cells by lowering the amount of energy (activation energy) needed to start these reactions. They do this by temporarily binding with the molecules upon which they act. This somehow strains or distorts the chemical structure of these molecules and thereby increases the likelihood of a chemical change. Thus, in the presence of enzymes, metabolic reactions are often speeded up by a factor of a million or more.

Enzymes are needed in very small quantities because as they function, they are not consumed and can, therefore, function repeatedly. Also, each enzyme acts only on a particular kind of substance, which is called its **substrate.** For example, the substrate of an enzyme called *catalase* (found in the peroxisomes of liver and kidney cells) is hydrogen peroxide, a toxic by-product of certain metabolic reactions. This enzyme's only function is to cause the decomposition of hydrogen peroxide into water and oxygen. In this way, it helps prevent an accumulation of hydrogen peroxide that might damage cells.

> You may have noticed the action of the enzyme catalase if you have ever used hydrogen peroxide to cleanse a wound. Injured cells release catalase, and when hydrogen peroxide contacts them, bubbles of oxygen are set free. This is useful because the resulting foam aids in removing debris from inaccessible parts of the wound.

Cellular metabolism includes hundreds of different chemical reactions, and each of these reactions is controlled by a specific kind of enzyme. Thus, hundreds of different kinds of enzymes must be present in every cell, and each enzyme must be able to "recognize" its specific substrate. This ability to identify a

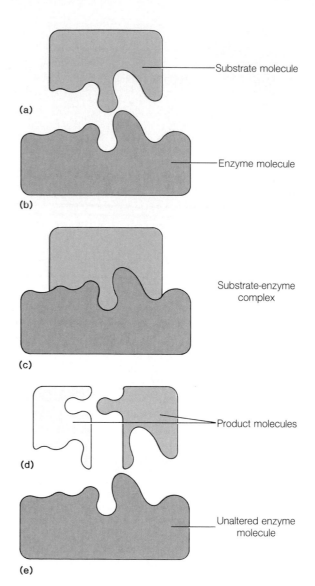

Figure 4.5 (*a*) The shape of a substrate molecule fits (*b*) the shape of the enzyme's active site. (*c*) When the substrate molecule combines temporarily with the enzyme, a chemical reaction occurs. The result is (*d*) product molecules and (*e*) an unaltered enzyme.

substrate depends upon the shape of an enzyme molecule. That is, each enzyme's polypeptide chain is twisted and coiled into a unique three-dimensional form that fits the special shape of its substrate molecule. In short, an enzyme molecule fits a molecule of its substrate much as a key fits a particular lock.

During an enzyme-controlled reaction, particular regions of the enzyme molecule called *active sites* temporarily combine with portions of the substrate, forming a substrate-enzyme complex. This interaction seems to strain chemical bonds within the substrate, increasing the likelihood of a change in the substrate molecule. When the substrate is changed, the product of the reaction appears, and the enzyme is released in its original form (figure 4.5).

This activity can be summarized as follows:

$$\text{Substrate molecule} + \text{Enzyme molecule} \rightarrow \text{Substrate-enzyme complex} \rightarrow \text{Product (changed substrate)} + \text{Enzyme molecule}$$

The speed of an enzyme-controlled reaction is related to the number of enzyme and substrate molecules present within the cell. Generally, the reaction occurs more rapidly if the concentration of the enzyme or the concentration of the substrate is increased. Also, the efficiency of different kinds of enzymes varies greatly. Thus, some enzymes seem able to process only a few substrate molecules per second, while others can process thousands or even millions of substrate molecules per second.

Commonly, enzyme names are derived from the names of the substrates they act upon, with the suffix *-ase* added. For example, a lipid-splitting enzyme is called *lipase,* a protein-splitting enzyme is *protease,* and a starch- (amylum) splitting enzyme is *amylase.* Similarly, *sucrase* is an enzyme that splits the sugar sucrose, *maltase* splits the sugar maltose, and *lactase* splits the sugar lactose.

Cofactors and Coenzymes

Often the protein portion of an enzyme molecule (apoenzyme) is inactive until it is combined with an additional substance. This nonprotein part is needed to complete the proper shape of the active site of the enzyme molecule or to help bind the enzyme to its substrate. Such a substance is called a **cofactor.** A cofactor may be an ion of an element, such as copper, iron, or zinc, or it may be a relatively small organic molecule, in which case it is called a **coenzyme.** Coenzymes are often composed of vitamin molecules or have vitamin molecules incorporated into their structures.

Vitamins are essential organic substances that cannot be synthesized (or cannot be synthesized in sufficient quantities) by human cells and therefore must be obtained in the diet. Since vitamins provide coenzymes that can, like enzymes, function again and again, vitamins are needed by cells in very small quantities. Vitamins are discussed in chapter 15.

Factors That Alter Enzymes

Except for their cofactor portions, enzymes are proteins, and like other proteins, they can be denatured by exposure to excessive heat, radiation, electricity, certain chemicals, or fluids with extreme pH values. For example, many enzymes become inactive at 45°C, and nearly all of them are denatured at 55°C. Some poisons are chemicals that cause enzymes to be denatured. Cyanides, for instance, can interfere with respiratory enzymes and damage cells by halting their energy-releasing processes.

The antibiotic drug penicillin acts by interfering with enzymes that function in the production of bacterial cell walls. These walls surround and protect normal bacterial cells. In the presence of penicillin, the cell walls are not properly produced, and the affected bacteria are less able to survive. Thus, penicillin may be used to control the growth of bacteria that might otherwise cause a serious bacterial infection.

1. What is an enzyme?
2. How can an enzyme control a metabolic reaction?
3. How does an enzyme "recognize" its substrate?
4. What is the role of a cofactor?
5. What factors are likely to denature enzymes?

Energy for Metabolic Reactions

Energy is the capacity to produce changes in matter or to move something; that is, it is the ability to do work. Therefore, energy is recognized by what it can do. Common forms of energy include heat, light, sound, electrical energy, mechanical energy, and chemical energy.

Although energy cannot be created or destroyed, it can be converted from one form to another. Sunlight is converted to heat when it is absorbed by skin; an ordinary incandescent light bulb changes electrical energy to heat and light; and an automobile engine converts the chemical energy in gasoline to heat and mechanical energy.

Whenever changes take place, energy is being transferred from one part to another. Thus, all metabolic reactions depend on the presence of energy in one form or another.

Release of Chemical Energy

Most metabolic processes use chemical energy. This form of energy is held in the bonds between the atoms of molecules and is released when these bonds are broken. For example, the chemical energy of many substances can be released by burning. Such a reaction must be started, and this is usually accomplished by applying heat. As the substance burns, molecular bonds are broken, and energy escapes as heat and light.

Similarly, glucose molecules are "burned" in cells; this process is more correctly called **oxidation.** The energy released by the oxidation of glucose is used to promote cellular metabolism. There are, however, some important differences between the oxidation of substances inside cells and the burning of substances outside them.

Burning usually requires a relatively large amount of energy to activate the process, and most of the energy released escapes as heat or light. In cells, the oxidation process is initiated by enzymes that reduce the amount of energy (activation energy) needed. Also, by transferring energy to special energy-carrying molecules, cells are able to capture in the form of chemical energy about half of the energy released. The rest escapes as heat, which helps maintain body temperature.

The process by which energy is released from molecules such as glucose and is transferred to other molecules is quite complex. The process is called **cellular respiration,** and it involves a number of chemical reactions that must occur in a particular sequence, each one controlled by a separate kind of enzyme. Some of these enzymes are in the cell's cytoplasm, while others are in the mitochondria.

1. What is meant by energy?
2. How does cellular oxidation differ from burning?
3. Define cellular respiration.

Anaerobic Respiration

When a 6-carbon glucose molecule is decomposed during cellular respiration, enzymes control a series of reactions that break glucose into two 3-carbon pyruvic acid molecules. This phase of the process occurs in the cytoplasm, and because it takes place in the absence of oxygen, it is called **anaerobic respiration** (glycolysis).

Although some energy is needed to activate the reactions of anaerobic respiration, more energy is released than is used. The excess is used to synthesize molecules of an energy-carrying substance called **ATP** (adenosine triphosphate) (figure 4.6).

Aerobic Respiration

Following the anaerobic phase of cellular respiration, oxygen must be available in order for the process to continue. For this reason, the second phase is called **aerobic respiration.** It takes place within the mitochondria, and as a result, considerably more energy is transferred to ATP molecules.

When the decomposition of a glucose molecule is complete, carbon dioxide molecules and hydrogen atoms remain. The carbon dioxide diffuses out of the cell as a waste, and the hydrogen atoms combine with oxygen to form water molecules. Thus, the final products of glucose oxidation are carbon dioxide, water, and energy.

ATP Molecules

For each glucose molecule that is decomposed completely, up to thirty-eight molecules of ATP can be produced. Two of these are the result of anaerobic respiration, but the rest are formed during the aerobic phase.

Each ATP molecule consists of three main parts—an adenine portion, a ribose portion, and three phos-

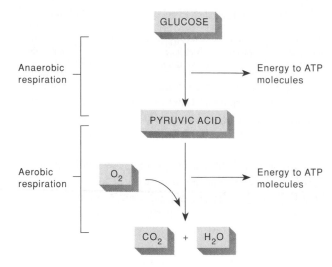

Figure 4.6 Anaerobic respiration occurs in the cytoplasm in the absence of oxygen, while aerobic respiration occurs in the mitochondria in the presence of oxygen.

phates in a chain (figure 4.7). (Note the similarity between the general structure of an ATP molecule and the structure of a nucleotide within a nucleic acid molecule described in chapter 2; see figure 2.28.)

As energy is released during cellular respiration, some of it is captured in the bond of the end phosphate (high-energy phosphate bond) of an ATP molecule. This stored chemical energy may be quickly transferred to another molecule involved in some metabolic process. When such an energy transfer occurs, the terminal, high-energy phosphate bond of the ATP molecule is broken, and the energy of this bond is released. Energy stored in ATP molecules is used whenever cellular work is performed, as when muscle cells contract, when membranes carry on active transport, or when cells synthesize substances.

An ATP molecule that has lost its terminal phosphate becomes an **ADP** (adenosine diphosphate) molecule. However, the ADP molecule can convert back into an ATP by capturing some energy and a phosphate. Thus, as figure 4.8 shows, ATP and ADP molecules shuttle back and forth between the energy-releasing reactions of cellular respiration and the energy-utilizing reactions of the cell.

Although ATP is not the only kind of energy-carrying molecule within a cell, it is the primary one. Without a source of ATP, most cells die quickly.

1. What is meant by anaerobic respiration? Aerobic respiration?
2. What happens to the energy released by cellular respiratory processes?
3. What are the final products of these cellular respiratory reactions?
4. What is the function of ATP molecules?

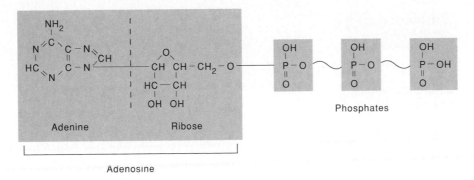

Figure 4.7 An ATP molecule consists of an adenine portion, a ribose portion, and three phosphates. The wavy lines connecting the last two phosphates represent high-energy chemical bonds from which energy can be released quickly.

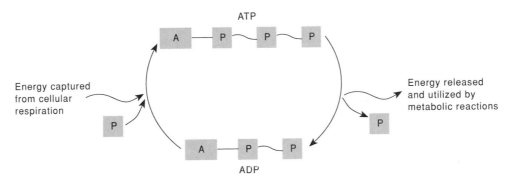

Figure 4.8 What is the significance of this cyclic process?

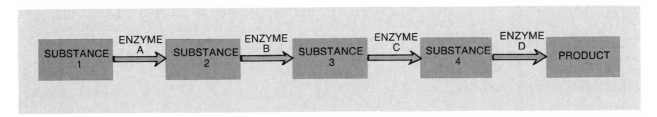

Figure 4.9 A metabolic pathway consists of a series of enzyme-controlled reactions leading to a product.

Metabolic Pathways

Anabolic and catabolic reactions that occur in cells usually involve a number of different steps that must occur in a particular sequence. For example, the anaerobic phase of cellular respiration by which glucose is converted to pyruvic acid, involves ten separate reactions. Since each reaction is controlled by a specific kind of enzyme, these enzymes must act in proper order. Such precision of activity suggests that the enzymes are positioned in the exact sequence as that of the reactions they control. The enzymes responsible for aerobic respiration are located in tiny, stalked particles on the membranes (cristae) within the mitochondria (chapter 3).

Such a sequence of enzyme-controlled reactions that leads to the production of particular products is called a **metabolic pathway** (figure 4.9). Typically these pathways are interconnected so that molecules of a certain kind of substance may enter more than one pathway as the substance is metabolized. For example, carbohydrate molecules from foods may enter catabolic pathways and be used to supply energy, or they may enter anabolic pathways and be stored as glycogen, fat, or other molecules (figure 4.10).

Carbohydrate Pathways

The average human diet consists largely of carbohydrates, which are changed by digestion into monosac-

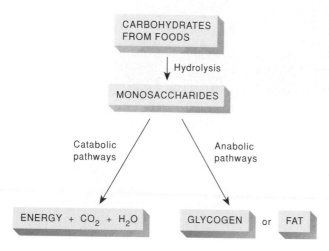

Figure 4.10 Carbohydrates from foods are changed into monosaccharides by hydrolysis. The resulting molecules may enter catabolic pathways and be used as energy sources, or they may enter anabolic pathways and be converted to glycogen or fat.

charides such as glucose. These substances are used primarily as cellular energy sources, which means they usually enter the catabolic pathways of cellular respiration. As discussed previously, the first phase of this process occurs in the cytoplasm and is anaerobic. This phase is called *glycolysis* and involves the conversion of a 6-carbon glucose into two 3-carbon pyruvic acid molecules. During the second phase of the process, the pyruvic acid (pyruvate) is transformed into a 2-carbon acetyl group that combines with a molecule of coenzyme A to form a substance called *acetyl coenzyme A*. It, in turn, is transported into a mitochondrion and is changed into a number of intermediate products by a complex series of chemical reactions known as the **citric acid cycle** (Krebs cycle). As these changes occur, energy is released and some of it is transferred to molecules of ATP, but the rest is lost as heat. The end products of the oxidation process are molecules of carbon dioxide and water.

Glycolysis

The chemical reactions of glycolysis are illustrated in figure 4.11. In the early steps of this metabolic pathway, the original glucose molecule is altered by the addition of phosphate groups (a process called *phosphorylation*) and by the rearrangement of its atoms. The phosphate groups and the energy needed for the reactions are supplied by ATP molecules. The result is a molecule of fructose combined with two phosphate groups (fructose-1,6-diphosphate). It is changed into two 3-carbon molecules (glyceraldehyde-3-phosphate), which in turn are converted to pyruvic acid, as follows:

a. An inorganic phosphate group is added to glyceraldehyde-3-phosphate to form 1,3-

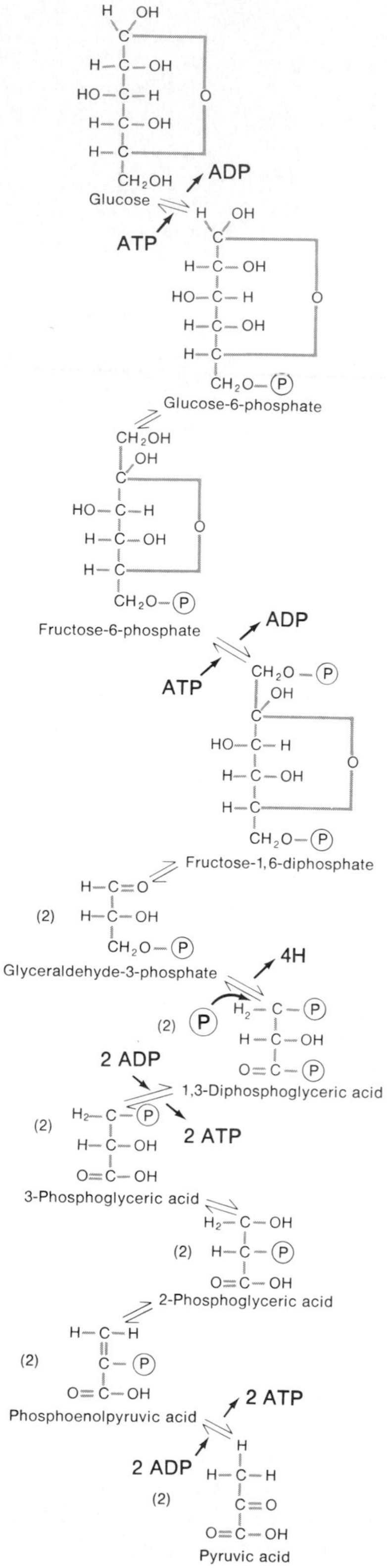

Figure 4.11 Chemical reactions of glycolysis.

diphosphoglyceric acid. At the same time, two hydrogen atoms are released.

b. 1,3-diphosphoglyceric acid is changed to 3-phosphoglyceric acid. As this occurs, some energy in the form of a high-energy phosphate is transferred from the 1,3-diphosphoglyceric acid to an ADP molecule, converting the ADP to ATP.

c. A slight alteration of 3-phosphoglyceric acid occurs to form 2-phosphoglyceric acid.

d. A change in 2-phosphoglyceric acid converts it into phosphoenolpyruvic acid.

e. Finally, a high-energy phosphate is transferred from the phosphoenolpyruvic acid to an ADP molecule, converting it to ATP. A molecule of pyruvic acid remains.

Since two 3-carbon molecules are involved originally, two molecules of pyruvic acid are produced by these reactions. Also, a total of four hydrogen atoms are released (step *a*), and four ATP molecules are formed (two in step *b* and two in step *e*). However, because two molecules of ATP are used early in glycolysis, there is a net gain of only two ATP molecules during this phase of cellular respiration.

In the absence of oxygen, the resulting pyruvic acid molecules may be converted into lactic acid. In the presence of oxygen, however, each pyruvic acid molecule is oxidized to an acetyl group, which is then combined with a molecule of coenzyme A (obtained from the vitamin called pantothenic acid) to form acetyl coenzyme A. As this occurs, two hydrogen atoms are released for each molecule of acetyl coenzyme A formed. The acetyl coenzyme A is then broken down by means of the citric acid cycle, which is illustrated in figure 4.12.

Citric Acid Cycle

An acetyl coenzyme A molecule enters the citric acid cycle (Krebs cycle) by combining with a molecule of oxaloacetic acid to form citric acid. The citric acid is then changed by a series of reactions back into oxaloacetic acid, and the cycle is completed.

As citric acid is produced, coenzyme A is released and thus can be used again and again in the formation of acetyl coenzyme A from pyruvic acid molecules.

During various steps in the citric acid cycle, carbon dioxide and hydrogen atoms are released. More specifically, for each glucose molecule metabolized in the presence of oxygen, two molecules of acetyl coenzyme A enter the citric acid cycle. As a result of the cycle, four carbon dioxide molecules and sixteen hydrogen atoms are released. At the same time, two more molecules of ATP are formed.

The released carbon dioxide dissolves in the cellular fluid and is transported away by the blood. Most of the hydrogen atoms released from the citric acid cycle, and those released during glycolysis and during the formation of acetyl coenzyme A, supply electrons that are involved with the production of ATP.

ATP Synthesis

Note in figure 4.12 that the hydrogen atoms released from various metabolic reactions are passed in pairs to hydrogen carriers. One of these carriers is NAD^+ (nicotinamide adenine dinucleotide). When NAD^+ accepts a pair of hydrogen atoms, one of the atoms becomes a hydrogen ion and the other bonds to NAD^+ to form NADH, as follows:

$$NAD^+ + 2H \rightarrow NADH + H^+$$

NAD^+ is a coenzyme obtained from a vitamin (niacin), and when it combines with hydrogen, it is said to be *reduced*. (*Reduction* results from the addition of hydrogen or the gain of electrons; it is the opposite of oxidation.) Another hydrogen acceptor, FAD, acts in a similar manner, combining with hydrogen to form $FADH_2$ (figure 4.12).

In their reduced states, the hydrogen carriers NADH and $FADH_2$ hold most of the energy once held by the original glucose molecule.

As shown in figure 4.13, when hydrogen is released from NADH, NAD^+ reappears and can function once again as a hydrogen acceptor. Since this reaction involves the removal of hydrogen, the NAD^+ is said to be *oxidized*. (*Oxidation* results from the removal of hydrogen or the loss of electrons; it is the opposite of reduction.) At the same time, two electrons from the original pair of hydrogen atoms are passed to a sequence of electron carriers.

The molecules that act as electron carriers comprise an *electron transport system* (cytochrome system). As electrons are passed from one carrier to another, the carriers are alternately reduced and oxidized as they accept or release electrons.

Among the members of the electron transport system are several proteins, including a set of iron-containing molecules called *cytochromes*. The cytochromes are located in the inner membranes of the mitochondria. (See chapter 3.)

The last cytochrome of the electron transport system (cytochrome oxidase) gives up a pair of electrons and causes two hydrogen ions (formed at the beginning of the sequence) to combine with an atom of oxygen. This process produces a water molecule:

$$2e^- + 2H^+ + \tfrac{1}{2}O_2 \rightarrow H_2O$$

Thus, oxygen is the final acceptor of the electrons.

Note in figure 4.13 that at the same time electrons are being passed through the electron transport

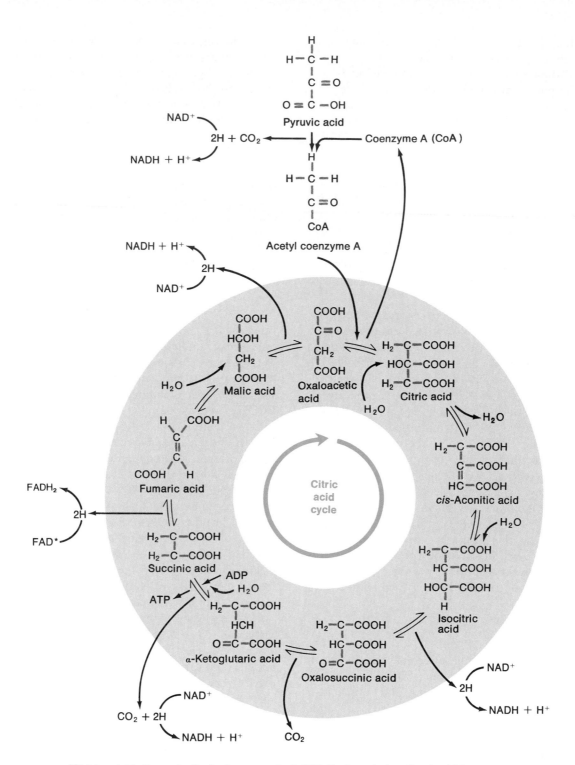

*At this point in the cycle, the hydrogen carrier is FAD (flavine adenine dinucleotide).

Figure 4.12   Chemical reactions of the citric acid cycle.

system, energy is being released. Some of this energy is used by the enzyme *ATP synthetase* to combine phosphate and ADP by a high-energy bond (phosphorylation), forming ATP.

Also note in figures 4.11 and 4.12 that twelve pairs of hydrogen atoms are released during the complete breakdown of one glucose molecule—two pairs from glycolysis, two pairs from the conversion of pyruvic acid to acetyl coenzyme A (one pair from each of two pyruvic acid molecules), and eight pairs from the citric acid cycle (four pairs for each of two acetyl coenzyme A molecules).

As a result of the oxidation (loss of electrons) of ten pairs of these hydrogen atoms, thirty ATP mole-

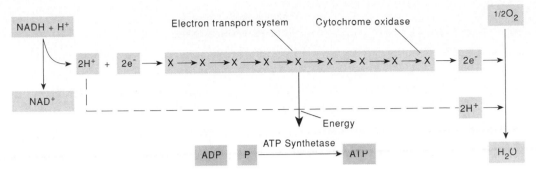

Figure 4.13 ATP synthesis by oxidative phosphorylation.

cules are produced, while metabolism of the other two pairs results in the formation of four ATP molecules. Also, there is a net gain of two ATP molecules during glycolysis, and two ATP molecules are formed as a result of the two acetyl coenzyme A molecules entering the citric acid cycle. Thus, a maximum of thirty-eight ATP molecules are formed for each glucose molecule that is metabolized.

Because this process of forming ATP involves both the oxidation of hydrogen atoms and the bonding of phosphate to ADP, it is called *oxidative phosphorylation.*

Storage of Carbohydrates

Whenever excess glucose is present, it may enter anabolic carbohydrate pathways and be converted into storage forms such as glycogen.

Although most cells can produce glycogen, the liver and muscle cells store the greatest amounts. Following a meal, when the blood glucose concentration is relatively high, liver cells obtain glucose from the blood and convert it to glycogen. Between meals, when the blood glucose concentration is lower, the reaction is reversed, and glucose is released into the blood. This mechanism ensures that various tissues will continue to have an adequate supply of glucose molecules to support their respiratory processes.

Glucose can also be converted into fat molecules, which are later deposited in fat tissues. This happens when a person takes in more carbohydrates than can be stored as glycogen or are needed for normal activities. The body has an almost unlimited capacity to perform this type of anabolic metabolism, so an excessive intake of nutrients can result in becoming overweight (obese).

Lipid Pathways

Although foods may contain lipids in the form of phospholipids or cholesterol, the most common dietary lipids are fats called *triglycerides.* As was explained in chapter 2, such a fat molecule consists of a glycerol portion and three fatty acids.

The metabolism of lipids is controlled mainly by the liver, which can remove them from the circulating blood and alter their molecular structures. For example, the liver can shorten or lengthen the carbon chains of fatty acid molecules or introduce double bonds into these chains, thus converting fatty acids from one form to another.

Lipids provide for a variety of physiological functions; however, fats are used mainly to supply energy. Gram for gram, fats contain more than twice as much chemical energy as carbohydrates or proteins.

Before energy can be released from a triglyceride molecule, the molecule must undergo hydrolysis. As shown in figure 4.14, some of the resulting fatty acid portions can then be converted into molecules of acetyl coenzyme A by a series of reactions called *beta oxidation.*

In the first phase of beta oxidation reactions, the fatty acids are converted into activated forms. This change requires a supply of energy from ATP molecules and the presence of a special group of enzymes (thiokinases). Each of the enzymes in this group can act upon a fatty acid with a particular carbon chain length. Once the fatty acid molecules have been activated, they are broken down by other enzymes called *fatty acid oxidases* that are located within mitochondria. In this phase of the reactions, segments of fatty acid chains (containing two carbon atoms each) are removed. Some of these segments are converted into acetyl coenzyme A molecules. Other segments are converted into compounds called *ketone bodies,* such as acetone, which later may be changed to acetyl coenzyme A. In either case, the resulting acetyl coenzyme A can be oxidized by means of the citric acid cycle. The glycerol portions of the triglyceride molecules can also enter metabolic pathways leading to the citric acid cycle, or they can be used to synthesize glucose.

It is also possible for glycerol and fatty acid molecules resulting from the hydrolysis of fats to be changed back into fat molecules by anabolic processes and to be stored in fat tissue. Additional fat molecules can be synthesized from excess molecules of glucose or amino acids.

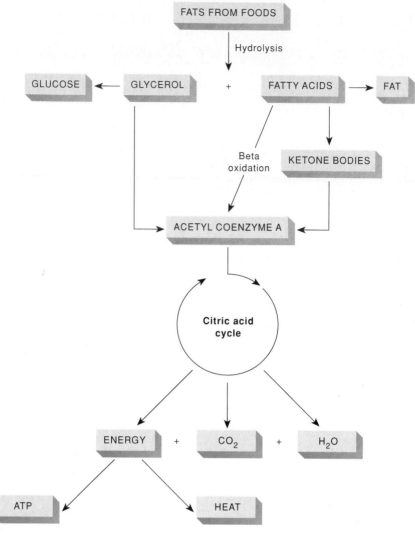

Figure 4.14 Fats from foods are digested into glycerol and fatty acids. These molecules may enter catabolic pathways and be used as energy sources.

When ketone bodies are formed more rapidly than they can be decomposed, some of them are eliminated through the lungs and kidneys. Consequently, the breath and urine may develop a fruity odor due to the presence of the ketone *acetone*. This sometimes happens when a person fasts, forcing body cells to metabolize fat, in order to lose weight. Persons suffering from *diabetes mellitus* also are likely to metabolize excessive amounts of fats, and they too may have acetone in the breath and urine. At the same time, they may develop a serious imbalance in pH called *acidosis,* which is due to an overaccumulation of still other acidic ketone bodies.

1. What is meant by a metabolic pathway?
2. How are carbohydrates used within cells?
3. What must happen to fat molecules before they can be used as energy sources?

Protein Pathways

When dietary proteins are digested, the resulting amino acids are absorbed and transported by the blood to various body cells. Many of these amino acids are reunited to form new protein molecules, as specified by DNA, which then may be incorporated into cell parts or serve as enzymes. Still others may be decomposed to supply energy.

When protein molecules are used as energy sources, they must first be broken down into amino acids. The amino acids undergo *deamination,* a process that occurs in the liver and involves removing the nitrogen-containing portions ($-NH_2$ groups) from the amino acids. These $-NH_2$ groups are converted later into a waste substance called *urea.*

Depending upon the amino acids involved, the remaining deaminated portions of the amino acid molecules are decomposed by one of several pathways. Some

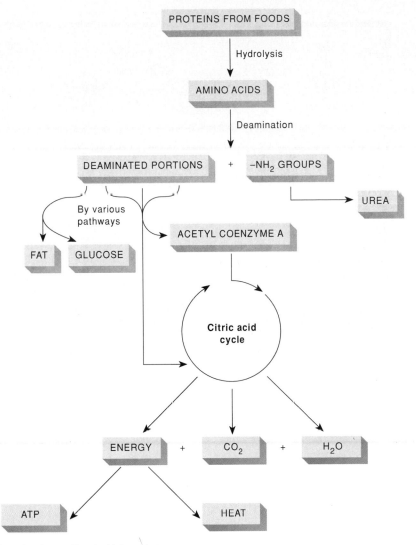

PROTEINS FROM FOODS

Hydrolysis

AMINO ACIDS

Deamination

DEAMINATED PORTIONS + –NH₂ GROUPS

By various pathways

UREA

FAT   GLUCOSE   ACETYL COENZYME A

Citric acid cycle

ENERGY + CO₂ + H₂O

ATP   HEAT

Figure 4.15   Proteins from foods are digested into amino acids, but before these smaller molecules can be used as energy sources, they must be deaminated.

of these pathways lead to the formation of acetyl coenzyme A, and others lead more directly to various steps of the citric acid cycle. As energy is released from the cycle, some of it is captured in molecules of ATP (figure 4.15). If energy is not needed immediately, the deaminated portions of the amino acids may be changed into glucose or fat molecules by still other metabolic pathways.

Urea is produced in the liver from amino groups that are formed by the deamination of amino acids. It is carried in the blood to the kidneys, which excrete the urea in urine. As a result of certain kidney disorders, the ability to remove urea from the blood is impaired, and the blood urea concentration rises. Thus, a blood test called *blood urea nitrogen* (BUN), which determines the blood urea concentration, is often used to evaluate kidney function.

Glucose can be changed back into some amino acids if certain nitrogen-containing molecules are available. However, about eight necessary amino acids cannot be synthesized in adequate amounts in human cells and so must be provided in the diet. For this reason, these are called *essential amino acids*. They are discussed in chapter 15.

Regulation of Metabolic Pathways

As mentioned previously, the rate at which an enzyme-controlled reaction occurs usually increases if either the number of substrate molecules or the number of enzyme molecules is increased. However, the rate at which a metabolic pathway functions is often determined by a regulatory enzyme responsible for one of its steps. This regulatory enzyme is present in limited quantity. Consequently, it can become saturated with substrate molecules whenever the substrate concentration increases

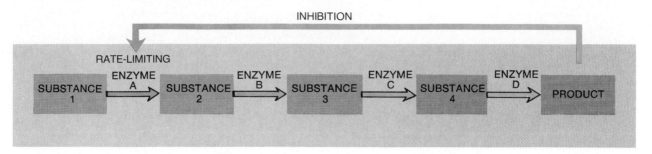

Figure 4.16   A rate-limiting enzyme in a metabolic pathway may be controlled by a negative feedback mechanism in which the product of the pathway inhibits the enzyme.

above a certain amount. Once the enzyme is saturated, increasing the amount of substrate will no longer affect the rate of reaction. In this way, a single enzyme in a pathway can control the whole pathway.

As a rule, such a *rate-limiting enzyme* is the first enzyme in a series. This position is important because some intermediate substance of the pathway might accumulate if an enzyme occupying another location in the sequence were involved.

Commonly, a rate-limiting enzyme is controlled, in turn, by a *negative feedback mechanism.* In this case, the final product of a metabolic pathway acts to inhibit the regulatory enzyme. Thus, as the product increases, the pathway is inhibited and synthesis of the product decreases; when the amount of product decreases, the pathway is not inhibited and more product is synthesized. Consequently, the rate at which the product is produced remains relatively stable (figure 4.16).

1. How do cells utilize proteins?
2. What is meant by a rate-limiting enzyme?
3. How might a metabolic pathway be controlled by a negative feedback mechanism?

Nucleic Acids and Protein Synthesis

Because enzymes control the metabolic processes that enable cells to survive, the cells must possess information for producing these specialized proteins. Such information is held in **DNA** (deoxyribonucleic acid) molecules in the form of a *genetic code.* This code "instructs" cells how to synthesize specific protein molecules.

Genetic Information

Children resemble their parents because of inherited traits; but what is actually passed from parents to a child is *genetic information.* This information is received in the form of DNA molecules from the parents'

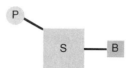

Figure 4.17   Each nucleotide of a nucleic acid consists of a 5–carbon sugar (S), a phosphate group (P), and an organic base (B).

sex cells, and as an offspring develops, the information is passed from cell to cell by mitosis. Genetic information "tells" the cells of the developing body how to construct specific protein molecules, which, in turn, function as structural materials, enzymes, or other vital substances.

The portion of a DNA molecule that contains the genetic information for making one kind of protein (polypeptide) is called a **gene.** Thus, inherited traits are determined by the genes contained in the parents' sex cells that fuse to form the first cell of an offspring's body. These genes instruct the cells to synthesize the particular enzymes needed to control metabolic pathways.

DNA Molecules

As described in chapter 2, the building blocks of nucleic acids (nucleotides) each contain a 5-carbon sugar (ribose or deoxyribose), a phosphate group, and one of several organic bases (figure 4.17).

These nucleotides are joined so that the sugar and phosphate portions alternate and form a long chain, or "backbone." In a DNA molecule, the organic bases project out from this backbone and are bound by weak hydrogen bonds to the bases of the second strand (figure 4.18). The resulting structure is something like a ladder in which the uprights represent the sugar and phosphate backbones of the two strands, and the crossbars represent the organic bases. In addition, the molecular ladder is twisted to form a *double helix,* which is somewhat like a spiral staircase (figure 4.19).

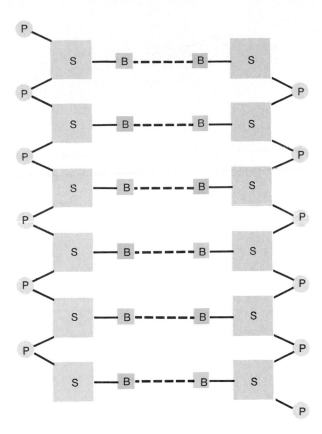

Figure 4.18 The nucleotides of a DNA strand are joined to form a sugar-phosphate backbone (S-P). The organic bases of the nucleotides (B) extend from this backbone and are weakly held to the bases of the second strand by hydrogen bonds (dashed lines).

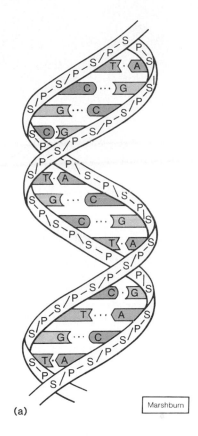

Marshburn

(a)

The organic base of a DNA nucleotide can be one of four kinds: *adenine* (A), *thymine* (T), *cytosine* (C), or *guanine* (G) (figure 4.20). Therefore, there are only four kinds of DNA nucleotides: adenine nucleotide, thymine nucleotide, cytosine nucleotide, and guanine nucleotide.

A strand of DNA consists of nucleotides arranged in a particular sequence (figure 4.21). Moreover, the nucleotides of one strand are paired in a special way with those of the other strand. Only certain organic bases have the molecular shapes needed to fit together so that their nucleotides can bond with one another. Specifically, because of their special molecular shapes, an adenine will bond only to a thymine, and a cytosine will bond only to a guanine. As a consequence of such base pairing, a DNA strand possessing the base sequence T, C, A, G would have to be joined to a second strand with the complementary base sequence A, G, T, C (figure 4.22). It is the particular sequence of base pairs that encodes the genetic information held in a DNA molecule.

(b)

Figure 4.19 (a) The molecular ladder of a double-stranded DNA molecule is twisted into the form of a double helix. Each strand contains only four kinds of organic bases—adenine (A), thymine (T), cytosine (C), and guanine (G). (b) Model of a portion of a DNA molecule.

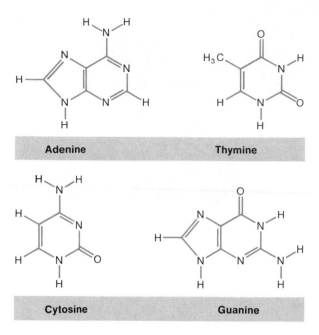

Adenine

Thymine

Cytosine

Guanine

Figure 4.20 Each nucleotide of a DNA molecule contains one of the four organic bases: adenine, thymine, cytosine, or guanine.

The patterns of tiny ridges in the skin of a person's fingertips are so distinctive it is possible to match a person with a fingerprint. Similarly, because the molecular sequence of the DNA in each person's cells is unique it is possible to match a person with a sample of body cells by analyzing the DNA in those cells. Such DNA sequences are sometimes used in criminal investigations to identify the individual source of blood, semen, or other cellular evidence.

Genetic Code

Genetic information contains the instructions for synthesizing proteins, and because proteins consist of twenty different amino acids joined in particular sequences, the genetic information must tell how to position the amino acids correctly in a polypeptide chain.

Each of the twenty different amino acids is represented in a DNA molecule by a particular sequence of three nucleotides. That is, the sequence C, G, T in a DNA strand represents one kind of amino acid; the sequence G, C, A represents another kind; and T, T, A still another kind (chart 4.1). Other sequences represent instructions for beginning or ending the synthesis of a protein molecule.

Thus, the sequence in which the nucleotide groups are arranged within a DNA molecule can denote the arrangement of amino acids within a protein molecule, as well as indicating how to start or stop the synthesis of a protein. This method of storing information used for the synthesis of particular protein molecules is the **genetic code.**

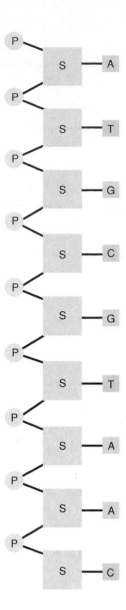

Figure 4.21 A single strand of DNA consists of a chain of nucleotides arranged in a particular sequence. Within the chain are four kinds of nucleotides: adenine nucleotide (A); thymine nucleotide (T); cytosine nucleotide (C); and guanine nucleotide (G).

Although DNA molecules are located in the chromatin within a cell's nucleus, protein synthesis occurs in the cytoplasm. Therefore, the genetic information must somehow be transferred from the nucleus into the cytoplasm. This transfer of information is the function of certain RNA molecules.

RNA Molecules

RNA (ribonucleic acid) molecules differ from DNA molecules in several ways. For example, RNA molecules are usually single-stranded, and their nucleotides contain ribose rather than deoxyribose sugar. Like DNA, RNA nucleotides each contain one of four or-

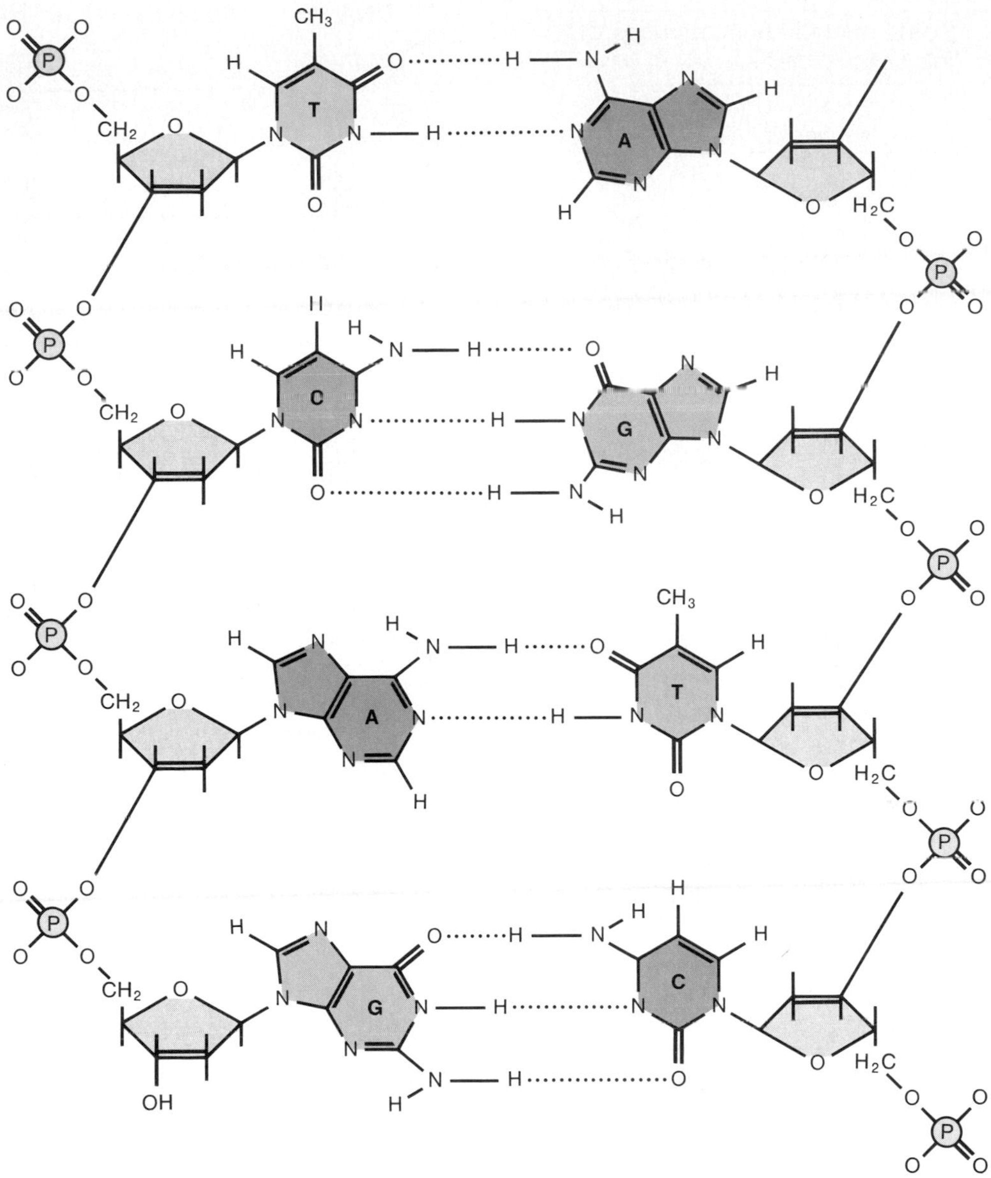

Figure 4.22 The nucleotides of a double-stranded molecule of DNA are paired so that an adenine nucleotide of one strand is joined to a thymine nucleotide of the other strand, and a guanine nucleotide of one is joined to a cytosine nucleotide of the other. The dotted lines represent weak hydrogen bonds between the paired bases of these nucleotides.

ganic bases, but while adenine, cytosine, and guanine nucleotides occur in both DNA and RNA, thymine nucleotides are found only in DNA. In place of thymine nucleotides, RNA molecules contain *uracil* (U) nucleotides (figure 4.23).

One step in the transfer of information from the nucleus to the cytoplasm involves the synthesis of a type of RNA called **messenger RNA** (mRNA).

In the process of producing messenger RNA, an enzyme called *RNA polymerase* becomes associated with the DNA base sequence that represents the beginning of a gene (the instructions for synthesizing a particular protein). Such a DNA base sequence is called a *promoter*. As a result of the enzymatic action, a part

of the double-stranded DNA molecule unwinds and pulls apart, exposing a portion of the gene. The RNA polymerase moves along the strand, causing other portions of the gene to be exposed. At the same time, a molecule of messenger RNA is formed. This new molecule is composed of nucleotides complementary to those arranged along the DNA strand. For example, if the sequence of DNA bases is A, T, G, C, G, T, A, A, C, the complementary bases in the developing messenger RNA molecule will be U, A, C, G, C, A, U, U, G, as shown in figure 4.24. (The other strand of DNA is not used in this process, but it is important in the replication of the DNA molecule, as is explained in a subsequent section of this chapter.)

CHART 4.1 Some nucleotide sequences of the genetic code

Amino acids	DNA sequence	RNA sequence
Alanine	CGT	GCA
Arginine	GCA	CGU
Asparagine	TTA	AAU
Aspartic acid	CTA	GAU
Cysteine	ACA	UGU
Glutamic acid	CTT	GAA
Glutamine	GTT	CAA
Glycine	CCG	GGC
Histidine	GTA	CAU
Isoleucine	TAG	AUC
Leucine	GAA	CUU
Lysine	TTT	AAA
Methionine	TAC	AUG
Phenylalanine	AAA	UUU
Proline	GGA	CCU
Serine	AGG	UCC
Threonine	TGC	ACG
Tryptophan	ACC	UGG
Tyrosine	ATA	UAU
Valine	CAA	GUU

Instructions		
Start protein synthesis	TAC	AUG
Stop protein synthesis	ATT	UAA

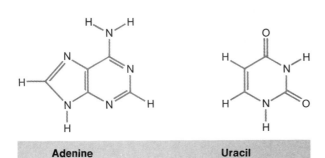

Adenine   **Uracil**

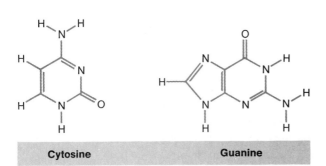

Cytosine   **Guanine**

Figure 4.23   Each nucleotide of an RNA molecule contains one of four organic bases: adenine, uracil, cytosine, or guanine. How are these nucleotides similar to those found in DNA molecules? How are they different?

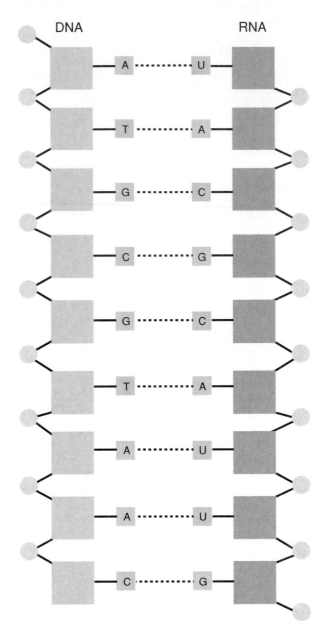

Figure 4.24   When an RNA molecule is synthesized beside a strand of DNA, complementary nucleotides bond as in a double-stranded molecule of DNA, with one exception: RNA contains uracil nucleotides (U) in place of thymine nucleotides (T).

The RNA polymerase continues to move along the DNA strand, exposing portions of the gene, until it reaches a special DNA base sequence (termination signal) that represents the end of the gene. At this point, the RNA polymerase releases the newly formed messenger RNA molecule and leaves the DNA. The DNA then rewinds and assumes its previous double-helix structure. This process of copying DNA information into the structure of a messenger RNA molecule is called *transcription*.

Because an amino acid is represented by a sequence of three nucleotides in a DNA molecule, the

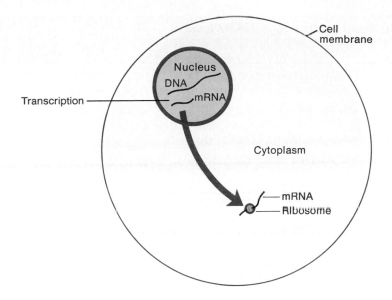

Figure 4.25 After a section of DNA information is transcribed, a messenger RNA molecule (mRNA) moves out of the nucleus and enters the cytoplasm where it becomes associated with a ribosome.

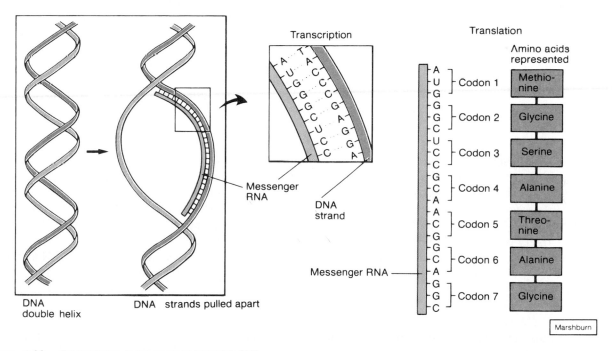

Figure 4.26 Transcription is the process by which DNA information is copied into the structure of messenger RNA. Translation is the process by which messenger RNA information is used in the synthesis of protein.

same amino acid will be represented in the transcribed messenger RNA by the complementary set of three nucleotides. Such a set of nucleotides in a messenger RNA molecule is called a *codon.*

Once they are formed, messenger RNA molecules (each of which consists of hundreds or even thou-

sands of nucleotides) can move out of the nucleus through the tiny pores in the nuclear envelope and enter the cytoplasm (figure 4.25). There they become associated with ribosomes and act as patterns or templates for the synthesis of protein molecules—a process called *translation.* (See figure 4.26.)

CHART 4.2 A comparison of DNA and RNA molecules

	DNA	RNA
Main location	Within chromosomes of the nucleus	Within the cytoplasm
5-carbon sugar	Deoxyribose	Ribose
Basic molecular structure	Double-stranded	Single-stranded
Organic bases included	Adenine, thymine, cytosine, guanine	Adenine, uracil, cytosine, guanine
Major functions	Contains genetic code for protein synthesis, replicates itself prior to cell division	Messenger RNA carries transcribed DNA information to cytoplasm, and acts as template for synthesis of protein molecules; transfer RNA carries amino acids to messenger RNA; ribosomal RNA provides structure for ribosomes

Recently it was discovered that RNA molecules in some organisms also act as enzymes; that is, they can catalyze changes in themselves and in other RNA molecules. Such RNA enzymes seem to play roles in tailoring the final structure of functional RNA molecules, and they may play some presently unknown role in protein synthesis.

Chart 4.2 provides a comparison of DNA and RNA molecules.

Protein Synthesis

Before a protein molecule can be synthesized, the correct amino acids must be present in the cytoplasm to serve as building blocks. Furthermore, these amino acids must be positioned in the proper locations along a strand of messenger RNA. The positioning of amino acid molecules is the function of a second kind of RNA molecule, which is synthesized in the nucleus and called **transfer RNA** (tRNA). A transfer RNA molecule, however, consists of only seventy to eighty nucleotides and has a complex three-dimensional shape.

Since twenty different kinds of amino acids are involved in protein synthesis, there must be at least twenty different kinds of transfer RNA molecules to serve as guides. However, before a transfer RNA molecule can pick up its particular kind of amino acid, the amino acid must be activated. This step is influenced by the presence of special enzymes and by ATP, which provides the energy needed to form a bond between the amino acid and its transfer RNA molecule.

Each type of transfer RNA includes a special region at one end that contains three nucleotides in a particular sequence. These nucleotides can only bond to the complementary set of three nucleotides of a messenger RNA molecule—a *codon*. Thus, the set of three nucleotides in the transfer RNA molecule is called an *anticodon*. In this way, the transfer RNA carries its amino acid to a correct position on a messenger RNA strand. This action occurs in close association with a ribosome.

A ribosome is a tiny particle of two unequal-sized subunits composed of *ribosomal RNA* (rRNA) and protein. By its smaller subunit, a ribosome binds to a molecule of messenger RNA near a codon at the beginning of the messenger strand. This action allows a transfer RNA molecule with the complementary anticodon to bring the amino acid it carries into position and become temporarily joined to the ribosome. Then the transfer RNA molecule releases its amino acid and returns to the cytoplasm (figure 4.27). This process is repeated again and again as the ribosome moves along the messenger RNA.

The amino acids, which are released by the transfer RNA molecules, are added one at a time to a developing protein molecule. The enzymes necessary for the bonding of these amino acids are found in the larger subunit of the ribosome. This subunit also seems to hold the growing chain of amino acids as the protein molecule is formed.

Actually, a molecule of messenger RNA is usually associated with several ribosomes at the same time. Thus, several protein molecules, each in a different stage of development, may be present at any given moment.

As the protein molecule develops, it folds into its unique shape, and when the process is completed, it is released as a separate functional molecule. The transfer RNA molecules can pick up other amino acids from the cytoplasm, and the messenger RNA molecules can function repeatedly.

ATP molecules provide the energy needed to form the various chemical bonds in this process. A complete protein molecule may consist of many hundreds of amino acids, and the energy from three ATP molecules is required to link each amino acid to the growing chain. Consequently, a large proportion of a cell's energy

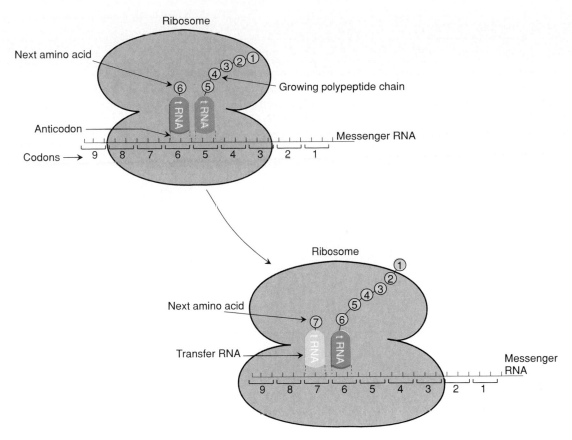

Figure 4.27 Molecules of transfer RNA (tRNA) bring specific amino acids and place them in the sequence determined by the codons of the messenger RNA molecule. These amino acids bond and form the polypeptide chain of a protein molecule.

supply is needed to support protein synthesis, especially if the cell is young and growing.

The quantity of a particular protein that a cell synthesizes is generally proportional to the quantity of the corresponding messenger RNA that is present. Control of such protein production involves the rate at which messenger RNA is synthesized in the nucleus and the rate at which the messenger RNA is destroyed by enzymes (ribonucleases) in the cytoplasm. Although this control mechanism is not completely understood, it is known that substances from outside a cell, such as hormones, sometimes provide stimuli that bring about the destruction of certain messenger RNA molecules and thus limit the production of specific proteins.

The process of protein synthesis is summarized in chart 4.3.

1. What is the function of DNA?
2. How is information carried from the nucleus to the cytoplasm?
3. How are protein molecules synthesized?

Each mitochondrion of a cell contains the enzymes necessary to carry on aerobic respiration and ATP formation. The genetic information needed to synthesize most of these enzymes is found within chromosomal DNA of the nucleus. The enzymes are produced in association with cytoplasmic ribosomes and are later moved into the mitochondria. The information for synthesizing the remainder of the respiratory enzyme set occurs in a circle of mitochondrial DNA, and these enzymes are manufactured using ribosomes found within each mitochondrion.

Replication of DNA

When a cell reproduces, each newly formed cell needs a copy of the parent cell's genetic information so it will be able to synthesize the proteins necessary to build cellular parts and carry on metabolism.

DNA molecules can be replicated, and this replication takes place during the interphase of the cell's life cycle.

As the *replication* process begins, bonds are broken between the complementary base pairs of the

CHART 4.3 Protein synthesis

Transcription (occurring within the nucleus)

1. RNA polymerase becomes associated with the base sequence of a gene within a DNA molecule.
2. This enzyme causes a portion of the DNA molecule to unwind and its strands to pull apart, thus exposing a portion of the gene.
3. The RNA polymerase moves along one strand of the exposed gene and causes the production of an mRNA molecule, whose nucleotides are complementary to those of the strand of the gene.
4. When the RNA polymerase reaches the end of the gene, the newly formed mRNA molecule is released.
5. The DNA molecule rewinds and assumes its previous structure.
6. The mRNA molecule passes through a pore in the nuclear envelope and enters the cytoplasm.

Translation (occurring within the cytoplasm)

1. A ribosome binds to the mRNA molecule near the codon at the beginning of the messenger strand.
2. A tRNA molecule that has the complementary anticodon becomes associated with the ribosome, and the tRNA molecule releases the amino acid it carries.
3. This process is repeated for each codon in the mRNA sequence as the ribosome moves along its length.
4. Enzymes associated with the ribosome cause the amino acids that are released from the tRNA molecules to bind together, thus forming a chain of amino acids.
5. As the chain of amino acids develops, it folds into the unique shape of a functional protein molecule.
6. The completed protein molecule is released. The mRNA molecule, ribosome, and tRNA molecules can function repeatedly to synthesize other protein molecules.

double strands in each DNA molecule. Then the double-stranded structure pulls apart and unwinds, exposing the organic bases of its nucleotides. New nucleotides (of the four types found in DNA) pair with the exposed bases, and enzymes cause the complementary bases to join together. In this way, a new strand of complementary nucleotides is constructed along each of the old strands. As a result, two complete DNA molecules are produced, each with one old strand of the original molecule and one new strand (figure 4.28).

During mitosis, the two DNA molecules that are held within the chromatids of each of the chromosomes are separated so that one of these DNA molecules passes to each of the new cells.

Changes in Genetic Information

The amount of genetic information held within a set of human chromosomes is very large. In fact, some investigators estimate that it may equal the amount of text contained in twenty sets of *Encyclopaedia Britannica*.

Since each of the trillions of cells in an adult body has resulted from mitosis, this large amount of information had to be replicated many times and with a high degree of accuracy. However, occasionally a mistake occurs or the DNA structure is damaged, and the genetic information is altered. Such a change is called a **mutation.** If the cell in which it occurs survives, the mutation is likely to be passed to future generations of cells.

Nature of Mutations

Mutations can originate in a number of ways. For example, during DNA replication, an organic base may be paired incorrectly within the newly forming strand, or some extra organic bases may be built into its structure. In other instances, sections of DNA strands may be deleted, moved to other regions of the molecule, or even attached to other chromosomes. In any case, the consequences are similar—the genetic information is changed. If a protein is constructed from this information, its molecular structure is likely to be faulty and the protein nonfunctional (figure 4.29).

Fortunately, cells can detect damage in their DNA molecules, and they possess enzymes that often can repair damage occurring in a single strand of DNA. These *repair enzymes* are able to clip out defective nucleotides and cause the resulting gap to be filled with nucleotides complementary to those on the remaining strand of DNA. In this way, the original structure of the double-stranded molecule can be restored.

On the other hand, both strands of the DNA molecule may be damaged in the same region. This type of change is unlikely to be repaired, and if the cell survives and reproduces, the newly formed cells are likely to receive copies of the mutation. Also, there is evidence that humans vary in their abilities to repair DNA damage. Those with impaired repair mechanisms seem to experience increased incidence of certain diseases, including cancers.

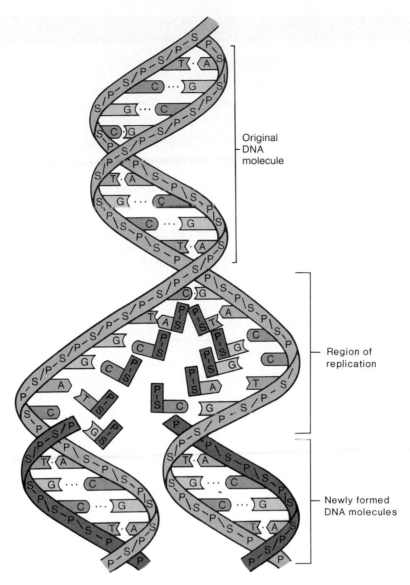

Figure 4.28   When a DNA molecule replicates, its strands pull apart, and a new strand of complementary nucleotides forms along each of the old strands.

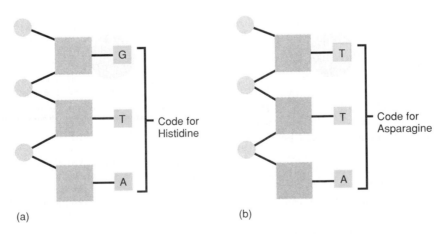

(a)

(b)

Figure 4.29   (a) The DNA code for the amino acid histidine is GTA. (b) If something happens to change the guanine in this section of the molecule to thymine, the DNA code is changed to TTA, which represents the amino acid asparagine. Such a change would be a mutation.

Mutagenic Factors

Although some mutations arise spontaneously without known cause, certain factors are known to increase the likelihood of mutations. These factors are called *mutagens,* and they include ionizing radiation and various chemicals.

Some common forms of ionizing radiation, such as ultraviolet light and X ray, are most likely to damage single strands of DNA by breaking them or by causing adjacent organic bases to link together. Since this type of damage may be repaired by repair enzymes, such radiation is considered to be a weaker mutagen than other more energetic forms. For example, radiation that includes neutrons may damage both strands of a DNA molecule, and that type of damage is unlikely to be repaired.

Ultraviolet light is not very penetrating, so its effect is usually confined to the surfaces of the skin and eyes. X ray, on the other hand, is very penetrating and can affect cells located in deep tissues and organs.

Chemical mutagens produce a variety of effects on DNA molecules, including removal of bases, production of cross-links between strands, and the linking of themselves to organic bases. Although some of this damage may be repairable, many chemicals are considered to be very powerful mutagens.

It is difficult to avoid contact with mutagenic factors in everyday living. Skin is exposed to ultraviolet light from sunlight; X rays and radioactive isotopes are often used in the diagnosis and treatment of diseases; and mutagenic chemicals occur in a variety of commonly used substances such as tobacco and various petroleum products. Yet, since we know that these factors are hazardous to cells of present and future generations, the only wise course is to limit exposure to them whenever possible.

It is estimated that there are as many as three billion base pairs in a complete set of human chromosomes. However, the number of mistakes made in matching nucleotides during the process of replicating DNA molecules seems to be only about one in ten billion. This high degree of accuracy is due to the presence of enzymes that "check" or "proofread" each nucleotide as it is selected and added to a growing strand of DNA, and that expel any mismatched nucleotides from that strand.

Effects of Mutations

The effects of mutations vary greatly. At one extreme, there is little or no effect. For example, if a mutant cell is unable to manufacture an enzyme that is relatively unimportant, the cell can continue to function effectively. At the other extreme, as a result of a mutation a cell may be unable to produce an enzyme needed for energy release, so it cannot survive. The mutations of most concern are those between these extremes. They bring about some decrease in cell efficiency, so that although an affected cell may live and reproduce, it has difficulty functioning normally. In time, the increasing numbers of these mutated cells may produce abnormal effects in the structure and function of the tissues and organs in which they occur.

It also makes a difference whether a mutation occurs in a body cell of an adult or in a cell that is part of a developing embryo. In an adult, a mutant cell might not be noticed because there are many normally functioning cells around it. In the embryo, however, the abnormal cell might become the ancestor of great numbers of cells that are forming the body of a child. In fact, all the cells of the child's body could be defective if the mutation were present in the fertilized egg.

Many human diseases are known to be related to defects in the genetic information of cells that have been caused by mutations. The majority of these involve an inability to produce specific enzymes. As an example, the disease called phenylketonuria (PKU) results from a defective enzyme. The normal enzyme functions in the metabolic breakdown of the amino acid phenylalanine. When the defective enzyme is present, phenylalanine is only partially metabolized, and a toxic substance accumulates. This substance interferes with the normal activity of the developing nervous system. If the condition is untreated in an affected newborn, it usually results in severe mental retardation. To be effective, treatment for PKU must begin during the first few days of a newborn's life. The treatment involves restricting the intake of foods containing phenylalanine, and must be continued for several years at least, and perhaps throughout the life of the patient.

1. How are DNA molecules replicated?
2. What is a mutation?
3. How do mutations occur?
4. What kinds of mutations are of greatest concern?

Chapter Summary

Introduction (page 98)

A cell continuously carries on metabolic processes.

Metabolic Processes (page 98)

1. Anabolic metabolism
   a. Anabolic metabolism consists of constructive processes in which smaller molecules are used to build larger ones.
   b. In dehydration synthesis, hydrogen atoms and hydroxyl groups are removed, water is formed, and smaller molecules become bound by shared atoms
   c. Complex carbohydrates are synthesized from monosaccharides, fats are synthesized from glycerol and fatty acids, and proteins are synthesized from amino acids.
2. Catabolic metabolism
   a. Catabolic metabolism consists of decomposition processes in which larger molecules are broken down into smaller ones.
   b. In hydrolysis, a water molecule is used to supply a hydrogen atom to one portion of a molecule and a hydroxyl group to a second portion; the bond between these two portions is broken.
   c. Complex carbohydrates are decomposed into monosaccharides, fats are decomposed into glycerol and fatty acids, and proteins are decomposed into amino acids.

Control of Metabolic Reactions (page 98)

Metabolic reactions are controlled by enzymes.

1. Enzymes and their actions
   a. Metabolic reactions require energy to get them started.
   b. Enzymes are proteins that promote metabolic reactions.
   c. An enzyme acts upon a molecule by temporarily combining with it and causing its chemical structure to become distorted.
   d. The shape of an enzyme molecule fits the shape of its substrate molecule.
   e. When an enzyme combines with its substrate, the substrate is changed, resulting in a product, but the enzyme is released in its original form.
   f. The rate of enzyme-controlled reactions depends upon the number of enzyme and substrate molecules present and the efficiency of the enzyme.
   g. Enzymes are usually named according to their substrates, with -ase added.
2. Cofactors and coenzymes
   a. Cofactors are necessary parts of some enzyme molecules.
   b. A cofactor may be an ion or a small organic molecule called a coenzyme.
   c. Vitamins, which provide coenzymes, usually cannot be synthesized by human cells in adequate amounts.

3. Factors that alter enzymes
   a. Enzymes are proteins and can be denatured.
   b. Factors that may denature enzymes include excessive heat, radiation, electricity, certain chemicals, and extreme pH values.

Energy for Metabolic Reactions (page 101)

Energy is a capacity to produce change or to do work. Common forms of energy include heat, light, sound, electrical energy, mechanical energy, and chemical energy. Whenever changes take place, energy is being transferred from one part to another.

1. Release of chemical energy
   a. Most metabolic processes utilize chemical energy that is released when molecular bonds are broken.
   b. The energy released from glucose during cellular respiration is used to promote metabolism.
   c. Cellular respiration is controlled by enzymes in the cytoplasm and mitochondria.
2. Anaerobic respiration
   a. The first phase of glucose decomposition occurs in the cytoplasm and does not require oxygen.
   b. Some of the energy released is transferred to molecules of ATP.
3. Aerobic respiration
   a. The second phase of glucose decomposition occurs within the mitochondria and requires oxygen.
   b. Considerably more energy is transferred to ATP molecules during this phase than during the anaerobic phase.
   c. The final products of glucose decomposition are carbon dioxide, water, and energy.
4. ATP molecules
   a. Thirty-eight molecules of ATP can be produced for each glucose molecule that is decomposed.
   b. Energy is captured in the bond of the terminal phosphate of each ATP molecule.
   c. Captured energy is released when the terminal phosphate bond of an ATP molecule is broken.
   d. An ATP molecule that loses its terminal phosphate becomes an ADP molecule.
   e. An ADP molecule can be converted to an ATP molecule by capturing some energy and a phosphate.

Metabolic Pathways (page 103)

Metabolic processes usually involve a number of steps that must occur in the correct sequence. A sequence of enzyme-controlled reactions is called a metabolic pathway. Typically, metabolic pathways are interconnected.

1. Carbohydrate pathways
   a. Carbohydrates may enter catabolic pathways and be used as energy sources.
   b. Glycolysis involves the conversion of a 6-carbon glucose molecule into two 3-carbon pyruvic acid molecules, resulting in a small net gain of ATP.

c. The citric acid cycle is a complex series of reactions in which molecules are decomposed, carbon dioxide and hydrogen atoms are released, and ATP molecules are formed.

d. Hydrogen atoms from the citric acid cycle are converted to hydrogen ions, which, in turn, combine with oxygen to form water molecules.

e. Electrons from hydrogen atoms enter an electron transport system, and as energy is released from the system, some of it is used to form ATP.

f. For each glucose molecule metabolized, a maximum of thirty-eight ATP molecules are formed.

g. When present in excess, carbohydrates may enter anabolic pathways and be converted into glycogen or fat.

2. Lipid pathways
   a. Most dietary fats are triglycerides.
   b. Before fats can be used as an energy source they must be converted into glycerol and fatty acids.
   c. Fatty acids are decomposed by beta oxidation.
      (1) Beta oxidation involves activating fatty acids and breaking them down into segments containing two carbon atoms each.
      (2) Fatty acid segments are converted into acetyl coenzyme A, which can be oxidized by means of the citric acid cycle.
   d. Fats can be synthesized from glycerol and fatty acids and from excess glucose or amino acids.

3. Protein pathways
   a. Proteins are used as building materials for cellular parts, as enzymes, and as energy sources.
   b. Before proteins can be used as energy sources, they must be decomposed into amino acids, and the amino acids must be deaminated.
   c. The deaminated portions of amino acids can be broken down into carbon dioxide and water, or converted into glucose or fat.
   d. Approximately eight essential amino acids cannot be synthesized in adequate amounts by human cells and must be obtained in foods.

4. Regulation of metabolic pathways
   a. A metabolic pathway may be regulated by a rate-limiting enzyme.
   b. The regulatory enzyme, in turn, may be controlled by a negative feedback mechanism in which the product of the pathway inhibits the enzyme.
   c. The rate of product formation usually remains stable.

Nucleic Acids and Protein Synthesis (page 110)

DNA molecules contain information that tells a cell how to synthesize the correct enzymes.

1. Genetic information
   a. Inherited traits result from DNA information that is passed from parents to offspring.
   b. A gene is a portion of a DNA molecule that contains the genetic information for making one kind of protein.

2. DNA molecules
   a. The nucleotides of a DNA strand are arranged in a particular sequence.
   b. The nucleotides are paired with those of the second strand in a complementary fashion.

3. Genetic code
   a. The sequence of nucleotides in a DNA molecule represents the sequence of amino acids in a protein molecule.
   b. Genetic information is transferred from the nucleus to the cytoplasm by RNA molecules.

4. RNA molecules
   a. RNA molecules are usually single-stranded, contain ribose instead of deoxyribose, and contain uracil nucleotides in place of thymine nucleotides.
   b. Messenger RNA molecules, which are synthesized in the nucleus, contain a nucleotide sequence that is complementary to that of an exposed strand of DNA.
   c. Messenger RNA molecules move into the cytoplasm, become associated with ribosomes, and act as patterns for the synthesis of protein molecules.

5. Protein synthesis
   a. Molecules of transfer RNA serve to position amino acids along a strand of messenger RNA.
   b. A ribosome binds to a messenger RNA molecule and allows a transfer RNA molecule to recognize its correct position on the messenger RNA.
   c. The ribosome contains enzymes needed for the synthesis of the developing protein and holds the protein until it is completed.
   d. As the protein develops, it folds into a unique shape.
   e. ATP provides the energy needed for these molecular changes.

6. Replication of DNA
   a. Each new cell needs a copy of the parent cell's genetic information.
   b. DNA molecules are replicated during interphase of the cell's life cycle.
   c. Each new DNA molecule contains one old strand and one new strand.

Changes in Genetic Information (page 118)

The amount of information within a DNA molecule is very large. Occasionally a change called a mutation occurs in the genetic information.

1. Nature of mutations
   a. Mutations involve a variety of kinds of changes within DNA molecules.
   b. A protein synthesized from damaged DNA information is likely to be nonfunctional.
   c. Repair enzymes can correct some forms of DNA damage.
2. Effects of mutations
   a. The mutations that cause a decrease in cell efficiency are of most concern.
   b. A mutation that occurs in a cell of a developing embryo may affect a great number of cells in the developing body.

Clinical Application of Knowledge

1. Because enzymes are proteins, they may become denatured. Relate this to the fact that changes in the pH value of body fluids during an illness may be life-threatening.
2. Some weight-reducing diets drastically limit the dieter's intake of carbohydrates but allow liberal use of fat and protein foods. What changes would such a diet cause in the cellular metabolism of the dieter? What changes could be noted in the urine of such a person?
3. How would you explain the fact that vitamins that function as coenzymes in cells are needed in extremely low concentrations?
4. Why do you think it is of particular importance for a pregnant woman to avoid exposure to forms of ionizing radiation or to chemicals that are known to cause mutations?
5. What changes in concentrations of oxygen and carbon dioxide would you expect to find in the blood of a person who is forced to exercise on a treadmill beyond his or her normal capacity? How might these changes affect the pH of the person's blood?
6. What mutagenic factors might healthcare workers encounter more frequently than other workers?

Review Activities

1. Define *anabolic* and *catabolic metabolism*.
2. Distinguish between dehydration synthesis and hydrolysis.
3. Explain what is meant by a *peptide bond*.
4. Define *enzyme*.
5. Describe how an enzyme interacts with its substrate.
6. List three factors that increase the rates of enzyme-controlled reactions.
7. Explain how enzymes are named.
8. Define *cofactor*.
9. Explain why humans need vitamins in their diets.
10. Explain how an enzyme may be denatured.
11. Define *energy*.
12. Explain how the oxidation of molecules inside cells differs from the burning of substances outside cells.
13. Define *cellular respiration*.
14. Distinguish between anaerobic and aerobic respiration.
15. Explain the importance of ATP to cellular processes.
16. Describe the relationship between ATP and ADP molecules.
17. Explain what is meant by a *metabolic pathway*.
18. Describe the process called *glycolysis*.
19. Review the major events of the citric acid cycle.
20. Explain how carbohydrates are stored.
21. Define *beta oxidation*.
22. Explain how fats may serve as energy sources.
23. Define *deamination* and explain its importance.
24. Explain why some amino acids are called essential.
25. Explain how one enzyme can regulate a metabolic pathway.
26. Describe how a negative feedback mechanism can help control a metabolic pathway.
27. Explain what is meant by *genetic information*.
28. Describe the relationship between a DNA molecule and a gene.
29. Describe the general structure of a DNA molecule.
30. Explain how genetic information is stored in a DNA molecule.
31. Distinguish between messenger RNA and transfer RNA.
32. Distinguish between transcription and translation.
33. Explain the function of a ribosome in protein synthesis.
34. Distinguish between a codon and an anticodon.
35. Explain how a DNA molecule is replicated.
36. Define *mutation*, and explain how mutations may originate.
37. Define *repair enzyme*.
38. Explain how a mutation may affect the cells of an organism.

5

Tissues

*C*ells, the basic units of structure and function within the human organism, are organized into groups and layers called *tissues*. Each type of tissue is composed of similar cells that are specialized to carry on particular functions. For example, epithelial tissues form protective coverings and function in secretion and absorption; connective tissues provide support for softer body parts and bind structures together; muscle tissues are responsible for producing body movements; and nervous tissue is specialized to conduct impulses that help control and coordinate body activities ■

After you have studied this chapter, you should be able to:

1. Describe the general characteristics and functions of epithelial tissue.

2. Name the types of epithelium and identify an organ in which each is found.

3. Explain how glands are classified.

4. Describe the general characteristics of connective tissue.

5. Describe the major cell types and fibers of connective tissue.

6. List the types of connective tissue that occur within the body.

7. Describe the major functions of each type of connective tissue.

8. Distinguish among the three types of muscle tissue.

9. Describe the general characteristics and functions of nervous tissue.

10. Complete the review activities at the end of this chapter. Note that the items are worded in the form of specific learning objectives. You may want to refer to them before reading the chapter.

adipose tissue (ad' ĭ-pōs tish'u)
cartilage (kar'tĭ-lij)
chondrocyte (kon'dro-sīt)
connective tissue (kŏ-nek'tiv tish'u)
epithelial tissue (ep''ĭ-the'le-al tish'u)
fibroblast (fi'bro-blast)
fibrous tissue (fi'brus tish'u)
macrophage (mak'ro-fāj)
muscle tissue (mus'el tish'u)
nervous tissue (ner'vus tish'u)
neuroglia (nu-rog'le-ah)
neuron (nu'ron)
osteocyte (os'te-o-sīt)
osteon (os'te-on)
reticuloendothelial tissue (rĕ-tik''u-lo-en''do-the'le-al tish'u)

adip-, fat: *adip*ose tissue—tissue that stores fat.

chondr-, cartilage: *chondr*ocyte—a cartilage cell.

-cyt, cell: osteo*cyte*—a bone cell.

epi-, upon: *epi*thelial tissue—tissue that covers all free body surfaces.

-glia, glue: neuro*glia*—cells that bind nervous tissue together.

inter-, between: *inter*calated disk—band located between the ends of adjacent cardiac muscle cells.

macro-, large: *macro*phage—a large phagocytic cell.

neuro-, nerve: *neuro*n—a nerve cell.

osseo-, bone: *osseo*us tissue—bone tissue.

phago-, to eat: *phago*cyte—a cell that engulfs and destroys foreign particles.

pseudo-, false: *pseudo*stratified epithelium—tissue whose cells appear to be arranged in layers, but are not.

squam-, scale: *squam*ous epithelium—tissue whose cells appear flattened or scalelike.

strat-, layer: *strat*ified epithelium—tissue whose cells occur in layers.

stria-, groove: *stria*ted muscle—tissue whose cells are characterized by alternating light and dark cross-markings.

I n all complex organisms, cells are organized into layers or groups called **tissues.** Although the cells of different tissues vary in size, shape, arrangement, and function, those within a tissue are quite similar.

Usually tissue cells are separated by nonliving, intercellular materials that the cells secrete. These intercellular materials vary in composition from one tissue to another and may take the form of solids, semisolids, or liquids. For example, bone tissue cells are separated by a solid substance, while blood tissue cells are separated by a liquid.

The tissues of the human body include four major types: *epithelial tissues, connective tissues, muscle tissues,* and *nervous tissue.* These tissues are organized into organs that have specialized functions.

This chapter is concerned primarily with various types of epithelial and connective tissues. Other kinds of tissues will be discussed in subsequent chapters.

1. What is a tissue?
2. List the four major types of tissue.

Epithelial Tissues

General Characteristics

Epithelial tissues are widespread throughout the body. They cover all body surfaces—inside and out—and are the major tissues of glands.

Since epithelium covers organs, forms the inner lining of body cavities, and lines hollow organs, it always has a free surface—one that is exposed to the outside or to an open space internally. The underside of this tissue is anchored to connective tissue by a thin, nonliving layer called the *basement membrane* (basement lamina).

As a rule, epithelial tissues lack blood vessels. However, they are nourished by substances that diffuse from underlying connective tissues, which are well supplied with blood vessels.

Although the cells of some tissues have limited abilities to reproduce, those of epithelium reproduce readily. Injuries to epithelium are likely to heal rapidly as new cells replace lost or damaged ones. For example, skin cells and the cells that line the stomach and intestines are continually being damaged and replaced.

Epithelial cells are tightly packed, and there is little intercellular material between them. Often they are attached to one another by desmosomes. (See chapter 3.) Consequently, these cells provide effective protective barriers in such structures as the outer layer of the skin and the inner lining of the mouth. Other epithelial functions include secretion, absorption, excretion, and sensory reception.

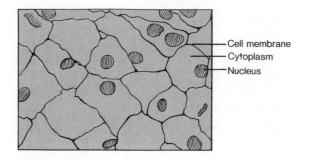

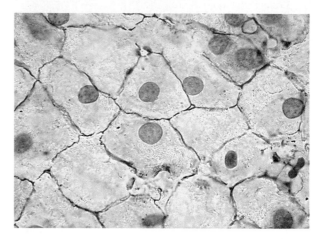

Figure 5.1 Simple squamous epithelium consists of a single layer of tightly packed flattened cells (×250). (Note: In the micrograph, the cells are viewed on the flattened surface of the tissue.)

Epithelial tissues are classified according to the specialized shapes, arrangements, and functions of their cells. For example, epithelial tissues that are composed of single layers of cells are called *simple;* those with many layers of cells are said to be *stratified;* those with thin, flattened cells are called *squamous;* those with cubelike cells are called *cuboidal;* and those with elongated cells are called *columnar.* In the following descriptions, note that the free surfaces of various epithelial cells are modified in ways that reflect their specialized functions.

Simple Squamous Epithelium

Simple squamous epithelium consists of a single layer of thin, flattened cells. These cells fit tightly together, somewhat like floor tiles, and their nuclei are usually broad and thin (figure 5.1).

As a rule, substances pass rather easily through this type of tissue, and it occurs commonly where diffusion and filtration are taking place. For instance, simple squamous epithelium lines the air sacs of the lungs where oxygen and carbon dioxide are exchanged. It also forms the walls of capillaries, lines the insides

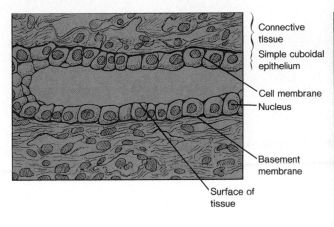

Figure 5.2 Simple cuboidal epithelium consists of a single layer of tightly packed cube-shaped cells (×250).

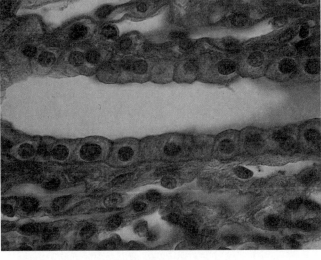

of blood and lymph vessels, and covers the membranes that line body cavities. However, because it is so thin and delicate, simple squamous epithelium can be damaged relatively easily.

Simple Cuboidal Epithelium

Simple cuboidal epithelium consists of a single layer of cube-shaped cells. These cells usually have centrally located, spherical nuclei (figure 5.2).

This tissue covers the ovaries, and lines many of the kidney tubules and the ducts of various glands, such as the salivary glands, pancreas, and liver. In the kidneys, it functions in secretion and absorption; in glands, it is concerned with the secretion of glandular products.

Simple Columnar Epithelium

The cells of **simple columnar epithelium** are elongated; that is, they are longer than they are wide. This tissue is composed of a single layer of cells whose nuclei are usually located at about the same level, near the basement membrane (figure 5.3).

Simple columnar epithelium occurs in the linings of the uterus and various organs of the digestive tract, including the stomach and intestines. Because its cells are elongated, this tissue is relatively thick, providing protection for underlying tissues. It also functions in the secretion of digestive fluids and in the absorption of nutrient molecules that result from the digestion of foods.

Columnar cells, whose principal function is absorption, often have numerous minute, cylindrical processes extending outward from their cell surfaces. These processes, called *microvilli,* are from 0.5 to 1.0 μm in length. They function to increase the surface of the cell membrane where it is exposed to the substances being absorbed (figure 5.4).

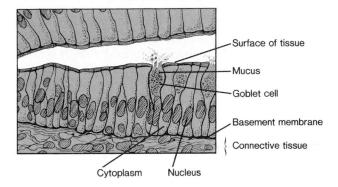

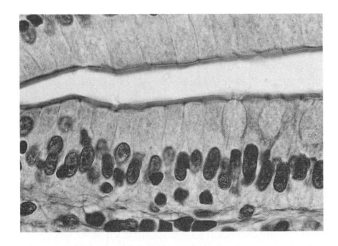

Figure 5.3 Simple columnar epithelium consists of a single layer of elongated cells (×400).

Typically, specialized, flask-shaped glandular cells are scattered among the columnar cells of this tissue. These cells, called *goblet cells,* secrete a protective fluid called *mucus* onto the free surface of the tissue (figure 5.3).

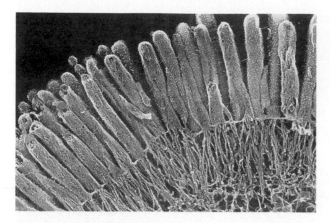

Figure 5.4 A scanning electron micrograph of microvilli, which occur on the exposed surfaces of some columnar epithelial cells.

Pseudostratified Columnar Epithelium

The cells of **pseudostratified columnar epithelium** appear stratified or layered, but they are not. The layered effect occurs because their nuclei are located at two or more levels within the cells of the tissue. However, the cells, which vary in shape, all reach the basement membrane, even though some of them may not contact the free surface.

These cells commonly possess *cilia,* which measure 7–10 μm in length. The cilia extend from the free surfaces of the cells, and they move constantly. Goblet cells also are scattered throughout this tissue, and the mucus they secrete is swept along by the cilia (figure 5.5).

Pseudostratified columnar epithelium lines the passages of the respiratory system and various tubes of the reproductive systems. In the respiratory passages, the mucus-covered linings are sticky and tend to trap particles of dust and microorganisms that enter with the air. The cilia move the mucus and its captured particles upward and out of the airways. In the reproductive tubes, the cilia aid in moving sex cells from one region to another.

Stratified Squamous Epithelium

Stratified squamous epithelium consists of many layers of cells, making this tissue relatively thick. Only the cells near the free surface, however, are likely to be flattened. Those in the deeper layers, where cellular reproduction occurs, are usually cuboidal or columnar. As the newer cells grow, older ones are pushed further and further outward, and they tend to become flattened (figure 5.6).

This tissue forms the outermost layer of the skin (epidermis). As the older cells are pushed outward, they accumulate a protein called *keratin,* become hardened, and die. This action produces a covering of dry, tough,

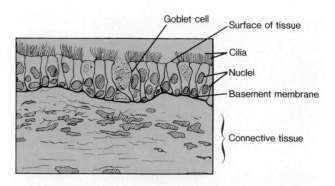

Figure 5.5 Pseudostratified columnar epithelium appears stratified because the nuclei of various cells are located at different levels (×500).

protective material that prevents water and other substances from escaping from underlying tissues and prevents various chemicals and microorganisms from entering.

Stratified squamous epithelium also lines the mouth cavity, throat, vagina, and anal canal. In these parts, the tissue is not keratinized; it stays moist, and the cells on its free surfaces remain alive.

Transitional Epithelium

Transitional epithelium (uroepithelium) is specialized to undergo changes in response to increased tension. It forms the inner lining of the urinary bladder and the passageways of the urinary system. When the wall of one of these organs is contracted, the tissue consists of several layers of cuboidal cells; however, when the organ is distended, the tissue is stretched, and the relationships among the cells are changed. In this distended condition, the tissue appears to contain only a few layers of cells (figure 5.7).

In addition to providing an expandable lining, transitional epithelium forms a barrier that helps prevent the contents of the urinary tract from diffusing out of its passageways.

Chart 5.1 summarizes the characteristics of the different types of epithelial tissue.

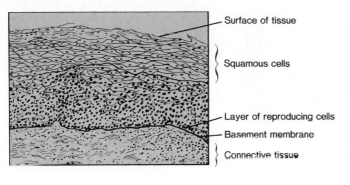

- Surface of tissue
- Squamous cells
- Layer of reproducing cells
- Basement membrane
- Connective tissue

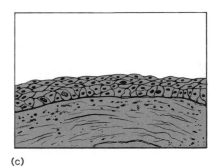

Figure 5.6   Stratified squamous epithelium consists of many layers of cells (×67).

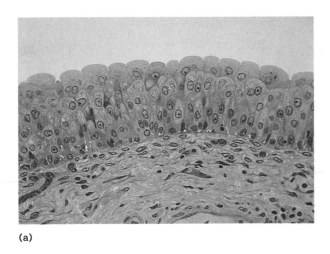

(a)

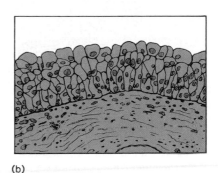

(b)

(c)

Figure 5.7   (a) Micrograph of transitional epithelium (×100). (b) This tissue consists of many layers when the organ wall is contracted. (c) The tissue is thinner when the wall is stretched.

CHART 5.1	Epithelial tissues	
Type	Function	Location
Simple squamous epithelium	Filtration, diffusion, osmosis	Air sacs of lungs, walls of capillaries, linings of blood and lymph vessels
Simple cuboidal epithelium	Secretion, absorption	Surface of ovaries, linings of kidney tubules, and linings of ducts of various glands
Simple columnar epithelium	Protection, secretion, absorption	Linings of uterus and tubes of the digestive tract
Pseudostratified columnar epithelium	Protection, secretion, movement of mucus and cells	Linings of respiratory passages and various tubes of the reproductive systems
Stratified squamous epithelium	Protection	Outer layer of skin, linings of mouth cavity, throat, vagina, and anal canal
Transitional epithelium	Distensibility, protection	Inner lining of urinary bladder and passageways of urinary tract

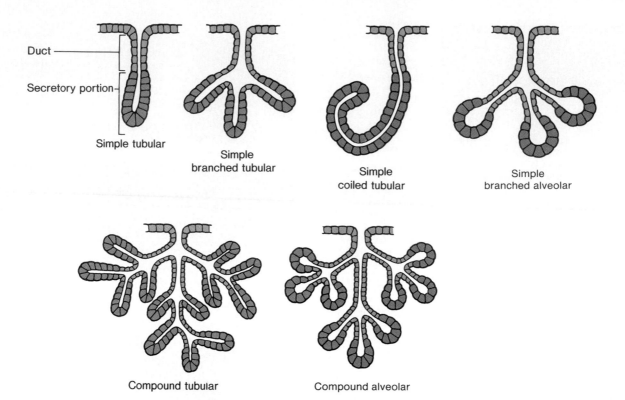

Figure 5.8 Structural types of exocrine glands.

Duct

Secretory portion

Simple tubular

Simple branched tubular

Simple coiled tubular

Simple branched alveolar

Compound tubular

Compound alveolar

A cancer originating in epithelium is called a *carcinoma*, and it is estimated that up to 90% of all human cancers are of this type. Most carcinomas begin on surfaces that contact the external environment, such as skin, linings of the airways in the respiratory tract, or linings of the stomach or intestines in the digestive tract. This observation suggests that the more common cancer-causing agents may not penetrate tissues very deeply.

1. List the general characteristics of epithelial tissue.
2. Explain how epithelial tissues are classified.
3. Describe the structure of each type of epithelium.
4. Describe the special functions of each type of epithelium.

Glandular Epithelium

Glandular epithelium is composed of cells that are specialized to produce and secrete various substances into ducts or into body fluids. Such cells are usually found within columnar or cuboidal epithelium, and one or more of these cells constitutes a *gland*. Glands that secrete their products into ducts that open onto some internal or external surface are called **exocrine glands.** Those that secrete their products into tissue fluid or blood are called **endocrine glands.** (Endocrine glands are discussed in chapter 13.)

An exocrine gland may consist of a single epithelial cell (unicellular gland), such as a mucus-secreting goblet cell, or it may be composed of many cells (multicellular gland). In turn, the multicellular forms can be subdivided structurally into two groups—simple and compound glands.

A *simple gland* communicates with the surface by means of an unbranched duct, and a *compound gland* has a branched duct. These two types of glands can be further classified according to the shapes of their secretory portions. Thus, those that consist of epithelial-lined tubes are called *tubular glands*; those whose terminal portions form saclike dilations are called *alveolar glands* (acinar glands). Figure 5.8 illustrates several types of exocrine glands.

Chart 5.2 summarizes the types of exocrine glands, lists their characteristics, and provides an example of each type.

Glandular secretions are classified according to whether they consist of cellular products or portions of the glandular cells. Glands that release fluid products through cell membranes without the loss of cytoplasm are called *merocrine* glands. Those of an intermediate type that lose small portions of their glandular cell

CHART 5.2 Types of exocrine glands

Type	Characteristics	Example
Unicellular glands	A single secretory cell	Mucus-secreting goblet cell (see figure 5.3)
Multicellular glands	Glands that consist of many cells	
Simple glands	Glands that communicate with surface by means of unbranched ducts	
1. Simple tubular gland	Straight tubelike gland that opens directly onto surface	Intestinal glands of small intestine (see figure 14.3)
2. Simple coiled tubular gland	Long, coiled, tubelike gland; long duct	Eccrine (sweat) glands of skin (see figure 6.11)
3. Simple branched tubular gland	Branched, tubelike gland; duct short or absent	Mucous glands in small intestine (see figure 14.3)
4. Simple branched alveolar gland	Secretory portions of gland expand into saclike compartments arranged along duct	Sebaceous gland of skin (see figure 6.9)
Compound glands	Glands that communicate with surface by means of branched ducts	
1. Compound tubular gland	Secretory portions are coiled tubules, usually branched	Bulbourethral glands of male (see figure 22.1)
2. Compound alveolar gland	Secretory portions are irregularly branched tubules with numerous saclike outgrowths	Salivary glands (see figure 14.12)

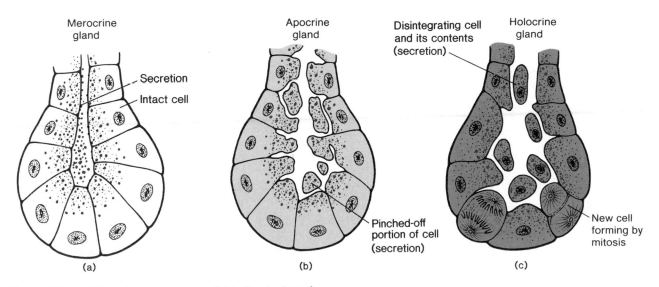

Figure 5.9 (a) Merocrine glands release fluid without a loss of cytoplasm; (b) apocrine glands lose small portions of their cell bodies during secretion; and (c) holocrine glands release entire cells filled with secretory products.

bodies during secretion are called *apocrine* glands. Glands that release entire cells filled with secretory products are called *holocrine* glands (figures 5.9 and 5.10). Chart 5.3 summarizes these glands and their secretions.

Most secretory cells are merocrine, and they can be further subdivided as either *serous cells* or *mucous cells*. The secretion of serous cells is typically watery, has a high concentration of enzymes, and is called *serous fluid*. Such cells are common in the glands of the digestive tract. Mucous cells secrete a thicker fluid called *mucus*. This substance is rich in the glycoprotein *mucin* and is secreted abundantly from the inner linings of the digestive and respiratory tubes.

1. Distinguish between exocrine and endocrine glands.
2. Explain how exocrine glands are classified.
3. Distinguish between a serous cell and a mucous cell.

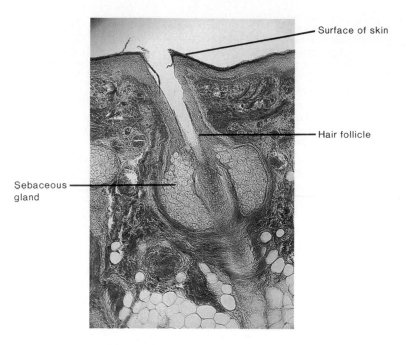

Figure 5.10 The sebaceous gland associated with a hair follicle is an example of a simple branched alveolar gland that secretes entire cells (×25).

CHART 5.3	Types of glandular secretions	
Type	Description of secretion	Example
Merocrine glands	A fluid product that is released through the cell membrane	Salivary glands, pancreatic glands, certain sweat glands of the skin
Apocrine glands	Cellular product and portions of the free ends of glandular cells that are pinched off during secretion	Mammary glands, certain sweat glands of the skin
Holocrine glands	Entire cells that are laden with secretory products	Sebaceous glands of the skin

Connective Tissues

General Characteristics

Connective tissues occur throughout the body and represent the most abundant type of tissue by weight. They bind structures together, provide support and protection, serve as frameworks, fill spaces, store fat, produce blood cells, provide protection against infections, and help repair tissue damage.

Connective tissue cells are usually further apart than epithelial cells, and they have an abundance of intercellular material, or *matrix,* between them. This matrix consists of *fibers* and a *ground substance* whose consistency varies from fluid to semisolid to solid.

Connective tissue cells are able to reproduce. In most cases, they have good blood supplies and are well nourished. Although some connective tissues, such as bone and cartilage, are quite rigid, loose connective tissue, adipose connective tissue, and fibrous connective tissue are more flexible.

Major Cell Types

Connective tissues contain a variety of cell types. Some of them are called *resident cells* because they are usually present in relatively stable numbers. These include fibroblasts, macrophages, and mast cells. Another group, known as *wandering cells,* temporarily appears in the tissues, usually in response to an injury or infection. The wandering cells include several types of white blood cells.

The **fibroblast** is the most common kind of resident cell in connective tissues. It is a relatively large cell and usually is star-shaped. Fibroblasts function to produce fibers by secreting protein into the matrix of connective tissues (figure 5.11).

Macrophages (histiocytes) are almost as numerous as fibroblasts in some connective tissues. They are usually attached to fibers, but can become detached and actively move about. Macrophages are specialized to carry on phagocytosis. Because they function as scavenger cells that can clear foreign particles from

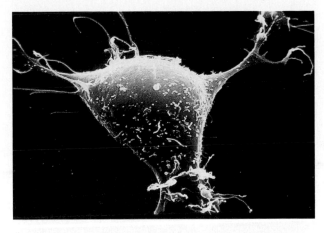

Figure 5.11 A scanning electron micrograph of a fibroblast (×6,000).

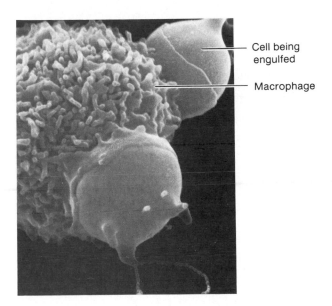

Cell being engulfed

Macrophage

Figure 5.12 Macrophages are common in connective tissues where they function as scavenger cells. In this scanning electron micrograph, a macrophage can be seen engulfing two cells (about ×5,600).

tissues, macrophages represent an important defense against infectious agents (figure 5.12). They also play a role in immunity, which is discussed in chapter 19.

Mast cells are relatively large cells that are widely distributed in connective tissues and are usually located near blood vessels (figure 5.13). Their function is not fully understood; however, they are known to release *heparin,* a compound that prevents blood clotting. They also release *histamine,* a substance that promotes some of the reactions associated with inflammation and allergies, such as asthma and hay fever. (See chapter 19.)

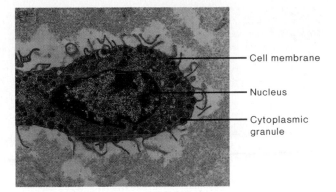

Cell membrane

Nucleus

Cytoplasmic granule

Figure 5.13 A transmission electron micrograph of a mast cell (×4,000).

Connective Tissue Fibers

Three types of connective tissue fibers are produced by fibroblasts: *collagenous fibers, elastic fibers,* and *reticular fibers.*

Collagenous fibers are relatively thick, threadlike parts composed of the protein *collagen,* which is the major structural protein of the body. (Collagen also is the main ingredient of leather; when this protein is denatured by boiling, it becomes gelatin.) These fibers are grouped in long, parallel bundles, and they are flexible but only slightly elastic. More importantly, they have great tensile strength—that is, they are capable of resisting considerable pulling force. Thus, collagenous fibers are important components of body parts that hold structures together (figure 5.14a).

When collagenous fibers are present in abundance, the tissue containing them is called *dense connective tissue.* Such tissue appears white, and for this reason collagenous fibers are sometimes called *white fibers. Loose connective tissue,* on the other hand, is sparsely supplied with collagenous fibers.

Synthesis of the most common form of collagen, which occurs as a major component in bone, ligaments, and tendons, is thought to require at least a dozen enzymes. Because mutations may cause an inability to produce particular enzymes, it is not surprising that a number of inherited connective tissue disorders exist. These include Marfan's syndrome, elastic skin syndrome, and familial joint instability.

Elastic fibers are composed of bundles of microfibrils embedded in a protein called *elastin.* These fibers tend to be branched and to form complex networks in various tissues. They have less strength than collagenous fibers, but they are very elastic. That is, they are easily stretched or deformed and will resume their original lengths and shapes when the force acting upon them is removed. They are common in body parts that are normally subjected to stretching, such as the vocal

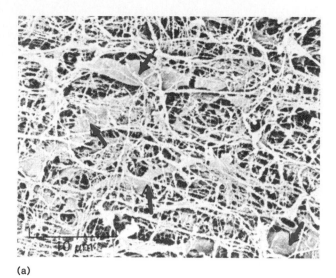

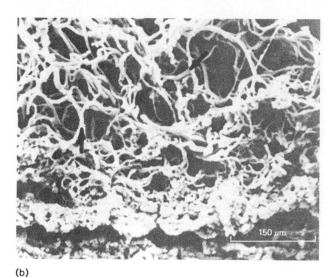

(a)

(b)

Figure 5.14 Scanning electron micrograph of (a) collagenous fibers (arrows) and (b) elastic fibers (arrows). How do these fibers differ?

CHART 5.4	Components of connective tissue	
Component	Characteristic	Function
Fibroblasts	Widely distributed, large, star-shaped cells	Secrete proteins that become fibers
Macrophages	Motile cells that are sometimes attached to fibers	Clear foreign particles from tissues by phagocytosis
Mast cells	Large cells, usually located near blood vessels	Release substances that may help prevent blood clotting and promote inflammation
Collagenous fibers (white fibers)	Thick, threadlike fibers of collagen with great tensile strength	Hold structures together
Elastic fibers (yellow fibers)	Bundles of microfibrils composed of elastin that is very elastic	Provide elastic quality to parts that stretch
Reticular fibers	Thin fibers of collagen	Form supportive networks within tissues

cords and various air passages of the respiratory system. Elastic fibers are sometimes called *yellow fibers,* because tissues amply supplied with them appear yellowish (figure 5.14b).

Reticular fibers are very thin fibers composed of collagen. They are highly branched and form delicate supporting networks in a variety of tissues.

Chart 5.4 summarizes the components of connective tissue.

When skin is excessively exposed to sunlight, its connective tissue fibers tend to lose their elasticity, and the skin becomes increasingly stiff and leathery. In time, it may sag and wrinkle. On the other hand, skin of a healthy, well-nourished person that remains covered usually shows much less change.

1. What are the general characteristics of connective tissue?
2. What are the major types of resident cells in connective tissue?
3. What is the primary function of fibroblasts?
4. What are the characteristics of collagen and elastin?

Loose Connective Tissue

Loose (areolar) connective tissue forms delicate, thin membranes throughout the body. The cells of this tissue, which are mainly fibroblasts, are located some distance apart and are separated by a gel-like ground substance that contains many collagenous and elastic fibers (figure 5.15).

This tissue binds the skin to the underlying organs and fills spaces between muscles. It lies beneath most layers of epithelium, where its numerous blood vessels provide nourishment for nearby epithelial cells.

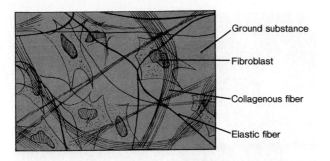

Ground substance
Fibroblast
Collagenous fiber
Elastic fiber

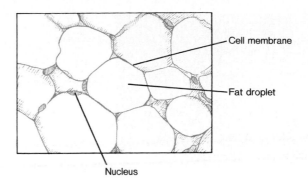

Cell membrane
Fat droplet
Nucleus

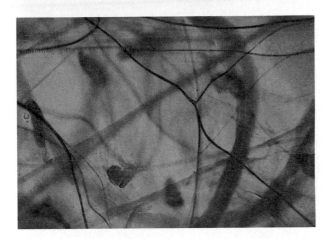

Figure 5.15 Loose connective tissue contains numerous fibroblasts that produce collagenous and elastic fibers (×250).

Figure 5.16 Adipose tissue cells contain large fat droplets that cause the nuclei to be pushed close to their cell membranes (×250).

Adipose Tissue

Adipose tissue, commonly called fat, is a specialized form of loose connective tissue. Certain cells within loose connective tissue (adipocytes) store fat in droplets within their cytoplasm. At first these cells resemble fibroblasts, but as they accumulate fat, they become enlarged, and their nuclei are pushed to one side (figure 5.16). When they occur in such large numbers that other cell types are crowded out, they form adipose tissue.

Adipose tissue is found beneath the skin and in spaces between muscles. It also occurs around the kidneys, behind the eyeballs, in certain abdominal membranes, on the surface of the heart, and around various joints.

Adipose tissue serves as a protective cushion for joints and some organs, such as the kidneys. It also functions as an insulator beneath the skin, and it stores energy in fat molecules.

Because excess food substances are likely to be converted to fat and stored, the amount of adipose tissue present in the body is usually related to a person's diet. During a period of fasting, adipose cells may lose their fat droplets, shrink in size, and become more like fibroblasts again.

Infants and young children usually have a continuous layer of adipose tissue just beneath the skin. This layer gives their bodies the rounded appearance of well-nourished youngsters. In adults, this subcutaneous fat tends to become thinner in some regions and thicker in others. For example, in males it usually thickens in the upper back, upper arms, lower back, and buttocks; in females it is more likely to develop in the breasts, buttocks, and upper legs.

Fibrous Connective Tissue

Fibrous connective tissue is a dense tissue that contains many closely packed, thick, collagenous fibers and a fine network of elastic fibers. It has relatively few cells, almost all of which are fibroblasts (figure 5.17).

Because collagenous fibers are very strong, this type of tissue can withstand pulling forces, and it often binds body parts together. For example, **tendons,** which connect muscles to bones, and **ligaments,** which connect bones to bones at joints, are composed largely of fibrous connective tissue. This type of tissue also occurs in the protective white layer of the eyeball and in the deeper portions of the skin.

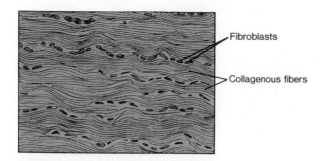

Fibroblasts

Collagenous fibers

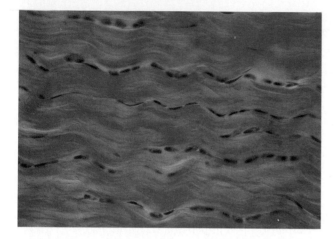

Figure 5.17 Fibrous connective tissue consists largely of tightly packed collagenous fibers (×100).

Figure 5.18 Elastic connective tissue contains many elastic fibers with collagenous fibers between them (×100).

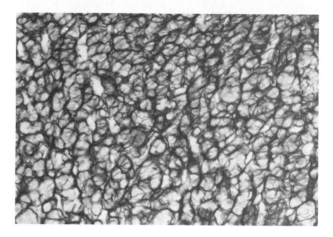

Figure 5.19 Reticular connective tissue is composed of thin collagenous fibers arranged in a network (×110).

The blood supply to fibrous connective tissue is relatively poor, so tissue repair occurs slowly. This is why a sprain, which involves damage to the tissues surrounding a joint, may take considerable time to heal.

Elastic Connective Tissue

Elastic connective tissue consists mainly of yellow, elastic fibers arranged in parallel strands or in branching networks. Spaces between the fibers contain collagenous fibers and fibroblasts.

This tissue is found in the attachments between adjacent vertebrae of the backbone (ligamenta flava). It also occurs in the layers within the walls of various hollow internal organs, including the larger arteries, some portions of the heart, and the larger airways. It imparts an elastic quality to these structures. (See figure 5.18.)

Reticular Connective Tissue

Reticular connective tissue is composed of thin, collagenous fibers arranged in a three-dimensional network. It functions as the supporting tissue in the walls of certain internal organs, such as the liver, spleen, and various lymphatic organs. (See figure 5.19.)

1. How is loose connective tissue related to adipose tissue?
2. What are the functions of adipose tissue?
3. How can fibrous, elastic, and reticular connective tissues be distinguished?

Cartilage

Cartilage is one of the rigid connective tissues. It supports parts, provides frameworks and attachments, protects underlying tissues, and forms structural models for many developing bones.

The matrix of cartilage is abundant and is composed largely of collagenous fibers embedded in a gel-like ground substance. This ground substance is rich in a protein-polysaccharide complex (chondromucoprotein) and contains an abundance of water. Cartilage cells, or **chondrocytes,** occupy small chambers called *lacunae,* and thus are completely surrounded by matrix.

A cartilaginous structure is enclosed in a covering of fibrous connective tissue called the *perichon-*

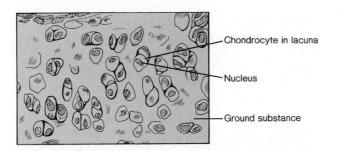

- Chondrocyte in lacuna
- Nucleus
- Ground substance

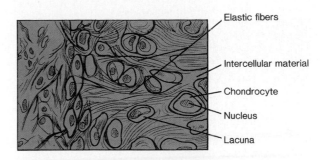

- Elastic fibers
- Intercellular material
- Chondrocyte
- Nucleus
- Lacuna

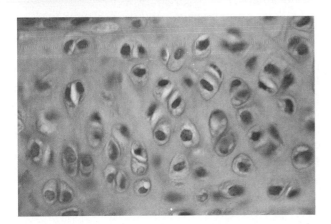

Figure 5.20 Hyaline cartilage cells (chondrocytes) are located in lacunae, which are in turn surrounded by intercellular material containing very fine collagenous fibers (×250).

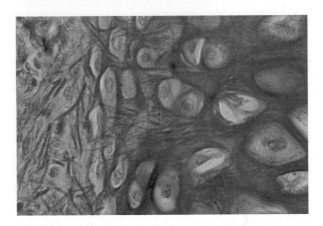

Figure 5.21 Elastic cartilage contains many elastic fibers in its intercellular material (×100).

drium. Although cartilage tissue lacks a direct blood supply, there are blood vessels in the perichondrium surrounding it. The cartilage cells obtain nutrients from these vessels by diffusion, which is aided by the water in the matrix. This lack of a direct blood supply is related to the slow rate of cellular reproduction and repair that is characteristic of cartilage.

There are three kinds of cartilage, and each kind contains a different type of intercellular material. Hyaline cartilage has very fine collagenous fibers in its matrix, elastic cartilage contains a dense network of elastic fibers, and fibrocartilage contains many large collagenous fibers.

Hyaline cartilage (figure 5.20), the most common type of cartilage, looks somewhat like milk glass. It occurs on the ends of bones in many joints, in the soft part of the nose, and in the supporting rings of the respiratory passages. In an embryo, many of the skeletal parts are formed at first of hyaline cartilage, which is later replaced by bone. Hyaline cartilage also plays an important role in the growth of most bones and in the repair of bone fractures. (See chapter 7.)

Elastic cartilage (figure 5.21), which is more flexible than hyaline cartilage because its matrix contains many elastic fibers, provides the framework for the external ears and parts of the larynx.

Fibrocartilage (figure 5.22), a very tough tissue, contains many collagenous fibers. It often serves as a shock absorber for structures that are subjected to pressure. For example, fibrocartilage forms pads (intervertebral disks) between the individual parts of the backbone. It also forms protective cushions between bones in the knees and between bones in the pelvic girdle.

Bone

Bone (osseous tissue) is the most rigid of the connective tissues. Its hardness is due largely to the presence of mineral salts, such as calcium phosphate and calcium carbonate, in its matrix. This intercellular material also contains a great amount of collagen, whose fibers provide flexible reinforcement for the mineral components of bone.

Bone provides an internal support for body structures. It protects vital parts in the cranial and thoracic cavities, and serves as an attachment for muscles. It also contains red marrow, which functions in forming blood cells, and it stores various inorganic salts.

Bone matrix is deposited in thin layers called *lamellae,* which most commonly are arranged in concentric patterns around tiny longitudinal tubes called

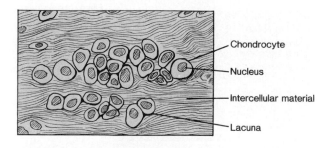

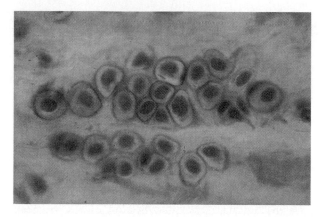

osteonic (Haversian) canals. Bone cells, or osteocytes, are located in lacunae, rather evenly spaced between the lamellae. Consequently, they too are arranged in patterns of concentric circles. (See figure 5.23.)

In a bone, the osteocytes and layers of intercellular material, which are clustered concentrically around an osteonic canal, form a cylinder-shaped unit called an **osteon** (Haversian system). Many of these units cemented together form the substance of bone. (See chapter 7.)

Each osteonic canal contains a blood vessel, so that every bone cell is fairly close to a nutrient supply. In addition, the bone cells have numerous cytoplasmic processes that extend outward and pass through minute tubes in the matrix called *canaliculi*. These cellular processes are attached to the membranes of nearby cells by gap junctions. (See chapter 3.) As a result, materials can move rapidly between blood vessels and bone cells. Thus, in spite of its inert appearance, bone is a very active tissue. When it is injured, bone heals much more rapidly than cartilage.

Chart 5.5 lists the characteristics of the major types of connective tissue.

Figure 5.22 Fibrocartilage contains many large collagenous fibers in its intercellular material (×195).

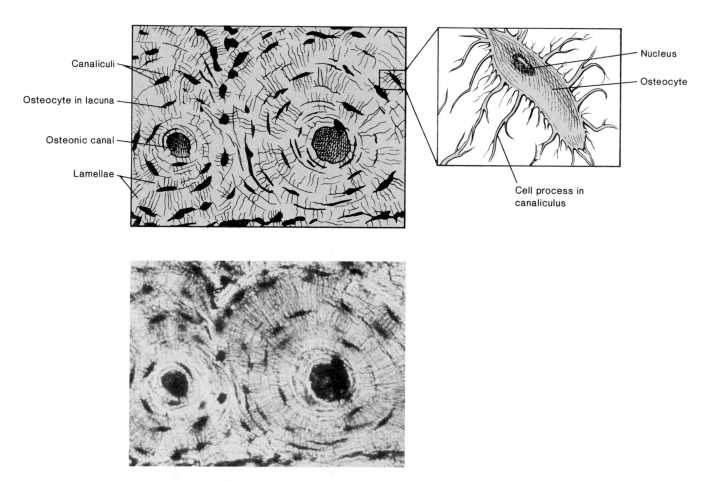

Figure 5.23 Bone matrix is deposited in concentric layers around osteonic canals (×160).

CHART 5.5 Connective tissues

Type	Function	Location
Loose connective tissue	Binds organs together, holds tissue fluids	Beneath the skin, between muscles, beneath epithelial tissues
Adipose tissue	Protection, insulation, and storage of fat	Beneath the skin, around the kidneys, behind the eyeballs, on the surface of the heart
Fibrous connective tissue	Binds organs together	Tendons, ligaments, deep layer of the skin
Elastic connective tissue	Provides elastic quality	Between adjacent vertebrae, in walls of arteries and airways
Reticular connective tissue	Support	Walls of liver, spleen, and lymphatic organs
Hyaline cartilage	Support, protection, provides framework	Ends of bones, nose, and rings in walls of respiratory passages
Elastic cartilage	Support, protection, provides flexible framework	Framework of external ear and part of larynx
Fibrocartilage	Support, protection, shock absorption	Between bony parts of backbone, pelvic girdle, and knee
Bone	Support, protection, provides framework	Bones of skeleton

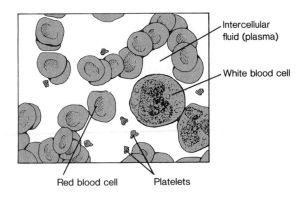

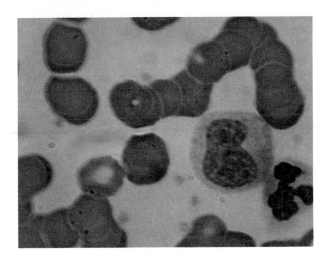

Figure 5.24 Blood tissue consists of an intercellular fluid in which red blood cells, white blood cells, and platelets are suspended (×640).

Other Connective Tissues

Other important connective tissues include blood (vascular tissue) and reticuloendothelial tissue.

Blood is composed of cells that are suspended in the fluid intercellular matrix called *blood plasma*. These cells include red blood cells, white blood cells, and certain cellular fragments called *platelets*. Most blood cells are formed by special tissues (hematopoietic tissues) in red marrow within the hollow parts of various bones (figure 5.24). Blood is described in chapter 17.

Of the blood cells, only the red cells function entirely within the blood vessels. The white blood cells typically leave the blood vessels by migrating through capillary walls. They enter connective tissues where they carry on their major activities, and they usually reside there until they die.

Reticuloendothelial tissue is composed of a variety of specialized cells that are widely scattered throughout the body. As a group, these cells are phagocytic; that is, they function to ingest and destroy foreign particles, such as microorganisms, that may invade the body. Thus, they are particularly important in defending the body against infection.

Reticuloendothelial cells include types found in the blood, lungs, brain, bone marrow, spleen, liver, and lymph glands. The most common ones, however, are *macrophages*. Typically, a macrophage remains in a fixed position until it "senses" a foreign particle. Then, the phagocyte becomes motile, moves toward the invader, and may engulf and destroy it. Once the invader has been destroyed, the macrophage becomes fixed again (figure 5.12).

Reticuloendothelial tissue is discussed in chapter 19.

Inflammation

Inflammation is a localized response to factors that are damaging or potentially damaging to tissues. These factors include traumatic injuries, excessive heat, chemical irritants, bacterial infections, and other stressful conditions.

The inflammatory reaction occurs mainly in connective tissues and involves various components of the circulatory system. It helps prevent the spread of infections, brings about the destruction of foreign substances entering the tissues, and promotes the healing process.

As an example, imagine that some bacteria have entered connective tissue through a break in the skin. In response, chemicals, such as histamine, are released from damaged and stressed cells. These substances, together with others from the invading bacteria, stimulate a variety of changes to take place. Nearby blood vessels become dilated, causing an increase in the blood supply to the affected tissues. Consequently, the region appears redder and feels warmer. At the same time, the permeability of the smaller blood vessels increases, and extra fluid filters into the intercellular spaces of the tissue. This produces swelling (a condition called *edema*) and puts pressure on local nerve endings, which causes the region to become painful.

The fluid escaping from the blood soon clots. This seals off the inflamed area by filling tissue spaces and lymphatic vessels, and it helps delay the spread of bacteria into surrounding tissues.

Macrophages residing in the connective tissue become mobilized and begin to phagocytize bacteria almost immediately. The inflamed tissue releases a substance (leukocytosis-inducing factor) that is carried away by the blood, and it stimulates the release of many white blood cells from the bone marrow. Within a few hours, large numbers of these cells (neutrophils) migrate through capillary walls and enter the inflamed tissues. They are attracted toward chemicals diffusing out from the affected tissues, and they act as phagocytes.

Eventually, a fluid-filled cavity containing a creamy, yellowish substance called *pus* may appear in the inflamed region. The pus represents an accumulation of tissue fluid, bacteria, and dead and dying white blood cells and macrophages. Sometimes such a cavity opens itself to the surface or into a body cavity, and the pus drains out.

As the inflammation subsides, macrophages clean up the cellular debris by phagocytosis, and the excess fluid is absorbed by surrounding tissues. The gap is filled with new connective tissue formed by fibroblasts that migrate into the area.

---

1. Describe the general characteristics of cartilage.
2. Explain why injured bone heals more rapidly than injured cartilage.
3. What are the major components of blood?
4. What do the cells of reticuloendothelial tissue have in common?

Muscle Tissues

General Characteristics

Muscle tissues are contractile; that is, their elongated cells, or *muscle fibers,* can change shape by becoming shorter and thicker. As they contract, the muscle fibers pull at their attached ends and cause body parts to move. The three types of muscle tissue are skeletal muscle, smooth muscle, and cardiac muscle.

Skeletal Muscle Tissue

Skeletal muscle tissue (figure 5.25) is found in muscles that usually are attached to bones and can be controlled by conscious effort. For this reason, it is often called *voluntary* muscle tissue. The cells of this tissue are long and threadlike, with alternating light and dark cross-markings called *striations*. Each cell, or muscle fiber, has many nuclei located just beneath its cell membrane. When the muscle fiber is stimulated by an

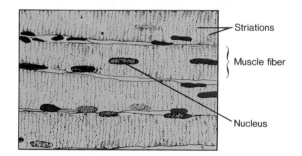

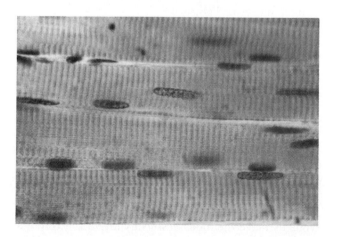

Figure 5.25   Skeletal muscle tissue is composed of striated muscle fibers that contain many nuclei (×250).

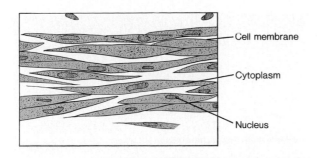

Cell membrane

Cytoplasm

Nucleus

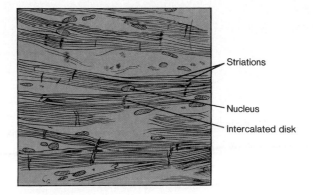

Striations

Nucleus

Intercalated disk

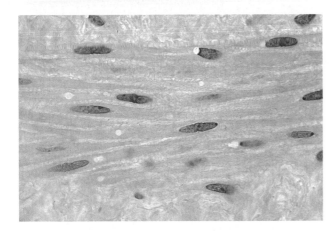

Figure 5.26 Smooth muscle tissue is formed of spindle-shaped cells, each containing a single nucleus (×250).

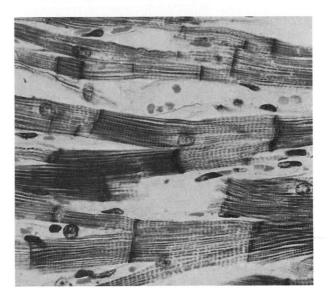

Figure 5.27 How is this cardiac muscle tissue (×400) similar to skeletal muscle? How is it similar to smooth muscle?

action of a nerve fiber, it contracts and then relaxes. If the connecting nerve fiber is damaged, it may not be possible to control the muscle fiber's action, and the muscle fiber is paralyzed.

Skeletal muscles, composed mainly of skeletal muscle tissue, are responsible for moving the head, trunk, and limbs. They also produce the movements involved with making facial expressions, writing, talking, and singing, and with such actions as chewing, swallowing, and breathing.

Smooth Muscle Tissue

Smooth muscle tissue (figure 5.26) is called smooth because its cells lack striations. This tissue is found in the walls of hollow internal organs, such as the stomach, intestines, urinary bladder, uterus, and blood vessels. Unlike skeletal muscle, smooth muscle usually cannot be stimulated to contract by conscious efforts. Thus, it is a type of *involuntary* muscle tissue.

Smooth muscle cells are shorter than those of skeletal muscle, and each has a single, centrally located nucleus. Smooth muscle tissue is responsible for moving food through the digestive tube, constricting blood vessels, and emptying the urinary bladder.

Cardiac Muscle Tissue

Cardiac muscle tissue (figure 5.27) occurs only in the heart. Its cells, which are striated, are joined end to

end. The resulting muscle fibers are branched and interconnected in complex networks. Each cell within a cardiac muscle fiber usually has a single nucleus. At its end, where it touches another cell, is a specialized intercellular junction called an *intercalated disk,* which occurs only in cardiac tissue.

The cells of different tissues vary greatly in their abilities to reproduce. For example, the epithelial cells of the skin and the inner lining of the digestive tube, and the connective tissue cells that form blood cells in red bone marrow reproduce continuously. However, striated and cardiac muscle cells and nerve cells do not seem to reproduce at all after becoming differentiated.

Fibroblasts respond rapidly to injuries by increasing in numbers and becoming active in fiber production. Therefore, they are often the principal agents of repair in tissues that have limited abilities to regenerate themselves. For instance, cardiac muscle tissue typically degenerates in the regions damaged by a heart attack. Such tissue may be replaced by connective tissue formed by fibroblasts, which later appears as a scar.

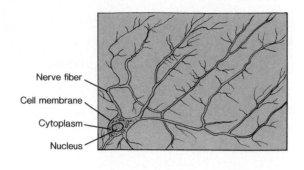

Nerve fiber
Cell membrane
Cytoplasm
Nucleus

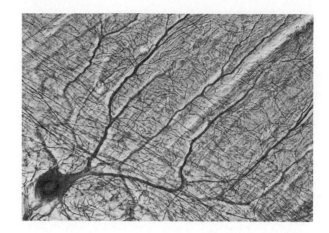

Figure 5.28 A nerve cell with nerve fibers extending into its surroundings (×450).

CHART 5.6	Muscle and nervous tissues				
Type	Function	Location	Type	Function	Location
Skeletal muscle tissue	Voluntary movements of skeletal parts	Muscles usually attached to bones	Cardiac muscle tissue	Heart movements	Heart muscle
Smooth muscle tissue	Involuntary movements of internal organs	Walls of hollow internal organs	Nervous tissue	Sensitivity and conduction of nerve impulses	Brain, spinal cord, and peripheral nerves

Cardiac muscle, like smooth muscle, is controlled involuntarily and, in fact, can continue to function without being stimulated by nerve impulses. This tissue makes up the bulk of the heart and is responsible for pumping blood through the heart chambers and into certain blood vessels.

1. List the general characteristics of muscle tissue.
2. Distinguish among skeletal, smooth, and cardiac muscle tissue.

Nervous Tissue

Nervous tissue is found in the brain, spinal cord, and peripheral nerves. The basic cells of this tissue are called *nerve cells,* or *neurons,* and they are among the more highly specialized body cells. Neurons are sensitive to certain types of changes in their surroundings. They respond by transmitting nerve impulses along cytoplasmic extensions (nerve fibers) to other neurons or to muscles or glands (figure 5.28).

As a result of the extremely complex patterns by which neurons are connected with each other and with various body parts, they are able to coordinate, regulate, and integrate many body functions.

In addition to neurons, nervous tissue contains **neuroglial cells.** These cells support and bind the components of nervous tissue together, carry on phagocy-

tosis, and help supply nutrients to neurons by connecting them to blood vessels. Nervous tissue is discussed in chapter 10.

Chart 5.6 summarizes the general characteristics of muscle and nervous tissues.

1. Describe the general characteristics of nervous tissue.
2. Distinguish between neurons and neuroglial cells.

Chapter Summary

Introduction (page 126)

Cells are arranged in layers or groups called tissues. Cells are separated by intercellular materials whose composition varies from solid to liquid. The four major types of human tissue are epithelial tissues, connective tissues, muscle tissues, and nervous tissue.

Epithelial Tissues (page 126)

1. General characteristics
   a. Epithelial tissue covers all free body surfaces and is the major tissue of glands.
   b. Epithelium is anchored to connective tissue by a basement membrane. Epithelial tissue lacks blood vessels, contains little intercellular material, and is replaced continuously.
   c. It functions in protection, secretion, absorption, excretion, and sensory reception.

2. Simple squamous epithelium
   a. This tissue consists of a single layer of thin, flattened cells through which substances pass relatively easily.
   b. It functions in the exchange of gases in the lungs and lines blood vessels, lymph vessels, and various membranes within the thorax and abdomen.
3. Simple cuboidal epithelium
   a. This tissue consists of a single layer of cube-shaped cells.
   b. It carries on secretion and absorption in the kidneys and various glands.
4. Simple columnar epithelium
   a. This tissue is composed of elongated cells whose nuclei are located near the basement membrane.
   b. It lines the uterus and digestive tract, where it functions in protection, secretion, and absorption.
   c. Absorbing cells often possess microvilli.
   d. This tissue usually contains goblet cells that secrete mucus.
5. Pseudostratified columnar epithelium
   a. This tissue appears stratified because the nuclei are located at two or more levels within the cells.
   b. Its cells may have cilia that function to move mucus or cells over the surface of the tissue.
   c. It lines various tubes of the respiratory and reproductive systems.
6. Stratified squamous epithelium
   a. This tissue is composed of many cell layers.
   b. It protects underlying cells from harmful environmental effects.
   c. It covers the skin and lines the mouth, throat, vagina, and anal canal.
7. Transitional epithelium
   a. This tissue is specialized to undergo distension.
   b. It occurs in the walls of various organs of the urinary tract.
   c. It helps prevent the contents of the urinary passageways from diffusing outward.
8. Glandular epithelium
   a. Glandular epithelium is composed of cells that are specialized to secrete substances.
   b. One or more cells can constitute a gland.
      (1) Exocrine glands secrete into ducts.
      (2) Endocrine glands secrete into tissue fluid or blood.
   c. Glands are classified according to the arrangement of their cells.
      (1) Simple glands have unbranched ducts.
      (2) Compound glands have branched ducts.
      (3) Tubular glands consist of simple epithelial-lined tubes.
      (4) Alveolar glands consist of saclike dilations connected to the surface by narrowed ducts.
   d. Exocrine glands are classified according to the composition of their secretions.
      (1) Merocrine glands secrete watery fluids without loss of cytoplasm. Most secretory cells are merocrine.
         (a) Serous cells secrete watery fluid with a high enzyme content.
         (b) Mucous cells secrete mucus.
      (2) Apocrine glands lose portions of their cells during secretion.
      (3) Holocrine glands release cells filled with secretory products.

Connective Tissues (page 132)

1. General characteristics
   a. Connective tissue connects, supports, protects, provides frameworks, fills spaces, stores fat, produces blood cells, provides protection against infection, and helps repair damaged tissues.
   b. Connective tissue cells are usually some distance apart, and they have considerable intercellular material between them.
   c. This intercellular matrix consists of fibers and a ground substance.
2. Major cell types
   a. Fibroblasts produce collagenous and elastic fibers.
   b. Macrophages function as phagocytes.
   c. Mast cells may release heparin and histamine, and usually are located near blood vessels.
3. Connective tissue fibers
   a. Collagenous fibers are composed of collagen and have great tensile strength.
   b. Elastic fibers are composed of microfibrils embedded in elastin and are very elastic.
   c. Reticular fibers are very fine collagenous fibers.
4. Loose connective tissue
   a. This tissue forms thin membranes between organs and binds them together.
   b. It is found beneath the skin and between muscles.
   c. Its intercellular spaces contain tissue fluid.
5. Adipose tissue
   a. Adipose tissue is a specialized form of loose connective tissue that stores fat, provides a protective cushion, and functions as an insulator.
   b. It is found beneath the skin, in certain abdominal membranes, and around the kidneys, heart, and various joints.
6. Fibrous connective tissue
   a. This tissue is composed largely of strong, collagenous fibers that bind parts together.
   b. It is found in tendons, ligaments, white portions of the eyes, and the deep layer of the skin.
7. Elastic connective tissue
   a. This tissue is composed mainly of elastic fibers.
   b. It imparts an elastic quality to the walls of certain hollow internal organs.

8. Reticular connective tissue
   a. This tissue consists largely of thin, branched collagenous fibers.
   b. It supports the walls of the liver, spleen, and lymphatic organs.
9. Cartilage
   a. Cartilage provides support and framework for various parts.
   b. Its intercellular material is largely composed of fibers and a gel-like ground substance.
   c. It lacks a direct blood supply and is slow to heal following an injury.
   d. Cartilaginous structures are enclosed in a perichondrium, which contains blood vessels.
   e. Major types are hyaline cartilage, elastic cartilage, and fibrocartilage.
   f. Cartilage occurs at the ends of various bones, in the ear, in the larynx, and in pads between bones of the backbone, pelvic girdle, and knees.
10. Bone
    a. The intercellular matrix of bone contains mineral salts and collagen.
    b. Its cells are usually arranged in concentric circles around osteonic canals and are interconnected by canaliculi.
    c. It is an active tissue that heals rapidly following an injury.
11. Other connective tissues
    a. Blood
       (1) Blood is composed of cells suspended in fluid.
       (2) Blood cells are formed by special tissue in the hollow parts of certain bones.
       (3) Blood transports white blood cells to connective tissues.
    b. Reticuloendothelial tissue
       (1) This tissue is composed of a variety of phagocytic cells that are widely distributed throughout the body.
       (2) It defends the body against invasion by microorganisms.

Muscle Tissues (page 140)

1. General characteristics
   a. Muscle tissue is contractile tissue that moves parts attached to it.
   b. Three types are skeletal, smooth, and cardiac muscle tissues.
2. Skeletal muscle tissue
   a. Muscles containing this tissue are usually attached to bones and are controlled by conscious effort.
   b. Cells or muscle fibers are long and threadlike with alternating light and dark cross-markings.
   c. Muscle fibers contract when stimulated by nerve action, then relax immediately.

3. Smooth muscle tissue
   a. This tissue is found in walls of hollow internal organs.
   b. Usually it is controlled by involuntary activity.
4. Cardiac muscle tissue
   a. This tissue is found only in the heart.
   b. Cells are joined by intercalated disks and arranged in branched, interconnecting networks.
   c. Cardiac muscle tissue is controlled by involuntary activity.

Nervous Tissue (page 142)

1. Nervous tissue is found in the brain, spinal cord, and peripheral nerves.
2. Neurons
   a. Neurons are sensitive to changes and respond by transmitting nerve impulses to other neurons or to other body parts.
   b. They function in coordinating, regulating, and integrating body activities.
3. Neuroglial cells
   a. Some forms of these cells bind and support nervous tissue.
   b. Others carry on phagocytosis.
   c. Still others connect neurons to blood vessels.

Clinical Application of Knowledge

1. Joints such as the elbow, shoulder, and knee contain considerable amounts of cartilage and fibrous connective tissue. How does this relate to the fact that joint injuries are often very slow to heal?
2. A group of disorders called collagenous diseases are characterized by deterioration of connective tissues. Why would you expect such diseases to produce widely varying symptoms?
3. The cardinal signs of inflammation are swelling, redness, pain, and warmth. How would you explain the reason for each of these signs to a patient?
4. Sometimes, in response to the presence of irritants, mucous cells secrete excessive amounts of mucus. What symptoms might this produce if it occurred in (a) the respiratory passageways; (b) the digestive tract?
5. As a result of continued and prolonged irritation by tobacco smoke, the pseudostratified columnar epithelium that forms the inner lining of the upper respiratory tubes may be replaced by stratified squamous epithelium. What might be the consequences of this tissue change?

Review Activities

1. Define *tissue*.
2. Name the four major types of tissue found in the human body.
3. Describe the general characteristics of epithelial tissues.
4. Distinguish between a simple epithelium and a stratified epithelium.
5. Explain how the structure of simple squamous epithelium is related to its function.
6. Name an organ in which each of the following tissues is found, and give the function of the tissue in each case:
   a. Simple squamous epithelium
   b. Simple cuboidal epithelium
   c. Simple columnar epithelium
   d. Pseudostratified columnar epithelium
   e. Stratified squamous epithelium
   f. Transitional epithelium
7. Define *gland*.
8. Distinguish between an exocrine gland and an endocrine gland.
9. Explain how glands are classified according to the structure of their ducts and the arrangement of their cells.
10. Explain how glands are classified according to the nature of their cellular secretions.
11. Distinguish between a serous cell and a mucous cell.
12. Describe the general characteristics of connective tissue.
13. Describe three major types of connective tissue cells.
14. Distinguish between collagen and elastin.
15. Explain the relationship between loose connective tissue and adipose tissue.
16. Explain how the quantity of adipose tissue in the body is related to diet.
17. Distinguish between elastic and reticular connective tissues.
18. Explain why injured fibrous connective tissue and cartilage are usually slow to heal.
19. Name the major types of cartilage, and describe their differences and similarities.
20. Describe how bone cells are arranged in bone tissue.
21. Explain how nutrients are supplied to bone cells.
22. Describe the solid components of blood.
23. Define *reticuloendothelial tissue*.
24. Describe the general characteristics of muscle tissues.
25. Distinguish between skeletal, smooth, and cardiac muscle tissues.
26. Describe the general characteristics of nervous tissue.
27. Distinguish between neurons and neuroglial cells.

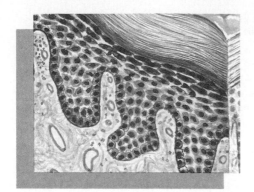

6

Skin and the Integumentary System

*T*he previous chapters dealt with the lower levels of organization within the human organism—tissues, cells, cellular organelles, and the chemical substances that form these parts.

Chapter 6 explains how tissues are grouped to form organs and how organs comprise organ systems. In this explanation, various membranes, including the skin, are used as examples of organs. Since the skin acts with hair follicles, sebaceous glands, and sweat glands to provide a variety of vital functions, these organs together constitute the integumentary organ system ■

After you have studied this chapter, you should be able to:

1. Describe the four major types of membranes.

2. Describe the structure of the various layers of the skin.

3. List the general functions of each layer of the skin.

4. Describe the accessory organs associated with the skin.

5. Explain the functions of each accessory organ.

6. Explain how the skin functions in regulating body temperature.

7. Summarize the factors that determine skin color.

8. Complete the review activities at the end of this chapter. Note that the items are worded in the form of specific learning objectives. You may want to refer to them before reading the chapter.

conduction (kon-duk'shun)

convection (kon-vek'shun)

cutaneous membrane (ku-ta'ne-us mem'brān)

dermis (der'mis)

epidermis (ep''ĭ-der'mis)

evaporation (e-vap''o-ra'shun)

hair follicle (hār fol' ĭ-kl)

integumentary (in-teg-u-men'tar-e)

keratinization (ker''ah-tin''ĭ-za'shun)

melanin (mel'ah-nin)

mucous membrane (mu'kus mem'brān)

sebaceous gland (se-ba'shus gland)

serous membrane (se'rus mem'brān)

subcutaneous (sub''ku-ta'ne-us)

sweat gland (swet gland)

alb-, white: *alb*inism—condition characterized by a lack of pigment.

cut-, skin: sub*cut*aneous—beneath the skin.

derm-, skin: *derm*is—inner layer of the skin.

epi-, upon: *epi*dermis—outer layer of the skin.

follic-, small bag: hair *follic*le—tubelike depression in which a hair develops.

kerat-, horn: *kerat*in—protein produced as epidermal cells die and harden.

melan-, black: *melan*in—dark pigment produced by certain cells.

por-, channel: *por*e—opening by which a sweat gland communicates to the skin's surface.

seb-, grease: *seb*aceous gland—gland that secretes an oily substance.

To review, two or more kinds of tissues grouped together and performing specialized functions constitute an organ. Thus, the thin, sheetlike structures called *membranes,* which are usually composed of epithelium and connective tissue and that cover body surfaces and line body cavities, are organs. The cutaneous membrane, one of these membranes, together with various accessory organs makes up the **integumentary organ system.**

Types of Membranes

The four major types of membranes are: *serous, mucous, cutaneous,* and *synovial.* Usually these structures are relatively thin. Serous, mucous, and cutaneous membranes are composed of epithelial tissue and some underlying connective tissue; synovial membranes are composed entirely of various connective tissues.

Serous membranes line the body cavities that lack openings to the outside. They form the inner linings of the thorax and abdomen, and they cover the organs within these cavities. A serous membrane consists of a layer of simple squamous epithelium (mesothelium) and a thin layer of loose connective tissue. Cells of a serous membrane secrete watery *serous fluid,* which helps lubricate the surfaces of the membrane. (See figures 1.7 and 1.8.)

Mucous membranes line the cavities and tubes that open to the outside of the body. These include the oral and nasal cavities, and the tubes of the digestive, respiratory, urinary, and reproductive systems. A mucous membrane consists of epithelium overlying a layer of loose connective tissue; however, the type of epithelium varies with the location of the membrane. For example, stratified squamous epithelium lines the oral cavity, pseudostratified columnar epithelium lines part of the nasal cavity, and simple columnar epithelium lines the small intestine. Specialized cells within a mucous membrane secrete *mucus.*

The **cutaneous membrane** is an organ of the integumentary organ system and is more commonly called *skin.* It is described in detail in this chapter.

Synovial membranes form the inner linings of joint cavities between the ends of bones at freely movable joints (synovial joints). These membranes usually include fibrous connective tissue overlying loose connective tissue and adipose tissue. Cells of a synovial membrane secrete a thick, colorless *synovial fluid* into the joint cavity, which lubricates the ends of the bones within the joint. (Synovial joints are described in chapter 8.)

Skin and Its Tissues

The skin is one of the larger and more versatile organs of the body, and it plays vital roles in the maintenance of homeostasis. For example, the skin functions as a protective covering that prevents many harmful substances, including microorganisms, from entering the body. At the same time, it retards the loss of water by diffusion from deeper tissues. Skin helps regulate body temperature, houses sensory receptors, synthesizes various chemicals, and excretes small quantities of waste substances. Also, people are recognized (and their ages often judged) by their skin characteristics.

The skin includes two distinct layers of tissues. The outer layer, called the **epidermis,** is composed of stratified squamous epithelium. The inner layer, or **dermis,** is thicker than the epidermis, and it contains a variety of tissues, including fibrous connective tissue, epithelial tissue, smooth muscle tissue, nervous tissue, and blood. These two layers of the skin are separated by a basement membrane that is anchored to the dermis by short fibrils (figure 6.1).

Beneath the dermis are masses of loose connective and adipose tissues that bind the skin to underlying organs. These tissues form the **subcutaneous layer.** (See figure 6.2.)

1. Name the four types of membranes, and explain how they differ.
2. List the general functions of the skin.
3. Name the tissue found in the outer layer of the skin.
4. Name the tissues found in the inner layer of the skin.

Epidermis

Since the epidermis is composed of stratified squamous epithelium, it lacks blood vessels. However, the deepest cells of the epidermis, which are close to the dermis, are nourished by dermal blood vessels and are capable of reproducing. These deep cells form a layer (stratum basale) in which the cells divide and grow. As the newly formed cells enlarge, they push the older epidermal cells away from the dermis toward the surface of the skin. The further the cells travel, the poorer their nutrient supply becomes, and, in time, they die.

Meanwhile, the cell membranes of the older cells (keratinocytes) thicken and develop numerous desmosomes that fasten them to adjacent cells (chapter 3). At the same time, the cells begin to undergo a hardening process called **keratinization,** during which strands of tough, fibrous, waterproof protein called *keratin* develop within the cell membranes. As a result,

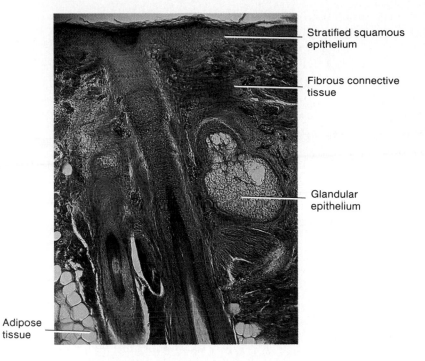

Stratified squamous epithelium

Fibrous connective tissue

Glandular epithelium

Adipose tissue

Figure 6.1 The skin is an organ that is composed of several kinds of tissues.

many layers of tough, tightly packed dead cells accumulate in the outer portion of the epidermis. This outermost layer is called the *stratum corneum,* and the dead cells that compose it are often rubbed away.

The structural organization of the epidermis varies from region to region. For example, it is most highly developed on the palms of the hands and the soles of the feet, where it may be 0.8–1.4 mm thick. In these areas, four layers can be distinguished. They are the *stratum basale* (stratum germinativum, or basal cell layer), which is the deepest layer; the *stratum spinosum,* a relatively thick layer; the *stratum granulosum,* a granular layer; and the *stratum corneum,* a fully keratinized layer (horny layer). The cells of these layers are characterized by the changes they undergo as they are pushed toward the surface (figure 6.3).

In other body regions, the epidermis is usually much thinner, averaging 0.07–0.12 mm in most areas. Also, where the epidermis is thin, the stratum granulosum may be missing. Chart 6.1 describes the characteristics of each layer of the epidermis.

In healthy skin, the production of epidermal cells is closely balanced with the loss of dead cells from the stratum corneum, so that skin seldom wears away completely. In fact, the rate of cellular reproduction tends to increase in regions where the skin is being rubbed or pressed regularly. This response causes the growth of

thickened areas called *calluses* on the palms and soles, as well as the development of horny, conical masses called *corns* on the toes when poorly fitting shoes rub the skin excessively.

Because blood vessels in the dermis supply nutrients to the epidermis, any interference with blood flow is likely to result in the death of epidermal cells. For example, when a person lies in one position for a prolonged period, the weight of the body pressing against the bed interferes with the skin's blood supply. If cells die, the tissues begin to break down (necrosis), and a *pressure ulcer* (also called a decubitus ulcer or bedsore) may appear.

Pressure ulcers usually occur in the skin overlying bony projections, such as on the hip, heel, elbow, or shoulder. These ulcers can often be prevented by changing the body position frequently or by massaging the skin to stimulate blood flow in regions associated with bony prominences. In the case of a paralyzed person who cannot feel pressure or respond to it by shifting position, special care must be taken to turn the body frequently in order to prevent pressure ulcers from developing.

A diet rich in proteins and other necessary nutrients, and an adequate intake of fluids to maintain blood volume also help prevent this condition.

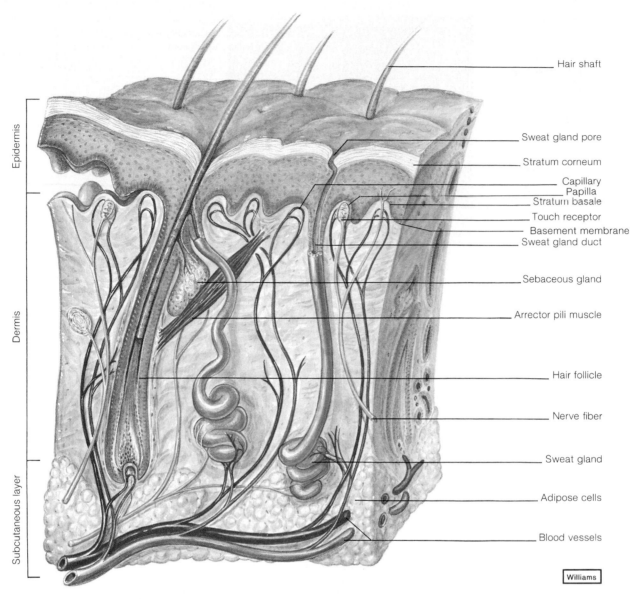

Figure 6.2 A section of skin.

Labels (top to bottom, left side): Epidermis, Dermis, Subcutaneous layer

Labels (right side, top to bottom): Hair shaft, Sweat gland pore, Stratum corneum, Capillary, Papilla, Stratum basale, Touch receptor, Basement membrane, Sweat gland duct, Sebaceous gland, Arrector pili muscle, Hair follicle, Nerve fiber, Sweat gland, Adipose cells, Blood vessels

Williams

CHART 6.1	Layers of the epidermis	
Layer	**Location**	**Characteristics**
Stratum corneum (horny layer)	Outermost layer	Many layers of keratinized, dead epithelial cells that are flattened and nonnucleated
Stratum granulosum	Beneath the stratum corneum	Three to five layers of flattened granular cells that contain shrunken fibers of keratin and shriveled nuclei
Stratum spinosum	Beneath the stratum granulosum	Many layers of cells with centrally located, large, oval nuclei and developing fibers of keratin; cells becoming flattened
Stratum basale (basal cell layer)	Deepest layer	A single row of cuboidal or columnar cells that undergo mitosis; this layer also includes pigment-producing melanocytes

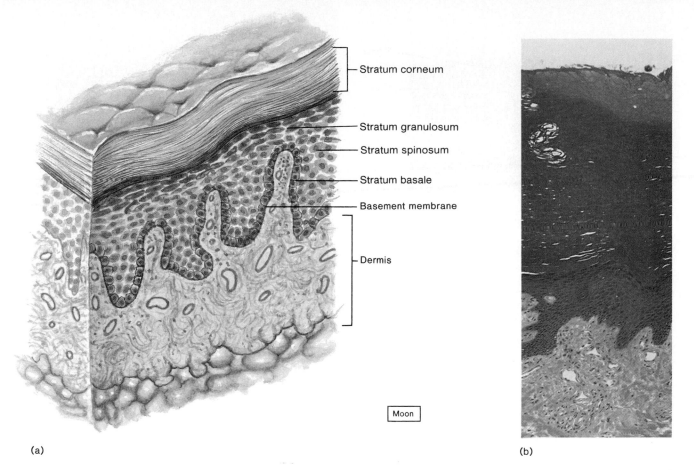

- Stratum corneum
- Stratum granulosum
- Stratum spinosum
- Stratum basale
- Basement membrane
- Dermis

Moon

(a)

(b)

Figure 6.3 (*a*) The various layers of the epidermis are characterized by changes that occur in cells as they are pushed toward the surface of the skin. (*b*) Which layers of the epidermis can you identify in this micrograph from the palm of the hand?

The epidermis has important protective functions. It shields the moist underlying tissues against excessive water loss, mechanical injury, and the effects of harmful chemicals. When it is unbroken, the epidermis also prevents the entrance of many disease-causing microorganisms. **Melanin** is a dark pigment that occurs in the epidermis and is produced by specialized cells known as *melanocytes* (figure 6.4). It absorbs light energy, and in this way, helps protect still deeper cells from the damaging effects of ultraviolet rays of sunlight.

Melanocytes are found in the deepest portion of the epidermis and in the underlying connective tissue of the dermis. Although they are the only cells that can produce melanin, the pigment also may occur in nearby epidermal cells. This happens because melanocytes have long, pigment-containing cellular extensions that pass upward between neighboring epidermal cells, and the

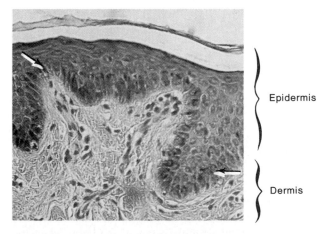

Epidermis

Dermis

Figure 6.4 Melanocytes (arrow) that occur mainly in the deeper layers of the epidermis produce the pigment called melanin (×160).

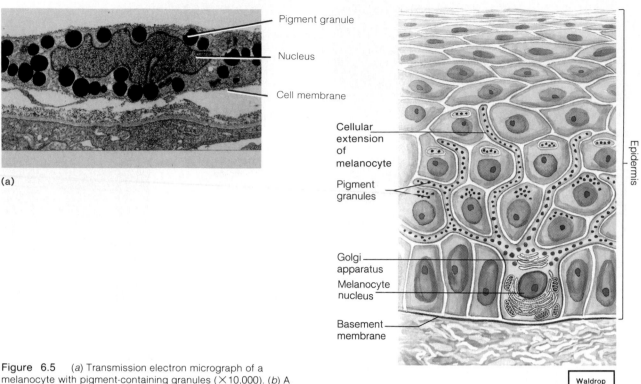

Pigment granule

Nucleus

Cell membrane

(a)

Cellular extension of melanocyte

Pigment granules

Golgi apparatus

Melanocyte nucleus

Basement membrane

Epidermis

Waldrop

(b)

Figure 6.5 (*a*) Transmission electron micrograph of a melanocyte with pigment-containing granules ($\times$10,000). (*b*) A melanocyte may have pigment-containing extensions that pass between neighboring epidermal cells.

extensions can transfer granules of melanin into these other cells by a process called *cytocrine secretion*. Consequently, nearby epidermal cells may contain more melanin than the melanocytes. (See figure 6.5.)

1. Explain how the epidermis is formed.
2. What factors help prevent the loss of body fluids through the skin?
3. What is the function of melanin?

Dermis

The surface between the epidermis and dermis is usually uneven, because the epidermis has ridges projecting inward and the dermis has fingerlike *papillae* passing into the spaces between the ridges. (See figure 6.2.)

The dermis binds the epidermis to the underlying tissues. It is composed largely of fibrous connective tissue that includes tough collagenous fibers and elastic fibers surrounded by a gel-like ground substance. Networks of these fibers give the skin toughness and elasticity. On the average, the dermis is 1.0–2.0 mm thick; however, it may be as thin as 0.5 mm or less on the eyelids, or as thick as 3.0 mm on the soles of the feet.

As a person ages, the skin is likely to undergo changes, particularly in those regions that are exposed to sunlight. The rate and degree of change vary from person to person. Typically, the skin becomes drier, scaly, and more translucent. At the same time, its thickness decreases. Also, there is some degeneration of connective tissue fibers, and this leads to loss of elasticity and tensile strength, followed by wrinkling.

The dermis also contains muscle fibers. Some regions, such as the skin that encloses the testes (scrotum), contain numerous smooth muscle cells that can cause the skin to wrinkle when they contract. Other smooth muscles in the dermis are associated with accessory organs such as hair follicles and various glands. Many striated muscle fibers are anchored to the dermis in the skin of the face. They help produce the voluntary movements associated with facial expressions.

As was discussed, blood vessels in the dermis supply nutrients to the deep, living layers of the epidermis, as well as to dermal cells. These vessels also play an important role in the regulation of body temperature—a function that is explained in a subsequent section of this chapter.

Skin Cancer

Skin cancer is most likely to arise from nonpigmented epithelial cells within the deep layer of the epidermis or from pigmented melanocytes. Skin cancers originating from epithelial cells are called *cutaneous carcinomas* (basal cell carcinoma or squamous cell carcinoma); those arising from melanocytes are known as *cutaneous melanomas* (melanocarcinomas or malignant melanomas).

Cutaneous carcinomas represent the most common type of skin cancer, and they occur most frequently in members of light-skinned populations who are over forty years of age. These cancers seem to be caused by the effects of the ultraviolet portion of sunlight on the DNA of epithelial cells, and they usually appear in persons who are exposed to sunlight regularly. Thus, the incidence of cutaneous carcinoma is increased in persons who have spent considerable amounts of time outdoors—farmers, sailors, athletes, sunbathers, and so forth.

Cutaneous carcinoma often develops from hard, dry, scaly growths (lesions) that have reddish bases. Such lesions may be either flat or raised above the surface, and they adhere firmly to the skin, appearing most often on the neck, face, and scalp. Fortunately, cutaneous carcinomas are typically slow growing and can usually be cured completely by surgical removal or radiation treatment.

Because cutaneous melanomas develop from melanocytes, they are pigmented with melanin. Such lesions often have a variety of colored areas—variegated brown, black, gray, or blue—that are arranged haphazardly. They usually have irregular margins rather than smooth, regular outlines (figure 6.6).

Cutaneous melanomas may appear in young adults as well as in older ones, and seem to be caused by relatively short, intermittent exposure to high-intensity sunlight. Thus, the incidence of melanoma is increased in persons who generally stay indoors but occasionally sustain blistering sunburns during weekend activities or vacations.

Cutaneous melanomas occur most often in light-skinned persons, particularly those whose skins tend to burn rather than tan. They usually appear in the skin of the trunk,

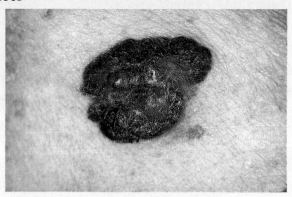

Figure 6.6   A cutaneous malignant melanoma.

especially the back, or in the skin of the limbs. Such a lesion may arise from normal-appearing skin or from a mole (nevus). Typically, the lesion enlarges by spreading through the skin horizontally, but eventually it may thicken and grow downward into the skin, invading the deeper tissues as well. If the melanoma is removed surgically while it is in its horizontal growth phase, its growth may be arrested. Once it has thickened and spread into the deeper tissues, however, it becomes difficult to treat, and the survival rate for persons with this form of cancer is very low.

For reasons that are not well understood, the incidence of melanoma within the U.S. population has been increasing rapidly for the past twenty years. To reduce the chances of occurrence, it is advisable to avoid exposing the skin to high-intensity sunlight and other sources of ultraviolet light. Whenever people are exposed to such radiation, they should wear protective clothing or use sunscreening lotions with high protective factors. In addition, such individuals should examine their skin regularly, and any unusual lesions—particularly those undergoing changes in color, shape, or surface texture—should be examined by a physician.

Numerous nerve fibers are scattered throughout the dermis. Some of them (motor fibers) carry impulses to dermal muscles and glands, causing these structures to react. Others (sensory fibers) carry impulses away from specialized sensory receptors, such as touch receptors, located within the dermis (figure 6.2).

One set of dermal sensory receptors (Pacinian corpuscles) is stimulated by heavy pressure, while another set (Meissner's corpuscles) is sensitive to light touch. Still other receptors are stimulated by temperature changes or by factors that can damage tissues. Sensory receptors are discussed in chapter 12.

Hair follicles, sebaceous glands, and sweat glands also occur at various depths in the dermis. These parts, which are composed largely of epithelial tissue, are discussed in subsequent sections of this chapter.

Skin cells can play an important role in the production of vitamin D, which is necessary for the normal development of bones and teeth. This vitamin can be formed from a substance (dehydrocholesterol) that is synthesized by cells in the digestive system or obtained in the diet. When it reaches the skin by means of the blood and is exposed to ultraviolet light from the sun, this compound is converted to a substance that becomes vitamin D. (See chapter 13.)

Certain skin cells (keratinocytes) also seem to function as part of the immune system by producing hormonelike substances that influence the development of a particular group of white blood cells. These white blood cells (T-lymphocytes) are of special importance in defending the body against invasions by disease-causing bacteria and viruses. (See chapter 19.)

Subcutaneous Layer

As was mentioned previously in this chapter, the subcutaneous layer (hypodermis) lies beneath the dermis and consists largely of loose connective and adipose tissues. (See figure 6.2.) The collagenous and elastic fibers of this layer are continuous with those of the dermis, and although most of these fibers run parallel to the surface of the skin, they travel in all directions. As a result, no sharp boundary exists between the dermis and the subcutaneous layer.

The adipose tissue of the subcutaneous layer functions as an insulator—helping to conserve body heat and impeding the entrance of heat from the outside. The amount of adipose tissue varies greatly with each individual's nutritional condition. It also varies in thickness from one region to another. For example, adipose tissue is usually quite thick over the abdomen, but absent altogether in the eyelids.

Subcutaneous injections are administered through a hollow needle into the subcutaneous layer beneath the skin. *Intradermal injections,* on the other hand, are injected into the tissues within the skin. Subcutaneous injections and intramuscular injections (administered into muscles) are sometimes called hypodermic injections.

In other instances, substances are administered through the skin by means of an adhesive transdermal patch. Such a skin patch includes a small reservoir that contains a certain quantity of a drug. The drug passes from the reservoir through a permeable membrane at a known rate. It then diffuses into the epidermis and enters the blood vessels of the dermis. Transdermal patches are used commonly to protect against motion sickness, chest pain associated with heart disease, and elevated blood pressure.

The subcutaneous layer also contains the major blood vessels that supply the skin. Branches of these vessels form a network (rete cutaneum) between the dermis and the subcutaneous layer. They, in turn, give off smaller vessels that supply the dermis above and the underlying adipose tissue.

1. What kinds of tissues make up the dermis?
2. What are the functions of these tissues?
3. What are the functions of the subcutaneous layer?

Accessory Organs of the Skin

Hair Follicles

Hair is present on all skin surfaces except the palms, soles, lips, nipples, and various parts of the external reproductive organs; however, it is not always well developed. For example, hair on the forehead and on the anterior surface of the arm is usually very fine.

Each hair develops from a group of epidermal cells at the base of a tubelike depression called a **hair follicle.** This follicle extends from the surface into the dermis and is occupied by the *root* of the hair. The epidermal cells at its base receive nourishment from dermal blood vessels that occur in a projection of connective tissue (dermal papilla) at the deep end of the follicle. As these epidermal cells divide and grow, older cells are pushed toward the surface. The cells that move upward and away from the nutrient supply become keratinized and die. Their remains constitute the structure of a developing hair, whose *shaft* extends away from the skin surface. In other words, a hair is composed of dead epidermal cells (figures 6.7 and 6.8).

Usually a hair grows for a time and then undergoes a resting period during which it remains anchored in its follicle. Later a new hair begins to grow from the base of the follicle, and the old hair is pushed outward and drops off. Sometimes, however, the hairs are not replaced. When this occurs in the scalp, the result is baldness (alopecia). Baldness is commonly due to heredity and is most likely to occur in males.

As with skin color, hair color is determined by the genes that direct the type and amount of pigment produced by epidermal melanocytes. In the case of hair color, the melanocytes are located at the deep ends of the follicles. For example, dark hair contains an abundance of melanin, while blond hair contains an intermediate quantity; the white hair of a person with albinism lacks melanin altogether. Bright red hair contains an iron pigment (trichosiderin) that does not occur in hair of any other color. A mixture of pigmented and unpigmented hair usually appears gray.

A bundle of smooth muscle cells, forming the *arrector pili muscle* (figure 6.2), is attached to each hair

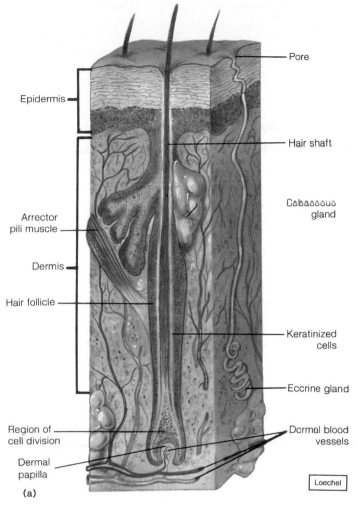

Epidermis

Arrector
pili muscle

Dermis

Hair follicle

Region of
cell division

Dermal
papilla

(a)

Pore

Hair shaft

Sebaceous
gland

Keratinized
cells

Eccrine gland

Dermal blood
vessels

Loechel

Dermal tissue

Hair follicle

Hair shaft

Region of cell
division
Dermal papilla

Adipose
tissue

(b)

Figure 6.7 (a) A hair grows from the base of a hair follicle when epidermal cells undergo cell division and older cells move outward and become keratinized. (b) A light micrograph of a hair follicle.

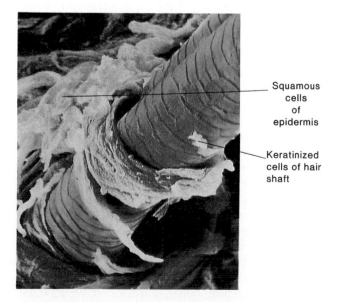

Squamous
cells
of
epidermis

Keratinized
cells of hair
shaft

Figure 6.8 A scanning electron micrograph of a hair emerging from the epidermis.

follicle. This muscle is positioned so that a short hair within the follicle stands on end when the muscle contracts. If a person is emotionally upset or very cold, nerve impulses may stimulate the arrector pili muscles to contract, causing gooseflesh or goose bumps. Each hair follicle also has one or more sebaceous glands associated with it.

Sebaceous Glands

Sebaceous glands (figure 6.2) contain groups of specialized epithelial cells and are usually associated with hair follicles. They are holocrine glands (chapter 5), and their cells produce globules of a fatty material that accumulate, causing the cells to swell and burst. The resulting mixture of fatty material and cellular debris is called *sebum*.

Sebum is secreted into hair follicles through short ducts and helps keep the hairs and the skin soft, pliable, and relatively waterproof (figure 6.9).

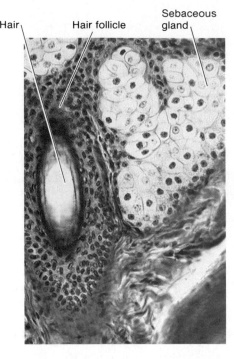

Figure 6.9   A sebaceous gland secretes sebum into a hair follicle (shown here in cross section; ×175).

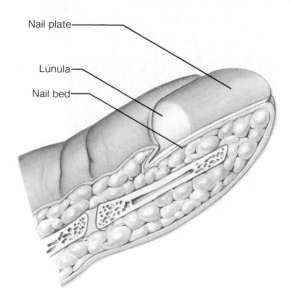

Figure 6.10   A nail is produced by epithelial cells that reproduce and undergo keratinization in the lunula of the nail.

Sebaceous glands are scattered throughout the skin, except on the palms and soles where they are lacking. In some regions, such as the lips, the corners of the mouth, and various parts of the external reproductive organs, sebaceous glands open directly to the surface of the skin rather than being connected to hair follicles.

> A disorder of the sebaceous glands is responsible for the condition called *acne* (acne vulgaris), which is common in adolescents. In this condition, the glands in some body regions become overactive and inflamed. At the same time, their ducts may become plugged, and the inflamed glands may be surrounded by small red elevations containing blackheads (comedones) or pimples (pustules).

Nails

Nails are protective coverings on the ends of the fingers and toes. Each nail consists of a *nail plate* that overlies a surface of skin called the *nail bed*. The nail plate is produced by specialized epithelial cells that are continuous with the epithelium of the skin. The whitish, thickened, half-moon-shaped region (lunula) at the base of a nail plate is its most active growing region. The epithelial cells in this region reproduce, and the newly formed cells undergo keratinization. This gives rise to tiny, horny scales that become part of the nail plate, pushing it forward over the nail bed. In time, the plate extends beyond the end of the nail bed and with normal use is gradually worn away (figure 6.10).

Sweat Glands

Sweat glands (sudoriferous glands) occur in nearly all regions of the skin, but are most numerous in the palms and soles. Each gland consists of a tiny tube that originates as a ball-shaped coil in the deeper dermis or superficial subcutaneous layer. The coiled portion of the gland is closed at its deep end and is lined with sweat-secreting epithelial cells (figure 6.2).

Certain sweat glands, known as the *apocrine glands,* respond to emotional stress. Apocrine secretions typically have odors, and the glands are considered to be scent glands. They begin to function as an individual becomes sexually mature (at puberty) and can cause some skin regions to become moist when a person is emotionally upset, frightened, or experiencing pain. They also become active when a person is sexually stimulated.

In adults, the apocrine glands are most numerous in the armpits (axillary regions), the groin, and the regions around the nipples. They are usually associated with hair follicles.

Other sweat glands, the *eccrine glands,* are not connected to hair follicles. They function throughout life by responding to body temperature that is elevated due to environmental heat or physical exercise. These glands are common on the forehead, neck, and back, where they produce profuse sweating on hot days or when a person is physically active. They are also responsible for the moisture that sometimes appears on the palms and soles when a person is emotionally stressed.

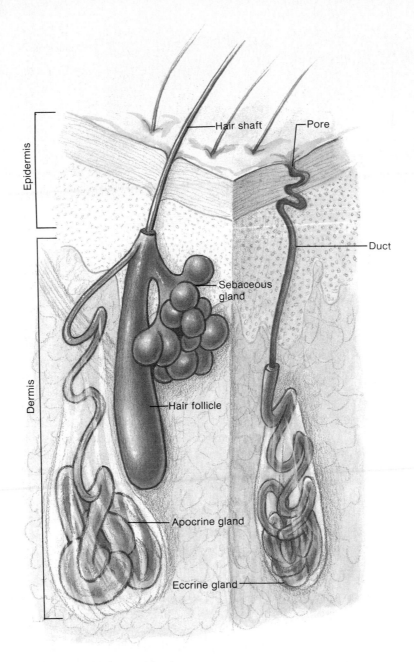

Hair shaft

Pore

Epidermis

Sebaceous gland

Duct

Dermis

Hair follicle

Apocrine gland

Eccrine gland

Figure 6.11   How do the functions of eccrine sweat glands and apocrine sweat glands differ?

With advancing age, there is a reduction in sweat gland activity, and in the very old, sweat glands may be replaced by fibrous tissues. As a result, the person becomes less able to control the body temperature effectively. There is also a decrease in sebaceous gland activity with age, so that the skin of elderly people tends to be dry and lacking oils.

The fluid secreted by the eccrine glands (sweat) is carried away by a tubular part that opens at the surface as a *pore*. Although it is mostly water, sweat con-tains small quantities of salts and certain wastes, such as urea and uric acid. Thus, the secretion of sweat is, to a limited degree, an excretory function (figures 6.11 and 6.12).

1. Explain how a hair forms.
2. What causes gooseflesh?
3. What is the function of the sebaceous glands?
4. How does the composition of a fingernail differ from that of a hair?
5. Describe the functions of the sweat glands.

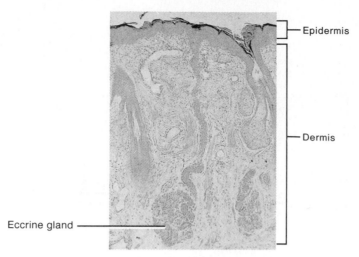

Epidermis

Dermis

Eccrine gland

Figure 6.12   Light micrograph of an eccrine gland.

Regulation of Body Temperature

The regulation of body temperature is vitally important because metabolic processes involve chemical reactions whose rates are affected by heat conditions. As a result, even slight shifts in body temperature can disrupt the rates of such reactions and produce metabolic disorders.

Normally, the temperature of deeper body parts remains close to 37°C (98.6°F). The maintenance of a stable temperature requires that the amount of heat lost by the body be balanced by the amount it produces. The skin plays a key role in the homeostatic mechanism that regulates body temperature.

Heat Production and Loss

Heat is a product of cellular metabolism. Thus, the most active cells are the major heat producers. These include the skeletal and heart muscle cells, and the cells of certain glands, such as the liver.

When heat production is excessive, nerve impulses stimulate structures in the skin and other organs to react in ways that promote heat loss. For example, during physical exercise, active muscles release heat, and the blood carries away the excess. When the warmed blood reaches the part of the brain that functions as the body's thermostat (the hypothalamus), this part signals muscles in the walls of specialized dermal blood vessels to relax. As the vessels dilate (vasodilation), more blood enters them and some of the heat the blood carries is likely to escape to the outside. At the same time, various deeper blood vessels are caused to contract (vasoconstriction), diverting blood to the surface, and the heart is stimulated to beat faster, causing more blood to move out of the deeper regions.

The primary means of body heat loss is **radiation,** by which heat in the form of *infrared heat rays* escapes from warmer surfaces to cooler surroundings. These rays radiate in all directions, much like those from the bulb of a heat lamp.

Lesser amounts of heat are lost by conduction and convection. In **conduction,** heat moves from the body directly into the molecules of cooler objects in contact with its surface. For example, heat is lost by conduction into the seat of a chair when a person sits down. The heat loss continues as long as the chair is cooler than the body surface touching it.

Heat is also lost by conduction to the air molecules that contact the body. As air becomes heated, it moves away from the body, carrying heat with it, and is replaced by cooler air moving toward the body. This type of continuous circulation of air over a warm surface is called **convection.**

Still another means of body heat loss is **evaporation.** When the body temperature is rising above normal, the nervous system stimulates eccrine sweat glands to release sweat onto the surface of the skin. As this fluid evaporates (changes from a liquid to a gas), it carries heat away from the surface, cooling the skin.

If heat is lost excessively, as may occur in a very cold environment, the brain triggers a different set of responses in the skin structures. For example, muscles in the walls of dermal blood vessels are stimulated to contract; this decreases the flow of heat-carrying blood through the skin and helps reduce heat loss by radiation, conduction, and convection. At the same time, the sweat glands remain inactive, decreasing heat loss by evaporation. If the body temperature continues to drop, the nervous system may stimulate muscle fibers in the skeletal muscles throughout the body to contract slightly. This action requires an increase in the rate of cellular respiration and produces heat as a by-product. If this response does not raise the body temperature to normal, small groups of muscles may be stimulated to

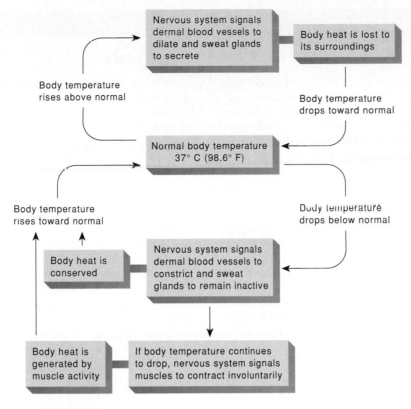

Figure 6.13 Actions involved in body temperature regulation.

contract rhythmically with still greater force, causing the person to shiver, and thus generating more heat.

Figure 6.13 summarizes some of the actions involved in the body's temperature-regulating mechanism.

Problems in Temperature Regulation

Unfortunately, the body's temperature-regulating mechanism does not always operate satisfactorily, and the consequences may be unpleasant and dangerous. For example, air can hold only a limited amount of water vapor, so that on a hot, humid day the air may become nearly saturated with water. At such times, the sweat glands may be activated, but the sweat cannot quickly evaporate. The skin becomes wet, but the person remains hot and uncomfortable. In addition, if the air temperature is high, heat is lost less effectively by radiation. In fact, if the air temperature is higher than the body temperature, the person may gain heat from the surroundings, and the body temperature may rise even higher.

1. Why is the regulation of body temperature so important?
2. How is body heat produced?
3. By what means is heat lost from the body when heat production is excessive?
4. How does the skin help regulate body temperature?

Because body temperature regulation depends, in large part, on evaporation of sweat from the skin's surface, and because high humidity hinders evaporation, athletes are advised to slow down their activities on hot, humid days. On these days, they should also stay out of the sunlight whenever possible and drink enough fluids to avoid the possibility of becoming dehydrated. Such precautions are necessary to prevent the symptoms of *heat exhaustion,* which include fatigue, dizziness, headache, muscle cramps, and nausea.

Skin Color

Skin color, like hair color, is due largely to the presence or absence of pigment produced by epidermal melanocytes. The amount of pigment synthesized by these cells is influenced by genetic, environmental, and physiological factors.

Genetic Factors

Regardless of racial origin, all humans have about the same concentrations of melanocytes in their skins. Differences in skin coloration are due largely to differences in the amount of melanin these cells produce, which is directed by the genes each person inherits. Those whose genes cause a relatively large amount of

Elevated Body Temperature

Elevated body temperature (fever), which is called *pyrexia* (pi-rek'se-ah), may be due to an infectious disease or to failure of the body's temperature-regulating mechanism. In the case of disease, substances from infectious agents, such as bacteria, or from diseased tissues cause phagocytic cells to release yet another substance, called *interleukin*.

Interleukin is carried by the blood to the brain and causes the set point of the brain's thermostat to be raised. At the same time, signals from the brain cause the skeletal muscles to increase heat production, the flow of blood into the skin to be reduced, and the sweat glands to decrease their secretion of fluid. As a consequence, the body temperature rises toward the new set point, and the person develops a fever.

If the temperature-regulating mechanism fails, body heat may accumulate faster than it is given off, and the body

temperature may become elevated even though the set point of the thermostat is normal. This sometimes happens to long distance runners, whose active muscles release great amounts of heat, especially in warm weather or when they become dehydrated.

In any case, a body temperature above 40° C (104° F) is of particular concern, especially in very young or very old persons. For example, an extremely elevated body temperature (hyperthermia) may cause seizures in a child under six years of age. Similarly, it may cause mental disturbances, low blood pressure, or heart problems in adults.

Treatment of hyperthermia may involve providing liquids to replace lost body fluids and electrolytes, administering agents (antipyretic drugs) that act to normalize the set point of the body's thermostat, sponging the skin with water to increase cooling by evaporation, or covering the person with a refrigerated blanket.

---

pigment to be produced have dark skins, while those whose genes cause less pigment to be formed have lighter complexions. Still others inherit mutant genes, and their cells are unable to manufacture melanin. As a consequence, their skins remain nonpigmented, producing the condition called *albinism* (figure 6.14).

> Skin color is also influenced by the distribution and the size of pigment granules within the melanocytes. The granules of very dark skin tend to occur singly and are relatively large; those in lighter skin tend to occur in clusters of two to four granules and are somewhat smaller.

Figure 6.14 Because of a mutant gene, a person with albinism lacks the ability to produce the skin pigment melanin.

Environmental Factors

Environmental factors such as sunlight, ultraviolet light from sunlamps, or X rays also affect skin color. These factors cause existing melanin to darken rapidly, and they stimulate melanocytes to produce more pigment and transfer it to nearby epidermal cells within a few days. This is why sunbathing results in skin tanning. Unless the exposure to sunlight is continued, however, the tan is eventually lost, as pigmented epidermal cells become keratinized and are worn away.

Physiological Factors

Blood in the dermal vessels causes additional skin color. For example, when the blood is well oxygenated, the blood pigment *hemoglobin* is bright red, making the skins of light-complexioned people and those with albinism appear pinkish. On the other hand, when the

blood oxygen concentration is low, hemoglobin is dark red, and the skin appears bluish—a condition called *cyanosis*.

The state of the blood vessels also has an effect on skin color. If the vessels are dilated, more blood enters the dermis, and the skin of a light-complexioned person becomes redder than usual. This may happen when a person is overheated, embarrassed, or under the influence of alcohol. Conversely, conditions that produce blood vessel constriction cause blanching of the skin. Thus, if the body temperature is dropping abnormally or if a person is frightened, the skin may appear pale.

The presence of a yellow-orange plant pigment called *carotene*, which is especially common in yellow vegetables, may cause the skin to appear yellowish. The yellowish color results from the accumulation of carotene in the adipose tissue of the subcutaneous layer.

Healing of Wounds and Burns

As described in chapter 5, **inflammation** is a normal response to injury or stress in which blood vessels in the affected tissues become dilated. At the same time, these vessels become more permeable, and excessive amounts of fluids tend to leave the blood vessels and enter the damaged tissues. Thus, when skin is injured, it may become reddened, swollen, warm, and painful to touch. On the other hand, the dilated blood vessels provide the tissues with more nutrients and oxygen, which aid the healing process.

The process by which damaged skin heals depends on the extent of the injury. If a break in the skin is shallow, epithelial cells along its margin are stimulated to reproduce more rapidly than usual, and the newly formed cells simply fill the gap.

If the injury extends into the dermis or subcutaneous layer, blood vessels are broken, and the blood that escapes forms a clot in the wound. Such a clot consists mainly of a fibrous protein (fibrin) that forms from the blood plasma, as well as blood cells and platelets that become entrapped in the protein fibers. Tissue fluids seep into the area and become dried. The blood clot and the dried fluids form a *scab* that covers and protects the underlying tissues. Before long, fibroblasts from connective tissue at the wound margins are attracted by various chemicals, which are released as a result of the injury. These cells migrate into the injured region and begin forming new collagenous fibers. These fibers tend to bind the edges of the wound together. The closer the edges of the wound, the sooner the gap can be filled; this is one reason for suturing or otherwise closing a large break in the skin. In addition, components of the connective tissue matrix release substances called *growth factors* that stimulate certain cell types to proliferate and regenerate the damaged tissue cells.

As the healing process continues, blood vessels send out new branches that grow into the area beneath the scab. Phagocytic cells remove dead cells and other debris. Eventually, the damaged tissues are replaced, and the scab sloughs off. If the wound is extensive, the newly formed connective tissue may appear on the surface as a *scar.*

In large, open wounds, the healing process may be accompanied by the formation of *granulations* that develop in the exposed tissues as small, rounded masses. Each of these granulations consists of a new branch of a blood vessel and a cluster of collagen-secreting fibroblasts that are being nourished by the vessel. In time, some of the blood vessels are resorbed, and the fibroblasts migrate away, leaving a scar that is composed largely of collagenous fibers. Figure 6.15 shows the stages in the healing of a wound.

As with other types of injuries, the skin's response to a burn depends upon the amount of damage it sustains. If the skin is only slightly burned, as in a minor sunburn, the skin may become warm and reddened (erythema) as a result of dermal blood vessel dilation. This response may be accompanied by mild edema, and, in time, the surface layer of skin may be shed. Such a burn that involves injury to the epidermis alone is called a *superficial partial-thickness* (first degree) *burn.* Healing usually occurs within a few days to two weeks, and there is no scarring of the skin.

A burn that involves destruction of some epidermis as well as some underlying dermis is called a *deep partial-thickness* (second degree) *burn.* In this condition, fluid escapes from damaged dermal capillaries, and as the fluid accumulates beneath the outer layer of epidermal cells, blisters appear. The injured region becomes moist and firm, and its color may vary from dark red to waxy white. Such a burn most commonly occurs as a result of exposure to hot objects, hot liquids, flames, or burning clothing.

The healing of a deep partial-thickness burn involves accessory organs of the skin that survive the injury because they are located in deeper portions of the dermis. These organs, which include hair follicles, sweat glands, and sebaceous glands, contain epithelial cells. During healing, these cells may grow out onto the surface of the dermis, spread over it, and form a new epidermis. In time, the skin usually recovers completely, and scar tissue does not develop unless an infection occurs.

A burn that destroys the epidermis, dermis, and the accessory organs of the skin is called a *full-thickness* (third degree) *burn.* In this instance, the injured skin becomes dry and leathery, and its color may vary from red to black to white.

A full-thickness burn usually occurs as a result of immersion in hot liquids or prolonged exposure to hot objects, flames, or corrosive chemicals. Since most of the epithelial cells in the affected region are likely to be destroyed, spontaneous healing can occur only by the growth of epithelial cells inward from the margin of the burn. If the injury is large, treatment may involve removing a thin layer of skin from an unburned region of the body and transplanting it in the injured area. This procedure is called an *autograft.*

If the burn is too extensive to allow removal of skin from the patient, cadaveric skin from a skin bank may be used to cover the injury. In this case, the transplant is called a *homograft,* and it serves as a temporary covering that decreases the size of the wound, helps prevent infection, and aids in preserving deeper tissues. In time, after healing has begun, the temporary covering may be removed and replaced with an autograft, as skin becomes available in areas that have healed. However, the healing of wounds using grafts is likely to be accompanied by extensive scarring.

Various skin substitutes also may be used to temporarily cover extensive burns. These include amniotic membrane that previously surrounded a human fetus before its birth and artificial membranes composed of silicone, polyurethane, or nylon together with a network of collagenous fibers. Still another type of skin substitute can be prepared by culturing human epithelial cells in a laboratory. In this instance, a relatively large, thin sheet of cells can be grown from a small bit of human skin.

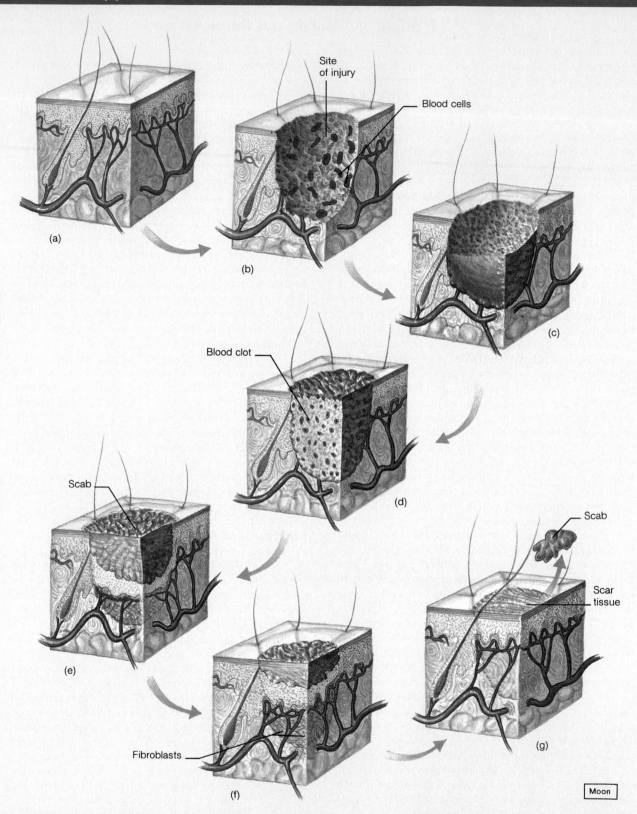

Figure 6.15 (*a*) If normal skin is (*b*) injured deeply, (*c*) blood escapes from dermal blood vessels, and (*d*) a blood clot soon forms. (*e*) The blood clot and dried tissue fluid form a scab that protects the damaged region. (*f*) Later, blood vessels send out branches, and fibroblasts migrate into the area. (*g*) The fibroblasts produce new connective tissue fibers, and when the skin is largely repaired, the scab sloughs off.

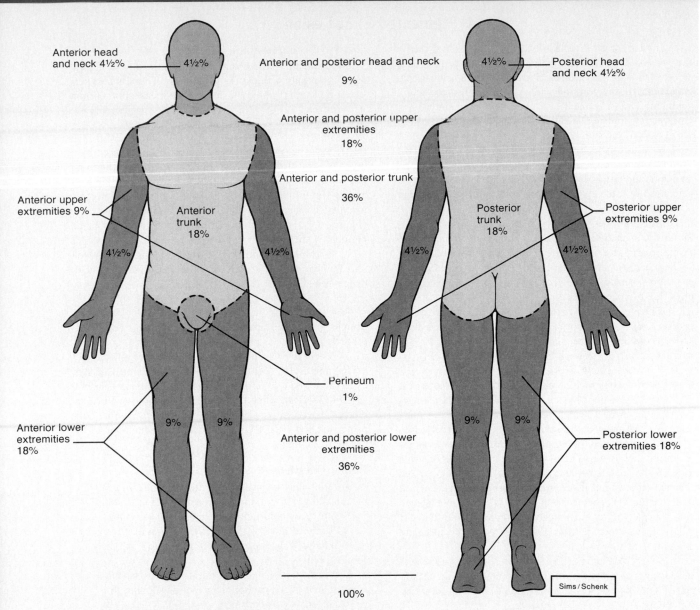

Anterior head and neck 4½%

4½%

Anterior and posterior head and neck
9%

Anterior and posterior upper extremities
18%

Anterior and posterior trunk
36%

Anterior upper extremities 9%

Anterior trunk
18%

Posterior trunk
18%

4½%

4½%

Posterior head and neck 4½%

4½%

Posterior upper extremities 9%

4½%

4½%

Perineum
1%

Anterior lower extremities
18%

9%

9%

Anterior and posterior lower extremities
36%

9%

9%

Posterior lower extremities 18%

Sims/Schenk

100%

Figure 6.16 As an aid for estimating the extent of damage caused by burning, the body is subdivided into regions, each of which represents 9% (or some multiple of 9%) of the total skin surface area.

The treatment of a burned patient usually involves estimating the extent to which the body's surface has been affected. Such an estimate can be made using the so-called "rule of nines," in which the skin's surface is subdivided into regions, each of which accounts for 9% (or some multiple of 9%) of the total surface area. (See figure 6.16.)

Calculating the percentage of body surface affected is important in planning for the replacement of body fluids and electrolytes lost from injured tissues, and for covering the burned area with skin or skin substitutes.

A Pigmented Skin Lesion

Margaret M., a healthy, light-complexioned thirty-two-year-old homemaker, spent many pleasant summer days and weekends at the seashore. She took pride in being deeply tanned most of the year and seldom applied sunscreen before going outdoors. She usually avoided becoming sunburned, however, by limiting the duration of her exposure to the direct rays of the sun.

Margaret was aware of the dangers associated with the development of pigmented skin cancers (cutaneous melanomas), and she knew that such cancers were thought to be caused by excessive exposure to the ultraviolet portion of sunlight.

She examined her skin periodically and was watchful for any changes that might occur. Her skin was marked by several pigmented moles (nevi) that had been present for as long as she could remember. One mole, located on her left thigh, seemed to be enlarging slowly, so she asked her physician to inspect it.

The physician observed the growth (lesion) and noted that its outline was smooth and regular, that it was elevated above the skin surface, and that it was uniformly pigmented. Its color was dark brown, almost black.

Although the lesion lacked some of the characteristics of a cutaneous melanoma, Margaret requested that it be removed, and her physician agreed to perform the necessary procedure.

The lesion and the surrounding skin were cleansed with an antiseptic solution, and a local anesthetic was injected into the tissues. The physician excised the lesion using a spoon-shaped instrument (curette). He then cauterized (burned) the wound using high-frequency electric current (electrocautery). The excised lesion was placed in a labeled vial of preservative and sent to a pathology laboratory for study. Margaret was cautioned to keep the wound clean and dry as it healed.

At the pathology laboratory, the specimen was sectioned, stained, and examined microscopically by a pathologist. The pathologist's report included the following comments:

Gross Description

The specimen consists of an ellipse of skin measuring 6.0 × 4.0 mm with a thickness of 2.0 mm. There is a hyperkeratotic (horny) lesion on the surface measuring 3.0 mm in diameter.

Microscopic Description

Sections of the specimen show wrinkled skin covered by a layer of stratified squamous epithelium (figure 5.6) and bearing a thick keratotic scale. The epithelial cells stain with average intensity, and the basement membrane (figure 5.6) is intact. The dermal tissue includes numerous melanocytes (figures 6.3 and 6.4) that contain dense, brownish black pigment in their cytoplasm.

Within the dermis are scattered nevus cells as well as a few lymphocytes (white blood cells). There is no evidence of malignancy (cancer).

Diagnosis

Simple pigmented papilloma (noncancerous epithelial tumor).

Margaret was advised that the best way to protect herself against the occurrence of skin cancer is to avoid skin exposure to sunlight, especially between the hours of 10:00 AM and 3:00 PM, when the light is most intense. Furthermore, she was warned always to apply sunscreen with a sun-protection factor (SPF) of at least 15 to her skin before exposure to sunlight.

Review Questions

1. What characteristics of a cutaneous melanoma were lacking in the lesion described in this case study?
2. Why is it important to have an excised lesion examined microscopically by a pathologist?
3. What cells are responsible for the brownish color of a nevus?
4. What additional advice might be given to this patient to reduce her chance of developing skin cancer?

Various diseases may affect skin color. For example, in certain forms of liver disease, bile pigments accumulate in the skin and cause it to appear yellowish (jaundiced). A person with anemia may have pale skin due to a decrease in the concentrations of hemoglobin in the red blood cells.

1. What factors influence skin color?
2. Which of these factors are genetic? Which are environmental?

Common Skin Disorders

acne (ak′ne) A disease of the sebaceous glands accompanied by blackheads and pimples.

alopecia (al″o-pe′she-ah) Loss of hair.

athlete's foot (ath′-lētz foot) A fungus infection (tinea pedis) usually involving the skin of the toes and soles.

birthmark (berth′ mark) Vascular or skin tumor involving the skin and subcutaneous tissues, visible at birth or soon after.

boil (boil) A bacterial infection (furuncle) involving a hair follicle and/or sebaceous glands.

carbuncle (kar′ bung-kl) A bacterial infection, similar to a boil, that spreads into the subcutaneous tissues.

cyst (sist) A liquid-filled sac or capsule.

dermatitis (der″mah-ti′tis) An inflammation of the skin.

eczema (ek′ze-mah) A noncontagious skin rash often accompanied by itching, blistering, and scaling.

erythema (er″i-the′mah) Reddening of the skin due to dilation of dermal blood vessels in response to injury or inflammation.

herpes (her′pez) An infectious disease of the skin usually caused by the virus called herpes simplex and characterized by recurring formations of small clusters of vesicles.

impetigo (im″pe-ti′go) A contagious disease of bacterial origin characterized by pustules that rupture and become covered with loosely held crusts.

keloid (ke′loid) An elevated, enlarging fibrous scar usually initiated by an injury.

mole (mol) A fleshy skin tumor (nevus) that is usually pigmented; colors range from brown to black.

pediculosis (pe-dik″u-lo′sis) A disease produced by an infestation of lice.

pruritus (proo-ri′tus) Itching of the skin.

psoriasis (so-ri′ah-sis) A chronic skin disease characterized by red patches covered with silvery scales.

pustule (pus′tul) An elevated, pus-filled area.

scabies (ska′bez) A disease resulting from an infestation of mites.

seborrhea (seb″o-re′ah) A disease characterized by hyperactivity of the sebaceous glands, accompanied by greasy skin and dandruff.

shingles (shing′g′lz) A disease caused by a viral infection (varicella zoster) of certain spinal nerves; characterized by severe localized pain and blisters in the skin areas supplied by the affected nerves.

ulcer (ul′ser) An open sore.

urticaria (ur″ti-ka′re-ah) An allergic reaction of the skin that produces reddish, elevated patches (hives).

wart (wort) A flesh-colored, raised area caused by a viral infection.

Chapter Summary

Introduction (page 148)

Organs are composed of two or more kinds of tissues. The skin, together with its accessory organs, constitute the integumentary organ system.

Types of Membranes (page 148)

1. Serous membranes
   a. Serous membranes are organs that line body cavities lacking openings to the outside.
   b. They are composed of epithelium and loose connective tissue.
   c. Cells of serous membranes secrete watery serous fluid that lubricates membrane surfaces.

2. Mucous membranes
   a. Mucous membranes are organs that line cavities and tubes opening to the outside.
   b. They are composed of various kinds of epithelium and loose connective tissue.
   c. Cells of mucous membranes secrete mucus.

3. The cutaneous membrane is the external body covering commonly called the skin.

4. Synovial membranes
   a. Synovial membranes are organs that line joint cavities.
   b. They are composed of fibrous connective tissue overlying loose connective tissue and adipose tissue.
   c. Cells of synovial membranes secrete synovial fluid that lubricates the ends of bones at joints.

Skin and Its Tissues (page 148)

Skin functions as a protective covering, aids in regulating body temperature, houses sensory receptors, synthesizes various chemicals, and excretes wastes. It is composed of an epidermis and a dermis separated by a basement membrane with a subcutaneous layer beneath.

1. Epidermis
   a. The epidermis is a layer of stratified squamous epithelium that lacks blood vessels.
   b. The deepest layer, called stratum basale, contains cells undergoing mitosis.
   c. Epidermal cells undergo keratinization as they are pushed toward the surface.
   d. The outermost layer, called stratum corneum, is composed of dead epidermal cells.
   e. The production of epidermal cells is balanced with the rate at which they are lost at the surface.
   f. Epidermis functions to protect underlying tissues against water loss, mechanical injury, and the effects of harmful chemicals.
   g. Melanin protects underlying cells from the effects of ultraviolet light.
   h. Melanocytes transfer melanin to nearby epidermal cells.

2. Dermis
   a. The dermis is a layer composed largely of fibrous connective tissue that binds the epidermis to underlying tissues.
   b. It also contains muscle fibers, blood vessels, and nerve fibers.
   c. Dermal blood vessels supply nutrients to all skin cells and help regulate body temperature.
   d. Nervous tissue is scattered through the dermis.
      (1) Some dermal nerve fibers carry impulses to muscles and glands of the skin.
      (2) Other dermal nerve fibers are associated with various sensory receptors in the skin.

3. Subcutaneous layer
   a. The subcutaneous layer is composed of loose connective tissue and adipose tissue.
   b. Adipose tissue helps conserve body heat.
   c. This layer contains blood vessels that supply the skin.

Accessory Organs of the Skin (page 154)

1. Hair follicles
   a. Hair occurs in nearly all regions of the skin.
   b. Each hair develops from epidermal cells at the base of a tubelike hair follicle.
   c. As newly formed cells develop and grow, older cells are pushed toward the surface and undergo keratinization.
   d. A hair usually grows for a while, undergoes a resting period, and then is replaced by a new hair.
   e. Hair color is determined by genes that direct the type and amount of pigment in its cells.
   f. A bundle of smooth muscle cells and one or more sebaceous glands are attached to each hair follicle.
2. Sebaceous glands
   a. Sebaceous glands secrete sebum, which helps keep skin and hair soft and waterproof.
   b. In some regions, they open directly to the skin surface.
3. Nails
   a. Nails are protective covers on the ends of fingers and toes.
   b. They are produced by epidermal cells that undergo keratinization.
4. Sweat glands
   a. Sweat glands are located in nearly all regions of the skin.
   b. Each gland consists of a coiled tube.
   c. Apocrine glands respond to emotional stress, while eccrine glands respond to elevated body temperature.
   d. Sweat is primarily water, but also contains salts and waste products.

Regulation of Body Temperature (page 158)

Regulation of body temperature is vital because heat affects the rates of metabolic reactions. Normal temperature of deeper body parts is about 37°C (98.6°F).

1. Heat production and loss
   a. Heat is a by-product of cellular respiration.
   b. When the body temperature rises above normal, more blood is caused to enter dermal blood vessels.
   c. Heat is lost to the outside by radiation, conduction, convection, and evaporation.
   d. Sweat gland activity increases heat loss by evaporation.
   e. If the body temperature drops below normal, dermal blood vessels constrict, and sweat glands become inactive.
   f. When heat is lost excessively, skeletal muscles are stimulated to contract involuntarily; this increases cellular respiration and produces additional heat.

2. Problems in temperature regulation
   a. Air can hold a limited amount of water vapor.
   b. When the air is saturated with water, sweat may fail to evaporate, and body temperature may remain elevated.

Skin Color (page 159)

All humans have about the same concentration of melanocytes. Skin color is due largely to the amount of melanin in the epidermis.

1. Genetic factors
   a. Each person inherits genes for melanin production.
   b. Dark skin is due to genes that cause large amounts of melanin to be produced; lighter skin is due to genes that cause lesser amounts of melanin to form.
   c. Mutant genes may cause a lack of melanin in the skin.
2. Environmental factors
   a. Environmental factors that influence skin color include sunlight, ultraviolet light, and X ray.
   b. These factors cause existing melanin to darken, and they stimulate additional melanin production.
3. Physiological factors
   a. The oxygen content of the blood in dermal vessels may cause the skin of light-complexioned persons to appear pinkish or bluish.
   b. Carotene in the subcutaneous layer may cause the skin to appear yellowish.
   c. Various diseases may affect skin color.

Clinical Application of Knowledge

1. What special problems would result from the loss of 50% of a person's functional skin surface? How might this person's environment be modified to compensate partially for such a loss?
2. A premature infant typically lacks subcutaneous adipose tissue. Also, the surface area of an infant's body is relatively large compared to its volume. How do you think these factors affect the ability of an infant to regulate its body temperature?
3. As a rule, a superficial partial-thickness burn is more painful than one involving deeper tissues. How would you explain this observation?
4. Which of the following would result in the more rapid absorption of a drug: a subcutaneous injection or an intradermal injection? Why do you think so?
5. What methods might be used to cool the skin of a child experiencing a high fever? For each method you list, identify the means by which it promotes heat loss—radiation, conduction, convection, or evaporation.
6. How would you explain to an athlete the importance of keeping the body hydrated when exercising in warm weather?

Review Activities

1. Explain why a membrane is an organ.
2. Define *integumentary organ system*.
3. Distinguish between serous and mucous membranes.
4. Explain the function of synovial membranes.
5. List six functions of skin.
6. Distinguish between the epidermis and the dermis.
7. Describe the subcutaneous layer.
8. Explain what happens to epidermal cells as they undergo keratinization.
9. List the layers of the epidermis.
10. Describe the function of melanocytes.
11. Describe the structure of the dermis.
12. Review the functions of dermal nervous tissue.
13. Explain the functions of the subcutaneous layer.
14. Distinguish between a hair and a hair follicle.
15. Review how hair color is determined.
16. Explain the function of sebaceous glands.
17. Describe how nails are formed.
18. Distinguish between apocrine and eccrine glands.
19. Explain the importance of body temperature regulation.
20. Describe the role of the skin in promoting the loss of excess body heat.
21. Explain how body heat is lost by radiation.
22. Distinguish between conduction and convection
23. Describe the body's responses to decreasing body temperature.
24. Review how air saturated with water vapor may interfere with body temperature regulation.
25. Explain how environmental factors affect skin color.
26. Describe three physiological factors that affect skin color.

UNIT

2

Support and Movement

The chapters of unit 2 deal with structures and functions of the skeletal and muscular systems. They describe how organs of the skeletal system support and protect other body parts, and how they function with organs of the muscular system to enable body parts to move. They also describe how skeletal structures participate in the formation of blood and in the storage of inorganic salts, and how muscular tissues act to produce body heat and to move body fluids.

7
Skeletal System

8
Joints of the Skeletal System

9
Muscular System

Athletic performance depends upon the conversion of chemical energy from nutrients into mechanical energy of contracting skeletal muscles.

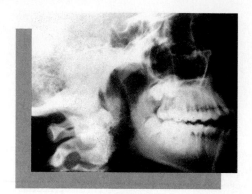

7

Skeletal System

*B*ones are composed of several kinds of tissues, and they are the organs of the *skeletal system.*

Because bones are relatively rigid structures, they provide support and protection for softer tissues, and they act together with skeletal muscles to make body movements possible. They also house the tissue that produces blood cells, and they store inorganic salts.

The shapes of individual bones are closely related to their functions. Projections provide places for the attachment of muscles, tendons, and ligaments; openings serve as passageways for blood vessels and nerves; and the ends of bones are modified to form joints with other bones ∎

Chapter Objectives

After you have studied this chapter, you should be able to:

1. Classify bones according to their shapes and name an example from each group.

2. Describe the general structure of a bone and list the functions of its parts.

3. Distinguish between intramembranous and endochondral bones, and explain how such bones grow and develop.

4. Describe the effects of sunlight, nutrition, hormonal secretions, and exercise on bone development.

5. Discuss the major functions of bones.

6. Distinguish between the axial and appendicular skeletons, and name the major parts of each.

7. Locate and identify the bones and the major features of the bones that comprise the skull, vertebral column, thoracic cage, pectoral girdle, upper limb, pelvic girdle, and lower limb.

8. Complete the review activities at the end of this chapter. Note that the items are worded in the form of specific learning objectives. You may want to refer to them before reading the chapter.

Key Terms

articular cartilage (ar-tik'u-lar kar'ti-lij)

compact bone (kom'pakt bōn)

diaphysis (di-af'i-sis)

endochondral bone (en'do-kon'dral bōn)

epiphyseal disk (ep''i-fiz'e-al disk)

epiphysis (e-pif'i-sis)

fontanel (fon'tah-nel)

hematopoiesis (hem''ah-to-poi-e'sis)

intramembranous bone (in''trah-mem'brah-nus bōn)

lever (lev'er)

marrow (mar'o)

medullary cavity (med'u-lār''e kav'i-te)

osteoblast (os'te-o-blast)

osteoclast (os'te-o-klast)

osteocyte (os'te-o-sīt)

osteon (os'te-on)

periosteum (per''e-os'te-um)

spongy bone (spun'je bōn)

Aids to Understanding Words

acetabul-, vinegar cup: acetabulum—depression in the coxal bone that articulates with the head of the femur.

ax-, axis: axial skeleton—upright portion of the skeleton that supports the head, neck, and trunk.

-blast, budding or developing: osteoblast—cell that forms bone tissue.

carp-, wrist: carpals—wrist bones.

-clast, broken: osteoclast—cell that breaks down bone tissue.

condyl-, knob: condyle—rounded, bony process.

corac-, beaklike: coracoid process—beaklike process of the scapula.

cribr-, sievelike: cribriform plate—portion of the ethmoid bone with many small openings.

crist-, ridge: crista galli—bony ridge that projects upward into the cranial cavity.

fov-, pit: fovea capitis—pit in the head of a femur.

gladi-, sword: gladiolus—middle portion of the bladelike sternum.

glen-, joint socket: glenoid cavity—depression in the scapula that articulates with the head of a humerus.

inter-, between: intervertebral disk—a structure located between adjacent vertebrae.

intra-, inside: intramembranous bone—bone that forms within sheetlike masses of connective tissue.

meat-, passage: auditory meatus—canal of the temporal bone that leads inward to parts of the ear.

odont-, tooth: odontoid process—toothlike process of the second cervical vertebra.

poie-, making: hematopoiesis—process by which blood cells are formed.

An individual bone is composed of a variety of tissues, including bone tissue, cartilage, fibrous connective tissue, blood, and nervous tissue. Because so much nonliving material is present in the matrix of bone tissue, the whole organ may appear to be inert. A bone, however, contains very active, living tissues.

Bone Structure

Although the various bones of the skeletal system differ greatly in size and shape, they are similar in their structure, development, and functions.

Classification of Bones

Bones can be classified according to their shapes—long, short, flat, or irregular (figure 7.1).

Long bones have long longitudinal axes and expanded ends. Examples are the arm and leg bones.

Short bones are somewhat cubelike, with their lengths and widths roughly equal. The bones of the wrists and ankles are examples of this type.

Flat bones are platelike structures with broad surfaces, such as the ribs, scapulae, and some bones of the skull.

Irregular bones have a variety of shapes and are usually connected to several other bones. Irregular bones include the vertebrae that comprise the backbone and many of the facial bones.

In addition to these four groups of bones, some authorities recognize a fifth group called *sesamoid* bones. The bones of this group are usually small and nodular, and they usually are embedded within tendons adjacent to joints, where the tendons undergo compression. The kneecap (patella) is an example of a very large sesamoid bone.

Parts of a Long Bone

To describe the structure of bone, the femur, a long bone in the upper leg, will be used as an example (figure 7.2).

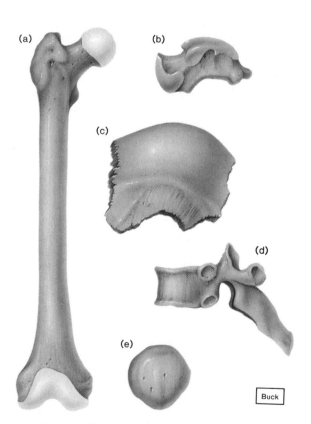

Figure 7.1 (*a*) The femur of the leg is a long bone; (*b*) a tarsal bone of the ankle is a short bone; (*c*) a parietal bone of the skull is a flat bone; (*d*) a vertebra of the backbone is an irregular bone; and (*e*) the patella of the knee is a round bone.

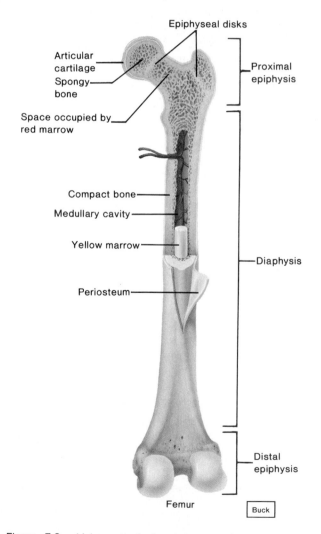

Figure 7.2 Major parts of a long bone.

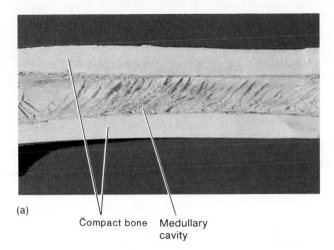

(a)

Compact bone Medullary
cavity

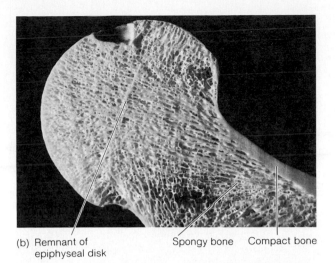

(b) Remnant of   Spongy bone   Compact bone
epiphyseal disk

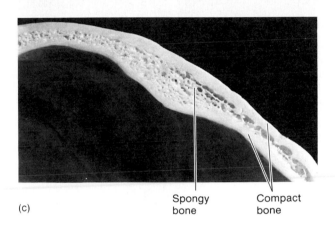

(c)   Spongy   Compact
                        bone   bone

Figure 7.3   (a) In a femur, the wall of the diaphysis consists of compact bone. (b) The epiphyses of the femur contain spongy bone enclosed by a thin layer of compact bone. (c) This skull bone contains a layer of spongy bone sandwiched between plates of compact bone.

At each end of such a bone is an expanded portion called an **epiphysis** (pl. *epiphyses*), which articulates (or forms a joint) with another bone. On its outer surface, the articulating portion of the epiphysis is coated with a layer of hyaline cartilage called **articular cartilage.** The shaft of the bone, which is located between the epiphyses, is called the **diaphysis.**

Except for the articular cartilage on its ends, the bone is completely enclosed by a tough, vascular covering of fibrous tissue called the **periosteum.** This membrane is firmly attached to the bone, and the periosteal fibers are continuous with various ligaments and tendons that are connected to the membrane. The periosteum also functions in the formation and repair of bone tissue.

Each bone has a shape closely related to its functions. Bony projections called *processes,* for example, provide sites for the attachment of ligaments and tendons; grooves and openings serve as passageways for blood vessels and nerves; a depression of one bone might articulate with a process of another.

The wall of the diaphysis is composed mainly of tightly packed tissue called **compact bone** (cortical bone). This type of bone is solid, strong, and resistant to bending (figure 7.3a).

The epiphyses, on the other hand, are composed largely of **spongy bone** (cancellous bone) with thin layers of compact bone on their surfaces (figure 7.3b). Spongy bone consists of numerous branching bony plates called **trabeculae.** Irregular interconnecting spaces occur between these plates and help reduce the weight of the bone. Spongy bone provides strength, and its bony plates are most highly developed in the regions of the epiphyses that are subjected to the forces of compression.

Both compact and spongy tissues are usually present in each bone. Short, flat, and irregular bones typically consist of a mass of spongy bone that is either covered by a layer of compact bone or sandwiched between plates of compact bone (figure 7.3c).

Compact bone in the diaphysis of a long bone forms a semirigid tube with a hollow chamber called the **medullary cavity.** This cavity is continuous with the spaces of the spongy bone. All of these areas are lined with a thin layer of squamous cells called **endosteum** and are filled with a specialized type of soft connective tissue called **marrow.**

Microscopic Structure

As was discussed in chapter 5, bone cells (osteocytes) are located in minute, bony chambers called *lacunae,* which are arranged in concentric circles around *osteonic canals* (Haversian canals). These cells communicate with nearby cells by means of cellular processes passing through canaliculi. The intercellular material

Figure 7.4 What features do you recognize in this scanning electron micrograph of a single osteon in compact bone (about ×1,300)? (*Tissues and Organs: A Text-Atlas of Scanning Electron Microscopy*, by R. G. Kessel and R. H. Kardon. © 1979 W.H. Freeman and Company.)

of bone tissue is largely collagen and inorganic salts. The collagen gives bone its strength and resilience, and the inorganic salts make it hard and resistant to crushing.

Compact Bone

In compact bone, the osteocytes and layers of intercellular material clustered concentrically around an osteonic canal form a cylinder-shaped unit called an **osteon** (Haversian system). Many of these units cemented together form the substance of compact bone (figure 7.4).

Each osteonic canal contains one or two small blood vessels (usually capillaries) and nerves surrounded by some loose connective tissue. Blood in these vessels provides nourishment for the bone cells associated with the osteonic canal.

Osteonic canals travel longitudinally through bone tissue. They are interconnected by transverse *communicating canals* (Volkmann's canals). These canals contain larger blood vessels and nerves by which the vessels and nerves in the osteonic canals communicate with the surface of the bone and the medullary cavity (figure 7.5).

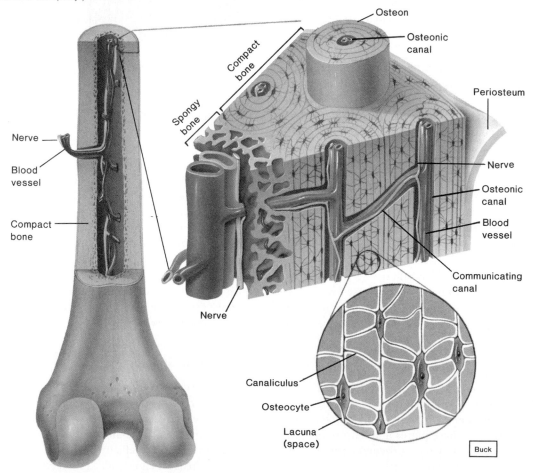

Figure 7.5 Compact bone is composed of osteons cemented together.

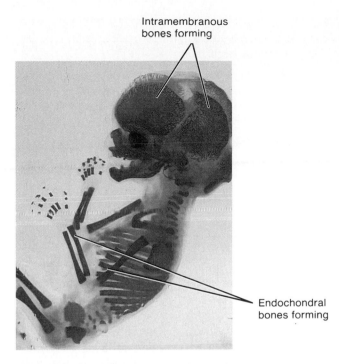

Intramembranous
bones forming

Endochondral
bones forming

Figure 7.6 The tissues of this miscarried fetus (about fourteen weeks old) have been cleared, and the developing bones have been stained selectively.

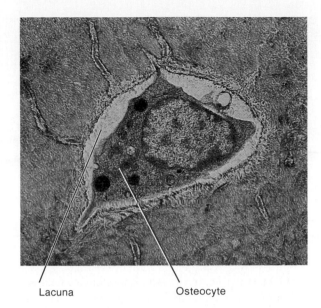

Lacuna   Osteocyte

Figure 7.7 Transmission electron micrograph of an osteocyte isolated within a lacuna.

Spongy Bone

Spongy bone is also composed of osteocytes and intercellular material. However, the bone cells are not arranged around osteonic canals. Instead, the cells occur within the trabeculae, and they are nourished by diffusion of substances into the canaliculi that lead from the bone cells to the surface of these thin, bony plates.

1. Explain how bones are classified.
2. List five major parts of a long bone.
3. How do compact and spongy bone differ in structure?
4. Describe the microscopic structure of a compact bone.

Bone Development and Growth

Parts of the skeletal system begin to form during the first few weeks of prenatal development, and bony structures continue to grow and develop into adulthood. Bones form by the replacement of existing connective tissue in one of two ways. Some first appear between sheetlike layers of connective tissues; they are called intramembranous bones. Others begin as masses of cartilage that are later replaced by bone tissue; they are called endochondral bones (figure 7.6).

Intramembranous Bones

Examples of **intramembranous bones** are the broad, flat bones of the skull. During their development (osteo-

genesis), membranelike layers of primitive connective tissues appear at the sites of the future bones. These layers are supplied with dense networks of blood vessels, and some of the connective tissue cells become arranged around these vessels. These primitive cells enlarge and differentiate into bone-forming cells called **osteoblasts,** which, in turn, deposit bony matrix around themselves. As a result, spongy bone is produced in all directions along the blood vessels within the layers of primitive connective tissues. Later, some of the spongy bone may be converted to compact bone, as spaces become filled with bone matrix.

As development continues, the osteoblasts may become completely surrounded by matrix, and in this manner they become secluded within lacunae. At the same time, matrix enclosing the cellular processes of the osteoblasts gives rise to canaliculi. Once they are isolated in lacunae, the bone cells are called **osteocytes.** (See figure 7.7.)

The cells of the primitive connective tissue that persist outside the developing bone give rise to the periosteum. Osteoblasts on the inside of the periosteum form a layer of compact bone over the surface of the newly formed spongy bone.

This process of forming an intramembranous bone by the replacement of connective tissue is called *intramembranous ossification*. Chart 7.1 lists the major steps of the process.

Endochondral Bones

Most of the bones of the skeleton are **endochondral bones.** Their development proceeds from masses of hyaline cartilage with shapes similar to future bony structures. These cartilaginous models grow rapidly for

CHART 7.1 Major steps in bone development

Intramembranous ossification	Endochondral ossification
1. Membranelike layers of primitive connective tissue appear at sites of future bones.	1. Masses of hyaline cartilage form models of future bones.
2. Primitive connective tissue cells become arranged around the blood vessels in these layers.	2. Cartilage tissue breaks down and disappears.
3. Connective tissue cells differentiate into osteoblasts, which form spongy bone.	3. Blood vessels and differentiating osteoblasts from the periosteum invade the disintegrating tissue.
4. Osteoblasts become osteocytes when they are completely surrounded by bony matrix.	4. Osteoblasts form spongy bone in the space occupied by cartilage.
5. Connective tissue on the surface of each developing structure forms a periosteum.	5. Compact bone is deposited around the spongy bone by osteoblasts beneath the periosteum.
6. Osteoblasts on the inside of the periosteum form compact bone over the spongy bone.	

a time, and then begin to undergo extensive changes. For example, cartilage cells enlarge and increase the sizes of their respective lacunae. This is accompanied by destruction of the surrounding matrix, and soon the cartilage cells die and degenerate.

About the same time, a periosteum forms from connective tissue that encircles the developing structure. As the cartilage breaks down, blood vessels and undifferentiated connective tissue cells invade the disintegrating tissue. Some of the invading cells differentiate into osteoblasts and begin to form spongy bone in the spaces previously occupied by the cartilage. Compact bone is deposited around the spongy bone by intramembranous ossification occurring beneath the periosteum.

This process of forming an endochondral bone by the replacement of hyaline cartilage is called *endochondral ossification.* Its major steps are listed in chart 7.1.

Growth of an Endochondral Bone

In a long bone, replacement of hyaline cartilage by bony tissue begins in the center of the diaphysis. This region is called the *primary ossification center,* and bone develops from it toward the ends of the cartilaginous structure. Meanwhile, osteoblasts from the periosteum deposit a thin layer of compact bone around the primary ossification center by intramembranous ossification. The epiphyses of the developing bone remain cartilaginous and continue to grow. Later, *secondary ossification centers* appear in the epiphyses, and spongy bone forms in all directions from them. As spongy bone is deposited in the diaphysis and in the epiphysis, a band of cartilage, called the **epiphyseal disk** (physis), is left between the two ossification centers. (See figures 7.2 and 7.3b.)

The cartilaginous cells of the epiphyseal disk are arranged in four layers, each of which may be several cells thick, as shown in figure 7.8. The first layer, closest to the end of the epiphysis, is composed of resting cells. Although these cells are not actively participating in the growing process, this layer anchors the epiphyseal disk to the bony tissue of the epiphysis.

The second layer contains rows of numerous young cells that are undergoing mitosis. As new cells are produced, and as intercellular material is formed around them, the cartilaginous disk thickens.

The rows of older cells, which are left behind when new cells appear, form the third layer. These cells enlarge and cause the epiphyseal disk to thicken still more. Consequently, the length of the entire bone increases. At the same time, calcium salts accumulate in the intercellular matrix adjacent to the oldest of the cartilaginous cells, and as the matrix becomes calcified, the cells begin to die.

The fourth layer of the epiphyseal disk is quite thin and is composed largely of dead cells and calcified intercellular substance.

In time, the calcified matrix is broken down by the action of large, multinucleated cells called **osteoclasts.** These large cells originate by the fusion of certain single-nucleated white blood cells (monocytes). (See chapter 17.) Osteoclasts secrete an acid that dissolves the inorganic component of the calcified substance, and at the same time, their lysosomal enzymes digest the organic components of the substance. After the matrix is removed, bone-building osteoblasts invade the region and deposit bone tissue in place of the calcified cartilage.

A long bone will continue to grow in length while the cartilaginous cells of the epiphyseal disks are active. However, once the ossification centers of the diaphysis and epiphyses come together and the epiphyseal disks become ossified, growth in length is no longer possible in that end of the bone.

A developing bone grows in thickness as compact bone is deposited on the outside, just beneath the periosteum. As this compact bone is forming on the surface, other bone tissue is being eroded away on the

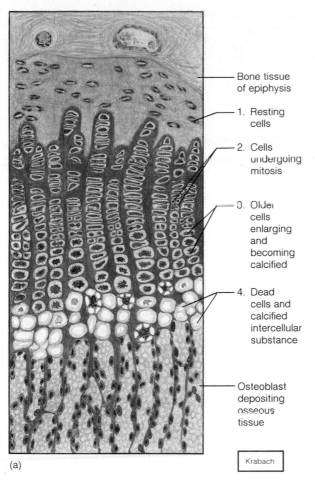

Bone tissue of epiphysis

1. Resting cells

2. Cells undergoing mitosis

3. Older cells enlarging and becoming calcified

4. Dead cells and calcified intercellular substance

Osteoblast depositing osseous tissue

(a)

Krabach

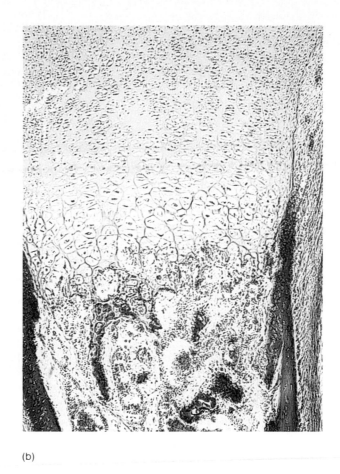

(b)

Figure 7.8 (*a*) The cartilaginous cells of an epiphyseal disk are arranged in four layers, each of which may be several cells thick. (*b*) A micrograph of an epiphyseal disk.

inside by osteoclasts (figure 7.9). The space that is produced becomes the medullary cavity of the diaphysis, which later fills with marrow.

The bone in the central regions of the epiphyses and diaphysis remains spongy, and hyaline cartilage on the ends of the epiphyses persists throughout life as articular cartilage. Figure 7.10 illustrates the stages in the development and growth of an endochondral bone. Chart 7.2 lists the ages at which various bones become ossified.

It is possible to determine whether a child's long bones are still growing by examining an X-ray photograph to see if the epiphyseal disks are present (figure 7.11). If a disk is damaged as a result of a fracture before it becomes ossified, elongation of the long bone may cease prematurely, or if growth continues, it may be uneven. For this reason, injuries to the epiphyses of a young person's bones are of special concern. On the other hand, an epiphysis is sometimes altered surgically in order to equalize growth of bones that are developing at very different rates.

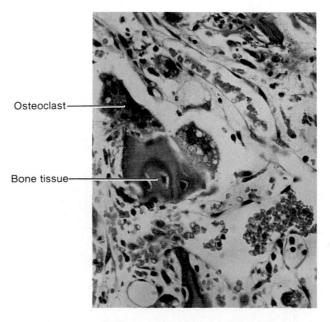

Osteoclast

Bone tissue

Figure 7.9 Micrograph of a bone-resorbing osteoclast.

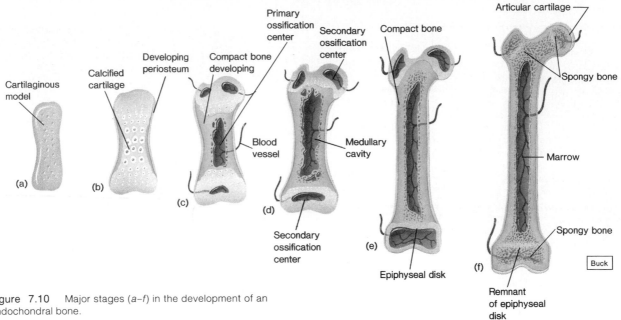

Figure 7.10 Major stages (a–f) in the development of an endochondral bone.

CHART 7.2 Ossification timetable

Age	Occurrence	Age	Occurrence
Third month of prenatal development	Ossification in long bones is beginning.	17 to 20 years	Bones of the upper limbs and scapulae become completely ossified.
Fourth month	Most primary ossification centers have appeared in the diaphyses of bones.	18 to 23 years	Bones of the lower limbs and coxal bones become completely ossified.
Birth to 5 years	Secondary ossification centers appear in the epiphyses.	23 to 25 years	Bones of the sternum, clavicles, and vertebrae become completely ossified.
5 to 12 years in females, or 5 to 14 years in males	Ossification spreads rapidly from the ossification centers, and various bones are becoming ossified.	By 25 years	Nearly all bones are completely ossified.

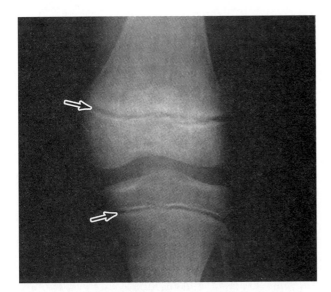

Figure 7.11 The presence of epiphyseal disks (arrows) in a child's femur is an indication that the bone is still growing in length.

Cancers involving bone are commonly accompanied by an abnormal increase in osteoclastic activity and a consequent loss of bone tissue. An exception to this is cancer originating in the male prostate gland. For reasons that are not well understood, prostate cancer cells stimulate the activity of osteoblasts. If such cancer cells reach the bone marrow, as they do in most cases of advanced prostatic cancer, these cells promote the formation of new bone on the surfaces of the bony trabeculae.

Homeostasis of Bone Tissue

After the intramembranous and endochondral bones have formed, they are continually remodeled by the activities of osteoclasts and osteoblasts. Thus, throughout life, osteoclasts are being stimulated to resorb bone tissue at specific sites, and osteoblasts are being activated to replace the bone. These opposing processes of resorption and replacement, however, are well regulated. Consequently, the total mass of bone tissue within an adult skeleton normally remains nearly constant, even though 3% to 5% of bone calcium is exchanged each year.

Factors Affecting Bone Development, Growth, and Repair

Bone development, growth, and repair are influenced by a number of factors, including nutrition, exposure to sunlight, hormonal secretions, and physical exercise. For example, vitamin D is necessary for proper absorption of calcium in the small intestine. In the absence of this vitamin, calcium is poorly absorbed, the inorganic salt portion of bone matrix is deficient in calcium, and the bones are likely to be deformed. In children, this condition is called *rickets,* and in adults it is called *osteomalacia.*

Vitamin D is relatively uncommon in natural foods, except for eggs. But it is readily available in milk and other dairy products to which vitamin D has been added. Vitamin D can also be formed from a substance (dehydrocholesterol) that is produced by cells in the digestive tract or obtained in the diet. This substance is carried by the blood to the skin, and when exposed to ultraviolet light from the sun, it is converted to a compound that becomes vitamin D.

Vitamins A and C are also needed for normal bone development and growth. Vitamin A is necessary for bone resorption that occurs during normal development. Thus, a deficiency of vitamin A may result in retardation of bone development. Vitamin C is needed for the synthesis of collagen, so its lack also may inhibit bone development. In this case, osteoblasts produce less collagen in the intercellular material of the bone tissue, and the resulting bones are abnormally slender and fragile.

Hormones that affect bone development and growth include those secreted by the pituitary gland, thyroid gland, parathyroid glands, and ovaries or testes. The pituitary gland, for instance, secretes a chemical called **growth hormone,** which stimulates reproduction of cartilage cells in the epiphyseal disks. In the absence of this hormone, the long bones of the limbs fail to develop normally, and the child may become a pituitary dwarf. Such a person is very short, but has normal body proportions. If excessive amounts of growth hormone are released before the epiphyseal disks are ossified, the affected person may attain a height of 8 feet or more—a condition called *pituitary giantism.*

In an adult, excessive secretion of growth hormone causes a condition called *acromegaly,* in which the hands, feet, and jaw enlarge. (See chapter 13.)

Thyroid hormone causes cartilage in the epiphyseal disks of long bones to be replaced by bone tissue. Thus, an excessive secretion of thyroid hormone can halt bone growth by causing premature ossification of the

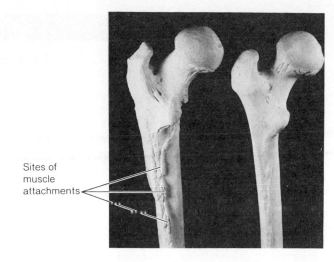

Sites of muscle attachments

Figure 7.12 Note the increased amount of bone at the sites of muscle attachments in the femur on the left. What factors might be responsible for such growth of bone tissue?

disks. A deficiency of thyroid hormone also may produce stunted growth, because without the normal effect of the thyroid hormone, the pituitary gland fails to secrete enough growth hormone. (See chapter 13.)

Both male and female sex hormones (called androgens and estrogens, respectively) from the testes, ovaries, and adrenal glands promote the formation of bone tissue. Beginning at puberty, these hormones are secreted abundantly, and the long bones tend to grow considerably. However, sex hormones also stimulate ossification of the epiphyseal disks, and consequently they cause bones to stop growing in length at a relatively early age. The effect of estrogens on the disks is somewhat stronger than that of androgens. For this reason, females typically reach their maximum heights earlier than males.

Physical stress also has a stimulating effect on bone growth. For example, when skeletal muscles contract, they pull at their attachments on bones, and the resulting stress stimulates the bone tissue to thicken and strengthen (hypertrophy). Conversely, with lack of exercise, the same bone tissue undergoes a wasting process and tends to become thinner and weaker (atrophy). This is why the bones of athletes are usually stronger and heavier than those of nonathletes. It is also the reason that the fractured bones of limbs that are immobilized by casting may decrease in size. (See figure 7.12.)

1. Describe the development of an intramembranous bone.
2. Explain how an endochondral bone develops.
3. List the steps in the growth of a long bone.
4. Explain how nutritional factors affect bone development.
5. What effects do hormones have on bone growth?
6. How does physical exercise affect bone structure?

Fractures

Although a **fracture** may involve injury to cartilaginous structures, it is usually defined as a break in a bone. A fracture can be classified according to its cause and the nature of the break sustained. For example, a break due to injury is a *traumatic* fracture, while one resulting from disease is a *spontaneous,* or *pathologic,* fracture.

If a broken bone is exposed to the outside by an opening in the skin, the injury is termed a *compound fracture.* Such a fracture is accompanied by the added danger of infection, since microorganisms almost surely enter through the broken skin. On the other hand, if the break is protected by uninjured skin, it is called a *simple fracture.* Figure 7.13 shows several types of traumatic fractures.

Repair of a Fracture

Whenever a bone is broken, blood vessels within the bone and its periosteum are ruptured, and the periosteum is likely to be torn. Blood escaping from the broken vessels spreads through the damaged area and soon forms a blood clot, or *hematoma.* As vessels in surrounding tissues dilate, those tissues become swollen and inflamed.

Within days or weeks, the hematoma is invaded by developing blood vessels and large numbers of osteoblasts originating from the periosteum. The osteoblasts multiply rapidly in the regions close to the new blood vessels, building spongy bone nearby. Granulation tissue develops, and in regions further from a blood supply, fibroblasts produce masses of fibrocartilage.

A *greenstick* fracture is incomplete, and the break occurs on the convex surface of the bend in the bone.

A *fissured* fracture involves an incomplete longitudinal break.

A *comminuted* fracture is complete and results in several bony fragments.

A *transverse* fracture is complete, and the break occurs at a right angle to the axis of the bone.

An *oblique* fracture occurs at an angle other than a right angle to the axis of the bone.

A *spiral* fracture is caused by twisting a bone excessively.

Figure 7.13 Various types of traumatic fractures.

Meanwhile, phagocytic cells begin to remove the blood clot as well as any dead or damaged cells in the affected area. Osteoclasts also appear and resorb bone fragments, thus aiding in "cleaning up" debris.

In time, a large amount of fibrocartilage fills the gap between the ends of the broken bone, and this mass is termed a cartilaginous *callus*. The callus is later replaced by bone tissue in much the same way that the hyaline cartilage of a developing endochondral bone is replaced. That is, the cartilaginous callus is broken down, the area is invaded by blood vessels and osteoblasts, and the space is filled with a bony callus.

Typically, more bone is produced at the site of a healing fracture than is needed to replace the damaged tissues. However, osteoclasts usually remove the excess, and the final result of the repair process is a bone shaped very much like the original one. Figure 7.14 shows the steps in the healing of a fracture.

The rate at which a fracture is repaired depends on several factors. For instance, if the ends of the broken bone are close together, healing is more rapid than if they are far apart. Thus, the setting of fractured bones and using casts or metal pins to keep the broken ends together help speed the healing process, as well as keeping the fractured parts aligned. Also, some bones naturally heal more rapidly than others. The long bones of the arms, for example, may heal in half the time required by the long bones of the legs. Furthermore, as age increases, so does the time required for healing.

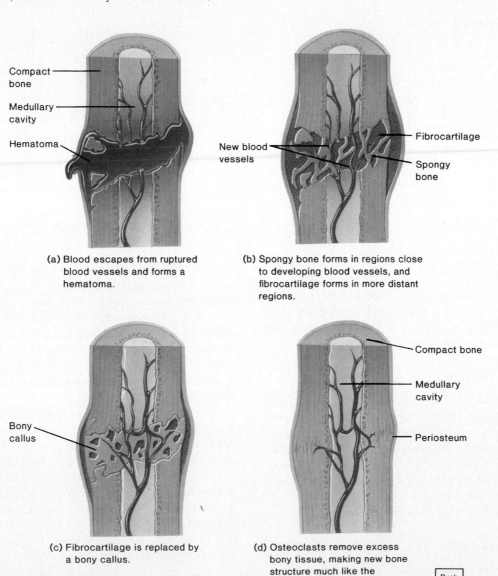

(a) Blood escapes from ruptured blood vessels and forms a hematoma.

(b) Spongy bone forms in regions close to developing blood vessels, and fibrocartilage forms in more distant regions.

(c) Fibrocartilage is replaced by a bony callus.

(d) Osteoclasts remove excess bony tissue, making new bone structure much like the original.

Figure 7.14   Major steps in the repair of a fracture.

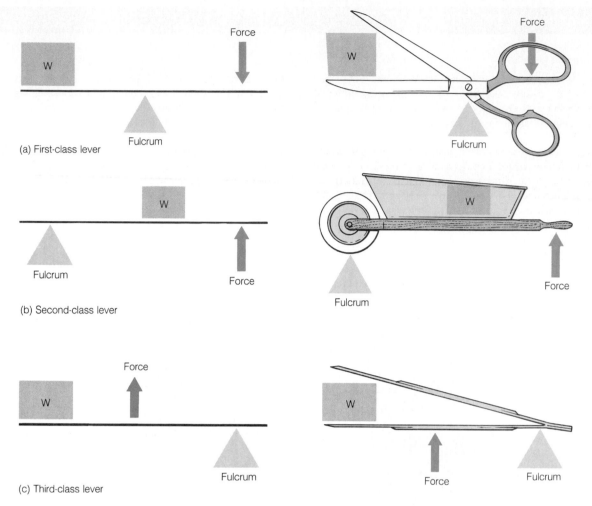

(a) First-class lever

(b) Second-class lever

(c) Third-class lever

Figure 7.15 Three types of levers: (*a*) A first-class lever is used in a pair of scissors; (*b*) a second-class lever is used in a wheelbarrow; and (*c*) a third-class lever is used in a pair of forceps.

Functions of Bones

Skeletal parts provide shape, support, and protection for body structures. They also act as levers that aid body movements, house tissues that produce blood cells, and store various inorganic salts.

Support and Protection

Bones give shape to structures such as the head, face, thorax, and limbs. They also provide support and protection. For example, the bones of the feet, legs, pelvis, and backbone support the weight of the body. The bones of the skull protect the eyes, ears, and brain. Those of the rib cage and shoulder girdle protect the heart and lungs, while bones of the pelvic girdle protect the lower abdominal and internal reproductive organs.

Body Movement

Whenever limbs or other body parts are moved, bones and muscles function together as simple mechanical devices called *levers*. Such a lever has four basic components: (a) a rigid bar or rod, (b) a pivot or fulcrum

on which the bar turns, (c) an object or weight (resistance) that is moved, and (d) a force that supplies energy for the movement of the bar.

A playground seesaw is a lever. The board of the seesaw serves as a rigid bar that rocks on a pivot near its center. The person on one end of the board represents the weight that is moved, while the person at the opposite end supplies the force needed for moving the board and its rider.

There are three kinds of levers, and they differ in the arrangements of their parts, as shown in figure 7.15. A first-class lever is one whose parts are arranged like those of the seesaw. Its pivot is located between the weight and the force, making the sequence of parts weight-pivot-force. Other examples of first-class levers are scissors and hemostats (devices used to clamp blood vessels closed).

The parts of a second-class lever are arranged in the sequence pivot-weight-force, as in a wheelbarrow.

The parts of a third-class lever are arranged in the sequence pivot-force-weight. This type of lever is employed when eyebrow tweezers or forceps are used to grasp an object.

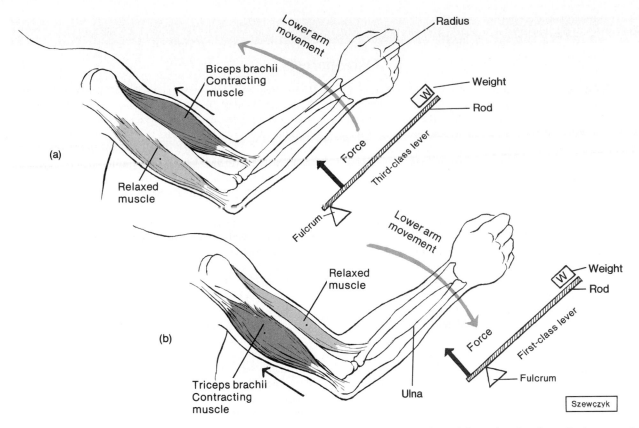

Figure 7.16 (a) When the arm is bent at the elbow, the bones and muscles involved work together as a third-class lever.

(b) When the arm is straightened at the elbow, the bones and muscles act as a first-class lever.

The actions of bending and straightening the arm at the elbow, for example, involve bones and muscles functioning together as levers, as illustrated in figure 7.16. When the arm is bent, the lower arm bones represent the rigid bar; the elbow joint is the pivot; the hand is the weight that is moved; and the force is supplied by muscles on the anterior side of the upper arm. One of these muscles, the *biceps brachii,* is attached by a tendon to a projection (radial tuberosity) on the *radius* bone in the lower arm, a short distance below the elbow. Since the parts of this lever are arranged in the sequence pivot-force-weight, it is an example of a third-class lever.

When the arm is straightened at the elbow, the lower arm bones again serve as the rigid bar, the hand as the weight, and the elbow joint as the pivot. However, this time the force is supplied by the *triceps brachii,* a muscle located on the posterior side of the upper arm. A tendon of this muscle is attached to a projection (olecranon process) of the *ulna* bone at the point of the elbow. Since the parts of the lever are arranged weight-pivot-force, it is an example of a first-class lever.

Although many lever arrangements occur throughout the skeletal-muscular systems, they are not always easy to identify. Nevertheless, these levers provide advantages in performing movements. The parts of some levers, such as those that function in moving the limbs, are arranged in ways that produce rapid motions, while others, such as those that move the head, aid in maintaining posture with minimal effort.

Blood Cell Formation

The process of blood cell formation is called **hematopoiesis.** Very early in life, it occurs in a structure called a *yolk sac,* which lies outside the body of a human embryo. (See chapter 23.) Later in development, blood cells are manufactured in the liver and spleen, and still later they are formed in the marrow of various bones.

Marrow is a soft, netlike mass of connective tissue found within the medullary cavities of long bones, in the irregular spaces of spongy bone, and in the larger osteonic canals of compact bone tissue.

There are two kinds of marrow—red marrow and yellow marrow. *Red marrow* functions in the formation of red blood cells (erythrocytes), certain white blood cells (leukocytes), and blood platelets (thrombocytes). It is red because of the red, oxygen-carrying pigment called **hemoglobin** that is contained within the red blood cells.

Red marrow occupies the cavities of most bones in an infant. With increasing age, however, more and more of it is replaced by *yellow marrow,* which functions as fat storage tissue and is inactive in blood cell production.

Osteoporosis

Osteoporosis is a disorder of the skeletal system in which there is excessive loss of bone volume and mineral content. This disorder is associated with the aging process. Within the affected bones, trabeculae tend to be lost. Consequently, the bones develop spaces and canals, and as these enlarge they fill with fibrous and fatty tissues. Such bones are easily fractured and may break spontaneously because they are no longer able to support body weight. For example, a person with osteoporosis may suffer a spontaneous fracture of the upper leg bone (femur) at the hip or the collapse of sections of the backbone (vertebrae). Similarly, the distal portion of a lower arm bone (radius) near the wrist may fracture as a result of a minor stress.

Osteoporosis is responsible for a large proportion of fractures occurring in persons over forty-five years of age. Although it may affect persons of either gender, it is most common in thin, light-complexioned females after menopause. (See chapter 22.)

Factors that increase the risk of osteoporosis include low intake of dietary calcium and lack of physical exercise (particularly during the early growing years), and in females, a decrease in the blood concentration of the sex hormone called *estrogen*. (This hormone is produced by the ovaries, which cease to secrete estrogen at menopause.) Heavy use of alcohol or tobacco also seems to increase the risk. In addition, some people may have a hereditary tendency for developing this condition.

Fortunately, osteoporosis may be prevented if steps are taken early enough. It is known, for example, that bone mass usually reaches a maximum at about age thirty-five.

Thereafter, bone loss may exceed bone formation in both males and females. To reduce such loss, people in their mid-twenties and older are advised to ensure that their dietary calcium intake reaches at least the U.S. government's recommended daily allowance of 800 milligrams. (Some nutritionists believe that 1,000–1,500 milligrams of calcium are needed daily to help control bone loss.) An 8-ounce glass of nonfat milk, for example, contains about 275 milligrams of calcium. It is also recommended that people engage in some type of regular physical exercise in which their bones support their body weight, such as walking or jogging. Additionally, postmenopausal women may require estrogen replacement therapy, which should be carried out under the supervision of a physician. As a rule, women have about 30% less bone than men; after menopause, women typically lose bone mass twice as fast as men do.

Confirming that a patient is developing osteoporosis is sometimes difficult. An X-ray film, for example, may not reveal a decrease in bone density until 20% to 30% of the bone tissue has been lost. Various noninvasive diagnostic techniques, however, are available for detecting rapid changes in bone mass. These include a scanner, called a *densimometer*, that measures the density of the wrist bones, and *quantitative computed tomography*, which can be used to visualize the density of other bones. (See chapter 1.)

In other cases, a bone sample may be removed, usually from a hipbone, in order to directly assess the condition of the bone tissue. Such a biopsy may also be used to judge the effectiveness of the treatment for bone disease.

In an adult, red marrow is found primarily in the spongy bone of the skull, ribs, sternum, clavicles, vertebrae, and pelvis. If the blood cell supply is deficient, some yellow marrow may change back into red marrow and become active in blood cell production. Blood cell formation is described in chapter 17.

Storage of Inorganic Salts

As mentioned previously, the intercellular matrix of bone tissue contains both collagen and inorganic mineral salts. Actually, the salts are responsible for about 70% of the matrix by weight, and are mostly in the form of tiny crystals of a type of calcium phosphate called *hydroxyapatite.*

Calcium is needed for a number of vital metabolic processes, including blood clot formation, nerve impulse conduction, and muscle cell contraction. When a low blood calcium ion concentration exists, osteoclasts are stimulated to break down bone tissue, and calcium salts are released from the intercellular matrix

into the blood. On the other hand, if the blood calcium ion concentration is excessively high, osteoclast activity is inhibited, and osteoblasts are stimulated to form bone tissue. As a result, the excessive calcium is stored in the matrix. The details of this homeostatic mechanism are presented in chapter 13.

In addition to storing calcium and phosphorus (as calcium phosphate), bone tissue contains lesser amounts of magnesium, sodium, potassium, and carbonate ions. Bones also tend to accumulate certain metallic elements such as lead, radium, and strontium, which are not normally present in the body but are sometimes ingested accidentally.

1. Name three major functions of bones.
2. Explain how parts of the arm form a first-class lever and a third-class lever.
3. Explain how the concentration of blood calcium ions is regulated.
4. List the substances normally stored in bone tissue.

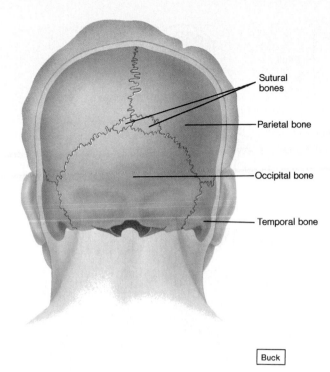

Sutural bones

Parietal bone

Occipital bone

Temporal bone

Buck

Figure 7.17 Wormian bones are extra bones that sometimes develop in sutures between the flat bones of the skull.

Organization of the Skeleton

Number of Bones

Although the number of bones in a human skeleton is often reported to be 206, the actual number varies from person to person. Some people lack certain bones, while others have extra ones. For example, the flat bones of the skull usually grow together and become tightly joined along irregular lines called **sutures.** Occasionally, extra bones called *sutural bones* (wormian bones) develop in these sutures (figure 7.17). Also, extra small, round sesamoid bones may develop in tendons, where they function to reduce friction in places where tendons pass over bony prominences. (See chart 7.3.)

Divisions of the Skeleton

For purposes of study, it is convenient to divide the skeleton into two major portions—an axial skeleton and an appendicular skeleton (figure 7.18).

The **axial skeleton** consists of the bony and cartilaginous parts that support and protect the organs of the head, neck, and trunk. These parts include the following:

1. **Skull.** The skull is composed of the *cranium* (brain case) and the *facial bones*.
2. **Hyoid bone.** The hyoid (hi′oid) bone is located in the neck between the lower jaw and the

CHART 7.3	Bones of the adult skeleton	
1. Axial Skeleton		
a. Skull		22 bones
8 cranial bones		
frontal 1		
parietal 2		
occipital 1		
temporal 2		
sphenoid 1		
ethmoid 1		
13 facial bones		
maxilla 2		
palatine 2		
zygomatic 2		
lacrimal 2		
nasal 2		
vomer 1		
inferior nasal concha 2		
1 mandible		
b. Middle ear bones		6 bones
malleus 2		
incus 2		
stapes 2		
c. Hyoid		1 bone
hyoid bone 1		
d. Vertebral column		26 bones
cervical vertebra 7		
thoracic vertebra 12		
lumbar vertebra 5		
sacrum 1		
coccyx 1		
e. Thoracic cage		25 bones
rib 24		
sternum 1		
2. Appendicular Skeleton		
a. Pectoral girdle		4 bones
scapula 2		
clavicle 2		
b. Upper limbs		60 bones
humerus 2		
radius 2		
ulna 2		
carpal 16		
metacarpal 10		
phalanx 28		
c. Pelvic girdle		2 bones
coxal bone 2		
d. Lower limbs		60 bones
femur 2		
tibia 2		
fibula 2		
patella 2		
tarsal 14		
metatarsal 10		
phalanx 28		
	Total	206 bones

larynx (figure 7.19). It does not articulate with any other bones, but is fixed in position by muscles and ligaments. The hyoid bone supports the tongue and serves as an attachment for certain muscles that help move the tongue and function in swallowing. It can

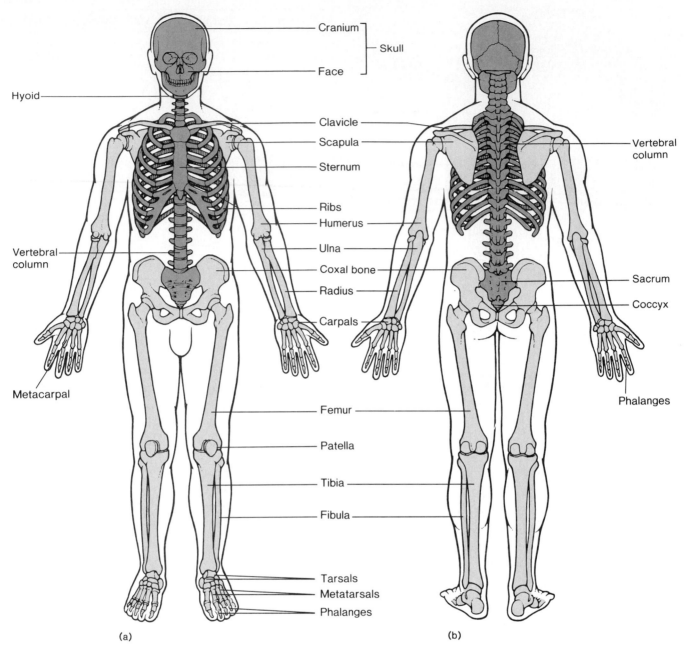

Cranium ⎤
          ⎦— **Skull**
Face ⎦

Hyoid

Clavicle

Scapula

Sternum

Ribs

Humerus

Ulna

Vertebral column

Coxal bone

Radius

Carpals

Metacarpal

Femur

Patella

Tibia

Fibula

Tarsals

Metatarsals

Phalanges

Vertebral column

Sacrum

Coccyx

Phalanges

(a)

(b)

Figure 7.18 Major bones of the skeleton. (*a*) Anterior view; (*b*) posterior view. (Note: The axial portions are shown in orange and the appendicular portions are shown in yellow.)

be felt approximately a finger's width above the anterior prominence of the larynx.

3. **Vertebral column.** The vertebral column, or backbone, consists of many vertebrae separated by cartilaginous *intervertebral disks*. This column forms the central axis of the skeleton. Near its distal end, several vertebrae are fused to form the **sacrum,** which is part of the pelvis.

A small, rudimentary tailbone called the **coccyx** is attached to the end of the sacrum.

4. **Thoracic cage.** The thoracic cage protects the organs of the thorax and the upper abdomen. It is composed of twelve pairs of **ribs,** which articulate posteriorly with thoracic vertebrae. It also includes the **sternum** (ster′num), or breastbone, to which most of the ribs are attached anteriorly.

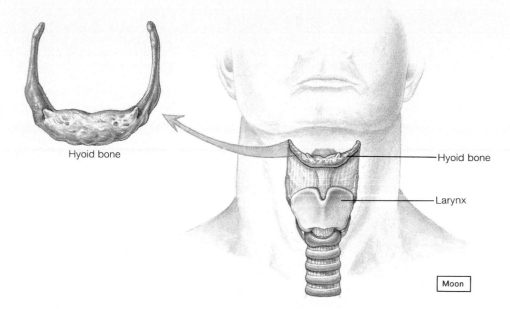

Hyoid bone

Hyoid bone

Larynx

Moon

Figure 7.19   The hyoid bone supports the tongue and serves as an attachment for muscles that move the tongue and function in swallowing.

The **appendicular skeleton** consists of the bones of the limbs and the bones that anchor the limbs to the axial skeleton. It includes the following:

1. **Pectoral girdle.** The pectoral girdle is formed by a **scapula** (scap'u-lah), or shoulder blade, and a **clavicle** (klav'i-k'l), or collarbone, on both sides of the body. The pectoral girdle connects the bones of the arms to the axial skeleton and aids in arm movements.
2. **Upper limbs** (arms). Each upper limb consists of a **humerus** (hu'mer-us), or upper arm bone, and two lower arm bones—a **radius** (ra'de-us) and an **ulna** (ul'nah). These three bones articulate with each other at the elbow joint. At the distal end of the radius and ulna, are eight **carpals** (kar'pals), or wrist bones. The bones of the palm are called **metacarpals,** and the finger bones are called **phalanges** (fah-lan'jēz).
3. **Pelvic girdle.** The pelvic girdle is formed by two **coxal** (kok'sal), or **innominate** (ĭ-nom'ĭ-nāt) bones (hipbones), which are attached to each other anteriorly and to the sacrum posteriorly. They connect the bones of the legs to the axial skeleton and, with the sacrum and coccyx, form the **pelvis,** which protects the lower abdominal and internal reproductive organs.
4. **Lower limbs** (legs). Each lower limb consists of a **femur** (fe'mur), or thighbone, and two lower

leg bones—a large **tibia** (tib'e-ah), or shinbone, and a slender **fibula** (fib'u-lah), or calf bone. These three bones articulate with each other at the knee joint, where the **patella** (pah-tel'ah), or kneecap, covers the anterior surface. At the distal ends of the tibia and fibula, are seven **tarsals** (tahr'sals), or anklebones. The bones of the foot are called **metatarsals,** and those of the toes (like the fingers) are called **phalanges.** Chart 7.4 defines some terms used to describe skeletal structures.

1. Distinguish between the axial and appendicular skeletons.
2. List the bones of the axial skeleton; of the appendicular skeleton.

Skull

A human skull usually consists of twenty-two bones that, except for the lower jaw, are firmly interlocked along lines called *sutures* (su'churz). Eight of these interlocked bones make up the cranium, and thirteen form the facial skeleton. The **mandible** (man'dĭ-b'l), or lower jawbone, is a movable bone held to the cranium by ligaments (figures 7.20, 7.21, 7.22, and 7.23).

A set of photographs of the human skull and its parts are presented in plates 8–36 on pages 228–43.

CHART 7.4 Terms used to describe skeletal structures

Term	Definition	Example
condyle (kon'dil)	A rounded process that usually articulates with another bone	Occipital condyle of the occipital bone (figure 7.23)
crest (krest)	A narrow, ridgelike projection	Iliac crest of the ilium (figure 7.55)
epicondyle (ep''ĭ-kon'dĭl)	A projection situated above a condyle	Medial epicondyle of the humerus (figure 7.50)
facet (fas'et)	A small, nearly flat surface	Facet of a thoracic vertebra (figure 7.40)
fissure (fish'-ur)	A cleft or groove	Inferior orbital fissure in the orbit of the eye (figure 7.21)
fontanel (fon''tah-nel')	A soft spot in the skull where membranes cover the space between bones	Anterior fontanel between the frontal and parietal bones (figure 7.35)
foramen (fo-ra'men)	An opening through a bone that usually serves as a passageway for blood vessels, nerves, or ligaments	Foramen magnum of the occipital bone (figure 7.23)
fossa (fos'ah)	A relatively deep pit or depression	Olecranon fossa of the humerus (figure 7.50)
fovea (fo've-ah)	A tiny pit or depression	Fovea capitis of the femur (figure 7.60)
head (hed)	An enlargement on the end of a bone	Head of the humerus (figure 7.50)
linea (lin'e-ah)	A narrow ridge	Linea aspera of the femur (figure 7.60)
meatus (me-a'tus)	A tubelike passageway within a bone	External auditory meatus of the ear (figure 7.22)
process (pros'es)	A prominent projection on a bone	Mastoid process of the temporal bone (figure 7.22)
ramus (ra'mus)	A structure given off from another larger one	Ramus of the mandible (figure 7.33)
sinus (si'nus)	A cavity within a bone	Frontal sinus of the frontal bone (figure 7.27)
spine (spīn)	A thornlike projection	Spine of the scapula (figure 7.47)
suture (soo'cher)	An interlocking line of union between bones	Lambdoidal suture between the occipital and parietal bones (figure 7.22)
trochanter (tro-kan'ter)	A relatively large process	Greater trochanter of the femur (figure 7.60)
tubercle (tu'ber-kl)	A small, knoblike process	Tubercle of a rib (figure 7.44)
tuberosity (tu''bĕ-ros' ĭ-te)	A knoblike process usually larger than a tubercle	Radial tuberosity of the radius (figure 7.51)

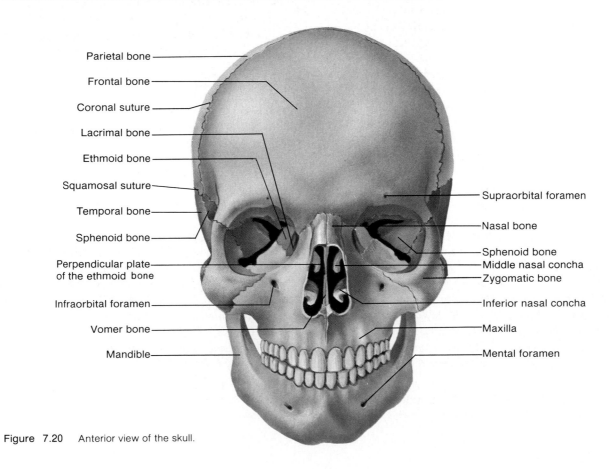

Figure 7.20 Anterior view of the skull.

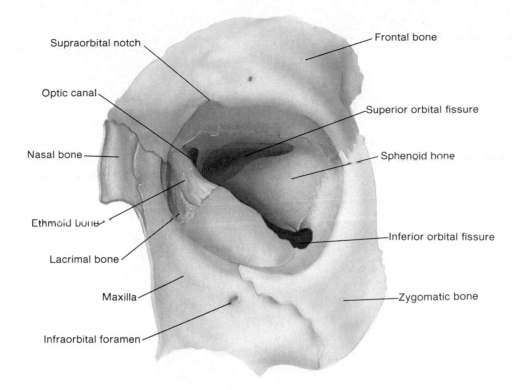

Supraorbital notch

Optic canal

Nasal bone

Ethmoid bone

Lacrimal bone

Maxilla

Infraorbital foramen

Frontal bone

Superior orbital fissure

Sphenoid bone

Inferior orbital fissure

Zygomatic bone

Figure 7.21   The orbit of the eye includes both cranial and facial bones.

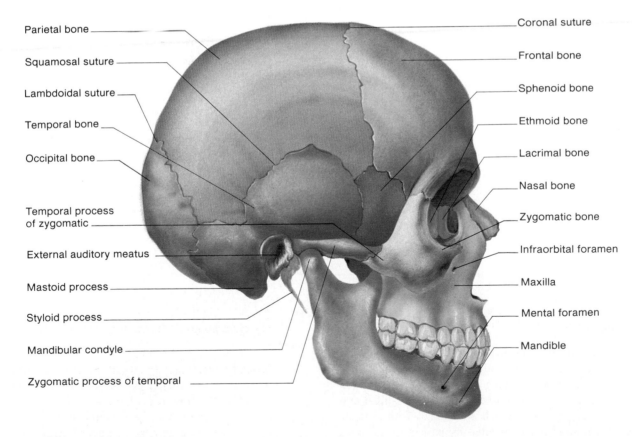

Parietal bone

Squamosal suture

Lambdoidal suture

Temporal bone

Occipital bone

Temporal process of zygomatic

External auditory meatus

Mastoid process

Styloid process

Mandibular condyle

Zygomatic process of temporal

Coronal suture

Frontal bone

Sphenoid bone

Ethmoid bone

Lacrimal bone

Nasal bone

Zygomatic bone

Infraorbital foramen

Maxilla

Mental foramen

Mandible

Figure 7.22   Lateral view of the skull.

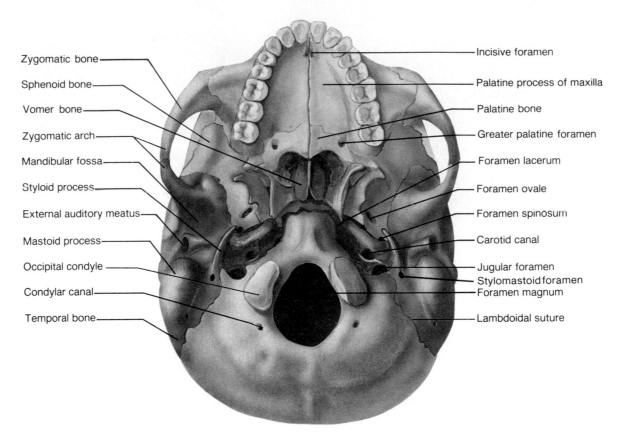

Zygomatic bone — Incisive foramen
Sphenoid bone — Palatine process of maxilla
Vomer bone — Palatine bone
Zygomatic arch — Greater palatine foramen
Mandibular fossa — Foramen lacerum
Styloid process — Foramen ovale
External auditory meatus — Foramen spinosum
Mastoid process — Carotid canal
Occipital condyle — Jugular foramen
Condylar canal — Stylomastoid foramen
— Foramen magnum
Temporal bone — Lambdoidal suture

Figure 7.23 Inferior view of the skull.

CHART 7.5	Cranial bones	
Name and number	Description	Special features
Frontal (1)	Forms forehead, roof of nasal cavity, and roofs of orbits	Supraorbital foramen, frontal sinuses
Parietal (2)	Form side walls and roof of cranium	Fused at midline along sagittal suture
Occipital (1)	Forms back of skull and base of cranium	Foramen magnum, occipital condyles
Temporal (2)	Form side walls and floor of cranium	External auditory meatus, mandibular fossa, mastoid process, styloid process, zygomatic process
Sphenoid (1)	Forms parts of base of cranium, sides of skull, and floors and sides of orbits	Sella turcica, sphenoidal sinuses
Ethmoid (1)	Forms parts of roof and walls of nasal cavity, floor of cranium, and walls of orbits	Cribriform plates, perpendicular plate, superior and middle nasal conchae, ethmoidal sinuses, crista galli

Cranium

The **cranium** (kra'ne-um) encloses and protects the brain, and its surface provides attachments for various muscles that make chewing and head movements possible. Some of the cranial bones contain air-filled cavities called *sinuses,* which are lined with mucous membranes and are connected by passageways to the nasal cavity. Sinuses reduce the weight of the skull and increase the intensity of the voice by serving as resonant sound chambers.

The eight bones of the cranium (chart 7.5) are as follows:

1. **Frontal bone.** The frontal (frun'tal) bone forms the anterior portion of the skull above the eyes, including the forehead, the roof of the nasal cavity, and the roofs of the orbits (bony sockets) of the eyes. On the upper margin of each orbit, the frontal bone is marked by a *supraorbital foramen* (or *supraorbital notch* in some skulls), through which blood vessels and

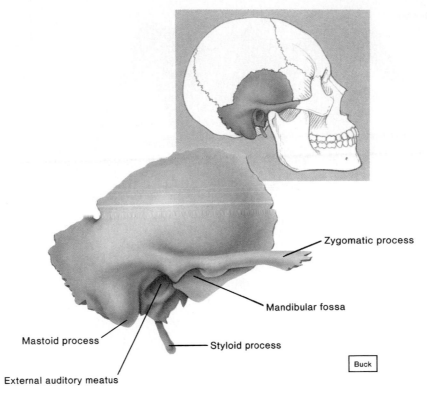

Figure 7.24 Lateral surface of the right temporal bone. What sensory structures are located within this bone?

Zygomatic process

Mandibular fossa

Styloid process

Mastoid process

External auditory meatus

Buck

nerves pass to the tissues of the forehead. Within the frontal bone are two *frontal sinuses,* one above each eye near the midline. Although the frontal bone is a single bone in adults, it develops in two parts. (See figure 7.35.) These halves grow together and are usually completely fused by the fifth or sixth year of age.

2. **Parietal bones.** One parietal (pah-ri′ĕ-tal) bone is located on each side of the skull just behind the frontal bone. Each is shaped like a curved plate and has four borders. Together, the parietal bones form the bulging sides and roof of the cranium. They are fused at the midline along the *sagittal suture,* and they meet the frontal bone along the *coronal suture.*

3. **Occipital bone.** The occipital (ok-sip′ĭ-tal) bone joins the parietal bones along the *lambdoidal* (lam′doid-al) *suture.* It forms the back of the skull and the base of the cranium. There is a large opening on its lower surface called the *foramen magnum,* through which nerve fibers from the brain pass and enter the vertebral canal to become part of the spinal cord. Rounded processes called *occipital condyles,* located on each side of the foramen magnum, articulate with the first vertebra (atlas) of the vertebral column.

4. **Temporal bones.** A temporal (tem′por-al) bone (figure 7.24) on each side of the skull joins the parietal bone along a *squamosal* (skwa-mo′sal) *suture.* The temporal bones form parts of the sides and the base of the cranium. Located near the inferior margin is an opening, the *external auditory meatus,* which leads inward to parts of the ear. The temporal bones also house the internal ear structures and have depressions called the *mandibular fossae* (glenoid fossae) that articulate with processes of the mandible. Below each external auditory meatus are two projections—a rounded *mastoid process* and a long, pointed *styloid process.* The mastoid process provides an attachment for certain muscles of the neck, while the styloid process serves as an anchorage for muscles associated with the tongue and pharynx. An opening near the mastoid process, the *carotid canal,* transmits a branch of the carotid artery, while an opening between the temporal and occipital bones, the *jugular foramen,* accommodates the jugular vein (figure 7.23).

A *zygomatic process* projects anteriorly from the temporal bone in the region of the external auditory meatus. It joins the *zygomatic bone* and helps form the prominence of the cheek.

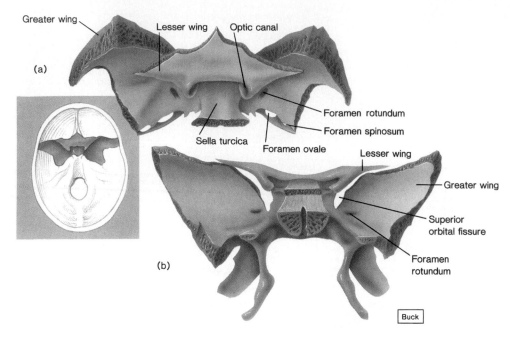

Greater wing
Lesser wing
Optic canal
(a)
Foramen rotundum
Foramen spinosum
Sella turcica
Foramen ovale
Lesser wing
Greater wing
Superior orbital fissure
(b)
Foramen rotundum
Buck

Figure 7.25 (*a*) The sphenoid bone viewed from above; (*b*) posterior view.

The mastoid process is of clinical interest because it may become infected. The tissues in this region of the temporal bone contain a number of interconnected air cells lined with mucous membranes that communicate with the middle ear. These spaces sometimes become inflamed when microorganisms spread into them from an infected middle ear (*otitis media*). The resulting mastoid infection, called *mastoiditis,* is of particular concern because the membranes that surround the brain are close by and may also become infected.

5. **Sphenoid bone.** The sphenoid (sfe′noid) bone (figure 7.25) is wedged between several other bones in the anterior portion of the cranium. It consists of a central part and two winglike structures that extend laterally toward each side of the skull. This bone helps form the base of the cranium, the sides of the skull, and the floors and sides of the orbits. Along the midline within the cranial cavity, a portion of the sphenoid bone rises up and forms a saddle-shaped mass called *sella turcica* (sel′ah tur′si-ka; Turk's saddle). The depression of this saddle is occupied by the pituitary gland, which hangs from the base of the brain by a stalk.

The sphenoid bone also contains two *sphenoidal sinuses,* which lie side by side and are separated by a bony septum that projects downward into the nasal cavity.

6. **Ethmoid bone.** The ethmoid (eth′moid) bone (figure 7.26) is located in front of the sphenoid bone. It consists of two masses, one on each side

of the nasal cavity, which are joined horizontally by thin *cribriform* (krib′ri-form) *plates.* These plates form part of the roof of the nasal cavity, and nerves associated with the sense of smell pass through tiny openings (olfactory foramina) in them. Portions of the ethmoid bone also form sections of the cranial floor, orbital walls, and nasal cavity walls. A *perpendicular plate* projects downward in the midline from the cribriform plates to form most of the nasal septum.

Delicate, scroll-shaped plates, called the *superior nasal concha* (kong′kah) and the *middle nasal concha,* project inward from the lateral portions of the ethmoid bone toward the perpendicular plate. These bones, which are also called *turbinate bones,* support mucous membranes that line the nasal cavity. The mucous membranes, in turn, begin the processes of moistening, warming, and filtering air as it enters the respiratory tract. The lateral portions of the ethmoid bone contain many small air spaces, the *ethmoidal sinuses.* Various structures in the nasal cavity are shown in figure 7.27.

Projecting upward into the cranial cavity between the cribriform plates is a triangular process of the ethmoid bone called the *crista galli* (kris′tă gal′li; cock's comb). This process serves as an attachment for membranes that enclose the brain. A view of the cranial cavity is shown in figure 7.28.

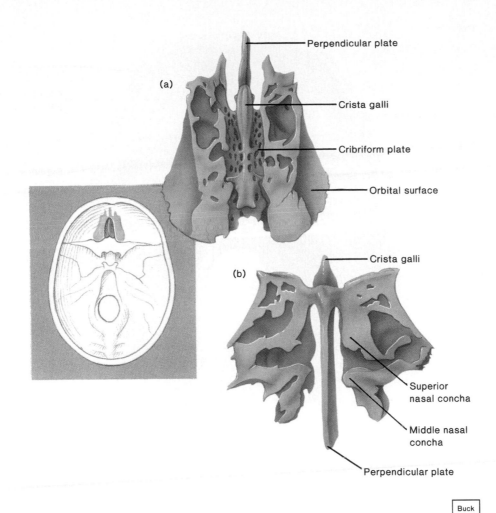

(a)

Perpendicular plate

Crista galli

Cribriform plate

Orbital surface

(b)

Crista galli

Superior nasal concha

Middle nasal concha

Perpendicular plate

Buck

Figure 7.26 (*a*) The ethmoid bone viewed from above and (*b*) from behind.

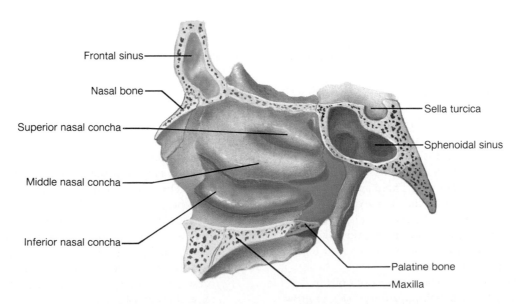

Frontal sinus

Nasal bone

Superior nasal concha

Middle nasal concha

Inferior nasal concha

Sella turcica

Sphenoidal sinus

Palatine bone

Maxilla

Figure 7.27 Lateral wall of the nasal cavity.

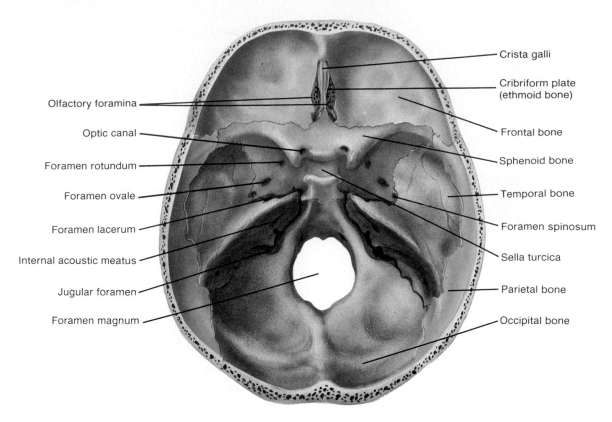

Olfactory foramina

Optic canal

Foramen rotundum

Foramen ovale

Foramen lacerum

Internal acoustic meatus

Jugular foramen

Foramen magnum

Crista galli

Cribriform plate (ethmoid bone)

Frontal bone

Sphenoid bone

Temporal bone

Foramen spinosum

Sella turcica

Parietal bone

Occipital bone

Figure 7.28   Floor of the cranial cavity viewed from above.

Facial Skeleton

The **facial skeleton** consists of thirteen immovable bones and a movable lower jawbone. In addition to forming the basic shape of the face, these bones provide attachments for various muscles that move the jaw and control facial expressions.

The bones of the facial skeleton are as follows:

1. **Maxillary bones.** The maxillary (mak'sĭ-ler″e) bones (pl. *maxillae,* mak-sĭl'e) form the upper jaw; together they form the keystone of the face, since all the other immovable facial bones articulate with them.

    Portions of these bones comprise the anterior roof of the mouth (*hard palate*), the floors of the orbits, and the sides and floor of the nasal cavity. They also contain the sockets of the upper teeth. Inside the maxillae, lateral to the nasal cavity, are *maxillary sinuses* (antrums of Highmore). These spaces are the largest of the sinuses, and they extend from the floor of the orbits to the roots of the upper teeth. Figure 7.29 shows the locations of the maxillary and other sinuses. See chart 7.6 for a summary of the sinuses.

    During development, portions of the maxillary bones called *palatine processes* grow

CHART 7.6	Sinuses of the cranial and facial bones	
Sinuses	Number	Location
Frontal sinuses	2	Frontal bone above each eye and near the midline
Sphenoidal sinuses	2	Sphenoid bone above the posterior portion of the nasal cavity
Ethmoidal sinuses	2 groups of small air cells	Ethmoid bone on either side of the upper portion of the nasal cavity
Maxillary sinuses	2	Maxillary bones lateral to the nasal cavity and extending from the floor of the orbits to the roots of the upper teeth

together and fuse along the midline to form the anterior section of the hard palate (figure 7.23).

The inferior border of each maxillary bone projects downward, forming an *alveolar* (al-ve'o-lar) *process.* Together these processes form a horseshoe-shaped *alveolar arch* (dental arch). Cavities in this arch (dental alveoli) are occupied by the teeth, which are attached to these bony sockets by fibrous connective tissue. (See chapter 14.)

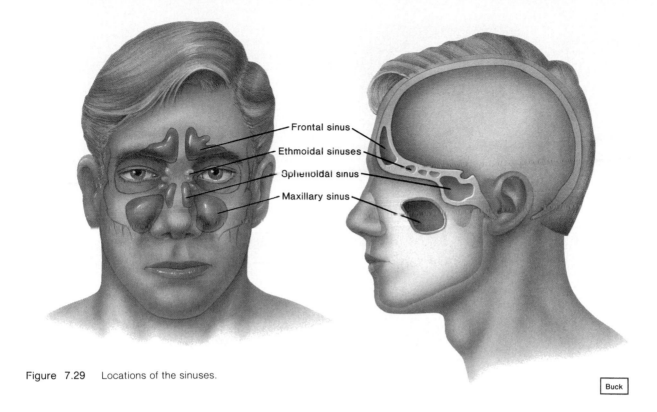

Figure 7.29 Locations of the sinuses.

Buck

2. **Palatine bones.** The palatine (pal'ah-tīn) bones (figure 7.30) are located behind the maxillae. Each bone is roughly L-shaped. The horizontal portions serve as both the posterior section of the hard palate and the floor of the nasal cavity. The perpendicular portions help form the lateral walls of the nasal cavity.

3. **Zygomatic bones.** The zygomatic (zi"go-mat'ik) bones (malar bones) are responsible for the prominences of the cheeks below and to the sides of the eyes. These bones also help form the lateral walls and the floors of the orbits. Each bone has a *temporal process,* which extends posteriorly to join the zygomatic process of a temporal bone. Together these processes form a *zygomatic arch* (figures 7.22 and 7.23).

4. **Lacrimal bones.** A lacrimal (lak'ri-mal) bone is a thin, scalelike structure located in the medial wall of each orbit between the ethmoid bone and the maxilla (figure 7.22). A groove in its anterior portion leads from the orbit to the nasal cavity, providing a pathway for a tube that carries tears from the eye to the nasal cavity.

5. **Nasal bones.** The nasal (na'zal) bones are long, thin, and nearly rectangular (figure 7.20). They lie side by side and are fused at the midline, where they form the bridge of the nose. These bones serve as attachments for the cartilaginous tissues that are largely responsible for the shape of the nose.

6. **Vomer bone.** The thin, flat vomer (vo'mer) bone is located along the midline within the nasal cavity. Posteriorly, it joins the perpendicular plate of the ethmoid bone, and together they form the nasal septum (figures 7.31 and 7.32).

7. **Inferior nasal conchae.** The inferior nasal conchae (kong'ke) are fragile, scroll-shaped bones attached to the lateral walls of the nasal cavity. They are the largest of the conchae and are positioned below the superior and middle nasal conchae of the ethmoid bone (figures 7.20 and 7.27). Like the ethmoidal conchae, the inferior conchae provide support for mucous membranes within the nasal cavity.

8. **Mandible.** The mandible (man'di-b'l), or lower jawbone, consists of a horizontal, horseshoe-shaped body with a flat *ramus* projecting upward at each end. The rami are divided into

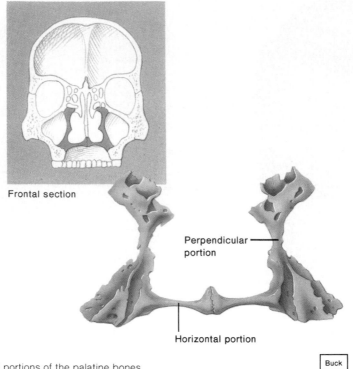

Frontal section

Perpendicular portion

Horizontal portion

Buck

Figure 7.30 The horizontal portions of the palatine bones form the posterior section of the hard palate, and the perpendicular portions help form the lateral walls of the nasal cavity.

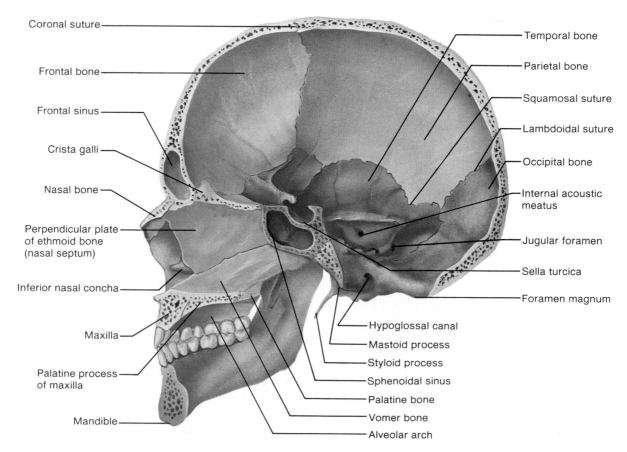

Coronal suture

Frontal bone

Frontal sinus

Crista galli

Nasal bone

Perpendicular plate of ethmoid bone (nasal septum)

Inferior nasal concha

Maxilla

Palatine process of maxilla

Mandible

Temporal bone

Parietal bone

Squamosal suture

Lambdoidal suture

Occipital bone

Internal acoustic meatus

Jugular foramen

Sella turcica

Foramen magnum

Hypoglossal canal

Mastoid process

Styloid process

Sphenoidal sinus

Palatine bone

Vomer bone

Alveolar arch

Figure 7.31 Sagittal section of the skull.

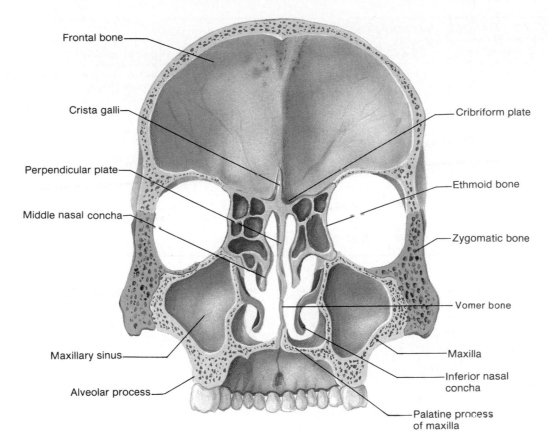

Figure 7.32 Frontal section of the skull (posterior view).

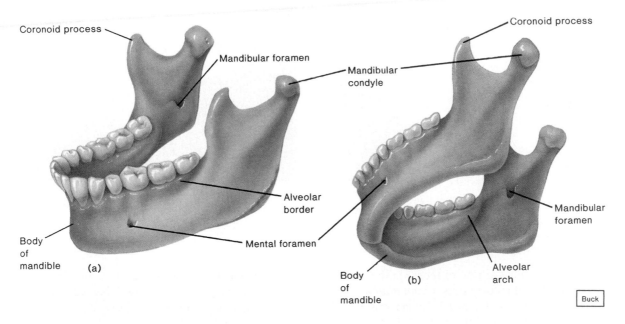

Figure 7.33 (a) Lateral view of the mandible; (b) inferior view.

two processes—a posterior *mandibular condyle* and an anterior *coronoid process* (figure 7.33). The mandibular condyles articulate with the mandibular fossae of the temporal bones, while the coronoid processes serve as attachments for muscles used in chewing. Other large chewing muscles are inserted on the lateral surfaces of the rami.

A curved bar of bone on the superior border of the mandible, the *alveolar border,* contains the hollow sockets (dental alveoli) that bear the lower teeth.

CHART 7.7 Bones of the facial skeleton

Name and number	Description	Special features
Maxillary (2)	Form upper jaw, anterior roof of mouth, floors of orbits, and sides and floor of nasal cavity	Alveolar processes, maxillary sinuses, palatine process
Palatine (2)	Form posterior roof of mouth, and floor and lateral walls of nasal cavity	
Zygomatic (2)	Form prominences of cheeks, and lateral walls and floors of orbits	Temporal process
Lacrimal (2)	Form part of medial walls of orbits	Groove that leads from orbit to nasal cavity
Nasal (2)	Form bridge of nose	
Vomer (1)	Forms inferior portion of nasal septum	
Inferior nasal conchae (2)	Extend into nasal cavity from its lateral walls	
Mandible (1)	Forms lower jaw	Body, ramus, mandibular condyle, coronoid process, alveolar process, mandibular foramen, mental foramen

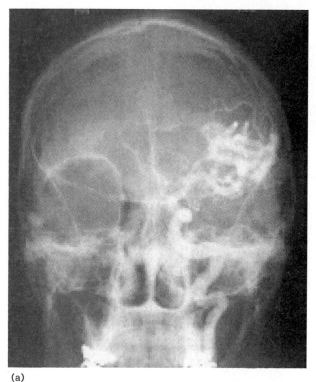
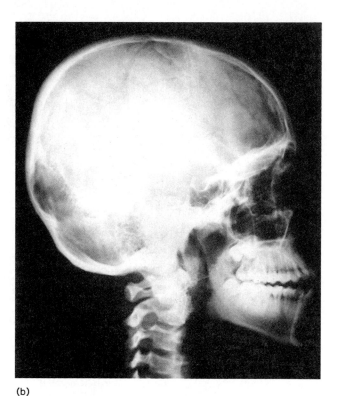

(a)

(b)

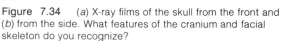

Figure 7.34 (a) X-ray films of the skull from the front and (b) from the side. What features of the cranium and facial skeleton do you recognize?

On the medial side of the mandible, near the center of each ramus, is a *mandibular foramen*. This opening admits blood vessels and a nerve, which supply the roots of the lower teeth. Dentists commonly inject anesthetic into the tissues near this foramen to temporarily block nerve impulse conduction and cause the teeth on that side of the jaw to become insensitive. Branches of the blood vessels and the nerve emerge from the mandible through the *mental foramen*, which opens on the outside near the point of the jaw. They supply the tissues of the chin and lower lip.

Chart 7.7 contains a descriptive summary of the fourteen facial bones. Various features of these bones can be seen in the X-ray films in figure 7.34.

Chart 7.8 lists the major openings (foramina) and passageways through bones of the skull, as well as their general locations and the structures they transmit.

CHART 7.8 Passageways through bones of the skull

Passageway	Location	Major structures transmitted
Carotid canal (figure 7.23)	Inferior surface of the temporal bone	Internal carotid artery, veins, and nerves
Condylar canal (figure 7.23)	Base of skull in occipital bone	Veins between the skull and the neck
Foramen lacerum (figure 7.23)	Floor of cranial cavity between temporal and sphenoid bones	Branch of pharyngeal artery (in life opening is largely covered by fibrocartilage)
Foramen magnum (figure 7.23)	Base of skull in occipital bone	Nerve fibers passing between the brain and spinal cord, and certain arteries
Foramen ovale (figure 7.23)	Floor of cranial cavity in sphenoid bone	Mandibular division of trigeminal nerve and veins
Foramen rotundum (figure 7.28)	Floor of cranial cavity in sphenoid bone	Maxillary division of trigeminal nerve
Foramen spinosum (figure 7.28)	Floor of cranial cavity in sphenoid bone	Middle meningeal blood vessels and branch of mandibular nerve
Greater palatine foramen (figure 7.23)	Posterior portion of hard palate in palatine bone	Palatine blood vessels and nerves
Hypoglossal canal (figure 7.31)	Near margin of foramen magnum in occipital bone	Hypoglossal nerve
Incisive foramen (figure 7.23)	Anterior portion of hard palate	Nasopalatine nerves
Inferior orbital fissure (figure 7.21)	Floor of the orbit	Maxillary nerve and blood vessels
Infraorbital foramen (figure 7.21)	Below the orbit in maxillary bone	Infraorbital blood vessels and nerves
Internal acoustic meatus (figure 7.28)	Floor of cranial cavity in temporal bone	Branches of facial, vestibular, and cochlear nerves and blood vessels
Jugular foramen (figure 7.23)	Base of the skull between temporal and occipital bones	Glossopharyngeal, vagus and accessory nerves, and blood vessels
Mandibular foramen (figure 7.33)	Inner surface of ramus of mandible	Inferior alveolar blood vessels and nerves
Mental foramen (figure 7.33)	Near point of jaw in mandible	Mental nerve and blood vessels
Optic canal (figure 7.21)	Posterior portion of orbit in sphenoid bone	Optic nerve and ophthalmic artery
Stylomastoid foramen (figure 7.23)	Between styloid and mastoid processes	Facial nerve and blood vessels
Superior orbital fissure (figure 7.21)	Lateral wall of orbit	Oculomotor, trochlear, abducent, and ophthalmic division of trigeminal nerves
Supraorbital foramen (figure 7.20)	Upper margin of orbit in frontal bone	Supraorbital blood vessels and nerves

Infantile Skull

At birth, the skull is incompletely developed, and the cranial bones are separated by fibrous membranes. These membranous areas are called **fontanels** (fon"tah-nel'z) or, more commonly, soft spots. They permit some movement between the bones, so that the developing skull is partially compressible and can change shape slightly. This action, called *molding,* enables an infant's skull to pass more easily through the birth canal. Eventually, the fontanels close as the cranial bones grow together. The posterior fontanel usually closes about two months after birth; the sphenoid fontanel closes at about three months; the mastoid fontanel closes near the end of the first year; and the anterior fontanel may not close until the middle or end of the second year.

Other characteristics of an infantile skull (figure 7.35) include a relatively small face with a prominent forehead and large orbits. The jaw and nasal cavity are small, the sinuses are incompletely formed, and the frontal bone is in two parts. The skull bones are thin, but they are also somewhat flexible and thus are less easily fractured than adult bones.

In the infantile skull, the two parts of the developing frontal bone are separated in the midline by a *frontal suture* (metopic suture). This suture usually closes before the sixth year; however, in a small proportion of adult skulls, the frontal suture remains visible.

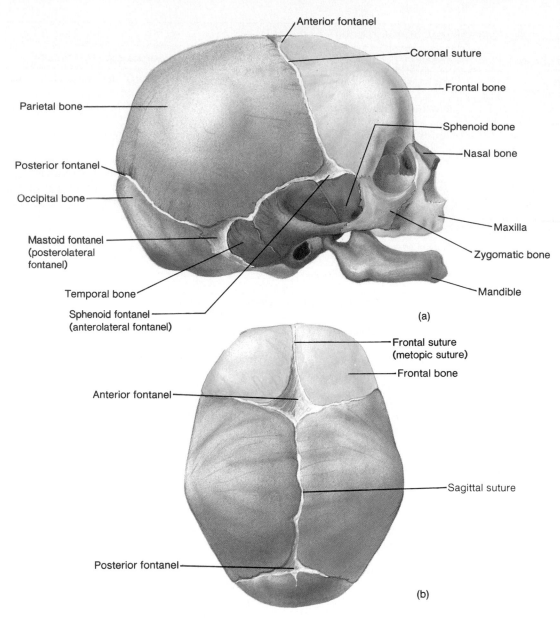

Figure 7.35 (a) Lateral view and (b) superior view of the infantile skull.

1. Locate and name each of the bones of the cranium.
2. Locate and name each of the facial bones.
3. Explain how an adult skull differs from that of an infant.

Vertebral Column

The **vertebral column** extends from the skull to the pelvis and forms the vertical axis of the skeleton (figure 7.36). It is composed of many bony parts called **vertebrae** (ver′tĕ-bre). These are separated by masses of fibrocartilage called *intervertebral disks* and are connected to one another by ligaments. The vertebral column supports the head and the trunk of the body, yet is flexible enough to permit movements, such as bending forward, backward, or to the side, and turning or rotating on the central axis. It also protects the spinal cord, which passes through a *vertebral canal* formed by openings in the vertebrae.

In an infant, there are thirty-three separate bones in the vertebral column. Five of these bones eventually fuse to form the sacrum, and four others join to become the coccyx. As a result, an adult vertebral column has twenty-six bones.

Normally, the vertebral column has four curvatures, which give it a degree of resiliency. The names of the curves correspond to the regions in which they occur, as shown in figure 7.36. The *thoracic* and *pelvic curvatures* are concave anteriorly and are called pri-

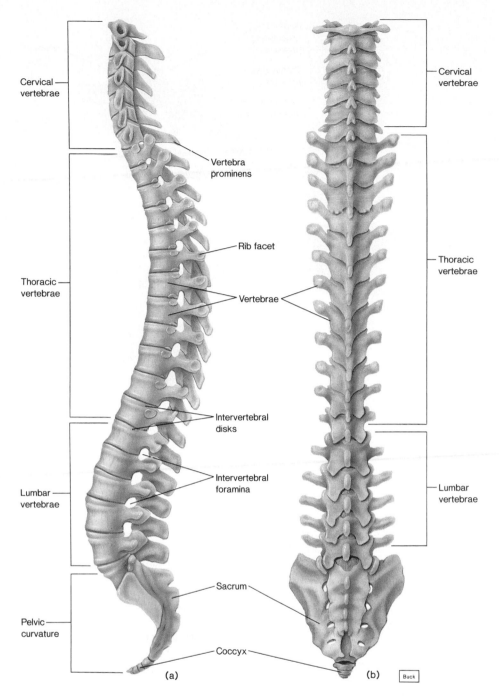

Cervical vertebrae

Vertebra prominens

Rib facet

Thoracic vertebrae

Vertebrae

Intervertebral disks

Lumbar vertebrae

Intervertebral foramina

Sacrum

Pelvic curvature

Coccyx

(a)

Cervical vertebrae

Thoracic vertebrae

Lumbar vertebrae

(b)

Buck

Figure 7.36 The curved vertebral column consists of many vertebrae separated by intervertebral disks. (*a*) Left lateral view; (*b*) posterior view.

mary curves. The *cervical curvature* in the neck and the *lumbar curvature* in the lower back are convex anteriorly and are called secondary curves.

A Typical Vertebra

Although the vertebrae in different regions of the vertebral column have special characteristics, they also have features in common. Thus, a typical vertebra (figure 7.37) has a drum-shaped *body* (centrum), which forms the thick, anterior portion of the bone. A longitudinal row of these vertebral bodies supports the

weight of the head and trunk. The intervertebral disks, which separate adjacent vertebrae, are fastened to the roughened upper and lower surfaces of the bodies. These disks cushion and soften the forces caused by such movements as walking and jumping, which might otherwise fracture vertebrae or jar the brain.

The bodies of adjacent vertebrae are joined on their anterior surfaces by *anterior ligaments* and on their posterior surfaces by *posterior ligaments*.

Projecting posteriorly from each vertebral body are two short stalks called *pedicles* (ped'ĭ-k'lz). They form the sides of the *vertebral foramen*. Two plates

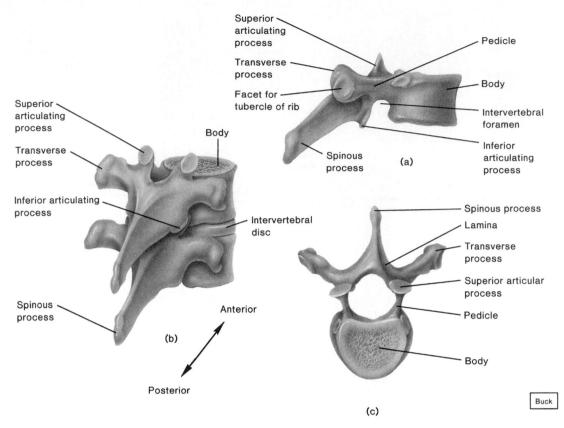

Figure 7.37 (*a*) Lateral view of a typical vertebra; (*b*) adjacent vertebrae are joined at their articulating processes; (*c*) superior view of a typical thoracic vertebra.

called *laminae* (lam'ı̆-ne) arise from the pedicles and fuse in the back to become a *spinous process.* The pedicles, laminae, and spinous process together complete a bony *vertebral arch* around the vertebral foramen, through which the spinal cord passes.

Between the pedicles and laminae of a typical vertebra is a *transverse process,* which projects laterally and toward the back. Various ligaments and muscles are attached to the dorsal spinous process and the transverse processes. Projecting upward and downward from each vertebral arch are *superior* and *inferior articulating processes.* These processes bear cartilage-covered facets by which each vertebra is joined to the one above and the one below it.

Athletes, such as gymnasts, jumpers, and pole vaulters, who repeatedly land on hard surfaces, sometimes develop breaks or cracks in their vertebrae. These fractures commonly involve the articulating processes of these bones. Such a fracturing of a vertebra is called *spondylolysis.* It may be prevented by limiting the number of landings experienced during individual practice periods and by padding the landing areas with floor mats.

On the lower surfaces of the vertebral pedicles are notches that align to help form openings called *intervertebral foramina* (in''ter-ver'tĕ-bral fo-ram'ı̆-nah). These openings provide passageways for spinal nerves that proceed between adjacent vertebrae and connect to the spinal cord.

Cervical Vertebrae

Seven **cervical vertebrae** comprise the bony axis of the neck. Although these are the smallest of the vertebrae, their bone tissues are denser than those in any other region of the vertebral column.

The transverse processes of the cervical vertebrae are distinctive because they have *transverse foramina,* which serve as passageways for arteries leading to the brain. Also, the spinous processes of the second through the fifth cervical vertebrae are uniquely forked (bifid). These processes provide attachments for various muscles.

The spinous process of the seventh vertebra is longer and protrudes beyond the other cervical spines. It is called the *vertebra prominens,* and because it can be felt through the skin, it is a useful landmark for locating other vertebral parts (figure 7.36).

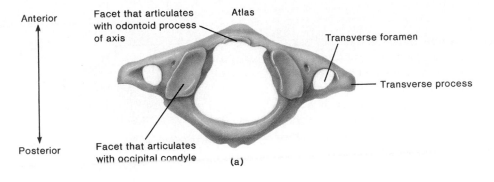

Anterior

Posterior

Facet that articulates with odontoid process of axis

Atlas

Transverse foramen

Transverse process

Facet that articulates with occipital condyle

(a)

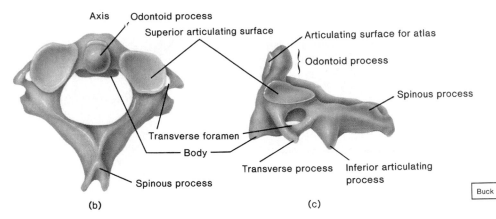

Axis   Odontoid process

Superior articulating surface

Articulating surface for atlas

Odontoid process

Spinous process

Transverse foramen

Body

Spinous process

Transverse process   Inferior articulating process

Buck

(b)

(c)

Figure 7.38   How do the structures of the (*a*) atlas and (*b*) axis function together to allow movement of the head? (*c*) Lateral view of the axis.

Two of the cervical vertebrae, shown in figure 7.38, are of special interest. The first vertebra, or **atlas** (at′las), supports and balances the head. It has practically no body or spine and appears as a bony ring with two transverse processes. On its upper surface, the atlas has two kidney-shaped *facets,* which articulate with the occipital condyles of the skull.

The second cervical vertebra, or **axis** (ak′sis), bears a toothlike *odontoid process* (dens) on its body. This process projects upward and lies in the ring of the atlas. As the head is turned from side to side, the atlas pivots around the odontoid process. (See figures 7.38 and 7.39.)

Thoracic Vertebrae

The twelve **thoracic vertebrae** are larger than those in the cervical region. Each vertebra has a long, pointed spinous process, which slopes downward, and facets on the sides of its body, which articulate with a rib.

Beginning with the third thoracic vertebra and moving downward, the bodies of these bones increase in size. Thus, they are adapted to the stress placed on them by the increasing amounts of body weight they bear.

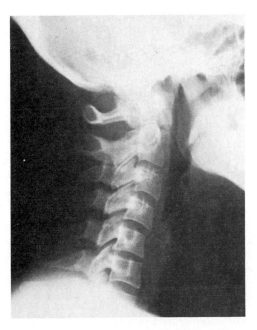

Figure 7.39   What features can you identify in this X-ray film of the neck?

Lumbar Vertebrae

There are five **lumbar vertebrae** in the small of the back (loins). The lumbars are adapted to support more weight than the vertebrae above them, and have larger and stronger bodies. The transverse processes of these vertebrae project backward at relatively sharp angles, while their short, thick spinous processes are directed nearly horizontally.

Figure 7.40 compares the structures of the cervical, thoracic, and lumbar vertebrae.

Gymnasts, football players, and others whose vertebral columns are bent excessively and forcefully may experience the slipping of one vertebra over the one below it. This painful condition is called *spondylolisthesis*. It usually involves the fifth lumbar vertebra sliding forward over the body of the sacrum. The condition may be prevented by doing exercises designed to strengthen the back muscles associated with the vertebral column.

Sacrum

The **sacrum** (sa′krum) is a triangular structure at the base of the vertebral column. It is composed of five vertebrae that develop as separate structures, but gradually become fused together between the eighteenth and thirtieth years. The spinous processes of these fused bones are represented by a ridge of tubercles that form the *median sacral crest*. To the sides of the tubercles are rows of openings, the *dorsal sacral foramina*, through which nerves and blood vessels pass (figure 7.41).

The sacrum is wedged between the coxal bones of the pelvis and is united to them at its *auricular surfaces* by fibrocartilage of the *sacroiliac joints*. The weight of the body is transmitted to the legs through the pelvic girdle at these joints (figure 7.18).

The sacrum forms the posterior wall of the pelvic cavity. The upper anterior margin of the sacrum, which represents the body of the first sacral vertebra, is called the *sacral promontory* (sa′kral prom′on-to″re). This projection can be felt during a vaginal examination and is used as a guide in determining the size of the pelvis. This measurement is helpful in estimating how easily an infant may be able to pass through a woman's pelvic cavity during childbirth.

The vertebral foramina of the sacral vertebrae form the *sacral canal*, which continues through the sacrum to an opening of variable size at the tip called the *sacral hiatus* (hi-a′tus). This foramen exists because the laminae of the last sacral vertebra are not fused. On the ventral surface of the sacrum, four pairs of *pelvic sacral foramina* provide passageways for nerves and blood vessels.

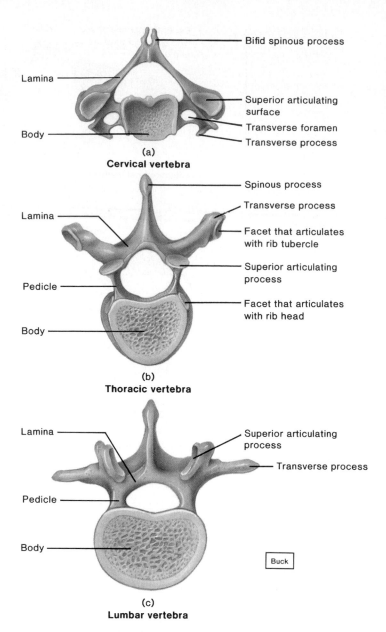

Figure 7.40 Superior view of (*a*) a cervical vertebra, (*b*) a thoracic vertebra, and (*c*) a lumbar vertebra.

Coccyx

The **coccyx** (kok′siks), or tailbone, is the lowest part of the vertebral column and is usually composed of four vertebrae that fuse together by the twenty-fifth year. It is attached by ligaments to the margins of the sacral hiatus (figure 7.41). When a person is sitting, pressure is exerted upon the coccyx, and it moves forward, acting somewhat like a shock absorber. Sitting down with great force sometimes causes the coccyx to be fractured or dislocated.

Chart 7.9 summarizes the bones of the vertebral column.

Disorders of the Vertebral Column

A common vertebral problem involves changes in the intervertebral disks. Each disk is composed of a tough, outer layer of fibrocartilage (annulus fibrosus) and an elastic central mass (nucleus pulposus). As a person ages, these disks tend to undergo degenerative changes in which the central masses lose their firmness and the outer layers become thinner and weaker, and develop cracks. Extra pressure, as when a person falls or lifts a heavy object, can break the outer layers of the disks and allow the central masses to squeeze out. Such a rupture may cause pressure on the spinal cord or on spinal nerves that branch from it. This condition, called a *ruptured,* or *herniated, disk,* may cause back pain and numbness or loss of muscular function in the parts innervated by the affected spinal nerves.

In order to relieve the pain associated with a herniated disk, a surgical procedure called a *laminectomy* may be performed. In this procedure, a portion of the posterior arch of a vertebra is removed to reduce the pressure on the affected nerve tissues. In other instances, a protein-digesting enzyme (chymopapain) may be injected into the injured disk so that its size is decreased.

Sometimes problems develop in the curvatures of the vertebral column because of poor posture, injury, or disease. For example, if an exaggerated thoracic curvature appears, the person develops rounded shoulders and a hunchback.

This condition, called *kyphosis,* occasionally develops in adolescents who are active in strenuous athletic activities. Unless the problem is corrected before bone growth is complete, the vertebral column may be permanently deformed.

Sometimes the vertebral column develops an abnormal lateral curvature, so that one hip or shoulder is lower than the other. At the same time, the thoracic and abdominal organs may be displaced or compressed. This condition is called *scoliosis.* Although it most commonly develops without known cause in adolescent females, it also may accompany such diseases as poliomyelitis, rickets, or tuberculosis.

If the vertebral column develops an accentuated lumbar curvature, the deformity is called *lordosis,* or swayback.

In addition to problems involving the intervertebral disks and curvatures, the vertebral column is subject to degenerative diseases. For example, as a person ages, the intervertebral disks tend to become smaller and more rigid, and the vertebral bodies are more likely to be fractured by compression. Consequently, the height of an elderly person may decrease, and the thoracic curvature of the vertebral column may be accentuated, causing the back to become bowed.

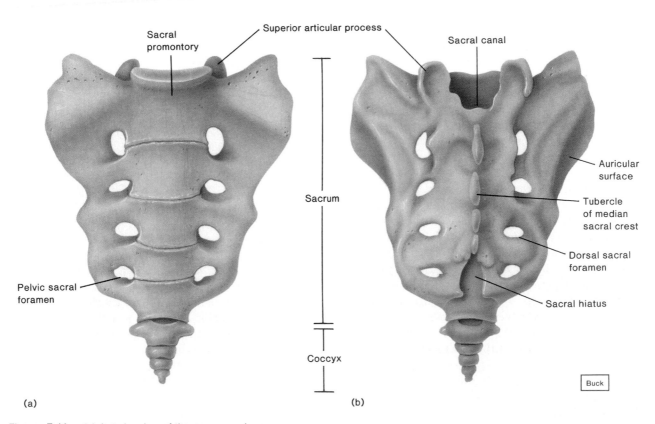

Figure 7.41   (a) Anterior view of the sacrum and coccyx; (b) posterior view.

CHART 7.9 Bones of the vertebral column

Bones	Number	Special features	Bones	Number	Special features
Cervical vertebrae	7	Transverse foramina; facets of atlas articulate with occipital condyles of skull; odontoid process of axis articulates with atlas; spinous processes of second through fifth vertebrae are bifid	Lumbar vertebrae	5	Large bodies; transverse processes that project backward at sharp angles; short, thick spinous processes directed nearly horizontally
Thoracic vertebrae	12	Pointed spinous processes that slope downward; facets that articulate with ribs	Sacrum	5 vertebrae fused into 1 bone	Dorsal sacral foramina, auricular surfaces, sacral promontory, sacral canal, sacral hiatus, pelvic sacral foramina
			Coccyx	4 vertebrae fused into 1 bone	Attached by ligaments to the margins of the sacral hiatus

1. Describe the structure of the vertebral column.
2. Explain the difference between the vertebral column of an adult and that of an infant.
3. Describe a typical vertebra.
4. How do the structures of cervical, thoracic, and lumbar vertebrae differ?

Thoracic Cage

The **thoracic cage** includes the ribs, the thoracic vertebrae, the sternum, and the costal cartilages by which the ribs are attached to the sternum. These parts support the shoulder girdle and arms, protect the visceral organs in the thoracic and upper abdominal cavities, and play a role in breathing (figures 7.42 and 7.43).

Ribs

Regardless of sex, each person usually has twelve pairs of ribs—one pair attached to each of the twelve thoracic vertebrae. Occasionally, however, some individuals develop extra ribs associated with their cervical or lumbar vertebrae.

The first seven rib pairs, which are called the *true ribs* (vertebrosternal ribs), join the sternum directly by their costal cartilages. The remaining five pairs are called *false ribs,* because their cartilages do not reach the sternum directly. Instead, the cartilages of the upper three false ribs (vertebrochondral ribs) join the cartilages of the ribs next above, while the last two rib pairs have no cartilaginous attachments to the sternum. These last two pairs (or sometimes the last three pairs) are called *floating ribs* (vertebral ribs).

A typical rib (figure 7.44) has a long, slender shaft, which curves around the chest and slopes down-

ward. On the posterior end is an enlarged *head* by which the rib articulates with a facet on the body of its own vertebra and with the body of the next higher vertebra. The neck of the rib is a flattened region, lateral to the head, to which various ligaments are attached. Near the neck is a *tubercle,* which articulates with the transverse process of the vertebra.

The costal cartilages are composed of hyaline cartilage. They are attached to the anterior ends of the ribs and continue in line with them toward the sternum.

Sternum

The **sternum** (ster′num), or breastbone, is located along the midline in the anterior portion of the thoracic cage. It is a flat, elongated bone that develops in three parts—an upper *manubrium* (mah-nu′bre-um), a middle *body* (gladiolus), and a lower *xiphoid* (zif′oid) *process* that projects downward (figure 7.42).

> The manubrium and body of the sternum lie in somewhat different planes, so that the line of union between them projects slightly forward. This projection, which occurs at the level of the second costal cartilage, is called the *sternal angle* (angle of Louis). It is commonly used as a clinical landmark when a particular rib must be located accurately (figure 7.42).

The sides of the manubrium and the body are notched where they articulate with costal cartilages. The manubrium also articulates with the clavicles by facets on its superior border. It usually remains as a separate bone until middle age or later, when it fuses to the body of the sternum.

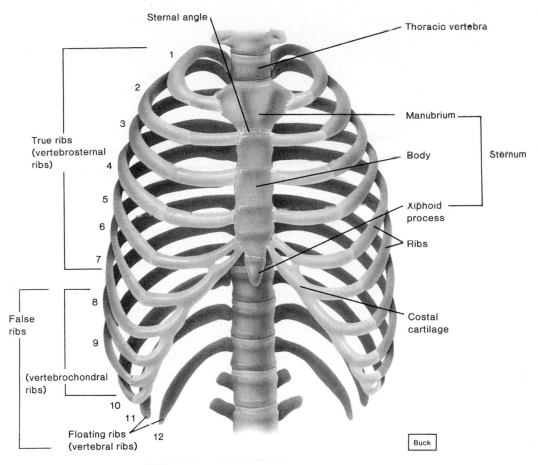

Figure 7.42   The thoracic cage includes the thoracic vertebrae, the sternum, the ribs, and the costal cartilages that attach the ribs to the sternum.

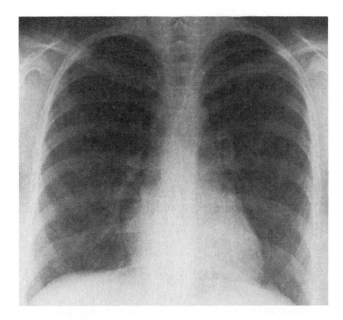

Figure 7.43   X-ray film of the thoracic cage viewed from the front. The light region behind the sternum and above the diaphragm is the heart.

The xiphoid process begins as a piece of cartilage. It slowly ossifies, and by middle life it is usually fused to the body of the sternum also.

The red marrow within the spongy bone of the sternum functions in blood cell formation into adulthood. Since the sternum has a thin covering of compact bone and is easy to reach, samples of its blood-cell-forming tissue may be removed for use in diagnosing diseases. This procedure, a *sternal puncture,* involves suctioning (aspirating) some marrow through a hollow needle. (Marrow may also be removed from the iliac crest of a coxal bone.)

1. What bones make up the thoracic cage?
2. Describe a typical rib.
3. What are the differences between true, false, and floating ribs?

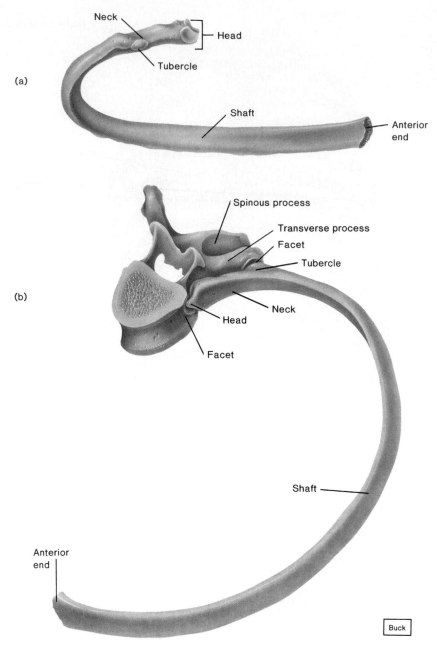

Neck

Head

Tubercle

(a)

Shaft

Anterior end

Spinous process

Transverse process

Facet

Tubercle

Neck

(b)

Head

Facet

Shaft

Anterior end

Buck

Figure 7.44 (*a*) A typical rib (posterior view); (*b*) articulations of a rib with a thoracic vertebra (superior view).

Pectoral Girdle

The **pectoral** (pek′to-ral) **girdle** (shoulder girdle) is composed of four parts—two *clavicles* (collarbones) and two *scapulae* (shoulder blades). Although the word *girdle* suggests a ring-shaped structure, the pectoral girdle is an incomplete ring. It is open in the back between the scapulae, and its bones are separated in front by the sternum. The pectoral girdle supports the arms and serves as an attachment for several muscles that move the arms (figures 7.45 and 7.46).

Clavicles

The **clavicles** are slender, rodlike bones with elongated S-shapes (figure 7.45). They are located at the base of the neck and run horizontally between the sternum and the shoulders. The medial (or sternal) ends of the clavicles articulate with the manubrium, while the lateral (or acromial) ends join processes of the scapulae.

The clavicles act as braces for the freely movable scapulae, and thus help hold the shoulders in place. They also provide attachments for muscles of the arms,

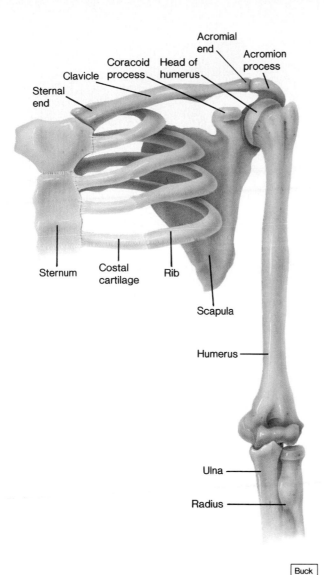

Figure 7.45   The pectoral girdle, to which the arms are attached, consists of a clavicle and a scapula on each side.

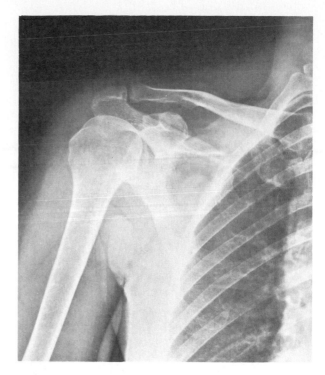

Figure 7.46   X-ray film of the right shoulder region viewed from the front. What features can you identify?

chest, and back. Because of its elongated double curve, the clavicle is structurally weak. If compressed lengthwise due to abnormal pressure on the shoulder, it is likely to fracture.

Scapulae

The **scapulae** are broad, somewhat triangular bones located on either side of the upper back. They have flat bodies with concave anterior surfaces. The posterior surface of each scapula is divided into unequal portions by a *spine*. This spine leads to a *head*, which bears two processes—an *acromion* (ah-kro'me-on) *process* that forms the tip of the shoulder and a *coracoid* (kor'ah-koid) *process* that curves forward and downward below the clavicle (figure 7.47). The acromion process artic-

ulates with the clavicle and provides attachments for muscles of the arm and chest. The coracoid process also provides attachments for arm and chest muscles.

On the head of the scapula between the processes is a depression called the *glenoid cavity*. It articulates with the head of the upper arm bone (humerus).

1. What bones form the pectoral girdle?
2. What is the function of the pectoral girdle?

Upper Limb

The bones of the upper limb form the framework of the arm, wrist, palm, and fingers. They also provide attachments for muscles, and they function in levers that move limb parts. These bones include a humerus, a radius, an ulna, and several carpals, metacarpals, and phalanges (figures 7.48 and 7.49).

Humerus

The **humerus** (figure 7.50) is a heavy bone that extends from the scapula to the elbow. At its upper end is a smooth, rounded *head* that fits into the glenoid cavity of the scapula. Just below the head are two processes—a *greater tubercle* on the lateral side and a *lesser tubercle* on the anterior side. These tubercles provide attachments for muscles that move the arm at the

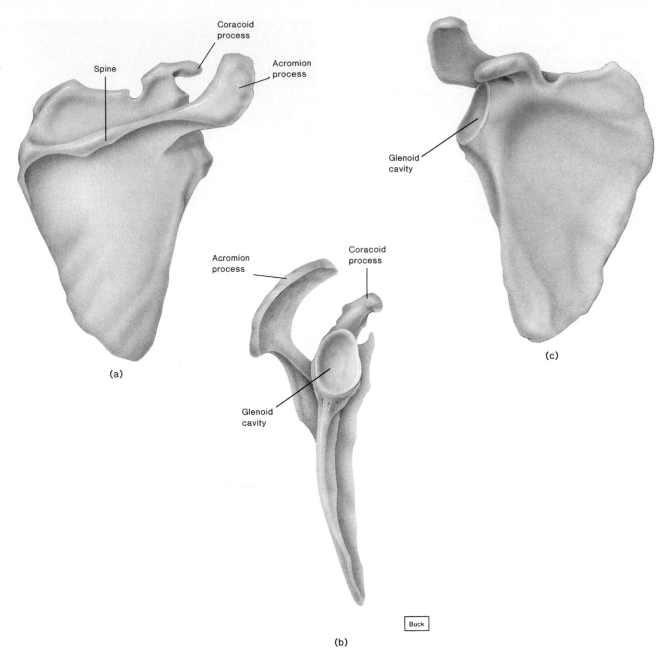

Spine

Coracoid
process

Acromion
process

(a)

Acromion
process

Coracoid
process

Glenoid
cavity

Buck

(b)

Glenoid
cavity

(c)

Figure 7.47 (a) Posterior surface of the right scapula;
(b) lateral view showing the glenoid cavity that articulates with
the head of the humerus (c) anterior surface.

shoulder. Between them is a narrow furrow, the *inter-tubercular groove,* through which a tendon passes from a muscle in the upper arm (biceps brachii) to the shoulder.

The narrow depression along the lower margin of the head that separates it from the tubercles is called the *anatomical neck*. Just below the head and the tu-

bercles of the humerus is a tapering region called the *surgical neck,* so named because fractures commonly occur there. Near the middle of the bony shaft on the lateral side is a rough V-shaped area called the *deltoid tuberosity*. It provides an attachment for the muscle (deltoid) that raises the arm horizontally to the side.

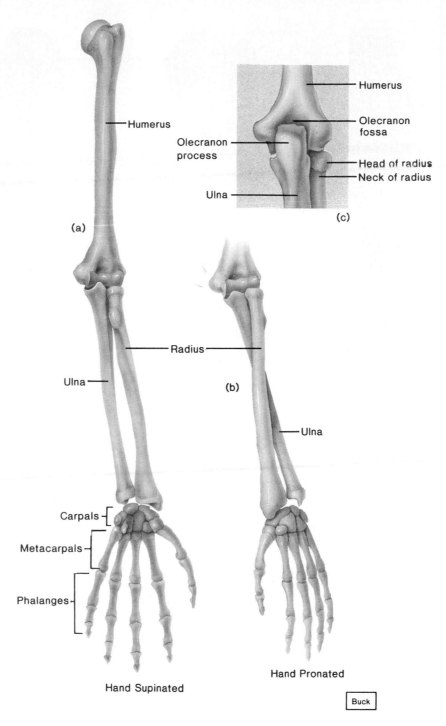

Humerus

Olecranon
process

Ulna

(a)

Ulna

Radius

(b)

Ulna

Carpals

Metacarpals

Phalanges

Hand Supinated

Humerus

Olecranon
fossa

Head of radius
Neck of radius

Ulna

(c)

Hand Pronated

Buck

Figure 7.48 (a) Frontal view of the left arm with the hand supinated; (b) with the hand pronated; (c) posterior view of the right elbow.

At the lower end of the humerus are two smooth *condyles*—a knoblike *capitulum* (kah-pit'u-lum) on the lateral side and a pulley-shaped *trochlea* (trok'le-ah) on the medial side. The capitulum articulates with the radius at the elbow, while the trochlea joins the ulna.

Above the condyles on either side are *epicondyles,* which provide attachments for muscles and ligaments of the elbow. Between the epicondyles anteriorly is a depression, the *coronoid* (kor'o-noid) *fossa,* that receives a process of the ulna (coronoid process) when the elbow is bent. Another depression on the posterior surface, the *olecranon* (o''lek'ra-non) *fossa,* receives an ulnar process (olecranon process) when the arm is straightened at the elbow.

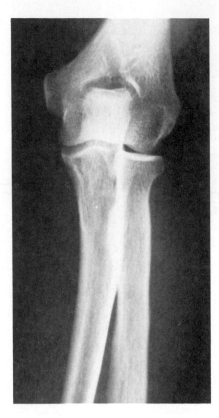

Figure 7.49 X-ray film of the left elbow and lower arm viewed from the front.

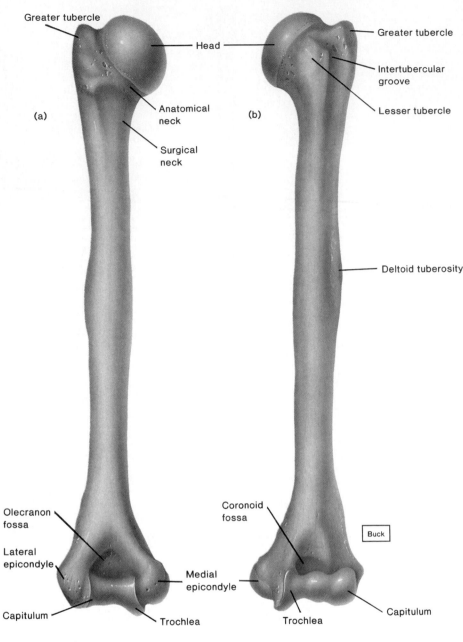

Greater tubercle

Head

Anatomical neck

Surgical neck

(a)

Greater tubercle

Intertubercular groove

Lesser tubercle

(b)

Deltoid tuberosity

Olecranon fossa

Lateral epicondyle

Capitulum

Medial epicondyle

Trochlea

Coronoid fossa

Buck

Trochlea

Capitulum

Figure 7.50 (a) Posterior surface and (b) anterior surface of the left humerus.

Radius

The **radius**, located on the thumb side of the lower arm, is somewhat shorter than its companion, the ulna (figure 7.51). The radius extends from the elbow to the wrist and crosses over the ulna when the hand is turned so that the palm faces backward.

A thick, disklike *head* at the upper end of the radius articulates with the capitulum of the humerus and a notch of the ulna (radial notch). This arrangement allows the radius to rotate freely.

On the radial shaft just below the head is a process called the *radial tuberosity*. It serves as an attachment for a muscle (biceps brachii) that bends the arm at the elbow. At the lower end of the radius, a lateral *styloid* (sti'loid) *process* provides attachments for ligaments of the wrist.

Ulna

The **ulna** is longer than the radius and overlaps the end of the humerus posteriorly. At its upper end, the ulna

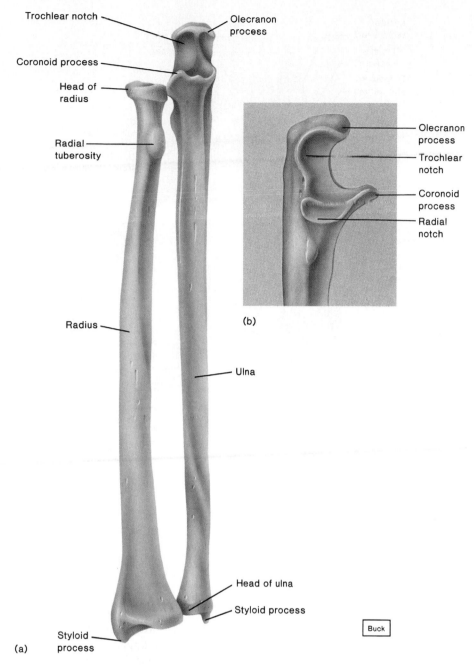

Trochlear notch

Olecranon process

Coronoid process

Head of radius

Radial tuberosity

Radius

Olecranon process

Trochlear notch

Coronoid process

Radial notch

(b)

Ulna

Head of ulna

Styloid process

Styloid process

(a)

Buck

Figure 7.51 (*a*) The head of the right radius articulates with the radial notch of the ulna, and the head of the ulna articulates with the ulnar notch of the radius; (*b*) lateral view of the proximal end of the ulna.

has a wrenchlike opening, the *trochlear notch* (semi-lunar notch), that articulates with the trochlea of the humerus. There is a process on either side of this notch. The *olecranon process,* located above the trochlear notch, provides an attachment for the muscle (triceps brachii) that straightens the arm at the elbow. During this movement, the olecranon process of the ulna fits into the olecranon fossa of the humerus. Similarly, the *coronoid process,* just below the trochlear notch, fits

into the coronoid fossa of the humerus when the elbow is bent.

At the lower end of the ulna, its knoblike *head* articulates with a notch of the radius (ulnar notch) laterally and with a disk of fibrocartilage inferiorly (figure 7.51). This disk, in turn, joins a wrist bone (triangular). A medial *styloid process* at the distal end of the ulna provides attachments for ligaments of the wrist.

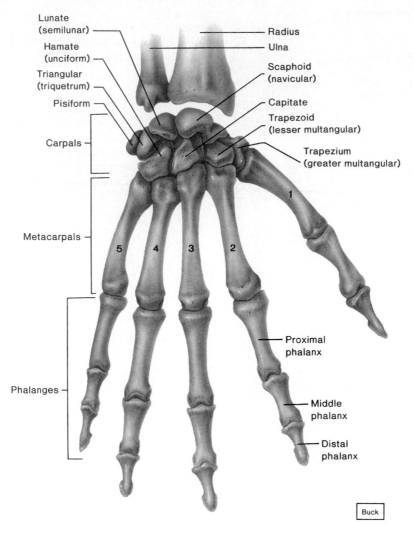

Figure 7.52   The right hand viewed from the back.

Hand

The hand is composed of a wrist, a palm, and five fingers (figures 7.52 and 7.53). The skeleton of the wrist consists of eight small **carpal bones** that are firmly bound in two rows of four bones each. The resulting compact mass is called a *carpus* (kar′pus).

The carpus is rounded on its proximal surface, where it articulates with the radius and with the fibrocartilaginous disk on the ulnar side. The carpus is concave anteriorly, forming a canal through which tendons and nerves extend to the palm. Its distal surface articulates with the metacarpal bones. The individual bones of the carpus are named in figure 7.52.

Five **metacarpal** (met″ah-kar′pal) **bones,** one in line with each finger, form the framework of the palm. These bones are cylindrical, with rounded distal ends that form the knuckles of a clenched fist. The metacarpals articulate proximally with the carpals and dis-

tally with the phalanges. The metacarpal on the lateral side is the most freely movable; it permits the thumb to oppose the fingers when something is grasped in the hand. These bones are numbered 1 to 5, beginning with the metacarpal of the thumb.

The **phalanges** are the bones of the fingers. There are three in each finger—a proximal, a middle, and a distal phalanx—and two in the thumb. (The thumb lacks a middle phalanx.) Thus, there are fourteen finger bones in each hand.

Chart 7.10 summarizes the bones of the pectoral girdle and upper limbs.

1. Locate and name each of the bones of the upper limb.
2. Explain how the bones of the upper limb articulate with one another.

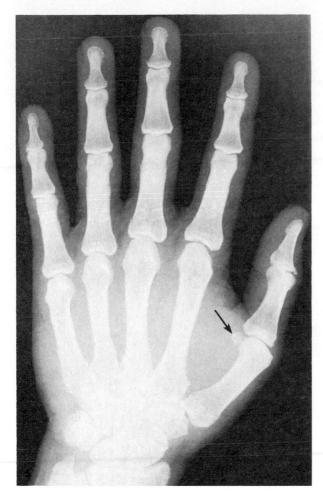

Figure 7.53 X-ray film of the left hand. Note the small sesamoid bone associated with the joint at the base of the thumb (arrow).

Pelvic Girdle

The **pelvic girdle** consists of the two coxal bones (hipbones), which articulate with each other anteriorly and with the sacrum posteriorly (figure 7.54). The sacrum, coccyx, and pelvic girdle together form the ringlike *pelvis,* which provides support for the trunk of the body and attachments for the legs. The weight of the body is transmitted through the pelvis to the legs and then onto the ground. The pelvis also protects the urinary bladder, the distal end of the large intestine, and the internal reproductive organs.

Coxal Bones

Each **coxal bone** (os coxa) develops from three parts— an ilium, an ischium, and a pubis. These parts fuse in the region of a cup-shaped cavity called the *acetabulum* (as″ĕ-tab′u-lum). This depression is on the lateral surface of the hipbone, and it receives the rounded head of the femur or thighbone (figure 7.55).

The **ilium** (il′e-um), which is the largest and uppermost portion of the coxal bone, flares outward to form the prominence of the hip. The margin of this prominence is called the *iliac crest.*

Posteriorly, the ilium joins the sacrum at the *sacroiliac* (sa″kro-il′e-ak) *joint.* Anteriorly, a projection of the ilium, the *anterior superior iliac spine,* can be felt lateral to the groin. This spine provides attachments for ligaments and muscles, and is an important surgical landmark.

A common injury that occurs in contact sports, such as football, involves bruising the soft tissues and bone associated with the anterior superior iliac spine. This painful injury is called a *hip pointer* and may be prevented by wearing protective padding.

CHART 7.10	Bones of the pectoral girdle and upper limbs	
Name and number	**Location**	**Special features**
Clavicle (2)	Base of neck between sternum and scapula	Sternal end, acromial end
Scapula (2)	Upper back, forming part of shoulder	Body, spine, head, acromion process, coracoid process, glenoid cavity
Humerus (2)	Upper arm, between scapula and elbow	Head, greater tubercle, lesser tubercle, intertubercular groove, surgical neck, deltoid tuberosity, capitulum, trochlea, medial epicondyle, lateral epicondyle, coronoid fossa, olecranon fossa
Radius (2)	Lateral side of lower arm, between elbow and wrist	Head, radial tuberosity, styloid process
Ulna (2)	Medial side of lower arm, between elbow and wrist	Trochlear notch, olecranon process, head, styloid process
Carpal (16)	Wrist	Arranged in two rows of four bones each
Metacarpal (10)	Palm	One in line with each finger
Phalanx (28)	Finger	Three in each finger; two in each thumb

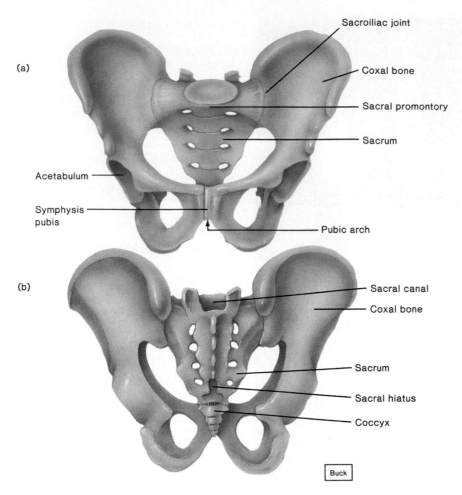

(a)

Sacroiliac joint

Coxal bone

Sacral promontory

Sacrum

Acetabulum

Symphysis pubis

Pubic arch

(b)

Sacral canal

Coxal bone

Sacrum

Sacral hiatus

Coccyx

Buck

Figure 7.54 (*a*) Anterior view and (*b*) posterior view of the pelvic girdle. This girdle provides an attachment for the legs, and together with the sacrum and coccyx forms the pelvis.

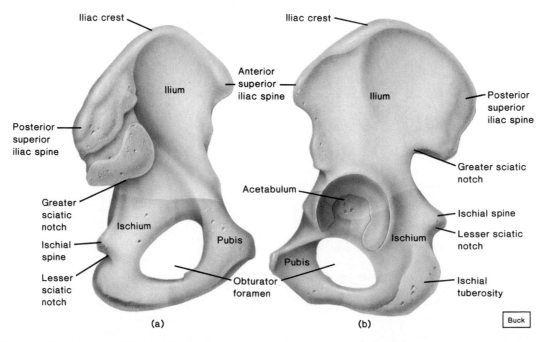

Iliac crest

Iliac crest

Ilium

Anterior superior iliac spine

Ilium

Posterior superior iliac spine

Posterior superior iliac spine

Greater sciatic notch

Greater sciatic notch

Acetabulum

Ischial spine

Ischium

Pubis

Ischium

Lesser sciatic notch

Ischial spine

Pubis

Ischial tuberosity

Lesser sciatic notch

Obturator foramen

(a)

(b)

Buck

Figure 7.55 (*a*) Medial surface of the left coxal bone; (*b*) lateral view.

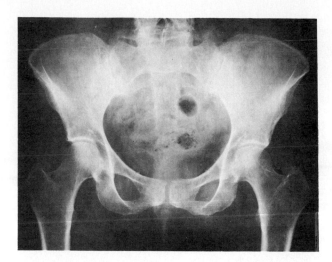

Figure 7.56 What features can you identify in this X-ray film of the pelvic girdle?

CHART 7.11	Differences between the male and female skeletons
Part	**Differences**
Skull	Female skull is relatively smaller and lighter, and its muscular attachments are less conspicuous. Female forehead is longer vertically, facial area is rounder, jaw smaller, and mastoid process less prominent than those of a male.
Pelvis	Female pelvic bones are lighter, thinner, and have less obvious muscular attachments. The obturator foramina and the acetabula are smaller and farther apart than those of a male.
Pelvic cavity	Female pelvic cavity is wider in all diameters, and is shorter, roomier, and less funnel-shaped. The distances between the ischial spines and between the ischial tuberosities are greater than in a male.
Sacrum	Female sacrum is relatively wider, sacral promontory projects forward to a lesser degree, and sacral curvature is bent more sharply posteriorly than in a male.
Coccyx	Female coccyx is more movable than that of a male.

On the posterior border of the ilium is a *posterior superior iliac spine*. Below this spine is a deep indentation, the *greater sciatic notch,* through which a number of nerves and blood vessels pass.

The **ischium** (is'ke-um), which forms the lowest portion of the coxal bone, is L-shaped, with its angle, the *ischial tuberosity,* pointing posteriorly and downward. This tuberosity has a rough surface that provides attachments for ligaments and leg muscles. It also supports the weight of the body when a person is sitting. Above the ischial tuberosity, near the junction of the ilium and ischium, is a sharp projection called the *ischial spine*. This spine, which can be felt during a vaginal examination, is used as a guide for determining the size of the pelvis. The distance between the ischial spines represents the shortest diameter of the pelvic outlet.

The **pubis** (pu'bis) constitutes the anterior portion of the coxal bone. The two pubic bones come together at the midline to form a joint called the *symphysis pubis* (sim'fi-sis pu'bis). The angle formed by these bones below the symphysis is the *pubic arch*.

A portion of each pubis passes posteriorly and downward to join an ischium. Between the bodies of these bones on either side is a large opening, the *obturator foramen,* which is the largest foramen in the skeleton. This foramen is covered and nearly closed by an obturator membrane. (See figures 7.55 and 7.56.)

Greater and Lesser Pelves

If a line were drawn along each side of the pelvis from the sacral promontory downward and anteriorly to the upper margin of the symphysis pubis, it would mark the *pelvic brim* (linea terminalis). This margin separates the lower, or lesser (true), pelvis from the upper, or greater (false), pelvis (figure 7.57).

The *greater pelvis* is bounded posteriorly by the lumbar vertebrae, laterally by the flared parts of the iliac bones, and anteriorly by the abdominal wall. The false pelvis helps support the abdominal organs.

The *lesser pelvis* is bounded posteriorly by the sacrum and coccyx, and laterally and anteriorly by the lower ilium, ischium, and pubis bones. This portion of the pelvis surrounds a short, canal-like cavity that has an upper inlet and a lower outlet. This cavity is of special interest because an infant passes through it during childbirth.

Differences between Male and Female Pelves

Some basic structural differences exist between the male and the female pelves, even though it may be difficult to find all of the "typical" characteristics in any one individual. These differences are related to the function of the female pelvis as a birth canal. Usually, the female iliac bones are more flared than those of the male, and consequently, the female hips are usually broader. The angle of the female pubic arch may be greater, there may be more distance between the ischial spines and the ischial tuberosities, and the sacral curvature may be shorter and flatter. Thus, the female pelvic cavity is usually wider in all diameters than that of the male. Also, the bones of the female pelvis are usually lighter, more delicate, and show less evidence of muscle attachments (figure 7.57). Chart 7.11 summarizes some of the differences between the male and female skeletons.

1. Locate and name each pelvic bone.
2. Explain what is meant by the greater pelvis and the lesser pelvis.
3. How can male and female pelves be distinguished?

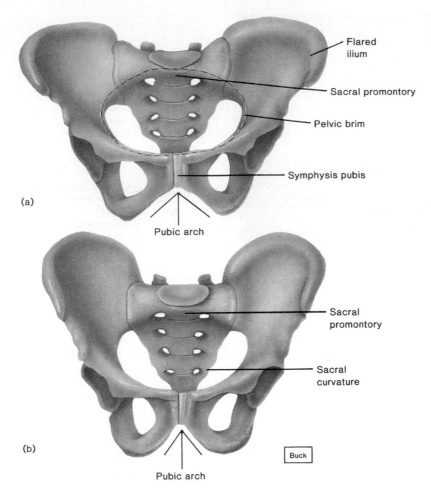

Flared ilium

Sacral promontory

Pelvic brim

Symphysis pubis

(a)

Pubic arch

Sacral promontory

Sacral curvature

(b)

Buck

Pubic arch

Figure 7.57 The female pelvis is usually wider in all diameters and roomier than that of the male. (a) Female pelvis; (b) male pelvis.

Lower Limb

The bones of the lower limb form the frameworks of the leg, ankle, instep, and toes. They include a femur, a tibia, a fibula, and several tarsals, metatarsals, and phalanges (figures 7.58 and 7.59).

Femur

The **femur,** or thighbone, is the longest bone in the body and extends from the hip to the knee. A large, rounded *head* at its upper end projects medially into the acetabulum of the coxal bone. On the head, a pit called the *fovea capitis* marks the attachment of a ligament. Just below the head are a constriction, or *neck,* and two large processes—an upper, lateral *greater trochanter* and a lower, medial *lesser trochanter.* These processes provide attachments for muscles of the legs and buttocks. On the posterior surface in the middle third of the shaft is a longitudinal crest called the *linea aspera.* This rough strip serves as an attachment for several muscles (figure 7.60).

At the lower end of the femur, two rounded processes, the *lateral* and *medial condyles,* articulate with the tibia of the lower leg. A patella also articulates with the femur on its distal anterior surface.

On the medial surface at its distal end is a prominent *medial epicondyle,* and on the lateral surface is a *lateral epicondyle.* These projections provide attachments for various muscles and ligaments.

Hip fracture is one of the more serious causes of hospitalization of elderly persons. The site of such a fracture is most commonly the neck of a femur or the region between the trochanters (intertrochanteric region). Hip fracture is most likely to result from a fall or from a bone disease, such as osteoporosis, in which calcified bone tissue has been lost excessively.

Patella

The **patella,** or kneecap, is a flat sesamoid bone located in a tendon that passes anteriorly over the knee. (See

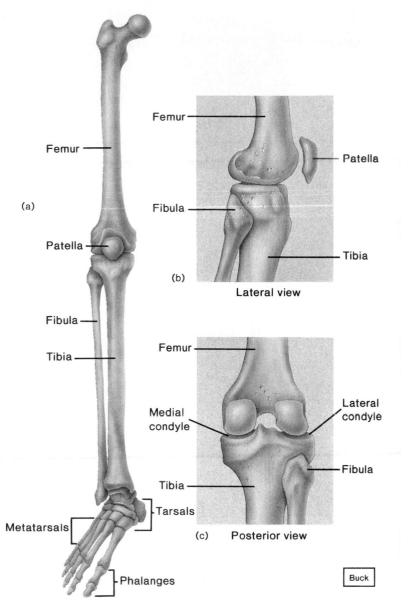

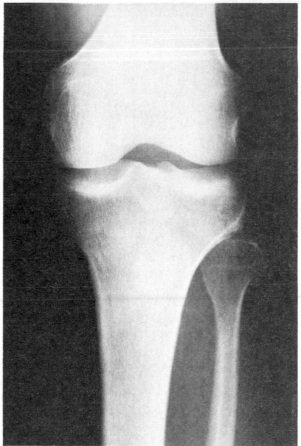

Figure 7.58 (a) Anterior view of the right leg; (b) lateral view of the right knee; (c) posterior view of the right knee.

Figure 7.59 X-ray film of the left knee, showing the ends of the femur, tibia, and fibula.

figure 7.58.) Because of its position, the patella controls the angle at which this tendon continues toward the tibia, and so it functions in lever actions associated with lower leg movements.

As a result of a blow to the knee or a forceful unnatural movement of the leg, the patella sometimes slips to one side. This painful condition is called a *patellar dislocation.* Such a displacement may be prevented by exercises that strengthen muscles associated with the knee and by wearing protective padding. Unfortunately, once the soft tissues that hold the patella in place have been stretched, patellar dislocation tends to recur.

Tibia

The **tibia,** or shinbone, is the larger of the two lower leg bones and is located on the medial side. Its upper end is expanded into *medial* and *lateral condyles,* which have concave surfaces and articulate with the condyles of the femur. Below the condyles, on the anterior surface, is a process called the *tibial tuberosity,* which provides an attachment for the *patellar ligament* (a continuation of the patella-bearing tendon). A prominent *anterior crest* extends downward from the tuberosity and serves as an attachment for connective tissues in the lower leg.

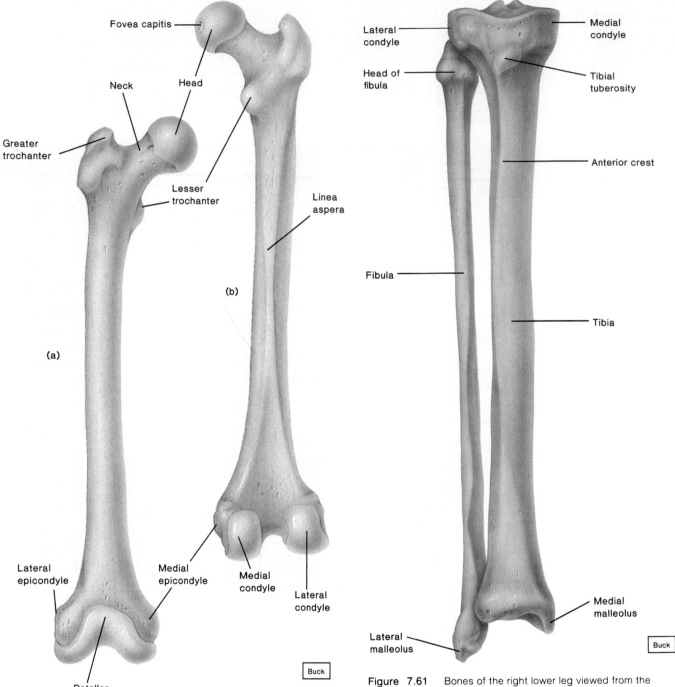

Figure 7.60 (*a*) Anterior surface and (*b*) posterior surface of the right femur.

Figure 7.61 Bones of the right lower leg viewed from the front.

At its lower end, the tibia expands to form a prominence on the inner ankle called the *medial malleolus* (mah-le'o-lus), which serves as an attachment for ligaments. On its lateral side is a depression that articulates with the fibula. The inferior surface of the tibia's distal end articulates with a large bone (the talus) in the foot (figure 7.61).

During a life-threatening emergency, if it is necessary to administer fluids or drugs to a young child, the red bone marrow is sometimes used as an entrance into the blood. In this procedure, called *intraosseous infusion*, the substances are introduced through a hollow needle that has been inserted into the marrow cavity of a long bone, such as the tibia. Once inside the marrow, the substances quickly enter the general blood circulation.

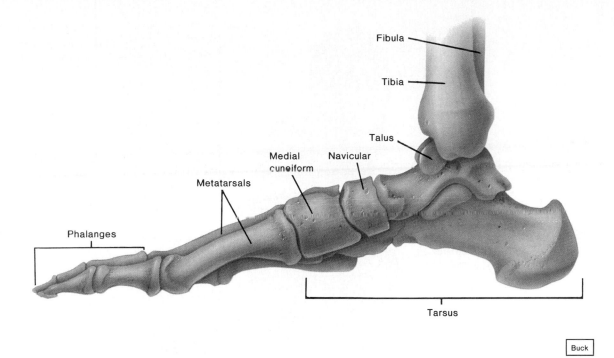

Labels on figure: Fibula, Tibia, Talus, Navicular, Medial cuneiform, Metatarsals, Phalanges, Tarsus, Buck

Figure 7.62 The talus moves freely where it articulates with the tibia and fibula.

Fibula

The **fibula** is a long, slender bone located on the lateral side of the tibia. Its ends are slightly enlarged into an upper *head* and a lower *lateral malleolus*. The head articulates with the tibia just below the lateral condyle; however, it does not enter into the knee joint and does not bear any body weight. The lateral malleolus articulates with the ankle and forms a prominence on the lateral side (figure 7.61).

Foot

The foot consists of an ankle, an instep, and five toes. The ankle is composed of seven **tarsal bones,** forming a group called the *tarsus* (tahr′sus). These bones are arranged so that one of them, the **talus** (ta′lus), can move freely where it joins the tibia and fibula. The remaining tarsal bones are bound firmly together, forming a mass on which the talus rests. The individual bones of the tarsus are named in figures 7.62 and 7.63.

The largest of the anklebones, the **calcaneus** (kal-ka′ne-us), or heel bone, is located below the talus where it projects backward to form the base of the heel. The calcaneus helps support the weight of the body and provides an attachment for muscles that move the foot.

The instep consists of five elongated **metatarsal** (met″ah-tar′sal) **bones,** which articulate with the tarsus. They are numbered 1 to 5, beginning on the medial side (figure 7.63). The heads at the distal ends of these bones form the ball of the foot. The tarsals and metatarsals are arranged and bound by ligaments to form the arches of the foot. A longitudinal arch extends from the heel to the toe, and a transverse arch stretches across the foot. These arches provide a stable, springy base for the body. Sometimes, however, the tissues that bind the metatarsals become weakened, producing fallen arches, or flat feet.

The **phalanges** of the toes are similar to those of the fingers. They are in line with the metatarsals and articulate with them. Each toe has three phalanges—a proximal, a middle, and a distal phalanx—except the great toe, which has only two because it lacks the middle phalanx. (See figures 7.63 and 7.64.) Chart 7.12 summarizes the bones of the pelvic girdle and lower limbs.

1. Locate and name each of the bones of the lower limb.
2. Explain how the bones of the lower limb articulate with one another.
3. Describe how the foot is adapted to support the body.

Clinical Terms Related to the Skeletal System

achondroplasia (a-kon″dro-pla′ze-ah) An inherited condition in which the formation of cartilaginous bone is retarded. The result is a type of dwarfism.

acromegaly (ak″ro-meg′ah-le) A condition caused by an overproduction of growth hormone in adults and characterized by abnormal enlargement of facial features, hands, and feet.

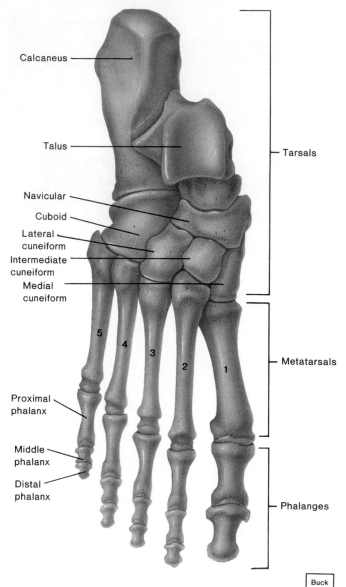

Calcaneus

Talus

Navicular

Cuboid

Lateral cuneiform

Intermediate cuneiform

Medial cuneiform

Tarsals

5 4 3 2 1

Metatarsals

Proximal phalanx

Middle phalanx

Distal phalanx

Phalanges

Buck

Figure 7.63 The right foot viewed from above.

Colles fracture (kol' ēz frak'tūre) A fracture at the distal end of the radius in which the smaller fragment is displaced posteriorly.

epiphysiolysis (ep''ĭ-fiz''e-ol'ĭ-sis) A separation or loosening of the epiphysis from the diaphysis of a bone.

laminectomy (lam''ĭ-nek'to-me) Surgical removal of the posterior arch of a vertebra, usually to relieve the symptoms of a ruptured intervertebral disk.

lumbago (lum-ba'go) A dull ache in the lumbar region of the back.

orthopedics (or''tho-pe'diks) The science of prevention, diagnosis, and treatment of diseases and abnormalities involving the skeletal and muscular systems.

ostalgia (os-tal'je-ah) Pain in a bone.

ostectomy (os-tek'to-me) Surgical removal of a bone.

osteitis (os''te-i'tis) Inflammation of bone tissue.

osteochondritis (os''te-o-kon-dri'tis) Inflammation of bone and cartilage tissues.

osteogenesis (os''te-o-jen'ĕ-sis) The development of bone.

osteogenesis imperfecta (os''te-o-jen'ĕ-sis im-per-fek'ta) A congenital condition characterized by the development of deformed and abnormally brittle bones.

osteoma (os''te-o'mah) A tumor composed of bone tissue.

osteomalacia (os''te-o-mah-la'she-ah) A softening of adult bone due to a disorder in calcium and phosphorus metabolism, usually caused by a deficiency of vitamin D.

osteomyelitis (os''te-o-mi''ĕ-li'tis) Inflammation of bone caused by the action of bacteria or fungi.

osteonecrosis (os''te-o-ne-kro'sis) Death of bone tissue. This condition occurs most commonly in the head of the femur in elderly persons and may be due to obstructions in arteries that supply the bone.

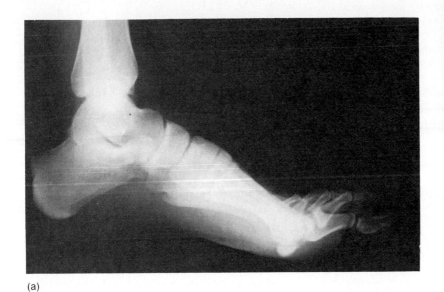

(a)

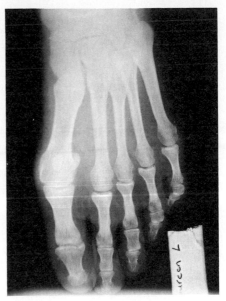

(b)

Figure 7.64 (a) X-ray film of the left foot viewed from the medial side; (b) viewed from above. What features can you identify?

CHART 7.12 Bones of the pelvic girdle and lower limbs

Name and number	Location	Special features
Coxal bone (2)	Hip, articulating with the other coxal bone anteriorly and with the sacrum posteriorly	Ilium, iliac crest, anterior superior iliac spine, ischium, ischial tuberosity, ischial spine, obturator foramen, acetabulum, pubis
Femur (2)	Upper leg, between hip and knee	Head, fovea capitis, neck, greater trochanter, lesser trochanter, linea aspera, lateral condyle, medial condyle
Patella (2)	Anterior surface of knee	A flat sesamoid bone located within a tendon
Tibia (2)	Medial side of lower leg, between knee and ankle	Medial condyle, lateral condyle, tibial tuberosity, anterior crest, medial malleolus
Fibula (2)	Lateral side of lower leg, between knee and ankle	Head, lateral malleolus
Tarsal (14)	Ankle	Freely movable talus that articulates with lower leg bones; six other tarsal bones bound firmly together
Metatarsal (10)	Instep	One in line with each toe, arranged and bound by ligaments to form arches
Phalanx (28)	Toe	Three in each toe, two in great toe

osteopathology (os″te-o-pah-thol′o-je) The study of bone diseases.

osteopenia (os″te-o-pe′ni-ah) Decrease in bone mass due to a reduction in the rate of bone tissue formation.

osteoporosis (os″te-o-po-ro′sis) Condition in which the mineral content of bone tissue is decreased.

osteotomy (os″te-ot′o-me) The cutting of a bone.

roentgenogram (rent-gen′o-gram″) An image obtained on film by using X rays.

Chapter Summary

Introduction (page 172)

Individual bones are the organs of the skeletal system. A bone contains very active tissues.

Bone Structure (page 172)

Bone structure reflects its function.

1. Classification of bones
   Bones are grouped according to their shapes—long, short, flat, irregular, or sesamoid.

2. Parts of a long bone
   a. Epiphyses at each end are covered with articular cartilage and articulate with other bones.
   b. The shaft of a bone is called the diaphysis.
   c. Except for the articular cartilage, a bone is covered by a periosteum.
   d. Compact bone provides strength and resistance to bending.
   e. Spongy bone provides strength where needed and reduces the weight of bone.
   f. The diaphysis contains a medullary cavity filled with marrow.
3. Microscopic structure
   a. Compact bone contains osteons cemented together.
   b. Osteonic canals contain blood vessels that nourish the cells of osteons.
   c. Communicating canals connect osteonic canals transversely and communicate with the bone's surface and the medullary cavity.
   d. Cells of spongy bone are nourished by diffusion from the surface of the thin bony plates.

Bone Development and Growth (page 175)

1. Intramembranous bones
   a. Certain flat bones of the skull are intramembranous bones.
   b. They develop from layers of connective tissues.
   c. Bone tissue is formed by osteoblasts within the membranous layers.
   d. Mature bone cells are called osteocytes.
   e. Primitive connective tissue gives rise to the periosteum.
2. Endochondral bones
   a. Most of the bones of the skeletal system are endochondral.
   b. They develop first as hyaline cartilage that is later replaced by bone tissue.
3. Growth of an endochondral bone
   a. Primary ossification center appears in the diaphysis, while secondary ossification centers appear in the epiphyses.
   b. An epiphyseal disk remains between the primary and secondary ossification centers.
   c. An epiphyseal disk consists of layers of cells: resting cells, young reproducing cells, older enlarging cells, and dying cells.
   d. The epiphyseal disk is responsible for growth in length.
   e. Long bones continue to grow in length until the epiphyseal disks are ossified.
   f. Growth in thickness is due to intramembranous ossification occurring beneath the periosteum.
   g. The medullary cavity is formed by the action of osteoclasts.
4. Homeostasis of bone tissue
   a. Bones are continually remodeled by osteoclasts and osteoblasts.
   b. The total mass of bone remains nearly constant.

5. Factors affecting bone development, growth, and repair
   a. Deficiencies of vitamin A, C, or D result in abnormal development.
   b. Insufficient secretion of pituitary growth hormone may result in dwarfism; excessive secretion may result in giantism, or acromegaly.
   c. Deficiency of thyroid hormone delays bone growth.
   d. Male and female sex hormones promote bone formation and stimulate ossification of the epiphyseal disks.

Functions of Bones (page 182)

1. Support and protection
   a. Skeletal parts provide the shape and form of body structures.
   b. Bones support and protect softer, underlying tissues.
2. Body movement
   a. Bones and muscles function together as levers.
   b. A lever consists of a rod, a pivot (fulcrum), the weight to be moved, and a force that supplies energy.
   c. Parts of a first-class lever are arranged weight-pivot-force; parts of a second-class lever are arranged pivot-weight-force; parts of a third-class lever are arranged pivot-force-weight.
3. Blood cell formation
   a. At different ages, hematopoiesis occurs in the yolk sac, the liver, the spleen, and the red bone marrow.
   b. Red marrow functions in the production of red blood cells, white blood cells, and blood platelets.
4. Storage of inorganic salts
   a. The intercellular material of bone tissue contains large quantities of calcium phosphate in the form of hydroxyapatite.
   b. When blood calcium ion concentration is low, osteoclasts resorb bone, thus releasing calcium salts.
   c. When blood calcium ion concentration is high, osteoblasts are stimulated to form bone tissue and store calcium salts.
   d. Bone stores small amounts of sodium, magnesium, potassium, and carbonate ions.
   e. Bone tissues may accumulate lead, radium, or strontium.

Organization of the Skeleton (page 185)

1. Number of bones
   a. Usually there are 206 bones in the human skeleton, but the number may vary.
   b. Extra bones in sutures are called sutural bones.

2. Divisions of the skeleton
   a. The skeleton can be divided into axial and appendicular portions.
   b. The axial skeleton consists of the skull, hyoid bone, vertebral column, and thoracic cage.
   c. The appendicular skeleton consists of the pectoral girdle, upper limbs, pelvic girdle, and lower limbs.

Skull (page 187)

The skull consists of twenty-two bones, which include eight cranial bones, thirteen facial bones, and one mandible.

1. Cranium
   a. The cranium encloses and protects the brain, and provides attachments for muscles.
   b. Some cranial bones contain air-filled sinuses that help reduce the weight of the skull.
   c. Cranial bones include the frontal bone, parietal bones, occipital bone, temporal bones, sphenoid bone, and ethmoid bone.
2. Facial skeleton
   a. Facial bones form the basic shape of the face and provide attachments for muscles.
   b. Facial bones include the maxillary bones, palatine bones, zygomatic bones, lacrimal bones, nasal bones, vomer bone, inferior nasal conchae, and mandible.
3. Infantile skull
   a. Incompletely developed bones, separated by fontanels, enable the infantile skull to change shape slightly during childbirth.
   b. The proportions of the infantile skull are different from those of an adult skull, and its bones are less easily fractured.

Vertebral Column (page 200)

The vertebral column extends from the skull to the pelvis and protects the spinal cord. It is composed of vertebrae separated by intervertebral disks. It consists of thirty-three bones in an infant and twenty-six bones in an adult. It has four curvatures—cervical, thoracic, lumbar, and pelvic.

1. A typical vertebra
   a. A typical vertebra consists of a body, pedicles, laminae, spinous process, transverse processes, and superior and inferior articulating processes.
   b. Notches on the lower surfaces of the pedicles form intervertebral foramina through which spinal nerves pass.
2. Cervical vertebrae
   a. Cervical vertebrae comprise the bones of the neck.
   b. Transverse processes bear transverse foramina.
   c. The atlas (first vertebra) supports and balances the head.
   d. The odontoid process of the axis (second vertebra) provides a pivot for the atlas when the head is turned from side to side.

3. Thoracic vertebrae
   a. Thoracic vertebrae are larger than cervical vertebrae.
   b. Their long spinous processes slope downward, and facets on the sides of bodies articulate with the ribs.
4. Lumbar vertebrae
   a. Vertebral bodies of lumbar vertebrae are large and strong.
   b. Their transverse processes project back at sharp angles, and their spinous processes are directed horizontally.
5. Sacrum
   a. The sacrum is a triangular structure that bears rows of dorsal sacral foramina.
   b. It is united with the coxal bones at the sacroiliac joints.
   c. The sacral promontory provides a guide for determining the size of the pelvis.
6. Coccyx
   a. The coccyx forms the lowest part of the vertebral column.
   b. It acts as a shock absorber when a person sits.

Thoracic Cage (page 206)

The thoracic cage includes the ribs, thoracic vertebrae, sternum, and costal cartilages. It supports the shoulder girdle and arms, protects visceral organs, and functions in breathing.

1. Ribs
   a. Twelve pairs of ribs are attached to the twelve thoracic vertebrae.
   b. Costal cartilages of the true ribs join the sternum directly; those of the false ribs join indirectly or not at all.
   c. A typical rib has a shaft, head, and tubercles that articulate with the vertebrae.
2. Sternum
   a. The sternum consists of a manubrium, body, and xiphoid process.
   b. It articulates with costal cartilages and clavicles.

Pectoral Girdle (page 208)

The pectoral girdle is composed of two clavicles and two scapulae. It forms an incomplete ring that supports the arms and provides attachments for muscles that move the arms.

1. Clavicles
   a. Clavicles are rodlike bones that run horizontally between the sternum and shoulders.
   b. They hold the shoulders in place and provide attachments for muscles.
2. Scapulae
   a. The scapulae are broad, triangular bones with bodies, spines, heads, acromion processes, coracoid processes, and glenoid cavities.
   b. They articulate with the humerus of each arm, and provide attachments for muscles of the arms and chest.

Upper Limb (page 209)
Bones of the limb provide the frameworks and attachments of muscles, and function in levers that move the limb and its parts.

1. Humerus
   a. The humerus extends from the scapula to the elbow.
   b. It bears a head, greater tubercle, lesser tubercle, intertubercular groove, anatomical neck, surgical neck, deltoid tuberosity, capitulum, trochlea, epicondyles, coronoid fossa, and olecranon fossa.
2. Radius
   a. The radius is located on the thumb side of the lower arm between the elbow and wrist.
   b. It has a head, radial tuberosity, and styloid process.
3. Ulna
   a. The ulna is longer than the radius and overlaps the humerus posteriorly.
   b. It bears a trochlear notch, olecranon process, head, and styloid process.
   c. It articulates with the radius laterally and with a disk of fibrocartilage inferiorly.
4. Hand
   a. The hand is composed of a wrist, palm, and five fingers.
   b. It includes eight carpals that form a carpus, five metacarpals, and fourteen phalanges.

Pelvic Girdle (page 215)
The pelvic girdle consists of two coxal bones that articulate with each other anteriorly and with the sacrum posteriorly. The sacrum, coccyx, and pelvic girdle form the pelvis. The girdle provides support for body weight and attachments for muscles, and protects visceral organs.

1. Coxal bones
   Each coxal bone consists of an ilium, ischium, and pubis, which are fused in the region of the acetabulum.
   a. Ilium
      (1) The ilium, the largest portion of the coxal bone, joins the sacrum at the sacroiliac joint.
      (2) It bears an iliac crest with anterior and posterior superior iliac spines.
   b. Ischium
      (1) The ischium is the lowest portion of the coxal bone.
      (2) It bears an ischial tuberosity and ischial spine.
   c. Pubis
      (1) The pubis is the anterior portion of the coxal bone.
      (2) Pubis bones are fused anteriorly at the symphysis pubis.

2. Greater and lesser pelves
   a. The lesser pelvis is below the pelvic brim; the greater pelvis is above it.
   b. The lesser pelvis functions as a birth canal; the greater pelvis helps support abdominal organs.
3. Differences between male and female pelves
   a. Differences between male and female pelves are related to the function of the female pelvis as a birth canal.
   b. Usually the female pelvis is more flared; pubic arch is broader; distance between the ischial spines and the ischial tuberosities is greater; and sacral curvature is shorter.

Lower Limb (page 218)
Bones of the lower limb provide the frameworks of the leg, ankle, instep, and toes.

1. Femur
   a. The femur extends from the knee to the hip.
   b. It bears a head, fovea capitis, neck, greater trochanter, lesser trochanter, linea aspera, lateral condyle, and medial condyle.
2. Patella
   a. The patella is a flat sesamoid bone in the tendon that passes anteriorly over the knee.
   b. It controls the angle of this tendon and functions in lever actions associated with lower leg movements.
3. Tibia
   a. The tibia is located on the medial side of the lower leg.
   b. It bears medial and lateral condyles, tibial tuberosity, anterior crest, and medial malleolus.
   c. It articulates with the talus of the ankle.
4. Fibula
   a. The fibula is located on the lateral side of the tibia.
   b. It bears a head and lateral malleolus that articulates with the ankle.
5. Foot
   a. The foot consists of an ankle, an instep, and five toes.
   b. It includes seven tarsals that form the tarsus, five metatarsals, and fourteen phalanges.

Clinical Application of Knowledge

1. What steps do you think should be taken to reduce the chances of a person accumulating abnormal metallic elements such as lead, radium, and strontium in bones?
2. Why do you think incomplete, longitudinal fractures of bone shafts (greenstick fractures) are more common in children than in adults?
3. When a child's bone is fractured, growth may be stimulated at the epiphyseal disk of that bone. What problems might this extra growth cause in an arm or leg before the growth of the other limb compensates for the difference in length?

4. How would you explain the observation that elderly persons often develop bowed backs and appear shorter than they were in earlier years?

5. How might the condition of an infant's fontanels be used to evaluate its development? How might the fontanels be used to estimate intracranial pressure?

6. Why are women more likely to develop osteoporosis than men? What steps might be taken to reduce the risks of developing this condition?

Review Activities

Part A

1. List four groups of bones based upon their shapes, and name an example from each group.

2. Sketch a typical long bone, and label its epiphyses, diaphysis, medullary cavity, periosteum, and articular cartilages.

3. Distinguish between spongy and compact bone.

4. Explain how osteonic canals and communicating canals are related.

5. Explain how the development of intramembranous bone differs from that of endochondral bone.

6. Distinguish between osteoblasts and osteocytes.

7. Explain the function of an epiphyseal disk.

8. Explain how a bone grows in thickness.

9. Define osteoclast.

10. Explain how osteoclasts and osteoblasts function in regulating bone mass.

11. Describe the effects of vitamin deficiencies on bone development.

12. Explain the causes of pituitary dwarfism and giantism.

13. Describe the effects of thyroid and sex hormones on bone development.

14. Explain the effects of exercise on bone structure.

15. Provide several examples to illustrate how bones support and protect body parts.

16. Describe a lever, and explain how its parts may be arranged to form first-, second-, and third-class levers.

17. Describe the functions of red and yellow bone marrow.

18. Explain the mechanism that regulates the concentration of blood calcium ions.

19. List three substances that may be stored in bone abnormally.

20. Distinguish between the axial and appendicular skeletons.

21. List the bones that form the pectoral and pelvic girdles.

22. Name the bones of the cranium and the facial skeleton.

23. Explain the importance of fontanels.

24. Describe a typical vertebra.

25. Explain the differences between cervical, thoracic, and lumbar vertebrae.

26. Describe the locations of the sacroiliac joint, the sacral promontory, and the sacral hiatus.

27. Name the bones that comprise the thoracic cage.

28. Name the bones of the upper limb.

29. Define coxal bone.

30. List the major differences that may occur between the male and female pelves.

31. List the bones of the lower limb.

Part B

Match the parts listed in column I with the bones listed in column II.

I	II
1. Coronoid process	A. Ethmoid bone
2. Cribriform plate	B. Frontal bone
3. Foramen magnum	C. Mandible
4. Mastoid process	D. Maxillary bone
5. Palatine process	E. Occipital bone
6. Sella turcica	F. Temporal bone
7. Supraorbital notch	G. Sphenoid bone
8. Temporal process	H. Zygomatic bone
9. Acromion process	I. Femur
10. Deltoid tuberosity	J. Fibula
11. Greater trochanter	K. Humerus
12. Lateral malleolus	L. Radius
13. Medial malleolus	M. Scapula
14. Olecranon process	N. Sternum
15. Radial tuberosity	O. Tibia
16. Xiphoid process	P. Ulna

Reference Plates
Human Skull

*T*he following set of reference plates is presented to help you locate some of the more prominent features of the human skull. As you study these photographs, it is important to remember that individual human skulls vary in every characteristic. Also, the photographs in this set depict bones from several different skulls.

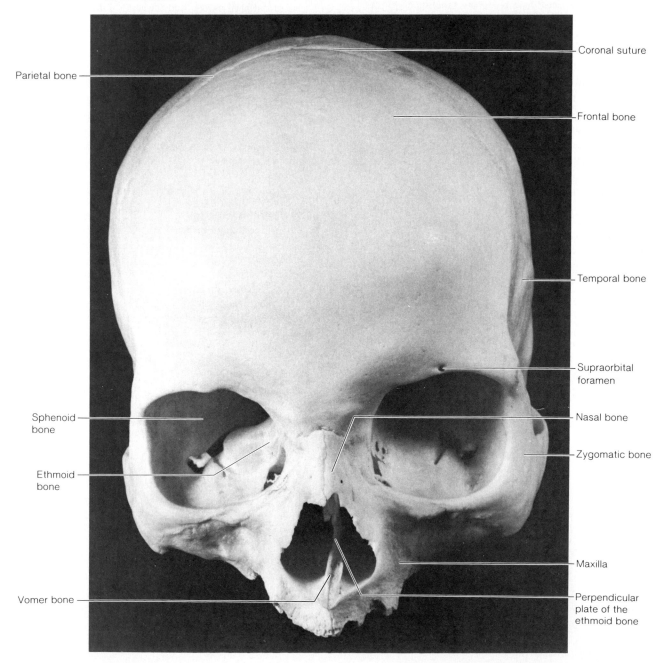

Parietal bone

Coronal suture

Frontal bone

Temporal bone

Supraorbital foramen

Sphenoid bone

Nasal bone

Ethmoid bone

Zygomatic bone

Vomer bone

Maxilla

Perpendicular plate of the ethmoid bone

Plate 8   The skull, frontal view.

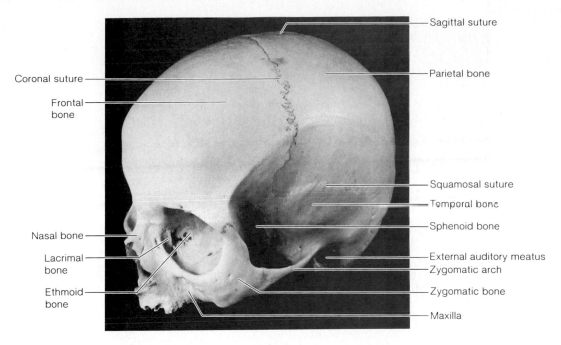

Coronal suture

Frontal bone

Nasal bone

Lacrimal bone

Ethmoid bone

Sagittal suture

Parietal bone

Squamosal suture

Temporal bone

Sphenoid bone

External auditory meatus

Zygomatic arch

Zygomatic bone

Maxilla

Plate 9   The skull, left lateral view.

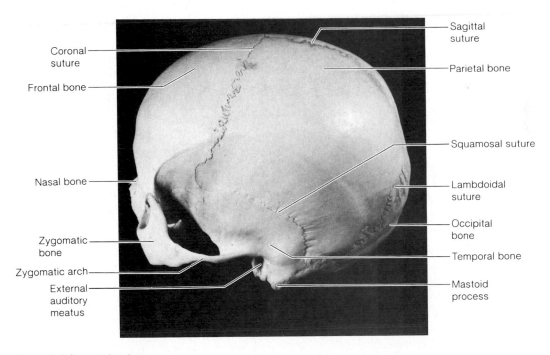

Coronal suture

Frontal bone

Nasal bone

Zygomatic bone

Zygomatic arch

External auditory meatus

Sagittal suture

Parietal bone

Squamosal suture

Lambdoidal suture

Occipital bone

Temporal bone

Mastoid process

Plate 10   The skull, left posterior view.

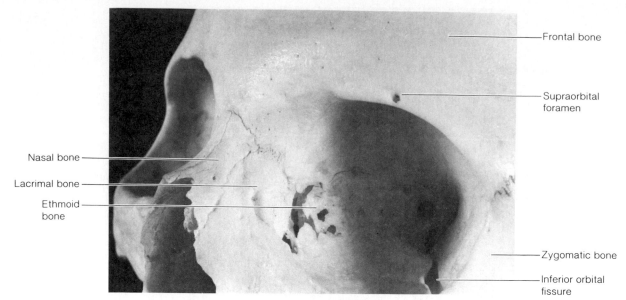

Frontal bone

Supraorbital
foramen

Nasal bone

Lacrimal bone

Ethmoid
bone

Zygomatic bone

Inferior orbital
fissure

Plate 11   Bones of the left orbital region.

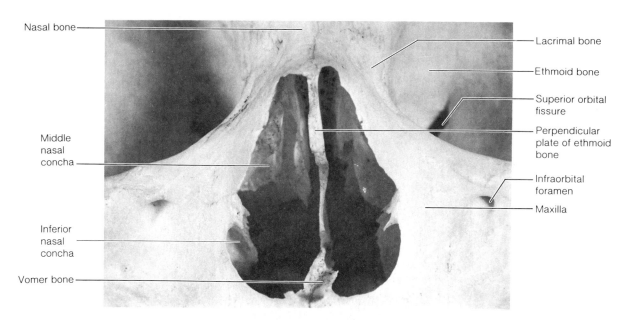

Nasal bone

Lacrimal bone

Ethmoid bone

Superior orbital
fissure

Perpendicular
plate of ethmoid
bone

Middle
nasal
concha

Infraorbital
foramen

Maxilla

Inferior
nasal
concha

Vomer bone

Plate 12   Bones of the anterior nasal region.

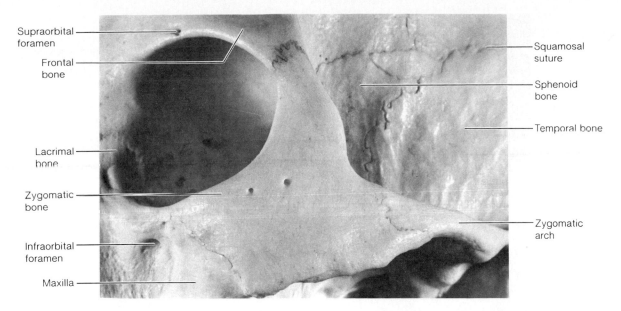

Supraorbital foramen

Frontal bone

Lacrimal bone

Zygomatic bone

Infraorbital foramen

Maxilla

Squamosal suture

Sphenoid bone

Temporal bone

Zygomatic arch

Plate 13 Bones of the left zygomatic region.

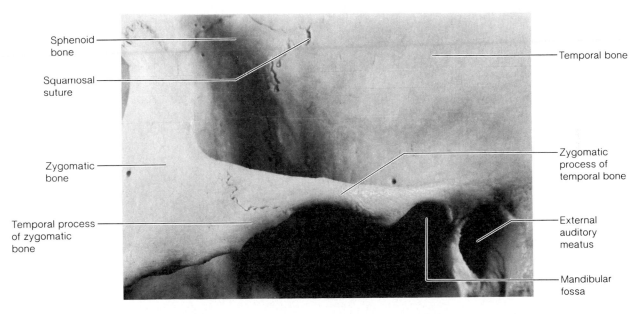

Sphenoid bone

Squamosal suture

Zygomatic bone

Temporal process of zygomatic bone

Temporal bone

Zygomatic process of temporal bone

External auditory meatus

Mandibular fossa

Plate 14 Bones of the left temporal region.

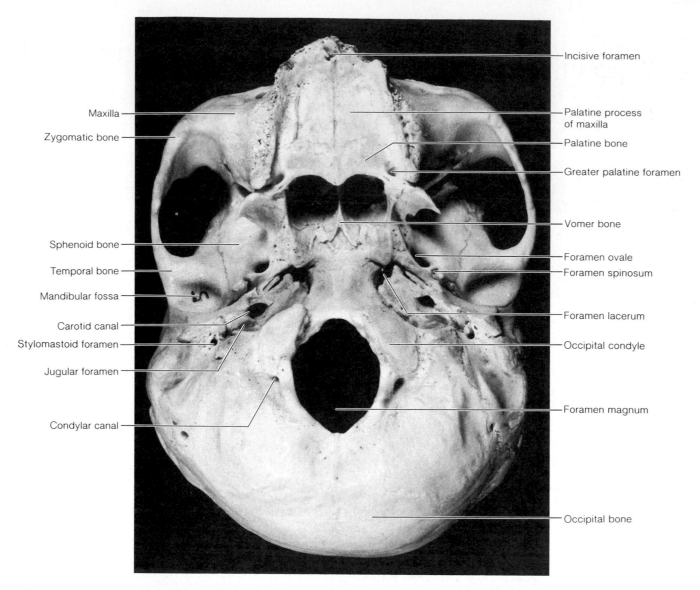

Maxilla

Zygomatic bone

Sphenoid bone

Temporal bone

Mandibular fossa

Carotid canal

Stylomastoid foramen

Jugular foramen

Condylar canal

Incisive foramen

Palatine process
of maxilla

Palatine bone

Greater palatine foramen

Vomer bone

Foramen ovale

Foramen spinosum

Foramen lacerum

Occipital condyle

Foramen magnum

Occipital bone

Plate 15 The skull, inferior view.

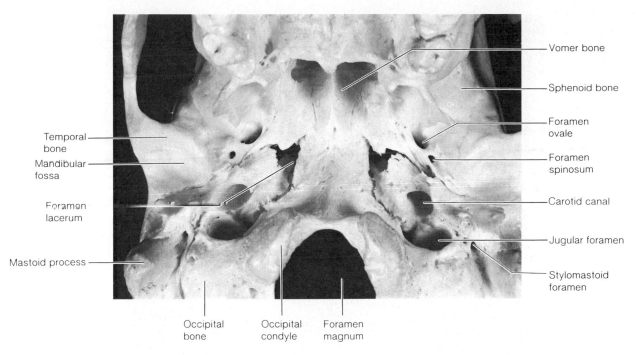

Temporal bone

Mandibular fossa

Foramen lacerum

Mastoid process

Vomer bone

Sphenoid bone

Foramen ovale

Foramen spinosum

Carotid canal

Jugular foramen

Stylomastoid foramen

Occipital bone

Occipital condyle

Foramen magnum

Plate 16 Base of the skull, sphenoidal region.

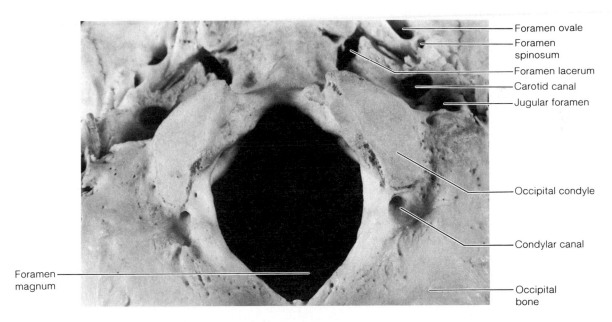

Foramen ovale

Foramen spinosum

Foramen lacerum

Carotid canal

Jugular foramen

Occipital condyle

Condylar canal

Foramen magnum

Occipital bone

Plate 17 Base of the skull, occipital region.

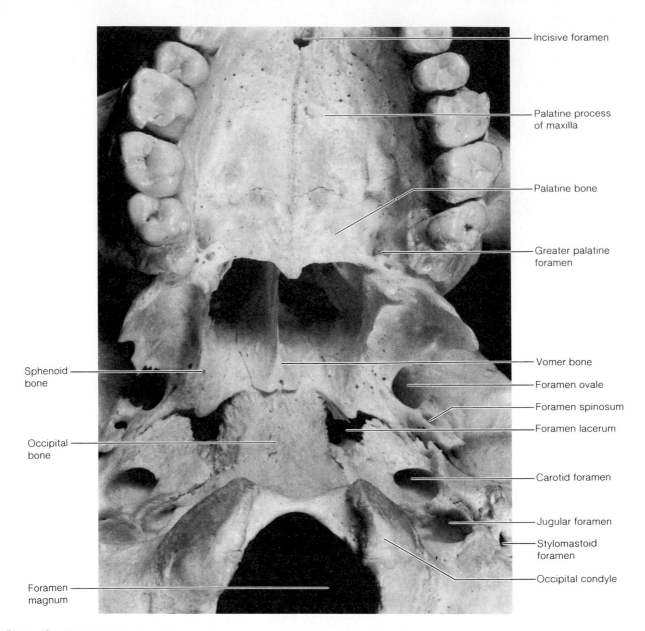

Incisive foramen

Palatine process
of maxilla

Palatine bone

Greater palatine
foramen

Vomer bone

Foramen ovale

Foramen spinosum

Foramen lacerum

Carotid foramen

Jugular foramen

Stylomastoid
foramen

Occipital condyle

Sphenoid
bone

Occipital
bone

Foramen
magnum

Plate 18   Base of the skull, maxillary region.

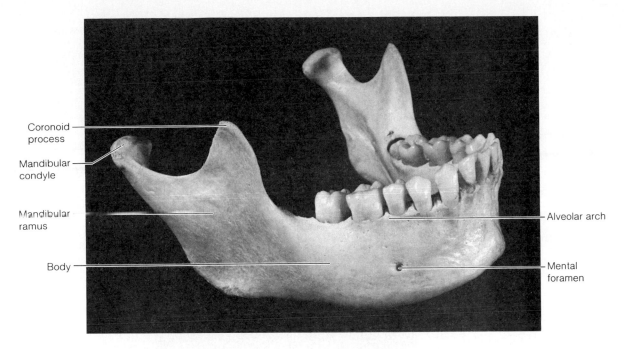

Coronoid process

Mandibular condyle

Mandibular ramus

Body

Alveolar arch

Mental foramen

Plate 19 Mandible, lateral view.

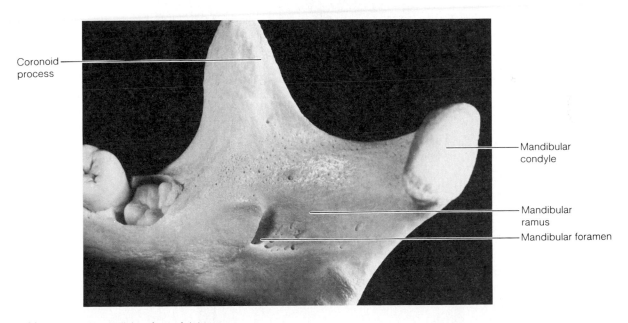

Coronoid process

Mandibular condyle

Mandibular ramus

Mandibular foramen

Plate 20 Mandible, medial surface of right ramus.

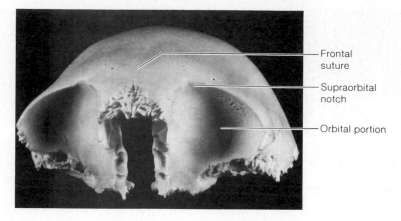

Plate 21   Frontal bone, anterior view.

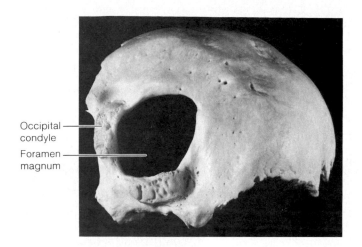

Plate 22   Occipital bone, inferior view.

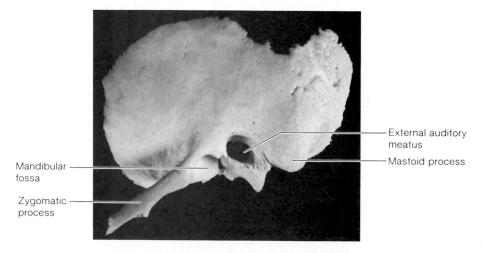

Plate 23   Temporal bone, left lateral view.

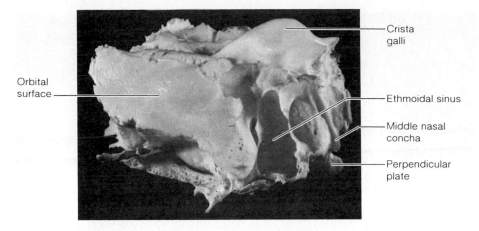

Crista galli

Orbital surface

Ethmoidal sinus

Middle nasal concha

Perpendicular plate

Plate 24 Ethmoid bone, right lateral view.

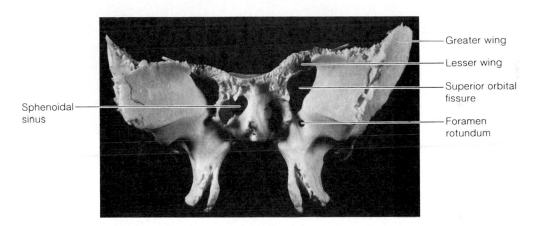

Greater wing

Lesser wing

Superior orbital fissure

Sphenoidal sinus

Foramen rotundum

Plate 25 Sphenoid bone, anterior view.

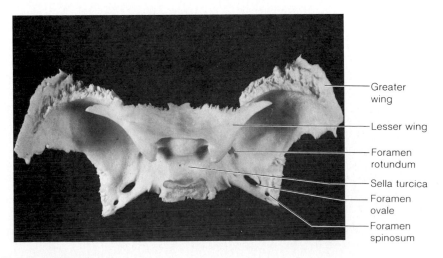

Greater wing

Lesser wing

Foramen rotundum

Sella turcica

Foramen ovale

Foramen spinosum

Plate 26 Sphenoid bone, posterior view.

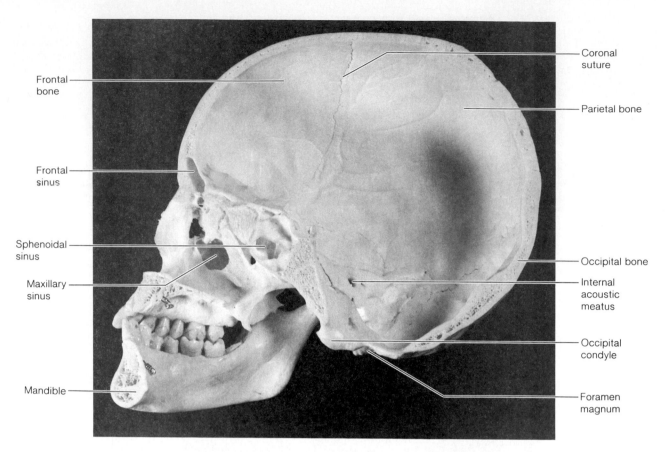

Frontal
bone

Frontal
sinus

Sphenoidal
sinus

Maxillary
sinus

Mandible

Coronal
suture

Parietal bone

Occipital bone

Internal
acoustic
meatus

Occipital
condyle

Foramen
magnum

Plate 27 The skull, sagittal section.

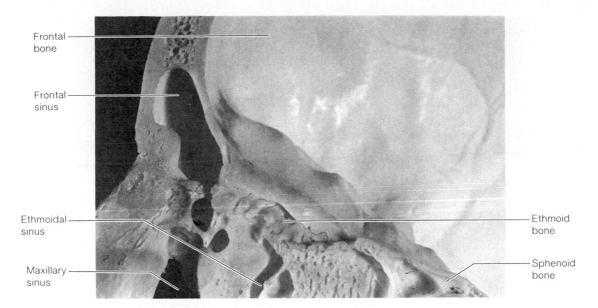

Frontal bone

Frontal sinus

Ethmoidal sinus

Maxillary sinus

Ethmoid bone

Sphenoid bone

Plate 28   Ethmoidal region, sagittal section.

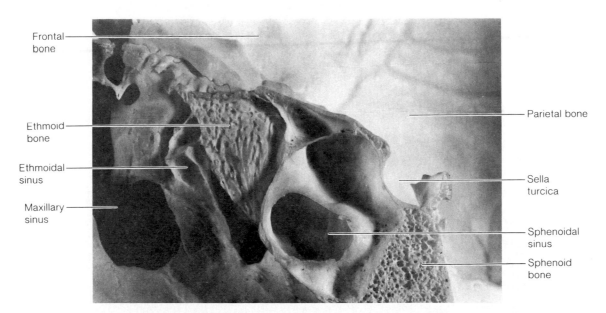

Frontal bone

Ethmoid bone

Ethmoidal sinus

Maxillary sinus

Parietal bone

Sella turcica

Sphenoidal sinus

Sphenoid bone

Plate 29   Sphenoidal region, sagittal section.

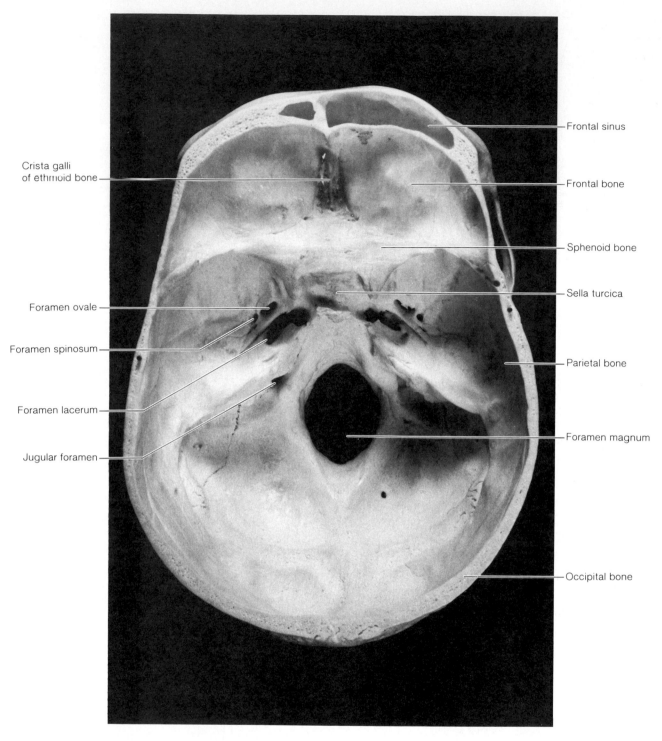

Crista galli
of ethmoid bone

Foramen ovale

Foramen spinosum

Foramen lacerum

Jugular foramen

Frontal sinus

Frontal bone

Sphenoid bone

Sella turcica

Parietal bone

Foramen magnum

Occipital bone

Plate 30   The skull, floor of the cranial cavity.

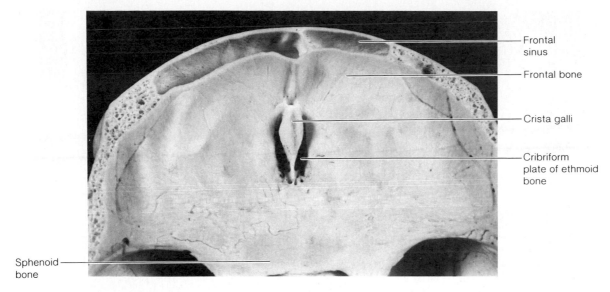

Frontal
sinus

Frontal bone

Crista galli

Cribriform
plate of ethmoid
bone

Sphenoid
bone

Plate 31 Frontal region, transverse section.

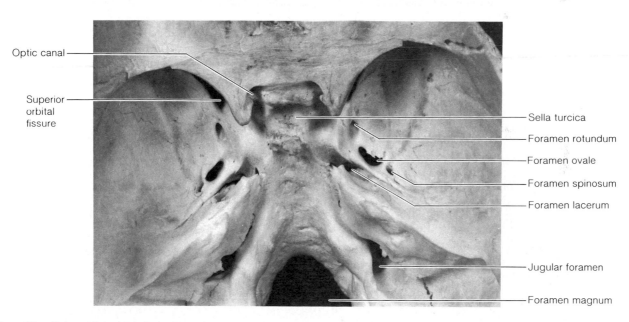

Optic canal

Superior
orbital
fissure

Sella turcica

Foramen rotundum

Foramen ovale

Foramen spinosum

Foramen lacerum

Jugular foramen

Foramen magnum

Plate 32 Sphenoidal region, floor of the cranial cavity.

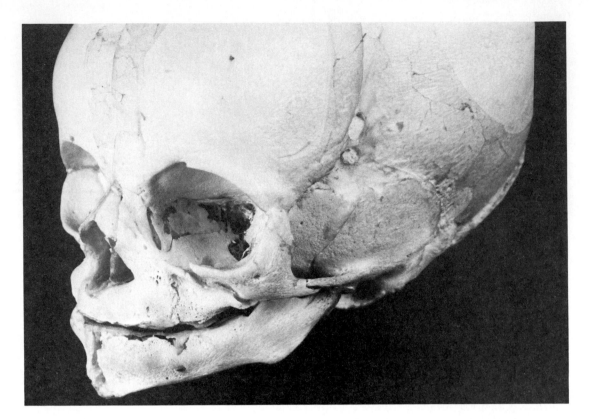

Plate 33   Skull of a fetus, left lateral view.

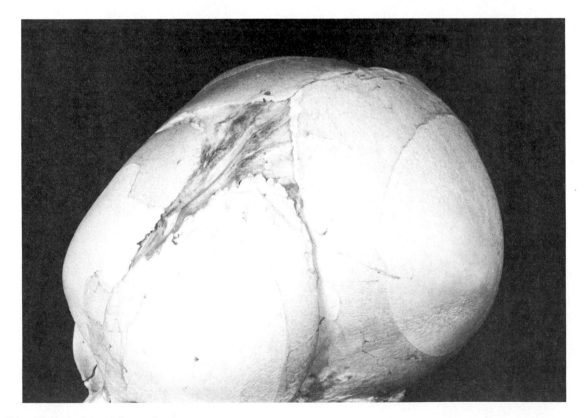

Plate 34   Skull of a fetus, left superior view.

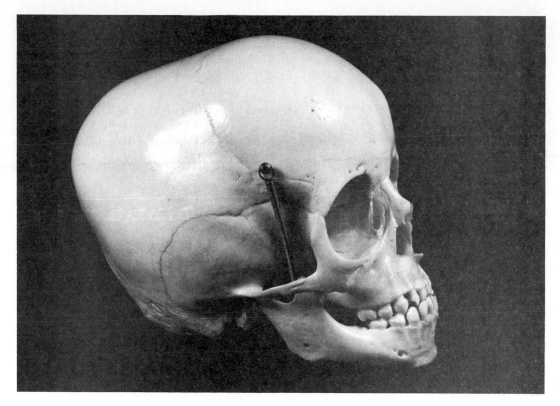

Plate 35   Skull of a child.

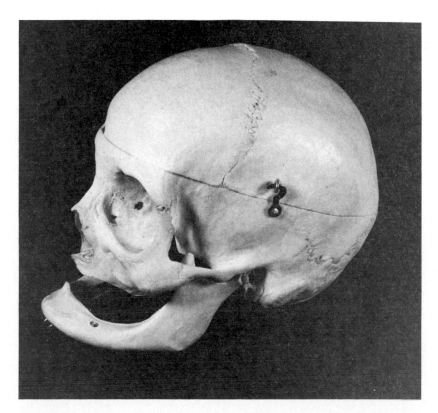

Plate 36   Skull of an aged person.

8

Joints of the Skeletal System

*J*oints are the junctions between the bones of the skeletal system. That is, a joint occurs wherever two or more bones come together. Most joints function in lever systems that make movements possible by bending or straightening. Others, however, are relatively rigid structures that help hold bones in place or enable bones to grow.

This chapter describes how various types of joints are classified, how they differ in structure, and how they permit different movements. It also describes in detail the structures of several large, freely movable joints ■

After you have studied this chapter, you should be able to:

1. Explain how joints can be classified according to the type of tissue that binds the bones together.

2. Describe how bones of fibrous joints are held together.

3. Describe how bones of cartilaginous joints are held together.

4. Describe the general structure of a synovial joint.

5. List six types of synovial joints and name an example of each type.

6. Explain how skeletal muscles produce movements at joints and identify several types of joint movements.

7. Describe the shoulder joint and explain how its articulating parts are held together.

8. Describe the elbow joint and explain how its articulating parts are held together.

9. Describe the hip joint and explain how its articulating parts are held together.

10. Describe the knee joint and explain how its articulating parts are held together.

11. Complete the review activities at the end of this chapter. Note that they are worded in the form of specific learning objectives. You may want to refer to them before reading the chapter.

articulation (ar-tik′u-la″shun)
bursa (ber′sah)
gomphosis (gom-fo′sis)
ligament (lig′ah-ment)
meniscus (mĕ-nis′kus)
suture (su′chur)
symphysis (sim′fĭ-sis)
synchondrosis (sin″kon-dro′sis)
syndesmosis (sin″des-mo′sis)
synovial (sĭ-no′ve-al)

acetabul-, vinegar cup: *acetabul*um—depression of the coxal bone that articulates with the head of the femur.

annul-, ring: *annul*ar ligament—ring-shaped band of connective tissue within the elbow that encircles the head of the radius.

burs-, pouch: prepatellar *burs*a—fluid-filled sac between the skin and the patella.

condyl-, knob: medial *condyl*e—rounded bony process at the distal end of the femur.

fov-, pit: *fov*ea capitis—pit in the head of the femur to which a ligament is attached.

glen-, joint socket: *glen*oid cavity—depression in the scapula that articulates with the head of the humerus.

labr-, lip: glenoidal *labr*um—rim of fibrocartilage attached to the margin of the glenoid cavity.

ov-, egglike: syn*ov*ial fluid—thick fluid within a joint cavity that resembles egg white.

sutur-, sewing: *sutur*e—type of joint in which flat bones are interlocked by a set of tiny bony processes.

syndesm-, binding together: *syndesm*osis—type of joint in which the bones are held together by long fibers of connective tissue.

Wherever two or more bones meet, a **joint** is formed. Such joints, or **articulations,** represent the functional junctions between bones. They bind various parts of the skeletal system together, allow bone growth to occur, permit certain parts of the skeleton to change shape during childbirth, and enable body parts to move in response to skeletal muscle contractions.

Classification of Joints

Although joints vary considerably in structure and function, they can be classified according to the type of tissue that binds the bones together at each junction. On this basis, three general groups can be identified—fibrous joints, cartilaginous joints, and synovial joints.

Joints can also be grouped according to the amount of movement possible at the bony junctions. In this scheme, joints are classified as immovable (synarthrotic), slightly movable (amphiarthrotic), and freely movable (diarthrotic).

Fibrous Joints

Fibrous joints are found between bones that come into close contact with one another. The bones at such joints are fastened tightly together by a thin layer of fibrous connective tissue. As a rule, no appreciable movement occurs between these bones; sometimes, however, a very small degree of motion is possible. The three types of fibrous joints are:

1. **Syndesmosis.** In this type of joint, the bones are bound together by relatively long fibers of connective tissue that form an *interosseous ligament.* Because this ligament is flexible and may be twisted, the joint may permit slight movement and thus is amphiarthrotic (am″fe-ar-thro′tik). An example of a syndesmosis is found at the distal ends of the tibia and fibula, where they join to form the tibiofibular articulation. (See figure 8.1.)

2. **Suture.** Sutures occur only between flat bones of the skull, where the relatively broad margins of adjacent bones grow together and become united by a thin layer of fibrous connective tissue called a *sutural ligament.* In the infantile skull, as described in chapter 7, the skull is incompletely developed, and several of the bones are separated by membranous areas called *fontanels.* (See figure 7.35.) These areas allow the skull to change shape slightly during childbirth, but as the bones continue to grow, the fontanels close and are replaced by sutures. With time, some of the bones at sutures become interlocked by a set of tiny bony processes, and the sutural ligament itself may be changed to

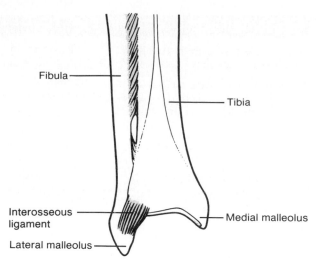

Figure 8.1 The articulation at the distal ends of the tibia and fibula provides an example of a syndesmosis.

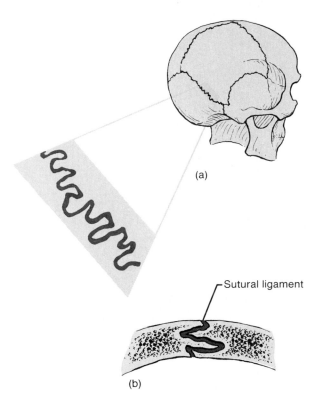

Figure 8.2 (*a*) The fibrous joints between the bones of the cranium are immovable and are called sutures. (*b*) The bones at a suture are separated by a sutural ligament.

bone. An example of such a suture can be found in the adult human skull where the parietal and occipital bones meet to form the lambdoidal suture. Because they are immovable, sutures are synarthrotic (sin″ar-thro′tik) joints. (See figures 8.2 and 8.3.)

3. **Gomphosis.** A gomphosis is a joint formed by the union of a cone-shaped bony process in a

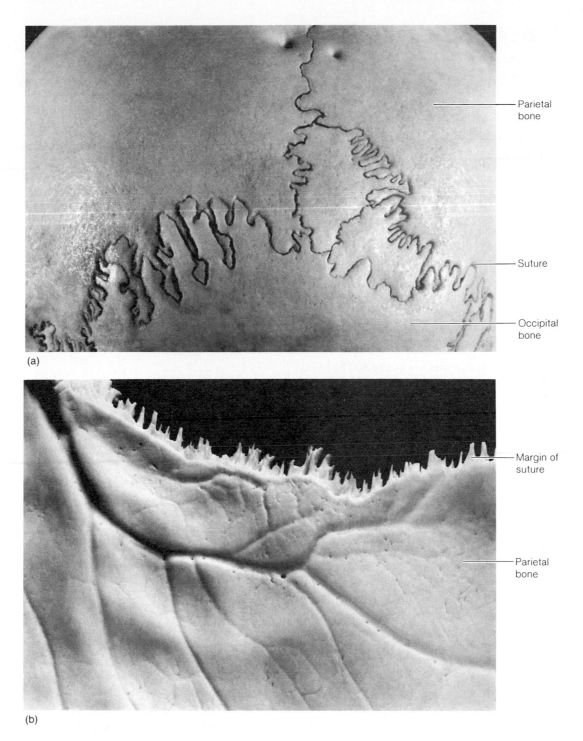

Parietal
bone

Suture

Occipital
bone

(a)

Margin of
suture

Parietal
bone

(b)

Figure 8.3 (*a*) Sutures between the parietal and occipital bones of the skull; (*b*) the inner margin of a parietal suture. (Note: The grooves on the inside of this parietal bone mark the paths of blood vessels located near the surface of the brain.)

bony socket. Such an articulation is formed by the peglike root of a tooth fastened to a jawbone by a *periodontal ligament*. This ligament surrounds the root and attaches it firmly to the jaw with bundles of thick collagenous fibers. A gomphosis is a synarthrotic joint. (See figure 8.4.)

1. What is a joint?
2. How are joints classified?
3. Describe three types of fibrous joints.
4. What is the function of the fontanels?

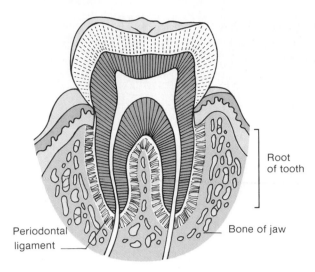

Figure 8.4 The articulation between the root of a tooth and the jawbone is a gomphosis.

Cartilaginous Joints

The bones of **cartilaginous joints** are connected by hyaline cartilage or fibrocartilage. The two types are:

1. **Synchondrosis.** In a synchondrosis, the bones are united by bands of hyaline cartilage. Many of these joints are temporary structures that disappear as a result of the growth process. An example is an immature long bone where an epiphysis is connected to a diaphysis by a band of hyaline cartilage (the epiphyseal disk). As described in chapter 7, this band functions in the growth in length of the bone, and in time is converted from cartilage into bone. When this ossification is complete, usually before the age of twenty-five years, movement no longer occurs at the joint. Thus, the joint is synarthrotic. (See figure 7.11.)

   Another example of a synchondrosis occurs between the sternum and the first rib, which are united directly by costal cartilage. (See figure 8.5.) In this instance, the joint is permanent. (Note: The joints between the costal cartilages and the sternum of ribs 2 through 7 are usually synovial joints.)

2. **Symphysis.** The articular surfaces of the bones at a symphysis are covered by a thin layer of hyaline cartilage, and the cartilage, in turn, is attached to a pad of springy fibrocartilage. A limited amount of movement occurs at such a

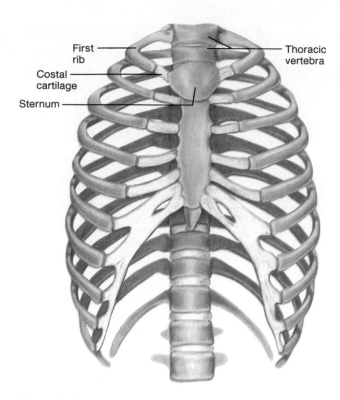

Figure 8.5 The articulation between the first rib and the sternum is an example of a synchondrosis.

joint whenever forces cause the cartilaginous pad to become compressed or deformed. The joint formed by the bodies of two adjacent vertebrae separated by an intervertebral disk is an example of a symphysis. (See figure 8.6 and reference plate 51.)

   Each intervertebral disk is composed of a band of fibrous fibrocartilage (annulus fibrosus) that surrounds a gelatinous core (nucleus pulposus). The disk acts as a shock absorber and helps equalize pressures between the vertebrae when the body moves. Since each disk is slightly flexible, the combined movement of many of the joints in the vertebral column allows the limited motion that occurs when the back is bent forward, bent to the side, or twisted. Another example of this type of joint is the symphysis pubis in the pelvis, which allows some movement of the maternal pelvic bones as an infant passes through the birth canal during childbirth. Because these joints allow slight movements, they are amphiarthrotic joints.

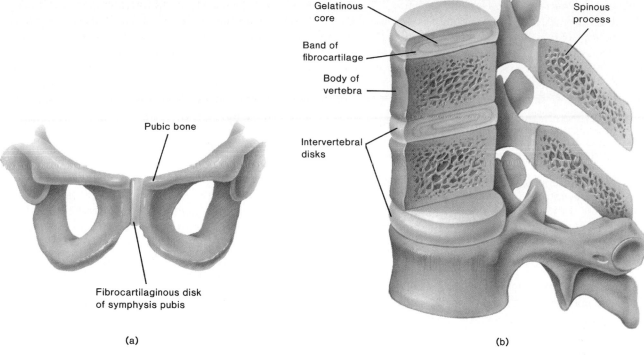

Gelatinous core

Band of fibrocartilage

Body of vertebra

Intervertebral disks

Spinous process

Pubic bone

Fibrocartilaginous disk of symphysis pubis

(a)

(b)

Figure 8.6 (a) The symphysis pubis that separates the pubic bones and (b) the intervertebral disks that separate adjacent vertebrae are composed of fibrocartilage.

Synovial Joints

Most joints of the skeletal system are **synovial joints**, and because they allow free movement, they are diarthrotic (di″ar-thro′tik). These joints have much more complex structures than the fibrous or cartilaginous types. For example, they contain articular cartilage, a joint capsule, and a synovial membrane.

General Structure of a Synovial Joint

The articular ends of the bones in a synovial joint are covered with a thin layer of hyaline cartilage. (See figure 8.7.) This layer, which is called the **articular cartilage,** is resistant to wear and produces a minimum of friction when it is compressed as the joint is moved. Articular cartilage may also transmit body weight onto an underlying bone.

Typically, the bone upon which articular cartilage lies (subchondral plate) contains cancellous bone, which is somewhat elastic (figure 8.7). This plate acts as a shock absorber that helps protect the joint from

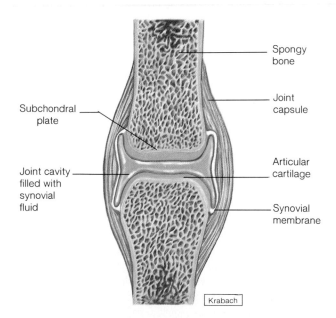

Subchondral plate

Spongy bone

Joint capsule

Joint cavity filled with synovial fluid

Articular cartilage

Synovial membrane

Krabach

Figure 8.7 The generalized structure of a synovial joint. What is the function of the synovial fluid within this type of joint?

stresses caused by the load of body weight and by forces produced by contracting muscles. (See figure 7.3.)

Excessive mechanical stress may cause tiny fractures to appear in a subchondral plate. Although such fractures usually heal, the bone that regenerates may be less elastic than the original, and its protective function may be reduced.

In addition to causing fractures in the subchondral plate, excessive mechanical stress due to obesity or certain athletic activities may damage the articular cartilage of various joints and cause it to deteriorate. Joints that are subjected to repeated overloading, as in the elbow of a baseball pitcher or the knee of a basketball player, are likely to develop the degenerative joint disease called *osteoarthritis.*

The bones of a synovial joint are held together by a tubular **joint capsule** that has two distinct layers. The outer layer consists largely of dense, white, fibrous connective tissue, whose fibers are attached to the periosteum around the circumference of each bone of the joint near its articular end. Thus, the outer fibrous layer of the capsule completely encloses the other parts of the joint. It is, however, flexible enough to permit movement and strong enough to help prevent the articular surfaces from being pulled apart.

Bundles of strong, tough collagenous fibers called **ligaments** reinforce the joint capsule and help bind the articular ends of the bones together. Some ligaments appear as thickenings in the fibrous layer of the capsule, while others are *accessory structures* located outside the capsule. In either case, these structures help prevent excessive movement at the joint. That is, the ligament is relatively inelastic, and it becomes tightly drawn whenever a normal limit of movement has been achieved in the joint.

The inner layer of the joint capsule consists of a shiny, vascular lining of loose connective tissue called the **synovial membrane.** This membrane, which is only a few cells thick, covers all of the surfaces within the joint capsule, except the areas covered by the articular cartilage. The synovial membrane surrounds a closed sac called the *synovial cavity,* and into this joint cavity the membrane secretes a clear, viscous fluid called **synovial fluid.** In some regions, the surface of the synovial membrane possesses villi as well as larger folds and projections that extend into the cavity. Besides filling spaces and irregularities of the joint cavity, these extensions increase the surface area of the synovial membrane. In addition to secreting fluid, the membrane may store adipose tissue and form movable fatty pads within the joint. The synovial membrane also functions to reabsorb fluid. Thus, it may help remove substances from a joint cavity that is injured or infected.

Synovial fluid has a consistency similar to egg white, and it moistens and lubricates the smooth cartilaginous surfaces within the joint. It also helps supply articular cartilage with nutrients that are obtained from blood vessels of the synovial membrane. However, the volume of synovial fluid present in a joint cavity is relatively small. Usually there is just enough to cover the articulating surfaces with a thin film of fluid. For example, the amount of synovial fluid in the cavity of the knee is 0.5 ml or less.

Some synovial joints are partially or completely divided into two compartments by disks of fibrocartilage called **menisci** (sing. *meniscus*) located between the articular surfaces. Each meniscus is attached to the fibrous layer of the joint capsule peripherally, and its free surface projects into the joint cavity. In the case of the knee joint, crescent-shaped menisci cushion the articulating surfaces and help distribute the body weight onto these surfaces. (See figure 8.8.)

Certain synovial joints also have closed, fluid-filled sacs called **bursae** associated with them. Each bursa has an inner lining of synovial membrane, which may be continuous with the synovial membrane of a nearby joint cavity. These sacs contain synovial fluid and are commonly located between the skin and underlying bony prominences, as in the case of the patella of the knee or the olecranon process of the elbow. Bursae act as cushions and aid the movement of tendons that glide over bony parts or over other tendons. The names of bursae indicate their locations. Thus, there is a *suprapatellar bursa,* a *prepatellar bursa,* and an *infrapatellar bursa,* as shown in figure 8.8.

Articular cartilage, like other cartilaginous structures, lacks a direct blood supply. (See chapter 5.) Instead, it depends upon the surrounding synovial fluid for its supply of oxygen, nutrients, and other vital substances. Normal body movements cause a forceful action within a joint that facilitates the passage of these substances into the cartilage. When a joint is immobilized or is not used for a prolonged time, lack of action may result in degeneration of the articular cartilage.

Although degeneration of the cartilage may be reversed when joint movements are resumed, it is important to avoid exercises that cause excessive compression of the tissue during the period of regeneration. Otherwise, chondrocytes in the thinned cartilage may be injured, and the repair process may be hindered.

1. Describe two types of cartilaginous joints.
2. What is the function of an intervertebral disk?
3. Describe the structure of a synovial joint.

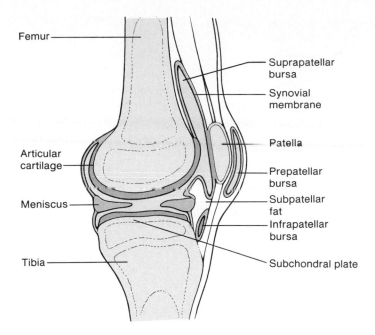

Figure 8.8 The articulating surfaces of the femur and tibia are separated by menisci. There are also several bursae associated with the knee joint.

(Labels on figure:)
Femur
Suprapatellar bursa
Synovial membrane
Articular cartilage
Patella
Prepatellar bursa
Subpatellar fat
Meniscus
Infrapatellar bursa
Tibia
Subchondral plate

Types of Synovial Joints

The articulating bones of synovial joints have a variety of shapes that allow different kinds of movement. Based upon the shapes of their parts and the movements they permit, these joints can be classified into six major types—ball-and-socket joints, condyloid joints, gliding joints, hinge joints, pivot joints, and saddle joints.

1. *Ball-and-socket joints.* A **ball-and-socket joint** (spheroidal joint) consists of a bone with a globular or slightly egg-shaped head that articulates with the cup-shaped cavity of another bone. Such a joint allows for a wider range of motion than does any other kind. Movements in all planes, as well as rotational movement around a central axis, are possible. The hip and shoulder contain joints of this type. (See figure 8.9a.)
2. *Condyloid joints.* In a **condyloid joint** (ellipsoidal joint), the ovoid condyle of one bone fits into the elliptical cavity of another bone, as in the joints between the metacarpals and phalanges. This type of joint permits a variety of movements in different planes; rotational movement, however, is not possible. (See figure 8.9b.)
3. *Gliding joints.* The articulating surfaces of **gliding joints** (arthrodial joints) are nearly flat or slightly curved. These joints allow only sliding or back-and-forth motion. Most of the joints within the wrist and ankle, as well as those between the articular processes of

adjacent vertebrae, belong to this group. (See figure 8.9c.)
4. *Hinge joints.* In a **hinge joint** (ginglymoidal joint), the convex surface of one bone fits into the concave surface of another, as in the elbow and the joints of the phalanges. Such a joint resembles the hinge of a door in that it permits movement in one plane only. (See figure 8.9d.)
5. *Pivot joints.* In a **pivot joint** (trochoidal joint), the cylindrical surface of one bone rotates within a ring formed of bone and the fibrous tissue of a ligament. Movement at such a joint is limited to rotation around a central axis. The joint between the proximal ends of the radius and the ulna, where the head of the radius rotates in a ring formed by the radial notch of the ulna and a ligament (annular ligament), is of this type. Similarly, a pivot joint functions in the neck as the head is turned from side to side. In this case, the ring formed by a ligament (transverse ligament) and the anterior arch of the atlas rotates around the odontoid process of the axis. (See figure 8.9e.)
6. *Saddle joints.* A **saddle joint** (sellar joint) is formed between bones whose articulating surfaces have both concave and convex regions. The surface of one bone fits the complementary surface of the other. This arrangement permits a variety of movements, as in the case of the joint between the carpal (trapezium) and the metacarpal of the thumb. (See figure 8.9f.)

Chart 8.1 summarizes the types of joints.

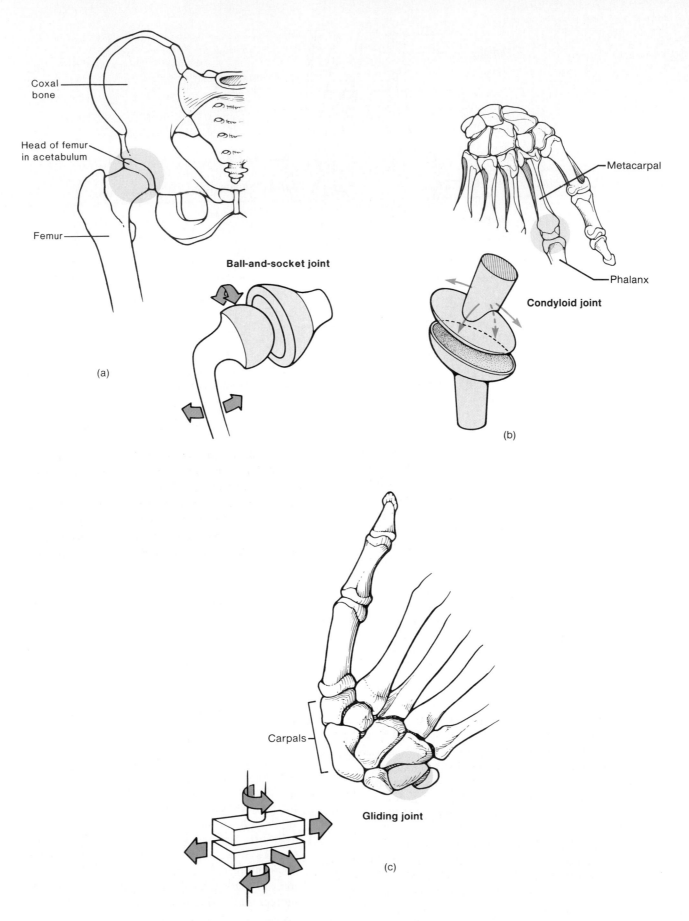

Coxal bone

Head of femur in acetabulum

Femur

Ball-and-socket joint

(a)

Metacarpal

Phalanx

Condyloid joint

(b)

Carpals

Gliding joint

(c)

Figure 8.9   (a)–(f) Types and examples of freely movable joints.

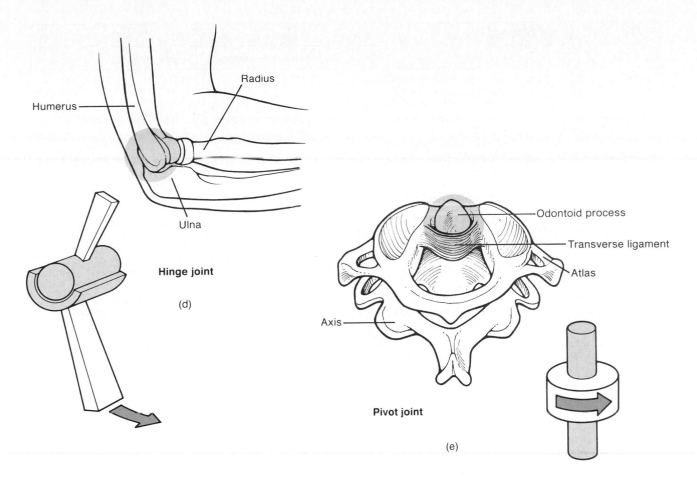

Radius
Humerus

Ulna

Hinge joint

(d)

Odontoid process

Transverse ligament

Atlas

Axis

Pivot joint

(e)

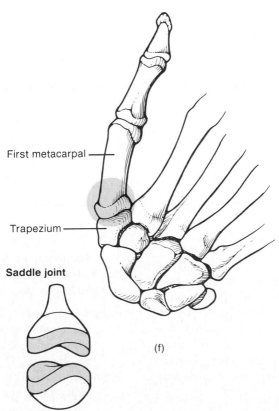

First metacarpal

Trapezium

Saddle joint

(f)

Figure 8.9 continued.

CHART 8.1 Types of joints

Type of joint	Description	Possible movements	Example
Fibrous	Articulating bones fastened together by thin layer of fibrous connective tissue		
1. *Syndesmosis* (amphiarthrotic)	Bones bound by interosseous ligament	Joint flexible and may be twisted	Tibiofibular articulation
2. *Suture* (synarthrotic)	Flat bones united by sutural ligament	None	Parietal bones articulate at sagittal suture of skull
3. *Gomphosis* (synarthrotic)	Cone-shaped process fastened in bony socket by periodontal ligament	None	Root of tooth united with jawbone
Cartilaginous	Articulating bones connected by hyaline cartilage or fibrocartilage		
1. *Synchondrosis* (synarthrotic)	Bones united by bands of hyaline cartilage	Movement occurs during growth process until ossification occurs	Joint between epiphysis and diaphysis of a long bone
2. *Symphysis* (amphiarthrotic)	Articular surfaces separated by thin layers of hyaline cartilage attached to band of fibrocartilage	Limited movement, as when back is bent or twisted	Joints between bodies of adjacent vertebrae
Synovial (diarthrotic)	Articulating bones surrounded by joint capsule of ligaments and synovial membranes; ends of articulating bones covered by hyaline cartilage and separated by synovial fluid		
1. *Ball-and-socket*	Ball-shaped head of one bone articulates with cup-shaped socket of another	Movements in all planes; rotation	Shoulder, hip
2. *Condyloid*	Oval-shaped condyle of one bone articulates with elliptical cavity of another	Variety of movements in different planes, but no rotation	Joints between metacarpals and phalanges
3. *Gliding*	Articulating surfaces are nearly flat or slightly curved	Sliding or twisting	Joints between various bones of wrist and ankle
4. *Hinge*	Convex surface of one bone articulates with concave surface of another	Flexion and extension	Elbow and joints of phalanges
5. *Pivot*	Cylindrical surface of one bone articulates with ring of bone and fibrous tissue	Rotation	Joint between proximal ends of radius and ulna
6. *Saddle*	Articulating surfaces have both concave and convex regions; surface of one bone fits complementary surface of another	Variety of movements	Joint between carpal and metacarpal of thumb

Types of Joint Movements

Movements at synovial joints are produced by actions of skeletal muscles. Typically, one end of a muscle is attached to a relatively immovable or fixed part on one side of a joint, and the other end of the muscle is fastened to a movable part on the other side. When the muscle contracts, its fibers pull its movable end (insertion) toward its fixed end (origin), and a movement occurs at the joint.

The following terms are used to describe movements at joints that occur in different directions and in different planes (figures 8.10, 8.11, and 8.12):

flexion (flek'shun)   Bending parts at a joint so that the angle between them is decreased and the parts come closer together (bending the leg at the knee).

extension (ek-sten'shun)   Straightening parts at a joint so that the angle between them is

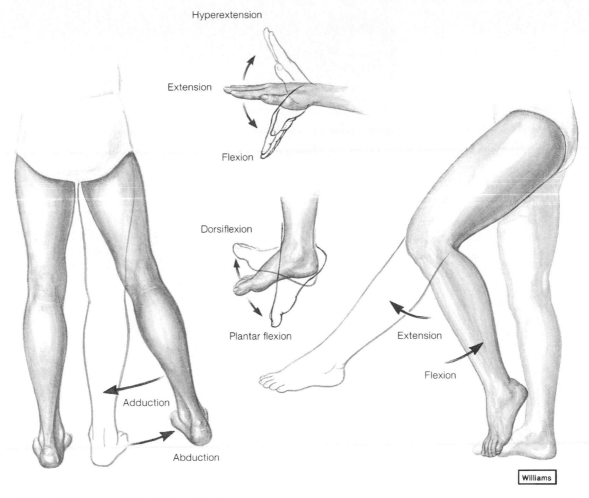

Figure 8.10 Joint movements illustrating adduction, abduction, hyperextension, dorsiflexion, plantar flexion, extension, and flexion.

increased and the parts move further apart (straightening the leg at the knee).

hyperextension (hi″per-ek-sten′shun)   Excessive extension of the parts at a joint, beyond the anatomical position (bending the head back beyond the upright position).

dorsiflexion (dor″sĭ-flek′shun)   Flexing the foot at the ankle (bending the foot upward).

plantar flexion (plan′tar flek′shun)   Extending the foot at the ankle (bending the foot downward).

abduction (ab-duk′shun)   Moving a part away from the midline (lifting the arm horizontally to form a right angle with the side of the body).

adduction (ah-duk′shun)   Moving a part toward the midline (returning the arm from the horizontal position to the side of the body).

rotation (ro-ta′shun)   Moving a part around an axis (twisting the head from side to side).

circumduction (ser″kum-duk′shun)   Moving a part so that its end follows a circular path (moving the finger in a circular motion without moving the hand).

supination (soo″pĭ-na′shun)   Turning the hand so the palm is upward or turning the foot so that the medial margin is raised.

pronation (pro-na′shun)   Turning the hand so the palm is downward or turning the foot so that the medial margin is lowered.

eversion (e-ver′zhun)   Turning the foot so the sole is outward.

inversion (in-ver′zhun)   Turning the foot so the sole is inward.

protraction (pro-trak′shun)   Moving a part forward (thrusting the chin forward).

retraction (re-trak′shun)   Moving a part backward (pulling the chin backward).

elevation (el″ĕ-va′shun)   Raising a part (shrugging the shoulders).

depression (de-presh′un)   Lowering a part (drooping the shoulders).

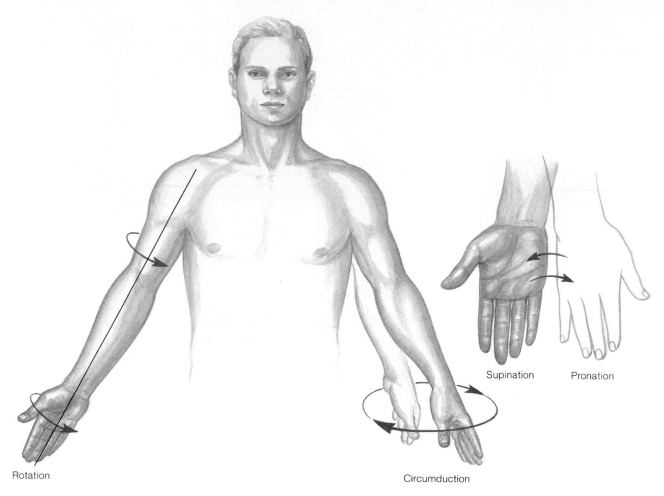

Rotation

Supination   Pronation

Circumduction

Figure 8.11 Joint movements illustrating rotation, circumduction, supination, and pronation.

Williams

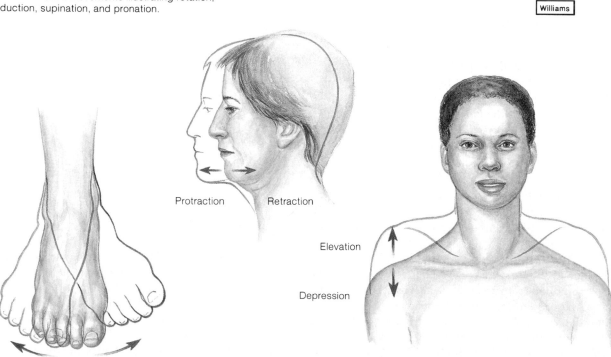

Protraction   Retraction

Elevation

Depression

Eversion   Inversion

Figure 8.12 Illustrations of several joint movements: eversion, inversion, protraction, retraction, elevation, and depression.

Williams

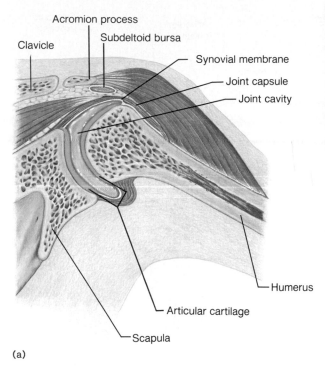

(a)

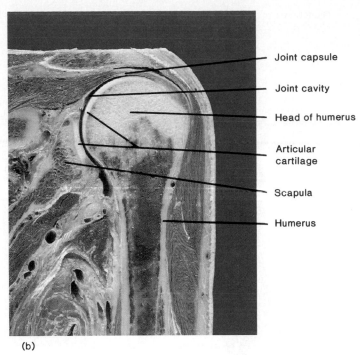

(b)

Figure 8.13 (a) The shoulder joint allows movements in all directions. (Note that there is a bursa associated with this joint.) (b) Photograph of the shoulder joint (frontal section).

1. Name six types of synovial joints.
2. Describe the structure of each type of synovial joint.
3. What terms are used to describe various movements that occur at synovial joints?

Examples of Synovial Joints

The shoulder, elbow, hip, and knee are examples of large, freely movable joints. Although these joints have much in common, each has a unique structure that is related to its specific function.

Shoulder Joint

The **shoulder joint** is a ball-and-socket joint that consists of the rounded head of the humerus and the shallow glenoid cavity of the scapula. These parts are protected above by the coracoid and acromion processes of the scapula, and they are held together by various fibrous connective tissues and muscles.

The **joint capsule** of the shoulder is attached along the circumference of the glenoid cavity and the anatomical neck of the humerus. Although it completely envelopes the joint, the capsule is very loose, and by itself is unable to keep the bones of the joint in close contact. However, the capsule is surrounded and reinforced by muscles and tendons, and these structures are largely responsible for keeping the articulating parts of the shoulder together. (See figure 8.13.)

The tendons of several muscles are intimately blended with the fibrous layer of the shoulder joint capsule, forming the *rotator cuff,* which reinforces and supports the shoulder joint. The rotator cuff is sometimes injured as a result of the centrifugal forces created when the shoulder joint is used in throwing.

The ligaments that help prevent displacement of the articulating surfaces of the shoulder joint include the following:

1. **Coracohumeral** (kor″ah-ko-hu′mer-al) **ligament.** This ligament is composed of a broad band of connective tissue that connects the coracoid process of the scapula to the greater tubercle of the humerus. It strengthens the superior portion of the joint capsule.
2. **Glenohumeral** (gle″no-hu′mer-al) **ligaments.** These include three bands of fibers that appear as thickenings in the ventral wall of the joint capsule. They extend from the edge of the glenoid cavity to the lesser tubercle and the anatomical neck of the humerus.
3. **Transverse humeral ligament.** This ligament consists of a narrow sheet of connective tissue fibers that runs between the lesser and the greater tubercles of the humerus. Together with the intertubercular groove of the humerus, the

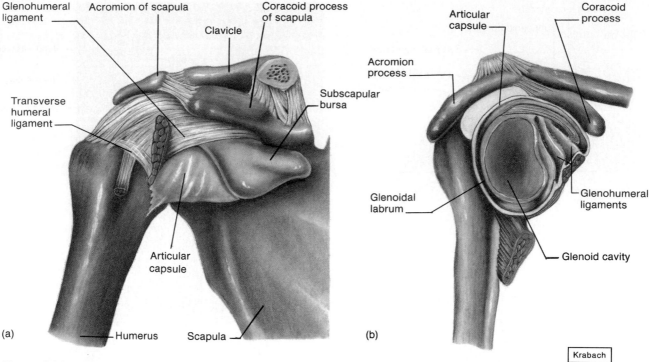

Figure 8.14 (a) The articulating surfaces of the shoulder are held together by ligaments. (b) The glenoidal labrum is a ligament composed of fibrocartilage.

ligament forms a canal (retinaculum) through which the long head of the biceps brachii muscle passes.

4. **Glenoidal labrum** (gle'noid-al la'brum). This ligament is composed of fibrocartilage. It is attached along the margin of the glenoid cavity and forms a rim with a thin, free edge that serves to deepen the cavity. (See figure 8.14.)

Several bursae are associated with the shoulder joint. The major ones include the *subscapular bursa* located between the joint capsule and the tendon of the subscapularis muscle, the *subdeltoid bursa* between the joint capsule and the deep surface of the deltoid muscle, the *subacromial bursa* between the joint capsule and the undersurface of the acromion process of the scapula, and the *subcoracoid bursa* between the joint capsule and the coracoid process of the scapula. Of these, the subscapular bursa is usually continuous with the synovial cavity of the joint cavity, and although the others do not communicate with the joint cavity, they may be connected to each other. (See figures 8.13 and 8.14.)

Due to the looseness of its attachments and the relatively large articular surface of the humerus compared to the shallow depth of the glenoid cavity, the shoulder joint is capable of a very wide range of movement. These include flexion, extension, abduction, adduction, rotation, and circumduction. Such movements may also be aided by motion occurring simultaneously in the joint formed between the scapula and the clavicle.

Because the bones of the shoulder joint are held together mainly by supporting muscles rather than by bony structures and strong ligaments, the joint is somewhat weak. Consequently, the articulating surfaces may become displaced or dislocated rather easily. Such a *dislocation* most commonly occurs during forceful abduction, as when a person falls on an outstretched arm. This movement may cause the head of the humerus to press against the lower part of the joint capsule where its wall is relatively thin and poorly supported by ligaments.

Elbow Joint

The **elbow joint** is a complex structure that includes two articulations—a hinge joint between the trochlea of the humerus and the trochlear notch of the ulna, and a gliding joint between the capitulum of the humerus and a shallow depression (fovea) on the head of the radius. These unions are completely enclosed and held together by a joint capsule, whose sides are thickened by ulnar and radial collateral ligaments and whose anterior surface is reinforced by fibers from a muscle (brachialis) in the upper arm. (See figure 8.15.)

The **ulnar collateral ligament,** which consists of a thick band of fibrous connective tissue, is located in the medial wall of the capsule. The anterior portion of this ligament connects the medial epicondyle of the humerus to the medial margin of the coronoid process of the ulna. Its posterior part is attached to the medial epicondyle of the humerus and to the olecranon process of the ulna.

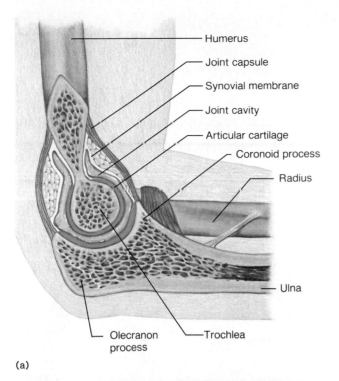

Humerus
Joint capsule
Synovial membrane
Joint cavity
Articular cartilage
Coronoid process
Radius
Ulna
Olecranon process
Trochlea

(a)

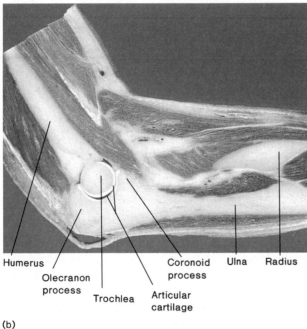

Humerus
Olecranon process
Trochlea
Coronoid process
Articular cartilage
Ulna
Radius

(b)

Figure 8.15 (a) The elbow joint allows hinge movements, as well as pronation and supination of the hand; (b) photograph of the elbow joint (sagittal section).

The **radial collateral ligament,** which strengthens the lateral wall of the joint capsule, is composed of a fibrous band extending between the lateral epicondyle of the humerus and the *annular ligament* of the radius. The annular ligament, in turn, is attached to the margin of the trochlear notch of the ulna, and it encircles the head of the radius, functioning to keep the head in contact with the radial notch of the ulna. The resulting radioulnar joint is also enclosed by the elbow joint cap-

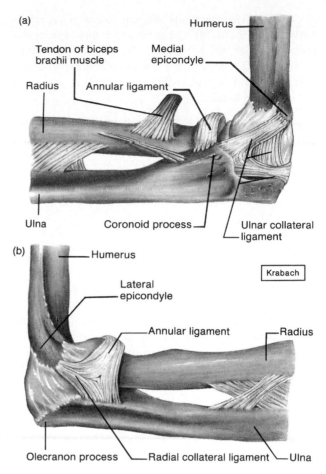

(a)
Tendon of biceps brachii muscle
Medial epicondyle
Radius
Annular ligament
Humerus
Ulna
Coronoid process
Ulnar collateral ligament

Krabach

(b)
Humerus
Lateral epicondyle
Annular ligament
Radius
Olecranon process
Radial collateral ligament
Ulna

Figure 8.16 (a) The ulnar collateral ligament and (b) the radial collateral ligament strengthen the capsular wall of the elbow joint.

sule, so that its function is closely associated with the elbow. (See figure 8.16.)

The *synovial membrane* that forms the inner lining of the elbow capsule projects into the joint cavity between the radius and ulna, and partially divides the joint into humerus-ulnar and humerus-radial portions. Also, varying amounts of adipose tissue form fatty pads between the synovial membrane and the fibrous layer of the joint capsule. These pads help protect nonarticular bony areas during joint movements.

Injuries to the elbow, shoulder, and knee are commonly diagnosed and treated using a procedure called *arthroscopy.* This procedure makes use of a thin, tubular instrument about 25 centimeters long called an *arthroscope.* The instrument, which contains optical fibers, can be inserted through a small incision in the joint capsule. A surgeon can then view the interior of the joint directly or observe an image of the joint on a television screen. In either case, the surgeon can use the arthroscope to explore the joint cavity and to guide other instruments inserted through the capsule wall in order to repair or remove injured parts.

The only movements that can occur at the elbow between the humerus and ulna are hinge-type movements—flexion and extension. The head of the radius, however, is free to rotate in the annular ligament, and this movement is responsible for pronation and supination of the hand.

1. What parts help keep the articulating surfaces of the shoulder joint together?
2. What factors allow an especially wide range of motion in the shoulder?
3. What parts make up the hinge joint of the elbow?
4. What parts of the elbow permit pronation and supination of the hand?

Hip Joint

The **hip joint** is a ball-and-socket joint that consists of the head of the femur and the cup-shaped acetabulum of the coxal bone. A ligament (ligamentum capitis) is attached to a pit (fovea capitis) on the head of the femur and to connective tissue within the acetabulum. This attachment, however, seems to have little importance in holding the articulating bones together. Instead, it serves to carry blood vessels to the head of the femur (figure 8.17).

A horseshoe-shaped ring of fibrocartilage (acetabular labrum) at the rim of the acetabulum deepens the cavity of the acetabulum. It encloses the head of the femur and helps hold it securely in place. In addition, a heavy, cylindrical joint capsule that is reinforced with still other ligaments surrounds the articulating structures and connects the neck of the femur to the margin of the acetabulum. (See figure 8.18.)

The major ligaments of the hip joint include the following:

1. **Iliofemoral** (il″e-o-fem′o-ral) **ligament.** This ligament consists of a Y-shaped band of very strong fibers that connects the anterior inferior iliac spine of the coxal bone to a bony line (intertrochanteric line) extending between the greater and lesser trochanters of the femur. The iliofemoral ligament is the strongest ligament in the body, and it helps prevent extension of the femur when the body is standing erect.
2. **Pubofemoral** (pu″bo-fem′o-ral) **ligament.** The pubofemoral ligament extends between the superior portion of the pubis and the iliofemoral ligament. Its fibers also blend with the fibers of the joint capsule.
3. **Ischiofemoral** (is″ke-o-fem′o-ral) **ligament.** This ligament consists of a band of strong fibers that originates on the ischium just posterior to the acetabulum and blends with the fibers of the joint capsule. (See figure 8.19.)

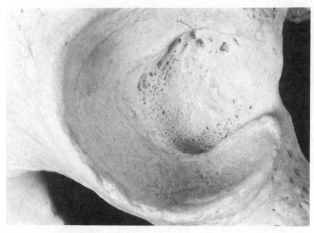

(a)

(b)

Figure 8.17 (a) The acetabulum provides the socket for the head of the femur in the hip joint. (b) The pit (fovea capitis) in the head of the femur marks the attachment of a ligament that carries blood vessels to the head of the femur.

As in the case of the shoulder, the joint capsule of the hip is surrounded by muscles. But the articulating parts of the hip are held more closely together than those of the shoulder. Thus, there is considerably less freedom of movement at the hip joint. The structure of the hip joint, however, still permits a wide variety of movements, including extension, flexion, abduction, adduction, rotation, and circumduction.

To correct problems caused by joint injury or disease, it is sometimes desirable to replace the articulating parts with artificial (prosthetic) devices. The hip joint, for example, can be totally replaced. In this procedure, called *total hip arthroplasty,* the acetabulum is replaced by a low-friction polyethylene socket that is fastened into the coxal bone with surgical bone cement. The head of the femur is replaced by a metallic, ball-shaped part.

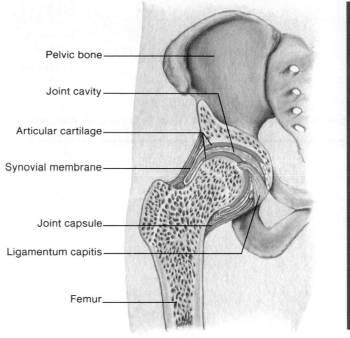

Pelvic bone

Joint cavity

Articular cartilage

Synovial membrane

Joint capsule

Ligamentum capitis

Femur

(a)

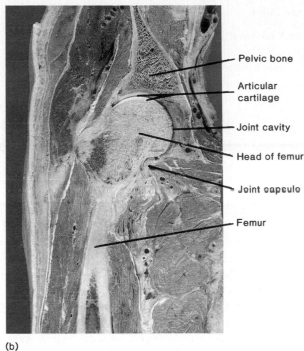

Pelvic bone

Articular cartilage

Joint cavity

Head of femur

Joint capsule

Femur

(b)

Figure 8.18 (*a*) The hip joint is held together by a ring of cartilage in the acetabulum and a joint capsule that is reinforced by ligaments; (*b*) photograph of the hip joint (frontal section).

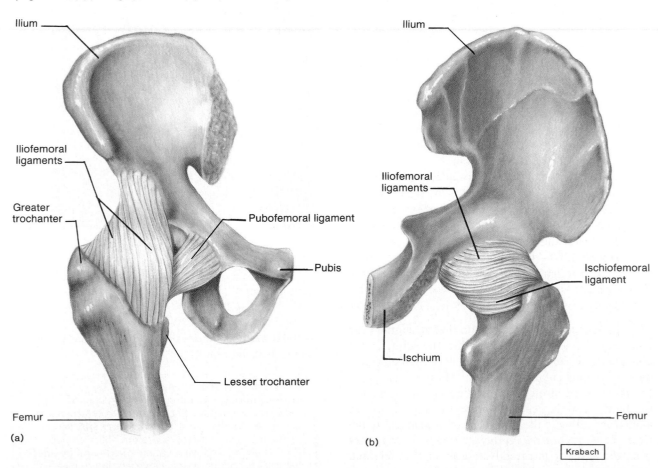

Ilium

Iliofemoral ligaments

Greater trochanter

Pubofemoral ligament

Pubis

Lesser trochanter

Femur

(a)

Ilium

Iliofemoral ligaments

Ischiofemoral ligament

Ischium

Femur

(b)

Krabach

Figure 8.19 The major ligaments of the right hip joint. (*a*) Anterior view; (*b*) posterior view.

Knee Joint

The **knee joint** is the largest and most complex of the synovial joints. It consists of the medial and lateral condyles at the distal end of the femur, and the medial and lateral condyles at the proximal end of the tibia. In addition, the femur articulates anteriorly with the patella. Although the knee is sometimes considered a modified hinge joint, the articulations between the femur and tibia are condyloid, and the joint between the femur and patella is a gliding joint.

The *joint capsule* of the knee is relatively thin, but it is greatly strengthened by ligaments and the tendons of several muscles. Anteriorly, for example, the capsule is covered by the fused tendons of several muscles in the thigh. Fibers from these tendons descend to the patella, partially enclose it, and continue downward to the tibia. The capsule is attached to the margins of the femoral and tibial condyles as well as to the areas between these condyles. (See figure 8.20.)

The ligaments associated with the joint capsule that help keep the articulating surfaces of the knee joint in contact include the following:

1. **Patellar** (pah-tel′ar) **ligament.** This ligament represents a continuation of a tendon from a large muscle group in the thigh (quadriceps femoris). It consists of a strong, flat band that extends from the margin of the patella to the tibial tuberosity.
2. **Oblique popliteal** (ŏ′blēk pop-lit′e-al) **ligament.** This ligament connects the lateral condyle of the femur to the margin of the head of the tibia.
3. **Arcuate** (ar′ku-āt) **popliteal ligament.** This ligament appears as a Y-shaped system of fibers that extends from the lateral condyle of the femur to the head of the fibula.
4. **Tibial collateral** (tib′e-al kŏ-lat′er-al) **ligament** (medial collateral ligament). This ligament is a broad, flat band of tissue that connects the medial condyle of the femur to the medial condyle of the tibia.
5. **Fibular** (fib′u-lar) **collateral ligament** (lateral collateral ligament). This ligament consists of a strong, round cord located between the lateral condyle of the femur and the head of the fibula.

In addition to the ligaments that strengthen the joint capsule, two ligaments within the joint, called **cruciate** (kroo′she-āt) **ligaments,** help prevent displacement of the articulating surfaces. These strong bands of fibrous tissue stretch upward between the tibia and the femur, crossing each other on the way. They are named according to their positions of attachment to the tibia. Thus, the *anterior cruciate ligament* originates from the anterior intercondylar area of the tibia and extends to the lateral condyle of the femur. The *posterior cruciate ligament* connects the posterior inter-

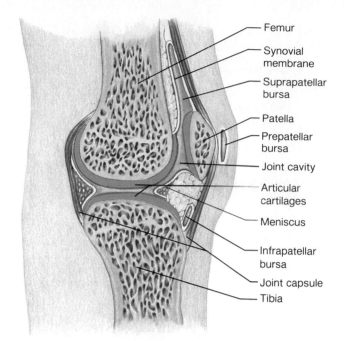

(a)

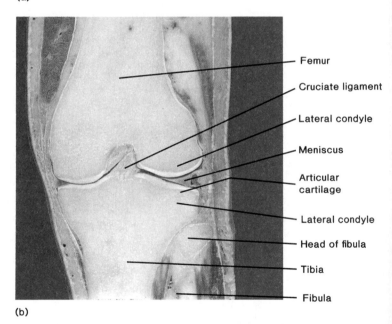

(b)

Figure 8.20 (*a*) The knee joint is the most complex of the synovial joints (sagittal section); (*b*) photograph of the knee joint (frontal section).

condylar area of the tibia to the medial condyle of the femur. (See figure 8.21.)

One of the more common serious knee injuries involves tearing the anterior cruciate ligament. Such a tear usually occurs when an athlete pivots quickly on one leg in order to change direction. At the moment of injury, the person often hears a "pop" from the knee, and within about two hours the knee typically becomes severely swollen due to bleeding within the joint.

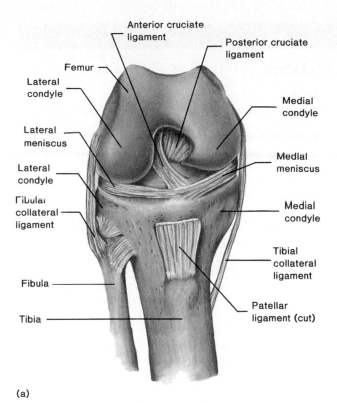

Anterior cruciate
ligament

Posterior cruciate
ligament

Femur

Lateral
condyle

Medial
condyle

Lateral
meniscus

Lateral
condyle

Medial
meniscus

Fibular
collateral
ligament

Medial
condyle

Fibula

Tibial
collateral
ligament

Tibia

Patellar
ligament (cut)

(a)

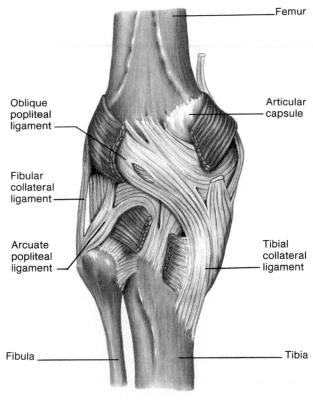

Femur

Oblique
popliteal
ligament

Articular
capsule

Fibular
collateral
ligament

Arcuate
popliteal
ligament

Tibial
collateral
ligament

Fibula

Tibia

Krabach

(b)

Figure 8.21 Ligaments within the knee joint help to
strengthen it. (a) Right anterior view; (b) left posterior view.

Separating the articulating surfaces of the femur
and tibia are two fibrocartilaginous *menisci*. Each meniscus is roughly C-shaped, with a thick rim and a
thinner center, and is attached to the head of the tibia.
The medial and lateral menisci form depressions that
fit the corresponding condyles of the femur, thus com-
pensating for the differences in shapes between the surfaces of the femur and tibia (figure 8.20).

Another common type of injury to the knee
involves tearing or displacing a meniscus. Usually
this occurs as a result of forcefully twisting the
knee when the leg is flexed. Since the meniscus is
composed of fibrocartilage, such an injury is likely
to heal very slowly. Also, if a torn and displaced
portion of cartilage becomes jammed between the
articulating surfaces, movement of the joint may be
impeded (ankylosis).

Following such a knee injury, the synovial
membrane may become inflamed (acute synovitis)
and secrete fluid excessively. As a result, the joint
cavity becomes distended so that the knee
appears enlarged above and on the sides of the
patella.

Several bursae are associated with the knee joint.
These include a large extension of the knee joint cavity
called the *suprapatellar bursa* located between the anterior surface of the lower part of the femur and the
muscle group (quadriceps femoris) above it; a large
prepatellar bursa between the patella and the skin; and
a smaller *infrapatellar bursa* between the upper part
of the tibia and the patellar ligament. (See figure 8.8.)

As with a hinge joint, the basic structure of the
knee joint permits flexion and extension. However, when
the knee is flexed, rotation is also possible.

A patient with severe pain and loss of knee joint
function as a result of rheumatoid arthritis or
osteoarthritis may receive a knee replacement.
During this surgical procedure called *total knee
arthroplasty,* the damaged surfaces of the distal
femur are removed and replaced with a smooth-
surfaced metallic device. Similarly, the proximal
surface of the tibia is removed and replaced by a
device consisting of low-friction polyethylene
attached to a metal plate. The metallic parts of
these devices are connected to the bones by
means of pegs, screws, or bone cement.

1. What parts help keep the articulating surfaces of
   the hip together?
2. What types of movement does the structure of the
   hip permit?
3. What types of joints are included within the knee?
4. What parts help hold the articulating surfaces of
   the knee together?

Disorders of Joints

Joints are subjected to considerable stress due to several factors. They are used frequently, and they must provide for a great variety of body movements. Some of them must support body weight. Injuries such as dislocations and sprains occur commonly during strenuous physical activity. In addition, joints may be affected by inflammation as well as by a number of degenerative diseases such as arthritis.

Dislocation

A *dislocation* (luxation) involves the displacement of the articulating bones of a joint. This condition usually occurs as the result of a fall or some other unusual body movement. The joints of the shoulders, knees, elbows, fingers, and jaw are common sites for such an injury. A dislocation is characterized by an obvious deformity of the joint, some loss of ability to move the parts involved, localized pain, and swelling. It presents a severe physical problem and requires medical attention.

Sprains

Sprains are the result of overstretching or tearing the connective tissues, ligaments, and tendons associated with a joint, but without dislocating the articular bones. Usually sprains are caused by forceful wrenching or twisting movements involving joints of the wrists or ankles. For example, the ankle may be sprained if it is inverted excessively when running or jumping, causing the ligaments on its lateral surface to be stretched. In severe injuries, these tissues may be pulled loose from their attachments. A sprained joint is likely to be painful and swollen, and movement at the joint may be restricted. The immediate treatment for a sprain is rest; more serious cases require medical attention. However, immobilization of a joint, even for a relatively short period, results in resorption of bone tissue and weakening of ligaments. Consequently, following such treatment, exercises may be needed to help strengthen the joint.

Bursitis

Bursitis is an inflammation of a bursa that may be caused by excessive use of a joint or by stress on a bursa. For example, the bursa between the heel bone (calcaneus) and the Achilles tendon may become inflamed as a result of a sudden increase in physical activity involving use of the feet. Similarly, a form of bursitis called "tennis elbow" involves the bursa between the olecranon process and the skin. Generally, bursitis is treated with rest, although severe cases may require medication.

Arthritis

Arthritis is a disease condition that causes inflamed, swollen, and painful joints. Although there are several types of arthritis, the most prevalent forms are rheumatoid arthritis and osteoarthritis.

Rheumatoid arthritis (RA) is an autoimmune disease (chapter 19) and is the most painful and potentially crippling of the arthritic conditions. In this type of arthritis, the synovial membrane of a joint becomes inflamed and grows thicker, forming a mass called a *pannus*. This change is usually followed by damage to the articular cartilages of the joint and by an invasion of the joint by fibrous tissues. These tissues interfere increasingly with joint movements, and, in time, they may become ossified so that the articulating bones become fused together (bony ankylosis).

Rheumatoid arthritis may affect many joints or only a few. It may be accompanied by the development of other disorders, including anemia, osteoporosis, and muscular atrophy as well as abnormal changes in the skin, eyes, lungs, blood vessels, and heart.

Although the mechanism by which joints are damaged in rheumatoid arthritis is not well understood, it seems to involve the presence of certain white blood cells (chronic inflammatory cells, such as lymphocytes and monocytes). These cells release substances (cytokines) that stimulate osteoclastic activity, which, in turn, can cause destruction of bone tissue.

Osteoarthritis (degenerative joint disease) is the most common type of arthritis. It occurs as a result of aging, as well as certain other factors, and affects a large proportion of the population over sixty years of age. In this condition, the articular cartilages soften and disintegrate gradually, so that the articular surfaces become roughened. Consequently, the joints involved are painful, and their movement is somewhat restricted. Osteoarthritis is most likely to affect joints that have received the greatest use over a person's lifetime, such as those of the fingers, hips, knees, and the lower regions of the vertebral column. Other factors that seem to increase the chance of developing osteoarthritis include joint injuries, excess body weight, and certain metabolic disorders.

As a rule, osteoarthritis develops relatively slowly, and its symptoms may be controlled by medications. In more severe cases, joint functions are disrupted, and parts of joints may need to be replaced surgically.

Clinical Terms Related to Joints

ankylosis (ang″ki-lo′sis) An abnormal stiffness of a joint or fusion of bones at a joint, often due to damage of the joint membranes from chronic rheumatoid arthritis.

arthralgia (ar-thral′je-ah) Pain in a joint.

arthrocentesis (ar″thro-sen-te′sis) Puncture and removal of fluid from a joint cavity.

arthrodesis (ar″thro-de′sis) Surgery performed to fuse the bones at a joint.

arthrogram (ar′thro-gram) X-ray film of a joint after an injection of radio-opaque fluid into the joint cavity.

arthrology (ar-throl′o-je) Study of joints and the diseases involving them.

arthropathy (ar-throp′ah-the) Any joint disease.

arthroplasty (ar′thro-plas″te) Surgery performed to make a joint more movable.

arthroscopy (ar-thros′ko-pe) Examination of the interior of a joint using a tubular instrument called an arthroscope.

arthrostomy (ar-thros′to-me) Surgical opening of a joint to allow fluid drainage.

arthrotomy (ar-throt′o-me) Surgical incision of a joint.

gout (gowt) Metabolic disease in which excessive uric acid in the blood is deposited in joints, causing them to become inflamed, swollen, and painful.

hemarthrosis (hem″ar-thros′sis) Blood in a joint cavity.

hydrarthrosis (hi″drar-thro′sis) Accumulation of fluid within a joint cavity.

luxation (luk-sa′shun) Dislocation of a joint.

subluxation (sub″luk-sa′shun) Partial dislocation of a joint.

synovectomy (sin″o-vek′to-me) Surgical removal of the synovial membrane of a joint.

Chapter Summary

Introduction (page 246)

A joint is formed wherever two or more bones meet.

Joints are the functional junctions between bones.

Classification of Joints (page 246)

Joints can be classified according to the type of tissue that binds the bones together.

1. Fibrous joints
   a. Bones at fibrous joints are fastened tightly together by a layer of fibrous connective tissue.
   b. Little or no movement occurs at a fibrous joint.
   c. There are three types of fibrous joints.
      (1) A syndesmosis is characterized by bones bound by relatively long fibers of connective tissue.
      (2) A suture occurs where flat bones are united by a thin layer of connective tissue and become interlocked by a set of bony processes.
      (3) A gomphosis is formed by the union of a cone-shaped bony process in a bony socket.

2. Cartilaginous joints
   a. Bones of cartilaginous joints are held together by a layer of cartilage.
   b. There are two types of cartilaginous joints.
      (1) A synchondrosis is characterized by bones united by hyaline cartilage that disappears as a result of growth.
      (2) A symphysis is a joint whose articular surfaces are covered by hyaline cartilage and attached to a pad of fibrocartilage.

3. Synovial joints
   a. Synovial joints have a more complex structure than other types of joints.
   b. These joints include articular cartilage, a joint capsule, and a synovial membrane.

General Structure of a Synovial Joint (page 249)

The articular ends of bones are covered by a layer of cartilage called *articular cartilage*.

1. The bones are held together by a joint capsule that is strengthened by ligaments.
2. The inner layer of the joint capsule is lined by a synovial membrane that secretes synovial fluid.
3. Synovial fluid moistens and lubricates the articular surfaces.
4. Some synovial joints are divided into compartments by menisci.
5. Some synovial joints have fluid-filled bursae.
   a. Bursae are usually located between the skin and underlying bony prominences.
   b. Bursae act as cushions and aid the movement of tendons over bony parts.
   c. Bursae are named according to their locations.

Types of Synovial Joints (page 251)

1. Ball-and-socket joints
   a. In a ball-and-socket joint, the globular head of a bone fits into the cup-shaped cavity of another.
   b. These joints permit a wide variety of movements.
   c. The hip and shoulder are ball-and-socket joints.

2. Condyloid joints
   a. A condyloid joint consists of an ovoid condyle of one bone fitting into an elliptical cavity of another.
   b. This joint permits a variety of movements.
   c. The joints between the metacarpals and phalanges are condyloid.

3. Gliding joints
   a. Articular surfaces of gliding joints are nearly flat.
   b. These joints permit the articular surfaces to slide back and forth.
   c. Most of the joints of the wrist and ankle are gliding joints.

4. Hinge joints
   a. In a hinge joint, the convex surface of one bone fits into the concave surface of another.
   b. This joint permits movement in one plane only.
   c. The elbow and the joints of the phalanges are the hinge type.
5. Pivot joints
   a. In a pivot joint, a cylindrical surface of one bone rotates within a ring of bone or fibrous tissue.
   b. This joint permits rotational movement.
   c. The articulation between the proximal ends of the radius and the ulna is a pivot joint.
6. Saddle joints
   a. A saddle joint is formed between bones that have complementary surfaces with both concave and convex regions.
   b. This joint permits a variety of movements.
   c. The articulation between the carpal and metacarpal of the thumb is a saddle joint.

Types of Joint Movements (page 254)

1. Muscles acting at synovial joints produce movements in different directions and in different planes.
2. Joint movements include flexion, extension, hyperextension, dorsiflexion, plantar flexion, abduction, adduction, rotation, circumduction, supination, pronation, eversion, inversion, elevation, depression, protraction, and retraction.

Examples of Synovial Joints (page 257)

1. Shoulder joint
   a. The shoulder joint is a ball-and-socket joint that consists of the head of the humerus and the glenoid cavity of the scapula.
   b. A cylindrical joint capsule envelops the joint. .
      (1) The capsule is loose and by itself cannot keep the articular surfaces together.
      (2) It is reinforced by surrounding muscles and tendons.
   c. Several ligaments help prevent displacement of the bones.
   d. Several bursae are associated with the shoulder joint.
   e. Because its parts are loosely attached, the shoulder joint permits a wide range of movements.
2. Elbow joint
   a. The elbow has a hinge joint between the humerus and the ulna, and a gliding joint between the humerus and the radius.
   b. The joint capsule is reinforced by collateral ligaments.
   c. A synovial membrane partially divides the joint cavity into two portions.
   d. The joint between the humerus and the ulna permits flexion and extension only.

3. Hip joint
   a. The hip joint is a ball-and-socket joint between the femur and the coxal bone.
   b. A ring of fibrocartilage deepens the cavity of the acetabulum.
   c. The articular surfaces are held together by a heavy joint capsule that is reinforced by ligaments.
   d. The hip joint permits a wide variety of movements.
4. Knee joint
   a. The knee joint includes two condyloid joints between the femur and the tibia, and a gliding joint between the femur and the patella.
   b. The joint capsule is relatively thin, but is strengthened by ligaments and tendons.
   c. Several ligaments, some of which are within the joint capsule, help keep the articular surfaces together.
   d. Two menisci separate the articulating surfaces of the femur and the tibia, and help compensate for differences in the shapes of these surfaces.
   e. Several bursae are associated with the knee joint.
   f. The knee joint permits flexion and extension; when the leg is flexed at the knee, some rotation is possible.

Clinical Application of Knowledge

1. How would you explain to an athlete why damaged joint ligaments and cartilages are so slow to heal following an injury?
2. Compared to the shoulder and hip joints, in what way is the knee joint poorly protected, and thus especially vulnerable to injuries?
3. Based upon your knowledge of joint structures, which do you think could be more satisfactorily replaced by a prosthetic device, a hip joint or a knee joint? Why?
4. If a patient's lower arm and elbow were immobilized by a cast for several weeks, what changes would you expect to occur in the bones of the arm?
5. Why is it important to encourage an inactive patient to keep all joints mobile, even if it is necessary to have another person or a device move the joints (passive movement)?
6. How would you explain to a person with a dislocated shoulder that the shoulder is likely to become dislocated more easily in the future?

Review Activities

Part A

1. Define *joint*.
2. Explain how joints are classified.
3. Compare the structure of a fibrous joint with that of a cartilaginous joint.
4. Distinguish between a syndesmosis and a suture.
5. Describe a gomphosis, and name an example.
6. Compare the structures of a synchondrosis and a symphysis.
7. Explain how the joints between adjacent vertebrae permit movement.
8. Describe the general structure of a synovial joint.
9. Describe how a joint capsule may be reinforced.
10. Explain the function of the synovial membrane.
11. Explain the function of synovial fluid.
12. Define *meniscus*.
13. Define *bursa*.
14. List six types of synovial joints, and name an example of each type.
15. Describe the movements permitted by each type of synovial joint.
16. Name the parts that comprise the shoulder joint.
17. Name the major ligaments associated with the shoulder joint.
18. Explain why the shoulder joint permits a wide range of movements.
19. Name the parts that comprise the elbow joint.
20. Describe the major ligaments associated with the elbow joint.
21. Name the movements permitted by the elbow joint.
22. Name the parts that comprise the hip joint.
23. Describe how the articular surfaces of the hip joint are held together.
24. Explain why there is less freedom of movement in the hip joint than in the shoulder joint.
25. Name the parts that comprise the knee joint.
26. Describe the major ligaments associated with the knee joint.
27. Explain the function of the menisci of the knee.
28. Describe the locations of the bursae associated with the knee joint.

Part B

Match the movements in column I with the descriptions in column II.

I	II
1. Rotation	A. Turning palm upward
2. Supination	B. Decreasing angle between parts
3. Extension	C. Moving part forward
4. Eversion	D. Moving part around an axis
5. Protraction	E. Turning sole of foot outward
6. Flexion	F. Increasing angle between parts
7. Pronation	G. Lowering a part
8. Abduction	H. Turning palm downward
9. Depression	I. Moving part away from midline

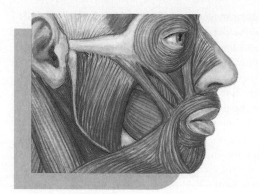

Muscular System

*M*uscles, the organs of the *muscular system,* consist largely of cells that are specialized to undergo contractions. During these contractions, chemical energy from nutrients is converted into mechanical energy, or movement.

When muscle cells contract, they pull on the body parts to which they are attached. This action usually causes movement, as when joints of the legs are flexed and extended during walking. But at other times, muscular contractions resist motion, as when they help hold body parts in postural positions. Muscles are also responsible for the movement of body fluids, such as blood and urine. In addition, they function in heat production, which helps maintain body temperature ∎

Chapter Objectives

After you have studied this chapter, you should be able to:

1. Describe how connective tissue is included in the structure of a skeletal muscle.

2. Name the major parts of a skeletal muscle fiber and describe the function of each part.

3. Explain the major events that occur during muscle fiber contraction.

4. Explain how energy is supplied to the muscle fiber contraction mechanism, how oxygen debt develops, and how a muscle may become fatigued.

5. Distinguish between fast and slow muscles.

6. Distinguish between a twitch and a sustained contraction.

7. Describe how skeletal muscles are affected by exercise.

8. Explain how various types of muscular contractions produce body movements and help maintain posture.

9. Distinguish between the structures and functions of a multiunit smooth muscle and a visceral smooth muscle.

10. Compare the fiber contraction mechanisms of skeletal, smooth, and cardiac muscles.

11. Explain how the locations of skeletal muscles are related to the movements they produce and how muscles interact to produce such movements.

12. Identify and describe the locations of the major skeletal muscles of each body region and describe the action of each muscle.

13. Complete the review activities at the end of this chapter. Note that the items are worded in the form of specific learning objectives. You may want to refer to them before reading the chapter.

Key Terms

actin (ak′tin)
antagonist (an-tag′o-nist)
aponeurosis (ap″o-nu-ro′sēz)
fascia (fash′e-ah)
insertion (in-ser′shun)
motor neuron (mo′tor nu′ron)
motor unit (mo′tor u′nit)
muscle impulse (mus′el im′puls)
myofibril (mi″o-fi′bril)
myogram (mi′o gram)
myosin (mi′o-sin)
neurotransmitter (nu″ro-trans′mit-er)
origin (or′ĭ-jin)
oxygen debt (ok′sĭ-jen det)
prime mover (prim moov′er)
recruitment (re-kroot′ment)
sarcomere (sar′ko-mēr)
synergist (sin′er-jist)
threshold stimulus (thresh′old stim′u-lus)

Aids to Understanding Words

calat-, something inserted: intercalated disk—membranous band that separates adjacent cardiac muscle cells.

erg-, work: synergist—muscle that works together with a prime mover to produce a movement.

fasc-, bundle. fasciculus—a bundle of muscle fibers.

-gram, something written: myogram—recording of a muscular contraction.

hyper-, over, more: muscular hypertrophy—enlargement of muscle fibers.

inter-, between: intercalated disk—membranous band that separates adjacent cardiac muscle cells.

iso-, equal: isotonic contraction—contraction during which the tension in a muscle remains unchanged.

laten-, hidden: latent period—the period between a stimulus and the beginning of a muscle contraction.

myo-, muscle: myofibril—contractile fiber of a muscle cell.

reticul-, a net: sarcoplasmic reticulum—network of membranous channels within a muscle fiber.

sarco-, flesh: sarcoplasm—substance (cytoplasm) within a muscle fiber.

syn-, together: synergist—muscle that works together with a prime mover to produce a movement.

tetan-, stiff: tetanic contraction—sustained muscular contraction.

-tonic, stretched: isotonic contraction—contraction during which the tension of a muscle remains unchanged.

-troph, well fed: muscular hypertrophy—enlargement of muscle fibers.

voluntar-, of one's free will: voluntary muscle—muscle that can be controlled by conscious effort.

The three types of muscle tissues are skeletal muscle, smooth muscle, and cardiac muscle, as described in chapter 5. This chapter is primarily concerned with skeletal muscle, the type found in muscles that are attached to bones and are under conscious control.

Structure of a Skeletal Muscle

A skeletal muscle is an organ of the muscular system and is composed of several kinds of tissue, including skeletal muscle tissue, nervous tissue, blood, and various connective tissues.

Connective Tissue Coverings

An individual skeletal muscle is separated from adjacent muscles and held in position by layers of fibrous connective tissue called **fascia.** This connective tissue surrounds each muscle and may project beyond the end of its muscle fibers to form a cordlike **tendon.** Fibers in a tendon intertwine with those in the periosteum of a bone, thus attaching the muscle to the bone. In other cases, the connective tissues associated with a muscle form broad, fibrous sheets called **aponeuroses,** which may be attached to the coverings of adjacent muscles (figure 9.1).

> A tendon, or the connective tissue sheath of a tendon (tenosynovium), may become painfully inflamed and swollen following an injury or the repeated stress of athletic activity. These conditions are called *tendinitis* and *tenosynovitis,* respectively. The tendons most commonly affected are those associated with the joint capsules of the shoulder, elbow, and hip, and those involved with moving the wrist, hand, thigh, and foot.

The layer of connective tissue that closely surrounds a skeletal muscle is called the *epimysium.* Other layers of connective tissue, called the *perimysium,* extend inward from the epimysium and separate the muscle tissue into small sections. These sections contain bundles of skeletal muscle fibers called *fascicles* (fasciculi). Each muscle fiber within a fascicle (fasciculus) is surrounded by a layer of connective tissue in the form of a thin covering called *endomysium* (figure 9.2).

Thus, all parts of a skeletal muscle are enclosed in layers of connective tissue. This arrangement allows the parts to move independently. Also, numerous blood vessels and nerves pass through these layers (figure 9.3).

The fascia associated with the individual organs of the muscular system is part of a complex network of

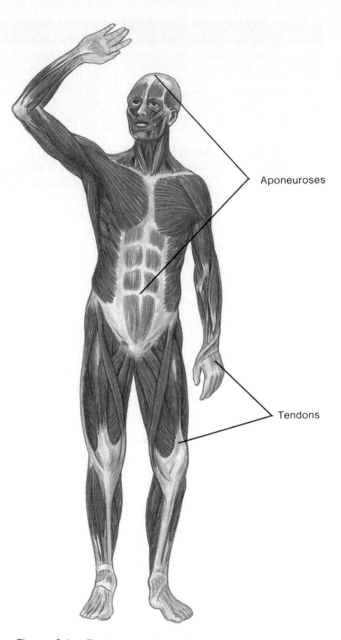

Figure 9.1 Tendons attach muscles to bones, while aponeuroses attach muscles to other muscles.

fasciae that extends throughout the body. The portion of the network that surrounds and penetrates the muscles is called *deep fascia.* It is continuous with the *subcutaneous fascia* that lies just beneath the skin, forming the subcutaneous layer described in chapter 6. The network is also continuous with the *subserous fascia* that forms the connective tissue layer of the serous membranes covering organs in various body cavities and lining those cavities (see chapter 6).

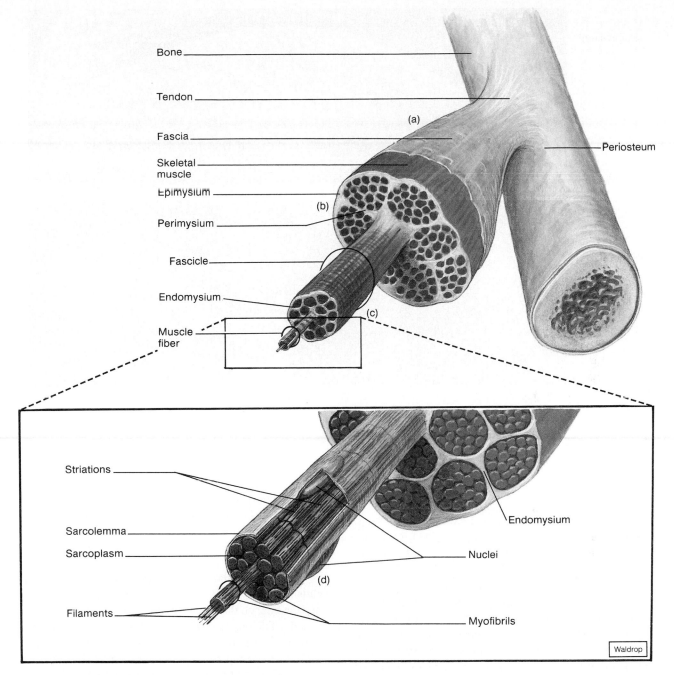

Bone

Tendon

Fascia

Skeletal muscle

Epimysium

Perimysium

Fascicle

Endomysium

Muscle fiber

(a)

(b)

(c)

Periosteum

Striations

Sarcolemma

Sarcoplasm

Filaments

(d)

Endomysium

Nuclei

Myofibrils

Waldrop

Figure 9.2 (*a*) A skeletal muscle is composed of a variety of tissues, including layers of connective tissue. (*b*) Fascia covers the surface of the muscle, epimysium lies beneath the fascia, and perimysium extends into the structure of the muscle where it separates muscle cells into fascicles. (*c*) Individual muscle fibers are separated by endomysium. (*d*) A single muscle fiber.

The space occupied by a particular group of muscles, blood vessels, and nerves, all tightly enclosed by relatively inelastic fascia, constitutes a *compartment*. There are many such spaces in the arms and legs. If an injury causes fluid, such as blood from an internal hemorrhage, to accumulate within a compartment, the pressure inside will rise. The increased pressure, in turn, may interfere with blood flow into the region, thus reducing the supply of oxygen and nutrients to the affected tissues. This condition, called *compartment syndrome*, often produces severe, unrelenting pain, and if the compartmental pressure remains elevated, the enclosed muscles and nerves may be irreversibly damaged.

Treatment for compartment syndrome may involve making a surgical incision through the fascia (fasciotomy) to relieve the excessive pressure and restore the circulation of blood.

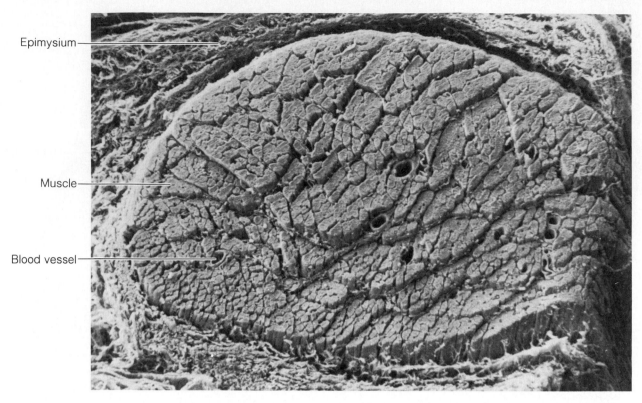

Epimysium

Muscle

Blood vessel

Figure 9.3 Scanning electron micrograph of a muscle surrounded by its connective tissue sheath, the epimysium (×215). (*Tissues and Organs: A Text-Atlas of Scanning Electron Microscopy,* by R. G. Kessel and R. H. Kardon. © 1979 W.H. Freeman and Company.)

Skeletal Muscle Fibers

As mentioned in chapter 5, a skeletal muscle fiber represents a single cell of a muscle (figure 5.25). This fiber responds to stimulation by contracting and then relaxing.

Each skeletal muscle fiber is a thin, elongated cylinder with rounded ends that are attached to connective tissues associated with a muscle. Just beneath its cell membrane (or *sarcolemma*), the cytoplasm (or *sarcoplasm*) of the fiber contains many small, oval nuclei and mitochondria. The sarcoplasm also contains numerous, threadlike **myofibrils** that lie parallel to one another (figure 9.2).

The myofibrils play a fundamental role in the muscle contraction mechanism. They contain two kinds of protein filaments—thick ones composed of the protein **myosin** and thin ones composed of the protein **actin.** The arrangement of these filaments produces the characteristic alternating light and dark *striations* of skeletal muscle fiber (figures 9.4 and 9.5).

Myosin filaments are located within the dark portions, or *A bands,* of the striations, and actin filaments occur primarily in the light areas, or *I bands.* The actin filaments, however, also extend into the A bands, and when the muscle fiber contracts, the actin filaments slide farther into these bands. Note in figure 9.4b that only myosin filaments occur between the ends of the actin filaments within the A bands. As a result, light H zones appear in the centers of the A bands. Within the H zone, thickenings of the myosin filaments produce an *M line.*

The actin filaments are attached to *Z lines* at the ends of the I bands. These Z lines extend across the muscle fiber so that those of adjacent myofibrils lie side by side. The segment of the myofibril between two successive Z lines is called a **sarcomere** (figures 9.4 and 9.5). The consequent regular arrangement of the sarcomeres causes the muscle fiber to appear striated.

Although muscle fibers and the connective tissues associated with them are flexible, they can be torn if they are overstretched. This type of injury is common in athletes and is called a *muscle strain* or *muscle pull.* The seriousness of the injury depends on the degree of damage sustained by the tissues. In a mild strain, for example, only a few muscle fibers are injured, the fascia remains intact, and there is little loss of function. In a severe strain, many muscle fibers as well as fascia are torn, and muscle function may be lost completely. A severe strain is very painful and is accompanied by discoloration and swelling of tissues due to ruptured blood vessels. Such an injury may require surgery to reconnect the separated tissues.

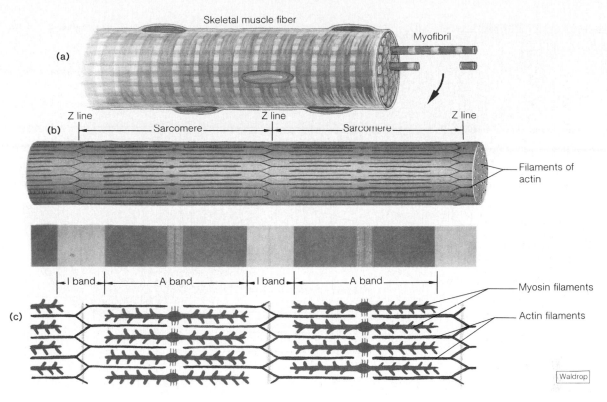

(a)

Skeletal muscle fiber

Myofibril

(b)

Z line   Z line   Z line

Sarcomere   Sarcomere

Filaments of actin

I band   A band   I band   A band

(c)

Myosin filaments

Actin filaments

Waldrop

Figure 9.4 (*a*) A skeletal muscle fiber contains numerous myofibrils, each consisting of (*b*) units called sarcomeres. (*c*) The characteristic striations of a sarcomere are due to the arrangement of actin and myosin filaments.

Figure 9.5 Identify the bands of the striations in this transmission electron micrograph of myofibrils (×20,000).

Within the cytoplasm of a muscle fiber is a network of membranous channels that surrounds each myofibril and runs parallel to it. These membranes form the **sarcoplasmic reticulum,** which corresponds to the endoplasmic reticulum of other cells. Another set of membranous channels called **transverse tubules** (T-tubules) extends inward, as invaginations from the fiber's membrane, and passes all the way through the fiber. Thus, each of these tubules opens to the outside of the muscle fiber and contains extracellular fluid. Furthermore, each transverse tubule lies between two enlarged portions of the sarcoplasmic reticulum called *cisternae* near the region where the actin and myosin filaments overlap. The sarcoplasmic reticulum and the transverse tubules activate the muscle contraction mechanism when the fiber is stimulated (figure 9.6).

1. Describe how connective tissue is associated with a skeletal muscle.
2. Describe the general structure of a skeletal muscle fiber.
3. Explain why skeletal muscle fibers appear striated.
4. Explain the relationship between the sarcoplasmic reticulum and the transverse tubules.

Neuromuscular Junction

Each skeletal muscle fiber is connected to a fiber of a nerve cell. Such a nerve fiber is an extension of a **motor neuron** that passes outward from the brain or spinal cord. Usually a skeletal muscle fiber will contract only when it is stimulated by the action of a motor neuron.

The site where the nerve fiber and muscle fiber meet is called a **neuromuscular junction** (myoneural junction). At this junction, the muscle fiber membrane is specialized to form a **motor end plate.** In this region of the muscle fiber, nuclei and mitochondria are abundant, and the sarcolemma is extensively folded (figure 9.7).

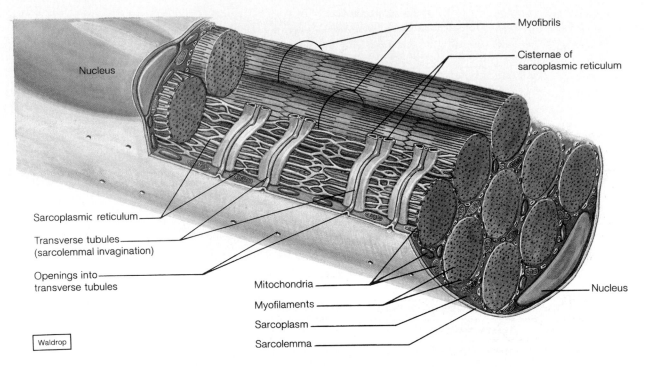

Figure 9.6   Within the sarcoplasm of a skeletal muscle fiber are a network of sarcoplasmic reticulum and a system of transverse tubules.

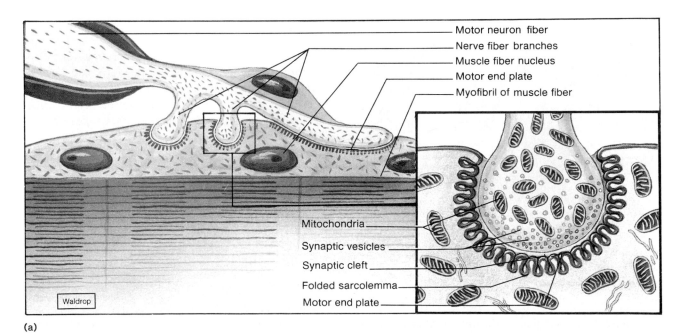

(a)

Figure 9.7   (a) A neuromuscular junction includes the end of a motor neuron and the motor end plate of a muscle fiber.

The end of the motor nerve fiber is branched, and the ends of these branches project into recesses (synaptic clefts) of the muscle fiber membrane. The cytoplasm at the distal ends of the nerve fibers is rich in mitochondria and contains many tiny vesicles (synaptic vesicles) that store chemicals called **neurotransmitters.**

When a nerve impulse traveling from the brain or spinal cord reaches the end of a motor nerve fiber, some of the vesicles release a neurotransmitter into the gap between the nerve and the motor end plate of the muscle fiber. This action stimulates the muscle fiber to contract.

Motor Units

Although a muscle fiber usually has a single motor end plate, motor nerve fibers are densely branched. By means of these fibers, one motor fiber may be connected to many muscle fibers. Furthermore, when a motor nerve fiber transmits an impulse, all of the muscle fibers connected to it are stimulated to contract simultaneously. Together, a motor neuron and the muscle fibers it controls constitute a **motor unit** (figure 9.8).

The number of muscle fibers in a motor unit varies considerably. The fewer muscle fibers in the motor units,

however, the finer the movements that can be produced in a particular muscle. For example, the motor units of the muscles that move the eyes may contain fewer than ten muscle fibers per motor unit and can produce very slight movements. Conversely, the motor units of the large muscles in the back may include a hundred or more muscle fibers. When these motor units are stimulated, the movements produced are coarse in comparison with those of the eye muscles.

Skeletal Muscle Contraction

A muscle fiber contraction is a complex action involving a number of cell parts and chemical substances. The final result is a sliding movement within the myofibrils in which the filaments of actin and myosin merge. When this happens, the muscle fiber is shortened, and it pulls on its attachments.

Role of Myosin and Actin

Myosin is the most abundant of the muscle proteins and accounts for about two-thirds of the protein within skeletal muscles. A myosin molecule is composed of two twisted protein strands with globular parts called crossbridges projecting outward along their lengths. Many

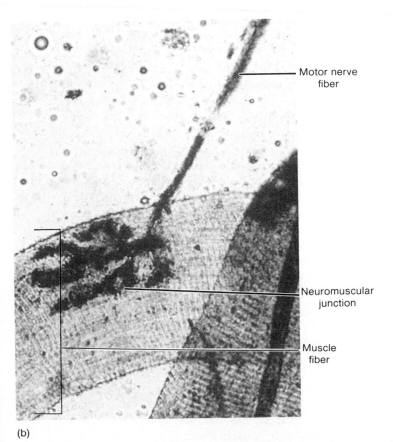

(b)

Figure 9.7 continued. (b) Micrograph of a neuromuscular junction.

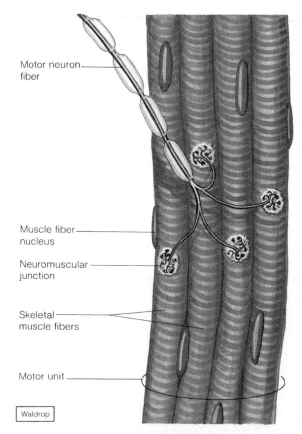

Figure 9.8 A motor unit consists of one motor neuron and all the muscle fibers with which it communicates.

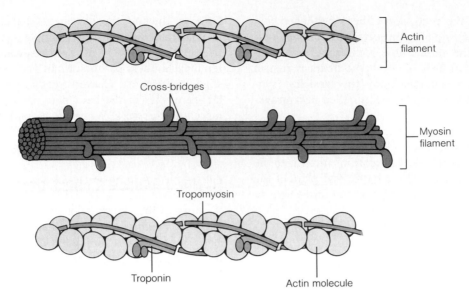

Figure 9.9 Myosin molecules have cross-bridges that extend toward nearby actin filaments.

of these molecules comprise a myosin filament. In the presence of calcium ions, the myosin cross-bridges can react with actin filaments and form linkages with them. This reaction between the myosin and actin filaments generates the force involved in shortening the myofibrils during a muscle contraction.

Actin accounts for about one-fourth of the total protein in skeletal muscle. An actin molecule is a globular structure with ADP molecules attached to its surface. These ADP molecules serve as active sites with which the cross-bridges of the myosin molecules interact to form linkages. Many of these actin molecules, arranged together in a double twisted strand (helix), form an actin filament.

Two other proteins, **tropomyosin** and **troponin,** are associated with an actin filament. The tropomyosin molecules are rod-shaped and occupy the longitudinal grooves of the actin helix. Each tropomyosin has a troponin molecule attached to its surface, forming a tropomyosin-troponin complex (figure 9.9).

When a muscle fiber is at rest, these tropomyosin-troponin complexes seem to inhibit the active sites on the actin molecules and thus prevent the formation of linkages. If a high concentration of *calcium ions* is present, however, the calcium ions bind to the troponin, and this apparently modifies the position of the tropomyosin. As the tropomyosin molecules move, the active sites on the actin filaments are exposed, and linkages can form between the actin and myosin filaments.

It is not completely understood how the formation of linkages results in shortening of the myofibrils. One theory (the ratchet theory) suggests that the head of a myosin cross-bridge can attach to an actin active site and bend slightly, pulling the actin filament with it. Then the head may release, straighten itself, and combine with another active site further down the actin filament. Presumably this cycle can be repeated again and again, and as the actin filament is moved toward the center of the sarcomere, the sarcomere shortens (figures 9.10 and 9.11).

Stimulus for Contraction

A skeletal muscle fiber normally does not contract until it is stimulated by a neurotransmitter. In skeletal muscle, the neurotransmitter is a compound called **acetylcholine.** This substance is synthesized in the cytoplasm of the motor neuron and is stored in vesicles near the distal end of its motor nerve fiber. When a nerve impulse reaches the end of the axon, some of these vesicles release their acetylcholine into the gap between the nerve fiber and the motor end plate (figure 9.7a).

The acetylcholine diffuses rapidly across the gap, combines with certain protein molecules (receptors) in the sarcolemma, and thus stimulates the muscle fiber membrane. As a result of this stimulus, a **muscle impulse** (action potential), very much like a nerve impulse (described in chapter 10), is generated, and the muscle impulse passes in all directions over the surface of the sarcolemma. It also travels through the transverse tubules, deep into the fiber, and reaches the sarcoplasmic reticulum.

The sarcoplasmic reticulum contains a high concentration of calcium ions. In response to a muscle impulse, the membranes of the cisternae become more permeable to these ions, and the ions diffuse into the sarcoplasm of the muscle fiber.

When a relatively high concentration of calcium ions is present in the sarcoplasm, linkages form between the actin and myosin filaments, and a muscle

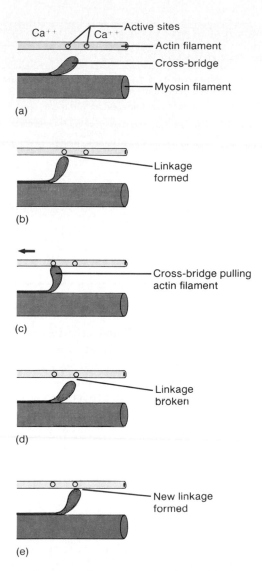

Figure 9.10 According to the ratchet theory, (a) when calcium ions are present, active sites on an actin filament are exposed. (b) Cross-bridges on a myosin filament form linkages at the active sites. (c) A myosin cross-bridge bends slightly, pulling an actin filament. (d) The linkage is broken, and (e) the myosin cross-bridge forms a linkage with the next active site.

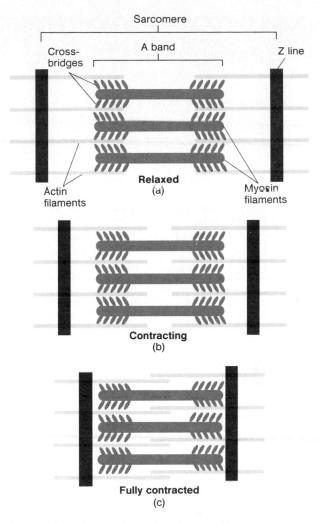

Figure 9.11 (a) As linkages form between filaments of actin and myosin, (b and c) the actin filaments are pulled toward the center of the A band, causing the myofibril to contract.

contraction occurs. The contraction continues while the calcium ions are present, but the ions are quickly moved back into the sarcoplasmic reticulum by an active transport mechanism (calcium pump). The calcium ion concentration of the sarcoplasm is decreased, the linkages are broken, and, once again, the interaction between the filaments is inhibited by the troponin and tropomyosin molecules.

Meanwhile, the acetylcholine that stimulated the muscle fiber in the first place is rapidly decomposed by the action of an enzyme called **cholinesterase.** This enzyme is present at the neuromuscular junction within the membranes of the motor end plate, and its action prevents a single nerve impulse from causing a continued stimulation of the muscle fiber.

When the acetylcholine is decomposed, the stimulus to the sarcolemma and membranes within the muscle fiber ceases. In response, these membranes quickly return to their original resting conditions. Consequently, the muscle fiber relaxes.

Chart 9.1 summarizes the major events leading to muscle contraction and relaxation.

A substance called botulinus toxin, produced by a bacterium (*Clostridium botulinum*), can prevent the release of acetylcholine from motor nerve fibers at the neuromuscular junctions. This bacterium is responsible for a very serious form of food poisoning called *botulism.* This condition is most likely to result from eating home-processed food that has not been heated enough to kill the bacteria present in it or to inactivate the toxin.

When botulinus toxin affects the body, muscle fibers fail to be stimulated, and muscles, including those responsible for breathing, may be paralyzed. Without prompt medical treatment, botulism may cause death.

CHART 9.1 Major events of muscle contraction and relaxation

Muscle fiber contraction	Muscle fiber relaxation
1. Acetylcholine is released from the distal end of a motor neuron.	1. Cholinesterase causes acetylcholine to decompose, and the muscle fiber membrane is no longer stimulated.
2. Acetylcholine diffuses across the gap at the neuromuscular junction.	2. Calcium ions are actively transported into the sarcoplasmic reticulum.
3. The sarcolemma is stimulated, and a muscle impulse travels over the surface of the muscle fiber and deep into the fiber through the transverse tubules and reaches the sarcoplasmic reticulum.	3. Linkages between actin and myosin filaments are broken.
4. Calcium ions diffuse from the sarcoplasmic reticulum into the sarcoplasm and bind to troponin molecules.	4. Troponin and tropomyosin molecules inhibit the interaction between myosin and actin filaments.
5. Tropomyosin molecules move and expose specific sites on actin filaments.	5. Actin and myosin filaments slide apart.
6. Linkages form between actin and myosin filaments.	6. Muscle fiber lengthens as it relaxes and its resting state is reestablished.
7. Actin filaments slide inward along the myosin filaments.	
8. Muscle fiber shortens as a contraction occurs.	

1. Describe a neuromuscular junction.
2. Define a motor unit.
3. List four proteins associated with myofibrils, and explain their relationships.
4. Explain how the filaments of a myofibril interact during muscle contraction.
5. Explain how a motor nerve impulse can trigger a muscle contraction.

Energy Sources for Contraction

The energy used during muscle fiber contraction comes from ATP molecules, which are supplied by numerous mitochondria that are positioned close to the myofibrils. The globular portions of the myosin filaments contain an enzyme called **ATPase.** This enzyme causes ATP to decompose into ADP and phosphate, and at the same time, to release energy. (See chapter 4.) This energy makes possible the reaction between the actin and myosin filaments. However, there is only enough ATP in a muscle fiber to operate the contraction mechanism for a very short time. Consequently, when a fiber is active, ATP must be regenerated.

The primary source of energy available to regenerate ATP from ADP and phosphate is a substance called **creatine phosphate.** Like ATP, creatine phosphate contains high-energy phosphate bonds, and it is actually four to six times more abundant in the muscle fibers than ATP. Creatine phosphate, however, cannot directly supply energy to a cell's energy-utilizing reactions. Instead, it stores excess energy released from mitochondria. Thus, whenever sufficient ATP is present, an enzyme in the mitochondria (creatine phosphokinase) promotes the synthesis of creatine phosphate, and the excess energy is stored in its bonds (figure 9.12).

When ATP is being decomposed, the energy from creatine phosphate can be transferred to ADP molecules, and they, in turn, are quickly converted back into ATP. The amount of ATP and creatine phosphate present in a skeletal muscle, however, is usually not sufficient to support maximal muscle activity for more than a few seconds. Consequently, the muscle fibers in an active muscle soon become dependent upon cellular respiration of glucose as a source of energy for synthesizing ATP. Typically, a muscle has a supply of glucose in the form of stored glycogen molecules.

Oxygen Supply and Cellular Respiration

As described in chapter 4, the early phase of cellular respiration occurs in the cytoplasm and is *anaerobic,* taking place in the absence of oxygen. This phase involves a partial breakdown of energy-supplying molecules such as glucose, and results in a gain of only a few ATP molecules. The complete breakdown of glucose occurs in the mitochondria and is *aerobic.* This process, which includes the complex series of reactions of the *citric acid cycle,* produces a relatively large number of ATP molecules.

The oxygen needed to support aerobic respiration is carried from the lungs to body cells by the blood. Oxygen is transported within the red blood cells, loosely bound to molecules of hemoglobin, the pigment responsible for the red color of the blood. In regions of the body where the oxygen concentration is relatively low, oxygen is released from hemoglobin and becomes available for cellular respiration.

Another pigment, **myoglobin,** is synthesized in the muscle cells and is responsible for the reddish brown color of skeletal muscle tissue. Myoglobin has properties similar to hemoglobin in that it can combine loosely with oxygen. In fact, myoglobin (muscle hemoglobin) has a greater attraction for oxygen than does hemoglobin, and it can store oxygen in muscle tissue, at least temporarily. This ability to store oxygen reduces a muscle's need for a continuous blood supply

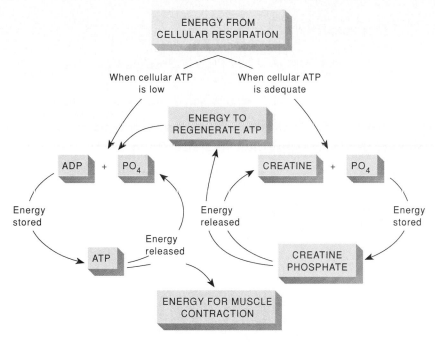

Figure 9.12 Energy released by cellular respiration may be used to promote the synthesis of ATP or to synthesize creatine phosphate. Later, energy from creatine phosphate may be used to promote ATP synthesis.

during muscular contraction, which is important because blood flow may decrease during muscular contraction, as blood vessels are compressed by the contracting muscle fibers nearby (figure 9.13).

Oxygen Debt

When a person is resting or moderately active, the respiratory and circulatory systems can usually supply sufficient oxygen to the skeletal muscles to support aerobic respiration. However, when skeletal muscles are used strenuously for even a minute or two, these systems usually cannot supply enough oxygen to meet the needs of aerobic respiration. Consequently, the muscle fibers must depend increasingly on the anaerobic phase of respiration for their energy.

In anaerobic respiration, glucose molecules are changed to *pyruvic acid*. (See chapter 4.) If the oxygen supply is low, however, the pyruvic acid is converted quickly to *lactic acid* (lactate), which diffuses out of the muscle fibers and is carried to the liver by the blood.

The liver cells can change the lactic acid into *glucose*, but this conversion also requires energy from ATP. During strenuous exercise, the available oxygen is used primarily to synthesize the ATP needed for the muscle fiber contraction mechanism rather than to make ATP for changing lactic acid into glucose. Consequently, as lactic acid accumulates, a person develops an **oxygen debt** that must be repaid at a later time. The amount of oxygen debt acquired is equal to the amount of oxygen needed by the liver cells to convert the accumulated lactic acid into glucose, plus the amount needed by the muscle cells to resynthesize ATP and creatine phosphate, and return these substances to their original concentrations (figure 9.14).

The conversion of lactic acid back into glucose is a relatively slow process. It may require several hours to repay an oxygen debt following strenuous exercise.

Muscle Fatigue

If a muscle is exercised strenuously for a prolonged period, it may lose its ability to contract, a condition called *fatigue*. This condition may result from an interruption in the muscle's blood supply or, rarely, from an exhaustion of the supply of acetylcholine in its motor nerve fibers. However, muscle fatigue is most likely to arise from an accumulation of lactic acid in the muscle as a result of anaerobic respiration. The lactic acid causes factors, such as pH, to change so that muscle fibers are no longer responsive to stimulation.

Occasionally a muscle becomes fatigued and develops cramps at the same time. A cramp is a painful condition in which the muscle contracts spasmodically, but does not relax completely. This condition seems to be due to a lack of ATP, which is needed to move calcium ions back into the sarcoplasmic reticulum and break the linkages between actin and myosin filaments before muscle fibers can relax.

Tolerance to lactic acid varies, so that some persons experience muscular fatigue more quickly than others. Resistance to the effects of lactic acid also varies

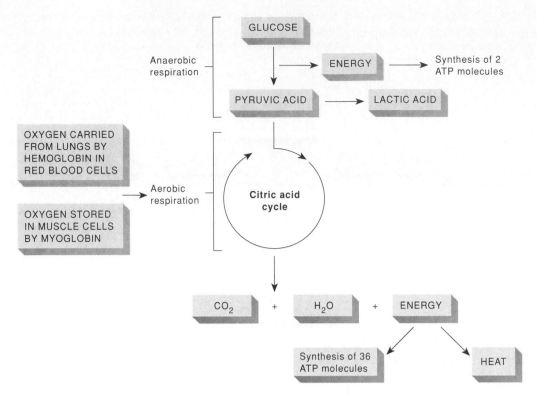

Figure 9.13 The oxygen needed to support aerobic respiration is carried in the blood and stored in myoglobin.

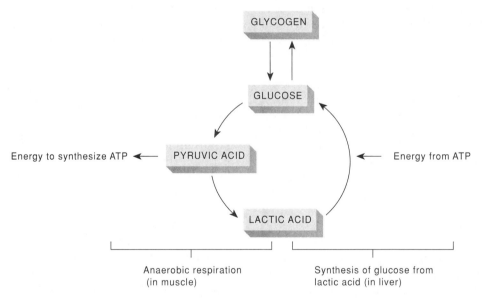

Figure 9.14 The lactic acid that accumulates in muscles as a result of anaerobic respiration can be converted back into glucose by liver cells.

among athletes, although in general, athletes can exercise and produce less lactic acid than nonathletes. In part, this is because the strenuous exercise of physical training stimulates new capillaries to grow within the muscles, allowing more oxygen and nutrients to be supplied to the muscle fibers of athletes. Such physical training also causes muscle fibers to produce additional mitochondria, increasing their ability to carry on cellular respiration.

> A few hours after death, the skeletal muscles undergo a partial contraction that causes the joints to become fixed. This condition, *rigor mortis*, may continue for seventy-two hours or more. It seems to result from an increase in membrane permeability to calcium ions, which promotes contraction, and a decrease in availability of ATP in the muscle fibers, which prevents relaxation. Thus, the actin and myosin filaments of the muscle fibers remain linked together until the muscles begin to decompose.

Fast and Slow Muscles

Muscles vary in the speeds at which they contract. Those that move the eyes contract about ten times faster than those responsible for maintaining posture, and the muscles that move the limbs contract at intermediate rates. Thus, the speed of contraction is related to the special function of the muscle.

The slow-contracting (slow-twitch) postural muscles, such as the long muscles of the back, are often called *red muscles* because most of their fibers contain the red, oxygen-storing pigment myoglobin. These fibers also are well supplied with blood containing the red, oxygen-carrying pigment hemoglobin. In addition, red muscle fibers contain many mitochondria, so they are well adapted to carry on aerobic respiration. These fibers have a high respiratory capacity in that they can generate ATP fast enough to keep up with the ATP breakdown that occurs when they contract. For this reason, they can contract for prolonged periods without undergoing fatigue.

Fast-contracting (fast-twitch) muscles are called *white muscles* because they contain less myoglobin and have a poorer blood supply than red muscles. They include certain hand muscles as well as those that move the eyes. Most of the fibers of these muscles have fewer mitochondria and thus have a reduced respiratory capacity. However, the white fibers have more extensive sarcoplasmic reticulum, which stores and reabsorbs calcium ions, and their ATPase has higher activity than that of red fibers. Because of these factors, white muscle fibers can contract rapidly, though they tend to fatigue as ATP and the substances needed to regenerate ATP are depleted and as lactic acid accumulates.

Although some muscles contain a predominate number of slow-contracting red fibers or fast-contracting white fibers, most skeletal muscles contain a mixture of the two types. Also, muscle fibers with intermediate characteristics have been detected.

Heat Production

Heat is produced as a by-product of cellular respiration, so it is generated by all active cells. Since muscle tissue represents such a large proportion of the total body mass, it is a major source of heat.

Only about 25% of the energy released by cellular respiration is available for use in metabolic processes; the rest becomes heat. Thus, whenever muscles are active, large amounts of heat are released. This heat is transported to other tissues by the blood and helps maintain body temperature. If heat is present in excess, mechanisms that promote heat loss (described in chapter 6) are activated, and homeostasis is maintained.

1. What substances provide the energy used to regenerate ATP?
2. What are the sources of the oxygen needed for aerobic respiration?
3. How are lactic acid, oxygen debt, and muscle fatigue related?
4. Distinguish between fast-contracting and slow-contracting muscles.
5. What is the relationship between cellular respiration and heat production?

Muscular Responses

One way to observe muscle contraction is to remove a single muscle fiber from a skeletal muscle and connect it to a device that senses and records changes in the fiber's length. In such experiments, the muscle fiber is usually stimulated with an electrical stimulator capable of producing stimuli of varying strengths and frequencies.

Threshold Stimulus

By exposing an isolated muscle fiber to a series of stimuli of increasing strength, it can be shown that the fiber remains unresponsive until a certain strength of stimulation is applied. This minimal strength needed to cause a contraction is called the **threshold stimulus.**

All-or-None Response

When a muscle fiber is exposed to a stimulus of threshold strength (or above), it responds to its fullest extent. Increasing the strength of the stimulus does not affect the degree to which the fiber contracts. In other words, there are no partial contractions of a muscle fiber—if it contracts at all, it contracts completely, even though it may not shorten completely. This phenomenon is called the **all-or-none response.**

Recording a Muscle Contraction

To record how a whole muscle responds to stimulation, a skeletal muscle can be removed from a frog or other small animal and mounted in a special laboratory apparatus. The muscle is then stimulated electrically, and when it contracts, it pulls on a lever. The lever's movement is recorded, and the resulting pattern is called a **myogram.**

If a muscle is exposed to a single stimulus of sufficient strength to activate some of its motor units, the muscle will contract and then relax. This action—a single contraction that lasts only a fraction of a second—is called a **twitch.** A twitch produces a myogram like that in figure 9.15. It is apparent from this recording that the muscle response did not begin immediately following the stimulation. There was a delay between the time the stimulus was applied and the time the muscle responded. This time lag is called the **latent period.** In a frog muscle, the latent period lasts for about 0.01 second, and in a human muscle, it is even shorter.

The latent period is followed by a *period of contraction* during which the muscle pulls at its attachments, and a *period of relaxation* during which it returns to its former length (figure 9.15).

If a muscle is exposed to two stimuli (of threshold strength or above) in quick succession, it may respond with a twitch to the first stimulus but not to the second. This is

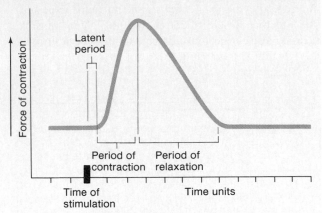

Figure 9.15 A myogram of a single muscle twitch.

because it takes an instant following a contraction for the muscle fibers to reestablish the electrolyte concentrations necessary to conduct another muscle impulse. Thus, there is a very brief moment following stimulation during which a muscle remains unresponsive. This time is called the **refractory period.**

Recruitment of Motor Units

Since the muscle fibers within a muscle are organized into motor units, and each motor unit is controlled by a single motor neuron, all the muscle fibers in a motor unit are stimulated at the same time. Consequently, a motor unit also responds in an all-or-none manner. A whole muscle, however, does not behave like this, because it is composed of many motor units controlled by different motor neurons, which respond to different thresholds of stimulation. Thus, if only the motor neurons with low thresholds of stimulation are stimulated, a relatively small number of motor units contract. At higher intensities of stimulation, other motor neurons respond and more motor units are activated. Such an increase in the number of motor units being activated is called **recruitment.** As the intensity of stimulation increases, recruitment of motor units continues until finally all possible motor units are activated and the muscle is contracting with maximal tension.

Staircase Effect

If a muscle that has been inactive is subjected to a series of stimuli, it undergoes a series of twitches (figure 9.16a). However, the strength of each successive con-

traction increases until a maximum is reached. This phenomenon is called the **staircase effect** (treppe). (See figure 9.16b.)

Although the cause of the staircase effect is poorly understood, it seems to involve a net increase in the concentration of calcium ions available in the sarcoplasm of the muscle fibers. This increase might occur if each stimulus in the series caused the release of calcium ions and if the sarcoplasmic reticulum failed to recapture those ions immediately.

Sustained Contractions

If a muscle is exposed to a series of stimuli of increasing frequency, a point is reached when the muscle is unable to complete its relaxation period before the next stimulus in the series arrives. When this happens, the individual contractions begin to combine and the muscle contraction becomes *sustained.* Such a combination of twitches is called *summation of twitches* or *wave summation* (figure 9.16c).

At the same time that twitches are combining, the strength of the contractions may be increasing. This is due to the recruitment of motor units. The smaller motor units, which have finer fibers, tend to respond earlier in the series of stimuli. The larger motor units,

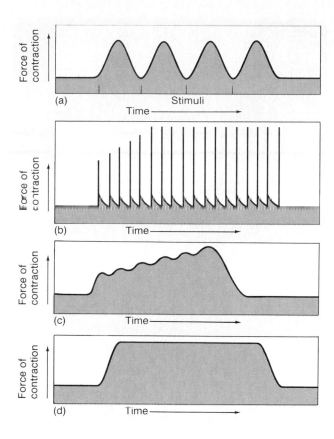

Figure 9.16 Myograms of (a) a series of twitches, (b) a staircase effect, (c) a combining of twitches, and (d) a tetanic contraction. (Note: The frequency of stimulation increases from one myogram to the next.)

which contain thicker fibers, respond later and produce more forceful contractions. The production of a sustained contraction of increasing strength is called *multiple motor unit summation*. When the resulting forceful, sustained contraction lacks even partial relaxation, it is termed a **tetanic** (tĕ-tan′ik) **contraction** (tetanus) (figure 9.16d).

Although twitches may occur occasionally in human skeletal muscles, as when an eyelid twitches, such contractions are of limited use. More commonly, muscular contractions are sustained, and they are smooth rather than irregular or jerky because a mechanism operates within the spinal cord to stimulate contractions in different sets of motor units at different moments. Thus, while some motor units are contracting, others are relaxing.

Tetanic contractions occur frequently in skeletal muscles during everyday activities. In many cases, the condition occurs in only a portion of a muscle. For example, when a person lifts a weight or walks, sustained contractions are maintained in the arm or leg muscles for varying lengths of time. These contractions are responses to a rapid series of stimuli transmitted from the brain and spinal cord on motor neuron fibers.

Actually, even when a muscle appears to be at rest, a certain amount of sustained contraction is occurring in its fibers. This is called **muscle tone** (tonus). Muscle tone is a response to nerve impulses originating repeatedly in the spinal cord and traveling to small numbers of muscle fibers. The result is a continuous state of partial contraction.

Muscle tone is particularly important in maintaining posture. Tautness in the muscles of the neck, trunk, and legs enables a person to hold the head upright, stand, or sit. If tone is suddenly lost, such as when a person loses consciousness, the body will collapse. Although muscle tone is maintained in health, it is lost if motor nerve fibers are cut or if diseases interfere with the conduction of nerve impulses.

> When skeletal muscles are contracted very forcefully, they may generate up to 50 pounds of pull for each square inch of muscle cross section. Consequently, large muscles such as those in the thigh can pull with several hundred pounds of force. Occasionally, this force is so great that the tendons of muscles are torn away from their attachments to the bones.

Isotonic and Isometric Contractions

Sometimes muscles shorten when they contract. For example, if a person lifts an object, the tautness in the muscles remains unchanged, their attached ends pull closer together, and the object is moved. This type of contraction is termed **isotonic.**

At other times, a skeletal muscle contracts, but the parts to which it is attached do not move. This happens, for instance, when a person pushes against the wall of a building. Tension within the muscles increases, but the wall does not move, and the muscles remain the same length. Contractions of this type are called **isometric.** Isometric contractions occur continuously in postural muscles that function to stabilize skeletal parts and hold the body upright. Figure 9.17 illustrates these two types of contractions.

Skeletal muscles can contract either isotonically or isometrically, and most body actions involve both types of contraction. In walking, for instance, certain leg muscles contract isometrically and keep the limb stiff as it touches the ground, while other muscles contract isotonically, causing the leg to bend and lift upward.

1. Define threshold stimulus.
2. What is meant by an all-or-none response?
3. Distinguish between a twitch and a sustained contraction.
4. Define muscle tone.
5. Explain the difference between isometric and isotonic contractions.

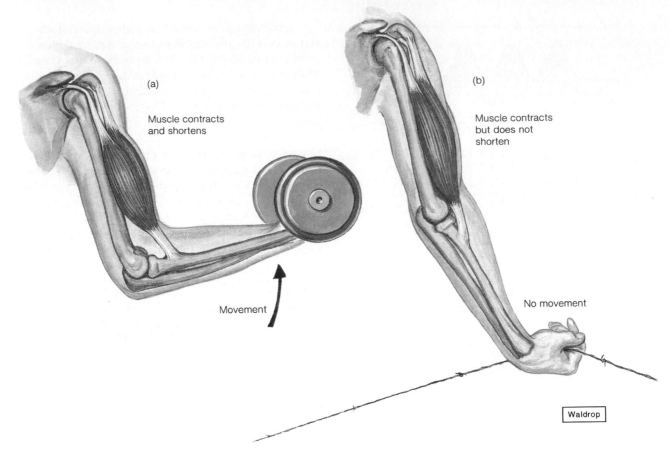

Figure 9.17 (a) Isotonic contractions occur when a muscle contracts and shortens. (b) Isometric contractions occur when a muscle contracts, but does not shorten.

Smooth Muscles

The contractile mechanisms of smooth and cardiac muscles are essentially the same as those of skeletal muscles. However, the cells of these tissues have important structural and functional differences.

Smooth Muscle Fibers

As discussed in chapter 5, smooth muscle cells are shorter than the fibers of skeletal muscle, and they have single, centrally located nuclei. These cells are elongated with tapering ends. Smooth muscle cells contain filaments of *actin* and *myosin* in myofibrils that extend the lengths of the cells. However, these filaments are very thin and more randomly arranged than those in skeletal muscle fibers. Consequently, smooth muscle cells lack striations. They also lack transverse tubules, and their sarcoplasmic reticula are not well developed.

The two major types of smooth muscles are multiunit and visceral. In **multiunit smooth muscle,** the muscle fibers are less well organized and occur as separate fibers rather than in sheets. Smooth muscle of this type is found in the irises of the eyes and in the walls of blood vessels. Typically, multiunit smooth muscle contracts only after stimulation by motor nerve impulses.

Visceral smooth muscle is composed of sheets of spindle-shaped cells that are in close contact and possess gap junctions between one another. These cells are positioned so that the thick portion of each cell is next to the thin parts of adjacent cells. Visceral smooth muscle is the more common type and is found in the walls of hollow visceral organs, such as the stomach, intestines, urinary bladder, and uterus. Usually there are two thicknesses of smooth muscle in the walls of these organs. The fibers of the outer coats are directed longitudinally, while those of the inner coats are arranged circularly. These muscular layers are responsible for the changes in size and shape that occur in visceral organs as they carry on their special functions.

The fibers of visceral smooth muscles are capable of stimulating each other. Consequently, when one fiber is stimulated, the impulse moving over its surface may excite adjacent fibers that, in turn, stimulate still others. Visceral smooth muscles also display *rhythmicity*—a

Use and Disuse of Skeletal Muscles

Skeletal muscles are very responsive to use and disuse. For example, those that are forcefully exercised tend to enlarge. This phenomenon is called *muscular hypertrophy.* Conversely, a muscle that is not used undergoes *atrophy*—that is, it decreases in size and strength.

The way a muscle responds to use also depends on the type of exercise involved. For instance, when a muscle contracts relatively weakly, as during swimming and running, its slow, fatigue-resistant red fibers are most likely to be activated. As a result, these fibers develop more mitochondria, and more extensive capillary networks develop around the fibers. Such changes increase the fibers' abilities to resist fatigue during prolonged periods of exercise, although their sizes and strengths may remain unchanged.

Forceful exercise, such as weightlifting, in which a muscle exerts more than 75% of its maximum tension, involves the muscle's fast, fatigable white fibers. In response, existing muscle fibers develop new filaments of actin and myosin, and as their diameters increase, the entire muscle enlarges. However, no new muscle fibers are produced during hypertrophy.

Since the strength of a contraction is directly related to the diameter of the muscle fibers, an enlarged muscle is capable of producing stronger contractions than before. However, such a change does not increase the muscle's ability to resist fatigue during activities such as running or swimming.

If regular exercise is discontinued, there is a reduction in the capillary networks and in the number of mitochondria within the muscle fibers. Also, the size of the actin and myosin filaments decreases, and the entire muscle atrophies. Such atrophy commonly occurs when injured limbs are immobilized by casts or when accidents or diseases interfere with motor nerve impulses. A muscle that cannot be exercised may decrease to less than one-half its usual size within a few months.

The fibers of muscles whose motor neurons are severed not only decrease in size, but also may become fragmented and, in time, be replaced by fat or fibrous connective tissue. However, if such a muscle is reinnervated within the first few months following an injury, its function may be restored.

---

pattern of repeated contractions. This phenomenon is due to the presence of self-exciting fibers from which spontaneous impulses travel periodically into the surrounding muscle tissue.

These two features of visceral smooth muscle—transmission of impulses from cell to cell and rhythmicity—are largely responsible for the wavelike motion called **peristalsis** that occurs in various tubular organs. (See chapter 14.) Peristalsis involves alternate contractions and relaxations of the longitudinal and circular muscles. These movements help force the contents of a tube along its length. In the intestines, for example, peristaltic waves move masses of food substances through these organs and mix them with digestive fluids. Similar activity in the ureters moves urine from the kidneys to the urinary bladder.

Smooth Muscle Contraction

Smooth muscle contraction resembles skeletal muscle contraction in a number of ways. Both mechanisms involve reactions of actin and myosin; both are triggered by membrane impulses and the release of calcium ions; and both use energy from ATP molecules. There are, however, significant differences. For example, smooth muscle fibers seem to lack troponin, the protein that binds to calcium ions in skeletal muscle. Instead, the smooth muscle uses a protein called *calmodulin,* which binds to calcium ions released when its fibers are stimulated, thus activating the actin-myosin contraction mechanism.

Although acetylcholine is the neurotransmitter substance in skeletal muscle, two neurotransmitters affect smooth muscle—*acetylcholine* and *norepinephrine.* Each of these substances stimulates contractions in some smooth muscles and inhibits contractions in others. These actions are described in more detail in the discussion of the autonomic nervous system in chapter 11.

Smooth muscles are also affected by a number of hormones that stimulate contraction in some cases and alter the amount of response to neurotransmitters in others. For example, during the later stages of the birth process, the smooth muscles in the wall of the uterus are stimulated to contract by the hormone called *oxytocin.* (See chapter 22.)

Stretching of smooth muscle fibers can also trigger contractions. This response is particularly important to the function of visceral smooth muscle in the walls of various hollow organs, such as the urinary bladder and the intestines. For example, when the wall of the intestine is stretched by its contents, an automatic contraction can move the contents away.

Smooth muscle is slower to contract and slower to relax than skeletal muscle. On the other hand, smooth muscle can maintain a forceful contraction for a longer time with the same amount of ATP. Unlike skeletal muscle, smooth muscle fibers can change length without

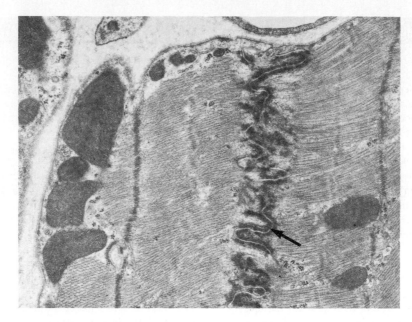

Figure 9.18 The intercalated disks of cardiac muscle, shown in this transmission electron micrograph, hold adjacent cells together.

changing tautness; because of this, smooth muscles in the stomach and intestinal walls can stretch as these organs become filled, while the pressure inside the organs remains unchanged.

1. Describe the two major types of smooth muscle.
2. What special characteristics of visceral smooth muscle make peristalsis possible?
3. How is smooth muscle contraction similar to skeletal muscle contraction?
4. How do the contraction mechanisms of smooth and skeletal muscles differ?

Cardiac Muscle

Cardiac muscle appears only in the heart. It is composed of striated cells joined end-to-end, forming fibers. These fibers are interconnected in branching, three-dimensional networks. Each cell contains a single nucleus and numerous filaments of actin and myosin similar to those in skeletal muscle. A cardiac muscle cell also has a well-developed sarcoplasmic reticulum, a system of transverse tubules, and many mitochondria. However, the cisternae of the sarcoplasmic reticulum of a cardiac muscle fiber are less well developed and store less calcium than those of a skeletal muscle fiber. On the other hand, the transverse tubules of cardiac muscle fibers are larger than those in skeletal muscle, and they release large quantities of calcium ions into the sarcoplasm in response to a single muscle impulse. This extra calcium from the transverse tubules enables cardiac muscle fibers to maintain a contraction longer than skeletal muscle fibers.

The calcium ions provided by transverse tubules are obtained from the fluid outside the muscle fiber. Consequently, the strength of the cardiac muscle contraction is controlled partially by the extracellular calcium ion concentration.

The opposing ends of cardiac muscle cells are separated by cross-bands called *intercalated disks*. These bands are the result of elaborate junctions of membranes. They help hold adjacent cells together and transmit the force of contraction from cell to cell. Intercellular junctions between the fused membranes of the intercalated disks allow diffusion of ions between the cells. This makes it possible for muscle impulses to travel rapidly from cell to cell. (See figures 9.18 and 5.27.)

When one portion of the cardiac muscle network is stimulated, the impulse passes to the other fibers of the network, and the whole structure contracts as a unit; that is, the network responds to stimulation in an all-or-none manner. Cardiac muscle is also self-exciting and rhythmic. Consequently, a pattern of contraction and relaxation is repeated again and again and is responsible for the rhythmic contractions of the heart. Also, unlike skeletal muscle, cardiac muscle remains refractory until a contraction is completed, so that sustained or tetanic contractions do not occur in the heart muscle.

Chart 9.2 summarizes the characteristics of the three types of muscles.

1. How is cardiac muscle similar to skeletal muscle?
2. How does cardiac muscle differ from skeletal muscle?
3. What is the function of intercalated disks?
4. What characteristic of cardiac muscle is responsible for the contraction of the heart as a unit?

CHART 9.2   Characteristics of muscle tissues

	Skeletal	Smooth	Cardiac
Major location	Skeletal muscles	Walls of hollow visceral organs	Wall of the heart
Major function	Movement of bones at joints; maintenance of posture	Movement of visceral organs; peristalsis	Pumping action of the heart
Cellular characteristics			
Striations	Present	Absent	Present
Nucleus	Multiple nuclei	Single nucleus	Single nucleus
Special features	Transverse tubule system is well developed	Lacks transverse tubules	Transverse tubule system is well developed; adjacent cells are separated by intercalated disks
Mode of control	Voluntary	Involuntary	Involuntary
Contraction characteristics	Contracts and relaxes relatively rapidly	Contracts and relaxes relatively slowly; self-exciting; rhythmic	Network of fibers contracts as a unit; self-exciting; rhythmic; remains refractory until contraction is completed

Skeletal Muscle Actions

Skeletal muscles are responsible for a great variety of body movements. The action of each muscle—that is, the movement it causes—depends largely upon the kind of joint it is associated with and the way the muscle is attached on either side of that joint.

Origin and Insertion

As mentioned in chapter 7, one end of a skeletal muscle is usually fastened to a relatively immovable or fixed part, and the other end is connected to a movable part on the other side of a joint. The immovable end is called the **origin** of the muscle, and the movable end is called its **insertion.** When a muscle contracts, its insertion is pulled toward its origin (figure 9.19).

Some muscles have more than one origin or insertion. The *biceps brachii* in the upper arm, for example, has two origins. This is reflected in its name *biceps,* meaning *two heads.* (Note: The head of a muscle is the part nearest its origin.) As figure 9.19 shows, one head of the muscle is attached to the coracoid process of the scapula, and the other head arises from a tubercle above the glenoid cavity of the scapula. The muscle extends along the front surface of the humerus and is inserted by means of a tendon on the radial tuberosity of the radius. When the biceps brachii contracts, its insertion is pulled toward its origin, and the arm bends at the elbow.

Interaction of Skeletal Muscles

Skeletal muscles almost always function in groups rather than singly. Consequently, when a particular body movement occurs, a person must do more than cause a single muscle to contract; instead, after learning to make a particular movement, the person wills the

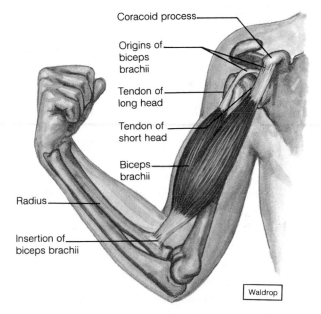

Figure 9.19   The biceps brachii has two heads that originate on the scapula. This muscle is inserted on the radius by means of a tendon. What movement results as this muscle contracts?

movement to occur, and the nervous system causes the appropriate group of muscles to respond.

By carefully observing body movements, it is possible to determine the special roles of various muscles. For instance, when the arm is lifted horizontally away from the side, a contracting *deltoid* muscle is responsible for most of the movement and so is said to be the **prime mover.** However, while a prime mover is acting, certain nearby muscles are also contracting. In the case of a contracting deltoid muscle, nearby muscles help hold the shoulder steady and in this way make the action of the prime mover more effective. Muscles that contract and assist a prime mover are called **synergists.**

Still other muscles act as **antagonists** to prime movers. These muscles are capable of resisting a prime mover's action and are responsible for movement in the opposite direction—the antagonist of the prime mover that raises the arm can lower the arm, or the antagonist of the prime mover that bends the arm can straighten it. If both a prime mover and its antagonist contract simultaneously, the part they act upon remains rigid. Consequently, smooth body movements depend upon the antagonists' relaxing and giving way to the prime movers whenever the prime movers are contracting. Once again, these complex actions are controlled by the nervous system, as described in chapter 11.

1. Distinguish between the origin and the insertion of a muscle.
2. Define prime mover.
3. What is the function of a synergist? An antagonist?

Major Skeletal Muscles

The following section concerns the locations, actions, origins, and insertions of some of the major skeletal muscles. The charts that summarize the information concerning groups of these muscles also include the names of nerves that supply the individual muscles within each group. The origins and pathways of these nerves are presented in chapter 11.

Figures 9.20 and 9.21 show the locations of superficial skeletal muscles—that is, those near the surface. Notice that the names of muscles often describe them in some way. A name may indicate a muscle's size, shape, location, action, number of attachments, or the direction of its fibers, as in the following examples:

pectoralis major A muscle of large size (*major*) located in the pectoral region (chest).
deltoid Shaped like a delta or triangle.
extensor digitorum Acts to extend the digits (fingers or toes).
biceps brachii A muscle with two heads (*biceps*), or points of origin, located in the brachium or arm.
sternocleidomastoid Attached to the sternum, clavicle, and mastoid process.
external oblique Located near the outside, with fibers that run obliquely or in a slanting direction.

Muscles of Facial Expression

A number of small muscles that lie beneath the skin of the face and scalp enable us to communicate feelings through facial expression. Many of these muscles are located around the eyes and mouth, and they are responsible for such expressions as surprise, sadness, anger, fear, disgust, and pain. As a group, the muscles of facial expression connect the bones of the skull to

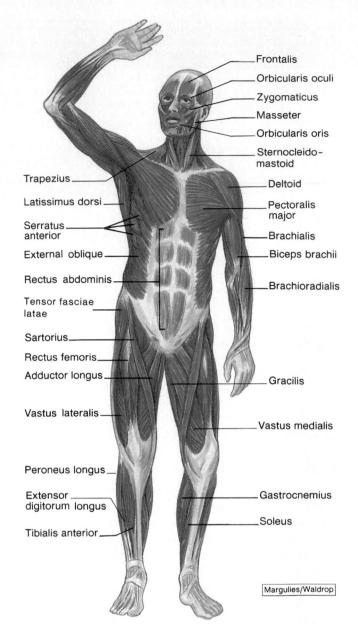

Figure 9.20   Anterior view of superficial skeletal muscles.

connective tissue in various regions of the overlying skin. They are shown in figure 9.22 and reference plate 61, and are listed in chart 9.3. The muscles of facial expression include the following:

Epicranius	*Buccinator*
Orbicularis oculi	*Zygomaticus*
Orbicularis oris	*Platysma*

The **epicranius** (ep″ĭ-kra′ne-us) covers the upper part of the cranium and consists of two muscular parts—the *frontalis* (frun-ta′lis), which lies over the frontal bone, and the *occipitalis* (ok-sip″ĭ-ta′lis), which lies over the occipital bone. These parts are united by a broad, tendinous membrane called the *epicranial*

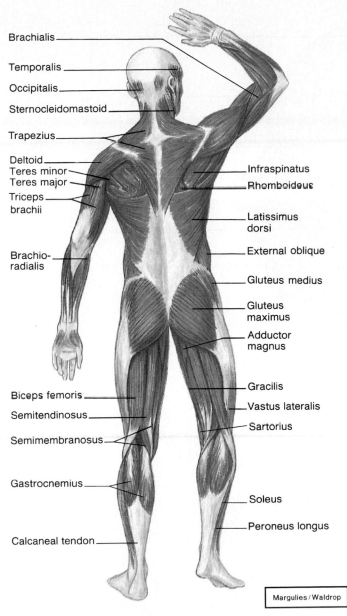

Brachialis
Temporalis
Occipitalis
Sternocleidomastoid
Trapezius
Deltoid
Teres minor
Teres major
Triceps brachii
Brachio-radialis
Biceps femoris
Semitendinosus
Semimembranosus
Gastrocnemius
Calcaneal tendon

Infraspinatus
Rhomboideus
Latissimus dorsi
External oblique
Gluteus medius
Gluteus maximus
Adductor magnus
Gracilis
Vastus lateralis
Sartorius
Soleus
Peroneus longus

Margulies/Waldrop

Figure 9.21   Posterior view of superficial skeletal muscles.

aponeurosis, which covers the cranium like a cap. Contraction of the epicranius raises the eyebrows and causes the skin of the forehead to wrinkle horizontally, as when a person expresses surprise. Headaches often result from sustained contraction of this muscle.

The **orbicularis oculi** (or-bik′u-la-rus ok′u-li) is a ringlike band of muscle, called a *sphincter muscle,* that surrounds the eye. It lies in the subcutaneous tissue of the eyelid and causes the eye to close or blink. At the same time, it compresses the nearby tear gland, or *lacrimal gland,* aiding the flow of tears over the surface of the eye. Contraction of the orbicularis oculi also causes the folds, or crow's feet, that radiate laterally from the corner of the eye.

The **orbicularis oris** (or-bik′u-la-rus o′ris) is a sphincter muscle that encircles the mouth. It lies between the skin and the mucous membranes of the lips, extending upward to the nose and downward to the region between the lower lip and chin. The orbicularis oris is sometimes called the kissing muscle because it causes the lips to close and pucker.

The **buccinator** (buk′si-na″tor) is located in the wall of the cheek. Its fibers are directed forward from the bones of the jaws to the angle of the mouth, and when they contract, the cheek is compressed inward. This action helps hold food in contact with the teeth when a person is chewing. The buccinator also aids in blowing air out of the mouth, and for this reason, it is sometimes called the trumpeter muscle.

The **zygomaticus** (zi″go-mat′ik-us) extends from the zygomatic arch downward to the corner of the mouth. When it contracts, the corner of the mouth is drawn up, as in smiling or laughing.

The **platysma** (plah-tiz′mah) is a thin, sheetlike muscle whose fibers extend from the chest upward over the neck to the face. It functions to pull the angle of the mouth downward, as in pouting. The platysma also helps lower the mandible.

CHART 9.3	Muscles of facial expression			
Muscle	Origin	Insertion	Action	Nerve supply
Epicranius	Occipital bone	Skin and muscles around eye	Raises eyebrow as when surprised	Facial n.
Orbicularis oculi	Maxillary and frontal bones	Skin around eye	Closes eye as in blinking	Facial n.
Orbicularis oris	Muscles near the mouth	Skin of central lip	Closes lips, protrudes lips as for kissing	Facial n.
Buccinator	Outer surfaces of maxilla and mandible	Orbicularis oris	Compresses cheeks inward as when blowing air	Facial n.
Zygomaticus	Zygomatic bone	Orbicularis oris	Raises corner of mouth as when smiling	Facial n.
Platysma	Fascia in upper chest	Lower border of mandible	Draws angle of mouth downward as when pouting	Facial n.

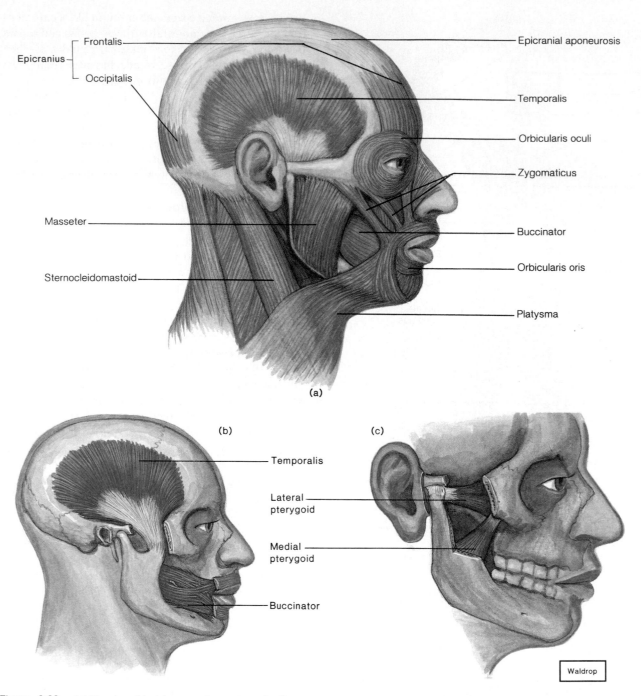

Figure 9.22 (a) Muscles of facial expression and mastication; isolated views of (b) the temporalis and buccinator muscles and (c) the lateral and medial pterygoid muscles.

CHART 9.4 Muscles of mastication

Muscle	Origin	Insertion	Action	Nerve supply
Masseter	Lower border of zygomatic arch	Lateral surface of mandible	Elevates mandible	Trigeminal n.
Temporalis	Temporal bone	Coronoid process and anterior ramus of mandible	Elevates mandible	Trigeminal n.
Medial pterygoid	Sphenoid, palatine, and maxillary bones	Medial surface of mandible	Elevates mandible	Trigeminal n.
Lateral pterygoid	Sphenoid bone	Anterior surface of mandibular condyle	Depresses and protracts mandible, and moves it from side to side	Trigeminal n.

The muscles that move the eye are described in chapter 12.

Muscles of Mastication

Chewing movements are produced by four pairs of muscles that are attached to the mandible. Three pairs of these muscles act to close the lower jaw, as in biting; the fourth pair can lower the jaw, cause side-to-side grinding motions of the mandible, and pull the mandible forward, causing it to protrude. The muscles of mastication are shown in figure 9.22 and reference plate 61, and are listed in chart 9.4. They include the following:

> Masseter
> Temporalis
> Medial pterygoid
> Lateral pterygoid

The **masseter** (mas-se'ter) is a thick, flattened muscle that can be felt just in front of the ear when the teeth are clenched. Its fibers extend downward from the zygomatic arch to the mandible. The masseter functions primarily to raise the jaw, but it can also control the rate at which the jaw falls open in response to gravity (figure 9.22a).

The **temporalis** (tem-po-ra'lis) is a fan-shaped muscle located on the side of the skull above and in front of the ear. Its fibers, which also raise the jaw, pass downward beneath the zygomatic arch to the mandible (figure 9.22a and b).

The **medial pterygoid** (ter'i-goid) extends back and downward from the sphenoid, palatine, and maxillary bones to the ramus of the mandible. It closes the jaw (figure 9.22c).

The fibers of the **lateral pterygoid** are directed forward from the region just below the mandibular condyle to the sphenoid bone. This muscle can open the mouth, pull the mandible forward to make it protrude, and move the mandible from side to side (figure 9.22c).

Sometimes an individual who is emotionally tense or has a poor bite (malocclusion) uses his or her jaw muscles to grind or clench the teeth excessively. This action may cause stress on the temporomandibular joint—the articulation between the mandibular condyle of the mandible and the mandibular fossa of the temporal bone. As a result, the person may experience a variety of unpleasant symptoms, including headache, earache, and pain in the jaw, neck, or shoulder. This condition is called *temporomandibular joint syndrome* (TMJ syndrome).

Muscles That Move the Head

Head movements result from the actions of paired muscles in the neck and upper back. These muscles are responsible for flexing, extending, and rotating the head. They are shown in figures 9.23 and 9.25, and listed in chart 9.5. Muscles that move the head include the following:

> Sternocleidomastoid
> Splenius capitis
> Semispinalis capitis
> Longissimus capitis

The **sternocleidomastoid** (ster''no-kli''do-mas'toid) is a long muscle in the side of the neck that extends upward from the thorax to the base of the skull behind the ear. When the sternocleidomastoid on one side contracts, the face is turned to the opposite side. When both muscles contract, the head is bent toward the chest. If the head is fixed in position by other muscles, the sternocleidomastoids can raise the sternum— an action that aids forceful inhalation (figure 9.25).

The **splenius capitis** (sple'ne-us kap'i-tis) is a broad, straplike muscle located in the back of the neck. It connects the base of the skull to the vertebrae in the neck and upper thorax. A splenius capitus acting singly causes the head to rotate and bend toward one side. Acting together, these muscles bring the head into an upright position (figure 9.23).

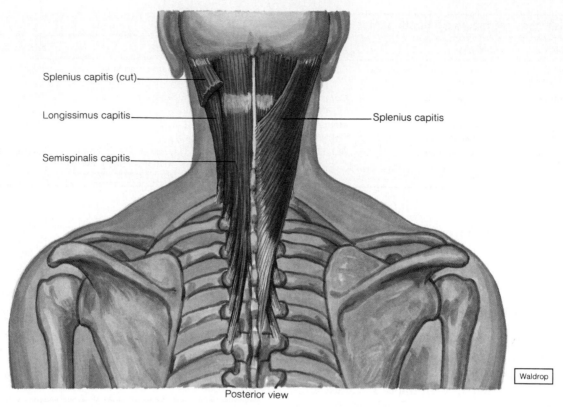

Splenius capitis (cut)

Longissimus capitis

Semispinalis capitis

Splenius capitis

Waldrop

Posterior view

Figure 9.23 Deep muscles in the back of the neck help move the head (posterior view). The splenius capitis is removed on the right to show the underlying muscles.

CHART 9.5	Muscles that move the head			
Muscle	Origin	Insertion	Action	Nerve supply
Sternocleido-mastoid	Anterior surface of sternum and upper surface of clavicle	Mastoid process of temporal bone	Pulls head to one side, flexes neck, or elevates sternum	Accessory, C2, and C3 nerves
Splenius capitis	Spinous processes of lower cervical and upper thoracic vertebrae	Mastoid process of temporal bone	Rotates head, bends head to one side, or extends neck	Cervical nerves
Semispinalis capitis	Processes of lower cervical and upper thoracic vertebrae	Occipital bone	Extends head, bends head to one side, or rotates head	Cervical and thoracic spinal nerves
Longissimus capitis	Processes of lower cervical and upper thoracic vertebrae	Mastoid process of temporal bone	Extends head, bends head to one side, or rotates head	Lower cervical, thoracic, and lumbar spinal nerves

The **semispinalis capitis** (sem″e-spi-na′lis kap′ĭ-tis) is a broad, sheetlike muscle extending upward from the vertebrae in the neck and thorax to the occipital bone. It functions to extend the head, bend it to one side, or rotate it (figure 9.23).

The **longissimus capitis** (lon-jis′ĭ-mus kap′ĭ-tis) is a narrow band of muscle that ascends from the vertebrae of the neck and thorax to the temporal bone. It also functions to extend the head, bend it to one side, or rotate it (figure 9.23).

Muscles That Move the Pectoral Girdle

The muscles that move the pectoral girdle are closely associated with those that move the upper arm. A number of these chest and shoulder muscles connect the scapula to nearby bones and act to move the scapula upward, downward, forward, and backward. These muscles are shown in figures 9.24, 9.25, 9.26, and in reference plates 63, 64, and are listed in chart 9.6.

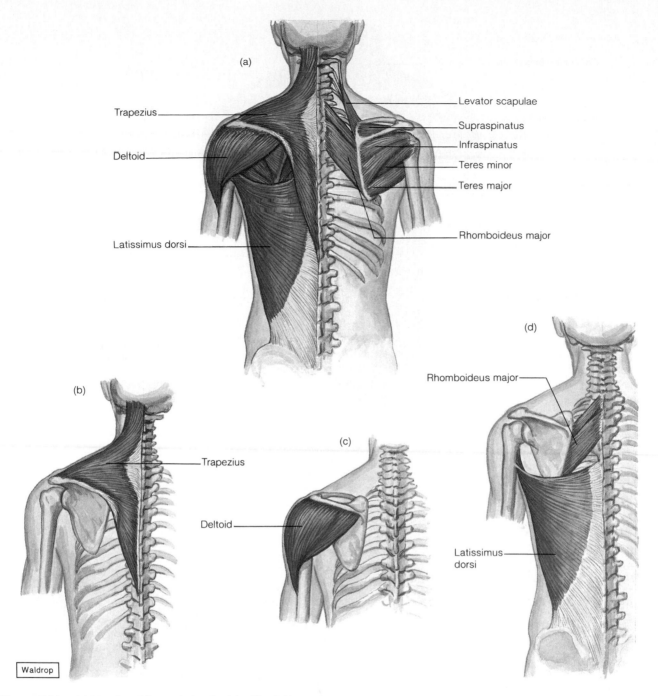

Figure 9.24 (a) Muscles of the posterior shoulder. The right trapezius is removed to show underlying muscles. Isolated views of (b) trapezius, (c) deltoid, and (d) rhomboideus and latissimus dorsi muscles.

Muscles that move the pectoral girdle include the following:

Trapezius
Rhomboideus major
Levator scapulae
Serratus anterior
Pectoralis minor

The **trapezius** (trah-pe'ze-us) is a large, triangular muscle in the upper back that extends horizontally from the base of the skull and the cervical and thoracic vertebrae to the shoulder. Its fibers are arranged in three groups—upper, middle, and lower. Together these fibers rotate the scapula. The upper fibers acting alone raise the scapula and shoulder, as when

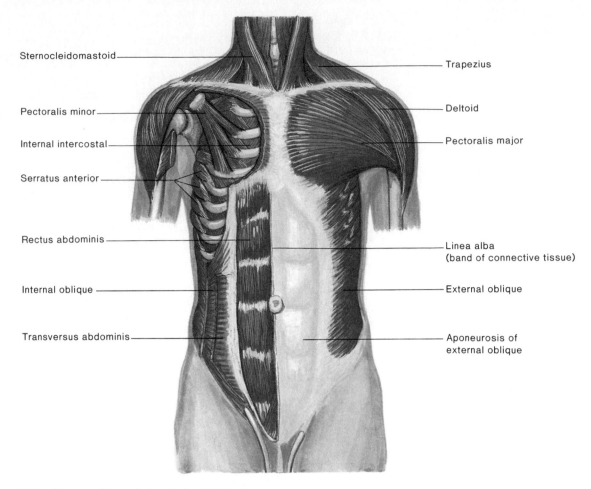

Sternocleidomastoid

Pectoralis minor

Internal intercostal

Serratus anterior

Rectus abdominis

Internal oblique

Transversus abdominis

Trapezius

Deltoid

Pectoralis major

Linea alba
(band of connective tissue)

External oblique

Aponeurosis of
external oblique

Figure 9.25 Muscles of the anterior chest and abdominal wall. (The left pectoralis major is removed to show the pectoralis minor.)

CHART 9.6	Muscles that move the pectoral girdle			
Muscle	Origin	Insertion	Action	Nerve supply
Trapezius	Occipital bone and spines of cervical and thoracic vertebrae	Clavicle, spine, and acromion process of scapula	Rotates scapula; various fibers raise scapula, pull scapula medially, or pull scapula and shoulder downward	Accessory n.
Rhomboideus major	Spines of upper thoracic vertebrae	Medial border of scapula	Raises and adducts scapula	Dorsal scapular n.
Levator scapulae	Transverse processes of cervical vertebrae	Medial margin of scapula	Elevates scapula	Dorsal scapular, and C3-C5 nerves
Serratus anterior	Outer surfaces of upper ribs	Ventral surface of scapula	Pulls scapula anteriorly and downward	Long thoracic n.
Pectoralis minor	Sternal ends of upper ribs	Coracoid process of scapula	Pulls scapula forward and downward or raises ribs	Pectoral n.

the shoulders are shrugged to express a feeling of indifference. The middle fibers pull the scapula toward the vertebral column, and the lower fibers draw the scapula and shoulder downward. When the shoulder is fixed in position by other muscles, the trapezius can pull the head backward or to one side (figure 9.24).

The **rhomboideus** (rom-boid′-e-us) **major** connects the upper thoracic vertebrae to the scapula. It acts to raise the scapula and to adduct it (figure 9.24).

The **levator scapulae** (le-va′tor scap′u-le) is a straplike muscle that runs almost vertically through the neck, connecting the cervical vertebrae to the scapula.

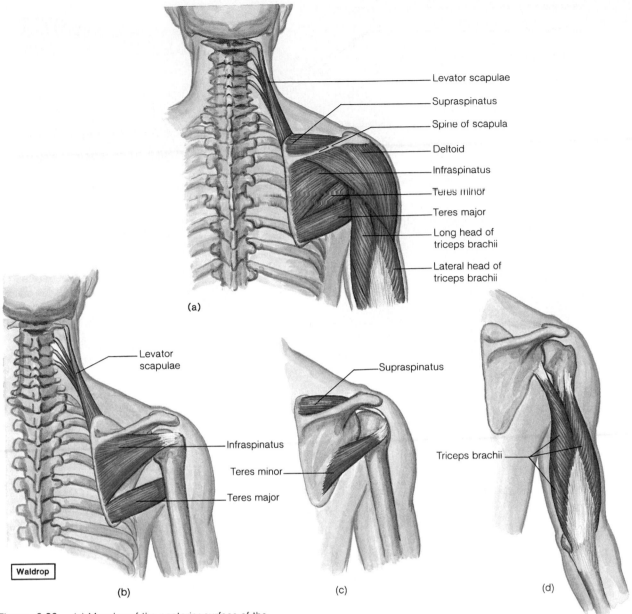

Labels in figure (a):
Levator scapulae
Supraspinatus
Spine of scapula
Deltoid
Infraspinatus
Teres minor
Teres major
Long head of triceps brachii
Lateral head of triceps brachii

(a)

Labels in figure (b):
Levator scapulae
Infraspinatus
Teres minor
Teres major

Waldrop

(b)

Labels in figure (c):
Supraspinatus
Teres minor

(c)

Labels in figure (d):
Triceps brachii

(d)

Figure 9.26 (a) Muscles of the posterior surface of the scapula and the upper arm; (b and c) muscles associated with the scapula. (d) Isolated view of the triceps brachii.

It functions to elevate the scapula (figures 9.24 and 9.26).

The **serratus anterior** (ser-ra′tus an-te′re-or) is a broad, curved muscle located on the side of the chest. It arises as fleshy, narrow strips on the upper ribs and extends along the medial wall of the axilla to the ventral surface of the scapula. It functions to pull the scapula downward and anteriorly, and is used to thrust the shoulder forward, as when pushing something (figure 9.25).

The **pectoralis** (pek″to-ra′lis) **minor** is a thin, flat muscle that lies beneath the larger pectoralis major. It extends laterally and upward from the ribs to the scapula and serves to pull the scapula forward and downward. When other muscles fix the scapula in position, the pectoralis minor can raise the ribs and thus aid forceful inhalation (figure 9.25).

A small, triangular region, called the *triangle of auscultation,* is located in the back where the trapezius overlaps the superior border of the latissimus dorsi and the underlying rhomboideus major. This area, which is near the medial border of the scapula, becomes enlarged when a person bends forward with the arms folded across the chest. By placing the bell of a stethoscope within the triangle of auscultation, a physician can usually hear clearly the sounds of the respiratory organs.

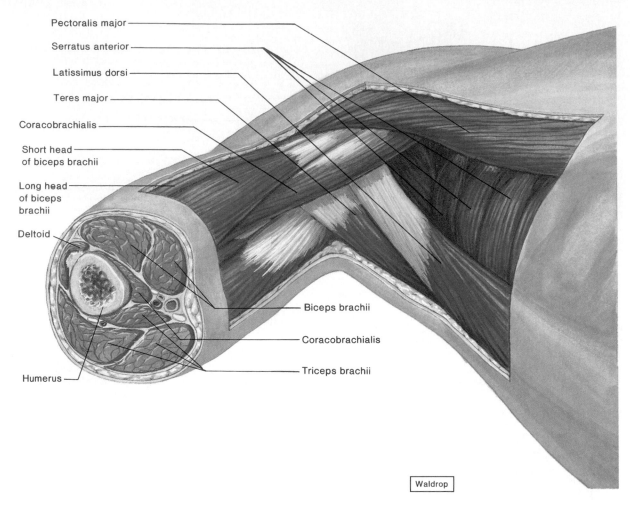

Pectoralis major
Serratus anterior
Latissimus dorsi
Teres major
Coracobrachialis
Short head of biceps brachii
Long head of biceps brachii
Deltoid
Humerus

Biceps brachii
Coracobrachialis
Triceps brachii

Waldrop

Figure 9.27 Cross section of the upper arm.

Muscles That Move the Upper Arm

The upper arm is one of the more freely movable parts of the body. Its many movements are made possible by muscles that connect the humerus to various regions of the pectoral girdle, ribs, and vertebral column. These muscles can be grouped according to their primary actions—flexion, extension, abduction, and rotation. They are shown in figures 9.26, 9.27, 9.28, and in reference plates 62, 63, 64, and are listed in chart 9.7. Muscles that move the upper arm include the following:

Flexors
   Coracobrachialis
   Pectoralis major
Extensors
   Teres major
   Latissimus dorsi

Abductors
   Supraspinatus
   Deltoid
Rotators
   Subscapularis
   Infraspinatus
   Teres minor

Flexors

The **coracobrachialis** (kor″ah-ko-bra′ke-al-is) extends from the scapula to the middle of the humerus along its medial surface. It acts to flex and adduct the upper arm (figures 9.27 and 9.28).

The **pectoralis major** is a thick, fan-shaped muscle located in the upper chest. Its fibers extend from the center of the thorax through the armpit to the humerus. This muscle functions primarily to pull the upper arm forward and across the chest. It can also rotate the humerus medially and adduct the arm from a raised position (figure 9.25).

Extensors

The **teres** (te′rēz) **major** connects the scapula to the humerus. It extends the humerus and can also adduct and rotate the upper arm medially (figures 9.24 and 9.26).

The **latissimus dorsi** (lah-tis′ĭ-mus dor′si) is a wide, triangular muscle that curves upward from the lower back, around the side, and to the armpit. It can extend and adduct the arm and rotate the humerus inwardly. It also pulls the shoulder downward and back. This muscle is used to pull the arm back in swimming, climbing, and rowing (figures 9.24 and 9.27).

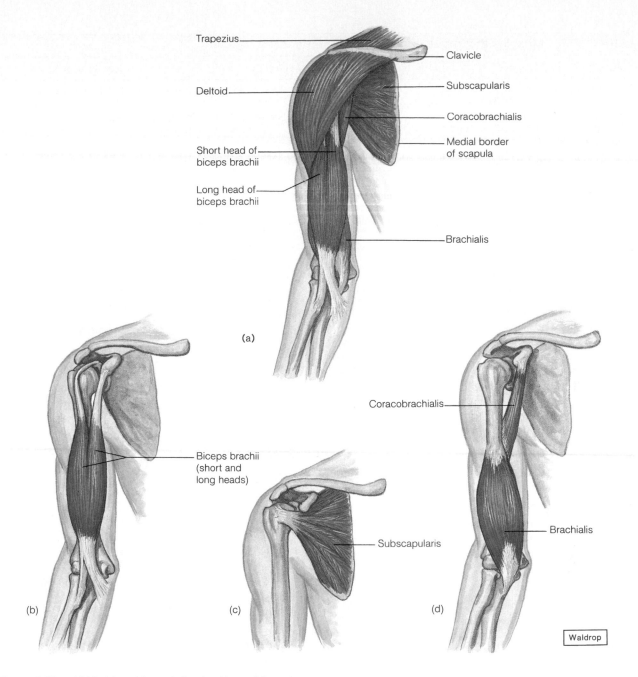

Figure 9.28 (a) Muscles of the anterior shoulder and the upper arm, with the rib cage removed. (b, c, and d) Isolated views of muscles associated with the upper arm.

Abductors

The **supraspinatus** (su″prah-spi′na-tus) is located in the depression above the spine of the scapula on its posterior surface. It connects the scapula to the greater tubercle of the humerus and acts to abduct the upper arm (figures 9.24 and 9.26).

The **deltoid** (del′toid) is a thick, triangular muscle that covers the shoulder joint. It connects the clavicle and scapula to the lateral side of the humerus and acts to abduct the upper arm. The deltoid's posterior fibers can extend the humerus, and its anterior fibers can flex the humerus (figure 9.24).

If the humerus is fractured at its surgical neck (figure 7.50), the nerve that supplies the deltoid muscle (axillary nerve) may be damaged. If this occurs, the muscle is likely to decrease in size and become weakened. In order to test the deltoid for such weakness, a physician may ask a patient to abduct the arm against some resistance and maintain that posture for a time.

CHART 9.7 Muscles that move the upper arm

Muscle	Origin	Insertion	Action	Nerve supply
Coracobrachialis	Coracoid process of scapula	Shaft of humerus	Flexes and adducts the upper arm	Musculocutaneous n.
Pectoralis major	Clavicle, sternum, and costal cartilages of upper ribs	Intertubercular groove of humerus	Flexes, adducts, and rotates humerus, or adducts arm	Pectoral n.
Teres major	Lateral border of scapula	Intertubercular groove of humerus	Extends humerus, or adducts and rotates arm medially	Lower subscapular n.
Latissimus dorsi	Spines of sacral, lumbar, and lower thoracic vertebrae, iliac crest, and lower ribs	Intertubercular groove of humerus	Extends and adducts the arm and rotates humerus inwardly, or pulls the shoulder downward and back	Thoracodorsal n.
Supraspinatus	Posterior surface of scapula	Greater tubercle of humerus	Abducts the upper arm	Suprascapular n.
Deltoid	Acromion process, spine of the scapula, and the clavicle	Deltoid tuberosity of humerus	Abducts upper arm, extends humerus, or flexes humerus	Axillary n.
Subscapularis	Anterior surface of scapula	Lesser tubercle of humerus	Rotates arm medially	Subscapular n.
Infraspinatus	Posterior surface of scapula	Greater tubercle of humerus	Rotates arm laterally	Suprascapular n.
Teres minor	Lateral border of scapula	Greater tubercle of humerus	Rotates arm laterally	Axillary n.

Rotators

The **subscapularis** (sub-scap′u-lar-is) is a large, triangular muscle that covers the anterior surface of the scapula. It connects the scapula to the humerus and functions to rotate the upper arm medially (figure 9.28).

The **infraspinatus** (in″frah-spi′na-tus) occupies the depression below the spine of the scapula on its posterior surface. The fibers of this muscle attach the scapula to the humerus and act to rotate the upper arm laterally (figure 9.26).

The **teres minor** is a small muscle connecting the scapula to the humerus. It rotates the upper arm laterally (figures 9.24 and 9.26).

Muscles That Move the Forearm

Most forearm movements are produced by muscles that connect the radius or ulna to the humerus or pectoral girdle. A group of muscles located along the anterior surface of the humerus acts to flex the elbow, while a single posterior muscle serves to extend this joint. Other muscles cause movements at the radioulnar joint and are responsible for rotating the forearm.

The muscles that move the forearm are shown in figures 9.28, 9.29, 9.30, 9.31, and in reference plates 63, 65, and are listed in chart 9.8. They include the following:

Flexors
Biceps brachii
Brachialis
Brachioradialis

Extensor
Triceps brachii
Rotators
Supinator
Pronator teres
Pronator quadratus

Flexors

The **biceps brachii** (bi′seps bra′ke-i) is a fleshy muscle that forms a long, rounded mass on the anterior side of the upper arm. It connects the scapula to the radius and functions to flex the arm at the elbow and to rotate the hand laterally (supination), as when a person turns a doorknob or screwdriver (figure 9.28).

The **brachialis** (bra′ke-al-is) is a large muscle beneath the biceps brachii. It connects the shaft of the humerus to the ulna and is the strongest flexor of the elbow (figure 9.28).

The **brachioradialis** (bra″ke-o-ra″de-a′lis) connects the humerus to the radius and aids in flexing the elbow (figure 9.29).

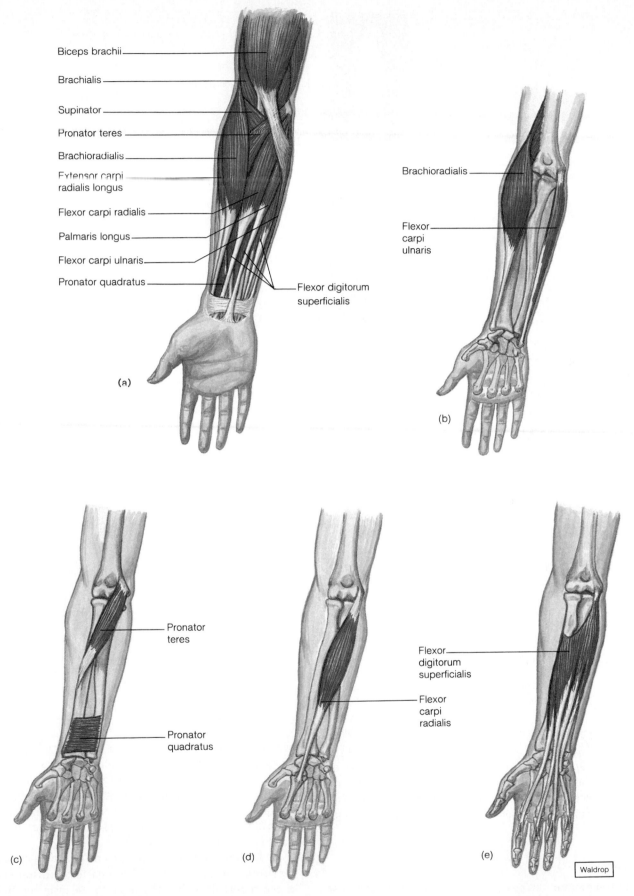

Biceps brachii

Brachialis

Supinator

Pronator teres

Brachioradialis

Extensor carpi radialis longus

Flexor carpi radialis

Palmaris longus

Flexor carpi ulnaris

Pronator quadratus

Flexor digitorum superficialis

(a)

Brachioradialis

Flexor carpi ulnaris

(b)

Pronator teres

Pronator quadratus

(c)

Flexor carpi radialis

(d)

Flexor digitorum superficialis

Flexor carpi radialis

(e)

Waldrop

Figure 9.29 (a) Muscles of the anterior forearm. (b–e) Isolated views of muscles associated with the anterior forearm.

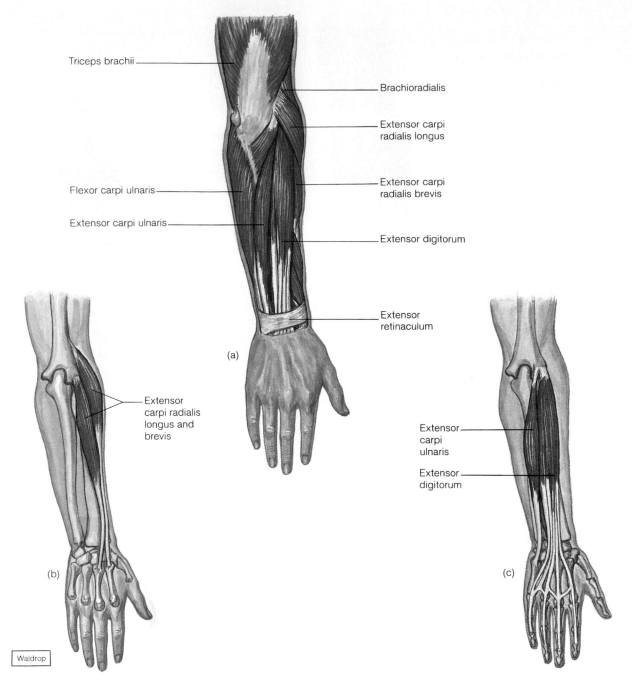

Triceps brachii

Flexor carpi ulnaris

Extensor carpi ulnaris

Brachioradialis

Extensor carpi
radialis longus

Extensor carpi
radialis brevis

Extensor digitorum

Extensor
retinaculum

(a)

Extensor
carpi radialis
longus and
brevis

(b)

Extensor
carpi
ulnaris

Extensor
digitorum

(c)

Waldrop

Figure 9.30   (a) Muscles of the posterior forearm. (b and c)
Isolated views of muscles associated with the posterior forearm.

Extensor

The **triceps brachii** (tri'seps bra'ke-i) has three heads
and is the only muscle on the back of the upper arm.
It connects the humerus and scapula to the ulna, and
is the primary extensor of the elbow (figures 9.26 and
9.27).

Rotators

The **supinator** (su'pĭ-na-tor) is a short muscle whose
fibers run from the ulna and the lateral end of the humerus to the radius. It assists the biceps brachii in rotating the forearm laterally (supination) (figure 9.29).

The **pronator teres** (pro-na'tor te'rēz) is a short
muscle connecting the ends of the humerus and ulna to
the radius. It functions to rotate the arm medially, as
when the hand is turned so the palm is facing downward (pronation) (figure 9.29).

The **pronator quadratus** (pro-na'tor kwod-ra'tus)
runs from the distal end of the ulna to the distal end
of the radius. It assists the pronator teres in rotating
the arm medially (figure 9.29).

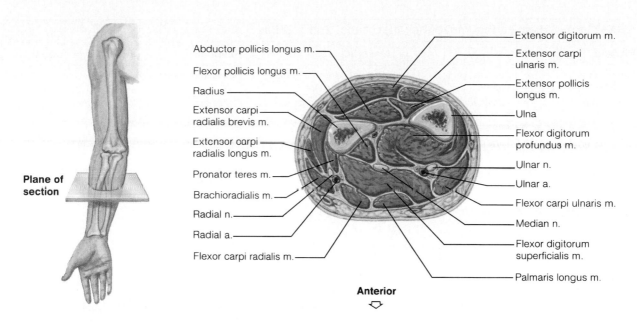

Figure 9.31 A cross section of the forearm.

Anterior

CHART 9.8	Muscles that move the forearm			
Muscle	Origin	Insertion	Action	Nerve supply
Biceps brachii	Coracoid process and tubercle above glenoid cavity of scapula	Radial tuberosity of radius	Flexes arm at elbow and rotates hand laterally	Musculocutaneous n.
Brachialis	Anterior shaft of humerus	Coronoid process of ulna	Flexes arm at elbow	Musculocutaneous, median, and radial nerves
Brachioradialis	Distal lateral end of humerus	Lateral surface of radius above styloid process	Flexes arm at elbow	Radial n.
Triceps brachii	Tubercle below glenoid cavity and lateral and medial surfaces of humerus	Olecranon process of ulna	Extends arm at elbow	Radial n.
Supinator	Lateral epicondyle of humerus and crest of ulna	Lateral surface of radius	Rotates forearm laterally	Radial n.
Pronator teres	Medial epicondyle of humerus and coronoid process of ulna	Lateral surface of radius	Rotates arm medially	Median n.
Pronator quadratus	Anterior distal end of ulna	Anterior distal end of radius	Rotates arm medially	Median n.

Muscles That Move the Wrist, Hand, and Fingers

Many muscles are responsible for wrist, hand, and finger movements. They originate from the distal end of the humerus and from the radius and ulna. The two major groups of these muscles are flexors on the anterior side of the forearm and extensors on the posterior side. These muscles are shown in figures 9.29, 9.30, 9.31, and in reference plate 65, and are listed in chart 9.9. Muscles that move the wrist, hand, and fingers include the following:

Flexors
 Flexor carpi radialis
 Flexor carpi ulnaris
 Palmaris longus
 Flexor digitorum profundus
 Flexor digitorum superficialis
Extensors
 Extensor carpi radialis longus
 Extensor carpi radialis brevis
 Extensor carpi ulnaris
 Extensor digitorum

CHART 9.9   Muscles that move the wrist, hand, and fingers

Muscle	Origin	Insertion	Action	Nerve supply
Flexor carpi radialis	Medial epicondyle of humerus	Base of second and third metacarpals	Flexes and abducts wrist	Median n.
Flexor carpi ulnaris	Medial epicondyle of humerus and olecranon process	Carpal and metacarpal bones	Flexes and adducts wrist	Ulnar n.
Palmaris longus	Medial epicondyle of humerus	Fascia of palm	Flexes wrist	Median n.
Flexor digitorum profundus	Anterior surface of ulna	Bases of distal phalanges in fingers 2-5	Flexes distal joints of fingers	Median and ulnar nerves
Flexor digitorum superficialis	Medial epicondyle of humerus, coronoid process of ulna, and radius	Tendons of fingers	Flexes fingers and hand	Median n.
Extensor carpi radialis longus	Distal end of humerus	Base of second metacarpal	Extends wrist and abducts hand	Radial n.
Extensor carpi radialis brevis	Lateral epicondyle of humerus	Base of second and third metacarpals	Extends wrist and abducts hand	Radial n.
Extensor carpi ulnaris	Lateral epicondyle of humerus	Base of fifth metacarpal	Extends and adducts wrist	Radial n.
Extensor digitorum	Lateral epicondyle of humerus	Posterior surface of phalanges in fingers 2-5	Extends fingers	Radial n.

Flexors

The **flexor carpi radialis** (flek′sor kar-pi′ ra″de-a′lis) is a fleshy muscle that runs medially on the anterior side of the forearm. It extends from the distal end of the humerus into the hand, where it is attached to metacarpal bones. The flexor carpi radialis functions to flex and abduct the wrist (figure 9.29).

The **flexor carpi ulnaris** (flek′sor kar-pi′ ul-na′ris) is located along the medial border of the forearm. It connects the distal end of the humerus and the proximal end of the ulna to carpal and metacarpal bones. It serves to flex and adduct the wrist (figure 9.29).

The **palmaris longus** (pal-ma′ris long′gus) is a slender muscle located on the medial side of the forearm between the flexor carpi radialis and the flexor carpi ulnaris. It connects the distal end of the humerus to fascia of the palm and functions to flex the wrist (figure 9.29).

The **flexor digitorum profundus** (flek′sor dij″ĭ-to′rum pro-fun′dus) is a large muscle that connects the ulna to the distal phalanges. It acts to flex the distal joints of the fingers, as when a fist is made (figure 9.29).

The **flexor digitorum superficialis** (flek′sor dij″ĭ-to′rum su″per-fish″e-a′-lis) is a large muscle located beneath the flexor carpi ulnaris. It arises by three heads—one from the medial epicondyle of the humerus, one from the medial side of the ulna, and one from the radius. It is inserted in the tendons of the fingers and acts to flex the fingers and, by a combined action, to flex the hand (figure 9.29).

Extensors

The **extensor carpi radialis longus** (eks-ten′sor kar-pi′ ra″de-a′lis long′gus) runs along the lateral side of the forearm, connecting the humerus to the hand. It functions to extend the wrist and assists in abducting the hand (figures 9.30 and 9.31).

The **extensor carpi radialis brevis** (eks-ten′sor kar-pi′ ra″de-a′lis brev′is) is a companion of the extensor carpi radialis longus and is located medially to it. This muscle runs from the humerus to metacarpal bones and functions to extend the wrist. It also assists in abducting the hand (figures 9.30 and 9.31).

The **extensor carpi ulnaris** (eks-ten′sor kar-pi′ ul-na′ris) is located along the posterior surface of the ulna and connects the humerus to the hand. It acts to extend the wrist and assists in adducting it (figures 9.30 and 9.31).

The **extensor digitorum** (eks-ten′sor dij″ĭ-to′rum) runs medially along the back of the forearm. It connects the humerus to the posterior surface of the phalanges and functions to extend the fingers (figures 9.30 and 9.31).

A structure called the *extensor retinaculum* consists of a group of heavy connective tissue fibers in the fascia of the wrist (figure 9.30). It connects the lateral margin of the radius with the medial border of the styloid process of the ulna and certain bones of the wrist. The retinaculum gives off branches of connective tissue to the underlying wrist bones, creating a series of sheathlike compartments through which the tendons of the extensor muscles pass to the wrist and fingers.

Muscles of the Abdominal Wall

Although the walls of the chest and pelvic regions are supported directly by bone, those of the abdomen are not. Instead, the anterior and lateral walls of the abdomen are composed of broad, flattened muscles arranged in layers. These muscles connect the rib cage and vertebral column to the pelvic girdle. A band of tough connective tissue, called the **linea alba,** extends from the xiphoid process of the sternum to the symphysis pubis. It serves as an attachment for some of the abdominal wall muscles.

Contraction of these muscles decreases the size of the abdominal cavity and increases the pressure inside. This action helps press air out of the lungs during forceful exhalation, and also aids in defecation, urination, vomiting, and childbirth.

The abdominal wall muscles are shown in figure 9.32 and reference plate 62, and are listed in chart 9.10. They include the following:

> *External oblique*
> *Internal oblique*
> *Transversus abdominis*
> *Rectus abdominis*

The **external oblique** (eks-ter'nal ō-blēk) is a broad, thin sheet of muscle whose fibers slant downward from the lower ribs to the pelvic girdle and the linea alba. When this muscle contracts, it tenses the abdominal wall and compresses the contents of the abdominal cavity.

Similarly, the **internal oblique** (in-ter'nal ō-blēk) is a broad, thin sheet of muscle located beneath the external oblique. Its fibers run up and forward from the pelvic girdle to the lower ribs. Its function is similar to that of the external oblique.

The **transversus abdominis** (trans-ver'sus abdom'ĭ-nis) forms a third layer of muscle beneath the external and internal obliques. Its fibers run horizontally from the lower ribs, lumbar vertebrae, and ilium to the linea alba and pubic bones. It functions in the same manner as the external and internal obliques.

The **rectus abdominis** (rek'tus ab-dom'ĭ-nis) is a long, straplike muscle that connects the pubic bones to the ribs and sternum. It is crossed transversely by three or more fibrous bands that give it a segmented appearance. The muscle functions with other abdominal wall muscles to compress the contents of the abdominal cavity, and it also helps to flex the vertebral column.

Muscles of the Pelvic Outlet

The outlet of the pelvis is spanned by two muscular sheets—a deeper **pelvic diaphragm** and a more superficial **urogenital diaphragm.** The pelvic diaphragm forms the floor of the pelvic cavity, and the urogenital diaphragm fills the space within the pubic arch. The muscles of the male and female pelvic outlets are shown in figure 9.33 and chart 9.11. They include the following:

Pelvic diaphragm
> *Levator ani*
> *Coccygeus*
Urogenital diaphragm
> *Superficial transversus perinei*
> *Bulbospongiosus*
> *Ischiocavernosus*
> *Sphincter urethrae*

Pelvic Diaphragm

The **levator ani** (le-va'tor ah-ni') muscles form a thin sheet across the pelvic outlet. They are connected at the midline posteriorly by a ligament that extends from the tip of the coccyx to the anal canal. Anteriorly, they are separated in the male by the urethra and the anal canal, and in the female by the urethra, vagina, and anal canal. These muscles help support the pelvic viscera and provide sphincterlike action in the anal canal and vagina.

An *external anal sphincter* that is under voluntary control and an *internal anal sphincter* that is formed of involuntary muscle fibers of the intestine encircle the anal canal and keep it closed.

The **coccygeus** (kok-sij'e-us) is a fan-shaped muscle that extends from the ischial spine to the coccyx and sacrum. It aids the levator ani in its functions.

Urogenital Diaphragm

The **superficial transversus perinei** (su"per-fish'al transver'sus per"ĭ-ne'i) consists of a small bundle of muscle fibers that passes medially from the ischial tuberosity along the posterior border of the urogenital diaphragm. It assists other muscles in supporting the pelvic viscera.

In males, the **bulbospongiosus** (bul"bo-spon"je-o'sus) muscles are united surrounding the base of the penis. They assist in emptying the urethra. In females, these muscles are separated medially by the vagina and act to constrict the vaginal opening. They can also retard the flow of blood in veins, which helps maintain an erection in the penis of the male and in the clitoris of the female.

The **ischiocavernosus** (is"ke-o-kav"er-no'sus) muscle is a tendinous structure that extends from the ischial tuberosity to the margin of the pubic arch. It assists the function of the bulbospongiosus muscle.

The **sphincter urethrae** (sfingk'ter u-re'thrē) are muscles that arise from the margins of the pubic and ischial bones. Each arches around the urethra and unites with the one on the other side. Together they act as a sphincter that closes the urethra by compression and opens it by relaxation, thus helping control the flow of urine.

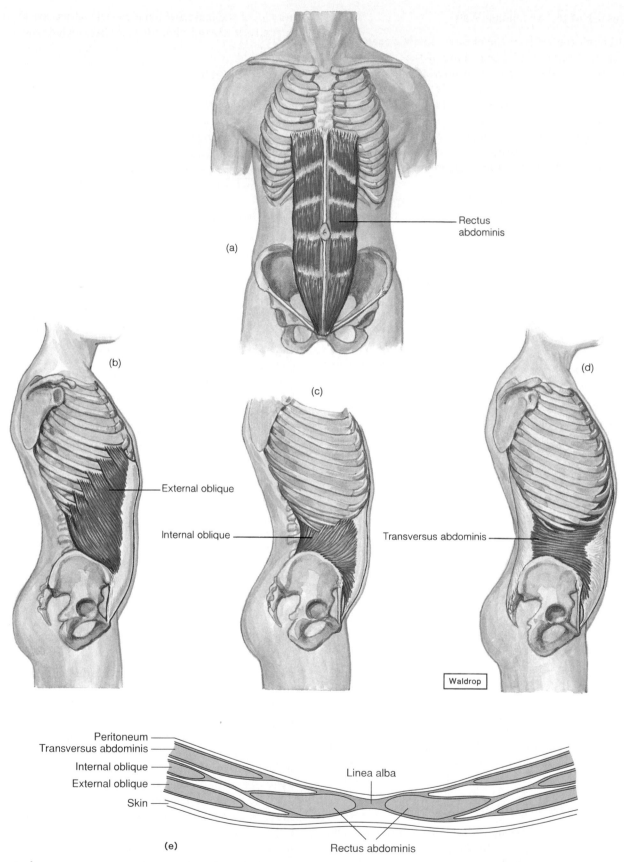

Rectus
abdominis

(a)

(b)

(c)

(d)

External oblique

Internal oblique

Transversus abdominis

Waldrop

Peritoneum
Transversus abdominis
Internal oblique
External oblique
Skin

Linea alba

Rectus abdominis

(e)

Figure 9.32 (*a–d*) Isolated muscles of the abdominal wall; (*e*) transverse section through the abdominal wall.

CHART 9.10 Muscles of the abdominal wall

Muscle	Origin	Insertion	Action	Nerve supply
External oblique	Outer surfaces of lower ribs	Outer lip of iliac crest and linea alba	Tenses abdominal wall and compresses abdominal contents	Intercostal nerves 7–12
Internal oblique	Crest of ilium and inguinal ligament	Cartilages of lower ribs, linea alba, and crest of pubis	Same as above	Intercostal nerves 7–12
Transversus abdominis	Costal cartilages of lower ribs, processes of lumbar vertebrae, lip of iliac crest, and inguinal ligament	Linea alba and crest of pubis	Same as above	Intercostal nerves 7–12
Rectus abdominis	Crest of pubis and symphysis pubis	Xiphoid process of sternum and costal cartilages	Same as above; also flexes vertebral column	Intercostal nerves 7–12

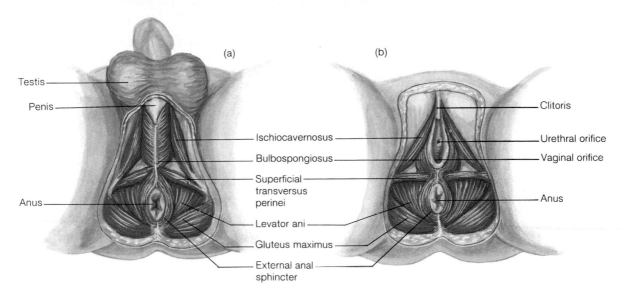

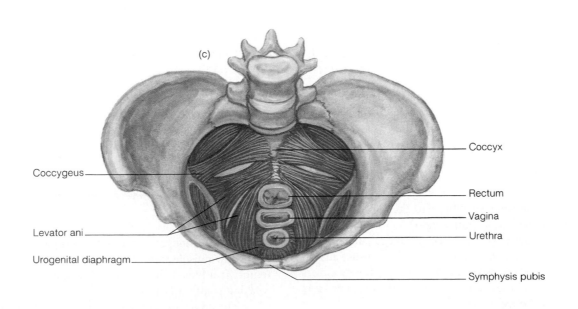

Figure 9.33 External view of muscles of (*a*) the male pelvic outlet, and (*b*) the female pelvic outlet. (*c*) Internal view of pelvic and urogenital diaphragms.

CHART 9.11 Muscles of the pelvic outlet

Muscle	Origin	Insertion	Action	Nerve supply
Levator ani	Pubic bone and ischial spine	Coccyx	Supports pelvic viscera, and provides sphincterlike action in anal canal and vagina	Pudendal n.
Coccygeus	Ischial spine	Sacrum and coccyx	Same as above	S4 and S5 nerves
Superficial transversus perinei	Ischial tuberosity	Central tendon	Supports pelvic viscera	Pudendal n.
Bulbospongiosus	Central tendon	Males: Urogenital diaphragm and fascia of penis Females: Pubic arch and root of clitoris	Males: Assists emptying of urethra Females: Constricts vagina	Pudendal n.
Ischiocavernosus	Ischial tuberosity	Pubic arch	Assists function of bulbospongiosus	Pudendal n.
Sphincter urethrae	Margins of pubis and ischium	Fibers of each unite with those from other side	Opens and closes urethra	Pudendal n.

Muscles That Move the Thigh

The muscles that move the thigh are attached to the femur and to some part of the pelvic girdle. They can be separated into anterior and posterior groups. The muscles of the anterior group act primarily to flex the thigh; those of the posterior group extend, abduct, or rotate it. The muscles in these groups are shown in figures 9.34, 9.35, 9.36, 9.37, and in reference plate 66, and are listed in chart 9.12. Muscles that move the thigh include the following:

Anterior group
  Psoas major
  Iliacus

Posterior group
  Gluteus maximus
  Gluteus medius
  Gluteus minimus
  Tensor fasciae latae

Still another group of muscles, attached to the femur and pelvic girdle, functions to adduct the thigh. This group includes:

  Adductor longus
  Adductor magnus
  Gracilis

Anterior Group

The **psoas** (so'as) **major** is a long, thick muscle that connects the lumbar vertebrae to the femur. It functions to flex the thigh (figure 9.34).

The **iliacus** (il'e-ak-us), a large, fan-shaped muscle, lies along the lateral side of the psoas major. The iliacus and the psoas major are the primary flexors of the thigh, and they serve to advance the leg in walking movements (figure 9.34).

Posterior Group

The **gluteus maximus** (gloo'te-us mak'si-mus) is the largest muscle in the body and covers a large part of each buttock. It connects the ilium, sacrum, and coccyx to the femur by fascia of the thigh and acts to extend the thigh. The gluteus maximus causes the leg to straighten at the hip when a person walks, runs, or climbs. It is also used to raise the body from a sitting position (figure 9.35).

The **gluteus medius** (gloo'te-us me'de-us) is partly covered by the gluteus maximus. Its fibers extend from the ilium to the femur, and they function to abduct the thigh and rotate it medially (figure 9.35).

The **gluteus minimus** (gloo'te-us min'i-mus) lies beneath the gluteus medius and is its companion in attachments and functions (figure 9.35).

The **tensor fasciae latae** (ten'sor fash'e-e lah-te) connects the ilium to the fascia of the thigh, which continues downward to the tibia. This muscle functions to abduct and flex the thigh and to rotate it medially (figure 9.35).

> The gluteus medius and gluteus minimus help support and maintain the normal position of the pelvis. If these muscles are paralyzed as a result of injury or disease, the pelvis tends to drop to one side whenever the foot on that side is raised. Consequently, the person walks with a waddling limp called the *gluteal gait*.

Thigh Adductors

The **adductor longus** (ah-duk'tor long'gus) is a long, triangular muscle that runs from the pubic bone to the femur. It adducts the thigh and assists in flexing and rotating it laterally (figure 9.34).

The **adductor magnus** (ah-duk'tor mag'nus) is the largest adductor of the thigh. It is a triangular muscle that connects the ischium to the femur. It adducts the thigh and assists in extending and rotating it laterally (figure 9.34).

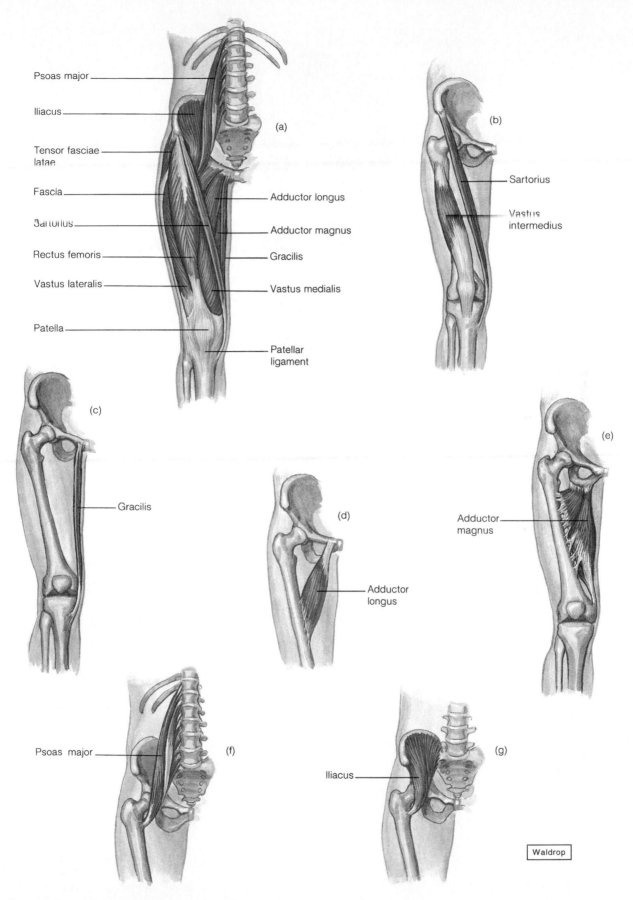

Psoas major

Iliacus

Tensor fasciae latae

Fascia

Sartorius

Rectus femoris

Vastus lateralis

Patella

(a)

Adductor longus

Adductor magnus

Gracilis

Vastus medialis

Patellar ligament

(b)

Sartorius

Vastus intermedius

(c)

Gracilis

(d)

Adductor longus

(e)

Adductor magnus

(f)

Psoas major

(g)

Iliacus

Waldrop

Figure 9.34 (*a*) Muscles of the anterior right thigh. Isolated views of (*b*) the vastus intermedius; (*c–e*) adductors of the thigh; (*f–g*) flexors of the thigh.

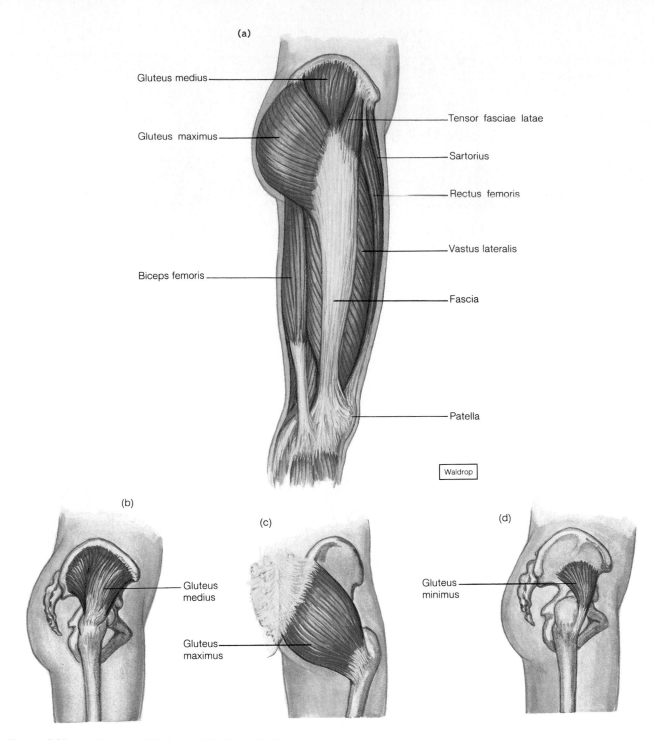

Figure 9.35 (*a*) Muscles of the lateral right thigh. (*b–d*) Isolated views of the gluteal muscles.

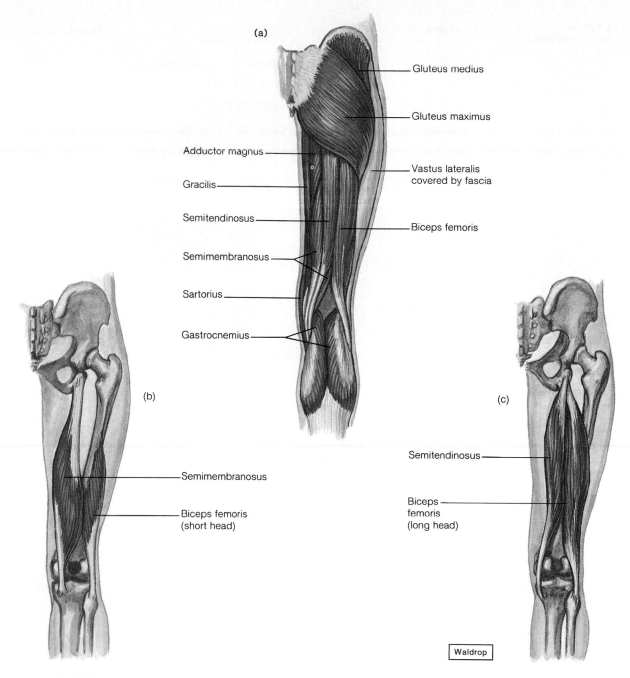

(a)

Gluteus medius

Gluteus maximus

Adductor magnus

Gracilis

Semitendinosus

Semimembranosus

Sartorius

Gastrocnemius

Vastus lateralis covered by fascia

Biceps femoris

(b)

Semimembranosus

Biceps femoris (short head)

(c)

Semitendinosus

Biceps femoris (long head)

Waldrop

Figure 9.36 (*a*) Muscles of the posterior right thigh. (*b* and *c*) Isolated views of muscles that flex the leg at the knee.

The **gracilis** (gras′il-is) is a long, straplike muscle that passes from the pubic bone to the tibia. It functions to adduct the thigh and to flex the leg at the knee (figure 9.34).

Muscles That Move the Lower Leg

The muscles that move the lower leg connect the tibia or fibula to the femur or to the pelvic girdle. They can be separated into two major groups—those that cause flexion at the knee and those that cause extension at the knee. The muscles of these groups are shown in fig-ures 9.34, 9.35, 9.36, 9.37, in reference plates 66 and 67, and are listed in chart 9.13. Muscles that move the lower leg include the following:

Flexors
  Biceps femoris
  Semitendinosus
  Semimembranosus
  Sartorius
Extensor
  Quadriceps femoris group

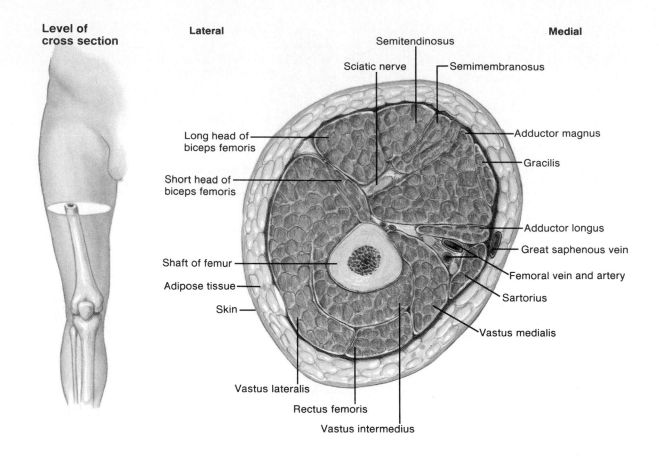

Level of cross section

Lateral

Medial

Semitendinosus

Sciatic nerve

Semimembranosus

Long head of biceps femoris

Adductor magnus

Gracilis

Short head of biceps femoris

Adductor longus

Great saphenous vein

Shaft of femur

Femoral vein and artery

Adipose tissue

Sartorius

Skin

Vastus medialis

Vastus lateralis

Rectus femoris

Vastus intermedius

Anterior

Figure 9.37   A cross section of the thigh.

CHART 9.12	Muscles that move the thigh			
Muscle	Origin	Insertion	Action	Nerve supply
Psoas major	Lumbar intervertebral disks; bodies and transverse processes of lumbar vertebrae	Lesser trochanter of femur	Flexes thigh	Branches of L1-3 nerves
Iliacus	Iliac fossa of ilium	Lesser trochanter of femur	Flexes thigh	Femoral n.
Gluteus maximus	Sacrum, coccyx, and posterior surface of ilium	Posterior surface of femur and fascia of thigh	Extends leg at hip	Inferior gluteal n.
Gluteus medius	Lateral surface of ilium	Greater trochanter of femur	Abducts and rotates thigh medially	Superior gluteal n.
Gluteus minimus	Lateral surface of ilium	Greater trochanter of femur	Same as gluteus medius	Superior gluteal n.
Tensor fasciae latae	Anterior iliac crest	Fascia of thigh	Abducts, flexes, and rotates thigh medially	Superior gluteal n.
Adductor longus	Pubic bone near symphysis pubis	Posterior surface of femur	Adducts, flexes, and rotates thigh laterally	Obturator n.
Adductor magnus	Ischial tuberosity	Posterior surface of femur	Adducts, extends, and rotates thigh laterally	Obturator and branch of sciatic n.
Gracilis	Lower edge of symphysis pubis	Medial surface of tibia	Adducts thigh and flexes leg at the knee	Obturator n.

CHART 9.13   Muscles that move the lower leg

Muscle	Origin	Insertion	Action	Nerve supply
Hamstring group				
Biceps femoris	Ischial tuberosity and linea aspera of femur	Head of fibula and lateral condyle of tibia	Flexes and rotates leg laterally and extends thigh	Tibial n.
Semitendinosus	Ischial tuberosity	Medial surface of tibia	Flexes and rotates leg medially and extends thigh	Tibial n.
Semimembranosus	Ischial tuberosity	Medial condyle of tibia	Flexes and rotates leg medially and extends thigh	Tibial n.
Sartorius	Anterior superior iliac spine	Medial surface of tibia	Flexes leg and thigh, abducts thigh, and rotates thigh laterally	Femoral n.
Quadriceps femoris group				
Rectus femoris	Spine of ilium and margin of acetabulum	Patella by common tendon, which continues as patellar ligament to tibial tuberosity	Extends leg at knee	Femoral n.
Vastus lateralis	Greater trochanter and posterior surface of femur			
Vastus medialis	Medial surface of femur			
Vastus intermedius	Anterior and lateral surfaces of femur			

Flexors

As its name implies, the **biceps femoris** (bi'seps fem'or-is) has two heads, one attached to the ischium and the other attached to the femur. This muscle passes along the back of the thigh on the lateral side and connects to the proximal ends of the fibula and tibia. The biceps femoris is one of the hamstring muscles, and its tendon (hamstring) can be felt as a lateral ridge behind the knee. This muscle functions to flex and rotate the leg laterally and to extend the thigh (figures 9.35 and 9.36).

The **semitendinosus** (sem"e-ten'di-no-sus) is another of the hamstring muscles. It is a long, bandlike muscle on the back of the thigh toward the medial side, connecting the ischium to the proximal end of the tibia. The semitendinosus is so named because it becomes tendinous in the middle of the thigh, continuing to its insertion as a long, cordlike tendon. It functions to flex and rotate the leg medially, and to extend the thigh (figure 9.36).

The **semimembranosus** (sem"e-mem'brah-no-sus) is the third hamstring muscle and is the most medially located muscle in the back of the thigh. It connects the ischium to the tibia, and functions to flex and rotate the leg medially and to extend the thigh (figure 9.36).

The **sartorius** (sar-to're-us) is an elongated, straplike muscle that passes obliquely across the front of the thigh and then descends over the medial side of the knee. It connects the ilium to the tibia, and functions to flex the leg and the thigh. It can also abduct the thigh and rotate it laterally (figures 9.34 and 9.35).

The tendinous attachments of the hamstring muscles to the ischial tuberosity are sometimes torn as a result of strenuous running or kicking motions. This painful injury is commonly called "pulled hamstrings," and is usually accompanied by internal bleeding from damaged blood vessels that supply the muscles.

Extensor

The large, fleshy muscle group called the **quadriceps femoris** (kwod'ri-seps fem'or-is) occupies the front and sides of the thigh, and is the primary extensor of the knee. It is composed of four parts—*rectus femoris, vastus lateralis, vastus medialis,* and *vastus intermedius* (figures 9.34 and 9.37). These parts connect the ilium and femur to a common *patellar tendon,* which passes over the front of the knee and attaches to the patella. This tendon then continues as the *patellar ligament* to the tibia.

Occasionally, as a result of traumatic injury in which a muscle such as the quadriceps femoris is compressed against an underlying bone, new bone tissue may begin to develop within the damaged muscle. This condition is called *myositis ossificans.* When the bone tissue has matured (several months following the injury), the newly formed bone can be removed surgically.

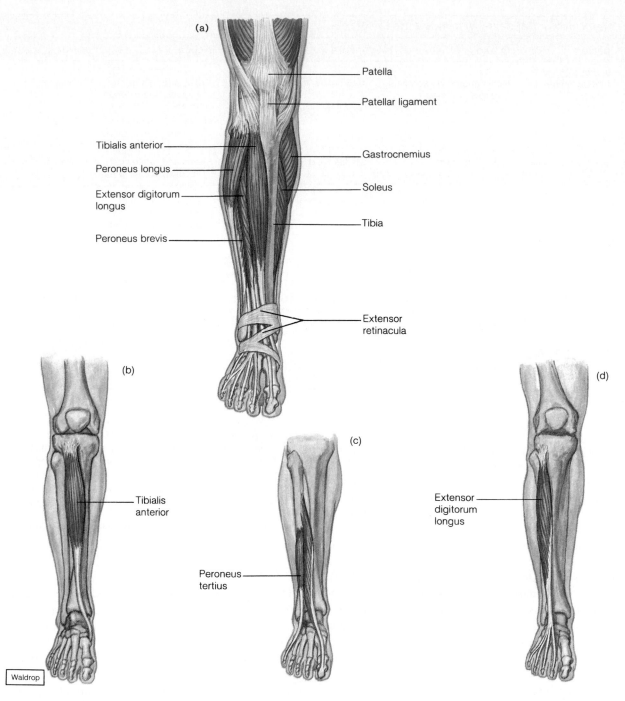

(a) Patella

Patellar ligament

Tibialis anterior

Peroneus longus

Extensor digitorum longus

Peroneus brevis

Gastrocnemius

Soleus

Tibia

Extensor retinacula

(b)

Tibialis anterior

(c)

Peroneus tertius

(d)

Extensor digitorum longus

Waldrop

Figure 9.38 (*a*) Muscles of the anterior right lower leg. (*b–d*) Isolated views of muscles associated with the anterior lower leg.

Muscles That Move the Ankle, Foot, and Toes

A number of muscles that function to move the ankle, foot, and toes are located in the lower leg. They attach the femur, tibia, and fibula to various bones of the foot and are responsible for a variety of movements—moving the foot upward (dorsiflexion) or downward (plantar flexion), and turning the sole of the foot inward (inversion) or outward (eversion). These muscles are shown in figures 9.38, 9.39, 9.40, 9.41, in reference plates 68, 69, 70, and are listed in chart 9.14. Muscles

that move the ankle, foot, and toes include the following:

Dorsal flexors
   Tibialis anterior
   Peroneus tertius
   Extensor digitorum longus
Plantar flexors
   Gastrocnemius
   Soleus
   Flexor digitorum longus

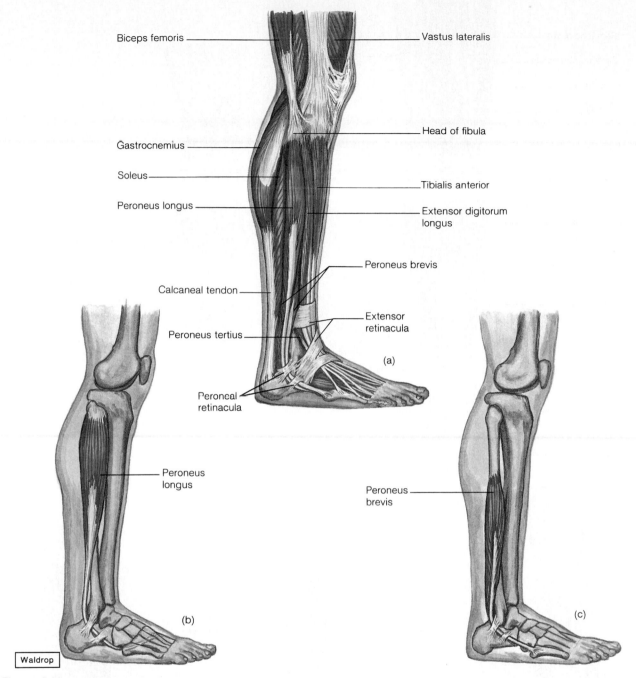

Figure 9.39 (*a*) Muscles of the lateral right lower leg. Isolated views of (*b*) peroneus longus and (*c*) peroneus brevis.

Invertor
  Tibialis posterior
Evertor
  Peroneus longus

Dorsal Flexors

The **tibialis anterior** (tib″e-a′lis an-te′re-or) is an elongated, spindle-shaped muscle located on the front of the lower leg. It arises from the surface of the tibia, passes medially over the distal end of the tibia, and attaches to bones of the ankle and foot. Contraction of the tibialis anterior causes dorsiflexion and inversion of the foot (figure 9.38).

The **peroneus tertius** (per″o-ne′us ter′shus) is a muscle of variable size that connects the fibula to the lateral side of the foot. It functions in dorsiflexion and eversion of the foot (figure 9.38).

The **extensor digitorum longus** (eks-ten′sor dij″i-to′rum long′gus) is situated along the lateral side of the lower leg just behind the tibialis anterior. It arises from the proximal end of the tibia and the shaft of the fibula.

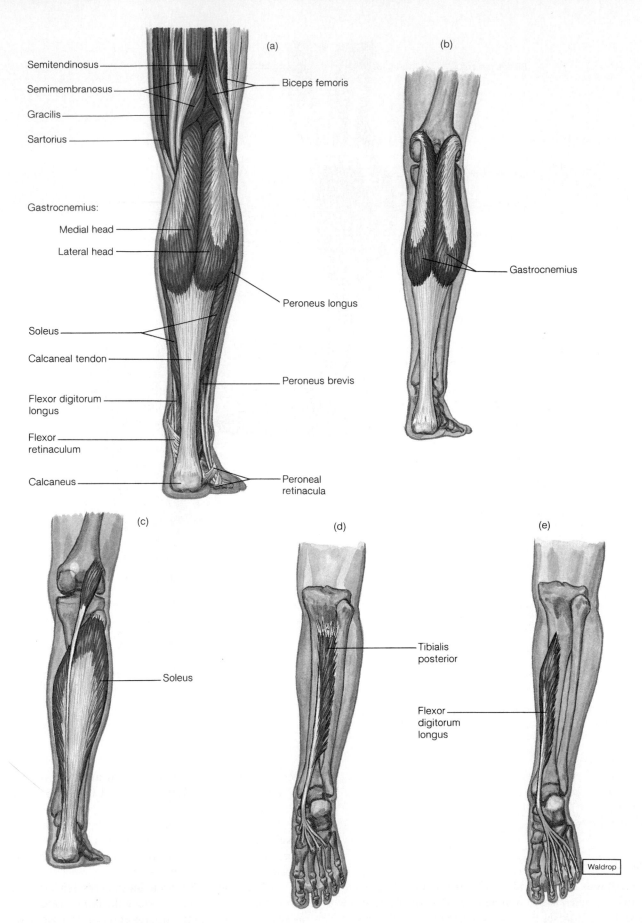

Figure 9.40 (*a*) Muscles of the posterior right lower leg. (*b–e*) Isolated views of muscles associated with the posterior right lower leg.

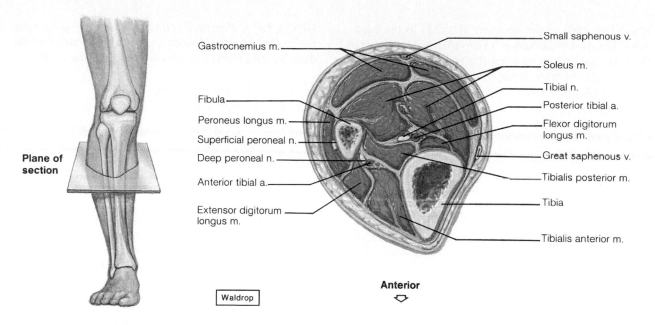

Figure 9.41   A cross section of the lower leg (superior view).

CHART 9.14	Muscles that move the ankle, foot, and toes			
Muscle	Origin	Insertion	Action	Nerve supply
Tibialis anterior	Lateral condyle and lateral surface of tibia	Tarsal bone (cuneiform) and first metatarsal	Dorsiflexion and inversion of foot	Deep peroneal n.
Peroneus tertius	Anterior surface of fibula	Dorsal surface of fifth metatarsal	Dorsiflexion and eversion of foot	Deep peroneal n.
Extensor digitorum longus	Lateral condyle of tibia and anterior surface of fibula	Dorsal surfaces of second and third phalanges of four lateral toes	Dorsiflexion and eversion of foot and extension of toes	Deep peroneal n.
Gastrocnemius	Lateral and medial condyles of femur	Posterior surface of calcaneus	Plantar flexion of foot and flexion of leg at knee	Tibial n.
Soleus	Head and shaft of fibula and posterior surface of tibia	Posterior surface of calcaneus	Plantar flexion of foot	Tibial n.
Flexor digitorum longus	Posterior surface of tibia	Distal phalanges of four lateral toes	Plantar flexion and inversion of foot, and flexion of four lateral toes	Tibial n.
Tibialis posterior	Lateral condyle and posterior surface of tibia, and posterior surface of fibula	Tarsal and metatarsal bones	Plantar flexion and inversion of foot	Tibial n.
Peroneus longus	Lateral condyle of tibia, and head and shaft of fibula	Tarsal and metatarsal bones	Plantar flexion and eversion of foot; also supports arch	Superficial peroneal n.

Its tendon divides into four parts as it passes over the front of the ankle. These parts continue over the surface of the foot and attach to the four lateral toes. The actions of the extensor digitorum longus include dorsiflexion of the foot, eversion of the foot, and extension of the toes (figures 9.38 and 9.40).

Plantar Flexors

The **gastrocnemius** (gas″trok-ne′me-us) on the back of the lower leg forms part of the calf. It arises by two heads from the femur. The distal end of this muscle joins the strong *calcaneal tendon* (Achilles tendon), which descends to the heel and attaches to the calcaneus. The gastrocnemius is a powerful plantar flexor of the foot that aids in pushing the body forward when a person walks or runs. It also functions to flex the leg at the knee (figure 9.39).

> As a result of strenuous athletic activity, the calcaneal (Achilles) tendon may be partially or completely torn. This injury occurs most frequently in middle-aged athletes who run or play sports that involve quick movements and directional changes. A torn calcaneal tendon usually requires surgical treatment.

The **soleus** (so′le-us) is a thick, flat muscle located beneath the gastrocnemius, and together these two muscles make up the calf of the leg. The soleus arises from the tibia and fibula, and it extends to the heel by way of the calcaneal tendon. It acts with the gastrocnemius to cause plantar flexion of the foot (figures 9.39 and 9.40).

The **flexor digitorum longus** (flek′sor dij″ĭ-to′rum long′gus) extends from the posterior surface of the tibia to the foot. Its tendon passes along the plantar surface of the foot. There it divides into four parts that attach to the terminal bones of the four lateral toes. This muscle assists in plantar flexion of the foot, flexion of the four lateral toes, and inversion of the foot (figure 9.40).

Invertor

The **tibialis posterior** (tib″e-a′lis pos-tēr′e-or) is the deepest of the muscles on the back of the lower leg. It connects the fibula and tibia to anklebones by means of a tendon that curves under the medial malleolus. This muscle assists in inversion and plantar flexion of the foot (figure 9.40).

Evertor

The **peroneus longus** (per″o-ne′us long′gus) is a long, straplike muscle located on the lateral side of the lower leg. It connects the tibia and the fibula to the foot by means of a stout tendon that passes behind the lateral malleolus. It functions in eversion of the foot, assists in plantar flexion, and helps support the arch of the foot (figures 9.39 and 9.41).

As in the wrist, fascia in various regions of the ankle is thickened to form retinacula. Anteriorly, for example, *extensor retinacula* connect the tibia and fibula as well as the calcaneus and fascia of the sole. These retinacula form sheaths for tendons crossing the front of the ankle (figure 9.39).

Posteriorly, on the inside, a *flexor retinaculum* runs between the medial malleolus and the calcaneus, and forms sheaths for tendons passing beneath the foot

(figure 9.40). *Peroneal retinacula* connect the lateral malleolus and the calcaneus, providing sheaths for tendons on the lateral side of the ankle (figure 9.39).

Clinical Terms Related to the Muscular System

contracture (kon-trak′tūr) A condition in which there is great resistance to the stretching of a muscle.

convulsion (kun-vul′shun) A series of involuntary contractions of various voluntary muscles.

electromyography (e-lek″tro-mi-og′rah-fe) A technique for recording the electrical changes that occur in muscle tissues.

fibrillation (fi″bri-la′shun) Spontaneous contractions of individual muscle fibers, producing rapid and uncoordinated activity within a muscle.

fibrosis (fi-bro′sis) A degenerative disease in which skeletal muscle tissue is replaced by fibrous connective tissue.

fibrositis (fi″bro-si′tis) An inflammatory condition of fibrous connective tissues, especially in the muscle fascia. This disease is also called muscular rheumatism.

muscular dystrophy (mus′ku-lar dis′tro-fe) A progressively crippling disease of unknown cause in which the muscles gradually weaken and atrophy.

myalgia (mi-al′je-ah) Pain resulting from any muscular disease or disorder.

myasthenia gravis (mi″as-the′ne-ah grav′is) A chronic disease characterized by muscles that are weak and easily fatigued. It results from a disorder at some of the neuromuscular junctions so that a stimulus is not transmitted from the motor neuron to the muscle fiber.

myokymia (mi″o-ki′me-ah) Persistent quivering of a muscle.

myology (mi-ol′o-je) The study of muscles.

myoma (mi-o′mah) A tumor composed of muscle tissue.

myopathy (mi-op′ah-the) Any muscular disease.

myositis (mi″o-si′tis) An inflammation of skeletal muscle tissue.

myotomy (mi-ot′o-me) The cutting of muscle tissue.

myotonia (mi″o-to′ne-ah) A prolonged muscular spasm.

paralysis (pah-ral′ĭ-sis) The loss of ability to move a body part.

paresis (pah-re′sis) A partial or slight paralysis of the muscles.

shin splints (shin′splints) A soreness on the front of the lower leg due to straining the flexor digitorum longus, often as a result of walking up and down hills.

torticollis (tor″ti-kol′is) A condition in which the neck muscles, such as the sternocleidomastoids, contract involuntarily. It is more commonly called wryneck.

Chapter Summary

Introduction (page 270)

The three types of muscle tissue are skeletal, smooth, and cardiac.

Structure of a Skeletal Muscle (page 270)

Skeletal muscles are composed of nervous, vascular, and various connective tissues, as well as skeletal muscle tissue.

1. Connective tissue coverings
   a. Skeletal muscles and their parts are covered with fascia.
   b. Other connective tissues surround cells and groups of cells within the muscle's structure.
   c. Fascia is part of a complex network of connective tissue that extends throughout the body.
2. Skeletal muscle fibers
   a. Each skeletal muscle fiber represents a single muscle cell, which is the unit of contraction.
   b. Muscle fibers are cylindrical cells with numerous nuclei.
   c. The cytoplasm contains mitochondria, a sarcoplasmic reticulum, and myofibrils composed of actin and myosin.
   d. Striations are produced by the arrangement of the actin and myosin filaments.
   e. Transverse tubules extend from the cell membrane into the cytoplasm and are associated with the cisternae of the sarcoplasmic reticulum.
3. Neuromuscular junction
   a. Muscle fibers are stimulated to contract by motor neurons.
   b. The motor end plate of a muscle fiber lies on one side of a neuromuscular junction.
   c. In response to a nerve impulse, the end of a motor nerve fiber secretes a neurotransmitter, which diffuses across the junction and stimulates the muscle fiber.
4. Motor units
   a. One motor neuron and the muscle fibers associated with it constitute a motor unit.
   b. Finer movements can be produced in muscles whose motor units contain small numbers of muscle fibers.

Skeletal Muscle Contraction (page 275)

Muscle fiber contraction results from a sliding movement within the myofibrils in which actin and myosin filaments merge.

1. Role of myosin and actin
   a. Cross-bridges of myosin filaments can form linkages with actin filaments.
   b. The reaction between actin and myosin filaments generates the force of contraction.
   c. When a fiber is at rest, troponin and tropomyosin molecules interfere with linkage formation.
   d. When calcium ions are present, the inhibition is removed.

2. Stimulus for contraction
   a. Muscle fiber is usually stimulated by acetylcholine released from the end of a motor nerve fiber.
   b. Acetylcholine causes muscle fiber to conduct an impulse that travels over the surface of the sarcolemma and reaches the deep parts of the fiber by means of the transverse tubules.
   c. A muscle impulse signals the sarcoplasmic reticulum to release calcium ions.
   d. Linkages form between myosin and actin, and the actin filaments move inward, shortening the sarcomere.
   e. The muscle fiber relaxes when calcium ions are transported back into the sarcoplasmic reticulum.
   f. Cholinesterase decomposes acetylcholine.
3. Energy sources for contraction
   a. ATP supplies the energy for muscle fiber contraction.
   b. Creatine phosphate stores energy that can be used to synthesize ATP as it is decomposed.
   c. Active muscles depend upon cellular respiration for energy.
4. Oxygen supply and cellular respiration
   a. Anaerobic respiration results in a gain of only a few ATP molecules, while aerobic respiration results in the gain of many ATP molecules.
   b. Oxygen is carried from the lungs to body cells by hemoglobin in red blood cells.
   c. Myoglobin in muscle cells stores some oxygen temporarily.
5. Oxygen debt
   a. During rest or moderate exercise, oxygen is supplied to muscles in sufficient concentration to support aerobic respiration.
   b. During strenuous exercise, an oxygen deficiency may develop, and lactic acid may accumulate as a result of anaerobic respiration.
   c. The amount of oxygen needed to convert accumulated lactic acid to glucose and to restore the supplies of ATP and creatine phosphate is called oxygen debt.
6. Muscle fatigue
   a. A fatigued muscle loses its ability to contract.
   b. Muscle fatigue is usually due to the effects of an accumulation of lactic acid.
   c. Athletes usually produce less lactic acid than nonathletes because of their increased ability to supply oxygen and nutrients to muscles.
7. Fast-contracting and slow-contracting muscles
   a. The speed of contraction is related to a muscle's special function.
   b. Slow-contracting, or red, muscles can generate ATP fast enough to keep up with ATP breakdown and can contract for long periods.
   c. Fast-contracting, or white, muscles have reduced ability to carry on aerobic respiration and tend to fatigue relatively rapidly.

8. Heat production
   a. Muscles represent an important source of body heat.
   b. Most of the energy released by cellular respiration is lost as heat.

Muscular Responses (page 281)

1. Threshold stimulus is the minimal stimulus needed to elicit a muscular contraction.
2. All-or-none response
   a. If a muscle fiber contracts at all, it will contract completely.
   b. Motor units respond in an all-or-none manner.
3. Recruitment of motor units
   a. At low intensity of stimulation, relatively small numbers of motor units contract.
   b. At increasing intensities of stimulation, other motor units are recruited until the muscle contracts with maximal tension.
4. Staircase effect
   a. An inactive muscle undergoes a series of contractions of increasing strength when subjected to a series of stimuli.
   b. This staircase effect seems to be due to failure to remove calcium ions from the sarcoplasm rapidly enough.
5. Sustained contractions
   a. A rapid series of stimuli may produce a summation of twitches and a sustained contraction.
   b. When contractions fuse, the strength of contraction may increase due to recruitment of fibers.
   c. Even when a muscle is at rest, its fibers usually maintain tone—that is, remain partially contracted.
6. Isotonic and isometric contractions
   a. When a muscle contracts and its ends are pulled closer together, the contraction is called isotonic.
   b. When a muscle contracts but its attachments do not move, the contraction is called isometric.
   c. Most body movements involve both isometric and isotonic contractions.

Smooth Muscles (page 284)

The contractile mechanisms of smooth and cardiac muscles are similar to those of skeletal muscle.

1. Smooth muscle fibers
   a. Smooth muscle cells contain filaments of myosin and actin.
   b. They lack transverse tubules, and the sarcoplasmic reticula are not well developed.
   c. Types include multiunit smooth muscle and visceral smooth muscle.
   d. Visceral smooth muscle displays rhythmicity.
   e. Movement of material through visceral organs is often aided by peristalsis.
2. Smooth muscle contraction
   a. In smooth muscles, calmodulin binds to calcium ions and activates the contraction mechanism.
   b. Both acetylcholine and norepinephrine serve as neurotransmitter substances for smooth muscles.
   c. Other factors that affect smooth muscle contractions include hormones and stretching.
   d. With a given amount of energy, smooth muscle can maintain a contraction for a longer time than can skeletal muscle.
   e. Smooth muscles can change their lengths without changing their tautness.

Cardiac Muscle (page 286)

1. Cardiac muscle remains contracted for a longer time than skeletal muscle because transverse tubules supply extra calcium ions.
2. The ends of adjacent cardiac muscle cells are separated by intercalated disks, which hold the cells together.
3. A network of fibers contracts as a unit and responds to stimulation in an all-or-none manner.
4. Cardiac muscle is self-exciting, rhythmic, and remains refractory until a contraction is completed.

Skeletal Muscle Actions (page 287)

1. Origin and insertion
   a. The movable end of a skeletal muscle is its insertion, and the immovable end is its origin.
   b. Some muscles have more than one origin or insertion.
2. Interaction of skeletal muscles
   a. Skeletal muscles function in groups.
   b. A prime mover is responsible for most of a movement; synergists aid prime movers; antagonists can resist the movement of a prime mover.
   c. Smooth movements depend upon antagonists giving way to the actions of prime movers.

Major Skeletal Muscles (page 288)

Muscle names often describe sizes, shapes, locations, actions, number of attachments, or direction of fibers.

1. Muscles of facial expression
   a. These muscles lie beneath the skin of the face and scalp, and are used to communicate feelings through facial expression.
   b. They include the epicranius, orbicularis oculi, orbicularis oris, buccinator, zygomaticus, and platysma.
2. Muscles of mastication
   a. These muscles are attached to the mandible and are used in chewing.
   b. They include the masseter, temporalis, medial pterygoid, and lateral pterygoid.
3. Muscles that move the head
   a. Head movements are produced by muscles in the neck and upper back.
   b. They include the sternocleidomastoid, splenius capitis, semispinalis capitis, and longissimus capitus.
4. Muscles that move the pectoral girdle
   a. Most of these muscles connect the scapula to nearby bones and are closely associated with muscles that move the upper arm.
   b. They include the trapezius, rhomboideus major, levator scapulae, serratus anterior, and pectoralis minor.
5. Muscles that move the upper arm
   a. These muscles connect the humerus to various regions of the pectoral girdle, ribs, and vertebral column.
   b. They include the coracobrachialis, pectoralis major, teres major, latissimus dorsi, supraspinatus, deltoid, subscapularis, infraspinatus, and teres minor.
6. Muscles that move the forearm
   a. These muscles connect the radius and ulna to the humerus and pectoral girdle.
   b. They include the biceps brachii, brachialis, brachioradialis, triceps brachii, supinator, pronator teres, and pronator quadratus.
7. Muscles that move the wrist, hand, and fingers
   a. These muscles arise from the distal end of the humerus and from the radius and ulna.
   b. They include the flexor carpi radialis, flexor carpi ulnaris, palmaris longus, flexor digitorum profundus, flexor digitorum superficialis, extensor carpi radialis longus, extensor carpi radialis brevis, extensor carpi ulnaris, and extensor digitorum.
   c. An extensor retinaculum forms sheaths for tendons of the extensor muscles.
8. Muscles of the abdominal wall
   a. These muscles connect the rib cage and vertebral column to the pelvic girdle.
   b. They include the external oblique, internal oblique, transversus abdominis, and rectus abdominis.
9. Muscles of the pelvic outlet
   a. These muscles form the floor of the pelvic cavity and fill the space of the pubic arch.
   b. They include the levator ani, coccygeus, superficial transversus perinei, bulbospongiosus, ischiocavernosus, and sphincter urethrae.
10. Muscles that move the thigh
    a. These muscles are attached to the femur and to some part of the pelvic girdle.
    b. They include the psoas major, iliacus, gluteus maximus, gluteus medius, gluteus minimus, tensor fasciae latae, adductor longus, adductor magnus, and gracilis.
11. Muscles that move the lower leg
    a. These muscles connect the tibia or fibula to the femur or pelvic girdle.
    b. They include the biceps femoris, semitendinosus, semimembranosus, sartorius, and the quadriceps femoris group.
12. Muscles that move the ankle, foot, and toes
    a. These muscles attach the femur, tibia, and fibula to various bones of the foot.
    b. They include the tibialis anterior, peroneus tertius, extensor digitorum longus, gastrocnemius, soleus, flexor digitorum longus, tibialis posterior, and peroneus longus.
    c. Retinacula form sheaths for tendons passing to the foot.

Clinical Application of Knowledge

1. Why do you think athletes generally perform better if they warm up by exercising lightly before a competitive event?

2. Following childbirth, a woman may experience a loss of urinary control (incontinence) when sneezing or coughing. What muscles of the pelvic floor should be strengthened by exercise to help control this problem?

3. What steps might be taken to minimize atrophy of skeletal muscles in patients who are confined to bed for prolonged times?

4. As lactic acid and other substances accumulate in an active muscle, they tend to stimulate pain receptors, and the muscle may feel sore. How might the application of heat or substances that cause blood vessels to dilate help relieve such soreness?

5. Several important nerves and blood vessels course through the muscles of the gluteal region. In order to avoid the possibility of damaging such parts, intramuscular injections into this region are usually made into the lateral, superior portion of the gluteus medius. What landmarks would help you locate this muscle in a patient?

6. Following an injury to a nerve, the muscles it supplies with motor nerve fibers may become paralyzed. How would you explain to a patient the importance of having the disabled muscles moved passively or causing them to contract by using electrical stimulation?

Review Activities
Part A

1. List the three types of muscle tissue.
2. Distinguish between a tendon and an aponeurosis.
3. Describe the connective tissue coverings of a skeletal muscle.
4. Distinguish between deep fascia, subcutaneous fascia, and subserous fascia.
5. List the major parts of a skeletal muscle fiber, and describe the function of each part.
6. Describe a neuromuscular junction.
7. Explain the function of a neurotransmitter substance.
8. Define *motor unit,* and explain how the number of fibers within a unit affects muscular contractions.
9. Describe the major events that occur when a muscle fiber contracts.
10. Explain how ATP and creatine phosphate are related and how these substances function in the muscle fiber contraction mechanism.
11. Describe how oxygen is supplied to skeletal muscles.
12. Describe how an oxygen debt may develop.
13. Explain how muscles may become fatigued and how a person's physical condition may affect tolerance to fatigue.
14. Distinguish between fast-contracting and slow-contracting muscles.
15. Explain how the maintenance of body temperature is related to the actions of skeletal muscles.
16. Define *threshold stimulus.*
17. Explain what is meant by an *all-or-none response.*
18. Explain what is meant by *motor unit recruitment.*
19. Describe the staircase effect.
20. Explain how a skeletal muscle can be stimulated to produce a sustained contraction.
21. Distinguish between a tetanic contraction and muscle tone.
22. Distinguish between isometric and isotonic contractions, and explain how each is used in body movements.
23. Compare the structures of smooth and skeletal muscle fibers.
24. Distinguish between multiunit and visceral smooth muscles.
25. Define *peristalsis* and explain its function.
26. Compare the characteristics of smooth and skeletal muscle contractions.
27. Compare the structures of cardiac and skeletal muscle fibers.
28. Compare the characteristics of cardiac and skeletal muscle contractions.
29. Distinguish between a muscle's origin and its insertion.
30. Define *prime mover, synergist,* and *antagonist.*

Part B

Match the muscles in column I with the descriptions and functions in column II.

I	II
1. Buccinator	A. Inserted on the coronoid process of the mandible.
2. Epicranius	B. Draws the corner of the mouth upward.
3. Lateral pterygoid	C. Can raise and adduct the scapula.
4. Platysma	D. Can pull the head into an upright position.
5. Rhomboideus major	E. Consists of two parts—the frontalis and the occipitalis.
6. Splenius capitis	F. Compresses the cheeks.
7. Temporalis	G. Extends over the neck from the chest to the face.
8. Zygomaticus	H. Pulls the jaw from side to side.

I	II
9. Biceps brachii	I. Primary extensor of the elbow.
10. Brachialis	J. Pulls the shoulder back and downward.
11. Deltoid	K. Abducts the arm.
12. Latissimus dorsi	L. Rotates the arm laterally.
13. Pectoralis major	M. Pulls the arm forward and across the chest.
14. Pronator teres	N. Rotates the arm medially.
15. Teres minor	O. Strongest flexor of the elbow.
16. Triceps brachii	P. Strongest supinator of the forearm.

I	II
17. Biceps femoris	Q. Inverts the foot.
18. External oblique	R. A member of the quadriceps femoris group.
19. Gastrocnemius	S. A plantar flexor of the foot.
20. Gluteus maximus	T. Compresses the contents of the abdominal cavity.
21. Gluteus medius	U. Largest muscle in the body.
22. Gracilis	V. A hamstring muscle.
23. Rectus femoris	W. Adducts the thigh.
24. Tibialis anterior	X. Abducts the thigh.

Part C
What muscles can you identify in the bodies of these models whose muscles are enlarged by exercise?

Reference Plates

Surface Anatomy

*T*he following set of reference plates is presented to help you locate some of the more prominent surface features in various regions of the body. For the most part, the labeled structures are easily seen or palpated through the skin. As a review, you may want to locate as many of these features as possible on your own body.

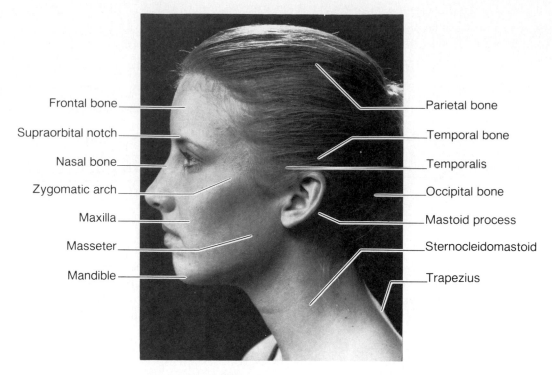

Frontal bone

Supraorbital notch

Nasal bone

Zygomatic arch

Maxilla

Masseter

Mandible

Parietal bone

Temporal bone

Temporalis

Occipital bone

Mastoid process

Sternocleidomastoid

Trapezius

Plate 37   Surface anatomy of head and neck, lateral view.

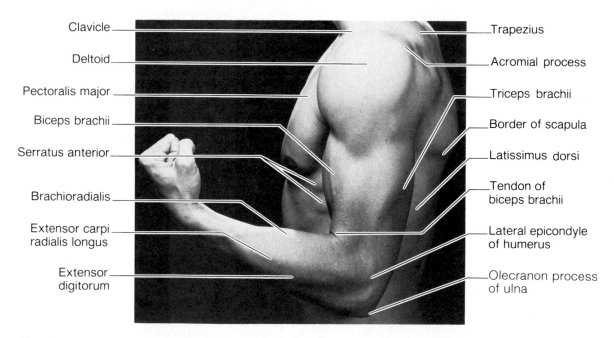

Clavicle

Deltoid

Pectoralis major

Biceps brachii

Serratus anterior

Brachioradialis

Extensor carpi
radialis longus

Extensor
digitorum

Trapezius

Acromial process

Triceps brachii

Border of scapula

Latissimus dorsi

Tendon of
biceps brachii

Lateral epicondyle
of humerus

Olecranon process
of ulna

Plate 38   Surface anatomy of arm and thorax, lateral view.

Trapezius

Teres major

Spinous processes
of vertebrae

Biceps brachii
Triceps brachii
Deltoid

Infraspinatus

Border of scapula

Vertebral spine

Latissimus dorsi

Plate 39 Surface anatomy of back and arms, posterior view.

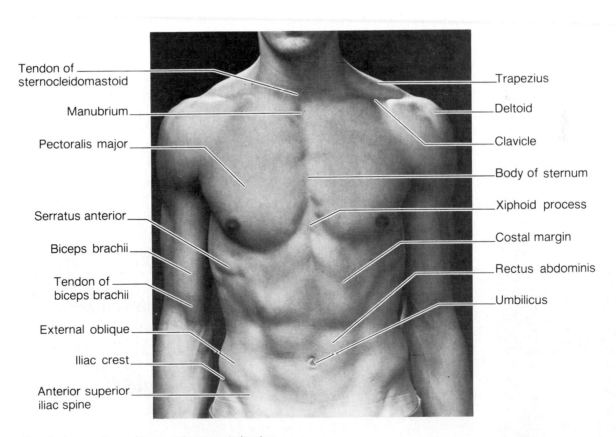

Tendon of
sternocleidomastoid

Manubrium

Pectoralis major

Serratus anterior

Biceps brachii

Tendon of
biceps brachii

External oblique

Iliac crest

Anterior superior
iliac spine

Trapezius

Deltoid

Clavicle

Body of sternum

Xiphoid process

Costal margin

Rectus abdominis

Umbilicus

Plate 40 Surface anatomy of torso and arms, anterior view.

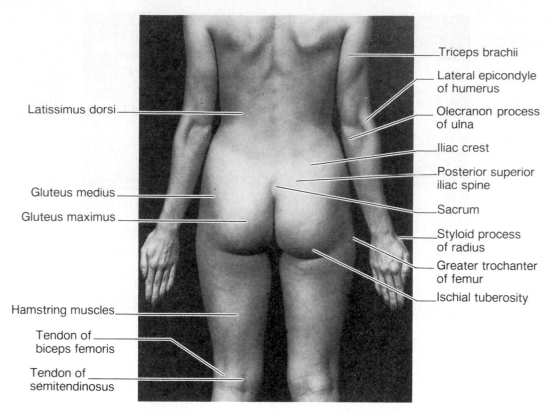

Triceps brachii

Lateral epicondyle
of humerus

Olecranon process
of ulna

Iliac crest

Posterior superior
iliac spine

Sacrum

Styloid process
of radius

Greater trochanter
of femur

Ischial tuberosity

Latissimus dorsi

Gluteus medius

Gluteus maximus

Hamstring muscles

Tendon of
biceps femoris

Tendon of
semitendinosus

Plate 41 Surface anatomy of torso and upper legs,
posterior view.

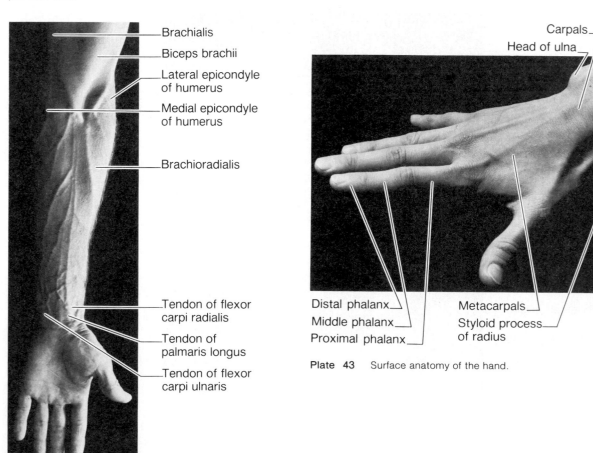

Brachialis

Biceps brachii

Lateral epicondyle
of humerus

Medial epicondyle
of humerus

Brachioradialis

Tendon of flexor
carpi radialis

Tendon of
palmaris longus

Tendon of flexor
carpi ulnaris

Plate 42 Surface anatomy of arm, anterior view.

Carpals

Head of ulna

Distal phalanx

Middle phalanx

Proximal phalanx

Metacarpals

Styloid process
of radius

Plate 43 Surface anatomy of the hand.

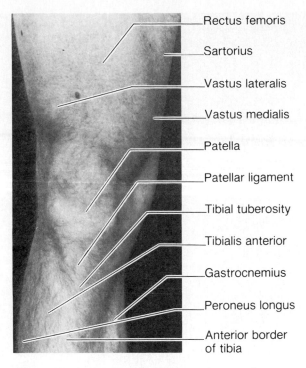

Rectus femoris

Sartorius

Vastus lateralis

Vastus medialis

Patella

Patellar ligament

Tibial tuberosity

Tibialis anterior

Gastrocnemius

Peroneus longus

Anterior border
of tibia

Plate 44 Surface anatomy of knee and surrounding area, anterior view.

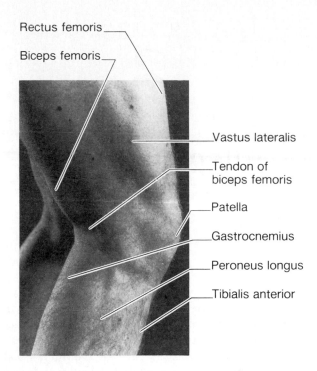

Rectus femoris

Biceps femoris

Vastus lateralis

Tendon of
biceps femoris

Patella

Gastrocnemius

Peroneus longus

Tibialis anterior

Plate 45 Surface anatomy of knee and surrounding area, lateral view.

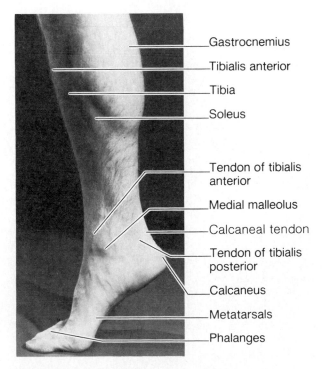

Gastrocnemius

Tibialis anterior

Tibia

Soleus

Tendon of tibialis
anterior

Medial malleolus

Calcaneal tendon

Tendon of tibialis
posterior

Calcaneus

Metatarsals

Phalanges

Plate 46 Surface anatomy of ankle and lower leg, medial view.

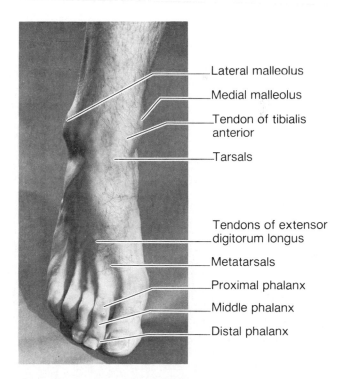

Lateral malleolus

Medial malleolus

Tendon of tibialis
anterior

Tarsals

Tendons of extensor
digitorum longus

Metatarsals

Proximal phalanx

Middle phalanx

Distal phalanx

Plate 47 Surface anatomy of ankle and foot.

Surface Anatomy • 327

UNIT

3

Integration and Coordination

T he chapters of unit 3 are concerned with the structures and functions of the nervous and endocrine systems. They describe how the organs of these systems keep the parts of the human body functioning together as a whole and how these organs help maintain a stable internal environment, which is vital to the survival of the organism.

10
Nervous System I: Basic Structure and Function

11
Nervous System II: Divisions of the Nervous System

12
Somatic and Special Senses

13
Endocrine System

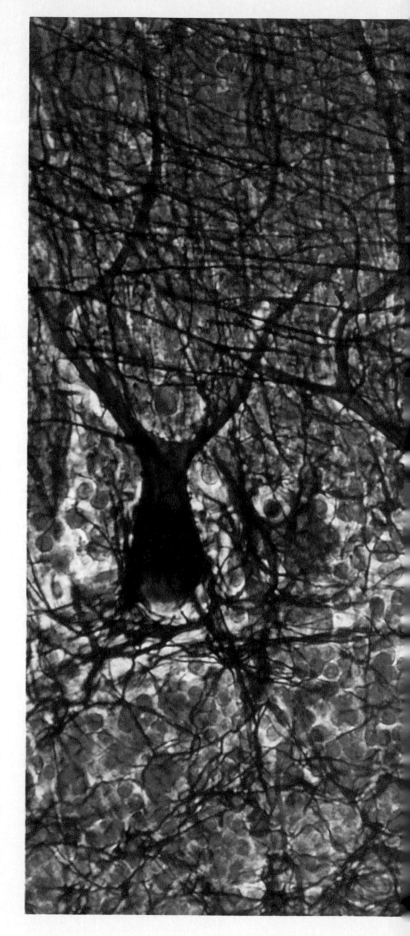

A light micrograph of nervous tissue from the brain (cerebellum) revealing a layer of darkly stained nerve cell bodies and branched nerve fibers (×250).

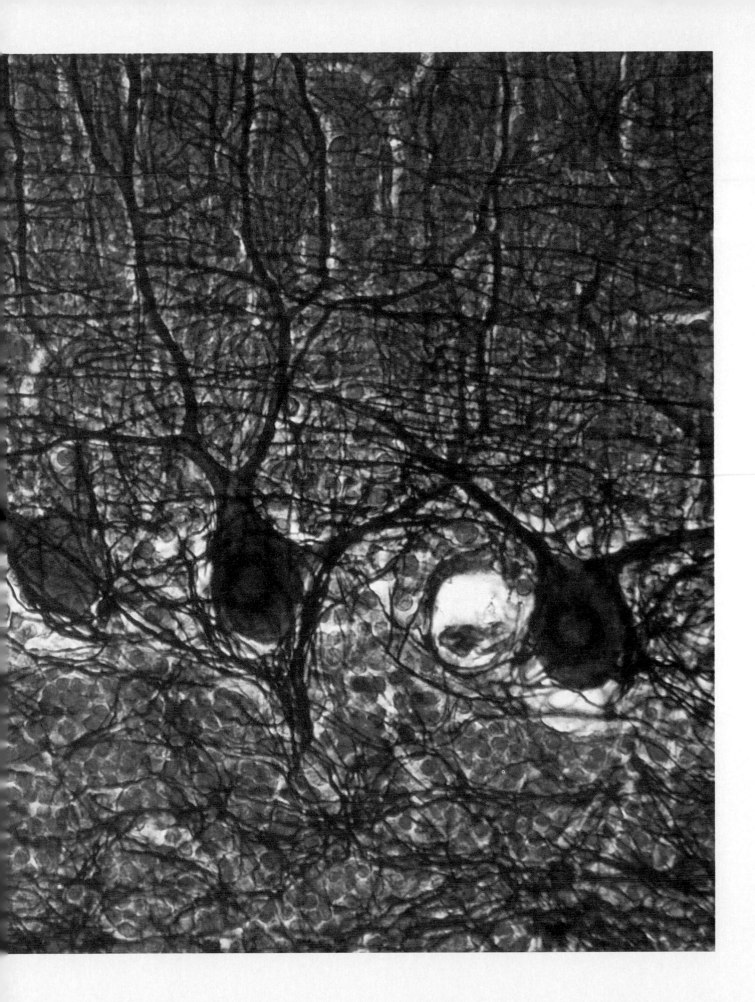

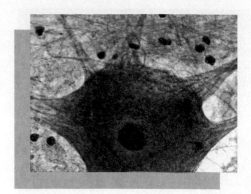

10

Nervous System I
Basic Structure
and Function

*N*eurons, the structural and functional units of the nervous system, communicate with one another by means of nerve impulses. Some neurons are specialized to detect changes that occur inside and outside the body and can transmit impulses to processing centers within the brain or spinal cord. Other neurons gather the incoming information, integrate it, and act on it in various ways. Still others transmit signals to muscles and glands, causing them to respond.

These functions are essential in the control and coordination of the body, enabling its parts to work together to maintain a stable internal environment ■

Chapter Objectives

After you have studied this chapter, you should be able to:

1. Explain the general functions of the nervous system.

2. Describe the general structure of a neuron.

3. Name four types of neuroglial cells and describe the functions of each.

4. Explain how an injured nerve fiber may regenerate.

5. Explain how a membrane becomes polarized.

6. Describe the events that lead to the conduction of a nerve impulse.

7. Explain how a nerve impulse is transmitted from one neuron to another.

8. Distinguish between excitatory and inhibitory postsynaptic potentials.

9. Explain two ways impulses are processed in neuronal pools.

10. Explain how neurons are classified.

11. Describe how nerve fibers in peripheral nerves are classified.

12. Describe a reflex arc.

13. Explain what is meant by reflex behavior.

14. Complete the review activities at the end of this chapter. Note that the items are worded in the form of specific learning objectives. You may want to refer to them before reading the chapter.

Key Terms

action potential (ak′shun po-ten′shal)

axon (ak′son)

central nervous system (sen′tral ner′vus sis′tem)

convergence (kon-ver′jens)

dendrite (den′drit)

divergence (di-ver′jens)

effector (e-fek′tor)

facilitation (fah-sil″ĭ-ta′shun)

myelin (mi′ĕ-lin)

neurilemma (nu″ri-lem′mah)

neuroglia (nu-rog′le-ah)

neuron (nu′ron)

neurotransmitter (nu″ro-trans-mit′er)

Nissl body (nis′l bod′e)

peripheral nervous system (pĕ-rif′er-al ner′vus sis′tem)

receptor (re-sep′tor)

reflex (re′fleks)

summation (sum-ma′shun)

synapse (sin′aps)

threshold (thresh′old)

Aids to Understanding Words

astr-, starlike: *astr*ocyte—a star-shaped neuroglial cell.

ax-, axle: *ax*on—a cylindrical nerve fiber that carries impulses away from a neuron cell body.

dendr-, tree: *dendr*ite—a branched nerve fiber that serves as the receptor surface of a neuron.

ependym-, tunic: *ependym*a—neuroglial cells that line spaces within the brain and spinal cord.

-lemm, rind or peel: neuri*lemm*a—a sheath that surrounds the myelin of a nerve fiber.

moto-, moving: *mot*or neuron—a neuron that stimulates a muscle to contract or a gland to release a secretion.

oligo-, few: *oligo*dendrocyte—a small neuroglial cell with few cellular processes.

peri-, all around: *peri*pheral nervous system—the portion of the nervous system that consists of the nerves branching from the brain and spinal cord.

saltator-, a dancer: *saltator*y conduction—nerve impulse conduction in which the impulse seems to jump from node to node along the nerve fiber.

sens-, feeling: *sens*ory neuron—a neuron that can be stimulated by a sensory receptor and conducts impulses into the brain or spinal cord.

syn-, together: *syn*apse—the junction between two neurons.

Organs of the nervous system, like other organs, are composed of various kinds of tissues, including nervous tissue, connective tissues, and blood. These organs can be divided into two groups. One group, consisting of the brain and spinal cord, forms the **central nervous system** (CNS), and the other, composed of the nerves (peripheral nerves) that connect the central nervous system to other body parts, is called the **peripheral nervous system** (PNS) (figure 10.1). Together these systems provide three general functions—a sensory function, an integrative function, and a motor function.

General Functions of the Nervous System

The sensory function of the nervous system involves sensory *receptors* at the ends of peripheral nerves. (See chapter 11.) These receptors are specialized to gather information by detecting changes that occur inside and outside the body. They monitor such external environmental factors as light and sound intensities as well as the temperature, oxygen concentration, and conditions of the body's internal fluids.

The information gathered by the sensory receptors is converted to signals in the form of *nerve impulses,* which are then transmitted over peripheral nerves to the central nervous system. There the signals are integrated—that is, they are brought together, creating sensations (perceptions), adding to memory, or helping produce thoughts. As a result of this integrative function, conscious or subconscious decisions are made and then acted upon by means of motor functions.

The motor functions of the nervous system employ peripheral nerves that carry impulses from the central nervous system to responsive parts called *effectors.* These effectors are outside the nervous system and include muscles that contract when they are stimulated by nerve impulses, and glands that produce a secretion when they are stimulated.

Thus, the nervous system can detect changes occurring in the body, make decisions on the basis of the information received, and cause muscles or glands to respond. Typically, these responses are directed toward counteracting the effects of the changes, and in this way, the nervous system helps maintain homeostasis.

Nervous Tissue

The nervous tissue of the brain and spinal cord consists of masses of nerve cells, or **neurons,** that are the structural and functional units of the nervous system. These cells are specialized to react to physical and chemical changes occurring in their surroundings. They also conduct nerve impulses to other neurons and to cells outside the nervous system. (See figure 10.2.)

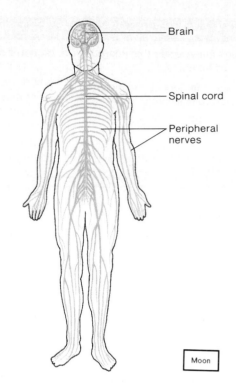

Figure 10.1 The nervous system consists of the brain, spinal cord, and peripheral nerves.

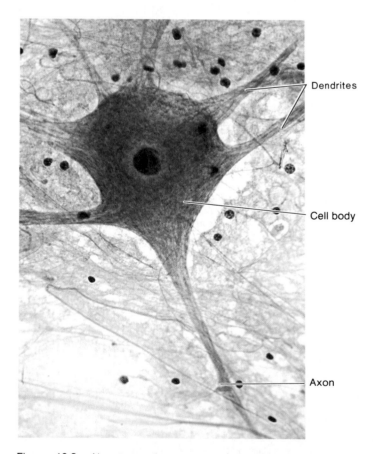

Figure 10.2 Neurons are the structural and functional units of the nervous system (×50). (Note: The dark spots in the area surrounding the neuron are nuclei of neuroglial cells.)

As mentioned in chapter 5, **neuroglial cells** are accessory cells within nervous tissue. They greatly outnumber the neurons, and, in addition to providing for certain physiological needs of the neurons, they function much like connective tissue cells in other body structures; that is, they fill spaces and surround or protect various parts.

Neuron Structure

Although neurons vary considerably in size and shape, they have certain features in common. For example, every neuron has a cell body and tubular processes filled with cytoplasm that conduct nerve impulses to or from the cell body (figure 10.3).

The **cell body** (soma or perikaryon) contains a mass of granular cytoplasm, a cell membrane, and various other organelles usually found in cells, although in nerve cells some of these parts have special names. Inside the cell body, for example, are mitochondria, lysosomes, a Golgi apparatus, and numerous microtubules. There is also a network of fine threads called **neurofibrils,** which extend into the cell processes and provide support for them. Scattered throughout the cytoplasm are many membranous sacs called **Nissl bodies,** which are similar to the rough endoplasmic reticulum in other cells. Ribosomes attached to the surfaces of these membranous parts help manufacture vital protein molecules. Cytoplasmic inclusions are common in neurons, and they contain substances such as glycogen, lipids, or pigments such as melanin.

Near the center of the neuron cell body is a large, spherical nucleus with a conspicuous nucleolus. This nucleus apparently does not undergo mitosis after the nervous system is developed, and consequently, mature neurons seem to be incapable of reproduction.

Two kinds of nerve fibers, **dendrites** and **axons,** extend from the cell bodies of most neurons. Although a neuron may have many dendrites, it has only one axon.

In most neurons, the dendrites are relatively short and highly branched. These branches, together with the membrane of the cell body, provide the main receptive surfaces of the neuron to which processes from other neurons communicate. Often the dendrites possess tiny, thornlike spines (dendritic spines) on their surfaces, which serve as contact points for parts of other neurons.

The axon, which usually arises from a slight elevation of the cell body (axonal hillock), is a slender, cylindrical process with a nearly smooth surface and uniform diameter. It is specialized to conduct nerve impulses away from the cell body. Many mitochondria, microtubules, and neurofibrils occur within the cytoplasm of the axon. Although it begins as a single fiber, an axon may give off branches, called *collaterals.* Near its end, it may have many fine extensions, each with a specialized ending called a *presynaptic terminal,* which contacts the receptive surface of another cell.

In addition to conducting nerve impulses, an axon conveys substances that are produced in the neuron cell body. This process, called *axonal transport,* involves carrying organelles, such as vesicles and mitochondria, as well as ions, molecules of nutrients, and some neurotransmitter molecules from the cell body to the ends of the axon.

Larger axons of the peripheral nerves are commonly enclosed in sheaths composed of many neuroglial cells called **Schwann cells.** These cells are tightly wound around the axons, somewhat like a bandage wrapped around a finger. As a result, such axons are coated with many layers of cell membranes that have little or no cytoplasm between them. These membranes are composed largely of a lipid-protein (lipoprotein) that has a higher proportion of lipid than other surface membranes. This lipoprotein is called **myelin,** and it forms a *myelin sheath* on the outside of an axon. In addition, the portions of the Schwann cells that contain most of the cytoplasm and the nuclei remain outside the myelin sheath and comprise a **neurilemma** (neurilemmal sheath), which surrounds the myelin sheath (figure 10.4). Narrow gaps in the myelin sheath between adjacent Schwann cells are called **nodes of Ranvier** (figures 10.3 and 10.4).

The smallest axons of peripheral nerves are also enclosed by Schwann cells, but the Schwann cells are not wound around these axons. Instead, the axon or a group of axons may lie in a longitudinal groove of Schwann cells. Consequently, such axons lack myelin sheaths.

Axons that possess myelin sheaths are called *myelinated* (medullated) nerve fibers, and those that lack these sheaths are *unmyelinated* nerve fibers (figure 10.5). Groups of myelinated fibers appear white, and masses of such fibers are responsible for the *white matter* in the brain and spinal cord. In this instance, however, the myelin is produced by another kind of neuroglial cell (oligodendrocyte) rather than by Schwann cells. Furthermore, nerve fibers in the brain and spinal cord lack neurilemmal sheaths.

Unmyelinated nerve tissue appears gray. Thus, the *gray matter* within the brain and spinal cord contains an abundance of unmyelinated nerve fibers and neuron cell bodies.

> When neurons are deprived of oxygen, they undergo a series of irreversible structural changes in which their shapes are altered and their nuclei shrink. This phenomenon is called *ischemic cell change,* and in time, the affected cells disintegrate. Such an oxygen deficiency can result from lack of blood flow through nerve tissue (ischemia), an abnormally low blood oxygen concentration (hypoxemia), or the presence of toxins that block aerobic respiration.

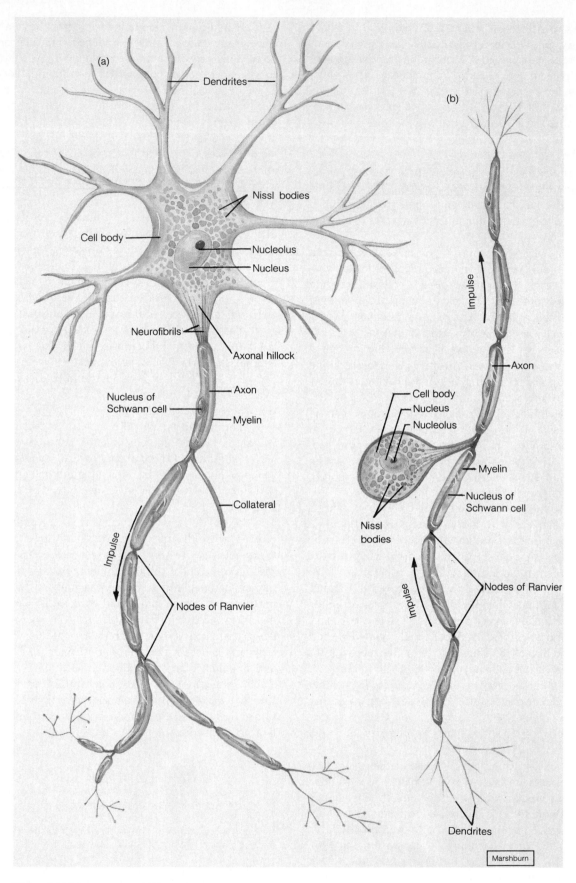

Figure 10.3 (a) Motor neuron and (b) sensory neuron.

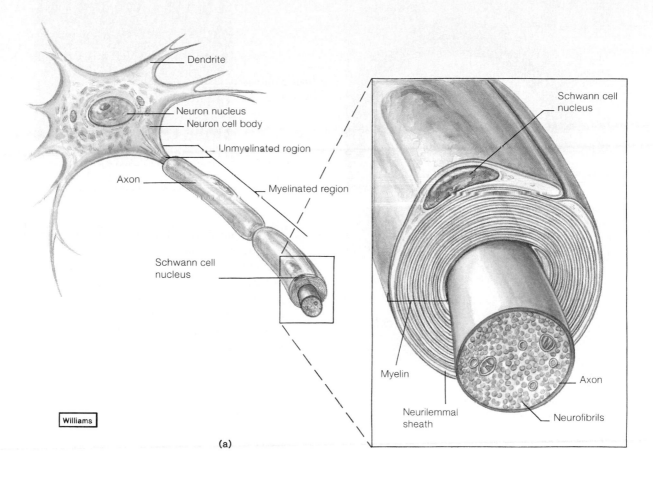

Williams

(a)

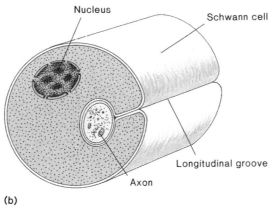

(b)

Figure 10.4 (a) The portion of a Schwann cell that is tightly wound around an axon forms the myelin sheath. The cytoplasm and nucleus of the Schwann cell, remaining on the outside, form the neurilemmal sheath. (b) An axon lying in a longitudinal groove of a Schwann cell is unmyelinated.

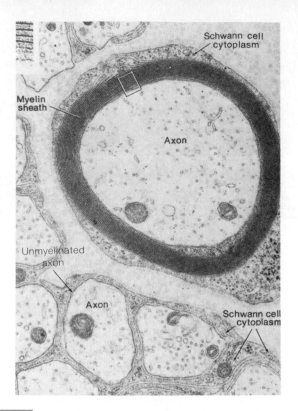

(a)

(b)

Figure 10.5 (a) A transmission electron micrograph of myelinated and unmyelinated axons in cross section; (b) light micrograph of a myelinated nerve fiber (longitudinal section) (×300).

---

1. List the general functions of the nervous system.
2. Describe a neuron.
3. Explain how myelin is formed in the peripheral nervous system.

Neuroglial Cells

As mentioned earlier, neuroglial cells (glial cells) fill spaces, support neurons, and provide frameworks within the organs of the nervous system. They also enclose neurons in ways that prevent contact between nearby cells except at particular sites (figures 10.6 and 10.7).

Within the peripheral nervous system, the neuroglial cells include the Schwann cells previously described. In the central nervous system, the following types are present:

1. **Astrocytes.** As their name implies, astrocytes are star-shaped cells. They are commonly found between neurons and blood vessels, where they provide structural support and hold parts together by means of numerous cellular processes. They seem to play a role in the metabolism of certain substances, such as glucose, and they may help regulate the concentrations of important ions, such as potassium ions, within the interstitial space of nervous tissue. Astrocytes also respond to injury of brain tissue and are responsible for the formation of scar tissue, which fills spaces and closes gaps following such injuries. They may also have a nutritive function involving the transport of substances from blood vessels to neurons.

2. **Oligodendrocytes.** Oligodendrocytes resemble astrocytes, but are smaller and have fewer processes. They are commonly arranged in rows along nerve fibers, and they function in the formation of myelin within the brain and spinal cord.

   Unlike the Schwann cells of the peripheral nervous system, oligodendrocytes can send out a number of cellular processes, each of which wraps tightly around a nearby axon. In this way, a single oligodendrocyte may provide myelin for many axons. However, these cells do not form neurilemmal sheaths.

CHART 10.1	Types of neuroglial cells of the central nervous system	
Type	Characteristics	Functions
Astrocyte	Star-shaped cell found between neurons and blood vessels	Structural support, formation of scar tissue, transport of substances between blood vessels and neurons
Oligodendrocyte	Shaped like astrocytes, but with fewer cellular processes, and arranged in rows along nerve fibers	Formation of myelin within the brain and spinal cord
Microglia	Small cells with few cellular processes and found throughout the CNS	Structural support and phagocytosis
Ependyma	Cuboidal and columnar cells in the inner lining of the ventricles of the brain and the central canal of the spinal cord	Form a porous layer through which substances diffuse between the interstitial fluid of the brain and spinal cord, and the cerebrospinal fluid

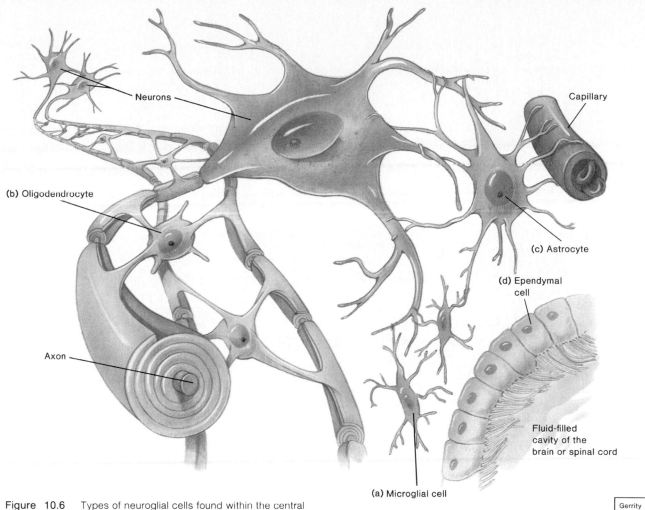

Figure 10.6 Types of neuroglial cells found within the central nervous system include (*a*) microglial cell, (*b*) oligodendrocyte, (*c*) astrocyte, and (*d*) ependymal cell.

Gerrity

3. **Microglia.** Microglial cells are relatively small and have fewer processes than other types of neuroglial cells. These cells are scattered throughout the central nervous system, where they help support neurons and phagocytize bacterial cells and cellular debris. They usually increase in number whenever the brain or spinal cord is inflamed because of injury or disease.

4. **Ependyma.** Ependymal cells are cuboidal or columnar in shape and may have cilia. They form an epithelial-like membrane that is one cell thick and covers the inside of spaces within the brain called *ventricles*. (See chapter 11.) They also form the inner lining of the *central canal* that extends downward through the spinal cord. Ependymal cells are loosely joined to one another by gap junctions. They form a porous layer through which substances diffuse freely between the interstitial fluid of the brain tissues and the fluid (cerebrospinal fluid) within the ventricles.

Chart 10.1 summarizes the characteristics of the neuroglial cells.

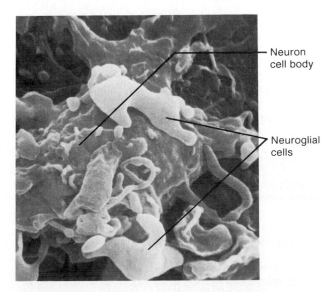

Figure 10.7 A scanning electron micrograph of a neuron cell body and some of the neuroglial cells associated with it (about ×10,000). (*Tissues and Organs: A Text-Atlas of Scanning Electron Microscopy*, by R. G. Kessel and R. H. Kardon. © 1979 W.H. Freeman and Company.)

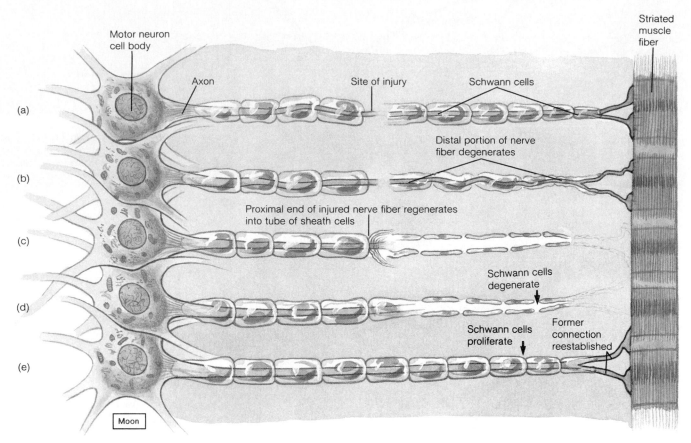

Figure 10.8 If a myelinated axon is injured, (a) the proximal portion of the fiber may survive, but (b) the portion distal to the injury degenerates. (c and d) In time, the proximal portion may develop extensions that grow into the tube of basement membrane and connective tissue cells previously occupied by the fiber, and (e) the former connection may be reestablished.

Although myelin begins to form on nerve fibers during the fourteenth week of development, many of the nerve fibers in a newborn infant are not completely myelinated. Consequently, the nervous system is unable to function as effectively as that of an older child or adult. An infant's responses to stimuli are therefore coarse and undifferentiated, and may involve its whole body. Essentially all myelinated fibers have begun to develop sheaths by the time a child starts to walk. Myelination continues into adolescence.

Any interference with the supply of essential nutrients during the developmental years may result in an insufficient amount of myelin formation. This, in turn, may cause impaired function of the nervous system.

Regeneration of Nerve Fibers

Injury to its cell body is most likely to cause the death of the neuron; however, a damaged axon may be regenerated. For example, if an axon in a peripheral nerve is separated from its cell body by injury or disease, the distal portion of the axon and its myelin sheath deteriorate within a few weeks. Macrophages remove the fragments of myelin and other cellular debris. Although some Schwann cells may also degenerate, a thin basement membrane and a layer of connective tissue (endoneurium) surrounding the Schwann cells probably will remain. These parts form a tube that leads back to the original connection of the axon.

The proximal end of the injured axon develops sprouts shortly after the injury, and one of these sprouts may grow into the tube formed by the basement membrane and connective tissue. At the same time, any remaining Schwann cells proliferate along the length of the degenerating fiber, and form new myelin around the growing axon.

The growth of such a regenerating fiber is slow (3 to 4 millimeters per day), but eventually the new fiber may reestablish the former connection (figure 10.8).

If an axon of a nerve fiber within the central nervous system is separated from its cell body, the distal portion of the axon will degenerate as before, although the rate of degeneration is somewhat slower. However, axons within the central nervous system lack connective tissue sheaths, and the myelin-producing oligodendrocytes fail to proliferate following an injury.

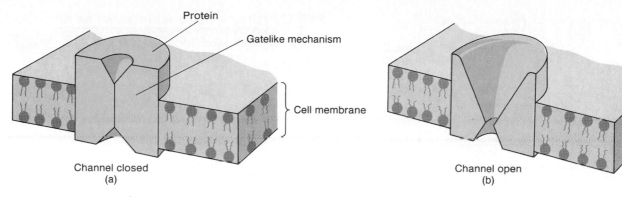

Protein

Gatelike mechanism

Cell membrane

Channel closed
(a)

Channel open
(b)

Figure 10.9 Some of the channels in cell membranes through which ions pass can be (a) closed or (b) opened by a gatelike mechanism.

Consequently, if the proximal end of a damaged axon begins to grow, there is no tube of sheath cells to guide it, and a functionally significant regeneration is unlikely to take place.

> If a peripheral nerve is severed, it is very important that the two cut ends be closely connected as soon as possible. As a result, the regenerating sprouts of the nerve fibers can more easily reach the tubes formed by the basement membranes and connective tissues on the distal side of the gap.
>
> When the gap exceeds 3 millimeters, the regenerating fibers tend to form a tangled mass called a *neuroma*. A neuroma, which is composed of sensory nerve fibers, is likely to be painfully sensitive to pressure. The development of such growths sometimes complicates a patient's recovery following the amputation of a limb.

1. What is a neuroglial cell?
2. Name and describe four types of neuroglial cells.
3. Explain how an injured peripheral nerve fiber might regenerate.
4. Explain why functionally significant regeneration is unlikely to occur in the central nervous system.

Cell Membrane Potential

A cell membrane is usually electrically charged, or *polarized,* so that the outside is positively charged with respect to the inside. This polarization is due to an unequal distribution of ions on either side of the membrane, and it is particularly important in the conduction of muscle and nerve impulses.

Distribution of Ions

The distribution of the ions associated with membranes is determined in part by the presence of pores or channels in those membranes, as discussed in chapter 3. The size and shape of these channels are related to the ways

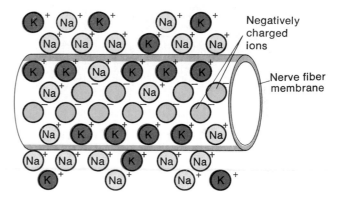

Figure 10.10 A nerve fiber at rest is polarized as a result of an unequal distribution of ions on either side of its membrane.

proteins and lipids are arranged in the membranes. Some channels are always open, and others can be opened or closed, as if by a gate. Furthermore, they can be selective; that is, a particular kind of channel may allow one kind of ion to pass through and exclude all other kinds. The effective size and electrical charge of an ion also affect the way it passes through a membrane (figure 10.9).

As a consequence of such factors, potassium ions tend to pass through cell membranes much more easily than sodium ions, and sodium ions pass through more easily than calcium ions. The relative ease with which potassium ions diffuse through membranes makes them a major contributor to membrane polarization.

Resting Potential

When nerve cells are at rest (that is, not conducting impulses), there is a relatively greater concentration of sodium ions (Na^+) outside their membranes and a relatively greater concentration of potassium ions (K^+) inside their membranes (figure 10.10). In the cytoplasm of these cells are large numbers of negatively

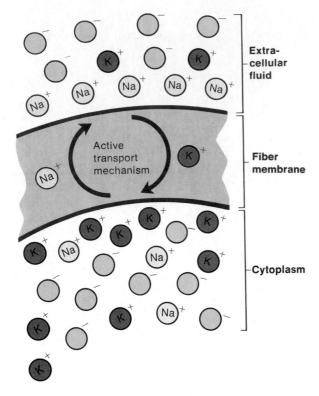

Figure 10.11 An active transport mechanism in the nerve fiber membrane moves sodium ions outward and potassium ions inward.

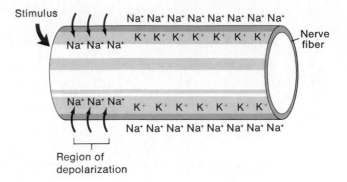

Figure 10.12 When a polarized nerve fiber is stimulated, sodium channels open, some sodium ions diffuse inward, and the membrane is depolarized.

charged ions, including those of phosphate, sulfate, and protein, that cannot diffuse through the cell membranes.

However, because a cell membrane is very permeable to potassium ions and only slightly permeable to sodium ions, potassium ions tend to diffuse freely through the membrane to the outside, and sodium ions diffuse inward more slowly. At the same time, the membrane expends energy to actively transport these ions in the opposite directions, which prevents them from reaching equilibrium by diffusion. Therefore, sodium ions are actively transported outward, and potassium ions are actively transported inward. Since potassium ions can readily diffuse out again and sodium can enter only with difficulty, the net effect is that more positively charged ions leave a cell than enter it. As a result, the outside of a cell membrane becomes positively charged with respect to the negatively charged inside (figure 10.11).

The difference in electrical charge between two points is called the *potential difference* and can be measured in units called *volts*. In the case of a resting nerve cell, the difference between the region inside and the region outside of the membrane is about −70 millivolts and is called the **resting potential.** As long as the nerve cell membrane is undisturbed, the membrane remains in this polarized state.

Local Potential Changes

Nerve cells are excitable; that is, they can respond to changes in their surroundings. Some nerve cells, for example, are specialized to detect changes in temperature, light, or pressure occurring outside the body, while others are responsive to signals coming from nearby neurons. In either case, such changes or stimuli usually affect the resting potential in a particular local region of the cell membrane. If, in response to a stimulus, the membrane's resting potential becomes less negative (moving toward zero), the membrane is said to be *depolarizing*; if the resting potential becomes more negative (moving away from zero), the membrane is *hyperpolarizing.*

Such changes in the local potential of a membrane are *graded.* This means the amount of change in potential is directly related to the intensity of the stimulation. Furthermore, if another stimulus of the same type is received before the effect of the first one subsides, the change in local membrane potential is greater. This additive phenomenon is called *summation,* and as a result of summated potentials, a level called the *threshold potential* may be reached. Thus, by this means, many subthreshold potential changes may be combined to reach threshold. At threshold, an *action potential* is produced in a nerve fiber.

Action Potentials

Once the threshold potential is reached, the portion of the cell membrane that is being stimulated undergoes a sudden change in permeability. Channels in the membrane that are highly selective for sodium ions open and allow the sodium ions to pass inward. This movement is aided by the fact that the sodium ions are attracted by the negative electrical condition on the other side of the membrane (figure 10.12).

As the sodium ions rush inward, the membrane loses its electrical charge and becomes *depolarized*. At almost the same time, channels open in the membrane that allow some of the potassium ions to pass through, and as they diffuse outward, the outside of the membrane becomes positively charged once more. The membrane then is said to become *repolarized,* and it remains in this state until it is stimulated again (figure 10.13).

This rapid sequence of changes, involving depolarization and repolarization, takes about 1/1,000th second or less and is called an **action potential.** Actually, only a small proportion of the sodium and potassium ions present move through the membrane during an action potential, so that many action potentials can occur before the concentrations of sodium ions and potassium ions on either side of the membrane change significantly. Eventually, these ion concentrations may change. However, an active transport mechanism in the membrane soon reestablishes the original concentrations of sodium and potassium, and the resting potential returns.

When an action potential occurs at one point in a nerve cell membrane, it causes an electric current to flow to adjacent portions of the membrane. This *local current* stimulates the membrane to its threshold level, triggering other action potentials. These in turn stimulate still other areas, and a wave of action potentials moves away in all directions from the point of stimulation, traveling to the end of the fiber without decreasing in amplitude. This propagation of action potentials along a nerve fiber constitutes a **nerve impulse** (figure 10.14).

A nerve impulse is similar to the muscle impulse mentioned in chapter 9. In the muscle fiber, stimulation at the motor end plate triggers an impulse to travel over the surface of the fiber and down into its transverse tubules. See chart 10.2 for a summary of the events leading to the conduction of a nerve impulse.

Refractory Period

For a moment following the passage of a nerve impulse, an ordinary stimulus will not trigger another impulse on a nerve fiber. This brief period, called the **refractory period,** has two parts. During the *absolute refractory period,* which lasts about 1/2,500 of a second, the fiber's membrane is changing in sodium permeability and

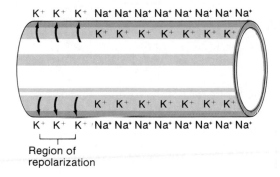

Figure 10.13 When the potassium channels open, potassium ions diffuse outward, and the membrane is repolarized.

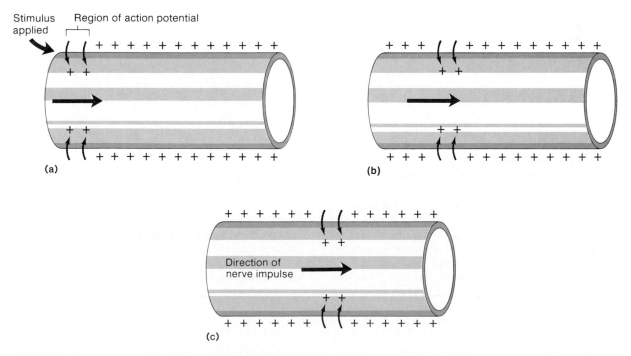

Figure 10.14 (*a*) An action potential in one region stimulates the adjacent region, and (*b* and *c*) a wave of action potentials (nerve impulse) moves along the fiber.

Recording an Action Potential

Since an action potential occurs in about 1/1,000 second, an instrument that responds rapidly is needed to record the sequence of changes. The instrument usually used is the **cathode ray oscilloscope.** This instrument includes a *cathode ray tube* that looks and functions somewhat like the picture tube of a television (figure 10.15). The cathode ray tube contains an *electron gun,* which fires a concentrated beam of electrons from the neck of the tube toward its face. The face is coated with a thin layer of fluorescent material that produces a spot of light where the electron beam strikes it.

Horizontal deflection plates are located on either side of the electron beam within the cathode ray tube. These plates can be electrically charged by the action of a *sweep circuit* of the oscilloscope. This circuit causes one of the horizontal deflection plates to become positively charged, while the other becomes negatively charged. When this happens, the beam of electrons (which are negatively charged) is drawn toward the positive plate and is repelled by the negative one. As a result, the electron beam moves across the face of the tube and produces a glowing horizontal line called a *trace.*

The sweep circuit can be adjusted to move the electron beam across the face of the cathode ray tube from left to right at a known velocity. Each time the beam reaches the right edge of the tube, it instantly jumps back to the left side and begins a new horizontal trace.

Other plates, called *vertical deflection plates,* are located above and below the electron beam within the cathode ray tube. If these plates become electrically charged, the electron beam moves up or down.

When the oscilloscope is being used to record the action potential of a nerve fiber, electrodes from an electronic *amplifier* are placed in contact with the nerve fiber, and the amplifier is connected to the vertical deflection plates. Any change in nerve fiber membrane potential is detected by the electrodes, and a signal is transmitted to the amplifier. The amplifier increases the intensity of the signal in direct proportion to the amount of change occurring in the membrane potential and causes the vertical deflection plates to become charged. If this happens when the electron beam is moving across the face of the cathode ray tube, its horizontal trace will be deflected up or down in direct proportion to the change in membrane potential detected by the electrodes.

Figure 10.16 illustrates an action potential of the type recorded on an oscilloscope when one electrode is placed on the surface of a nerve fiber membrane and the other electrode is inserted into the cytoplasm of the fiber. Note that

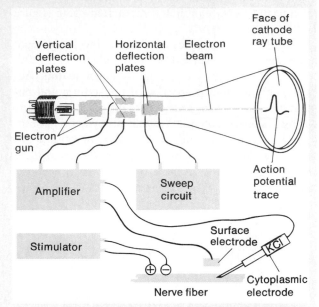

Figure 10.15 A cathode ray oscilloscope can be used to record the action potential of a nerve fiber.

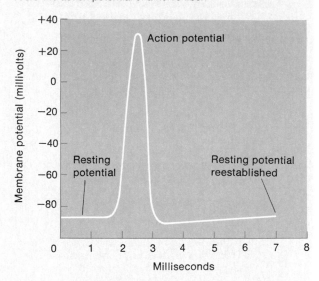

Figure 10.16 An action potential as it might be recorded on the face of a cathode ray tube.

the fiber's resting potential is about −85 millivolts. As the action potential occurs following stimulation of the fiber, the potential momentarily becomes positive, and then the negative resting potential is reestablished.

CHART 10.2   Events leading to the conduction of a nerve impulse

1. Nerve fiber membrane develops resting potential.	6. Potassium ions diffuse outward, causing the membrane to repolarize.
2. Threshold stimulus is received.	7. The resulting action potential causes a local electric current that stimulates adjacent portions of the membrane.
3. Sodium channels in a local region of the membrane open.	
4. Sodium ions diffuse inward, causing the membrane to depolarize.	8. Wave of action potentials travels the length of the nerve fiber as a nerve impulse.
5. Potassium channels in the membrane open.	

cannot be stimulated. This is followed by a *relative refractory period,* during which the membrane is reestablishing its resting potential. While the membrane is in the relative refractory period, even though polarization is incomplete, an impulse may be triggered by a stimulus of high intensity.

As time passes, however, the fiber gradually becomes more excitable. Simultaneously, the intensity of stimulation needed to trigger an impulse decreases until the fiber's original excitability is restored. This return to the resting state usually takes from 10 to 30 milliseconds.

Because of the refractory period, a nerve fiber cannot be stimulated continuously. Thus, the refractory period limits the rate at which nerve impulses can be conducted. In other words, the time between impulses cannot be less than the absolute refractory period, and the maximum rate of nerve impulses is about one impulse per millisecond.

All-or-None Response

Like muscle fiber contraction, nerve impulse conduction is an *all-or-none response.* In other words, if a nerve fiber responds at all, it responds completely. Thus, a nerve impulse is conducted whenever a stimulus of threshold intensity or above is applied to a nerve fiber, and all impulses carried on that fiber will be of the same strength. A greater intensity of stimulation does not produce a stronger impulse.

Impulse Conduction

An unmyelinated nerve fiber conducts an impulse over its entire surface. A myelinated fiber functions differently. Myelin contains a high proportion of lipid that excludes water and water soluble substances. Thus, it serves as an insulator and prevents almost all flow of ions through the membrane that is enclosed in myelin.

Considering this, it might seem that the myelin sheath would prevent the conduction of a nerve impulse altogether, and this would be true if the sheath were continuous. It is, however, interrupted by the constrictions called **nodes of Ranvier,** which occur between adjacent Schwann cells (figure 10.3). At these nodes, the nerve fiber membrane is especially permeable to sodium and potassium ions.

When such a myelinated nerve fiber is stimulated, an action potential occurs at a node. This causes an electric current to flow away through the cytoplasm of the fiber. As this local current reaches the next node, it stimulates the membrane to its threshold level, and an action potential occurs there. Consequently, a nerve impulse traveling along a myelinated nerve fiber appears to jump from node to node. This type of impulse conduction, called **saltatory conduction,** is many times faster than conduction on an unmyelinated nerve fiber (figure 10.17).

The speed of nerve impulse conduction is also related to the diameter of the fiber—the greater the diameter, the faster the impulse. For example, an impulse on a relatively thick, myelinated nerve fiber, such as a motor fiber associated with a skeletal muscle, might travel 120 meters per second, while an impulse on a thin, unmyelinated nerve fiber, such as a sensory fiber associated with the skin, might move only 0.5 meter per second.

1. Summarize how a resting potential is achieved.
2. Explain how a polarized nerve fiber responds to stimulation.
3. List the major events that occur during an action potential.
4. Define refractory period.
5. Explain how impulse conduction differs in myelinated and unmyelinated nerve fibers.

The Synapse

Within the nervous system, nerve impulses travel from neuron to neuron along complex nerve pathways. The junction between the parts of two communicating neurons is called a **synapse.** Actually, two such cells, called *presynaptic* and *postsynaptic neurons,* are not in direct contact at the synapse. There is a gap called a *synaptic cleft* between them, and for an impulse to continue along a nerve pathway it must cross this gap (figure 10.18).

Synaptic Transmission

As was mentioned, a nerve impulse travels in both directions away from the point of stimulation. Within a neuron (presynaptic neuron), however, an impulse travels from a dendrite or cell body and then moves

Factors Affecting Impulse Conduction

A number of substances affect nerve fiber membranes by influencing their permeability to ions. For example, calcium ions are needed for the closure of the sodium channels in the nerve fiber membrane during an action potential. Consequently, if calcium is deficient, the sodium channels may remain open, and sodium ions may diffuse through the membrane again and again so that impulses are transmitted repeatedly. These spontaneous impulses may travel along the nerve fibers to skeletal muscle fibers. When this happens, the muscles may undergo continuous spasms (tetanus or tetany). Such a decrease in the level of blood calcium, accompanied by tetany, sometimes occurs in women during pregnancy as the developing fetus makes increasing demands for calcium. Tetanic contraction may also occur in persons who have an inadequate supply of calcium or vitamin D in their diets, or who lose calcium ions excessively due to prolonged diarrhea.

A small increase in the concentration of extracellular potassium ions causes the resting potential of nerve fibers to be less negative (partially depolarized). As a result, the threshold potential can be reached with a less intense stimulus than usual. The affected fibers are very excitable, and the person may experience convulsions.

If the extracellular potassium ion concentration is greatly increased, the resting potentials of the nerve fibers may remain so low that action potentials cannot occur. In this case, impulses are not triggered, and muscles become paralyzed.

Certain drugs, such as procaine, produce special effects by decreasing membrane permeability to sodium ions. When one of these drugs is present in the tissue fluids surrounding a nerve fiber, impulses are prevented from passing through the affected region. Consequently, the drugs are useful as local anesthetics because they keep impulses from reaching the brain and thus prevent the perception of touch and pain.

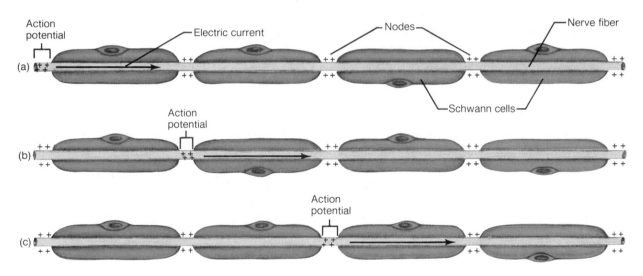

Figure 10.17 On a myelinated fiber, a nerve impulse appears to jump from node to node.

along the axon to the presynaptic terminal at its end. There the impulse crosses a synapse and continues to the dendrite or cell body of another neuron (postsynaptic neuron). The process of crossing the gap at a synapse is called *synaptic transmission.*

The typical one-way transmission from axon to dendrite or cell body is due to the fact that axons usually have several rounded *synaptic knobs* at their presynaptic terminals, which dendrites lack. These knobs contain numerous membranous sacs, called *synaptic*

vesicles, and when a nerve impulse reaches a synaptic knob, some of the synaptic vesicles respond by releasing a **neurotransmitter** (neurohumor).

As figure 10.19 shows, the released neurotransmitter diffuses across the synaptic cleft and reacts with specific receptors of the postsynaptic neuron membrane. If enough neurotransmitter is released, the postsynaptic membrane is stimulated to the threshold level, and a nerve impulse is triggered (figure 10.20).

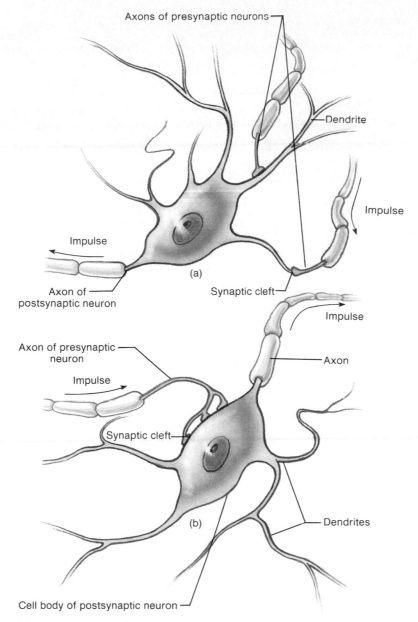

Figure 10.18 For an impulse to continue from one neuron to another, it must cross the synaptic cleft at a synapse. A synapse may occur (*a*) between an axon and a dendrite, or (*b*) between an axon and a cell body.

Neurotransmitter Substances

Many different neurotransmitter substances are produced in the nervous system, and although some neurons seem to release only one kind, other neurons produce two or three kinds of neurotransmitters. These neurotransmitters include *acetylcholine,* which also stimulates skeletal muscle contractions (see chapter 9); a group of compounds called *monoamines* (such as epinephrine, norepinephrine, dopamine, and serotonin), which are formed by modifying amino acid molecules;

a group of unmodified *amino acids* (such as glycine, glutamic acid, aspartic acid, and gamma-aminobutyric acid—GABA); and a large group of *peptides* (such as enkephalins and substance P), each of which consists of a relatively short chain of amino acids.

These neurotransmitter substances are usually synthesized in the cytoplasm of the synaptic knobs and stored in the synaptic vesicles.

When an action potential passes over the membrane of a synaptic knob, it produces an increase in the

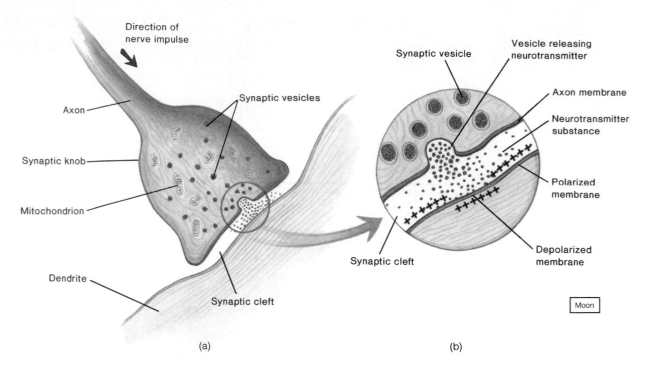

Figure 10.19 (a) When a nerve impulse reaches the synaptic knob at the end of an axon, (b) synaptic vesicles release a neurotransmitter substance that diffuses across the synaptic cleft.

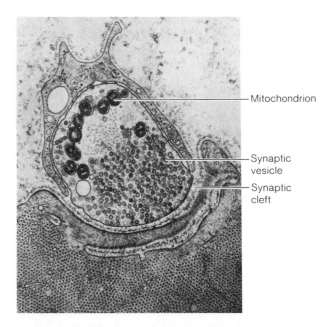

Figure 10.20 A transmission electron micrograph of a synaptic knob filled with synaptic vesicles.

membrane's permeability to calcium ions by causing its calcium ion channels to open. Calcium ions diffuse inward, and in response to their presence, some of the synaptic vesicles fuse with the membrane and release their contents into the synaptic cleft. The quantity of neurotransmitter released by the vesicles is related directly to the quantity of calcium that has entered the synaptic knob.

Once a vesicle has released its neurotransmitter, it breaks away from the membrane and reenters the cytoplasm where it is quickly resupplied with neurotransmitter substance. Vesicles that are damaged or lost are replaced by new ones formed in the neuron cell body and transported through the axon to the synaptic knob. (Chart 10.3 summarizes this activity.)

Some neurotransmitters are rapidly decomposed by enzymes present in the synaptic clefts. Others are removed from the cleft by being transported back into the synaptic knob of the neuron that released them or into nearby neurons or neuroglial cells. Acetylcholine, for example, is decomposed by the enzyme *cholinesterase,* which is present in the membranes at synapses that release this neurotransmitter. Similarly, the monoamine neurotransmitters epinephrine and norepinephrine are inactivated by the enzyme *monoamine oxidase* found in mitochondria. Such destruction or removal of the neurotransmitter prevents a continuous stimulation of the postsynaptic neuron.

Neuropeptides

Neuropeptides are peptides that are synthesized by neurons in the brain or spinal cord. They act as neurotransmitters or as *neuromodulators*—substances that alter a neuron's response to a neurotransmitter or block the release of a neurotransmitter.

Among the neuropeptides is a form called *enkephalin* that occur throughout the brain and spinal

CHART 10.3   Events leading to the release of a neurotransmitter

1. Action potential passes along a nerve fiber and over the surface of its synaptic knob.
2. Synaptic knob membrane becomes more permeable to calcium ions, and they diffuse inward.
3. In the presence of calcium ions, synaptic vesicles fuse to synaptic knob membrane.
4. Synaptic vesicles release their neurotransmitter into the synaptic cleft.
5. Synaptic vesicles reenter the cytoplasm of the nerve fiber and are refilled with neurotransmitter.

cord. Each enkephalin molecule consists of five amino acids in a chain. The synthesis of these substances seems to increase during periods of painful stress, and they are known to bind to the same receptors in the brain (opiate receptors) as the narcotic morphine. Although they are thought to play a role in relieving pain sensations, enkephalins probably have other functions as well.

Another morphinelike peptide, called *betaendorphin,* is found mainly in the pituitary gland. It seems to be decomposed less rapidly and to have a much more potent pain-relieving property than the enkephalins.

Still another neuropeptide, which consists of eleven amino acids and is widely distributed throughout the nervous system, is called *substance P.* It seems to function as a neurotransmitter (or perhaps as a neuromodulator) in the neurons that transmit pain impulses into the spinal cord and on to the brain. Some investigators think the enkephalins and endorphins may relieve pain by inhibiting the release of substance P from pain-transmitting neurons.

As discussed in chapter 13, some neurotransmitters also function as hormones.

Synaptic Potentials

The local membrane potentials at a synapse are called *synaptic potentials,* and they provide a means by which one neuron can influence another. As mentioned before, the changes in such potentials are graded and can be depolarizing or hyperpolarizing.

Although the kinds of neurotransmitters released by the synaptic knobs of a particular neuron are always the same, other neurons may secrete different neurotransmitters. Furthermore, the effects of the neurotransmitters vary. Some cause various ion channels to open; others cause ion channels to close. The effects on synaptic potentials also vary—some effects are excitatory and others are inhibitory.

For example, if a neurotransmitter acts on a postsynaptic membrane by combining with specialized receptors and causes its sodium ion channels to open, sodium ions will diffuse inward and cause the membrane to depolarize. As a result, an action potential may be triggered. This kind of membrane change (hypopolarization) is called an *excitatory postsynaptic potential* (EPSP), and it lasts for about 15 milliseconds.

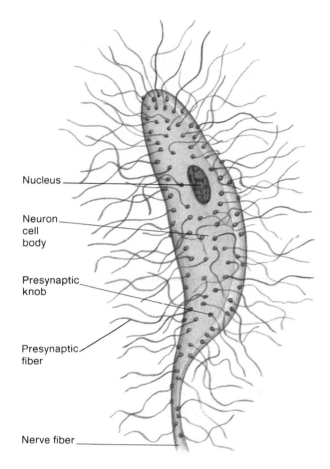

Nucleus

Neuron cell body

Presynaptic knob

Presynaptic fiber

Nerve fiber

Figure 10.21   The synaptic knobs of many dendrites may communicate with the cell body of a neuron.

If, instead, a different neurotransmitter combines with other receptors and causes the permeability of potassium to increase, these ions will diffuse outward, and the membrane will become *hyperpolarized.* Since an action potential is now less likely to occur, this change is called an *inhibitory postsynaptic potential* (IPSP).

Within the brain and spinal cord, the synaptic knobs of a thousand or more nerve fibers may communicate with the dendrites and cell body of each neuron. Furthermore, at any moment, some of the postsynaptic potentials may be producing excitatory effects on each neuron, while others are producing inhibitory effects (figure 10.21).

Factors Affecting Synaptic Transmission

If nerve impulses reach synaptic knobs at rapid rates, the supplies of neurotransmitters may become exhausted. Impulses cannot be transferred between the neurons involved until more neurotransmitters are synthesized.

Such a condition seems to occur during an epileptic seizure when abnormal and excessive discharges of impulses originate from certain brain cells. Some of these impulses reach skeletal muscle fibers and stimulate violent contractions. In time, the synaptic knobs seem to run out of neurotransmitters, and the seizure subsides.

A drug called Dilantin (diphenylhydantoin) is sometimes used to treat epilepsy. It seems to have a stabilizing effect upon excitable neuron membranes, by increasing the effectiveness of the sodium active transport mechanism. Consequently, as more sodium ions move through the membranes to the outside, thresholds of the neuron membranes are stabilized against excessive stimulation.

Other factors that affect synaptic transmission include various drugs and chemicals. For example, caffeine, which is found in coffee, tea, and cola drinks, stimulates activity in the nervous system. It does this by lowering the thresholds at synapses. When caffeine is present, certain neurons are more easily excited than usual.

A number of drugs produce their special effects by interfering with the normal actions of neurotransmitters. In fact, nearly all of the drugs used to treat functional disorders of the nervous system act by blocking or promoting the effects of some neurotransmitter substance. For example, the stimulants called *amphetamines* cause a person to feel more alert and energetic by enhancing the release of norepinephrine and dopamine from the axonal ends of certain neurons. The *tricyclic antidepressant drugs* (as well as cocaine) produce similar results by blocking the normal uptake of norepinephrine and dopamine, thus leaving these substances in synaptic clefts where they can continue producing their effects longer than usual.

Another group of antidepressant drugs inhibits the enzyme monoamine oxidase. Because this enzyme functions to inactivate monoamines, these drugs, *monoamine inhibitors,* allow prolonged activity of monoamine neurotransmitters, such as norepinephrine and serotonin.

Whether or not an action potential is triggered on a neuron depends upon the integrated sum of the postsynaptic potential effects. That is, if the net effect is more excitatory than inhibitory, threshold may be reached, and an action potential will occur. Conversely, if the net effect is inhibitory, no impulse will be transmitted.

This summation of the excitatory and inhibitory effects of the postsynaptic potentials commonly takes place at a specialized "trigger zone" in the membrane of a dendrite or in a proximal region of the axon. This region has an especially low threshold for triggering an action potential; thus, it serves as a decision-making part of the neuron.

1. Describe a synapse.
2. Explain the function of a neurotransmitter.
3. Define neuropeptide.
4. Distinguish between EPSP and IPSP.

Processing of Impulses

The way the nervous system processes nerve impulses and acts upon them reflects, in part, the organization of neurons and their nerve fibers within the brain and spinal cord.

Neuronal Pools

The neurons within the central nervous system are organized into groups called *neuronal pools* that have varying numbers of cells. Each neuronal pool receives impulses from input (afferent) nerve fibers. These impulses are processed according to the special characteristics of the pool, and any resulting impulses are conducted away on output (efferent) fibers.

Each input fiber divides many times as it enters, and its branches spread over a certain region of the neuronal pool. The branches give off smaller branches, and their terminals form hundreds of synapses with the dendrites and cell bodies of the neurons in the pool.

Facilitation

As a result of incoming impulses and the release of neurotransmitters, a particular neuron of the neuronal pool may receive excitatory and inhibitory stimulation. As before, if the net effect of the stimulation is excitatory, threshold may be reached, and an outgoing impulse will be triggered. If the net effect is excitatory but subthreshold, an impulse will not be triggered; however, in this case the neuron is more excitable to incoming stimulation than before and is said to be *facilitated*.

Convergence

Any single neuron in a neuronal pool may receive impulses from two or more incoming fibers. Furthermore, these fibers may originate from different parts of the nervous system, and they are said to *converge* when they lead to the same neuron.

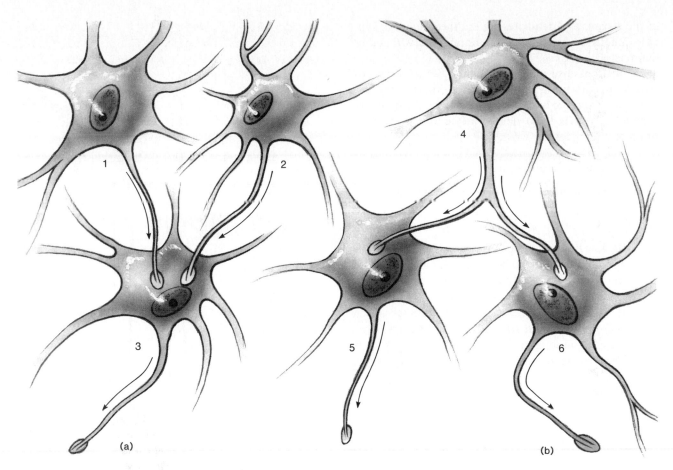

Figure 10.22 (a) Nerve fibers of neurons 1 and 2 converge to the cell body of neuron 3. (b) The nerve fiber of neuron 4 diverges to the cell bodies of neurons 5 and 6.

Convergence makes it possible for a neuron to summate impulses from different sources. For example, if a neuron is facilitated by receiving sub-threshold stimulation from one input fiber, its threshold may be reached if it receives additional stimulation from a second input fiber. Thus, if an output impulse is triggered from this neuron, it reflects a summation of impulses from two different sources. Such an output impulse may travel to a particular effector and cause a response (figure 10.22a).

Incoming impulses often represent information from various sensory receptors that have detected changes taking place. Convergence allows the nervous system to bring together a variety of information, process it, and respond to it in a special way.

Divergence

Impulses leaving a neuron of a neuronal pool often *diverge* by passing into several other output fibers. For example, an impulse from one neuron may stimulate two others; each of these, in turn, may stimulate several others, and so forth. Such an arrangement of diverging nerve fibers can cause an impulse to be

amplified—that is, to spread to increasing numbers of neurons within the pool (figure 10.22b).

As a result of divergence, an impulse originating from a single neuron in the central nervous system may be amplified so that enough impulses reach the motor units within a skeletal muscle to cause a forceful contraction.

Similarly, an impulse originating from a sensory receptor may diverge and reach several different regions of the central nervous system, where the resulting impulses can be processed and acted upon.

1. What is a neuronal pool?
2. Define facilitation.
3. What is meant by convergence?
4. What is the relationship between divergence and amplification?

Classification of Neurons and Nerve Fibers

Neurons differ in the structure, size, and shape of their cell bodies. Likewise, they vary in the length and size of their axons and dendrites, and in the number of synaptic knobs by which they communicate with other neurons.

Neurons also vary in function. Some carry impulses into the brain or spinal cord; others carry impulses out from the brain or spinal cord; and still others conduct impulses from neuron to neuron within the brain or spinal cord.

Classification of Neurons

On the basis of *structural differences,* neurons can be classified into three major groups, as shown in figure 10.23:

1. **Bipolar neurons.** The cell body of a bipolar neuron has only two nerve fibers, one arising from either end. Although these fibers have similar structural characteristics, one serves as an axon and the other as a dendrite. Such neurons are found within specialized parts of the eyes, nose, and ears.

2. **Unipolar neurons.** Each unipolar neuron has a single nerve fiber extending from its cell body. A short distance from the cell body, this fiber divides into two branches: One branch is connected to a peripheral body part and serves as a dendrite, and the other enters the brain or spinal cord and serves as an axon. The cell bodies of some unipolar neurons occur in specialized masses of nerve tissue called *ganglia,* which are located outside the brain and spinal cord. (Note: The sensory neuron illustrated in figure 10.3 is unipolar.)

3. **Multipolar neurons.** Multipolar neurons have many nerve fibers arising from their cell bodies. Only one fiber of each neuron is an axon; the rest are dendrites. Most neurons whose cell bodies lie within the brain or spinal cord are of this type. For example, the motor neuron illustrated in figure 10.3 is multipolar.

Neurons can also be classified on the basis of their *functional differences* into the following groups:

1. **Sensory neurons** (afferent neurons) carry nerve impulses from peripheral body parts into the brain or spinal cord. These neurons have specialized *receptor ends* at the tips of their dendrites, or they have dendrites that are closely associated with *receptor cells* located in the skin or in various sensory organs.

    Changes that occur inside or outside the body are likely to stimulate receptor ends or receptor cells, triggering sensory nerve impulses. The impulses travel along the sensory neuron fibers that lead to the brain or spinal cord, and they are then processed in these parts

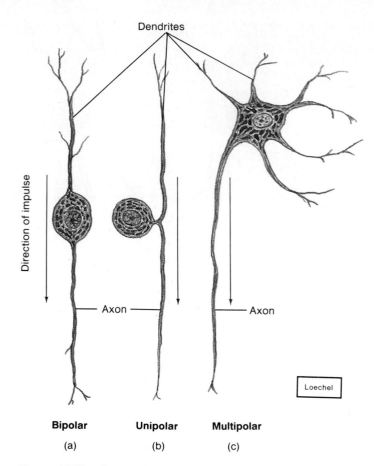

Figure 10.23 Structural types of neurons include (*a*) the bipolar neuron, (*b*) the unipolar neuron, and (*c*) the multipolar neuron. Where can examples of each of these be found in the body?

by other neurons. Most sensory neurons are unipolar, although some are bipolar (figure 10.23b).

2. **Interneurons** (also called association, intercalated, or internuncial neurons) lie within the brain or spinal cord. They are multipolar and form links between other neurons (figure 10.23c). Interneurons transmit impulses from one part of the brain or spinal cord to another. That is, they may direct incoming sensory impulses to appropriate parts for processing and interpreting. Other incoming impulses are transferred to motor neurons.

3. **Motor neurons** (efferent neurons) are multipolar and carry nerve impulses out of the brain or spinal cord to effectors—parts of the body capable of responding, such as muscles or glands. For example, when motor impulses reach muscles, these effectors are stimulated to contract; when motor impulses reach glands, these glands are stimulated to release secretions.

CHART 10.4 Types of neurons

A. Classified by structure

Type	Structural characteristics	Location
1. Bipolar neuron	Cell body with a nerve fiber arising from each end	In specialized parts of the eyes, nose, and ears
2. Unipolar neuron	Cell body with a single nerve fiber that divides into two branches	In ganglia outside the brain or spinal cord
3. Multipolar neuron	Cell body with many nerve fibers, one of which is an axon	Most common type of neuron in the brain and spinal cord

B. Classified by function

Type	Functional characteristics	Structural characteristics
1. Sensory neuron	Conducts nerve impulses from receptors in peripheral body parts into the brain or spinal cord	Most unipolar; some bipolar
2. Interneuron	Transmits nerve impulses between neurons within the brain and spinal cord	Multipolar
3. Motor neuron	Conducts nerve impulses from the brain or spinal cord out to effectors—muscles or glands	Multipolar

Two specialized groups of motor neurons, accelerator and inhibitory neurons, supply impulses to smooth and cardiac muscles. The *accelerator neurons* cause an increase in muscular activities, while the *inhibitory neurons* cause such actions to decrease.

Chart 10.4 summarizes the classification of neurons.

Types of Nerves and Nerve Fibers

Although a nerve fiber is an extension of a neuron, a **nerve** is a cordlike bundle (or group of bundles) of nerve fibers held together by layers of connective tissue (figure 10.24). Nerve structure is described in chapter 11.

Like nerve fibers, nerves that conduct impulses into the brain or spinal cord are called **sensory nerves,** and those that carry impulses to muscles or glands are termed **motor nerves.** Most nerves, however, include both sensory and motor fibers, and they are called **mixed nerves.**

Nerves originating from the spinal cord that communicate with other body parts are called *spinal nerves,* while those originating from the brain that communicate with other body parts are called *cranial nerves.* The nerve fibers within these structures can be subdivided further into four groups as follows:

1. **General somatic efferent fibers** carry motor impulses outward from the brain or spinal cord to skeletal muscles and cause them to contract.
2. **General visceral efferent fibers** carry motor impulses outward to various smooth muscles and glands associated with internal organs,

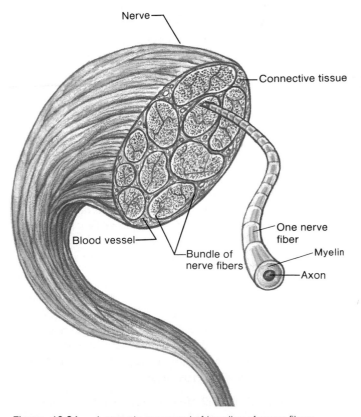

Figure 10.24 A nerve is composed of bundles of nerve fibers held together by connective tissue.

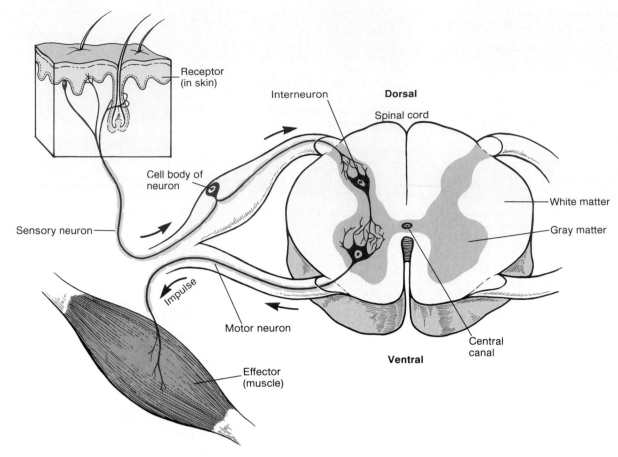

Figure 10.25 A reflex arc usually includes a receptor, a sensory neuron, interneurons, a motor neuron, and an effector.

causing certain muscles to contract or glands to release their secretions.

3. **General somatic afferent fibers** carry sensory impulses inward to the brain or spinal cord from receptors in the skin and skeletal muscles.

4. **General visceral afferent fibers** carry sensory impulses to the central nervous system from blood vessels and internal organs.

The term *general* in each of the categories just listed indicates that the fibers are associated with general structures such as the skin, skeletal muscles, glands, and visceral organs. Three other groups of fibers, found only in cranial nerves, are associated with more specialized, or *special,* structures:

1. **Special visceral efferent fibers** carry motor impulses outward from the brain to the muscles involved with chewing, swallowing, speaking, and forming facial expressions.

2. **Special visceral afferent fibers** carry sensory impulses inward to the brain from the olfactory and taste receptors.

3. **Special somatic afferent fibers** carry sensory impulses inward from the receptors of sight, hearing, and equilibrium.

1. Explain how neurons are classified according to structure and according to function.
2. How is a neuron related to a nerve?
3. What is a mixed nerve?
4. Explain how cranial and spinal nerve fibers are grouped.

Nerve Pathways

The routes followed by nerve impulses as they travel through the nervous system are called *nerve pathways.* The simplest of these pathways includes only a few neurons and is called a reflex arc.

Reflex Arcs

A **reflex arc** begins with a receptor at the end of a sensory nerve fiber. This fiber usually leads to several interneurons within the central nervous system, which serve as a processing center, or *reflex center.* Fibers from these interneurons may connect with interneurons in other parts of the nervous system. They also communicate with motor neurons, whose fibers pass outward from the central nervous system to effectors. (See figure 10.25.)

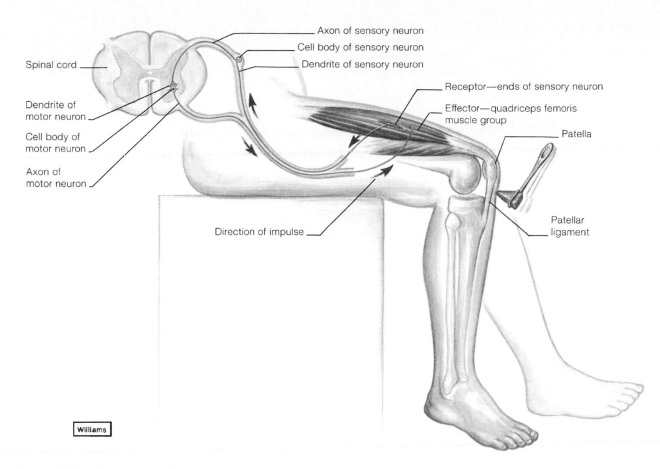

Spinal cord

Dendrite of motor neuron

Cell body of motor neuron

Axon of motor neuron

Axon of sensory neuron

Cell body of sensory neuron

Dendrite of sensory neuron

Receptor—ends of sensory neuron

Effector—quadriceps femoris muscle group

Patella

Patellar ligament

Direction of impulse

Williams

Figure 10.26 The knee-jerk reflex involves two neurons—a sensory neuron and a motor neuron.

Such a reflex arc represents the behavioral unit of the nervous system. That is, it constitutes the structural and functional basis for the simplest acts—the reflexes.

Reflex Behavior

Reflexes are automatic, unconscious responses to changes (stimuli) occurring within or outside the body. Such responses to stimuli are mechanisms that help maintain homeostasis by controlling many involuntary processes such as heart rate, breathing rate, blood pressure, and digestive activities. Reflexes are also involved in the automatic actions of swallowing, sneezing, coughing, and vomiting.

The *knee-jerk reflex* (patellar tendon reflex) is an example of a simple reflex that employs only two neurons—a sensory neuron communicating directly to a motor neuron. This reflex is initiated by striking the patellar ligament just below the patella. As a result, the quadriceps femoris group of muscles, which is attached to the patella by a tendon, is pulled slightly, and stretch receptors located within the muscle group are stimulated. These receptors, in turn, trigger impulses that pass along the fiber of a sensory neuron into the lumbar region of the spinal cord. Within the spinal cord,

the sensory axon forms a synapse with a dendrite of a motor neuron. The impulse then continues along the axon of the motor neuron and travels back to the quadriceps femoris. The muscles respond by contracting, and the reflex is completed as the lower leg extends (figure 10.26).

This reflex is helpful in maintaining an upright posture. For example, if a person is standing still and the knee begins to bend as a result of gravity, the quadriceps femoris is stretched, the reflex is triggered, and the leg straightens again.

Another type of reflex, called a *withdrawal reflex* (figure 10.27), occurs when a person unexpectedly touches a finger to something painful. As this happens, skin receptors are activated, and sensory impulses travel to the spinal cord. There the impulses pass on to interneurons of a reflex center and are directed to motor neurons. The motor neurons transmit signals to the flexor muscles in the arm, and the muscles contract in response, causing the part to be pulled away from the painful stimulus.

At the same time, some of the incoming impulses stimulate interneurons that inhibit the action of the antagonistic extensor muscles. This inhibition (reciprocal innervation) allows flexor muscles to effectively withdraw the hand.

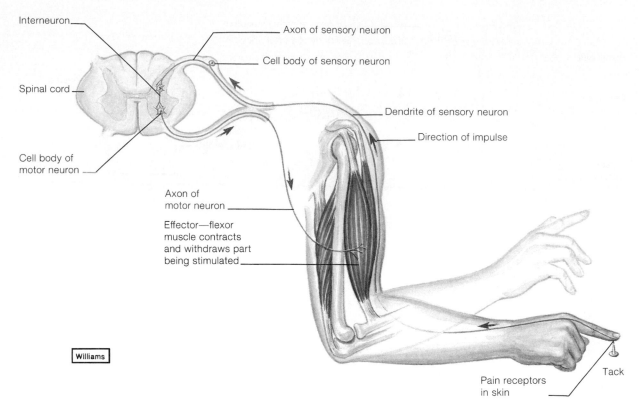

Interneuron

Axon of sensory neuron

Cell body of sensory neuron

Spinal cord

Dendrite of sensory neuron

Direction of impulse

Cell body of
motor neuron

Axon of
motor neuron

Effector—flexor
muscle contracts
and withdraws part
being stimulated

Williams

Pain receptors
in skin

Tack

Figure 10.27   A withdrawal reflex involves a sensory neuron,
an interneuron, and a motor neuron.

While flexor muscles of the stimulated side (ip-silateral side) are caused to contract, the flexor muscles of the other arm (contralateral side) are inhibited. Furthermore, the extensor muscles on this other side are caused to contract. This phenomenon, which is called a *crossed extensor reflex,* is due to interneuron pathways within the reflex center of the spinal cord that allow sensory impulses arriving on one side of the cord to pass across to the other side and produce an opposite effect (figure 10.28).

Concurrent with the withdrawal reflex, other interneurons in the spinal cord carry sensory impulses to the brain, and the person becomes aware of the experience and may feel pain.

A withdrawal reflex is, of course, protective because it prevents excessive tissue damage when a body part touches something potentially harmful. Chart 10.5 summarizes the parts of a reflex arc.

1. What is a nerve pathway?
2. Describe a reflex arc.
3. Define a reflex.
4. Describe the actions that occur during a withdrawal reflex.

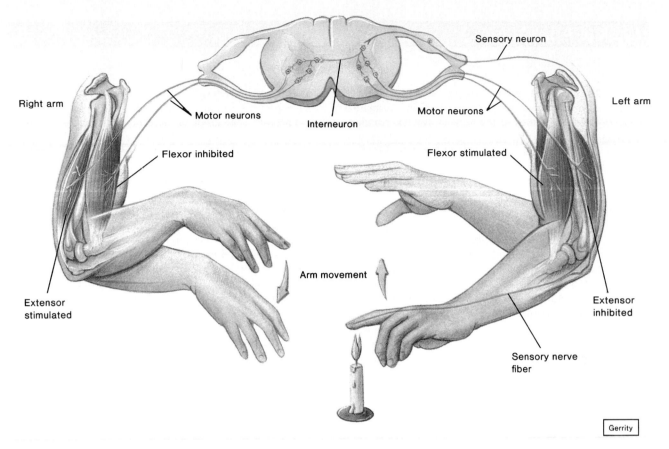

Right arm

Sensory neuron

Left arm

Motor neurons

Interneuron

Motor neurons

Flexor inhibited

Flexor stimulated

Extensor
stimulated

Extensor
inhibited

Arm movement

Sensory nerve
fiber

Gerrity

Figure 10.28 When the flexor muscle on one side is stimulated to contract in a withdrawal reflex, the extensor muscle on the opposite side also contracts.

CHART 10.5	Parts of a reflex arc	
Part	**Description**	**Function**
Receptor	The receptor end of a dendrite or a specialized receptor cell in a sensory organ	Sensitive to a specific type of internal or external change
Sensory neuron	Dendrite, cell body, and axon of a sensory neuron	Transmits nerve impulse from the receptor into the brain or spinal cord
Interneuron	Dendrite, cell body, and axon of a neuron within the brain or spinal cord	Serves as processing center; conducts nerve impulse from the sensory neuron to a motor neuron
Motor neuron	Dendrite, cell body, and axon of a motor neuron	Transmits nerve impulse from the brain or spinal cord out to an effector
Effector	A muscle or gland outside the nervous system	Responds to stimulation by the motor neuron and produces the reflex or behavioral action

Uses of Reflexes

Since normal reflexes depend on normal neuron functions, reflexes are commonly used to obtain information concerning the condition of the nervous system. An anesthesiologist, for instance, may try to initiate a reflex in a patient who is being anesthetized in order to determine how the anesthetic drug is affecting nerve functions. Also, in the case of injury to some part of the nervous system, various reflexes may be tested to discover the location and extent of the damage.

If any portion of a reflex arc is injured, the normal characteristics of that arc are likely to be altered. For example, a *plantar reflex* is normally initiated by stroking the sole of the foot, and the usual response is flexion of the foot and toes. However, in persons who have suffered damage to certain nerve pathways (corticospinal tract) there may be an abnormal response called the *Babinski reflex*. In this case, the reflex response is dorsiflexion, in which the great toe extends upward and the smaller toes fan apart. If the injury is minor, the response may consist of plantar flexion with failure of the great toe to flex, or plantar flexion followed by dorsiflexion. The Babinski reflex is present normally in infants up to the age of twelve months and is thought to reflect a degree of immaturity in their corticospinal tracts.

Other reflexes that may be tested during a neurological examination include the following:

1. **Biceps-jerk reflex.** This reflex can be elicited by bending a person's arm at the elbow. The examiner's finger is placed on the inside of the bent elbow over the tendon of the biceps muscle, and the finger is tapped. The biceps contracts in response, and the forearm is flexed at the elbow.
2. **Triceps-jerk reflex.** This reflex can be caused by flexing a person's arm at the elbow and tapping the short tendon of the triceps muscle close to its insertion near the tip of the elbow. The muscle contracts in response, and the forearm is extended slightly.
3. **Abdominal reflexes.** These reflexes occur when the examiner strokes the skin of the abdomen. For example, a dull pin may be drawn from the sides of the abdomen upward toward the midline and above the umbilicus. Normally, the abdominal muscles underlying the skin contract in response, and the umbilicus is moved toward the region that was stimulated.
4. **Ankle-jerk reflex.** This reflex is elicited by tapping the calcaneal tendon just above its insertion on the calcaneus. The response is plantar flexion, produced by contraction of the gastrocnemius and soleus muscles.
5. **Cremasteric reflex.** This reflex is obtained in males by stroking the upper inside of the thigh. As a result, the testis on the same side is elevated by contracting muscles.

Chapter Summary

Introduction (page 332)

Organs of the nervous system are divided into the central and peripheral nervous systems. These parts provide sensory, integrative, and motor functions.

General Functions of the Nervous System (page 332)

1. Sensory functions employ receptors that detect changes in internal and external body conditions.
2. Integrative functions bring sensory information together and make decisions that are acted upon by using motor functions.
3. Motor functions use effectors that respond when they are stimulated by motor impulses.

Nervous Tissue (page 332)

1. Neuron structure
   a. A neuron includes a cell body, cell processes, and the organelles usually found in cells.
   b. Dendrites and the cell body provide receptive surfaces.
   c. A single axon arises from the cell body and may be enclosed in a myelin sheath and a neurilemma.
2. Neuroglial cells
   a. Neuroglial cells are accessory cells.
   b. They fill spaces, support neurons, hold nervous tissue together, play a role in the metabolism of glucose, help regulate potassium ion concentration, produce myelin, and carry on phagocytosis.
   c. They include Schwann cells, astrocytes, oligodendrocytes, microglia, and ependymal cells.
3. Regeneration of nerve fibers
   a. If a neuron cell body is injured, the neuron is likely to die.
   b. If a peripheral nerve fiber is severed, its distal portion will die, but the proximal portion may regenerate and reestablish its former connections, provided it has a tube of connective tissue to guide it.
   c. Significant regeneration is unlikely to take place in the central nervous system.

Cell Membrane Potential (page 339)

A cell membrane is usually polarized as a result of an unequal distribution of ions on either side. This distribution of ions is due to the presence of pores and channels in the membrane that allow passage of some ions but not others.

1. Resting potential
   a. There is a high concentration of sodium ions on the outside of the membrane and a high concentration of potassium ions on the inside.
   b. There are large numbers of negatively charged ions on the inside of the cell.
   c. In a resting cell, more positive ions leave the cell than enter it, so the outside of the cell membrane develops a positive charge with respect to the inside.
2. Local potential changes
   a. Stimulation of a membrane affects its resting potential in a local region.
   b. The membrane is depolarized if it becomes less negative; it is hyperpolarized if it becomes more negative.
   c. Local potential changes are graded and subject to summation.
   d. If threshold potential is reached, an action potential is triggered.
3. Action potentials
   a. At threshold, sodium channels open and sodium ions diffuse inward, causing depolarization.
   b. About the same time, potassium channels open and potassium ions diffuse outward, causing repolarization.
   c. This rapid change in potential is an action potential.
   d. Many action potentials can occur before an active transport mechanism reestablishes the original resting potential.
   e. The propagation of action potentials along a nerve fiber is an impulse.
4. Refractory period
   a. The refractory period is a brief time following the passage of a nerve impulse when the membrane is unresponsive to an ordinary stimulus.
   b. During the absolute refractory period, the membrane cannot be stimulated; during the relative refractory period, the membrane can be stimulated with a stimulus of high intensity.
5. All-or-none response
   a. A nerve impulse is conducted in an all-or-none manner whenever a stimulus of threshold intensity is applied to a fiber.
   b. All the impulses conducted on a fiber are the same strength.
6. Impulse conduction
   a. Unmyelinated fibers conduct impulses that travel over their entire surfaces.
   b. Myelinated fibers conduct impulses that travel from node to node.
   c. Impulse conduction is more rapid on myelinated fibers with large diameters.

The Synapse (page 343)

A synapse is a junction between two neurons. A synaptic cleft is the gap between parts of two neurons at a synapse.

1. Synaptic transmission
   a. Impulses usually travel from dendrite or cell body, then along the axon to a synapse.
   b. Axons have synaptic knobs at their distal ends that secrete neurotransmitters.
   c. The neurotransmitter is released when a nerve impulse reaches the end of an axon, and the neurotransmitter diffuses across the synaptic cleft.
   d. When the neurotransmitter reaches the nerve fiber on the distal side of the cleft, a nerve impulse is triggered.
2. Neurotransmitter substances
   a. About thirty neurotransmitters are thought to be produced in the nervous system.
   b. Calcium ions diffuse into synaptic knobs in response to action potentials, causing the release of neurotransmitters.
   c. Neurotransmitters are quickly decomposed or removed from synaptic clefts.
3. Neuropeptides
   a. Neuropeptides are composed of amino acids in chains.
   b. Some of them serve as neurotransmitters or neuromodulators.
   c. They include enkephalins, endorphins, and substance P.
4. Synaptic potentials
   a. Some neurotransmitters can cause postsynaptic membranes to depolarize and trigger an action potential.
   b. Others cause the membranes to become hyperpolarized and inhibit action potentials.
   c. Summation of stimuli occurs in a trigger zone of the neuron.

Processing of Impulses (page 348)

The way impulses are processed reflects the organization of the neurons in the brain and spinal cord.

1. Neuronal pools
   a. Neurons are organized into pools within the central nervous system.
   b. Each pool receives impulses, processes them, and conducts impulses away.
2. Facilitation
   a. Each neuron in a pool may receive excitatory and inhibitory stimuli.
   b. A neuron is facilitated when it receives subthreshold stimuli and becomes more excitable.

3. Convergence
   a. Impulses from two or more incoming fibers may converge on a single neuron.
   b. Convergence makes it possible for a neuron to summate impulses from different sources.
4. Divergence
   a. Impulses leaving a pool may diverge by passing onto several output fibers.
   b. Divergence allows impulses to be amplified.

Classification of Neurons and Nerve Fibers (page 349)

Neurons differ in structure and function.

1. Classification of neurons
   a. On the basis of structure, neurons can be classified as bipolar neurons, unipolar neurons, and multipolar neurons.
   b. On the basis of function, neurons can be classified as sensory neurons, interneurons, or motor neurons.
2. Types of nerves and nerve fibers
   a. Nerves are cordlike bundles of nerve fibers.
   b. Nerves can be classified as sensory nerves, motor nerves, or mixed nerves, depending on which type of fibers they contain.
   c. Nerve fibers within the central nervous system can be subdivided into groups with general and special functions.

Nerve Pathways (page 352)

A nerve pathway is the route followed by an impulse as it travels through the nervous system.

1. Reflex arcs
   a. A reflex arc usually includes a sensory neuron, a reflex center composed of interneurons, and a motor neuron.
   b. The reflex arc is the behavioral unit of the nervous system.
2. Reflex behavior
   a. Reflexes are automatic, unconscious responses to changes.
   b. They help maintain homeostasis.
   c. The knee-jerk reflex employs only two neurons.
   d. Withdrawal reflexes are protective actions.

Clinical Application of Knowledge

1. How would you explain the following observations?
   a. When motor nerve fibers in the leg are severed, the muscles they innervate become paralyzed; however, in time, control over the muscles often returns.
   b. When motor nerve fibers in the spinal cord are severed, the muscles they control become paralyzed permanently.
2. People who have a hereditary disease called familial periodic paralysis often develop very low blood potassium concentrations. How would you explain the fact that the paralysis may disappear quickly when potassium ions are administered intravenously?
3. Multiple sclerosis is a disease in which nerve fibers in the central nervous system lose their myelin. Why would this loss be likely to affect the person's ability to control skeletal muscles?
4. What might be deficient in the diet of a pregnant woman who is complaining of leg muscle cramping? How would you explain this to her?
5. Why are rapidly growing cancers that originate in nervous tissue more likely to be composed of neuroglial cells than of neurons?
6. The biceps-jerk reflex employs motor neurons that exit from the spinal cord in the 5th spinal nerve (C5), that is, fifth from the top of the cord. The triceps-jerk reflex involves motor neurons in the 7th spinal nerve (C7). How might these reflexes be used to help locate the site of damage in a patient with a neck injury?

Review Activities

1. Explain the relationship between the central nervous system and the peripheral nervous system.
2. List three general functions of the nervous system.
3. Distinguish between neurons and neuroglial cells.
4. Describe the generalized structure of a neuron.
5. Define *myelin*.
6. Distinguish between myelinated and unmyelinated nerve fibers.
7. Describe how an injured nerve fiber may regenerate.
8. Discuss the functions of each type of neuroglial cell.
9. Explain how a membrane may become polarized.
10. Define *resting potential*.
11. Distinguish between depolarizing and hyperpolarizing.
12. List the changes that occur during an action potential.
13. Explain how action potentials are related to nerve impulses.
14. Define *refractory period*.
15. Define *saltatory conduction*.
16. Define *synapse*.
17. Explain how a nerve impulse is transmitted from one neuron to another.
18. Explain the role of calcium in the release of neurotransmitters.
19. Define *neuropeptide*.
20. Distinguish between excitatory and inhibitory postsynaptic potentials.
21. Explain what is meant by the "trigger zone" of a neuron.
22. Describe the relationship between an input nerve fiber and its neuronal pool.
23. Define *facilitation*.
24. Distinguish between convergence and divergence.
25. Explain how nerve impulses can be amplified.
26. Explain how neurons can be classified on the basis of their structure.
27. Explain how neurons can be classified on the basis of their function.
28. Distinguish between sensory, motor, and mixed nerves.
29. List four general types of nerve fibers.
30. Describe a reflex arc.
31. Define *reflex*.
32. Describe a withdrawal reflex.

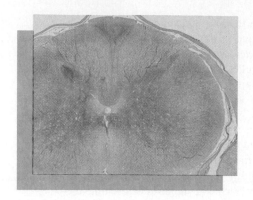

11

Nervous System II Divisions of the Nervous System

*T*he organs of the nervous system can be divided into two major groups. The brain and spinal cord comprise the central nervous system, while nerves that connect the brain and spinal cord to other body parts form the peripheral nervous system. The peripheral nervous system can be subdivided further into a somatic system, composed of the peripheral nerves that communicate with skin and skeletal muscles, and an autonomic system, composed of the peripheral nerves that communicate with visceral organs ∎

Chapter Objectives

After you have studied this chapter, you should be able to:

1. Describe the coverings of the brain and spinal cord.

2. Describe the structure of the spinal cord and its major functions.

3. Name the major parts of the brain and describe the functions of each.

4. Distinguish among motor, sensory, and association areas of the cerebral cortex.

5. Explain what is meant by hemisphere dominance.

6. Explain the stages in memory storage.

7. Describe the formation and function of cerebrospinal fluid.

8. Explain the functions of the limbic system and the reticular formation.

9. List the major parts of the peripheral nervous system.

10. Describe the structure of a peripheral nerve.

11. Name the cranial nerves and list their major functions.

12. Explain how spinal nerves are named.

13. Explain the function of a spinal nerve.

14. Describe the general characteristics of the autonomic nervous system.

15. Distinguish between the sympathetic and the parasympathetic divisions of the autonomic nervous system.

16. Describe a sympathetic and a parasympathetic nerve pathway.

17. Explain how the autonomic neurotransmitters produce different effects on visceral effectors.

18. Complete the review activities at the end of this chapter. Note that the items are worded in the form of specific learning objectives. You may want to refer to them before reading the chapter.

Key Terms

adrenergic (ad″ren-er′jik)

autonomic nervous system (aw″to-nom′ik ner′vus sis′tem)

brain stem (brān stem)

cerebellum (ser″ĕ-bel′um)

cerebral cortex (ser′ĕ-bral kor′teks)

cerebral hemisphere (ser′ĕ-bral hem′ĭ-sfēr)

cerebrospinal fluid (ser″ĕ-bro-spi′nal floo′id)

cerebrum (ser′ĕ-brum)

cholinergic (ko″lin-er′jik)

choroid plexus (ko′roid plek′sus)

diencephalon (di″en-sef′ah-lon)

hypothalamus (hi″po-thal′ah-mus)

medulla oblongata (mĕ-dul′ah ob″long-ga′tah)

meninges (mĕ-nin′jēz)

midbrain (mid′brān)

parasympathetic (par″ah-sim″pah-thet′ik)

postganglionic (pōst″gang-gle-on′ik)

preganglionic (pre″gang-gle-on′ik)

reticular formation (rĕ-tik′u-lar fōr-ma′shun)

sympathetic (sim″pah-thet′ik)

thalamus (thal′ah-mus)

ventricle (ven′trĭ-kl)

Aids to Understanding Words

chiasm-, cross: optic *chiasm*a—an X-shaped structure produced by the crossing over of optic nerve fibers.

flacc-, flabby: *flacc*id paralysis—paralysis characterized by loss of tone in muscles innervated by damaged nerve fibers.

funi-, small cord or fiber: *funi*culus—a major nerve tract or bundle of myelinated nerve fibers within the spinal cord.

gangli-, swelling: *gangli*on—a mass of neuron cell bodies.

mening-, membrane: *mening*es—membranous coverings of the brain and spinal cord.

plex-, interweaving: choroid *plex*us—a mass of specialized capillaries associated with spaces in the brain.

The organs of the central nervous system (CNS) are surrounded by bones, membranes, and fluid. More specifically, the brain lies within the cranial cavity of the skull, while the spinal cord, which is continuous with the brain, occupies the vertebral canal within the vertebral column. Beneath these bony coverings, the brain and spinal cord are protected by membranes called *meninges* located between the bone and the soft tissues of the nervous system (figure 11.1a).

Meninges

The **meninges** (singular, *meninx*) have three layers—dura mater, arachnoid mater, and pia mater (figure 11.1b).

The **dura mater** is the outermost layer. It is composed primarily of tough, white fibrous connective tissue and contains many blood vessels and nerves. It is attached to the inside of the cranial cavity and forms the internal periosteum of the surrounding skull bones. (See reference plate 53.)

In some regions, the dura mater extends inward between lobes of the brain and forms partitions that support and protect these parts (chart 11.1). In other areas, the dura mater splits into two layers, forming channels called *dural sinuses,* shown in figure 11.1b. Venous blood flows through these channels as it returns from the brain to vessels leading to the heart.

The dura mater continues into the vertebral canal as a strong, tubular sheath that surrounds the spinal cord. It is attached to the cord at regular intervals by a band of pia mater (denticulate ligaments) that ex-

tends the length of the spinal cord on either side. The dural sheath terminates as a blind sac at the level of the second sacral vertebra, below the end of the spinal cord. The sheath around the spinal cord is not attached directly to the vertebrae, but is separated by an *epidural space,* which lies between the dural sheath and the bony walls (figure 11.2). This space contains blood vessels, loose connective tissue, and adipose tissue that provide a protective pad around the spinal cord.

A blow to the head may cause some blood vessels associated with the brain to rupture, and the escaping blood may collect in the space beneath the dura mater. This condition, called *subdural hematoma,* can create increasing pressure between the rigid bones of the skull and the soft tissues of the brain. Unless the accumulating blood is evacuated promptly, the resulting compression of the brain may lead to functional losses or even death.

CHART 11.1	Partitions of the dura mater
Partition	**Location**
Falx cerebelli	Separates the right and left cerebellar hemispheres
Falx cerebri	Extends downward into the longitudinal fissure, and separates the right and left cerebral hemispheres
Tentorium cerebelli	Separates the occipital lobes of the cerebrum from the cerebellum

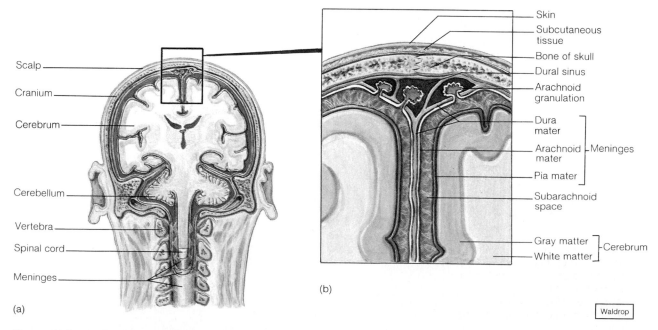

(a)

(b)

Waldrop

Figure 11.1 (a) The brain and spinal cord are enclosed by bone and by membranes called meninges. (b) The meninges include three layers: dura mater, arachnoid mater, and pia mater.

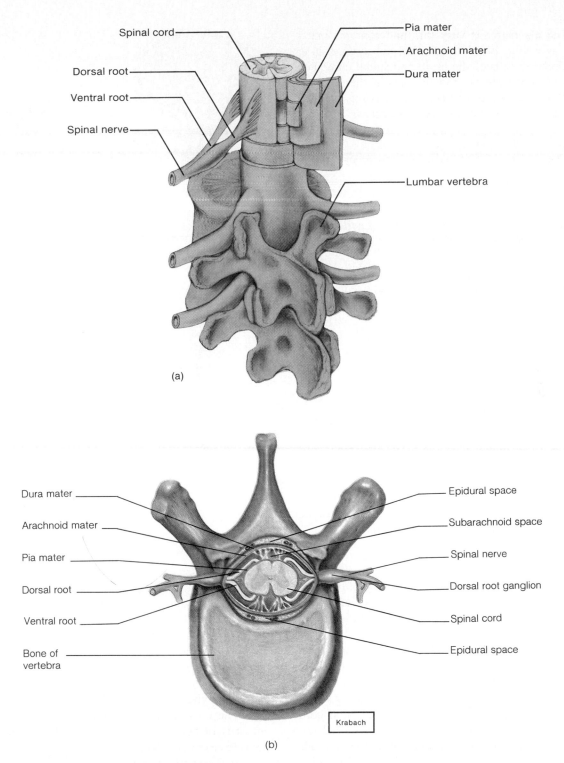

Spinal cord
Dorsal root
Ventral root
Spinal nerve

Pia mater
Arachnoid mater
Dura mater

Lumbar vertebra

(a)

Dura mater
Arachnoid mater
Pia mater
Dorsal root
Ventral root
Bone of
vertebra

Epidural space
Subarachnoid space
Spinal nerve
Dorsal root ganglion
Spinal cord
Epidural space

Krabach

(b)

Figure 11.2 (a) The dura mater forms a tubular sheath around the spinal cord. (b) The epidural space between the dural sheath and the bone of the vertebra is filled with tissues that provide a protective pad around the cord.

The **arachnoid mater** is a thin, weblike membrane that lacks blood vessels and is located between the dura and pia maters. It spreads over the brain and spinal cord, but generally does not dip into the grooves and depressions on their surfaces. Many thin strands extend from its undersurface and are attached to the pia mater.

Between the arachnoid and pia maters is a *subarachnoid space,* which contains the clear, watery **cerebrospinal fluid.**

The **pia mater** is very thin and contains many nerves, as well as blood vessels that aid in nourishing the underlying cells of the brain and spinal cord. The pia mater is attached to the surfaces of these organs and follows their irregular contours, passing over the high areas and dipping into the depressions.

An inflammation of the meninges is called *meningitis.* This condition is usually caused by certain bacteria or viruses that invade the cerebrospinal fluid. Although meningitis may involve the dura mater, it is more commonly limited to the arachnoid and pia maters.

Meningitis occurs most often in infants and children, and is considered one of the more serious childhood infections. Possible complications of this disease include loss of vision or hearing, paralysis, mental retardation, or death.

1. Describe the meninges.
2. Name the layers of the meninges.
3. Explain where cerebrospinal fluid occurs.

Spinal Cord

The **spinal cord** is a slender nerve column that passes downward from the brain into the vertebral canal. Although it is continuous with the brain, the spinal cord is said to begin where nervous tissue leaves the cranial cavity at the level of the foramen magnum. (See reference plate 55.) The cord tapers to a point and terminates near the intervertebral disk that separates the first and second lumbar vertebrae (figure 11.3a).

Structure of the Spinal Cord

The spinal cord consists of thirty-one segments, each of which gives rise to a pair of **spinal nerves.** These nerves branch out to various body parts and connect them with the central nervous system.

In the neck region, a thickening in the spinal cord, called the *cervical enlargement,* gives off nerves to the arms. A similar thickening in the lower back, the *lumbar enlargement,* gives off nerves to the legs. Just inferior to the lumbar enlargement, the spinal cord narrows to a sharp tip called the *conus medullaris.* From this tip, a thin cord of connective tissue descends to the upper surface of the coccyx. This cord is called the *filum terminale* (figure 11.3b).

Two grooves, a deep *anterior median fissure* and a shallow *posterior median sulcus,* extend the length of the spinal cord, dividing it into right and left halves. A cross section of the cord (figure 11.4) reveals that it consists of a core of gray matter surrounded by white matter. The pattern produced by the gray matter roughly resembles a butterfly with its wings outspread. The upper and lower wings of gray matter are called the *posterior horns* and the *anterior horns,* respec-

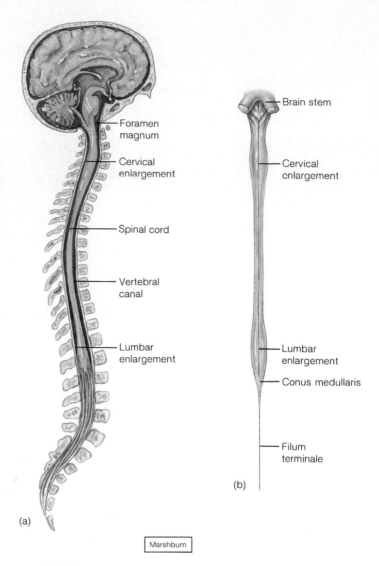

Marshburn

Figure 11.3 (*a*) The spinal cord begins at the level of the foramen magnum. At what level does it terminate? (*b*) Posterior view of the spinal cord with the spinal nerves removed.

tively. Between them on either side is a protrusion of gray matter called the *lateral horn.* Neurons with relatively large cell bodies in the anterior horns (anterior horn cells) give rise to motor fibers that pass out through spinal nerves to various skeletal muscles. However, the majority of neurons in the gray matter are interneurons. (See chapter 10.)

A horizontal bar of gray matter in the middle of the spinal cord, the *gray commissure,* connects the wings of the gray matter on the right and left sides. This bar surrounds the **central canal,** which is continuous with the ventricles of the brain and contains cerebrospinal fluid. Although the central canal is prominent during embryonic development, it becomes almost microscopic in an adult.

The white matter of the spinal cord is divided by the gray matter into three regions on each side. These regions are known as the *anterior, lateral,* and *poste-*

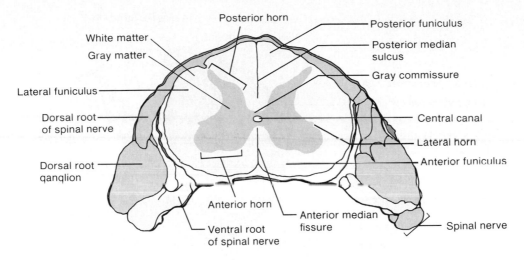

(a)

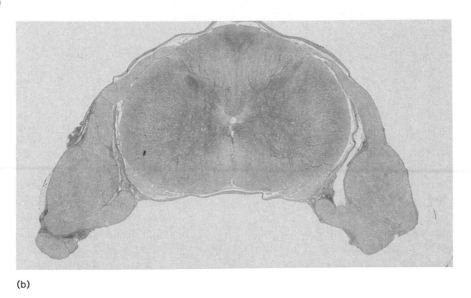

(b)

Figure 11.4 (*a*) A cross section of the spinal cord. (*b*) Identify the parts of the spinal cord in this micrograph.

rior funiculi. Each funiculus consists of longitudinal bundles of myelinated nerve fibers that comprise major nerve pathways called **nerve tracts** (figure 11.4).

Functions of the Spinal Cord

The spinal cord has two major functions: it conducts nerve impulses, and it serves as a center for spinal reflexes.

The nerve tracts of the spinal cord provide a two-way communication system between the brain and body parts outside the nervous system. The tracts that conduct impulses from body parts and carry sensory information to the brain are called **ascending tracts;** those that conduct motor impulses from the brain to muscles and glands are called **descending tracts.**

The nerve fibers within these tracts are axons. Typically, all the axons within a given tract originate from neuron cell bodies located in the same part of the nervous system and end together in some other part. The names that identify nerve tracts often reflect these common origins and terminations. For example, a *spinothalamic tract* begins in the spinal cord and carries sensory impulses associated with the sensations of pain and touch to the thalamus of the brain. A *corticospinal tract* originates in the cortex of the brain and carries motor impulses downward through the spinal cord and spinal nerves. These impulses function in the control of skeletal muscle movements.

Ascending Tracts

Among the major ascending tracts of the spinal cord are the following (figure 11.5):

1. **Fasciculus gracilis** (fah-sik′u-lus gras′il-is) and **fasciculus cuneatus** (ku′ne-at-us). These tracts are located in the posterior funiculi of the spinal cord. Their fibers conduct sensory

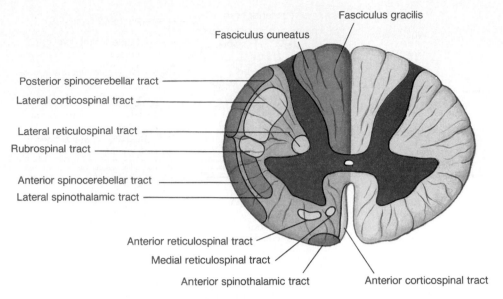

Figure 11.5 Major ascending and descending tracts within a cross section of the spinal cord. (Note: Ascending tracts are shown in darker tones; descending tracts are in lighter tones.)

impulses from the skin, muscles, tendons, and joints to the brain, where they are interpreted as sensations of touch, pressure, and body movement.

At the base of the brain in an area called the *medulla oblongata,* most of these fibers cross over (decussate) from one side to the other—that is, those ascending on the left side of the spinal cord pass across to the right side and vice versa. As a result, the impulses originating from sensory receptors on the left side of the body reach the right side of the brain, and those originating on the right side of the body reach the left side of the brain (figure 11.6).

2. **Spinothalamic** (spi″no-thah-lam′ik) **tracts.** The *lateral* and *anterior spinothalamic tracts* are located in the lateral and anterior funiculi, respectively. Impulses in these tracts cross over in the spinal cord. The lateral tracts conduct impulses from various body regions to the brain, and give rise to sensations of pain and temperature. Impulses carried on fibers of the anterior tracts are interpreted as touch and pressure.

> Persons suffering from severe, unremitting pain are sometimes treated with a surgical operation called a *cordotomy.* This is done only as a last resort. In this procedure, fibers in the lateral spinothalamic tracts on the side opposite the affected area are severed above the level of the pain receptors that are being stimulated.

3. **Spinocerebellar** (spi″no-ser″ĕ-bel′ar) **tracts.** The *posterior* and *anterior spinocerebellar tracts* are found near the surface in the lateral funiculi of the spinal cord. Fibers in the posterior tracts remain uncrossed, while those in the anterior tracts cross over in the cord. Impulses conducted on their fibers originate in the muscles of the legs and trunk, and then travel to the cerebellum of the brain. These impulses are necessary for the coordination of muscular movements.

Descending Tracts

The major descending tracts of the spinal cord are shown in figure 11.5. They include the following:

1. **Corticospinal** (kor″ti-ko-spi′nal) **tracts.** The *lateral* and *anterior corticospinal tracts* occupy the lateral and anterior funiculi, respectively. Most of the fibers of the lateral tracts cross over in the lower portion of the medulla oblongata. Those of the anterior tracts descend uncrossed. The corticospinal tracts conduct motor impulses from the brain to spinal nerves and outward to various skeletal muscles. Thus, they function in the control of voluntary movements (figure 11.7).

    The corticospinal tracts are also called *pyramidal tracts* after the pyramid-shaped regions in the medulla oblongata through which they pass. The other descending tracts are called *extrapyramidal tracts,* and they include the reticulospinal and rubrospinal tracts.

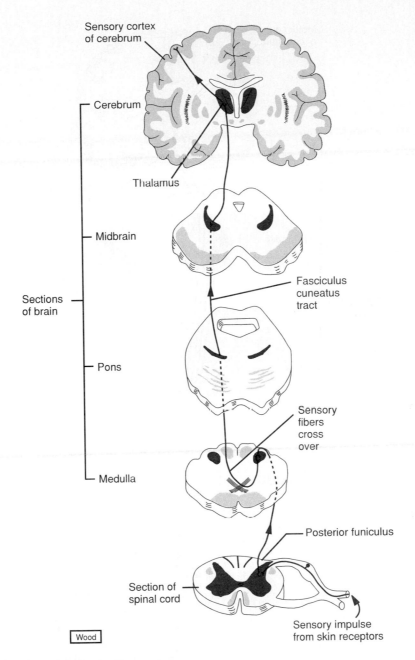

Sensory cortex
of cerebrum

Cerebrum

Thalamus

Sections
of brain

Midbrain

Fasciculus
cuneatus
tract

Pons

Sensory
fibers
cross
over

Medulla

Posterior funiculus

Section of
spinal cord

Sensory impulse
from skin receptors

Wood

Figure 11.6 Sensory impulses originating in skin receptors ascend in the fasciculus cuneatus tract and cross over in the medulla of the brain.

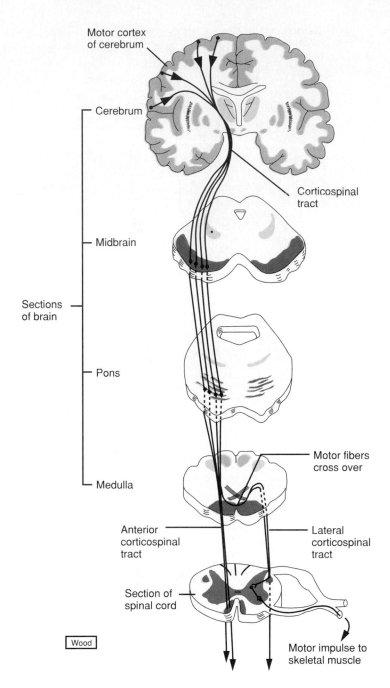

Motor cortex of cerebrum

Cerebrum

Corticospinal tract

Midbrain

Sections of brain

Pons

Motor fibers cross over

Medulla

Anterior corticospinal tract

Lateral corticospinal tract

Section of spinal cord

Wood

Motor impulse to skeletal muscle

Figure 11.7 Motor fibers of the corticospinal tract begin in the cerebral cortex, cross over in the medulla, and descend in the spinal cord. There they synapse with neurons that have fibers that lead to spinal nerves supplying skeletal muscles.

CHART 11.2 Nerve tracts of the spinal cord

Tract	Location	Function
Ascending tracts		
1. Fasciculus gracilis and fasciculus cuneatus	Posterior funiculi	Conduct sensory impulses associated with the senses of touch, pressure, and body movement from skin, muscles, tendons, and joints to the brain
2. Spinothalamic tracts (anterior and posterior)	Lateral and anterior funiculi	Conduct sensory impulses associated with the senses of pain, temperature, touch, and pressure from various body regions to the brain
3. Spinocerebellar tracts (posterior and anterior)	Lateral funiculi	Conduct sensory impulses needed for the coordination of muscle movements from muscles of the legs and trunk to the cerebellum
Descending tracts		
1. Corticospinal tracts (lateral and anterior)	Lateral and anterior funiculi	Conduct motor impulses associated with voluntary movements from the brain to various skeletal muscles
2. Reticulospinal tracts (lateral, anterior, and medial)	Lateral and anterior funiculi	Conduct motor impulses associated with the maintenance of muscle tone and the activity of sweat glands from the brain
3. Rubrospinal tracts	Lateral funiculi	Conduct motor impulses associated with muscular coordination and the maintenance of posture from the brain

2. **Reticulospinal** (rĕ-tik″u-lo-spi′nal) **tracts.** The *lateral reticulospinal tracts* are located in the lateral funiculi, while the *anterior* and *medial reticulospinal tracts* are in the anterior funiculi. Some fibers in the lateral tracts cross over, while others remain uncrossed. Those of the anterior and medial tracts remain uncrossed. Motor impulses transmitted on the reticulospinal tracts originate in the brain and function in the control of muscular tone and in the activity of the sweat glands.

3. **Rubrospinal** (roo″bro-spi′nal) **tracts.** The fibers of the rubrospinal tracts cross over in the brain and pass through the lateral funiculi. They carry motor impulses from the brain to skeletal muscles, and are involved with muscular coordination and posture control.

The nerve tracts of the spinal cord are summarized in chart 11.2.

In addition to serving as a pathway for various nerve tracts, the spinal cord functions in many reflexes, like the knee-jerk and withdrawal reflexes described in chapter 10. Such reflexes are called **spinal reflexes,** because their reflex arcs pass through the cord.

The disease called *poliomyelitis* is caused by viruses that sometimes affect motor neurons, including those in the anterior horns of the spinal cord. When this happens, the person may develop paralysis in the muscles innervated by the affected neurons, but may not suffer any sensory losses in these parts.

Another condition involving motor neurons is called *amyotrophic lateral sclerosis* (ALS). In this disorder, motor neurons degenerate within the spinal cord, brain stem, and cerebral cortex. Later, these neurons are replaced by fibrous tissue. The early signs of ALS include loss of dexterity (particularly in fine finger movements), wasting of hand muscles, severe muscle cramps, difficulty in walking, and problems with swallowing. The cause of ALS, which is also called Lou Gehrig's disease, is unknown.

1. Describe the structure of the spinal cord.
2. What is meant by ascending and descending tracts?
3. What is the consequence of fibers crossing over?
4. Name the major tracts of the spinal cord, and list the kinds of impulses each conducts.

Spinal Cord Injuries

Injuries to the spinal cord may be caused indirectly, as by a blow to the head or a fall, or they may be caused by forces applied directly to the cord. The consequences will depend on the amount of damage sustained by the cord.

Normal spinal reflexes depend on two-way communication between the spinal cord and the brain. When nerve pathways are injured, the cord's reflex activities in sites below the injury are usually depressed. At the same time, there is a lessening of sensations and muscular tone in the parts innervated by the affected fibers. This condition is called *spinal shock,* and it may last for days or weeks; although normal reflex activity may return eventually. However, if nerve fibers are severed, some of the cord's functions are likely to be permanently lost.

Less severe injuries to the spinal cord, as from a blow to the head, whiplash in an automobile accident, or rupture of an intervertebral disk, cause the cord to be compressed or distorted. Such injuries are often accompanied by pain, weakness, and muscular atrophy in the regions supplied by the damaged nerve fibers.

Among the more common causes of severe direct injury to the spinal cord are gunshot wounds, stabbings, and fractures or dislocations of vertebrae during vehicular accidents (figure 11.8). Regardless of the cause, if nerve fibers in ascending tracts are cut, sensations arising from receptors below the level of the injury will be lost. If descending tracts are damaged, the result will be a loss of motor functions.

For example, if the right lateral corticospinal tract is severed in the neck near the first cervical vertebra, control of the voluntary muscles in the right arm and leg is lost, and the limbs become paralyzed (hemiplegia). Problems of

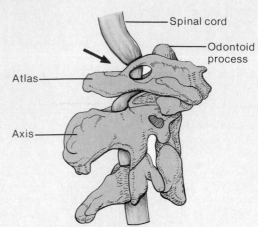

Figure 11.8 A dislocation of the atlas may cause a compression injury to the spinal cord.

this type, involving fibers of the descending tracts, are said to produce the *upper motor neuron syndrome.* This condition is characterized by *spastic paralysis* in which muscle tone increases, with very little atrophy of the muscles involved. However, uncoordinated reflex activity (hyperreflexia) usually occurs, during which the flexor and extensor muscles of affected limbs alternately undergo spasms.

If, on the other hand, motor neurons or their fibers in the horns of the spinal cord are injured, the resulting condition is called *lower motor neuron syndrome.* It is characterized by *flaccid paralysis,* in which muscle tone and reflex activity are totally lost, and the muscles undergo extreme atrophy.

Brain

The **brain** is the largest and most complex part of the nervous system. It occupies the cranial cavity and is composed of about one hundred billion (10^{11}) multipolar neurons and innumerable nerve fibers, by which these neurons communicate with one another and with neurons in other parts of the system.

The brain contains nerve centers associated with sensory functions, and is responsible for sensations and perceptions. It issues motor commands to skeletal muscles, and carries on higher mental functions, such as memory and reasoning. It also contains centers associated with the coordination of muscular movements, as well as centers and nerve pathways involved in regulating visceral activities.

Development of the Brain

The basic structure of the brain reflects the way it forms during early (embryonic) development. It begins as a tubelike structure (neural tube) that gives rise to the

central nervous system. The portion that becomes the brain has three major cavities, or vesicles, at one end. They are called the *forebrain* (prosencephalon), *midbrain* (mesencephalon), and *hindbrain* (rhombencephalon) (figure 11.9). Later, the forebrain divides into anterior and posterior portions (telencephalon and diencephalon, respectively), and the hindbrain partially divides into two parts (metencephalon and myelencephalon). The resulting five cavities persist in the mature brain as fluid-filled spaces called *ventricles* and the tubes that connect them. The tissue surrounding the spaces differentiates into various structural and functional regions of the brain.

The wall of the anterior portion of the forebrain gives rise to the *cerebrum* and *basal ganglia,* while the posterior portion forms a section of the brain stem called the *diencephalon.* The region produced by the midbrain continues to be called the *midbrain* in the adult structure, and the hindbrain gives rise to the *cerebellum, pons,* and *medulla oblongata* (figure 11.10 and chart 11.3).

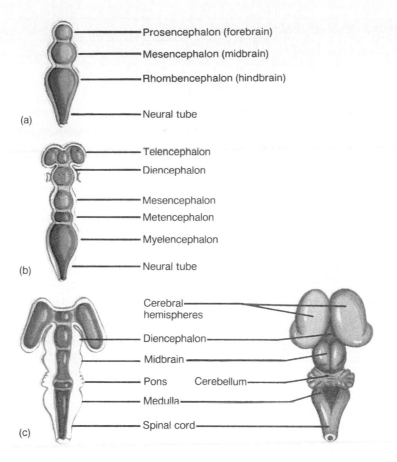

(a)

- Prosencephalon (forebrain)
- Mesencephalon (midbrain)
- Rhombencephalon (hindbrain)
- Neural tube

(b)

- Telencephalon
- Diencephalon
- Mesencephalon
- Metencephalon
- Myelencephalon
- Neural tube

(c)

- Cerebral hemispheres
- Diencephalon
- Midbrain
- Pons
- Cerebellum
- Medulla
- Spinal cord

Figure 11.9 (a) The brain develops from a tubular part with three cavities. (b) The cavities persist as the ventricles and their interconnections. (c) The wall of the tube gives rise to various regions of the brain.

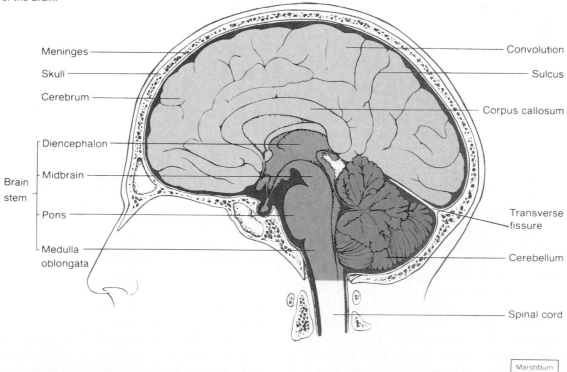

Meninges

Skull

Cerebrum

Brain stem

- Diencephalon
- Midbrain
- Pons
- Medulla oblongata

Convolution

Sulcus

Corpus callosum

Transverse fissure

Cerebellum

Spinal cord

Marshburn

Figure 11.10 The major portions of the brain include the cerebrum, the cerebellum, and the brain stem.

CHART 11.3 Structural development of the brain

Embryonic vesicle	Spaces produced	Regions of the brain produced
Forebrain (prosencephalon)		
Anterior portion (telencephalon)	Lateral ventricles	Cerebrum Basal ganglia
Posterior portion (diencephalon)	Third ventricle	Thalamus Hypothalamus Posterior pituitary gland Pineal gland
Midbrain (mesencephalon)	Cerebral aqueduct	Midbrain
Hindbrain (rhombencephalon)		
Anterior portion (metencephalon)	Fourth ventricle	Cerebellum Pons
Posterior portion (myelencephalon)	Fourth ventricle	Medulla oblongata

Occasionally during early development, the neural tube does not close completely. If this happens in the anterior end, the brain fails to develop, or only a rudimentary one forms. This condition is called *anencephaly,* and it usually results in the death of the developing offspring.

When the neural tube defect occurs in the posterior portion, the vertebral column remains open, and the spinal cord is exposed to some degree. This condition is called *spina bifida.* If the spinal cord is covered by the developing skin, the problem may be corrected surgically, and the child may have only minor functional problems. However, if the nerve tissue is protruding from the site of the defect at birth, the child is likely to suffer irreversible brain damage and some degree of paralysis, and will require extensive medical and surgical treatment.

Structure of the Cerebrum

The cerebrum, which develops from the anterior portion of the forebrain, is the largest part of the mature brain. It consists of two large masses called **cerebral hemispheres,** which are essentially mirror images of each other (figure 11.11 and reference plate 49). These hemispheres are connected by a deep bridge of nerve fibers called the **corpus callosum** and are separated by a layer of dura mater called the *falx cerebri.*

The surface of the cerebrum is marked by numerous ridges or **convolutions** (sing. *gyrus;* pl. *gyri*), which are separated by grooves. A shallow groove is called a **sulcus,** and a very deep groove is called a **fissure.** Although the arrangement of these elevations and depressions is complex, they form fairly distinct patterns in all normal brains. For example, a *longitudinal fissure* separates the right and left cerebral hemispheres; a *transverse fissure* separates the cerebrum from the cerebellum; and various sulci divide each hemisphere into lobes (figures 11.10 and 11.11).

The lobes of the cerebral hemispheres (figure 11.11) are named after the skull bones that they underlie. They include the following:

1. **Frontal lobe.** The frontal lobe forms the anterior portion of each cerebral hemisphere. It is bordered posteriorly by a *central sulcus* (fissure of Rolando), which passes out from the longitudinal fissure at a right angle, and inferiorly by a *lateral sulcus* (fissure of Sylvius), which passes out from the undersurface of the brain along its sides.
2. **Parietal lobe.** The parietal lobe is posterior to the frontal lobe and is separated from it by the central sulcus.
3. **Temporal lobe.** The temporal lobe lies below the frontal and parietal lobes and is separated from them by the lateral sulcus.
4. **Occipital lobe.** The occipital lobe forms the posterior portion of each cerebral hemisphere and is separated from the cerebellum by a shelflike extension of dura mater called the *tentorium cerebelli.* There is no distinct boundary between the occipital lobe and the parietal and temporal lobes.
5. **Insula.** The insula (island of Reil) is located deep within the lateral sulcus and is covered by parts of the frontal, parietal, and temporal lobes. It is separated from them by a *circular sulcus.*

A thin layer of gray matter (2 to 5 millimeters thick) called the **cerebral cortex** constitutes the outermost portion of the cerebrum. This layer covers the convolutions, and dips into the sulci and fissures. It is estimated to contain nearly 75% of all the neuron cell bodies in the nervous system.

Just beneath the cerebral cortex is a mass of white matter that makes up the bulk of the cerebrum. This mass contains bundles of myelinated nerve fibers that connect neuron cell bodies of the cortex with other parts

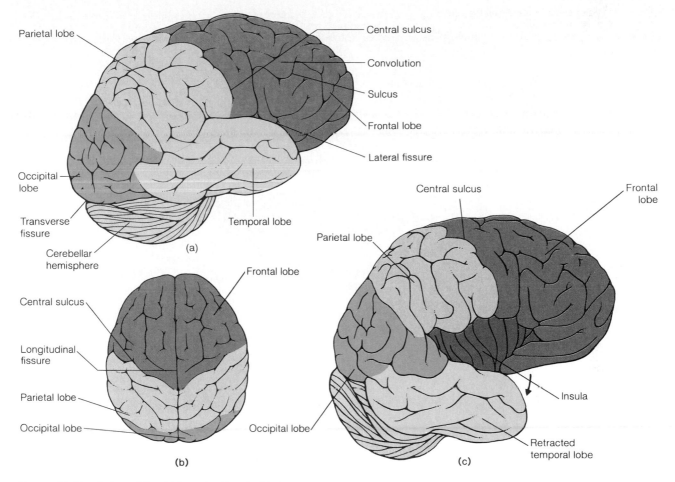

Figure 11.11 Lobes of the right cerebral hemisphere are distinguished by color. (*a*) Lateral view of the right hemisphere; (*b*) hemispheres viewed from above; (*c*) lateral view of the left hemisphere with the insula exposed.

of the nervous system. Some of these fibers pass from one cerebral hemisphere to the other by way of the corpus callosum, and others carry sensory or motor impulses from portions of the cortex to nerve centers in the brain or spinal cord.

Functions of the Cerebrum

The cerebrum is concerned with higher brain functions; that is, it contains centers for interpreting sensory impulses arriving from various sense organs as well as centers for initiating voluntary muscular movements. It stores the information of memory and utilizes this information in reasoning processes. It also functions in determining a person's intelligence and personality.

Functional Regions of the Cortex

The regions of the cerebral cortex that perform specific functions have been located using a variety of techniques. For example, persons who have suffered brain injuries or have had portions of their brains removed surgically have been studied. Their impaired abilities provide clues as to the functions of the specific parts that were injured or removed.

In other studies, areas of cortices have been exposed surgically and stimulated mechanically or electrically. Then the responses in certain muscles or the specific sensations that result have been observed.

As a result of such investigations, it is possible to map the functional regions of the cerebral cortex. Although there is considerable overlap in these areas, the cortex can be divided into sections known as motor, sensory, and association areas.

Motor Areas

The *primary motor areas* of the cerebral cortex lie in the frontal lobes just in front of the central sulcus (precentral gyrus), and in the anterior wall of this sulcus (figure 11.12). The nervous tissue in these regions contains numerous, large *pyramidal cells,* so named because of their pyramid-shaped cell bodies.

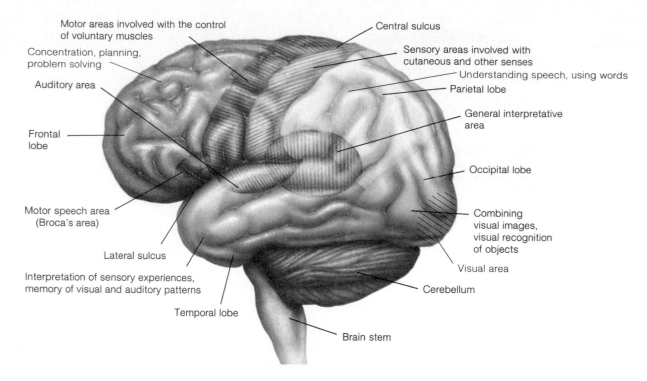

Figure 11.12 Some motor, sensory, and association areas of the left cerebral cortex.

Labels on figure:
- Motor areas involved with the control of voluntary muscles
- Concentration, planning, problem solving
- Auditory area
- Frontal lobe
- Motor speech area (Broca's area)
- Lateral sulcus
- Interpretation of sensory experiences, memory of visual and auditory patterns
- Temporal lobe
- Brain stem
- Central sulcus
- Sensory areas involved with cutaneous and other senses
- Understanding speech, using words
- Parietal lobe
- General interpretative area
- Occipital lobe
- Combining visual images, visual recognition of objects
- Visual area
- Cerebellum

Impulses from these pyramidal cells travel downward through the brain stem and into the spinal cord on the *corticospinal tracts*. Most of the nerve fibers in these tracts cross over from one side of the brain to the other within the brain stem and descend as lateral corticospinal tracts. (See figure 11.7.)

Within the spinal cord, the corticospinal fibers synapse with motor neurons in the gray matter of the anterior horns. Axons of the motor neurons lead outward through peripheral nerves to various voluntary muscles. Impulses transmitted on these pathways in special patterns and frequencies are responsible for fine movements in skeletal muscles. More specifically, as figure 11.13 shows, cells in the upper portions of the motor areas send impulses to muscles in the legs and thighs; those in the middle portions control muscles in the shoulders and arms; and those in lower portions activate muscles of the head, face, and tongue.

The *reticulospinal* and *rubrospinal tracts* coordinate and control motor functions involved with maintaining balance and posture. Many of these fibers pass into the basal ganglia on the way to the spinal cord. Some of the impulses conducted on these pathways normally inhibit muscular actions.

In addition to the primary motor areas, certain other regions of the frontal lobe are involved with motor functions. For example, a region called *Broca's area* is just anterior to the primary motor cortex and above the lateral sulcus (figure 11.12). This area usually occurs in the left cerebral hemisphere. It coordinates the complex muscular actions of the mouth, tongue, and larynx, which make speech possible. A person with an injury to this area may be able to understand spoken words, but may be unable to speak.

Above Broca's area is a region called the *frontal eye field*. The motor cortex in this area controls the voluntary movements of the eyes and eyelids. Nearby is the cortex responsible for movements of the head that direct the eyes. Another region just in front of the primary motor area controls the muscular movements of the hands and fingers that make such skills as writing possible (figure 11.12).

An injury to the motor system may result in a loss of the ability to produce purposeful muscular movements. Such a condition that impairs use of the hands, arms, legs, head, or eyes is called *apraxia*. Apraxia involving the speech muscles, accompanied by loss of the ability to speak is called *aphasia*.

Sensory Areas

Sensory areas, which occur in several lobes of the cerebrum, interpret impulses that arrive from various

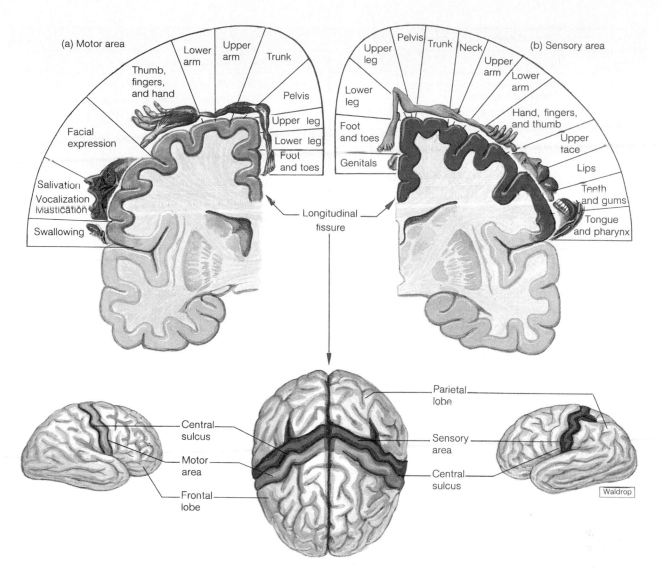

Figure 11.13 (*a*) Motor areas that control voluntary muscles; (*b*) sensory areas involved with cutaneous and other senses.

sensory receptors. These interpretations give rise to feelings or sensations. For example, the sensations of temperature, touch, pressure, and pain involving the skin arise in the anterior portions of the parietal lobes along the central sulcus (postcentral gyrus) and in the posterior wall of this sulcus (figure 11.12). The posterior parts of the occipital lobes are concerned with vision, while the posterior, dorsal portions of the temporal lobes contain the centers for hearing. The sensory areas for taste are located near the bases of the central sulci along the lateral sulci, and the sense of smell arises from centers deep within the cerebrum.

Like motor fibers, sensory fibers, such as those in the *fasciculus cuneatus tract,* cross over. (See figure 11.6.) Thus, the centers in the right cerebral hemisphere interpret impulses originating from the left side of the body, and vice versa. However, the sensory areas concerned with vision receive impulses from both eyes, and those concerned with hearing receive impulses from both ears.

Association Areas

Association areas are those areas of the cerebral cortex that are not primarily sensory or motor in function and are interconnected with each other and with other brain structures. These areas occupy the anterior portions of the frontal lobes and are widespread in the lateral portions of the parietal, temporal, and occipital lobes. They analyze and interpret sensory experiences and are involved with memory, reasoning, verbalizing, judgment, and emotional feelings (figure 11.12).

The association areas of the frontal lobes are concerned with a number of higher intellectual processes, including those necessary for concentrating, planning,

CHART 11.4   Functions of the cerebral lobes

Lobe	Functions	Lobe	Functions
Frontal lobes	Motor areas control movements of voluntary skeletal muscles. Association areas carry on higher intellectual processes such as those required for concentrating, planning, complex problem solving, and judging the consequences of behavior.	Temporal lobes	Sensory areas are responsible for hearing. Association areas function in interpreting sensory experiences and in remembering visual scenes, music, and other complex sensory patterns.
Parietal lobes	Sensory areas are responsible for the sensations of temperature, touch, pressure, and pain involving the skin. Association areas function in understanding speech and in using words to express thoughts and feelings.	Occipital lobes	Sensory areas are responsible for vision. Association areas function in combining visual images with other sensory experiences.

and complex problem solving. The anterior and inferior portions of these lobes (prefrontal areas) control emotional behavior and produce awareness of the possible consequences of behavior.

The association areas of the parietal lobes help interpret sensory information, and aid in understanding speech and choosing words to express thoughts and feelings. Awareness of the form of objects, including one's own body parts, stems from the posterior regions of these lobes.

The association areas of the temporal lobes and the regions at the posterior ends of the lateral fissures are concerned with the interpretation of complex sensory experiences, such as those needed to understand speech and to read printed words. These regions also are involved with the memory of visual scenes, music, and other complex sensory patterns.

The association areas of the occipital lobes, adjacent to the visual centers, are important in analyzing visual patterns and combining visual images with other sensory experiences—as when one recognizes another person.

Of particular importance is the region where the parietal, temporal, and occipital association areas come together near the posterior end of the lateral sulcus. This region is called the *general interpretative area,* and it plays the primary role in complex thought processing. It makes it possible for a person to recognize words and arrange them to express a thought, and to read and understand ideas presented in writing. Chart 11.4 summarizes the functions of the cerebral lobes.

1. How does the brain form during early development?
2. Describe the cerebrum.
3. List the general functions of the cerebrum.
4. Where in the brain are the primary motor and sensory regions located?
5. Explain the functions of association areas.

Hemisphere Dominance

Both cerebral hemispheres participate in basic functions, such as receiving and analyzing sensory impulses, controlling skeletal muscles on opposite sides of the body, and storing memory. However, in most persons one side acts as a **dominant hemisphere** for certain other functions.

In over 90% of the population, for example, the left hemisphere is dominant for the language-related activities of speech, writing, and reading. It is also dominant for complex intellectual functions requiring verbal, analytical, and computational skills. In other persons, the right hemisphere is dominant, and in some, the hemispheres are equally dominant.

Various tests indicate that the left hemisphere is dominant in 90% of right-handed adults and in 64% of left-handed ones. The right hemisphere is dominant in 10% of right-handed adults and in 20% of left-handed ones. The hemispheres are equally dominant in the remaining 16% of left-handed persons. As a consequence of hemisphere dominance, Broca's area on one side almost completely controls the motor activities associated with speech. For this reason, over 90% of patients with language impairment involving the cerebrum have disorders in the left hemisphere.

In addition to carrying on basic functions, the nondominant hemisphere seems to specialize in nonverbal functions, such as those involving motor tasks that require orientation of the body in surrounding space, understanding and interpreting musical patterns, and visual experiences. It is also concerned with emotional and intuitive thought processes. For example, although the region in the nondominant hemisphere that corresponds to Broca's area is incapable of controlling speech, it seems to influence the emotional aspects of the language expressed in speech.

Cerebral Injuries

When the brain is jarred intensely, as a result of a violent blow to the head, and the person loses consciousness, this condition is called a *concussion*. If the injury is mild, the unconsciousness may last only a few seconds, but if it is severe, the loss of consciousness may be prolonged.

Following a concussion, a person may experience a variety of symptoms, including headache, impairment of memory, and lessened ability to focus attention. These symptoms can continue for varying lengths of time, depending upon the degree of damage sustained by the cerebrum.

In another type of brain injury, the cerebrum is damaged when blood flow in a vessel that supplies brain tissues is suddenly interrupted. This malady is called a "stroke," or *cerebrovascular accident* (CVA). It most commonly occurs following the rupture of a blood vessel, accompanied by a hemorrhage into the brain, or following a prolonged obstruction in a vessel due to the presence of a blood clot. (See chapter 17.) In either case, the brain tissues downstream from the vascular accident may die (tissue infarction), causing a permanent loss of function.

If the interruption of blood flow to brain tissue is only temporary, the condition is called a *transient ischemic attack* (TIA). A TIA may occur as the result of a blood clot that soon disintegrates, so that its effect on blood flow is limited to a few minutes.

The lasting effects of various injuries to the cerebral cortex depend on which areas are damaged and to what extent. When particular portions of the cortex are injured, the special functions of these portions are likely to be lost or at least depressed.

It is often possible to deduce the location and extent of a brain injury by determining what abilities the patient is missing. For example, if the motor area of one frontal lobe has been damaged, the person is likely to be partially or completely paralyzed on the opposite side of the body. Similarly, if the visual cortex of one occipital lobe is injured, the person may suffer partial blindness in both eyes; if both visual cortices are damaged, the blindness may be total.

A person with damage to the association areas of the frontal lobes may have difficulty concentrating on complex mental tasks. Such an individual usually appears disorganized and is easily distracted.

If the general interpretative area of the dominant hemisphere is injured, the person may be unable to interpret sounds as words or to understand ideas presented in writing. However, the dominance of one hemisphere usually does not become established until after five or six years of age. Consequently, if the general interpretative area is destroyed in a child, the corresponding region on the other side of the brain may be able to take over the functions, and the child's language abilities may develop normally. If such an injury occurs in an adult, the nondominant hemisphere may develop only limited interpretative functions, and the person is likely to have a severe intellectual disability.

Brain damage may also interfere with memory. For example, if the portion of the temporal lobe that functions in processing recent memory is lost, the person will be unable to acquire new recent memory or subsequently to store it as long-term memory. Generally, the amount of stored memory that is lost as a result of injury is roughly proportional to the amount of cerebral cortex damaged.

Nerve fibers of the *corpus callosum,* which connect the cerebral hemispheres, make it possible for the dominant hemisphere to control the motor cortex of the nondominant hemisphere. These fibers also allow sensory information reaching the nondominant hemisphere to be transferred to the general interpretative area of the dominant one, where the information can be used in decision making.

Storage of Memory

Memory refers to the storage of information and the ability to recall this information when it is needed. Memory is required for all higher brain functions, such as learning, reasoning, and adapting behavior to immediate needs.

Although most memory occurs in the cerebral cortex, certain other parts of the nervous system can also store information and are involved with sorting and retrieving it.

The establishment of memory occurs in stages. In the first stage, sensory information reaches the brain and is held by ongoing neuronal activity for a few seconds or so. This type of temporary storage is called **short-term memory,** and although it may not last long, it is useful for remembering a small amount of information, such as a telephone number, long enough to use it.

Short-term memory seems to be processed mainly in the anterior portion of the frontal lobe and in a part of the brain stem called the *thalamus*. This memory is easily disrupted and can be displaced and lost as new sensory information arrives. However, it can also be transferred into the next phase of memory storage, which is more permanent.

Information in short-term memory may be recalled quickly, and if it is consciously reviewed, it may be moved into a form called **recent memory.** This phase of processing involves a portion of the temporal lobe

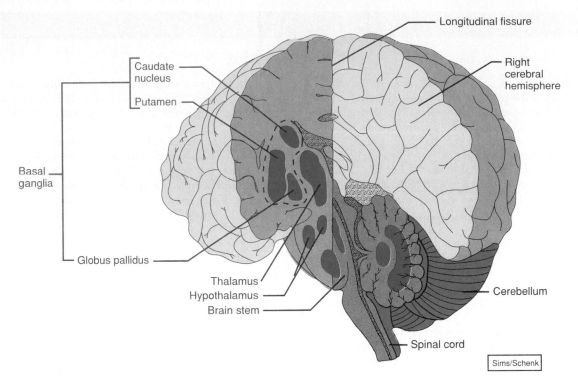

Longitudinal fissure

Right cerebral hemisphere

Caudate nucleus

Putamen

Basal ganglia

Globus pallidus

Thalamus

Hypothalamus

Brain stem

Cerebellum

Spinal cord

Sims/Schenk

Figure 11.14 A frontal section of the left cerebral hemisphere reveals some of the basal ganglia.

called the *hippocampus* and produces memory that may last for minutes or days. However, recent memory is usually stored relatively weakly and is subject to loss as a result of disuse.

The most stable form of memory is called **long-term memory.** It results from repeated recall of new information or from repeated sensory experiences. Processing this form of memory takes place over large areas of the cerebral cortex, and the information is stored diffusely in both hemispheres. However, bits of information seem to be sorted into groups, so that each new bit is stored in association with other information of the same kind.

The brain has a vast capacity to store long-term memory. The stored information, such as one's name, one's date of birth, letters of the alphabet, and words commonly used in speech, usually can be recalled quickly.

The way in which long-term memory is processed is poorly understood, but some researchers believe it involves neuronal activity that is intense and repetitive enough to cause stable changes in nerve pathways. For

example, structural or chemical changes might occur that produce persistent facilitation of certain synapses (chapter 10). If all of the synapses of a particular pathway remain facilitated, an incoming impulse at a later time might excite this pathway, causing impulses to flow through it, thus recalling a bit of memory.

Basal Ganglia

The **basal ganglia** (basal nuclei) are masses of gray matter located deep within the cerebral hemispheres. They are called the *caudate nucleus,* the *putamen,* and the *globus pallidus,* and they develop from the anterior portion of the forebrain (figure 11.14). Although the precise function of these parts is not completely understood, the neuron cell bodies they contain are known to serve as relay stations for motor impulses originating in the cerebral cortex and passing into the brain stem and spinal cord. It is also known that most of the inhibitory neurotransmitter called *dopamine* is produced in the basal ganglia. Impulses from these parts normally inhibit motor functions and thus aid in controlling various muscular activities.

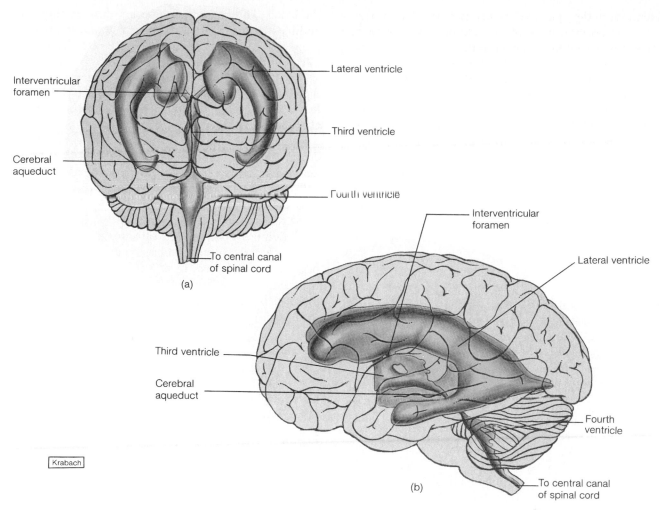

Interventricular foramen

Cerebral aqueduct

Lateral ventricle

Third ventricle

Fourth ventricle

To central canal of spinal cord

(a)

Interventricular foramen

Lateral ventricle

Third ventricle

Cerebral aqueduct

Fourth ventricle

To central canal of spinal cord

(b)

Krabach

Figure 11.15 (a) Anterior view of the ventricles within the cerebral hemispheres and brain stem; (b) lateral view.

Because the basal ganglia function to inhibit muscular activity, disorders in this portion of the brain may be accompanied by reduced mobility (hypokinesia) due to an excessive discharge of impulses from the ganglia or by involuntary movements (hyperkinesia) due to lack of inhibiting impulses from the ganglia.

Parkinson's disease, for example, is thought to be caused by degeneration of the neurons that synthesize dopamine in the basal ganglia. This condition is characterized by slow movements, difficulty in initiating voluntary muscular actions, and tremors. It is often treated with a drug called L-dopa, which can be converted into dopamine by cellular enzymes. Although this treatment may control some of the symptoms of Parkinson's disease for a time, it usually does not alter the course of the disease. Huntington's chorea is another example of a condition that results from lesions in the basal ganglia.

1. What is meant by hemisphere dominance?
2. What are the functions of the nondominant hemisphere?
3. Distinguish between short-term, recent, and long-term memory.
4. What is the function of the basal ganglia?

Ventricles and Cerebrospinal Fluid

Within the cerebral hemispheres and brain stem is a series of interconnected cavities called **ventricles,** shown in figure 11.15 and reference plates 53 and 54. These spaces are continuous with the central canal of the spinal cord and, like it, they are filled with cerebrospinal fluid.

The largest ventricles are the *lateral ventricles,* which are the first and second ventricles, (the first ventricle in the left cerebral hemisphere and the second ventricle in the right cerebral hemisphere). They extend into the cerebral hemispheres and occupy portions of the frontal, temporal, and occipital lobes.

A narrow space that constitutes the *third ventricle* is located in the midline of the brain, beneath the corpus callosum. This ventricle communicates with the lateral ventricles through openings *(interventricular foramina)* in its anterior end.

The *fourth ventricle* is located in the brain stem just in front of the cerebellum. It is connected to the third ventricle by a narrow canal, the *cerebral aqueduct* (aqueduct of Sylvius), which passes lengthwise through the brain stem. This ventricle is continuous with the central canal of the spinal cord and has openings in its roof that lead into the subarachnoid space of the meninges.

Cerebrospinal Fluid

The *cerebrospinal fluid* (CSF) is secreted by tiny, reddish cauliflower-like masses of specialized capillaries from the pia mater called **choroid plexuses.** These structures project into the cavities of the ventricles (figure 11.16). They are covered by a single layer of specialized ependymal cells (chapter 10) that are joined closely to one another by tight junctions. These cells provide a barrier that hinders the passage of water-soluble substances between the blood and the cerebrospinal fluid. At the same time, the cells selectively transfer certain substances from the blood into the cerebrospinal fluid by facilitated diffusion and transfer other substances by active transport (chapter 3), thus regulating the composition of the cerebrospinal fluid.

Although choroid plexuses occur in the medial walls of the lateral ventricles and in the roofs of the third and fourth ventricles, most of the cerebrospinal fluid seems to arise in the lateral ventricles. From there it circulates slowly into the third and fourth ventricles, and into the central canal of the spinal cord. It also enters the subarachnoid space of the meninges by passing through the wall of the fourth ventricle near the cerebellum.

Nearly 500 milliliters of cerebrospinal fluid are secreted daily. However, only about 140 milliliters are present in the nervous system at any time, because cerebrospinal fluid is continuously being reabsorbed into the blood. This reabsorption occurs through tiny, fingerlike structures called *arachnoid granulations* that project from the subarachnoid space into the blood-filled dural sinuses (figure 11.16).

Cerebrospinal fluid is a clear, somewhat viscid liquid that differs in composition from the fluid that leaves the capillaries in other parts of the body. Specifically, it contains a greater concentration of sodium, and lesser concentrations of glucose and potassium than do other extracellular fluids. Its function seems to be nutritive as well as protective. Cerebrospinal fluid also helps maintain a stable ionic concentration in the central nervous system and provides a pathway to the blood for waste substances.

Because it occupies the subarachnoid space of the meninges, cerebrospinal fluid completely surrounds the brain and spinal cord. In effect, these organs float in the fluid that supports and protects them by absorbing forces that might otherwise jar and damage their delicate tissues.

In order to treat certain cancers developing within the central nervous system, anticancer (antineoplastic) drugs are sometimes injected directly into the cerebrospinal fluid. This procedure is called an *intrathecal injection,* and the drug is usually administered into the subarachnoid space of the lumbar region. Such an injection may also be made into a silicone reservoir that has been surgically implanted beneath the scalp and communicates with a lateral ventricle by means of a tube. In either case, because the volume of cerebrospinal fluid is relatively small and is in close contact with brain tissues, intrathecal drug concentrations must be very carefully monitored to prevent serious toxic reactions.

Various pain-relieving drugs, such as the opioids sometimes used for patients with chronic pain, may also be administered into the cerebrospinal fluid through a thin catheter placed into the epidural or subarachnoid space.

1. Where are the ventricles of the brain located?
2. How is cerebrospinal fluid formed?
3. Describe the pattern of cerebrospinal fluid circulation.

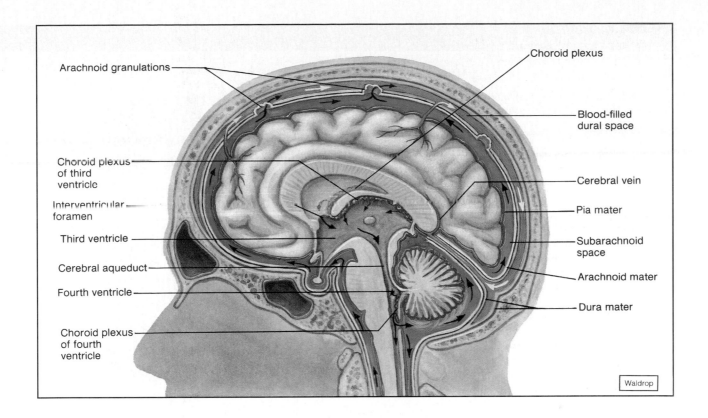

Arachnoid granulations

Choroid plexus

Choroid plexus
of third
ventricle

Interventricular
foramen

Third ventricle

Cerebral aqueduct

Fourth ventricle

Choroid plexus
of fourth
ventricle

Blood-filled
dural space

Cerebral vein

Pia mater

Subarachnoid
space

Arachnoid mater

Dura mater

Waldrop

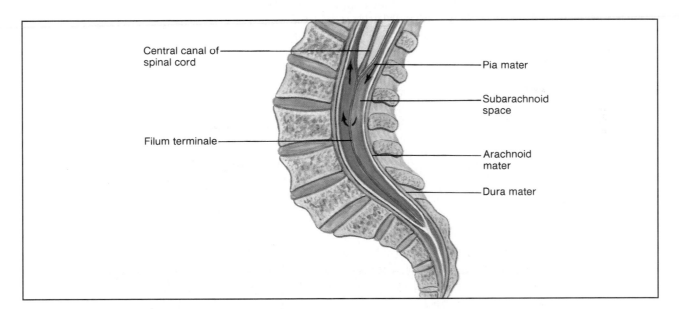

Central canal of
spinal cord

Filum terminale

Pia mater

Subarachnoid
space

Arachnoid
mater

Dura mater

Figure 11.16 Cerebrospinal fluid is secreted by choroid plexuses in the walls of the ventricles. The fluid circulates through the ventricles and central canal, enters the subarachnoid space, and is reabsorbed into the blood of the dural sinuses through arachnoid granulations.

Cerebrospinal Fluid Pressure

Because cerebrospinal fluid is secreted and reabsorbed continuously, the fluid pressure in the ventricles remains relatively constant. However, sometimes an infection, a tumor, or a blood clot interferes with the circulation of this fluid, and the pressure within the ventricles (intracranial pressure) increases. As the pressure increases, there is danger that cerebral blood vessels will collapse, retarding blood flow, and that the brain tissues may be injured by being forced against the skull.

The pressure of cerebrospinal fluid can be determined by performing a *lumbar puncture* (spinal tap). This procedure involves inserting a fine, hollow needle into the subarachnoid space between the third and fourth or between the fourth and fifth lumbar vertebrae—a level below the end of the spinal cord, as shown in figure 11.17. The pressure of the fluid, which is usually about 130 millimeters of water (10 millimeters of mercury), is measured using an instrument called a *manometer.* At the same time, samples of cerebrospinal fluid may be withdrawn and tested for the presence of abnormal constituents. The presence of red blood cells, for example, is abnormal and may indicate the existence of a hemorrhage somewhere in the central nervous system.

If the pressure is found to be excessive, a temporary drain may be used to decrease the fluid volume and thus lower the pressure. Such a drain may be inserted into the subarachnoid space between the fourth and fifth lumbar vertebrae. In an infant whose cranial sutures have not yet united, increasing intracranial pressure (ICP) may cause an enlargement of the cranium called *hydrocephalus,* or "water on the brain." Hydrocephalus is often treated by inserting a shunt that drains fluid away from the cranial cavity and into some other body region where it can either be reabsorbed into the blood or excreted (figure 11.18).

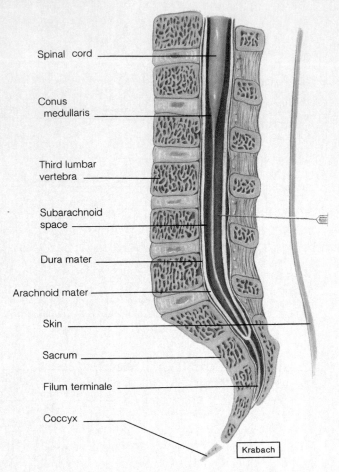

Figure 11.17 A lumbar puncture is performed by inserting a fine needle between the third and fourth lumbar vertebrae and withdrawing a sample of cerebrospinal fluid from the subarachnoid space.

Brain Stem

The region of the brain that connects the cerebrum to the spinal cord is called the *brain stem.* It consists of the diencephalon, midbrain, pons, and medulla oblongata. These parts include numerous tracts of nerve fibers and several masses of gray matter called *nuclei* (figures 11.10 and 11.19).

Diencephalon

The *diencephalon* develops from the posterior portion of the forebrain and is located between the cerebral hemispheres and above the midbrain. It surrounds the third ventricle and is composed largely of gray matter. Within the diencephalon, a dense mass, called the *thalamus,* bulges into the third ventricle from each side. Another region of the diencephalon that includes many nuclei is the *hypothalamus.* It lies below the thalamic nuclei and forms the lower walls and floor of the third ventricle. (See reference plates 49 and 53.)

Other parts of the diencephalon include: (a) the **optic tracts** and the **optic chiasma** that is formed by the optic nerve fibers crossing over; (b) the **infundibulum,** a conical process behind the optic chiasma to which the

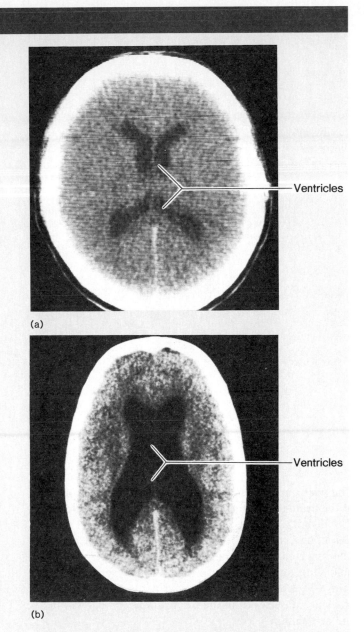

(a)

(b)

Figure 11.18 CT scans of the human brain. (a) Normal ventricles. (b) Ventricles enlarged due to excessive accumulation of fluid within the ventricles.

pituitary gland is attached; (c) the **posterior pituitary gland,** which hangs from the floor of the hypothalamus; (d) the **mammillary** (mam′i-ler″e) **bodies,** which appear as two rounded structures behind the infundibulum; and (e) the **pineal gland,** which forms as a cone-shaped evagination from the roof of the diencephalon. (See chapter 13.)

The **thalamus** serves as a central relay station for sensory impulses ascending from other parts of the nervous system to the cerebral cortex. It receives all sensory impulses (except those associated with the sense of smell) and channels them to appropriate regions of the cortex for interpretation. In addition, all regions of the cerebral cortex can communicate with the thalamus by means of descending fibers.

Although the cerebral cortex pinpoints the origin of sensory stimulation, the thalamus produces a general awareness of certain sensations such as pain, touch, and temperature.

The **hypothalamus** is interconnected by nerve fibers to the cerebral cortex, thalamus, and other parts of the brain stem so that it can receive impulses from them and send impulses to them. The hypothalamus plays key roles in maintaining homeostasis by regulating a variety of visceral activities and by serving as a link between the nervous and endocrine systems.

Among the many important functions of the hypothalamus are the following:

1. Regulation of heart rate and arterial blood pressure.
2. Regulation of body temperature.
3. Regulation of water and electrolyte balance.
4. Control of hunger and regulation of body weight.
5. Control of movements and glandular secretions of the stomach and intestines.
6. Production of neurosecretory substances that stimulate the pituitary gland to release various hormones. These hormones help regulate growth, control various glands, and influence reproductive physiology.
7. Regulation of sleep and wakefulness.

Structures in the general region of the diencephalon also play important roles in controlling emotional responses. For example, portions of the cerebral cortex in the medial parts of the frontal and temporal lobes are interconnected with the hypothalamus, thalamus, basal ganglia, and other deep nuclei. Together, these structures comprise a complex called the *limbic system.*

The **limbic system** is involved in emotional experience and expression, and can modify the way a person acts. It produces such feelings as fear, anger, pleasure, and sorrow. More specifically, the limbic system seems to recognize upsets in a person's physical or psychological condition that might threaten life. By causing pleasant or unpleasant feelings about experiences, the limbic system guides a person into behavior that may increase the chance of survival. In addition, portions of the limbic system interpret sensory impulses from the receptors associated with the sense of smell (olfactory receptors).

Midbrain

The **midbrain** (mesencephalon) is a short section of the brain stem located between the diencephalon and the pons. It contains bundles of myelinated nerve fibers that

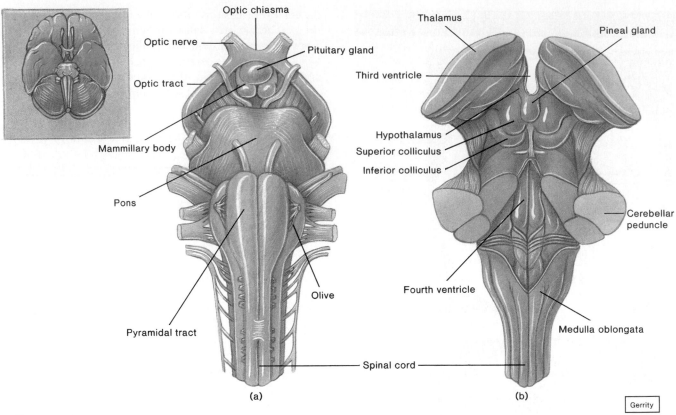

Figure 11.19 (a) Ventral view of the brain stem; (b) dorsal view of the brain stem with the cerebellum removed, exposing the fourth ventricle.

join lower parts of the brain stem and spinal cord with higher parts of the brain. The midbrain includes several masses of gray matter that serve as reflex centers. It also contains the *cerebral aqueduct* that connects the third and fourth ventricles (figure 11.19).

Two prominent bundles of nerve fibers on the underside of the midbrain comprise the *cerebral peduncles*. These fibers include the corticospinal tracts and are the main motor pathways between the cerebrum and lower parts of the nervous system. Beneath the cerebral peduncles are some large bundles of sensory fibers that carry impulses upward to the thalamus.

Two pairs of rounded knobs on the upper surface of the midbrain mark the location of four nuclei known collectively as *corpora quadrigemina* (figure 11.19). The upper masses (superior colliculi) contain the centers for certain visual reflexes, such as those responsible for moving the eyes to view something as the head is turned. The lower ones (inferior colliculi) contain the auditory reflex centers that operate when it is necessary to move the head so that sounds can be heard more distinctly.

Near the center of the midbrain is a mass of gray matter called the *red nucleus*. This nucleus communicates with the cerebellum and with centers of the spinal cord, and it functions in reflexes concerned with maintaining posture.

Pons

The **pons** appears as a rounded bulge on the underside of the brain stem where it separates the midbrain from the medulla oblongata (figure 11.19). The dorsal portion of the pons consists largely of longitudinal nerve fibers, which relay impulses to and from the medulla oblongata and the cerebrum. Its ventral portion contains large bundles of transverse nerve fibers, which transmit impulses from the cerebrum to centers within the cerebellum.

Several nuclei of the pons relay sensory impulses from peripheral nerves to higher brain centers. Other nuclei function with centers of the medulla oblongata to regulate the rate and depth of breathing.

Medulla Oblongata

The **medulla oblongata** is an enlarged continuation of the spinal cord extending from the level of the foramen magnum to the pons (figure 11.19). Its dorsal surface is flattened to form the floor of the fourth ventricle, and its ventral surface is marked by the corticospinal tracts, most of whose fibers cross over at this level. On each side of the medulla oblongata is an oval swelling called the *olive,* from which a large bundle of nerve fibers arises and passes to the cerebellum.

Because of its location, all the ascending and descending nerve fibers connecting the brain and spinal cord must pass through the medulla oblongata. As in the spinal cord, the white matter of the medulla surrounds a central mass of gray matter. Here, however, the gray matter is broken up into nuclei that are separated by nerve fibers. Some of these nuclei relay ascending impulses to the other side of the brain stem and then on to higher brain centers. The *nucleus gracilis* and the *nucleus cuneatus,* for example, receive sensory impulses from fibers of the fasciculus gracilis and the fasciculus cuneatus, and pass them on to the thalamus or the cerebellum.

Other nuclei within the medulla oblongata function as control centers for vital visceral activities. These centers include the following:

1. **Cardiac center.** Impulses originating in the cardiac center are transmitted to the heart on peripheral nerves. Impulses on these nerves can cause the heart rate to increase or decrease.

2. **Vasomotor center.** Certain cells of the vasomotor center initiate impulses that travel to smooth muscles in the walls of blood vessels and stimulate them to contract. This action causes constriction of the vessels (vasoconstriction) and a consequent rise in blood pressure. Other cells of the vasomotor center can produce the opposite effect—dilation of the blood vessels (vasodilation) and a consequent drop in the blood pressure.

3. **Respiratory center.** The respiratory center acts with centers in the pons to regulate the rate, rhythm, and depth of breathing.

Still other nuclei within the medulla oblongata function as centers for certain nonvital reflexes, including those associated with coughing, sneezing, swallowing, and vomiting. Since the medulla contains vital reflex centers, injuries to this part of the brain stem are often fatal.

Reticular Formation

Scattered throughout the medulla oblongata, pons, and midbrain is a complex network of nerve fibers associated with tiny islands of gray matter. This network, the **reticular formation** (reticular activating system), extends from the upper portion of the spinal cord into the diencephalon (figure 11.20). Its intricate system of nerve fibers interconnects centers of the hypothalamus, basal ganglia, cerebellum, and cerebrum with fibers in all the major ascending and descending tracts.

When sensory impulses reach the reticular formation, it responds by activating the cerebral cortex into a state of wakefulness. Without this arousal, the cortex remains unaware of stimulation and cannot interpret sensory information or carry on thought processes. Thus, decreased activity in the reticular formation results in sleep. If the reticular formation

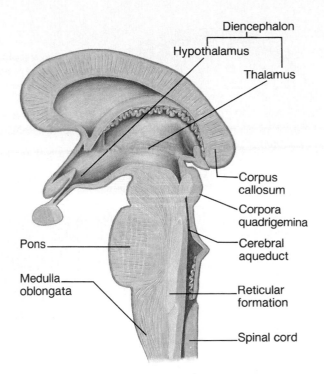

Figure 11.20 The reticular formation (shown in green) extends from the upper portion of the spinal cord into the diencephalon. What is the function of this complex network of nerve fibers?

ceases to function, as in certain injuries, the person remains unconscious and cannot be aroused even with strong stimulation (a comatose state).

The reticular formation also seems to act as a filter for incoming sensory impulses. Impulses that are judged to be important, such as those originating in pain receptors, are passed on to the cerebral cortex, while others are disregarded. This selective action of the reticular formation frees the cortex from what would otherwise be a continual bombardment of sensory stimulation, and allows the cortex to concentrate on more significant information. The cerebral cortex can also activate the reticular system, so that intense cerebral activity tends to keep a person awake.

In addition, the reticular formation plays a role in regulating motor activities so that various skeletal muscles move together evenly, and it influences spinal reflexes so that some are inhibited and others are enhanced.

Types of Sleep

The two types of sleep are normal sleep and paradoxical sleep. **Normal sleep** (slow wave sleep) occurs when a person is particularly tired. It results from decreasing activity of the reticular formation, and is restful, dreamless, and accompanied by reduced blood pressure and reduced respiratory rate.

Paradoxical sleep is so named because it occurs when some areas of the brain are active and others are

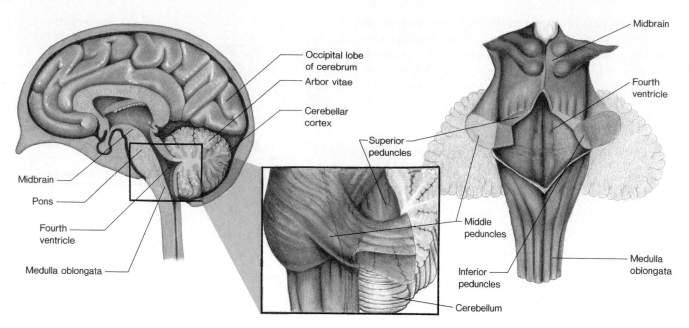

Figure 11.21 The cerebellum, which is located below the occipital lobes of the cerebrum, communicates with other parts of the nervous system by means of the cerebellar peduncles.

asleep. Episodes of this type of sleep usually last from five to twenty minutes and occur about every ninety minutes during normal sleep. However, the incidence of these episodes tends to increase as a person becomes rested. This form of sleep seems to occur because activating impulses are channeled to some regions of the brain and not to others. It is accompanied by dreams, irregular heart and respiratory rates, and rapid eye movements (REM).

1. What are the major functions of the thalamus? Of the hypothalamus?
2. How may the limbic system influence a person's behavior?
3. What vital reflex centers are located in the brain stem?
4. What is the function of the reticular formation?
5. Describe two types of sleep.

Cerebellum

The **cerebellum** is a large mass of tissue located below the occipital lobes of the cerebrum and posterior to the pons and medulla oblongata (figures 11.10 and 11.19). It consists of two lateral hemispheres partially separated by a layer of dura mater called the *falx cerebelli.* These hemispheres are connected in the midline by a structure called the *vermis.*

Like the cerebrum, the cerebellum is composed primarily of white matter with a thin layer of gray

matter, the **cerebellar cortex,** on its surface. This cortex is doubled over on itself in a series of complex folds that have myelinated nerve fibers branching into them. A cut into the cerebellum reveals a treelike pattern of white matter, called the *arbor vitae,* that is surrounded by gray matter. A number of nuclei lie deep within each cerebellar hemisphere, the largest and most important of which is the *dentate nucleus.*

The cerebellum communicates with other parts of the central nervous system by means of three pairs of nerve tracts called **cerebellar peduncles** (ser″ĕ-bel′ar pe-dung′k′ls). One pair, the *inferior peduncles,* brings sensory information concerning the actual position of body parts such as limbs and joints to the cerebellum via the spinal cord and medulla oblongata. The *middle peduncles* transmit information from the cerebral cortex about the desired position of these body parts. After integrating and analyzing the information from these two sources, the cerebellum sends impulses from the dentate nucleus via the *superior peduncles* to the midbrain (figure 11.21). In response, motor impulses travel downward through the pons, medulla oblongata, and spinal cord, and stimulate or inhibit skeletal muscles to cause desired body movements. This activity of the cerebellum makes rapid and complex muscular movements possible.

Thus, the cerebellum is the center for integrating sensory information concerning the position of body parts and for coordinating skeletal muscle activity and

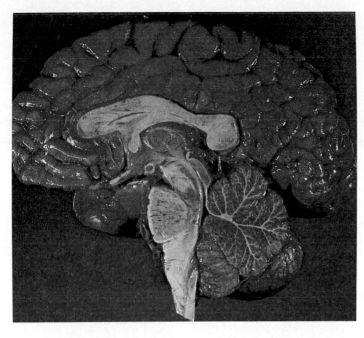

maintaining posture. The sensory impulses it receives come from receptors in muscles, tendons, and joints (proprioceptors) and from special sense organs, such as the eyes and ears. (See chapter 12.) For example, the cerebellum receives sensory information from the semicircular canals of the inner ears. This information concerns the motion and position of the head, and is used by the cerebellum to help maintain equilibrium. (See figure 11.22.)

Damage to the cerebellum is likely to result in tremors, inaccurate movements of voluntary muscles, loss of muscle tone, a reeling walk, and loss of equilibrium.

The characteristics and functions of the major parts of the brain are summarized in chart 11.5.

1. Where is the cerebellum located?
2. What are the major functions of the cerebellum?
3. What kinds of receptors provide information to the cerebellum?

Figure 11.22 What structures can you identify in this sagittal section of the brain?

CHART 11.5	Major parts of the brain	
Part	Characteristics	Functions
1. Cerebrum	Largest part of the brain; consists of two hemispheres connected by the corpus callosum	Controls higher brain functions, including interpreting sensory impulses, initiating muscular movements, storing memory, reasoning, and determining intelligence
2. Basal ganglia	Masses of gray matter located deep within the cerebral hemispheres	Act as relay stations for motor impulses originating in the cerebral cortex and passing into the brain stem and spinal cord
3. Brain stem	Region that connects the cerebrum to the spinal cord	
a. Diencephalon of the brain stem includes masses of gray matter called the thalamus and hypothalamus		The thalamus serves as a relay station for sensory impulses ascending from other parts of the nervous system to the cerebral cortex; the hypothalamus helps maintain homeostasis by regulating various visceral activities and by serving as a link between the nervous and endocrine systems
b. Midbrain contains masses of gray matter and bundles of nerve fibers that join the spinal cord to higher regions of the brain		Contains reflex centers that function in moving the eyes and head, and in maintaining posture
c. Pons appears as a bulge on the underside of the brain stem and contains masses of gray matter and nerve fibers		Relays nerve impulses to and from the medulla oblongata and cerebrum; helps regulate the rate and depth of breathing
d. Medulla oblongata is an enlarged continuation of the spinal cord that extends from the foramen magnum to the pons, and contains masses of gray matter and nerve fibers		Conducts ascending and descending impulses between the brain and spinal cord; contains cardiac, vasomotor, and respiratory control centers and various nonvital reflex control centers
4. Cerebellum	A large mass of tissue located below the cerebrum and posterior to the brain stem; includes two lateral hemispheres connected by the vermis	Communicates with other parts of the CNS by nerve tracts; acts as a center for integrating sensory information concerning the position of body parts, and for coordinating muscle activities and maintaining posture

Brain Waves

Brain waves are recordings of fluctuating electrical changes that occur in the brain. To obtain such a recording, electrodes are positioned on the surface of a surgically exposed brain (an electrocorticogram, ECoG) or on the outer surface of the head (an electroencephalogram, EEG). These electrodes detect electrical changes taking place in the extracellular fluid of the brain in response to changes in potential occurring among large groups of neurons. The resulting signals from the electrodes are amplified and recorded.

The source of brain waves seems to be the neurons in the cerebral cortex. However, the changes that occur in these waves are also related to activities in other parts of the brain that influence the cortex, such as the reticular formation.

In general, the intensity of electrical changes is directly related to the degree of neuronal activity. Thus, brain waves vary markedly in amplitude and frequency between sleep and wakefulness.

Brain waves are classified as alpha, beta, theta, and delta waves. *Alpha waves* are recorded most easily from the posterior regions of the head and have a frequency of 8–13 cycles per second. They occur when a person is awake but resting, with the eyes closed. These waves disappear during sleep, and if the wakeful person's eyes are opened, the alpha waves are replaced by higher frequency beta waves.

Beta waves have a frequency of more than 13 cycles per second and are usually recorded in the anterior region of the head. They occur when a person is actively engaged in mental activity or is under tension.

Theta waves have a frequency of 4–7 cycles per second and occur mainly in the parietal and temporal regions of the cerebrum. They are produced normally in children. Theta waves usually do not occur in adults, although some adults produce them in early stages of sleep or at times of emotional stress.

Delta waves have a frequency below 4 cycles per second and occur during sleep. They seem to come from the cerebral cortex when it is not being activated by the reticular formation (figure 11.23).

The patterns of brain waves are sometimes useful for diagnosing disease conditions. For example, they are used to distinguish various types of seizure disorders (epilepsy) and to locate brain tumors. (Because a tumor either takes up space previously occupied by normal nerve tissue, or

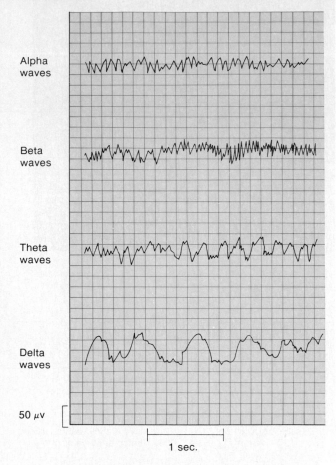

Figure 11.23 Brain waves are recordings of fluctuating electrical changes that occur in the brain.

compresses the normal tissues around itself, abnormal brain waves produced in the affected area may reveal the location of the tumor.)

Brain waves are also used to determine when *brain death* has occurred. Brain death, characterized by the cessation of neuronal activity, may be verified by an EEG that lacks waves (isoelectric EEG). However, because certain drugs can greatly depress brain functions, such drugs must be excluded as the cause of the flat EEG pattern before brain death can be confirmed.

Peripheral Nervous System

The **peripheral nervous system** (PNS) consists of the nerves that branch out from the central nervous system (CNS) and connect it to other body parts. The PNS includes the *cranial nerves* that arise from the brain and the *spinal nerves* that arise from the spinal cord.

The peripheral nervous system can also be subdivided into somatic and autonomic nervous systems.

Generally, the **somatic nervous system** consists of the cranial and spinal nerve fibers that connect the CNS to the skin and skeletal muscles, so it is involved with conscious activities. The **autonomic nervous system** includes those fibers that connect the CNS to visceral organs such as the heart, stomach, intestines, and various glands. Thus, the autonomic nervous system is concerned with unconscious actions. Chart 11.6 outlines the subdivisions of the nervous system.

Structure of Peripheral Nerves

As described in chapter 10, a peripheral nerve consists of bundles of nerve fibers surrounded by connective tissue. The outermost layer of the connective tissue, called the *epineurium,* is dense and includes many collagenous fibers. Each bundle of nerve fibers (fascicle) is, in turn, enclosed in a sleeve of less dense connective tissue called the *perineurium*. The individual nerve fibers are surrounded by a small amount of loose connective tissue called *endoneurium* (figures 11.24 and 11.25).

Blood vessels are usually found in the epineurium and perineurium, and they give rise to a network of capillaries in the endoneurium.

Cranial Nerves

Twelve pairs of **cranial nerves** arise from various locations on the underside of the brain. With the exception of the first pair, which begins within the cerebrum, these nerves originate from the brain stem. They pass from their sites of origin through various foramina of the skull and lead to parts of the head, neck, and trunk.

Although most of the cranial nerves are mixed nerves, some of those associated with special senses,

CHART 11.6	Subdivisions of the nervous system

1. Central nervous system (CNS)
   a. Brain
   b. Spinal cord
2. Peripheral nervous system (PNS)
   a. Cranial nerves arising from the brain
      (1) Somatic fibers connecting to the skin and skeletal muscles
      (2) Autonomic fibers connecting to visceral organs
   b. Spinal nerves arising from the spinal cord
      (1) Somatic fibers connecting to the skin and skeletal muscles
      (2) Autonomic fibers connecting to visceral organs

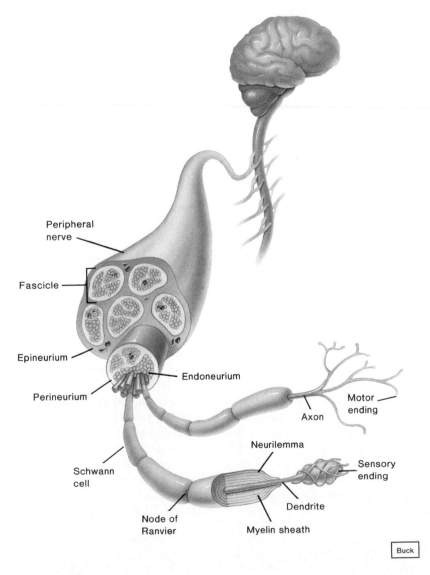

Buck

Figure 11.24 The structure of a peripheral nerve.

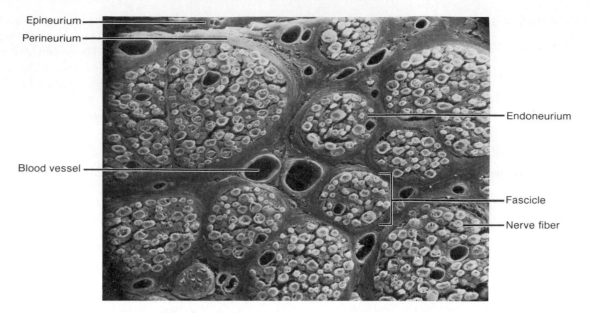

Epineurium

Perineurium

Endoneurium

Blood vessel

Fascicle

Nerve fiber

Figure 11.25 Scanning electron micrograph of a peripheral nerve in cross section (about ×1,000). Note the bundles or fascicles of nerve fibers. (*Tissues and Organs: A Text-Atlas of Scanning Electron Microscopy,* by R. G. Kessel and R. H. Kardon. © 1979 W.H. Freeman and Company.)

such as smell and vision, contain only sensory fibers. Others that are closely involved with the activities of muscles and glands are composed primarily of motor fibers and have only limited sensory functions.

When sensory fibers are present in the cranial nerves, the neuron cell bodies to which the fibers are attached are located outside the brain and are usually in groups called *ganglia* (sing. *ganglion*). On the other hand, the motor neuron cell bodies are typically located within the gray matter of the brain.

The cranial nerves are designated either by numbers or by names. The numbers indicate the order in which the nerves arise from the brain, going from anterior to posterior. The names describe their primary functions or the general distribution of their fibers (figure 11.26).

The first pair of cranial nerves, the **olfactory nerves** (I), are associated with the sense of smell and contain only sensory neurons. These neurons are located in the lining of the upper nasal cavity where they serve as *olfactory receptor cells.* Axons from these receptors pass upward through the cribriform plates of the ethmoid bone and into *olfactory bulbs,* which are extensions of the cerebral cortex, located just beneath the frontal lobes. Sensory impulses travel from the olfactory bulbs along *olfactory tracts* to cerebral centers where they are interpreted. The result of this interpretation is the sensation of smell.

The second pair, the **optic nerves** (II), lead from the eyes to the brain and are associated with the sense of vision. The sensory cell bodies of these nerve fibers occur in ganglion cell layers within the eyes, and their axons pass through the *optic foramina* of the orbits and continue into the visual nerve pathways of the brain (chapter 12).

The third pair, the **oculomotor** (ok″u-lo-mo′tor) **nerves** (III), arise from the midbrain and pass into the orbits of the eyes. One component of each nerve connects to a number of voluntary muscles. These include the muscles that function to raise the eyelids, as well as most of the muscles attached to the eye surfaces that cause them to move.

A second portion of each oculomotor nerve is part of the autonomic nervous system, supplying involuntary muscles inside the eyes. These muscles help adjust the amount of light that enters the eyes and help focus the lenses of the eyes.

Although the fibers of the oculomotor nerves are primarily motor, some sensory fibers are present. These transmit sensory information to the brain concerning the condition of various muscles.

The fourth pair, the **trochlear** (trok′le-ar) **nerves** (IV), are the smallest cranial nerves. They arise from the midbrain and carry motor impulses to a pair of external eye muscles, the *superior oblique muscles,* which are not supplied by the oculomotor nerves. The troch-

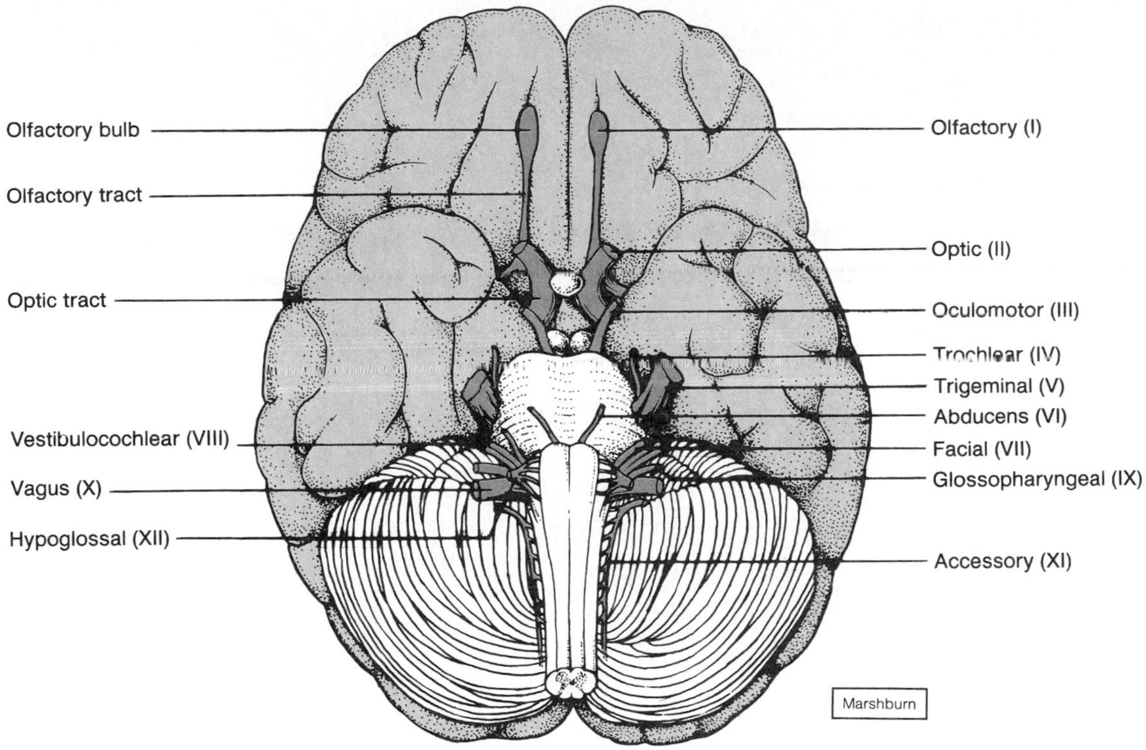
Olfactory bulb
Olfactory tract
Optic tract
Vestibulocochlear (VIII)
Vagus (X)
Hypoglossal (XII)
Olfactory (I)
Optic (II)
Oculomotor (III)
Trochlear (IV)
Trigeminal (V)
Abducens (VI)
Facial (VII)
Glossopharyngeal (IX)
Accessory (XI)
Marshburn

Figure 11.26 Except for the first pair, the cranial nerves arise from the brain stem. They are identified either by numbers indicating their order, or by names describing their function or the general distribution of their fibers.

lear nerves also contain sensory fibers that transmit information about the condition of certain muscles.

The fifth pair, the **trigeminal** (tri-jem′ĭ-nal) **nerves** (V), are the largest of the cranial nerves and arise from the pons. They are mixed nerves, but their sensory portions are more extensive than their motor portions. Each sensory component includes three large branches, called the ophthalmic, maxillary, and mandibular divisions (figure 11.27).

The *ophthalmic division* consists of sensory fibers that bring impulses to the brain from the surface of the eye, the tear gland, and the skin of the anterior scalp, forehead, and upper eyelid. The fibers of the *maxillary division* carry sensory impulses from the upper teeth, upper gum, and upper lip, as well as from the mucous lining of the palate and the skin of the face. The *mandibular division* includes both motor and sensory fibers. The sensory branches transmit impulses from the scalp behind the ears, the skin of the jaw, the lower teeth, the lower gum, and the lower lip. The motor branches supply the muscles of mastication and certain muscles in the floor of the mouth.

A disorder of the trigeminal nerve called *trigeminal neuralgia* (tic douloureux) is characterized by severe recurring pain in the face and forehead on the affected side. If the pain cannot be controlled with drugs, the sensory portion of the nerve is sometimes severed surgically. Although this procedure may relieve the pain, the patient loses other sensations in the parts supplied by the sensory branch. Consequently, persons who have had such surgery must be cautious when eating or drinking hot foods or liquids, because they may not sense burning. They should also inspect their mouths daily for the presence of food particles or damage to their cheeks from biting, which they may not feel.

The sixth pair, the **abducens** (ab-du′senz) **nerves** (VI), are quite small, and originate from the pons near the medulla oblongata. They enter the orbits of the eyes and supply motor impulses to a pair of external eye muscles, the *lateral rectus muscles*. The sensory fibers of these nerves provide the brain with information concerning the condition of certain muscles.

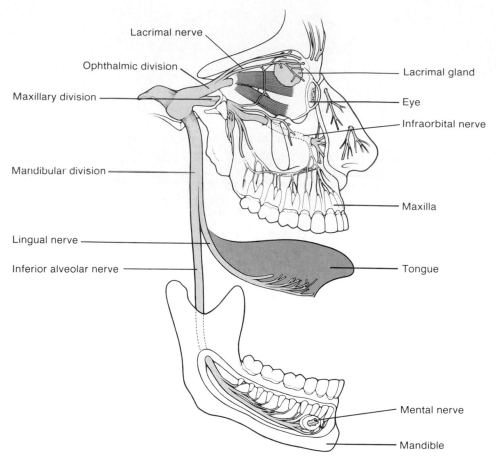

Figure 11.27 Each trigeminal nerve has three large branches that supply various parts of the head and face: the ophthalmic division, the maxillary division, and the mandibular division.

The seventh pair, the **facial** (fa'shal) **nerves** (VII), arise from the lower part of the pons and emerge on the sides of the face. Their sensory branches are associated with taste receptors on the anterior two-thirds of the tongue, and some of their motor fibers transmit impulses to muscles of facial expression. Still other motor fibers of these nerves function in the autonomic nervous system by stimulating secretions from tear glands and certain salivary glands (submandibular and sublingual glands). (See figure 11.28.)

The eighth pair, the **vestibulocochlear** (ves-tib"u-lo-kok'le-ar) **nerves** (VIII, acoustic, or auditory, nerves), are sensory nerves that arise from the medulla oblongata. Each of these nerves has two distinct parts—a vestibular branch and a cochlear branch.

The neuron cell bodies of the *vestibular branch* fibers are located in ganglia near the vestibule and semicircular canals of the inner ear. These structures contain receptors that are sensitive to changes in the position of the head. The impulses they initiate pass into the cerebellum, where they are used in reflexes associated with maintaining equilibrium.

The neuron cell bodies of the *cochlear branch* fibers are located in a ganglion of the cochlea, a part of the inner ear that houses the hearing receptors. Impulses from this branch pass through the medulla oblongata and midbrain on their way to the temporal lobe where they are interpreted.

The ninth pair, the **glossopharyngeal** (glos"o-fah-rin'je-al) **nerves** (IX), are associated with the tongue and pharynx. These nerves arise from the medulla oblongata, and, although they are mixed nerves, their predominant fibers are sensory. These fibers carry impulses from the lining of the pharynx, tonsils, and posterior third of the tongue to the brain. Fibers in the motor component of the glossopharyngeal nerves innervate constrictor muscles in the wall of the pharynx that function in swallowing.

The tenth pair, the **vagus** (va'gus) **nerves** (X), originate in the medulla oblongata and extend downward through the neck into the chest and abdomen. These nerves are mixed, and although they contain both somatic and autonomic branches, the autonomic fibers are the predominant branches.

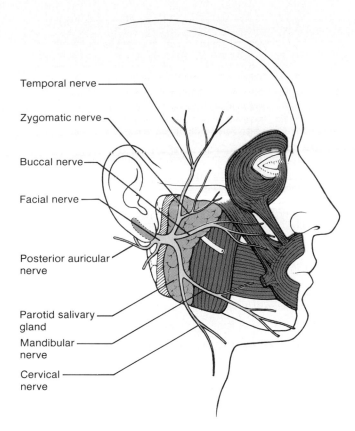

Temporal nerve

Zygomatic nerve

Buccal nerve

Facial nerve

Posterior auricular
nerve

Parotid salivary
gland

Mandibular
nerve

Cervical
nerve

Figure 11.28 The facial nerves are associated with the taste receptors on the tongue and with the muscles of facial expression.

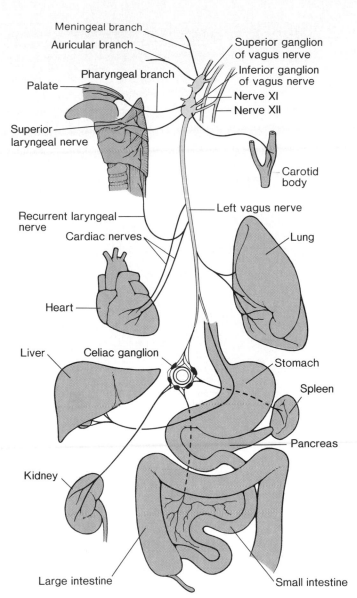

Meningeal branch

Auricular branch

Pharyngeal branch

Palate

Superior ganglion of vagus nerve

Inferior ganglion of vagus nerve

Nerve XI

Nerve XII

Superior laryngeal nerve

Carotid body

Recurrent laryngeal nerve

Left vagus nerve

Cardiac nerves

Lung

Heart

Liver

Celiac ganglion

Stomach

Spleen

Pancreas

Kidney

Large intestine

Small intestine

Figure 11.29 The vagus nerves extend from the medulla oblongata downward into the chest and abdomen to supply many visceral organs.

Among the somatic components of the vagus nerves are motor fibers that carry impulses to muscles of the larynx. These fibers are associated with speech and swallowing reflexes that employ muscles in the soft palate and pharynx. Vagal sensory fibers carry impulses from the linings of the pharynx, larynx, and esophagus, and from the viscera of the thorax and abdomen to the brain.

Autonomic motor fibers of the vagus nerves supply the heart, and a variety of smooth muscles and glands in the visceral organs of the thorax and abdomen. (See figure 11.29.)

The eleventh pair, the **accessory** (ak-ses'o-re) **nerves** (XI, spinal accessory), originate in the medulla oblongata and the spinal cord; thus, they have both cranial and spinal branches.

Each *cranial branch* of an accessory nerve joins a vagus nerve and carries impulses to muscles of the soft palate, pharynx, and larynx. The *spinal branch* descends into the neck and supplies motor fibers to the trapezius and sternocleidomastoid muscles.

The twelfth pair, the **hypoglossal** (hi"po-glos'al) **nerves** (XII), arise from the medulla oblongata and pass into the tongue. They consist primarily of motor fibers that carry impulses to muscles that move the tongue in speaking, chewing, and swallowing.

The functions of the cranial nerves are summarized in chart 11.7.

The consequences of cranial nerve injuries depend on the location and extent of the injuries. For example, if only one member of a nerve pair is damaged, loss of function is limited to the affected side, but if both nerves are injured, losses occur on both sides. Also, if a nerve is severed completely, the functional loss is total; if the cut is incomplete, the loss may be partial.

CHART 11.7   Functions of cranial nerves

Nerve	Type	Function
I Olfactory	Sensory	Sensory fibers transmit impulses associated with the sense of smell.
II Optic	Sensory	Sensory fibers transmit impulses associated with the sense of vision.
III Oculomotor	Primarily motor	Motor fibers transmit impulses to muscles that raise the eyelids, move the eyes, adjust the amount of light entering the eyes, and focus the lenses. Some sensory fibers transmit impulses associated with the condition of muscles.
IV Trochlear	Primarily motor	Motor fibers transmit impulses to muscles that move the eyes. Some sensory fibers transmit impulses associated with the condition of muscles.
V Trigeminal	Mixed	
Ophthalmic division		Sensory fibers transmit impulses from the surface of the eyes, tear glands, scalp, forehead, and upper eyelids.
Maxillary division		Sensory fibers transmit impulses from the upper teeth, upper gum, upper lip, lining of the palate, and skin of the face.
Mandibular division		Sensory fibers transmit impulses from the scalp, skin of the jaw, lower teeth, lower gum, and lower lip. Motor fibers transmit impulses to muscles of mastication and to muscles in the floor of the mouth.
VI Abducens	Primarily motor	Motor fibers transmit impulses to muscles that move the eyes. Some sensory fibers transmit impulses associated with the condition of muscles.
VII Facial	Mixed	Sensory fibers transmit impulses associated with taste receptors of the anterior tongue. Motor fibers transmit impulses to muscles of facial expression, tear glands, and salivary glands.
VIII Vestibulocochlear	Sensory	
Vestibular branch		Sensory fibers transmit impulses associated with the sense of equilibrium.
Cochlear branch		Sensory fibers transmit impulses associated with the sense of hearing.
IX Glossopharyngeal	Mixed	Sensory fibers transmit impulses from the pharynx, tonsils, posterior tongue, and carotid arteries. Motor fibers transmit impulses to salivary glands and to muscles of the pharynx used in swallowing.
X Vagus	Mixed	Somatic motor fibers transmit impulses to muscles associated with speech and swallowing; autonomic motor fibers transmit impulses to the heart and to smooth muscles and glands of visceral organs in the thorax and abdomen. Sensory fibers transmit impulses from the pharynx, larynx, esophagus, and visceral organs of the thorax and abdomen.
XI Accessory	Primarily motor	
Cranial branch		Motor fibers transmit impulses to muscles of the soft palate, pharynx, and larynx.
Spinal branch		Motor fibers transmit impulses to muscles of the neck and back.
XII Hypoglossal	Primarily motor	Motor fibers transmit impulses to muscles that move the tongue.

1. Define peripheral nervous system.
2. Distinguish between somatic and autonomic nerve fibers.
3. Describe the structure of a peripheral nerve.
4. Name the cranial nerves, and list the major functions of each.

Spinal Nerves

Thirty-one pairs of **spinal nerves** originate from the spinal cord. They are mixed nerves, and they provide two-way communication between the spinal cord and parts of the arms, legs, neck, and trunk.

Although the spinal nerves are not named individually, they are grouped according to the level from which they arise, and each nerve is numbered in sequence (figure 11.30). Thus, there are eight pairs of *cervical nerves* (numbered C1 to C8), twelve pairs of *thoracic nerves* (numbered T1 to T12), five pairs of *lumbar nerves* (numbered L1 to L5), five pairs of *sacral nerves* (numbered S1 to S5), and one pair of *coccygeal nerves* (Co).

The nerves arising from the upper part of the spinal cord pass outward almost horizontally, while those from the lower portions of the spinal cord descend at sharp angles. This arrangement is a consequence of growth. In early life, the spinal cord extends the entire length of the vertebral column, but with age, the column grows more rapidly than the cord. Thus, the adult spinal cord ends at the level between the first

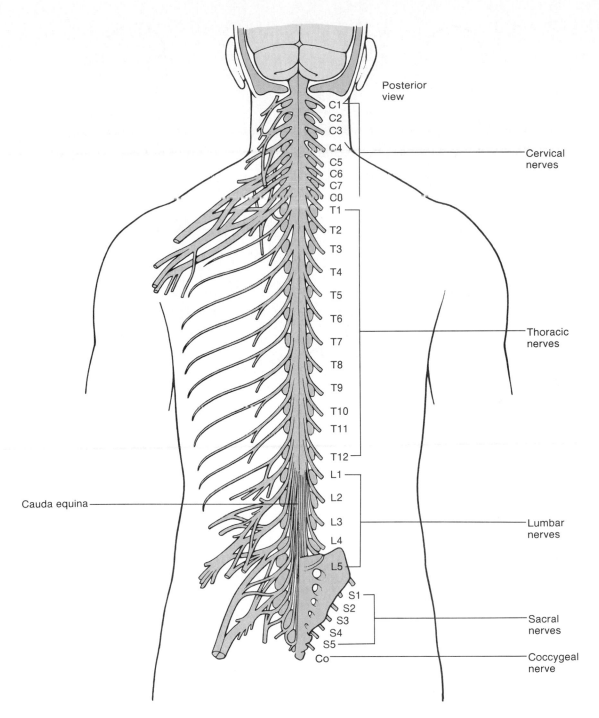

Posterior view

C1
C2
C3
C4
C5
C6
C7
C0

T1
T2
T3
T4
T5
T6
T7
T8
T9
T10
T11
T12

L1
L2
L3
L4
L5

S1
S2
S3
S4
S5
Co

Cervical nerves

Thoracic nerves

Lumbar nerves

Sacral nerves

Coccygeal nerve

Cauda equina

Figure 11.30 The thirty-one pairs of spinal nerves are grouped according to the level from which they arise, and are numbered in sequence.

and second lumbar vertebrae, so the lumbar, sacral, and coccygeal nerves descend to their exits beyond the end of the cord. These descending nerves form a structure called the *cauda equina* (horse's tail). (See figure 11.30.)

Each spinal nerve emerges from the cord by two short branches, or *roots,* which lie within the vertebral column. The **dorsal root** (posterior, or sensory, root) can

be identified by an enlargement called the *dorsal root ganglion.* This ganglion contains the cell bodies of the sensory neurons whose dendrites conduct impulses inward from the peripheral body parts. The axons of these neurons extend through the dorsal root and into the spinal cord, where they form synapses with dendrites of other neurons.

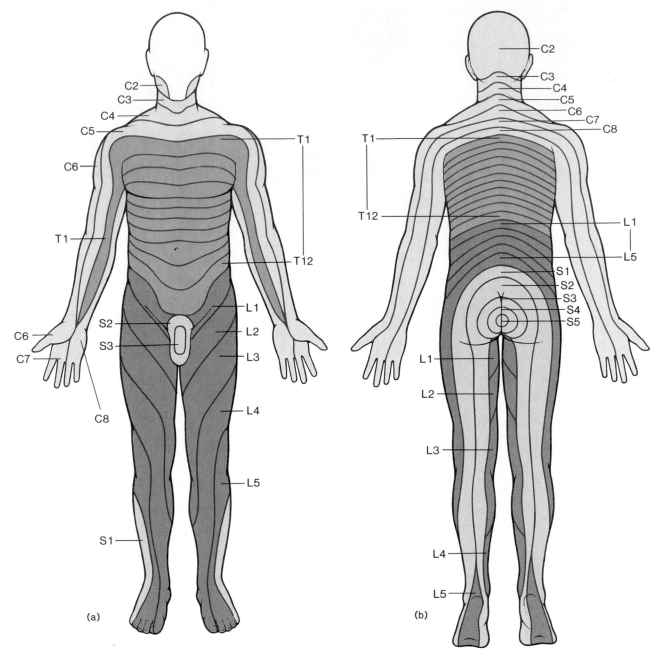

Figure 11.31 (a) Dermatomes on the anterior body surface and (b) on the posterior surface. (Note: Spinal nerve C1 does not supply any skin area.)

An area of skin that is innervated by the sensory nerve fibers of a particular spinal nerve is called a *dermatome*. Dermatomes are highly organized, but they vary considerably in size and shape, as indicated in figure 11.31. A map of the dermatomes is often useful in localizing the sites of injuries to dorsal roots or the spinal cord.

The **ventral root** (anterior, or motor, root) of each spinal nerve consists of axons from the motor neurons whose cell bodies are located within the gray matter of the cord.

A ventral root and a dorsal root unite to form a spinal nerve, which extends outward from the vertebral canal through an *intervertebral foramen*. Just beyond its foramen, each spinal nerve divides into several parts.

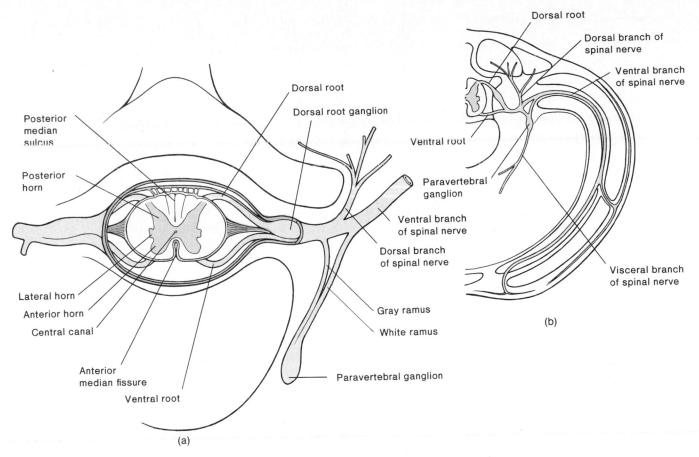

Figure 11.32 (a) Each spinal nerve has a posterior and an anterior branch. (b) The thoracic and lumbar spinal nerves also have a visceral branch.

One of these parts, the small *meningeal branch,* reenters the vertebral canal through the intervertebral foramen and supplies the meninges and blood vessels of the cord, as well as the intervertebral ligaments and the vertebrae.

As figure 11.32 shows, a *posterior branch* (posterior ramus) of each spinal nerve turns posteriorly and innervates the muscles and skin of the back. The main portion of the nerve, the *anterior branch* (anterior ramus), continues forward to supply muscles and skin on the front and sides of the trunk and limbs.

The spinal nerves in the thoracic and lumbar regions have a fourth, or *visceral branch,* which is part of the autonomic nervous system.

Except in the thoracic region, anterior branches of the spinal nerves combine to form complex networks called **plexuses** instead of continuing directly to the peripheral body parts. In a plexus, the fibers of various spinal nerves are sorted and recombined, so that fibers associated with a particular peripheral body part reach it in the same nerve, even though the fibers originate from different spinal nerves (figure 11.33).

Cervical Plexuses

The **cervical plexuses** lie deep in the neck on either side. They are formed by the anterior branches of the first four cervical nerves. Fibers from these plexuses supply the muscles and skin of the neck. In addition, fibers from the third, fourth, and fifth cervical nerves pass into the right and left **phrenic** (fren'ik) **nerves,** which conduct motor impulses to the muscle fibers of the diaphragm.

Brachial Plexuses

The anterior branches of the lower four cervical nerves and the first thoracic nerve give rise to **brachial plexuses.** These networks of nerve fibers are located deep within the shoulders between the neck and the axillae (armpits). The major branches emerging from the brachial plexuses include the following:

1. *Musculocutaneous nerves* supply muscles of the arms on the anterior sides, and the skin of the forearms.
2. *Ulnar nerves* supply muscles of the forearms and hands, and the skin of the hands.

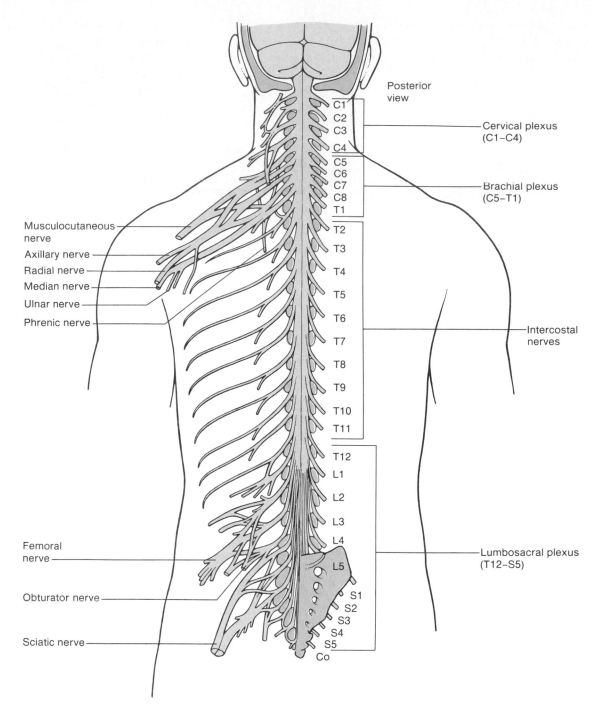

Posterior view

Cervical plexus (C1–C4)

C1
C2
C3
C4
C5
C6
C7
C8
T1

Brachial plexus (C5–T1)

Musculocutaneous nerve
Axillary nerve
Radial nerve
Median nerve
Ulnar nerve
Phrenic nerve

T2
T3
T4
T5
T6
T7
T8
T9
T10
T11

Intercostal nerves

T12
L1
L2
L3
L4
L5

Femoral nerve

Obturator nerve

Sciatic nerve

S1
S2
S3
S4
S5
Co

Lumbosacral plexus (T12–S5)

Figure 11.33 The anterior branches of the spinal nerves in the thoracic region give rise to intercostal nerves. Those in other regions combine to form complex networks called plexuses.

3. *Median nerves* supply muscles of the forearms, and muscles and skin of the hands.
4. *Radial nerves* supply muscles of the arms on the posterior sides, and the skin of the forearms and hands.
5. *Axillary nerves* supply muscles and skin of the upper, lateral, and posterior regions of the arm.

Other nerves associated with the brachial plexus that innervate various skeletal muscles include:

1. The *lateral* and *medial pectoral nerves* supply the pectoralis major and pectoralis minor muscles.
2. The *dorsal scapular nerve* supplies the rhomboideus major and levator scapulae muscles.

3. The *lower subscapular nerve* supplies the subscapularis and teres major muscles.
4. The *thoracodorsal nerve* supplies the latissimus dorsi muscle.
5. The *suprascapular nerve* supplies the supraspinatus and infraspinatus muscles. (See figure 11.34.)

In contact sports such as football, a player's shoulder girdle is sometimes depressed at the same time that the head and neck are pushed away from the depressed side. This may result in a painful stretching of the brachial plexus, which athletes call a "stinger."

Lumbosacral Plexuses

The **lumbosacral** (lum″bo-sa′kral) **plexuses** are formed by the last thoracic nerve and the lumbar, sacral, and coccygeal nerves. These networks of nerve fibers extend from the lumbar region of the back into the pelvic cavity, giving rise to a number of motor and sensory fibers associated with the lower abdominal wall, external genitalia, buttocks, thighs, legs, and feet. The major branches of these plexuses include the following:

1. The *obturator nerves* supply the adductor muscles of the thighs.
2. The *femoral nerves* divide into many branches, supplying motor impulses to muscles of the thighs and legs, and receiving sensory impulses from the skin of the thighs and lower legs.
3. The *sciatic nerves* are the largest and longest nerves in the body. They pass downward into the buttocks and descend into the thighs, where they divide into *tibial* and *common peroneal nerves*. The many branches of these nerves supply muscles and skin in the thighs, legs, and feet.

Other nerves associated with the lumbosacral plexus that innervate various skeletal muscles include:

1. The *pudendal nerve* supplies the muscles of the perineum.
2. The *inferior* and *superior gluteal nerves* supply the gluteal muscles and the tensor fasciae latae muscle. (See figure 11.35.)

The anterior branches of the thoracic spinal nerves do not enter a plexus. Instead, they travel into spaces between the ribs and become **intercostal** (in″ter-kos′tal) **nerves.** These nerves supply motor impulses to the intercostal muscles and the upper abdominal wall muscles. They also receive sensory impulses from the skin of the thorax and abdomen.

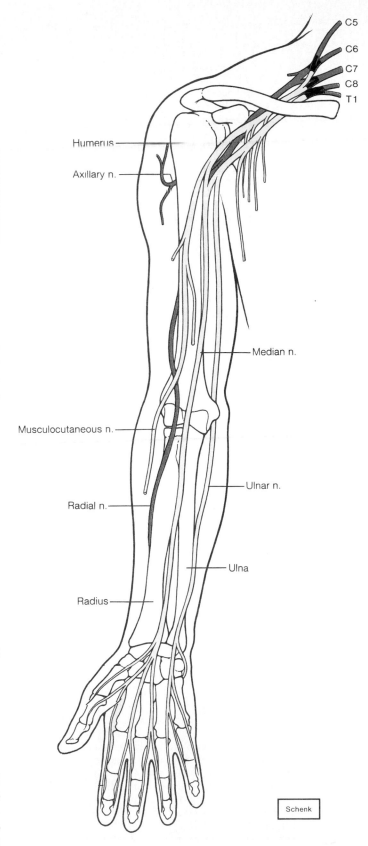

Figure 11.34 Nerves of the brachial plexus.

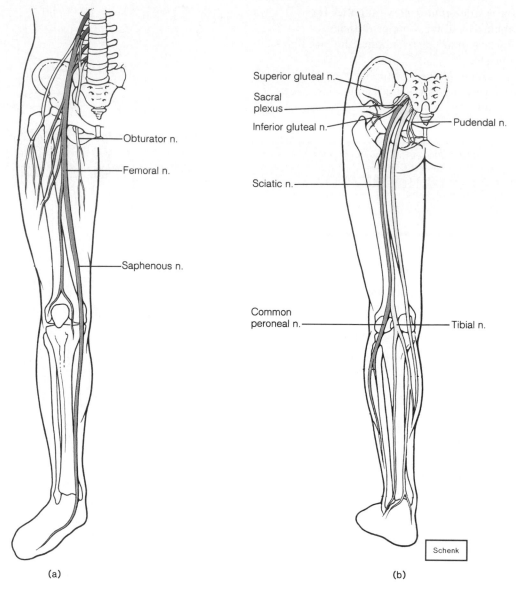

Labels on figure:
(a): Obturator n., Femoral n., Saphenous n.
(b): Superior gluteal n., Sacral plexus, Inferior gluteal n., Pudendal n., Sciatic n., Common peroneal n., Tibial n., Schenk

Figure 11.35 Nerves of the lumbosacral plexus. (*a*) Anterior view; (*b*) posterior view.

In order to control pain or to prepare patients for various clinical procedures, a drug is sometimes injected into the pathway of a sensory nerve to produce a *nerve block*. In some cases, a short-acting local anesthetic substance may achieve the desired effect. In other instances, a substance, such as alcohol or phenol that causes tissue destruction, may be used to produce a more permanent effect. If the injection involves a destructive agent, care must be taken that only sensory fibers are affected, because damage to motor fibers may cause permanent motor impairment.

1. How are spinal nerves grouped?
2. Describe how a spinal nerve joins the spinal cord.
3. Name and locate the major nerve plexuses.

Autonomic Nervous System

The **autonomic nervous system** is the portion of the peripheral nervous system that functions independently (autonomously) and continuously, without conscious effort. This system controls visceral activities by regulating the actions of smooth muscles, cardiac muscles,

Spinal Nerve Injuries

Spinal nerves may be injured in a variety of ways, including birth injuries, dislocations and fractures of vertebrae, stabs, gunshot wounds, and pressure from tumors in surrounding tissues.

The nerves of the cervical plexuses are sometimes compressed by the sudden bending of the neck, called *whiplash,* which may occur during rear end automobile collisions. A victim of such an injury may suffer continuing headache and pain in the neck and skin, which are supplied by the cervical nerves.

If the phrenic nerves associated with the cervical plexuses are severed or damaged by a broken or dislocated vertebra, partial or complete paralysis of the diaphragm may result.

Sometimes people whose occupations require prolonged abduction of the arm, as in painting or typing, develop intermittent or constant pain in the neck, shoulder, or a region of the arm due to excessive pressure on the brachial plexus. This condition, called *thoracic outlet syndrome,* may also be caused by congenital malformations of skeletal parts that compress the plexus during arm and shoulder movements.

As a result of degenerative changes, an intervertebral disk in the lumbar region may be compressed. (See chapter 8.) This may produce a condition called *sciatica,* which is characterized by pain in the lower back and gluteal region that sometimes radiates into the thigh, calf, ankle, and foot. Sciatica is most common in persons forty to fifty years of age. It usually involves compression of spinal nerve roots between L2 and S1, some of which contain fibers of the sciatic nerve.

Although most persons with sciatica recover spontaneously after a period of rest, others require treatment with drugs or surgery. In some cases, a surgical procedure called a *laminectomy* may be performed. In this procedure, the bony lamina of a vertebra is removed so that the surgeon can reach the displaced portion of intervertebral disk and relieve the pressure on the affected spinal nerve.

---

and various glands. It is concerned with regulating heart rate, blood pressure, breathing rate, body temperature, and other visceral activities that aid in maintaining homeostasis. Portions of the autonomic nervous system are also responsive during times of emotional stress, and they prepare the body to meet the demands of strenuous physical activity.

General Characteristics

Autonomic activities are regulated largely by reflexes in which sensory signals originate from receptors within the visceral organs and the skin. These signals are transmitted on afferent nerve fibers to nerve centers within the brain or spinal cord. In response, motor impulses travel out from these centers on efferent nerve fibers within cranial and spinal nerves.

Typically, these efferent fibers lead to ganglia outside the central nervous system. The impulses they carry are integrated within the ganglia and are relayed to various visceral organs (muscles or glands) that respond by contracting, secreting, or being inhibited. The integrative function of the ganglia provides the autonomic nervous system with some degree of independence from the brain and spinal cord, and the visceral efferent nerve fibers associated with these ganglia comprise the autonomic nervous system.

The autonomic nervous system includes two divisions, called the **sympathetic** and **parasympathetic divisions** that act together. For example, some visceral organs are supplied by nerve fibers from each of the divisions. In such cases, impulses on one set of fibers may activate an organ, while impulses on the other set inhibit it. Thus, the divisions may act antagonistically, so that the actions of some visceral organs are regulated by alternately being activated or inhibited.

The functions of the autonomic divisions are mixed; that is, each activates some organs and inhibits others. However, the divisions have important functional differences. The sympathetic division is concerned primarily with preparing the body for energy-expending, stressful, or emergency situations. Conversely, the parasympathetic division is most active under ordinary, restful conditions. It also counterbalances the effects of the sympathetic division, and restores the body to a resting state following a stressful experience. For example, during an emergency, the sympathetic division will cause the heart and breathing rates to increase; following the emergency, the parasympathetic division will decrease these activities.

Autonomic Nerve Fibers

All of the nerve fibers of the autonomic nervous system are efferent, or motor, fibers. Unlike the motor pathways of the somatic nervous system, however, which usually include a single neuron between the brain or spinal cord and a skeletal muscle, those of the auto-

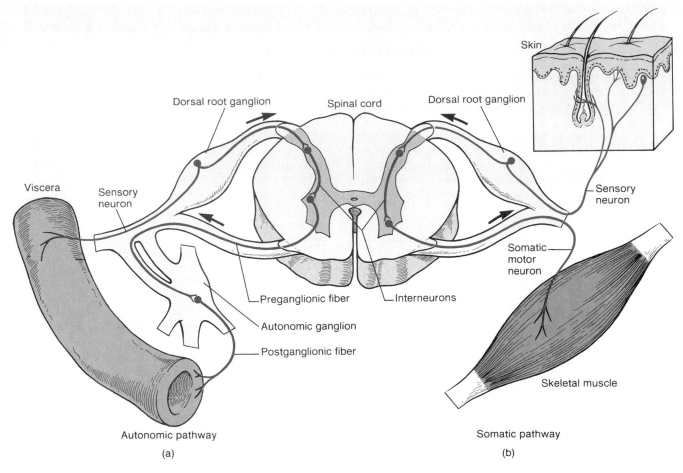

Figure 11.36 (a) Autonomic neuron pathways involve two interneurons between the central nervous system and an effector. (b) Somatic neuron pathways usually have a single neuron between the central nervous system and an effector.

nomic system involve two neurons, as shown in figure 11.36. The cell body of one neuron is located in the brain or spinal cord. Its axon, the **preganglionic fiber,** leaves the CNS and forms a synapse with one or more nerve fibers whose cell bodies are housed within an autonomic ganglion. The axon of such a second neuron is called a **postganglionic fiber,** and it extends to a visceral effector.

Sympathetic Division

Within the sympathetic division (thoracolumbar division), the preganglionic fibers originate from neurons within the lateral horn of the spinal cord. These neurons are found in all of the thoracic segments and in the upper two or three lumbar segments of the cord. Their axons exit through the ventral roots of spinal nerves along with various somatic motor fibers.

After traveling a short distance, preganglionic fibers leave the spinal nerves through branches called *white rami* (sing. *ramus*) and enter sympathetic ganglia. Two groups of such ganglia, called **paravertebral ganglia,** are located in chains along the sides of the ver-

tebral column. These ganglia, together with the fibers that interconnect them, comprise the **sympathetic trunks.** (See figure 11.37.)

The paravertebral ganglia are found just beneath the parietal pleura in the thorax and beneath the parietal peritoneum in the abdomen. (See chapter 1.) Although these ganglia are located some distance from the visceral organs they help control, other sympathetic ganglia are positioned nearer to the viscera. The *collateral ganglia,* for example, are found within the abdomen, closely associated with certain large blood vessels. (See figure 11.38.)

Some of the preganglionic fibers that enter paravertebral ganglia synapse with neurons within these ganglia. Other fibers travel through the ganglia and pass up or down the sympathetic trunk and synapse with neurons in ganglia at higher or lower levels within the chain. Still other fibers pass through to collateral ganglia before they synapse. Typically, a preganglionic axon will synapse with several other neurons within a sympathetic ganglion.

The axons of the second neurons in sympathetic pathways, the postganglionic fibers, extend out from the

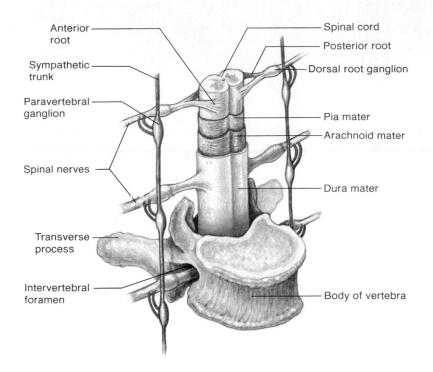

Figure 11.37 A chain of paravertebral ganglia extends along each side of the vertebral column.

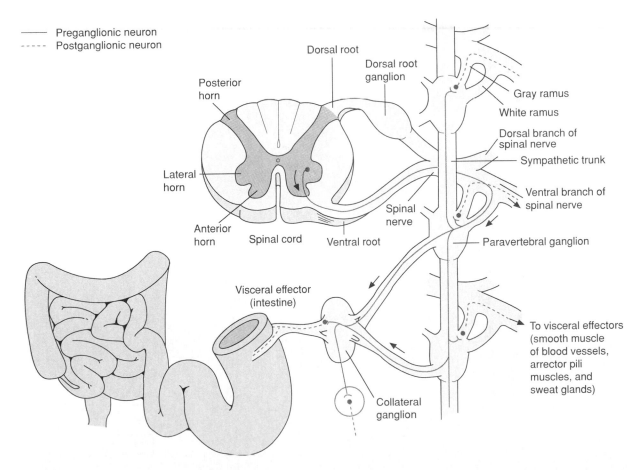

Figure 11.38 Sympathetic fibers leave the spinal cord in the ventral roots of spinal nerves, enter sympathetic ganglia, and synapse with other neurons that extend to visceral effectors.

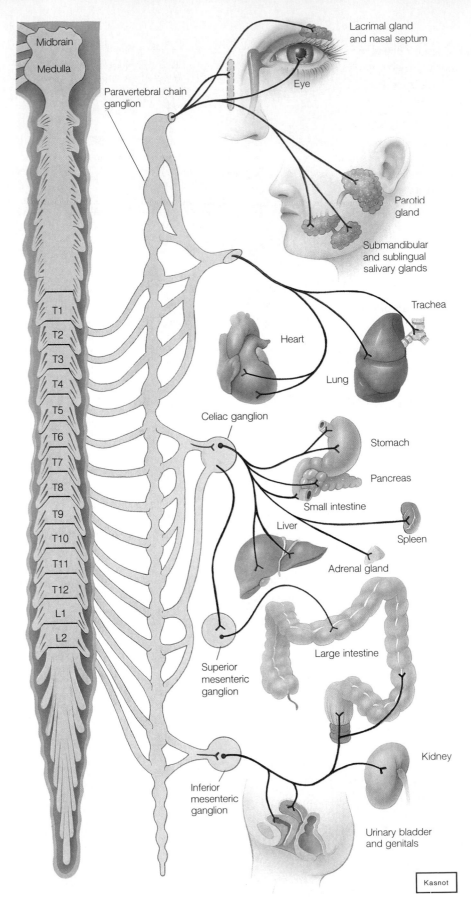

Figure 11.39 The preganglionic fibers of the sympathetic division of the autonomic nervous system arise from the thoracic and lumbar regions of the spinal cord.

sympathetic ganglia to visceral effectors. Those leaving paravertebral ganglia usually pass through branches called **gray rami** and return to a spinal nerve before proceeding to an effector. (See figure 11.38.) These branches appear gray because the postganglionic axons generally are unmyelinated, whereas the preganglionic axons in the white rami are nearly all myelinated. An important exception to the usual arrangement of sympathetic fibers occurs in a set of preganglionic fibers that pass through the sympathetic ganglia and extend out to the medulla of each adrenal gland. These fibers terminate within the glands on special hormone-secreting cells that release norepinephrine or epinephrine when they are stimulated. The functions of the adrenal medulla gland and its hormones are discussed in chapter 13. The sympathetic division is shown in figure 11.39.

Parasympathetic Division

The preganglionic fibers of the parasympathetic division (craniosacral division) arise from neurons in the midbrain, pons, and medulla oblongata of the brain stem and from the sacral region of the spinal cord. From there, they lead outward on cranial or sacral nerves to ganglia located near or within various visceral organs (terminal ganglia). The relatively short postganglionic fibers continue from the ganglia to specific muscles or glands within these visceral organs. As in the case of the sympathetic fibers, the parasympathetic preganglionic axons are usually myelinated, and the parasympathetic postganglionic fibers are unmyelinated.

The parasympathetic preganglionic fibers associated with parts of the head are included in the oculomotor, facial, and glossopharyngeal nerves. Those that innervate organs of the thorax and upper abdomen are parts of the vagus nerves. (The vagus nerves carry about 75% of all parasympathetic fibers.) Preganglionic fibers arising from the sacral region of the spinal cord are found within the branches of the second through the fourth sacral spinal nerves, and they carry impulses to visceral organs within the pelvis. (See figure 11.40.)

1. What is the general function of the autonomic nervous system?
2. How are the divisions of the autonomic system distinguished?
3. Describe a sympathetic nerve pathway and a parasympathetic nerve pathway.

Autonomic Neurotransmitters

The preganglionic fibers of the sympathetic and parasympathetic divisions all secrete *acetylcholine*. The parasympathetic postganglionic fibers also secrete acetylcholine, and for this reason they are called **cholinergic fibers.** Most sympathetic postganglionic fibers, however, secrete norepinephrine (noradrenalin) and are

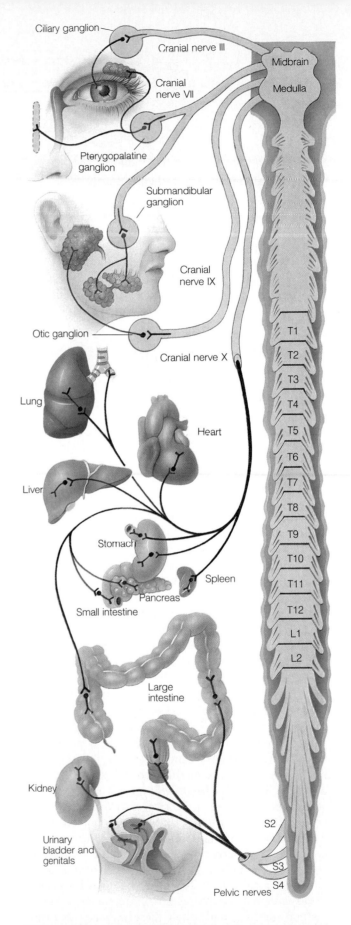

Figure 11.40 The preganglionic fibers of the parasympathetic division of the autonomic nervous system arise from the brain and sacral region of the spinal cord.

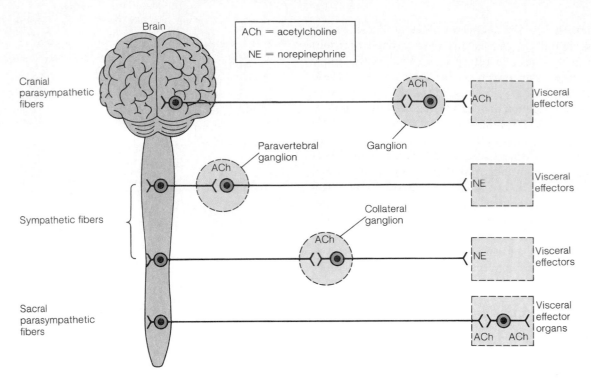

Brain

ACh = acetylcholine

NE = norepinephrine

Cranial parasympathetic fibers

ACh

ACh

Visceral effectors

Ganglion

Paravertebral ganglion

ACh

NE

Visceral effectors

Sympathetic fibers

Collateral ganglion

ACh

NE

Visceral effectors

Sacral parasympathetic fibers

Visceral effector organs

ACh   ACh

Figure 11.41 Most sympathetic fibers are adrenergic and secrete norepinephrine at the ends of the postganglionic fibers; parasympathetic fibers are cholinergic and secrete acetylcholine at the ends of the postganglionic fibers.

called **adrenergic fibers** (figure 11.41). Exceptions to this include the sympathetic postganglionic fibers leading to sweat glands and to blood vessels in muscles; these fibers secrete acetylcholine, and therefore, are cholinergic.

The different postganglionic neurotransmitters (mediators) are responsible for the different effects that the sympathetic and parasympathetic divisions have on visceral organs.

Although each division can activate some effectors and inhibit others, most visceral organs are controlled primarily by one division. In other words, the divisions usually are not actively antagonistic. For example, the diameter of most blood vessels, which lack parasympathetic innervation, is regulated by the sympathetic division. Smooth muscles in the walls of these vessels are continuously stimulated by sympathetic impulses; thus, they are maintained in a state of partial contraction called *sympathetic tone.* The diameter of such a blood vessel can be increased (dilated) by decreasing the sympathetic stimulation, which allows the muscular wall to relax. Conversely, the vessel can be constricted by increasing the sympathetic stimulation.

Similarly, the parasympathetic division is dominant in controlling movement in the digestive system. Parasympathetic impulses stimulate stomach and intestinal motility, and when these impulses decrease, movement is reduced.

The effects of adrenergic and cholinergic fibers on various visceral effectors are summarized in chart 11.8.

Actions of Autonomic Neurotransmitters

As in the case of stimulation occurring at neuromuscular junctions (chapter 9) and at synapses (chapter 10), the actions of autonomic neurotransmitters result from the combining of the neurotransmitters with receptors. These receptors are proteins located in the membranes of effector cells, and when a neurotransmitter combines with a receptor, the membrane changes. For example, the membrane's permeability to certain ions may increase, and in smooth muscle cells, an action potential followed by muscular contraction may result. Similarly, a gland cell may respond to a change in its membrane by secreting a special product.

Acetylcholine is known to combine with two types of cholinergic receptors, called muscarinic receptors and nicotinic receptors. (Note: These receptor names come from *muscarine,* a toxin from a fungus that can activate muscarinic receptors, and *nicotine,* the toxin of tobacco that can activate nicotinic receptors.) The *muscarinic receptors* are located in the membranes of effector cells at the ends of all postganglionic parasympathetic nerve fibers and at the ends of the cholinergic sympathetic fibers. Responses from these receptors are

CHART 11.8 Effects of autonomic stimulation on various effectors

Effector location	Response to sympathetic stimulation	Response to parasympathetic stimulation
Integumentary system		
Apocrine glands	Increased secretion	No action
Eccrine glands	Increased secretion (cholinergic effect)	No action
Special senses		
Iris of eye	Dilation	Constriction
Tear gland	Slightly increased secretion	Greatly increased secretion
Endocrine system		
Adrenal cortex	Increased secretion	No action
Adrenal medulla	Increased secretion	No action
Digestive system		
Muscle of gallbladder wall	Relaxation	Contraction
Muscle of intestinal wall	Decreased peristaltic action	Increased peristaltic action
Muscle of internal anal sphincter	Contraction	Relaxation
Pancreatic glands	Reduced secretion	Greatly increased secretion
Salivary glands	Reduced secretion	Greatly increased secretion
Respiratory system		
Muscles in walls of bronchioles	Dilation	Constriction
Cardiovascular system		
Blood vessels supplying muscles	Constriction (alpha adrenergic) Dilation (beta adrenergic) Dilation (cholinergic)	No action
Blood vessels supplying skin	Constricted	No action
Blood vessels supplying heart (coronary arteries)	Dilation (beta adrenergic) Constriction (alpha adrenergic)	Dilation
Muscles in wall of heart	Increased contraction rate	Decreased contraction rate
Urinary system		
Muscle of bladder wall	Relaxation	Contraction
Muscle of internal urethral sphincter	Contraction	Relaxation
Reproductive systems		
Blood vessels to clitoris and penis	No action	Dilation leading to erection of clitoris and penis
Muscles associated with internal reproductive organs	Male ejaculation, female orgasm	

excitatory and occur relatively slowly. The *nicotinic receptors* occur in the synapses between the preganglionic and postganglionic neurons of the parasympathetic and sympathetic pathways. They produce rapid, excitatory responses (figure 11.42). (Receptors at neuromuscular junctions of skeletal muscles are also nicotinic.)

Both norepinephrine and epinephrine from the adrenal glands combine with adrenergic receptors of effector cells. The two major types of these receptors are called alpha adrenergic and beta adrenergic receptors. When norepinephrine is present, it stimulates the effector cells by combining mainly with the *alpha receptors,* although it has only a slight effect on the *beta receptors.* However, epinephrine can combine with either type of receptor. Consequently, the way each of these adrenergic substances influences effector cells depends upon the relative numbers of alpha and beta receptors present in the cell membranes. For example, an effector cell that contains both types of receptors will be affected by both substances, but one with only

beta receptors will be influenced mainly by epinephrine (figure 11.43).

The actions resulting from adrenergic receptors are sometimes excitatory and sometimes inhibitory. Therefore, when norepinephrine combines with the alpha receptors of smooth muscles, the action is usually excitatory and the cells contract, but when epinephrine combines with the beta receptors of these cells, the action is inhibitory and the cells relax.

The acetylcholine released by cholinergic fibers is rapidly decomposed by the enzyme *cholinesterase.* (This decomposition also occurs in the neuromuscular junctions of skeletal muscle.) Thus, acetylcholine usually produces an effect for only a fraction of a second. Much of the norepinephrine released from adrenergic fibers, however, is taken back into the nerve endings by an active transport mechanism. Norepinephrine can be inactivated by the enzyme *monoamine oxidase* found in mitochondria. This may take a few seconds, and during that time some molecules may diffuse into nearby tissues and be decomposed by other enzymes.

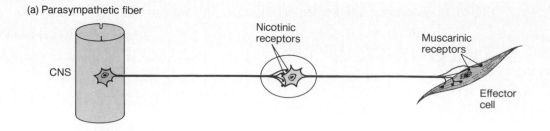

(a) Parasympathetic fiber

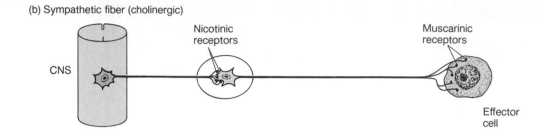

(b) Sympathetic fiber (cholinergic)

(c) Somatic motor fiber

Figure 11.42 (*a* and *b*) Muscarinic receptors occur in the membranes of effector cells at the ends of cholinergic fibers. (*c*) Nicotinic receptors are found in the membranes of skeletal muscle fibers.

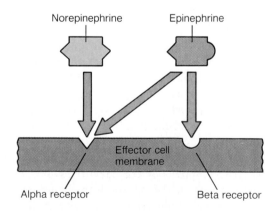

Figure 11.43 Norepinephrine combines mainly with alpha adrenergic receptors, while epinephrine combines with both alpha and beta adrenergic receptors.

On the other hand, some norepinephrine molecules may enter the blood and remain active until they diffuse into tissues containing the inactivating enzymes. For these reasons, norepinephrine is likely to produce a more prolonged effect than acetylcholine. In fact, when the adrenal medulla releases norepinephrine and epinephrine into the blood in response to sympathetic stimulation, these substances may trigger sympathetic responses in organs throughout the body that last up to thirty seconds.

Control of Autonomic Activity

Although the autonomic nervous system has some degree of independence resulting from the integration of impulses within its ganglia, it is controlled largely by the brain and spinal cord. For example, as discussed

Biofeedback

It was once believed that the activities of the autonomic nervous system were beyond conscious control. It is now known that some people can learn to alter their visceral responses consciously. To accomplish this, a person needs to know the state of the visceral action to be controlled. For example, a person is normally unaware of blood pressure, but electronic equipment can measure blood pressure, and continuously feed back this information to the person by means of a signal. Then the person can consciously try to produce a visceral response that will alter the blood pressure in a desired way.

This procedure, called **biofeedback,** has been used successfully to control heart rates and blood pressures, and to alter the diameters of blood vessels and the amount of blood flowing into certain regions of the body. It has also been used to control migraine headaches, which are associated with changes in the muscular tone of certain blood vessels, and to control epileptic seizures.

previously, there are control centers in the medulla oblongata for cardiac, vasomotor, and respiratory activities. These reflex centers receive sensory impulses from visceral organs by means of vagus nerve fibers, and they use autonomic nerve pathways to stimulate motor responses in various muscles and glands. Thus, they exert control over the autonomic nervous system. Similarly, the hypothalamus helps regulate body temperature, hunger, thirst, and water and electrolyte balance by employing and controlling autonomic pathways.

Still higher levels within the brain, including the limbic system and the cerebral cortex, control the autonomic nervous system when a person is stressed emotionally. These parts use the autonomic pathways to regulate a person's emotional expression and behavior.

The regulation of particular visceral organs is discussed in subsequent chapters that deal with individual organs and organ systems.

Many drugs influence autonomic functions. Some, like ephedrine, enhance sympathetic effects by stimulating the release of norepinephrine from postganglionic sympathetic nerve endings. Others, like reserpine, inhibit sympathetic activity by preventing the synthesis of norepinephrine. Another group of drugs, which includes pilocarpine, produce parasympathetic effects, and some, like atropine, block the action of acetylcholine in visceral effectors.

---

1. Distinguish between cholinergic and adrenergic fibers.
2. Explain how the fibers of one autonomic division can control the actions of a particular organ.
3. What neurotransmitter substances are used in the autonomic nervous system?
4. Describe two types of cholinergic receptors and two types of adrenergic receptors.

Clinical Terms Related to the Nervous System

analgesia (an″al-je′ze-ah) Loss or reduction in the ability to sense pain, but without loss of consciousness.

analgesic (an″al-je′sik) A pain-relieving drug.

anesthesia (an″es-the′ze-ah) A loss of feeling.

aphasia (ah-fa′ze-ah) A disturbance or loss in the ability to use words or to understand them, usually due to damage to cerebral association areas.

apraxia (ah-prak′se-ah) An impairment in a person's ability to make correct use of objects.

ataxia (ah-tak′se-ah) A partial or complete inability to coordinate voluntary movements.

cerebral palsy (ser′ĕ-bral pawl′ze) A condition characterized by partial paralysis and lack of muscular coordination.

coma (ko′mah) An unconscious condition in which a person does not respond to stimulation.

cordotomy (kor-dot′o-me) A surgical procedure in which a nerve tract within the spinal cord is severed, usually to relieve intractable pain.

craniotomy (kra″ne-ot′o-me) A surgical procedure in which part of the skull is opened.

electroencephalogram (EEG) (e-lek″tro-en-sef ′ah-lo-gram″) A recording of the electrical activity of the brain.

encephalitis (en″sef-ah-li′tis) An inflammation of the brain and meninges characterized by drowsiness and apathy.

epilepsy (ep′ĭ-lep″se) A disorder of the central nervous system that is characterized by temporary disturbances in normal brain impulses; it may be accompanied by convulsive seizures and loss of consciousness.

hemiplegia (hem″ĭ-ple′je-ah) Paralysis on one side of the body and the limbs on that side.

Huntington's chorea (hunt'ing-tunz ko-re'ah) A rare hereditary disorder of the brain characterized by involuntary convulsive movements and mental deterioration.

laminectomy (lam''i-nek'to-me) Surgical removal of the posterior arch of a vertebra, usually to relieve the symptoms of a ruptured intervertebral disk that is pressing on a spinal nerve.

monoplegia (mon''o-ple'je-ah) Paralysis of a single limb.

multiple sclerosis (mul'ti-pl skle-ro'sis) A disease of the central nervous system characterized by loss of myelin and the appearance of scarlike patches throughout either the brain or the spinal cord, or throughout both.

neuralgia (nu-ral'je-ah) A sharp, recurring pain associated with a nerve, usually caused by inflammation or injury.

neuritis (nu-ri'tis) An inflammation of a nerve.

paraplegia (par''ah-ple'je-ah) Paralysis of both legs.

quadriplegia (kwod''ri-ple'je-ah) Paralysis of all four limbs.

vagotomy (va-got'o-me) Surgical severing of a vagus nerve.

Chapter Summary

Introduction (page 362)

The brain and spinal cord are surrounded by bone and protective membranes called meninges.

Meninges (page 362)

1. The meninges consist of a dura mater, arachnoid mater, and pia mater.
2. Cerebrospinal fluid occupies the space between the arachnoid and pia maters.

Spinal Cord (page 364)

The spinal cord is a nerve column that extends from the brain into the vertebral canal. It terminates at the level between the first and second lumbar vertebrae.

1. Structure of the spinal cord
   a. The spinal cord is composed of thirty-one segments, each of which gives rise to a pair of spinal nerves.
   b. It is characterized by a cervical enlargement, a lumbar enlargement, and two deep longitudinal grooves that divide it into right and left halves.
   c. It has a central core of gray matter that is surrounded by white matter.
   d. The white matter is composed of bundles of myelinated nerve fibers.
2. Functions of the spinal cord
   a. The cord provides a two-way communication system between the brain and body parts outside the nervous system.
   b. Ascending tracts carry sensory impulses to the brain; descending tracts carry motor impulses to muscles and glands.
   c. Many of the fibers in the ascending and descending tracts cross over in the spinal cord or brain.

Brain (page 370)

The brain is the largest and most complex part of the nervous system. It contains nerve centers that are associated with sensations. The brain issues motor commands and carries on higher mental functions.

1. Development of the brain
   a. Brain structure reflects the way it forms.
   b. The brain develops from a tubular part with three cavities—the forebrain, midbrain, and hindbrain.
   c. The cavities persist as ventricles, and the walls give rise to structural and functional regions.
2. Structure of the cerebrum
   a. The cerebrum consists of two cerebral hemispheres connected by the corpus callosum.
   b. Its surface is marked by ridges and grooves; sulci divide each hemisphere into lobes.
   c. The cerebral cortex is a thin layer of gray matter near the surface.
   d. White matter consists of myelinated nerve fibers that interconnect neurons within the nervous system and communicate with other body parts.
3. Functions of the cerebrum
   a. The cerebrum is concerned with higher brain functions, such as thought, reasoning, interpretation of sensory impulses, control of voluntary muscles, and storage of memory.
   b. The cerebral cortex can be subdivided into sensory, motor, and association areas.
   c. The primary motor regions lie in the frontal lobes near the central sulcus and are aided by other areas of the frontal lobes that control special motor functions.
   d. Areas responsible for interpreting sensory impulses from the skin are located in the parietal lobes near the central sulcus; other specialized sensory areas occur in the temporal and occipital lobes.
   e. Association areas analyze and interpret sensory impulses and are involved with memory, reasoning, verbalizing, judgment, and emotional feelings.
   f. In most persons, one cerebral hemisphere is dominant for certain intellectual functions.
   g. Memory is established in phases and is stored diffusely in both hemispheres.
4. Basal ganglia
   a. Basal ganglia are masses of gray matter located deep within the cerebral hemispheres.
   b. They function as relay stations for motor impulses originating in the cerebral cortex, and they aid in the control of motor activities.

5. Ventricles and cerebrospinal fluid
   a. Ventricles are interconnected cavities within the cerebral hemispheres and brain stem.
   b. These spaces are filled with cerebrospinal fluid.
   c. Cerebrospinal fluid is secreted by choroid plexuses in the walls of the ventricles.
   d. Ependymal cells of the choroid plexus regulate the composition of cerebrospinal fluid.
   e. Cerebrospinal fluid circulates through the ventricles and is reabsorbed into the blood of the dural sinuses.
6. Brain stem
   a. The brain stem extends from the base of the cerebrum to the spinal cord.
   b. The brain stem consists of the diencephalon, midbrain, pons, and medulla oblongata.
   c. The diencephalon contains the thalamus, which serves as a central relay station for incoming sensory impulses, and the hypothalamus, which plays important roles in maintaining homeostasis.
   d. The limbic system functions to produce emotional feelings and to modify behavior.
   e. The midbrain contains reflex centers associated with eye and head movements.
   f. The pons transmits impulses between the cerebrum and other parts of the nervous system, and contains centers that help regulate the rate and depth of breathing.
   g. The medulla oblongata transmits all ascending and descending impulses, and contains several vital and nonvital reflex centers.
   h. The reticular formation filters incoming sensory impulses, arousing the cerebral cortex into wakefulness whenever significant impulses are received.
   i. Normal sleep results from decreasing activity of the reticular formation, and paradoxical sleep occurs when activating impulses are received by some parts of the brain, but not by others.
7. Cerebellum
   a. The cerebellum consists of two hemispheres connected by the vermis.
   b. It is composed of white matter surrounded by a thin cortex of gray matter.
   c. The cerebellum functions primarily as a reflex center in the coordination of skeletal muscle movements and the maintenance of equilibrium.

Peripheral Nervous System (page 388)

The peripheral nervous system consists of cranial and spinal nerves that branch out from the brain and spinal cord to all body parts. It can be subdivided into somatic and autonomic portions.

1. Structure of peripheral nerves
   a. A nerve consists of a bundle of nerve fibers surrounded by connective tissues.
   b. The connective tissues form an outer epineurium, a perineurium enclosing bundles of nerve fibers, and an endoneurium surrounding each fiber.

2. Cranial nerves
   a. Twelve pairs of cranial nerves connect the brain to parts in the head, neck, and trunk.
   b. Although most cranial nerves are mixed, some are pure sensory, and others are primarily motor.
   c. The names of cranial nerves indicate their primary functions or the general distributions of their fibers.
   d. Some cranial nerve fibers are somatic and others are autonomic.
3. Spinal nerves
   a. Thirty-one pairs of spinal nerves originate from the spinal cord.
   b. These mixed nerves provide a two-way communication system between the spinal cord and parts in the arms, legs, neck, and trunk.
   c. Spinal nerves are grouped according to the levels from which they arise, and they are numbered in sequence.
   d. Each nerve emerges by a dorsal and a ventral root.
      (1) A dorsal root contains sensory fibers and is characterized by the presence of a dorsal root ganglion.
      (2) A ventral root contains motor fibers.
   e. Just beyond its foramen, each spinal nerve divides into several branches.
   f. Most spinal nerves combine to form plexuses in which nerve fibers are sorted and recombined so that those fibers associated with a particular part reach it together.

Autonomic Nervous System (page 400)

The autonomic nervous system consists of the portions of the nervous system that function without conscious effort. It is concerned primarily with the regulation of visceral activities that aid in maintaining homeostasis.

1. General characteristics
   a. Autonomic functions operate as reflex actions controlled from centers in the hypothalamus, brain stem, and spinal cord.
   b. Autonomic nerve fibers are associated with ganglia in which impulses are integrated before passing out to effectors.
   c. The integrative function of the ganglia provides a degree of independence from the central nervous system.
   d. The autonomic nervous system consists of the visceral efferent fibers associated with these ganglia.
   e. The autonomic nervous system is subdivided into two divisions—sympathetic and parasympathetic.
   f. The sympathetic division prepares the body for stressful and emergency conditions.
   g. The parasympathetic division is most active under ordinary conditions.

2. Autonomic nerve fibers
   a. The autonomic fibers are efferent, or motor.
   b. Sympathetic fibers leave the spinal cord and synapse in ganglia.
      (1) Preganglionic fibers pass through white rami to reach paravertebral ganglia.
      (2) Paravertebral ganglia and interconnecting fibers comprise the sympathetic trunks.
      (3) Preganglionic fibers synapse within paravertebral or collateral ganglia.
      (4) Postganglionic fibers usually pass through gray rami to reach spinal nerves before passing to effectors.
      (5) A special set of sympathetic preganglionic fibers passes through ganglia and extends to the adrenal medulla.
   c. Parasympathetic fibers begin in the brain stem and sacral region of the spinal cord, and synapse in ganglia near various visceral organs.
3. Autonomic neurotransmitters
   a. Sympathetic and parasympathetic preganglionic fibers secrete acetylcholine.
   b. Most sympathetic postganglionic fibers secrete norepinephrine and are adrenergic; postganglionic parasympathetic fibers secrete acetylcholine and are cholinergic.
   c. The different effects of the autonomic divisions are due to different transmitter substances released by the postganglionic fibers.
   d. Most visceral organs are controlled mainly by one division.
4. Actions of autonomic neurotransmitters
   a. Neurotransmitters combine with receptors and cause changes in cell membranes.
   b. There are two types of cholinergic receptors and two types of adrenergic receptors.
   c. How cells respond to neurotransmitters depends upon the number and type of receptors present in their membranes.
   d. Acetylcholine acts very briefly; norepinephrine and epinephrine may have more prolonged effects.
5. Control of autonomic activity
   a. The autonomic nervous system is controlled largely by the central nervous system.
   b. The medulla oblongata employs autonomic fibers to regulate cardiac, vasomotor, and respiratory activities.
   c. The hypothalamus employs autonomic fibers in the regulation of various visceral functions.
   d. The limbic system and cerebral cortex control emotional responses through the autonomic nervous system.

Clinical Application of Knowledge

1. If a physician plans to obtain a sample of spinal fluid from a patient, what anatomical site can be safely used, and how should the patient be positioned to facilitate this procedure?
2. What functional losses would you expect to observe in a patient who has suffered injury to the right occipital lobe of the cerebral cortex? To the right temporal lobe?
3. The Brown-Sequard syndrome is due to an injury on one side of the spinal cord. It is characterized by paralysis below the injury and on the same side as the injury, and by loss of sensations of temperature and pain on the opposite side. How would you explain these symptoms?
4. Substances used by intravenous drug abusers are sometimes obtained in tablet form and must be crushed and dissolved before they can be injected. Such tablets may contain fillers, such as talc or cornstarch, that were added during the manufacturing process. Either talc or cornstarch may cause obstructions in tiny blood vessels of the cerebrum. What problems might these obstructions create?
5. In planning treatment for a patient who has had a cerebrovascular accident, why would it be important to know whether the CVA involved a ruptured blood vessel or an obstructed blood vessel?
6. What symptoms might be produced by the sympathetic division of the autonomic nervous system in a patient who is experiencing stress?

Review Activities

1. Name the layers of the meninges, and explain their functions.
2. Describe the location of cerebrospinal fluid within the meninges.
3. Describe the structure of the spinal cord.
4. Name the major ascending and descending tracts of the spinal cord, and list the functions of each.
5. Explain the consequences of nerve fibers crossing over.
6. Describe how the brain develops.
7. Describe the structure of the cerebrum.
8. Define *cerebral cortex.*
9. Describe the location and function of the primary motor areas of the cortex.
10. Describe the location and function of Broca's area.
11. Describe the location and function of the sensory areas of the cortex.

12. Explain the function of the association areas of the lobes of the cerebrum.
13. Define *hemisphere dominance.*
14. Explain the function of the corpus callosum.
15. Distinguish between short-term, recent, and long-term memory.
16. Describe the location and function of the basal ganglia.
17. Describe the location of the ventricles of the brain.
18. Explain how cerebrospinal fluid is produced and how it functions.
19. Name the parts of the brain stem, and describe the general functions of each.
20. Define the limbic system, and explain its functions.
21. Name the parts of the midbrain, and describe the general functions of each.
22. Describe the pons and its functions.
23. Describe the medulla oblongata and its functions.
24. Describe the location and function of the reticular formation.
25. Distinguish between normal and paradoxical sleep.
26. Describe the functions of the cerebellum.
27. Distinguish between the somatic and autonomic nervous systems.
28. Describe the structure of a peripheral nerve.
29. Name, locate, and describe the major functions of each pair of cranial nerves.
30. Explain how the spinal nerves are grouped and numbered.
31. Define *cauda equina.*
32. Describe the structure of a spinal nerve.
33. Define *plexus,* and locate the major plexuses of the spinal nerves.
34. Distinguish between the sympathetic and parasympathetic divisions of the autonomic nervous system.
35. Explain how autonomic ganglia provide a degree of independence from the central nervous system.
36. Distinguish between a preganglionic fiber and a postganglionic fiber.
37. Define *paravertebral ganglion.*
38. Trace a sympathetic nerve pathway through a ganglion to an effector.
39. Explain why the effects of the sympathetic and parasympathetic autonomic divisions differ.
40. Distinguish between cholinergic and adrenergic nerve fibers.
41. Define *sympathetic tone.*
42. Explain how autonomic transmitters influence the actions of effector cells.
43. Distinguish between alpha adrenergic and beta adrenergic receptors.
44. Describe three examples in which the central nervous system employs autonomic nerve pathways.

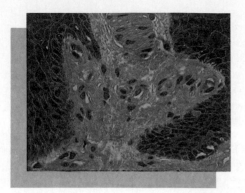

12

Somatic and Special Senses

*B*efore the nervous system can act to control body functions, it must detect what is occurring inside and outside the body. This information is gathered by sensory receptors that are sensitive to changes occurring in their surroundings.

Although sensory receptors vary greatly in their individual characteristics, they can be grouped into two major categories. The receptors of one group are widely distributed throughout the skin and deeper tissues, and generally have simple forms. These receptors are associated with the *somatic senses* of touch, pressure, temperature, and pain. Receptors of the second group are parts of complex, specialized sensory organs that are responsible for the *special senses* of smell, taste, hearing, equilibrium, and vision ■

Chapter Objectives	Key Terms	Aids to Understanding Words

Chapter Objectives

After you have studied this chapter, you should be able to:

1. Name five kinds of receptors and explain the function of each kind.

2. Explain how receptors stimulate sensory impulses.

3. Explain how a sensation is produced.

4. Distinguish between somatic and special senses.

5. Describe the receptors associated with the senses of touch and pressure, temperature, and pain.

6. Describe how the sense of pain is produced.

7. Explain the importance of stretch receptors in muscles and tendons.

8. Explain the relationship between the senses of smell and taste.

9. Name the parts of the ear and explain the function of each part.

10. Distinguish between static and dynamic equilibrium.

11. Name the parts of the eye and explain the function of each part.

12. Explain how light is refracted by the eye.

13. Explain how depth and distance are perceived.

14. Describe the visual nerve pathway.

15. Complete the review activities at the end of this chapter. Note that the items are worded in the form of specific learning objectives. You may want to refer to them before reading the chapter.

Key Terms

accommodation (ah-kom″o-da′shun)

ampulla (am-pul′lah)

auditory (aw′di-to″re)

chemoreceptor (ke″mo-re-sep′tor)

cochlea (kok′le-ah)

cornea (kor′ne-ah)

dynamic equilibrium (di-nam′ik e″kwi-lib′re-um)

labyrinth (lab′i-rinth)

macula (mak′u-lah)

mechanoreceptor (mek″ah-no-re-sep′tor)

olfactory (ol-fak′to-re)

optic (op′tik)

photoreceptor (fo″to-re-sep′tor)

projection (pro-jek′shun)

proprioceptor (pro″pre-o-sep′tor)

referred pain (re-furd′ pān)

refraction (re-frak′shun)

retina (ret′i-nah)

rhodopsin (ro-dop′sin)

sclera (skle′rah)

sensory adaptation (sen′so-re ad″ap-ta′shun)

static equilibrium (stat′ik e″kwi-lib′re-um)

thermoreceptor (ther″mo-re-sep′tor)

Aids to Understanding Words

aud-, to hear: auditory—pertaining to hearing.

choroid, skinlike: choroid coat—the middle, vascular layer of the eye.

cochlea, snail: cochlea—the coiled tube within the inner ear.

corn-, horn: cornea—the transparent outer layer in the anterior portion of the eye.

iris, rainbow: iris—the colored, muscular part of the eye.

labyrinth, maze: labyrinth—a complex system of interconnecting chambers and tubes of the inner ear.

lacri-, tears: lacrimal gland—a tear gland.

lut-, yellow: macula lutea—a yellowish spot on the retina.

macula, spot: macula lutea—a yellowish spot on the retina.

oculi-, eye: orbicularis oculi—a muscle associated with the eyelid.

olfact-, to smell: olfactory—pertaining to the sense of smell.

palpebra, eyelid: levator palpebrae superioris—a muscle associated with the eyelid.

scler-, hard: sclera—the tough, outer protective layer of the eye.

therm-, heat: thermoreceptor—a receptor sensitive to changes in temperature.

tympan-, drum: tympanic membrane—the eardrum.

vitre-, glass: vitreous humor—a clear, jellylike substance within the eye.

415

As changes occur within the body and in its surroundings, *sensory receptors* are stimulated, and they, in turn, trigger nerve impulses. These impulses travel on sensory pathways into the central nervous system to be processed and interpreted. As a result, a person experiences (perceives) a particular feeling or sensation (perception).

Receptors and Sensations

Although there are many kinds of sensory receptors, they have certain features in common. For example, each type of receptor is particularly sensitive to a distinct kind of environmental change, and is much less sensitive to other forms of stimulation.

Types of Receptors

Five general groups of sensory receptors have been identified, based on their sensitivities to changes in one of the following factors:

1. *Chemical concentration.* Receptors that are stimulated by changes in the concentration of chemical substances are called **chemoreceptors.** Receptors associated with the senses of smell and taste are of this type. Chemoreceptors in various internal organs can detect changes in the blood concentrations of such substances as oxygen, hydrogen ions, and glucose.
2. *Tissue damage.* Whenever tissues are damaged in any way, **pain receptors** (nociceptors) are likely to be stimulated. These receptors may be triggered by exposure to excessive mechanical, electrical, thermal, or chemical energy.
3. *Temperature change.* Receptors sensitive to temperature change are called **thermoreceptors,** and there are two different sets. One set, the *heat receptors,* responds when the amount of heat energy increases above a certain level; the other set, the *cold receptors,* responds when the amount of heat energy decreases below a certain level.
4. *Mechanical forces.* A number of sensory receptors are sensitive to mechanical forces, such as changes in the pressure or movement of fluids. As a rule, these **mechanoreceptors** detect changes that cause the receptors to become deformed. For example, those called *proprioceptors* are sensitive to changes in the tensions of muscles and tendons, while mechanoreceptors called *pressoreceptors* (baroreceptors) in certain blood vessels can detect changes in blood pressure. Similarly, in the lungs mechanoreceptors called *stretch receptors* are sensitive to the degree of inflation.

5. *Light intensity.* Light receptors, or **photoreceptors,** occur only in the eyes, and they respond whenever they are exposed to light energy of sufficient intensity.

Sensory Impulses

Although the ends of nerve fibers often serve as sensory receptors, other kinds of cells located close to nerve fiber endings may also serve as receptors. In either case, when the receptors are stimulated, local changes occur in their membrane potentials (receptor potentials). As a result, an electric current is generated that is graded, and thus reflects the intensity of stimulation. (See chapter 10.)

If a receptor is a nerve fiber and the change in its membrane potential is sufficient to reach threshold, an action potential is generated, and a sensory impulse travels away on the nerve fiber. However, if the receptor is another type of cell, its receptor potential must be transferred to a nerve fiber before an action potential can be triggered.

Such sensory impulses enter the central nervous system by means of peripheral nerves and are analyzed and interpreted within the brain. (See chapter 11.)

Sensations

A **sensation** (perception) is a feeling that occurs when sensory impulses are interpreted by the brain. Because all the nerve impulses that travel away from sensory receptors into the central nervous system are alike, different kinds of sensations must be due to the way the brain interprets the impulses rather than to differences in the receptors. In other words, when a receptor is stimulated, the resulting sensation depends on what region of the cerebral cortex receives the impulse. For example, impulses reaching one region are always interpreted as sounds, and those reaching another portion are always sensed as touch.

It makes little difference how the receptors are stimulated. Pain receptors, for example, can be stimulated by excessive heat, cold, or pressure, but the sensation is always the same because in each case the resulting nerve impulses are interpreted by the same part of the brain. Similarly, nerve impulses are sometimes triggered in visual receptors by factors other than light. When this happens, the person may "see" lights, even though no light is entering the eye, since any impulses reaching the visual cortex are interpreted as light.

At the same time that a sensation is created, the cerebral cortex causes it to seem to come from the receptors being stimulated. This process is called **projection,** because the brain projects the sensation back to its apparent source. Projection allows a person to pinpoint the region of stimulation. Thus, the eyes seem to see and the ears seem to hear.

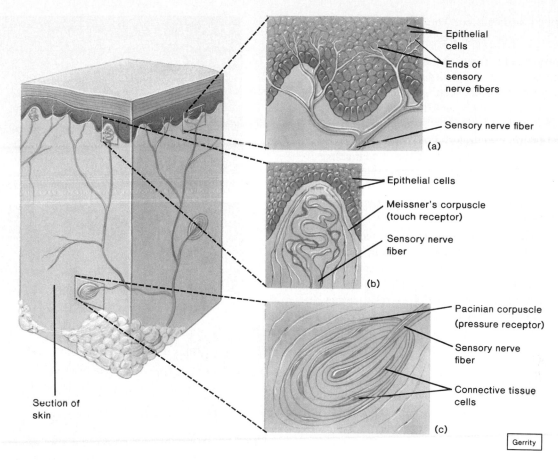

Figure 12.1 Touch and pressure receptors include (*a*) free ends of sensory nerve fibers, (*b*) Meissner's corpuscles, and (*c*) Pacinian corpuscles.

Sensory Adaptation

When sensory receptors are subjected to continuous stimulation, many of them undergo an adjustment called **sensory adaptation.** As the receptors adapt, impulses leave them at decreasing rates, until finally these receptors may completely fail to send signals. Once receptors have adapted, impulses can be triggered only if the strength of the stimulus is changed.

Sensory adaptation is experienced when a person enters a room where there is a strong odor. At first the odor seems intense, but it becomes less and less noticeable as the smell (olfactory) receptors adapt. If the person remains in the room for a minute or more, he or she may become totally adapted to the odor and quite unaware of its presence. However, if the person leaves the room and then reenters it, the receptors will probably be stimulated once again.

1. List the five general types of sensory receptors.
2. What do all types of receptors have in common?
3. Explain how a sensation occurs.
4. What is meant by sensory adaptation?

Somatic Senses

Somatic senses involve receptors associated with the skin, muscles, joints, and visceral organs. These senses can be divided into three groups, as follows:

1. Senses associated with changes occurring at the body surface (exteroceptive senses), which include the senses of touch, pressure, and temperature.
2. Senses associated with changes occurring in muscles and tendons, and in body position (proprioceptive senses).
3. Senses associated with changes occurring in visceral organs (visceroceptive senses). Except for visceral pain, these senses will be discussed in subsequent chapters.

Touch and Pressure Senses

The senses of touch and pressure employ three kinds of receptors (figure 12.1). As a group, these receptors are sensitive to mechanical forces that cause tissues to be deformed or displaced. The touch and pressure receptors include the following:

1. **Sensory nerve fibers.** These receptors are common in epithelial tissues where their free ends are found between epithelial cells. They are associated with sensations of touch and pressure.

2. **Meissner's corpuscles.** These structures consist of small, oval masses of flattened connective tissue cells surrounded by connective tissue sheaths. Two or more sensory nerve fibers branch into each corpuscle and end within it as tiny knobs.

    Meissner's corpuscles are especially numerous in the hairless portions of the skin, such as the lips, fingertips, palms, soles, nipples, and external genital organs. They are sensitive to the motion of objects that barely contact the skin, and impulses from them are interpreted as the sensation of light touch. They are also used when a person touches something to judge its texture. (See figure 12.2.)

3. **Pacinian corpuscles.** These sensory bodies are relatively large, ellipsoidal structures composed of connective tissue fibers and cells. They are common in the deeper subcutaneous tissues of such organs as the hands, feet, penis, clitoris, urethra, and breasts, and also occur in tendons of muscles and ligaments of joints.

    Pacinian corpuscles are stimulated by heavy pressure and are thought to be associated with the sensation of deep pressure. They may also detect vibrations in tissues.

Temperature Senses

As mentioned previously, the temperature senses employ two kinds of thermoreceptors. Although there is some question about the identity of these receptors, they seem to include two types of *free nerve endings* located in the skin called *heat receptors* and *cold receptors*. The heat receptors are most sensitive to temperatures above 25°C (77°F) and become unresponsive at temperatures above 45°C (113°F). As 45°C is approached, pain receptors are also triggered, producing a *burning sensation*.

Cold receptors are most sensitive to temperatures between 10°C (50°F) and 20°C (68°F). If the temperature drops below 10°C, pain receptors are stimulated, and the person feels a *freezing sensation*.

Both heat and cold receptors demonstrate rapid adaptation, so that within about a minute of continuous stimulation, the sensation of heat or cold begins to fade. A person experiences this form of sensory adaptation upon entering a tub of hot water or wading into a cold lake or stream. At first the temperature sensation may

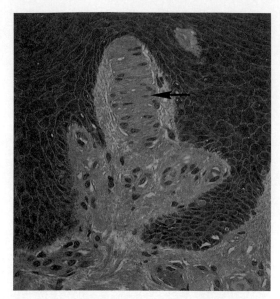

(a)

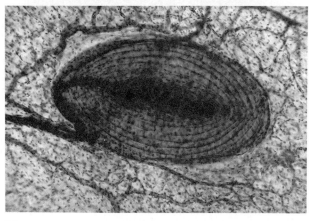

(b)

Figure 12.2 (*a*) Light micrograph of a Meissner's corpuscle from the skin of the palm (×400); (*b*) light micrograph of a Pacinian corpuscle (×25).

be unpleasant, but within a few moments the temperature receptors adapt, at least partially, and the person feels more comfortable.

Sense of Pain

The sense of pain also involves receptors that consist of *free nerve endings*. These receptors are widely distributed throughout the skin and internal tissues, except in the nervous tissue of the brain, which lacks pain receptors.

Pain Receptors

The pain receptors have a protective function in that they are stimulated whenever tissues are being damaged. The pain sensation is usually perceived as unpleasant, and it serves as a signal that something should be done to remove the source of the stimulation.

Headache

One of the more common forms of pain is headache. Although the nervous tissue of the brain lacks pain receptors, nearly all other tissues of the head, including the meninges and blood vessels associated with the brain, are well supplied with them.

Most headaches seem to be related to stressful life situations that result in fatigue, emotional tension, anxiety, or frustration. These conditions are reflected in various physiological changes. For example, they may cause prolonged contraction of skeletal muscles in the forehead, sides of the head, or back of the neck. Such contractions stimulate pain receptors and produce what is often called a *tension headache*. In other cases, people suffer more severe *vascular headaches,* which accompany constriction or dilation of cranial blood vessels. The throbbing headache of a hangover following excessive consumption of alcohol may be due to blood pulsating through cranial vessels. Similar headaches may be experienced by persons who are sensitive to certain food additives, such as monosodium glutamate (MSG) that is sometimes used to enhance flavor.

Still another form of vascular headache is called *migraine*. In this disorder, certain cranial blood vessels seem to constrict, producing a localized cerebral blood deficiency. As a result, the person may experience a variety of symptoms, such as seeing patterns of bright light that interfere with vision or feeling numbness in the limbs or face. Typically, the vasoconstriction is followed by vasodilation of the cranial vessels and a severe headache, which is usually limited to one side of the head and lasts for several hours. Migraine headaches tend to run in families and generally begin in adolescence. They are more common in women than in men.

Other causes of headaches include high blood pressure, increased intracranial pressure due to tumor or blood escaping from a ruptured vessel, temporomandibular joint syndrome (see chapter 9), decreased cerebrospinal fluid pressure following a lumbar puncture (see chapter 11), or sensitivity to or withdrawal from various drugs.

Although most pain receptors can be stimulated by more than one type of change, some are most sensitive to mechanical damage while others are particularly sensitive to extremes in temperature. Still other pain receptors are most responsive to various chemicals such as hydrogen ions, potassium ions, or polypeptides (kinins) from the breakdown of proteins, histamine, and acetylcholine. A deficiency of blood (ischemia), and thus a deficiency of oxygen (hypoxia) in a tissue, or the stimulation of mechanical-sensitive receptors also triggers the pain sensation. The pain elicited during a muscle cramp, for example, seems to be related to an interruption of blood flow that occurs as the sustained contraction squeezes capillaries, as well as to the stimulation of mechanical-sensitive pain receptors. Also, when blood flow is interrupted, pain-stimulating chemicals accumulate excessively. Increasing the blood flow through the sore tissue may relieve the resulting pain, and this is why heat is sometimes applied to reduce muscle soreness. The heat causes blood vessels to dilate and thus promotes blood flow, which helps reduce the concentration of the pain-stimulating substances. In some conditions, accumulating chemicals cause the thresholds of pain receptors to be lowered. Consequently, inflamed tissues may become more sensitive to heat or pressure than before.

Pain receptors adapt very little, if at all, and once such a receptor has been activated, even by a single stimulus, it may continue to send impulses into the central nervous system for some time.

The stimulation of pain receptors associated with injuries to bones, tendons, or ligaments may also cause nearby skeletal muscles to undergo contraction. As the muscles contract, they may become ischemic, and this condition may trigger still other pain receptors within the muscle tissue. This additional stimulation of pain receptors may lead to increased muscle contraction, thus establishing a "vicious circle."

Visceral Pain

As a rule, pain receptors are the only receptors in visceral organs whose stimulation produces sensations. Pain receptors in these organs seem to respond differently to stimulation than those associated with surface tissues. For example, localized damage to intestinal tissue during surgical procedures may not elicit any pain sensations, even in a conscious person. However, when visceral tissues are subjected to more widespread stimulation, as when intestinal tissues are stretched or when the smooth muscles in the intestinal walls undergo spasms, a strong pain sensation may follow. Once again, the resulting pain seems to be related to the stimulation of mechanical-sensitive receptors and to a decreased blood flow accompanied by a lower tissue oxygen concentration and an accumulation of pain-stimulating chemicals.

Another characteristic of visceral pain is that it may feel as if it is coming from some part of the body

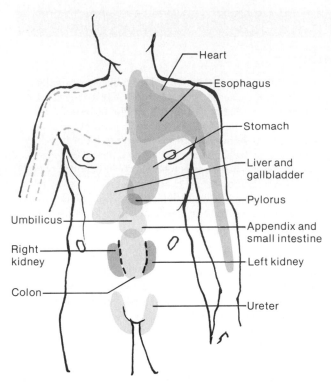

Figure 12.3 Surface regions to which visceral pain may be referred.

other than the part being stimulated—a phenomenon called **referred pain.** For example, pain originating in the heart may be referred to the left shoulder or the inside of the left arm. Pain from the lower esophagus, stomach, or small intestine may seem to be coming from the upper central (epigastric) region of the abdomen. Pain from the urogenital tract may be referred to the lower central (hypogastric) region of the abdomen or to the sides between the ribs and the hip (figure 12.3).

The occurrence of referred pain seems to be related to *common nerve pathways* that are used by sensory impulses coming both from skin areas and from visceral organs. (See chapter 11.) In other words, pain impulses from the heart seem to be conducted over the same nerve paths as those conducting impulses from the skin of the left shoulder and the inside of the left arm, as shown in figure 12.4. Consequently, the cerebral cortex may incorrectly interpret the source of the impulses as the shoulder or arm rather than the heart.

Pain originating in the parietal layers of thoracic and abdominal membranes—parietal pleura, parietal pericardium, or parietal peritoneum—is usually not referred; instead, such pain is likely to be felt directly over the area being stimulated.

Pain Nerve Pathways

The nerve fibers that conduct impulses away from pain receptors include two main types: acute pain fibers and chronic pain fibers.

The *acute pain fibers* (also known as A-delta fibers) are relatively thin, myelinated nerve fibers. They

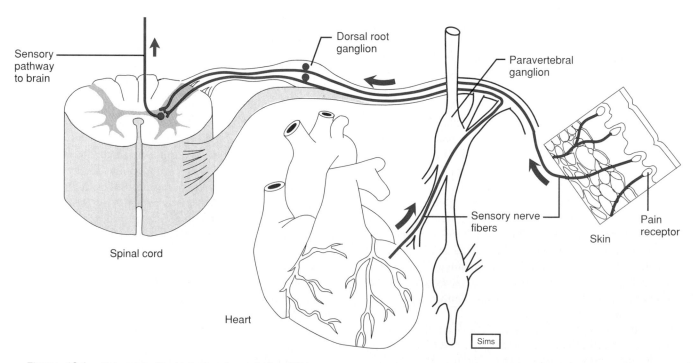

Figure 12.4 Pain originating in the heart may feel as if it is coming from the skin because sensory impulses from those two regions follow common nerve pathways to the brain.

Control of Pain

Most people suffer pain arising from injuries and diseases, and few are able to remain indifferent to it.

Interestingly, the intensity of stimulation needed to trigger pain receptors seems to be about the same for all individuals, yet people vary greatly in the ways they perceive pain and respond to it. These differences seem to be related to past experiences, cultural conditioning, and the circumstances under which the painful stimuli are encountered. For example, fatigue, depression, and anxiety can increase a person's susceptibility to pain.

Acute pain, which is short-term and often due to a known cause, can usually be endured. However, *chronic pain,* which persists for a long period, tends to weaken its victim and become the dominant feature of that person's life. Such pain may also trigger reflexes that cause changes in blood pressure and heart rate, as well as nausea, insomnia, loss of appetite, and loss of weight.

When pain is persistent and interferes with a person's ability to function effectively, he or she may try to control it by taking a pain-relieving drug (analgesic) or by some other means. Methods employed by physical therapists may help reduce pain by improving the flexibility of joints, relieving the stress on spinal musculature, or decompressing the nerve roots associated with vertebral joints. Massage of superficial or deep soft tissues may also reduce pain by improving the movement of body fluids.

Other methods of relieving pain include applying heat, applying a liniment that produces a local irritation of the skin, or treatment with acupuncture—the insertion of needles into various tissues. Although acupuncture is not well understood, it is known to trigger impulses on large sensory peripheral nerve fibers, and they, in turn, may inhibit the transmission of pain impulses by causing the release of enkephalins or endorphins within the central nervous system.

Treatment of chronic pain often requires the use of narcotic drugs. In some difficult cases, electrodes are applied to the skin or implanted in appropriate regions of the spinal cord and are then used along with a battery-powered device to stimulate certain nerve fibers. Such electrical stimulation produces an analgesic effect over relatively large areas of the body. In other cases, various sensory nerve pathways may be sectioned surgically. This procedure prevents pain impulses from reaching the pain-interpreting centers of the brain, but it also results in a permanent loss of sensation from the body parts innervated by the sectioned nerve fibers.

---

conduct nerve impulses rapidly, at velocities up to 30 meters per second. These impulses are associated with the sensation of sharp pain, which typically seems to originate in the skin and is restricted to a local area. This type of pain seldom continues after the pain-producing stimulus is discontinued.

The *chronic pain fibers* (C fibers) are thin, unmyelinated nerve fibers. They conduct impulses more slowly than the acute pain fibers, at velocities up to 2 meters per second. These impulses are related to the dull, aching pain sensation that may be widespread and difficult to pinpoint. Such pain may continue for some time after the original stimulus has been eliminated. Although acute pain is usually sensed as coming from the surface, chronic pain is likely to be felt in deeper tissues as well as in the skin. Visceral pain impulses also are usually carried on C fibers.

Commonly, an event that stimulates pain receptors will trigger impulses on both types of pain fibers. This causes a dual sensation—a sharp, pricking pain, followed shortly by a dull, aching one. The aching pain is usually more intense and may become even more painful with time. This chronic type of pain sometimes gives rise to prolonged suffering that is very resistant to relief and control.

Pain impulses that originate from tissues of the head reach the brain on sensory fibers of the fifth, seventh, ninth, and tenth cranial nerves. All other pain impulses travel on sensory fibers of spinal nerves, and they pass into the spinal cord by way of the dorsal roots of these spinal nerves.

Upon reaching the spinal cord, pain impulses enter the gray matter of the dorsal horn, where they are processed. (See chapters 10 and 11.) The fast-conducting fibers synapse with long nerve fibers that cross over to the opposite side of the spinal cord and ascend in the lateral spinothalamic tracts. The impulses carried on the slow-conducting fibers pass through one or more interneurons before reaching the long fibers that cross over and ascend to the brain.

Within the brain, most of the pain fibers terminate in the reticular formation (chapter 11), and from there are conducted on the fibers of still other neurons to the thalamus, hypothalamus, and cerebral cortex.

Regulation of Pain Impulses

The awareness of pain seems to occur when the pain impulses reach the level of the thalamus—that is, even before they reach the cerebral cortex. However, the cerebral cortex is needed to judge the intensity of the pain and to locate its source. The cerebral cortex is also responsible for various emotional and motor responses to pain.

Still other parts of the brain regulate the flow of pain impulses up from the spinal cord. These parts include areas of gray matter in the midbrain, pons, and medulla oblongata. Impulses from special neurons in these areas descend in the lateral funiculus (chapter 11) to various levels within the spinal cord. The impulses cause the ends of certain nerve fibers to release substances that can block pain signals by inhibiting presynaptic nerve fibers in the dorsal horn of the spinal cord.

Among the inhibiting substances released in the dorsal horn are a group of neuropeptides called *enkephalins* and the monoamine called *serotonin* (chapter 10). Enkephalins can suppress both acute and chronic pain impulses; thus they can relieve relatively strong pain sensations, much as morphine and other opiate drugs do. In fact, enkephalins seem to bind to the same receptor sites on neuron membranes as does morphine. Serotonin is thought to act by stimulating other neurons to release enkephalins.

Another group of neuropeptides that has pain-suppressing, morphinelike actions are the *endorphins*. They occur in the pituitary gland and in various regions of the nervous system, such as the hypothalamus, that are involved with the transmission of pain impulses.

It is believed that enkephalins and endorphins are released when pain impulses are triggered excessively, and that in this way they provide natural pain control. Such pain-relieving substances may also be released in response to indirect stimulation by transcutaneous electrical impulses or acupuncture.

1. Describe three types of touch and pressure receptors.
2. Describe the thermoreceptors.
3. What types of stimuli excite pain receptors?
4. What is referred pain?
5. Explain how neuropeptides may help control pain.

Stretch Receptors

Stretch receptors are proprioceptors that provide information to the spinal cord and brain concerning the lengths and tensions of muscles. The two main kinds of stretch receptors are muscle spindles and Golgi tendon organs; however, no sensation results when they are stimulated.

Muscle spindles are located in skeletal muscles near their junctions with tendons. Each spindle consists of one or more small, modified skeletal muscle fibers (intrafusal fibers) enclosed in a connective tissue sheath.

Near its center, each fiber has a specialized nonstriated region with the end of a sensory nerve fiber wrapped around it (figure 12.5a).

The striated portions of the muscle fiber can contract, and they keep the spindle taut. However, if the whole muscle becomes relaxed and is stretched longer than usual, the muscle spindle detects this change, and sensory nerve impulses may be triggered on its nerve fiber. Such sensory impulses travel into the spinal cord and onto motor fibers leading back to the same muscle, causing it to contract. This action, called a **stretch reflex,** opposes the lengthening of the muscle and helps a person maintain the desired position of a limb in spite of gravitational or other forces tending to move it. (See chapter 10.)

Golgi tendon organs are found in tendons close to their attachments to muscles. Each is connected to a set of skeletal muscle fibers and is innervated by a sensory neuron (figure 12.5b).

These receptors have relatively high thresholds and are stimulated by increased tension. Sensory impulses from them produce a reflex that inhibits contraction of the muscle whose tendon they occupy. Thus, the Golgi tendon organs stimulate a reflex with an effect opposite that of a stretch reflex. This reflex also helps maintain posture, and it protects muscle attachments from being pulled away from their insertions by excessive tension. Chart 12.1 summarizes the somatic receptors and their functions.

1. Describe a muscle spindle.
2. Explain how muscle spindles help maintain posture.
3. Where are Golgi tendon organs located?
4. What is the function of Golgi tendon organs?

Special Senses

Special senses are those whose sensory receptors occur within relatively large, complex sensory organs in the head. These organs include the olfactory organs, taste buds, ears, organs of equilibrium, and eyes, which are associated with the senses of smell, taste, hearing, static equilibrium, dynamic equilibrium, and sight, respectively.

Sense of Smell

The sense of smell is associated with sensory structures in the upper region of the nasal cavity.

Olfactory Receptors

The smell, or olfactory, receptors are similar to those for taste (discussed in a subsequent section of this

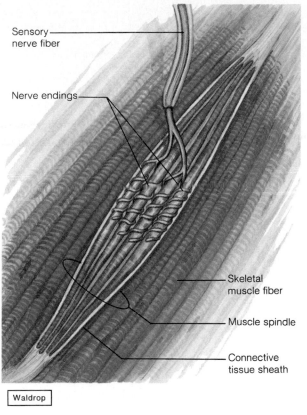

Sensory nerve fiber

Nerve endings

Skeletal muscle fiber

Muscle spindle

Connective tissue sheath

Waldrop

(a)

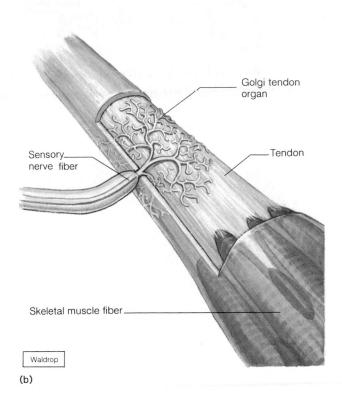

Golgi tendon organ

Sensory nerve fiber

Tendon

Skeletal muscle fiber

Waldrop

(b)

Figure 12.5 (a) Muscle spindles, which are modified muscle fibers, are stimulated by changes in muscle length; (b) Golgi tendon organs inhibit the contraction of muscles whose tendons they occupy.

CHART 12.1	Somatic receptors	
Type	Function	Sensation
Free nerve endings (mechanoreceptors)	Detect changes in pressure	Touch, pressure
Meissner's corpuscles (mechanoreceptors)	Detect objects moving over the skin	Touch, texture
Pacinian corpuscles (mechanoreceptors)	Detect changes in pressure	Deep pressure, vibrations
Free nerve endings (thermoreceptors for heat)	Detect changes in temperature	Heat
Free nerve endings (thermoreceptors for cold)	Detect changes in temperature	Cold
Free nerve endings (pain receptors)	Detect tissue damage	Pain
Muscle spindle (mechanoreceptor)	Detect changes in muscle length	None
Golgi tendon organ (mechanoreceptor)	Detect changes in muscle tension	None

chapter) in that they are chemoreceptors stimulated by chemicals dissolved in liquids. These two senses function closely together and aid in food selection, because food is usually smelled at the same time it is tasted. In fact, it is often difficult to tell what part of a food sensation is due to smell and what part is due to taste. For this reason, an onion tastes quite different when sampled with the nostrils closed, because much of the usual onion sensation is due to odor. Similarly, if excessive mucous secretions cover the olfactory receptors when a person has a head cold, food may seem to lose its taste.

Olfactory Organs

The **olfactory organs,** which contain the olfactory receptors, also include various epithelial supporting cells. These organs appear as yellowish brown masses surrounded by pinkish mucous membrane. They cover the upper parts of the nasal cavity, the superior nasal conchae, and a portion of the nasal septum (figure 12.6).

The **olfactory receptor cells** are bipolar neurons surrounded by columnar epithelial cells. These neurons have knobs at the distal ends of their dendrites that are

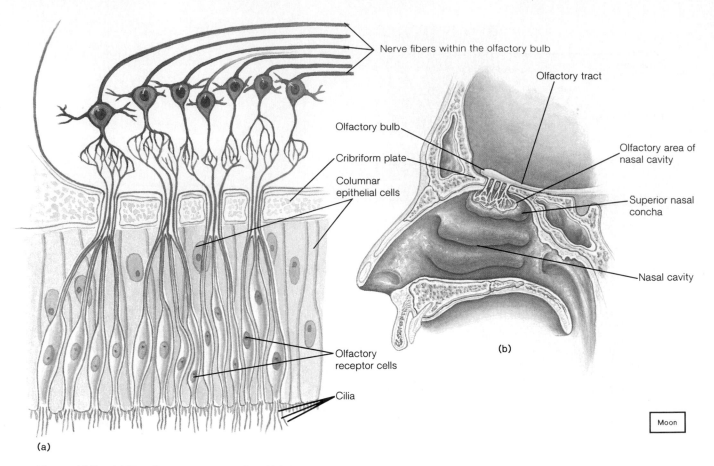

Nerve fibers within the olfactory bulb

Olfactory tract

Olfactory bulb

Cribriform plate

Columnar epithelial cells

Olfactory area of nasal cavity

Superior nasal concha

Nasal cavity

Olfactory receptor cells

Cilia

(a)

(b)

Moon

Figure 12.6 (*a*) The olfactory receptor cells, which have cilia at their distal ends, are supported by columnar epithelial cells. (*b*) The olfactory area is associated with the superior nasal concha.

covered by hairlike cilia. The cilia project into the nasal cavity and are thought to be the sensitive portions of the receptors (figure 12.7).

Chemicals that stimulate olfactory receptors enter the nasal cavity as gases, but they must dissolve at least partially in the watery fluids that surround the cilia before they can be detected. It is believed that such chemicals must also be soluble in lipids before they can stimulate the receptors, because cilia are composed primarily of lipid materials.

Olfactory Nerve Pathways

Once the olfactory receptors have been stimulated, nerve impulses travel along the axons of the receptor cells, which are the fibers of the olfactory nerves. These nerve fibers pass through tiny openings in the cribriform plates of the ethmoid bone and lead to neurons located in the enlargements of the **olfactory bulbs.** These structures lie on either side of the crista galli of the ethmoid bone. (See figure 7.26.) Within the olfactory bulbs, the sensory impulses are analyzed, and as a result, additional impulses travel along the **olfactory tracts** to portions of the limbic system (chapter 11). The

main interpreting areas for the olfactory impulses (olfactory cortex) are located deep within the temporal lobes and at the bases of the frontal lobes, anterior to the hypothalamus.

Olfactory Stimulation

The mechanism by which various substances stimulate the olfactory receptors is poorly understood. One hypothesis suggests that the shapes of gaseous molecules may fit the complementary shapes of receptor sites on the cilia. A nerve impulse, according to this idea, is triggered when a molecule binds to its particular receptor site.

Attempts to classify the sensations associated with odors have met with only limited success. However, many researchers agree that there are at least seven groups of primary odors:

1. *Camphoraceous,* like the scent of camphor.
2. *Musky,* like the scent of musk.
3. *Floral,* like the scent of flowers.
4. *Pepperminty,* like the scent of oil of peppermint.

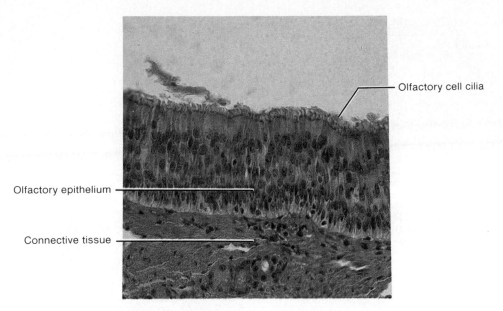

Olfactory cell cilia

Olfactory epithelium

Connective tissue

Figure 12.7 Light micrograph of the olfactory epithelium (×250).

5. *Ethereal,* like the scent of ether.
6. *Pungent,* like the scent of spices.
7. *Putrid,* like the scent of decaying meat.

Any particular odor may be described as one of the primary odors or as some combination of two or more of them.

Because the olfactory organs are located high in the nasal cavity above the usual pathway of inhaled air, a person may have to sniff and force air over the receptor areas in order to smell a faint odor. Also, olfactory receptors undergo sensory adaptation rather rapidly, so the intensity of an olfactory sensation drops about 50% within a second following the stimulation. Within a minute, the receptors may become almost insensitive to a given odor, but even though they have adapted to one scent, their sensitivity to other odors remains unchanged.

Partial or complete loss of smell is called *anosmia.* This condition may be caused by a variety of factors, including inflammation of the nasal cavity lining during a head cold. It may also result from excessive tobacco smoking or from the use of drugs, such as epinephrine or cocaine. Persons who have injured their olfactory bulbs, olfactory tracts, or cerebral olfactory interpreting centers experience some degree of anosmia.

Elderly persons may require special stimulation to encourage eating because their olfactory sensitivities have diminished. For this reason, it is important that their foods be visibly attractive and that mealtimes be pleasant. Otherwise, disinterest in eating may contribute to malnutrition.

The olfactory receptor neurons are the only parts of the nervous system that are in direct contact with the outside environment. Because of their exposed positions, these neurons are subject to damage; and because damaged neurons usually are not replaced, persons are likely to experience a progressive diminishing of olfactory sense with age. In fact, some investigators have estimated that a person loses about 1% of the olfactory receptors every year.

1. Where are the olfactory receptors located?
2. Trace the pathway of an olfactory impulse from a receptor to the cerebrum.
3. Why is the sense of smell likely to decrease with age?

Sense of Taste

Taste buds are the special organs of taste. They occur primarily on the surface of the tongue where they are associated with tiny elevations called **papillae** (figures 12.8 and 12.9). They also are found in smaller numbers in the roof of the mouth and the walls of the pharynx.

Taste Receptors

Each taste bud includes a group of modified epithelial cells, which are the **taste cells** (gustatory cells) that function as receptors. The taste bud also includes a number of epithelial supporting cells. The entire structure is somewhat spherical, with an opening, the **taste pore,** on its free surface. Tiny projections (microvilli), called **taste hairs,** protrude from the outer ends of the

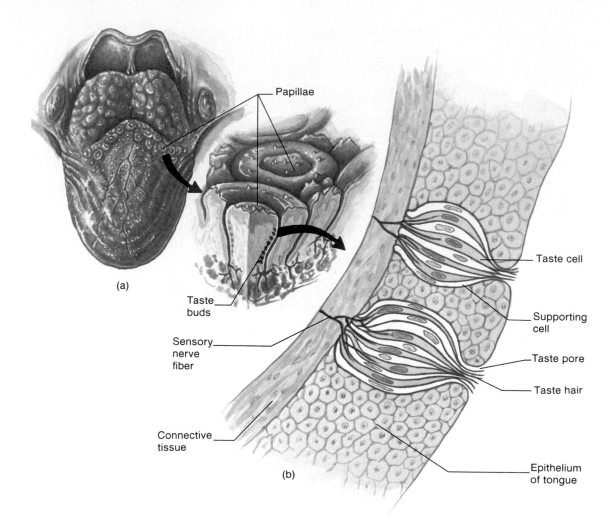

Papillae

Taste cell

Supporting cell

Taste pore

Taste hair

Epithelium of tongue

Taste buds

Sensory nerve fiber

Connective tissue

(a)

(b)

Figure 12.8 (a) Taste buds on the surface of the tongue are associated with nipplelike elevations called papillae. (b) A taste bud contains taste cells and has an opening, the taste pore, at its free surface.

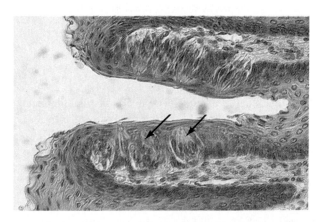

Figure 12.9 A light micrograph of some taste buds (arrows).

taste cells and jut out through the taste pore. It is believed that these taste hairs are the sensitive parts of the receptor cells.

Interwoven among and wrapped around the taste cells is a network of nerve fibers. The ends of these fibers

are in close contact with the receptor cell membranes. When a receptor cell is stimulated, an impulse is triggered on a nearby nerve fiber and travels into the brain.

Before the taste of a particular chemical can be detected, the chemical must be dissolved in the watery fluid surrounding the taste buds. This fluid is supplied by the salivary glands. To demonstrate its importance, blot your tongue and try to taste some dry food; then repeat the test after moistening your tongue with saliva.

The mechanism by which various substances stimulate taste cells is not well understood. It is known, however, that taste cell membranes are normally polarized and that application of certain chemicals causes changes in their polarization. The amount of change is directly proportional to the concentration of the stimulating substance. One possible explanation holds that various substances combine with specific receptor sites on the taste hair surfaces. Such a combination is thought to be responsible for the change in membrane polarization and the generation of sensory impulses on nearby nerve fibers.

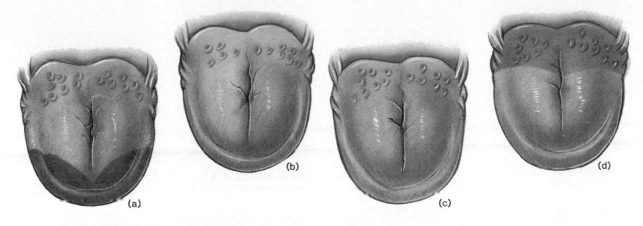

Figure 12.10 Patterns of taste receptor distribution are indicated by color: (*a*) sweet receptors; (*b*) sour receptors; (*c*) salt receptors; and (*d*) bitter receptors.

Gerrity

Although the taste cells in all taste buds appear very much alike microscopically, there are at least four types. Each type is most sensitive to a particular kind of chemical stimulus; consequently, there are at least four primary taste (gustatory) sensations.

Taste Sensations

The four *primary taste sensations* are:

1. *Sweet,* as produced by table sugar.
2. *Sour,* as produced by vinegar.
3. *Salty,* as produced by table salt.
4. *Bitter,* as produced by caffeine or quinine.

In addition to these four, some investigators recognize two other taste sensations, *alkaline* and *metallic.* Each of the many flavors we experience daily is believed to result from one of the primary sensations or from some combination of two or more of them. The way we experience flavors may also involve the concentration of chemicals as well as the sensations of odor, texture (touch), and temperature. Furthermore, chemicals in some foods—chili peppers and ginger, for instance—may stimulate pain receptors that cause the tongue to burn.

Each of the four major types of taste receptors is most highly concentrated in certain regions of the tongue's surface (figure 12.10). For example, *sweet receptors* are most plentiful near the tip of the tongue. This is why a child may prefer to lick a candy sucker rather than chew it. The chemicals that stimulate these receptors are usually *organic compounds,* such as sugars and polysaccharides. A few inorganic substances, including some salts of lead and beryllium, also elicit sweet sensations.

Sour receptors occur primarily along the margins of the tongue. They are stimulated mainly by acids. The intensity of a sour sensation is roughly propor-

tional to the concentration of the *hydrogen ions* in the substance being tasted.

Salt receptors are most common in the tip and the upper front portion of the tongue. They are stimulated mainly by ionized *inorganic salts.* The quality of the sensation each produces is related to the kind of positively charged ion, such as Na^+ from table salt, that it releases into solution.

Bitter receptors are located toward the back of the tongue. They are stimulated by a variety of chemical substances, most of which are organic compounds, although some inorganic salts of magnesium and calcium produce bitter sensations, too.

One group of bitter compounds of particular interest are the *alkaloids,* which include a number of poisons such as strychnine, nicotine, and morphine. The fact that persons commonly reject bitter substances may be related to a protective mechanism that causes them to avoid alkaloids in foods.

Taste receptors, like olfactory receptors, undergo sensory adaptation relatively rapidly. The resulting loss of taste can be avoided by moving bits of food over the surface of the tongue to stimulate different receptors at different moments.

Although the taste cells are located very close to the surface of the tongue and are somewhat exposed to damage by environmental factors, the sense of taste is not as likely to diminish with age as the sense of smell. This is because taste cells are reproduced continually, so that any one of these cells functions for only about a week before it is replaced.

Taste Nerve Pathways

Sensory impulses from the taste receptors located in the anterior two-thirds of the tongue travel on fibers of the facial nerve (VII); impulses from receptors in the posterior one-third of the tongue and the back of the mouth pass along the glossopharyngeal nerve (IX); and

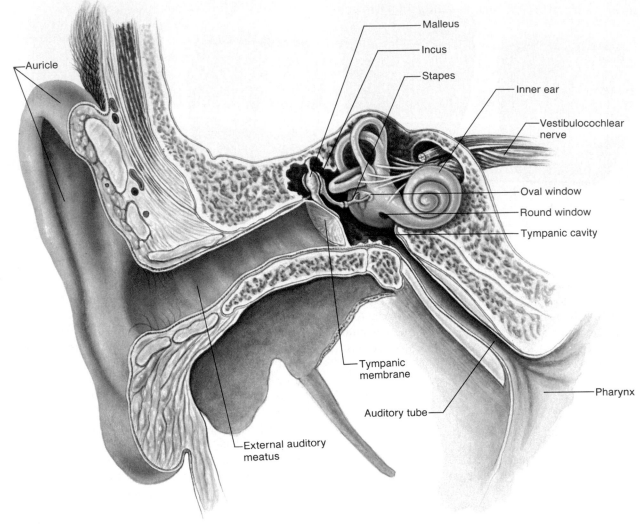

Figure 12.11 Major parts of the ear.

impulses from receptors at the base of the tongue and the pharynx travel on the vagus nerve (X). (See chapter 11.)

These cranial nerves conduct the impulses into the medulla oblongata. From there, the impulses ascend to the thalamus and are directed to the gustatory cortex of the cerebrum, located in the parietal lobe along a deep portion of the lateral sulcus.

1. Why is saliva necessary for a sense of taste?
2. Name the four primary taste sensations.
3. What characteristic of taste receptors helps maintain a sense of taste with age?
4. Trace a sensory impulse from a taste receptor to the cerebral cortex.

Sense of Hearing

The organ of hearing, the **ear,** has external, middle, and inner parts. In addition to making hearing possible, the ear functions in the sense of equilibrium, which will be discussed in a subsequent section of this chapter.

External Ear

The external ear consists of two parts: an outer, funnel-like structure called the **auricle** (pinna), and an S-shaped tube, the **external auditory meatus** (external auditory canal) that leads inward for about 2.5 centimeters (figure 12.11).

The external auditory meatus passes into the temporal bone. Near its opening, the tube is guarded by hairs. It is lined with skin that contains numerous modified sweat glands called *ceruminous glands,* which secrete wax (cerumen). The hairs and wax help keep relatively large foreign objects, such as insects, from entering the ear.

Sounds are generally produced by vibrating objects, and the vibrations are transmitted through matter in the form of sound waves. For example, the sounds of some musical instruments are produced by vibrating strings or reeds, and the sounds of the voice are caused by vibrating vocal folds in the larynx. The auricle of the ear helps collect sound waves traveling through air and directs them into the external auditory meatus.

After entering the meatus, the sound waves pass to the end of the tube and cause pressure changes on the eardrum. The eardrum moves back and forth in response and thus reproduces the vibrations of the sound wave source.

Middle Ear

The middle ear is comprised of the tympanic cavity, the tympanic membrane (eardrum), and three small bones called auditory ossicles.

The **tympanic cavity** is an air-filled space in the temporal bone that separates the external and internal ears.

The **tympanic membrane** (eardrum) is a semi-transparent membrane covered by a thin layer of skin on its outer surface and by mucous membrane on the inside. It has an oval margin and is cone-shaped, with the apex of the cone directed inward. Its cone shape is maintained by the attachment of one of the auditory ossicles (malleus).

The three **auditory ossicles** are called the *malleus*, the *incus*, and the *stapes*. They are attached to the wall of the tympanic cavity by tiny ligaments and are covered by mucous membrane. These bones form a bridge connecting the eardrum to the inner ear and function to transmit vibrations between these parts. Specifically, the malleus is attached to the eardrum, and when the eardrum vibrates, the malleus vibrates in unison with it. The malleus causes the incus to vibrate, and the incus passes the movement onto the stapes. The stapes is held by ligaments to an opening in the wall of the tympanic cavity called the **oval window.** Vibration of the stapes, which acts like a piston at the oval window, causes motion of a fluid within the inner ear. These vibrations of the fluid are responsible for stimulating the hearing receptors (figure 12.11).

In addition to transmitting vibrations, the auditory ossicles form a lever system that helps increase (amplify) the force of the vibrations as they are passed from the eardrum to the oval window. Also, because the ossicles transmit vibrations from the relatively large surface of the eardrum to a much smaller area at the oval window, the vibrational force becomes concentrated as it travels from the external to the inner ear. As a result of these two factors, the pressure (per square millimeter) applied by the stapes at the oval window is about twenty-two times greater than that exerted on the eardrum by sound waves.

The middle ear also contains two small skeletal muscles that are attached to the auditory ossicles and controlled involuntarily. One of them, the *tensor tympani,* is inserted on the medial surface of the malleus

and is anchored to the cartilaginous wall of the auditory tube. When it contracts, it pulls the malleus inward. The other muscle, the *stapedius,* is attached to the posterior side of the stapes and the inner wall of the tympanic cavity. It serves to pull the stapes outward (figure 12.12). These muscles are the effectors in the **tympanic reflex,** which is elicited in about one-tenth second following a loud, external sound. When the reflex occurs, the muscles contract, and the malleus and stapes are moved. As a result, the bridge of ossicles in the middle ear becomes more rigid, and its effectiveness in transmitting vibrations to the inner ear is reduced.

In this instance, the tympanic reflex is a protective mechanism that reduces pressure from loud sounds that might otherwise damage the hearing receptors. The tympanic reflex is also elicited by ordinary vocal sounds, as when a person speaks or sings, and this action muffles the lower frequencies of such sounds. As a result, the hearing of higher frequency sounds, which are common in human vocal sounds, is improved. In addition, the tensor tympani muscle also maintains a steady pull on the eardrum. This is important because a loose tympanic membrane would not be able to transmit vibrations effectively to the auditory ossicles.

The middle ear muscles require from 100 to 200 milliseconds to contract. For this reason, the tympanic reflex cannot protect the hearing receptors from the effects of loud sounds that occur very rapidly, such as those from an explosion or a gunshot. On the other hand, this protective mechanism can reduce the effects of intense sounds that arise relatively slowly, such as the roar of thunder.

Auditory Tube

An **auditory tube** (eustachian tube) connects each middle ear to the throat. This tube allows air to pass between the tympanic cavity and the outside of the body by way of the throat (nasopharynx) and mouth. It helps maintain equal air pressure on both sides of the eardrum, which is necessary for normal hearing (figure 12.11).

The function of the auditory tube can be experienced during rapid change in altitude. For example, as a person moves from a high altitude to a lower one, the air pressure on the outside of the eardrum becomes greater and greater. As a result, the eardrum may be pushed inward, out of its normal position, and hearing may be impaired.

When the air pressure difference is great enough, some air may force its way up through the auditory tube into the middle ear. This allows the pressure on both sides of the eardrum to equalize, and the eardrum moves back into its regular position. The person usually hears a popping sound at this moment, and normal

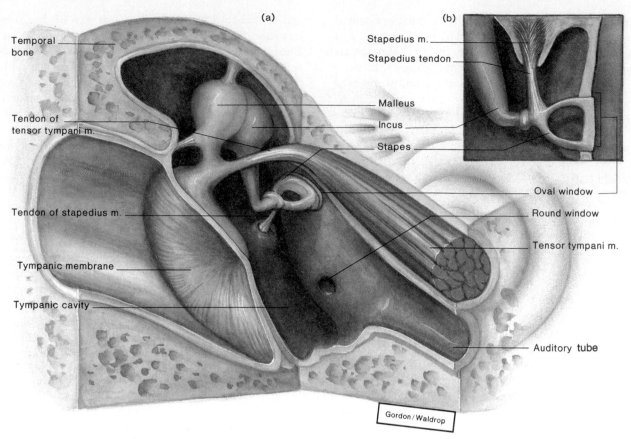

(a)

Temporal bone

Tendon of tensor tympani m.

Tendon of stapedius m.

Tympanic membrane

Tympanic cavity

Malleus

Incus

Stapes

(b)

Stapedius m.

Stapedius tendon

Oval window

Round window

Tensor tympani m.

Auditory tube

Gordon/Waldrop

Figure 12.12 Two small muscles attached to the (a) malleus and (b) stapes, the tensor tympani and the stapedius, serve as effectors in the tympanic reflex.

hearing is restored. A reverse movement of air ordinarily occurs when a person moves from a low altitude into a higher one.

The auditory tube is usually closed by valvelike flaps in the throat, which may inhibit air movements into the middle ear. Swallowing, yawning, or chewing aid in opening the valves, and can hasten the equalization of air pressure if discomfort is experienced during altitude changes.

The mucous membranes that line the auditory tubes are continuous with the linings of the middle ears. Consequently, the tubes provide a route by which mucous membrane infections of the throat (pharyngitis) may spread and cause an infection of the middle ear (otitis media). For this reason, it is poor practice to pinch a nostril when blowing the nose, because pressure in the nasal cavity may then force material from the throat up the auditory tube and into the middle ear.

Inner Ear

The inner ear consists of a complex system of intercommunicating chambers and tubes called a **labyrinth.** In fact, there are two such structures in each ear—the osseous labyrinth and the membranous labyrinth.

The *osseous labyrinth* is a bony canal in the temporal bone; the *membranous labyrinth* is a tube that lies within the osseous labyrinth and has a similar shape (figure 12.13). Between the osseous and membranous labyrinths is a fluid called *perilymph,* which is secreted by cells in the wall of the bony canal. The membranous labyrinth contains another fluid called *endolymph,* whose composition is slightly different.

The parts of the labyrinths include a **cochlea** that functions in hearing, and three **semicircular canals** that function in providing a sense of equilibrium. A bony chamber called the **vestibule,** which is located between the cochlea and the semicircular canals, contains membranous structures that serve both hearing and equilibrium.

The **cochlea** is shaped like the coiled shell of a snail. Inside, it contains a bony core (modiolus) and a thin, bony shelf (spiral lamina) that winds around the core like the threads of a screw. The shelf divides the bony labyrinth of the cochlea into upper and lower compartments. The upper compartment, called the *scala vestibuli,* leads from the oval window to the apex of the spiral. The lower compartment, the *scala tympani,* extends from the apex of the cochlea to a membrane-covered opening in the wall of the inner ear called the **round window.** These compartments constitute the

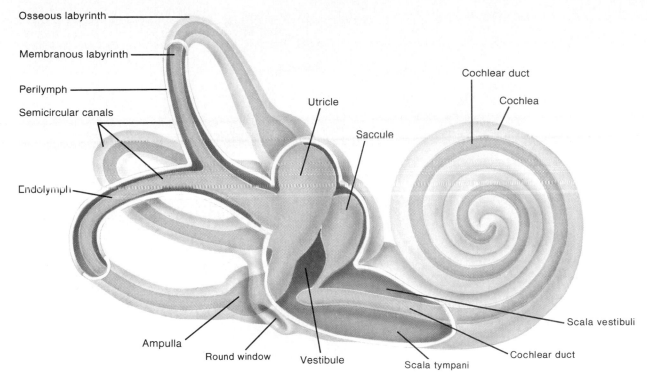

Figure 12.13 The osseous labyrinth of the inner ear is separated from the membranous labyrinth by perilymph. The membranous labyrinth contains endolymph.

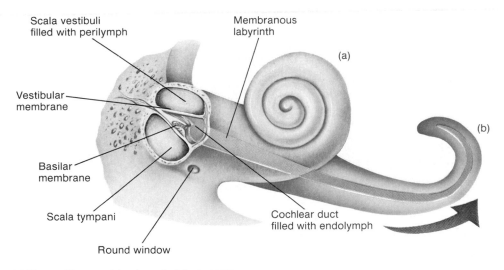

Figure 12.14 (a) The cochlea consists of a coiled, bony canal with a membranous tube inside. (b) If the cochlea could be unwound, the membranous tube would be seen ending as a closed sac at the apex of the bony canal.

bony labyrinth of the cochlea, and they are filled with perilymph. At the apex of the cochlea, the fluids in the chambers can flow together through a small opening (helicotrema).

A portion of the membranous labyrinth within the cochlea is called the *cochlear duct* (scala media). It is filled with endolymph and lies between the two bony compartments. The cochlear duct ends as a closed sac at the apex of the cochlea. The duct is separated from

the scala vestibuli by a *vestibular membrane* (Reissner's membrane) and from the scala tympani by a *basilar membrane* (figure 12.14).

The basilar membrane extends from the bony shelf of the cochlea and forms the floor of the cochlear duct. It contains many thousands of stiff, elastic fibers whose lengths vary, becoming progressively longer from the base of the cochlea to its apex. Vibrations entering the perilymph at the oval window travel along the scala

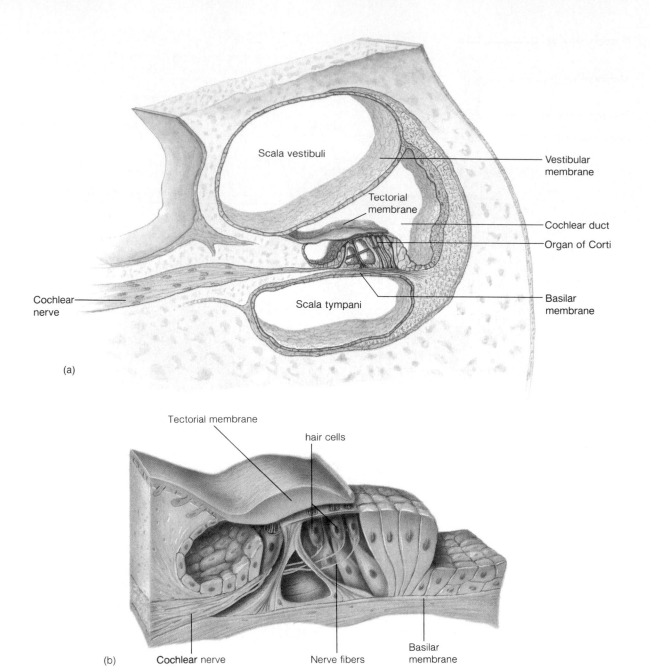

Scala vestibuli

Vestibular
membrane

Tectorial
membrane

Cochlear duct

Organ of Corti

Cochlear
nerve

Scala tympani

Basilar
membrane

(a)

Tectorial membrane

hair cells

(b) **Cochlear** nerve

Nerve fibers

Basilar
membrane

Figure 12.15 (*a*) Cross section of the cochlea; (*b*) organ of
Corti.

vestibuli and pass through the vestibular membrane to
enter the endolymph of the cochlear duct, where they
cause movements in the basilar membrane.

After passing through the basilar membrane, the
vibrations enter the perilymph of the scala tympani, and
their forces are dissipated into the air in the tympanic
cavity by movement of the membrane covering the
round window.

The **organ of Corti,** which contains about 16,000
hearing receptor cells, is located on the upper surface
of the basilar membrane and stretches from the apex
to the base of the cochlea. The receptor cells, which are

called **hair cells,** are arranged in four parallel rows, and
they possess numerous hairlike processes (stereocilia)
that extend into the endolymph of the cochlear duct.
Above these hair cells is a *tectorial membrane,* which
is attached to the bony shelf of the cochlea and passes
like a roof over the receptor cells, making contact with
the tips of their hairs. (See figures 12.15 and 12.16.)

Because of its structure, different portions of the
basilar membrane are moved by different frequencies
of vibration. Thus, a particular sound frequency causes
the hairs of a specific group of receptor cells to shear
back and forth against the tectorial membrane more

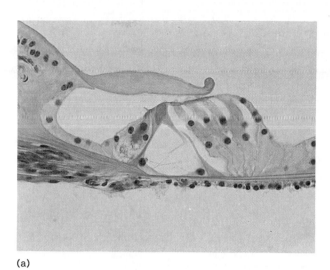

(a)

(b)

Figure 12.16   (*a*) A micrograph of the organ of Corti. (*b*) A scanning electron micrograph of hair cells in the organ of Corti.

intensely, while each of the other frequencies deflects another set of receptor cells.

Although these receptor cells are epithelial cells, they act somewhat like neurons. (See chapter 10.) For example, when such a cell is at rest, its membrane is polarized. When it is stimulated by having its hairs moved mechanically, its cell membrane becomes depolarized, ion channels open, and the membrane becomes more permeable to calcium ions. Although the receptor cell has no axon or dendrites, it has neurotransmitter-containing vesicles in the cytoplasm near its base. In the presence of calcium ions, some of these vesicles fuse with the cell membrane and release neurotransmitter substance to the outside. The neurotransmitter stimulates the ends of nearby sensory nerve fibers, and in response they transmit nerve impulses along the cochlear branch of the vestibulocochlear nerve to the brain.

Although the ear of a young person with normal hearing is able to detect sound waves with frequencies varying from about 20 to 20,000 or more vibrations per second, the range of greatest sensitivity is between 2,000 and 3,000 vibrations per second. (See figure 12.17.)

Auditory Nerve Pathways

The cochlear branches of the vestibulocochlear nerves enter the auditory nerve pathways that extend into the medulla oblongata and proceed through the midbrain to the thalamic region. From there they pass into the auditory cortices of the temporal lobes of the cerebrum where they are interpreted. On the way, some of these fibers cross over, so that impulses arising from each ear are interpreted on both sides of the brain. Consequently, damage to a temporal lobe on one side of the brain is not necessarily accompanied by complete hearing loss in the ear on that side. (See figure 12.18.)

Chart 12.2 summarizes the pathway of vibrations through the parts of the middle and inner ears.

The intensity of sound is measured in units called *decibels* (dB). The decibel scale begins at 0 dB, which is the intensity of the sound that is least perceptible by a normal human ear. The decibel scale is logarithmic, so that a sound of 10 dB is 10 times as intense as the least perceptible sound; a sound of 20 dB is 100 times as intense; and a sound of 30 dB is 1,000 times as intense.

On this scale, a whisper has an intensity of about 40 dB, normal conversation measures 60-70 dB, and heavy traffic produces about 80 dB. A sound of 120 dB, such as that commonly produced by the amplified sound at a rock concert, produces discomfort,and a sound of 140 dB, such as that emitted by a jet plane at takeoff, causes pain. It is important to note that frequent or prolonged exposure to sounds with intensities above 90 dB is likely to cause permanent hearing loss.

1. Describe the external, middle, and inner ears.
2. Explain how sound waves are transmitted through the parts of the ear.
3. Describe the tympanic reflex.
4. Distinguish between the osseous and membranous labyrinths.
5. Explain the function of the organ of Corti.

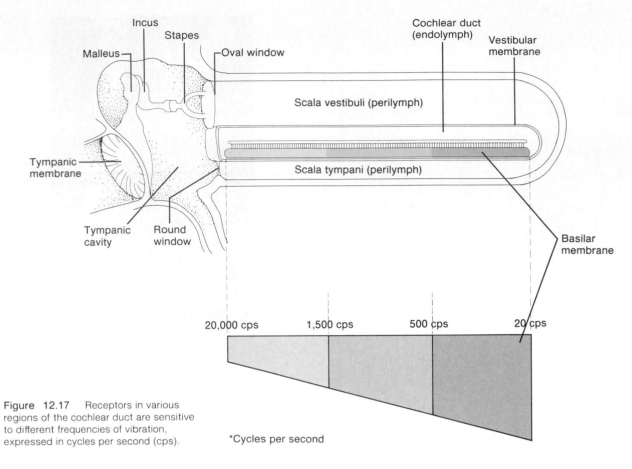

Figure 12.17 Receptors in various regions of the cochlear duct are sensitive to different frequencies of vibration, expressed in cycles per second (cps).

20,000 cps   1,500 cps   500 cps   20 cps

*Cycles per second

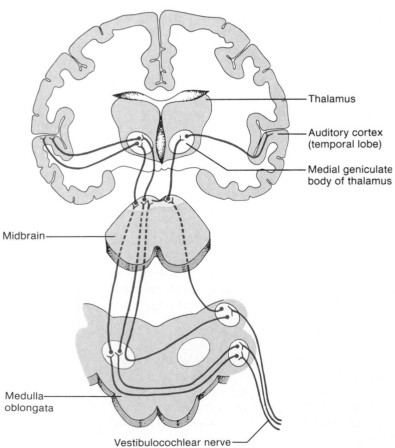

Figure 12.18 The auditory nerve pathway extends into the medulla oblongata, proceeds through the midbrain to the thalamus, and passes into the auditory cortex of the cerebrum.

Deafness

Partial or complete hearing loss can be caused by a variety of factors, including interference with the transmission of vibrations to the inner ear (conductive deafness) or damage to the cochlea, auditory nerve, or auditory nerve pathways (sensorineural deafness).

Conductive deafness may occur if the external auditory meatus becomes plugged by an accumulation of dry wax or by a foreign object. In other instances, the hearing impairment is related to changes in the eardrum or in the auditory ossicles. The eardrum, for example, may harden as a result of disease and thus be less responsive to sound waves, or it may be torn or perforated by disease or injury. The auditory ossicles may be damaged or destroyed by disease or may lose their mobility.

One of the more common diseases involving the auditory ossicles is *otosclerosis*. In this condition, new bone is deposited abnormally around the base of the stapes, interfering with the motion of the ossicle that is needed to transmit vibrations effectively to the inner ear. Although the cause of otosclerosis is unknown, a victim's hearing often can be restored by surgical procedures. For instance, the base of the stapes may be freed by chipping away the bone that holds it fixed in position, or the stapes may be removed and replaced with a wire or plastic substitute (prosthesis).

Two tests used to diagnose conductive deafness are the Weber test and the Rinne test. In the Weber test, the handle of a vibrating tuning fork is pressed against the forehead. A person with normal hearing perceives the sound coming from directly in front, while a person with sound conduction blockage in one middle ear hears the sound coming from the impaired side.

In the Rinne test, a vibrating tuning fork is held against the bone behind the ear. After the sound is no longer heard by conduction through the bones of the skull, the fork is moved to just in front of the external auditory meatus. In middle ear conductive deafness, the vibrating fork can no longer be heard, but a normal ear will continue to hear its tone.

Sensorineural deafness can be caused by excessively loud sounds. If the exposure is of short duration, the hearing loss may be temporary, but when exposure is repeated and prolonged, such as occurs in foundries, near jackhammers, or on a firing range, the impairment may be permanent.

Other causes of sensorineural deafness include tumors in the central nervous system, brain damage as a result of vascular accidents, and the use of certain drugs.

A prosthetic device called the *cochlear implant* has been developed to aid persons with sensorineural deafness. It consists of a microphone that senses external sounds, a tiny electronic processor that converts these sounds into electrical signals, and a cordlike electrode that is surgically threaded into the cochlea. In response to sounds detected by the microphone, the electrode stimulates fibers of the auditory nerve that remain functional. As a result, the person may be able to hear and interpret the information represented by some sounds.

In persons over age sixty-five, the ability to hear high-frequency sounds is often lost, and the ability to discriminate speech sounds is decreased—a condition called *presbycusis*. Thus, it is important to use lower voice tones in speaking to an elderly person. Also, the speaker should face the listener so that lip-reading is possible.

The degree of a person's hearing impairment can be measured with an *audiometer*. This device produces vibrations of known frequencies that are transmitted to the test subject through earphones. During the test, it is possible to determine the percentage of hearing loss a person has suffered at each vibration frequency.

CHART 12.2 Steps in the generation of sensory impulses from the ear

1. Sound waves enter the external auditory meatus.
2. Waves of changing pressures cause the eardrum to reproduce the vibrations coming from the sound wave source.
3. Auditory ossicles amplify and transmit vibrations to the end of the stapes.
4. Movement of the stapes at the oval window transmits vibrations to the perilymph in the scala vestibuli.
5. Vibrations pass through the vestibular membrane and enter the endolymph of the cochlear duct.
6. Different frequencies of vibration in endolymph stimulate different sets of receptor cells.
7. A receptor cell becomes depolarized; its membrane becomes more permeable to calcium ions.
8. In the presence of calcium ions, vesicles at the base of the receptor cell release neurotransmitter.
9. Neurotransmitter stimulates the ends of nearby sensory neurons.
10. Sensory impulses are triggered on fibers of the cochlear branch of the vestibulocochlear nerve.
11. The auditory cortex of the temporal lobe interprets the sensory impulses.

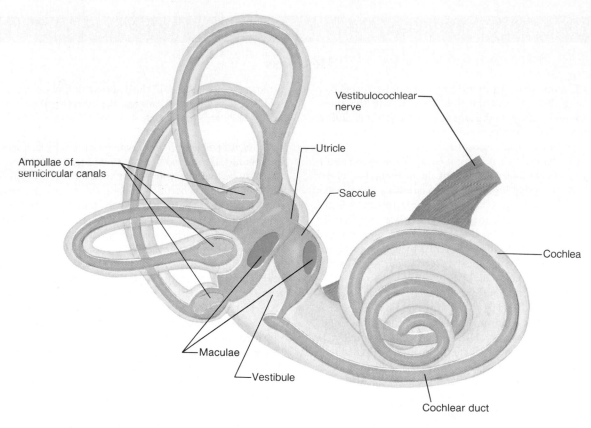

Figure 12.19 The saccule and utricle, which are expanded portions of the membranous labyrinth, are located within the bony chamber of the vestibule.

Sense of Equilibrium

The sense of equilibrium actually involves two senses—a sense of *static equilibrium* and a sense of *dynamic equilibrium*—that result from the actions of different sensory organs. The organs associated with static equilibrium function to sense the position of the head. They help maintain the stability and posture of the head and body when these parts are motionless. When the head and body are suddenly moved or rotated, the organs of dynamic equilibrium detect such motion and aid in maintaining balance.

Static Equilibrium

The organs of **static equilibrium** are located within the **vestibule,** a bony chamber between the semicircular canals and the cochlea. More specifically, the membranous labyrinth inside the vestibule consists of two expanded chambers—a **utricle** and a **saccule.** The larger utricle communicates with the saccule and the membranous portions of the semicircular canals; the saccule, in turn, communicates with the cochlear duct (figure 12.19).

Each of these chambers has a small patch of hair cells and supporting cells called a **macula** on its wall. When the head is upright, the hairs of the macula in the utricle project vertically, while those in the saccule project horizontally. In each case, the hairs are in contact with a sheet of gelatinous material (otolithic membrane) that has crystals of calcium carbonate (otoliths) embedded on its surface. These particles increase the weight of the gelatinous sheet, making it more responsive to changes in position. The hair cells, which serve as sensory receptors, have nerve fibers wrapped around their bases. These fibers are associated with the vestibular portion of the vestibulocochlear nerve.

The usual stimulus to the hair cells occurs when the head is bent forward, backward, or to one side. Such movements cause the gelatinous mass of one or more maculae to be tilted, and as the gelatinous material sags in response to gravity, the hairs projecting into it are bent. This action stimulates the hair cells, and they signal the nerve fibers associated with them in a manner similar to that of the hearing receptors. The resulting nerve impulses travel into the central nervous system by means of the vestibular branch of the vestibulocochlear nerve. These impulses inform the brain of the position of the head. The brain acts on this information by sending motor impulses to skeletal muscles, and they may contract or relax appropriately so that balance is maintained (figures 12.20 and 12.21).

The maculae also function in the sense of dynamic equilibrium. For example, if the head or body is

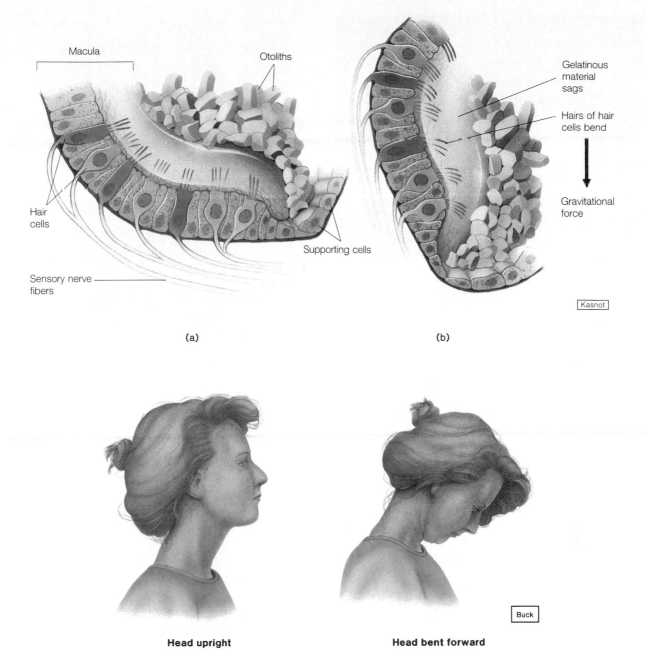

Macula

Otoliths

Hair
cells

Supporting cells

Sensory nerve
fibers

(a)

Gelatinous
material
sags

Hairs of hair
cells bend

Gravitational
force

Kasnot

(b)

Head upright

Head bent forward

Buck

Figure 12.20 The macula is responsive to changes in the position of the head. (*a*) Macula with the head in an upright position; (*b*) macula with the head bent forward.

thrust forward or backward abruptly, the gelatinous mass of the maculae lags slightly behind, and the hair cells are stimulated. In this way, the maculae aid the brain in detecting movements such as falling, and in maintaining posture while walking.

Dynamic Equilibrium

Each semicircular canal follows a circular path about 6 millimeters in diameter. The three bony **semicircular canals** lie at right angles to each other, and each occupies a different plane in space. Two of them, the *su-*

perior canal and the *posterior canal,* stand vertically, while the third, the *lateral canal,* is horizontal.

Suspended in the perilymph of each bony canal is a membranous semicircular canal that ends in a swelling called an **ampulla.** The ampullae communicate with the utricle of the vestibule.

An ampulla contains a septum that crosses the tube and houses a sensory organ. Each of these organs, called a **crista ampullaris,** contains a number of sensory hair cells and supporting cells. As in the maculae, the hairs of the hair cells extend upward into a dome-shaped gelatinous mass called the *cupula.* Also, the hair cells

Figure 12.21   Scanning electron micrograph of hairs of hair cells, such as those found in the utricle and saccule.

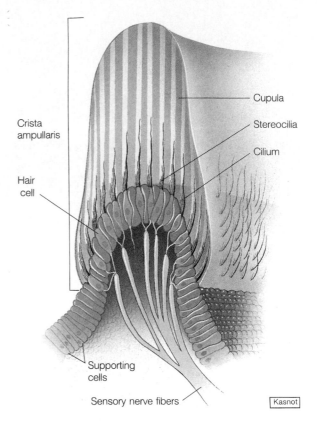

Figure 12.22   A crista ampullaris is located within the ampulla of each semicircular canal.

are connected at their bases to nerve fibers that make up part of the vestibular branch of the vestibulocochlear nerve (figure 12.22).

The hair cells of the crista are ordinarily stimulated by rapid turns of the head or body. At such times, the semicircular canals move with the head or body, but the fluid inside the membranous canals tends to remain stationary because of inertia. This causes the cupula in one or more of the canals to bend in a direction opposite that of the head or body movement, and the hairs embedded in it also bend. This bending of the hairs stimulates the hair cells to signal their associated nerve fibers, and, as a result, impulses travel to the brain (figure 12.23).

Parts of the cerebellum are particularly important in interpreting impulses from the semicircular canals. Analysis of such information allows the brain to predict the consequences of rapid body movements, and by stimulating appropriate skeletal muscles, the brain can prevent the loss of balance.

> Some people experience a discomfort called *motion sickness* when they are in a moving boat, airplane, or automobile. This discomfort seems to be caused by abnormal and irregular body motions that disturb the organs of equilibrium. Symptoms of motion sickness include nausea, vomiting, dizziness, headache, and prostration.

Other sensory structures also aid in maintaining equilibrium. For example, various proprioceptors, particularly those associated with the joints of the neck, supply the brain with information concerning the position of body parts. In addition, the eyes can detect changes in posture that result from body movements. Such visual information is so important that even though a person has suffered damage to the organs of equilibrium, he or she may be able to maintain normal balance by keeping the eyes open and moving slowly.

1. Distinguish between the senses of static and dynamic equilibrium.
2. What structures function to provide the sense of static equilibrium? Of dynamic equilibrium?
3. How does sensory information from other receptors help maintain equilibrium?

Sense of Sight

Although the eye contains the visual receptors, its functions are assisted by a number of *accessory organs*. These include the eyelids and lacrimal apparatus that help protect the eye, and a set of extrinsic muscles that move it.

Visual Accessory Organs

The eye, lacrimal gland, and extrinsic muscles are housed within the pear-shaped orbital cavity of the skull. The orbit, which is lined with the periosteums of various bones, also contains fat, blood vessels, nerves, and a variety of connective tissues.

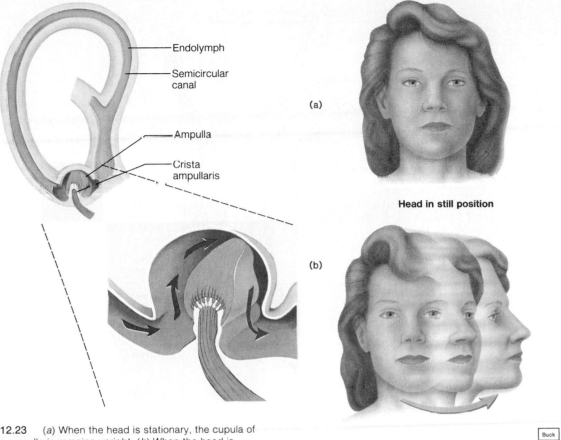

Figure 12.23 (*a*) When the head is stationary, the cupula of the crista ampullaris remains upright. (*b*) When the head is moving rapidly, the cupula is bent, and sensory receptors are stimulated.

Labels on figure: Endolymph; Semicircular canal; Ampulla; Crista ampullaris; (a) Head in still position; (b) Head rotating; Buck

Each **eyelid** (palpebra) is composed of four layers—skin, muscle, connective tissue, and conjunctiva. The skin of the eyelid, which is the thinnest skin of the body, covers the lid's outer surface and fuses with its inner lining near the margin of the lid (figure 12.24).

The muscles that move the eyelids include the *orbicularis oculi* and the *levator palpebrae superioris.* Fibers of the orbicularis oculi encircle the opening between the lids and spread out onto the cheek and forehead. This muscle acts as a sphincter that closes the lids when it contracts.

Fibers of the levator palpebrae superioris muscle arise from the roof of the orbit and are inserted in the connective tissue of the upper lid. When these fibers contract, the upper lid is raised and the eye opens.

The connective tissue layer of the eyelid, which helps give it form, contains numerous modified sebaceous glands (tarsal glands). The oily secretions of these glands are carried by ducts to openings along the borders of the lids. This secretion helps keep the lids from sticking together.

The **conjunctiva** is a mucous membrane that lines the inner surfaces of the eyelids and folds back to cover the anterior surface of the eyeball, except for its central portion (cornea). Although the portion that lines the eyelids is relatively thick, the conjunctiva that covers the eyeball is very thin. It is also freely movable and quite transparent, so that blood vessels are clearly visible beneath it.

The *lacrimal apparatus* (figure 12.25) consists of the **lacrimal gland,** which secretes tears, and a series of *ducts,* which carry the tears into the nasal cavity. The gland is located in the orbit, above and to the lateral side of the eye. It secretes tears continuously, and they pass out through tiny tubules, and flow downward and medially across the eye.

Tears are collected by two small ducts (superior and inferior canaliculi) whose openings (puncta) can be seen on the medial borders of the eyelids. From these ducts, the fluid moves into the *lacrimal sac,* which lies in a deep groove of the lacrimal bone, and then into the *nasolacrimal duct,* which empties into the nasal cavity.

Glandular cells of the conjunctiva also secrete a tearlike liquid that, together with the secretion of the lacrimal gland, keeps the surface of the eye and the lining of the lids moist and lubricated. Tears contain

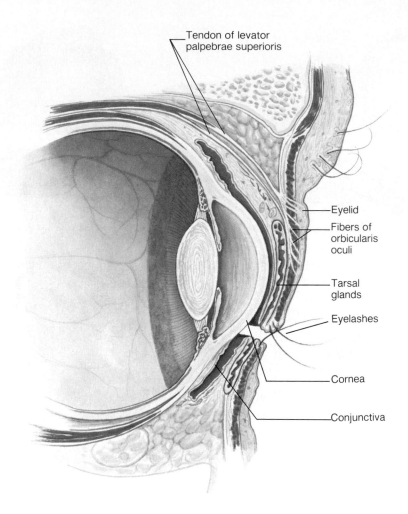

Figure 12.24 Sagittal section of the closed eyelids and the anterior portion of the eye.

Tendon of levator palpebrae superioris

Eyelid
Fibers of orbicularis oculi
Tarsal glands
Eyelashes
Cornea
Conjunctiva

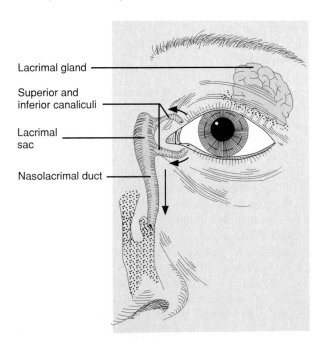

Lacrimal gland
Superior and inferior canaliculi
Lacrimal sac
Nasolacrimal duct

Figure 12.25 The lacrimal apparatus consists of a tear-secreting gland and a series of ducts.

an enzyme, called *lysozyme,* that functions as an antibacterial agent, reducing the chance of eye infections.

When a person is emotionally upset by grief or disappointment, or when the conjunctiva is irritated, the tear glands are likely to secrete excessive fluids. Tears may spill over the edges of the eyelids, and the nose may fill with fluid. This response involves motor impulses carried to the lacrimal glands on parasympathetic nerve fibers.

The **extrinsic muscles** of the eye arise from the bones of the orbit and are inserted by broad tendons on the eye's tough outer surface. Six such muscles function to move the eye in various directions (figure 12.26). Although any given eye movement may involve more than one of them, each muscle is associated with one primary action, as follows:

1. **Superior rectus**—rotates the eye upward and toward the midline.
2. **Inferior rectus**—rotates the eye downward and toward the midline.
3. **Medial rectus**—rotates the eye toward the midline.

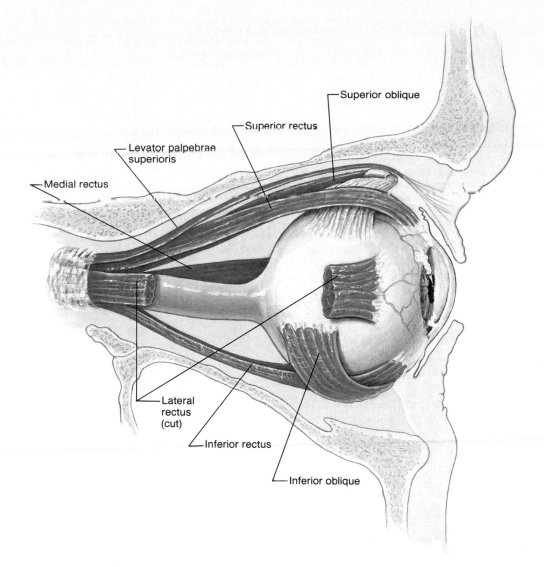

Figure 12.26 The extrinsic muscles of the right eye (lateral view).

Labels: Superior oblique, Superior rectus, Levator palpebrae superioris, Medial rectus, Lateral rectus (cut), Inferior rectus, Inferior oblique

4. **Lateral rectus**—rotates the eye away from the midline.
5. **Superior oblique**—rotates the eye downward and away from the midline.
6. **Inferior oblique**—rotates the eye upward and away from the midline.

The motor units of the extrinsic eye muscles contain the smallest number of muscle fibers (five to ten) of any muscles in the body. Because of this, the eyes can be moved with great precision. Also, the eyes move together so that they are aligned when looking at something. Such alignment involves complex motor adjustments that result in the contraction of certain eye muscles while their antagonists are relaxed. For example, when the eyes move to the right, the lateral rectus of the right eye and the medial rectus of the left eye must contract. At the same time, the medial rectus of the right eye and the lateral rectus of the left eye must relax. A person whose eyes are not coordinated well enough to produce alignment is said to have *strabismus,* or squint.

Chart 12.3 summarizes the muscles associated with the eyelids and eye.

When one eye deviates from the line of vision, the person has double vision (diplopia). If this condition persists, there is danger that changes will occur in the brain to suppress the image from the deviated eye. As a result, the turning eye may become blind (suppression amblyopia). Such monocular blindness can often be prevented if the eye deviation is treated early in life with exercises, eyeglasses, and surgery. For this reason, vision screening programs for preschool children are very important.

CHART 12.3 Muscles associated with the eyelids and eyes

Skeletal muscles			Smooth muscles		
Name	*Innervation*	*Function*	*Name*	*Innervation*	*Function*
Muscles of the eyelids			Ciliary muscles	Oculomotor nerve (III) parasympathetic fibers	Causes suspensory ligaments to relax
Orbicularis oculi	Facial nerve (VII)	Closes eye			
Levator palpebrae superioris	Oculomotor nerve (III)	Opens eye	Iris, circular muscles	Oculomotor nerve (III) parasympathetic fibers	Causes size of pupil to decrease
Extrinsic muscles of the eyes			Iris, radial muscles	Sympathetic fibers	Causes size of pupil to increase
Superior rectus	Oculomotor nerve (III)	Rotates eye upward and toward midline			
Inferior rectus	Oculomotor nerve (III)	Rotates eye downward and toward midline			
Medial rectus	Oculomotor nerve (III)	Rotates eye toward midline			
Lateral rectus	Abducens nerve (VI)	Rotates eye away from midline			
Superior oblique	Trochlear nerve (IV)	Rotates eye downward and away from midline			
Inferior oblique	Oculomotor nerve (III)	Rotates eye upward and away from midline			

1. Explain how the eyelid is moved.
2. Describe the conjunctiva.
3. What is the function of the lacrimal apparatus?
4. Describe the function of each extrinsic eye muscle.

Structure of the Eye

The eye is a hollow, spherical structure about 2.5 centimeters in diameter. Its wall has three distinct layers—an outer *fibrous tunic,* a middle *vascular tunic,* and an inner *nervous tunic.* The spaces within the eye are filled with fluids that provide support for its wall and internal parts, and help maintain its shape. Figure 12.27 shows the major parts of the eye.

The Outer Tunic

The anterior sixth of the outer tunic bulges forward as the transparent **cornea,** which serves as the window of the eye and helps focus entering light rays. It is composed largely of connective tissue with a thin layer of epithelium on its surface. The transparency of the cornea is due to the fact that it contains relatively few cells and no blood vessels. Also, the cells and collagenous fibers are arranged in unusually regular patterns.

On the other hand, the cornea is well supplied with nerve fibers that enter its margin and radiate toward its center. These fibers are associated with numerous pain receptors that have very low thresholds. Cold receptors are also abundant in the cornea, but heat and touch receptors seem to be lacking.

Along its circumference, the cornea is continuous with the **sclera,** the white portion of the eye. This part makes up the posterior five-sixths of the outer tunic and is opaque due to the presence of many large, haphazardly arranged collagenous and elastic fibers. The sclera provides protection and serves as an attachment for the extrinsic muscles of the eye.

In the back of the eye, the sclera is pierced by the **optic nerve** and certain blood vessels, and is attached to the dura mater that encloses these structures as well as other parts of the central nervous system.

Worldwide, the most common cause of blindness is corneal disease in which the transparency of the cornea is lost. Treatment for this condition may involve *corneal transplantation* (penetrating keratoplasty). In this procedure, the central two-thirds of the defective cornea is removed and replaced by a similar-sized portion of cornea from a donor eye. Because corneal tissues lack blood vessels, the transplanted tissue is usually not rejected by the recipient's immune system (chapter 19), and the success rate of this procedure is very high. Unfortunately for many persons in need of corneal transplantation, the supply of donor eyes remains inadequate.

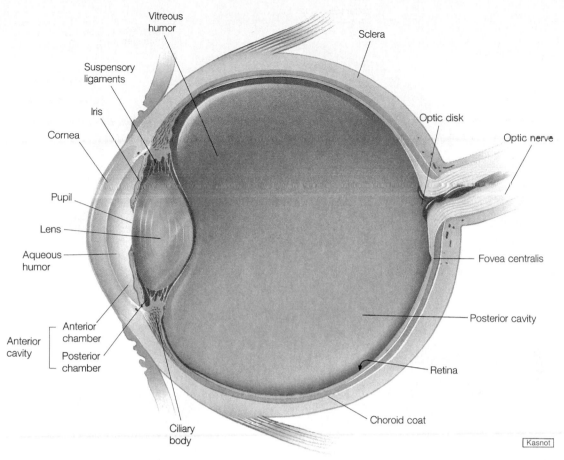

Figure 12.27 Transverse section of the eye (superior view).

The Middle Tunic

The middle, or vascular, tunic of the eyeball (uveal layer) includes the choroid coat, the ciliary body, and the iris.

The **choroid coat,** in the posterior five-sixths of the globe of the eye, is loosely joined to the sclera and is honeycombed with blood vessels that nourish the surrounding tissues. The choroid coat also contains numerous pigment-producing melanocytes that give it a brownish black appearance. The melanin of these cells absorbs excess light and helps keep the inside of the eye dark.

The **ciliary body,** which is the thickest part of the middle tunic, extends forward from the choroid coat and forms an internal ring around the front of the eye. Within the ciliary body are many radiating folds called *ciliary processes* and two distinct groups of muscle fibers that constitute the *ciliary muscles.* These structures are shown in figure 12.28.

The transparent **lens** is held in position by a large number of strong but delicate fibers, called *suspensory ligaments* (zonular fibers), which extend inward from the ciliary processes. The distal ends of these fibers are attached along the margin of a thin capsule that surrounds the lens. The body of the lens, which contains no blood vessels, lies directly behind the iris and pupil and is composed of specialized epithelial cells.

These epithelial cells originate from a single layer of epithelium located beneath the anterior portion of the lens capsule. The cells divide, and the newly formed cells on the surface of the lens capsule differentiate into columnar cells called *lens fibers.* These lens fibers constitute the substance of the lens. This process of lens fiber production continues slowly throughout life, so that the lens becomes progressively thicker from front to back. Simultaneously, the deeper lens fibers are compressed toward the center of the structure (figure 12.29).

The lens capsule is a clear, membranelike structure composed largely of intercellular material. It is quite elastic, and this quality keeps it under constant tension. As a result, the lens can assume a globular shape. However, the suspensory ligaments attached to the margin of the capsule are also under tension, and

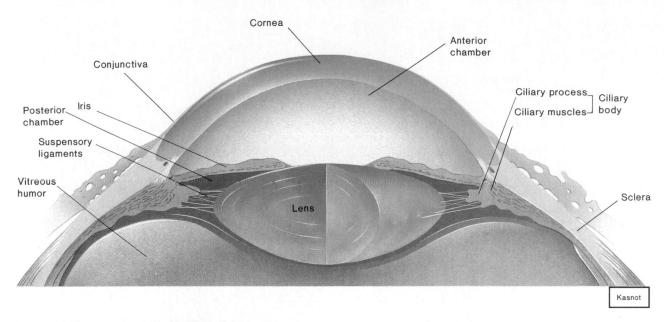

Figure 12.28   Anterior portion of the eye.

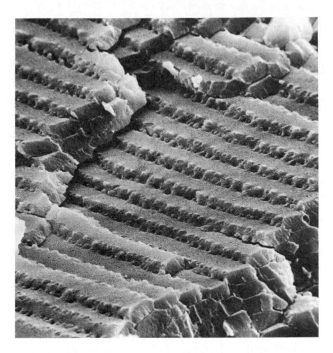

Figure 12.29   A scanning electron micrograph of the long, flattened lens fibers (×4,800). (Note the fingerlike junctions by which one fiber joins to another.)

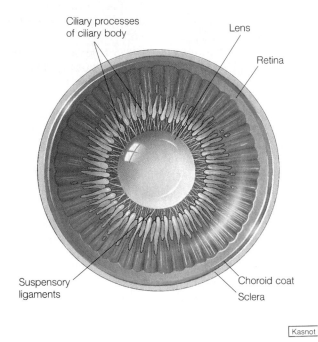

Figure 12.30   The lens and ciliary body viewed from behind.

because they pull outward, the capsule and the lens inside are kept somewhat flattened (figure 12.30).

If the tension on the suspensory ligaments is relaxed, the elastic capsule rebounds, and the lens surface becomes more convex. This change occurs in the lens when the eye is focused to view a close object. This adjustment is called **accommodation.**

The relaxation of the suspensory ligaments during accommodation is a function of the ciliary muscles. One

set of these muscle fibers is arranged in a circular pattern, forming a sphincterlike structure around the ciliary processes. The fibers of the other set extend back from fixed points in the sclera to the choroid coat. When the circular muscle fibers contract, the diameter of the ring formed by the ciliary processes is lessened; when the other fibers contract, the choroid coat is pulled forward and the ciliary body is shortened. Each of these actions causes the suspensory ligaments to become relaxed, and the lens thickens in response. In this thickened state, the lens is focused for viewing closer objects than before (figure 12.31).

To focus on a more distant object, the ciliary muscles are relaxed, tension on the suspensory ligaments increases, and the lens becomes thinner again.

A relatively common eye disorder, particularly in older people, is called *cataract*. In this condition, the lens or its capsule slowly loses its transparency and becomes cloudy, opaque, and discolored. As a result, clear images cannot be focused on the retina, and, in time, the person may become blind.

Cataract is often treated by surgically removing the lens and replacing it with an artificial one (intraocular lens implant). Sometimes the loss of refractive power following removal of the lens may also be corrected with eyeglasses or contact lenses.

1. Describe the outer and middle tunics of the eye.
2. What factors contribute to the transparency of the cornea?
3. How does the shape of the lens change during accommodation?

The **iris** is a thin diaphragm composed largely of connective tissue and smooth muscle fibers that is seen from the outside as the colored portion of the eye. It extends forward from the periphery of the ciliary body and lies between the cornea and the lens. The iris divides the space separating these parts, which is called the *anterior cavity,* into an *anterior chamber* (between the cornea and the iris) and a *posterior chamber* (between the iris and the vitreous humor, and occupied by the lens).

The epithelium on the inner surface of the ciliary body continuously secretes a watery fluid called **aqueous humor** into the posterior chamber. The fluid circulates from this chamber through the **pupil,** a circular opening in the center of the iris, and into the anterior chamber (figure 12.32). Aqueous humor fills the space between the cornea and the lens, helps nourish these parts, and aids in maintaining the shape of the front of the eye. It subsequently leaves the anterior chamber through veins and a special drainage canal (canal of Schlemm) located in its wall.

The smooth muscle fibers of the iris are arranged into two groups, a *circular set* and a *radial set*. These muscles control the size of the pupil, which is the opening that light passes through as it enters the eye. The circular set of muscle fibers acts as a sphincter, and when it contracts, the pupil gets smaller and the intensity of the light entering decreases. When the radial muscle fibers contract, the diameter of the pupil increases and the intensity of the light entering increases.

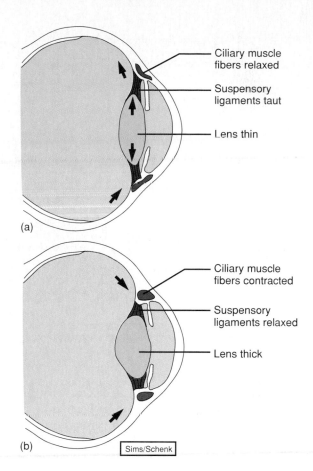

(a)

(b) Sims/Schenk

Figure 12.31 (a) How is the focus of the eye affected when the lens becomes thin as the ciliary muscle fibers relax? (b) How is focus affected when the lens thickens as the ciliary muscle fibers contract?

A disorder called *glaucoma* sometimes develops in the eyes as a person ages. This condition occurs when the rate of aqueous humor formation exceeds the rate of its removal. As a result, fluid accumulates in the anterior chamber of the eye, and the fluid pressure rises.

Because liquids cannot be compressed, the increasing pressure from the anterior chamber is transmitted to all parts of the eye, and, in time, the blood vessels that supply the receptor cells of the retina may be squeezed closed. If this happens, cells that fail to receive needed nutrients and oxygen may die, and permanent blindness may result.

When it is diagnosed early, glaucoma can usually be treated successfully with drugs, laser therapy, or surgery that will promote the outflow of aqueous humor. However, since glaucoma in its early stages typically produces no symptoms, discovery of the condition depends largely on measurement of the intraocular pressure using an instrument called a *tonometer.*

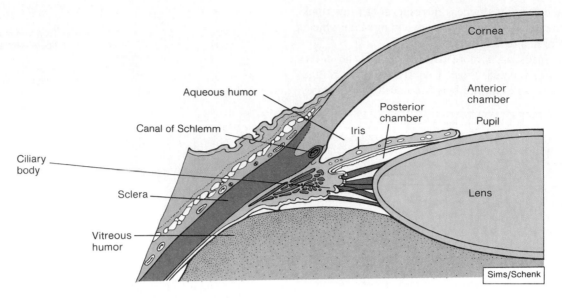

Figure 12.32 Aqueous humor, which is secreted into the posterior chamber, circulates into the anterior chamber and leaves it through the canal of Schlemm.

The sizes of the pupils change constantly in response to pupillary reflexes that are triggered by such factors as light intensity, gaze, and variations in emotional state. For example, bright light elicits a reflex, and impulses travel along parasympathetic nerve fibers to the *circular muscles* of the irises. The pupils become constricted in response. Conversely, in dim light, impulses travel on sympathetic nerve fibers to the *radial muscles* of the irises, and the pupils become dilated (figure 12.33).

The color of the eyes is determined largely by the amount and distribution of melanin in the irises and by the density of the tissue within the body of the iris. For example, if melanin is present only in the epithelial cells that cover an iris's posterior surface, the iris appears blue. When this condition exists together with denser than usual tissue within the body of the iris, it looks gray. And when melanin is present within the body of the iris as well as in the epithelial covering, the iris appears brown.

The Inner Tunic

The inner tunic of the eye consists of the **retina,** which contains the visual receptor cells (photoreceptors). This nearly transparent sheet of tissue is continuous with the optic nerve in the back of the eye and extends forward as the inner lining of the eyeball. It ends just behind the margin of the ciliary body.

Although the retina is thin and delicate, its structure is quite complex. It has a number of distinct layers, including pigmented epithelium, neurons, nerve fibers, and limiting membranes (figures 12.34 and 12.35).

There are five major groups of retinal neurons. The nerve fibers of three of these groups—the *receptor*

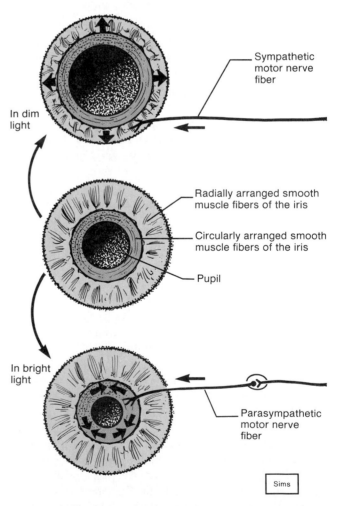

Figure 12.33 In dim light, the radial muscles of the iris are stimulated to contract, and the pupil dilates. In bright light, the circular muscles of the iris are stimulated to contract, and the pupil becomes smaller.

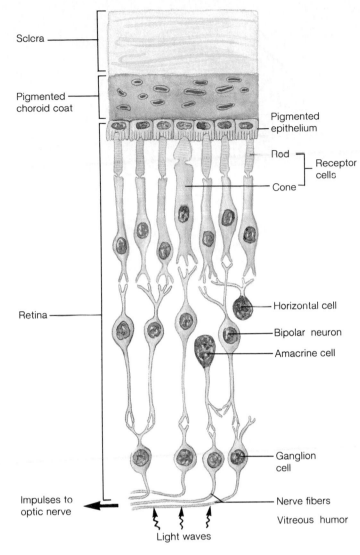

Sclera

Pigmented
choroid coat

Pigmented
epithelium

Rod — Receptor
cells

Cone

Retina

Horizontal cell

Bipolar neuron

Amacrine cell

Ganglion
cell

Impulses to
optic nerve

Nerve fibers

Vitreous humor

Light waves

Figure 12.34   The retina consists of several cell layers.

cells, bipolar neurons, and ganglion cells—provide a direct pathway for impulses triggered in the receptors to the optic nerve and brain. The nerve fibers of the other two groups of retinal cells, called *horizontal cells* and *amacrine cells,* pass laterally between retinal cells (figure 12.34). The horizontal and amacrine cells modify the impulses transmitted on the fibers of the direct pathway.

In the central region of the retina is a yellowish spot called the **macula lutea** that occupies about 1 square millimeter. A depression in its center, called the **fovea centralis,** is in the region of the retina that produces the sharpest vision and is explained in a subsequent section of this chapter.

Just medial to the fovea centralis is an area called the **optic disk** (figure 12.36). Here the nerve fibers from the retina leave the eye and become parts of the optic nerve. A central artery and vein also pass through at the optic disk. These vessels are continuous with capillary networks of the retina, and together with vessels in the underlying choroid coat, they supply blood to the cells of the inner tunic. Because there are no receptor cells in the region of the optic disk, it is commonly referred to as the *blind spot* of the eye.

The space bounded by the lens, ciliary body, and retina is the largest compartment of the eye and is called the *posterior cavity.* It is filled with a transparent, jellylike fluid called **vitreous humor,** which together with some collagenous fibers comprise the **vitreous body.** The vitreous body supports the internal parts of the eye and helps maintain its shape.

In summary, light waves entering the eye must pass through the cornea, aqueous humor, lens, vitreous humor, and several layers of the retina before they reach the photoreceptors. The layers of the eye are summarized in chart 12.4.

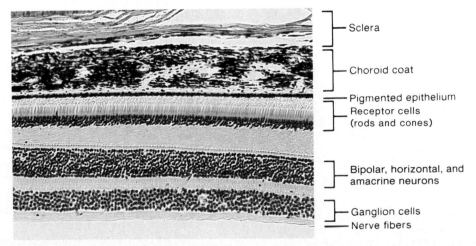

Sclera

Choroid coat

Pigmented epithelium
Receptor cells
(rods and cones)

Bipolar, horizontal, and
amacrine neurons

Ganglion cells
Nerve fibers

Figure 12.35   Note the layers of cells and nerve fibers in this light micrograph of the retina (×200).

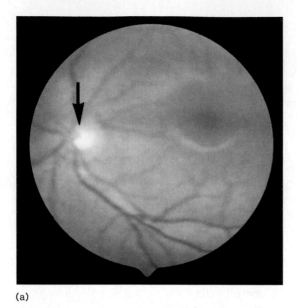

(a)

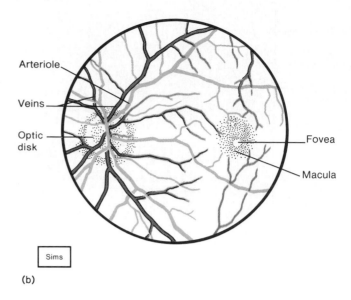

Arteriole

Veins

Optic disk

Fovea

Macula

Sims

(b)

Figure 12.36 (a) Nerve fibers leave the eye in the area of the optic disk (arrow) to form the optic nerve; (b) major features of the retina.

CHART 12.4	Layers of the eye			
Tunic	Posterior portion	Function	Anterior portion	Function
Outer layer	Sclera	Protection	Cornea	Light transmission and refraction
Middle layer	Choroid coat	Blood supply; pigment prevents reflection	Ciliary body, iris	Accommodation; controls light intensity
Inner layer	Retina	Photoreception, impulse transmission	None	

Commonly, as a person ages, tiny dense clumps of gel or deposits of crystal-like substances form in the vitreous humor. These clumps may cast shadows on the retina, and as a result, the person sees small moving specks in the field of vision. Such specks are known as *floaters,* and they are most apparent when looking at a plain background, such as the sky or a blank wall.

Also with age, the vitreous humor tends to shrink and pull away from the retina. As this occurs, receptor cells of the retina may be mechanically stimulated, and the person may see flashes of light. The presence of floaters or light flashes usually does not indicate a serious eye problem; however, if they develop suddenly or seem to be increasing in number or frequency, the eye should be examined by an ophthalmologist.

1. Explain the origin of aqueous humor and trace its path through the eye.
2. How is the size of the pupil regulated?
3. Describe the structure of the retina.

Refraction of Light

When a person sees something, either the object is giving off light, or light waves are being reflected from it. These light waves enter the eye, and an image of what is seen becomes focused upon the retina. This focusing process involves the bending of the light waves—a phenomenon called **refraction.**

Refraction occurs when light waves pass at an oblique angle from a medium of one optical density into a medium of a different optical density. For example, as figure 12.37 shows, when light passes obliquely from a less dense medium such as air into a denser medium such as glass, or from air into the cornea of the eye, the light is bent toward a line perpendicular to the surface between these substances. When the surface between such refracting media is curved, a lens is formed. A lens with a *convex* surface causes light waves to converge, and a lens with a *concave* surface causes light waves to diverge (figure 12.38).

When light arrives from objects outside the eye, the light waves are refracted primarily by the convex surface of the cornea. This surface is responsible for about 75% of the total refractive power of the eye. The light is refracted again by the convex surface of the lens and to a lesser extent by the surfaces of the fluids within the chambers of the eye.

If the shape of the eye is normal, light waves are focused sharply upon the retina, much as a motion picture image is focused on a screen for viewing. Unlike the motion picture image, however, the one formed on the retina is upside down and reversed from left to right (figure 12.39). When the visual cortex of the cerebrum interprets such an image, it somehow corrects this, and things are seen in their proper positions.

Light waves coming from objects more than 20 feet away are traveling in nearly parallel lines, and they are focused on the retina by the cornea and by the lens in its more flattened or "at rest" condition. Light waves arriving from objects less than 20 feet away, however, reach the eye along more divergent lines—in fact, the closer the object, the more divergent the lines.

Divergent light waves tend to come into focus behind the retina unless something is done to increase the refracting power of the eye. This increase is accomplished by *accommodation,* which results in a thickening of the lens. As the lens thickens, light waves are converged more strongly, so that diverging light waves coming from close objects are focused on the retina.

1. What is meant by refraction?
2. What parts of the eye provide refracting surfaces?
3. Why is it necessary to accommodate for viewing close objects?

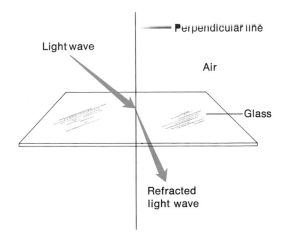

Figure 12.37 When light passes at an oblique angle from air into glass, the light waves are bent toward a line perpendicular to the surface of the glass.

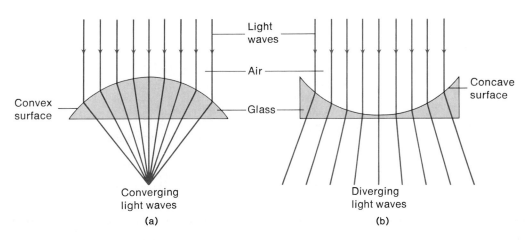

(a)   (b)

Figure 12.38 (a) A lens with a convex surface causes light waves to converge; (b) a lens with a concave surface causes them to diverge.

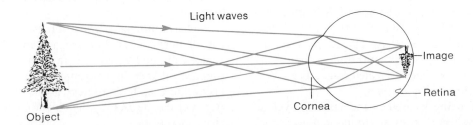

Figure 12.39 The image of an object is formed on the retina upside down.

Refraction Disorders

Unfortunately, the elastic quality of the lens capsule tends to lessen with time, and persons over forty-five years of age are often unable to accommodate sufficiently to read the fine print in books and newspapers. Their eyes remain focused for distant vision. This condition is termed *presbyopia,* or farsightedness of age, and it can usually be corrected using eyeglasses or contact lenses that make up for the eye's loss of refracting power.

Other persons suffer from problems that result when their eyeballs are too short or too long for sharp focusing. For example, if an eye is too short, light waves are not focused sharply on the retina because their point of focus lies some distance behind it. A person with this condition may be able to bring the image of distant objects into focus by accommodation, but this requires contraction of the ciliary muscles at times when these muscles are at rest in a normal eye. Still more accommodation is needed to view closer objects, and the person may suffer from ciliary muscle fatigue, pain, and headache when doing close work.

Since people with short eyeballs are usually unable to accommodate enough to focus on very close objects, they are said to be *farsighted.* This condition (hyperopia) is remedied by eyeglasses or contact lenses with *convex* surfaces that cause images to be focused closer to the front of the eye.

If an eyeball is too long, light waves tend to be focused in front of the retina, and the image produced on it is blurred. In other words, the refracting power of the eye, even when the lens is flattened, is too great. Although a person with this problem may be able to focus on close objects by accommodation, distance vision is invariably poor. For this reason, the person is said to be *nearsighted*. Treatment for nearsightedness (myopia) involves eyeglasses or contact lenses with *concave* surfaces that cause images to be focused further from the front of the eye (figures 12.40 and 12.41).

Still another refraction problem is termed *astigmatism.* This condition is characterized by a defect in the curvature of the cornea, or sometimes, in the curvature of the lens. The normal cornea has a spherical curvature, like the

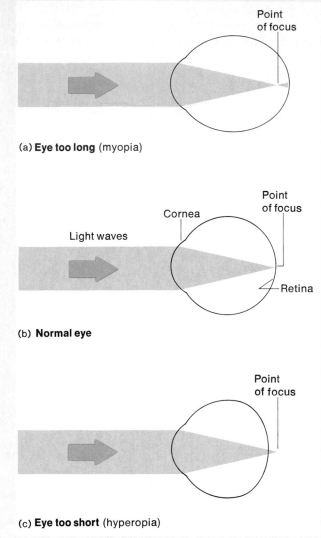

(a) **Eye too long** (myopia)

(b) **Normal eye**

(c) **Eye too short** (hyperopia)

Figure 12.40 (*a*) If an eye is too long, the focus point of images lies in front of the retina. (*b*) In a normal eye, the focus point is on the retina. (*c*) If an eye is too short, the focus point lies behind the retina.

Visual Receptors

The photoreceptors of the eye are actually modified neurons, and there are two distinct kinds. One group of receptor cells have long, thin projections at their terminal ends and are called **rods.** The retina contains about 100 million rods. The cells of the other group have short, blunt projections and are called **cones.** The retina contains about three million cones.

Instead of being located in the surface layer of the retina, the rods and cones are found in a deep portion, closely associated with a layer of pigmented epithelium. The projections from the receptors extend into the pigmented layer and contain light-sensitive visual pigments.

The epithelial pigment of the retina functions to absorb light waves that are not absorbed by the receptor cells, and together with the pigment of the choroid coat, it keeps light from reflecting off the surfaces inside the eye. The pigment layer also stores vitamin A, which can be used by the receptor cells to synthesize visual pigments.

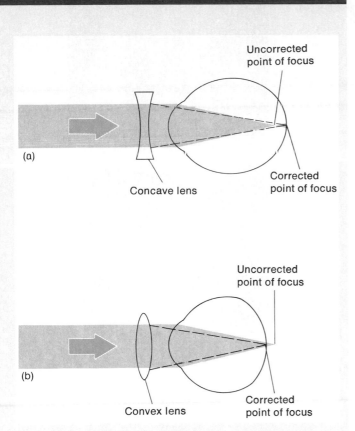

(a)

Uncorrected
point of focus

Concave lens

Corrected
point of focus

(b)

Uncorrected
point of focus

Convex lens

Corrected
point of focus

Figure 12.41 (a) Nearsightedness is corrected with a concave lens; (b) farsightedness is corrected with a convex lens.

inside of a ball; an astigmatic cornea usually has an elliptical curvature, like the bowl of a spoon. As a result, some portions of an image are in focus on the retina, but other portions are blurred, and vision is distorted.

Without corrective lenses, astigmatic eyes tend to accommodate back and forth reflexly in an attempt to sharpen focus. The consequence of this continual action is likely to be ciliary muscle fatigue and headache.

Persons with albinism, who lack melanin in all parts of their bodies including their eyes, usually have impaired visual sharpness (acuity) and an intolerance for light (photophobia). This is because light reflects from inside their eyes and stimulates their visual receptors excessively.

The visual receptors are only stimulated when light reaches them. Thus, when a light image is focused on an area of the retina, some receptors are stimulated, and impulses travel away from them to the brain. How-

ever, the impulse leaving each activated receptor provides only a fragment of the information needed for the brain to interpret a total scene.

Rods and cones function differently. For example, rods are hundreds of times more sensitive to light than cones, and as a result, rods enable persons to see in relatively dim light. In addition, rods produce colorless vision, while cones can detect colors of light.

Still another difference involves the sharpness of the images perceived. Cones allow people to see sharp images, while rods enable them to see more general outlines of objects. This characteristic is related to the fact that nerve fibers from many rods may converge, and their impulses may be transmitted to the brain on the same nerve fiber (see chapter 10). Thus, if a point of light stimulates a rod, the brain cannot tell which one of many receptors has actually been stimulated. Such a convergence of impulses occurs to a much lesser degree among cones, so when a cone is stimulated, the brain is able to pinpoint the stimulation more accurately (figure 12.42).

As was mentioned, the area of sharpest vision is the fovea centralis in the macula lutea. This area lacks rods, but contains densely packed cones with few or no converging fibers. Also, the overlying layers of the retina, as well as the retinal blood vessels, are displaced to the sides in the fovea. This displacement more fully exposes the receptors to incoming light. Consequently, to view something in detail, a person moves the eyes so that the important part of an image falls upon the fovea centralis.

The concentration of cones decreases in areas further away from the macula lutea, while the concentration of rods increases in these areas. Also, the amount of convergence among the rods and cones increases toward the periphery of the retina. As a result, the visual sensations from images focused on the sides of the retina tend to be blurred compared with those focused on the central portion of the retina.

Although the eye is well protected by the bony orbit, sometimes it is injured by a forceful blow. If the eye is jarred sufficiently, some of its contents may be displaced. For example, the suspensory ligaments may be torn and the lens dislocated into the posterior cavity. Similarly, a blow to the eye (or to the head) may cause the retina to pull away from the underlying vascular choroid coat. Once the retina is detached, there is danger that photoreceptor cells will die because of lack of oxygen and nutrients. Unless a *detached retina* can be repaired surgically, this injury may result in varying degrees of visual loss or blindness. Such injuries to the eye often occur in sports, and usually can be prevented by wearing protective devices, such as face masks or goggles.

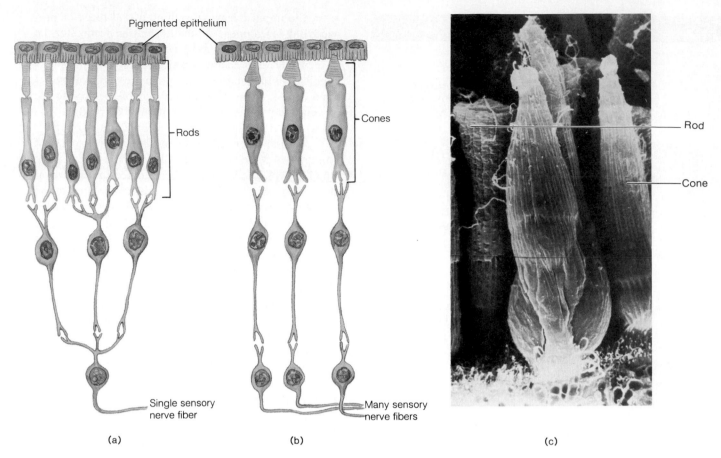

Figure 12.42 (a) Impulses from several rods may be transmitted to the brain on a single sensory nerve fiber. (b) Impulses from cones are often transmitted to the brain on separate sensory nerve fibers. (c) A scanning electron micrograph of rods and cones.

(a)   (b)   (c)

Visual Pigments

Both rods and cones contain light-sensitive pigments that decompose when they absorb light energy. The light-sensitive substance in rods is called **rhodopsin,** or visual purple, and it is embedded in disks of membrane that are stacked within these receptor cells. (See figure 12.43.) In the presence of light, rhodopsin molecules break down into molecules of a colorless protein called *opsin* and a yellowish organic substance called *retinal* (retinene) that is synthesized from vitamin A.

In darkness, sodium channels in portions of the receptor cell membranes are kept open by the presence of a nucleotide called *cyclic guanosine monophosphate* (c-GMP). When rhodopsin molecules absorb light, the opsin that is released becomes an active enzyme. This enzyme activates another enzyme (transducin), which, in turn, activates still another one (phosphodiesterase). The third enzyme of this series causes the breakdown of c-GMP, and as the concentration of c-GMP decreases, the sodium channels close, and the receptor cell membrane becomes hyperpolarized (chapter 10). The degree of hyperpolarization achieved by this process is directly proportional to the intensity of the light that is stimulating the receptor cells.

The hyperpolarization moves as a wave over the surface of the receptor cell membrane, and when it reaches the synaptic end of the cell, nerve impulses (action potentials) are triggered in nearby retinal neurons. Consequently, nerve impulses travel away from the retina, through the optic nerve, and into the brain. (Note: The transmission of hyperpolarization by photoreceptor cell membranes differs from the depolarization that occurs in most other sensory receptor cells.)

In bright light, nearly all of the rhodopsin in the rods of the retina is decomposed, and the sensitivity of these receptors is greatly reduced. In dim light, however, rhodopsin can be regenerated from scotopsin and retinal faster than it is broken down. This regeneration process requires cellular energy, which is provided by energy-carrying molecules of ATP (chapter 4).

The light sensitivity of an eye whose rods have converted the available scotopsin and retinal to rhodopsin increases about 100,000 times, and the eye is said to be *dark adapted.*

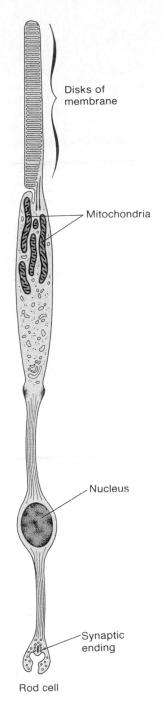

Disks of membrane

Mitochondria

Nucleus

Synaptic ending

Rod cell

Figure 12.43 Rhodopsin is embedded in disks of membrane that are stacked within the rod cells.

A person needs a dark-adapted eye to see in dim light. For example, when going from daylight into a darkened theater, it may be difficult to see well enough to locate a seat, but soon the eyes adapt to the dim light, and the vision improves. Later, when leaving the theater and entering the sunlight, the person may feel discomfort or even pain. This occurs at the moment that most of the rhodopsin decomposes in response to the

bright light. At the same time, the light sensitivity of the eyes decreases greatly, and they become *light adapted*.

Many people, particularly children, suffer from vitamin A deficiency due to improper diet. In such cases, the quantity of retinal available for manufacturing rhodopsin may be reduced, and consequently, the sensitivity of these individuals' rods may be low. This condition, called *night blindness*, is characterized by poor vision in dim light. Fortunately, the problem is usually easy to correct by adding vitamin A to the diet or by providing the patient vitamin A injections. Vitamin A deficiency remains one of the principal nutritional problems throughout the world, even though most countries have ample supplies of the plant foods that provide this needed substance.

The light-sensitive pigments of cones, called *iodopsins*, are similar to rhodopsin in that they are composed of retinal combined with a protein; the protein, however, differs from the protein in the rods. In fact, there are three sets of cones within the retina, each containing an abundance of one of the three different visual pigments.

The wavelength of a particular kind of light determines the color perceived from it. For example, the shortest wavelengths of visible light are perceived as violet, while the longest wavelengths of visible light are seen as red. As far as cone pigments are concerned, one type (erythrolabe) is most sensitive to red light waves, another (chlorolabe) to green light waves, and a third (cyanolabe) to blue light waves. There is, however, some overlap in the sensitivities of these pigments. For example, both red and green light pigments are sensitive to orange light waves. On the other hand, the red light pigment absorbs the orange light waves more effectively.

The color a person perceives depends upon which set of cones or combination of sets of cones is stimulated by the light in a given image. If all three sets of cones are stimulated, the person senses the light as white, and if none are stimulated, the person senses black.

Some persons have defective color vision, due to a decreased sensitivity in one or more cone sets. Such people perceive colors differently from those with normal vision and are said to be *color-blind*.

Stereoscopic Vision

Stereoscopic vision (stereopsis) is vision that involves perceiving the distance and depth as well as the height and width of objects. Such vision is due largely to the fact that the pupils of the eyes are 6–7 centimeters apart. Consequently, objects that are relatively close

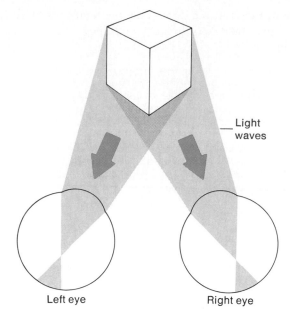

Figure 12.44 Stereoscopic vision results from the formation of two slightly different retinal images.

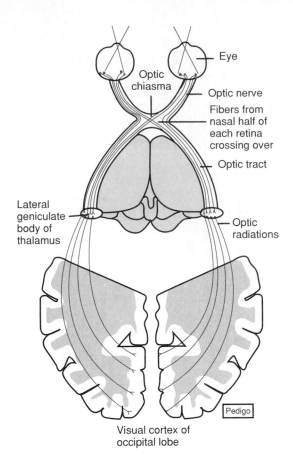

Figure 12.45 The visual pathway includes the optic nerve, optic chiasma, optic tract, and optic radiations.

(less than 20 feet away) produce slightly different retinal images. That is, the right eye sees a little more of one side of an object, while the left eye sees a little more of the other side. These two images are somehow superimposed and interpreted by the visual cortex of the brain, and the result is the perception of a single object in three dimensions (figure 12.44).

Because this type of depth perception depends on vision with two eyes (binocular vision), it follows that a one-eyed person has a lessened ability to judge distance and depth accurately. To compensate, a person with one eye can use clues provided by the relative sizes and positions of familiar objects.

Visual Nerve Pathways

As mentioned in chapter 11, the axons of the ganglion cells in the retina leave the eyes to form the *optic nerves.* Just anterior to the pituitary gland, these nerves give rise to the X-shaped *optic chiasma,* and within the chiasma some of the fibers cross over. More specifically, the fibers from the nasal (medial) half of each retina cross over, while those from the temporal (lateral) sides do not. Thus, fibers from the nasal half of the left eye and the temporal half of the right eye form the right *optic tract*; and fibers from the nasal half of the right eye and the temporal half of the left eye form the left optic tract.

The nerve fibers continue in the optic tracts, and just before they reach the thalamus, a few of them leave to enter nuclei that function in various visual reflexes. Most of the fibers, however, enter the thalamus and synapse in its posterior portion (lateral geniculate body). From this region the visual impulses enter nerve

pathways called *optic radiations,* and the pathways lead to the visual cortex of the occipital lobes (figure 12.45).

Other fibers conducting visual impulses pass downward into various regions of the brain stem. These impulses are important for controlling head and eye movements associated with tracking an object visually, for controlling the simultaneous movements of both eyes, and for controlling certain visual reflexes, such as those involved with movements of the iris muscles.

Since each visual cortex receives impulses from each eye, a person may develop partial blindness in both eyes if either visual cortex is injured. For example, if the right visual cortex (or the right optic tract) is injured, sight may be lost in the temporal side of the right eye and the nasal side of the left eye. Similarly, if the central portion of the optic chiasma, where fibers from the nasal sides of the eyes cross over, is damaged, the nasal sides of both eyes are blinded.

1. Distinguish between the rods and the cones of the retina.
2. Explain the roles of visual pigments.
3. What factors make stereoscopic vision possible?
4. Trace the pathway of visual impulses from the retina to the occipital cortex.

Clinical Terms Related to the Senses

amblyopia (am″ble-o′pe-ah) Dim vision due to a cause other than a refractive disorder or lesion.

anopia (an-o′pe-ah) Absence of an eye.

audiometry (aw″de-om′ĕ-tre) The measurement of auditory acuity for various frequencies of sound waves.

blepharitis (blef″ah-ri′tis) An inflammation of the margins of the eyelids.

causalgia (kaw-zal′je-ah) A persistent, burning pain usually associated with injury to a limb.

conjunctivitis (kon-junk″ti-vi′tis) An inflammation of the conjunctiva.

diplopia (di-plo′pe-ah) Double vision, or the sensation of seeing two objects when only one is viewed.

emmetropia (em″ĕ-tro′pe-ah) Normal condition of the eyes; eyes with no refractive defects.

enucleation (e-nu″kle-a′shun) Removal of the eyeball.

exophthalmos (ek″sof-thal′mos) Condition in which the eyes protrude abnormally.

hemianopsia (hem″e-an-op′se-ah) Defective vision affecting half of the visual field.

hyperalgesia (hi″per-al-je′ze-ah) An abnormally increased sensitivity to pain.

iridectomy (ir″ĭ-dek′to-me) The surgical removal of part of the iris.

iritis (i-ri′tis) An inflammation of the iris.

keratitis (ker″ah-ti′tis) An inflammation of the cornea.

labyrinthectomy (lab″ĭ-rin-thek′to-me) The surgical removal of the labyrinth.

labyrinthitis (lab″ĭ-rin-thi′tis) An inflammation of the labyrinth.

Ménière's disease (men″e-ārz′ di-zez) An inner ear disorder characterized by ringing in the ears, increased sensitivity to sounds, dizziness, and loss of hearing.

neuralgia (nu-ral′je-ah) Pain resulting from inflammation of a nerve or a group of nerves.

neuritis (nu-ri′tis) An inflammation of a nerve.

nystagmus (nis-tag′mus) An involuntary oscillation of the eyes.

otitis media (o-ti′tis me′de-ah) An inflammation of the middle ear.

otosclerosis (o″to-skle-ro′sis) A formation of spongy bone in the inner ear, which often causes deafness by fixing the stapes to the oval window.

pterygium (tĕ-rij′e-um) An abnormally thickened patch of conjunctiva that extends over part of the cornea.

retinitis pigmentosa (ret″ĭ-ni′tis pig″men-to′sa) A progressive retinal sclerosis characterized by deposits of pigment in the retina and by atrophy of the retina.

retinoblastoma (ret″ĭ-no-blas-to′mah) An inherited, highly malignant tumor arising from immature retinal cells.

tinnitus (ti-ni′tus) A ringing or buzzing noise in the ears.

tonometry (to-nom′ĕ-tre) The measurement of fluid pressure within the eyeball.

trachoma (trah-ko′mah) A bacterial disease of the eye, characterized by conjunctivitis, which may lead to blindness.

tympanoplasty (tim″pah-no-plas′te) The surgical reconstruction of the middle ear bones and establishment of continuity from the tympanic membrane to the oval window.

uveitis (u″ve-i′tis) An inflammation of the uvea, the region of the eye that includes the iris, the ciliary body, and the choroid coat.

vertigo (ver′ti-go) A sensation of dizziness.

Chapter Summary

Introduction (page 416)

Sensory receptors are sensitive to environmental changes and initiate impulses to the brain and spinal cord.

Receptors and Sensations (page 416)

1. Types of receptors
   a. Each type of receptor is sensitive to a distinct type of stimulus.
   b. The major types of receptors include the following:
      (1) Chemoreceptors, sensitive to changes in chemical concentration.
      (2) Pain receptors, sensitive to tissue damage.
      (3) Thermoreceptors, sensitive to temperature changes.
      (4) Mechanoreceptors, sensitive to mechanical forces.
      (5) Photoreceptors, sensitive to light.
2. Sensory impulses
   a. When receptors are stimulated, changes occur in their membrane potentials.
   b. Receptor potentials are transferred to nerve fibers, and action potentials are triggered.
3. Sensations
   a. Sensations are feelings resulting from sensory stimulation.
   b. A particular part of the sensory cortex interprets every impulse that reaches it in the same way.
   c. The cerebral cortex projects a sensation back to the region of stimulation.
4. Sensory adaptations are adjustments made by sensory receptors to continuous stimulation in which impulses are triggered at slower and slower rates.

Somatic Senses (page 417)

Somatic senses are those involved with receptors in skin, muscles, joints, and visceral organs. They can be grouped as exteroceptive, proprioceptive, and visceroceptive senses.

1. Touch and pressure senses
   a. Free ends of sensory nerve fibers are the receptors for the sensations of touch and pressure.
   b. Meissner's corpuscles are the receptors for the sensations of light touch.
   c. Pacinian corpuscles are the receptors for the sensations of heavy pressure and vibrations.
2. Thermoreceptors include two sets of free nerve endings that serve as heat and cold receptors.

3. Sense of pain
   a. Pain receptors
      (1) Pain receptors are free nerve endings that are stimulated by tissue damage.
      (2) Pain receptors provide a protective function, do not adapt rapidly, and can be stimulated by changes in temperature, mechanical force, and chemical concentration.
   b. The only receptors in visceral organs that provide sensations are pain receptors. These receptors are most sensitive to lack of blood flow and the presence of certain chemicals. The sensations they produce are likely to feel as if they were coming from some other part of the body (referred pain).
   c. Pain nerve pathways
      (1) The two main types of pain fibers are acute pain fibers and chronic pain fibers.
      (2) Acute pain fibers are fast conducting; chronic pain fibers are slower conducting.
      (3) Pain impulses are processed in the dorsal horn of the spinal cord, and they ascend in the spinothalamic tracts.
      (4) Within the brain, pain impulses pass through the reticular formation before being conducted to the cerebral cortex.
   d. Regulation of pain impulses
      (1) Awareness of pain occurs when impulses reach the thalamus.
      (2) The cerebral cortex functions to judge the intensity of pain and to locate its source.
      (3) Impulses descending from the brain cause neurons to release pain-relieving substances, such as enkephalins and serotonin.
      (4) Endorphin is a pain-relieving substance produced in the brain.
   e. Certain neuropeptides synthesized in the brain and spinal cord seem to relieve pain by inhibiting pain impulses.
4. Stretch receptors
   a. Stretch receptors provide information about the condition of muscles and tendons.
   b. Muscle spindles are stimulated when a muscle is relaxed, and they initiate a reflex that causes the muscle to contract.
   c. Golgi tendon organs are stimulated when muscle tension increases, and they initiate a reflex that causes muscle relaxation.

Special Senses (page 422)

Special senses are those whose receptors occur in relatively large, complex sensory organs of the head.

Sense of Smell (page 422)

1. Olfactory receptors
   a. Olfactory receptors are chemoreceptors that are stimulated by chemicals dissolved in liquid.
   b. Olfactory receptors function together with taste receptors and aid in food selection.

2. Olfactory organs
   a. The olfactory organs consist of receptors and supporting cells in the nasal cavity.
   b. Olfactory receptors are neurons with cilia that are sensitive to lipid-soluble chemicals.
3. Olfactory nerve pathways. Nerve impulses travel from the olfactory receptors through the olfactory nerves, olfactory bulbs, and olfactory tracts to interpreting centers in the limbic system of the brain.
4. Olfactory stimulation
   a. Olfactory impulses may result when various gaseous molecules combine with specific sites on the cilia of the receptor cells.
   b. Primary odors include camphoraceous, musky, floral, pepperminty, ethereal, pungent, and putrid.
   c. Olfactory receptors adapt rapidly.
   d. Olfactory receptors are often damaged by environmental factors, but are not replaced.

Sense of Taste (page 425)

1. Taste receptors
   a. Taste buds consist of receptor cells and supporting cells.
   b. Taste cells have taste hairs that are sensitive to particular chemicals dissolved in water.
   c. Taste hair surfaces have receptor sites to which chemicals combine and trigger impulses to the brain.
   d. There are four primary kinds of taste cells, each particularly sensitive to a certain group of chemicals.
2. Taste sensations
   a. The four primary taste sensations are sweet, sour, salty, and bitter.
   b. Various taste sensations result from the stimulation of one or more sets of taste receptors.
   c. Sweet receptors are most plentiful near the tip of the tongue, sour receptors along the margins, salt receptors in the tip and upper front, and bitter receptors toward the back.
3. Taste nerve pathways
   a. Sensory impulses from taste receptors travel on fibers of the facial, glossopharyngeal, and vagus nerves.
   b. These impulses are carried to the medulla and ascend to the thalamus, from which they are directed to the gustatory cortex in the parietal lobes.

Sense of Hearing (page 428)

1. The external ear collects sound waves created by vibrating objects.
2. Middle ear
   a. Auditory ossicles of the middle ear conduct sound waves from the tympanic membrane to the oval window of the inner ear. They also increase the force of these waves.
   b. Skeletal muscles attached to the auditory ossicles act in the tympanic reflex to protect the inner ear from the effects of loud sounds.
3. Auditory tubes connect the middle ears to the throat and help maintain equal air pressure on both sides of the eardrums.
4. Inner ear
   a. The inner ear consists of a complex system of interconnected tubes and chambers—the osseous and membranous labyrinths. It includes the cochlea, which in turn houses the organ of Corti.
   b. The organ of Corti contains the hearing receptors that are stimulated by vibrations in the fluids of the inner ear.
   c. Different frequencies of vibrations stimulate different sets of receptor cells; the human ear 'can detect sound frequencies from about 20 to 20,000 vibrations per second.
5. Auditory nerve pathways
   a. The nerve fibers from hearing receptors travel in the cochlear branch of the vestibulocochlear nerves.
   b. Auditory impulses travel into the medulla oblongata, midbrain, and thalamus, and are interpreted in the temporal lobes of the cerebrum.

Sense of Equilibrium (page 436)

1. Static equilibrium is concerned with maintaining the stability of the head and body when they are motionless. The organs of static equilibrium are located in the vestibule.
2. Dynamic equilibrium is concerned with balancing the head and body when they are moved or rotated suddenly. The organs of this sense are located in the ampullae of the semicircular canals.
3. Other parts that help maintain equilibrium include the eyes and the proprioceptors associated with certain joints.

Sense of Sight (page 438)

1. Visual accessory organs include the eyelids and lacrimal apparatus that protect the eye, and the extrinsic muscles that move the eye.
2. Structure of the eye
   a. The wall of the eye has an outer, a middle, and an inner tunic that function as follows:
      (1) The outer layer (sclera) is protective, and its transparent anterior portion (cornea) refracts light entering the eye.
      (2) The middle layer (choroid coat) is vascular and contains pigments that help keep the inside of the eye dark.
      (3) The inner layer (retina) contains the visual receptor cells.
   b. The lens is a transparent, elastic structure whose shape is controlled by the action of ciliary muscles.
   c. The iris is a muscular diaphragm that controls the amount of light entering the eye; the pupil is an opening in the iris.
   d. Spaces within the eye are filled with fluids (aqueous and vitreous humors) that help maintain its shape.
3. Refraction of light
   a. Light waves are refracted primarily by the cornea and lens to focus an image on the retina.
   b. The lens must be thickened to focus on close objects.
4. Visual receptors
   a. The visual receptors are called rods and cones.
   b. Rods are responsible for colorless vision in relatively dim light, and cones are responsible for color vision.
5. Visual pigments
   a. A light-sensitive pigment in rods (rhodopsin) decomposes in the presence of light and triggers a complex series of reactions that initiate nerve impulses on the optic nerve.
   b. Color vision is related to the presence of cones that are subdivided into three sets. Each set contains a different light-sensitive pigment, and each set is sensitive to a different wavelength of light; the color perceived depends on which set or sets of cones are stimulated.
6. Stereoscopic vision
   a. Stereoscopic vision involves the perception of distance and depth.
   b. Stereoscopic vision occurs because of the formation of two slightly different retinal images that the brain superimposes and interprets as one image in three dimensions.
   c. A one-eyed person uses relative sizes and positions of familiar objects to judge distance and depth.
7. Visual nerve pathways
   a. Nerve fibers from the retina form the optic nerves.
   b. Some fibers cross over in the optic chiasma.
   c. Most of the fibers enter the thalamus and synapse with others that continue to the visual cortex of the occipital lobes.
   d. Other impulses pass into the brain stem and function in various visual reflexes.

Clinical Application of Knowledge

1. How would you interpret the following observation? A person enters a tub of water and reports that it is uncomfortably warm, yet a few moments later says the water feels comfortable, even though the water temperature has remained unchanged.
2. Why are some serious injuries, such as those produced by a bullet entering the abdomen, relatively painless, while others, such as those involving a crushing of the skin, produce considerable discomfort?
3. Labyrinthitis is a condition in which the tissues of the inner ear are inflamed. What symptoms would you expect to observe in a patient with this disorder?
4. Sometimes, as a result of an injury to the eye, the retina becomes detached from its pigmented epithelium. Assuming that the retinal tissues remain functional, what is likely to happen to the person's vision if the retina moves unevenly toward the interior of the eye?
5. The auditory tubes of a child are shorter and directed more horizontally than those of an adult. How might this be related to the observation that children have middle ear infections more frequently than do adults?
6. A patient with heart disease experiences pain at the base of the neck and in the left shoulder and arm after exercise. How would you explain to the patient the origin of this pain?

Review Activities

1. List five groups of sensory receptors, and name the kind of change to which each is sensitive.
2. Explain how sensory impulses are stimulated by sensory receptors.
3. Define *sensation*.
4. Explain what is meant by the projection of a sensation.
5. Define *sensory adaptation*.
6. Explain how somatic senses can be grouped.
7. Describe the functions of free nerve endings, Meissner's corpuscles, and pacinian corpuscles.
8. Explain how thermoreceptors function.
9. Compare pain receptors with other types of somatic receptors.
10. List the factors that are likely to stimulate visceral pain receptors.
11. Define *referred pain*.
12. Explain how neuropeptides function to relieve pain.
13. Distinguish between muscle spindles and Golgi tendon organs.
14. Explain how the senses of smell and taste function together to create the flavors of foods.
15. Describe the olfactory organ and its function.
16. Trace a nerve impulse from the olfactory receptor to the interpreting centers of the brain.
17. List the seven primary olfactory sensations.
18. Explain how the salivary glands aid the function of the taste receptors.
19. Name the four primary taste sensations, and describe the patterns in which the taste receptors are distributed on the tongue.
20. Explain why taste sensation is less likely to diminish with age than olfactory sensation.
21. Trace the pathway of a taste impulse from the receptor to the cerebral cortex.
22. Distinguish between the external, middle, and inner ears.
23. Trace the path of a sound vibration from the tympanic membrane to the hearing receptors.
24. Describe the functions of the auditory ossicles.
25. Describe the tympanic reflex, and explain its importance.
26. Explain the function of the auditory tube.
27. Distinguish between the osseous and the membranous labyrinths.
28. Describe the cochlea and its function.
29. Describe a hearing receptor.
30. Explain how a hearing receptor stimulates a sensory neuron.
31. Trace a nerve impulse from the organ of Corti to the interpreting centers of the cerebrum.
32. Describe the organs of static and dynamic equilibrium and their functions.
33. Explain how the sense of vision helps maintain equilibrium.
34. List the visual accessory organs, and describe the functions of each.
35. Name the three layers of the eye wall, and describe the functions of each.
36. Describe how accommodation is accomplished.
37. Explain how the iris functions.
38. Distinguish between aqueous humor and vitreous humor.
39. Distinguish between the macula lutea and the optic disk.
40. Explain how light waves are focused on the retina.
41. Distinguish between rods and cones.
42. Explain why cone vision is generally more acute than rod vision.
43. Describe the function of rhodopsin.
44. Explain how the eye becomes adapted to light and dark.
45. Describe the relationship between light wavelengths and color vision.
46. Define *stereoscopic vision*.
47. Explain why a person with binocular vision is able to judge distance and depth of close objects more accurately than a one-eyed person.
48. Trace a nerve impulse from the retina to the visual cortex.

Intraocular Lens Implant

Mrs. T., a 76-year-old retired realtor, had worn eyeglasses to correct nearsightedness since she was a teenager. During the past several years, she had needed frequent refraction changes in her glasses, and she had noticed that her ability to focus sharply was decreasing. Also, her vision was becoming increasingly clouded, especially in bright light.

During an appointment with her ophthalmologist, she asked about these problems. The doctor told her that she was developing cataracts—that is, the lenses of her eyes were becoming sclerotic (hardened) and opaque as she grew older.

In order to inspect Mrs. T.'s lenses, the ophthalmologist instilled drops into her eyes that caused short-term paralysis of the ciliary muscles (cycloplegia) and dilation of the pupils (mydriasis). The doctor then examined each lens by using a slit-lamp biomicroscope that allowed him to study the layers of the lenses at varying magnifications.

The doctor explained that in each eye the cataract was located close to the posterior portion of the lens capsule (posterior subcapsular cataract), and that this type of cataract usually progresses rapidly. Because he expected her cataracts to be causing significant visual impairment within a relatively short time, and because there is presently no way to slow or reverse the progress of such changes, he suggested that the lenses be surgically removed and replaced with intraocular lens (IOL) implants. Mrs. T. agreed to have the lens of her weakest eye replaced.

In preparation for surgery, antibiotic solution was instilled into Mrs. T.'s left eye to prevent infection, and a drug (topical mydriatic) that caused her pupil to dilate and remain dilated was instilled. The ophthalmologist performed an extracapsular cataract extraction (ECCE) and an intraocular lens implant on the eye. In this procedure, an incision is made along the superior margin of the cornea, thus opening the anterior chamber of the eye. The lens is removed with the aid of a loop-shaped instrument, and any residual portions of the lens are aspirated from the lens chamber. A plastic lens with predetermined optical qualities that are correct for the eye is slipped through the incision and into the chamber previously occupied by the normal lens. The implanted lens is supported by the posterior portion of the lens capsule that was left intact during the cataract extraction.

Mrs. T. was instructed to instill drops into her eye periodically. These drops contained antibiotic solution to prevent infections and steroid solution to help control inflammation. She was also advised to wear dark glasses for about a month following the surgery, because she would experience an increased sensitivity to light. Her recovery was uneventful, and after a few weeks, her visual acuity in the left eye was nearly normal.

Review Questions

1. Why is cataract most common in older persons?
2. Why might a cataract cause greater visual impairment in bright light than in dim light?
3. Why is it desirable to have the pupil dilated when examining the interior of an eye?
4. What quality of a normal lens would be lacking in a plastic one?
5. What advantage is gained by leaving the posterior portion of the lens capsule intact during cataract extraction?
6. What options, other than intraocular lens implantation, are available to restore visual acuity following cataract extraction?

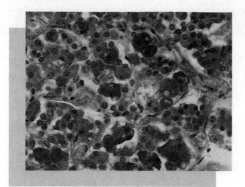

13

Endocrine System

*T*he endocrine system consists of a variety of cells, tissues, and organs that act together with parts of the nervous system to control body activities and maintain homeostasis. The endocrine system and the nervous system each provide a means by which the body parts can communicate with one another and adjust to changing needs. Whereas the parts of the nervous system communicate with various cells by means of nerve impulses carried on nerve fibers, the parts of the endocrine system use hormones that act as chemical messengers to their target cells ∎

Chapter Objectives

After you have studied this chapter, you should be able to:

1. Distinguish between endocrine and exocrine glands.

2. Describe how hormones can be classified according to their chemical composition.

3. Explain how steroid and nonsteroid hormones produce effects on target cells.

4. Discuss how hormonal secretions are regulated by negative feedback mechanisms.

5. Explain how hormonal secretions may be controlled by the nervous system.

6. Name and describe the locations of the major endocrine glands of the body, and list the hormones they secrete.

7. Describe the general functions of the hormones secreted by the endocrine glands.

8. Explain how the secretion of each hormone is regulated.

9. Distinguish between physical and psychological stress.

10. Describe the general stress response.

11. Complete the review activities at the end of this chapter. Note that the items are worded in the form of specific learning objectives. You may want to refer to them before reading the chapter.

Key Terms

adenylate cyclase (ah-den'ĭ-lat si'klās)

adrenal cortex (ah-dre'nal kor'teks)

adrenal medulla (ah-dre'nal me-dul'ah)

anterior pituitary (an-ter'e-or pĭ-tu'ĭ-tar''e)

cyclic AMP (sik'lik AMP)

hormone (hor'mōn)

kinase (kī'nās)

metabolic rate (met''ah-bol'ik rāt)

negative feedback (neg'ah-tiv fēd'bak)

pancreas (pan'kre-as)

parathyroid gland (par''ah-thi'roid gland)

pineal gland (pin'e-al gland)

posterior pituitary (pos-ter'e-or pĭ-tu'ĭ-tar''e)

prostaglandin (pros''tah-glan'din)

steroid (ste'roid)

target cell (tar'get sel)

thymus gland (thi'mus gland)

thyroid gland (thi'roid gland)

Aids to Understanding Words

-crin, to secrete: endo*crine*—pertaining to internal secretions.

diuret-, to pass urine: *diuret*ic—a substance that promotes the production of urine.

endo-, within: *endo*crine gland—a gland that releases its secretion internally into a body fluid.

exo-, outside: *exo*crine gland—a gland that releases its secretion to the outside through a duct.

hyper-, above: *hyper*thyroidism—a condition resulting from an above-normal secretion of thyroid hormone.

hypo-, below: *hypo*thyroidism—a condition resulting from a below-normal secretion of thyroid hormone.

lact-, milk: pro*lact*in—a hormone that promotes milk production.

para-, beside: *para*thyroid glands—a set of glands located near the surface of the thyroid gland.

toc-, birth: oxy*toc*in—a hormone that stimulates the uterine muscles to contract during childbirth.

-tropic, influencing: adrenocortico*tropic* hormone—a hormone secreted by the anterior pituitary gland that stimulates the adrenal cortex.

vas-, vessel: *vas*opressin—a substance that causes blood vessel walls to contract.

The term *endocrine* describes the cells, tissues, and organs that secrete hormones into the body fluids. By contrast, the term *exocrine* refers to those parts whose secretions are carried by tubes or ducts to internal or external body surfaces. Thus, the thyroid and parathyroid glands that secrete hormones into the blood are endocrine glands (ductless glands), and the sweat glands and salivary glands are exocrine glands (figure 13.1).

General Characteristics of the Endocrine System

As a group, endocrine glands and their hormones help regulate metabolic processes. They control the rates of certain chemical reactions, aid in the transport of substances through membranes, and help regulate water and electrolyte balances. They also play vital roles in the reproductive processes, and in development and growth.

Although some hormones are produced by specialized groups of cells, those described in this chapter are produced by the larger endocrine glands—the pituitary gland, thyroid gland, parathyroid glands, adrenal glands, and pancreas. Several other hormone-secreting glands and tissues, including those involved in the processes of digestion and reproduction, are discussed in subsequent chapters (figure 13.2).

Hormones and Their Actions

A **hormone** is a substance secreted by a cell that has an effect on the functions of another cell. Such substances are released into the extracellular spaces surrounding the hormone-secreting cells. Some hormones travel only short distances and produce their effects in nearby cells (paracrine secretion). Other hormones are transported in the blood to all parts of the body and may produce rather general effects. In either case, the physiological action of a particular hormone is restricted to its *target cells*—those cells that possess specific receptors for the hormone molecules. In other words, a hormone's target cells possess receptors that other cells lack.

Chemistry of Hormones

Hormones are organic substances with special molecular structures. They are very potent substances, and thus can stimulate cellular changes in target cells even though the hormones are present in extremely low concentrations. Chemically, most hormones are steroids or steroidlike substances that are synthesized from cholesterol (see chapter 2), or they are amines, peptides, proteins, or glycoproteins that are synthesized from amino acids.

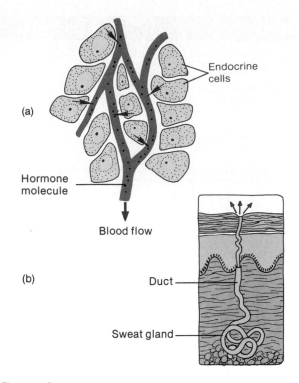

Figure 13.1 Endocrine glands, such as (*a*) the thyroid gland, release hormones into body fluids. Exocrine glands, such as (*b*) sweat glands, release their secretions into ducts that lead to body surfaces.

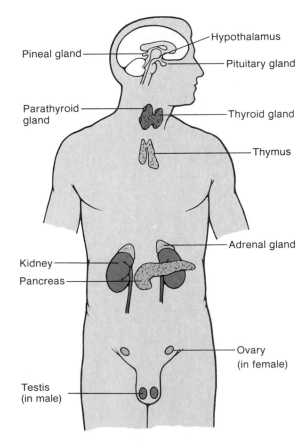

Figure 13.2 Locations of major endocrine glands.

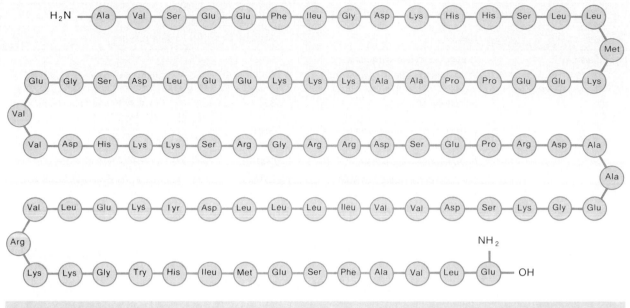

(a) Parathyroid hormone (PTH)

(b) Cortisol

(c) Norepinephrine

Figure 13.3 (a) Amino acid sequence of a protein hormone (PTH). (b and c) Structural formulas of a steroid hormone (cortisol) and an amine hormone (norepinephrine).

Steroids are compounds whose molecules contain complex rings of carbon and hydrogen atoms. The difference between one steroid and another is due to the types and numbers of atoms attached to these rings and the ways they are joined. The steroid hormones include sex hormones (such as estrogens and testosterone) and those secreted by the adrenal cortex (including aldosterone and cortisol). The substance commonly called vitamin D is a modified steroid and can be converted into a hormone, as is discussed later in this chapter. (See also chapter 15.)

Hormones that are *amines* include norepinephrine and epinephrine. As discussed in chapter 11, these substances are produced by certain neurons. They are also synthesized in the adrenal medulla (the inner portion of the adrenal gland) from a single amino acid (tyrosine).

The *peptide* hormones are composed of relatively short chains of amino acids. This group includes hormones associated with the posterior pituitary gland (ADH and OT) and some produced in the hypothalamus (including TRH, GnRH, and SS).

Protein hormones are composed of many amino acids united to form intricate molecular structures. (See chapter 2.) They include the hormone secreted by the parathyroid gland (PTH) and some of those secreted by the anterior pituitary gland (such as GH and PRL). (See figure 13.3.) Certain other hormones from the anterior pituitary gland (FSH, LH, and TSH) are *glycoproteins*. Their molecules include a protein joined to a carbohydrate portion.

Chart 13.1 lists the names and abbreviations of some of the hormones discussed in this chapter. Chart 13.2 summarizes the chemical composition of various hormones.

CHART 13.1 Hormone names and abbreviations

Source	Name	Abbreviation	Synonym
Hypothalamus	Corticotropin-releasing hormone	CRH	
	Gonadotropin-releasing hormone	GnRH	Luteinizing hormone-releasing hormone (LHRH)
	Somatostatin	SS	Growth hormone release-inhibiting hormone (GIH)
	Growth hormone-releasing hormone	GRH	
	Prolactin release-inhibiting factor	PIF	
	Prolactin-releasing factor	PRF	
	Thyrotropin-releasing hormone	TRH	
Anterior pituitary gland	Adrenocorticotropic hormone	ACTH	Corticotropin
	Follicle-stimulating hormone	FSH	Follitropin
	Growth hormone	GH	Somatotropin (STH)
	Luteinizing hormone	LH	Lutropin, interstitial cell-stimulating hormone (ICSH)
	Prolactin	PRL	
	Thyroid-stimulating hormone	TSH	Thyrotropin
Posterior pituitary gland	Antidiuretic hormone	ADH	Vasopressin
	Oxytocin	OT	
Thyroid gland	Calcitonin		
	Thyroxine	T_4	
	Triiodothyronine	T_3	
Parathyroid gland	Parathyroid hormone	PTH	Parathormone
Adrenal medulla	Epinephrine		Adrenaline
	Norepinephrine		Noradrenalin
Adrenal cortex	Aldosterone		
	Cortisol		Hydrocortisone
Pancreas	Glucagon		
	Insulin		
	Somatostatin	SS	

CHART 13.2 Types of hormones

Type of compound	Formed from	Examples
Amines	Amino acids	Norepinephrine, epinephrine
Peptides	Amino acids	ADH, OT, TRH, SS, GnRH
Proteins	Amino acids	PTH, GH, PRL
Glycoproteins	Protein and carbohydrate	FSH, LH, TSH
Steroids	Cholesterol	Estrogen, testosterone, aldosterone, cortisol

1. What is the difference between an endocrine gland and an exocrine gland?
2. What is a hormone?
3. How are hormones chemically classified?

Actions of Hormones

Hormones exert their effects by altering various metabolic processes. For example, some hormones alter the activities of enzymes that, in turn, cause substances to be synthesized; other hormones alter the rates at which chemicals are transported through cell membranes.

Steroid Hormones

Steroids are compounds whose molecules contain complex rings of carbon and hydrogen atoms. The difference between one type of steroid and another is due to the kinds and numbers of atoms attached to these rings and the ways they are joined. (See chapter 2.)

Steroid hormones, unlike amines, peptides, and proteins, are soluble in the lipids that make up the bulk of cell membranes. For this reason, steroid molecules can enter target cells relatively easily by diffusion. Once inside a target cell, they enter the nucleus and combine with specific nuclear protein molecules—the receptors. The steroid-protein complex thus formed binds to a

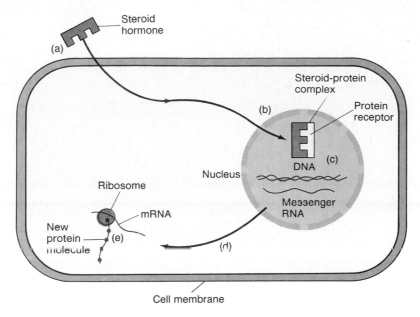

Figure 13.4 (*a*) A steroid hormone passes through a cell membrane and (*b*) combines with a protein receptor in the nucleus. (*c*) The steroid-protein complex activates the synthesis of messenger RNA. (*d*) The messenger RNA leaves the nucleus and (*e*) functions in the manufacture of protein molecules.

particular region of the target cell's DNA molecules, and activates specific genes. The activated genes, in turn, bring about the synthesis of particular kinds of messenger RNA (mRNA) molecules.

As discussed in chapter 4, messenger RNA can leave the nucleus and enter the cytoplasm, where it functions in manufacturing specific proteins. Thus, steroid hormones influence their target cells by causing special proteins to be synthesized—proteins that act as enzymes and alter the rates of cellular processes, act as parts of membrane transport systems, or cause enzymes to be activated or inhibited (figure 13.4 and chart 13.3).

The degree of cellular response induced by a steroid hormone is proportional to the number of hormone-receptor complexes formed.

Nonsteroid Hormones

The nonsteroid hormones, such as amines, peptides, and proteins, usually act by combining with specific receptors located in the target-cell membranes. Each receptor is a protein molecule that has a *binding site* and an *activity site*. A hormone delivers its message to the cell by uniting with the binding site of its receptor. This combination stimulates the receptor's activity site to interact with other membrane proteins. As a result, the actions of membrane-bound enzymes or membrane transport mechanisms are altered, causing changes in the concentrations of still other cellular substances. These substances serve as "second messengers" in the

CHART 13.3 Sequence of steroid hormone action

1. Endocrine gland secretes steroid hormone.
2. Steroid hormone enters target cell by diffusing through the cell membrane and then enters the nucleus.
3. Hormone combines with a nuclear receptor molecule.
4. Steroid hormone-protein complex binds to DNA and promotes the synthesis of messenger RNA.
5. Messenger RNA enters the cytoplasm and functions in manufacturing protein molecules.
6. Newly synthesized proteins may function as enzymes and cause reactions that are recognized as the hormone's action.

hormonal stimulation mechanism because they, in turn, induce cellular changes that appear as responses to the original hormone.

The second messenger employed by one group of hormones is **cyclic adenosine monophosphate** (cyclic AMP, or cAMP). In this mechanism, a hormone binds to its receptor, and the resulting hormone-receptor complex activates an enzyme called **adenylate cyclase** that is bound to the inside of the cell membrane. The activated enzyme causes ATP molecules within the cytoplasm to become molecules of cAMP (figure 13.5). The cAMP, in turn, activates another set of enzymes called **protein kinases.** Protein kinases bring about the transfer of phosphate groups from ATP molecules to various protein substrate molecules. This phosphorylation alters the shapes of the substrate molecules and

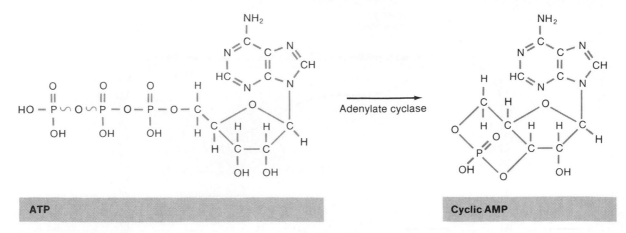

ATP

Cyclic AMP

Adenylate cyclase

Figure 13.5 ATP molecules are converted into cyclic AMP by the action of the enzyme called adenylate cyclase. What cellular changes may take place as a result of this reaction?

converts some of them from inactive forms into active ones. The activated protein molecules then induce changes in various cellular processes. (See figure 13.6.) Thus, the response of any particular cell to such a hormone is determined not only by the type of membrane receptors present, but also by the kinds of protein substrate molecules the cell contains. Chart 13.4 summarizes these actions.

Cellular responses to this second messenger mechanism include altering membrane permeabilities, activating various enzymes, promoting the synthesis of certain proteins, stimulating or inhibiting specific metabolic pathways, promoting cellular movements, and initiating the secretion of hormones and other substances.

The cyclic AMP produced by this mechanism is quickly inactivated by still another enzyme (phosphodiesterase) so that the cAMP action is short-lived. For this reason, a continuing response in a target cell will depend upon a continuing signal produced by hormone molecules combining with receptors in the target-cell membrane.

Hormones whose actions involve the cyclic AMP mechanism include releasing hormones from the hypothalamus; TSH, ACTH, FSH, and LH from the anterior pituitary gland; ADH from the posterior pituitary gland; PTH from the parathyroid glands; norepinephrine and epinephrine from the adrenal glands; calcitonin from the thyroid gland; and glucagon from the pancreas.

Another group of nonsteroid hormones employs second messengers other than cAMP. These hormones include GnRH and SS from the hypothalamus, GH and PRL from the anterior pituitary gland, OT from the posterior pituitary gland, and insulin from the pancreas. The mechanisms that produce responses to these

hormones are less well understood than the cAMP mechanism.

In one such mechanism, the binding of a hormone to its receptor causes an increase in calcium ion concentration within the target cell by stimulating the transport of calcium ions inward through the cell membrane or by inducing a release of calcium ions from cellular storage sites. The calcium ions combine with the protein *calmodulin* (see chapter 9), causing the calmodulin's molecular structure to be changed and activated. The activated calmodulin can interact with various cellular enzymes, altering their activities and thus producing a variety of responses.

Still another mechanism involves a substance called *cyclic guanosine monophosphate* (cyclic GMP, or cGMP). This substance is similar to cAMP, and it is thought to function as a second messenger in much the same manner as cAMP. Thus, a hormone-receptor combination activates the enzyme guanylate cyclase, which is responsible for the formation of cGMP, and the cGMP, in turn, induces changes in protein kinases.

As was mentioned, the cellular response to a steroid hormone is usually proportional to the number of hormone-receptor complexes formed; the response to a hormone operating through a second messenger, however, is greatly amplified. This amplification occurs when even a small number of hormone-receptor combinations is produced, because each combination can activate a membrane-bound enzyme. This enzyme, in turn, can cause a large number of second messenger molecules to appear, thus inducing a relatively strong cellular response. As a consequence of such amplification, cells are highly sensitive to changes in the concentrations of nonsteroid hormones.

Some protein hormones enter cells, although it is not known how they accomplish this. The significance

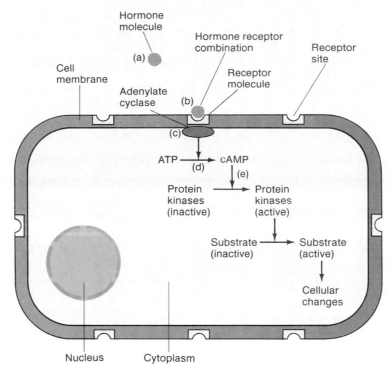

Figure 13.6 (a) Nonsteroid hormone molecules reach the target cell by means of body fluids and (b) combine with receptor sites on the cell membrane. (c) As a result, molecules of adenylate cyclase are activated, and (d) cause the change of ATP into cyclic AMP. (e) Cyclic AMP promotes a series of reactions leading to various cellular changes.

CHART 13.4	Sequence of nonsteroid hormone actions

1. Endocrine gland secretes hormone.
2. Hormone is carried to its target cell by body fluid.
3. Hormone combines with receptor site on membrane of its target cell.
4. Adenylate cyclase molecules are activated within target cell's membrane.
5. Adenylate cyclase causes ATP molecules to be converted into cyclic AMP molecules.
6. Cyclic AMP activates various protein kinases.
7. These enzymes activate protein substrates in the cell that induce changes in metabolic processes.
8. Cellular changes are recognized as the hormone's action.

of such movement is not clear. The act of such a protein hormone binding to receptor sites produces an immediate, short-term effect on a cell; however, long-lasting effects may be produced by actions taking place after the hormone, often with its receptor, has entered the cell.

Among the protein hormones that enter cells are growth hormone, prolactin, and gonadotropins from the anterior pituitary gland; PTH from the parathyroid glands; and insulin from the pancreas.

1. How does a steroid hormone act on its target cells?
2. How does a nonsteroid hormone act on its target cells?
3. What is meant by a "second messenger"?

Prostaglandins

Another group of compounds, called **prostaglandins,** has regulating effects on cells. These substances are lipids (20-carbon fatty acids that include 5-carbon rings) and are synthesized from a fatty acid (arachidonic acid) found in cell membranes. Prostaglandins occur in a wide variety of cells, including those of the liver, kidneys, heart, lungs, thymus gland, pancreas, brain, and various reproductive organs (figure 13.7).

Prostaglandins are very potent substances and are present in very small quantities. They are not stored in cells. Instead, they are synthesized just before they are released, and are rapidly inactivated.

Some prostaglandins regulate cellular responses to hormones. For example, prostaglandins influence the

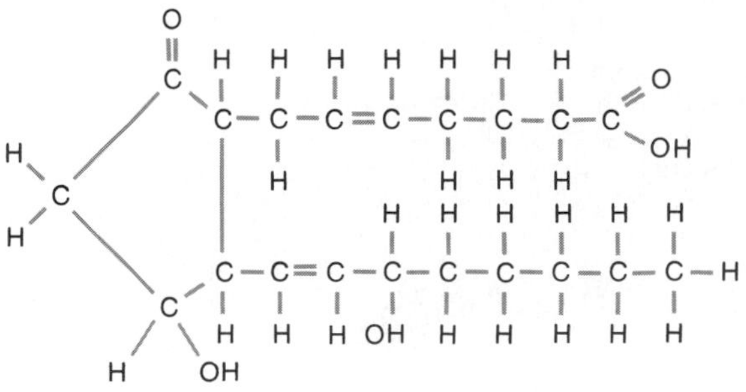

Prostaglandin E₂ (PGE₂)

Figure 13.7 A prostaglandin molecule contains a 20–carbon fatty acid with a 5–carbon ring.

activation of adenylate cyclase in cell membranes, and thus control the formation of cyclic AMP. In some cases, this action increases the formation of cyclic AMP, and in others it decreases its formation. In either instance, by controlling the amount of cyclic AMP produced, the prostaglandins are able to alter the cell's response to a hormone.

Prostaglandins produce a variety of effects. For example, some can cause the smooth muscles in the airways of the lungs and in the blood vessels to relax, so that these passageways dilate. Prostaglandins can also cause smooth muscle in the walls of the uterus and intestine to contract. They stimulate the secretions of hormones from the adrenal cortex and inhibit the secretion of hydrochloric acid from the wall of the stomach. Prostaglandins also influence the movements of sodium ions and water in the kidneys, help regulate blood pressure, and have powerful effects on both male and female reproductive physiology. When tissues are injured, they promote the inflammation response. (See chapter 5.)

> Prostaglandins are usually present in relatively high concentration in the synovial fluid of joints affected by rheumatoid arthritis. Various anti-inflammatory drugs, including aspirin and certain steroids, inhibit the production of these prostaglandins, and thus the drugs are useful for treating the symptoms of arthritis.

1. What are prostaglandins?
2. Describe one possible function of prostaglandins.
3. What kinds of effects do prostaglandins produce?

Control of Hormonal Secretions

Because hormones are very potent substances, their release by endocrine cells must be regulated precisely, so that the amounts released are balanced with the amounts used by the body. One mechanism that functions to control hormonal secretion is a *feedback system.*

Negative Feedback Systems

In a feedback system, a body part continuously receives information (feedback) in the form of signals concerning the cellular process it controls. If conditions change, the part can act upon the information and adjust its activity.

Commonly, the control of hormonal secretions involves a **negative feedback system,** as described in chapter 1. In such a system, an endocrine gland is sensitive to the concentration of a substance it regulates or to the concentration of a product from a process it controls. Whenever this concentration reaches a certain level, the endocrine gland is inhibited (a negative effect), and its secretory activity decreases. Then, as the concentration of the gland's hormone decreases, the concentration of the regulated substance decreases too, and the inhibition of the gland ceases. When the gland is no longer inhibited, it begins to secrete its hormone again (figure 13.8).

As a result of such negative feedback systems, the concentrations of some hormones remain relatively stable, although they may fluctuate slightly within normal ranges (figure 13.9).

Nerve Control

Another type of hormone control mechanism involves the nervous system. For example, some endocrine glands, such as the adrenal medulla, secrete their hormones in response to nerve impulses, and no other stimulus seems able to cause a secretion. In the case of the adrenal medulla, hormones (norepinephrine and epinephrine) are secreted in response to stimulation from preganglionic sympathetic nerve impulses originating in the hypothalamus of the brain. (See chapter 11.)

Still another kind of control system involves interaction between an endocrine gland and the hypothalamus. In this system, neurosecretory cells in the hypothalamus secrete substances, called releasing (or inhibiting) hormones (or factors). The target cells of these **releasing hormones** are in the anterior pituitary gland. The anterior pituitary gland responds to a releasing hormone by secreting its own hormone. Then, as the gland's hormone reaches a certain concentration in the body fluids, a negative feedback system inhibits the hypothalamus and its secretion of the releasing hormone decreases.

1. What is a feedback system?
2. How does a negative feedback system control hormonal secretions?
3. How does the nervous system help regulate hormonal secretions?

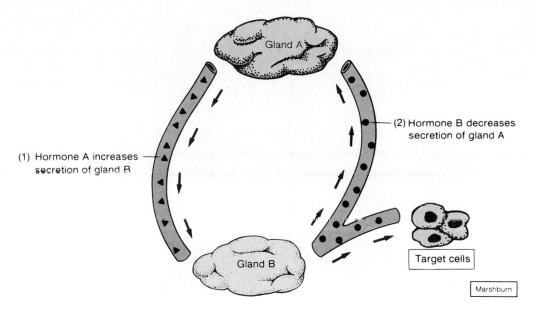

(1) Hormone A increases secretion of gland B

(2) Hormone B decreases secretion of gland A

Gland A

Gland B

Target cells

Marshburn

Figure 13.8 An example of a negative feedback system: (*1*) Gland A secretes a hormone that stimulates gland B to increase the secretion of another hormone. (*2*) The hormone from gland B causes changes in its target cells and inhibits the activity of gland A.

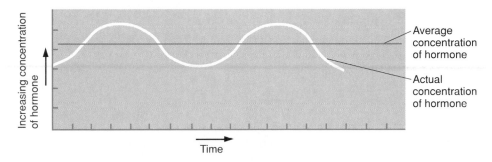

Increasing concentration of hormone

Time

Average concentration of hormone

Actual concentration of hormone

Figure 13.9 As a result of negative feedback systems, some hormone concentrations remain relatively stable, although they may fluctuate slightly above and below the average concentrations.

Pituitary Gland

The **pituitary gland** (hypophysis) is about 1 centimeter in diameter and is located at the base of the brain. It is attached to the hypothalamus by the pituitary stalk, or *infundibulum,* and lies in the sella turcica of the sphenoid bone, as shown in figure 13.10.

The pituitary gland consists of two distinct portions: an anterior lobe (adenohypophysis) and a posterior lobe (neurohypophysis). The *anterior lobe* secretes a number of hormones, including growth hormone (GH), thyroid-stimulating hormone (TSH), adrenocorticotropic hormone (ACTH), follicle-stimulating hormone (FSH), luteinizing hormone (LH), and prolactin (PRL). Although the cells of the

posterior lobe (pituicytes) do not synthesize any hormones, two important ones, antidiuretic hormone (ADH) and oxytocin (OT), are secreted by nerve cells (neurosecretory cells) within the tissues of the posterior lobe.

During fetal development, a narrow region appears between the anterior and posterior lobes of the pituitary gland. It is called the *intermediate lobe* (pars intermedia). It produces melanocyte-stimulating hormone (MSH), which regulates the formation of melanin—the pigment found in the skin, and in portions of the eyes and brain. This region atrophies during prenatal development, and appears only as a vestige in adults.

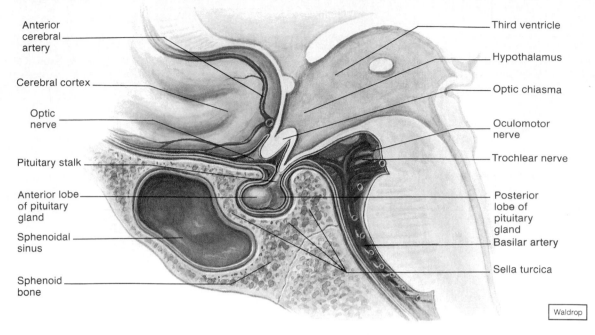

Anterior cerebral artery

Cerebral cortex

Optic nerve

Pituitary stalk

Anterior lobe of pituitary gland

Sphenoidal sinus

Sphenoid bone

Third ventricle

Hypothalamus

Optic chiasma

Oculomotor nerve

Trochlear nerve

Posterior lobe of pituitary gland

Basilar artery

Sella turcica

Waldrop

Figure 13.10 The pituitary gland is attached to the hypothalamus and lies in the sella turcica of the sphenoid bone.

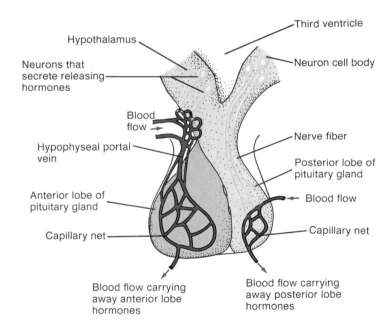

Hypothalamus

Neurons that secrete releasing hormones

Blood flow

Hypophyseal portal vein

Anterior lobe of pituitary gland

Capillary net

Blood flow carrying away anterior lobe hormones

Third ventricle

Neuron cell body

Nerve fiber

Posterior lobe of pituitary gland

Blood flow

Capillary net

Blood flow carrying away posterior lobe hormones

Figure 13.11 Axons in the posterior lobe of the pituitary gland are stimulated to release hormones by nerve impulses originating in the hypothalamus; cells of the anterior lobe are stimulated by releasing hormones secreted by hypothalamic neurons.

Most of the pituitary gland's activities are controlled by the brain. For example, the release of hormones from its posterior lobe occurs when nerve impulses from the hypothalamus signal the axon ends of neurosecretory cells in the posterior lobe. On the other hand, secretions from the anterior lobe typically are controlled by releasing hormones produced by the hypothalamus (figure 13.11). These releasing hormones are transmitted by blood in the vessels of a capillary net associated with the hypothalamus. These vessels merge to form the **hypophyseal portal veins** that pass downward along the pituitary stalk and give rise

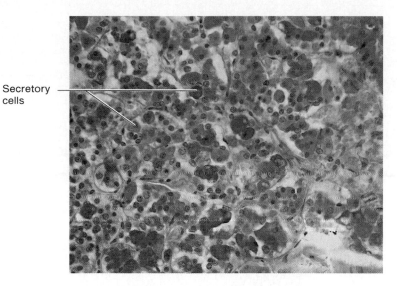

Secretory cells

Figure 13.12 Light micrograph of the anterior pituitary gland.

to a capillary net in the anterior lobe. Thus, substances released into the blood from the hypothalamus are carried directly to the anterior lobe.

The hypothalamus (described in chapter 11) receives information from nearly all parts of the nervous system. This information includes data concerning a person's emotional state, body temperature, blood nutrient concentrations, and so forth. The hypothalamus sometimes acts on such information by signaling the pituitary gland to release hormones.

1. Where is the pituitary gland located?
2. List the hormones secreted by the anterior lobe and the posterior lobe of the pituitary gland.
3. Explain how the hypothalamus controls the actions of the pituitary gland.

Anterior Pituitary Hormones

The anterior lobe of the pituitary gland is enclosed in a dense capsule of collagenous connective tissue, and consists largely of epithelial tissue arranged in blocks around many thin-walled blood vessels. Within the epithelial tissue, five types of secretory cells have been identified. They are *somatotropes* that secrete GH, *mammatropes* that secrete PRL, *thyrotropes* that secrete TSH, *corticotropes* that secrete ACTH, and *gonadotropes* that secrete FSH and LH. In males, LH (luteinizing hormone) is known as ICSH (interstitial cell-stimulating hormone). (See figures 13.12 and 13.13.)

Growth hormone (GH), which is also called *somatotropin* (STH), is a protein that stimulates the body cells to increase in size and undergo more rapid cell division than usual. It enhances the movement of amino acids through the cell membranes, and causes an in-

crease in the rate at which the cells convert these molecules into proteins. GH also causes cells to decrease the rate at which they utilize carbohydrates and to increase the rate at which they use fats. The hormone's effect on amino acids, however, seems to be the more important one.

Normally, growth hormone is secreted throughout the life cycle, although its blood concentration rises considerably at puberty and levels off following adolescence. In adult males, the concentration of growth hormone remains fairly constant at all ages. In adult females, growth hormone concentration is usually constant until menopause, when the concentration decreases sharply and remains relatively low for the remainder of life. The role of growth hormone in adults is not completely understood.

Growth hormone is secreted in rhythmic pulses (pulsatile secretion), especially during sleep. The mechanism for controlling its secretion involves two substances from the hypothalamus called *growth hormone-releasing hormone* (GRH) and *somatostatin* (SS). These two peptides are released alternately and have opposite effects on the secretion of growth hormone. Thus, when GRH is released, a pulse of GH is secreted; then, when SS is released, GH secretion is inhibited.

A person's nutritional state also seems to play a role in the control of GH. For example, more GH is released during periods of protein deficiency and abnormally low blood glucose concentration. Conversely, when blood protein and glucose concentrations increase, growth hormone secretion decreases. Apparently the hypothalamus is able to sense changes in the

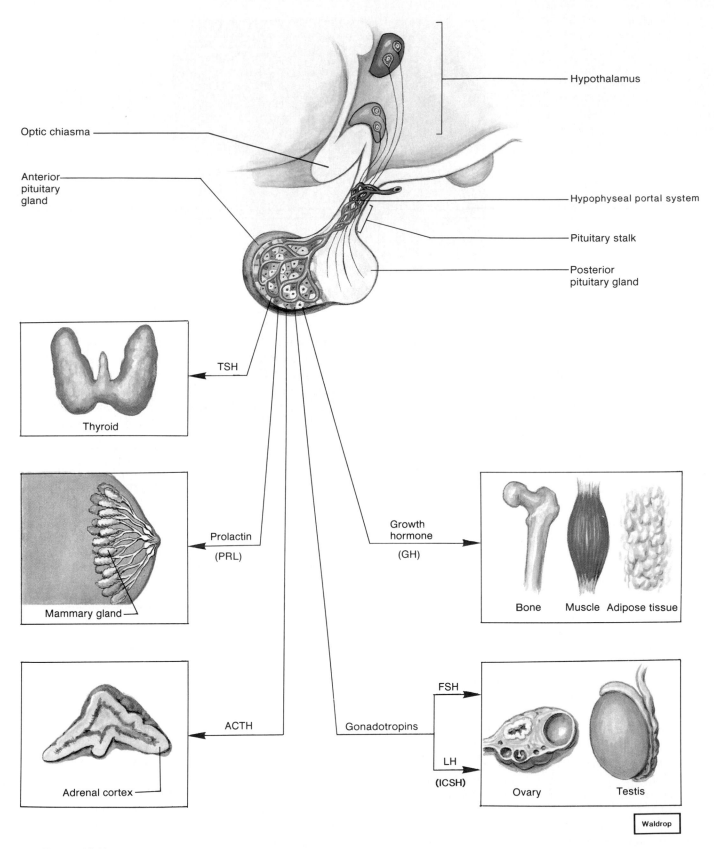

Hypothalamus

Optic chiasma

Anterior
pituitary
gland

Hypophyseal portal system

Pituitary stalk

Posterior
pituitary gland

TSH

Thyroid

Prolactin

(PRL)

Mammary gland

Growth
hormone

(GH)

Bone   Muscle Adipose tissue

ACTH

Gonadotropins

FSH

LH

(ICSH)

Adrenal cortex

Ovary   Testis

Waldrop

Figure 13.13   Hormones released from the anterior lobe of
the pituitary gland and their target organs.

concentrations of certain blood nutrients, and it releases GRH in response to some of them.

Although growth hormone is able to stimulate the elongation of bone tissue directly, its effect on cartilage requires the presence of a mediator substance. This substance, called *somatomedin* (insulin-like growth factor) is released from the liver in response to GH, and the somatomedin, in turn, promotes the growth of cartilage.

If an insufficient amount of growth hormone is secreted during childhood, body growth is limited and a type of *dwarfism* (hypopituitary dwarfism) results. In this condition, the body parts are usually correctly proportioned and mental development is normal. However, because an abnormally low secretion of growth hormone is usually accompanied by lessened secretions of other anterior pituitary hormones, additional hormone deficiency symptoms may appear. For example, a hypopituitary dwarf often fails to develop adult sexual features unless hormone therapy is provided.

Hypopituitary dwarfism is sometimes treated by administering growth hormone, and this treatment may stimulate a rapid increase in height. The procedure, however, must be started before the epiphyseal disks of the person's long bones have become ossified. Otherwise growth in height is not possible.

An oversecretion of growth hormone during childhood may result in *gigantism*—a condition in which the person's height may exceed 8 feet. Gigantism, which is relatively rare, is usually accompanied by a tumor of the pituitary gland. In such cases, various pituitary hormones in addition to GH are likely to be secreted excessively, so that a giant often suffers from a variety of metabolic disturbances.

If growth hormone is secreted excessively in an adult after the epiphyses of the long bones have ossified, the person does not grow taller. The soft tissues, however, may continue to enlarge and the bones may become thicker. As a consequence, an affected individual may develop greatly enlarged hands and feet, a protruding jaw, and a large tongue and nose. This condition is called *acromegaly,* and, like gigantism, it is often associated with a pituitary tumor (figure 13.14).

Prolactin (PRL) is a protein, and as its name suggests, it promotes milk production. More specifically, prolactin stimulates and sustains maternal milk production following childbirth.

Although prolactin seems to have little effect in males by itself, it causes a decrease in the secretion of the hormone LH. Because LH (ICSH) is necessary for the production of male sex hormones (androgens), excessive prolactin secretion may cause a male to develop a deficiency of sex hormones and become infertile. (See chapter 22.)

(a) (b) (c) (d)

Figure 13.14 Acromegaly is caused by an oversecretion of growth hormone in adulthood. Note the changes in facial features of this individual from ages (*a*) nine, (*b*) sixteen, (*c*) thirty-three, and (*d*) fifty-two.

Two substances from the hypothalamus are involved in the regulation of prolactin secretion. One of these, *prolactin release-inhibiting factor* (PIF), acts to restrain the secretion of prolactin, while the other, *prolactin-releasing factor* (PRF), stimulates its secretion.

Thyroid-stimulating hormone (TSH), also called *thyrotropin,* is a glycoprotein—a protein bound to a carbohydrate. Its major function is to control the secretion of certain hormones from the thyroid gland, which are described in a subsequent section of this chapter.

TSH secretion is partially regulated by the hypothalamus, which produces *thyrotropin-releasing hormone* (TRH). TSH secretion is also regulated by circulating thyroid hormones that exert an inhibiting effect on the release of TRH and TSH; therefore, as the blood concentration of thyroid hormones increases, the secretions of TRH and TSH are reduced (figure 13.15).

Certain external factors influence the release of TRH and TSH. These factors include exposure to ex-

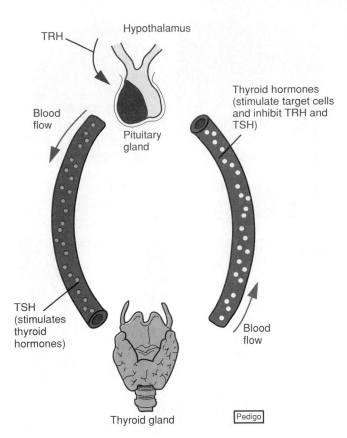

Figure 13.15 TRH from the hypothalamus stimulates the anterior pituitary gland to release TSH. TSH stimulates the thyroid gland to release hormones, which in turn cause the hypothalamus to reduce its secretion of TRH.

treme cold, which is accompanied by increased hormonal secretions, and emotional stress, which sometimes triggers increased hormonal secretions and other times causes decreased secretions.

1. How does growth hormone affect the cellular metabolism of carbohydrates, fats, and proteins?
2. What are the functions of prolactin in females? In males?
3. How is TSH secretion regulated?

Adrenocorticotropic hormone (ACTH) is a peptide that controls the manufacture and secretion of certain hormones from the outer layer (cortex) of the adrenal gland. These adrenal cortical hormones are discussed in a subsequent section of this chapter.

The secretion of ACTH may be regulated in part by *corticotropin-releasing hormone* (CRH), which is released from the hypothalamus in response to decreased concentrations of adrenal cortical hormones. Also, as in the case of TSH, various forms of stress result in increased secretion of ACTH by stimulating the release of CRH.

Both **follicle-stimulating hormone** (FSH) and **luteinizing hormone** (LH) are glycoproteins and are called *gonadotropins,* which means they exert their actions on the gonads or reproductive organs. FSH, for example, is responsible for growth and development of egg-cell-containing follicles in the female ovaries. It also stimulates the follicular cells to secrete a group of female sex hormones, collectively called *estrogen.*

In males, FSH stimulates the initial production of sperm cells in the testes at puberty. LH, which in males is also called *interstitial cell-stimulating hormone* (ICSH) because it acts upon the interstitial cells of the testes, promotes the secretion of sex hormones in both males and females, and is essential for the release of egg cells from the female ovaries. Other functions of the gonadotropins and the ways they interact are discussed in chapter 22.

The mechanism that regulates the secretion of gonadotropins is not well understood. However, it is known that the hypothalamus secretes a *gonadotropin-releasing hormone* (GnRH). The hypothalamus apparently does not secrete this hormone until puberty, because gonadotropins are virtually absent in the body fluids of infants and children.

1. What is the function of ACTH?
2. Describe the functions of FSH and LH in a male and in a female.
3. What is a gonadotropin?

Posterior Pituitary Hormones

Unlike the anterior lobe of the pituitary gland, which is composed primarily of glandular epithelial cells, the posterior lobe consists largely of nerve fibers and neuroglial cells (*pituicytes*). The neuroglial cells function to support the nerve fibers that originate in the hypothalamus.

As mentioned earlier, the two hormones associated with the posterior pituitary—antidiuretic hormone (ADH) and oxytocin (OT)—are produced by specialized neurons in the hypothalamus (figure 13.11). These hormones travel down axons through the pituitary stalk to the posterior pituitary and are stored in vesicles (secretory granules) near the ends of the axons. The hormones are released into the blood in response to nerve impulses coming from the hypothalamus.

Antidiuretic hormone (ADH) is a polypeptide consisting of a relatively short chain of amino acids. This is also true of oxytocin, and as figure 13.16 shows, the molecules of these substances are similar except for amino acids in two locations.

A *diuretic* is a substance that increases urine production. An *antidiuretic,* then, is a chemical that decreases urine formation. ADH produces its antidiuretic effect by acting on the kidneys and causing them to reduce the amount of water they excrete. In this way, ADH plays an important role in regulating the water concentration of the body fluids. (See chapter 20.)

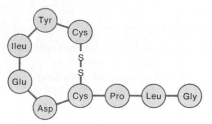

Oxytocin

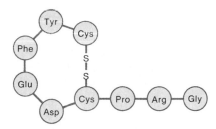

Antidiuretic hormone

Figure 13.16 What differences exist in the amino acid composition of these hormones?

Beverage alcohol (ethyl alcohol) is thought to inhibit the normal secretion of ADH. Consequently, an abnormally large volume of urine may be produced after a person drinks alcoholic beverages. The lost body fluid must later be replaced if normal water balance is to be maintained.

When ADH is present in high concentrations, it causes contractions of certain smooth muscles, including those in the walls of some blood vessels. As a result, the blood pressure in these vessels increases. For this reason, ADH is also called *vasopressin*. Although ADH is seldom present in quantities sufficient to elevate blood pressure, it may be released following a severe loss of blood. Blood pressure may drop as a consequence of excessive bleeding, and in this situation ADH's vasopressor effect may help restore normal blood pressure.

The secretion of ADH is regulated by the hypothalamus. Certain neurons in this part of the brain, called *osmoreceptors,* are sensitive to changes in the water concentration of body fluids. For example, if a person is dehydrating due to a lack of water intake, the solutes in blood become more and more concentrated. The osmoreceptors can sense the resulting increase in osmotic pressure, and their signals cause the posterior pituitary to release ADH, which is transported in the blood to the kidneys. As a result of the effect of ADH on kidney functions, less urine is produced. This action conserves water.

On the other hand, if a person drinks an excessive amount of water, the body fluids become more dilute and the release of ADH is inhibited. Consequently, the kidneys excrete more dilute urine until the water concentration of the body fluids returns to normal.

Another factor that affects ADH secretion is blood volume. Normal blood volume causes blood pressure that stretches the walls of certain blood vessels. Pressure receptors in some of these blood vessel walls are stimulated by such stretching, and they signal the hypothalamus with impulses that inhibit the release of ADH. However, if the blood volume is decreased due to a hemorrhage, the blood pressure drops, and the receptors send fewer inhibiting impulses. As a result, the secretion of ADH increases, and as before, the ADH causes the kidneys to conserve water. This action helps compensate for some of the decreased blood volume.

If any parts involved in the ADH regulating mechanism are damaged due to an injury or a tumor, insufficient quantities of the hormone may be synthesized or released. The resulting ADH deficiency produces a condition called *diabetes insipidus*. This condition is characterized by an output of as much as 25 to 30 liters of very dilute urine per day (polyuria) and by a rise in the concentration of solutes in the body fluids. It is also accompanied by a sensation of great thirst (polydipsia).

Oxytocin (OT) also has an antidiuretic action, but it is weaker in this respect than ADH. In addition, oxytocin can cause contractions of the smooth muscles in the uterine wall and plays a role in the later stages of childbirth by stimulating uterine contractions. The mechanism that triggers the release of oxytocin during childbirth is not clearly understood, but the uterus does become more and more sensitive to oxytocin's effects during pregnancy. It is believed that stretching of uterine and vaginal tissues late in pregnancy, caused by the growing fetus, may initiate nerve impulses to the hypothalamus. The hypothalamus may then signal the posterior pituitary to release oxytocin, which, in turn, may enhance uterine wall contractions during labor.

Oxytocin has an effect on the breasts, causing contractions in certain cells associated with the milk-producing glands and their ducts. In lactating breasts, this action forces liquid from the milk glands into the milk ducts and causes the milk to be ejected from the breasts—an effect that is necessary for breast-feeding.

The mechanical stimulation provided by suckling the nipple of a breast initiates nerve impulses that travel to the mother's hypothalamus. The hypothalamus responds by signaling the posterior pituitary to release oxytocin (sometimes called *milk let-down factor*), which, in turn, stimulates the release of milk. Thus, milk is normally not ejected from the milk glands and ducts

CHART 13.5   Hormones of the pituitary gland

Anterior lobe

Hormone	Action	Source of control
Growth hormone (GH)	Stimulates increase in size and rate of reproduction of body cells; enhances movement of amino acids through membranes; promotes growth of long bones	Growth hormone-releasing hormone (GRH) and somatostatin (SS) from the hypothalamus
Prolactin (PRL)	Sustains milk production after birth; amplifies effect of LH in males	Secretion restrained by prolactin release-inhibiting factor (PIF) and stimulated by prolactin-releasing factor (PRF) from the hypothalamus
Thyroid-stimulating hormone (TSH)	Controls secretion of hormones from the thyroid gland	Thyrotropin-releasing hormone (TRH) from the hypothalamus
Adrenocorticotropic hormone (ACTH)	Controls secretion of certain hormones from the adrenal cortex	Corticotropin-releasing hormone (CRH) from the hypothalamus
Follicle-stimulating hormone (FSH)	Responsible for development of egg-containing follicles in ovaries; stimulates follicular cells to secrete estrogen; in males, stimulates production of sperm cells	Gonadotropin-releasing hormone (GnRH) from the hypothalamus
Luteinizing hormone (LH or ICSH in males)	Promotes secretion of sex hormones; plays role in release of egg cell in females	Gonadotropin-releasing hormone (GnRH) from the hypothalamus

Posterior lobe

Hormone	Action	Source of control
Antidiuretic hormone (ADH)	Causes kidneys to reduce water excretion; in high concentration, causes blood pressure to rise	Hypothalamus in response to changes in blood water concentration and blood volume
Oxytocin (OT)	Causes contractions of muscles in uterine wall; causes muscles associated with milk-secreting glands to contract	Hypothalamus in response to stretch in uterine and vaginal walls, and stimulation of breasts

until it is needed. Oxytocin has no known function in males, although it is present in the male posterior pituitary.

Chart 13.5 reviews the hormones of the pituitary gland.

> If the uterus is not contracting sufficiently to expel a fully developed fetus, commercial preparations of oxytocin are sometimes used to stimulate uterine contractions, thus inducing labor. Also, such preparations are often administered to the mother following childbirth to ensure that the uterine muscles will contract enough to squeeze broken blood vessels closed, minimizing the danger of hemorrhage.

1. What is the function of ADH?
2. How is the secretion of ADH controlled?
3. What effects does oxytocin produce in females?

Thyroid Gland

The **thyroid gland,** shown in figure 13.17, is a very vascular structure that consists of two large lateral lobes connected by a broad isthmus. It is located just below the larynx on either side and in front of the trachea. It has a special ability to remove iodine from the blood.

Structure of the Gland

The thyroid gland is covered by a capsule of connective tissue and is made up of many secretory parts called *follicles.* The cavities within these follicles are lined with a single layer of cuboidal epithelial cells, and are filled with a clear, viscous glycoprotein called *colloid.* The follicular cells produce and secrete hormones that may either be stored in the colloid or released into the blood of nearby capillaries (figure 13.18). Other hormone-secreting cells, called *extrafollicular cells* (C cells), occur outside the follicles.

Thyroid Hormones

Three important hormones are produced by the thyroid gland. Two of these are synthesized by the follicular cells and have marked effects on the metabolic rates of body cells. Another hormone, which is produced by the extrafollicular cells, influences the blood concentrations of calcium and phosphate ions.

Of the thyroid hormones that affect cellular metabolic rates, the most important are **thyroxine,** or te-

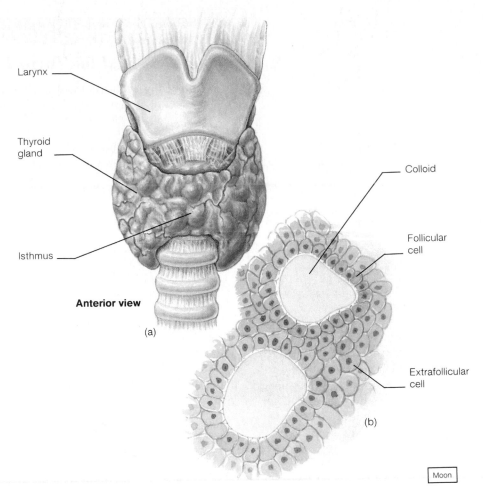

Anterior view

(a)

(b)

Moon

Figure 13.17 (a) The thyroid gland consists of two lobes connected anteriorly by an isthmus. (b) Thyroid hormones are secreted by follicular cells.

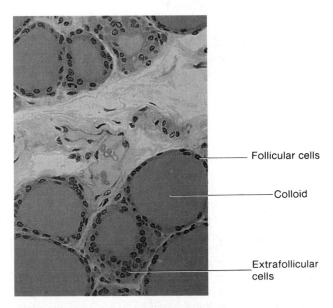

Figure 13.18 A light micrograph of thyroid gland tissue (×40). The open spaces surrounded by follicular cells are filled with colloid.

traiodothyronine (also called T$_4$, because its molecule includes four atoms of iodine), and **triiodothyronine** (also called T$_3$, because its molecule includes three atoms of iodine). They help regulate the metabolism of carbohydrates, lipids, and proteins. For example, they increase the rate at which the cells release energy from carbohydrates, enhance the rate of protein synthesis, and stimulate the breakdown and mobilization of lipids. As a result of their actions, these hormones are needed for normal growth and development, and they are essential for the maturation of the nervous system (figure 13.19).

Before follicular cells can produce thyroxine and triiodothyronine, they must be supplied with iodine salts (iodides). Such salts are normally obtained from foods, and after they have been absorbed from the intestine, some are carried by the blood to the thyroid gland. An efficient active transport mechanism called the *iodine pump* moves the iodides into the follicular cells, where they are concentrated. The iodides, together with an amino acid (tyrosine), are used to synthesize these thyroid hormones.

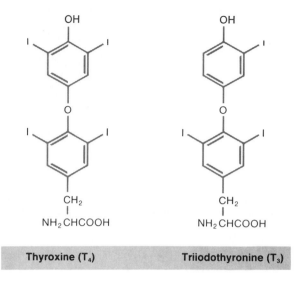

Figure 13.19 The hormones thyroxine and triiodothyronine have very similar molecular structures.

Follicular cells also secrete a protein called *thyroglobulin,* which is the main ingredient of thyroid colloid. Thyroglobulin is used to store thyroid hormones whenever they are produced in excess. The stored hormones are bound to the thyroglobulin until the hormone concentration of the body fluids drops below a certain level, then enzymes cause the hormones to be released from the colloid, and they diffuse into the blood. Once they are in the blood, thyroid hormones combine with blood proteins (alpha globulins) and are transported to body cells. Although triiodothyronine is nearly five times more potent, thyroxine accounts for at least 95% of the circulating thyroid hormones.

The thyroid hormone that influences blood calcium and phosphate ion concentrations is a polypeptide called **calcitonin.** This substance helps regulate the concentrations of calcium and phosphate ions by inhibiting the rate at which they leave the bones and enter the extracellular fluids. This is accomplished by inhibiting the bone-resorbing activity of osteoclasts. (See chapter 7.) At the same time, calcitonin causes an increase in the rate at which calcium and phosphate ions are deposited in bone matrix by stimulating the activity of osteoblasts. It also increases the excretion of calcium ions and phosphate ions by the kidneys. Thus, calcitonin acts to lower the blood calcium and phosphate ion concentrations.

The secretion of calcitonin is stimulated by a high blood calcium ion concentration, as may occur following the absorption of calcium ions from a recent meal. Its secretion is also stimulated by certain hormones, such as gastrin, that are released from digestive organs when they are active. For these reasons, it is

Disorders of the Thyroid Gland

Many functional disorders of the thyroid gland are characterized by *overactivity* (hyperthyroidism) or *underactivity* (hypothyroidism) of the gland cells.

A variety of laboratory tests are available to help a physician judge the functional condition of a patient's thyroid gland. These include tests that measure the blood concentration of triiodothyronine (T_3 test), the blood concentration of thyroxine (T_4 test), and the rate at which iodine is taken up by the gland (iodine uptake test).

In the iodine uptake test, the patient drinks distilled water containing a small amount of *radioactive iodine* (I-131). Several hours later, the radioactivity of the thyroid gland is measured using a scintillation counter. (See figure 2.2.) The results indicate how effectively the gland is able to remove iodine from the blood and concentrate it, thus giving an indication of the gland's condition (figure 2.3a).

A thyroid disorder may develop at any time during a person's life as the result of a developmental problem, injury, disease, or dietary deficiency. Such disorders may also arise from dysfunctions of the immune system in which thyroid tissues are destroyed or substances that stimulate thyroid secretions are produced.

One form of **hypothyroidism** appears in infants when their thyroid glands fail to function normally. An affected child may appear normal at birth, because it has received an adequate supply of thyroid hormones from its mother during pregnancy. When its own thyroid gland fails to produce sufficient quantities of these hormones, the child soon develops a condition called *cretinism*. Cretinism is characterized by severe symptoms including stunted growth, abnormal bone formation, retarded mental development, low body temperature, and sluggishness. Without treatment, within a month or so following birth the child is likely to suffer permanent mental retardation (figure 13.20).

If a person develops hypothyroidism later in life, the symptoms include an abnormally low metabolic rate, abnormal sensitivity to cold, physical sluggishness, and poor appetite. The person also may appear mentally dull and may develop swollen tissues due to an accumulation of body fluid in the subcutaneous tissue—a condition called *myxedema.*

On the other hand, a person with **hyperthyroidism** has an abnormally high metabolic rate, is sensitive to heat, is restless or overactive, loses weight in spite of a good appetite, and appears mentally alert. Also, the person's eyes are likely to protrude (exophthalmos) because of edematous swelling in the tissues behind them (figure 13.21). At the same time, the thyroid gland may enlarge, producing a bulge in the neck called a *goiter.* A goiter associated with hyperthyroidism is said to be a *toxic goiter.*

The most common form of hyperthyroidism is called *Graves disease* (thyrotoxicosis). It occurs most often in young women and is a type of autoimmune disease. (See chapter 19.) In Graves disease, the body's immune system

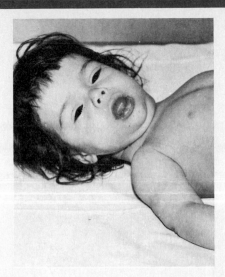

Figure 13.20 Cretinism is due to an underactive thyroid gland during infancy and childhood. How can this condition be prevented?

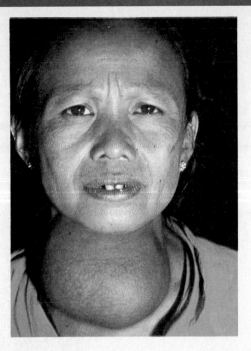

Figure 13.22 Endemic goiter is caused by an iodine deficiency.

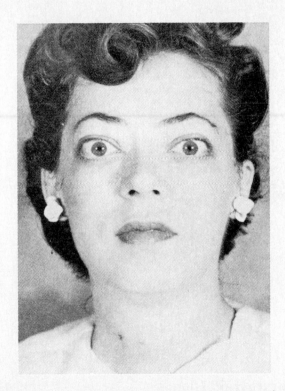

Figure 13.21 Hyperthyroidism may produce protrusion of the eyes.

produces molecules (antibodies) that bind to the TSH receptors of thyroid cell membranes and mimic the action of TSH. Consequently, the glandular cells are stimulated to secrete excessive amounts of thyroid hormones.

Treatment for hyperthyroidism often involves surgery in which some of the overactive thyroid tissue is removed, or it may involve exposure of these tissues to radioactive iodine-131 (chapter 2). After a dose of I-131 has been ingested and absorbed, the thyroid gland rapidly removes the iodine from the blood. Because the radioactive iodine is thus concentrated in the thyroid gland, the ionizing radiation that it releases can cause some of the surrounding glandular tissue to be destroyed.

Another disorder of the thyroid gland is called *simple goiter,* or *endemic goiter.* It may affect persons who live in regions where iodine is lacking in the soil and drinking water. Such a person is likely to develop an iodine deficiency, which is reflected in an inability to produce thyroid hormones. These hormones normally exert an inhibiting effect on the secretion of TSH, and without such inhibition, the anterior pituitary gland releases TSH excessively. The resulting overstimulation of the thyroid gland causes it to enlarge, but because the gland is unable to manufacture hormones, the condition is accompanied by the symptoms of hypothyroidism (figure 13.22).

Simple goiter can usually be prevented in regions where iodine deficiencies occur if the inhabitants supplement their diets with table salt containing iodides (*iodized salt*).

CHART 13.6   Hormones of the thyroid gland

Hormone	Action	Source of control
Thyroxine (T₄)	Increases rate of energy release from carbohydrates; increases rate of protein synthesis; accelerates growth; stimulates activity in the nervous system	TSH from the anterior pituitary gland
Triiodothyronine (T₃)	Same as above, but five times more potent than thyroxine	Same as above
Calcitonin	Lowers blood calcium and phosphate ion concentrations by inhibiting the release of calcium and phosphate ions from bones, and by increasing the excretion of these ions by the kidneys	Elevated blood calcium ion concentration, digestive hormones

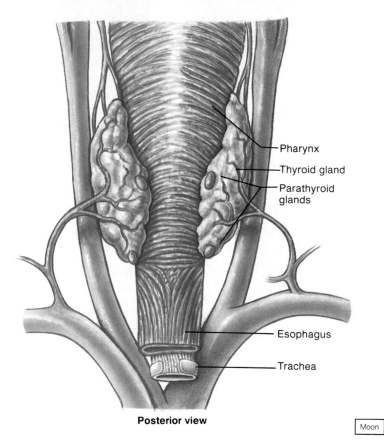

Posterior view

Moon

Figure 13.23   The parathyroid glands are embedded in the posterior surface of the thyroid gland.

thought that calcitonin may prevent prolonged elevation in blood calcium ion concentration after eating.

Chart 13.6 summarizes the actions and sources of control of the thyroid hormones.

1. Where is the thyroid gland located?
2. What hormones of the thyroid gland affect carbohydrate metabolism and protein synthesis?
3. What substance is essential for the production of thyroxine and triiodothyronine?
4. How does calcitonin influence the concentrations of blood calcium and phosphate ions?

Parathyroid Glands

The **parathyroid glands** are located on the posterior surface of the thyroid gland, as shown in figure 13.23. Usually there are four of them—a superior and an inferior gland associated with each of the thyroid's lateral lobes. The parathyroid glands secrete a hormone that functions in regulating the concentrations of calcium and phosphate ions in the blood.

Disorders of the Parathyroid Glands

As with the thyroid gland, disorders of the parathyroids usually involve excessive hormone secretions (hyperparathyroidism) or inadequate secretions (hypoparathyroidism).

Hyperparathyroidism is most common in females, and its incidence increases with age. Its symptoms include fatigue, muscular weakness, painful joints (arthralgia), altered mental functions, depression, and loss of weight. Hyperparathyroidism is usually caused by a tumor within the tissues of a parathyroid gland. The resulting increase in PTH secretion stimulates excessive osteoclastic activity, and as bone tissue is resorbed, the bones become soft, deformed, and subject to spontaneous fractures. In extreme cases, portions of bones may disappear altogether, leaving large holes in place of the former solid structure, a condition known as *osteoporosis*. Later these cavities may become filled with fibrous tissue, producing a condition called *osteitis fibrosa cystica*.

The excessive calcium and phosphate ions released into the body fluids, as a result of increased PTH activity, may be deposited in abnormal places, causing new problems such as kidney stones.

Treatment for *hyperparathyroidism* may include surgical removal of a tumor and orthopedic procedures to correct severe bone deformities.

Hypoparathyroidism can result if the parathyroid glands are injured or surgically removed. In either case, decreased PTH secretion is reflected in reduced osteoclastic activity, and although the bones remain strong, the blood calcium ion concentration drops, producing the symptoms of hypocalcemia, which include muscle cramps and seizures.

Treatment of *hypoparathyroidism* usually does not involve PTH because it can produce undesirable side effects. Instead, the patient may be given injections of calcium salts and massive doses of vitamin D, which promotes the absorption of calcium ions from the intestine and causes the blood calcium ion concentration to rise.

Structure of the Glands

Each parathyroid gland is a small, yellowish brown structure covered by a thin capsule of connective tissue. The body of the gland consists of numerous tightly packed secretory cells that are closely associated with capillary networks. (See figure 13.24.)

Parathyroid Hormone

The hormone secreted by the parathyroid glands is a protein called **parathyroid hormone** (PTH), or *parathormone* (figure 13.3). This substance causes an increase in blood calcium ion concentration and a decrease in blood phosphate ion concentration through actions in the bones, kidneys, and intestines.

The intercellular matrix of bone tissue contains a considerable amount of calcium phosphate and calcium carbonate. PTH stimulates bone resorption by osteocytes and osteoclasts, and inhibits the activity of osteoblasts. (See chapter 7.) As a result of increased resorption of bone, calcium and phosphate ions are released from the bones, and the blood concentrations of these ions increase. At the same time, PTH causes the kidneys to conserve blood calcium ions and to excrete more phosphate ions in the urine. It also indirectly stimulates the absorption of calcium ions from food in the intestine by influencing the metabolism of vitamin D.

Vitamin D (cholecalciferol) can be obtained in some foods. It can also be synthesized in the body from dietary cholesterol, which is converted into provitamin

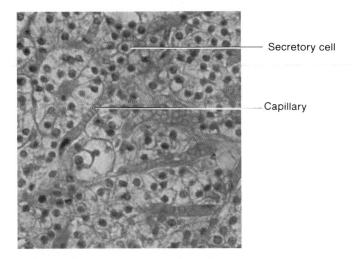

Figure 13.24 Light micrograph of the parathyroid gland.

D (7-dehydrocholesterol) by intestinal enzymes. This provitamin is stored largely in the skin, and it can be changed to vitamin D by exposure to ultraviolet light, as from sunlight. The resulting vitamin D molecules (hydroxycholecalciferol) can be transported away by the blood or stored in various tissues.

When PTH is present, hydroxycholecalciferol can be changed by the kidneys into an active form of vitamin D (dihydroxycholecalciferol). This active form controls the transport mechanism by which calcium ions are absorbed from the contents of the intestine. Thus, PTH indirectly regulates calcium ion absorption in the

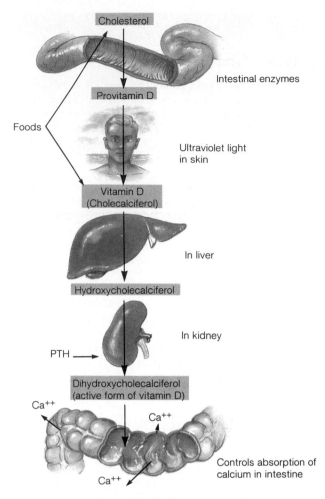

Figure 13.25 Mechanism by which PTH promotes the absorption of calcium in the intestine.

Cholesterol

Intestinal enzymes

Provitamin D

Foods

Ultraviolet light in skin

Vitamin D (Cholecalciferol)

In liver

Hydroxycholecalciferol

In kidney

PTH

Dihydroxycholecalciferol (active form of vitamin D)

Ca++

Ca++

Ca++

Controls absorption of calcium in intestine

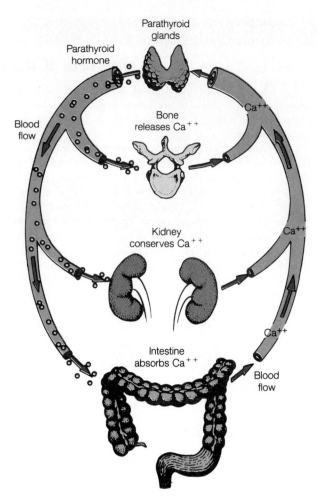

Figure 13.26 Parathyroid hormone stimulates the release of calcium from bone and the conservation of calcium by the kidneys. It indirectly stimulates the absorption of calcium by the intestine. The resulting increase in blood calcium concentration inhibits the secretion of parathyroid hormone.

Parathyroid glands

Parathyroid hormone

Blood flow

Bone releases Ca++

Ca++

Kidney conserves Ca++

Ca++

Intestine absorbs Ca++

Ca++

Blood flow

intestine by causing the kidneys to activate vitamin D molecules (figure 13.25).

The parathyroid glands seem to be free of control by the hypothalamus. Instead, the secretion of PTH is regulated by a negative feedback mechanism operating between the glands and the blood calcium ion concentration (figure 13.26). As the concentration of blood calcium ions rises, less PTH is secreted; as the concentration of blood calcium ions drops, more PTH is released.

To summarize, a stable concentration of blood calcium ions is maintained by the action of calcitonin, which tends to decrease the blood calcium ion concentration when it is relatively high, and by the action of PTH, which tends to increase the blood calcium ion concentration when it is relatively low.

The homeostasis of calcium ions is important in a number of physiological processes. For example, as the blood calcium ion concentration drops (hypocalcemia), the nervous system becomes abnormally excitable, and impulses may be triggered spontaneously. As a result, muscles, including the respiratory muscles,

may undergo tetanic contractions, and the person may die due to a failure to breathe. With an abnormally high concentration of blood calcium ions (hypercalcemia), the nervous system becomes depressed. Consequently, muscle contractions are weak and reflexes are sluggish.

1. Where are the parathyroid glands located?
2. How does parathyroid hormone help regulate the concentrations of blood calcium and phosphate ions?
3. How does the negative feedback system of the parathyroid glands differ from that of the thyroid gland?

Adrenal Glands

The **adrenal glands** (suprarenal glands) are closely associated with the kidneys. A gland sits atop each kidney like a cap and is embedded in the mass of adipose tissue that encloses the kidney.

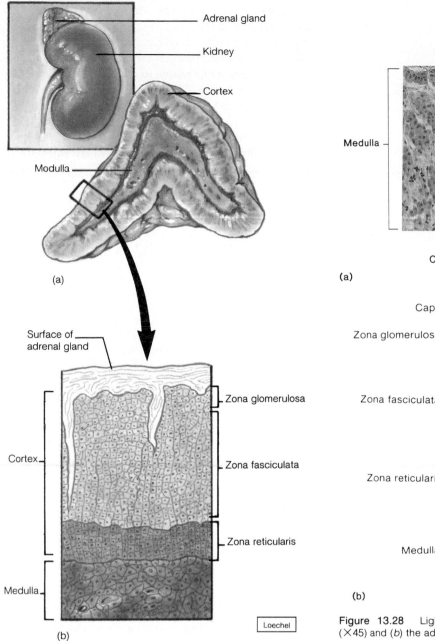

(a)

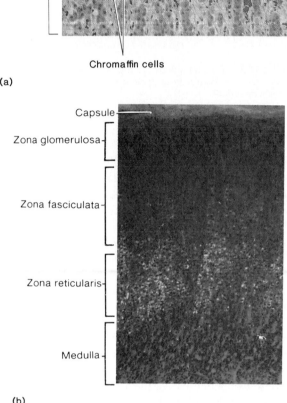

(a)

(b)

Figure 13.28 Light micrograph of (*a*) the adrenal medulla (×45) and (*b*) the adrenal cortex (×100).

(b)

Figure 13.27 (*a*) An adrenal gland consists of an outer cortex and an inner medulla. (*b*) The cortex consists of three layers, or zones, of cells.

Structure of the Glands

Although the adrenal glands may differ somewhat in size and shape, they are generally pyramidal. Each adrenal gland is very vascular and consists of two parts, as shown in figure 13.27. The central portion is the adrenal medulla, and the outer part is the adrenal cortex. Although these regions are not sharply divided, they represent distinct glands that secrete different hormones.

The **adrenal medulla** consists of irregularly shaped cells that are arranged in groups around blood vessels.

These cells are intimately connected with the sympathetic division of the autonomic nervous system. In fact, these adrenal medullary cells are modified postganglionic neurons, and preganglionic autonomic nerve fibers lead to them directly from the central nervous system. (See chapter 11.)

The **adrenal cortex,** which makes up the bulk of the adrenal gland, is composed of closely packed masses of epithelial cells arranged in layers. These layers form an outer, a middle, and an inner zone of the cortex—the zona glomerulosa, the zona fasciculata, and the zona reticularis, respectively. As in the case of the cells of the adrenal medulla, cells of the adrenal cortex are well supplied with blood vessels. (See figure 13.28.)

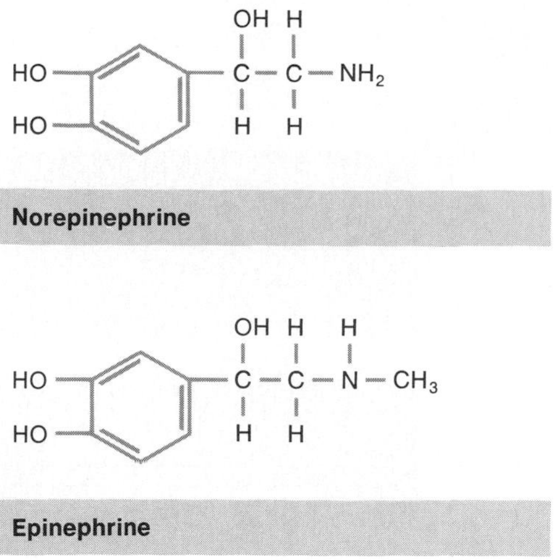

Norepinephrine

Epinephrine

Figure 13.29 Epinephrine and norepinephrine have similar molecular structures and similar functions. How are these molecules different?

Hormones of the Adrenal Medulla

The cells of the adrenal medulla (chromaffin cells) produce, store, and secrete two closely related hormones, **epinephrine** (adrenaline) and **norepinephrine** (noradrenalin). Both of these substances are amines (catecholamines), and they have similar molecular structures and physiological functions (figure 13.29). In fact, epinephrine is synthesized from norepinephrine.

The synthesis of these hormones begins with the amino acid *tyrosine*. In the first step of the process, tyrosine is converted into a substance called *dopa* by an enzyme (tyrosine hydroxylase) in the secretory cells. Dopa is converted to *dopamine* by a second enzyme (dopa decarboxylase), and dopamine is converted to norepinephrine by a third enzyme (dopamine betahydroxylase). In about 10% of the adrenal medullary cells, the process ends with norepinephrine; in the rest, still another enzyme (phenylethanolamine N-methyltransferase) converts the norepinephrine to epinephrine. The hormones are stored in tiny vesicles (chromaffin granules), much like neurotransmitters are stored in neurons.

The effects of the adrenal medullary hormones generally resemble those that result when sympathetic nerve fibers stimulate their effectors. However, the hormonal effects last up to ten times longer because the hormones are removed from the tissues relatively slowly. These effects include increased heart rate and increased force of cardiac muscle contraction, elevated blood pressure, increased breathing rate, and decreased activity in the digestive system. (See chart 11.8.)

The ratio of the two hormones in the adrenal medullary secretion varies with different physiological conditions, but usually the secretion is about 80% epinephrine and 20% norepinephrine. Although their effects are generally similar, certain effectors respond differently to the hormones. These differences are due to the relative numbers of alpha and beta receptors in the membranes of the effector cells. As described in chapter 11, epinephrine combines with either alpha or beta receptors, while norepinephrine combines mainly with alpha receptors. Chart 13.7 compares some of the differences in the effects of these hormones.

Impulses arriving by way of sympathetic nerve fibers stimulate the adrenal medulla to release its hormones at the same time other effectors are being stimulated by sympathetic impulses. As a rule, these impulses originate in the hypothalamus in response to various types of stress. Thus, the medullary secretions function together with the sympathetic division of the autonomic nervous system in preparing the body for energy-expending action—"fight or flight."

Although a lack of adrenal medullary hormones produces no significant disorder, tumors in the adrenal medulla (pheochromocytomas) are sometimes accompanied by excessive hormonal secretions. In such cases, the secretion of norepinephrine predominates, and affected persons show signs of prolonged sympathetic responses—high blood pressure, increased heart rate, elevated blood sugar, and so forth. The treatment of this condition generally involves surgical removal of the tumorous growth.

1. Describe the location and structure of the adrenal glands.
2. Name the hormones secreted by the adrenal medulla.
3. What general effects are produced by the hormones secreted by the adrenal medulla?
4. What usually stimulates the release of hormones from the adrenal medulla?

Hormones of the Adrenal Cortex

The cells of the adrenal cortex produce over thirty different steroids, including several hormones (corticosteroids). Unlike the adrenal medullary hormones, without which a person can survive, some of those released by the cortex are vital. In fact, in the absence of adrenal cortical secretions, a person usually dies within a week unless extensive electrolyte therapy is provided. The most important adrenal cortical hormones are aldosterone, cortisol, and certain sex hormones.

Aldosterone

Aldosterone is synthesized by cells in the outer zone (zona glomerulosa) of the adrenal cortex. It is called a *mineralocorticoid* because it helps regulate the concentration of mineral electrolytes, such as sodium and potassium ions. More specifically, aldosterone causes sodium ions to be reabsorbed, and thus conserved, and causes potassium ions to be excreted. As a consequence of the sodium ion reabsorption and an increasing blood

CHART 13.7 Comparative effects of epinephrine and norepinephrine

Part or function affected	Epinephrine	Norepinephrine
Heart	Rate increases	Less effect (rate may slow)
	Force of contraction increases	Force of contraction increases
Blood vessels	Vessels in skeletal muscle vasodilate, decreasing resistance to blood flow	Blood flow to skeletal muscles increases, resulting from constriction of blood vessels in skin and viscera
Systemic blood pressure	Some increase due to increased cardiac output	Great increase due to vasoconstriction
Airways	Dilated	Less effect
Reticular formation of brain	Activated	Little effect
Liver	Promotes change of glycogen to glucose, increasing blood sugar	Little effect on blood sugar
Metabolic rate	Increases	Increases

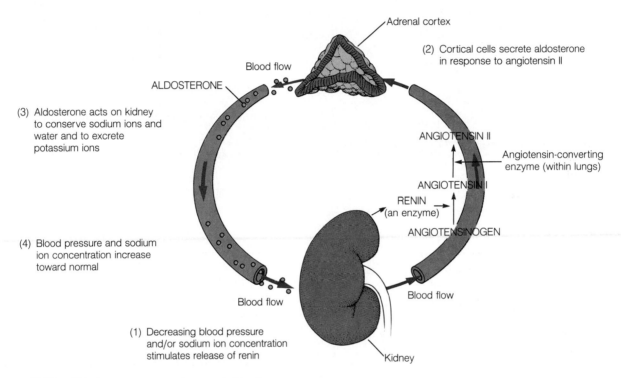

Figure 13.30 Aldosterone causes an increase in blood volume and blood pressure by promoting the conservation of sodium ions and water (as indicated by steps 1–4).

sodium ion concentration, ADH is released from the posterior pituitary gland. The action of ADH promotes the conservation of water and reduces urine output.

The cells that secrete aldosterone are stimulated by a decrease in the blood concentration of sodium ions or by an increase in the blood concentration of potassium ions. They are also responsive to a mechanism that involves groups of specialized cells in the kidneys (juxtaglomerular cells). These specialized kidney cells are influenced by changes in blood pressure and blood sodium ion concentration. If either of these factors is decreased, the cells release an enzyme called *renin,* which causes a blood protein (angiotensinogen) to decompose. A peptide product of this decomposition, *an-giotensin I,* is then transformed by an enzyme (angiotensin-converting enzyme, or ACE) in the lungs into another form called *angiotensin II.* The angiotensin II is carried by the blood to the adrenal cortex, where it stimulates the release of aldosterone. ACTH also stimulates the secretion of aldosterone.

By promoting the conservation of sodium ions and indirectly promoting water conservation, aldosterone increases the blood sodium ion concentration and the blood volume. This increase in blood volume is accompanied by an increase in blood pressure. When angiotensin II is present in relatively high concentrations, it also increases the systemic blood pressure by acting as a potent vasoconstrictor (figure 13.30).

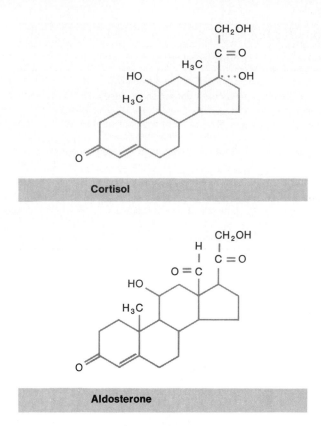

Figure 13.31 Cortisol and aldosterone are steroids with similar molecular structures.

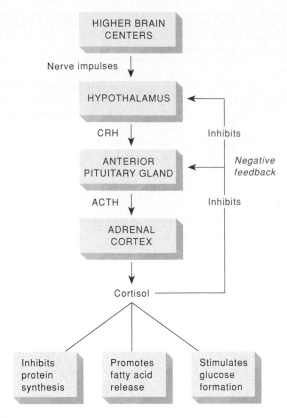

Figure 13.32 Cortisol secretion is regulated by a negative feedback mechanism.

Cortisol

Cortisol (hydrocortisone) is a *glucocorticoid,* which means it affects glucose metabolism. It is produced in the middle zone (zona fasciculata) of the adrenal cortex and has a molecular structure similar to aldosterone (figure 13.31). In addition to affecting glucose, cortisol influences protein and fat metabolism. Among the more important actions of cortisol are the following:

1. It inhibits the synthesis of protein in various tissues, causing an increase in the blood concentration of amino acids.
2. It promotes the release of fatty acids from adipose tissue, causing an increase in the use of fatty acids as an energy source and a decrease in the use of glucose as an energy source.
3. It stimulates liver cells to form glucose from noncarbohydrates (gluconeogenesis), such as circulating amino acids and glycerol, thus promoting an increase in the blood glucose concentration.

These actions help keep the blood glucose concentration within the normal range between meals, because the supply of glycogen used to provide glucose from the liver can be exhausted in a few hours without food.

The release of cortisol is controlled by a negative feedback mechanism that involves the hypothalamus, anterior pituitary gland, and adrenal cortex. More specifically, the hypothalamus of the brain secretes CRH (corticotropin-releasing hormone) into the hypophyseal portal veins, which were described previously in this chapter. These vessels carry the CRH to the anterior pituitary gland, stimulating it to secrete ACTH. In turn, the ACTH causes the adrenal cortex to release cortisol.

Cortisol has an inhibiting effect on the release of both CRH and ACTH, and as these substances decrease in concentration, less cortisol is produced.

Such a mechanism might act to maintain a relatively stable blood concentration of a hormone, but in the case of cortisol, the set point of the mechanism is changed from time to time. (See chapter 1.) In this way, the output of hormone can be altered to meet the demands of changing conditions. For example, when the body is subjected to stressful conditions—injury, disease, extreme temperature, extreme emotional feelings, and so forth—nerve impulses provide the brain with information concerning the stressful condition. In response, brain centers acting through the hypothalamus can cause the release of more cortisol and can maintain a higher concentration of the hormone until the cause of the stress is reduced (figure 13.32).

Disorders of the Adrenal Cortex

Hyposecretion of cortical hormones, which can be caused by an autoimmune disease or by an infectious disease such as tuberculosis, leads to *Addison's disease*. This condition is characterized by decreased blood sodium, increased blood potassium, low blood glucose level (hypoglycemia), dehydration, low blood pressure, and increased skin pigmentation. The treatment for Addison's disease involves the use of mineralocorticoids and glucocorticoids. An untreated person is likely to live only a few days because of severe disturbances in electrolyte balance.

Hypersecretion of glucocorticoids (primarily cortisol), which may be associated with an adrenal tumor or with an oversecretion of ACTH by the anterior pituitary gland, results in *Cushing's syndrome*. A person with this condition may have a great decrease in tissue protein, accompanied by wasting of muscles and loss of bone tissue. The blood glucose level tends to remain elevated, and sodium is retained excessively. As a result, tissue fluid may increase, and the skin may appear puffy and thin, and may bruise easily. Adipose tissue is often deposited in the face and in the back, between the shoulders, producing a characteristic "moon face" and a "buffalo hump." At the same time, an increase in adrenal sex hormones may produce masculinizing effects in a female, such as growth of a beard or development of a deeper voice.

The treatment of Cushing's syndrome is usually directed toward reducing an oversecretion of ACTH. For example, a tumor associated with the pituitary gland may be removed surgically or destroyed by radiation treatment. In other cases, the adrenal glands may be partially or completely removed (adrenalectomy). If the patient develops symptoms of cortical insufficiency after treatment, mineralocorticoids and glucocorticoids must be administered.

When cortisol is present in abnormally high concentrations, it can have an important effect on certain blood vessels. Tissues subjected to stress sometimes become inflamed—red, swollen, and painful. (See chapter 5.) This reaction is due in part to an increase in the permeability of the capillaries in the affected area, which allows fluids to leak from these blood vessels and cause swelling in the surrounding tissues. Also, lysosomes of injured cells may release enzymes that produce irritation and add to the inflammation. A high concentration of cortisol, however, acts to block such inflammation, and although it is not clear how it accomplishes this, the cortisol may decrease the permeability of the capillaries and thus inhibit the leakage of fluids. Cortisol may also stabilize lysosomal membranes so that their enzymes are not released, and cortisol may inhibit the synthesis of prostaglandins.

Cortisol and related compounds are commonly used to treat patients with inflammatory diseases, such as rheumatoid arthritis and allergies. Unfortunately, when used in the concentrations needed to produce anti-inflammatory effects, these compounds may cause toxic reactions. Consequently, their usefulness is limited. They are also used to depress the inflammatory responses that accompany organ transplants and tissue grafts, and in this way help prevent the rejection of donor tissues.

Sex Hormones

Adrenal sex hormones are produced by cells in the inner zone (zona reticularis) of the cortex. Although the hormones of this group are male types (adrenal androgens), some of them are converted into female hormones (estrogens) by the skin, liver, and adipose tissues. The normal functions of these hormones are not well understood, but they may supplement the supply of sex hormones from the gonads and stimulate early development of the reproductive organs. Also, there is some evidence that adrenal androgens play a role in controlling the female sex drive. Chart 13.8 summarizes the actions of the cortical hormones.

---

1 Name the most important hormones of the adrenal cortex.
2 What is the function of aldosterone?
3 What actions does cortisol produce?
4 How are the blood concentrations of aldosterone and cortisol regulated?

Pancreas

The **pancreas** contains two major types of secretory tissues, reflecting its dual function as an exocrine gland that secretes digestive juice through a duct, and an endocrine gland that releases hormones into body fluids.

Structure of the Gland

The pancreas is an elongated, somewhat flattened organ that is posterior to the stomach and behind the parietal peritoneum (figure 13.33). It is attached to the first section of the small intestine (duodenum) by a duct, which transports its digestive juice into the intestine.

The endocrine portion of the pancreas consists of cells arranged in groups, closely associated with blood

CHART 13.8	Hormones of the adrenal cortex	
Hormone	Action	Factors regulating secretion
Aldosterone	Helps regulate the concentration of extracellular electrolytes by causing sodium ions to be conserved and potassium ions to be excreted	Electrolyte concentrations in body fluids, and renin-angiotensin mechanism
Cortisol	Decreases protein synthesis, increases fatty acid release, and stimulates the formation of glucose from noncarbohydrates	CRH from the hypothalamus, and ACTH from the anterior pituitary gland
Adrenal androgens	Supplement the sex hormones from the gonads, may be converted into estrogens	

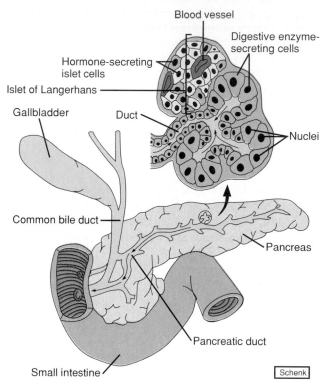

Figure 13.33 The hormone-secreting cells of the pancreas are arranged in clusters, or islets, that are closely associated with blood vessels. Other pancreatic cells secrete digestive enzymes into ducts.

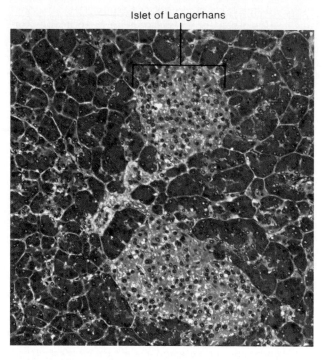

Figure 13.34 Light micrograph of an islet of Langerhans within the pancreas (×50).

vessels. These groups, called *islets of Langerhans,* include three distinct types of hormone-secreting cells—*alpha cells,* which secrete glucagon; *beta cells,* which secrete insulin; and *delta cells,* which secrete somatostatin (figure 13.34). The digestive functions of the pancreas are discussed in chapter 14.

Hormones of the Islets of Langerhans

Glucagon is a protein that stimulates the liver to convert glycogen into glucose (glycogenolysis) and to convert noncarbohydrates, such as amino acids, into glucose (gluconeogenesis). Glucagon also stimulates the breakdown of fats into fatty acids and glycerol.

The secretion of glucagon is regulated by a negative feedback system in which a low concentration of blood sugar stimulates the release of this hormone from the alpha cells. When the blood sugar concentration rises, glucagon is no longer secreted. This mechanism prevents hypoglycemia from occurring at times when glucose concentration is relatively low, such as between meals, or when glucose is being used rapidly—during periods of exercise, for example.

The hormone **insulin** is also a protein, and its main effect is exactly opposite that of glucagon. Insulin acts on the liver to stimulate the formation of glycogen from glucose and to inhibit the conversion of noncarbohydrates into glucose. Insulin also has the special effect of promoting the facilitated diffusion (see chapter 3) of glucose through the membranes of cells that possess insulin receptors. These cells include those of the skeletal muscles, cardiac muscles, and adipose tissues. As a result of its actions, insulin causes a decrease in the concentration of blood glucose. In addition, it promotes the transport of amino acids into the cells and increases the synthesis of proteins; and it stimulates the adipose cells to synthesize and store fat.

As in the case of glucagon secretion, insulin secretion is regulated by a negative feedback system that is sensitive to the concentration of blood glucose. When the glucose concentration is relatively high, as may occur following a meal, insulin is released from the beta cells. By promoting the formation of glycogen in the liver and the entrance of glucose into adipose and muscle cells, this hormone helps prevent an excessive rise in the blood glucose concentration (hyperglycemia). Then, when the glucose concentration is relatively low, between meals or during the night, less insulin is released (figure 13.35).

As the insulin concentration decreases, less and less glucose enters the adipose and muscle cells, and the glucose remaining in the blood is available for use by cells that lack insulin receptors, such as nerve cells. Neurons are dependent on a continuous supply of glucose for ATP production.

At the same time that the insulin concentration is decreasing, glucagon secretion is increasing, so that these hormones function together to help maintain a relatively constant blood glucose concentration, despite great variations in the amounts of carbohydrates that are eaten.

Because nerve cells, including those of the brain, lack insulin receptors, they must obtain glucose by simple diffusion. For this reason, nerve cells are particularly sensitive to changes in the blood glucose concentration, and conditions that cause such changes—excessive insulin secretion, for example—are likely to affect brain functions.

Somatostatin, which is released by the delta cells, helps regulate carbohydrates by inhibiting the secretion of glucagon.

Chart 13.9 summarizes the hormones of the islets of Langerhans.

Cancer cells that develop from nonendocrine tissues sometimes synthesize and secrete great amounts of peptide hormones or peptide hormonelike substances inappropriately. Consequently, a cancer patient may develop an endocrine disorder that seems unrelated to the cancer (endocrine paraneoplastic syndrome). Most commonly, such disorders involve excessive production of ADH, ACTH, a PTH-like substance, or an insulin-like substance.

1. What is the name of the endocrine portion of the pancreas?
2. What is the function of glucagon?
3. What is the function of insulin?
4. How are the secretions of glucagon and insulin controlled?
5. Why are nerve cells particularly sensitive to changes in the blood glucose concentration?

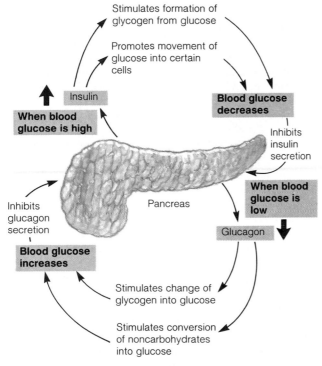

Figure 13.35 Insulin and glucagon function together to help maintain a relatively stable blood glucose concentration.

CHART 13.9	Hormones of the islets of Langerhans	
Hormone	Action	Source of control
Glucagon	Stimulates the liver to convert glycogen and noncarbohydrates into glucose; stimulates breakdown of fats	Blood glucose concentration
Insulin	Promotes formation of glycogen from glucose, inhibits conversion of noncarbohydrates into glucose, and enhances movement of glucose through adipose and muscle cell membranes, causing the blood glucose concentration to fall; promotes transport of amino acids into cells; enhances synthesis of proteins and fats	Blood glucose concentration
Somatostatin	Helps regulate carbohydrates	

Diabetes Mellitus

Diabetes mellitus is characterized by a deficiency of insulin (hypoinsulinism) and by severe disturbances in carbohydrate metabolism as well as by disorders in protein and fat metabolism. More specifically, the movement of glucose into adipose and skeletal muscle cells decreases, glycogen formation decreases, and consequently the concentration of blood sugar rises (hyperglycemia). When the blood sugar reaches a certain high concentration, the kidneys begin to excrete the excess, and glucose appears in the urine (glycosuria). The presence of sugar raises the osmotic pressure of the urine, and more water and electrolytes are excreted than usual. As a result of excessive urine output (diuresis, or polyuria), the affected person becomes dehydrated and experiences great thirst (polydipsia).

The name of this condition is derived from "diabetes," referring to the increased output of urine, and "mellitus" (honey in Latin), referring to its sugar content.

In addition to the above symptoms, diabetes mellitus is accompanied by decreases in the synthesis of protein and fats. As a result of decreased protein synthesis, and increased use of protein as an energy source by glucose-starved cells, tissues tend to waste away. The affected person is likely to lose weight rapidly and have a decreased ability to grow or repair damaged tissues.

As a result of decreased fat synthesis and storage, fatty acids tend to accumulate in the blood abnormally. At the same time, ketone bodies increase in the blood (ketosis). As explained in chapter 4, ketones are a by-product of fat metabolism, and they can be excreted in the urine (ketonuria) in the form of sodium salts. However, when ketones are excreted, large quantities of water follow them osmotically. This water loss intensifies the dehydration of the tissues, and the loss of sodium salts causes the sodium ion concentration to decrease.

The accumulation of acidic ketone bodies in the blood and the loss of sodium ions leads to *metabolic acidosis* (see chapter 21), a condition in which the pH of the body fluids decreases. As a result of acidosis and dehydration, neurons in the brain are adversely affected, and without treatment, the person may become disoriented, comatose (diabetic coma), and die.

The two common forms of diabetes mellitus are *insulin-dependent diabetes mellitus* (also called type I, or juvenile onset diabetes mellitus) and *noninsulin-dependent diabetes mellitus* (type II, or maturity onset diabetes mellitus). Insulin-dependent diabetes mellitus (IDDM) usually appears before the age of twenty. It is an autoimmune disease (chapter 19) in which the body's immune system slowly causes the beta cells of the pancreas to be damaged and finally destroyed, so that the ability to secrete insulin is lost.

The usual treatment for this form of diabetes includes administering enough insulin to control carbohydrate metabolism. However, such therapy almost never achieves the blood glucose stability found in a healthy person. Since insulin is a protein and is broken down by digestive enzymes, it must be administered parenterally—that is, by some route other than the alimentary canal, usually by injection. The amount and type of insulin needed varies with individuals, depending on such factors as diet, physical activities, and health.

Noninsulin-dependent diabetes mellitus (NIDDM) is found in 70%–80% of diabetic patients. It usually appears gradually during adulthood (after the age of forty) and produces milder symptoms than the other form. It is thought to be caused by an inherited disorder, although most people who are affected are overweight when they first experience the symptoms. In this condition, the beta cells continue to function; in fact, the patient may have an excessive blood insulin concentration. Instead of resulting from a lack of insulin, the symptoms of noninsulin-dependent diabetes mellitus seem to involve a loss of sensitivity to the hormone, apparently due to a decreased number of insulin receptors in the target cells.

Treatment for this form of diabetes usually involves carefully controlling the diet, participating in an exercise program, avoiding substances that stimulate insulin secretion, and maintaining a desirable body weight.

A test commonly used to diagnose diabetes mellitus is the *glucose-tolerance test*. To perform the test, the patient ingests a specific amount of glucose, and the blood glucose concentration is measured at intervals to determine how the glucose is utilized. In a person with diabetes, the blood glucose concentration will rise excessively and remain elevated for several hours. In a normal person, the glucose rises to a lesser degree and returns to its normal concentration in about one and one-half hours.

Because a major goal in the treatment of a diabetic patient is to keep the blood glucose concentration under control, regular monitoring of blood glucose is usually necessary. Such testing may be done by a patient at home using a drop of capillary blood and a strip of blood glucose test paper or a blood glucose meter. The results of these tests indicate the blood glucose concentration at the time of the test. Another more complex method for determining a patient's blood glucose concentration involves a laboratory blood test for *hemoglobin A* (glycosylated hemoglobin). This test measures a patient's average blood glucose concentration for the preceding sixty days.

Other Endocrine Glands

Other organs that produce hormones and, therefore, are parts of the endocrine system include the pineal gland, thymus gland, reproductive glands, and certain glands of the digestive tract, as well as the heart and kidneys.

Pineal Gland

The **pineal gland** is a small, oval structure located deep between the cerebral hemispheres, where it is attached to the upper portion of the thalamus near the roof of the third ventricle. Its body consists largely of specialized *pineal cells* and *neuroglial cells* that provide support for the pineal cells. (See figure 11.19b.)

The pineal gland secretes a hormone called **melatonin,** which is synthesized from serotonin. The gland seems to be controlled by varying light conditions outside the body. Information concerning these conditions reaches the gland by means of nerve impulses. In the presence of light, the nerve impulses originate in the retinas of the eyes. Some of these impulses travel to the hypothalamus. From the hypothalamus, they enter the reticular formation and then pass downward into the spinal cord. In the spinal cord, the impulses are transmitted by means of sympathetic nerve fibers back into the brain, and finally they reach the pineal gland. In response to impulses that were triggered originally by light, the secretion of melatonin by the pineal gland decreases.

In the absence of light, nerve impulses from the eyes are decreased, and the secretion of melatonin by the pineal gland increases. This mechanism, by which hormonal secretion is responsive to varying light conditions, is sometimes involved in the regulation of **circadian rhythms.** Circadian rhythms are patterns of repeated activity that are associated with the environmental cycles of day and night, such as sleep/wake rhythm and the seasonal cycles of fertility and infertility that occur in many mammals.

Although the process by which melatonin acts is poorly understood, it seems to inhibit the secretion of gonadotropins from the anterior pituitary gland and is thought to help regulate the female reproductive cycle (menstrual cycle). It may also play a role in reproductive development by controlling the onset of puberty. (See chapter 22.)

Thymus Gland

The **thymus gland,** which lies in the mediastinum behind the sternum and between the lungs, is relatively large in young children but diminishes in size with age. It is believed that this gland secretes a group of hormones, including one called **thymosin** that affects the production of certain white blood cells (lymphocytes). This gland plays an important role in the mechanism of immunity, which is discussed in chapter 19.

Reproductive Glands

The reproductive organs that secrete important hormones include the **ovaries,** which produce estrogens and progesterone; the **placenta,** which produces estrogens, progesterone, and a gonadotropin; and the **testes,** which produce testosterone. These glands and their secretions are discussed in chapter 22.

Digestive Glands

The digestive glands that secrete hormones are generally associated with the linings of the stomach and small intestine. These structures and their secretions are described in chapter 14.

Other Hormone-producing Organs

Other organs that produce hormones include the heart, which secretes *atrial natriuretic factor* (chapter 18), and the kidneys, which secrete a red blood cell growth hormone called *erythropoietin* (chapter 17).

1. Where is the pineal gland located?
2. What seems to be the function of the pineal gland?
3. Where is the thymus gland located?

Stress and Its Effects

Because survival depends upon the maintenance of homeostasis, factors that cause changes in the body's internal environment are potentially life threatening. When such dangers are sensed, nerve impulses are directed to the hypothalamus, and physiological responses are triggered that tend to resist a loss of homeostasis. These responses often include increased activity in the sympathetic division of the autonomic nervous system and increased secretion of adrenal hormones. A factor capable of stimulating such a response is called a **stressor,** and the condition it produces in the body is called **stress.**

Types of Stress

The factors that produce stress are usually nonspecific forms of stimuli and may be physical, psychological, or a combination of both.

Physical stress is caused by exposure to stressors that are harmful or potentially harmful to tissues. These include extreme heat or cold, decreased oxygen concentration, infections, injuries, prolonged heavy exercise, and loud sounds. Often physical stress is accompanied by unpleasant or painful sensations.

Psychological stress results from thoughts about real or imagined dangers, personal losses, unpleasant social interactions (or lack of social interactions), or factors that in any way seem to threaten a person. Feelings of anger, fear, grief, anxiety, depression, and guilt are involved with this type of stress. In other instances,

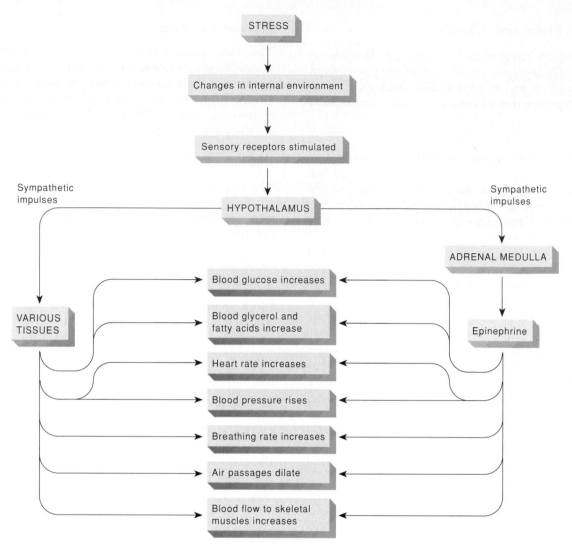

Figure 13.36 At times of stress, the hypothalamus helps prepare the body for fight or flight by triggering sympathetic impulses to various organs. It also stimulates the release of epinephrine, which intensifies the sympathetic responses.

psychological stress may stem from pleasant stimuli, such as friendly social contact, feelings of joy or happiness, or sexual arousal. The factors that produce psychological stress vary greatly from person to person. Thus, a situation that is stressful to one person may not affect another, and what is stressful at one time may not be at another time.

Responses to Stress

Regardless of its cause, the physiological responses to stress are directed toward maintaining homeostasis, and they usually involve a set of reactions called the *general stress syndrome* (general adaptation syndrome), which is largely controlled by the hypothalamus.

As discussed in chapter 11, the hypothalamus receives information from nearly all body parts, in-

cluding visceral receptors and various regions of the brain, such as the cerebral cortex, reticular formation, and limbic system. At times of stress, the hypothalamus may respond to incoming impulses by activating mechanisms that generally prepare the body for fight or flight. More specifically, sympathetic impulses from the hypothalamus cause a rise in the blood glucose concentration, an increase in the level of blood glycerol and fatty acids, an increase in the heart rate, a rise in blood pressure, an increase in the breathing rate, dilation of the air passages, a shunting of blood from the skin and digestive organs into the skeletal muscles, and an increase in the secretion of *epinephrine* from the adrenal medulla. The epinephrine, in turn, intensifies these sympathetic responses and prolongs their effects (figure 13.36).

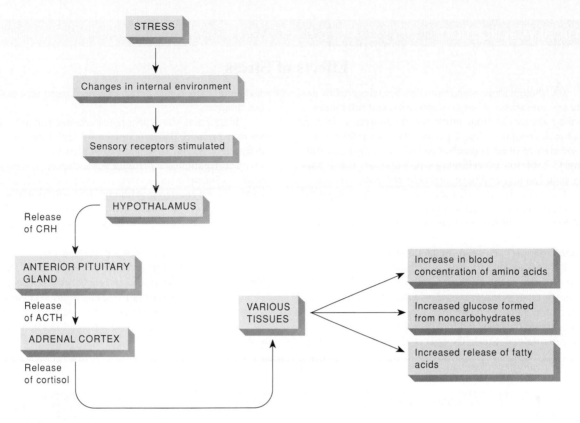

Figure 13.37 As a result of stress, the hypothalamus causes the adrenal cortex to release cortisol, which promotes reactions that resist the effects of stress.

At the same time, the hypothalamus releases *corticotropin-releasing hormone* (CRH), which stimulates the anterior pituitary gland to secrete ACTH. The ACTH causes the adrenal cortex to increase its secretion of *cortisol.*

As mentioned earlier, cortisol promotes increases in the blood concentration of amino acids, the release of fatty acids, and the formation of glucose from noncarbohydrates. Thus, while the sympathetic responses are preparing the body for physical activities that may be necessary for survival, the actions of cortisol supply cells with amino acids and extra energy sources that may be needed during times of stress. This effect of cortisol may also allow available glucose to be diverted from skeletal muscles to brain tissue (figure 13.37).

Other hormones whose secretions increase with stress include *glucagon* from the pancreas, *growth hormone* (GH) from the anterior pituitary, and *antidiuretic hormone* (ADH) from the posterior pituitary gland.

The general effect of glucagon and growth hormone is to aid in mobilizing energy sources, such as glucose, glycerol, and fatty acids, and to stimulate the uptake of amino acids by cells, facilitating the repair of injured tissues. ADH promotes the retention of sodium ions and water by the kidneys. This action decreases urine output and increases blood volume—an action of particular importance if a person is bleeding excessively or sweating heavily. Chart 13.10 summarizes the body's reactions to stress.

1. What is meant by stress?
2. Distinguish between physical stress and psychological stress.
3. Describe the general stress syndrome.

Clinical Terms Related to the Endocrine System

adrenalectomy (ah-dre″nah-lek′to-me) Surgical removal of the adrenal glands.

adrenogenital syndrome (ah-dre″no-jen′ĭ-tal sin′drōm) A group of symptoms associated with changes in sexual characteristics as a result of increased secretion of adrenal androgens.

diabetes mellitus (di″ah-be′tēz mel′ĭ-tus) A condition due to a deficiency of insulin and characterized by severe disturbances in carbohydrate, protein, and lipid metabolism.

Effects of Stress

Although stress responses are often triggered by extreme environmental changes or intense social interactions, they may also result from subtle social contacts or from a lack of social contact. Thus, a person who is socially isolated may experience stress as much as one who is anxious, fearful, or angry. A person experiencing stress may appear abnormally pale (pallor) and have cool skin and a dry mouth.

The intensity of a stress response can be judged by observing changes in a person's heart rate, respiratory rate, blood pressure, and perspiration on the palms of the hands. It can also be measured by determining the concentration of epinephrine or cortisol in the blood or urine.

For reasons that are poorly understood, increased secretion of cortisol may be accompanied by a decrease in the activities of the *lymphatic organs,* which include the thymus, lymph nodes, and spleen. At the same time, the number of *lymphocytes* in the blood tends to decrease. Since these white blood cells defend the body against infections, a person who is under stress may have a lowered resistance to infectious diseases.

In addition, the exposure to excessive amounts of cortisol may promote the development of high blood pressure, atherosclerosis, and gastrointestinal ulcers.

Most patients experience some degree of stress as a result of disease, pain, anxiety, or even the medical treatments they are receiving. Although it is usually not possible to prevent stress completely, psychological stress may often be reduced by discussing problems and sources of anxiety with patients and by explaining clinical procedures so that they understand what is taking place and why.

Physical stress in patients can be reduced by providing pleasant, comfortable external surroundings and by helping each patient maintain a stable internal environment by controlling pain, administering necessary fluids and electrolytes, providing drug treatments for infections, and so forth.

CHART 13.10 Major events in the general stress syndrome

1. As a result of stress, nerve impulses are transmitted to the hypothalamus.

2. Sympathetic impulses arising from the hypothalamus cause increases in blood glucose concentration, blood glycerol concentration, blood fatty acid concentration, heart rate, blood pressure, and breathing rate, as well as dilation of air passages, shunting of blood into skeletal muscles, and increased secretion of epinephrine from the adrenal medulla.

3. Epinephrine intensifies and prolongs sympathetic actions.

4. The hypothalamus secretes CRH, which stimulates the secretion of ACTH by the anterior pituitary gland.

5. ACTH stimulates the release of cortisol by the adrenal cortex.

6. Cortisol promotes increases in the concentration of blood amino acids, release of fatty acids, and formation of glucose from noncarbohydrate sources.

7. Secretions of glucagon from the pancreas and growth hormone from the anterior pituitary increase.

8. Glucagon and growth hormone aid the mobilization of energy sources and stimulate the uptake of amino acids by cells.

9. Secretion of ADH from the posterior pituitary increases.

10. ADH promotes the retention of sodium ions and water by the kidneys, which increases blood volume.

11. As a result of these actions, the body is better able to resist the effects of stress.

---

diabetes insipidus (di″ah-be′tēz in-sip′ĭ-dus) A metabolic disorder characterized by a large output of dilute urine containing no sugar, and caused by a decreased secretion of ADH from the posterior pituitary gland.

exophthalmos (ek″sof-thal′mos) An abnormal protrusion of the eyes.

hirsutism (her′sut-izm) Excessive growth of hair, especially in women.

hypercalcemia (hi″per-kal-se′me-ah) An excess of blood calcium.

hyperglycemia (hi″per-gli-se′me-ah) An excess of blood glucose.

hypocalcemia (hi″po-kal-se′me-ah) A deficiency of blood calcium.

hypoglycemia (hi″po-gli-se′me-ah) A deficiency of blood glucose.

hypophysectomy (hi-pof″ĭ-sek′to-me) Surgical removal of the pituitary gland.

parathyroidectomy (par″ah-thi″roi-dek′to-me) Surgical removal of the parathyroid glands.

pheochromocytoma (fe-o-kro″mo-si-to′mah) A type of tumor found in the adrenal medulla and usually accompanied by high blood pressure.

polyphagia (pol″e-fa′je-ah) Excessive eating.

thymectomy (thi-mek'to-me) Surgical removal of the thymus gland.

thyroidectomy (thi''roi-dek'to-me) Surgical removal of the thyroid gland.

thyroiditis (thi''roi-di'tis) Inflammation of the thyroid gland.

virilism (vir'ĭ-lizm) Masculinization of a female.

Chapter Summary

Introduction (page 462)

Endocrine glands secrete their products into body fluids; exocrine glands secrete their products into ducts that lead to the outside of the body.

General Characteristics of the Endocrine System (page 462)

As a group, endocrine glands are concerned with the regulation of metabolic processes.

Hormones and Their Actions (page 462)

Endocrine glands secrete hormones that affect target cells possessing specific receptors.

1. Chemistry of hormones
   a. Each kind of hormone has a special molecular structure and is very potent.
   b. Chemically, hormones are steroids, amines, peptides, proteins, or glycoproteins.
2. Actions of hormones
   a. Steroid hormones
      (1) Steroid hormones enter target cells and combine with receptors in the nucleus to form complexes.
      (2) These complexes activate specific genes, which, in turn, cause special proteins to be synthesized.
      (3) The degree of cellular response is proportional to the number of hormone-receptor complexes formed.
   b. Nonsteroid hormones
      (1) Nonsteroid hormones combine with receptors in the target-cell membrane.
      (2) A hormone-receptor combination stimulates membrane proteins, such as adenylate cyclase, to induce the formation of second messenger molecules.
      (3) A second messenger, such as cAMP, activates protein kinases.
      (4) Protein kinases activate certain protein substrate molecules, which, in turn, cause changes in cellular processes.
      (5) The cellular response to a nonsteroid hormone is amplified because the enzymes induced by a small number of hormone-receptor complexes can cause a large number of second messenger molecules to be formed.

3. Prostaglandins
   a. Prostaglandins are substances present in small quantities that have powerful hormonelike effects.
   b. Prostaglandins seem to function as modulators of hormones that regulate the formation of cyclic AMP.

Control of Hormonal Secretions (page 468)

The concentration of each hormone in the body fluids must be regulated precisely.

1. Negative feedback systems
   a. In a negative feedback system, a gland is sensitive to the concentration of a substance it regulates.
   b. When the concentration of the regulated substance reaches a certain concentration, it inhibits the gland.
   c. As the gland secretes less hormone, the controlled substance also decreases.
2. Nerve control
   a. Some endocrine glands secrete their hormones in response to nerve impulses.
   b. Other glands secrete hormones in response to releasing hormones secreted by the hypothalamus.

Pituitary Gland (page 469)

The pituitary gland, which is attached to the base of the brain, has an anterior lobe and a posterior lobe. Most pituitary secretions are controlled by the hypothalamus.

1. Anterior pituitary hormones
   a. The anterior pituitary consists largely of epithelial cells, and it secretes GH, PRL, TSH, ACTH, FSH, and LH.
   b. Growth hormone (GH)
      (1) Growth hormone stimulates body cells to increase in size and rate of reproduction.
      (2) The secretion of GH is controlled by growth hormone-releasing hormone and somatostatin from the hypothalamus.
   c. Prolactin (PRL)
      (1) PRL promotes breast development and stimulates milk production.
      (2) In males, prolactin decreases the secretion of LH (ICSH).
      (3) Prolactin release-inhibiting factor from the hypothalamus restrains the secretion of prolactin, while prolactin-releasing factor promotes its secretion.
   d. Thyroid-stimulating hormone (TSH)
      (1) TSH controls the secretion of hormones from the thyroid gland.
      (2) TSH secretion is regulated by the hypothalamus, which secretes thyrotropin-releasing hormone.

e. Adrenocorticotropic hormone (ACTH)
   (1) ACTH controls the secretion of certain hormones from the adrenal cortex.
   (2) The secretion of ACTH is regulated by the hypothalamus, which secretes corticotropin-releasing hormone.
f. Follicle-stimulating hormone (FSH) and luteinizing hormone (LH) are gonadotropins that influence the reproductive organs.

2. Posterior pituitary hormones
   a. The posterior lobe of the pituitary gland consists largely of neuroglial cells and nerve fibers that originate in the hypothalamus.
   b. The two hormones of the posterior pituitary are produced in the hypothalamus.
   c. Antidiuretic hormone (ADH)
      (1) ADH causes the kidneys to reduce the amount of water they excrete.
      (2) In high concentration, ADH causes blood vessel walls to constrict, thus raising the blood pressure.
      (3) The secretion of ADH is regulated by the hypothalamus.
   d. Oxytocin (OT)
      (1) Oxytocin has an antidiuretic effect and can cause muscles in the uterine wall to contract, thus playing a role in childbirth.
      (2) OT also causes contraction of certain cells associated with the production and ejection of milk from the milk glands of the breasts.

Thyroid Gland (page 476)

The thyroid gland is located in the neck and consists of two lateral lobes.

1. Structure of the gland
   a. The thyroid gland consists of many hollow secretory parts called follicles.
   b. The follicles are fluid filled and store the hormones secreted by the follicle cells.
   c. Extrafollicular cells secrete calcitonin.
2. Thyroid hormones
   a. Thyroxine and triiodothyronine
      (1) These hormones increase the rate of metabolism, enhance protein synthesis, and stimulate the breakdown of lipids.
      (2) These hormones are needed for normal growth and development, and for the maturation of the nervous system.
   b. Calcitonin
      (1) Calcitonin lowers the blood calcium and phosphate ion concentrations.
      (2) This hormone prevents prolonged elevation of calcium after a meal.

Parathyroid Glands (page 480)

The parathyroid glands are located on the posterior surface of the thyroid.

1. Structure of the glands
   a. Each gland consists of secretory cells that are well supplied with capillaries.
2. Parathyroid hormone (PTH)
   a. PTH causes an increase in blood calcium ion concentration and a decrease in blood phosphate ion concentration.
   b. PTH stimulates the resorption of bone tissue, causes the kidneys to conserve calcium ions and excrete phosphate ions, and indirectly stimulates the absorption of calcium ions from the intestine.
   c. The parathyroid glands seem to be regulated by a negative feedback mechanism that operates between the glands and the blood.

Adrenal Glands (page 482)

The adrenal glands are located atop the kidneys.

1. Structure of the glands
   a. Each adrenal gland consists of a medulla and a cortex.
   b. The adrenal medulla and adrenal cortex represent distinct glands that secrete different hormones.
2. Hormones of the adrenal medulla
   a. The adrenal medulla secretes epinephrine and norepinephrine.
   b. These hormones are synthesized from tyrosine and are closely related to each other chemically.
   c. These hormones produce effects similar to those of the sympathetic nervous system.
   d. The secretion of these hormones is stimulated by sympathetic impulses originating from the hypothalamus.
3. Hormones of the adrenal cortex
   a. The cortex produces a variety of steroids that include several hormones.
   b. Aldosterone
      (1) It causes the kidneys to conserve sodium ions and water, and to excrete potassium ions.
      (2) It is secreted in response to decreased sodium ion concentration, increased potassium ion concentration, or the presence of angiotensin II.
      (3) By promoting the conservation of sodium ions and water, it helps maintain blood volume and pressure.

c. Cortisol
   (1) It inhibits the synthesis of proteins, promotes the release of fatty acids, and stimulates the formation of glucose from noncarbohydrates.
   (2) It is controlled by a negative feedback mechanism involving the secretion of CRH from the hypothalamus and ACTH from the anterior pituitary gland.
d. Adrenal sex hormones
   (1) These hormones are of the male type although some can be converted into female hormones.
   (2) They are thought to supplement the sex hormones produced by the gonads.

Pancreas (page 487)

The pancreas secretes digestive juices as well as hormones.

1. Structure of the gland
   a. The pancreas is posterior to the stomach and is attached to the small intestine.
   b. The endocrine portion, which is called the islets of Langerhans, secretes glucagon, insulin, and somatostatin.
2. Hormones of the islets of Langerhans
   a. Glucagon stimulates the liver to produce glucose, causing an increase in the concentration of blood glucose. It also promotes the breakdown of fats.
   b. Insulin promotes the movement of glucose through cell membranes, stimulates the storage of glucose, promotes the synthesis of proteins, and stimulates the storage of fats.
   c. Nerve cells lack insulin receptors and depend upon diffusion for a glucose supply.
   d. Somatostatin may inhibit glucagon release.

Other Endocrine Glands (page 491)

1. Pineal gland
   a. The pineal gland is attached to the thalamus near the roof of the third ventricle.
   b. It is innervated by postganglionic sympathetic nerve fibers.
   c. It secretes melatonin, which seems to inhibit the secretion of gonadotropins from the anterior pituitary gland.
   d. It may help regulate the female reproductive cycle.
2. Thymus gland
   a. The thymus gland lies behind the sternum and between the lungs.
   b. Its size diminishes with age.
   c. It secretes thymosin, which affects the production of certain lymphocytes that, in turn, play important roles in immunity.

3. Reproductive glands
   a. The ovaries secrete estrogens and progesterone.
   b. The placenta secretes estrogens, progesterone, and a gonadotropin.
   c. The testes secrete testosterone.
4. The digestive glands include certain glands of the stomach and small intestine that secrete hormones.
5. Other hormone-producing organs include the heart and kidneys.

Stress and Its Effects (page 491)

Stress is produced when the body responds to stressors that threaten the maintenance of homeostasis. Stress responses involve increased activity of the sympathetic nervous system and increased secretion of adrenal hormones.

1. Types of stress
   a. Physical stress results from environmental factors that are harmful or potentially harmful to tissues.
   b. Psychological stress results from thoughts about real or imagined dangers.
   c. Factors that produce psychological stress vary from person to person and from situation to situation.
2. Responses to stress
   a. Responses to stress are directed toward maintaining homeostasis.
   b. Usually these responses involve a general stress syndrome that is controlled by the hypothalamus.

Clinical Application of Knowledge

1. Based on your understanding of the actions of glucagon and insulin, would a person with diabetes mellitus be likely to require more insulin or more sugar following strenuous exercise? Why?
2. What problems might result from the prolonged administration of cortisol to a person with a severe inflammatory disease?
3. How might the environment of a patient with hyperthyroidism be modified to minimize the drain on body energy resources?
4. What hormones should be administered to an adult whose anterior pituitary gland has been removed? Why?
5. A patient who has lost a relatively large volume of blood will have an increased secretion of aldosterone from the adrenal cortex. What effect will this increased secretion have on the patient's blood concentrations of sodium and potassium ions?
6. Both growth hormone and growth hormone-releasing hormone have been successfully used to promote growth in children with short statures. How would you explain the difference in the ways these hormones produce their effects?

Review Activities

1. Explain what is meant by an *endocrine gland*.
2. Define *hormone* and *target cell*.
3. Explain how hormones can be grouped on the basis of their chemical compositions.
4. Explain how steroid hormones exert their influence on cells.
5. Distinguish between the binding site and the activity site of a receptor molecule.
6. Explain how nonsteroid hormones may function through the formation of cAMP.
7. Explain how nonsteroid hormones may function through an increase in intracellular calcium ion concentration.
8. Explain how the cellular response to a hormone operating through a second messenger is amplified.
9. Define *prostaglandins*, and explain their general function.
10. Describe a negative feedback system.
11. Define *releasing hormone*, and provide an example of such a substance.
12. Describe the location and structure of the pituitary gland.
13. List the hormones secreted by the anterior pituitary gland.
14. Explain how pituitary gland activity is controlled by the brain.
15. Explain how growth hormone produces its effects.
16. List the major factors that affect the secretion of growth hormone.
17. Summarize the functions of prolactin.
18. Describe the mechanism that regulates the concentrations of circulating thyroid hormones.
19. Explain how the secretion of ACTH is controlled.
20. List the major gonadotropins, and explain the general functions of each.
21. Compare the cellular structures of the anterior and posterior lobes of the pituitary gland.
22. Name the hormones associated with the posterior pituitary, and explain their functions.
23. Explain how the release of ADH is regulated.
24. Describe the location and structure of the thyroid gland.
25. Name the hormones secreted by the thyroid gland, and list the general functions of each.
26. Define *iodine pump*.
27. Describe the location and structure of the parathyroid glands.
28. Explain the general functions of parathyroid hormone.
29. Describe the mechanism that regulates the secretion of parathyroid hormone.
30. Distinguish between the adrenal medulla and the adrenal cortex.
31. List the hormones produced by the adrenal medulla, and describe their general functions.
32. List the steps in the synthesis of adrenal medullary hormones.
33. Name the most important hormones of the adrenal cortex, and describe the general functions of each.
34. Describe how the secretion of aldosterone is regulated.
35. Describe how the secretion of cortisol is regulated.
36. Describe the location and structure of the pancreas.
37. List the hormones secreted by the islets of Langerhans, and describe the general functions of each.
38. Summarize how the secretion of hormones from the pancreas is regulated.
39. Describe the location and general function of the pineal gland.
40. Describe the location and general function of the thymus gland.
41. Distinguish between a stressor and stress.
42. List several factors that cause physical and psychological stress.
43. Describe the general stress syndrome.

UNIT
4

Processing and Transporting

T he chapters of unit 4 are concerned with the digestive, respiratory, circulatory, lymphatic, and urinary systems. They describe how organs of these systems obtain nutrients and oxygen from outside the body, how the nutrients are altered chemically and absorbed into body fluids, and how the nutrients and oxygen are transported to body cells. They also explain how these substances are utilized by cells, how resulting wastes are transported and excreted, and how stable concentrations of various substances are maintained in body fluids.

Innumerable microscopic projections (villi) extend into the lumen of the small intestine and greatly increase the surface area of the intestinal lining (×32).

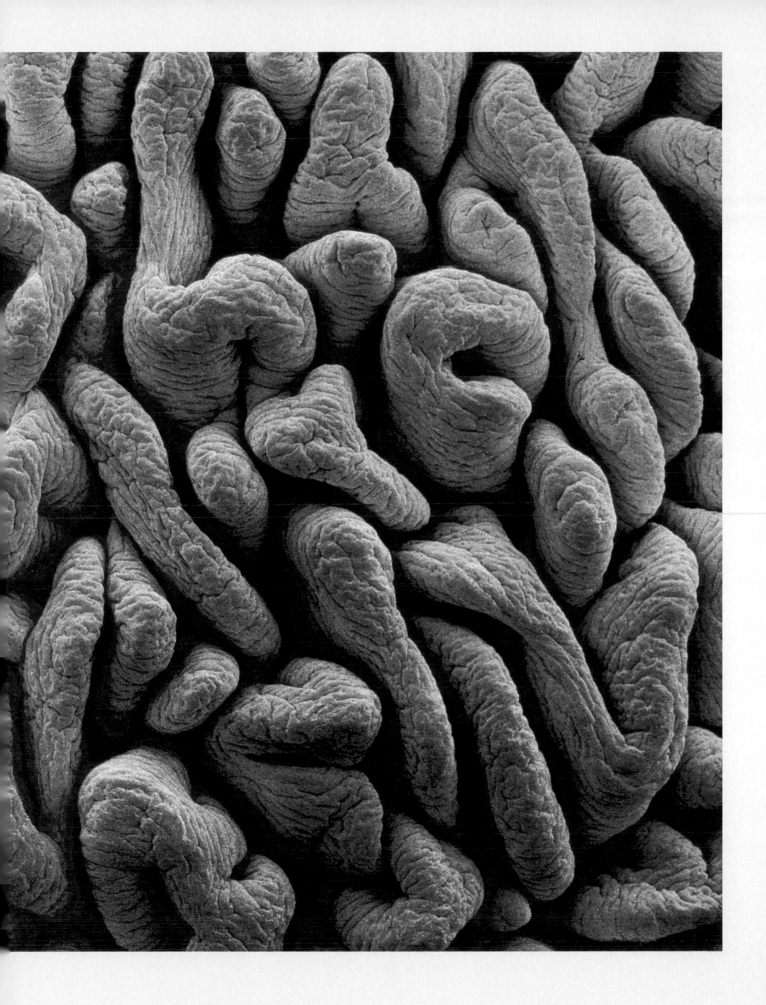

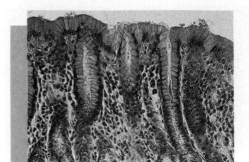

Digestive System

*M*ost food substances are composed of chemicals whose molecules cannot pass easily through membranes and so cannot be absorbed effectively by cells. The *digestive system* solves this problem. Its parts are adapted to ingest foods, to break large particles into smaller ones, to secrete enzymes that decompose food molecules, to absorb the products of this digestive action, and to eliminate the unused residues.

Foods are moved through the digestive tract by muscular contractions in the wall of the tubular alimentary canal. Digestive juices, which contain digestive enzymes, are secreted into the canal by various glands and accessory organs. These functions are controlled largely by interactions between the digestive, nervous, and endocrine systems ■

Chapter Objectives

After you have studied this chapter, you should be able to:

1. Name and describe the locations of the organs of the digestive system and their major parts.

2. Describe the general functions of each digestive organ and of the liver.

3. Describe the structure of the wall of the alimentary canal.

4. Explain how the contents of the alimentary canal are mixed and moved.

5. List the enzymes secreted by the various digestive organs and describe the function of each enzyme.

6. Describe how digestive secretions are regulated.

7. Explain how digestive reflexes function to control the movement of material through the alimentary canal.

8. Describe the mechanisms of swallowing, vomiting, and defecating.

9. Explain how the products of digestion are absorbed.

10. Complete the review activities at the end of this chapter. Note that the items are worded in the form of specific learning objectives. You may want to refer to them before reading the chapter.

Key Terms

absorption (ab-sorp'shun)

accessory organ (ak-ses'o-re or'gan)

alimentary canal (al''ĭ-men'tar-e kah-nal')

bile (bīl)

chyme (kīm)

circular muscle (ser'ku-lar mus'el)

deciduous (de-sid'u-us)

feces (fe'sez)

gastric juice (gas'trik jōōs)

intestinal juice (in-tes'tĭ-nal jōōs)

intrinsic (in-trin'sik)

longitudinal muscle (lon''jĭ-tu'dĭ-nal mus'el)

mesentery (mes'en-ter''e)

mucous membrane (mu'kus mem'brān)

pancreatic juice (pan''kre-at'ik jōōs)

peristalsis (per''ĭ-stal'sis)

serous layer (se'rus la'er)

sphincter muscle (sfingk'ter mus'el)

villi; singular, villus (vil'i, vil'us)

Aids to Understanding Words

aliment-, food: *aliment*ary canal—the tubelike portion of the digestive system.

cari-, decay: dental *cari*es—tooth decay.

cec-, blindness: *cec*um—blind-ended sac at the beginning of the large intestine.

chym-, juice: *chym*e—semifluid paste of food particles and gastric juice formed in the stomach.

decidu-, falling off: *decidu*ous teeth—teeth that are shed during childhood.

frenul-, a restraint: *frenul*um—membranous fold that anchors the tongue to the floor of the mouth.

gastr-, stomach: *gastr*ic gland—a portion of the stomach that secretes gastric juice.

hepat-, liver: *hepat*ic duct—duct that carries bile from the liver to the common bile duct.

hiat-, an opening: esophageal *hiat*us—opening through which the esophagus penetrates the diaphragm.

lingu-, the tongue: *lingu*al tonsil—mass of lymphatic tissue at the root of the tongue.

peri-, around: *peri*stalsis—a wavelike ring of contraction that moves material along the alimentary canal.

pylor-, gatekeeper: *pylor*ic sphincter—a muscle that serves as a valve between the stomach and small intestine.

vill-, hairy: *vill*i—tiny projections of mucous membrane in the small intestine.

igestion is the process by which food substances are changed into forms that can be absorbed through cell membranes. The **digestive system** includes the organs that promote digestion and absorb the products of this process. It consists of an *alimentary canal,* which extends from the mouth to the anus, and several *accessory organs,* which release secretions into the canal. The alimentary canal includes the mouth, pharynx, esophagus, stomach, small intestine, and large intestine. The accessory organs include the salivary glands, liver, gallbladder, and pancreas. The major organs of the digestive system are shown in figure 14.1.

General Characteristics of the Alimentary Canal

The **alimentary canal** is a muscular tube about 9 meters long that passes through the body's ventral cavity. Although it is specialized in various regions to carry on particular functions, the structure of its wall, the method by which it moves food, and its type of innervation are similar throughout its length (figure 14.2).

Structure of the Wall

The wall of the alimentary canal consists of four distinct layers, although the degree to which they are developed varies from region to region. Beginning with the innermost tissues, these layers, shown in figure 14.3, include the following:

1. **Mucous membrane** (mucosa). This layer is formed of surface epithelium, underlying connective tissue (lamina propria), and a small amount of smooth muscle. In some regions, it develops folds and tiny projections, which extend into the lumen of the digestive tube and increase its absorptive surface area. It may also contain glands that are tubular invaginations into which the lining cells secrete mucus and digestive enzymes. The mucosa protects the tissues beneath it and carries on absorption and secretion.
2. **Submucosa.** The submucosa contains considerable loose connective tissue as well as glands, blood vessels, lymphatic vessels, and nerves. Its vessels nourish the surrounding tissues and carry away absorbed materials.
3. **Muscular layer.** This layer, which is responsible for the movements of the tube, consists of two

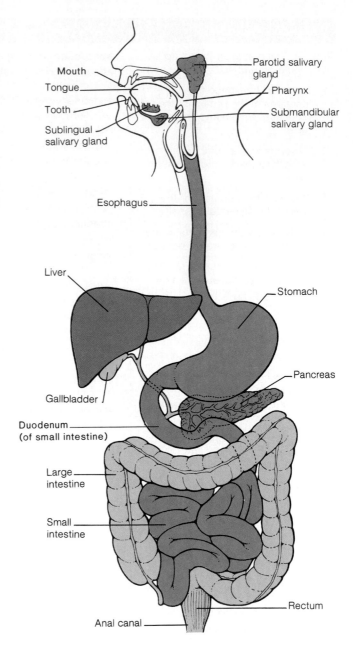

Figure 14.1 Major organs of the digestive system. What are the general functions of this system?

coats of smooth muscle tissue. The fibers of the inner coat are arranged so that they encircle the tube, and when these *circular fibers* contract, the diameter of the tube is decreased. The fibers of the outer muscular coat run lengthwise, and when these *longitudinal fibers* contract, the tube is shortened.

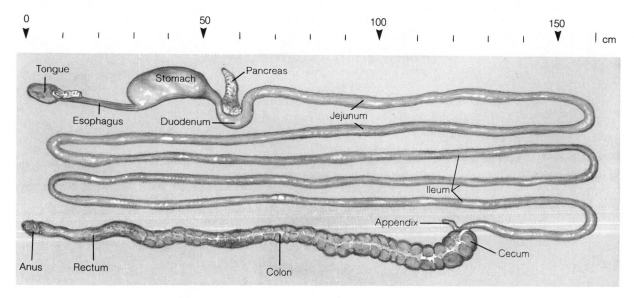

Figure 14.2 The alimentary canal is a muscular tube about nine meters long.

CHART 14.1	Layers of the wall of the alimentary canal	
Layer	**Composition**	**Function**
Mucous membrane	Epithelium, connective tissue, smooth muscle	Protection, absorption, secretion
Submucosa	Loose connective tissue, blood vessels, lymphatic vessels, nerves	Nourishes surrounding tissues, transports absorbed materials
Muscular layer	Smooth muscle fibers arranged in circular and longitudinal groups	Movements of the tube and its contents
Serous layer	Epithelium, connective tissue	Protection, lubrication

4. **Serous layer** (serosa). The serous layer, or outer covering of the tube, is composed of the *visceral peritoneum,* which is formed of epithelium on the outside and connective tissue beneath. The cells of the serosa protect underlying tissues and secrete serous fluid, which keeps the tube's outer surface moist. This lubricates the surface so that the organs within the abdominal cavity slide freely against one another.

The characteristics of these layers are summarized in chart 14.1.

Movements of the Tube

The motor functions of the alimentary canal are of two basic types—*mixing movements* and *propelling move-*ments (figure 14.4). Mixing occurs when smooth muscles in relatively small segments of the tube undergo rhythmic contractions. For example, when the stomach is full, waves of muscular contractions move along its wall from one end to the other. These waves occur every twenty seconds or so, and their action mixes food substances with the digestive juices secreted by the mucosa.

Propelling movements include a wavelike motion called **peristalsis.** When peristalsis occurs, a ring of contraction appears in the wall of the tube. At the same time, the muscular wall just ahead of the ring relaxes—a phenomenon called *receptive relaxation.* As the wave moves along, it pushes the tubular contents ahead of it. The usual stimulus for peristalsis is an expansion of the tube due to accumulation of food inside. Such movements cause sounds that can be heard through a stethoscope applied to the abdominal wall.

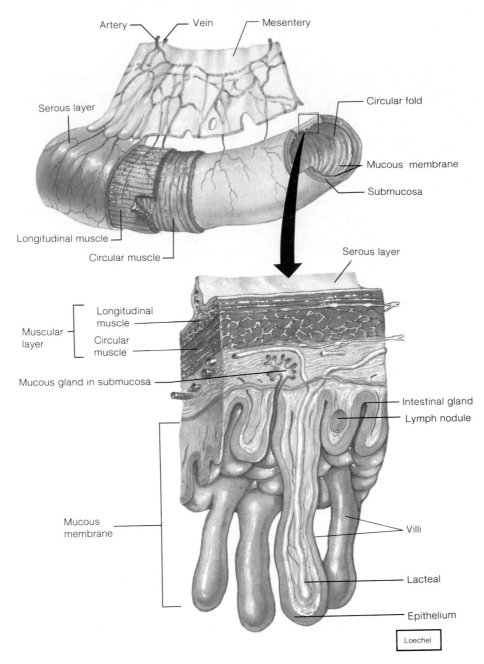

Artery — Vein — Mesentery

Serous layer

Circular fold

Mucous membrane

Submucosa

Longitudinal muscle

Circular muscle

Serous layer

Muscular layer
Longitudinal muscle
Circular muscle

Mucous gland in submucosa

Intestinal gland

Lymph nodule

Mucous membrane

Villi

Lacteal

Epithelium

Loechel

Figure 14.3 The wall of the small intestine, as in other portions of the alimentary canal, includes four layers: an inner mucous membrane, a submucosa, a muscular layer, and an outer serous layer.

Innervation of the Tube

The alimentary canal is innervated extensively by branches of the sympathetic and parasympathetic divisions of the autonomic nervous system. These nerve fibers are associated mainly with the tube's muscular layer, and they are responsible for maintaining muscle tone and for regulating the strength, rate, and velocity of muscular contractions.

Parasympathetic impulses generally cause an increase in the activities of the digestive system. Some of these impulses originate in the brain and are conducted on branches of the vagus nerves to the esophagus, stomach, pancreas, gallbladder, small intestine, and proximal half of the large intestine. Other parasympathetic impulses arise in the sacral region of the spinal cord and supply the distal half of the large intestine.

The effects on digestive actions produced by sympathetic nerve impulses usually are opposite those of the parasympathetic division. In other words, sympathetic impulses inhibit various digestive actions. Such impulses are responsible for the contraction of certain

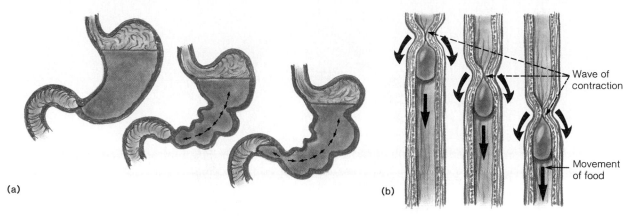

(a)

(b)

Wave of
contraction

Movement
of food

Figure 14.4 (a) Mixing movements occur when small segments of the muscular wall of the alimentary canal undergo rhythmic contractions. (b) Peristaltic waves cause the contents to be moved along the canal.

sphincter muscles in the wall of the alimentary canal, and when they are contracted, these muscles effectively block the movement of materials through the tube.

1. What organs constitute the digestive system?
2. Describe the wall of the alimentary canal.
3. Name the two types of movements that occur in the alimentary canal.
4. What effect do parasympathetic nerve impulses have on digestive actions? What effect do sympathetic nerve impulses have?

Mouth

The **mouth,** which is the first portion of the alimentary canal, is adapted to receive food and prepare it for digestion by mechanically reducing the size of solid particles and mixing them with saliva (mastication). The mouth also functions as an organ of speech and sensory reception. It is surrounded by the lips, cheeks, tongue, and palate, and includes a chamber between the palate and tongue called the *oral cavity,* as well as a narrow space between the teeth, cheeks, and lips called the *vestibule* (figure 14.5 and reference plates 49 and 55).

Cheeks and Lips

The **cheeks** form the lateral walls of the mouth. They consist of outer layers of skin, pads of subcutaneous fat, certain muscles associated with expression and chewing, and inner linings of moist, stratified squamous epithelium.

The **lips** are highly mobile structures that surround the mouth opening. They contain skeletal muscles and a variety of sensory receptors, which are useful in judging the temperature and texture of foods. Their normal reddish color is due to an abundance of blood vessels near their surfaces. The external borders of the lips mark the boundaries between the skin of the face and the mucous membrane that lines the alimentary canal.

Tongue

The **tongue** is a thick, muscular organ that occupies the floor of the mouth and nearly fills the oral cavity when the mouth is closed. It is covered by mucous membrane and is connected in the midline to the floor of the mouth by a membranous fold called the **frenulum.**

The *body* of the tongue is composed largely of skeletal muscle whose fibers run in several directions. These muscles aid in mixing food particles with saliva during chewing and in moving food toward the pharynx during swallowing. Rough projections, called **papillae,** on the surface of the tongue provide friction, which is useful in handling food. These papillae also contain taste buds (figure 14.6).

The posterior region, or *root,* of the tongue is anchored to the hyoid bone and is covered with rounded masses of lymphatic tissue called **lingual tonsils.**

Palate

The **palate** forms the roof of the oral cavity and consists of a hard anterior part and a soft posterior part. The *hard palate* is formed by the palatine processes of the maxillary bones in front and the horizontal portions of the palatine bones in back. The *soft palate* forms a muscular arch, which extends posteriorly and downward as a cone-shaped projection called the **uvula.**

During swallowing, muscles draw the soft palate and the uvula upward. This action closes the opening between the nasal cavity and the pharynx, preventing food from entering the nasal cavity.

In the back of the mouth, on either side of the tongue and closely associated with the palate, are masses of lymphatic tissue called **palatine tonsils.** These

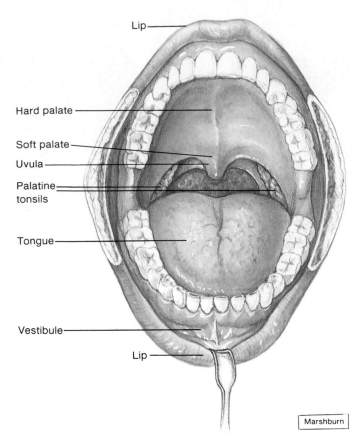

Figure 14.5 The mouth is adapted for ingesting food and preparing it for digestion.

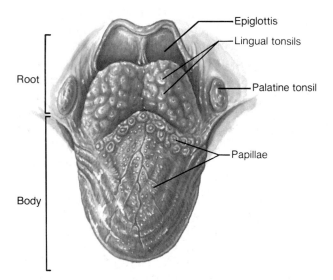

Figure 14.6 The surface of the tongue, viewed from above.

The palatine tonsils themselves are common sites of infections, and if they become inflamed, the condition is termed *tonsillitis.* Infected tonsils may become so swollen that they block the passageways of the pharynx and interfere with breathing and swallowing. Because the mucous membranes of the pharynx, auditory tubes, and middle ears are continuous, there is danger that such an infection may travel from the throat into the middle ears (otitis media).

When tonsillitis occurs repeatedly and remains unresponsive to antibiotic treatment, the tonsils are often removed. This surgical procedure is called *tonsillectomy.*

Still other masses of lymphatic tissue, called **pharyngeal tonsils,** or *adenoids,* occur on the posterior wall of the pharynx, above the border of the soft palate. If these parts become enlarged and block the passage between the pharynx and the nasal cavity, they also may be removed surgically (figure 14.7).

structures lie beneath the epithelial lining of the mouth and, like other lymphatic tissues, help protect the body against infections. (See chapter 19 for more detail concerning lymphatic functions.)

1. What is the function of the mouth?
2. How does the tongue function as part of the digestive system?
3. What is the role of the soft palate in swallowing?
4. Where are the tonsils located?

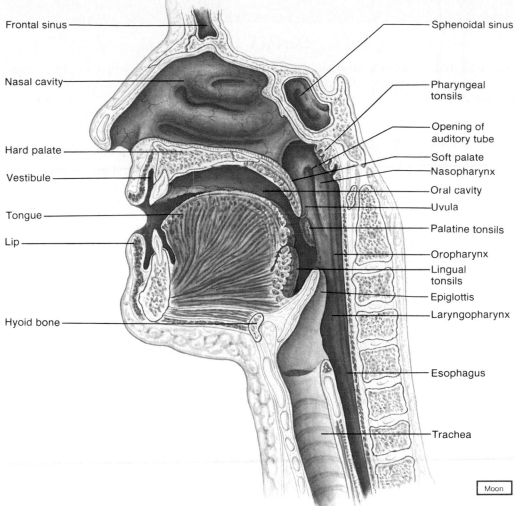

Figure 14.7   A sagittal section of the mouth, nasal cavity, and pharynx.

Teeth

The **teeth** develop in sockets within the alveolar processes of the mandibular and maxillary bones. Teeth are unique structures in that two sets form during development. The members of the first set, the *primary teeth* (deciduous teeth), usually erupt through the gums (gingiva) at regular intervals between the ages of six months and two to four years. There are twenty primary teeth—ten in each jaw—and they occur from the midline toward the sides in the following sequence: central incisor, lateral incisor, cuspid (canine), first molar, and second molar. (See figure 14.8.)

The primary teeth are usually shed in the same order they appeared. Before this happens, though, their roots are resorbed. The teeth are then pushed out of their sockets by pressure from the developing *secondary teeth* (permanent teeth). This secondary set consists of thirty-two teeth—sixteen in each jaw—and they are arranged from the midline as follows: central incisor, lateral incisor, cuspid (canine), first bicuspid

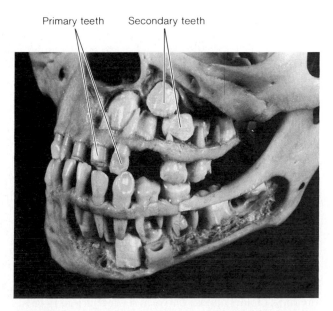

Figure 14.8   The primary and secondary teeth developing in the maxilla and mandible are revealed in this skull of a child.

Dental Caries

Dental caries (decay) involves decalcification of tooth enamel and is usually followed by destruction of the enamel and its underlying dentin. The result is a cavity, which must be cleaned and filled to prevent further erosion of the tooth.

Although the cause or causes of dental caries are not completely understood, lack of dental cleanliness and a diet high in sugar and starch contribute to the problem. Accumulations of food particles on the surfaces and between the teeth aid the growth of certain kinds of bacteria, especially a form called *Streptococcus mutans*. These microorganisms utilize carbohydrates in food particles and produce an abundance of acid by-products and some sticky substances that help keep the bacteria attached to the teeth. The acids then begin the process of destroying tooth enamel and dentin.

Preventing dental caries requires brushing the teeth at least once a day, using dental floss or tape regularly to remove debris from between the teeth, and limiting the intake of sugar and starch, especially between meals. Drinking fluoridated water or applying a fluoride solution to teeth also helps prevent dental decay.

Loss of teeth is most commonly associated with diseases of the gums (gingivitis) and the dental pulp (endodontitis). Such diseases can usually be avoided by practicing good oral hygiene and obtaining regular dental treatment.

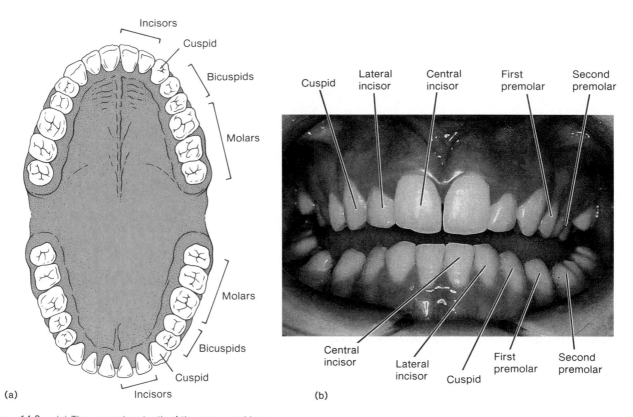

Figure 14.9   (a) The secondary teeth of the upper and lower jaws; (b) anterior view of the secondary teeth.

(premolar), second bicuspid, first molar, second molar, and third molar (figure 14.9).

Chart 14.2 summarizes the types and numbers of primary and secondary teeth.

The permanent teeth usually begin to appear at six years of age, but the set may not be completed until the third molars appear between seventeen and twenty-five years of age. Sometimes these third molars, which are also called wisdom teeth, become wedged in ab-

normal positions within the jaws and fail to erupt. Such teeth are said to be *impacted*.

The teeth break pieces of food into smaller pieces. This action increases the surface area of the food particles and thus makes it possible for digestive enzymes to react more effectively with the food molecules.

Different teeth are adapted to handle food in different ways. The *incisors* are chisel-shaped, and their sharp edges bite off relatively large pieces of food. The *cuspids* are cone-shaped, and they are useful in grasping or tearing food. The *bicuspids* and *molars* have somewhat flattened surfaces and are specialized for grinding food particles.

Each tooth consists of two main portions—the *crown*, which projects beyond the gum, and the *root*, which is anchored to the alveolar process of the jaw. The region where these portions meet is called the *neck* of the tooth. The crown is covered by glossy, white *enamel*. Enamel consists mainly of calcium salts and is the hardest substance in the body. Unfortunately, if enamel is damaged by abrasive action or injury, it is not replaced. It also tends to wear away with age.

The bulk of a tooth beneath the enamel is composed of *dentin*, a substance much like bone, but somewhat harder. The dentin, in turn, surrounds the tooth's central cavity (pulp cavity), which contains blood vessels, nerves, and connective tissue (pulp). The blood vessels and nerves reach this cavity through tubular *root canals*, which extend upward into the root.

The root is enclosed by a thin layer of bonelike material called *cementum*, which is surrounded by a *periodontal ligament* (periodontal membrane). This ligament contains bundles of thick collagenous fibers, which pass between the cementum and the bone of the alveolar process, firmly attaching the tooth to the jaw. It also contains blood vessels and nerves near the surface of the cementum-covered root (figure 14.10).

The mouth parts and their functions are summarized in chart 14.3.

1. How do primary teeth differ from secondary teeth?
2. How are various types of teeth adapted to provide specialized functions?
3. Describe the structure of a tooth.
4. Explain how a tooth is attached to the bone of the jaw.

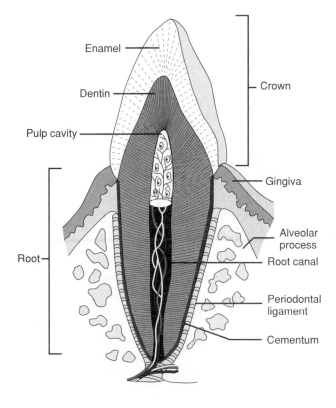

Figure 14.10   A section of a cuspid tooth.

CHART 14.2	Primary and secondary teeth		
Primary teeth (deciduous)		**Secondary teeth (permanent)**	
Type	*Number*	*Type*	*Number*
Incisor		Incisor	
Central	4	Central	4
Lateral	4	Lateral	4
Cuspid	4	Cuspid	4
		Bicuspid	
		First	4
		Second	4
Molar		Molar	
First	4	First	4
Second	4	Second	4
		Third	4
Total	20	Total	32

CHART 14.3	Mouth parts and their functions				
Part	**Location**	**Function**	**Part**	**Location**	**Function**
Cheeks	Form lateral walls of mouth	Hold food in mouth; muscles function in chewing	Tongue	Occupies floor of mouth	Aids in mixing food with saliva, moves food toward pharynx, contains taste receptors
Lips	Surround mouth opening	Contain sensory receptors used to judge characteristics of foods	Palate	Forms roof of mouth	Holds food in mouth; directs food to pharynx
			Teeth	In sockets of mandibular and maxillary bones	Break food particles into smaller pieces, help mix food with saliva during chewing

Salivary Glands

The **salivary glands** secrete saliva. This fluid moistens food particles, helps bind them together, and begins the digestion of carbohydrates. Saliva also acts as a solvent, dissolving various food chemicals so that they can be tasted, and it helps cleanse the mouth and teeth. Bicarbonate ions (HCO_3^-) in saliva help regulate (buffer) its acid concentration so that the pH of saliva usually remains near neutral, between 6.5 and 7.5. This is a favorable range for the action of the salivary enzyme and protects the teeth from dissolving in an excessively acid environment.

Many small salivary glands are scattered throughout the mucosa of the tongue, palate, and cheeks. They secrete fluid continuously so that the lining of the mouth remains moist. In addition, there are three pairs of major salivary glands: the parotid glands, the submandibular glands, and the sublingual glands.

Salivary Secretions

Within a salivary gland are two types of secretory cells, *serous cells* and *mucous cells*. These cells occur in varying proportions within different glands. The serous cells produce a watery fluid that contains a digestive enzyme called **amylase.** This enzyme splits starch and glycogen molecules into disaccharides—the first step in the digestion of carbohydrates. Mucous cells secrete the thick, stringy liquid called **mucus,** which binds food particles together and acts as a lubricant during swallowing.

Like other digestive structures, the salivary glands are innervated by branches of both sympathetic and parasympathetic nerves. Impulses arriving on sympathetic fibers stimulate the gland cells to secrete a small quantity of viscous saliva. Parasympathetic impulses, on the other hand, elicit the secretion of a large volume of watery saliva. Such parasympathetic impulses are activated reflexly when a person sees, smells, tastes, or even thinks about pleasant foods. Conversely, if food looks, smells, or tastes unpleasant, parasympathetic activity is inhibited, so that less saliva is produced and swallowing may become difficult.

Major Salivary Glands

The **parotid glands** are the largest of the major salivary glands. One lies in front of and somewhat below each ear, between the skin of the cheek and the masseter muscle. A *parotid duct* (Stensen's duct) passes from the gland inward through the buccinator muscle, entering the mouth just opposite the upper second molar on either side of the jaw. The parotid glands secrete a clear, watery fluid that is rich in amylase (figure 14.11).

The **submandibular glands** are located in the floor of the mouth on the inside surface of the lower jaw. The secretory cells of these glands are predominantly serous, although some mucous cells are present. Consequently, the submandibular glands secrete a more viscous fluid than the parotid glands. The ducts of the submandibular glands (Wharton's ducts) open under the tongue, near the frenulum (figure 14.11).

The **sublingual glands** are the smallest of the major salivary glands. They are found on the floor of the mouth under the tongue. Because their cells are primarily the mucous type, their secretions, which enter the mouth through many separate ducts, tend to be thick and stringy. (See figure 14.12.)

Chart 14.4 summarizes the characteristics of the major salivary glands.

1. What is the function of saliva?
2. What stimulates the salivary glands to secrete saliva?
3. Where are the major salivary glands located?

CHART 14.4	The major salivary glands		
Gland	Location	Duct	Type of secretion
Parotid glands	In front of and somewhat below the ears, between the skin of the cheeks and the masseter muscles	Parotid ducts pass through the buccinator muscles and enter the mouth opposite the upper second molars	Clear, watery serous fluid, rich in amylase
Submandibular glands	In the floor of the mouth on the inside surface of the mandible	Ducts open beneath the tongue near the frenulum	Primarily serous fluid, but with some mucus; more viscous than parotid secretion
Sublingual glands	In the floor of the mouth beneath the tongue	Many separate ducts	Primarily thick, stringy mucus

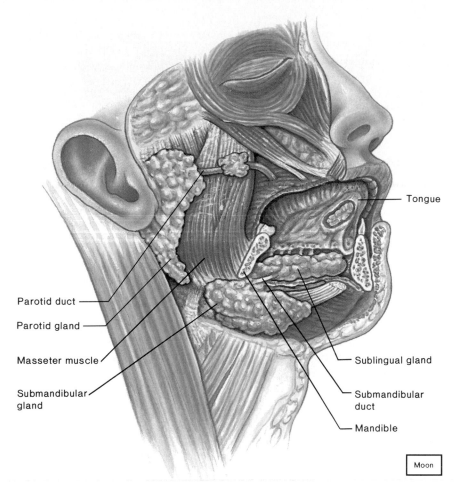

Parotid duct

Parotid gland

Masseter muscle

Submandibular gland

Tongue

Sublingual gland

Submandibular duct

Mandible

Moon

Figure 14.11 Locations of the major salivary glands.

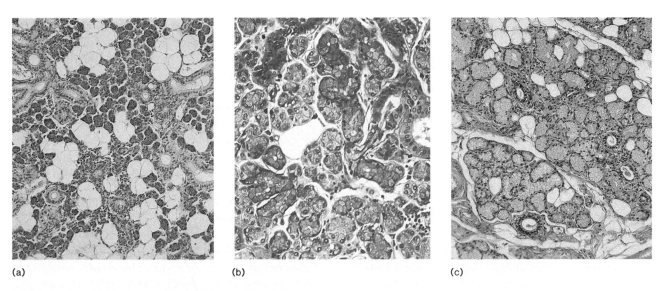

(a)

(b)

(c)

Figure 14.12 Light micrographs of (a) the parotid salivary gland (×200), (b) the submandibular salivary gland (×150), and (c) the sublingual salivary gland (×200).

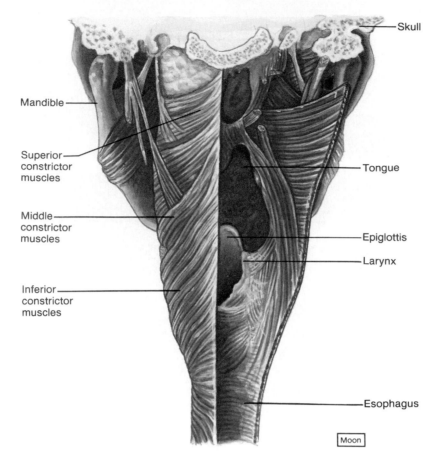

Figure 14.13 Muscles of the pharyngeal wall, viewed from behind.

Pharynx and Esophagus

The pharynx is a cavity behind the mouth from which the tubular esophagus leads to the stomach. Although neither the pharynx nor the esophagus contributes to the digestive process, both are important passageways, and their muscular walls function in swallowing.

Structure of the Pharynx

The **pharynx** connects the nasal and oral cavities with the larynx and esophagus. (See figure 14.7.) It can be divided into the following parts:

1. The **nasopharynx** is located above the soft palate. It communicates with the nasal cavity and provides a passageway for air during breathing. The auditory tubes, which connect the pharynx with the middle ears, open through the walls of the nasopharynx. (See chapter 12.)
2. The **oropharynx** is behind the mouth. It opens behind the soft palate into the nasopharynx and projects downward to the upper border of the epiglottis. This portion functions as a passageway for food moving downward from the mouth, and for air moving to and from the nasal cavity.
3. The **laryngopharynx** is located just below the oropharynx. It extends from the upper border of the epiglottis downward to the lower border of the cricoid cartilage of the larynx, where it is continuous with the esophagus.

The muscles in the walls of the pharynx are arranged in inner circular and outer longitudinal groups (figure 14.13). The circular muscles, called *constrictor muscles,* pull the walls inward during swallowing. The *superior constrictor muscles,* which are attached to bony processes of the skull and mandible, curve around the upper part of the pharynx. The *middle constrictor muscles* arise from projections on the hyoid bone and fan around the middle of the pharynx. The *inferior constrictor muscles* originate from cartilage of the larynx and pass around the lower portion of the pharyngeal cavity. Some of the lower inferior constrictor muscle fibers remain contracted most of the time, and this action prevents air from entering the esophagus during breathing.

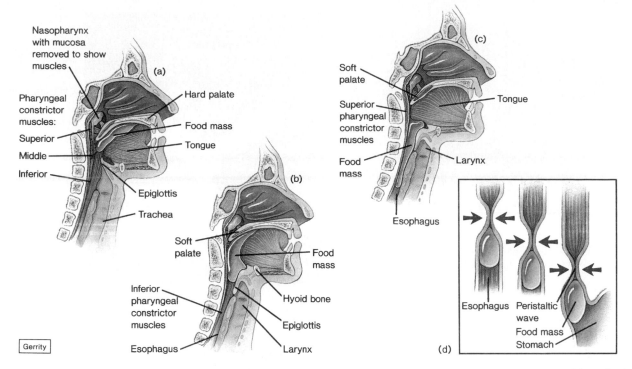

Figure 14.14   Steps in the swallowing reflex: (*a*) The tongue forces food into the pharynx. (*b*) The soft palate, hyoid bone, and larynx are raised, the tongue is pressed against the palate, and the inferior constrictor muscles relax so that the esophagus opens. (*c*) Superior constrictor muscles contract and force food into the esophagus. (*d*) Peristaltic waves move food through the esophagus to the stomach.

Although the pharyngeal muscles are skeletal muscles, they generally are not under voluntary control. Instead, they function involuntarily in the swallowing mechanism.

Swallowing Mechanism

The act of swallowing (deglutition) involves a set of complex reflexes and can be divided into three stages. In the first stage, which is initiated voluntarily, food is chewed and mixed with saliva. Then, this mixture is rolled into a mass (bolus) and forced into the pharynx by the tongue. The second stage begins as the food reaches the pharynx and stimulates sensory receptors located around the pharyngeal opening. This stimulation triggers the swallowing reflex, illustrated in figure 14.14, which includes the following actions:

1. The soft palate is raised, preventing food from entering the nasal cavity.
2. The hyoid bone and the larynx are elevated, so that food is less likely to enter the trachea.
3. The tongue is pressed against the soft palate, sealing off the oral cavity from the pharynx.
4. The longitudinal muscles in the pharyngeal wall contract, pulling the pharynx upward toward the food.
5. The lower portion of the inferior constrictor muscles relaxes, opening the esophagus.

6. The superior constrictor muscles contract, stimulating a peristaltic wave to begin in other pharyngeal muscles, and this wave forces the food into the esophagus.

As the swallowing reflex occurs, breathing is momentarily inhibited. Then, during the third stage of swallowing, the food is transported in the esophagus to the stomach by peristalsis.

Esophagus

The **esophagus** is a straight, collapsible tube about 25 centimeters long. It provides a passageway for substances from the pharynx to the stomach. It descends through the thorax behind the trachea, passing through the mediastinum. The esophagus penetrates the diaphragm through an opening, the *esophageal hiatus,* and is continuous with the stomach on the abdominal side of the diaphragm (figures 14.15, 14.16 and reference plates 57, 73).

Mucous glands are scattered throughout the submucosa of the esophagus, and their secretions keep the inner lining of the tube moist and lubricated.

Just above the point where the esophagus joins the stomach, some of the circular muscle fibers in its wall are thickened. These fibers usually remain contracted, and they function to close the entrance to the stomach. In this way, they help prevent regurgitation of the stomach contents into the esophagus.

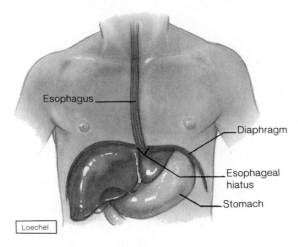

Figure 14.15 The esophagus functions as a passageway between the pharynx and the stomach.

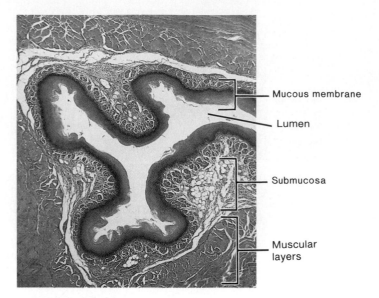

Figure 14.16 This cross section of the esophagus shows its muscular wall (×10).

Occasionally there is a weak place in the diaphragm due to a congenital defect or an injury. As a result, a portion of the stomach, large intestine, or some other abdominal organ may protrude upward through the esophageal hiatus and into the thorax. This condition is called *hiatal hernia.*

If gastric juice from the stomach enters the esophagus as a result of such a hiatal hernia or from regurgitation (reflux), the esophageal mucosa may become inflamed. This may lead to the discomfort commonly called "heartburn," to difficulty in swallowing, or to ulceration, which may be accompanied by loss of blood.

In response to the destructive action of gastric juice, the squamous epithelium that normally lines the esophagus may be replaced by columnar epithelium (chapter 5). This condition, called *Barrett's esophagus,* is associated with an increased risk of developing an esophageal cancer.

When peristaltic waves reach the stomach, the muscle fibers that guard its entrance relax and allow the food to enter.

1. Describe the regions of the pharynx.
2. List the major events that occur during swallowing.
3. What is the function of the esophagus?

Stomach

The **stomach** is a J-shaped, pouchlike organ, about 25–30 centimeters long, which hangs under the diaphragm in the upper left portion of the abdominal cavity. It has a capacity of about one liter or more, and its inner lining is marked by thick folds (rugae) that

tend to disappear when its wall is distended. The stomach receives food from the esophagus, mixes it with gastric juice, initiates the digestion of proteins, carries on a limited amount of absorption, and moves food into the small intestine.

In addition to the two layers of smooth muscle—an inner circular layer and an outer longitudinal layer—found in other regions of the alimentary canal, some parts of the stomach have another inner layer of oblique fibers. This third muscular layer is most highly developed near the opening of the esophagus and in the body of the stomach (figure 14.17).

Parts of the Stomach

The stomach, shown in figures 14.18 and 14.19 and reference plate 51, can be divided into the cardiac, fundic, body, and pyloric regions. The *cardiac region* is a small area near the esophageal opening (cardia). The *fundic region,* which balloons above the cardiac portion, acts as a temporary storage area and sometimes becomes filled with swallowed air. This produces a gastric air bubble, which may be used as a landmark on an X-ray film of the abdomen. The dilated *body region,* which is the main part of the stomach, is located between the fundic and pyloric portions. The *pyloric region* (antrum) narrows and becomes the *pyloric canal* as it approaches the small intestine.

At the end of the pyloric canal, the circular layer of fibers in its muscular wall is thickened, forming a powerful muscle, called the **pyloric sphincter** (pylorus). This muscle serves as a valve that prevents regurgitation of food from the intestine back into the stomach.

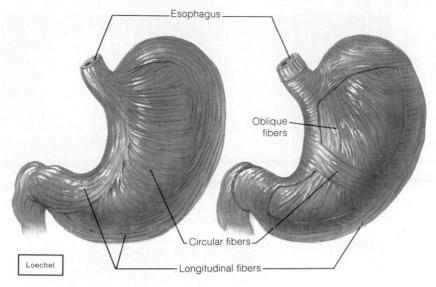

Figure 14.17 Some parts of the stomach have three layers of muscle fibers.

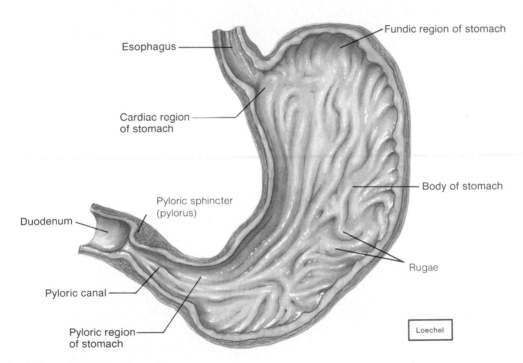

Figure 14.18 Major regions of the stomach.

Gastric Secretions

The mucous membrane that forms the inner lining of the stomach is relatively thick, and its surface is studded with many small openings. These openings, called *gastric pits,* are located at the ends of tubular **gastric glands** (oxyntic glands) (figure 14.20). Although their structure and the composition of their secretions vary in different parts of the stomach, gastric glands generally contain three types of secretory cells. One type, the *mucous cell* (goblet cell), occurs in the necks of the glands near the openings of the gastric pits. The other types, *chief cells* (peptic cells) and *parietal cells* (oxyntic cells), are found in the deeper parts of the glands (figures 14.20 and 14.21). The chief cells secrete *digestive enzymes,* and the parietal cells release a solution containing *hydrochloric acid.* The products of the mucous cells, chief cells, and parietal cells together form **gastric juice.**

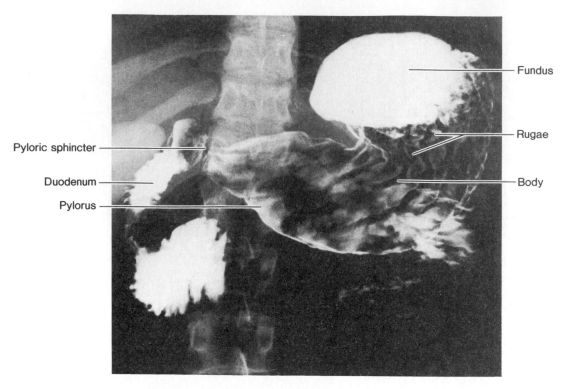

Figure 14.19 X-ray film of a stomach. (Note: A radiopaque substance that was swallowed by the patient appears white in the X-ray film.)

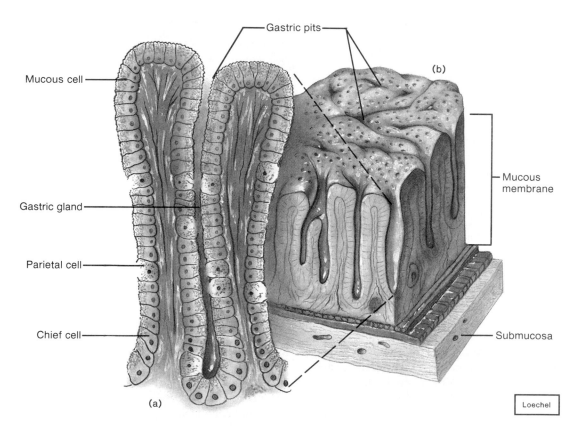

Figure 14.20 (a) Gastric glands include mucous cells, parietal cells, and chief cells. (b) The mucosa of the stomach is studded with gastric pits that are the openings of the gastric glands.

Although gastric juice contains several digestive enzymes, **pepsin** is by far the most important of them. It is secreted by the chief cells as an inactive, nonerosive enzyme precursor called **pepsinogen.** When pepsinogen contacts the hydrochloric acid from the parietal cells, however, it changes rapidly into pepsin.

Pepsin is a protein-splitting enzyme capable of beginning the digestion of nearly all types of dietary protein. This enzyme is most active in an acid environment, and the hydrochloric acid in gastric juice provides such an environment.

Small quantities of a fat-splitting enzyme, *gastric lipase,* also occur in gastric juice. However, its action is relatively weak due in part to the low pH of gastric juice. Gastric lipase acts mainly on butterfat.

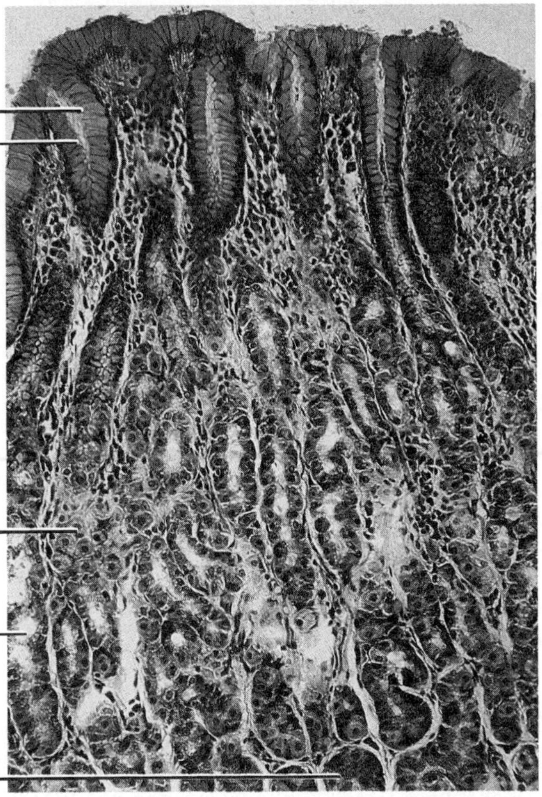

The mucous cells of the gastric glands secrete large quantities of thin mucus. In addition, the cells of the mucous membrane, associated with the inner lining of the stomach and between the gastric glands, release a more viscous and alkaline secretion, which forms a protective coating on the inside of the stomach wall. This coating is especially important because pepsin is capable of digesting the proteins of the stomach tissues, as well as those in foods. Thus, the coating normally prevents the stomach from digesting itself.

Still another component of gastric juice is **intrinsic factor.** This substance, which is secreted by the parietal cells of the gastric glands, aids in the absorption of vitamin B_{12} from the small intestine, as is explained in chapter 15.

The substances in gastric juice are summarized in chart 14.5.

1. Where is the stomach located?
2. What substance is secreted by the chief cells of the gastric glands? By the parietal cells?
3. What is the most important digestive enzyme in gastric juice?
4. How is the stomach prevented from digesting itself?

Regulation of Gastric Secretions

Although gastric juice is produced continuously, the rate of its production varies considerably from time to time and is under the control of neural and hormonal mechanisms. More specifically, within the gastric glands there are specialized cells closely associated with the parietal cells. These specialized cells secrete the hormone *somatostatin,* which inhibits the secretion of acid by the parietal cells. However, acetylcholine released from nerve fibers in response to parasympathetic impulses arriving on the vagus nerves suppresses the secretion of somatostatin and stimulates the gastric glands to secrete large amounts of gastric juice, which is rich

Figure 14.21 A light micrograph of cells associated with the gastric glands (×50).

Mucous cell
Gastric pit
Parietal cell
Gastric gland
Chief cell

CHART 14.5	Major components of gastric juice	
Component	Source	Function
Pepsinogen	Chief cells of the gastric glands	Inactive form of pepsin
Pepsin	Formed from pepsinogen in the presence of hydrochloric acid	A protein-splitting enzyme capable of digesting nearly all types of dietary protein
Hydrochloric acid	Parietal cells of the gastric glands	Provides the acid environment needed for the conversion of pepsinogen into pepsin and for the action of pepsin
Mucus	Goblet cells and mucous glands	Provides a viscous, alkaline protective layer on the stomach wall
Intrinsic factor	Parietal cells of the gastric glands	Aids in the absorption of vitamin B_{12}

in hydrochloric acid and pepsin. These parasympathetic impulses also stimulate certain stomach cells, found mainly in the pyloric region, to release a peptide hormone called **gastrin** that causes the gastric glands to increase their secretory activity (figure 14.22). Furthermore, the parasympathetic impulses and the presence of gastrin promotes the release of *histamine* from gastric mucosal cells, which, in turn, stimulates additional gastric secretion.

Histamine is very effective in promoting the secretion of gastric acid. Consequently, drugs that block the histamine receptors of gastric mucosal cells (H$_2$–blockers) are often used to inhibit gastric secretions in patients whose gastric acid content is excessive.

As a result of the interaction of these factors, three stages of gastric secretion can be recognized. They are called the cephalic, gastric, and intestinal phases.

The *cephalic phase* of gastric secretion begins before any food reaches the stomach and may sometimes begin before any food is eaten. In this stage, gastric secretions are stimulated by parasympathetic reflexes operating through the vagus nerves whenever a person tastes, smells, sees, or even thinks about food. Furthermore, the hungrier the person is, the greater the amount of gastric secretion. The cephalic phase of secretion is responsible for 30%–50% of the secretory response to a meal.

The *gastric phase* of gastric secretion, which accounts for 40%–50% of the secretory activity, starts when food enters the stomach. The presence of food chemicals and the distension of the stomach wall trigger the release of gastrin from the stomach. The gastrin, in turn, stimulates the production of still more gastric juice. Actually, the release of gastrin is promoted by several factors, including parasympathetic impulses, distension of the stomach wall, and the presence of certain food substances (secretagogues) in the stomach. For example, meat extracts, partially digested proteins, various spices, caffeine, and alcohol are especially stimulating to gastrin release. Food in the stomach also elicits additional parasympathetic impulses that promote the secretion of still more gastric juice.

Another important factor in regulating gastric secretion is the pH of the stomach contents. As food enters the stomach and becomes mixed with gastric juice, the pH of the contents typically rises, and, in response, the secretion of gastrin and acid is enhanced. Consequently, the pH of the stomach contents drops. As the pH approaches 3.0, the secretion of gastrin begins to be inhibited, and when the pH reaches 1.5, gastrin secretion ceases.

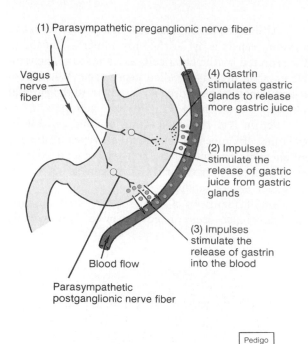

Figure 14.22 The secretion of gastric juice is regulated in part by parasympathetic nerve impulses that stimulate the release of gastric juice and gastrin.

In addition to its effect on gastric secretion, gastrin stimulates cell growth within the mucosa of the stomach and intestines, except in the region of the stomach where gastrin is produced. This effect is important in the repair of mucosal cells that are damaged as a result of disease or medical treatments.

In the process of secreting hydrochloric acid, hydrogen ions are removed from the blood, and an equivalent amount of alkaline bicarbonate ions is released into the blood. Consequently, following a meal, the blood concentration of bicarbonate ions increases, and an elevated amount of bicarbonate ions is excreted in the urine. This phenomenon is called the *alkaline tide*.

The *intestinal phase* of gastric secretion, which accounts for about 5% of the total secretory response to a meal, begins when food leaves the stomach and enters the small intestine. When food first contacts the intestinal wall, it stimulates intestinal cells to release a hormone that again enhances gastric gland secretions. Although the actual nature of this hormone is unknown, some investigators believe it is identical to gastrin, and they call it *intestinal gastrin*.

As more food moves into the small intestine, however, the secretion of gastric juice from the stomach wall is inhibited. This inhibition results from a sympathetic reflex triggered by the presence of acid in the upper part of the small intestine. Also, the presence of

CHART 14.6	Phases of gastric secretion
Phase	Action
Cephalic phase	Parasympathetic reflexes are triggered by the sight, taste, smell, or thought of food; gastric juice is secreted in response
Gastric phase	Food in stomach chemically and mechanically stimulates the release of gastrin, which, in turn, stimulates the secretion of gastric juice; reflex responses also stimulate gastric juice secretion
Intestinal phase	As food enters the small intestine, it stimulates intestinal cells to release intestinal gastrin, which, in turn, promotes the secretion of gastric juice from the stomach wall

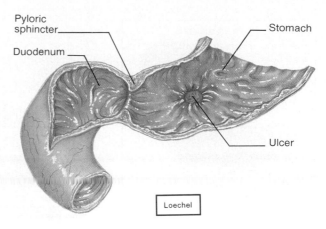

Figure 14.23   Gastric ulcers are usually caused by the digestive action of pepsin.

proteins and fats in this region of the intestine causes the release of the peptide hormone *cholecystokinin* from the intestinal wall, which causes a decrease in gastric motility. Similarly, the presence of fats in the small intestine stimulates intestinal cells to release *intestinal somatostatin,* which inhibits the release of gastric acid. Thus, these actions bring about a decrease in gastric secretion and a decrease in motility, as the small intestine fills with food.

Chart 14.6 summarizes the phases of gastric secretion.

1. How is the secretion of gastric juice controlled?
2. Distinguish between the cephalic, gastric, and intestinal phases of gastric secretion.
3. What is the function of cholecystokinin?

An *ulcer* is an open sore on the surface of an organ that results from a localized breakdown of the tissues. Although ulcers may occur in various parts of the alimentary canal, they often develop in the stomach and are called *gastric ulcers* (figure 14.23).

Ulcers are also common in the first portion of the small intestine, the duodenum. Such *duodenal ulcers* occur in regions that are exposed to pepsin as the contents of the stomach enter the intestine. Because gastric ulcers and duodenal ulcers are usually caused by pepsin, both are commonly called *peptic ulcers.* Ulcers often develop in people who are emotionally stressed and whose stomachs secrete increased amounts of acidic gastric juice between meals, when the stomach is essentially empty. There is also an increased frequency of gastric and duodenal ulcers among smokers.

Gastric Absorption

Although gastric enzymes begin the breakdown of proteins, the stomach wall is not well adapted to absorb digestive products. However, small quantities of water, glucose, certain salts, alcohol, and various lipid-soluble drugs may be absorbed by the stomach.

Mixing and Emptying Actions

As more and more food enters the stomach, the smooth muscles in its wall become stretched. Although the stomach may enlarge, its muscles maintain their tone, and internal pressure of the stomach normally remains unchanged. A person may eat more than the stomach can comfortably hold, and when this happens, the internal pressure may rise enough that pain receptors are stimulated. The result is a stomachache.

Following a meal, the mixing movements of the stomach wall aid in producing a semifluid paste of food particles and gastric juice called **chyme.** Peristaltic waves push the chyme toward the pyloric region of the stomach, and as chyme accumulates near the pyloric sphincter, this muscle begins to relax. The muscular pyloric region then pumps the chyme a little (5–15 milliliters) at a time into the small intestine. This process is illustrated in figure 14.24.

The rate at which the stomach empties depends on several factors, including the fluidity of the chyme and the type of food present. For example, liquids usually pass through the stomach quite rapidly, but solids remain until they are well mixed with gastric juice. Fatty foods may remain in the stomach three to six hours; foods high in proteins tend to be moved through more quickly; and carbohydrates usually pass through more rapidly than either fats or proteins.

When the duodenum becomes filled with chyme, its internal pressure increases, and the intestinal wall is stretched. This action stimulates sensory receptors in the wall, and an **enterogastric reflex** is triggered. The name of this reflex, like those of other digestive reflexes, describes the origin and termination of reflex

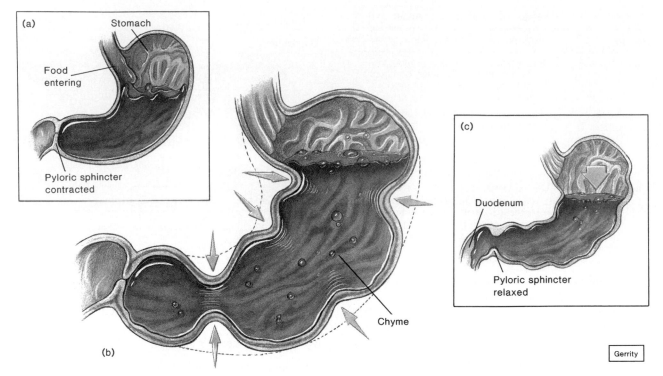

(a)
Stomach
Food
entering
Pyloric sphincter
contracted

(b)

(c)
Duodenum
Pyloric sphincter
relaxed

Chyme

Gerrity

Figure 14.24 Movements of the stomach. (*a*) As the stomach fills, its muscular wall becomes stretched, but the pyloric sphincter remains closed. (*b*) Mixing movements mix food and gastric juice, creating chyme. (*c*) Peristaltic waves move the chyme toward the pyloric sphincter, which relaxes and allows some chyme to enter the duodenum.

impulses. Thus, the enterogastric reflex begins in the small intestine (*entero*) and ends in the stomach (*gastro*).

As a result of the enterogastric reflex, fewer parasympathetic impulses travel to the stomach, and peristaltic waves are inhibited. Consequently, the intestine is filled less rapidly. Also, if the chyme entering the intestine has a high fat content, the hormone **cholecystokinin** is released from the intestinal wall, and it causes peristalsis to be inhibited even more (figure 14.25).

As the stomach contents enter the duodenum, accessory organs add their secretions to the chyme. These organs include the pancreas, liver, and gallbladder.

Vomiting results from a complex reflex that empties the stomach another way. This action is usually triggered by irritation or distension in some part of the alimentary canal, such as the stomach or intestines. Sensory impulses travel from the site of stimulation to the *vomiting center* in the medulla oblongata, and a number of motor responses follow. These include taking a deep breath, raising the soft palate and thus closing the nasal cavity, closing the opening to the trachea (glottis), relaxing the circular muscle fibers at the base of the esophagus, contracting the diaphragm so it moves downward over the stomach, and contracting the abdominal wall muscles so that pressure inside the abdominal cavity is increased. As a result, the stomach is squeezed from all sides, and its contents are forced upward and out through the esophagus, pharynx, and mouth.

Activity in the vomiting center can be stimulated by various drugs (emetics), by toxins in contaminated foods, and sometimes by rapid changes in body motion. In this last situation, sensory impulses from the labyrinths of the inner ears apparently reach the vomiting center, and in some people this produces motion sickness.

The vomiting center can also be activated by stimulation of higher brain centers through sights, sounds, odors, tastes, emotional feelings, or by mechanical stimulation of the back of the pharynx.

The feeling called *nausea* seems to be due to activity in the vomiting center or in nerve centers closely associated with it. During nausea, stomach movements usually are diminished or absent, and duodenal contents may be moved back into the stomach.

1. How is chyme produced?
2. What factors influence the speed with which chyme leaves the stomach?
3. Describe the enterogastric reflex.
4. Describe the vomiting reflex.
5. What factors may stimulate the vomiting reflex?

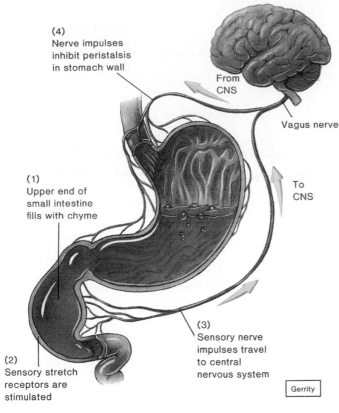

(4)
Nerve impulses
inhibit peristalsis
in stomach wall

From
CNS

Vagus nerve

(1)
Upper end of
small intestine
fills with chyme

To
CNS

(3)
Sensory nerve
impulses travel
to central
nervous system

(2)
Sensory stretch
receptors are
stimulated

Gerrity

Figure 14.25 The rate at which chyme leaves the stomach is regulated in part by the enterogastric reflex.

Pancreas

The shape of the **pancreas** and its general location are described in chapter 13, as are its endocrine functions. The pancreas also has an exocrine function—the secretion of a digestive juice.

Structure of the Pancreas

The pancreas is closely associated with the small intestine and is located behind the parietal peritoneum. It extends horizontally across the posterior abdominal wall, with its head in the C-shaped curve of the duodenum and its tail against the spleen (figure 14.26 and reference plate 59).

The cells that produce pancreatic juice are called *pancreatic acinar cells,* and they make up the bulk of the pancreas. These cells are clustered around tiny tubes, into which they release their secretions. The smaller tubes unite to form larger ones, which, in turn, give rise to a *pancreatic duct* extending the length of the pancreas. The pancreatic duct usually connects with the duodenum at the same place where the bile duct from the liver and gallbladder joins the duodenum (figures 13.33 and 14.26).

The pancreatic and bile ducts are joined at a short, dilated tube called the *hepatopancreatic ampulla* (ampulla of Vater). This ampulla is surrounded by a band of smooth muscle, called the *hepatopancreatic sphincter* (sphincter of Oddi).

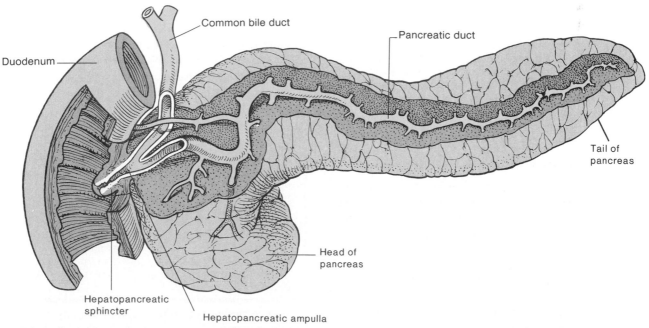

Common bile duct

Pancreatic duct

Duodenum

Tail of
pancreas

Head of
pancreas

Hepatopancreatic
sphincter

Hepatopancreatic ampulla

Figure 14.26 The pancreas is closely associated with the duodenum.

Pancreatic Juice

Pancreatic juice contains enzymes capable of digesting carbohydrates, fats, proteins, and nucleic acids.

The carbohydrate-digesting enzyme is called **pancreatic amylase.** It splits molecules of starch or glycogen into double sugars (disaccharides); the fat-digesting enzyme, **pancreatic lipase,** breaks triglyceride molecules into fatty acids and monoglycerides. (A monoglyceride molecule consists of one fatty acid bound to glycerol.)

The protein-splitting (proteolytic) enzymes are **trypsin, chymotrypsin,** and **carboxypeptidase.** Each of these enzymes splits the bonds between particular combinations of amino acids in proteins. Because no single enzyme can split all of the possible combinations of amino acids, several enzymes are necessary for the complete digestion of protein molecules.

The protein-splitting enzymes are stored in inactive forms within tiny cellular structures called *zymogen granules.* These enzymes, like gastric pepsin, are secreted in inactive forms and must be activated by other enzymes after they reach the small intestine. For example, the pancreatic cells release inactive **trypsinogen.** This substance becomes active trypsin when it contacts an enzyme called *enterokinase,* which is secreted by the mucosa of the small intestine. Chymotrypsin and carboxypeptidase are activated, in turn, by the presence of trypsin. This mechanism prevents the enzymatic digestion of proteins within the secreting cells and the pancreatic ducts.

> If the release of pancreatic juice is blocked, it may accumulate in the duct system of the pancreas, and the trypsinogen may become activated. As a result, portions of the pancreas may become digested, causing a painful condition called *acute pancreatitis.*
>
> The symptoms of acute pancreatitis include severe and steady pain centered in the epigastric region of the abdomen, accompanied by nausea, vomiting, and abdominal distension. Most commonly, this condition occurs in persons who use alcohol excessively or who have gallstones. It may also result from various infectious diseases, as a side effect of certain drugs, or from a traumatic injury.

Pancreatic juice also contains two **nucleases,** which are enzymes that break nucleic acid molecules into nucleotides, as well as a high concentration of bicarbonate ions that makes it alkaline. The alkaline content of pancreatic juice provides a favorable environment for the actions of the digestive enzymes and helps neutralize the acidic chyme as it arrives from the stomach. At the same time, the alkaline condition in the small intestine blocks the action of pepsin, which might otherwise damage the duodenal wall.

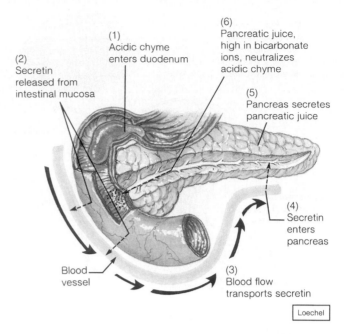

Figure 14.27 Acidic chyme entering the duodenum from the stomach stimulates the release of secretin, which in turn stimulates the release of pancreatic juice.

Regulation of Pancreatic Secretion

As with gastric and small intestinal secretions, the release of pancreatic juice is regulated by nerve actions as well as by hormones. For example, during the cephalic and gastric phases of gastric secretion, parasympathetic impulses reach the pancreas and stimulate it to release digestive enzymes. Also, when acidic chyme enters the duodenum, the pancreas is stimulated to secrete a large quantity of fluid. This stimulus, however, comes from a peptide hormone called **secretin,** which is released into the blood from the duodenal mucous membrane in response to the acid in chyme. The pancreatic juice secreted at this time contains few, if any, digestive enzymes. Instead, it has a high concentration of bicarbonate ions that neutralizes the acid in chyme (figure 14.27).

> *Cystic fibrosis* is an inherited condition characterized by the production of very thick, sticky mucus that adversely affects various exocrine glands. For example, the viscid mucus tends to clog the ducts of the pancreas. This interferes with the secretion of pancreatic juice, prevents the pancreatic digestive enzymes from reaching the duodenum, and leaves the person vulnerable to malnutrition. Cystic fibrosis is also commonly accompanied by excessive secretion from mucous glands associated with the respiratory tract, which may lead to chronic obstruction of the airways.

The presence of proteins and fats in chyme within the duodenum also stimulates the release of **cholecys-**

tokinin from the intestinal wall. As in the case of secretin, cholecystokinin reaches the pancreas by way of the blood. Cholecystokinin causes the secretion of pancreatic juice with a high concentration of digestive enzymes.

1. Where is the pancreas located?
2. List the enzymes found in pancreatic juice.
3. What are the functions of the enzymes in pancreatic juice?
4. How is the secretion of pancreatic juice regulated?

Liver

The **liver** is located in the upper right and central portions of the abdominal cavity, just below the diaphragm. It is partially surrounded by the ribs, and extends from the level of the fifth intercostal space to the lower margin of the ribs. It is reddish brown in color and well supplied with blood vessels (figures 14.28, 14.29, 14.30 and reference plates 51, 57, 74).

Functions of the Liver

The liver carries on many important metabolic activities. For example, it plays a key role in carbohydrate metabolism by helping maintain the normal concentration of blood glucose. As described in chapter 13, liver cells responding to various hormones can decrease blood glucose by converting glucose to glycogen; they can also increase blood glucose by changing glycogen to glucose or by converting noncarbohydrates into glucose.

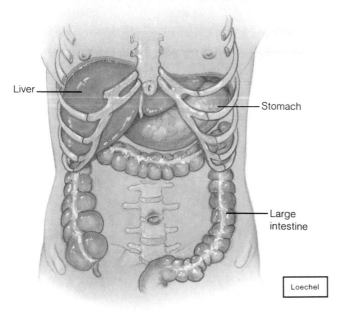

Figure 14.28 The liver is partially surrounded by the ribs.

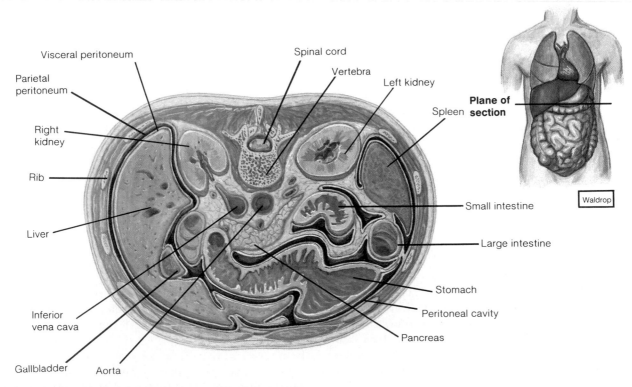

Figure 14.29 This transverse section of the abdomen reveals the liver and other organs within the upper portion of the abdominal cavity.

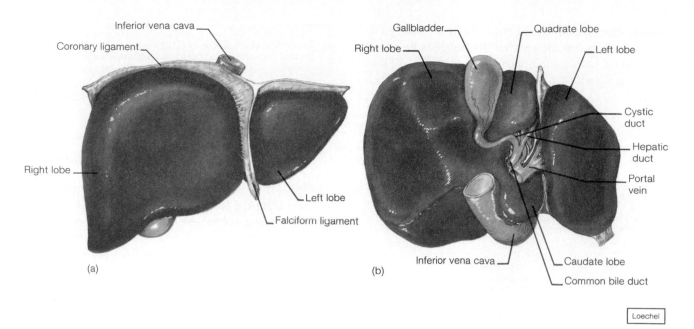

Figure 14.30 Lobes of the liver, viewed (a) from the front and (b) from below.

The liver's effects on lipid metabolism include oxidizing fatty acids at an especially high rate (see chapter 4); synthesizing lipoproteins, phospholipids, and cholesterol; and converting portions of carbohydrate and protein molecules into fat molecules. Fats synthesized in the liver are transported by the blood to adipose tissue for storage.

The most vital liver functions are probably those related to protein metabolism. They include deaminating amino acids; forming urea (see chapter 4); synthesizing various blood proteins, including several that are necessary for blood clotting (see chapter 17); and converting various amino acids to other amino acids.

> Ammonia formed by bacteria in the intestine is normally removed from the blood by liver cells and changed into urea. In the absence of this liver function, the concentration of blood ammonia rises excessively, causing *hepatic coma,* a condition that can lead to death.

The liver also stores a variety of substances, including glycogen, iron, and vitamins A, D, and B_{12}. (See chapter 15.) Iron storage occurs when the concentration of blood iron is excessive. The extra iron is combined with a protein (apoferritin) in liver cells, and as a result, a substance called *ferritin* is formed. The iron remains stored in this form until the blood iron concentration reaches a certain low level. Then some of the iron is released. Thus, the liver plays an important role in the homeostasis of iron.

In addition to the above functions, various liver cells help destroy damaged red blood cells and foreign substances by phagocytosis, alter the composition of toxic substances such as alcohol (detoxification), and secrete bile. The liver can also store from 200 to 400 milliliters of blood, thus it serves as a blood reservoir.

Because many of these liver functions are not directly related to the digestive system, they are discussed in other chapters. Bile secretion, however, is important to digestion and is explained in a subsequent section of this chapter.

Chart 14.7 summarizes the major functions of the liver.

Structure of the Liver

The liver is enclosed in a fibrous capsule and is divided by connective tissue into *lobes*—a large *right lobe* and a smaller *left lobe.* These major lobes are separated by the *falciform ligament,* a fold of visceral peritoneum that also fastens the liver to the abdominal wall anteriorly. As figure 14.30 shows, in addition to the two major lobes, there are two minor ones, the *quadrate lobe* near the gallbladder, and the *caudate lobe* close to the vena cava.

On its superior surface, the liver is attached to the diaphragm by a fold of visceral peritoneum called the *coronary ligament.* Each lobe is separated into numerous tiny **hepatic lobules,** which are the functional units of the gland (figures 14.31 and 14.32). A lobule consists of numerous *hepatic cells* that radiate outward from a *central vein.* Platelike groups of these cells are separated from each other by vascular channels called **hepatic sinusoids.** Blood from the digestive tract, which is carried in *portal veins* (see chapter 18), brings newly absorbed nutrients into the sinusoids and nourishes the hepatic cells (figure 14.33).

Hepatitis

*H*epatitis is an inflammation of the liver and is most commonly caused by a viral infection.

One form of viral hepatitis, called *type A hepatitis,* usually occurs in children or young adults. It is spread by contact with food or objects, such as eating utensils or toys, that have been contaminated with virus-containing feces. This form of hepatitis is often mild, although it may be accompanied by weakness, abdominal discomfort, nausea, and jaundice. Usually the person recovers completely, with no lasting damage to the liver.

Type B hepatitis produces symptoms similar to those of type A, but the effects may last for a much longer time. This form of the disease is spread by contact with virus-containing body fluids, such as blood, saliva, or semen. Thus, it may be transmitted by means of blood transfusions, hypodermic needles, or sexual activity. Most patients recover completely from type B hepatitis; however, some persons continue to harbor live viruses and become "carriers" who may seem healthy, but can transmit the condition to others.

Another type of viral hepatitis is called *non-A, non-B hepatitis.* This form of the disease is thought to be caused by at least two viruses, one resembling the hepatitis A virus and the other resembling the hepatitis B virus. Non-A, non-B hepatitis is usually transmitted by means of blood or blood products and is responsible for 80%–90% of the cases of hepatitis that develop following blood transfusions.

People who exhibit the symptoms of hepatitis for six months or more are said to have *chronic hepatitis.* In such cases, there is danger that the liver will be permanently damaged and its functions impaired. In addition to being caused by viral infections, chronic hepatitis may be caused by the effects of certain drugs, alcohol, or by autoimmune reactions. (See chapter 19.)

CHART 14.7 Major functions of the liver

General function	Specific function	General function	Specific function
Carbohydrate metabolism	Converts glucose to glycogen, glycogen to glucose, and noncarbohydrates to glucose	Protein metabolism	Deaminates amino acids; forms urea; synthesizes blood proteins; interconverts amino acids
Lipid metabolism	Oxidizes fatty acids; synthesizes lipoproteins, phospholipids, and cholesterol; converts portions of carbohydrate and protein molecules into fats	Storage	Stores glycogen, vitamins A, D, and B_{12}, iron, and blood
		Blood filtering	Removes damaged red blood cells and foreign substances by phagocytosis
		Detoxification	Alters composition of toxic substances
		Secretion	Secretes bile

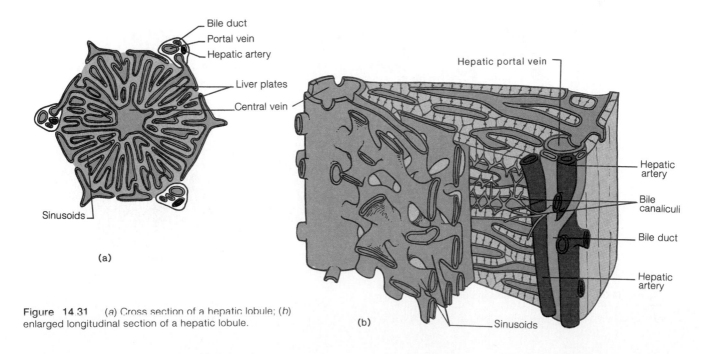

Figure 14.31 (*a*) Cross section of a hepatic lobule; (*b*) enlarged longitudinal section of a hepatic lobule.

Often the blood in the portal veins contains some bacterial cells that have entered through the intestinal wall. However, large **Kupffer cells,** which are fixed to the inner lining (endothelium) of the hepatic sinusoids, remove most of the bacteria from the blood by phagocytosis. Then the blood passes into the *central veins* of the hepatic lobules and moves out of the liver.

Within the liver lobules, are many fine *bile canals,* which receive secretions from the hepatic cells. The canals of neighboring lobules unite to form larger ducts, and then converge to become the **hepatic ducts.** These ducts merge, in turn, to form the *common hepatic duct.*

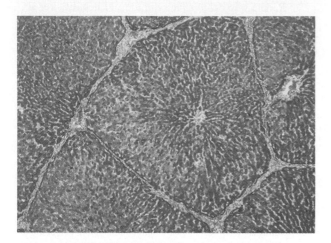

Figure 14.32 A light micrograph of hepatic lobules. What features can you identify?

1. Describe the location of the liver.
2. Review the functions of the liver.
3. What liver function is directly related to digestion?
4. Describe an hepatic lobule.

Composition of Bile

Bile is a yellowish green liquid that is secreted continuously by the hepatic cells. In addition to water, it contains *bile salts, bile pigments, cholesterol,* and various *electrolytes.* Of these, bile salts are the most abundant, and they are the only substances in bile that have a digestive function.

Hepatic cells use cholesterol to produce bile salts, and in the process of secreting these salts, they release some cholesterol into the bile. Cholesterol has no special function in bile or in the alimentary canal.

If the excretion of bile pigments is prevented due to obstructed bile ducts, the pigments tend to accumulate in the blood (hyperbilirubinemia) and tissues, causing jaundice (icterus) that is characterized by a yellowish tinge in the sclera of the eye and other light-colored tissues. This condition is called *obstructive jaundice.* Jaundice may also accompany liver diseases in which the hepatic cells are unable to secrete ordinary amounts of bile pigments. This disorder is called *hepatocellular jaundice.* An excessive destruction of red blood cells accompanied by rapid release of pigments produces *hemolytic jaundice.*

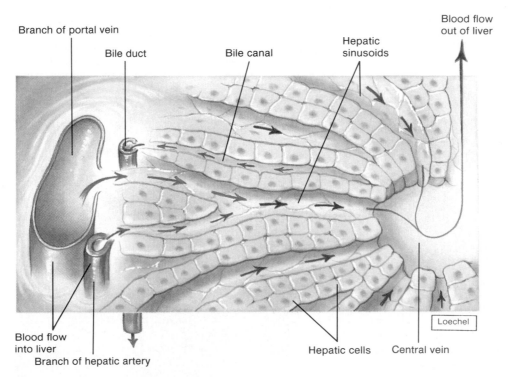

Figure 14.33 The paths of blood and bile within a hepatic lobule.

The bile pigments (bilirubin and biliverdin) are products of the breakdown of hemoglobin from red blood cells, and these pigments are normally excreted in the bile. (See chapter 17.)

Gallbladder and Its Functions

The **gallbladder** is a pear-shaped sac located in a depression on the inferior surface of the liver. It is connected to the **cystic duct,** which, in turn, joins the hepatic duct. The gallbladder has a capacity of 30–50 milliliters, is lined with columnar epithelial cells, and has a strong muscular layer in its wall (figure 14.30 and reference plate 59). It stores bile between meals, concentrates bile by reabsorbing water, and releases bile into the duodenum when stimulated by *cholecystokinin* from the small intestine.

The **common bile duct** is formed by the union of the hepatic and cystic ducts. It leads to the duodenum, where its exit is guarded by the hepatopancreatic sphincter muscle (figure 14.26). This sphincter normally remains contracted, so that bile collects in the common bile duct and backs up into the cystic duct. When this happens, the bile flows into the gallbladder and is stored there.

While the bile is in the gallbladder, its composition is altered because the lining of the gallbladder reabsorbs some of the water and electrolytes. As these substances are removed, the bile salts, bile pigments, and cholesterol become increasingly concentrated. Although the cholesterol normally remains in solution, under certain conditions it may precipitate and form solid crystals. If cholesterol continues to come out of solution, these crystals become larger and larger, forming *gallstones.*

Gallstones may form if the bile is concentrated excessively, if the hepatic cells secrete too much cholesterol, or if the gallbladder is inflamed (cholecystitis). If such stones get into the bile duct, they may block the flow of bile, causing obstructive jaundice and considerable pain (figure 14.34). Generally, gallstones that cause obstructions are surgically removed. At the same time, the gallbladder is removed by a surgical procedure called *cholecystectomy.* Following such surgery, the person is unable to produce gallstones or store bile. However, bile continues to reach the intestine by means of the hepatic and common bile ducts.

Regulation of Bile Release

Normally, bile does not enter the duodenum until the gallbladder is stimulated to contract by *cholecystokinin.* This hormone is released from the intestinal mucosa in response to the presence of proteins and fats in the contents of the small intestine. The hepatopancreatic sphincter usually remains contracted until a

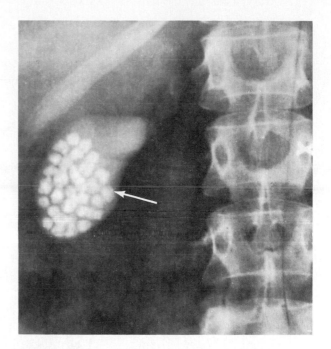

Figure 14.34 X-ray film of a gallbladder that contains gallstones (arrow). What other anatomic features can you identify?

peristaltic wave in the duodenal wall approaches it. Then, just before the wave reaches it, the sphincter undergoes receptive relaxation and a squirt of bile enters the duodenum (figure 14.35).

The hormones that help control digestive functions are summarized in chart 14.8.

Functions of Bile Salts

Although they do not act as digestive enzymes, bile salts aid the actions of digestive enzymes and enhance the absorption of fatty acids and certain fat-soluble vitamins.

Molecules of fats tend to clump together, forming masses called *fat globules.* Bile salts affect fat globules much like a soap or detergent would affect them. That is, bile salts cause fat globules to break up into smaller droplets, an action called **emulsification.** Emulsification is also aided by the presence of monoglycerides, resulting from the action of pancreatic lipase on triglyceride molecules. As a result of emulsification, the total surface area of the fatty substance is greatly increased, and the tiny droplets mix with water. The fat-splitting enzymes (lipases) can then digest the fat molecules more effectively.

Bile salts aid in the absorption of fatty acids and cholesterol by forming complexes (micelles) that are very soluble in chyme and more easily absorbed by epithelial cells. Along with these lipids, various fat-soluble vitamins, such as vitamins A, D, E, and K, are also absorbed. Thus, if bile salts are lacking, lipids may

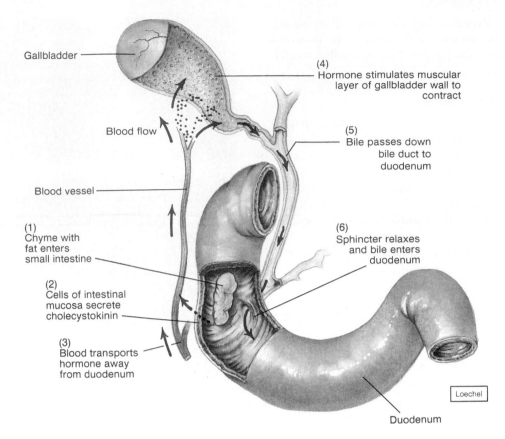

Gallbladder

(4)
Hormone stimulates muscular
layer of gallbladder wall to
contract

Blood flow

(5)
Bile passes down
bile duct to
duodenum

Blood vessel

(1)
Chyme with
fat enters
small intestine

(6)
Sphincter relaxes
and bile enters
duodenum

(2)
Cells of intestinal
mucosa secrete
cholecystokinin

(3)
Blood transports
hormone away
from duodenum

Loechel

Duodenum

Figure 14.35 The gallbladder is stimulated to release bile when fat-containing chyme enters the duodenum.

CHART 14.8 Hormones of the digestive tract

Hormone	Source	Function
Gastrin	Gastric cells, in response to the presence of food	Causes gastric glands to increase their secretory activity
Intestinal gastrin	Cells of small intestine, in response to the presence of chyme	Causes gastric glands to increase their secretory activity
Somatostatin	Gastric cells	Inhibits secretion of acid by parietal cells
Intestinal somatostatin	Intestinal wall cells in response to the presence of fats	Inhibits secretion of acid by parietal cells
Cholecystokinin	Intestinal wall cells, in response to the presence of proteins and fats in the small intestine	Causes gastric glands to decrease their secretory activity and inhibits gastric motility; stimulates pancreas to secrete fluid with a high digestive enzyme concentration; stimulates gallbladder to contract and release bile
Secretin	Cells in the duodenal wall, in response to acidic chyme entering the small intestine	Stimulates pancreas to secrete fluid with a high bicarbonate ion concentration

be poorly absorbed and the person is likely to develop vitamin deficiencies.

Nearly all the bile salts are reabsorbed by the mucous membrane of the small intestine along with the fatty acids. They enter the blood and are carried to the liver where they are resecreted into the bile ducts by the hepatic cells. The small quantities that are lost in the feces are replaced by synthesis of bile salts in liver cells.

1. Explain how bile originates.
2. Describe the function of the gallbladder.
3. How is the secretion of bile regulated?
4. How does bile function in digestion?

Small Intestine

The **small intestine** is a tubular organ that extends from the pyloric sphincter to the beginning of the large in-

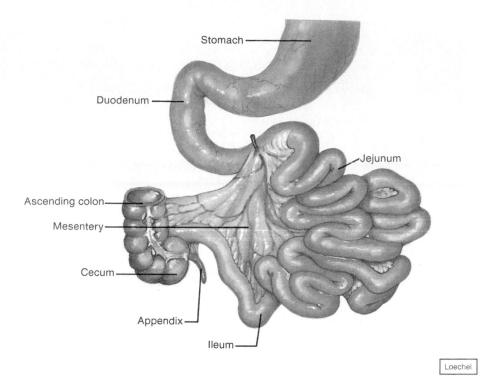

Figure 14.36 The three parts of the small intestine are the duodenum, the jejunum, and the ileum.

testine. With its many loops and coils, it fills much of the abdominal cavity. Although it is 5.5–6.0 meters (18–20 feet) long in a cadaver when the muscular wall lacks tone, the small intestine may be only half this long in a living person.

As mentioned, this portion of the alimentary canal receives secretions from the pancreas and liver. It also completes the digestion of the nutrients in chyme, absorbs the various products of digestion, and transports the remaining residues to the large intestine.

Parts of the Small Intestine

The small intestine, shown in figures 14.36 and 14.37, and in reference plates 51, 58, 74, and 75, consists of three portions: the duodenum, the jejunum, and the ileum.

The **duodenum,** which is about 25 centimeters long and 5 centimeters in diameter, lies behind the parietal peritoneum (retroperitoneal). It is the shortest and most fixed portion of the small intestine. The duodenum follows a C-shaped path as it passes in front of the right kidney and the upper three lumbar vertebrae.

The remainder of the small intestine is mobile and lies free in the peritoneal cavity. The proximal two-fifths of this portion is the **jejunum,** and the remainder is the **ileum.** Although there is no distinct separation between the jejunum and ileum, the diameter of the jejunum tends to be greater, and its wall is thicker, more vascular, and more active than that of the ileum.

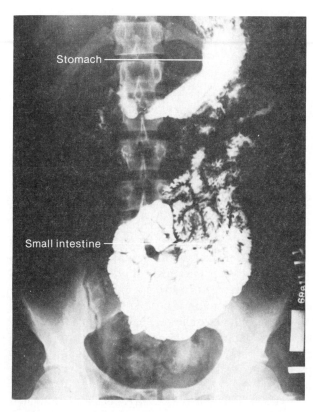

Figure 14.37 X-ray film showing a normal small intestine.

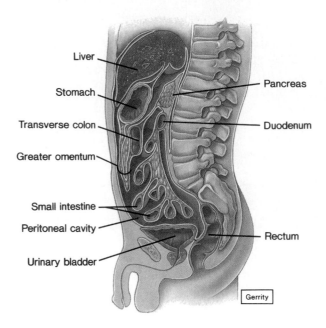

Figure 14.38 Portions of the small intestine are suspended from the posterior abdominal wall by mesentery formed by folds of the peritoneal membrane.

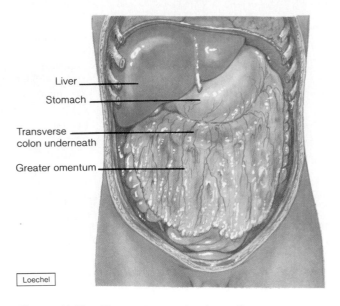

Figure 14.39 The greater omentum hangs like an apron over the abdominal organs.

The jejunum and ileum are suspended from the posterior abdominal wall by a double-layered fold of peritoneum called **mesentery** (figure 14.38). This supporting tissue contains the blood vessels, nerves, and lymphatic vessels that supply the intestinal wall.

A filmy fold of peritoneal membrane called the *greater omentum* drapes like an apron from the stomach over the transverse colon and the folds of the small intestine. If infections occur in the wall of the alimentary canal, cells from the omentum may adhere to the inflamed region and help wall it off so that the infection is less likely to enter the peritoneal cavity. (See figure 14.39.)

Structure of the Small Intestinal Wall

Throughout its length, the inner wall of the small intestine has a velvety appearance. This is due to the presence of innumerable tiny projections of mucous membrane called **intestinal villi** (figures 14.40 and 14.41). These structures are most numerous in the duodenum and the proximal portion of the jejunum. They project into the passageway, or **lumen,** of the alimentary canal, contacting the intestinal contents. Villi increase the surface area of the intestinal lining, and they play an important role in the absorption of digestive products.

Each villus consists of a layer of simple columnar epithelium and a core of connective tissue containing blood capillaries, a lymphatic capillary called a **lacteal,** and nerve fibers. At their free surfaces, the epithelial cells possess many fine extensions called *microvilli* that

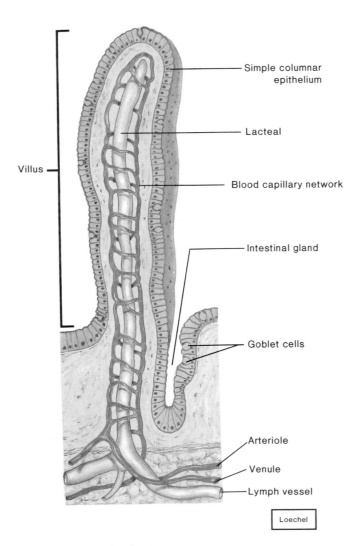

Figure 14.40 Structure of a single intestinal villus. What is the function of such a projection?

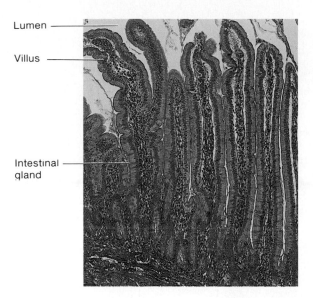

Figure 14.41 Light micrograph of intestinal villi from the wall of the duodenum.

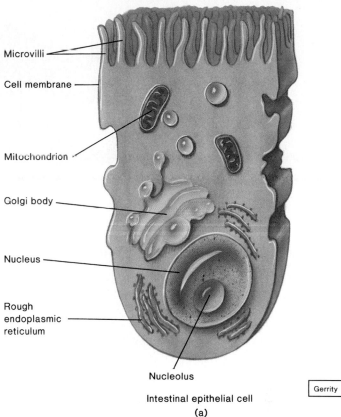

Intestinal epithelial cell
(a)

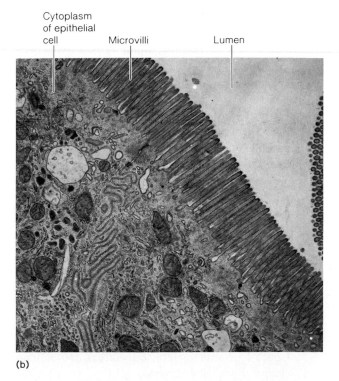

(b)

Figure 14.42 (a) Microvilli increase the surface area of intestinal epithelial cells; (b) transmission electron micrograph of microvilli.

form a brushlike border and greatly increase the surface area of the intestinal cells (figure 14.42). The presence of microvilli enhances the process of absorption.

The blood and lymph capillaries carry away substances absorbed by a villus, and impulses transmitted on the nerve fibers stimulate or inhibit its activities.

Between the bases of adjacent villi are tubular **intestinal glands** (crypts of Lieberkühn), which extend downward into the mucous membrane. The deeper layers of the small intestinal wall are much like those of other parts of the alimentary canal in that they include a submucosa, a muscular layer, and a serous layer.

The lining of the small intestine is also characterized by the presence of numerous circular folds of mucosa, called *plicae circulares,* that are especially well developed in the lower duodenum and upper jejunum. Together with the villi and microvilli, these folds help increase the surface area of the intestinal lining (figure 14.43).

> The epithelial cells that form the lining of the small intestine are continually being replaced. The new cells are formed within the intestinal glands by mitosis, and they migrate outward onto the surface of a villus. When the migrating cells reach the tip of the villus, they are shed. As a result of this process, which is called *cellular turnover,* the epithelial lining of the small intestine is renewed every three to six days.

Secretions of the Small Intestine

In addition to the mucus-secreting goblet cells, which occur extensively throughout the mucosa of the small intestine, many specialized *mucus-secreting glands*

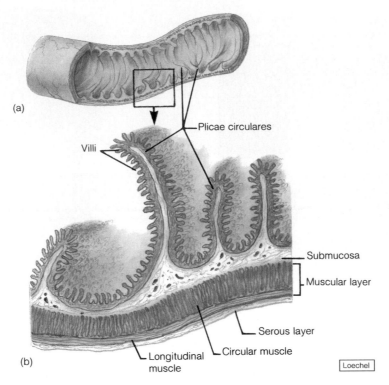

Figure 14.43 (a) The inner lining of the small intestine contains many circular folds, the plicae circulares; (b) a longitudinal section through some of these folds.

(Brunner's glands) occur in the submucosa within the proximal portion of the duodenum. These glands secrete large quantities of viscid, alkaline mucus in response to various stimuli.

The intestinal glands at the bases of the villi secrete relatively large amounts of a watery fluid (figure 14.40). The villi rapidly reabsorb this fluid, and it provides a vehicle for moving digestive products into the villi. The fluid secreted by the intestinal glands has a pH that is nearly neutral (6.5–7.5), and it seems to lack digestive enzymes. However, the epithelial cells of the intestinal mucosa have digestive enzymes embedded in the surfaces of their microvilli. These enzymes can break down food molecules just before absorption takes place. They include **peptidases,** which split peptides into amino acids; **sucrase, maltase,** and **lactase,** which split the double sugars (disaccharides) sucrose, maltose, and lactose into the simple sugars (monosaccharides) glucose, fructose, and galactose; and **intestinal lipase,** which splits fats into fatty acids and glycerol.

Chart 14.9 summarizes the sources and actions of the major digestive enzymes.

The amount of lactase produced in the small intestine usually reaches a maximum shortly after birth and thereafter tends to decrease. As a result, some adults produce insufficient quantities of lactase to break down the lactose (milk sugar) in their diets, a condition called *lactose intolerance*. When this happens, the lactose from milk and various milk products remains undigested and causes an increase in the osmotic pressure of the intestinal contents. Consequently, water is drawn from the tissues into the intestine. At the same time, intestinal bacteria may act upon the undigested sugar and produce organic acids and gases. As a result, the person may feel bloated and suffer from intestinal cramps and diarrhea.

Regulation of Small Intestinal Secretions

Since mucus protects the intestinal wall in the same way it protects the stomach lining, it is not surprising that mucus secretions are increased in response to mechanical stimulation and the presence of irritants, such as gastric juice. Consequently, as the stomach contents enter the small intestine, the duodenal mucous glands are stimulated to release large quantities of mucus.

Secretions from goblet cells and intestinal glands are similarly stimulated by direct contact with chyme, which provides both chemical and mechanical stimuli, and by reflexes triggered by distension of the intestinal

CHART 14.9 Summary of the major digestive enzymes

Enzyme	Source	Digestive action
Salivary enzyme		
Amylase	Salivary glands	Begins carbohydrate digestion by converting starch and glycogen to disaccharides
Gastric enzymes		
Pepsin	Gastric glands	Begins the digestion of proteins
Lipase	Gastric glands	Begins the digestion of butterfat
Pancreatic enzymes		
Amylase	Pancreas	Converts starch and glycogen into disaccharides
Lipase	Pancreas	Converts fats into fatty acids and glycerol
Proteinases a. Trypsin b. Chymotrypsin c. Carboxypeptidase	Pancreas	Convert proteins or partially digested proteins into peptides
Nucleases	Pancreas	Convert nucleic acids into nucleotides
Intestinal enzymes		
Peptidase	Mucosal cells	Converts peptides into amino acids
Sucrase, maltase, lactase	Mucosal cells	Convert disaccharides into monosaccharides
Lipase	Mucosal cells	Converts fats into fatty acids and glycerol
Enterokinase	Mucosal cells	Converts trypsinogen into trypsin

Figure 14.44   Digestion changes complex carbohydrates into disaccharides. The disaccharides are then converted into monosaccharides that are absorbed by intestinal villi and enter the blood.

wall. These reflex actions involve parasympathetic motor impulses that cause secretory cells to increase their activities.

1. Describe the parts of the small intestine.
2. Distinguish between intestinal villi and microvilli.
3. What is the function of the intestinal glands?
4. List the digestive enzymes formed by intestinal cells.

Absorption in the Small Intestine

Because villi greatly increase the surface area of the intestinal mucosa, the small intestine is the most important absorbing organ of the alimentary canal. In fact, the small intestine is so effective in absorbing digestive products, water, and electrolytes, that very little absorbable material reaches its distal end.

Carbohydrate digestion begins in the mouth with the activity of salivary amylase, and is completed in the small intestine by enzymes from the intestinal mucosa and pancreas (figure 14.44). The resulting monosaccharides are absorbed by the villi and enter blood capillaries. Although small quantities may pass into the villi by diffusion, most simple sugars are absorbed by active transport or facilitated diffusion. (See chapter 3.) The exact mechanisms, however, are poorly understood.

Protein digestion begins in the stomach as a result of pepsin activity, and is completed in the small intestine by enzymes from the intestinal mucosa and the

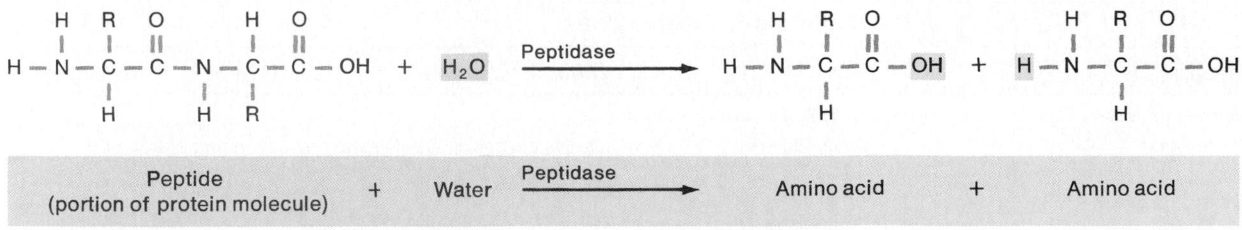

Figure 14.45 The amino acids that result from protein digestion are absorbed by intestinal villi and enter the blood.

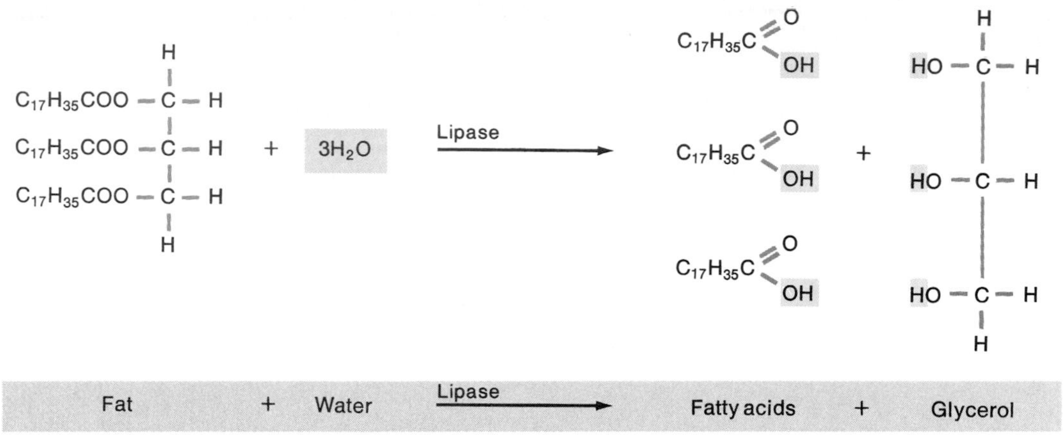

Figure 14.46 Fatty acids and glycerol result from fat digestion. They are absorbed by intestinal villi, and most are resynthesized into fat molecules before they enter the blood or lymph.

pancreas. During this process, large protein molecules are converted into amino acids, as shown in figure 14.45. These smaller particles are then absorbed into the villi by active transport and are carried away by the blood.

Fat molecules are digested almost entirely by enzymes from the intestinal mucosa and pancreas (figure 14.46). The mechanism by which the resulting fatty acid molecules are absorbed involves the following steps: (1) The fatty acid molecules are dissolved in the epithelial cell membranes of the villi and diffuse into them. (2) The endoplasmic reticula of the cells use the fatty acids to resynthesize fat molecules similar to those previously digested. (3) These fats collect in clusters that become encased in protein. (4) The resulting large molecules of lipoprotein are called *chylomicrons*, and they make their way to the lacteal of the villus. (5) Periodic contractions of smooth muscles in the villus help empty the lacteal, and the lymph carries the chylomicrons to the blood. (See figure 14.47.)

Chylomicrons are transported by the blood to capillaries of muscle and adipose tissues. A specific type of protein (apoprotein) associated with the surface of chylomicrons activates an enzyme (lipoprotein lipase) that is attached to the inner lining of such capillaries. As a result of the enzyme's action, fatty acids and monoglycerides are released from the chylomicrons, and

they enter muscle or adipose cells to be utilized as energy sources or to be stored. The remnants of the chylomicrons travel in the blood to the liver, where they bind to receptors on the surface of liver cells. The remnants quickly enter liver cells by receptor-mediated endocytosis (see chapter 3) and are destroyed by lysosomal activity.

Some fatty acids with relatively short carbon chains may be absorbed directly into the blood capillary of the villus without being converted back into fat.

In addition to absorbing the products of carbohydrate, protein, and fat digestion, the intestinal villi absorb various electrolytes and water. Certain ions, such as those of sodium, potassium, chloride, nitrate, and bicarbonate, are readily absorbed; but others, including ions of calcium, magnesium, and sulfate, are poorly absorbed.

The absorption of electrolytes usually involves active transport, and water is absorbed by osmosis. Thus, even though the intestinal contents may be hypertonic to the epithelial cells at first, as nutrients and electrolytes are absorbed, the intestinal contents tend to become hypotonic to the cells. Then, water follows the nutrients and electrolytes into the villi by osmosis.

The absorption process is summarized in chart 14.10.

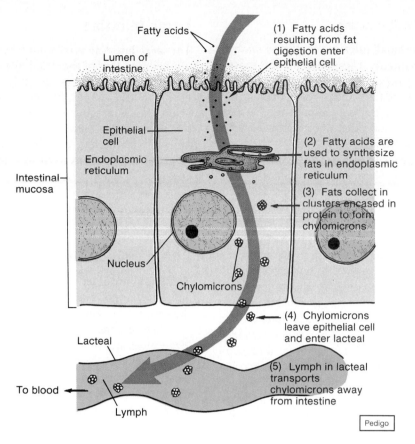

Fatty acids

Lumen of
intestine

Epithelial
cell

Endoplasmic
reticulum

Intestinal
mucosa

Nucleus

Chylomicrons

Lacteal

To blood

Lymph

(1) Fatty acids
resulting from fat
digestion enter
epithelial cell

(2) Fatty acids are
used to synthesize
fats in endoplasmic
reticulum

(3) Fats collect in
clusters encased in
protein to form
chylomicrons

(4) Chylomicrons
leave epithelial cell
and enter lacteal

(5) Lymph in lacteal
transports
chylomicrons away
from intestine

Pedigo

Figure 14.47 Fatty acid absorption involves several steps.

CHART 14.10	Intestinal absorption of nutrients	
Nutrient	Absorption mechanism	Means of transport
Monosaccharides	Diffusion, facilitated diffusion, or active transport	Blood in capillaries
Amino acids	Active transport	Blood in capillaries
Fatty acids and glycerol	Diffusion into cells (a) Most fatty acids are converted back into fats and incorporated in chylomicrons for transport.	Lymph in lacteals
	(b) Some fatty acids with relatively short carbon chains are transported without being converted back into fats.	Blood in capillaries
Electrolytes	Diffusion and active transport	Blood in capillaries
Water	Osmosis	Blood in capillaries

When some nutrients are digested but not absorbed by the small intestine, the condition is called *malabsorption.* Causes of malabsorption include surgical removal of a portion of the small intestine, obstruction of lymphatic vessels due to a tumor, or interference with the production and release of bile as a result of liver disease.

Another cause of malabsorption involves a reaction to the presence of *gluten,* a substance found in certain grains, especially wheat and rye. This condition is called *Celiac disease.* As a result of the reaction to gluten, microvilli are damaged and, in severe cases, villi may be destroyed. Both of these effects reduce the absorptive surface of the small intestine, and consequently some nutrients fail to be absorbed. The symptoms of malabsorption include diarrhea, weight loss, weakness, vitamin deficiencies, anemia, and demineralization of the bones.

1. What substances resulting from the digestion of carbohydrate, protein, and fat molecules are absorbed by the small intestine?
2. What ions are absorbed by the small intestine?
3. What transport mechanisms are used by intestinal villi?
4. Describe how fatty acids are absorbed and transported.

Movements of the Small Intestine

Like the stomach, the small intestine carries on *mixing movements* and *peristalsis*. The major mixing movement is called *segmentation*. It involves the formation of small, ringlike contractions that occur periodically, cutting the chyme into segments and moving it to and fro. Segmentation also slows the movement of chyme through the small intestine.

The chyme is propelled through the small intestine by peristaltic waves. These waves are usually weak, and they stop after pushing the chyme a short distance. Consequently, food materials move relatively slowly through the small intestine, taking from three to ten hours to travel its length.

As might be expected, parasympathetic impulses enhance both mixing and peristaltic movements, and sympathetic impulses inhibit them. Reflexes involving parasympathetic impulses to the small intestine sometimes originate in the stomach. For example, as the stomach fills with food and its wall becomes distended, a reflex (gastroenteric reflex) is triggered, and peristaltic activity in the small intestine is greatly increased. Another reflex is initiated when the duodenum is filled with chyme and its wall is stretched. This reflex causes the chyme to be moved through the small intestine more rapidly.

Stimulation of the small intestinal wall by overdistension or by severe irritation may elicit a strong *peristaltic rush* that passes along the entire length of the small intestine. This type of movement sweeps the contents of the small intestine into the large intestine relatively rapidly and helps relieve the small intestine of its problem. This rapid movement of chyme may prevent the normal absorption of water, nutrients, and electrolytes from the intestinal contents. The result is *diarrhea,* a condition in which defecation becomes more frequent and the stools become watery. If the diarrhea continues for a prolonged time, imbalances in water and electrolyte concentrations are likely to develop.

At the distal end of the small intestine, where the ileum joins the cecum of the large intestine, is a sphincter muscle called the **ileocecal valve.** Normally, this sphincter remains constricted, preventing the contents of the small intestine from entering the large intestine and at the same time, preventing the contents of the large intestine from backing up into the ileum. However, after a meal, a reflex (gastroileal reflex) is elicited, and peristalsis in the ileum is increased. This action forces some of the contents of the small intestine into the cecum.

1. Describe the movements of the small intestine.
2. How are the movements of the small intestine initiated?
3. What is meant by a peristaltic rush?
4. What stimulus causes the ileocecal valve to relax?

Large Intestine

The **large intestine** is so named because its diameter is greater than that of the small intestine. This portion of the alimentary canal is about 1.5 meters long, and it begins in the lower right side of the abdominal cavity where the ileum joins the cecum. From there, the large intestine travels upward on the right side, crosses obliquely to the left, and descends into the pelvis. At its distal end, it opens to the outside of the body as the anus.

The large intestine reabsorbs water and electrolytes from the chyme remaining in the alimentary canal. It also forms and stores the feces until defecation occurs.

Parts of the Large Intestine

The large intestine, shown in figures 14.48 and 14.49, and in reference plates 51, 52, 58, and 75, consists of the cecum, the colon, the rectum, and the anal canal.

The **cecum,** which represents the beginning of the large intestine, is a dilated, pouchlike structure that hangs slightly below the ileocecal opening. Projecting downward from it is a narrow tube with a closed end called the **vermiform appendix.** Although the human appendix has no known digestive function, it does contain lymphatic tissue that can serve to resist infections.

> Occasionally, the appendix may become inflamed and infected, a condition called *appendicitis.* When this happens, the appendix is often removed surgically to prevent it from rupturing. If it does break open, the contents of the large intestine may enter the abdominal cavity and cause a serious infection of the peritoneum called *peritonitis.*

The large intestine, or **colon,** is divided into four portions—the ascending, transverse, descending, and sigmoid colons. The **ascending colon** begins at the cecum and travels upward against the posterior abdominal wall to a point just below the liver. There it turns sharply to the left (as the right colic, or hepatic, flexure) and becomes the **transverse colon.** The transverse colon is the longest and most mobile part of the large intestine. It is suspended by a fold of peritoneum and tends to sag in the middle below the stomach. As the transverse colon approaches the spleen, it turns abruptly downward (as the left colic, or splenic, flexure) and becomes the **descending colon.** At the brim of the pelvis, the descending colon makes an S-shaped curve, called the **sigmoid colon,** and then becomes the rectum.

The **rectum** lies next to the sacrum and generally follows its curvature. It is firmly attached to the sacrum by the peritoneum, and it ends about 5 centimeters below the tip of the coccyx, where it becomes the anal canal (figure 14.50).

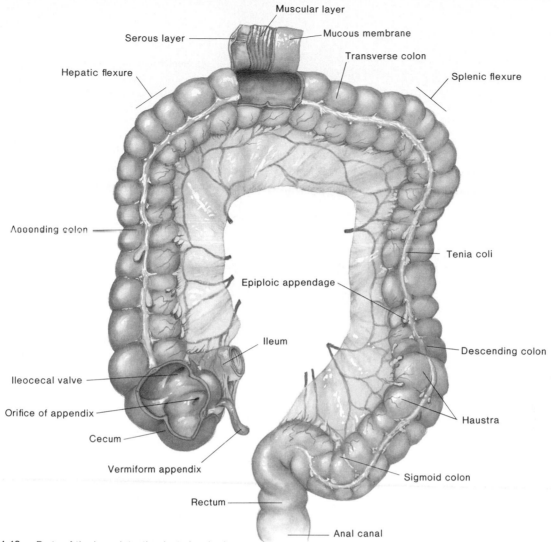

Figure 14.48   Parts of the large intestine (anterior view).

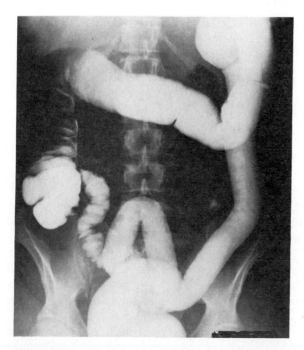

Figure 14.49   X-ray film of the large intestine. What portions can you identify?

The **anal canal** is formed by the last 2.5 to 4.0 centimeters of the large intestine. The mucous membrane in the canal is folded into a series of six to eight longitudinal *anal columns*. At its distal end, the canal opens to the outside as the **anus**. This opening is guarded by two sphincter muscles—an *internal anal sphincter muscle* composed of smooth muscle under involuntary control, and an *external anal sphincter muscle* composed of skeletal muscle under voluntary control.

> Each anal column contains a branch of the rectal vein; if something interferes with the blood flow in these vessels, the anal columns may become enlarged and inflamed. This condition, called *hemorrhoids,* may be aggravated by bowel movements and may be accompanied by discomfort and bleeding.

1. What is the general function of the large intestine?
2. Describe the parts of the large intestine.
3. Distinguish between the internal sphincter muscle and the external sphincter muscle of the anus.

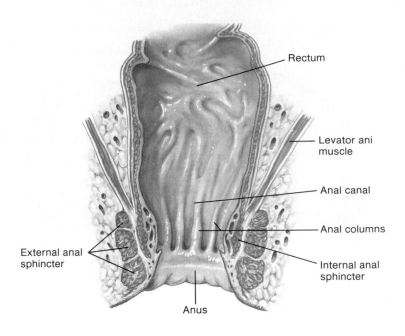

Figure 14.50 The rectum and the anal canal are located at the distal end of the alimentary canal.

Structure of the Large Intestinal Wall

Although the wall of the large intestine includes the same types of tissues found in other parts of the alimentary canal, it has some unique features. For example, it lacks the villi that are characteristic of the small intestine. Also, the layer of longitudinal muscle fibers does not cover its wall uniformly. Instead, the fibers are arranged in three distinct bands (teniae coli) that extend the entire length of the colon. These bands exert tension on the wall, creating a series of pouches (haustra). The large intestinal wall is also characterized by small collections of fat (epiploic appendages) in the serous layer on its outer surface (figure 14.44).

Functions of the Large Intestine

Unlike the small intestine, which secretes digestive enzymes and absorbs the products of digestion, the large intestine has little or no digestive function. However, the mucous membrane that forms the inner lining of the large intestine contains many tubular glands. Structurally, these glands are similar to those of the small intestine, but they are composed almost entirely of goblet cells. Consequently, mucus is the only significant secretion of this portion of the alimentary canal (figures 14.51 and 14.52).

The rate of mucus secretion is controlled largely by mechanical stimulation from chyme and by parasympathetic impulses. In both cases, the goblet cells respond by increasing their production of mucus, which, in turn, protects the intestinal wall against the abrasive

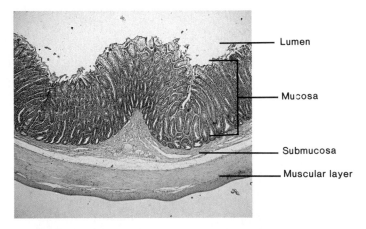

Figure 14.51 Light micrograph of the large intestinal wall (×10).

action of materials passing through it. Mucus also aids in holding particles of fecal matter together, and, because it is alkaline, mucus helps control the pH of the large intestinal contents. This is important because acids are sometimes released from the feces as a result of bacterial activity.

The chyme entering the large intestine usually has few nutrients remaining in it. It contains materials that were not digested or absorbed by the small intestine. It also contains water, various electrolytes, mucus, and bacteria.

Absorption in the large intestine is normally limited to water and electrolytes, and this usually takes place in the proximal half of the tube. Electrolytes, such

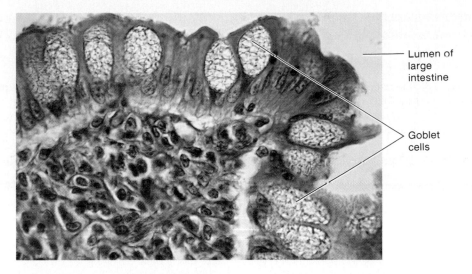

Figure 14.52 Light micrograph of the large intestinal mucosa (×250).

as sodium ions, can be absorbed by active transport, while the water follows, passively entering the mucosa by osmosis. As a result, about 90% of the water that enters the large intestine is absorbed, and little sodium or water is lost in the feces.

Many bacteria normally inhabit the large intestine, and these organisms can break down some of the substances that escape the actions of the human digestive enzymes. For instance, cellulose, a complex carbohydrate found in food of plant origin, passes through the alimentary canal almost unchanged, but colon bacteria can break down cellulose and use it as an energy source. At the same time, these bacteria synthesize certain vitamins that can be absorbed by the intestinal mucosa and used by body cells. Vitamins formed in this manner include vitamins K, B_{12}, thiamine, and riboflavin. Bacterial actions in the large intestine also may give rise to gas (flatus), which sometimes causes discomfort and pain.

1. How does the structure of the large intestine differ from that of the small intestine?
2. What substances can be absorbed by the large intestine?
3. What useful substances are produced by bacteria inhabiting the large intestine?

Movements of the Large Intestine

The movements of the large intestine—mixing and peristalsis—are similar to those of the small intestine, although they are usually more sluggish. The mixing movements break the fecal matter into segments and turn it so that all portions are exposed to the intestinal mucosa. This helps in water and electrolyte absorption.

The peristaltic waves of the large intestine are different from those of the small intestine. Instead of occurring frequently, they happen only two or three times each day. These waves produce *mass movements* in which a relatively large section of the intestinal wall constricts vigorously, forcing the intestinal contents to move toward the rectum. Typically, mass movements occur following a meal, as a result of a reflex that is initiated in the small intestine (duodenocolic reflex). Abnormal irritations of the intestinal mucosa can also trigger such movements. For instance, a person suffering from an inflamed colon (colitis) may experience frequent mass movements.

When it is appropriate to defecate, a person usually can initiate a *defecation reflex* by holding a deep breath and contracting the abdominal wall muscles. This action increases the internal abdominal pressure and forces feces into the rectum. As the rectum fills, its wall is distended and the defecation reflex is triggered. As a result, peristaltic waves in the descending colon are stimulated, and the internal anal sphincter relaxes. At the same time, other reflexes involving the sacral region of the spinal cord cause the peristaltic waves to strengthen, the diaphragm to lower, the glottis to close, and the abdominal wall muscles to contract. These actions cause an additional increase in the internal abdominal pressure and assist in squeezing the rectum. The external anal sphincter is signaled to relax, and the feces are forced to the outside. A person can inhibit defecation voluntarily by keeping the external anal sphincter contracted.

The action of taking a deep breath, closing the glottis, and forcibly contracting the abdominal muscles is called the *Valsalva maneuver*. This action causes the internal abdominal pressure to increase and aids defecation. It also causes the internal thoracic pressure to rise and may interfere with the return of blood to the heart. For this reason, the Valsalva maneuver may present a hazard to persons with heart disease.

Disorders of the Large Intestine

If feces are allowed to remain in the colon too long, the material becomes hard and dry, as more and more water is removed from it. This may produce a condition called *constipation*, in which the feces become so dry that defecation is difficult.

The causes of constipation include prolonged inhibition of the defecation reflex, lack of tone in the muscular layer of the large intestine, and muscular spasms in the intestinal wall that block the movement of feces. Constipation is sometimes treated with laxatives (cathartics), which are substances that can stimulate fecal movements.

Some laxatives act by causing an irritating effect on the intestinal mucosa, which stimulates peristaltic waves. Others prevent reabsorption of water by coating the feces with oil or by keeping the feces moist by increasing the osmotic pressure within the contents of the large intestine.

Constipation may also be treated by enemas, by increasing the fluid intake, by adding bulk foods to the diet, or by resensitizing the person to the defecation reflex.

Diarrhea causes an effect opposite that of constipation, because the feces move more rapidly than usual through the colon and remain very liquid. Among the more common causes of diarrhea are abnormally frequent peristaltic waves and hypersecretion of intestinal mucus that adds to the volume and increases the electrolyte concentrations of fluid within the colon. Increased peristalsis may accompany emotional stress or irritations produced by infections in the intestines. Hypersecretion may be induced by the presence of certain hormones, bacterial toxins, or dietary nutrients.

Sometimes the mucosa of the large intestine becomes extensively ulcerated, a condition called *ulcerative colitis*. The cause of this condition is unknown, but it is often associated with prolonged periods of emotional stress. In any case, the colon becomes very active, mass movements occur frequently, and the person has diarrhea most of the time.

Treatment for ulcerative colitis involves reducing nervous tension, changing the diet, and taking medication. If these measures are unsuccessful, a surgeon may perform an *ileostomy*, in which the colon is removed and a portion of the ileum is connected to an opening in the abdominal wall. As a result, the small intestinal contents can move to the outside of the body without passing through the colon.

A related procedure, called *colostomy*, is sometimes performed in cases of colon cancer, abscesses of the colon, intestinal obstructions, or removal of the rectum. In this procedure, the colon is connected to an opening on the surface of the abdomen so that the rectum and anus are bypassed.

Feces

As was mentioned, the **feces** are composed largely of materials that were not digested or absorbed, together with water, electrolytes, mucus, and bacteria. Usually the feces are about 75% water, and their color is normally due to the presence of bile pigments that have been altered somewhat by bacterial action.

The pungent odor of the feces results from a variety of compounds produced by bacteria. These compounds include phenol, hydrogen sulfide, indole, skatole, and ammonia.

1. How does peristalsis in the large intestine differ from peristalsis in the small intestine?
2. List the major events that occur during defecation.
3. Describe the composition of feces.

Clinical Terms Related to the Digestive System

achalasia (ak″ah-la′ze-ah) Failure of the smooth muscle to relax at some junction in the digestive tube, such as that between the esophagus and stomach.

achlorhydria (ah″klor-hi′dre-ah) A lack of hydrochloric acid in gastric secretions.

aphagia (ah-fa′je-ah) Inability to swallow.

cholecystitis (ko″le-sis-ti′tis) Inflammation of the gallbladder.

cholelithiasis (ko″le-li-thi′ah-sis) The presence of stones in the gallbladder.

cholestasis (ko″le-sta′sis) Condition in which the flow of bile from the gallbladder is prevented.

cirrhosis (si-ro′sis) A liver condition in which the hepatic cells degenerate and the surrounding connective tissues thicken.

diverticulitis (di″ver-tik″u-li′tis) Inflammation of small pouches (diverticula) that sometimes form in the lining and wall of the colon.

dumping syndrome (dum′ping sin′drōm) A set of symptoms, including diarrhea, that often occur following a gastrectomy.

dysentery (dis′en-ter″e) An intestinal infection, caused by viruses, bacteria, or protozoans, that is accompanied by diarrhea and cramps.

dyspepsia (dis-pep′se-ah) Indigestion; difficulty in digesting a meal.

dysphagia (dis-fa′ze-ah) Difficulty in swallowing.

enteritis (en″tĕ-ri′tis) Inflammation of the intestine.

esophagitis (e-sof″ah-ji′tis) Inflammation of the esophagus.

gastrectomy (gas-trek′to-me) Partial or complete removal of the stomach.

gastritis (gas-tri′tis) Inflammation of the stomach lining.

gastrostomy (gas-tros′to-me) The creation of an opening in the stomach wall through which food and liquids may be administered when swallowing is not possible.

gingivitis (jin″ji-vi′tis) Inflammation of the gums.

glossitis (glŏ-si'tis) Inflammation of the tongue.

hemorrhoidectomy (hem''ŏ-roi-dek'to-me) Removal of hemorrhoids.

hepatitis (hep''ah-ti'tis) Inflammation of the liver.

ileitis (il''e-i'tis) Inflammation of the ileum.

ileus (il'e-us) An obstruction of the intestine due to an inhibition of motility or a mechanical cause.

pharyngitis (far''in-ji'tis) Inflammation of the pharynx.

pyloric stenosis (pi-lor'ik stĕ-no'sis) A congenital obstruction at the pyloric sphincter due to an enlarged pyloric muscle.

pylorospasm (pi-lor'o-spazm) A spasm of the pyloric portion of the stomach or of the pyloric sphincter.

pyorrhea (pi''o-re'ah) Inflammation of the dental periosteum, accompanied by the formation of pus.

stomatitis (sto''mah-ti'tis) Inflammation of the lining of the mouth.

vagotomy (va-got'o-me) Sectioning of the vagus nerve fibers.

Chapter Summary

Introduction (page 504)

Digestion is the process of changing food substances into forms that can be absorbed. The digestive system consists of an alimentary canal and several accessory organs.

General Characteristics of the Alimentary Canal (page 504)

Various regions of the canal are specialized to perform specific functions.

1. Structure of the wall
   a. The wall consists of four layers.
   b. These layers include the mucous membrane, submucosa, muscular layer, and serous layer.
2. Movements of the tube
   a. Motor functions include mixing and propelling movements.
   b. Peristalsis is responsible for propelling movements.
   c. The wall of the tube undergoes receptive relaxation just ahead of a peristaltic wave.
3. Innervation of the tube
   a. The tube is innervated by branches of the sympathetic and parasympathetic divisions of the autonomic nervous system.
   b. Parasympathetic impulses generally cause an increase in digestive activities; sympathetic impulses generally inhibit digestive activities.
   c. Sympathetic impulses are responsible for the contraction of certain sphincter muscles that control movement through the alimentary canal.

Mouth (page 507)

The mouth is adapted to receive food and begin preparing it for digestion. It also serves as an organ of speech and sensory perception.

1. Cheeks and lips
   a. Cheeks form the lateral walls of the mouth.
   b. Lips are highly mobile and possess a variety of sensory receptors useful in judging the characteristics of food.
2. Tongue
   a. The tongue is a thick, muscular organ that aids in mixing food with saliva and moving it toward the pharynx.
   b. The rough surface of the tongue aids in handling food and contains taste buds.
   c. Lingual tonsils are located on the root of the tongue.
3. Palate
   a. The palate comprises the roof of the mouth and includes hard and soft portions.
   b. The soft palate closes the opening to the nasal cavity during swallowing.
   c. Palatine tonsils are located on either side of the tongue in the back of the mouth.
   d. Tonsils consist of lymphatic tissues.
4. Teeth
   a. Two sets of teeth develop in sockets of the mandibular and maxillary bones.
   b. There are twenty primary and thirty-two secondary teeth.
   c. Teeth break food into smaller pieces, increasing the surface area of food that is exposed to digestive actions.
   d. Different kinds of teeth are adapted to handle foods in different ways, such as biting, grasping, or grinding.
   e. Each tooth consists of a crown and root, and is composed of enamel, dentin, pulp, nerves, and blood vessels.
   f. A tooth is attached to the alveolar process by collagenous fibers of the periodontal ligament.

Salivary Glands (page 512)

Salivary glands secrete saliva, which moistens food, helps bind food particles together, begins digestion of carbohydrates, makes taste possible, helps cleanse the mouth, and regulates pH in the mouth.

1. Salivary secretions
   a. Salivary glands include serous cells that secrete digestive enzymes and mucous cells that secrete mucus.
   b. Parasympathetic impulses stimulate the secretion of serous fluid.
2. Major salivary glands
   a. The parotid glands are the largest, and they secrete saliva that is rich in amylase.
   b. The submandibular glands in the floor of the mouth produce viscid saliva.
   c. The sublingual glands in the floor of the mouth primarily secrete mucus.

Pharynx and Esophagus (page 514)

The pharynx and esophagus serve as passageways.

1. Structure of the pharynx
   a. The pharynx is divided into a nasopharynx, oropharynx, and laryngopharynx.
   b. The muscular walls of the pharynx contain fibers arranged in circular and longitudinal groups.
2. Swallowing mechanism
   a. The act of swallowing occurs in three stages.
      (1) Food is mixed with saliva and forced into the pharynx.
      (2) Involuntary reflex actions move the food into the esophagus.
      (3) Food is transported to the stomach.
3. Esophagus
   a. The esophagus passes through the mediastinum and penetrates the diaphragm.
   b. Circular muscle fibers at the distal end of the esophagus help prevent the regurgitation of food from the stomach.

Stomach (page 516)

The stomach receives food, mixes it with gastric juice, carries on a limited amount of absorption, and moves food into the small intestine.

1. Parts of the stomach
   a. The stomach is divided into cardiac, fundic, body, and pyloric regions.
   b. The pyloric sphincter serves as a valve between the stomach and the small intestine.
2. Gastric secretions
   a. Gastric glands secrete gastric juice.
   b. Gastric juice contains pepsin, hydrochloric acid, lipase, and intrinsic factor.
3. Regulation of gastric secretions
   a. Gastric secretions are enhanced by parasympathetic impulses and by gastrin, a hormone.
   b. The three stages of gastric secretion are the cephalic, gastric, and intestinal phases.
   c. The presence of food in the small intestine reflexly inhibits gastric secretions.
4. Gastric absorption
   a. The stomach is not well adapted for absorption.
   b. A few substances such as water and other small molecules may be absorbed through the stomach wall.

5. Mixing and emptying actions
   a. As the stomach fills, its wall stretches, but its internal pressure remains unchanged.
   b. Mixing movements aid in producing chyme; peristaltic waves move the chyme into the pyloric region.
   c. The muscular wall of the pyloric region pumps chyme into the small intestine.
   d. The rate of emptying depends on the fluidity of the chyme and the type of food present.
   e. The upper part of the small intestine fills, and an enterogastric reflex causes the peristaltic waves in the stomach to be inhibited.
   f. Vomiting results from a complex reflex that can be stimulated by a variety of factors.

Pancreas (page 523)

1. Structure of the pancreas
   a. The pancreas is closely associated with the duodenum.
   b. It produces pancreatic juice that is secreted into a pancreatic duct.
   c. The pancreatic duct leads to the duodenum.
2. Pancreatic juice
   a. Pancreatic juice contains enzymes that can split carbohydrates, proteins, fats, and nucleic acids.
   b. Pancreatic juice has a high bicarbonate ion concentration that helps neutralize chyme and causes the intestinal contents to be alkaline.
3. Regulation of pancreatic secretion
   a. Secretin from the duodenum stimulates the release of pancreatic juice that contains few digestive enzymes but has a high bicarbonate ion concentration.
   b. Cholecystokinin from the intestinal wall stimulates the release of pancreatic juice that has a high concentration of digestive enzymes.

Liver (page 525)

1. Functions of the liver
   a. The liver is located in the upper right and central portion of the abdominal cavity.
   b. It carries on a variety of important functions involving the metabolism of carbohydrates, lipids, and proteins; the storage of substances; the filtering of blood; the destruction of toxic chemicals; and the secretion of bile.
   c. Bile is the only liver secretion that directly affects digestion.
2. Structure of the liver
   a. The liver is a highly vascular organ, enclosed in a fibrous capsule, and divided into lobes.
   b. Each lobe contains hepatic lobules, the functional units of the liver.
   c. Bile from the lobules is carried by bile canals to hepatic ducts that unite to form the common bile duct.

3. Composition of bile
   a. Bile contains bile salts, bile pigments, cholesterol, and various electrolytes.
   b. Only the bile salts have digestive functions.
   c. Bile pigments are products of red blood cell breakdown.
4. Gallbladder and its functions
   a. The gallbladder stores bile between meals.
   b. Release of bile from the common bile duct is controlled by a sphincter muscle.
   c. Gallstones may sometimes form within the gallbladder.
5. Regulation of bile release
   a. Release of bile is stimulated by cholecystokinin from the small intestine.
   b. The sphincter muscle at the base of the common bile duct relaxes as a peristaltic wave in the duodenal wall approaches.
6. Functions of bile salts
   a. Bile salts emulsify fats and aid in the absorption of fatty acids, cholesterol, and certain vitamins.
   b. Bile salts are reabsorbed in the small intestine.

Small Intestine (page 530)

The small intestine extends from the pyloric sphincter to the large intestine. It receives secretions from the pancreas and liver, completes the digestion of nutrients, absorbs the products of digestion, and transports the residues to the large intestine.

1. Parts of the small intestine
   a. The small intestine consists of the duodenum, jejunum, and ileum.
   b. The small intestine is suspended from the posterior abdominal wall by mesentery.
2. Structure of the small intestinal wall
   a. The wall is lined with villi that increase the surface area and aid in mixing and absorption.
   b. Microvilli on the free ends of epithelial cells greatly increase the surface area.
   c. Intestinal glands are located between the villi.
   d. Circular folds in the lining of the intestinal wall also increase its surface area.
3. Secretions of the small intestine
   a. Intestinal glands primarily secrete a watery fluid that lacks digestive enzymes but provides a vehicle for moving food substances to the villi.
   b. Digestive enzymes embedded in the surfaces of microvilli can split molecules of sugars, proteins, and fats.
4. Regulation of small intestinal secretions
   a. Secretions are enhanced by the presence of gastric juice and chyme.
   b. Reflexes stimulated by distension of the small intestinal wall also increase intestinal secretions.

5. Absorption in the small intestine
   a. Monosaccharides, amino acids, fatty acids, and glycerol are absorbed by the villi.
   b. Villi also absorb water and electrolytes.
   c. Fat molecules with longer chains of carbon atoms enter the lacteals of the villi; fatty acids with relatively short carbon chains enter the blood capillaries of the villi.
6. Movements of the small intestine
   a. Movements include mixing by segmentation and peristalsis.
   b. Overdistension or irritation may stimulate a peristaltic rush and result in diarrhea.
   c. The ileocecal valve controls movement of the intestinal contents from the small intestine into the large intestine.

Large Intestine (page 538)

The large intestine reabsorbs water and electrolytes, and forms and stores feces.

1. Parts of the large intestine
   a. The large intestine consists of the cecum, colon, rectum, and anal canal.
   b. The colon is divided into ascending, transverse, descending, and sigmoid portions.
2. Structure of the large intestinal wall
   a. Basically, the large intestine wall is like the wall in other parts of the alimentary canal.
   b. Unique features include a layer of longitudinal muscle fibers that are arranged in distinct bands.
3. Functions of the large intestine
   a. The large intestine has little or no digestive function, although it secretes mucus.
   b. The rate of mucus secretion is controlled by mechanical stimulation and parasympathetic impulses.
   c. Absorption in the large intestine is generally limited to water and electrolytes.
   d. Many bacteria inhabit the large intestine and may aid the body by synthesizing certain vitamins.
4. Movements of the large intestine
   a. Movements are similar to those in the small intestine.
   b. Mass movements occur two to three times each day.
   c. Defecation is stimulated by a defecation reflex.
5. Feces
   a. Feces are formed and stored in the large intestine.
   b. Feces consist largely of water, undigested material, mucus, and bacteria.
   c. The color of feces is due to bile salts that have been altered by bacterial action.

Clinical Application of Knowledge

1. If a patient has 95% of the stomach removed (subtotal gastrectomy) as treatment for severe ulcers or cancer, how would the digestion and absorption of foods be affected? How would the patient's eating habits have to be altered? Why?

2. Why may a person with inflammation of the gallbladder (cholecystitis) also develop an inflammation of the pancreas (pancreatitis)?

3. What effect is a before-dinner cocktail likely to have on digestion? Why are such beverages inadvisable for persons with ulcers?

4. What type of acid-alkaline disorder is likely to develop if the stomach contents are lost repeatedly by vomiting over a prolonged period? What acid-alkaline disorder is likely to develop as a result of prolonged diarrhea?

5. In infants, the pyloric sphincter is sometimes abnormally thickened (hypertrophic pyloric stenosis) and obstructs the opening into the duodenum. What symptoms might accompany this condition?

6. How would you explain to a patient with malabsorption due to gluten sensitivity the importance of avoiding wheat and rye flour in the diet?

Review Activities

1. List and describe the locations of the major parts of the alimentary canal.
2. List and describe the locations of the accessory organs of the digestive system.
3. Name the four layers of the wall of the alimentary canal.
4. Distinguish between mixing movements and propelling movements.
5. Define *peristalsis*.
6. Explain the relationship between peristalsis and receptive relaxation.
7. Describe the general effects of parasympathetic and sympathetic impulses on the alimentary canal.
8. Discuss the functions of the mouth and its parts.
9. Distinguish between lingual, palatine, and pharyngeal tonsils.
10. Compare the primary and secondary teeth.
11. Explain how the various types of teeth are adapted to perform specialized functions.
12. Describe the structure of a tooth.
13. Explain how a tooth is anchored in its socket.
14. List and describe the locations of the major salivary glands.
15. Explain how the secretions of the salivary glands differ.
16. Discuss the digestive functions of saliva.
17. Name and locate the three major regions of the pharynx.
18. Describe the mechanism of swallowing.
19. Explain the function of the esophagus.
20. Describe the structure of the stomach.
21. List the enzymes in gastric juice, and explain the function of each enzyme.
22. Explain how gastric secretions are regulated.
23. Describe the mechanism that controls the emptying of the stomach.
24. Describe the enterogastric reflex.
25. Explain the mechanism of vomiting.
26. Describe the locations of the pancreas and the pancreatic duct.
27. List the enzymes found in pancreatic juice, and explain the function of each enzyme.
28. Explain how pancreatic secretions are regulated.
29. List the major functions of the liver.
30. Describe the structure of the liver.
31. Describe the composition of bile.
32. Trace the path of bile from a bile canal to the small intestine.
33. Explain how gallstones form.
34. Define *cholecystokinin*.
35. Explain the functions of bile salts.
36. List and describe the locations of the parts of the small intestine.
37. Name the enzymes of the intestinal mucosa, and explain the function of each enzyme.
38. Explain how the secretions of the small intestine are regulated.
39. Describe the functions of the intestinal villi.
40. Summarize how each major type of digestive product is absorbed.
41. Explain how the movement of the intestinal contents is controlled.
42. List and describe the locations of the parts of the large intestine.
43. Explain the general functions of the large intestine.
44. Describe the defecation reflex.

Gallbladder Disease

Molly G., an overweight, 47-year-old college administrator and mother of four, had been feeling healthy until recently. Then she regularly began to feel pain in the upper right quadrant of her abdomen (figure 1.19). Sometimes the discomfort seemed to radiate around to her back and move upward into her right shoulder. Most commonly, she felt this pain after her evening meal; occasionally it also occurred during the night, causing her to awaken. After an episode of particularly severe pain accompanied by sweating (diaphoresis) and nausea, Molly decided to speak to her physician.

During an examination of Molly's abdomen, the physician discovered tenderness in the epigastric region (figure 1.18). She decided that Molly might be experiencing the symptoms of *acute cholecystitis*—an inflammation of the gallbladder. The physician recommended that Molly have a *cholecystogram*—an X ray of the gallbladder.

Molly was given some tablets containing a contrast medium to swallow the night before the X-ray procedure. This schedule allowed time for the substance to be absorbed by the small intestine, carried to the liver, and excreted into the bile. Later, the bile and contrast medium would be stored and concentrated in the gallbladder and would make the contents of the gallbladder opaque to X rays.

Molly's cholecystogram (figure 14.34) revealed several stones (calculi) in her gallbladder, a condition called *cholelithiasis*. Because Molly was experiencing increasingly severe symptoms of gallbladder disease, her physician recommended that she consult with an abdominal surgeon about undergoing a *cholecystectomy*—surgical removal of the gallbladder.

During the surgical procedure, an incision was made in Molly's right subcostal region. Her gallbladder was excised from the liver. Then the cystic duct (figure 14.30) and hepatic ducts were explored for the presence of stones, but none were found.

Unfortunately, following her recovery from surgery, Molly's symptoms persisted. Her surgeon ordered a *cholangiogram*—an X-ray series of the bile ducts. This study showed a residual stone at the distal end of Molly's common bile duct (figure 14.26).

The surgeon decided to extract the residual stone using a *fiberoptic endoscope,* a long, flexible tube that can be passed through the patient's esophagus and stomach and into the duodenum. This instrument enables a surgeon to observe features of the gastrointestinal tract by viewing them directly through the eyepiece of the endoscope or by watching an image on the screen of a television monitor. A surgeon can also perform various manipulations using specialized parts that are passed through the endoscope to its distal end.

In Molly's case, the surgeon performed an *endoscopic papillotomy*—an incision of the hepatopancreatic sphincter (figure 14.26). This was done by applying an electric current to a wire extending from the end of the endoscope. He then removed the exposed stone by manipulating a tiny basket at the tip of the endoscope.

Review Questions

1. Why is the pain of gallbladder disease likely to occur following a large meal?
2. What substance most commonly forms gallstones?
3. What functions might be lost by a patient after removal of the gallbladder?
4. How would you explain to a patient the difference between a cholecystogram and a cholangiogram?
5. What would be indicated if a patient with gallbladder disease developed a yellowing of the skin and eyes (jaundice)?

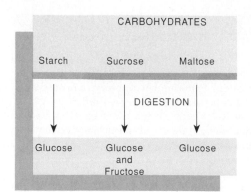

CARBOHYDRATES

Starch   Sucrose   Maltose

DIGESTION

Glucose   Glucose and Fructose   Glucose

C H A P T E R

15

Nutrition and Metabolism

*T*o maintain health, the human body must obtain a variety of chemicals from its environment in adequate amounts. These substances, which include carbohydrates, lipids, proteins, vitamins, and minerals, are called nutrients. They supply energy for cellular processes, play vital roles in metabolic reactions, and provide building materials for growth, maintenance, and repair of tissues. Taken together, the processes by which nutrients are ingested, assimilated, and utilized by the body are called nutrition ∎

Chapter Objectives	Key Terms	Aids to Understanding Words

After you have studied this chapter, you should be able to:

1. Define *nutrition, nutrients,* and *essential nutrients.*

2. List the major sources of carbohydrates, lipids, and proteins.

3. Describe how carbohydrates are utilized by cells.

4. Describe how lipids are utilized by cells.

5. Describe how amino acids are utilized by cells.

6. Define *nitrogen balance.*

7. Explain how the energy values of foods are determined.

8. Explain how various factors affect an individual's energy needs.

9. Define *energy balance.*

10. Explain what is meant by desirable weight.

11. List the fat-soluble and water-soluble vitamins.

12. Describe the general functions of each vitamin.

13. Distinguish between a vitamin and a mineral.

14. List the major minerals and trace elements.

15. Describe the general functions of each mineral and trace element.

16. Describe an adequate diet.

17. Distinguish between primary and secondary malnutrition.

18. Complete the review activities at the end of this chapter. Note that the items are worded in the form of specific learning objectives. You may want to refer to them before reading the chapter.

---

acetyl coenzyme A (as'ĕ-til ko-en'zīm)

antioxidant (an''te-ok'sĭ-dant)

basal metabolic rate (ba'sal met''ah-bol'ik rāt)

calorie (kal'o-re)

calorimeter (kal''o-rim'ĕ-ter)

dynamic equilibrium (di-nam'ik e''kwĭ-lib're-um)

energy balance (en'er-je bal'ans)

malnutrition (mal''nu-trish'un)

mineral (min'er-al)

nitrogen balance (ni'tro-jen bal'ans)

nutrient (nu'tre-ent)

triglyceride (tri-glis'er-īd)

vitamin (vi'tah-min)

---

bas-, base: *bas*al metabolic rate—metabolic rate of body under resting (basal) conditions.

calor-, heat: *calor*ie—unit used in measurement of heat or energy content of foods.

carot-, carrot: *carot*ene—yellowish plant pigment responsible for the color of carrots and other yellowish plant parts.

mal-, bad: *mal*nutrition—poor nutrition resulting from lack of food or failure to use available foods to best advantage.

-meter, measure: calori*meter*—instrument used to measure the caloric content of food.

nutri-, nourishing: *nutri*ent—substance needed to nourish body cells.

obes-, fat: *obes*ity—condition in which the body contains excessive fat.

pell-, skin: *pell*agra—vitamin deficiency condition characterized by inflammation of the skin and other symptoms.

The **nutrients** needed by the body include the carbohydrates, lipids, proteins, vitamins, minerals, and water introduced in chapters 2 and 4.

As a group, nutrients are obtained from foods, and some of them are altered by digestive processes after being ingested. Once they are processed by the digestive system, they are absorbed and transported to body cells by the blood. The ways that nutrients are changed chemically, and used anabolically and catabolically to support life processes constitute **metabolism.** (The metabolic pathways of carbohydrates, lipids, and proteins are discussed in chapter 4.)

Some nutrients, such as certain amino acids and fatty acids, are particularly important because they cannot be synthesized in adequate amounts by human cells. These are called **essential nutrients** because it is essential that they be provided in the diet.

Carbohydrates

Carbohydrates are organic compounds, such as sugars and starch, that are used primarily to supply energy for cellular processes.

Sources of Carbohydrates

Carbohydrates are ingested in a variety of forms, including starch from grains and certain vegetables; glycogen from meats and seafoods; disaccharides from cane sugar, beet sugar, and molasses; and simple sugars from honey and various fruits. During digestion, the more complex carbohydrates, such as starch and glycogen, are changed to monosaccharides—a form that can be absorbed and transported to body cells.

An exception to this digestive processing is the polysaccharide called *cellulose,* which is abundant in plant foods. Although cellulose is composed of glucose units arranged in chains, somewhat like plant starch, cellulose cannot be broken down by human digestive enzymes; consequently, it passes through the alimentary canal largely unchanged. However, the presence of cellulose in foods is important to the function of the digestive system. It provides bulk (sometimes called fiber or roughage) in the digestive tube against which the muscular wall of the tube can push. Thus, cellulose facilitates the movement of food substances through the digestive tract. (Other carbohydrates that provide fiber include hemicellulose, pectin, and lignin.)

Utilization of Carbohydrates

The monosaccharides that are absorbed from the digestive tract include *fructose, galactose,* and *glucose.* Fructose and galactose are normally converted into glucose by the enzymes of the liver (figure 15.1), and glucose is the form of carbohydrate that is most commonly oxidized by the cells as fuel.

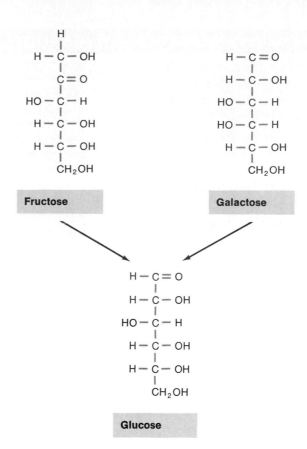

Figure 15.1 The monosaccharides fructose and galactose are converted into glucose by the liver.

If glucose is present in excessive amounts, some of it is converted to *glycogen* (glycogenesis) and stored as a glucose reserve in the liver and muscles. Glucose can be mobilized rapidly from glycogen (glycogenolysis) when it is needed to supply energy. However, only a certain amount of glycogen can be stored, and excess glucose is usually converted into fat and stored in adipose tissue (figure 15.2).

Although most carbohydrates are used to supply cellular energy, some are used by cells to synthesize vital substances. These vital substances include the 5-carbon sugars *ribose* and *deoxyribose,* which are needed for the production of the nucleic acids RNA and DNA, as well as the disaccharide *lactose* (milk sugar), which is synthesized when the breasts are actively secreting milk.

Many cells can also obtain energy by oxidizing fatty acids. However, some cells, such as neurons, are dependent on a continuous supply of glucose for survival. Even a temporary decrease in the glucose supply may result in a serious functional disorder of the nervous system or in the death of nerve cells. Consequently, the presence of a minimum amount of carbohydrates in the body is essential. If an adequate supply of carbohydrates is not received in foods, the liver may convert some noncarbohydrates, such as amino acids from proteins or glycerol from fats, into glucose—a process called *gluconeogenesis* (figure 15.3).

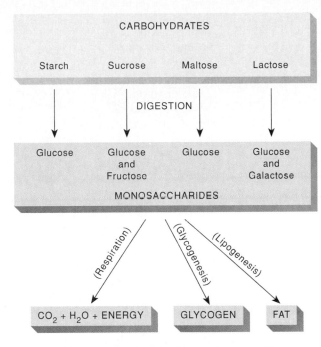

Figure 15.2 Monosaccharides from foods are used for energy, stored as glycogen, or converted to fat.

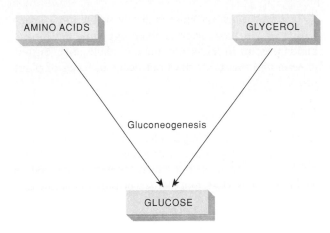

Figure 15.3 Whenever the supply of glucose is inadequate, liver cells may convert noncarbohydrates, such as amino acids, into glucose.

Thus, the need for glucose has physiological priority over the need to synthesize certain other substances, such as proteins, from the available amino acids.

It is estimated that about 100 grams of glycogen can be stored in an adult liver and that another 200 grams can be stored in various muscle tissues. This amount of glycogen is sufficient to meet the energy demands of body cells for about twelve hours when the body is at rest.

Humans survive on widely varying amounts of carbohydrates, as a result of differences in the carbohydrate content of available or preferred foods. In the United States, for example, it is estimated that a typical adult diet supplies about 50% of total body energy in the form of carbohydrates. Since carbohydrate foods are generally inexpensive, this percentage is likely to be higher among members of lower economic groups. However, it is possible to maintain adequate nutrition with relatively high or relatively low carbohydrate intake, provided the needs for all essential nutrients are met. Furthermore, poor nutrition among people with limited economic resources is more likely to be related to low intake of essential amino acids, vitamins, and minerals than to an excessive use of carbohydrate foods.

1. List several common sources of carbohydrates.
2. In what form are carbohydrates utilized as a cellular fuel?
3. Explain what happens to excessive glucose in the body.
4. Name two uses of carbohydrates other than supplying energy.
5. How does the body obtain glucose when its food supply of carbohydrate is insufficient?

Carbohydrate Requirements

Because carbohydrates provide the primary source of fuel for cellular processes, the need for carbohydrates varies with individual energy requirements. Therefore, persons who are physically active require more fuel than those who are sedentary. The minimal requirement for carbohydrates in the human diet is unknown. It is estimated, however, that an intake of at least 100–125 grams daily is necessary to spare protein (that is, to avoid an excessive breakdown of protein) and to avoid metabolic disorders that sometimes accompany excessive utilization of fats. Persons in the United States typically include 200–300 grams of carbohydrates in their daily diets.

1. Why do the daily requirements for carbohydrates vary from person to person?
2. What is the daily minimum requirement of carbohydrates?

Lipids

Lipids comprise the group of organic compounds that includes fats, oils, and fatlike substances (chapter 2). They are used to supply energy for cellular processes and to build structures, such as cell membranes. Although lipids include fats, phospholipids, and cholesterol, the most common dietary lipids are the fats called *triglycerides,* such as those described in chapter 2. (See figure 2.19.)

Sources of Lipids

Triglycerides are contained in foods of both plant and animal origin. They occur in meats, eggs, milk, and lard, as well as in various nuts and in plant oils, such as corn oil, peanut oil, and olive oil.

Cholesterol is obtained in relatively high concentrations from foods such as liver, egg yolk, and brain, and is present in lesser amounts in whole milk, butter, cheese, and meats. It does not occur in foods of plant origin.

Utilization of Lipids

During digestion, triglycerides are broken down into fatty acids and glycerol. After being absorbed, these products are transported by the lymph and blood to various tissues. The metabolism of these substances is controlled mainly by the liver and adipose tissue.

As explained in chapter 4, the liver can convert fatty acids from one form to another. However, it cannot synthesize certain of the fatty acids, such as the one called *linoleic acid*. This **essential fatty acid** is needed for the synthesis of phospholipids, which, in turn, are necessary for the formation of cell membranes and the transport of circulating lipids. Good sources of linoleic acid include corn oil, cottonseed oil, and soy oil. Other essential fatty acids are *linolenic acid* and *arachidonic acid*.

The liver uses free fatty acids to synthesize triglycerides, phospholipids, and lipoproteins that may then be released into the blood. Thus, the liver is largely responsible for the control of circulating lipids (figure 15.4). In addition, it regulates the total amount of cholesterol in the body by synthesizing cholesterol and releasing it into the blood, or by removing cholesterol from the blood and excreting it into the bile. The liver also uses cholesterol in the production of bile salts.

Cholesterol is not used as an energy source, but it does provide structural material for a variety of cell parts, including cell membranes, and it furnishes molecular components for the synthesis of various sex hormones and hormones produced by the adrenal cortex.

Excessive triglycerides are stored in adipose tissue, and if the blood lipid concentration drops (in response to fasting, for example), some of these triglycerides are hydrolyzed into free fatty acids and glycerol, and then released into the blood.

Lipid Requirements

As with carbohydrates, the lipid content of human diets varies widely. One person's diet may provide 10% of the body's energy in the form of fats, while another person's provides 40%.

The amounts and types of fats needed for health are unknown. However, linoleic acid is an essential fatty acid, and to prevent deficiency conditions from developing, nutritionists recommend that infants receive formulas in which 3% of the energy intake is in the form of linoleic acid. Fatty acid deficiencies have not been observed in adults, so it has been concluded that a typ-

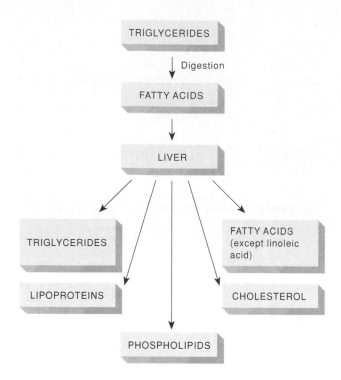

Figure 15.4 Fatty acids are used by the liver to synthesize a variety of lipids.

ical adult diet consisting of a variety of foods provides an adequate supply of this essential nutrient.

Because fats contain fat-soluble vitamins, the intake of fats must also be sufficient to supply the needed amounts of these vitamins.

> A report from the Surgeon General of the United States recommends that to improve health most people should reduce their intake of saturated fats by selecting leaner meats and fish for their meals and by using low-fat dairy products. The report also suggests that people should increase their intake of fruits, vegetables, and whole grains.
>
> The American Heart Association recommends that fats make up no more than 30% of the calories in a person's diet.

1. What foods commonly supply lipids?
2. Which fatty acid is an essential nutrient?
3. What is the role of the liver in the utilization of lipids?
4. What is the function of cholesterol?

Proteins

Proteins are organic compounds that serve as structural materials in cells, act as enzymes that regulate metabolic reactions, and may be used to supply energy. (See chapter 2.)

Dietary Fat and Colon Cancer

Experiments with laboratory animals and studies comparing cancer patients with healthy persons indicate that some food substances may promote the development of certain cancers. Conversely, there is evidence that other food substances may inhibit the growth of some cancers.

The food substance most clearly associated with cancer formation is *dietary fat,* which has been linked to cancer of the colon (as well as to cancer of the breasts, prostate gland, and ovaries).

Although its possible cancer-causing action is not well understood, some investigators believe that dietary fat may promote colon cancer by increasing the secretion of bile salts (chapter 14). These bile salts may then be converted by intestinal bacteria into substances that, in turn, can stimulate cancer to develop in the tissues of the colon.

On the other hand, *dietary fiber* may inhibit the appearance of colon cancers. Dietary fiber consists of complex carbohydrates, including cellulose, hemicellulose, pectin, and lignin. These carbohydrates comprise parts of plant cell walls, and they cannot be broken down by human digestive enzymes. When plant foods with high fiber content, such as cereal and grain products, are included in the diet, the fiber absorbs water and adds bulk to the intestinal contents, thereby diluting the concentration of bile salts. At the same time, the presence of the fiber facilitates the movement of substances through the intestine. More rapid movement of the intestinal contents is thought to reduce the time that the tissues of the colon are exposed to the potential cancer-causing action of bile salts or other cancer-promoting agents and thus reduces the chances of cancer development.

Sources of Proteins

Foods rich in proteins include meats, fish, poultry, cheese, nuts, milk, eggs, and cereals. Various legumes, such as beans and peas, contain lesser amounts of protein.

During digestion, proteins are broken down into their component amino acids, and these smaller molecules are absorbed by intestinal tissues and transported to the body cells in the blood.

As mentioned in chapter 4, the human body can synthesize many amino acids (nonessential amino acids). However, eight amino acids needed by the adult body (ten needed by growing children) cannot be synthesized, or are not synthesized in adequate amounts. They are called **essential amino acids** because it is essential that they be present in the diet.

> Plant proteins typically contain less than adequate amounts of one or more of the essential amino acids. However, protein-containing plant foods can be combined in a meal so that one supplies the amino acids that another lacks. For example, beans are deficient in the essential amino acid *methionine,* but they contain adequate amounts of *lysine.* Rice is deficient in lysine but contains adequate amounts of methionine. Thus, if beans and rice are eaten together, they complement one another, and the meal provides a suitable combination of essential amino acids.

All the essential amino acids must be present in the body at the same time for growth and tissue repair to occur. In other words, if one essential molecule is missing, the process of protein synthesis cannot take place. Since essential amino acids are not stored, those

CHART 15.1	Amino acids found in foods
Alanine	Leucine (e)
Arginine (ch)	Lysine (e)
Asparagine	Methionine (e)
Aspartic acid	Phenylalanine (e)
Cysteine	Proline
Glutamic acid	Serine
Glutamine	Threonine (e)
Glycine	Tryptophan (e)
Histidine (ch)	Tyrosine
Hydroxyproline	Valine (e)
Isoleucine (e)	

Eight essential amino acids (e) cannot be synthesized by human cells and must be provided in the diet. Two additional amino acids (ch) are essential in growing children.

that are present and are not used in protein synthesis are oxidized as energy sources or are converted into carbohydrates or fats.

Chart 15.1 lists the amino acids found in foods and indicates those that are essential.

On the basis of the kinds of amino acids that they provide, proteins can be classified as complete or incomplete. The **complete proteins,** which include those available in milk, meats, fish, poultry, and eggs, contain adequate amounts of the essential amino acids to maintain body tissues, and promote normal growth and development. **Incomplete proteins,** such as *zein* in corn, which lacks adequate amounts of the essential amino acids tryptophan and lysine, and *gelatin,* which lacks adequate tryptophan, are unable by themselves to support tissue maintenance or normal growth and development.

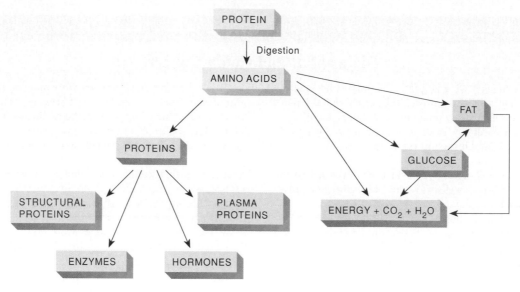

Figure 15.5 Amino acids resulting from the digestion of proteins are used to synthesize nonessential amino acids and proteins, and to supply energy.

A protein called *gliadin,* which occurs in wheat, is an example of a *partially complete protein.* It is deficient in the essential amino acid lysine, and although it does not contain enough lysine to promote growth, it contains enough to maintain life.

1. What foods provide rich sources of protein?
2. Why are some amino acids called essential?
3. Distinguish between a complete protein and an incomplete protein.

Utilization of Amino Acids

As discussed in chapter 4, amino acids are used by body cells in a variety of ways. For example, some amino acids are incorporated into protein molecules that provide cellular structure, as in the *actin* and *myosin* of muscle fibers, the *collagen* of connective tissue fibers, and the *keratin* of the skin. Other amino acids are used to synthesize proteins that function in various body processes: hemoglobin, which transports oxygen; plasma proteins, which help regulate water balance and control pH; enzymes, which catalyze chemical reactions; and certain hormones, which help regulate metabolic activities.

Amino acids also represent potential sources of energy. When they are present in excess or when supplies of carbohydrates and fats are insufficient to provide needed energy, amino acids are oxidized by various metabolic pathways (figure 15.5).

Body cells use carbohydrates preferentially as an energy source, thus a diet that supplies an adequate amount of carbohydrate is said to have a *protein-sparing effect.* In other words, when sufficient carbohydrates are present, proteins remain available for tissue building and repair rather than being converted into carbohydrates for use as an energy source.

A relatively high proportion of tissue proteins may be used as energy sources during times of starvation. These molecules are drawn from structural parts of cells, and consequently, such utilization of structural proteins is accompanied by the wasting of tissues.

Nitrogen Balance

In a healthy adult, proteins are continuously being built up and broken down. Although these processes occur at different rates in different tissues, the overall gain of body proteins equals the loss, so that a state of *dynamic equilibrium* exists. Since proteins contain a relatively high percentage of nitrogen, dynamic equilibrium is also characterized by a **nitrogen balance**—a condition in which the amount of nitrogen taken in is equal to the amount excreted.

A person who is starving, however, will have a *negative nitrogen balance* because the amount of nitrogen excreted as a result of amino acid oxidation will exceed the amount replaced by the diet. A growing

CHART 15.2 Protein, lipid, and carbohydrate nutrients

Nutrient	Sources	Calories per gram	Utilization	Adult daily requirement
Protein	Meats, fish, poultry, cheese, nuts, milk, eggs, cereals, legumes	4.1	Production of protein molecules used to build cell structure and to function as enzymes or hormones; used in the transport of oxygen, regulation of water balance, control ot pH; amino acids may be broken down and oxidized for energy or converted to carbohydrates or fats for storage	0.8 grams per kilogram of body weight; diet must include the essential amino acids
Lipid	Meats, eggs, milk, lard, plant oils	9.5	Oxidized for energy; production of triglycerides, phospholipids, lipoproteins, and cholesterol, stored in adipose tissue; glycerol portions of fat molecules may be used to synthesize glucose	Amount and type needed for health unknown; diet must supply an adequate amount of essential fatty acids and fat-soluble vitamins
Carbohydrate	Primarily from starch and sugars in foods of plant origin, and from glycogen in meats and seafoods	4.1	Oxidized for energy; used in production of ribose, deoxyribose, and lactose; stored in liver and muscles as glycogen; converted to fats and stored in adipose tissue	Varies with individual energy needs; at least 100 grams per day to prevent excessive breakdown of proteins

child, a pregnant woman, or an athlete in training is likely to have a *positive nitrogen balance,* since the amount of protein being built into new tissue probably exceeds the amount being used for energy.

Protein Requirements

In addition to supplying essential amino acids, proteins are needed to provide molecular parts and nitrogen for the synthesis of nonessential amino acids and various nonprotein nitrogenous substances. Consequently, the amount of protein required by individuals varies according to body size, metabolic rate, and nitrogen balance condition.

For an average adult, nutritionists recommend a daily protein intake of about 0.8 grams per kilogram (0.4 grams per pound) of body weight. For a pregnant woman, who needs to maintain a positive nitrogen balance, the recommendation is increased by an additional 30 grams of protein per day. Similarly, a nursing mother requires an additional 20 grams of protein per day to maintain a high level of milk production.

The consequences of protein deficiencies are severe, particularly in growing children. As mentioned previously, a decreased protein intake is likely to result in a negative nitrogen balance and tissue wasting. It may also be accompanied by a decrease in the level of plasma proteins, and, if this occurs, the osmotic pressure of the blood decreases. As a result of this osmotic change, fluids collect excessively in the tissues, producing a condition called *nutritional edema.*

Chart 15.2 summarizes the sources, uses, and requirements of protein, lipid, and carbohydrate nutrients.

> Protein deficiencies in pregnant women are likely to result in anemia, miscarriages, or premature births. Children whose diets are inadequate in proteins tend to develop a condition called *kwashiorkor,* which is characterized by nutritional edema and failure to grow (figure 15.6). The nervous systems of such children may also fail to develop, so they may be mentally retarded.

1. What are the physiological functions of proteins?
2. What is meant by a negative nitrogen balance? A positive nitrogen balance?
3. How can edema result from inadequate nutrition?

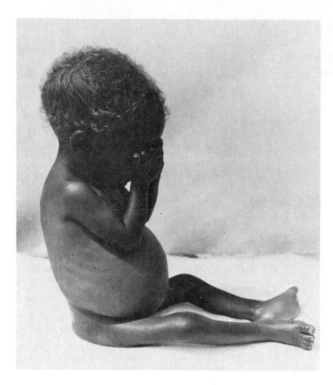

Figure 15.6 How can the condition known as kwashiorkor be prevented?

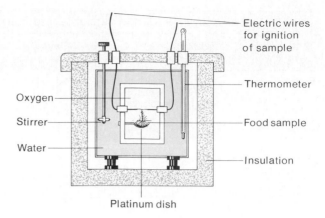

Figure 15.7 A bomb calorimeter can be used to measure the caloric content of a food sample.

Energy Expenditures

Energy can be obtained from carbohydrates, fats, and proteins. Since it is required for all metabolic processes, the supply of energy is of prime importance to cell survival. If the diet is deficient in energy-supplying materials, structural molecules may gradually be consumed, and if this continues, death may result.

Energy Values of Foods

The amount of potential energy contained in a food can be expressed as **calories,** which are units of heat.

Although a calorie is defined as the amount of heat needed to raise the temperature of a gram of water by 1 degree Celsius (°C), the calorie used to measure food energy is 1,000 times greater. This *large calorie* (Cal.) is equal to the amount of heat needed to raise the temperature of a kilogram (1,000 grams) of water by 1°C (actually from 14.5°C to 15.5°C). This unit is also called a *kilocalorie,* but it is customary in nutritional studies to refer to it simply as a calorie.

The caloric contents of various foods can be determined using an instrument called a *bomb calorimeter,* shown in figure 15.7, that consists of a metal chamber submerged in a known volume of water. A food sample is dried, weighed, and placed inside the chamber. The chamber is filled with oxygen gas and submerged in the water. Then, the food is ignited and allowed to oxidize completely. As heat is released from the food, it causes the temperature of the surrounding water to rise, and the change in temperature is measured. Since the volume of the water is known, the amount of heat released from the food can be calculated in calories.

Caloric values determined this way are usually somewhat higher than the amount of energy actually released by metabolic oxidation, because nutrients generally are not completely absorbed from the digestive tract. Also, molecules of amino acids in proteins are not completely oxidized in the body. Portions of these molecules, which still contain some energy, are excreted in urea or are transformed into other nitrogenous substances. When such losses are taken into account, cellular oxidation yields on the average about 4.1 calories from 1 gram of carbohydrates, 4.1 calories from 1 gram of protein, and 9.5 calories from 1 gram of fat.

1. What term is used to designate the potential energy in a food substance?
2. How can the energy value of a food be determined?
3. What is the energy value of a gram of carbohydrate? A gram of protein? A gram of fat?

Energy Requirements

The amount of energy required to support metabolic activities for twenty-four hours varies from person to person. The factors that influence energy needs include the individual's basal metabolic rate, degree of muscular activity, body temperature, and rate of growth.

The **basal metabolic rate** (BMR) is a measurement of the rate at which the body expends energy under *basal conditions*—the conditions that exist when a person is awake and at rest, after an overnight fast, and in a comfortable, controlled environment. The apparatus used to determine this rate measures the amount of oxygen taken in by a resting person as well as the amount of carbon dioxide given off during a certain time period.

The amount of oxygen consumed by the body is directly proportional to the amount of energy released by cellular respiration. The BMR, therefore, reveals the total amount of energy expended in a given time period to support the activities of such organs as the brain, heart, lungs, liver, and kidneys.

Although the average adult basal metabolic rate indicates a need for approximately one calorie of energy per hour for each kilogram of body weight, this need varies with such factors as sex, body size, body temperature, and level of endocrine gland activity. For example, since heat loss is directly proportional to the body surface area, and a smaller person has a relatively larger surface area, such a person will have a higher BMR. Males tend to have higher metabolic rates than females. As body temperature increases, the BMR increases, and as the blood level of thyroxine or epinephrine increases, so does the BMR.

Maintaining the basal metabolic rate usually requires the body's greatest expenditure of energy. The energy required to support voluntary muscular activity comes next, though this amount varies greatly with the type of activity (chart 15.3). For example, energy needed to maintain posture while sitting at a desk might amount to 100 Cal. per hour above the basal need, while running or swimming might require 500–600 Cal. per hour.

The maintenance of body temperature may require additional energy expenditure, particularly in cold weather. In this case, extra energy may be released by involuntary muscular contractions, such as shivering, or through voluntary muscular actions, such as walking.

As mentioned in the discussion of nitrogen balance, growing children and pregnant women have special needs for nutrients because their bodies are actively engaged in producing new tissues. They also require increased caloric intake, since tissue building uses energy.

Energy Balance

A state of **energy balance** exists when the caloric intake in the form of foods is equal to the caloric output resulting from the basal metabolic rate, muscular activities, and so forth. Under these conditions, the body weight remains constant, except perhaps for slight variations due to changes in water content.

If, on the other hand, the caloric intake exceeds the output, a *positive energy balance* occurs, and the excessive nutrients are likely to be stored in the tissues. At the same time, the body weight increases, since an excess of 3,500 Cal. can be stored as a pound of fat. Conversely, if the caloric output exceeds the input, the energy balance is *negative,* and stored materials are mobilized from the tissues for oxidation, causing body weight loss. Chart 15.4 gives the number of Cal. recommended for maintaining energy balance according to height, weight, and age.

CHART 15.3	Calories used during various activities
Kinds of activity	Calories (per hour)
Walking up stairs	1,100
Running (jogging)	570
Swimming	500
Vigorous exercise	450
Slow walking	200
Dressing and undressing	118
Sitting at rest	100

CHART 15.4	Recommended daily dietary allowances of calories for maintenance of good nutrition					
	Years (from–to)	Weight (kg)	Weight (lb)	Height (cm)	Height (in)	Energy (kcal/day)
Infants	0–6 mos	6	13	60	24	320
	6 mos–1 yr	9	20	71	28	500
Children	1–3	13	29	90	35	740
	4–6	20	44	112	44	950
	7–10	28	62	132	52	1,130
Males	11–14	45	99	157	62	1,440
	15–18	66	145	176	69	1,760
	19–24	72	160	177	70	1,780
	25–50	79	174	176	70	1,800
	51+	77	170	173	68	1,530
Females	11–14	46	101	157	62	1,310
	15–18	55	120	163	64	1,370
	19–24	58	128	164	65	1,350
	25–50	63	138	163	64	1,380
	51+	65	143	160	63	1,280
Pregnant						+300
Lactating						+500

Reprinted with permission from *Recommended Dietary Allowances,* 10th edition, © 1989 by the National Academy of Sciences. Published by National Academy Press, Washington, DC.

1. What is meant by basal metabolic rate?
2. What factors influence the BMR?
3. What is meant by *energy balance?*

Desirable Weight

The most obvious and common nutritional disorders involve caloric imbalances. Persons in underdeveloped countries, for example, are often undernourished, and obesity is rare. Conversely, in highly developed countries, such as the United States, people are likely to be overnourished and overweight.

It is difficult to determine a desirable body weight. In the past, weight standards were based on *average weights* and *heights* within a certain population, and the degrees of underweight and overweight were expressed as percentage deviations from these averages. These standards reflected the gradual gain in weight that usually occurs in members of the United States population as they grow older. Later, it was recognized

CHART 15.5   Desirable weights for women 25 years of age and over*
Weight in pounds according to frame (in indoor clothing and two-inch heels)

Height Feet	Inches	Small frame	Medium frame	Large frame
4	10	92–98	96–107	104–119
4	11	94–101	98–110	106–122
5	0	96–104	101–113	109–125
5	1	99–107	104–116	112–128
5	2	102–110	107–119	115–131
5	3	105–113	110–122	118–134
5	4	108–116	113–126	121–138
5	5	111–119	116–130	125–142
5	6	114–123	120–135	129–146
5	7	118–127	124–139	133–150
5	8	122–131	128–143	137–154
5	9	126–135	132–147	141–158
5	10	130–140	136–151	145–163
5	11	134–144	140–155	149–168
6	0	138–148	144–159	153–173

*For women between 18 and 25, subtract 1 pound for each year under 25.
Source: Courtesy Metropolitan Life Insurance Company.

CHART 15.6   Desirable weights for men 25 years of age and over
Weight in pounds according to frame (in indoor clothing and one-inch heels)

Height Feet	Inches	Small frame	Medium frame	Large frame
5	2	112–120	118–129	126–141
5	3	115–123	121–133	129–144
5	4	118–126	124–136	132–148
5	5	121–129	127–139	135–152
5	6	124–133	130–143	138–156
5	7	128–137	134–147	142–161
5	8	132–141	138–152	147–166
5	9	136–145	142–156	151–170
5	10	140–150	146–160	155–174
5	11	144–154	150–165	159–179
6	0	148–158	154–170	164–184
6	1	152–162	158–175	168–189
6	2	156–167	162–180	173–194
6	3	160–171	167–185	178–199
6	4	164–175	172–190	182–204

Source: Courtesy Metropolitan Life Insurance Company.

that such an increase in weight after the age of twenty-five to thirty years is not necessary and may not be conducive to health. Consequently, standards of *desirable weights* were prepared, as shown in charts 15.5 and 15.6.

These figures were based on the assumption that the average body weights for persons twenty-five to thirty years of age are desirable weights for all older persons, and it was recommended that each individual maintain this desirable weight throughout life.

More recently, height-weight guidelines have been prepared that are based upon the characteristics of people who live the longest. The weights in charts 15.7 and 15.8 tend to be somewhat more lenient than those in the desirable weight charts.

Using desirable weight as the standard, *over-weight* can be defined as the condition in which a person exceeds the desirable weight by 10% to 20%. A person who exceeds this standard by more than 20% is usually *obese*, though **obesity** is more correctly defined as the

CHART 15.7 Height-weight guidelines for adult women based on longevity
Weight in pounds according to frame (in 3 pounds of clothes and shoes with one-inch heels)

Height Feet	Inches	Small frame	Medium frame	Large frame
4	10	102–111	109–121	118–131
4	11	103–113	111–123	120–134
5	0	104–115	113–126	122–137
5	1	106–118	115–129	125–140
5	2	108–121	118–132	128–143
5	3	111–124	121–135	131–147
5	4	114–127	124–138	134–151
5	5	117–130	127–141	137–155
5	6	120–133	130–144	140–159
5	7	123–136	133–147	143–163
5	8	126–139	136–150	146–167
5	9	129–142	139–153	149–170
5	10	132–145	142–156	152–173
5	11	135–148	145–159	155–176
6	0	138–151	148–162	158–179

Source: Courtesy Metropolitan Life Insurance Company.

CHART 15.8 Height-weight guidelines for adult men based on longevity
Weight in pounds according to frame (in 5 pounds of clothes and shoes with one-inch heels)

Height Feet	Inches	Small frame	Medium frame	Large frame
5	2	128–134	131–141	138–150
5	3	130–136	133–143	140–153
5	4	132–138	135–145	142–156
5	5	134–140	137–148	144–160
5	6	136–142	139–151	146–164
5	7	138–145	142–154	149–168
5	8	140–148	145–157	152–172
5	9	142–151	148–160	155–176
5	10	144–154	151–163	158–180
5	11	146–157	154–166	161–184
6	0	149–160	157–170	164–188
6	1	152–164	160–174	168–192
6	2	155–168	164–178	172–197
6	3	158–172	167–182	176–202
6	4	162–176	171–187	181–207

Source: Courtesy Metropolitan Life Insurance Company.

Obesity

One of the more serious and growing nutritional problems in the United States is obesity—the accumulation of excessive body fat. Although the indirect causes of this condition are complex and may involve physiological as well as psychological and genetic factors, the direct cause is usually overeating. That is, the person has maintained a *positive energy balance* by taking in more nutrients than were needed to supply the body's energy requirements. The excessive nutrients were stored as fat, and, over a given time period, the person became obese.

Successful treatment of obesity involves maintaining a *negative energy balance*—taking in less food than is needed to supply the body's energy requirements. Since a pound of fat represents the storage of about 3,500 Cal., the person's diet must have a deficit of 3,500 Cal. in order for a pound of fat to be lost. Thus, if the daily intake provides a deficit of 500 Cal., one pound of fat should be lost in a week.

Some fad diets cause rapid reduction of weight by promoting the loss of body water. For example, certain weight-reducing drugs act as diuretics (chapter 13), which cause the kidneys to increase the production of urine; strenuous physical activity and exposure to heat are accompanied by increased water loss due to sweating. Also, as the body's store of glycogen is utilized in response to fasting, a quantity of water (obligatory water) is lost. Consequently, a person entering some weight reduction programs may lose weight rapidly at first. Unfortunately for the obese person, such a weight loss has little or no effect on stored fat.

Obesity is associated with a number of adverse health conditions. These include heart disease, high blood pressure, strokes, noninsulin-dependent diabetes mellitus, damage to weight-bearing bones and associated joints, arthritis, gallstones, and certain types of cancer. Obese persons also tend to have shortened life spans.

(a)

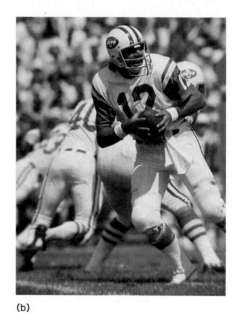

(b)

Figure 15.8 (*a*) An obese person is overweight and has excessive adipose tissue. (*b*) An athlete may be overweight due to muscular hypertrophy, but would not be considered obese.

presence of excessive adipose tissue. In other words, being overweight and being obese are not the same. For example, as figure 15.8 shows, an athlete or a person whose work involves heavy muscular activity may be overweight, but not obese. Conversely, a sedentary person may be excessively fat, and therefore obese, but not overweight.

Thus, determining what body weight a particular person should maintain for optimal health is a complex problem, and, in fact, the average weights and heights for clinically healthy persons in the United States are not known.

1. What is meant by desirable weight?
2. Distinguish between overweight and obesity.

Vitamins

Vitamins are organic compounds (other than carbohydrates, lipids, and proteins) that must be present in small amounts for normal metabolic processes, but cannot be synthesized in adequate amounts by body cells. Thus, they are essential nutrients that must be supplied in foods.

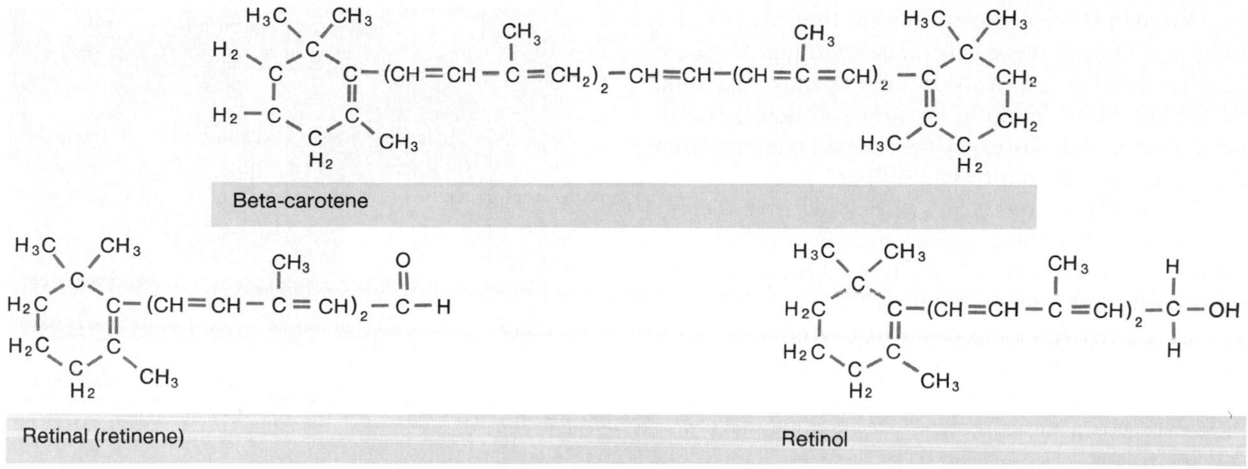

Beta-carotene

Retinal (retinene)

Retinol

Figure 15.9 A molecule of beta-carotene can be converted into two molecules of retinal, which in turn can be changed into retinol.

Vitamins can be classified on the basis of their solubilities, since some are soluble in fats (or fat solvents) and others are soluble in water. Those that are *fat-soluble* include vitamins A, D, E, and K; the *water-soluble* group includes the B vitamins and vitamin C.

Fat-Soluble Vitamins

Since the fat-soluble vitamins dissolve in fats, they occur in association with lipids and are influenced by the same factors that affect lipid absorption. For example, the presence of bile salts in the intestine promotes the absorption of these vitamins. As a group, the fat-soluble vitamins are stored in moderate quantities within various tissues, and because they are fairly resistant to the effects of heat, they are usually not destroyed by cooking or food processing.

1. What are vitamins?
2. Distinguish between fat-soluble vitamins and water-soluble vitamins.
3. How do bile salts affect the absorption of vitamins?

Vitamin A occurs in several forms, including retinol and retinal (retinene). This vitamin can be synthesized by body cells from a group of yellowish plant pigments called *carotenes*. Whenever vitamin A or its precursors are present in excess, they are stored mainly in the liver, which regulates the concentrations of these substances within the body (figure 15.9).

It is estimated that the liver of a healthy adult stores enough vitamin A to supply body needs for a year. Infants and children, on the other hand, usually lack such reserves, and consequently they are more likely to develop vitamin A deficiencies if their diets are inadequate.

Vitamin A is relatively stable to the effects of heat, acids, and alkalis, but it is readily destroyed by oxidation and is unstable in the presence of light.

Although the functions of vitamin A are not fully understood, it is known that retinal is used in the synthesis of *rhodopsin* (visual purple) in the rods of the retina. Vitamin A is thought to be required for the production of light-sensitive pigments in the cones as well. Thus, this vitamin is needed to maintain normal vision. It is also believed to function in the synthesis of mucoproteins and mucopolysaccharides of mucus, in the development of normal bones and teeth, and in the maintenance of epithelial cells in skin and mucous membranes.

Vitamin A as such is obtained only from foods of animal origin such as liver, fish, whole milk, butter, and eggs. However, its precursor, carotene, is widespread in leafy green vegetables and in yellow or orange vegetables and fruits.

A deficiency of vitamin A is characterized by a condition called *night blindness,* in which a person is unable to see normally in dim light. Also as a result of vitamin A deficiency, various epithelial tissues tend to undergo degenerative changes, and the body becomes more susceptible to infections by microorganisms.

An excessive intake of vitamin A may produce a serious condition called *hypervitaminosis A,* which is characterized by peeling skin, loss of hair, nausea, headache, and dizziness. In chronic cases, growth may be inhibited, and the bones and joints may undergo degenerative changes.

1. What substance is used by body cells to synthesize vitamin A?
2. What conditions will destroy vitamin A?
3. What foods are good sources of vitamin A?

Vitamin D is a group of steroids that have similar properties. One of these substances, vitamin D_3 (cholecalciferol), is found in foods such as milk, egg yolk, and fish liver oils. Vitamin D_2 (ergocalciferol) is produced commercially by exposing a steroid obtained from yeasts (ergosterol) to ultraviolet light.

Vitamin D can also be synthesized from dietary cholesterol that has been converted to provitamin D by intestinal enzymes, then stored in the skin and exposed to ultraviolet light. (See chapter 13.)

Like other fat-soluble vitamins, vitamin D is resistant to the effects of heat and to oxidation, acids, and alkalis. It is stored primarily in the liver and occurs to a lesser extent in the skin, brain, spleen, and bones.

As it is needed, vitamin D stored in the form of hydroxycholecalciferol is released into the blood. When parathyroid hormone is present, this form of vitamin D is converted in the kidneys into an active form of the vitamin (dihydroxycholecalciferol). This substance, in turn, is carried as a hormone in the blood to the intestine where it promotes the absorption of calcium and phosphorus, and, thus, helps ensure that adequate amounts of these minerals are available in the blood for tooth and bone formation and various metabolic processes.

Although natural foods are generally poor sources of vitamin D, it is often added to food during processing. For example, homogenized milk, nonfat milk, and evaporated milk are typically fortified with vitamin D. (Note: *Fortified* means essential nutrients have been added to a food in which they originally were absent or present in lesser amounts; *enriched* means essential nutrients have been at least partially replaced in a food that has lost nutrients as a result of processing.)

Figure 15.10 Rickets, which is characterized by failure of the bones and teeth to develop normally, is caused by a deficiency of vitamin D.

The presence of excessive amounts of vitamin D produces a toxic condition called *hypervitaminosis D,* which is characterized by diarrhea, nausea, and weight loss. Over prolonged periods, excessive intake of vitamin D can cause calcification of various soft tissues and irreversible damage to the kidneys.

In children, a deficiency of vitamin D results in the condition called *rickets,* in which the bones and teeth fail to develop normally (figure 15.10). In adults or in elderly persons who have a minimal exposure to sunlight, such a deficiency may be accompanied by *osteomalacia,* a condition in which the bones undergo decalcification and become weakened due to disturbances in calcium and phosphorus metabolism.

1. Where is vitamin D stored?
2. What are the functions of vitamin D?
3. What foods are good sources of vitamin D?

Vitamin E includes a group of compounds. The most active of these is *alphatocopherol.* This substance is resistant to the effects of heat, acids, and visible light, but is unstable in alkalis and in the presence of ultraviolet light or oxygen. In fact, it combines so readily with oxygen that its presence may prevent the oxidation of other compounds, and for this reason it is said to be an **antioxidant.**

Although vitamin E is found in all tissues, it is stored primarily in the muscles and adipose tissues. It also occurs in relatively high concentrations in the pituitary and adrenal glands.

The precise functions of vitamin E are unknown, but it apparently plays a role as an antioxidant by inhibiting the oxidation of vitamin A in the digestive tract and of polyunsaturated fatty acids in the tissues. It may also help maintain the stability of cell membranes.

While vitamin E is widely distributed among foods, its richest sources are oils from cereal seeds such as wheat germ oil. Other good sources are salad oils, margarine, shortenings, fruits, nuts, and vegetables. Since this vitamin is so easily obtained, deficiency conditions are rare.

1. Where is vitamin E stored?
2. What are the functions of vitamin E?
3. What foods are good sources of vitamin E?

CHART 15.9	Fat-soluble vitamins		
Vitamin	Characteristics	Functions	Sources
Vitamin A	Occurs in several forms; synthesized from carotenes; stored in liver; stable in heat, acids, and alkalis; unstable in light	Necessary for synthesis of visual pigments, mucoproteins, and mucopolysaccharides; for normal development of bones and teeth; and for maintenance of epithelial cells	Liver, fish, whole milk, butter, eggs, leafy green vegetables, yellow and orange vegetables and fruits
Vitamin D	A group of steroids; resistant to heat, oxidation, acids, and alkalis; stored in liver, skin, brain, spleen, and bones	Promotes absorption of calcium and phosphorus; promotes development of teeth and bones	Produced in skin exposed to ultraviolet light; in milk, egg yolk, fish liver oils, fortified foods
Vitamin E	A group of compounds; resistant to heat and visible light; unstable in presence of oxygen and ultraviolet light; stored in muscles and adipose tissue	An antioxidant; prevents oxidation of vitamin A and polyunsaturated fatty acids; may help maintain stability of cell membranes	Oils from cereal seeds, salad oils, margarine, shortenings, fruits, nuts, and vegetables
Vitamin K	Occurs in several forms; resistant to heat, but destroyed by acids, alkalis, and light; stored in liver	Needed for synthesis of prothrombin, which functions in blood clotting	Leafy green vegetables, egg yolk, pork liver, soy oil, tomatoes, cauliflower

Vitamin K, like the other fat-soluble vitamins, occurs in several chemical forms. One of these, vitamin K_1 (phylloquinone), is found in foods, while another, vitamin K_2, is produced by bacteria (*Escherichia coli*) that normally inhabit the human intestinal tract.

These vitamins are resistant to the effects of heat, but are destroyed by oxidation or by exposure to acids, alkalis, or light. They are stored to a limited degree in the liver.

Vitamin K functions primarily in the liver, where it is necessary for the formation of several proteins needed for blood clotting, including *prothrombin* (chapter 17). Consequently, a deficiency of vitamin K is characterized by prolonged blood clotting time and a tendency to hemorrhage.

The richest sources of vitamin K are leafy green vegetables. Other good sources are egg yolk, pork liver, soy oil, tomatoes, and cauliflower.

The fat-soluble vitamins and their properties are summarized in chart 15.9.

Newborns sometimes develop vitamin K deficiencies and experience abnormal bleeding (hemorrhagic disease of the newborn) because their intestines lack bacteria that can synthesize this vitamin. Similarly, adults who have been treated with antibiotic drugs may develop a vitamin K deficiency because such drugs often interfere with the activities of intestinal bacteria.

1. Where in the body is vitamin K synthesized?
2. What is the function of vitamin K?
3. What foods are good sources of vitamin K?

Water-Soluble Vitamins

The water-soluble vitamins include the B vitamins and vitamin C. The **B vitamins** consist of several compounds that are essential for normal cellular metabolism and are involved with the oxidation of carbohydrates, lipids, and proteins. Since the B vitamins often occur together in foods, they are usually referred to as a group called the *vitamin B complex.* Members of this group differ chemically and have unique functions.

The B-complex vitamins include the following substances:

1. **Thiamine, or vitamin B_1.** In its pure form, thiamine is a crystalline compound called thiamine hydrochloride. It is destroyed by exposure to heat and oxygen, especially in alkaline environments. (Its molecular structure is shown in figure 15.16.)

Thiamine functions as part of a coenzyme called *cocarboxylase,* which acts in the oxidation of carbohydrates. More specifically, thiamine is required for pyruvic acid to enter the citric acid cycle (chapter 4); in the absence of this vitamin, pyruvic acid accumulates in the blood. Thiamine also functions as a coenzyme in the synthesis of sugars, such as ribose, that, in turn, are used in the production of the nucleotides of nucleic acids.

Thiamine is absorbed primarily through the wall of the duodenum and is transported by the blood to body cells. Only small amounts are stored in the tissues, and excesses are excreted in the urine.

Since this vitamin functions in the oxidation of carbohydrates, the quantity required by the cells varies with the caloric intake. It is recommended that an adult diet contain 0.5 milligram (mg) of thiamine for every 1,000 calories ingested each day.

Good sources of thiamine are lean pork, other lean meats, fish, liver, eggs, whole-grain cereals, leafy green vegetables, and legumes.

> A mild deficiency of thiamine is characterized by loss of appetite, fatigue, and nausea, while a prolonged deficiency leads to a disease called *beriberi*. This condition may be accompanied by gastrointestinal disturbances, mental confusion, muscular weakness and paralysis, and enlargement of the heart. In severe cases, death may occur as a result of heart failure.
> In the United States, beriberi occurs mainly in chronic alcoholics who have substituted alcohol for needed foods. Moreover, since thiamine is required for the metabolic oxidation of alcohol, alcoholics are particularly likely to develop a thiamine deficiency.

2. **Riboflavin,** or **vitamin B$_2$.** Riboflavin is a yellowish brown crystalline substance that is relatively stable to the effects of heat, acids, and oxidation, but is destroyed by exposure to alkalis and ultraviolet light.

This vitamin functions as a part of several enzymes and coenzymes, such as FAD, that are known as *flavoproteins*. These substances are essential for the oxidation of glucose and fatty acids, and for cellular growth. The absorption of riboflavin seems to be regulated by an active transport system that controls the amount taken in by the intestinal mucosa. It is carried in the blood combined with blood proteins called *albumins*. Any excessive amounts of riboflavin in the blood are excreted in the urine, and any that remain unabsorbed in the intestine are lost in the feces.

The amount of riboflavin required by the body varies with the caloric intake, and it is estimated that 0.6 mg of riboflavin per 1,000 calories is sufficient to meet daily cellular needs.

Riboflavin is widely distributed in foods, and rich sources include meats and dairy products. Leafy green vegetables, whole-grain cereals, and enriched cereals provide lesser amounts.

Although vitamin B$_2$ is essential for growth, a deficiency does not cause a clearly defined set of symptoms.

3. **Niacin.** Niacin, which is also known as *nicotinic acid,* occurs in plant tissues, and is stable in the presence of heat, acids, and alkalis. After ingestion, it is converted to a physiologically active form called *niacinamide* (figure 15.11). Niacinamide is the form of niacin that is present in foods of animal origin.

Niacin functions as part of two coenzymes (coenzyme I, also called NAD, and coenzyme II, called NADP) that play essential roles in the oxidation of glucose and in the synthesis of proteins and fats (figure 15.12). These coenzymes are also needed for the synthesis of the sugars used in the production of nucleic acids.

Niacin can be readily absorbed from foods, and can be synthesized by human cells from the essential amino acid *tryptophan.* Consequently, the daily requirement for niacin varies with the intake of tryptophan.

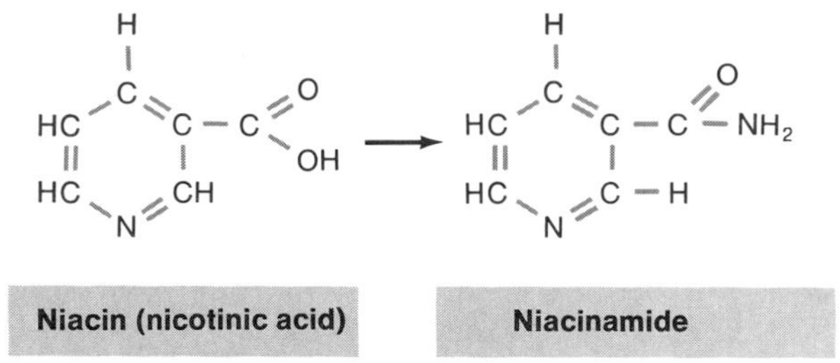

Niacin (nicotinic acid)   **Niacinamide**

Figure 15.11 Niacin from foods is converted into physiologically active niacinamide by body cells.

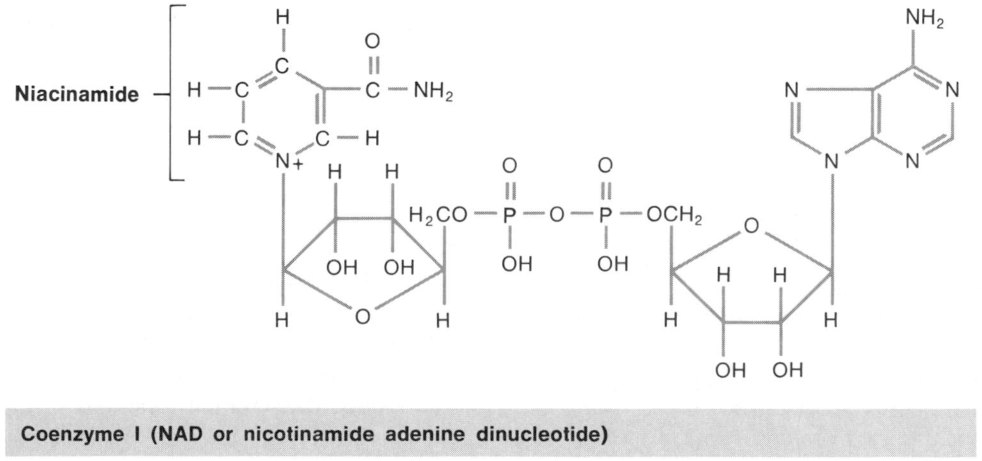

Coenzyme I (NAD or nicotinamide adenine dinucleotide)

Figure 15.12 Niacinamide is incorporated into molecules of coenzyme I.

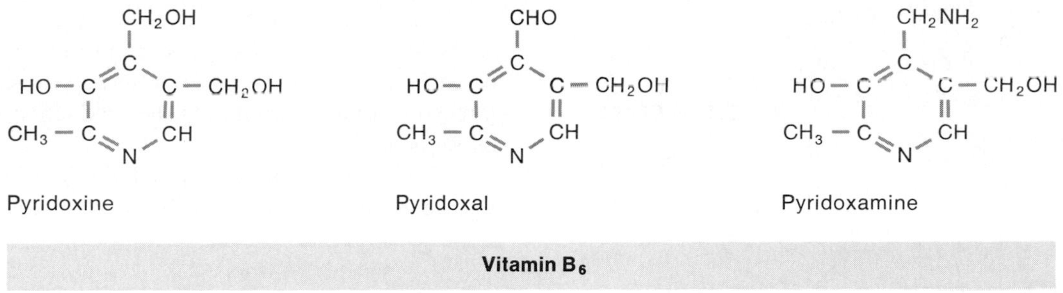

Pyridoxine

Pyridoxal

Pyridoxamine

Vitamin B₆

Figure 15.13 Vitamin B₆ consists of three closely related chemical compounds.

It is believed that 60 mg of tryptophan are equivalent to 1 mg of niacin, but this equivalence seems to vary from one individual to another. With these factors in mind, nutritionists recommend a daily niacin (or niacin equivalent) intake of 6.6 mg per 1,000 calories.

Rich sources of niacin (and tryptophan) include liver, lean meats, poultry, peanut butter, and legumes. Although milk is a poor source of niacin, it is a good source of tryptophan.

Historically, niacin deficiencies have been associated with diets consisting largely of corn and corn products, which are very low in niacin and lack adequate tryptophan. Such a deficiency causes a disease called *pellagra* that is characterized by dermatitis, inflammation of the digestive tract, diarrhea, and various mental disorders.

Although pellagra is relatively rare in the United States today, it was a serious problem in the rural South in the early 1900s. It sometimes occurs in chronic alcoholics who have substituted alcohol for needed foods over prolonged time periods.

4. **Vitamin B₆.** Vitamin B₆ is a group of three compounds that are closely related chemically, as figure 15.13 shows. They are called *pyridoxine, pyridoxal,* and *pyridoxamine.* These substances have similar actions and are fairly stable in the presence of heat and acids; but they are destroyed by oxidation, or by exposure to alkalis or ultraviolet light.

The vitamin B₆ compounds function as coenzymes that play essential roles in a wide variety of metabolic pathways, including those associated with the synthesis of proteins and various amino acids, the conversion of tryptophan to niacin, the production of antibodies, and the synthesis of nucleic acids.

Since the functions of vitamin B₆ are generally related to the metabolism of nitrogen-containing substances, the requirement for this vitamin varies with the protein content of the diet rather than with the caloric intake.

The recommended daily allowance of vitamin B₆ is 2.0 mg, but because this substance is so widespread in foods, deficiency conditions are quite rare.

Good sources of vitamin B₆ include liver, meats, fish, poultry, bananas, avocados, beans, peanuts, whole-grain cereals, and egg yolk.

5. **Pantothenic acid.** Pantothenic acid is a yellowish oil that is destroyed by heat, acids, and alkalis. It functions as part of a complex molecule called *coenzyme A,* which, in turn, reacts with intermediate products of carbohydrate and fat metabolism to become *acetyl coenzyme A.* Pantothenic acid is therefore essential to the energy-releasing mechanisms of cells.

A daily adult intake of 4–7 mg seems to be adequate. Most diets apparently provide sufficient amounts, since deficiencies are rare, and no clearly defined set of deficiency symptoms is known.

Good sources of pantothenic acid include meats, fish, whole-grain cereals, legumes, milk, fruits, and vegetables.

6. **Cyanocobalamin,** or **vitamin B₁₂.** Cyanocobalamin has a complex molecular structure that contains a single atom of the element *cobalt* (figure 15.14). In its pure form, this vitamin is red. It is stable to the effects of heat, but is inactivated by exposure to light, or by the presence of strong acids or alkalis.

The absorption of cyanocobalamin is regulated by the secretion of *intrinsic factor* from the parietal cells of the gastric glands, as explained in chapter 14. Intrinsic factor is thought to combine with cyanocobalamin and to facilitate its transfer through the epithelial lining of the small intestine and into the blood. Although the details of this transport mechanism are not well understood, it is known that calcium ions must be present for the process to take place.

Cyanocobalamin is stored in various tissues, particularly those of the liver, and an average adult body is thought to contain a reserve sufficient to supply cellular needs for three to five years.

This vitamin is essential for the functions of all cells. It serves as a part of coenzymes needed for the synthesis of nucleic acids and the metabolism of carbohydrates and fats. It also seems to play a role in the formation of myelin within the central nervous system.

Cyanocobalamin is found only in foods of animal origin; good sources include liver, meats, poultry, fish,

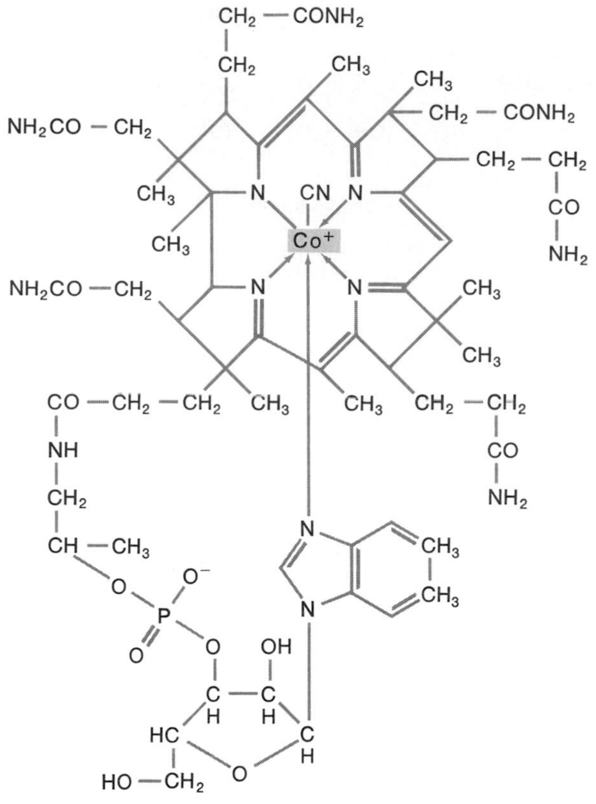

Vitamin B₁₂ (cyanocobalamin)

Figure 15.14 Vitamin B₁₂, which has the most complex molecular structure of the vitamins, contains cobalt (Co). What is the function of this vitamin?

milk, cheese, and eggs. Because persons in the United States consume relatively large amounts of such foods, a dietary lack of this vitamin seldom occurs, although strict vegetarians (vegans) may develop a deficiency.

A cyanocobalamin deficiency may develop in association with a disorder of the absorption mechanism. Specifically, the gastric glands of some individuals fail to secrete adequate amounts of intrinsic factor. As a result, the vitamin is poorly absorbed. This leads to a condition called *pernicious anemia* that is characterized by the presence of abnormally large red blood cells called *macrocytes*. These cells are produced when bone marrow cells fail to divide properly because of defective synthesis of DNA.

7. **Folacin,** or **folic acid.** Folacin is a yellow crystalline compound that exists in several forms. It is easily oxidized in an acid environment and is destroyed by heat in alkaline solutions; consequently, this vitamin may be lost in foods that are stored or cooked.

Folacin is readily absorbed from the digestive tract and is stored in the liver, where it is converted to a physiologically active substance called *folinic acid.*

Folinic acid functions as a coenzyme that is necessary for the metabolism of certain amino acids and for the synthesis of DNA. It also acts together with cyanocobalamin in promoting the production of normal red blood cells.

Good sources of folacin include liver, leafy green vegetables, whole-grain cereals, and legumes.

A deficiency of folacin leads to *megaloblastic anemia,* which is characterized by a reduced number of normal red blood cells and the presence of large, nucleated red cells.

8. **Biotin.** Biotin is a relatively simple compound that is stable to the effects of heat, acids, and light, but may be destroyed by oxidation or alkalis. (The molecular structure of biotin is shown in figure 15.16.)

Biotin functions as a necessary coenzyme in a number of metabolic pathways, including those involved with the metabolism of amino acids and fatty acids. It also plays a role in the synthesis of certain organic bases (purines) which are essential in the synthesis of nucleic acids.

Small quantities of biotin are stored in the tissues of metabolically active organs such as the brain, liver, and kidneys. It is believed to be manufactured by bacteria that inhabit the intestinal tract.

Biotin is widely distributed in foods, and dietary deficiencies are relatively rare. Good sources include liver, egg yolk, nuts, legumes, and mushrooms.

1. What substances make up the vitamin B complex?
2. What foods are good sources of vitamin B complex?
3. Which of the B-complex vitamins can be synthesized from tryptophan?
4. What is the general function of each member of the B complex?

Vitamin C, or **ascorbic acid.** Ascorbic acid is a crystalline compound that contains six carbon atoms. Chemically, it is closely related to the monosaccharides (figure 15.15). It is one of the least stable of the vitamins, in that it can be destroyed by oxidation, heat, light, or alkalis. It is fairly stable in acids.

The functions of ascorbic acid are poorly understood. However, it is known to be necessary for the production of the protein *collagen,* which is a vital part of various connective tissues such as bone, cartilage, and the fibrous connective tissue of skin. Ascorbic acid is needed in the conversion of folacin to folinic acid and in the metabolism of certain amino acids. It also promotes the absorption of iron and the synthesis of various hormones from cholesterol.

Although vitamin C is not stored in any great amount, tissues such as the adrenal cortex, pituitary gland, and intestinal glands contain relatively high

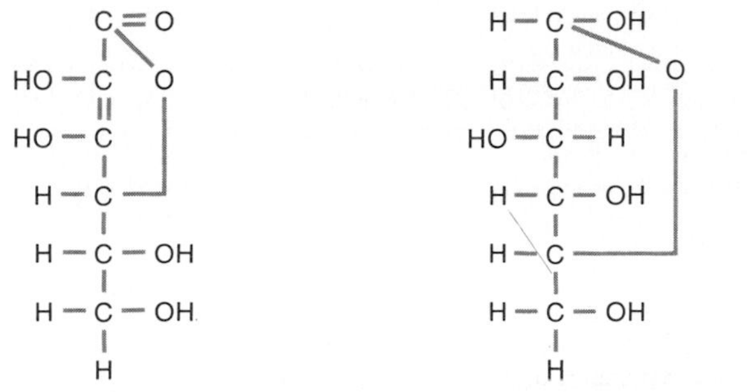

C = O
HO — C   O
 ‖
HO — C
H — C
H — C — OH
H — C — OH.
         H

H — C — OH
H — C — OH   O
HO — C — H
H — C — OH
H — C
H — C — OH
         H

**Vitamin C
(ascorbic acid)**

**Glucose
(a monosaccharide)**

Figure 15.15   Vitamin C is closely related chemically to the monosaccharides.

concentrations. Excess quantities are either excreted in the urine or oxidized.

There is considerable controversy as to how much ascorbic acid is required to maintain optimal health, and individual needs may vary. It is believed that 10 mg per day is sufficient to prevent deficiency symptoms and that 80 mg per day will saturate the tissues within a few weeks. Consequently, many nutritionists generally recommend a daily adult intake of 60 mg, which is thought to be enough to replenish normal losses and to provide a satisfactory level for cellular needs.

Ascorbic acid is fairly widespread in plant foods; particularly high concentrations are found in citrus fruits, citrus juices, tomatoes, and cabbage. Potatoes, leafy green vegetables, and fresh fruits are also good sources.

Chart 15.10 summarizes the water-soluble vitamins and their characteristics.

A prolonged deficiency of ascorbic acid leads to a disease called *scurvy* that occurs more frequently in infants and children than in adults. This condition is characterized by abnormal bone development and swollen, painful joints. There is also a tendency for cells to pull apart, and consequently, the gums may swell and bleed easily, resistance to infection is lowered, and wounds heal slowly.

1. What factors cause the destruction of vitamin C?
2. What is the function of vitamin C?
3. What foods are good sources of vitamin C?

Minerals

The nutrients that have been discussed—carbohydrates, lipids, proteins, and vitamins—are all organic substances. However, dietary **minerals** are inorganic elements that play essential roles in human metabo-

lism. These elements are usually extracted from the soil by plants. Humans, in turn, obtain them from plant foods or from animals that have eaten plants.

Characteristics of Minerals

Minerals are responsible for about 4% of the body weight and are most concentrated in the bones and teeth. In fact, the minerals *calcium* and *phosphorus,* which are very abundant in these tissues, account for nearly 75% of the body's minerals.

Minerals are usually incorporated into organic molecules. For example, phosphorus occurs in phospholipids, iron in hemoglobin, and iodine in thyroxine. However, some occur in inorganic compounds, such as the calcium phosphate of bone; others occur as free ions, such as the sodium, chloride, and calcium ions in the blood.

Minerals are present in all body cells, where they comprise parts of the structural materials. They also function as portions of enzyme molecules, help create the osmotic pressure of body fluids, and play vital roles in the conduction of nerve impulses, the contraction of muscle fibers, the coagulation of blood, and the maintenance of pH.

The concentrations of various minerals in body fluids are regulated by homeostatic mechanisms, which ensure that the excretion of these substances is balanced with the dietary intake. Thus, toxic excesses are avoided, while minerals that are present in limited amounts are conserved.

1. How do minerals differ from other nutrients?
2. What are the major functions of minerals?
3. What are the most abundant minerals in the body?

Major Minerals

Calcium and phosphorus account for nearly 75% of the mineral elements in the body; thus, they are **major minerals** (macrominerals). Other major minerals, each of which accounts for 0.05% or more of the body weight, include potassium, sulfur, sodium, chlorine, and magnesium. Descriptions of the major minerals follow:

1. **Calcium.** Calcium (Ca) is widely distributed in cells and body fluids, even though 99% of the body's supply occurs in the inorganic salts of the bones and teeth. It is essential for nerve impulse conduction, muscle fiber contraction, and blood coagulation. It also serves to decrease the permeability of cell membranes and to activate certain enzymes.

The amount of calcium that is absorbed varies with a number of factors. For example, the proportion of calcium absorbed increases as the body's need for calcium increases. Vitamin D and high protein intake promote calcium absorption; increased motility of the

CHART 15.10 Water-soluble vitamins

Vitamin	Characteristics	Functions	Sources
Thiamine (Vitamin B₁)	Destroyed by heat and oxygen, especially in alkaline environment	Part of coenzyme needed for oxidation of carbohydrates; coenzyme needed in synthesis of ribose	Lean meats, liver, eggs, whole-grain cereals, leafy green vegetables, legumes
Riboflavin (Vitamin B₂)	Stable to heat, acids, and oxidation; destroyed by alkalis and light	Part of enzymes and coenzymes, such as FAD, needed for oxidation of glucose and fatty acids and for cellular growth	Meats, dairy products, leafy green vegetables, whole-grain cereals
Niacin (Nicotinic Acid)	Stable to heat, acids, and alkalis; converted to niacinamide by cells, synthesized from tryptophan	Part of coenzymes NAD and NADP needed for oxidation of glucose and synthesis of proteins, fats, and nucleic acids	Liver, lean meats, poultry, peanuts, legumes
Vitamin B₆	Group of three compounds; stable to heat and acids; destroyed by oxidation, alkalis, and ultraviolet light	Coenzyme needed for synthesis of proteins and various amino acids, for conversion of tryptophan to niacin, for production of antibodies, and for synthesis of nucleic acids	Liver, meats, fish, poultry, bananas, avocados, beans, peanuts, whole-grain cereals, egg yolk
Pantothenic Acid	Destroyed by heat, acids, and alkalis	Part of coenzyme A needed for oxidation of carbohydrates and fats	Meats, fish, whole-grain cereals, legumes, milk, fruits, vegetables
Cyanocobalamin (Vitamin B₁₂)	Complex, cobalt-containing compound; stable to heat; inactivated by light, strong acids, and strong alkalis; absorption regulated by intrinsic factor from gastric glands; stored in liver	Part of coenzyme needed for synthesis of nucleic acids and for metabolism of carbohydrates; plays role in synthesis of myelin	Liver, meats, poultry, fish, milk, cheese, eggs
Folacin (Folic Acid)	Occurs in several forms; destroyed by oxidation in acid environment or by heat in alkaline environment; stored in liver where it is converted into folinic acid	Coenzyme needed for metabolism of certain amino acids and for synthesis of DNA; promotes production of normal red blood cells	Liver, leafy green vegetables, whole-grain cereals, legumes
Biotin	Stable to heat, acids, and light; destroyed by oxidation and alkalis	Coenzyme needed for metabolism of amino acids and fatty acids, and for synthesis of nucleic acids	Liver, egg yolk, nuts, legumes, mushrooms
Ascorbic Acid (Vitamin C)	Closely related to monosaccharides; stable in acids, but destroyed by oxidation, heat, light, and alkalis	Needed for production of collagen, conversion of folacin to folinic acid, and metabolism of certain amino acids; promotes absorption of iron and synthesis of hormones from cholesterol	Citrus fruits, citrus juices, tomatoes, cabbage, potatoes, leafy green vegetables, fresh fruits

digestive tract or an excessive intake of fats causes a decreased absorption. Consequently, the amount of calcium needed to provide an adequate cellular supply may vary from time to time. However, nutritionists believe a daily intake of 800 mg is sufficient to cover adult needs in spite of variations in absorption.

Unfortunately, only a few foods contain significant amounts of calcium. Milk and milk products and fish with bones, such as salmon or sardines, are the richest sources. Leafy green vegetables, such as mustard greens, turnip greens, and kale, are good sources.

Since these vegetables are not very popular foods in the United States, it is difficult for most people to maintain an adequate intake of calcium unless milk or milk products, such as cheese, cottage cheese, and ice cream, are regularly incorporated into the diet.

A deficiency of calcium in children may be accompanied by stunted growth, misshapen bones, and enlarged wrists and ankles. In adults, such a deficiency may result in removal of calcium from the bones. As this occurs, the bones become thinner and more fragile, and may fracture easily.

2. **Phosphorus.** Phosphorus (P) is responsible for about 1% of the total body weight, and most of it is incorporated in the calcium phosphate of the bones and teeth. The remainder is distributed throughout the body cells, where it serves as a structural component and plays important roles in nearly all metabolic reactions. More specifically, phosphorus is a constituent of nucleic acids, many proteins, some enzymes, and some vitamins. It also occurs in the phospholipids of cell membranes, the energy-carrying molecules of ATP, and the phosphates of body fluids that function in the regulation of pH. (The molecular structure of ATP is shown in figure 4.7.)

The recommended daily adult intake of phosphorus is 800 mg, and since this mineral is abundant in protein foods, diets adequate in proteins are also adequate in phosphorus. In addition to being found in protein foods such as meats, poultry, fish, cheese, and nuts, phosphorus occurs in whole-grain cereals, milk, and legumes.

1. What are the functions of calcium?
2. What are the functions of phosphorus?
3. What foods are good sources of calcium and phosphorus?

3. **Potassium.** Potassium (K) is widely distributed throughout the body and tends to be concentrated inside cells rather than in extracellular fluids. On the other hand, sodium, which has similar chemical properties, tends to be concentrated outside the cells. More specifically, the ratio of potassium to sodium within a cell is 10:1, while the ratio outside the cell is 1:28.

Within cells, potassium helps maintain the intercellular osmotic pressure and regulate pH. It promotes reactions involved in carbohydrate and protein metabolism, and plays a vital role in the membrane polarization that occurs in nerve impulse conduction and muscle fiber contraction.

Nutritionists recommend a daily adult intake of 2.5 grams (2,500 mg) of potassium, and because this mineral is widely distributed in foods, it is estimated that a typical adult diet provides between 2 and 6 grams each day. Although deficiencies of potassium due to an inadequate diet are relatively rare, they may occur for other reasons. For example, when a person has diarrhea, the intestinal contents may pass through the digestive tract so rapidly that potassium absorption is greatly reduced. Vomiting or using diuretic drugs may also lead to potassium depletion. The consequences of such losses may include muscular weakness, cardiac abnormalities, and edema.

Foods rich in potassium are avocados, dried apricots, meats, milk, peanut butter, potatoes, and bananas. Citrus fruits, citrus juices, apples, carrots, and tomatoes provide lesser amounts.

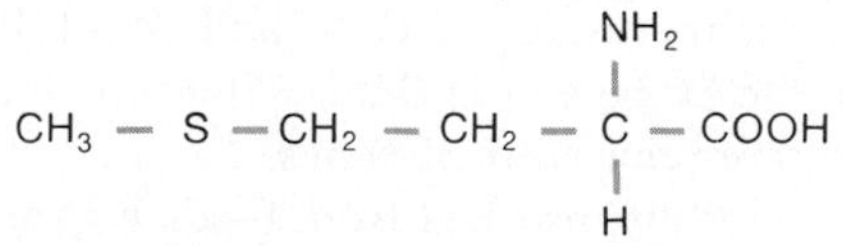

Methionine

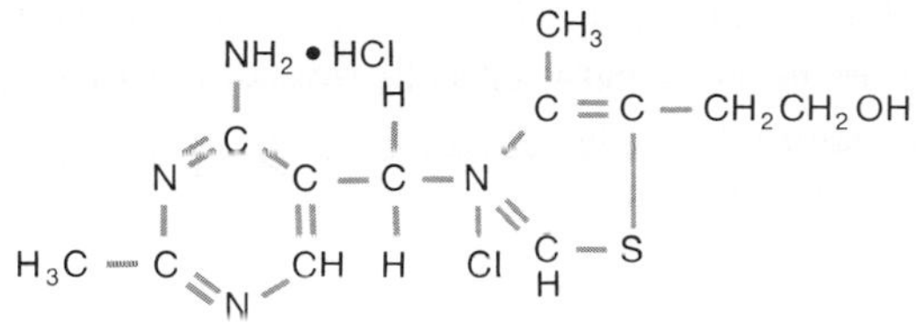

Thiamine hydrochloride (vitamin B₁)

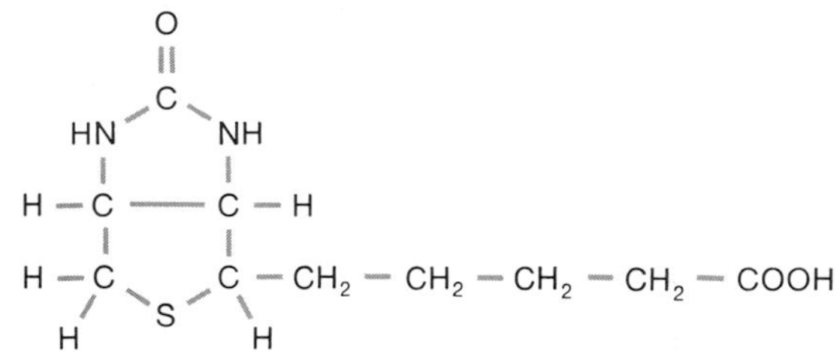

Biotin

Figure 15.16 Three examples of essential sulfur-containing nutrients.

1. How is potassium distributed in the body?
2. What is the function of potassium?
3. What foods are good sources of potassium?

4. **Sulfur.** Sulfur (S) is responsible for about 0.25% of the body weight and is widely distributed through the tissues. It is particularly abundant in the skin, hair, and nails. Most sulfur is incorporated in molecules of the amino acids *methionine* and *cysteine*. Other sulfur-containing compounds include thiamine, insulin, and biotin (figure 15.16). In addition, sulfur is a constituent of mucopolysaccharides that are found in cartilage, tendons, and bones, and of sulfolipids that occur in the liver, kidneys, salivary glands, and brain.

No daily requirement for sulfur has been established. It is thought, however, that a diet providing adequate amounts of protein will also meet the body's need for sulfur. Good food sources of this mineral include meats, milk, eggs, and legumes.

5. **Sodium.** It is estimated that about 0.15% of the adult body weight is due to sodium (Na), which is widely distributed throughout the body. Only about 10%

of this mineral occurs inside the cells, and about 40% is found within the extracellular fluids. The remainder is bound to the inorganic salts of bones.

Sodium is readily absorbed from foods by active transport, and the blood concentration of this element is regulated by the kidneys under the influence of the adrenal cortical hormone *aldosterone.*

Sodium helps maintain the osmotic concentrations of extracellular fluids and thus serves to regulate the water balance between cells and their surroundings. Its presence is necessary for nerve impulse conduction and for muscle fiber contraction, and it plays a role in the movement of substances, such as chloride ions, through cell membranes (chapter 20).

Although a daily requirement for sodium has not been established, it is believed that the usual human diet provides more than enough to meet the body's needs. On the other hand, sodium may be lost as a result of diarrhea, vomiting, kidney disorders, sweating, or using diuretics. Such losses may cause a variety of symptoms including edema, muscular cramps, and convulsions.

The amount of sodium naturally present in foods varies greatly, and it is commonly added to foods in the form of table salt (sodium chloride). In some geographic regions, there are significant concentrations of sodium in the drinking water.

Foods that have relatively high sodium contents include cured ham, sauerkraut, cheese, and graham crackers.

1. In what compounds and tissues of the body is sulfur found?
2. What hormone regulates the blood concentration of sodium?
3. What are the functions of sodium?

6. **Chlorine.** Chlorine (Cl) in the form of chloride ions is closely associated with sodium. Like sodium, it is widely distributed throughout the body, although it is most highly concentrated in cerebrospinal fluid and in gastric juice.

Together with sodium, chlorine helps to maintain the osmotic concentrations of extracellular fluids, regulate pH, and maintain electrolyte balance. It is also essential for the formation of hydrochloric acid in gastric juice, and it functions in the transport of carbon dioxide by red blood cells.

Chlorine and sodium are usually obtained together in the form of table salt (sodium chloride), and as in the case of sodium, an ordinary diet usually provides considerably more chlorine than the body needs. It too can be lost excessively as a result of vomiting, diarrhea, kidney disorders, sweating, or using diuretics.

7. **Magnesium.** Magnesium (Mg) is responsible for about 0.05% of the body weight and is found in all cells. It is particularly abundant in bones in the form of phosphates and carbonates.

This mineral functions in a number of metabolic reactions including many that occur within mitochondria and are associated with the production of ATP. It also plays a role in the conversion of ATP to ADP; therefore, it is important in providing energy for cellular processes.

When the intake of magnesium is high, a smaller percentage is absorbed from the intestinal tract, and when the intake is low, a larger percentage is absorbed. Absorption increases as protein intake increases, and decreases as calcium and vitamin D intake increase. A reserve supply of magnesium is stored in bone tissue, and excess amounts are excreted in the urine under the influence of *aldosterone.*

The recommended daily allowance of magnesium is 300 mg for females and 350 mg for males. A typical diet usually provides only about 120 mg of magnesium for every 1,000 calories, thus it may barely meet the body's needs.

Good sources of magnesium include milk and dairy products (except butter), legumes, nuts, and leafy green vegetables.

The major minerals are summarized in chart 15.11.

1. Where are chloride ions most highly concentrated in the body?
2. Where is magnesium stored?
3. What factors influence the absorption of magnesium from the intestinal tract?

Trace Elements

Trace elements (microminerals) are essential minerals that occur in minute amounts, each making up less than 0.005% of the adult body weight. They include iron, manganese, copper, iodine, cobalt, and zinc.

Iron (Fe) is found primarily in the blood, although some is present in all body cells, and a reserve supply is stored in the liver, spleen, and bone marrow. It functions as part of the *hemoglobin molecules* in red blood cells and enables these molecules to carry oxygen (figure 15.17). Iron also occurs in *myoglobin molecules,* which are synthesized in muscle cells and act to store oxygen temporarily. In addition, iron functions to catalyze the formation of vitamin A, is incorporated into a number of enzymes, and is included in the cytochrome molecules that function in the synthesis of ATP (chapter 4).

It has been estimated that an adult male requires from 0.7 to 1 mg of iron daily, while a female needs 1.2 to 2 mg. It is also believed that although a typical diet supplies about 10 to 18 mg of iron each day, only 2% to 10% of this iron is absorbed. (See chapter 17.) Such a diet provides a narrow margin of safety as far as iron

CHART 15.11 Major minerals

Mineral	Distribution	Functions	Sources
Calcium (Ca)	Mostly in the inorganic salts of bones and teeth	Structure of bones and teeth; essential for nerve impulse conduction, muscle fiber contraction, and blood coagulation; increases permeability of cell membranes; activates certain enzymes	Milk, milk products, leafy green vegetables
Phosphorus (P)	Mostly in the inorganic salts of bones and teeth	Structure of bones and teeth; component in nearly all metabolic reactions; constituent of nucleic acids, many proteins, some enzymes, and some vitamins; occurs in cell membrane, ATP, and phosphates of body fluids	Meats, poultry, fish, cheese, nuts, whole-grain cereals, milk, legumes
Potassium (K)	Widely distributed; tends to be concentrated inside cells	Helps maintain intracellular osmotic pressure and regulate pH; promotes metabolism; needed for nerve impulse conduction and muscle fiber contraction	Avocados, dried apricots, meats, nuts, potatoes, bananas
Sulfur (S)	Widely distributed	Essential part of various amino acids, thiamine, insulin, biotin, and mucopolysaccharides	Meats, milk, eggs, legumes
Sodium (Na)	Widely distributed; large proportion occurs in extracellular fluids and bonded to inorganic salts of bone	Helps maintain osmotic pressure of extracellular fluids and regulate water balance; needed for conduction of nerve impulses and contraction of muscle fibers; aids in regulation of pH and in transport of substances across cell membranes	Table salt, cured ham, sauerkraut, cheese, graham crackers
Chlorine (Cl)	Closely associated with sodium; most highly concentrated in cerebrospinal fluid and gastric juice	Helps maintain osmotic pressure of extracellular fluids, regulate pH, and maintain electrolyte balance; essential in formation of hydrochloric acid; aids transport of carbon dioxide by red blood cells	Same as for sodium
Magnesium (Mg)	Abundant in bones	Needed in metabolic reactions that occur in mitochondria and are associated with the production of ATP; plays role in conversion of ATP to ADP	Milk, dairy products, legumes, nuts, leafy green vegetables

intake is concerned, and it may fail to provide the amount needed by some individuals.

> Pregnant women require additional quantities of iron in order to support the formation of a placenta, and the growth and development of a fetus. Iron is also needed for the synthesis of hemoglobin in the fetus as well as in the mother, whose blood volume increases during pregnancy.

Liver is the only really rich source of dietary iron, and since liver is not a very popular food, iron is one of the more difficult nutrients to obtain from natural sources in adequate amounts. Foods that contain some iron include lean meats; dried apricots, raisins, and prunes; enriched whole-grain cereals; legumes; and molasses.

Manganese (Mn) is most concentrated in the liver, kidneys, and pancreas. It is necessary for normal growth and development of skeletal structures and other connective tissues. Manganese occurs in enzymes that are essential for the synthesis of fatty acids and cholesterol, for the formation of urea, and for the normal functions of the nervous system.

The daily requirement for manganese is unknown. The richest sources include nuts, legumes, and whole-grain cereals; leafy green vegetables and fruits are good sources.

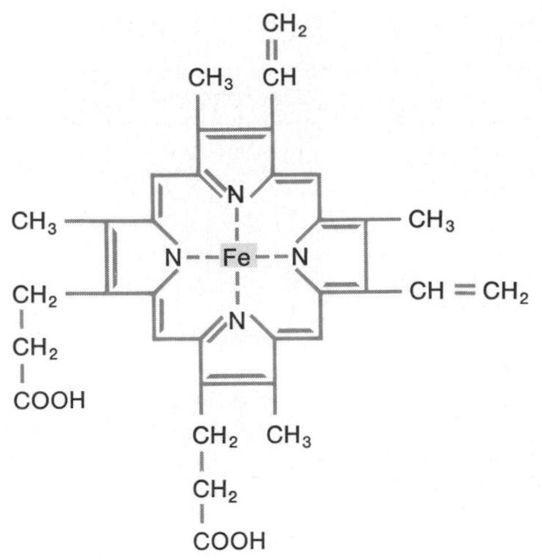

Heme

Figure 15.17 A hemoglobin molecule contains four heme portions, each of which contains a single iron atom (Fe) that can combine with oxygen.

1. What is the primary function of iron?
2. Why does the usual diet provide only a narrow margin of safety in supplying iron?
3. How is manganese utilized?
4. What foods are good sources of manganese?

Copper (Cu) is found in all body tissues, but is most highly concentrated in the liver, heart, and brain. It is essential for the synthesis of hemoglobin, the normal development of bone, the production of melanin, and the formation of myelin within the nervous system.

It is believed that a daily intake of 2 mg of copper is sufficient to provide the needs of body cells, and that a typical adult diet supplies about 2–5 mg of this mineral. Adults seldom develop copper deficiencies.

Foods rich in copper include liver, oysters, crabmeat, nuts, whole-grain cereals, and legumes.

Iodine (I) occurs in minute quantities in all tissues, but is highly concentrated within the thyroid gland. Its only known function is to provide an essential component for the synthesis of *thyroid hormones*. (The molecular structures of two of these hormones, thyroxine and triiodothyronine, are shown in figure 13.19.)

A daily intake of 1 microgram (.001 mg) of iodine per kilogram of body weight is thought to provide an adequate amount for most adults. Since the iodine content of foods varies with the iodine content of soils in different geographic regions, it is sometimes necessary to artificially increase the iodine in foods to prevent deficiencies from developing. The use of iodized table salt, which contains added *potassium iodide* (KI), has proved to be an effective means of accomplishing this.

Cobalt (Co) is widely distributed throughout the body, since it is an essential part of *cyanocobalamin molecules* (vitamin B_{12}). It is also thought to be necessary for the synthesis of several important enzymes.

The amount of cobalt required in the daily diet is unknown. This mineral is found in a great variety of foods, and the quantity present in the average diet is apparently sufficient to meet the body's need. Good sources of cobalt include liver, lean meats, poultry, fish, and milk.

Zinc (Zn) generally occurs in body tissues, but is most concentrated in the liver, kidneys, and brain. It is a constituent of a large number of enzymes involved in digestion, respiration, bone metabolism, and liver metabolism. It is also necessary for normal wound healing and for maintaining the integrity of the skin.

The daily requirement for zinc is believed to be about 15 mg, and most diets are thought to provide 10–15 mg. Since only a portion of this amount may be absorbed, zinc deficiencies may be relatively common.

The richest sources of zinc are seafoods and meats; cereals, legumes, nuts, and vegetables provide lesser amounts.

Characteristics of these trace elements are summarized in chart 15.12.

1. How is copper used?
2. What is the function of iodine?
3. Why are zinc deficiencies thought to be relatively common?

Adequate Diets

An adequate diet provides sufficient *energy* (calories), *essential fatty acids, essential amino acids, vitamins,* and *minerals* to support optimal growth and to maintain and repair body tissues. However, because individual needs for nutrients vary greatly with age, sex, growth rate, amount of physical activity, and level of stress, as well as with genetic and environmental factors, it is not possible to design a diet that is adequate for everyone.

On the other hand, nutrients are so widely distributed in foods that satisfactory amounts and combinations of essential substances can usually be obtained in spite of individual food preferences that are related to cultural backgrounds, life-styles, and emotional attitudes. Chart 15.13 lists the recommended daily allowance for each of the major nutrients.

Malnutrition

Malnutrition is poor nutrition that results from a lack of essential nutrients or a failure to use available foods to best advantage. It may result from *undernutrition*

CHART 15.12   Trace elements

Trace element	Distribution	Functions	Sources
Iron (Fe)	Primarily in blood; stored in liver, spleen, and bone marrow	Part of hemoglobin molecule; catalyzes formation of vitamin A; incorporated into a number of enzymes	Liver, lean meats, dried apricots, raisins, enriched whole-grain cereals, legumes, molasses
Manganese (Mn)	Most concentrated in liver, kidneys, and pancreas	Occurs in enzymes needed for synthesis of fatty acids and cholesterol, formation of urea, and normal functioning of the nervous system	Nuts, legumes, whole-grain cereals, leafy green vegetables, fruits
Copper (Cu)	Most highly concentrated in liver, heart, and brain	Essential for synthesis of hemoglobin, development of bone, production of melanin, and formation of myelin	Liver, oysters, crabmeat, nuts, whole-grain cereals, legumes
Iodine (I)	Concentrated in thyroid gland	Essential component for synthesis of thyroid hormones	Food content varies with soil content in different geographic regions; iodized table salt
Cobalt (Co)	Widely distributed	Component of cyanocobalamin; needed for synthesis of several enzymes	Liver, lean meats, poultry, fish, milk
Zinc (Zn)	Most concentrated in liver, kidneys, and brain	Constituent of several enzymes involved in digestion, respiration, bone metabolism, liver metabolism; necessary for normal wound healing and maintaining integrity of the skin	Seafoods, meats, cereals, legumes, nuts, vegetables

and produce the symptoms of deficiency diseases, or it may be due to *overnutrition* arising from an excessive intake of nutrients.

The factors leading to malnutrition are varied. For example, a deficiency condition may stem from lack of availability or poor quality of food. On the other hand, malnutrition may result from excessive caloric intake or from excessive intake of vitamin supplements, which causes hypervitaminosis. Malnutrition from causes that involve the diet alone is called *primary malnutrition.*

Secondary malnutrition occurs when a person's individual characteristics make a normally adequate diet insufficient. For example, a person who secretes low quantities of bile salts may develop a deficiency of fat-soluble vitamins because the presence of bile salts promotes the absorption of these substances. Likewise, severe and prolonged emotional stress may lead to secondary malnutrition, because stress can cause changes in hormonal concentrations, and such changes may result in abnormal breakdown of amino acids or in excessive excretion of various nutrients.

1. What is meant by an adequate diet?
2. What factors influence individual needs for nutrients?
3. What is meant by primary malnutrition? By secondary malnutrition?

In response to emotional problems and an intense fear of becoming overweight, teenagers (particularly girls) sometimes develop a disorder called *anorexia nervosa*. In this condition, which also occurs in adults, though less commonly, the person severely restricts food intake and may exercise excessively. As a result, 25% or more of the body weight is lost. The person also may engage in episodes of secretive, unrestrained eating (bulimia) followed by self-induced vomiting and the use of laxatives and diuretics (binge-purge behavior). Symptoms of anorexia nervosa that accompany the malnutrition associated with the chronic starvation include cessation of menstruation, decreased heart rate, inability to maintain normal body temperature, impaired judgment, and hallucinations. In 3% or more of the cases, anorexics die suddenly, usually as a result of electrolyte imbalances and cardiac disorders.

Clinical Terms Related to Nutrients and Nutrition

anorexia (an″o-rek′se-ah) A loss of appetite.
cachexia (kah-kek′se-ah) A state of chronic malnutrition and physical wasting.
casein (ka′se-in) The primary protein found in milk.
celiac disease (se′le-ak dĭ-zēz′) A digestive disorder characterized by the inability to digest or use fats and carbohydrates.

CHART 15.13 Summary of Recommended Dietary Allowances (RDA), 1989*

Age (years)	kg	lb.	cm	in.	Protein (g)	(µg RE) Vitamin A	(µg) Vitamin D	(mg α –TE) Vitamin E	(µg) Vitamin K	(mg) Vitamin C	(mg) Thiamin	(mg) Riboflavin	(mg NE) Niacin	(mg) Vitamin B$_6$	(µg) Folate	(µg) Vitamin B$_{12}$	(mg) Calcium	(mg) Phosphorus	(mg) Magnesium	(mg) Iron	(mg) Zinc	(µg) Iodine	(µg) Selenium
Infants																							
0.0–0.5	6	13	60	24	13	375	7.5	3	5	30	0.3	0.4	5	0.3	25	0.3	400	300	40	6	5	40	10
0.5–1.0	9	20	71	28	14	375	10	4	10	35	0.4	0.5	6	0.6	35	0.5	600	500	60	10	5	50	15
Children																							
1–3	13	29	90	35	16	400	10	6	15	40	0.7	0.8	9	1.0	50	0.7	800	800	80	10	10	70	20
4–6	20	44	112	44	24	500	10	7	20	45	0.9	1.1	12	1.1	75	1.0	800	800	120	10	10	90	20
7–10	28	62	132	52	28	700	10	7	30	45	1.0	1.2	13	1.4	100	1.4	800	800	170	10	10	120	30
Males																							
11–14	45	99	157	62	45	1,000	10	10	45	50	1.3	1.5	17	1.7	150	2.0	1,200	1,200	270	12	15	150	40
15–18	66	145	176	69	59	1,000	10	10	65	60	1.5	1.8	20	2.0	200	2.0	1,200	1,200	400	12	15	150	50
19–24	72	160	177	70	58	1,000	10	10	70	60	1.5	1.7	19	2.0	200	2.0	1,200	1,200	350	10	15	150	70
25–50	79	174	176	70	63	1,000	5	10	80	60	1.5	1.7	19	2.0	200	2.0	800	800	350	10	15	150	70
51+	77	170	173	68	63	1,000	5	10	80	60	1.2	1.4	15	2.0	200	2.0	800	800	350	10	15	150	70
Females																							
11–14	46	101	157	62	46	800	10	8	45	50	1.1	1.3	15	1.4	150	2.0	1,200	1,200	280	15	12	150	45
15–18	55	120	163	64	44	800	10	8	55	60	1.1	1.3	15	1.5	180	2.0	1,200	1,200	300	15	12	150	50
19–24	58	128	164	65	46	800	10	8	60	60	1.1	1.3	15	1.6	180	2.0	1,200	1,200	280	15	12	150	55
25–50	63	138	163	64	50	800	5	8	65	60	1.1	1.3	15	1.6	180	2.0	800	800	280	15	12	150	55
51+	65	143	160	63	50	800	5	8	65	60	1.0	1.2	13	1.6	180	2.0	800	800	280	10	12	150	55
Pregnant					60	800	10	10	65	70	1.5	1.6	17	2.2	400	2.2	1,200	1,200	300	30	15	175	65
Lactating																							
1st 6 mo					65	1,300	10	12	65	95	1.6	1.8	20	2.1	280	2.6	1,200	1,200	355	15	19	200	75
2nd 6 mo					62	1,200	10	11	65	90	1.6	1.7	20	2.1	260	2.6	1,200	1,200	340	15	16	200	75

Reprinted with permission from *Recommended Dietary Allowances*, 10th Edition, © 1989 by the National Academy of Sciences. Published by National Academy Press, Washington, DC.

emaciation (e-ma″se-a'shun) Excessive leanness of the body due to tissue wasting.

extrinsic factor (ek-strin'sik fak'tor) Vitamin B$_{12}$.

hyperalimentation (hi″per-al″ĭ men-ta'shun) Prolonged intravenous nutrition provided for patients with severe digestive disorders.

hypercalcemia (hi″per-kal-se'me-ah) An excessive level of calcium in the blood.

hypercalciuria (hi″per-kal-se-u're-ah) An excessive excretion of calcium in the urine.

hyperglycemia (hi″per-glis-e'me-ah) An excessive level of glucose in the blood.

hyperkalemia (hi″per-kah-le'me-ah) An excessive level of potassium in the blood.

hypernatremia (hi″per-nah-tre'me-ah) An excessive level of sodium in the blood.

hypoalbuminemia (hi″po-al-bu″mĭ-ne'me-ah) A low level of albumin in the blood.

hypoglycemia (hi″po-gli-se'me-ah) A low level of glucose in the blood.

hypokalemia (hi″po-kah-le'me-ah) A low level of potassium in the blood.

hyponatremia (hi″po-nah-tre'me-ah) A low level of sodium in the blood.

isocaloric (i″so-kah-lo'rik) Containing equal amounts of heat energy.

lipogenesis (lip″o-jen'ĕ-sis) The formation of fat.

nyctalopia (nik″tah-lo'pe-ah) Night blindness.

pica (pi'kah) A hunger for substances that are not suitable foods.

polyphagia (pol″e-fa'je-ah) Excessive intake of food.

proteinuria (pro″te-ĭ-nu're-ah) Presence of protein in the urine.

provitamin (pro-vi'tah-min) A substance used as a precursor in vitamin synthesis.

Food Selection

Although recommended daily allowances of essential nutrients are often used by dietitians as guides in planning adequate diets, these values are usually of limited use to the average person. A more useful aid is the idea of *basic food groups*.

A basic food group is a class of foods that will supply sufficient amounts of certain essential nutrients when a given quantity of food from the group is included in the diet. One basic food plan is based on four food groups, each of which provides a unique contribution toward achieving an adequate diet. In this plan, the food groups and the quantities of food required from each are as follows:

Group 1: Milk and Cheese Products. This group includes milk, cottage cheese, cream cheese, natural and processed cheese, and ice cream. It provides calcium, phosphorus, magnesium, protein, vitamin B_6, vitamin B_{12}, and riboflavin. It is recommended that an adult have two to four servings of these foods each day.

Group 2: Fruits and Vegetables. This group includes all fruits and vegetables, and it provides vitamin C, vitamin A, iron, magnesium, and vitamin B_6. It is recommended that an adult diet contain four servings from this group each day, and that one of these be citrus fruit or some other fruit or vegetable that is particularly high in vitamin C content. At least every other day, one of the servings should be a dark green, yellow, or orange vegetable.

Group 3: Meat, Poultry, and Fish. This group includes all meats, poultry, and fish as well as meat substitutes such as eggs, beans, peas, and nuts. It provides protein, vitamin A, thiamine, niacin, vitamin B_6, vitamin B_{12}, riboflavin, phosphorus, magnesium, and iron. For an adult, two servings from this group are recommended each day.

Group 4: Breads and Cereals. This group includes all whole grain or enriched breads and cereals. It provides iron, thiamine, niacin, riboflavin, protein, phosphorus, and magnesium. Four servings per day from this group are recommended for adults.

Chapter Summary

Introduction (page 550)

Nutrients include carbohydrates, lipids, proteins, vitamins, and minerals. The ways nutrients are used to support life processes constitute metabolism. Essential nutrients are needed for health and cannot be synthesized by body cells.

Carbohydrates (page 550)

Carbohydrates are organic compounds that are used primarily to supply cellular energy.

1. Sources of carbohydrates
   a. Carbohydrates are ingested in a variety of forms.
   b. Starch, glycogen, disaccharides, and monosaccharides are carbohydrates.
   c. Cellulose is a polysaccharide that cannot be digested by human enzymes, but is important in providing bulk that facilitates the movement of intestinal contents.
2. Utilization of carbohydrates
   a. Carbohydrates are absorbed as monosaccharides.
   b. Fructose and galactose are converted to glucose by the liver.
   c. Energy is released from glucose by oxidation.
   d. Excessive glucose is stored as glycogen or converted to fat.
3. Carbohydrate requirements
   a. Most carbohydrates are used to supply energy, although some are used to produce important sugars.
   b. Some cells depend on a continuous supply of glucose to survive.

   c. If inadequate amounts of glucose are available, amino acids may be converted to glucose.
   d. Humans survive with a wide range of carbohydrate intakes.
   e. Poor nutritional status is usually related to low intake of nutrients other than carbohydrates.

Lipids (page 551)

Lipids are organic compounds that supply energy and are used to build cell structures. They include fats, phospholipids, and cholesterol.

1. Sources of lipids
   a. Triglycerides are obtained from foods of plant and animal origins.
   b. Cholesterol is obtained in foods of animal origin only.
2. Utilization of lipids
   a. Metabolism of triglycerides is controlled mainly by the liver and adipose tissue.
   b. The liver can alter the molecular structures of fatty acids.
   c. Linoleic acid is an essential fatty acid.
   d. The liver also regulates the amount of cholesterol by synthesizing or excreting it.
3. Lipid requirements
   a. Humans survive with a wide range of lipid intakes.
   b. The amounts and types of lipids needed for health are unknown.
   c. Some fats contain fat-soluble vitamins, and the intake of fats must be sufficient to supply these essential nutrients.

Proteins (page 552)

Proteins are organic compounds that serve as structural materials, act as enzymes, and provide energy.

1. Sources of proteins
   a. Proteins are obtained mainly from meats, dairy products, cereals, and legumes.
   b. During digestion, proteins are broken down into amino acids.
   c. Eight amino acids are essential for adults, while ten are essential for growing children.
   d. All essential amino acids must be present at the same time in order for growth and repair of tissues to take place.
   e. Complete proteins contain adequate amounts of all the essential amino acids needed to maintain the tissues and promote growth.
   f. Incomplete proteins lack adequate amounts of one or more essential amino acids.
2. Utilization of amino acids
   a. Amino acids are incorporated into various structural and functional proteins, including enzymes.
   b. During starvation, tissue proteins may be used as energy sources; thus, the tissues waste away.
3. Nitrogen balance
   a. In healthy adults, the gain of protein equals the loss of protein, and a nitrogen balance exists.
   b. A starving person has a negative nitrogen balance; a growing child, a pregnant woman, or an athlete in training usually has a positive nitrogen balance.
4. Protein requirements
   a. Proteins and amino acids are needed to supply essential amino acids and nitrogen for the synthesis of various nitrogen-containing molecules.
   b. The consequences of protein deficiencies are particularly severe among growing children.

Energy Expenditures (page 556)

Energy is of prime importance to survival and may be obtained from carbohydrates, fats, or proteins.

1. Energy values of foods
   a. The potential energy values of foods are expressed in calories.
   b. When energy losses due to incomplete absorption and incomplete oxidation are taken into account, 1 gram of carbohydrate or 1 gram of protein yields about 4 calories, while 1 gram of fat yields about 9 calories.
2. Energy requirements
   a. The amount of energy required varies from person to person.
   b. Factors that influence energy requirements include basal metabolic rate, muscular activity, body temperature, and nitrogen balance.
3. Energy balance
   a. Energy balance exists when caloric intake equals caloric output.
   b. If energy balance is positive, body weight increases; if energy balance is negative, body weight decreases.
4. Desirable weight
   a. The most common nutritional disorders involve caloric imbalances.
   b. Average weights of persons 25–30 years of age are thought to be desirable for older persons as well.
   c. Recently, height-weight guidelines have been prepared based on longevity.
   d. A person who exceeds the desirable weight by 10%–20% is called overweight.
   e. A person whose body contains an excess of fatty tissue is said to be obese.

Vitamins (page 560)

Vitamins are organic compounds (other than carbohydrates, lipids, and proteins) that are essential for normal metabolic processes and cannot be synthesized by body cells in adequate amounts.

1. Fat-soluble vitamins
   a. General characteristics
      (1) Fat-soluble vitamins occur in association with lipids and are influenced by the same factors that affect lipid absorption.
      (2) They are fairly resistant to the effects of heat; thus, they are not destroyed by cooking or food processing.
   b. Vitamin A
      (1) Vitamin A occurs in several forms, is synthesized from carotenes, and is stored in the liver.
      (2) It functions in the production of pigments necessary for vision.
   c. Vitamin D
      (1) Vitamin D is represented by a group of related steroids.
      (2) It is found in certain foods and is produced commercially; it can also be synthesized in the skin.
      (3) When needed, vitamin D is converted by the kidneys to an active form that functions as a hormone and promotes the absorption of calcium and phosphorus by the intestine.
   d. Vitamin E
      (1) Vitamin E is represented by a group of compounds that are antioxidants.
      (2) It is stored in muscles and adipose tissue.
      (3) Its precise functions are unknown, but it seems to prevent oxidation of vitamin A and polyunsaturated fatty acids, and to stabilize cell membranes.

e. Vitamin K
   (1) Vitamin K_1 occurs in foods; vitamin K_2 is produced by certain bacteria that normally inhabit the intestinal tract.
   (2) It is stored to a limited degree in the liver.
   (3) It is used in the production of prothrombin that is needed for normal blood clotting.
2. Water-soluble vitamins
   a. General characteristics
      (1) Water-soluble vitamins include the B vitamins and vitamin C.
      (2) B vitamins make up a group called the vitamin B complex and are generally involved with the oxidation of carbohydrates, lipids, and proteins.
   b. Vitamin B complex
      (1) Thiamine
         (a) Thiamine functions as part of coenzymes that act in the oxidation of carbohydrates and in the synthesis of essential sugars.
         (b) Small amounts are stored in the tissues; excess is excreted in the urine.
         (c) Quantities needed vary with caloric intake.
      (2) Riboflavin
         (a) Riboflavin functions as part of several enzymes and coenzymes that are essential to the oxidation of glucose and fatty acids.
         (b) Its absorption is regulated by an active transport system; excess is excreted in the urine.
         (c) Quantities needed vary with caloric intake.
      (3) Niacin
         (a) Niacin functions as part of coenzymes needed for the oxidation of glucose and for the synthesis of proteins and fats.
         (b) It can be synthesized from tryptophan; daily requirement varies with the tryptophan intake.
      (4) Vitamin B_6
         (a) Vitamin B_6 is a group of compounds that function as coenzymes needed by a variety of metabolic pathways involved in the synthesis of proteins, various amino acids, antibodies, and nucleic acids.
         (b) Its requirement varies with protein intake.
      (5) Pantothenic acid
         (a) Pantothenic acid functions as part of coenzyme A; thus, it is essential for energy-releasing mechanisms.
         (b) Its daily requirement is not known.
      (6) Cyanocobalamin
         (a) The cyanocobalamin molecule contains cobalt.
         (b) Its absorption is regulated by the secretion of intrinsic factor from the gastric glands.
         (c) It functions as part of coenzymes needed for the synthesis of nucleic acids and for the metabolism of carbohydrates and fats.
      (7) Folacin
         (a) Folacin is converted by the liver to physiologically active folinic acid.
         (b) It functions as a coenzyme needed for the metabolism of certain amino acids, the synthesis of DNA, and the normal production of red blood cells.
      (8) Biotin
         (a) Biotin functions as a coenzyme needed for the metabolism of amino acids and fatty acids, and for the synthesis of nucleic acids.
         (b) It is stored in metabolically active organs.
   c. Ascorbic acid (vitamin C)
      (1) Vitamin C is closely related chemically to monosaccharides.
      (2) Its functions are poorly understood, but it is thought to be needed for the production of collagen, the metabolism of certain amino acids, and the absorption of iron.
      (3) It is not stored in large amounts; excess is excreted in the urine.

Minerals (page 567)

1. Characteristics of minerals
   a. Minerals are responsible for about 4% of body weight.
   b. About 75% of the minerals are found in bones and teeth as calcium and phosphorus.
   c. Minerals are usually incorporated into organic molecules, although some occur in inorganic compounds or as free ions.
   d. They comprise structural materials, function in enzymes, and play vital roles in various metabolic processes.
   e. Mineral concentrations are generally regulated by homeostatic mechanisms.

2. Major minerals
   a. Calcium
      (1) Calcium is essential for the formation of bones and teeth, the conduction of nerve impulses, the contraction of muscle fibers, the coagulation of blood, and the activation of various enzymes.
      (2) Its absorption is affected by existing calcium concentration, vitamin D, protein intake, and motility of the digestive tract.
   b. Phosphorus
      (1) Phosphorus is incorporated for the most part into the salts of bones and teeth.
      (2) It plays roles in nearly all metabolic reactions as a constituent of nucleic acids, proteins, enzymes, and some vitamins.
      (3) It also occurs in the phospholipids of cell membranes, in ATP, and in phosphates of body fluids.
   c. Potassium
      (1) Potassium tends to be concentrated inside cells.
      (2) It functions in maintenance of osmotic pressure, regulation of pH, metabolism of carbohydrates and proteins, conduction of nerve impulses, and contraction of muscle fibers.
   d. Sulfur
      (1) Sulfur is incorporated for the most part into the molecular structures of certain amino acids.
      (2) It is also included in thiamine, insulin, biotin, and mucopolysaccharides.
   e. Sodium
      (1) Most sodium occurs in extracellular fluids or is bound to the inorganic salts of bone.
      (2) The blood concentration of sodium is regulated by the kidneys under the influence of aldosterone.
      (3) Sodium helps maintain osmotic concentrations and regulates water balance.
      (4) It is essential for the conduction of nerve impulses, the contraction of muscle fibers, and the movement of substances through cell membranes.
   f. Chlorine
      (1) Chlorine is closely associated with sodium in the form of chloride ions.
      (2) It acts with sodium to help maintain osmotic pressure, regulate pH, and maintain electrolyte balance.
      (3) It is essential for the formation of hydrochloric acid and for the transport of carbon dioxide by red blood cells.
   g. Magnesium
      (1) Magnesium is particularly abundant in the bones as phosphates and carbonates.
      (2) It functions in the production of ATP and in the conversion of ATP to ADP.
      (3) A reserve supply of magnesium is stored in the bones; excesses are excreted in the urine.

3. Trace elements
   a. Iron
      (1) Iron occurs primarily in hemoglobin of red blood cells and in myoglobin of muscles.
      (2) A reserve supply of iron is stored in the liver, spleen, and bone marrow.
      (3) It is needed to catalyze the formation of vitamin A; it is also incorporated into various enzymes and the cytochrome molecules.
   b. Manganese
      (1) Most manganese is concentrated in the liver, kidneys, and pancreas.
      (2) It is necessary for normal growth and development of skeletal structures and other connective tissues; it is essential for the synthesis of fatty acids, cholesterol, and urea.
   c. Copper
      (1) Most copper is concentrated in the liver, heart, and brain.
      (2) It is needed for synthesis of hemoglobin, development of bones, production of melanin, and formation of myelin.
   d. Iodine
      (1) Iodine is most highly concentrated in the thyroid gland.
      (2) It provides an essential component for the synthesis of thyroid hormones.
      (3) It is often added to foods in the form of iodized table salt.
   e. Cobalt
      (1) Cobalt is widely distributed throughout the body.
      (2) It is an essential part of cyanocobalamin and is probably needed for the synthesis of several important enzymes.
   f. Zinc
      (1) Zinc is most concentrated in the liver, kidneys, and brain.
      (2) It is a constituent of several enzymes involved with digestion, respiration, bone metabolism, and liver metabolism.

Adequate Diets (page 572)

1. An adequate diet provides sufficient energy and essential nutrients to support optimal growth, as well as maintenance and repair, of tissues.
2. Individual needs vary so greatly that it is not possible to design a diet that is adequate for everyone.
3. Malnutrition
   a. Poor nutrition is due to lack of foods or failure to make the best use of available foods.
   b. Primary malnutrition is due to poor diet.
   c. Secondary malnutrition is due to an individual characteristic that makes a normal diet inadequate.

Clinical Application of Knowledge

1. How would you explain the fact that the blood sugar concentration of a person whose diet is relatively low in carbohydrates remains stable?
2. With the aid of nutrient tables available in reference books, calculate the carbohydrate, lipid, and protein content of your diet in grams for a twenty-four-hour period. Also calculate the total calories represented by these foods. Assuming that this twenty-four-hour sample is representative of your normal eating habits, what improvements could be made in its composition?
3. Examine the label information on the packages of a variety of dry breakfast cereals. Which types of cereals provide the best sources of vitamins and minerals? Which major nutrients are lacking in these cereals?
4. If a person decided to avoid eating meat and other animal products, such as milk, cheese, and eggs, what foods might be included in the diet to provide essential amino acids?
5. How might a diet be modified in order to limit the intake of cholesterol?
6. How do you think the nutritional requirements of a healthy 12-year-old boy and a healthy 60-year-old man would differ?

Review Activities

1. Define *essential nutrient*.
2. List some common sources of carbohydrates.
3. Explain the importance of cellulose in the diet.
4. Explain what happens to excessive amounts of glucose in the body.
5. Explain why a temporary drop in the blood glucose concentration may produce functional disorders of the nervous system.
6. List some of the factors that affect an individual's need for carbohydrates.
7. Define *triglyceride*.
8. List some common sources of lipids.
9. Describe the role of the liver in fat metabolism.
10. Discuss the functions of cholesterol.
11. List some common sources of protein.
12. Distinguish between essential and nonessential amino acids.
13. Explain why all of the essential amino acids must be present before growth can occur.
14. Distinguish between complete and incomplete proteins.
15. Review the major functions of amino acids.
16. Define *nitrogen balance*.
17. Explain why a protein deficiency may be accompanied by edema.
18. Define *calorie*.
19. Explain how the caloric values of foods are determined.
20. Define *basal metabolic rate*.
21. List some of the factors that affect the BMR.
22. Define *energy balance*.
23. Explain what is meant by *desirable weight*.
24. Distinguish between overweight and obesity.
25. Discuss the general characteristics of fat-soluble vitamins.
26. List the fat-soluble vitamins, and describe the major functions of each vitamin.
27. List some good sources for each of the fat-soluble vitamins.
28. Explain what is meant by the *vitamin B complex*.
29. List the water-soluble vitamins, and describe the major functions of each vitamin.
30. List some good sources for each of the water soluble vitamins.
31. Discuss the general characteristics of the mineral nutrients.
32. List the major minerals, and describe the major functions of each mineral.
33. List some good sources for each of the major minerals.
34. Distinguish between a major mineral and a trace element.
35. List the trace elements, and describe the major functions of each trace element.
36. List some good sources of each of the trace elements.
37. Define *adequate diet*.
38. Define *malnutrition*.
39. Distinguish between primary and secondary malnutrition.

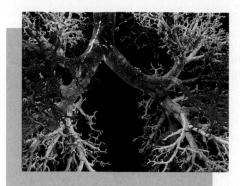

16

Respiratory System

*B*efore the body cells can oxidize nutrients and release energy, they must be supplied with oxygen, and the carbon dioxide that results from oxidation must be excreted. These two general processes—obtaining oxygen and removing carbon dioxide—are the primary functions of the *respiratory system.*

In addition, the respiratory organs filter particles from the incoming air, help control the temperature and water content of the incoming air, aid in producing vocal sounds, and play important roles in the sense of smell and in the regulation of the pH of blood ■

Chapter Objectives	Key Terms	Aids to Understanding Words

After you have studied this chapter, you should be able to:

1. List the general functions of the respiratory system.

2. Name and describe the locations of the organs of the respiratory system.

3. Describe the functions of each organ of the respiratory system.

4. Explain how inspiration and expiration are accomplished.

5. Name and define each of the respiratory air volumes and capacities.

6. Explain how the alveolar ventilation rate is calculated.

7. List several nonrespiratory air movements and explain how each occurs.

8. Locate the respiratory center and explain how it controls normal breathing.

9. Discuss how various factors affect the respiratory center.

10. Describe the structure and function of the respiratory membrane.

11. Explain how oxygen and carbon dioxide are transported in the blood.

12. Review the major events that occur during cellular respiration.

13. Explain how oxygen is used by cells.

14. Complete the review activities at the end of this chapter. Note that the items are worded in the form of specific learning objectives. You may want to refer to them before reading the chapter.

---

alveolus (al-ve′o-lus)

bronchial tree (brong′ke-al trē)

carbaminohemoglobin (kar-bam″ĭ-no-he″mo-glo′bin)

carbonic anhydrase (kar-bon′ik an-hi′drās)

cellular respiration (sel′u-lar res″pi-ra′shun)

citric acid cycle (sit′rik as′id si′kl)

expiration (ek″spi-ra′shun)

glottis (glot′is)

hemoglobin (he″mo-glo′bin)

hyperventilation (hi″per-ven″tĭ-la′shun)

inspiration (in″spĭ-ra′shun)

oxyhemoglobin (ok″se-he″mo-glo′bin)

partial pressure (par′shil presh′ur)

pleural cavity (ploo′ral kav′ĭ-te)

respiratory center (re-spi′rah-to″re sen′ter)

respiratory membrane (re-spi′rah-to″re mem′brān)

respiratory volume (re-spi′rah-to″re vol′ūm)

surface tension (ser′fas ten′shun)

surfactant (ser-fak′tant)

---

alveol-, small cavity: *alveol*us—a microscopic air sac within a lung.

bronch-, windpipe: *bronch*us—a primary branch of the trachea.

carcin-, spreading sore: *carcin*oma—a type of cancer.

carin-, keel-like: *carin*a—a ridge of cartilage between the right and left bronchi.

cric-, ring: *cric*oid cartilage—a ring-shaped mass of cartilage at the base of the larynx.

epi-, upon: *epi*glottis—a flaplike structure that partially covers the opening into the larynx during swallowing.

hem-, blood: *hem*oglobin—the pigment in red blood cells.

tuber-, swelling: *tuber*culosis—a disease characterized by the formation of fibrous masses within the lungs.

The **respiratory system** consists of a group of passages that filter incoming air and transport it from outside the body into the lungs, and numerous microscopic air sacs in which gas exchanges take place. The entire process of exchanging gases between the atmosphere and the body cells is called **respiration**, and it involves several events: the movement of air in and out of the lungs—commonly called *breathing,* or *pulmonary ventilation*; the exchange of gases between the air in the lungs and the blood; the transport of gases by the blood between the lungs and body cells; and the exchange of gases between the blood and the body cells. The utilization of oxygen and the production of carbon dioxide by the body cells is called **cellular respiration.**

Organs of the Respiratory System

The organs of the respiratory system include the nose, nasal cavity, sinuses, pharynx, larynx, trachea, bronchial tree, and lungs.

The parts of the respiratory system, shown in figure 16.1, can be divided into two groups, or tracts. Those organs outside the thorax constitute the *upper respiratory tract,* and those within the thorax comprise the *lower respiratory tract.*

Nose

The nose is covered with skin and is supported internally by bone and cartilage. Its two *nostrils* (external nares) provide openings through which air can enter and leave the nasal cavity. These openings are guarded by numerous internal hairs, which help prevent the entrance of relatively large particles sometimes carried in the air.

Nasal Cavity

The **nasal cavity,** a hollow space behind the nose, is divided medially into right and left portions by the **nasal septum.** This cavity is separated from the cranial cavity by the cribriform plate of the ethmoid bone and from the mouth by the hard palate.

> The nasal septum is usually straight at birth, although it is sometimes bent as the result of a birth injury. It remains straight throughout early childhood, but as a person ages, the septum tends to bend toward one side or the other. Such a *deviated septum* may cause an obstruction in the nasal cavity that makes breathing difficult.

As figure 16.2 shows, **nasal conchae** (turbinate bones) curl out from the lateral walls of the nasal cavity on each side, dividing the cavity into passageways called the *superior, middle,* and *inferior meatuses.* (See

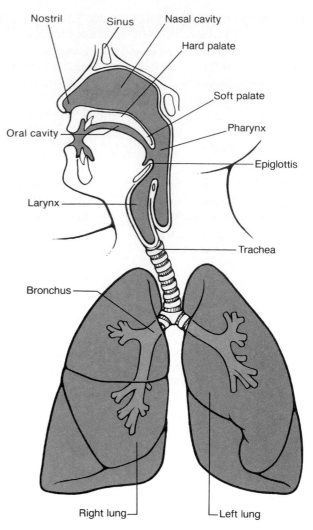

Figure 16.1   Organs of the respiratory system.

chapter 7.) They also support the mucous membrane that lines the nasal cavity and help increase its surface area.

The upper posterior portion of the nasal cavity, below the cribriform plate, is slitlike, and its lining contains the olfactory receptors that function in the sense of smell. The remainder of the cavity serves to conduct air to and from the nasopharynx.

The mucous membrane lining the nasal cavity contains pseudostratified ciliated epithelium that is rich in mucus-secreting goblet cells. (See chapter 5.) It also includes an extensive network of blood vessels, and normally appears pinkish. As air passes over the membrane, heat radiates from the blood and warms the air. In this way, the temperature of the incoming air quickly adjusts to that of the body. In addition, the incoming air becomes moistened by evaporation of water from the mucous lining. The sticky mucus secreted by the mucous membrane entraps dust and other small particles entering with the air.

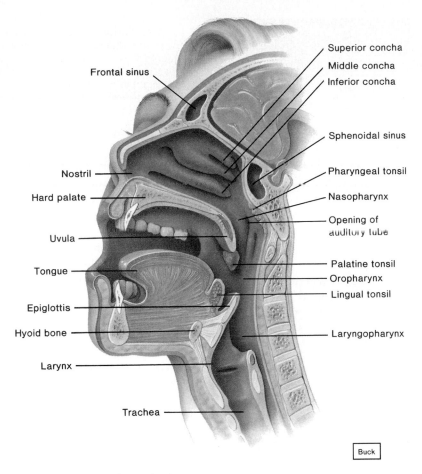

Superior concha
Middle concha
Inferior concha

Frontal sinus

Sphenoidal sinus

Pharyngeal tonsil

Nostril

Nasopharynx

Hard palate

Opening of
auditory tube

Uvula

Palatine tonsil

Tongue

Oropharynx

Lingual tonsil

Epiglottis

Hyoid bone

Laryngopharynx

Larynx

Trachea

Buck

Figure 16.2   Major features of the upper respiratory tract.

As the cilia of the epithelial cells move, a thin layer of mucus and any entrapped particles are pushed toward the pharynx (figure 16.3). When the mucus reaches the pharynx, it is swallowed. In the stomach, any microorganisms in the mucus, including disease-causing forms, are likely to be destroyed by the action of gastric juice. Thus, the filtering mechanism provided by the mucous membrane not only prevents particles from reaching the lower air passages, but also helps prevent respiratory infections.

1. What is meant by respiration?
2. What organs constitute the respiratory system?
3. What is the function of the mucous membrane that lines the nasal cavity?
4. What is the function of the cilia in the cells that line the nasal cavity?

Sinuses

As discussed in chapter 7, the **sinuses** (paranasal sinuses) are air-filled spaces located within the *maxillary, frontal, ethmoid,* and *sphenoid bones* of the skull (figure 16.4). These spaces open into the nasal cavity and are lined with mucous membranes that are contin-

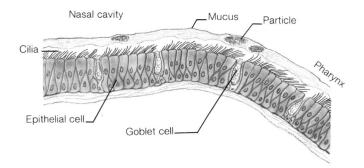

Nasal cavity

Mucus

Particle

Cilia

Pharynx

Epithelial cell

Goblet cell

Figure 16.3   Cilia move mucus and trapped particles from the nasal cavity to the pharynx.

uous with the lining of the nasal cavity. Consequently, mucus secretions can drain from the sinuses into the nasal cavity. If this drainage is blocked by membranes that are inflamed and swollen because of nasal infections or allergic reactions (sinusitis), the accumulating fluids may cause increasing pressure within a sinus and a painful sinus headache.

Although the sinuses function mainly to reduce the weight of the skull, they also serve as resonant chambers that affect the quality of the voice.

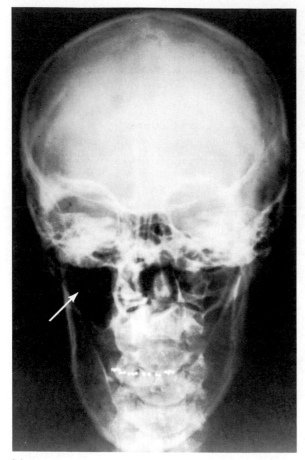

(a)

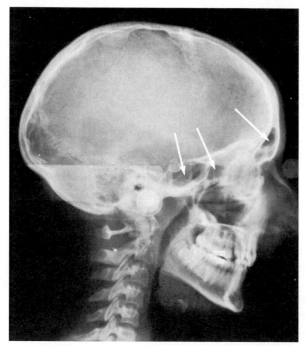

(b)

Figure 16.4 · (a) X-ray film of a skull from the front and (b) from the side, showing air-filled sinuses within the bones (arrows). Which of these sinuses can you identify?

1. Where are the sinuses located?
2. What are the functions of the sinuses?

Pharynx

The **pharynx** (throat) is located behind the oral cavity and between the nasal cavity and the larynx. It functions as a passageway for food traveling from the oral cavity to the esophagus and for air passing between the nasal cavity and the larynx (figure 16.2). It also aids in producing the sounds of speech. The subdivisions of the pharynx—the nasopharynx, oropharynx, and laryngopharynx—are described in chapter 14.

Larynx

The **larynx** is an enlargement in the airway at the top of the trachea and below the pharynx. It serves as a passageway for air moving in and out of the trachea and prevents foreign objects from entering the trachea. In addition, it houses the *vocal cords*. (See reference plates 49 and 71.)

The larynx is composed primarily of muscles and cartilages, which form the framework of the larynx and are bound together by elastic tissue. The largest of the cartilages (shown in figure 16.5) are the thyroid, cricoid, and epiglottic cartilages. These structures occur singly, and the other laryngeal cartilages—the arytenoid, corniculate, and cuneiform cartilages—are paired.

The **thyroid cartilage** was named for the thyroid gland that covers its lower part. This cartilage is the shieldlike structure that protrudes in the front of the neck and is sometimes called the Adam's apple. The protrusion typically is more prominent in males than in females because of an effect of male sex hormones on the development of the larynx.

The **cricoid cartilage** lies below the thyroid cartilage and marks the lowermost portion of the larynx.

The **epiglottic cartilage** is attached to the upper border of the thyroid cartilage and supports a flaplike structure called the **epiglottis.** The epiglottis usually stands upright and allows air to enter the larynx. During swallowing, however, the larynx is raised by muscular contractions, and the epiglottis is pressed downward by the base of the tongue. As a result, the epiglottis partially covers the opening into the larynx, helping prevent foods and liquids from entering the air passages.

The pyramid-shaped **arytenoid cartilages** are located above and on either side of the cricoid cartilage.

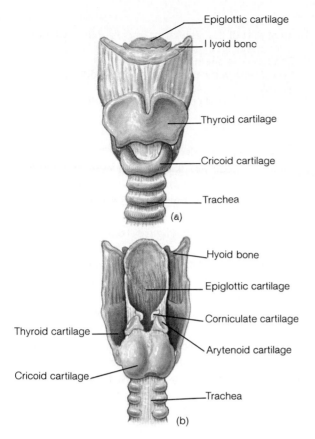

Epiglottic cartilage

Hyoid bone

Thyroid cartilage

Cricoid cartilage

Trachea

(a)

Hyoid bone

Epiglottic cartilage

Corniculate cartilage

Thyroid cartilage

Arytenoid cartilage

Cricoid cartilage

Trachea

(b)

Figure 16.5 (a) Anterior view and (b) posterior view of the larynx.

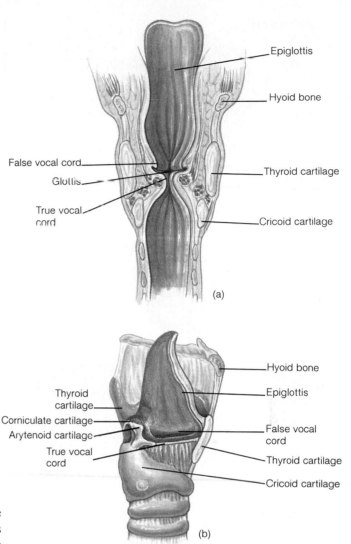

Epiglottis

Hyoid bone

False vocal cord

Thyroid cartilage

Glottis

True vocal cord

Cricoid cartilage

(a)

Thyroid cartilage

Hyoid bone

Epiglottis

Corniculate cartilage

Arytenoid cartilage

False vocal cord

True vocal cord

Thyroid cartilage

Cricoid cartilage

(b)

Figure 16.6 (a) Frontal section and (b) sagittal section of the larynx.

Attached to the tips of the arytenoid cartilages are the tiny, conelike **corniculate cartilages.** These cartilages serve as attachments for muscles that help regulate tension on the vocal cords during speech and aid in closing the larynx during swallowing.

The **cuneiform cartilages** are small, cylindrical parts found in the mucous membrane between the epiglottic and the arytenoid cartilages. They function to stiffen the soft tissues in this region.

Inside the larynx, two pairs of horizontal folds in the mucous membrane extend inward from the lateral walls. The upper folds (vestibular folds) are called *false vocal cords* because they do not function in the production of sounds. Muscle fibers within these folds help close the larynx during swallowing.

The lower folds are the *true vocal cords.* They contain elastic fibers and are responsible for vocal sounds, which are created when air is forced between these folds, causing them to vibrate from side to side. This action generates sound waves, which can be formed into words by changing the shapes of the pharynx and oral cavity and by using the tongue and lips. Both pairs of folds are shown in figure 16.6.

The *pitch* (musical tone) of the vocal sound is controlled by changing the tension on the cords. This

is accomplished by contracting or relaxing various laryngeal muscles. Increasing the tension produces a higher pitch, and decreasing the tension creates a lower pitch.

The *intensity* (loudness) of a vocal sound is related to the force of the air passing over the vocal cords. Louder sounds are produced by using stronger blasts of air to cause the vocal cords to vibrate.

During normal breathing, the vocal cords remain relaxed, and the opening between them, called the **glottis,** appears as a triangular slit. However, when food or liquid is swallowed the glottis is closed by muscles within the false vocal folds, and this prevents the food or liquid from entering the trachea (figure 16.7).

The mucous membrane that lines the larynx continues to filter the incoming air by entrapping particles and moving them toward the pharynx by ciliary action.

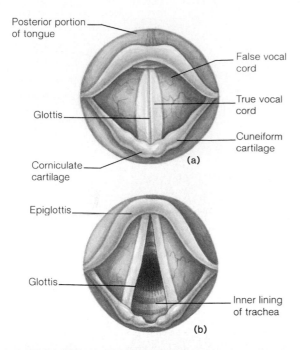

Posterior portion of tongue

False vocal cord

True vocal cord

Glottis

Cuneiform cartilage

Corniculate cartilage

(a)

Epiglottis

Glottis

Inner lining of trachea

(b)

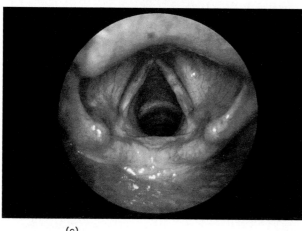

(c)

Figure 16.7 Vocal cords as viewed from above (a) with the glottis closed and (b) with the glottis open. (c) A photograph of the glottis and vocal folds.

Occasionally, the mucous membrane of the larynx becomes inflamed and swollen as the result of an infection or an irritation from inhaled vapors. When this happens, the vocal cords may not vibrate as freely as before, and the voice may sound harsh (hoarse). This condition is called *laryngitis,* and although it is usually mild, laryngitis is potentially dangerous because the swollen tissues may obstruct the airway and interfere with breathing. In such cases, it may be necessary to provide a passageway by inserting a tube (endotracheal tube) into the trachea through the nose or mouth.

1. What part of the respiratory tract is shared with the alimentary canal?
2. Describe the structure of the larynx.
3. How do the vocal cords function to produce sounds?
4. What is the function of the glottis? Of the epiglottis?

Trachea

The **trachea** (windpipe) is a flexible cylindrical tube about 2.5 centimeters in diameter and 12.5 centimeters in length. It extends downward in front of the esophagus and into the thoracic cavity, where it splits into right and left bronchi (figure 16.8 and reference plate 50).

The inner wall of the trachea is lined with a ciliated mucous membrane that contains many goblet cells. This membrane continues to filter the incoming air and to move entrapped particles upward into the pharynx.

Within the tracheal wall are about twenty C-shaped pieces of hyaline cartilage, arranged one above the other. The open ends of these incomplete rings are directed posteriorly, and the gaps between their ends are filled with smooth muscle and connective tissues (figures 16.9 and 16.10). These cartilaginous rings prevent the trachea from collapsing and blocking the airway. At the same time, the soft tissues that complete the rings in the back allow the nearby esophagus to expand as food moves through it on the way to the stomach.

If the trachea becomes obstructed by swollen tissues, excessive secretions, or a foreign object, it may be necessary to create a temporary, external opening in the tube so that air can bypass the obstruction (figure 16.11). This procedure is called a *tracheostomy.* It may be a lifesaving procedure because if the trachea is blocked, asphyxiation can occur within a few minutes.

Bronchial Tree

The **bronchial tree** consists of the branched airways leading from the trachea to the microscopic air sacs in the lungs. It begins with the right and left **primary bronchi,** which arise from the trachea at the level of the fifth thoracic vertebrae. The openings of the primary bronchi are separated by a ridge of cartilage called the *carina* (figure 16.8). Each bronchus, accompanied by large blood vessels, enters its respective lung.

Branches of the Bronchial Tree

A short distance from its origin, each primary bronchus divides into **secondary,** or **lobar, bronchi** (two on

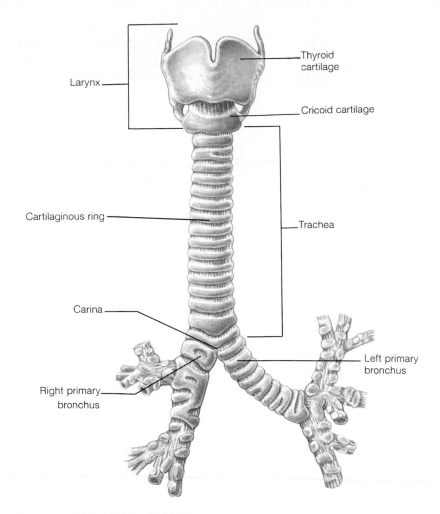

Figure 16.8   The trachea transports air between the larynx and the bronchi.

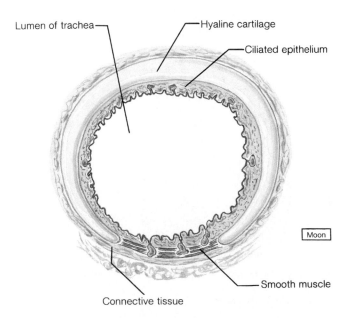

Figure 16.9   Cross section of the trachea. What is the significance of the C-shaped rings of hyaline cartilage in its wall?

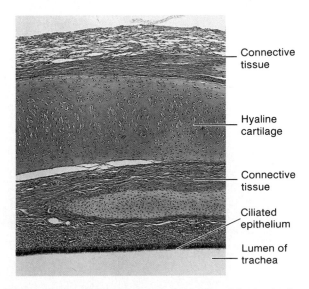

Figure 16.10   Light micrograph of a section of the tracheal wall (×63).

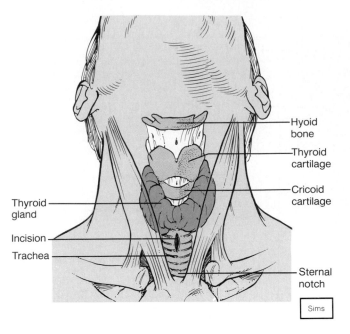

Figure 16.11 A tracheostomy may be performed to allow air to bypass an obstruction within the larynx.

Hyoid bone
Thyroid cartilage
Cricoid cartilage
Thyroid gland
Incision
Trachea
Sternal notch
Sims

the left and three on the right), that, in turn, branch again and again into finer and finer tubes (figures 16.12 and 16.13). The successive divisions of these branches from the lobar bronchus to the microscopic air sacs follow:

1. *Tertiary,* or *segmental, bronchi.* Each of these branches supplies a portion of the lung called a *bronchopulmonary segment.* Usually there are ten such segments in the right lung and eight in the left lung.
2. *Bronchioles.* These small branches of the segmental bronchi enter the basic units of the lung—the *lobules.*
3. *Terminal bronchioles.* These tubes branch from a bronchiole. There are fifty to eighty terminal bronchioles within a lobule of the lung.
4. *Respiratory bronchioles.* Two or more respiratory bronchioles branch from each terminal bronchiole. They are relatively short and have diameters of about 0.5 millimeter. They are called "respiratory" because a few air sacs bud from their sides; thus, they are the first structures in the sequence that can engage in gas exchange.

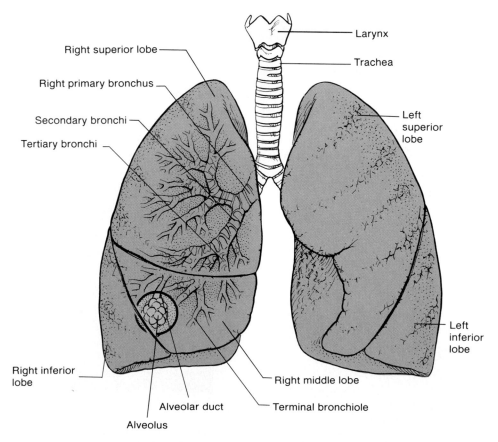

Right superior lobe
Right primary bronchus
Secondary bronchi
Tertiary bronchi
Larynx
Trachea
Left superior lobe
Left inferior lobe
Right inferior lobe
Alveolar duct
Alveolus
Right middle lobe
Terminal bronchiole

Figure 16.12 The bronchial tree consists of the passageways that connect the trachea and the alveoli.

5. *Alveolar ducts*. Two to ten long, branching alveolar ducts extend from each respiratory bronchiole (figure 16.14).
6. *Alveolar sacs*. Alveolar sacs are thin-walled, closely packed outpouchings of the alveolar ducts.
7. *Alveoli*. Alveoli are thin-walled, microscopic air sacs that open only on the side communicating with an alveolar sac. Thus, air can diffuse

freely from the alveolar ducts, through the alveolar sacs, and into the alveoli (figure 16.15).

Structure of the Respiratory Tubes

The structure of a bronchus is similar to that of the trachea, but the C-shaped cartilaginous rings are replaced with cartilaginous plates where the bronchus enters the lung. These plates are irregularly shaped, and they completely surround the tube, giving it a cylindrical form. However, as finer and finer branch tubes appear, the amount of cartilage decreases, and it finally disappears in the bronchioles, which have diameters of about 1 millimeter.

As the amount of cartilage decreases, a layer of smooth muscle that surrounds the tube just beneath the mucosa becomes more prominent. This muscular layer remains in the wall to the ends of the respiratory bronchioles, and only a few muscle fibers occur in the walls of the alveolar ducts.

Elastic fibers are scattered among the smooth muscle cells and are abundant in the connective tissue that surrounds the respiratory tubes. These fibers play an important role in breathing, as explained in a subsequent section of this chapter.

As the tubes become smaller, a change occurs in the type of cells that line them. For example, the lining of the larger tubes consists of pseudostratified, ciliated columnar epithelium and mucus-secreting goblet cells. In the finer tubes, beginning with the respiratory bronchioles, the lining is cuboidal epithelium; and in the alveoli, it is simple squamous epithelium closely

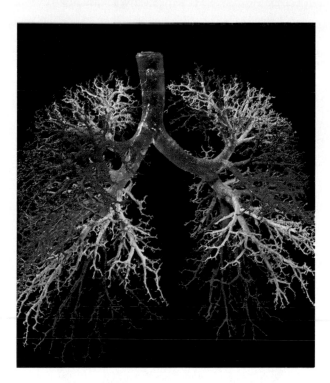

Figure 16.13   A plastic cast of the bronchial tree.

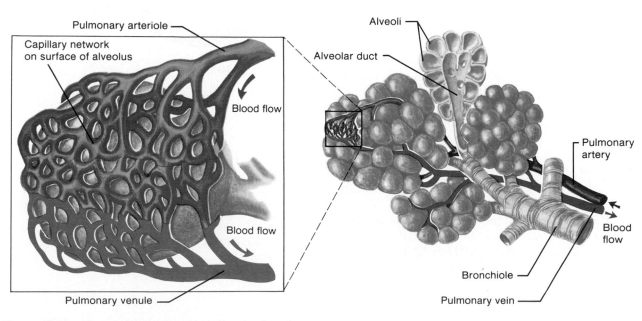

Figure 16.14   The respiratory tubes end in tiny alveoli, each of which is surrounded by a capillary network.

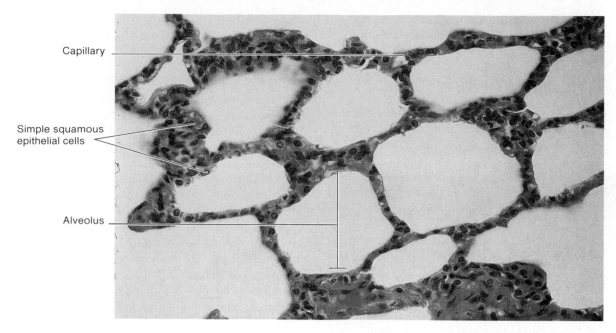

Capillary

Simple squamous
epithelial cells

Alveolus

Figure 16.15 Light micrograph of alveoli (×250).

associated with a dense network of capillaries. However, along the way the number of goblet cells and the
height of the other epithelial cells decline, and the
abundance of cilia decreases.

> The trachea and bronchial tree can be examined
> directly using a flexible optical instrument called a
> *fiberoptic bronchoscope*. This procedure
> (bronchoscopy) is sometimes used in diagnosing
> tumors or other pulmonary diseases. It may also be
> employed to locate and remove aspirated foreign
> bodies in the air passages.

Functions of the Respiratory Tubes and Alveoli

The branches of the bronchial tree serve as air passages, which continue to filter the incoming air and distribute it to the alveoli in all parts of the lungs. The
alveoli, in turn, provide a large surface area of thin epithelial cells through which gas exchanges can occur.
During these exchanges, oxygen diffuses through the
alveolar walls and enters the blood in nearby capillaries, and carbon dioxide diffuses from the blood
through these walls and enters the alveoli (figures 16.16,
16.17, and 16.18).

It is estimated that there are about three hundred
million alveoli in an adult lung and that these spaces
provide a total surface area of between 70 and 80 square
meters.

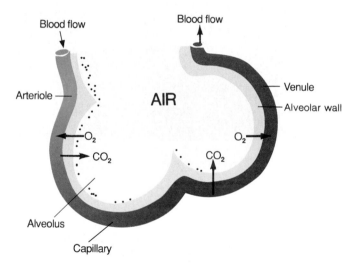

Blood flow   Blood flow

Arteriole   AIR   Venule
                                 Alveolar wall

O_2 O_2
CO_2 CO_2

Alveolus

Capillary

Figure 16.16 Oxygen diffuses from the air within the alveolus
into the capillary, while carbon dioxide diffuses from the blood
within the capillary into the alveolus.

1. What is the function of the cartilaginous rings in the
   tracheal wall?
2. How do the right and left bronchi differ in structure?
3. List the branches of the bronchial tree.
4. Describe the changes in structure that occur in the
   respiratory tubes as they become smaller and
   smaller.
5. How are gases exchanged in the alveoli?

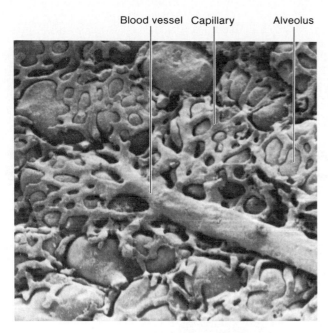

Blood vessel   Capillary   Alveolus

Figure 16.17 Scanning electron micrograph of casts of alveoli and associated capillary networks. These casts were prepared by filling the alveoli and blood vessels with resin and later removing the soft tissues by digestion, leaving only the resin casts (×415). (*Tissues and Organs: A Text-Atlas of Scanning Electron Microscopy*, by Richard D. Kessel and Randy Kardon. © 1979 W.H. Freeman and Company.)

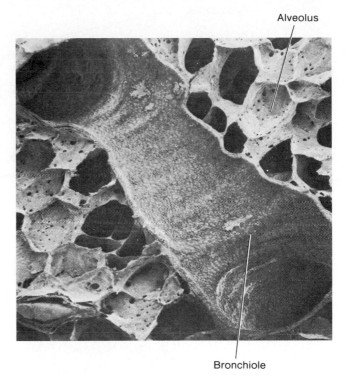

Alveolus

Bronchiole

Figure 16.18 Scanning electron micrograph of lung alveoli and a bronchiole.

Lungs

The lungs are soft, spongy, cone-shaped organs located in the thoracic cavity. The right and left lungs are separated medially by the heart and the mediastinum, and they are enclosed by the diaphragm and the thoracic cage. (See figures 1.8, 16.19, and reference plates 56, 57, and 71.)

Each lung occupies most of the thoracic space on its side and is suspended in the cavity by its attachments, which include a bronchus and some large blood vessels. These tubular parts enter the lung on its medial surface through a region called the **hilus.** A layer of serous membrane, the **visceral pleura,** is firmly attached to the surface of each lung, and this membrane folds back at the hilus to become the **parietal pleura.** The parietal pleura, in turn, forms part of the mediastinum and lines the inner wall of the thoracic cavity.

The potential space between the visceral and parietal pleurae is called the **pleural cavity,** and it contains a thin film of serous fluid. This fluid lubricates the adjacent pleural surfaces, reducing friction as they move against one another during breathing. It also helps hold the pleural membranes together, as explained in the next section of this chapter.

The right lung is larger than the left lung, and it is divided into three parts, called the superior, middle, and inferior **lobes.** The left lung consists of two parts, a superior and an inferior lobe.

Each lobe is supplied by a lobar bronchus of the bronchial tree. A lobe also has connections to blood and lymphatic vessels, and is enclosed by connective tissues. A lobe is further subdivided by connective tissue into **lobules,** and each of these units contains terminal bronchioles together with their alveolar ducts, alveolar sacs, alveoli, nerves, and associated blood and lymphatic vessels. Thus, the substance of a lung includes air passages, alveoli, blood vessels, connective tissues, lymphatic vessels, and nerves.

Chart 16.1 summarizes the characteristics of the major parts of the respiratory system.

1. Where are the lungs located?
2. What is the function of the serous fluid within the pleural cavity?
3. How does the structure of the right lung differ from that of the left lung?
4. What kinds of structures make up a lung?

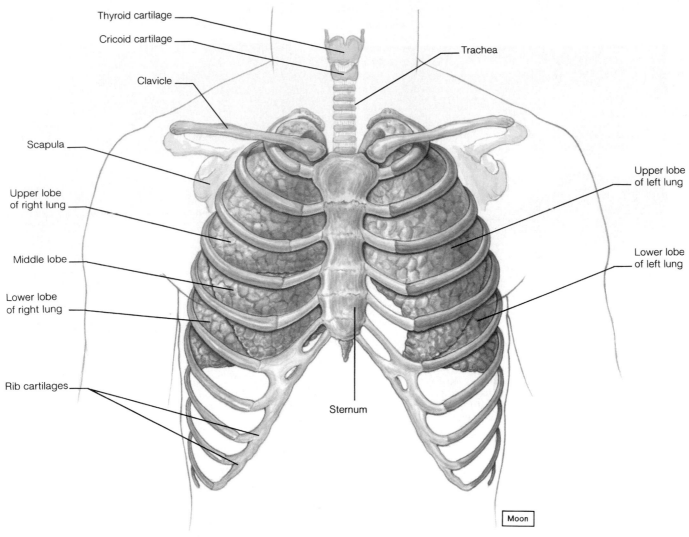

Figure 16.19 Locations of the lungs within the thoracic cavity.

CHART 16.1	Parts of the respiratory system	
Part	**Description**	**Function**
Nose	Part of face centered above the mouth and below the space between the eyes	Nostrils provide entrance to nasal cavity; internal hairs begin to filter incoming air
Nasal cavity	Hollow space behind nose	Conducts air to pharynx; mucous lining filters, warms, and moistens incoming air
Sinuses	Hollow spaces in various bones of the skull	Reduce weight of the skull; serve as resonant chambers
Pharynx	Chamber behind mouth cavity and between nasal cavity and larynx	Passageway for air moving from nasal cavity to larynx and for food moving from mouth cavity to esophagus
Larynx	Enlargement at the top of the trachea	Passageway for air; prevents foreign objects from entering trachea; houses vocal cords
Trachea	Flexible tube that connects larynx with bronchial tree	Passageway for air; mucous lining continues to filter air
Bronchial tree	Branched tubes that lead from the trachea to the alveoli	Conducts air from the trachea to the alveoli; mucous lining continues to filter incoming air
Lungs	Soft, cone-shaped organs that occupy a large portion of the thoracic cavity	Contain the air passages, alveoli, blood vessels, connective tissues, lymphatic vessels, and nerves of the lower respiratory tract

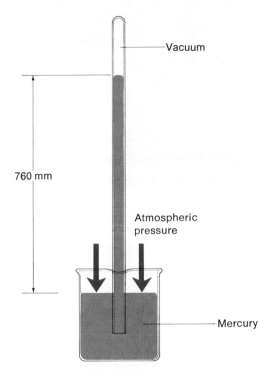

Figure 16.20 Atmospheric pressure is sufficient to support a column of mercury 760 millimeters high at sea level.

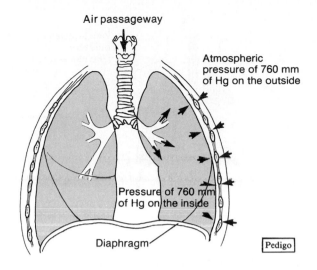

Figure 16.21 When the lungs are at rest, the pressure on the inside of the lungs is equal to the pressure on the outside of the thorax.

Breathing Mechanism

Breathing, which is also called *pulmonary ventilation,* is the movement of air from outside the body into the bronchial tree and alveoli, followed by a reversal of this air movement. The actions responsible for these air movements are termed **inspiration** (inhalation) and **expiration** (exhalation).

Inspiration

Atmospheric pressure due to the weight of the air is the force that causes air to move into the lungs. At sea level, this pressure is sufficient to support a column of mercury about 760 millimeters high in a tube (figure 16.20). Thus, normal air pressure is equal to 760 millimeters (mm) of mercury (Hg).

Air pressure is exerted on all surfaces in contact with the air, and because people breathe air, the inside surfaces of their lungs are also subjected to pressure. In other words, the pressures on the inside of the lungs and alveoli and on the outside of the thoracic wall are about the same (figure 16.21).

If the pressure inside the lungs and alveoli (intra-alveolar pressure) decreases, outside air will then be pushed into the airways by atmospheric pressure. This is what happens during normal inspiration, and it involves the action of muscle fibers within the dome-shaped *diaphragm.*

The diaphragm is located just below the lungs. It consists of an anterior group of skeletal muscle fibers (costal fibers), which originate from the ribs and sternum, and a posterior group (crural fibers), which originate from the vertebrae. Both groups of muscle fibers are inserted on a tendinous central portion of the diaphragm (reference plate 71).

The muscle fibers of the diaphragm are stimulated to contract by impulses carried on the *phrenic nerves,* which are associated with the cervical plexuses. When this occurs, the diaphragm moves downward, the size of the thoracic cavity is enlarged, and the intra-alveolar pressure is reduced about 2 mm Hg below that of atmospheric pressure. In response to this decreased pressure, air is forced into the airways by atmospheric pressure, and the lungs expand (figure 16.22).

While the diaphragm is contracting and moving downward, the *external intercostal muscles* between the ribs and certain thoracic muscles may be stimulated to contract. This action raises the ribs and elevates the sternum, so that the size of the thoracic cavity increases even more. As a result, the intra-alveolar pressure is further reduced, and more air is forced into the airways by the greater atmospheric pressure.

The expansion of the lungs is aided by the fact that the parietal pleura, on the inner wall of the thoracic cavity, and the visceral pleura, attached to the surface of the lungs, are separated only by a thin film of serous fluid. The *water molecules* in this fluid have a great attraction to one another, creating a force called **surface tension.** This force holds the moist surfaces of the pleural membranes tightly together. Consequently, when the thoracic wall is moved upward and outward by the action of the intercostal muscles, the parietal

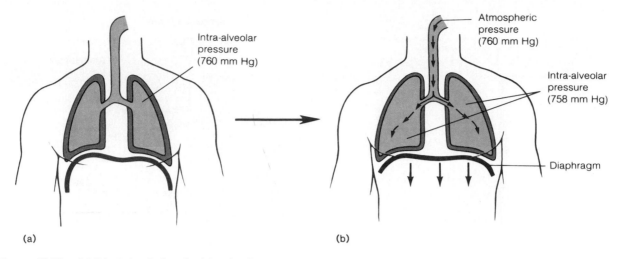

Figure 16.22 (a) Prior to inspiration, the intra-alveolar pressure is 760 mm Hg. (b) The intra-alveolar pressure decreases to about 758 mm Hg as the thoracic cavity enlarges and air is forced into the airways by the atmospheric pressure.

pleura is moved too, and the visceral pleura follows it. This helps expand the lungs in all directions. The steps in inspiration are summarized in chart 16.2.

The surface tension between the adjacent moist membranes is sufficient to cause the collapse of the alveoli, which have moist inner surfaces. However, certain alveolar cells (alveolar type II cells) synthesize a mixture of lipoproteins, called **surfactant.** Surfactant, which is secreted continuously into the alveolar air spaces, acts to reduce the surface tension and thus decreases the tendency of the alveoli to collapse when the lung volume is low.

Sometimes the lungs of a premature newborn fail to produce enough surfactant, and the newborn's breathing mechanism is unable to overcome the force of surface tension. Consequently, the lungs cannot easily be ventilated and the newborn is likely to die of suffocation. This condition is called *respiratory distress syndrome* (hyaline membrane disease), and it is the primary cause of respiratory difficulty in premature newborns and those born to mothers with diabetes mellitus.

If a person needs to take a deeper than normal breath, the diaphragm and external intercostal muscles may be contracted to an even greater extent. Additional muscles, such as the pectoralis minors and sternocleidomastoids, can also be used to pull the thoracic cage further upward and outward, enlarging the thoracic cavity and decreasing the internal pressure still more (figure 16.23).

The ease with which the lungs can be expanded as a result of pressure changes occurring during breathing is called *compliance* (distensibility). In a normal lung, compliance decreases as lung volume in-

CHART 16.2 Major events in inspiration

1. Nerve impulses travel on phrenic nerves to muscle fibers in the diaphragm, causing them to contract.

2. As the dome-shaped diaphragm moves downward, the size of the thoracic cavity increases.

3. At the same time, the external intercostal muscles may contract, raising the ribs and causing the size of the thoracic cavity to increase still more.

4. As the size of the thoracic cavity increases, the intra-alveolar pressure decreases.

5. Atmospheric pressure, which is relatively greater on the outside, forces air into the respiratory tract through the air passages.

6. The lungs become inflated.

creases, because an inflated lung is more difficult to expand than a deflated one. Conditions that tend to obstruct air passages, destroy lung tissue, or impede lung expansion in other ways also decrease compliance.

Expiration

The forces responsible for normal expiration come from *elastic recoil* of tissues and from surface tension. For example, the lungs and thoracic wall contain a considerable amount of elastic tissue, and as the lungs expand during inspiration, these tissues are stretched. When the diaphragm lowers, the abdominal organs beneath it are compressed. As the diaphragm and the external intercostal muscles relax following inspiration, the elastic tissues cause the lungs and thoracic cage to recoil, and they return to their original shapes. Similarly, elastic tissues within the abdominal organs cause them to spring back into their previous shapes, pushing the diaphragm upward. At the same time, surface ten-

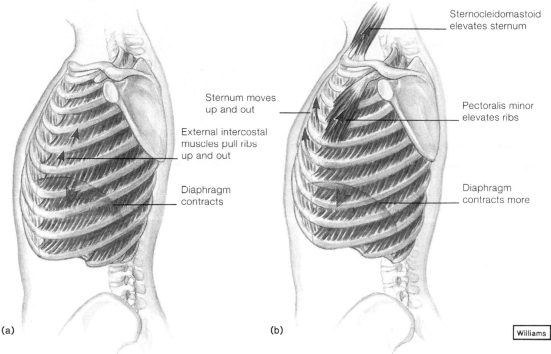

Sternocleidomastoid
elevates sternum

Sternum moves
up and out

External intercostal
muscles pull ribs
up and out

Pectoralis minor
elevates ribs

Diaphragm
contracts

Diaphragm
contracts more

(a)

(b)

Williams

Figure 16.23 (a) Shape of the thorax at the end of normal inspiration. (b) Shape of the thorax at the end of maximal inspiration, aided by contraction of the sternocleidomastoid and pectoralis minor muscles.

sion that develops between the moist surfaces of the alveolar linings tends to cause alveoli to collapse. Each of these factors tends to increase the intra-alveolar pressure about 1 mm Hg above atmospheric pressure, so that the air inside the lungs is forced out through the respiratory passages. Thus, normal expiration is a passive process.

The recoil of the elastic fibers within the lung tissues tends to reduce the pressure in the pleural cavity. Consequently, the pressure between the pleural membranes (intrapleural pressure) is usually about 4 mm Hg less than atmospheric pressure.

Because the visceral and parietal pleural membranes are held together by surface tension, no actual space normally exists in the pleural cavity between them. However, if the thoracic wall is punctured, atmospheric air may enter the pleural cavity and create a real space between the membranes. This condition is called *pneumothorax,* and when it occurs, the lung on the affected side may collapse because of the lung's elasticity.

Pneumothorax may be treated by covering the chest wound with an impermeable bandage, passing a tube (chest tube) through the thoracic wall into the pleural cavity, and applying suction to the tube. In response to the suction, negative pressure is reestablished within the cavity, and the collapsed lung expands rapidly.

If a person needs to exhale more air than normal, the posterior *internal intercostal muscles* can be contracted. These muscles pull the ribs and sternum downward and inward, increasing the pressure in the lungs. Also, the *abdominal wall muscles,* including the external and internal obliques, the transversus abdominis, and the rectus abdominis, can be used to squeeze the abdominal organs inward. Thus, the abdominal wall muscles can cause the pressure in the abdominal cavity to increase and force the diaphragm still higher against the lungs (figure 16.24). As a result of these actions, additional air may be squeezed out of the lungs.

Chart 16.3 summarizes the steps in expiration.

1. Describe the events in inspiration.
2. How does surface tension aid in expanding the lungs during inspiration?
3. What forces are responsible for normal expiration?

Respiratory Air Volumes and Capacities

Different degrees of effort in breathing result in different volumes of air being moved in or out of the lungs. The measurement of such air volumes is called *spirometry,* and these measurements reveal the presence of four distinct *respiratory volumes.* For example, the amount of air that enters the lungs during a normal, quiet inspiration is about 500 cubic centimeters (cc).

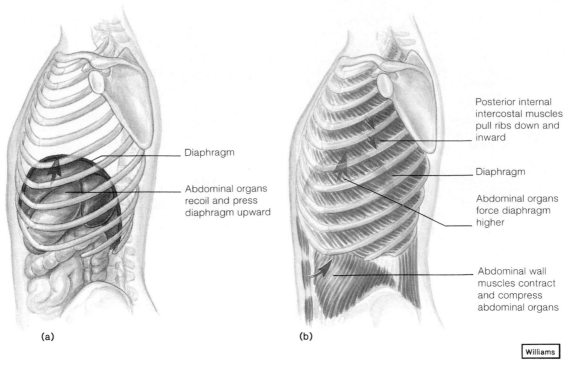

(a)   (b)

Williams

Figure 16.24 (a) Normal expiration is due to elastic recoil of the thoracic wall and abdominal organs. (b) Maximal expiration is aided by contraction of the abdominal wall muscles and posterior internal intercostal muscles.

In figure (a), labels read: Diaphragm; Abdominal organs recoil and press diaphragm upward.

In figure (b), labels read: Posterior internal intercostal muscles pull ribs down and inward; Diaphragm; Abdominal organs force diaphragm higher; Abdominal wall muscles contract and compress abdominal organs.

CHART 16.3 Major events in expiration

1. The diaphragm and external respiratory muscles relax.

2. Elastic tissues of the lungs, thoracic cage, and abdominal organs, which were stretched during inspiration, suddenly recoil, and surface tension causes alveolar walls to collapse.

3. Tissues recoiling around the lungs cause the intra-alveolar pressure to increase.

4. Air is squeezed out of the lungs and into the air passages.

Approximately the same amount leaves during a normal expiration. This amount of air that enters or leaves during normal breathing is termed the **tidal volume.** (See figure 16.25.)

During forced inspiration, a quantity of air in addition to the tidal volume enters the lungs. This additional volume is called the **inspiratory reserve volume** (complemental air), and it equals about 3,000 cc.

During forced expiration, about 1,100 cc of air in addition to the tidal volume can be expelled from the lungs. This quantity is called the **expiratory reserve volume** (supplemental air). However, even after the most forceful expiration, about 1,200 cc of air remains in the lungs. This volume is called the **residual volume.**

Residual air remains in the lungs at all times, and consequently, newly inhaled air is always mixed with air that is already in the lungs. This prevents the oxygen and carbon dioxide concentrations in the lungs from fluctuating excessively with each breath.

Once the respiratory volumes are known, four *respiratory capacities* can be calculated by combining two or more of the volumes. Thus, if the inspiratory reserve volume (3,000 cc) is combined with the tidal volume (500 cc) and the expiratory reserve volume (1,100 cc), the total is termed the **vital capacity** (4,600 cc). This capacity is the maximum amount of air a person can exhale after taking the deepest breath possible.

The tidal volume (500 cc) plus the inspiratory reserve volume (3,000 cc) gives the **inspiratory capacity** (3,500 cc), which is the maximum volume of air a person can inhale following the exhalation of the tidal volume. Similarly, the expiratory reserve volume (1,100 cc) plus the residual volume (1,200 cc) equals the **functional residual capacity** (2,300 cc), which is the volume of air that remains in the lungs following the exhalation of the tidal volume.

The vital capacity plus the residual volume equals the **total lung capacity** (about 5,800 cc). This total varies with age, sex, and body size.

Some of the air that enters the respiratory tract during breathing fails to reach the alveoli. This volume (about 150 cc) remains in the passageways of the trachea, bronchi, and bronchioles. Since gas exchanges do not occur through the walls of these passages, this air is said to occupy *anatomic dead space.*

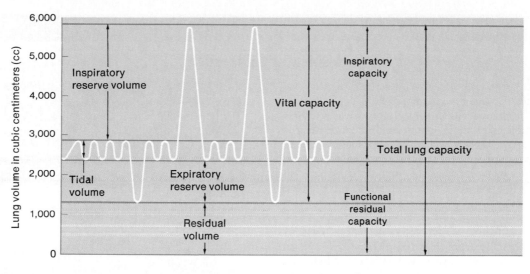

Figure 16.25   Respiratory air volumes and capacities.

CHART 16.4	Respiratory air volumes and capacities	
Name	Volume (average)	Description
Tidal volume (TV)	500 cc	Volume moved in or out of the lungs during quiet breathing
Inspiratory reserve volume (IRV)	3,000 cc	Volume that can be inhaled during forced breathing in addition to tidal volume
Expiratory reserve volume (ERV)	1,100 cc	Volume that can be exhaled during forced breathing in addition to tidal volume
Residual volume (RV)	1,200 cc	Volume that remains in the lungs at all times
Inspiratory capacity (IC)	3,500 cc	Maximum volume of air that can be inhaled following exhalation of tidal volume: IC=TV+IRV
Functional residual capacity (FRC)	2,300 cc	Volume of air that remains in the lungs following exhalation of tidal volume: FRC = ERV + RV
Vital capacity (VC)	4,600 cc	Maximum volume of air that can be exhaled after taking the deepest breath possible: VC = TV + IRV + ERV
Total lung capacity (TLC)	5,800 cc	Total volume of air that the lungs can hold: TLC = VC + RV

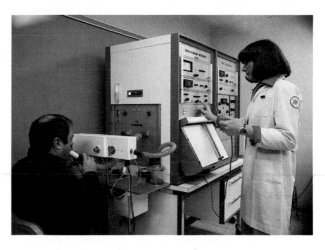

Figure 16.26   A spirometer can be used to measure respiratory air volumes.

combined, the total is called *physiologic dead space*. In a normal lung, however, the anatomic and physiologic dead spaces are essentially the same (about 150 cc).

The respiratory air volumes and capacities are summarized in chart 16.4.

With the exception of the residual volume, which is measured using special techniques, respiratory air volumes can be determined using an instrument called a *spirometer*, shown in figure 16.26. Such measurements may be useful in evaluating the courses of diseases like emphysema, pneumonia, or lung cancer in which functional lung tissue is lost. These measurements may also be valuable in studying the progress of bronchial asthma and other diseases that involve obstructions of the air passages.

Occasionally, air sacs in some regions of the lungs are nonfunctional due to poor blood flow in the adjacent capillaries. This creates *alveolar dead space*. If the anatomic and alveolar dead space volumes are

Respiratory Disorders

Although respiratory disorders are caused by a variety of factors, some are characterized by decreased ventilation. This group includes paralysis of various breathing muscles, bronchial asthma, emphysema, and lung cancer.

Paralysis of breathing muscles is sometimes caused by injuries to the respiratory center or to spinal nerve tracts that transmit motor impulses. In other instances, paralysis may be due to a disease, such as *poliomyelitis,* that affects parts of the central nervous system and injures motor neurons. The consequences of such paralysis depend on which muscles are affected. Sometimes by increasing their responses, other muscles are able to compensate for functional losses of a paralyzed muscle. If unaffected muscles are unable to ventilate the lungs adequately, a person must be provided with some type of mechanical breathing device in order to survive.

Bronchial asthma is a condition commonly caused by an *allergic reaction* to foreign substances in the respiratory tract. Typically, the foreign substance is a plant pollen that enters with inhaled air. As a result of this allergic reaction, the walls of the small bronchioles become edematous, the cells lining the respiratory tubes secrete abnormally large amounts of thick mucus, and the smooth muscles in these tubes contract. These muscles cause the bronchioles to constrict, reducing the diameters of the air passages. As these changes occur, the person finds it increasingly difficult to breathe and produces a characteristic wheezing sound as air moves through narrowed passages.

An asthmatic person usually finds it harder to force air out of the lungs than to bring it in. This is because inspiration involves powerful breathing muscles, and, as they contract, the lungs expand. This expansion of the tissues helps open the air passages. Expiration, on the other hand, is usually a passive process due to elastic recoil of stretched tissues. Also, it causes compression of the tissues and decreases the diameters of the bronchioles, and this adds to the problem of moving air through the narrowed air passages.

Emphysema is a progressive, degenerative disease characterized by the destruction of many alveolar walls. As a result, clusters of small air sacs merge to form larger chambers, so that the total surface area of the alveolar walls decreases. At the same time, the alveolar walls tend to lose their elasticity, and the capillary networks associated with the alveoli become less abundant (figure 16.27).

Because of the loss of tissue elasticity, a person with emphysema finds it increasingly difficult to force air out of the lungs. As was mentioned, normal expiration involves the passive elastic recoil of inflated tissues; in emphysema, abnormal muscular efforts are required to produce this movement.

The cause of emphysema is not well understood, but some investigators believe it develops in response to prolonged exposure to respiratory irritants, such as those in tobacco smoke and polluted air. Emphysema is becoming one

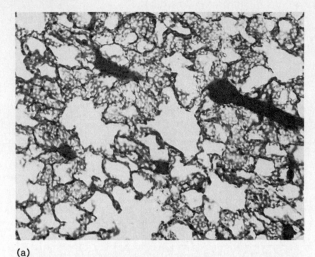

(a)

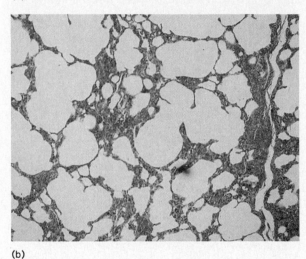

(b)

Figure 16.27 (*a*) Normal lung tissue. (*b*) As emphysema develops, the alveoli tend to merge, forming larger chambers.

of the more common respiratory disorders among older persons, although it is not limited to this group.

Lung cancer, like other cancers, involves an uncontrolled growth of abnormal cells. These cells develop in and around the normal tissues, and deprive them of nutrients. In effect, the cancer cells cause the death of normal cells by crowding them out.

Some cancerous growths in the lungs result from cancer cells that spread (metastasize) from other parts of the body, such as the breasts, alimentary tract, liver, or kidneys.

Primary pulmonary cancer, which begins in the lungs, is the most common form of cancer in males today. It is also becoming increasingly common among females and is rapidly replacing breast cancer as the leading cause of death from cancer among women. Primary pulmonary cancer may

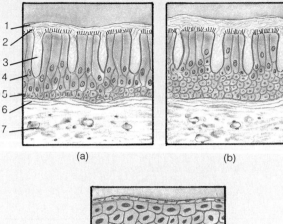

(a)   (b)

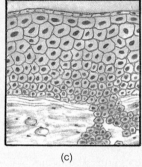

(c)

Figure 16.28 Lung cancer usually starts in the lining of the bronchus. (a) The normal epithelial lining shows (4) columnar cells with (2) hairlike cilia, (3) goblet cells that secrete (1) mucus, and (5) basal cells from which new columnar cells arise. (6) A basement membrane separates the epithelial cells from (7) the underlying connective tissue. (b) In the first stage of lung cancer, the basal cells divide repeatedly. The goblet cells secrete excessive mucus, and the cilia function less efficiently in moving the heavy mucus secretion. (c) With the continued multiplication of basal cells, the columnar and goblet cells are displaced. The basal cells penetrate the basement membrane and invade the deeper connective tissue.

arise from epithelial cells, connective tissue cells, or various blood cells. The most common form arises from epithelium and is called *bronchogenic carcinoma,* which means it originates from the cells that line the tubes of the bronchial tree. This type of cancer seems to occur in response to excessive irritation such as that produced by prolonged exposure to tobacco smoke (figure 16.28).

Once lung cancer cells have appeared, they are likely to produce masses that obstruct air passages and reduce the amount of alveolar surface available for gas exchange. Furthermore, bronchogenic carcinoma is likely to spread to other tissues relatively quickly and establish secondary cancers. Common sites of such secondary cancers include the lymph nodes, liver, bones, brain, and kidneys.

Lung cancer is often difficult to control. Usually it is treated by surgical removal of the diseased portions of the lungs, exposure to ionizing radiation, and the use of drugs (chemotherapy). Despite these treatments, the survival rate among lung cancer patients remains extremely low.

1. What is meant by tidal volume?
2. Distinguish between inspiratory and expiratory reserve volumes.
3. How is vital capacity measured?
4. How is the total lung capacity calculated?

Alveolar Ventilation

The amount of new atmospheric air that is moved into the respiratory passages each minute is called the *minute respiratory volume.* This volume can be calculated by multiplying the tidal volume by the breathing rate. Thus, if the tidal volume is 500 cc and the breathing rate is 12 breaths per minute, the minute respiratory volume is 500 cc × 12, or 6,000 cc per minute. However, not all of this new air reaches the alveoli. Instead, much of it remains in the air passages, occupying the physiologic dead space.

The amount of new air that does reach the alveoli and is available for gas exchange is calculated by subtracting the physiologic dead space (150 cc) from the tidal volume (500 cc). If the resulting volume (350 cc) is multiplied by the breathing rate (12 breaths per minute), the *alveolar ventilation rate* (4,200 cc per minute) is obtained. This alveolar ventilation rate is a major factor affecting the concentrations of oxygen and carbon dioxide in the alveoli; thus, it affects the exchange of gases between the alveolar air and the blood.

Nonrespiratory Air Movements

Air movements that occur in addition to breathing are called *nonrespiratory movements.* They are used to clear air passages, as in coughing and sneezing, or to express emotional feelings, as in laughing and crying.

Nonrespiratory movements usually result from *reflexes,* although sometimes they are initiated voluntarily. A cough, for example, can be produced through conscious effort or may be triggered by the presence of a foreign object in an air passage.

The act of *coughing* involves taking a deep breath, closing the glottis, and forcing air upward from the lungs against the closure. Then the glottis is suddenly opened, and a blast of air is forced upward from the lower respiratory tract. Usually this rapid rush of air will remove the substance that triggered the reflex.

The most sensitive areas of the air passages are in the larynx and in regions near the branches of the major bronchi. The distal portions of the bronchioles (respiratory bronchioles), alveolar ducts, and alveoli lack a nerve supply. Consequently, before any material in these parts can trigger a cough reflex, it must be moved into the larger passages of the respiratory tract.

A *sneeze* is much like a cough, but it clears the upper respiratory passages rather than the lower ones.

CHART 16.5 Nonrespiratory air movements

Air movement	Mechanism	Function
Coughing	Deep breath is taken, glottis is closed, and air is forced against the closure; suddenly the glottis is opened and a blast of air passes upward	Clears lower respiratory passages
Sneezing	Same as coughing, except air moving upward is directed into the nasal cavity by depressing the uvula	Clears upper respiratory passages
Laughing	Deep breath is released in a series of short expirations	Expresses emotional happiness
Crying	Same as laughing	Expresses emotional sadness
Hiccuping	Diaphragm contracts spasmodically while glottis is closed	No useful function
Yawning	Deep breath is taken	Ventilates a larger proportion of the alveoli and aids oxygenation of the blood
Speech	Air is forced through the larynx, causing vocal cords to vibrate; words are formed by lips, tongue, and soft palate	Vocal communication

This reflex is usually initiated by a mild irritation in the lining of the nasal cavity, and, in response, a blast of air is forced up through the glottis. This time, the air is directed into the nasal passages by depressing the uvula, thus closing the opening between the pharynx and the oral cavity.

Laughing involves taking a breath and releasing it in a series of short expirations. *Crying* consists of very similar movements, and sometimes it is necessary to note a person's facial expression in order to distinguish laughing from crying.

A *hiccup* is caused by sudden inspiration due to a spasmodic contraction of the diaphragm while the glottis is closed. The sound of the hiccup is caused by air striking the vocal folds. The reason for hiccups is not well understood, and they apparently serve no useful function.

Yawning is thought to aid respiration by providing an occasional deep breath. During normal, quiet breathing, not all of the alveoli are ventilated and some blood may pass through the lungs without becoming well oxygenated. It is believed that this low blood oxygen concentration somehow triggers the yawn reflex, and, in response, a very deep breath is taken. The deep breath ventilates a larger proportion of the alveoli, and the blood oxygen concentration rises.

As described previously, *speech* results from the sounds that are produced when air is forced through the larynx, causing the vocal cords to vibrate. Words are formed from these sounds by actions of the lips, tongue, and soft palate.

Chart 16.5 summarizes the characteristics of the nonrespiratory air movements.

1. How is the minute respiratory volume calculated? The alveolar ventilation rate?
2. What nonrespiratory air movements help clear the air passages?
3. What nonrespiratory air movements are used to express emotions?
4. What seems to be the function of a yawn?

Control of Breathing

Although the respiratory muscles can be controlled voluntarily, normal breathing is a rhythmic, involuntary act that continues when a person is unconscious.

Respiratory Center

Breathing is controlled by a poorly defined group of neurons in the brain stem called the **respiratory center.** This center periodically initiates impulses that travel on cranial and spinal nerves to various breathing muscles, causing inspiration and expiration. The respiratory center is also able to adjust the rate and depth of breathing. As a result, cellular needs for a supply of oxygen and for the removal of carbon dioxide are met, even during periods of strenuous physical exercise.

The components of the respiratory center are widely scattered throughout the pons and medulla oblongata. However, two areas of the respiratory center are of special interest. They are the rhythmicity area of the medulla and the pneumotaxic area of the pons, shown in figure 16.29.

The **medullary rhythmicity area** includes two groups of neurons that extend throughout the length of

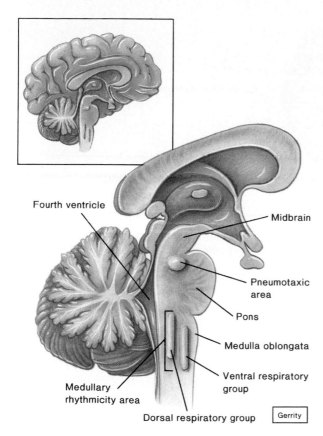

Figure 16.29   The respiratory center is located in the pons and the medulla oblongata.

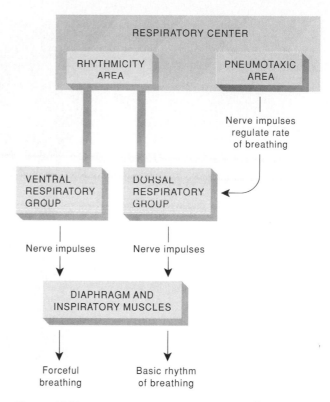

Figure 16.30   Breathing is controlled by the medullary rhythmicity and pneumotaxic areas of the respiratory center.

the medulla oblongata. They are called the dorsal respiratory group and the ventral respiratory group.

The *dorsal respiratory group* is responsible for the basic rhythm of breathing. The neurons of this group emit bursts of impulses that signal the diaphragm and other inspiratory muscles to contract. The impulses of each burst begin weakly, increase in strength for about two seconds, and cease abruptly. The breathing muscles that contract in response to the impulses cause the volume of air entering the lungs to increase steadily. The neurons remain inactive while expiration occurs passively, and then they emit another burst of inspiratory impulses so that the inspiration-expiration cycle is repeated.

The *ventral respiratory group* is quiescent during normal breathing. However, when more forceful breathing is necessary, the neurons in this group generate impulses that increase inspiratory movement. Other neurons in the group activate the muscles associated with forceful expiration as well. (See figure 16.29.)

The neurons in the **pneumotaxic area** transmit impulses to the dorsal respiratory group continuously

and regulate the duration of inspiratory bursts originating from the dorsal group. In this way the pneumotaxic neurons control the rate of breathing. More specifically, when the pneumotaxic signals are strong, the inspiratory bursts have shorter durations, and the rate of breathing is increased; when the pneumotaxic signals are weak, the inspiratory bursts have longer durations, and the rate of breathing is decreased (figure 16.30).

1. Where is the respiratory center located?
2. Describe how the respiratory center functions to maintain a normal breathing pattern.
3. Explain how the breathing pattern may be changed.

Factors Affecting Breathing

In addition to the controls exerted by the respiratory center, the breathing rate and depth are influenced by a variety of other factors. These include the presence of certain chemicals in body fluids, the degree to which the lung tissues are stretched, and the person's emotional state. For example, there are *chemosensitive areas* within the respiratory center. These areas are located in the ventral portion of the medulla oblongata near the origins of the vagus nerves, and they are very

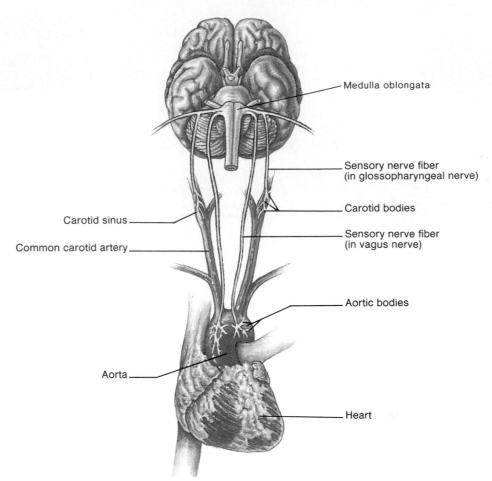

Figure 16.31 Chemoreceptors in the carotid and aortic bodies are stimulated by decreased blood oxygen concentration.

Labels in figure:
- Medulla oblongata
- Sensory nerve fiber (in glossopharyngeal nerve)
- Carotid bodies
- Sensory nerve fiber (in vagus nerve)
- Carotid sinus
- Common carotid artery
- Aortic bodies
- Aorta
- Heart

sensitive to changes in the blood concentrations of carbon dioxide and hydrogen ions. Thus, if the concentration of carbon dioxide or hydrogen ions rises, the chemosensitive areas signal the respiratory center, and the rate of breathing is increased. The similarity of the effects of these two substances is related to the fact that carbon dioxide combines with water in the blood or cerebrospinal fluid to form carbonic acid (H_2CO_3):

$$CO_2 + H_2O \rightleftharpoons H_2CO_3$$

The carbonic acid thus formed soon becomes ionized, releasing *hydrogen ions* (H^+) and *bicarbonate ions* (HCO_3^-):

$$H_2CO_3 \rightleftharpoons H^+ + HCO_3^-$$

Apparently, it is the presence of hydrogen ions that influences the chemosensitive areas rather than the presence of carbon dioxide molecules. In any event, a person's breathing rate increases when air rich in carbon dioxide is inhaled. As a result of the increased breathing rate, more carbon dioxide is lost in exhaled air, and the blood concentrations of carbon dioxide and hydrogen ions are reduced.

Air to which additional carbon dioxide has been added is sometimes used to stimulate the rate and depth of breathing. Ordinary air is about 0.04% carbon dioxide. If a patient inhales air containing 4% carbon dioxide, the breathing rate usually doubles.

Low blood oxygen concentration seems to have little direct effect on the chemosensitive areas associated with the respiratory center. Instead, changes in the blood oxygen concentration are sensed by *chemoreceptors* in specialized structures called the *carotid* and *aortic bodies,* which are located in the walls of certain large arteries (carotid arteries and aorta) in the neck and thorax. (See figure 16.31.) When these receptors are stimulated by decreased oxygen concentration, impulses are transmitted to the respiratory center, and the breathing rate is increased. This mechanism is usually not triggered until the blood oxygen concentration reaches a very low level; thus, oxygen seems to play only a minor role in the control of normal respiration.

Although the chemoreceptors of the carotid and aortic bodies are also stimulated by changes in the blood

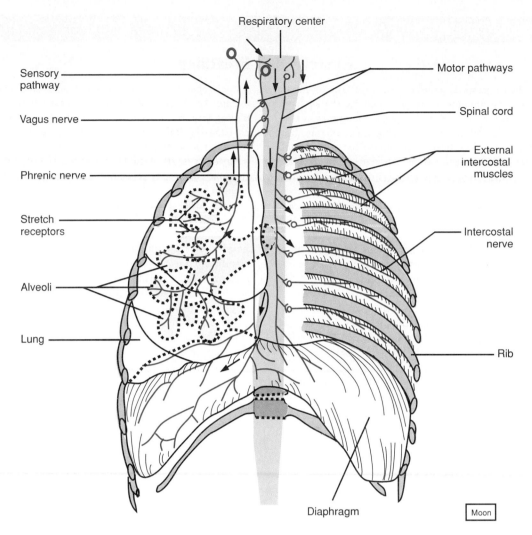

Respiratory center

Sensory pathway

Vagus nerve

Phrenic nerve

Stretch receptors

Alveoli

Lung

Motor pathways

Spinal cord

External intercostal muscles

Intercostal nerve

Rib

Diaphragm

Moon

Figure 16.32 In the process of inspiration, motor impulses travel from the respiratory center to the diaphragm and external intercostal muscles, which contract and cause the lungs to expand. This expansion stimulates stretch receptors in the lungs to send inhibiting impulses to the respiratory center, thus preventing overinflation.

concentrations of carbon dioxide and hydrogen ions, these substances have a much more powerful effect when they act on the chemosensitive areas of the respiratory center. Thus, the effects of carbon dioxide and hydrogen ions on the carotid and aortic bodies are relatively unimportant.

An exception to the normal pattern of chemical control may occur in patients who have chronic obstructive pulmonary diseases (COPD), such as asthma, bronchitis, and emphysema. Over a period of time, these patients seem to adapt to high concentrations of carbon dioxide, and for them, low oxygen concentrations may serve as effective respiratory stimuli.

An *inflation reflex* (Hering-Breuer reflex) helps regulate the depth of breathing. This reflex occurs when

stretch receptors in the visceral pleura, bronchioles, and alveoli are stimulated as a result of lung tissues being stretched. The sensory impulses of the reflex travel via the vagus nerves to the pneumotaxic area of the respiratory center, and cause the duration of inspiratory movements to shorten. This action prevents overinflation of the lungs during forceful breathing. (See figure 16.32.)

The normal breathing pattern also may be altered if a person is emotionally upset. Fear, for example, typically causes an increased breathing rate, as does pain; thus, a person may gasp in response to a sudden fright or the chill of a cold shower.

In addition, because the respiratory muscles are voluntary, the breathing pattern can be altered consciously. In fact, breathing can be stopped altogether for a time.

Exercise and Breathing

When a person engages in moderate to heavy physical exercise, the amount of oxygen used by the skeletal muscles increases greatly. For example, a young man at rest will utilize about 250 milliliters of oxygen per minute, but he may require 3,600 milliliters per minute during maximal exercise. While oxygen utilization is increasing, the volume of carbon dioxide produced increases also. Since decreased blood oxygen and increased blood carbon dioxide concentration are stimulating to the respiratory center, it is not surprising that exercise is accompanied by an increased breathing rate. However, studies have revealed that blood oxygen and carbon dioxide concentrations usually remain nearly unchanged during exercise—a reflection of the respiratory system's effectiveness in obtaining oxygen and releasing carbon dioxide to the outside.

This observation has led investigators to search for other factors that might increase the breathing rate during exercise. The mechanism that seems to be responsible for most of the increase involves the *cerebral cortex* and the *proprioceptors* associated with muscles and joints. (See chapter 12.) Specifically, the cortex seems to transmit stimulating impulses to the respiratory center whenever it signals the skeletal muscles to contract. At the same time, muscular movements stimulate the proprioceptors, and a *joint reflex* is triggered. In this reflex, sensory impulses are transmitted from the proprioceptors to the respiratory center, and the breathing rate increases.

Whenever an increase in the breathing rate occurs during exercise, an increase in the blood flow is also needed in order to meet the needs of the skeletal muscles. Thus, physical exercise places an increased demand on both the respiratory and the circulatory systems. If either of these systems fails to keep up with cellular demands, the person will begin to feel out of breath. This feeling, however, is usually due to the inability of the heart and circulatory system to move enough blood between the lungs and the body cells, rather than to the inability of the respiratory system to provide enough air.

CHART 16.6 Factors affecting breathing

Factor	Receptors stimulated	Response	Effect
Stretch of tissues	Stretch receptors in visceral pleura, bronchioles, and alveoli	Inhibits inspiration	Prevents overinflation of lungs during forceful breathing
Low blood oxygen	Chemoreceptors in carotid and aortic bodies	Increases breathing rate	Increases blood oxygen concentration
High blood carbon dioxide	Chemosensitive areas of the respiratory center	Increases breathing rate	Decreases blood carbon dioxide concentration
High blood hydrogen ion	Chemosensitive areas of the respiratory center	Increases breathing rate	Decreases blood hydrogen ion concentration

If a person decides to stop breathing, the blood concentrations of carbon dioxide and hydrogen ions begin to rise, and the concentration of oxygen falls. These changes stimulate the respiratory center, and soon the need to inhale overpowers the desire to hold the breath. On the other hand, a person can increase the breath-holding time by breathing rapidly and deeply in advance. This action, termed **hyperventilation,** causes a lowering of the blood carbon dioxide concentration. Following hyperventilation, it takes longer than usual for the carbon dioxide concentration to reach the level needed to produce an overwhelming effect upon the respiratory center.

Factors affecting breathing are summarized in chart 16.6.

Sometimes a person who is emotionally upset may hyperventilate, become dizzy, and lose consciousness. This condition seems to be due to a lowered carbon dioxide concentration followed by a rise in pH (respiratory alkalosis). This change in pH is accompanied by a localized vasoconstriction of cerebral blood arterioles and, consequently, a decreased blood flow to nearby brain cells. As a result, the oxygen supply to the brain cells may be inadequate, and the person may faint.

Hyperventilation should never be used to help hold the breath while swimming, because the person who has hyperventilated may lose consciousness under water and drown.

1. Describe the inflation reflex.
2. What chemical factors affect breathing?
3. How does hyperventilation result in a decreased respiratory rate?

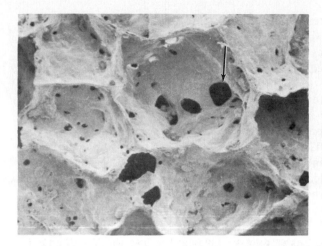

Figure 16.33 Alveolar pores (arrow) allow air to pass from one alveolus to another.

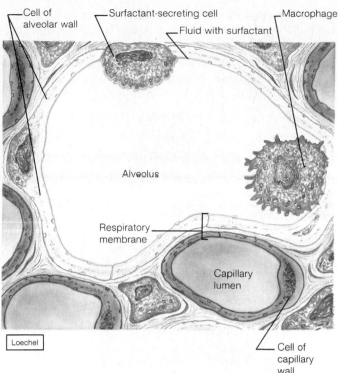

Figure 16.34 The respiratory membrane consists of the walls of the alveolus and the capillary.

Alveolar Gas Exchanges

While other parts of the respiratory system conduct air and move it in and out of the air passages, the alveoli carry on the vital process of exchanging gases between the air and the blood.

Alveoli

The **alveoli** are microscopic air sacs clustered at the distal ends of the finest respiratory tubes—the alveolar ducts. Each alveolus consists of a tiny space surrounded by a thin wall that separates it from adjacent alveoli. There are minute openings, called **alveolar pores,** in the walls of some alveoli, and although the function of these pores is not well understood, they may permit air to pass from one alveolus to another (figure 16.33). This arrangement may provide alternate air pathways if the passages in some portions of the lung become obstructed.

Alveolar pores may also allow phagocytic cells, called *alveolar macrophages,* to move between adjacent alveoli (figure 16.34). These macrophages, which normally inhabit the alveoli, help keep the alveoli clean by phagocytizing various airborne agents, such as bacterial cells.

Respiratory Membrane

The wall of an alveolus consists of an inner lining of simple squamous epithelium and a dense network of capillaries, which are also lined with simple squamous epithelial cells. Thin basement membranes separate the layers of these flattened cells, and in the spaces between them are elastic and collagenous fibers that help support the alveolar wall. There are at least two thicknesses of epithelial cells and basement membranes be-

tween the air in an alveolus and the blood in a capillary. These layers make up the **respiratory membrane** (alveolar-capillary membrane), which is of vital importance because through this membrane gas exchange occurs between the alveolar air and the blood. (See figures 16.34 and 16.35.)

Diffusion through the Respiratory Membrane

As described in chapter 3, gas molecules diffuse from regions where they are in higher concentration toward regions where they are in lower concentration. Similarly, gases move from regions of higher pressure toward regions of lower pressure, and the *pressure* of a gas determines the rate at which it will diffuse from one region to another.

Measured by volume, ordinary air is about 78% nitrogen, 21% oxygen, and 0.04% carbon dioxide. Air also contains small amounts of other gases that have little or no physiological importance.

In a mixture of gases, such as the air, each gas is responsible for a portion of the total weight or pressure produced by the mixture. The amount of pressure each gas contributes is called the **partial pressure,** and this pressure is directly related to the concentration of the gas in the mixture. For example, because air is 21% oxygen, this gas is responsible for 21% of the atmospheric pressure. Because 21% of 760 mm Hg is equal to 160 mm Hg, the partial pressure of oxygen, symbolized P_{O_2}, in atmospheric air is said to be 160 mm Hg. Similarly, the partial pressure of carbon dioxide (P_{CO_2}) in air can be calculated as 0.3 mm Hg.

When a mixture of gases dissolves in the blood, each gas exerts its own partial pressure in proportion to its dissolved concentration. Furthermore, each gas will diffuse between the blood and its surroundings, and as figure 16.36 shows, this movement will tend to equalize its partial pressures in the two regions.

For example, the P_{CO_2} in capillary blood is 45 mm Hg, but the P_{CO_2} in alveolar air is 40 mm Hg. As a consequence of the difference between these partial pressures, carbon dioxide diffuses from the blood, where its partial pressure is higher, through the respiratory membrane, and into the alveolar air. When the blood leaves the lungs, its P_{CO_2} is 40 mm Hg, which is about the same as the P_{CO_2} of the alveolar air.

Similarly, the P_{O_2} of capillary blood is 40 mm Hg, but that of alveolar air is 104 mm Hg. Thus, oxygen diffuses from the alveolar air into the blood, and the blood leaves the lungs with a P_{O_2} of 100 mm Hg.

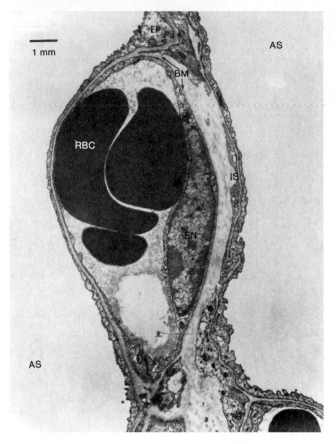

Figure 16.35 Electron micrograph of a capillary located between two adjacent alveoli. (*AS*, alveolar space; *RBC*, red blood cell; *BM*, basement membrane; *IS*, interstitial connective tissue; *EN*, epithelial nucleus.)

If the lungs are exposed to a high oxygen concentration (hyperoxia) for a prolonged time, tissues may be damaged. The lung tissues most likely to suffer are those that form the capillary walls, and as a result of such damage, excessive amounts of fluid may escape and flood the alveolar air spaces. Fluid in the alveoli interferes with gas exchange, and this may lead to death.

Similarly, the retinal capillaries of premature infants are very sensitive to damage by hyperoxia. The eyes of infants who are exposed to excessive oxygen may develop *retrolental fibroplasia* (RLF), a condition that may lead to blindness.

1. Describe the structure of the respiratory membrane.
2. What is meant by partial pressure?
3. What causes oxygen and carbon dioxide to move across the respiratory membrane?

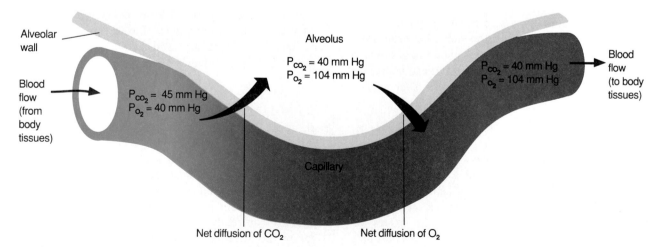

Figure 16.36 Gas exchanges occur between the air of the alveolus and the blood of the capillary as a result of differences in partial pressures.

Disorders Involving Gas Exchange

As mentioned previously, the disorder called *emphysema* is accompanied by a progressive decrease in the surface area available for gas exchanges, because the alveoli tend to merge and form larger chambers. The functional surface of an emphysematous lung may be reduced to one fourth that of a normal lung. This greatly decreases the lung's ability to carry on gas exchanges by diffusion, and its ability to oxygenate blood is further decreased by a loss of capillaries. Similar problems develop in pneumonia, tuberculosis, and a condition called atelectasis.

When a person has *pneumonia,* the alveoli become filled with fluids and blood cells. Pneumonia is most commonly due to an acute infection by bacteria called *Streptococcus pneumoniae* or to the presence of certain viruses.

In response to such infections, the linings of the alveoli become edematous and abnormally permeable. As a result, fluids and white blood cells enter the air sacs in abnormal quantities. As the alveoli fill up, the surface area available for gas exchanges is greatly reduced.

This type of infection is likely to spread to other regions of the respiratory tract and may involve lobes (lobar pneumonia) or an entire lung. The death rate in untreated cases of pneumonia is relatively high.

Tuberculosis is a disease caused by the bacteria called *Mycobacterium tuberculosis* that usually infects the lung tissues but may spread to other organs as well. In this condition, fibrous connective tissue develops around the sites of infection, forming structures called *tubercles.* By walling off the bacteria, the tubercles help inhibit their spread. Sometimes the protective mechanism fails, and the bacteria become widely distributed throughout the lungs.

In the later stages of tuberculosis, other kinds of bacteria are likely to appear and cause secondary infections in the lungs. There may be extensive destruction of lung tissues that greatly reduces the surface area of the respiratory membrane. In addition, the widespread development of fibrous tissue within a tubercular lung causes an increase in the thickness of the respiratory membrane, which further restricts gas exchange.

Atelectasis refers to the collapse of a lung or some part of it, together with the collapse of the blood vessels that supply the affected region.

Although atelectasis can be caused by a variety of factors, it is commonly due to an obstruction of a respiratory tube. Such a blockage may result from an abnormal secretion of mucus or the presence of a foreign object. In either case, the air in the alveoli beyond the obstruction tends to be absorbed by the lung tissues. As the air pressure in the alveoli decreases, their elastic walls cause them to collapse, and they become nonfunctional.

Fortunately, after a portion of a lung collapses, the functional regions that remain are often able to carry on enough gas exchange to sustain the body cells.

Respiratory distress syndrome (RDS) is a special form of atelectasis that is characterized by the collapse of the alveoli. In premature infants, RDS is due to a lack of surfactant, as was described previously. A related condition, called *adult respiratory distress syndrome* (ARDS) sometimes occurs when lung tissues are damaged. Common sources of such damage include respiratory infections, the aspiration of the stomach contents into the lungs, and physical trauma to the lungs from an injury or surgical procedure. As a result of such damage, the respiratory membrane that separates the air of the alveoli and the blood of the pulmonary capillaries is disrupted. This allows fluid to escape from the capillaries excessively, and it floods the alveoli, which collapse in response. In ARDS, surfactant is nonfunctional.

Transport of Gases

The transport of oxygen and carbon dioxide between the lungs and the body cells is a function of the blood. As these gases enter the blood, they dissolve in the liquid portion of the blood (plasma). They combine chemically with various blood components, and most are carried in combination with other atoms or molecules.

Oxygen Transport

Almost all the oxygen (over 98%) carried in the blood is combined with the compound **hemoglobin** that occurs within the red blood cells. Hemoglobin is responsible for the color of these blood cells. The remainder of the oxygen is dissolved in the blood plasma.

Hemoglobin has a complex molecular structure that consists of two subunits, called *heme* and *globin.* (See chapter 17.) Globin is a protein that contains 574 amino acids arranged in four polypeptide chains. Each chain is associated with a heme group, and each heme group contains an atom of iron. Each iron atom can combine loosely with an oxygen molecule. Thus, as oxygen dissolves in blood, it combines rapidly with hemoglobin, and the result of this chemical reaction is a new substance called **oxyhemoglobin.**

The amount of oxygen that combines with hemoglobin is determined by the P_{O_2}. The greater the P_{O_2}, the more oxygen will combine with hemoglobin, until the hemoglobin molecules are saturated with oxygen.

The chemical bonds that form between oxygen and hemoglobin molecules are relatively unstable, and as the P_{O_2} decreases, oxygen is released from oxyhemoglobin molecules (figure 16.37). This happens in tissues where cells have used oxygen in their respiratory processes and the free oxygen diffuses from the blood into nearby cells, as shown in figure 16.38.

The amount of oxygen that is released from oxyhemoglobin is affected by several other factors, including the blood concentration of carbon dioxide, the pH, and the temperature of the blood. Thus, as the concentration of carbon dioxide (P_{CO_2}) increases, oxyhemoglobin tends to release more oxygen (figure 16.39). Also, as the blood becomes more acidic or as the blood temperature increases, more oxygen is released (figures 16.40 and 16.41).

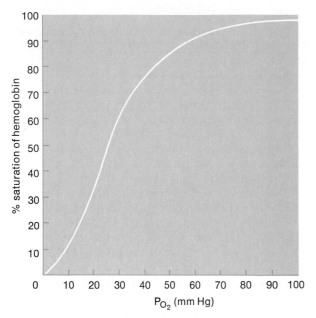

Oxyhemoglobin dissociation at 38°C

Figure 16.37   The oxyhemoglobin dissociation curve indicates that increased quantities of oxygen are released as the P_{O_2} decreases.

At high altitudes, the proportion of oxygen in air remains the same (about 21%), but the P_{O_2} decreases. Consequently, if a person breathes high altitude air, oxygen diffuses less rapidly from the alveoli into the blood, and the hemoglobin is less likely to become saturated with oxygen. Such a person may develop the symptoms of oxygen deficiency (hypoxia). For example, at 8,000 feet a person may experience anxiety, restlessness, an increased breathing rate, and a rapid pulse. At 12,500 feet, where the P_{O_2} is only 100 mm Hg rather than 160 mm Hg at sea level, the symptoms may include drowsiness, mental fatigue, headache, and nausea. As altitude increases, these conditions are likely to increase in intensity until the person loses consciousness and may die at about 23,000 feet.

Due to these factors, more oxygen is released from the blood to the skeletal muscles during periods of exercise because the increased muscular activity accompanied by an increased use of oxygen causes an increase in the P_{CO_2}, a decrease in the pH value, and a rise in the local temperature. At the same time, less active cells receive relatively smaller amounts of oxygen.

Carbon Monoxide

Carbon monoxide (CO) is a toxic gas produced in gasoline engines and some stoves as a result of incomplete combustion of fuels. It is also a component of tobacco smoke. The toxic effect of carbon monoxide occurs because it combines with hemoglobin more effectively than oxygen. Furthermore, carbon monoxide does not dissociate readily from hemoglobin. Thus, when a person breathes carbon monoxide, increasing quan-

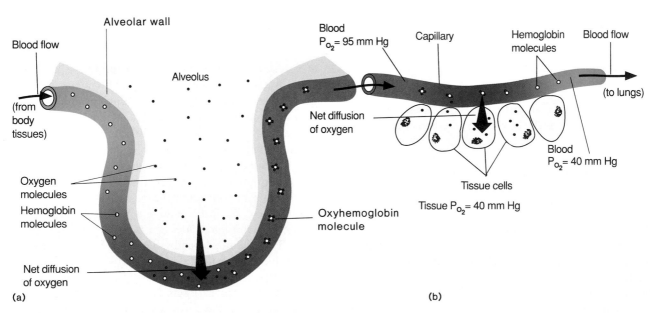

Figure 16.38   (a) Oxygen molecules, entering the blood from the alveolus, bond to hemoglobin, forming oxyhemoglobin. (b) In the regions of the body cells, oxyhemoglobin releases oxygen.

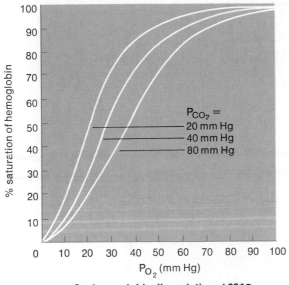

Oxyhemoglobin dissociation at 38°C

Figure 16.39 The amount of oxygen released from oxyhemoglobin increases as the PCO₂ increases.

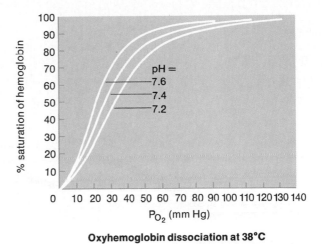

Oxyhemoglobin dissociation at 38°C

Figure 16.40 The amount of oxygen released from oxyhemoglobin increases as the blood pH decreases.

tities of hemoglobin become unavailable for oxygen transport, and the body cells soon begin to suffer from oxygen deficiency.

A patient with carbon monoxide poisoning may be treated by administering oxygen in high concentration to replace some of the carbon monoxide bound to hemoglobin molecules. Carbon dioxide is usually administered simultaneously to stimulate the respiratory center which, in turn, increases the breathing rate. Rapid breathing is desirable because it helps reduce the concentration of carbon monoxide in the alveoli.

1. How is oxygen transported from the lungs to the body cells?
2. What factors affect the release of oxygen from oxyhemoglobin?
3. Why is carbon monoxide toxic?

Carbon Dioxide Transport

Blood flowing through the capillaries of the body tissues gains carbon dioxide because the tissues have a relatively high PCO₂. This carbon dioxide is transported to the lungs in one of three forms: as carbon dioxide dissolved in the blood, as part of a compound formed by bonding to hemoglobin, or as part of a bicarbonate ion (figure 16.42).

The amount of carbon dioxide that dissolves in the blood is determined by its partial pressure. The higher the PCO₂ of the tissues, the more carbon dioxide will go into solution. However, only about 7% of the carbon dioxide is transported in this form.

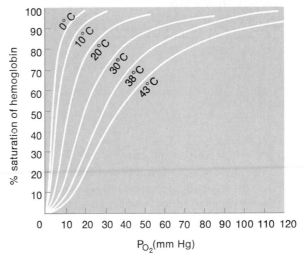

Oxyhemoglobin dissociation at various temperatures

Figure 16.41 The amount of oxygen released from oxyhemoglobin increases as the blood temperature increases.

Unlike oxygen, which combines with the iron atoms of hemoglobin molecules, carbon dioxide bonds with the amino groups ($-NH_2$) of these molecules. Consequently, oxygen and carbon dioxide do not compete for bonding sites, and both gases can be transported by a hemoglobin molecule at the same time.

When carbon dioxide combines with hemoglobin, a loosely bound compound called **carbaminohemoglobin** is formed. This substance decomposes readily in regions where the PCO₂ is low, and thus releases its carbon dioxide. Although this method of transporting carbon dioxide is theoretically quite effective, carbaminohemoglobin forms relatively slowly. It is believed that about 15%–25% of the total carbon dioxide is carried this way.

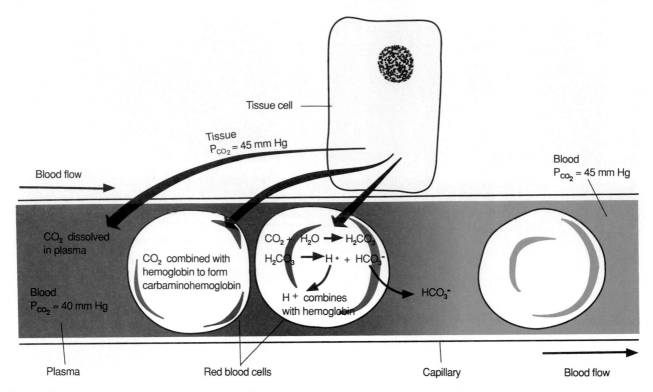

Figure 16.42 Carbon dioxide produced by tissue cells is transported in the blood in a dissolved state, either combined with hemoglobin or in the form of bicarbonate ions (HCO_3^-).

The most important carbon dioxide transport mechanism involves the formation of **bicarbonate ions** (HCO_3^-). As mentioned previously, carbon dioxide reacts with water to form carbonic acid (H_2CO_3). Although this reaction occurs slowly in the blood plasma, much of the carbon dioxide diffuses into the red blood cells, and these cells contain an enzyme called **carbonic anhydrase,** which speeds the reaction between carbon dioxide and water.

The resulting carbonic acid dissociates almost immediately, releasing hydrogen ions (H^+) and bicarbonate ions (HCO_3^-):

$$H_2CO_3 \rightarrow H^+ + HCO_3^-$$

Most of the hydrogen ions combine quickly with hemoglobin molecules; thus, hydrogen ions are prevented from accumulating and causing a great change in the blood pH. The bicarbonate ions tend to diffuse out of the red blood cells and enter the blood plasma. It is estimated that at least 70% of the carbon dioxide transported in the blood is carried in this form.

As the bicarbonate ions leave the red blood cells and enter the plasma, *chloride ions,* which also have negative charges, are repelled electrically, and they move from the plasma into the red blood cells. This exchange in position of the two negatively charged ions, shown in figure 16.43, maintains the ionic balance be-

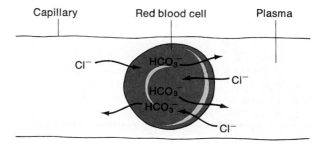

Figure 16.43 As bicarbonate ions (HCO_3^-) diffuse out of the red blood cell, chloride ions (Cl^-) from the plasma diffuse into the cell, thus maintaining the electrical balance between ions. This exchange of ions is called the chloride shift.

tween the red blood cells and the plasma. It is termed the **chloride shift.**

When blood passes through the capillaries of the lungs, it loses its dissolved carbon dioxide by diffusion into the alveoli. This occurs in response to the relatively low P_{CO_2} of the alveolar air. At the same time, hydrogen ions and bicarbonate ions in the red blood cells recombine to form carbonic acid molecules, and, under the influence of carbonic anhydrase, the carbonic acid gives rise to carbon dioxide and water:

$$H^+ + HCO_3^- \rightarrow H_2CO_3 \rightarrow CO_2 + H_2O$$

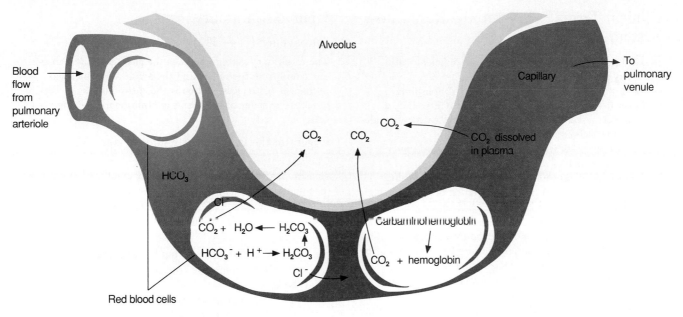

Figure 16.44 In the lungs, carbon dioxide diffuses from the blood into the alveoli.

Carbaminohemoglobin also releases its carbon dioxide, and, as carbon dioxide continues to diffuse out of the blood, an equilibrium is established between the P_{CO_2} of the blood and the P_{CO_2} of the alveolar air. This process is summarized in figure 16.44.

Chart 16.7 summarizes the transport of blood gases.

1. Describe three ways carbon dioxide can be transported from body cells to the lungs.
2. How is it possible for hemoglobin to carry oxygen and carbon dioxide at the same time?
3. What is meant by the chloride shift?
4. How is carbon dioxide released from the blood into the lungs?

Utilization of Oxygen

The most important use of oxygen by body cells occurs during cellular respiration, even though certain cells may use oxygen for other purposes.

Cellular Respiration

As discussed in chapter 4, **cellular respiration** involves the molecular breakdown of substances, such as glucose, and the release of energy from them.

To review, this process involves two phases. The first phase of cellular respiration takes place in the cytoplasm of cells. It occurs in the absence of oxygen and is therefore *anaerobic*. For glucose, this anaerobic phase includes the conversion of glucose to pyruvic acid and is accompanied by the transfer of a relatively small amount of energy to ATP molecules.

CHART 16.7	Gases transported in blood	
Gas	Reaction involved	Substance transported
Oxygen	Combines with iron atoms of hemoglobin molecules	Oxyhemoglobin
Carbon dioxide	About 7% dissolves in plasma	Carbon dioxide
	About 23% combines with the amino groups of hemoglobin molecules	Carbamino-hemoglobin
	About 70% reacts with water to form carbonic acid; the carbonic acid then dissociates to release hydrogen ions and bicarbonate ions	Bicarbonate ions

The second phase of cellular respiration takes place within the mitochondria and is *aerobic*, because it requires oxygen. During this phase, the pyruvic acid is converted into acetyl coenzyme A, and these molecules enter a complex series of chemical reactions that constitute the *citric acid cycle*. (See figure 4.12.)

1. What is meant by cellular respiration?
2. Distinguish between the anaerobic and aerobic phases of cellular respiration.

Clinical Terms Related to the Respiratory System

anoxia (ah-nok′se-ah) An absence or a deficiency of oxygen within tissues.

apnea (ap-ne′ah) A temporary absence of breathing.

asphyxia (as-fik′se-ah) A condition characterized by a deficiency of oxygen and an excess of carbon dioxide in the blood and tissues.

atelectasis (at″e-lek′tah-sis) The collapse of a lung or some portion of it.

bradypnea (brad″e-ne′ah) Abnormally slow breathing.

bronchitis (brong-ki′tis) An inflammation of the bronchial lining.

Cheyne-Stokes respiration (chān stōks res″pĭ-ra′shun) Irregular breathing characterized by a series of shallow breaths that increase in depth and rate, followed by breaths that decrease in depth and rate.

dyspnea (disp′ne-ah) Difficulty in breathing.

eupnea (up-ne′ah) Normal breathing.

hemothorax (he″mo-tho′raks) The presence of blood in the pleural cavity.

hypercapnia (hi″per-kap′ne-ah) Excessive carbon dioxide in the blood.

hyperoxia (hi″per-ok′se-ah) An excess in the oxygenation of the blood.

hyperpnea (hi″perp-ne′ah) An increase in the depth and rate of breathing.

hyperventilation (hi″per-ven″ti-la′-shun) Prolonged, rapid, and deep breathing.

hypoxemia (hi″pok-se′me-ah) A deficiency in the oxygenation of the blood.

hypoxia (hi-pok′se-ah) A diminished availability of oxygen in the tissues.

lobar pneumonia (lo′ber nu-mo′ne-ah) Pneumonia that affects an entire lobe of a lung.

pleurisy (ploo′ri-se) An inflammation of the pleural membranes.

pneumoconiosis (nu″mo-ko″ne-o′sis) A condition characterized by the accumulation of particles from the environment in the lungs and the reaction of the tissues to their presence.

pneumothorax (nu″mo-tho′raks) The entrance of air into the space between the pleural membranes, followed by collapse of the lung.

rhinitis (ri-ni′tis) An inflammation of the nasal cavity lining.

sinusitis (si″nŭ-si′tis) An inflammation of the sinus cavity lining.

tachypnea (tak″ip-ne′ah) Rapid, shallow breathing.

tracheotomy (tra″ke-ot′o-me) An incision in the trachea for exploration or the removal of a foreign object.

Chapter Summary

Introduction (page 582)

The respiratory system includes the passages that transport air to and from the lungs, and the air sacs in which gas exchanges occur. Respiration is the entire process by which gases are exchanged between the atmosphere and the body cells.

Organs of the Respiratory System (page 582)

The respiratory system includes the nose, nasal cavity, sinuses, pharynx, larynx, trachea, bronchial tree, and lungs. The upper respiratory tract includes the respiratory organs outside the thorax; the lower respiratory tract includes those respiratory organs within the thorax.

1. Nose
   a. The nose is supported by bone and cartilage.
   b. Nostrils provide entrances for air.
2. Nasal cavity
   a. The nasal cavity is a space behind the nose.
   b. It is divided medially by the nasal septum.
   c. Nasal conchae divide the cavity into passageways and help increase the surface area of the mucous membrane.
   d. Mucous membrane filters, warms, and moistens incoming air.
   e. Particles trapped in the mucus are carried to the pharynx by ciliary action and are swallowed.
3. Sinuses
   a. Sinuses are spaces in the bones of the skull that open into the nasal cavity.
   b. They are lined with mucous membrane that is continuous with the lining of the nasal cavity.
4. Pharynx
   a. The pharynx is located behind the mouth and between the nasal cavity and the larynx.
   b. It functions as a common passage for air and food.
   c. It aids in creating vocal sounds.
5. Larynx
   a. The larynx is an enlargement at the top of the trachea.
   b. It serves as a passageway for air and helps prevent foreign objects from entering the trachea.
   c. It is composed of muscles and cartilages; some of these cartilages are single while others are paired.
   d. It contains the vocal cords, which produce sounds by vibrating as air passes over them.
      (1) The pitch of a sound is related to the tension on the cords.
      (2) The intensity of a sound is related to the force of the air passing over the cords.
   e. The glottis and epiglottis help prevent food and liquid from entering the trachea.

6. Trachea
   a. The trachea extends into the thoracic cavity in front of the esophagus.
   b. It divides into the right and left bronchi.
   c. The mucous lining continues to filter incoming air.
   d. The wall is supported by incomplete cartilaginous rings.
7. Bronchial tree
   a. The bronchial tree consists of branched air passages that lead from the trachea to the air sacs.
   b. The branches of the bronchial tree include primary bronchi, lobar bronchi, segmental bronchi, bronchioles, terminal bronchioles, respiratory bronchioles, alveolar ducts, alveolar sacs, and alveoli.
   c. Structure of the respiratory tubes
      (1) As tubes branch, the amount of cartilage in the walls decreases and the muscular layer becomes more prominent.
      (2) Elastic fibers in the walls aid the breathing mechanism.
      (3) The epithelial lining changes from pseudostratified and ciliated to cuboidal and simple squamous as the tubes become progressively smaller.
   d. Functions of the respiratory tubes include distribution of air and exchange of gases between the alveolar air and the blood.
8. Lungs
   a. The left and right lungs are separated by the mediastinum and are enclosed by the diaphragm and the thoracic cage.
   b. The visceral pleura is attached to the surface of the lungs; parietal pleura lines the thoracic cavity.
   c. The right lung has three lobes, and the left lung has two.
   d. Each lobe is composed of lobules that contain alveolar ducts, alveolar sacs, alveoli, nerves, blood vessels, lymphatic vessels, and connective tissues.

Breathing Mechanism (page 593)

Inspiration and expiration movements are accompanied by changes in the size of the thoracic cavity.

1. Inspiration
   a. Air is forced into the lungs by atmospheric pressure.
   b. Inspiration occurs when the intra-alveolar pressure is reduced.
   c. The intra-alveolar pressure is reduced when the diaphragm moves downward, and the thoracic cage moves upward and outward.
   d. Expansion of the lungs is aided by surface tension that holds the pleural membranes together.
   e. Surfactant reduces surface tension within the alveoli.

2. Expiration
   a. The forces of expiration come from the elastic recoil of tissues and from surface tension within the alveoli.
   b. Expiration can be aided by thoracic and abdominal wall muscles that pull the thoracic cage downward and inward, and compress the abdominal organs inward.
3. Respiratory air volumes and capacities
   a. The amount of air that normally moves in and out during quiet breathing is the tidal volume.
   b. Additional air that can be inhaled is the inspiratory reserve volume; additional air that can be exhaled is the expiratory reserve volume.
   c. Residual air remains in the lungs and is mixed with newly inhaled air.
   d. The inspiratory capacity is the maximum volume of air a person can inhale following exhalation of the tidal volume.
   e. The functional residual capacity is the volume of air that remains in the lungs following the exhalation of the tidal volume.
   f. The vital capacity is the maximum amount of air a person can exhale after taking the deepest breath possible.
   g. The total lung capacity is equal to the vital capacity plus the residual air volume.
   h. Air in the anatomic and alveolar dead spaces is not available for gas exchange.
4. Alveolar ventilation
   a. Minute respiratory volume is calculated by multiplying the tidal volume by the breathing rate.
   b. Alveolar ventilation rate is calculated by subtracting the physiologic dead space from the tidal volume and multiplying the result by the breathing rate.
   c. The alveolar ventilation rate is a major factor affecting the exchange of gases between the alveolar air and the blood.
5. Nonrespiratory air movements
   a. Nonrespiratory air movements are air movements that occur in addition to breathing.
   b. They include coughing, sneezing, laughing, crying, hiccuping, and yawning.

Control of Breathing (page 600)

Normal breathing is rhythmic and involuntary, although the respiratory muscles can be controlled voluntarily.

1. Respiratory center
    a. The respiratory center is located in the brain stem, and includes parts of the medulla oblongata and pons.
    b. The medullary rhythmicity area includes two groups of neurons.
        (1) The dorsal respiratory group is responsible for the basic rhythm of breathing.
        (2) The ventral respiratory group increases inspiratory and expiratory movements during forceful breathing.
    c. The pneumotaxic area regulates the rate of breathing.
2. Factors affecting breathing
    a. Breathing is affected by certain chemicals, the stretching of lung tissues, and the person's emotional state.
    b. Chemosensitive areas are associated with the respiratory center.
        (1) Carbon dioxide combines with water to form carbonic acid, which, in turn, releases hydrogen ions.
        (2) Chemosensitive areas are influenced mainly by hydrogen ions.
        (3) Stimulation of these areas causes the breathing rate to increase.
    c. There are chemoreceptors in the carotid and aortic bodies of certain arteries.
        (1) These chemoreceptors are sensitive to low oxygen concentration.
        (2) When oxygen concentration is low, breathing rate increases.
    d. An inflation reflex is triggered by stretching the lung tissues.
        (1) This reflex reduces the duration of inspiratory movements.
        (2) This prevents overinflation of the lungs during forceful breathing.
    e. Hyperventilation causes the carbon dioxide concentration to decrease, but *this is very dangerous when associated with underwater swimming.*

Alveolar Gas Exchanges (page 605)

Gas exchanges between the air and the blood occur within the alveoli.

1. Alveoli
    a. The alveoli are tiny air sacs clustered at the distal ends of the alveolar ducts.
    b. Some alveoli have openings into adjacent air sacs that provide alternate pathways for air when passages are obstructed.
2. Respiratory membrane
    a. The respiratory membrane consists of the alveolar and capillary walls.
    b. Gas exchanges take place through these walls.
3. Diffusion through the respiratory membrane
    a. The partial pressure of a gas is determined by the concentration of that gas in a mixture of gases or the concentration of gas dissolved in a liquid.
    b. Gases diffuse from regions of higher partial pressure toward regions of lower partial pressure.
    c. Oxygen diffuses from the alveolar air into the blood; carbon dioxide diffuses from the blood into the alveolar air.

Transport of Gases (page 607)

Blood transports gases between the lungs and the body cells.

1. Oxygen transport
    a. Oxygen is mainly transported in combination with hemoglobin molecules.
    b. The resulting oxyhemoglobin is relatively unstable and releases its oxygen in regions where the P_{O_2} is low.
    c. More oxygen is released as the blood concentration of carbon dioxide increases, as the blood becomes more acidic, and as the blood temperature increases.
2. Carbon monoxide
    a. Carbon monoxide forms as a result of incomplete combustion of fuels.
    b. It combines with hemoglobin more readily than oxygen and forms a stable compound.
    c. Carbon monoxide is toxic because the hemoglobin with which it combines is no longer available for oxygen transport.
3. Carbon dioxide transport
    a. Carbon dioxide may be carried in solution, either bound to hemoglobin or as a bicarbonate ion.
    b. Most carbon dioxide is transported in the form of bicarbonate ions.
    c. Carbonic anhydrase, an enzyme, speeds the reaction between carbon dioxide and water to form carbonic acid.
    d. Carbonic acid dissociates to release hydrogen ions and bicarbonate ions.

Utilization of Oxygen (page 611)

1. Cellular respiration
   a. Cellular respiration occurs in two phases.
      (1) The anaerobic phase takes place in the cytoplasm.
      (2) The aerobic phase takes place in the mitochondria.
   b. During aerobic respiration, pyruvic acid is converted to acetyl coenzyme A, which enters the citric acid cycle.

Clinical Application of Knowledge

1. If the upper respiratory passages are bypassed with a tracheostomy, how might the air entering the trachea be different from air normally passing through this canal? What problems might this cause for the patient?

2. In certain respiratory disorders, such as emphysema, the capacity of the lungs to recoil elastically is reduced. Which respiratory air volumes will be affected by this condition? Why?

3. What changes would you expect to occur in the relative concentrations of blood oxygen and carbon dioxide in a patient who breathes rapidly and deeply for a prolonged time? Why?

4. If a person has stopped breathing and is receiving pulmonary resuscitation, would it be better to administer pure oxygen or a mixture of oxygen and carbon dioxide? Why?

5. The air pressure within the passenger compartment of a commercial aircraft may be equivalent to an altitude of 8,000 feet. What problem might this create for a person with a serious respiratory disorder?

6. Patients experiencing asthma attacks are often advised to breathe through pursed (puckered) lips. How might this help reduce the symptoms of the asthma?

Review Activities

1. Describe the general functions of the respiratory system.
2. Distinguish between the upper and lower respiratory tracts.
3. Explain how the nose and nasal cavity function in filtering incoming air.
4. Name and describe the locations of the major sinuses, and explain how a sinus headache may occur.
5. Distinguish between the pharynx and the larynx.
6. Name and describe the locations and functions of the cartilages of the larynx.
7. Distinguish between the false vocal cords and the true vocal cords.
8. Compare the structure of the trachea with the structure of the branches of the bronchial tree.
9. List the successive branches of the bronchial tree, from the primary bronchi to the alveoli.
10. Describe how the structure of the respiratory tube changes as the branches become finer.
11. Explain the functions of the respiratory tubes.
12. Distinguish between visceral pleura and parietal pleura.
13. Name and describe the locations of the lobes of the lungs.
14. Explain how normal inspiration and forced inspiration are accomplished.
15. Define *surface tension,* and explain how it aids the breathing mechanism.
16. Define *surfactant,* and explain its function.
17. Define *compliance.*
18. Explain how normal expiration and forced expiration are accomplished.
19. Distinguish between vital capacity and total lung capacity.
20. Distinguish between anatomic, alveolar, and physiologic dead spaces.
21. Distinguish between minute respiratory volume and alveolar ventilation rate.
22. Compare the mechanisms of coughing and sneezing, and explain the function of each.
23. Explain the function of yawning.
24. Describe the location of the respiratory center, and name its major components.
25. Describe how the basic rhythm of breathing is controlled.
26. Explain the function of the pneumotaxic area of the respiratory center.
27. Explain why increasing blood concentrations of carbon dioxide and hydrogen ions have similar effects on the respiratory center.
28. Describe the function of the chemoreceptors in the carotid and aortic bodies of certain arteries.

29. Describe the inflation reflex.
30. Discuss the effects of emotions on breathing.
31. Define *hyperventilation,* and explain how it affects the respiratory center.
32. Define *respiratory membrane,* and explain its function.
33. Explain the relationship between the partial pressure of a gas and its rate of diffusion.
34. Summarize the gas exchanges that occur through the respiratory membrane.

35. Describe how oxygen is transported in blood.
36. List three factors that cause an increased release of oxygen from the blood.
37. Explain why carbon monoxide is toxic.
38. List three ways that carbon dioxide is transported in blood.
39. Explain the function of carbonic anhydrase.
40. Define *chloride shift.*
41. Distinguish between the anaerobic phase and the aerobic phase of cellular respiration.

Chronic Obstructive Pulmonary Disease

George L., a slight, 64-year-old automobile mechanic, was looking forward to retirement. He had smoked about one pack of cigarettes per day for the past thirty-five years. Smoking had always caused him to cough, but over the past few years the coughing had become increasingly troublesome, particularly in the morning. He seemed to be coughing at times unrelated to his smoking, and he was experiencing shortness of breath (dyspnea) at work. Also, George was finding it increasingly difficult to exhale; in fact, he sometimes found it necessary to bend forward to help force air out of his lungs. His wife convinced him that he should talk to their family physician about his breathing problems.

After listening to George's description of his coughing and breathing difficulty, the physician examined his chest. He noted that the chest was slightly enlarged, so that the intercostal spaces were clearly visible, and the sternocleidomastoid muscles (figure 9.22) were hypertrophied. George's neck veins (figure 18.65) seemed swollen, and the physician heard wheezing within the lungs.

An X-ray film of George's chest (figure 7.43) failed to reveal any abnormalities; however, *pulmonary function tests* demonstrated reduced vital and total lung capacities (figure 16.27), as well as a forced expiratory volume (FEV) that was well below normal. Laboratory tests of *blood gases* disclosed a reduced PO_2 and an increased PCO_2. A *sputum test* was performed to check out the possibility of tuberculosis. The results of this test were negative.

On the bases of the physical examination and laboratory tests, the physician concluded that George had *chronic obstructive pulmonary disease* (COPD) and that the condition was probably caused by *emphysema*. (See figure 16.29.)

Hoping to prevent further development of the disease, the physician advised George to give up smoking immediately, to avoid breathing air containing particles that might irritate the lung tissues, and to have vaccinations to prevent respiratory infections, such as influenza and pneumonia. The physician also explained to George that persons with COPD have a high risk of early death and that this risk is closely related to the number of cigarettes smoked each day as well as to how deeply the smoke is inhaled. Although stopping smoking does not reverse the development of COPD, it produces a slow decrease in the risk of early death.

George was treated with drugs that help to liquify bronchial secretions (mucolytic agents) and drugs that cause bronchioles to dilate (bronchodilators) to make it easier for him to move air through his air passages. He was advised to increase his fluid intake to 2.5–3.0 liters per day to keep the bronchial secretions well hydrated and to report any changes in color or consistency of his sputum that might indicate the presence of a respiratory infection.

Review Questions

1. What is it that is obstructed in COPD caused by emphysema?
2. Why would a patient with COPD be likely to have an enlarged chest?
3. Why might the patient's neck veins be swollen?
4. How would you explain to a patient the difference between vital capacity and total lung capacity?
5. What does an FEV test measure?

CHAPTER

17

Blood

*T*he blood, heart, and blood vessels constitute the *circulatory system,* and provide a link between the body's internal parts and its external environment. More specifically, the blood transports nutrients from the digestive tract to body cells, moves oxygen from respiratory organs to the body cells, and carries wastes from these cells to the respiratory and excretory organs. It transports hormones from the endocrine glands to target tissues and bathes the body cells in a liquid of relatively stable composition. It also aids in body temperature control by distributing heat from the skeletal muscles and other active organs to all the body parts. Thus, the blood provides vital support for cellular activities and aids in maintaining a favorable cellular environment.

The heart and the closed system of blood vessels comprise the apparatus that moves blood throughout the body. These organs constitute the *cardiovascular system,* which is discussed in chapter 18 ■

Chapter Objectives	Key Terms	Aids to Understanding Words
After you have studied this chapter, you should be able to: 1. Describe the general characteristics of the blood and discuss its major functions. 2. Distinguish between the various types of blood cells. 3. Explain how blood cell counts are made and how they are used. 4. Discuss the life cycle of a red blood cell. 5. Explain how red blood cell production is controlled. 6. List the major components of blood plasma and describe the functions of each. 7. Define *hemostasis*, and explain the mechanisms that help to achieve it. 8. Review the major steps in blood coagulation. 9. Explain how coagulation can be prevented. 10. Explain the basis for blood typing. 11. Describe how blood reactions may occur between the fetal and maternal tissues. 12. Complete the review activities at the end of this chapter. Note that the items are worded in the form of specific learning objectives. You may want to refer to them before reading the chapter.	agglutinin (ah-gloo′tĭ-nin) agglutinogen (ag″loo-tin′o-jen) albumin (al-bu′min) antibody (an′tĭ-bod″e) basophil (ba′so-fil) coagulation (ko-ag″u-la′shun) embolus (em′bo-lus) eosinophil (e″o-sin′o-fil) erythrocyte (ĕ-rith′ro-sīt) erythropoietin (ĕ-rith″ro-poi′ĕ-tin) fibrinogen (fi-brin′o-jen) globulin (glob′u-lin) hemostasis (he″mo-sta′sis) leukocyte (lu′ko-sīt) lymphocyte (lim′fo-sīt) macrophage (mak′ro-fāj) monocyte (mon′o-sīt) neutrophil (nu′tro-fil) plasma (plaz′mah) platelet (plāt′let) thrombus (throm′bus)	**agglutin-**, to glue together: *agglutin*ation—the clumping together of red blood cells. **bil-**, bile: *bil*irubin— a pigment excreted in the bile. **-crit**, to separate: hemato*crit*— percentage of cells in a blood sample, determined by separating the cells from the plasma. **embol-**, stopper: *embol*us an obstruction in a blood vessel. **erythr-**, red: *erythr*ocyte—a red blood cell. **hemo-**, blood: *hemo*globin— the red pigment responsible for the color of blood. **hepar-**, liver: *hepar*in—anticoagulant secreted by liver cells. **leuko-**, white: *leuko*cyte—a white blood cell. **-lys**, to break up: fibrino*lys*in— protein-splitting enzyme that can digest fibrin. **macro-**, large: *macro*phage—a large phagocytic cell. **-osis**, abnormal increase in production: leukocyt*osis*— condition in which white blood cells are produced in abnormally large numbers. **-poiet**, making: erythro*poiet*in—a hormone that stimulates the production of red blood cells. **poly-**, many: *poly*cythemia— condition in which red blood cells appear in abnormally large numbers. **-stas**, halting: hemo*stas*is—arrest of bleeding from damaged blood vessels. **thromb-**, clot: *thromb*ocyte—a blood platelet that plays a role in the formation of a blood clot.

A s mentioned in chapter 5, blood is a type of connective tissue whose cells are suspended in a liquid intercellular material. It plays vital roles in transporting substances between the body cells and the external environment, and it aids in maintaining a stable cellular environment.

Blood and Blood Cells

Whole blood is slightly heavier and three to four times more viscous than water. Its cells, which are formed mostly in red bone marrow, include *red blood cells* and *white blood cells*. The blood also contains some cellular fragments called *blood platelets* (figure 17.1).

Volume and Composition of Blood

The volume of blood varies with body size. It also varies with changes in the fluid and electrolyte concentrations and the amount of adipose tissue present. However, an average-sized adult will have a blood volume of about 5 liters.

If a blood sample is allowed to stand in a tube for a while, the cells will become separated from the liquid portion of the blood and settle to the bottom. This separation can be speeded by centrifuging the sample so that the cells quickly become packed into the lower part of the centrifuge tube, as shown in figure 17.2. When the amounts of cells and liquid are measured, the percentage of each in the blood sample can be calculated.

A blood sample is usually about 45% cells. This percentage is called the **hematocrit** (HCT), or **packed cell volume** (PCV). The remaining 55% of a blood sample consists of clear, straw-colored **plasma.** The normal values for this and other common blood tests are provided in Appendix C, page 933.

In addition to red blood cells, which comprise over 99% of the blood cells, the solids (formed elements) of the blood include the white blood cells and blood platelets. The plasma is composed of a complex mixture that includes water, amino acids, proteins, carbohydrates, lipids, vitamins, hormones, electrolytes, and cellular wastes. The normal concentrations of plasma components are also listed in Appendix C.

1. What are the major components of blood?
2. What factors affect blood volume?
3. How is the hematocrit determined?

Characteristics of Red Blood Cells

Red blood cells, or **erythrocytes,** are tiny, biconcave disks that are thin near their centers and thicker around their rims. This special shape is related to the red cell's function of transporting gases. The shape provides an increased surface area through which gases can diffuse, and it places the cell membrane closer to various in-

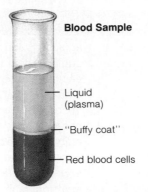

Blood Sample

- Liquid (plasma)
- "Buffy coat"
- Red blood cells

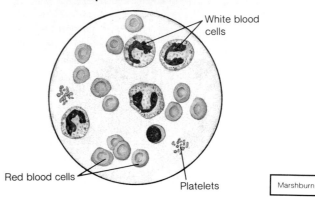

Peripheral Blood Smear

White blood cells

Red blood cells

Platelets

Marshburn

Figure 17.1 Blood consists of a liquid portion called plasma and a solid portion that includes red blood cells, white blood cells, and platelets. (Note: When blood components are separated, the white blood cells and platelets form a thin layer, called the "buffy coat," between the plasma and the red blood cells.)

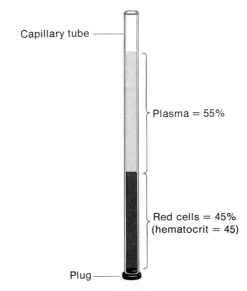

Capillary tube

Plasma = 55%

Red cells = 45% (hematocrit = 45)

Plug

Figure 17.2 If a blood-filled capillary tube is centrifuged, the red cells become packed in the lower portion, and the percentage of red cells (hematocrit) can be determined.

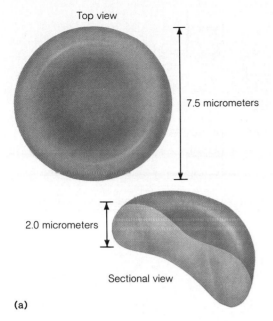

Top view

7.5 micrometers

2.0 micrometers

Sectional view

(a)

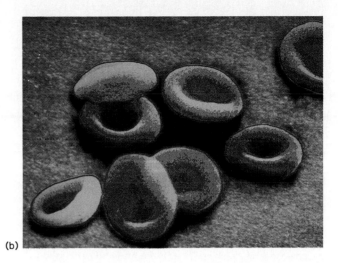

(b)

Figure 17.3 (a) How is the biconcave shape of a red blood cell related to its function? (b) Falsely colored scanning electron micrograph of human red blood cells (falsely colored).

terior parts where oxygen-carrying *hemoglobin* is found. Because of its shape, a red blood cell can also be deformed readily as it squeezes through the narrow passages of capillaries (figure 17.3).

Each red blood cell is about one-third hemoglobin by volume, and this substance is responsible for the color of the blood. The rest of the cell consists mainly of membrane, water, electrolytes, enzymes, and components of cellular respiratory mechanisms. When the hemoglobin is combined with oxygen, the resulting *oxyhemoglobin* is bright red, and when the oxygen is released, the resulting *deoxyhemoglobin* is darker. In fact, blood rich in deoxyhemoglobin may appear bluish when it is viewed through blood vessel walls.

A person experiencing a prolonged oxygen deficiency (hypoxia) may develop a symptom called *cyanosis*. In this condition, the skin and mucous membranes appear bluish due to an abnormally high concentration of deoxyhemoglobin in the blood. Cyanosis may also occur as a result of exposure to low temperature. In this case, the superficial blood vessels become constricted, the blood flow is slowed, and more oxygen than usual is removed from the blood flowing through the vessels.

Although the red blood cells have nuclei during their early stages of development, these nuclei are lost as the cells mature. This characteristic, like the shape of a red blood cell, seems to be related to the function of transporting oxygen, since the space previously occupied by the nucleus is available to hold hemoglobin.

Since their nuclei are missing, red blood cells cannot carry on DNA-directed protein synthesis (see chapter 4) and are unable to reproduce. On the other hand, their cytoplasmic enzymes remain functional, and the blood cells are able to carry on vital energy-releasing processes. With time, however, the red blood cells become less and less active.

Some individuals produce an abnormal type of hemoglobin molecule, which tends to form long chains when exposed to low oxygen concentrations. This causes the red blood cells containing such molecules to become distorted, or sickle-shaped, and the person is said to have *sickle-cell anemia*.

In this condition, the distorted cells may block capillaries and thus interrupt the flow of blood to tissues. The resulting symptoms may include severe joint and abdominal pain, skin ulceration, and chronic kidney disease.

1. Describe a red blood cell.
2. What is the function of hemoglobin?
3. What changes occur in a red blood cell as it matures?

Red Blood Cell Counts

The number of red blood cells in a cubic millimeter (mm^3) of blood is called the *red blood cell count* (RBC or RCC). Although this number varies from time to time even in healthy individuals, the normal range for adult males is 4,600,000 6,200,000 cells per mm^3, and

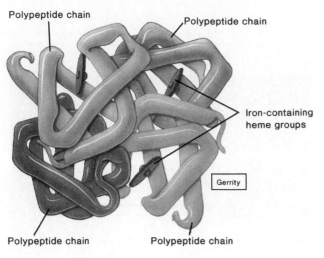

Polypeptide chain

Polypeptide chain

Iron-containing
heme groups

Gerrity

Polypeptide chain

Polypeptide chain

(a) Hebroglobin molecule

Figure 17.4 When a hemoglobin molecule (*a*) is decomposed, the heme portions (*b*) are broken down into iron (Fe) and biliverdin (*c*). Most of the biliverdin is then converted to bilirubin (*d*).

(b) **Heme**

(c) **Biliverdin (C$_{33}$H$_{34}$O$_6$N$_4$)** + Fe

(d) **Bilirubin (C$_{33}$H$_{36}$O$_6$N$_4$)**

that for adult females is 4,200,000–5,400,000 cells per mm^3. For children, the normal range is 4,500,000–5,100,000 cells per mm^3. (Note: These normal values may vary slightly with the hospital, physician, and type of equipment used to make blood cell counts.) The number of red blood cells generally increases following exercise, a large meal, a rise in temperature, or an increase in altitude.

Since the number of circulating red blood cells is closely related to the blood's *oxygen-carrying capacity,* any changes in this number may be significant. For this reason, red cell counts are routinely made to help diagnose and evaluate the courses of various diseases.

Blood cell counts are sometimes made using a special glass slide called a *hemocytometer.* This device is designed so that the cells in a small, but known volume of blood can be observed and counted with the aid of a microscope. Clinical laboratories are often equipped with electronic counters that detect and record the number of blood cells forced through a tiny opening in the instrument.

Destruction of Red Blood Cells

Red blood cells are quite elastic and flexible, and they readily change shape as they pass through small blood vessels. With age, however, these cells become more fragile, and they are often damaged. For example, they may be injured when they squeeze through capillaries, particularly those in active muscles or when they are bent as they pass through narrow, slitlike passages in the spleen.

Damaged red cells are phagocytized and destroyed primarily in the liver and spleen by large reticuloendothelial cells called **macrophages** (chapter 5). The fragments and contents of ruptured cells are also phagocytized and digested by macrophages.

The hemoglobin molecules from the red blood cells undergoing destruction are broken down into molecular subunits of *heme,* an iron-containing portion, and *globin,* a protein. The heme is further decomposed into iron and a greenish pigment called **biliverdin.** The iron, combined with a protein called *transferrin,* may be carried by the blood to the hematopoietic tissue in the red bone marrow and reused in the synthesis of a new hemoglobin. Otherwise the iron may be stored in the liver cells in the form of an iron-protein complex called *ferritin.* In time, the biliverdin is converted to an orange pigment called **bilirubin.** Biliverdin and bilirubin are excreted in the bile as bile pigments (figure 17.4). Chart 17.1 summarizes the process of red cell destruction.

1. Cells are damaged while squeezing through the capillaries of active tissues.
2. Damaged red cells are phagocytized by macrophages in the liver and spleen.
3. Hemoglobin from the red cells is decomposed into heme and globin.
4. Heme is decomposed into iron and biliverdin.
5. Iron is made available for reuse in the synthesis of new hemoglobin or is stored in the liver as ferritin.
6. Some biliverdin is converted into bilirubin.
7. Biliverdin and bilirubin are excreted in bile as bile pigments.

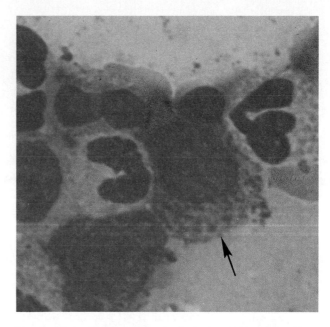

Figure 17.5 Light micrograph of a hemocytoblast (arrow) in red bone marrow (×640).

About one-third of all newborns develop a mild disorder called *physiologic jaundice* within a few days following birth. In this condition, as in other forms of jaundice, the skin and eyes become yellowish due to an accumulation of bilirubin in the tissues.

Physiologic jaundice is thought to be the result of immature liver cells, which are somewhat ineffective in excreting bilirubin into the bile. The condition is treated by feedings that promote bowel movements and by exposure to fluorescent light, which reduces the concentration of bilirubin in the tissues.

1. What is the normal red blood cell count for an adult male? For an adult female?
2. What happens to damaged red blood cells?
3. What are the end products of hemoglobin breakdown?

Red Blood Cell Production and Its Control

As mentioned in chapter 7, red blood cell formation (hematopoiesis) occurs initially in the yolk sac, liver, and spleen; but after an infant is born, these cells are produced almost exclusively by the tissue that lines the spaces within the red bone marrow.

Within the red marrow, cells called **hemocytoblasts** give rise to **erythroblasts** that can synthesize hemoglobin (figure 17.5). The erythroblasts also reproduce and give rise to many daughter cells. The nuclei of these newly formed cells soon shrink and are extruded by being pinched off in thin coverings of cytoplasm and cell membrane. The resulting cells are **erythrocytes.** Some of these young red cells may contain a netlike structure (reticulum) for a day or two. This network represents the endoplasmic reticulum, and such cells are called **reticulocytes.** (See figure 17.6.)

The average life span of a red blood cell is about 120 days, and a large number of these cells are removed from circulation each day. Yet, the number of cells in the circulating blood remains relatively stable. This observation suggests that a *homeostatic mechanism* regulates the rate of red cell production.

The rate of red blood cell formation is controlled by a *negative feedback mechanism* involving a hormone called **erythropoietin.** In response to prolonged oxygen deficiency, erythropoietin is released, primarily from the kidneys and to a lesser extent from the liver. (Note: In a fetus, the liver is the main site of erythropoietin production.)

At high altitudes, for example, where the P_{O_2} is reduced, the amount of oxygen delivered to the tissues decreases. As figure 17.7 shows, this drop in oxygen concentration triggers the release of erythropoietin, which travels via the blood to the red bone marrow and stimulates increased erythrocyte production.

After a few days, large numbers of newly formed red cells begin to appear in the circulating blood, and the increased rate of production continues as long as the kidney and liver tissues experience an oxygen deficiency. Eventually, the number of erythrocytes in circulation may be sufficient to supply these tissues with their oxygen needs. When this happens, or when the P_{O_2} returns to normal, the release of erythropoietin ceases, and the rate of red cell production is reduced.

Other conditions that produce a low P_{O_2} and stimulate erythropoietin release include loss of blood, which results in a decreased oxygen-carrying capacity of the circulatory system, and chronic lung diseases, which cause a decrease in the respiratory surface area available for gas exchanges.

Chart 17.2 summarizes the steps in the regulation of red blood cell production.

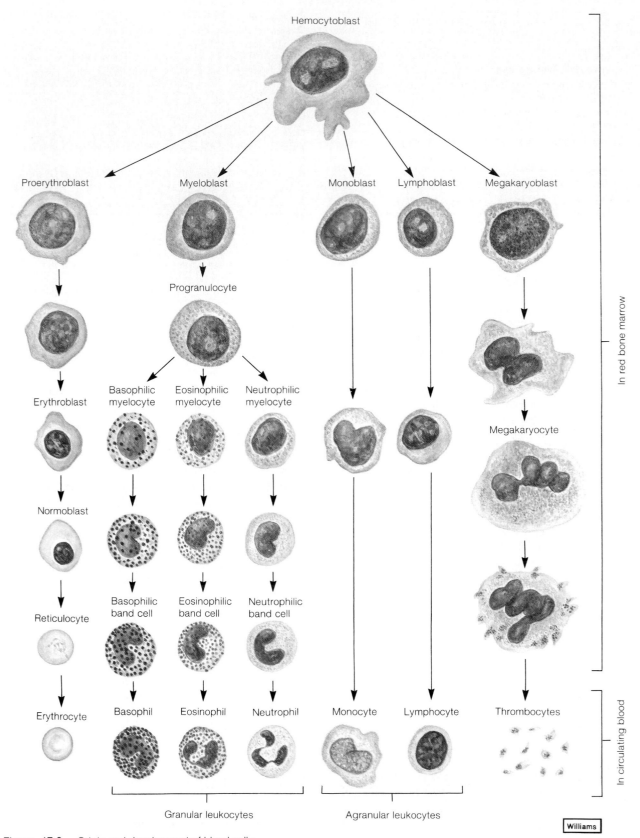

Hemocytoblast

Proerythroblast

Myeloblast

Monoblast

Lymphoblast

Megakaryoblast

Progranulocyte

Erythroblast

Basophilic
myelocyte

Eosinophilic
myelocyte

Neutrophilic
myelocyte

Normoblast

Megakaryocyte

Reticulocyte

Basophilic
band cell

Eosinophilic
band cell

Neutrophilic
band cell

Erythrocyte

Basophil

Eosinophil

Neutrophil

Monocyte

Lymphocyte

Thrombocytes

Granular leukocytes

Agranular leukocytes

In red bone marrow

In circulating blood

Williams

Figure 17.6   Origin and development of blood cells.

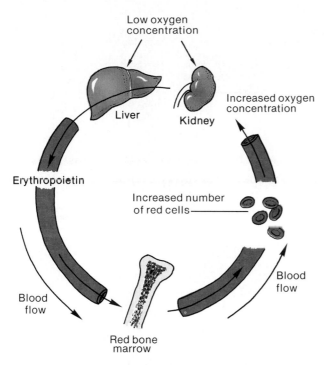

Figure 17.7 Low P_{O_2} causes the release of erythropoietin from the kidneys and liver. This substance stimulates the production of red blood cells, and these cells carry additional oxygen to the tissues.

Low oxygen concentration

Liver

Kidney

Increased oxygen concentration

Erythropoietin

Increased number of red cells

Blood flow

Blood flow

Red bone marrow

CHART 17.2	Major events in the hormonal control of red blood cell production

1. The kidney and liver tissues experience an oxygen deficiency.
2. These tissues release the hormone erythropoietin.
3. Erythropoietin travels to the red bone marrow and stimulates an increase in the production of red cells.
4. As increasing numbers of red blood cells are released into the circulation, the oxygen-carrying capacity of the blood rises.
5. The oxygen concentration in the kidney and liver tissues increases, and the release of erythropoietin decreases.

---

1. Where are red blood cells produced?
2. How is the production of red blood cells controlled?

Dietary Factors Affecting Red Blood Cell Production

Red blood cell production is significantly influenced by the availability of two of the B-complex vitamins—*vitamin B₁₂* and folic acid. These substances are required for the synthesis of DNA molecules, so they are needed by all cells for growth and reproduction. Since cellular reproduction occurs at a particularly high rate in hematopoietic tissue, this tissue is especially affected by a lack of either of these vitamins.

A lack of vitamin B_{12} is usually due to a disorder in the stomach lining rather than to a dietary deficiency. As mentioned in chapters 14 and 15, the parietal cells of the gastric glands normally secrete a substance called *intrinsic factor* that promotes the absorption of vitamin B_{12}.

> In the absence of intrinsic factor and the consequent decrease in vitamin B_{12} absorption, the red bone marrow forms red blood cells that are abnormally large, irregularly shaped, and have particularly thin membranes. These cells tend to be damaged more easily than normal red cells, and consequently they have shorter lives. The result is a condition called *pernicious anemia*.
>
> Because vitamin B_{12} is absorbed poorly by persons lacking intrinsic factor, pernicious anemia is usually treated with injections of the deficient vitamin rather than with oral ingestion.

As mentioned previously, iron is also needed for the synthesis of hemoglobin. Although much of the iron that is released during the decomposition of hemoglobin is available for reuse, some iron is lost each day and must be replaced.

Absorption of iron from food occurs mainly in the duodenum of the small intestine. Although the absorption mechanism is poorly understood, it is known to involve active transport and to occur relatively slowly. Because the absorption is slow, only a small percentage of iron is absorbed even when food contains an abundance of this mineral. However, the rate of absorption varies somewhat with the total amount of iron in the body. Thus, when the amount of iron is low, the rate of absorption is increased; and when the tissues are becoming saturated with iron, the rate is greatly decreased. Figure 17.8 summarizes the life cycle of a red blood cell.

The dietary factors that affect red blood cell production are summarized in chart 17.3.

> During pregnancy, the enlarging uterus and its developing vascular system create a demand for more blood. Consequently, the maternal blood volume usually increases about 30% above normal. Because this increase requires accelerated production of red blood cells, a pregnant woman must take in additional iron to meet the needs of her own hematopoietic tissues as well as those of the developing fetus.

1. What vitamins are necessary for red blood cell production?
2. Why is iron needed for the normal development of red blood cells?

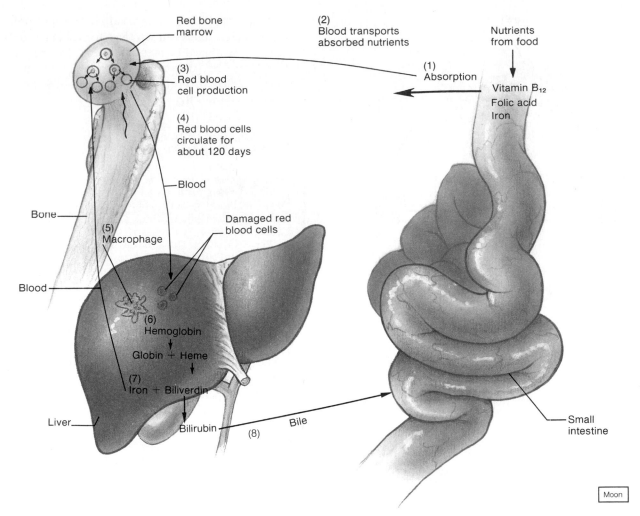

Red bone marrow

(2) Blood transports absorbed nutrients

Nutrients from food

(3) Red blood cell production

(1) Absorption

Vitamin B$_{12}$
Folic acid
Iron

(4) Red blood cells circulate for about 120 days

Blood

Bone

(5) Macrophage

Damaged red blood cells

Blood

(6) Hemoglobin

Globin + Heme

(7) Iron + Biliverdin

Liver

Bilirubin (8)

Bile

Small intestine

Moon

Figure 17.8 Life cycle of a red blood cell. (*1*) Essential nutrients are absorbed from the intestine; (*2*) nutrients are transported by blood to red bone marrow; (*3*) red blood cells are produced in red bone marrow by mitosis; (*4*) mature red blood cells are released into blood where they circulate for about 120 days; (*5*) damaged red blood cells are destroyed in liver by macrophages; (*6*) hemoglobin from red blood cells is decomposed into heme and globin; (*7*) iron from heme is returned to red bone marrow and reused; (*8*) biliverdin is excreted in the bile.

CHART 17.3	Dietary factors affecting red blood cell production	
Substance	**Source**	**Function**
Vitamin B$_{12}$	Absorbed from small intestine in the presence of intrinsic factor	Necessary for the synthesis of DNA
Iron	Absorbed from small intestine; conserved during red cell destruction and made available for reuse	Necessary for the synthesis of hemoglobin
Folic acid	Absorbed from small intestine	Necessary for the synthesis of DNA

Types of White Blood Cells

White blood cells, or **leukocytes,** function primarily to control various disease conditions. Although these cells do most of their work outside the circulatory system, they use the blood for transportation to sites of infection.

Normally, five types of white cells can be found in the circulating blood. They are distinguished by their size, the nature of their cytoplasm, the shape of their nucleus, and their staining characteristics. For example, some types of leukocytes have granular cytoplasm and make up a group called *granulocytes,* while others lack cytoplasmic granules and are called *agranulocytes.*

A typical **granulocyte** is about twice the size of a red cell. The members of this group include three types of white cells—neutrophils, eosinophils, and basophils.

Red Blood Cell Disorders

A deficiency of red blood cells or a reduction in the quantity of the hemoglobin they contain results in a condition called **anemia.** In disorders of this type, the oxygen-carrying capacity of the blood is reduced, and the person may appear pale and lack energy. Among the more common types of anemia are hemorrhagic anemia, hypochromic anemia, aplastic anemia, and hemolytic anemia.

Hemorrhagic anemia is caused by an abnormal loss of blood and may be mild or severe, depending on the amount of bleeding that occurs. Increased production of red cells usually makes up for a blood loss within a few weeks. If the blood loss is prolonged, however, the person may not be able to obtain enough iron from food to replace the lost hemoglobin. As a result, even though the normal number of red cells is reestablished, they may lack a normal concentration of hemoglobin. This condition is called *hypochromic anemia* (iron-deficiency anemia). It is characterized by the presence of small, pale red blood cells with relatively low hemoglobin content (figure 17.9).

Aplastic anemia is due to a malfunction of the red bone marrow and is characterized by decreased production of red blood cells. This condition is sometimes caused by an overexposure to X rays or to some other form of *ionizing radiation.* It can also be caused by the toxic effects of various chemicals or by adverse drug reactions.

Hemolytic anemia is characterized by an abnormally high rate of red blood cell rupture (hemolysis). Such cellular breakdown may be caused by a variety of factors including hereditary defects, parasitic infections, and adverse drug reactions.

Another type of red blood cell disorder is called **polycythemia.** In this condition, there is an abnormal increase in the number of circulating erythrocytes. Polycythemia may result from stimulation of the erythrocyte-forming tissues, as occurs when a person remains at a high altitude or engages in prolonged physical activity. It may occur when the blood becomes concentrated due to the excessive loss of body fluids that accompanies heavy sweating or shock (secondary polycythemia). Polycythemia may also be caused by an abnormal condition in the hematopoietic tissues, in which abnormally large numbers of erythrocytes are produced (polycythemia vera).

A general effect of polycythemia is to increase the viscosity of the blood, which depends largely on the concentration of red blood cells. (Viscosity is a property related to the ease at which a fluid flows—syrup has a high viscosity and flows sluggishly, while water has a low viscosity and flows easily.) As a result of increased viscosity, blood flow through the blood vessels is decreased, an abnormal amount of hemoglobin may be deoxygenated, and the skin may appear bluish (cyanotic).

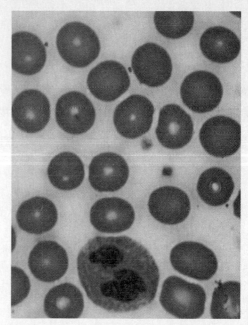

(a)

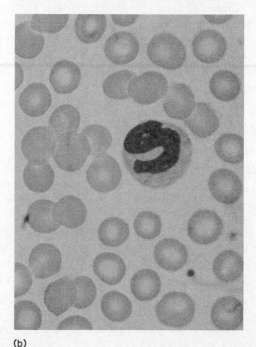

(b)

Figure 17.9 (a) Light micrograph of normal human erythrocytes; (b) light micrograph of erythrocytes from a person with hypochromic anemia (×800).

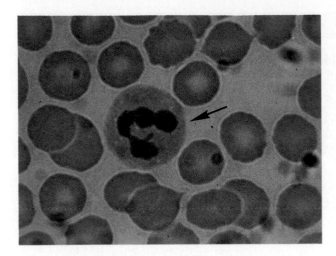

Figure 17.10 A neutrophil has a lobed nucleus with two to five parts (×400).

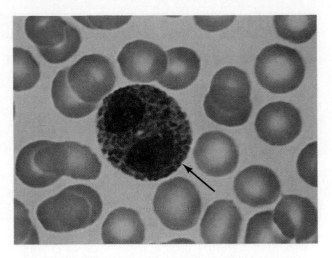

Figure 17.11 An eosinophil is characterized by the presence of red-staining cytoplasmic granules (×500).

These cells develop in the red bone marrow and are produced from *hemocytoblasts* in much the same manner as red cells. However, they have a relatively short life span, averaging about twelve hours. Figure 17.6 illustrates the stages in the development of these and other blood cells.

Neutrophils are characterized by the presence of fine cytoplasmic granules that stain pinkish in neutral stain. The nucleus of a neutrophil is lobed and consists of two to five parts connected by thin strands of chromatin. For this reason, neutrophils are also called *polymorphonuclear leukocytes,* which means their nuclei take many forms. Neutrophils account for 54%–62% of the leukocytes in a normal blood sample (figure 17.10).

Eosinophils contain coarse, uniformly sized cytoplasmic granules that stain deep red in acid stain. The nucleus usually has only two lobes (bilobed). These cells make up 1%–3% of the total number of circulating leukocytes (figure 17.11).

Basophils are similar to eosinophils in size and in the shape of their nuclei. However, they have fewer, more irregularly shaped cytoplasmic granules that stain deep blue in basic stain. This type of leukocyte usually accounts for less than 1% of the leukocytes (figure 17.12).

The leukocytes of the **agranulocyte** group include monocytes and lymphocytes. While monocytes generally arise from red bone marrow, lymphocytes are formed in the organs of the lymphatic system as well as in the red bone marrow.

Monocytes are the largest cells found in the blood, having diameters two to three times greater than those of red blood cells. Their nuclei vary in shape and are described as round, kidney-shaped, oval, or lobed. They

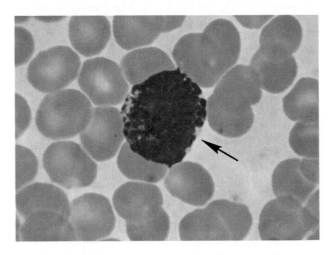

Figure 17.12 A basophil has cytoplasmic granules that stain deep blue (×563).

usually make up 3%–9% of the leukocytes in a blood sample and live for several weeks or even months (figure 17.13).

Although large **lymphocytes** are sometimes found in the blood, usually they are only slightly larger than the erythrocytes. A typical lymphocyte contains a relatively large, round nucleus surrounded by a thin rim of cytoplasm. These cells account for 25%–33% of the circulating leukocytes. They have relatively long life spans that may extend for years (figure 17.14).

1. Distinguish between granulocytes and agranulocytes.
2. List five types of white blood cells, and explain how they differ from one another.

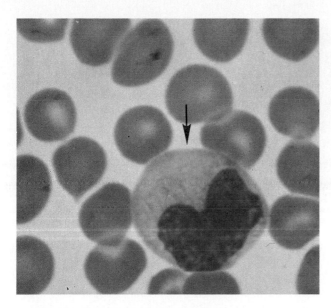

Figure 17.13 A monocyte is the largest type of blood cell (×640).

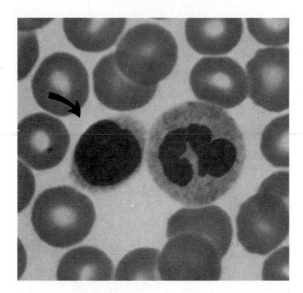

Figure 17.14 The lymphocyte (arrow) contains a large, round nucleus (×640). A neutrophil appears to the right of the lymphocyte.

White Blood Cell Counts

The procedure used to count white blood cells is similar to that used for counting red cells. However, before a *white cell count* is made the red cells in the blood sample are destroyed so they will not be mistaken for white cells. Normally, there are from 5,000 to 10,000 white cells per mm³ of human blood.

Since the total number of white blood cells may change in response to abnormal conditions, white blood cell counts are of clinical interest. For example, some

infectious diseases are accompanied by a rise in the number of circulating white cells, and, if the total number of white cells exceeds 10,000 per mm³ of blood, the person is said to have **leukocytosis.** This condition occurs during certain acute infections, such as appendicitis. It may also appear following vigorous exercise, emotional disturbances, or excessive loss of body fluids.

If the total white cell count drops below 5,000 per mm³ of blood, the condition is called **leukopenia.** Such a deficiency may accompany typhoid fever, influenza, measles, mumps, chicken pox, or poliomyelitis. It may also result from various anemias or from lead, arsenic, or mercury poisoning.

A *differential white blood cell count* (DIFF) is one in which the percentages of the various types of leukocytes in a blood sample are determined. This test is useful because the relative proportions of white cells may change in particular diseases. Neutrophils, for instance, usually increase during bacterial infections, and eosinophils may increase during certain parasitic infections and allergic reactions.

1. What is the normal human white blood cell count?
2. Distinguish between leukocytosis and leukopenia.
3. What is a differential white blood cell count?

Functions of White Blood Cells

As mentioned previously, white blood cells function to control disease conditions, such as infections caused by microorganisms. For example, some leukocytes phagocytize bacterial cells that get into the body, and others produce antibodies that react with such foreign particles to destroy or disable them.

Leukocytes can squeeze between the cells that form blood vessel walls. This movement, called **diapedesis,** allows the white cells to leave the circulation. (See figure 17.15.) Once outside the blood, they move through interstitial spaces using a form of self-propulsion called *ameboid motion.*

The most mobile and active phagocytic leukocytes are the *neutrophils* and the *monocytes.* Although the neutrophils are unable to ingest particles much larger than bacterial cells, monocytes can engulf relatively large objects. Both of these types of phagocytes contain numerous *lysosomes,* which are filled with digestive enzymes capable of breaking down various organic molecules, such as those in bacterial cell membranes. As a result of their activities, neutrophils and monocytes often become so engorged with digestive products and bacterial toxins that they also die.

When tissues are invaded by microorganisms, certain body cells respond by releasing substances that cause local blood vessels to dilate. One such substance,

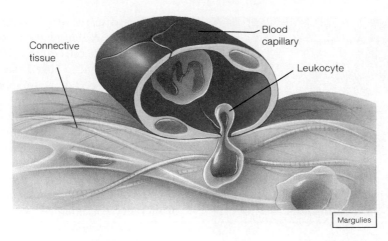

Connective
tissue

Blood
capillary

Leukocyte

Margulies

Figure 17.15 Leukocytes can squeeze between the cells of a capillary wall and enter the tissue space outside the blood system.

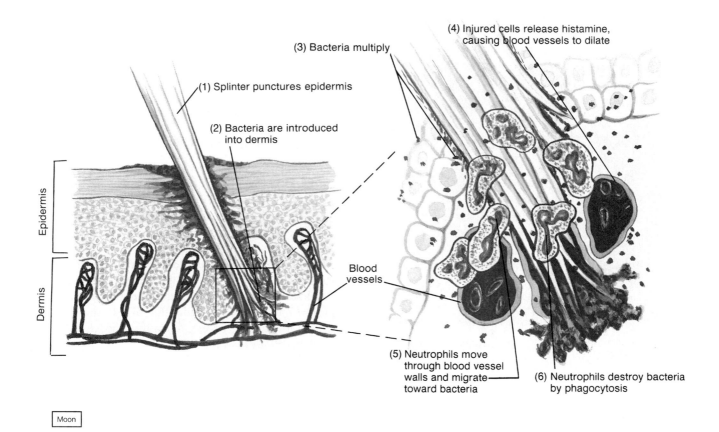

(3) Bacteria multiply

(4) Injured cells release histamine, causing blood vessels to dilate

(1) Splinter punctures epidermis

(2) Bacteria are introduced into dermis

Epidermis

Dermis

Blood vessels

(5) Neutrophils move through blood vessel walls and migrate toward bacteria

(6) Neutrophils destroy bacteria by phagocytosis

Moon

Figure 17.16 When bacteria invade the tissues, leukocytes migrate into the region and destroy the bacteria by phagocytosis.

histamine, also causes an increase in the permeability of nearby capillaries and the walls of small veins (venules). As these changes occur, the tissues become reddened and large quantities of fluids leak into the interstitial spaces. The swelling produced by this *inflammatory reaction* tends to delay the spread of the invading microorganisms into other regions. (See chapter 5.) At the same time, leukocytes are attracted by chemicals released from damaged cells, and they move toward these substances. This phenomenon is called **positive chemotaxis,** and it results in large numbers of white blood cells moving quickly into inflamed areas (figure 17.16).

Leukemia

Leukemia is a form of cancer characterized by an uncontrolled production of specific types of leukocytes (figure 17.17). There are two major types of leukemia. *Myeloid leukemia* results from an abnormal production of granulocytes by the red bone marrow, while *lymphoid leukemia* is accompanied by increased formation of lymphocytes from lymph nodes. In both types, the newly-formed leukocytes fail to mature into functional cells. Thus, even though large numbers of neutrophils may be formed in myeloid leukemia, these immature cells have little ability to phagocytize bacteria, and the patient has a lowered resistance to infections.

Eventually, the cells responsible for the overproduction of leukocytes tend to spread (metastasize) from the bone marrow or lymph nodes to other parts, and, as a result, white blood cells are produced abnormally in tissues throughout the body. As with other forms of cancer, the leukemic cells finally appear in such great numbers that they crowd out the normal, functioning cells. For example, leukemic cells originating in red bone marrow may invade other regions of the bone, weakening its structure and stimulating pain receptors. Also, as the normal red marrow is crowded out, the patient is likely to become anemic and develop a deficiency of blood platelets (thrombocytopenia). The lack of platelets is usually reflected in an increasing tendency to bleed.

Leukemias are also classified as either *acute* or *chronic*. An acute condition appears suddenly, the symptoms progress rapidly, and death occurs in a few months if the condition is untreated. Chronic forms begin more slowly and may remain undetected for many months. Without treatment, life expectancy is about three years.

The greatest success in treatment has been achieved with acute lymphoid leukemia, which is the most common cancerous condition in children. This treatment usually involves counteracting the side effects of the condition, such as anemia, hemorrhaging, and an increased susceptibility to infections, as well as administering chemotherapeutic drugs.

Although acute lymphoid leukemia may occur at any age, the chronic form usually occurs after fifty years of age. Acute myeloid leukemia also may occur at any age, but it is more frequent in adults; chronic myeloid leukemia is primarily a disease of adults between twenty and fifty years of age.

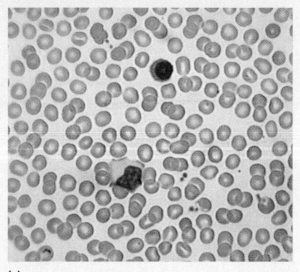

(a)

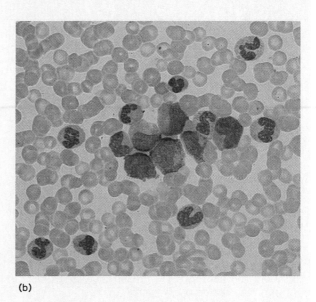

(b)

Figure 17.17 (*a*) Normal blood cells; (*b*) blood cells from a person with leukemia. (Note the increased number of leukocytes in *b*.)

As bacteria, leukocytes, and damaged cells accumulate in an inflamed area, a thick fluid called *pus* often forms, and continues to be present while the invading microorganisms are active. If the pus cannot escape to the outside of the body or into a body cavity, it may remain walled in the tissues for some time. Eventually it may be absorbed by the surrounding cells.

Eosinophils are only weakly phagocytic. However, they are attracted to various parasites that enter the body, and they can kill some of these invaders. Eosinophils also help control inflammation and allergic reactions by removing certain substances associated with these reactions.

Some of the cytoplasmic granules of *basophils* contain a blood clot inhibiting substance called *hep-*

CHART 17.4 Cellular components of blood

Component	Description	Number present	Function
Red blood cell (erythrocyte)	Biconcave disk without a nucleus; about one-third hemoglobin	4,000,000 to 6,000,000 per mm³	Transports oxygen and carbon dioxide
White blood cell (leukocyte)		5,000 to 10,000 per mm³	Aids in defense against infections by microorganisms
Granulocytes	About twice the size of red cells; cytoplasmic granules are present		
1. Neutrophil	Nucleus with two to five lobes; its cytoplasmic granules stain pink in neutral stain	54%–62% of white cells present	Destroys relatively small particles by phagocytosis
2. Eosinophil	Nucleus bilobed, its cytoplasmic granules stain red in acid stain	1%–3% of white cells present	Kills parasites and helps control inflammation and allergic reactions
3. Basophil	Nucleus lobed; cytoplasmic granules stain blue in basic stain	Less than 1% of white cells present	Releases anticoagulant, heparin, and histamine
Agranulocytes	Cytoplasmic granules are absent		
1. Monocyte	Two to three times larger than a red cell; nuclear shape varies from round to lobed	3%–9% of white cells present	Destroys relatively large particles by phagocytosis
2. Lymphocyte	Only slightly larger than a red cell; its nucleus nearly fills cell	25%–33% of white cells present	Mechanism of immunity
Platelet (thrombocyte)	Cytoplasmic fragment	130,000 to 360,000 per mm³	Helps control blood loss from broken vessels

arin, while other granules contain *histamine*. Basophils may help prevent intravascular blood clot formation by releasing heparin, and may cause an increase in blood flow to injured tissues by releasing histamine. They may also play major roles in certain allergic reactions. (See chapter 19.)

Lymphocytes play an important role in the mechanism of *immunity*. For example, some can form **antibodies** that act against various foreign substances entering the body. This function is discussed in chapter 19.

---

1. What are the primary functions of white blood cells?
2. How do white blood cells reach microorganisms that are outside blood vessels?
3. Which white blood cells are the most active phagocytes?
4. What are the functions of eosinophils and basophils?

Blood Platelets

Platelets, or **thrombocytes,** are not complete cells. They arise from very large cells in the red bone marrow called **megakaryocytes.** These cells release cytoplasmic fragments of themselves, and, as the fragments detach and enter the circulation, the smaller ones become platelets. The larger fragments of the megakaryocytes are reduced in size and become platelets as they pass through the blood vessels of the lungs (figure 17.6).

Each platelet is a round disk that lacks a nucleus and is less than half the size of a red blood cell. It is capable of ameboid movement and may live for about ten days. In normal blood, the *platelet count* will vary from 130,000 to 360,000 platelets per mm³.

Platelets help close breaks in damaged blood vessels by sticking to the broken surfaces. In the presence of such breaks, platelets release a substance called **serotonin,** which causes smooth muscles in the vessel walls to contract—an action that reduces blood flow. In addition, platelets initiate the formation of blood clots, as is explained in a subsequent section of this chapter.

Chart 17.4 summarizes the characteristics of blood cells and platelets.

---

1. What is leukemia?
2. Distinguish between myeloid and lymphoid leukemia.
3. What is the function of blood platelets?
4. What is the normal human blood platelet count?

Blood Plasma

Plasma is the clear, straw-colored, liquid portion of the blood in which the cells and platelets are suspended. It is approximately 92% water and contains a complex mixture of organic and inorganic substances that function in a variety of ways. These functions include transporting nutrients, gases, and vitamins; regulating fluid and electrolyte balances; and maintaining a favorable pH. Figure 17.18 shows the chemical makeup of plasma.

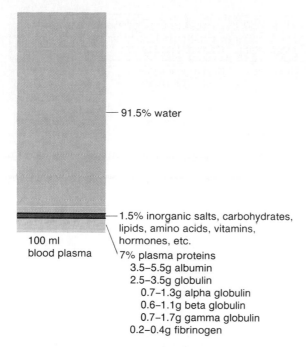

— 91.5% water

— 1.5% inorganic salts, carbohydrates, lipids, amino acids, vitamins, hormones, etc.

7% plasma proteins
3.5–5.5g albumin
2.5–3.5g globulin
0.7–1.3g alpha globulin
0.6–1.1g beta globulin
0.7–1.7g gamma globulin
0.2–0.4g fibrinogen

100 ml blood plasma

Figure 17.18 Plasma, which is primarily water, contains about 7% protein and 1.5% other substances.

Plasma Proteins

The most abundant of the dissolved substances (solutes) in plasma are the **plasma proteins.** These proteins remain in the blood and interstitial fluids, and ordinarily are not used as energy sources. The three main groups are albumins, globulins, and fibrinogen. The members of each group differ in their chemical compositions and in their physiological functions.

> The concentration of plasma proteins may decrease significantly if a person is starving or has a protein-deficient diet and is forced to use body protein as an energy source. Similarly, plasma protein concentration may drop if liver disease interferes with the synthesis of these proteins. In either case, as the blood protein concentration decreases, the osmotic pressure of the blood decreases, and water tends to accumulate in the intercellular spaces, causing tissues to swell (edema).

Albumins account for about 60% of the plasma proteins, and their molecules are the smallest of these proteins. They are synthesized in the liver, and because they are so plentiful, albumins together with other solutes help maintain the *osmotic pressure* of the blood.

As explained in chapter 3, whenever the concentration of solutes changes on either side of a cell membrane, water is likely to move through the membrane toward the region where the solutes are in higher concentration. For this reason, it is important that the concentration of solutes in the plasma remains relatively

CHART 17.5	Plasma proteins		
Protein	Percentage of total	Origin	Function
Albumin	60%	Liver	Helps maintain blood osmotic pressure
Globulin	36%		
Alpha globulins		Liver	Transports lipids and fat-soluble vitamins
Beta globulins		Liver	Transports lipids and fat-soluble vitamins
Gamma globulins		Lymphatic tissues	Constitute the antibodies of immunity
Fibrinogen	4%	Liver	Plays a key role in blood clot formation

stable. Otherwise, water tends to leave the blood and enter the tissues, or leave the tissues and enter the blood, by *osmosis*. Because the presence of albumins (and other plasma proteins) adds to the osmotic pressure of the plasma, albumins aid in regulating the water balance between the blood and the tissues. At the same time, they help control blood volume, which, in turn, is directly related to blood pressure. Albumins also bind and transport certain molecules, such as bilirubin and free fatty acids.

The **globulins,** which make up about 36% of the plasma proteins, can be further separated into fractions called *alpha globulins, beta globulins,* and *gamma globulins*. The alpha and beta globulins are synthesized in the liver, and they have a variety of functions including the transport of lipids and fat-soluble vitamins. The gamma globulins are produced in lymphatic tissues, and they include the proteins that function as *antibodies of immunity*. (See chapter 19.)

Fibrinogen, which constitutes about 4% of the plasma protein, plays a primary role in the blood clotting mechanism. It is synthesized in the liver and has the largest molecules of the plasma proteins. The function of fibrinogen is discussed in a subsequent section of this chapter.

Chart 17.5 summarizes the characteristics of the plasma proteins.

1. List three types of plasma proteins.
2. How does albumin help maintain a water balance between the blood and the tissues?
3. Which of the globulins functions in immunity?
4. What is the role of fibrinogen?

Nutrients and Gases

The *plasma nutrients* include amino acids, simple sugars, and various lipids that have been absorbed from the digestive tract. Glucose, for example, is transported by the plasma from the small intestine to the liver, where it may be stored as glycogen or changed into fat. If the blood glucose concentration drops below the normal range, glycogen may be converted back into glucose, as described in chapter 13.

Recently absorbed amino acids are also carried to the liver, where they may be used in the manufacture of proteins such as those found in the plasma, or deaminated and used as an energy source. (See chapter 4.)

The lipids of plasma include fats (triglycerides), phospholipids, and cholesterol. Generally, these lipids are combined with proteins in complexes called **lipoproteins.** Lipoprotein molecules are relatively large and consist of a core of triglyceride surrounded by a surface layer composed of phospholipid, cholesterol, and protein. Within the protein portion of this layer are specialized molecules called *apoproteins* that can combine with receptors on the membranes of specific target cells. Lipoprotein molecules also vary in the proportions of the various lipids they contain.

Because fats are less dense than proteins, as the proportion of triglycerides in a lipoprotein increases, the density of the particle decreases. Conversely, as the proportion of triglycerides decreases, the density increases.

On the basis of their densities, which reflect their composition, lipoproteins can be classified as *chylomicrons,* which consist mainly of triglycerides absorbed from the small intestine (see chapter 14); *very low-density lipoproteins* (VLDL), which have a relatively high concentration of triglycerides; *low-density lipoproteins* (LDL), which have a relatively high concentration of cholesterol and are the major cholesterol-carrying lipoproteins; and *high-density lipoproteins* (HDL), which have a relatively high concentration of protein and a lower concentration of lipids.

As discussed in chapter 14, chylomicrons are lipoproteins that transport dietary fats to muscle and adipose cells. Similarly, very low-density lipoproteins, which are produced by liver cells, are used to transport triglycerides that the liver has synthesized from excess dietary carbohydrates. After VLDL molecules have delivered their loads of triglycerides to adipose cells, the remnants of the VLDL molecules are converted to low-density lipoproteins by the action of an enzyme called *lipoprotein lipase.* Because most of the triglycerides have been removed, LDL molecules have a relatively higher cholesterol content than do the original VLDL molecules. Various cells, including liver cells, have surface receptors that combine with apoproteins associated with LDL molecules, and these cells slowly

CHART 17.6	Plasma Lipoproteins	
Lipoprotein	Characteristics	Functions
Chylomicron	High concentration of triglycerides	Transports dietary fats to muscle and adipose cells
Very low-density lipoprotein (VLDL)	Relatively high concentration of triglycerides; produced in the liver	Transports triglycerides that have been synthesized in the liver from carbohydrates to adipose cells
Low-density lipoprotein (LDL)	Relatively high concentration of cholesterol; formed from remnants of VLDL molecules that have given up their triglycerides	Delivers cholesterol to various cells, including liver cells
High-density lipoprotein (HDL)	Relatively high concentration of protein and low concentration of lipid	Transports to the liver remnants of chylomicrons that have given up their triglycerides

remove LDL from plasma by receptor-mediated endocytosis. (See chapter 3.) This process provides cells with a supply of cholesterol.

> Some individuals have an inherited condition in which their cells lack LDL receptors. As a consequence, LDL molecules accumulate in the plasma, and the blood cholesterol concentration becomes two to three times normal. This condition, which is called *familial hypercholesteremia,* is characterized by increased deposition of cholesterol over various joints and in the walls of arteries with accelerated development of the disease *atherosclerosis.* Persons with this condition typically have heart attacks when relatively young.

After chylomicrons have delivered their triglycerides to cells, the remnants of these molecules are transferred to high-density lipoproteins. These HDL molecules, which are formed in the liver and small intestine, transport chylomicron remnants to the liver, where they enter cells rapidly by receptor-mediated endocytosis. The liver disposes of the cholesterol it obtains in this manner by secreting it into bile or by using it to synthesize bile salts.

Chart 17.6 summarizes the characteristics and functions of these lipoproteins.

Much of the cholesterol and bile salts in bile are later reabsorbed by the small intestine and transported back to the liver, and the secretion-reabsorption cycle is repeated. During each cycle, some of the cholesterol and bile salts escape reabsorption, reach the large intestine, and are eliminated with the feces.

The most important *blood gases* are oxygen and carbon dioxide. Although plasma also contains a considerable amount of dissolved nitrogen, this gas ordinarily has no physiological function. The blood gases and their transport are discussed in chapter 16.

As a rule, blood gases are evaluated using a fresh sample of whole blood obtained from an artery. This blood is cooled to decrease the rate of the metabolic processes, and it is prevented from clotting by the addition of an anticoagulant. In the laboratory, the partial pressures of oxygen and carbon dioxide of the blood are determined, the blood pH is measured, and the plasma bicarbonate concentration is calculated. Such information is commonly used to diagnose and treat disorders of circulation, respiration, and electrolyte balance. The normal values for these laboratory tests are provided in Appendix C.

1. What nutrients are found in blood plasma?
2. How are triglycerides transported in plasma?
3. How is cholesterol eliminated from the liver?
4. What gases occur in plasma?

Nonprotein Nitrogenous Substances

Molecules that contain nitrogen atoms but are not proteins comprise a group called **nonprotein nitrogenous substances.** Within the plasma, this group includes amino acids, urea, uric acid, creatine, and creatinine. The *amino acids* are present as a result of protein digestion and amino acid absorption. *Urea* and *uric acid* are the products of protein and nucleic acid catabolism, respectively, and *creatinine* results from the metabolism of *creatine.*

As discussed in chapter 9, creatine occurs as **creatine phosphate** in muscle and brain tissues as well as in the blood. This substance functions to store high-energy phosphate bonds, much like those of ATP molecules.

Normally, the concentration of nonprotein nitrogenous (NPN) substances remains relatively stable because protein intake and utilization are balanced with the excretion of nitrogenous wastes. Because about half of the NPN is urea, which is ordinarily excreted by the kidneys, a rise in the plasma NPN concentration may suggest a kidney disorder. However, such an increase may also occur as a result of excessive protein catabolism or the presence of an infection.

Plasma Electrolytes

Blood plasma contains a variety of *electrolytes,* which have been absorbed from the intestine or have been released as by-products of cellular metabolism. They include sodium, potassium, calcium, magnesium,

chloride, bicarbonate, phosphate, and sulfate ions. Of these, sodium and chloride ions are the most abundant.

Such ions are important in maintaining the osmotic pressure and the pH of the plasma, and like other plasma constituents, they are regulated so that their blood concentrations remain relatively stable. These electrolytes are discussed in chapter 21 in connection with water and electrolyte balance.

1. What is meant by a nonprotein nitrogenous substance?
2. Why does kidney disease cause an increase in the blood concentration of these substances?
3. What are the sources of plasma electrolytes?

Hemostasis

The term **hemostasis** refers to the stoppage of bleeding, which is vitally important when blood vessels are damaged. Following an injury to the blood vessels, several actions may help prevent excessive blood loss. These include blood vessel spasm, platelet plug formation, and blood coagulation.

These mechanisms are most effective in minimizing blood losses from relatively small vessels, such as arterioles, capillaries, and venules. Injury to a larger vessel, particularly an artery, may result in a severe hemorrhage that requires special treatment in order to be controlled.

Blood Vessel Spasm

When an arteriole or a venule is cut or broken, the smooth muscles in its wall are stimulated to contract, and blood loss is decreased almost immediately. In fact, the ends of a severed vessel may be closed completely by such a *spasm.* This effect seems to result from direct stimulation of the vessel wall as well as from reflexes elicited by pain receptors in the injured tissues.

Although the reflex response may last only a few minutes, the effect of the direct stimulation usually continues for about thirty minutes, and, by then, a platelet plug and blood coagulation mechanisms are normally operating. Also, as the platelet plug forms, the platelets release a substance called *serotonin,* which causes smooth muscles in the blood vessel walls to contract. This vasoconstricting action helps maintain a prolonged vascular spasm.

Platelet Plug Formation

Platelets tend to stick to the exposed ends of injured blood vessels. Actually, they adhere to any rough surface, particularly to the *collagen* in connective tissue that underlies the endothelial lining of blood vessels.

When platelets contact collagen, their shapes change drastically, and numerous spiny processes begin

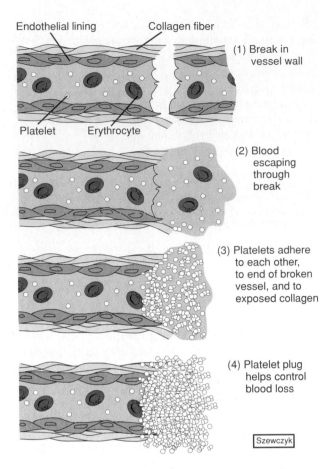

Endothelial lining   Collagen fiber

(1) Break in vessel wall

Platelet   Erythrocyte

(2) Blood escaping through break

(3) Platelets adhere to each other, to end of broken vessel, and to exposed collagen

(4) Platelet plug helps control blood loss

Szewczyk

Figure 17.19   Steps in platelet plug formation.

to protrude from their membranes. At the same time, they tend to stick to each other, forming a *platelet plug* in the vascular break. Such a plug may be able to control blood loss if the break is relatively small, but a larger one may require the aid of a blood clot to halt bleeding.

The steps in platelet plug formation are shown in figure 17.19.

1. What is meant by hemostasis?
2. How does a blood vessel spasm help control bleeding?
3. Describe the formation of a platelet plug.

Blood Coagulation

Coagulation, which is the most effective of the hemostatic mechanisms, causes the formation of a *blood clot*. This process may be initiated by a variety of substances, most of which are synthesized by liver cells, and it may involve an extrinsic or an intrinsic clotting mechanism.

The *extrinsic clotting mechanism* is triggered by the release of chemical substances from broken blood vessels or damaged tissues. The *intrinsic clotting*

CHART 17.7	Hemostatic mechanisms	
Mechanism	Stimulus	Effect
Blood vessel spasm	Direct stimulus to vessel wall or to pain receptors; serotonin released from platelets	Smooth muscles in vessel wall contract reflexly; vasoconstriction helps maintain prolonged vessel spasm
Platelet plug formation	Exposure of platelets to rough surfaces or to collagen of connective tissue	Platelets adhere to rough surfaces and to each other, forming a plug
Blood coagulation	Cellular damage and blood contact with foreign surfaces result in the production of substances that favor coagulation	Blood clot forms as a result of a series of reactions, terminating in the conversion of fibrinogen into fibrin

mechanism is stimulated by blood contact with foreign surfaces in the absence of tissue damage.

The mechanism by which the blood coagulates is very complex and involves many substances called *clotting factors.* Some of these factors promote coagulation (procoagulants) and others inhibit it (anticoagulants). Whether or not the blood coagulates depends on the balance that exists between these two groups of factors. Normally, the anticoagulants prevail, and the blood does not clot. However, as a result of injury (trauma), substances that favor coagulation may increase in concentration, and the blood may coagulate.

The basic event in blood clot formation is the conversion of the soluble plasma protein *fibrinogen* (factor I) into relatively insoluble threads of the protein **fibrin.** This change is triggered by the activation of certain plasma proteins that remain inactive until they are affected by the presence of still other protein factors.

Chart 17.7 and figure 17.20 summarize the three primary hemostatic mechanisms.

Extrinsic Clotting Mechanism

The extrinsic clotting mechanism is triggered when blood contacts damaged blood vessel walls or tissues outside blood vessels. Such damaged tissues release a complex of substances that is associated with disrupted cell membranes and is called *tissue thromboplastin.* The presence of tissue thromboplastin initiates a series of reactions involving several clotting factors. These reactions, which also depend on the presence of calcium ions for their completion, lead to the production of *prothrombin activator.*

Do Well on Tests and Exams
Get Your Own Tutor!

Wm. C. Brown Publishers has now developed a take-home tutor that will help you achieve your goals in the classroom and prepare for careers in the nursing/allied health sciences. The *Biology Learning Support Package* is a complete set of additional study aids designed to facilitate learning and achievement. The package includes:

STUDENT STUDY GUIDE
ISBN 0-697-12280-8

For each text chapter there is a corresponding study guide chapter that includes suggestions for review, objective questions, and essay questions that will review the major chapter concepts.

• • •

STUDY CARDS FOR HUMAN ANATOMY AND PHYSIOLOGY
ISBN 0-697-09731-5

This boxed set of 300 two-sided study cards provides a complete description of terms, clearly labeled drawings, pronunciation guides, and clinical information on diseases. They will accompany any anatomy and physiology textbook.

• • •

ATLAS OF THE SKELETAL MUSCLES
ISBN 0-697-10618-7

This inexpensive atlas depicts all the human skeletal muscles in precise black-and-white drawings that show their origin, insertion, action, innervation. Each muscle is presented on a separate page, and the spinal cord level of the nerve fibers is included in parentheses. These features aid in the visual understanding of the action of the muscles.

• • •

COLORING REVIEW GUIDE TO HUMAN ANATOMY
ISBN 0-697-03150-0

This helpful manual serves as a framework for learning human anatomy. It emphasizes the connection between facts and structures through the specific student activity of color association.

• • •

HUMAN SKELETAL MUSCULATURE VIDEOTAPE
ISBN 0-697-15210-0

This videotape provides an in-depth introduction to the major groups of skeletal muscles by using a combination of film clips, X rays, line art, and human cadaver photographs. This is an excellent introduction to how the muscles provide movement to the human body.

To get your own take-home tutor, just ask your bookstore manager, or if unavailable through your bookstore, call toll-free 1-800-338-5578 and order using your credit card.

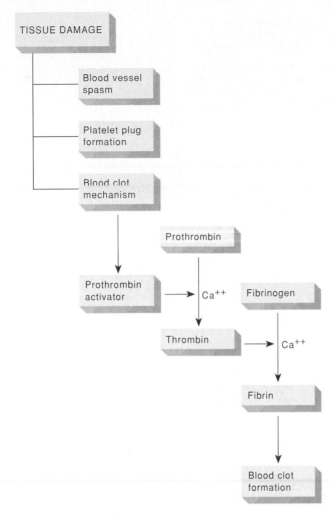

TISSUE DAMAGE

Blood vessel spasm

Platelet plug formation

Blood clot mechanism

Prothrombin

Prothrombin activator → Ca^{++} Fibrinogen

Thrombin → Ca^{++}

Fibrin

Blood clot formation

Figure 17.20 Hemostasis following tissue damage is likely to involve blood vessel spasm, platelet plug formation, and the extrinsic blood clotting mechanism.

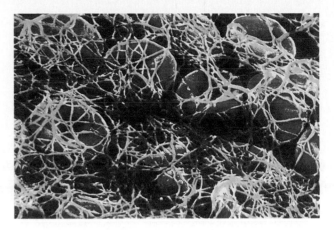

Figure 17.21 A scanning electron micrograph of fibrin threads ($\times$2,000). What factors initiate the formation of fibrin?

Prothrombin (factor II) is an alpha globulin that is continually produced by the liver and thus is normally present in the plasma. In the presence of calcium ions, prothrombin is converted into **thrombin** (factor IIa) by the action of prothrombin activator. Thrombin, in turn, acts as an enzyme and causes a reaction in the molecules of fibrinogen. In this reaction, small portions of the fibrinogen molecules are split off, and the remaining pieces develop attractions for other similar molecules. These activated fibrinogen molecules join end to end, forming long threads of *fibrin.* The production of fibrin threads is also enhanced by the presence of calcium ions and certain proteins. (See figure 17.20.)

Once the threads of fibrin have formed, they tend to stick to the exposed surfaces of the damaged blood vessels and create a meshwork in which various blood cells and platelets become entangled (figure 17.21). The resulting mass is a *blood clot,* which may effectively block a vascular break and prevent further loss of blood.

The amount of prothrombin activator that appears in the blood is directly proportional to the degree of tissue damage. Once a blood clot begins to form, it promotes still more clotting. This happens because thrombin also acts directly on blood clotting factors other than fibrinogen, and it can cause prothrombin to form still more thrombin. This type of self-initiating action is an example of a **positive feedback system,** in which the original action stimulates more of the same type of action. Such a mechanism produces unstable conditions and can operate for only a short time in a living system, because life depends on the maintenance of a stable internal environment (chapter 1).

Normally, the formation of a massive clot throughout the blood system is prevented by blood movement, which rapidly carries excessive thrombin away and thus keeps its concentration too low to enhance further clotting. Also, a naturally occurring substance called *antithrombin,* which is present in the blood and on the surface membranes of the endothelial cells, limits the formation of thrombin. Consequently, blood clot formation is usually limited to blood that is standing still (or moving relatively slowly), and clotting ceases where a clot comes in contact with circulating blood.

Sometimes the clotting mechanism is abnormally activated in widespread regions of the circulatory system. This condition, called *disseminated intravascular clotting* (DIC), is usually associated with the presence of bacteria or bacterial toxins in the blood or with a disorder causing widespread tissue damage. As a result, many small clots may appear and obstruct blood flow into various tissues and organs, particularly the kidneys. As the plasma clotting factors and platelets become depleted, the patient may develop a serious tendency to bleed.

Intrinsic Clotting Mechanism

Unlike the extrinsic clotting mechanism, all of the components necessary for the intrinsic clotting mechanism occur in the blood. The intrinsic clotting mechanism is initiated by the activation of a substance called the *Hageman factor* (factor XII) when blood is exposed to a foreign surface such as collagen or when blood is stored in a glass container. In the presence of *calcium ions,* the activated factor triggers a complex series of changes that involve several clotting factors and lead to the formation of *prothrombin activator.* The subsequent steps of blood clot formation are the same as those described for the extrinsic mechanism (figure 17.22).

> Laboratory tests commonly used to evaluate the blood coagulation mechanisms include *prothrombin time* (PT) and *partial thromboplastin time* (PTT). Both of these tests measure the time it takes for fibrin threads to form in a sample of blood plasma. However, the prothrombin time test is designed to check the extrinsic clotting mechanism, while the partial thromboplastin test evaluates the intrinsic clotting mechanism.

Fate of Blood Clots

After a blood clot has formed, it soon begins to retract due to platelet activity. The tiny processes extending from the platelet membranes adhere to strands of fibrin within the clot, and these processes contract. The blood clot becomes smaller, and the edges of the broken vessel are pulled closer together. At the same time, a fluid called **serum** is squeezed from the clot. Serum is essentially plasma minus all of its fibrinogen and most of the other factors involved in the clotting mechanism. Platelets associated with a blood clot also release a substance called *platelet-derived growth factor* (PDGF) that stimulates smooth muscle cells and fibroblasts to repair damaged blood vessel walls.

Blood clots that form in ruptured vessels are soon invaded by *fibroblasts.* These cells produce fibrous connective tissue throughout the clots, which helps strengthen and seal vascular breaks. Many clots, including those that form in tissues as a result of blood leakage (hematomas), disappear in time. This dissolution involves a plasma protein called *plasminogen* (profibrinolysin) that is absorbed in fibrin threads. Plasminogen is converted to *plasmin* by a substance (plasminogen activator) that is released from the lysosomes of damaged tissue cells. The plasmin thus formed is a protein-splitting enzyme that can digest fibrin threads and other proteins associated with blood clots. This action may cause a whole clot to dissolve; however, clots that fill large blood vessels are seldom removed by natural processes.

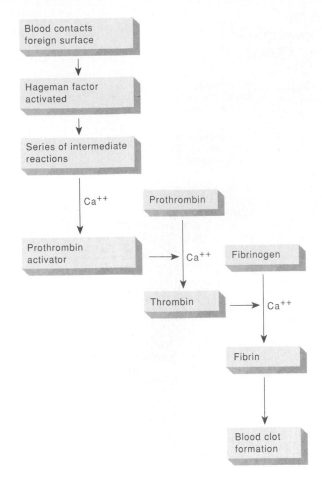

Figure 17.22 The intrinsic blood clotting mechanism may be initiated when blood contacts a foreign surface.

> Lysosomes of certain cells within the kidneys produce an abundance of a plasminogen activator called *urokinase.* This enzyme helps keep the functional units of the kidneys (nephrons) free of blood clots. (See chapter 20.) Urokinase or a bacterial enzyme called *streptokinase,* which can also cause plasmin formation, are sometimes used to dissolve blood clots that have formed abnormally in blood vessels.

If a blood clot forms in a vessel abnormally, it is termed a **thrombus.** If the clot becomes dislodged or if a fragment of it breaks loose and is carried away by the blood flow, it is called an **embolus.** Generally, emboli continue to move until they reach narrow places in vessels where they become lodged and may interfere with blood flow.

Such abnormal clot formations are often associated with conditions that cause changes in the endothelial linings of vessels. For example, in the disease *atherosclerosis,* arterial linings are changed by accumulations of fatty deposits. Such changes may initiate

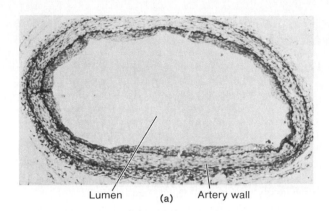

Lumen   **(a)**   Artery wall

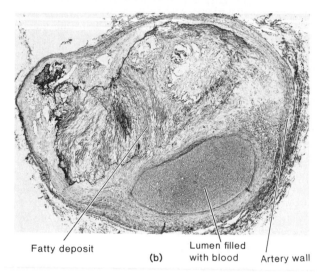

Fatty deposit   Lumen filled
          (b)   with blood   Artery wall

Figure 17.23   (a) Light micrograph of a normal artery (×60); (b) the inner wall of an artery changed as a result of atherosclerosis (×60). How might this condition promote the formation of blood clots?

the clotting mechanism and are the most common causes of blood clot formation (thrombosis) in medium-sized arteries. (figure 17.23).

Coagulation may also occur in blood that is flowing too slowly. In this instance, the concentration of clot-promoting substances may increase to a critical level instead of being carried away by more rapidly moving blood, and a clot may form. This mechanism is the usual cause of thrombosis occurring in veins.

Sometimes a blood clot forms in a vessel supplying a vital organ, such as the heart (coronary thrombosis) or the brain (cerebral thrombosis). As a consequence of the obstruction, tissues may die (infarction), and this may lead to the death of the individual. The blood vessels of the lungs are most likely to be obstructed by emboli arriving through the pulmonary arteries (pulmonary embolism). If a major branch of a pulmonary artery is obstructed, a portion of the lung will become nonfunctional, and the life of the patient may be threatened.

1. Distinguish between extrinsic and intrinsic clotting mechanisms.
2. What is the basic event in blood clot formation?
3. What prevents the formation of massive clots throughout the blood system?
4. Distinguish between a thrombus and an embolus.

Prevention of Coagulation

In a normal vascular system, spontaneous formation of blood clots is prevented in part by the endothelium of the blood vessels. This smooth lining carries a negative electrical charge that tends to repel platelets and various clotting factors. The endothelial cells also produce a prostaglandin (chapter 13) called *prostacyclin* (PGI_2), which inhibits the adherence of platelets to the inner surface of healthy blood vessel walls.

When a clot is forming, fibrin threads adsorb thrombin, thus helping prevent the spread of the clotting reaction. Any additional thrombin is likely to be inactivated by an alpha globulin called *antithrombin,* which is normally present in the plasma. This substance binds to thrombin and blocks its action on fibrinogen.

In addition, certain cells including basophils and mast cells, that are common in the connective tissue surrounding capillaries, secrete the anticoagulant *heparin.* This substance interferes with the formation of prothrombin activator, prevents the action of thrombin on fibrinogen, and promotes the removal of thrombin by antithrombin and fibrin adsorption.

Heparin-secreting cells are particularly abundant in the liver and lungs, where capillaries are likely to trap small blood clots that commonly occur in the slow-moving blood of veins. These cells are thought to secrete heparin continually, and thus they help prevent additional clotting in the circulatory system. Chart 17.8 summarizes the clot-inhibiting factors.

Because of its strong anticoagulant action, heparin is widely used to prevent blood clot formation in deep veins. Administration of heparin by injection almost immediately increases the time required for blood coagulation. Substances that are used to prevent blood clot formation in arteries usually act by inhibiting the formation of platelet plugs.

1. How does the lining of a blood vessel help prevent blood clot formation?
2. What is the function of antithrombin?
3. How does heparin help prevent blood clot formation?

Coagulation Disorders

Coagulation disorders fall into two main groups—those that result in excessive bleeding and those that produce abnormal blood clotting.

Because the liver plays an important role in the synthesis of various plasma proteins, such as prothrombin, it is not surprising that liver diseases are often accompanied by a tendency to bleed. Also, bile salts from the liver are necessary for the efficient absorption of *vitamin K* from the intestine, and this vitamin is essential for the synthesis of prothrombin, as well as certain coagulation factors. If the liver fails to produce enough bile, or if the bile ducts become obstructed, a vitamin K deficiency is likely to develop, and the ability to form blood clots may be diminished. For this reason, vitamin K is often administered to patients with liver diseases or bile duct obstructions before they are treated surgically.

Newborns sometimes have a tendency to hemorrhage because they have relatively low blood concentrations of certain coagulation factors. These factors normally rise in concentration during the first few weeks of life. Meanwhile, such infants are usually given preventive treatment with vitamin K, which reduces the chance of bleeding.

In other cases, a tendency to bleed is related to abnormally low platelet counts. This condition, called **thrombocytopenia,** is said to occur whenever the platelet count drops below 100,000 platelets per mm^3 of blood.

The cause of thrombocytopenia is usually unknown, but various factors that damage red bone marrow, such as excessive exposure to ionizing radiation or adverse drug reactions, may produce this condition. Affected persons bleed easily, experience numerous capillary hemorrhages throughout their body tissues, and typically exhibit many small, bruiselike spots on their skin.

The adherence of platelets to damaged blood vessel walls, which usually precedes blood clotting, requires the presence of a plasma protein called *von Willebrand factor* (factor VIII). This factor is synthesized by the endothelial lining of blood vessels and is present in the underlying collagen. Although their platelet counts may be normal, persons lacking the von Willebrand factor have a tendency to bleed easily and are subject to episodes of spontaneous bleeding from the mucous membranes of their gastrointestinal and urinary tracts.

Hemophilia is a hereditary disease that appears almost exclusively in males. Hemophiliacs lack a blood factor that is necessary for coagulation. Consequently, they typically experience repeated episodes of serious bleeding.

There are several types of hemophilia, and each is caused by the lack of a different blood factor. In addition to the tendency to hemorrhage severely following minor injuries, hemophilia is characterized by frequent nosebleeds, large intramuscular hematomas, blood in the urine (hematuria), and severe pain and disability caused by bleeding into joints and body cavities.

Treatment for hemophilia may involve applying pressure or packing accessible bleeding sites in an effort to control blood loss. Also, transfusions of fresh plasma, fresh-frozen plasma, or plasma concentrates (cryoprecipitates) that contain the missing blood factor (most commonly, factor VIII) are often used.

Certain drugs, including aspirin, ibuprofen, and penicillin, inhibit platelet activity. Consequently, prolonged use of such drugs is sometimes associated with a tendency to bruise easily or to bleed excessively following injuries, menstruation, or dental work.

CHART 17.8	Factors that inhibit blood clot formation		
Factor	Action	Factor	Action
Smooth lining of blood vessel	Prevents activation of intrinsic blood clotting mechanism	Fibrin threads	Adsorb thrombin
Prostacyclin	Inhibits adherence of platelets to blood vessel wall	Antithrombin in plasma	Interferes with the action of thrombin
Negative charge of blood vessel lining	Repels platelets	Heparin from mast cells and basophils	Interferes with the formation of prothrombin activator

Blood Groups and Transfusions

Early attempts to transfer blood from one person to another produced varied results. Sometimes the person receiving the transfusion was aided by the procedure. At other times, the recipient suffered a blood reaction in which the red blood cells clumped together, obstructing vessels and producing other serious consequences.

Eventually, it was discovered that each individual has a particular combination of substances in his or her blood. Some of these substances react with those in an-

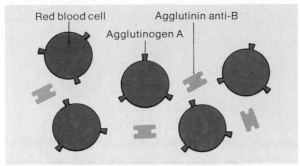

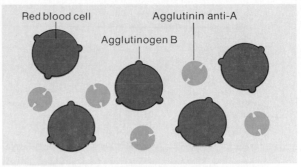

Type A blood

Type B blood

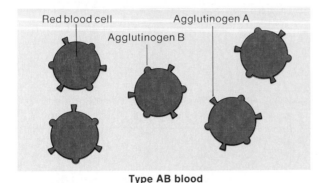

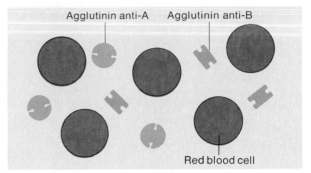

Type AB blood

Type O blood

Figure 17.24 Each blood type is characterized by a different combination of agglutinogens and agglutinins.

other person's blood. These discoveries led to the development of procedures for typing blood. It is now known that safe transfusions of whole blood depend upon properly matching the blood types of the donors and recipients.

Agglutinogens and Agglutinins

The clumping of red blood cells following a transfusion reaction is called **agglutination.** This phenomenon is due to the presence of substances called **agglutinogens** (antigens) in the red cell membranes and substances called **agglutinins** (antibodies) dissolved in the plasma.

Blood typing involves identifying the agglutinogens that are present in a person's red cells. Although many different agglutinogens are associated with human erythrocytes, only a few of them are likely to produce serious transfusion reactions. These include the agglutinogens of the ABO group and those of the Rh group.

Avoiding the mixture of certain kinds of agglutinogens and agglutinins prevents adverse transfusion reactions. Such reactions are described in a subsequent section of this chapter.

ABO Blood Group

The *ABO blood group* is based on the presence (or absence) of two major agglutinogens in red cell membranes—*agglutinogen A* and *agglutinogen B*—which

are present at birth as a result of inheritance. The erythrocytes of each person contain one of the four following combinations of agglutinogens: only A, only B, both A and B, or neither A nor B.

A person with only agglutinogen A is said to have *type A blood;* a person with only agglutinogen B has *type B blood;* one with both agglutinogen A and B has *type AB blood;* and one with neither agglutinogen A nor B has *type O blood* (also called type H). Thus, all humans have one of four possible blood types—A, B, AB, or O.

Certain agglutinins develop spontaneously in the plasma about two to eight months following birth. Specifically, whenever agglutinogen A is absent in the red blood cells, an agglutinin called *anti-A* develops, and whenever agglutinogen B is absent, an agglutinin called *anti-B* develops. Therefore, persons with type A blood also have agglutinin anti-B in their plasma; those with type B blood have agglutinin anti-A; those with type AB blood have neither agglutinin; and those with type O blood have both agglutinin anti-A and anti-B (figure 17.24).

Chart 17.9 summarizes the agglutinogens and agglutinins of the ABO blood group.

Because an agglutinin of one kind will react with an agglutinogen of the same kind and cause red blood cells to clump together, such combinations must be avoided. The major concern in blood transfusion procedures is that the cells in the *transfused donor blood*

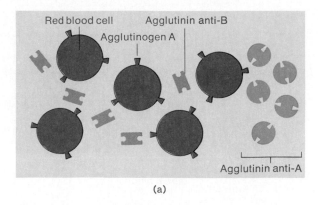

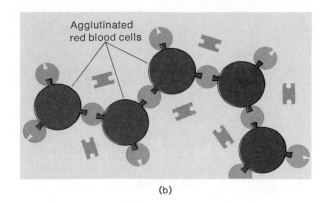

(a)   (b)

Figure 17.25   (a) If red blood cells with agglutinogen A are added to blood containing agglutinin anti-A, (b) the agglutinins react with the agglutinogens of the red blood cells and cause them to clump together.

CHART 17.9	Agglutinogens and agglutinins of the ABO blood group	
Blood type	Agglutinogen	Agglutinin
A	A	anti-B
B	B	anti-A
AB	A and B	Neither anti-A nor anti-B
O	Neither A nor B	Both anti-A and anti-B

CHART 17.10	Preferred and permissible blood types for transfusions	
Blood type of recipient	Preferred blood type of donor	Permissible blood type of donor (in an extreme emergency)
A	A	A,O
B	B	B,O
AB	AB	AB, A, B, O
O	O	O

not be agglutinated by the agglutinins in the recipient's plasma. For this reason, a person with type A (anti-B) blood must not be given blood of type B or AB, because the red cells of both types would be agglutinated by the anti-B in the recipient's type A blood. Likewise, a person with type B (anti-A) blood must not be given type A or AB blood, and a person with type O (anti-A and anti-B) blood must not be given type A, B, or AB blood (figure 17.25).

Because type AB blood lacks both anti-A and anti-B agglutinins, it would appear that an AB person could receive a transfusion of blood of any other type. For this reason, type AB persons are sometimes called *universal recipients*. It should be noted, however, that type A (anti-B) blood, type B (anti-A) blood, and type O (anti-A and anti-B) blood still contain agglutinins (either anti-A or anti-B) that could cause agglutination of type AB cells. Consequently, even for AB individuals, it is always best to use donor blood of the same type as the recipient blood. If the matching type is not available and type A, B, or O is used, it should be transfused slowly so that the donor blood is well diluted by the recipient's larger blood volume. This precaution usually avoids serious reactions between the donor's agglutinins and the recipient's agglutinogens.

Similarly, because type O blood lacks agglutinogens A and B, it would seem that this type could be

transfused into persons with blood of any other type. Therefore, persons with type O blood are sometimes called *universal donors*. Type O blood, however, does contain both anti-A and anti-B agglutinins, and if it is given to a person with blood type A, B, or AB, it too should be transfused slowly to minimize the chance of an adverse reaction.

Chart 17.10 summarizes preferred blood types for normal transfusions and permissible blood types for emergency transfusions.

The agglutinogens of the ABO group are *inherited factors* that are present in the red cell membranes at birth. The plasma agglutinins begin to appear spontaneously, for unknown reasons, about two to eight months after birth, and they reach a maximum concentration between eight and ten years of age. A person's blood type cannot be changed by a transfusion or by any other procedure.

Fluid replacements, such as isotonic saline solution or isotonic glucose solution, are sometimes administered to persons during emergencies. These replacements increase the blood volume only temporarily. If a patient has suffered a severe hemorrhage, a transfusion of whole blood may be necessary for survival.

Blood Component Therapy and Agglutination Reactions

In the past, a blood transfusion often involved infusing whole blood into a patient. Today, however, whole blood is often separated into its component parts—red blood cells, white blood cells, platelets, and plasma—and only the component needed to treat a particular condition is used.

For example, a patient with anemia or an acute blood loss might be given concentrated (packed) red blood cells; a person whose platelet function has been depressed by a drug reaction might be given platelets; and one whose white blood cells have been damaged by an anticancer treatment might be given a white blood cell preparation. Similarly, blood plasma might be used to replace lost blood volume, or to provide certain blood clotting factors to patients who lack them.

Before donor blood is used to obtain blood components, it is tested for the presence of particular viruses, including those that cause the diseases hepatitis B and AIDS (chapter 19). Then the blood components must be stored properly. For example, packed red blood cells can be stored for several years if they are frozen, but can be used for only about a month if they are not frozen. After thawing, however, such cells must be used within about a day. Platelet preparations must be used within five days, and white blood cell concentrates must be used immediately.

In blood reactions involving agglutinogens and agglutinins, the agglutinated red cells usually degenerate or are phagocytized by reticuloendothelial cells. At the same time, hemoglobin and other red cell contents are released, and the blood concentration of *free hemoglobin* increases greatly. Some of this hemoglobin may be phagocytized by macrophages and converted into bilirubin. As a result, the tissues may develop a yellowish stain—a characteristic of the condition called *jaundice* (icterus).

Free hemoglobin may also pass into the kidneys and interfere with their vital functions, so that a person with a blood transfusion reaction may develop kidney (renal) failure later.

The severity of a blood transfusion reaction depends on a number of factors, including the degree of incompatibility between the blood of the donor and that of the recipient, the quantity of blood that is transfused, and the rate at which the transfusion is administered.

If a reaction occurs, the patient is likely to experience anxiety, breathing difficulty, facial flushing, headache, and severe pain in the neck, chest, and lumbar area.

Infants with *erythroblastosis fetalis* are usually jaundiced and severely anemic. As their hematopoietic tissues respond to the need for more red cells, various immature erythrocytes, including *erythroblasts,* are released into the blood. (The presence of these immature cells is related to the name of the disease.)

An affected infant may suffer permanent brain damage as a result of bilirubin precipitating in the brain tissues and injuring neurons. This condition is called *kernicterus,* and, if the infant survives, he or she may have motor or sensory losses and exhibit mental deficiencies.

Treatment for erythroblastosis fetalis usually involves exposing the affected infant to bright blue or white *fluorescent light.* Bilirubin is a light-sensitive substance, and this exposure (phototherapy) causes a decrease in the blood bilirubin concentration.

In more severe cases, the infant's Rh-positive blood may be replaced with Rh-negative blood. This procedure is called an *exchange transfusion,* and it reduces the concentration of bilirubin in the infant's tissues in addition to removing the agglutinating red cells, the anti-Rh agglutinins, the free hemoglobin, and the other products of erythrocyte destruction. Such a transfusion also provides a temporary supply of red cells that will not be agglutinated by any remaining anti-Rh agglutinins. In time, the infant's hematopoietic tissues will replace the donor's blood cells with Rh-positive cells, but by then, the maternal agglutinins will have disappeared.

Erythroblastosis fetalis can be prevented in future offspring by treating Rh-negative mothers with a special blood serum within seventy-two hours after the birth of each Rh-positive child. This serum is obtained from the blood of another Rh-negative person who has formed anti-Rh agglutinin. The serum antibodies are able to inactivate any Rh-positive cells that may have entered the maternal blood at the time of the infant's birth. Consequently, the maternal tissues are not stimulated to manufacture anti-Rh agglutinins.

Every person's blood contains a great variety of substances, and although the agglutinogens of the ABO and Rh groups are the most likely to produce serious blood reactions, other factors sometimes cause problems. For this reason, it is important to determine whether samples of recipient and donor blood will produce agglutination of red cells before administering a transfusion to a patient. This procedure, called *crossmatching,* involves mixing a suspension of donor cells in a sample of recipient serum, and then mixing a suspension of recipient cells in a sample of donor serum. If the red cells do not agglutinate in either case, it is probably safe to give the transfusion.

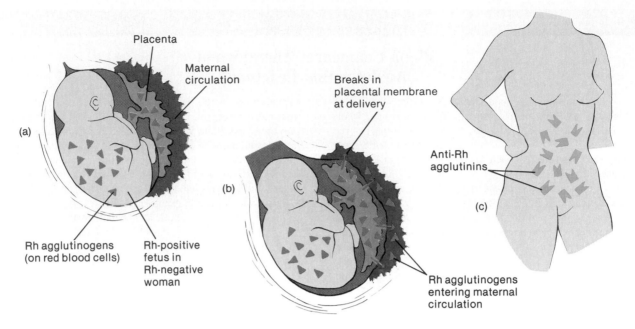

Placenta

Maternal
circulation

(a)

Rh agglutinogens
(on red blood cells)

Rh-positive
fetus in
Rh-negative
woman

Breaks in
placental membrane
at delivery

(b)

Anti-Rh
agglutinins

(c)

Rh agglutinogens
entering maternal
circulation

Figure 17.26 (a) If an Rh-negative woman is pregnant with an Rh-positive fetus, (b) some of the fetal red blood cells with Rh agglutinogens may enter the maternal blood at the time of birth. (c) As a result, the woman's cells may produce anti-Rh agglutinins.

1. Distinguish between agglutinogens and agglutinins.
2. What is meant by blood type?
3. What is the main concern when blood is transfused from one individual to another?
4. Why is a type AB person called a universal recipient?

Rh Blood Group

The *Rh blood group* was named after the *rhesus monkey* in which it was first studied. In humans, this group includes several Rh agglutinogens (factors). The most important of these is *agglutinogen D;* however, if any of the Rh factors are present in the red cell membranes, the blood is said to be *Rh positive.* Conversely, if the red cells lack Rh agglutinogens, the blood is called *Rh negative.*

As in the case of agglutinogens A and B, the presence (or absence) of an Rh agglutinogen is an inherited trait. Unlike anti-A and anti-B, agglutinins for Rh (*anti-Rh*) do not appear spontaneously. Instead, they form only in Rh-negative persons in response to special stimulation.

If an Rh-negative person receives a transfusion of Rh-positive blood, the recipient's antibody-producing cells are stimulated by the presence of the Rh agglutinogen and will begin producing *anti-Rh agglutinin.* Generally, no serious consequences result from this initial transfusion, but if the Rh-negative person—who is now sensitized to Rh-positive blood—receives

another transfusion of Rh-positive blood some months later, the donor's red cells are likely to agglutinate.

A related condition may occur when an Rh-negative woman is pregnant with an Rh-positive fetus for the first time. Such a pregnancy may be uneventful; however, at the time of this infant's birth (or if a miscarriage occurs), the placental membranes that separated the maternal blood from the fetal blood during the pregnancy may be broken, and some of the infant's Rh-positive blood cells may get into the maternal circulation. These Rh-positive cells may then stimulate the maternal tissues to begin producing anti-Rh agglutinins (figure 17.26).

If a mother who has already developed anti-Rh agglutinin becomes pregnant with a second Rh-positive fetus, these anti-Rh agglutinins can pass through the placental membrane and react with the fetal red cells, causing them to agglutinate. The fetus then develops a condition called **erythroblastosis fetalis** (hemolytic disease of the newborn) (figure 17.27). Chart 17.11 reviews the events that can lead to this condition.

Clinical Terms Related to the Blood

anisocytosis (an-i″so-si-to′sis) A condition characterized by an abnormal variation in the size of erythrocytes.
antihemophilic plasma (an″ti-he″mo-fil′ik plaz′mah) Normal blood plasma that has been processed to preserve an antihemophilic factor.

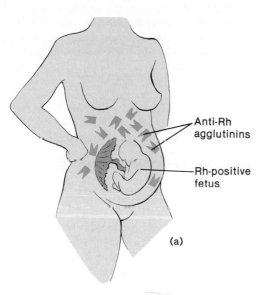

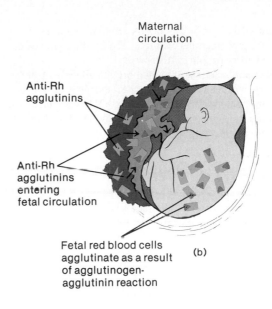

Figure 17.27 (a) If a woman who has developed anti-Rh agglutinins is pregnant with an Rh-positive fetus, (b) agglutinins may pass through the placental membrane and cause the fetal red blood cells to agglutinate.

CHART 17.11   Possible events leading to erythroblastosis fetalis

1. Rh-negative woman becomes pregnant with her first Rh-positive child.

2. Pregnancy is uneventful, but at the time of birth, some Rh-positive red cells enter the maternal circulation through damaged placental tissues.

3. Maternal tissues produce anti-Rh agglutinins.

4. A second Rh-positive child is conceived.

5. Anti-Rh agglutinins from the maternal circulation pass through the placental membranes and enter the fetal blood.

6. The fetus develops erythroblastosis fetalis as the maternal anti-Rh agglutinins react with the Rh agglutinogens of the fetal red blood cells and cause them to agglutinate.

Christmas disease (kris'mas dĭ-zēz') A hereditary bleeding disease that is due to a deficiency in a particular clotting factor (factor IX); also called *hemophilia B*.

citrated whole blood (sit'rāt-ed hōl blud) Normal blood to which a solution of acid citrate has been added to prevent coagulation.

dried plasma (drīd plaz'mah) Normal blood plasma that has been vacuum dried to prevent the growth of microorganisms.

hemorrhagic telangiectasia (hem″o-raj'ik tel-an″je-ek-ta'ze-ah) A hereditary disorder characterized by a tendency to bleed from localized lesions of the capillaries.

heparinized whole blood (hep'er-i-nīzed″ hōl blud) Normal blood to which a solution of heparin has been added to prevent coagulation.

macrocytosis (mak″ro-si-to'sis) A condition characterized by the presence of abnormally large erythrocytes.

microcytosis (mi″kro-si-to'sis) A condition characterized by the presence of abnormally small erythrocytes.

neutrophilia (nu″tro-fil'e-ah) A condition in which the number of circulating neutrophils has increased.

normal plasma (nor'mal plaz'mah) Plasma from which the blood cells have been removed by centrifugation or sedimentation.

packed red cells A concentrated suspension of red blood cells from which the plasma has been removed.

pancytopenia (pan″si-to-pe'ne-ah) A condition characterized by an abnormal depression of all the cellular components of blood.

poikilocytosis (poi″ki-lo-si-to'sis) A condition in which the erythrocytes are irregularly shaped.

pseudoagglutination (su″do-ah-gloo″ti-na'shun) The clumping of erythrocytes due to some factor other than an agglutinogen-agglutinin reaction.

purpura (per'pu-rah) A disease characterized by spontaneous bleeding into the tissues and through the mucous membrane.

septicemia (sep″tĭ-se'me-ah) A condition in which disease-causing microorganisms or their toxins are present in the blood.

spherocytosis (sfēr″o-si-to'sis) A hereditary form of hemolytic anemia characterized by the presence of spherical erythrocytes (spherocytes).

thalassemia (thal''ah-se'me-ah) A group of hereditary hemolytic anemias characterized by the presence of very thin, fragile erythrocytes.

von Willebrand's disease (fon vil'ĕ-brandz di-zēz') A hereditary condition caused by the deficiency of an antihemophilic blood factor (factor VIII) combined with capillary defects; characterized by bleeding from the nose, gums, and genitalia.

Chapter Summary

Introduction (page 620)

Blood is a type of connective tissue whose cells are suspended in a liquid intercellular material. It transports substances between the body cells and the external environment, and helps maintain a stable cellular environment.

Blood and Blood Cells (page 620)

Blood contains red blood cells, white blood cells, and platelets.

1. Volume and composition of blood
   a. Blood volume varies with body size, fluid and electrolyte balance, and adipose tissue content.
   b. Blood can be separated into solid and liquid portions.
      (1) The solid, cellular portion is mostly red blood cells.
      (2) The liquid plasma includes water, nutrients, hormones, electrolytes, and cellular wastes.
2. Characteristics of red blood cells
   a. Red blood cells are biconcave disks with shapes that provide increased surface area and place their cell membranes close to internal parts.
   b. They contain hemoglobin, which combines loosely with oxygen.
   c. The mature forms lack nuclei, but contain enzymes needed for energy-releasing processes.
3. Red blood cell counts
   a. The red blood cell count equals the number of cells per mm³ of blood.
   b. The average count may range from approximately 4,000,000 to 6,000,000 cells per mm³.
   c. Red cell count is related to the oxygen-carrying capacity of the blood and is used in diagnosing and evaluating the courses of diseases.
4. Destruction of red blood cells
   a. Red cells are fragile and are damaged while moving through capillaries.
   b. Damaged red cells are phagocytized by macrophages in the liver and spleen.
   c. Hemoglobin molecules are decomposed, and the iron they contain is conserved.
   d. Biliverdin and bilirubin are pigments resulting from hemoglobin breakdown.

5. Red blood cell production and its control
   a. During fetal development, red cells are formed in the yolk sac, liver, and spleen; later, red cells are produced by the red bone marrow.
   b. The number of red blood cells remains relatively stable.
   c. The rate of red cell production is controlled by a negative feedback mechanism that involves erythropoietin from the kidneys and liver.
      (1) Erythropoietin is released in response to low oxygen levels.
      (2) Low oxygen concentrations may be caused by high altitude, loss of blood, or chronic lung disease.
6. Dietary factors affecting red blood cell production
   a. Production of red blood cells is affected by the availability of vitamin B_{12} and folic acid that are needed for DNA synthesis.
   b. Iron is also needed for hemoglobin synthesis.
   c. The rate of iron absorption varies with the amount of iron in the body.
7. Types of white blood cells
   a. White blood cells function to control disease conditions.
   b. Granulocytes include neutrophils, eosinophils, and basophils.
   c. Agranulocytes include monocytes and lymphocytes.
8. White blood cell counts
   a. Normal total white blood cell counts vary from 5,000 to 10,000 cells per mm³ of blood.
   b. The number of white cells may change in abnormal conditions such as infections, emotional disturbances, or excessive loss of body fluids.
   c. A differential white cell count indicates the percentages of various types of leukocytes present.
9. Functions of white blood cells
   a. Neutrophils and monocytes phagocytize foreign particles.
   b. Leukocytes may be stimulated by the presence of chemicals released by damaged cells and may move toward these chemicals.
   c. Eosinophils kill parasites and help control inflammation and allergic reactions.
   d. Basophils release heparin, which inhibits blood clotting.
   e. Lymphocytes produce antibodies that act against specific foreign substances.
10. Blood platelets
    a. Blood platelets are fragments of giant cells that become detached and enter the circulation.
    b. The normal count varies from 130,000 to 360,000 platelets per mm³.
    c. They help close breaks in blood vessels.

Blood Plasma (page 632)

Plasma is the liquid part of the blood that is composed of water and a mixture of organic and inorganic substances. It transports nutrients and gases, regulates fluid and electrolyte balance, and helps maintain stable pH.

1. Plasma proteins
    a. Plasma proteins remain in blood and interstitial fluids, and are not normally used as energy sources.
    b. Three major groups exist.
        (1) Albumins help maintain the osmotic pressure of blood.
        (2) Globulins include the antibodies of immunity. They also transport lipids and fat-soluble vitamins.
        (3) Fibrinogen functions in blood clotting.
2. Nutrients and gases
    a. Plasma nutrients include amino acids, simple sugars, and lipids.
        (1) Glucose is stored in the liver as glycogen and is released whenever the blood glucose concentration falls.
        (2) Amino acids are used to synthesize proteins and are deaminated for use as energy sources.
        (3) Lipoproteins function in the transport of lipids.
    b. Gases in plasma include oxygen, carbon dioxide, and nitrogen.
3. Nonprotein nitrogenous substances
    a. Nonprotein nitrogenous substances are composed of molecules that contain nitrogen atoms but are not proteins.
    b. They include amino acids, urea, uric acid, creatine, and creatinine.
        (1) Urea and uric acid are products of catabolic metabolism.
        (2) Creatinine results from the metabolism of creatine.
    c. These substances usually remain stable; an increase may indicate a kidney disorder.
4. Plasma electrolytes
    a. Plasma electrolytes are obtained by absorption from the intestines and are released as by-products of cellular metabolism.
    b. They include ions of sodium, potassium, calcium, magnesium, chlorine, bicarbonate, phosphate, and sulfate.
    c. They are important in the maintenance of osmotic pressure and pH.

Hemostasis (page 635)

Hemostasis refers to the stoppage of bleeding. Hemostatic mechanisms are most effective in controlling blood loss from small vessels.

1. Blood vessel spasm
    a. Smooth muscles in walls of arterioles and arteries contract reflexly following injury.
    b. Platelets release serotonin that stimulates vasoconstriction and helps maintain vessel spasm.
2. Platelet plug formation
    a. Platelets adhere to rough surfaces and exposed collagen.
    b. Platelets stick together at the sites of injuries and form platelet plugs in broken vessels.
3. Blood coagulation
    a. Blood clotting is the most effective means of hemostasis and may be initiated by extrinsic or intrinsic mechanisms.
    b. Clot formation depends on the balance between clotting factors that promote clotting and those that inhibit clotting.
    c. The basic event of coagulation is the conversion of soluble fibrinogen into insoluble fibrin.
    d. Substances that promote clotting include the presence of prothrombin activator, prothrombin, and calcium ions.
    e. After forming, the clot retracts and pulls the edges of a broken vessel closer together.
    f. A thrombus is a blood clot in a vessel; an embolus is a clot or fragment of a clot that has moved in a vessel.
    g. A clot is invaded by fibroblasts that form connective tissue throughout the clot.
    h. The clot eventually may be destroyed by the action of protein-splitting enzymes.
4. Prevention of coagulation
    a. The smooth lining of blood vessels repels platelets.
    b. As a clot forms, fibrin adsorbs thrombin and prevents the reaction from spreading.
    c. Antithrombin interferes with the action of excessive thrombin.
    d. Some cells secrete heparin, an anticoagulant.

Blood Groups and Transfusions (page 640)

Blood can be typed on the basis of the substances it contains. Blood substances of certain types react adversely with other types.

1. Agglutinogens and agglutinins
   a. Red blood cell membranes may contain agglutinogens, and blood plasma may contain agglutinins.
   b. Blood typing involves identifying the agglutinogens present in the red cell membranes.
2. ABO blood group
   a. Blood can be grouped according to the presence or absence of agglutinogens A and B.
   b. Whenever agglutinogen A is absent, agglutinin anti-A is present; whenever agglutinogen B is absent, agglutinin anti-B is present.
   c. Adverse transfusion reactions are avoided by preventing the mixing of red cells that contain an agglutinogen with plasma that contains the corresponding agglutinin.
   d. Adverse reactions involve agglutination (clumping) of the red blood cells.
3. Rh blood group
   a. Rh agglutinogens are present in the red cell membranes of Rh-positive blood; they are absent in Rh-negative blood.
   b. If an Rh-negative person is exposed to Rh-positive blood, anti-Rh agglutinins are produced in response.
   c. Mixing Rh-positive red cells with plasma that contains anti-Rh agglutinins results in agglutination of the positive cells.
   d. If an Rh-negative female is pregnant with an Rh-positive fetus, some of the positive cells may enter the maternal blood at the time of birth and stimulate the maternal tissues to produce anti-Rh agglutinins.
   e. Anti-Rh agglutinins in maternal blood may pass through the placental tissues and react with the red cells of an Rh-positive fetus.

Clinical Application of Knowledge

1. What changes would you expect to occur in the hematocrit of a person who is dehydrated? Why?
2. If a patient with an inoperable cancer is treated using a drug that reduces the rate of cell division, what changes might occur in the patient's white blood cell count? How might the patient's environment be modified to compensate for the effects of these changes?
3. Hypochromic (iron-deficiency) anemia is relatively common among aging persons who are admitted to hospitals for other conditions. What environmental and sociological factors might promote this form of anemia?
4. Why do patients with liver diseases commonly develop blood clotting disorders?
5. How would you explain to a patient with leukemia, who has a greatly elevated white blood cell count, the importance of avoiding bacterial infections?
6. If a woman whose blood is Rh-negative and contains anti-Rh agglutinins is carrying a fetus with Rh-negative blood, will the fetus be in danger of developing erythroblastosis fetalis? Why?

Review Activities

1. List the major components of blood.
2. Define *hematocrit,* and explain how it is determined.
3. Describe a red blood cell.
4. Distinguish between oxyhemoglobin and deoxyhemoglobin.
5. Explain what is meant by a *red blood cell count.*
6. Describe the life cycle of a red blood cell.
7. Distinguish between biliverdin and bilirubin.
8. Define *erythropoietin,* and explain its function.
9. Explain how vitamin B_{12} and folic acid deficiencies affect red blood cell production.
10. List two sources of iron that can be used for the synthesis of hemoglobin.
11. Distinguish between granulocytes and agranulocytes.
12. Name five types of leukocytes, and list the major functions of each type.
13. Explain the significance of white blood cell counts as aids to diagnosing diseases.
14. Describe a blood platelet, and explain its functions.

15. Name three types of plasma proteins, and list the major functions of each type.
16. Define *lipoprotein*.
17. Define *apoprotein*.
18. Distinguish between low-density lipoprotein and high-density lipoprotein.
19. Name the sources of VLDL, LDL, HDL, and chylomicrons.
20. Describe how lipoproteins are removed from plasma.
21. Explain how cholesterol is eliminated from plasma and from the body.
22. Define *nonprotein nitrogenous substances,* and name those commonly present in plasma.
23. Name several plasma electrolytes.
24. Define *hemostasis.*
25. Explain how blood vessel spasms are stimulated following an injury.
26. Explain how a platelet plug forms.
27. List the major steps leading to the formation of a blood clot.
28. Distinguish between fibrinogen and fibrin.

29. Provide an example of a positive feedback system that operates during blood clotting.
30. Define *serum.*
31. Distinguish between a thrombus and an embolus.
32. Explain how a blood clot may be removed naturally from a blood vessel.
33. Describe how blood coagulation may be prevented.
34. Review the function of vitamin K.
35. Distinguish between agglutinogen and agglutinin.
36. Explain the basis of ABO blood types.
37. Explain why a person with blood type AB is sometimes called a universal recipient.
38. Explain why a person with blood type O is sometimes called a universal donor.
39. Distinguish between Rh-positive and Rh-negative blood.
40. Describe how a person may become sensitized to Rh-positive blood.
41. Define *erythroblastosis fetalis,* and explain how this condition may develop.

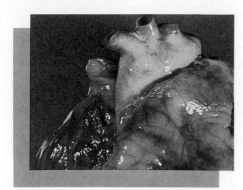

18

Cardiovascular System

*T*he *cardiovascular system* is the portion of the circulatory system that includes the heart and blood vessels. It moves the blood between the body cells and the organs of the integumentary, digestive, respiratory, and urinary systems that communicate with the external environment. In performing this function, the heart acts as a pump that forces blood through the blood vessels. The blood vessels, in turn, form a closed system of ducts, which transports the blood and allows exchanges of gases, nutrients, and wastes between the blood and the body cells ■

Chapter Objectives

After you have studied this chapter, you should be able to:

1. Name the organs of the cardiovascular system and discuss their functions.

2. Name and describe the locations of the major parts of the heart and discuss the function of each part.

3. Trace the pathway of the blood through the heart and the vessels of the coronary circulation.

4. Discuss the cardiac cycle and explain how it is controlled.

5. Compare the structures and functions of the major types of blood vessels.

6. Describe the mechanism that aids in returning venous blood to the heart.

7. Explain how blood pressure is produced and controlled.

8. Compare the pulmonary and systemic circuits of the cardiovascular system.

9. Identify and locate the major arteries and veins of the pulmonary and systemic circuits.

10. Complete the review activities at the end of this chapter. Note that the items are worded in the form of specific learning objectives. You may want to refer to them before reading the chapter.

Key Terms

arteriole (ar-te're-ōl)

atrium (a'tre-um)

cardiac conduction system (kar'de-ak kon-duk'shun sis'tem)

cardiac cycle (kar'de-ak si'kl)

cardiac output (kar'de-ak owt'poot)

diastolic pressure (di''ah-stol'ik presh'ur)

electrocardiogram (e-lek''tro-kar'de-o-gram'')

epicardium (ep''ĭ-kar'de-um)

functional syncytium (funk'shun-al sin-sish'e-um)

myocardium (mi''o-kar'de-um)

pacemaker (pas'māk-er)

pericardium (per''ĭ-kar'de-um)

peripheral resistance (pe-rif'er-al re-zis'tans)

pulmonary circuit (pul'mo-ner''e sur'kit)

sphygmomanometer (sfig''mo-mah-nom'ĕ-ter)

systemic circuit (sis-tem'ik sur'kit)

systolic pressure (sis-tol'ik presh'ur)

vasoconstriction (vas''o-kon-strik'shun)

vasodilation (vas''o-di-la'shun)

ventricle (ven'tri-kl)

venule (ven'ūl)

viscosity (vis-kos' ĭ-te)

Aids to Understanding Words

angio-, vessel: *angio*tensin—a substance that causes blood vessels to constrict.

brady-, slow: *brady*cardia—an abnormally slow heartbeat.

diastol-, dilation: *diastol*ic pressure—the blood pressure that occurs when the ventricle of the heart is relaxed (thus dilated).

ectop-, out of place: *ectop*ic beat—a heartbeat that occurs before it is expected in a normal series of cardiac cycles.

edem-, swelling: *edem*a—a condition in which fluids accumulate in the tissues and cause them to swell.

-gram, something written: electrocardio*gram*—a recording of the electrical changes that occur in the myocardium during a cardiac cycle.

myo-, muscle: *myo*cardium—the muscle tissue within the wall of the heart.

papill-, nipple: *papill*ary muscle—a small mound of muscle within a ventricular chamber of the heart.

phleb-, vein: *phleb*itis—an inflammation of a vein.

scler-, hard: *scler*osis—condition in which a blood vessel wall loses its elasticity and becomes hard.

syn-, together: *syn*cytium—a mass of merging cells that act together.

systol-, contraction: *systol*ic pressure—the blood pressure that occurs during a ventricular contraction.

tachy-, rapid: *tachy*cardia—an abnormally fast heartbeat.

A functional cardiovascular system is vital for survival, because without circulation, the tissues lack a supply of oxygen and nutrients, and waste substances accumulate. Under such conditions, the cells soon begin to undergo irreversible changes, which quickly lead to death of the organism. The general pattern of the cardiovascular system is shown in figure 18.1.

Structure of the Heart

The heart is a hollow, cone-shaped, muscular pump located within the mediastinum of the thorax and resting upon the diaphragm.

Size and Location of the Heart

Although heart size varies with body size, the heart of an average adult is generally about 14 centimeters long and 9 centimeters wide (figure 18.2).

The heart is located within the mediastinum, and is bordered laterally by the lungs, posteriorly by the backbone, and anteriorly by the sternum (figure 18.3 and reference plates 50, 56, 71, and 72). Its *base,* which is attached to several large blood vessels, lies beneath the second rib. Its distal end extends downward and to the left, terminating as a bluntly pointed *apex* at the

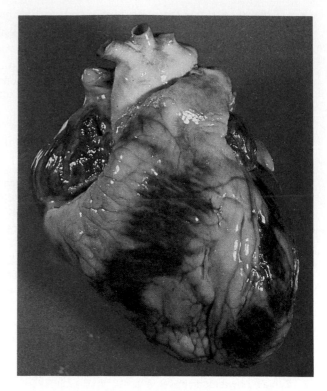

Figure 18.2 Anterior view of a human heart.

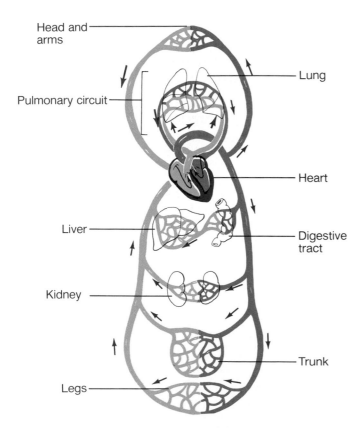

Figure 18.1 The cardiovascular system transports blood between the body cells and the organs that communicate with the external environment.

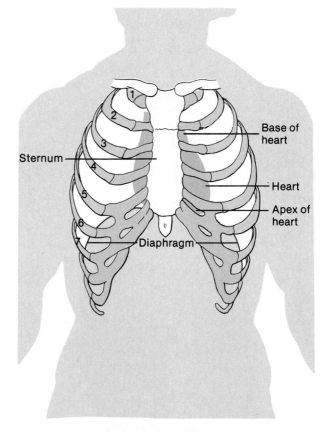

Figure 18.3 The heart is located behind the sternum, where it lies upon the diaphragm.

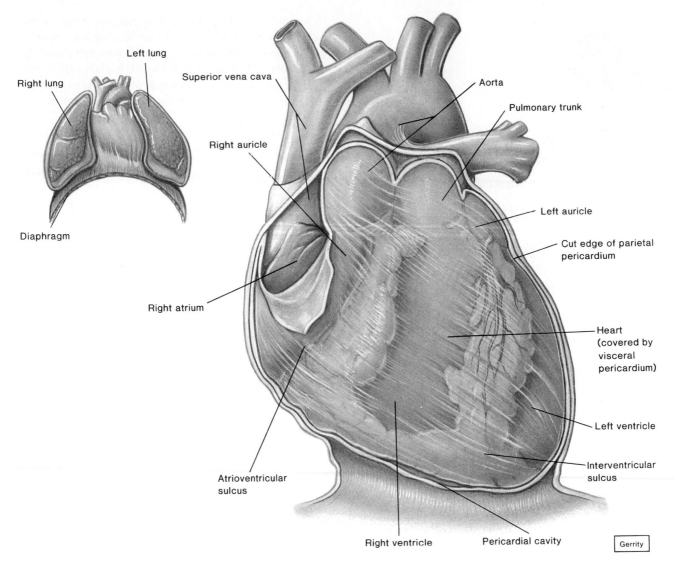

Right lung

Left lung

Superior vena cava

Aorta

Pulmonary trunk

Right auricle

Left auricle

Cut edge of parietal pericardium

Right atrium

Heart (covered by visceral pericardium)

Diaphragm

Left ventricle

Atrioventricular sulcus

Interventricular sulcus

Right ventricle

Pericardial cavity

Gerrity

Figure 18.4   The heart is within the mediastinum and is enclosed by a layered pericardium.

level of the fifth intercostal space. For this reason, it is possible to sense the *apical heartbeat* by feeling or listening to the chest wall between the fifth and sixth ribs, about 7.5 centimeters to the left of the midline. (See figure 7.43 for an X-ray view showing the position of the heart.)

Coverings of the Heart

The heart and the proximal ends of the large blood vessels, to which the heart is attached, are enclosed by the **pericardium.** The pericardium consists of an outer fibrous bag, the *fibrous pericardium,* that surrounds a more delicate, double-layered sac. The inner layer of this sac, the *visceral pericardium* (epicardium), covers the heart. At the base of the heart, the visceral pericardium turns back upon itself to become the *parietal pericardium.* The parietal pericardium, in turn, forms the inner lining of the fibrous pericardium.

The fibrous pericardium is a tough, protective sac composed largely of white fibrous connective tissue. It is attached to the central portion of the diaphragm, behind the sternum, the vertebral column, and the large blood vessels emerging from the heart. (See figures 1.8, 18.4 and reference plates 56 and 57.) Between the parietal and visceral layers of the pericardium is a potential space, the *pericardial cavity,* that contains a small amount of serous fluid. This fluid reduces friction between the pericardial membranes as the heart moves within them (figure 18.5).

If the pericardium becomes inflamed due to a bacterial or viral infection, the condition is called *pericarditis.* As a result of this inflammation, the layers of the pericardium sometimes become stuck together by adhesions. This may cause considerable pain and interfere with heart movements.

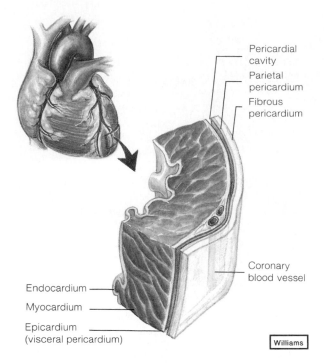

Endocardium —
Myocardium —
Epicardium —
(visceral pericardium)

Pericardial cavity
Parietal pericardium
Fibrous pericardium

Coronary blood vessel

Williams

Figure 18.5 The wall of the heart consists of three layers: an endocardium, a myocardium, and an epicardium.

1. Where is the heart located?
2. Where would you listen to hear the apical heartbeat?
3. Distinguish between the visceral pericardium and the parietal pericardium.
4. What is the function of the fluid in the pericardial cavity?

Wall of the Heart

The wall of the heart is composed of three distinct layers: an outer epicardium, a middle myocardium, and an inner endocardium (figure 18.5).

The **epicardium,** which corresponds to the visceral pericardium, functions as an outer protective layer. It is a serous membrane that consists of connective tissue covered by epithelium, and it includes blood capillaries, lymph capillaries, and nerve fibers. The deeper portion of the epicardium often contains fat, particularly along the paths of larger blood vessels.

The middle layer, or **myocardium,** is relatively thick and consists largely of the cardiac muscle tissue responsible for forcing the blood out of the heart chambers. The muscle fibers are arranged in planes, separated by connective tissues that are richly supplied with blood capillaries, lymph capillaries, and nerve fibers.

The inner layer, or **endocardium,** consists of epithelium and connective tissue that contains many elastic and collagenous fibers. The connective tissue also contains blood vessels and some specialized cardiac muscle

fibers called *Purkinje fibers.* The function of these fibers is described in a subsequent section of this chapter.

The endocardium lines all of the heart chambers and covers the structures, such as the heart valves, that project into them. This inner lining is also continuous with the inner lining of the blood vessels (endothelium) attached to the heart.

Chart 18.1 summarizes the characteristics of the three layers of the heart wall.

An inflammation of the endocardium is termed *endocarditis.* This condition sometimes accompanies bacterial diseases, such as scarlet fever or syphilis, and may produce lasting effects by damaging the valves of the heart.

Heart Chambers and Valves

Internally, the heart is divided into four hollow chambers, two on the left and two on the right. The upper chambers, called **atria** (sing. *atrium*), have relatively thin walls and receive the blood from veins. Small, ear-like projections, called **auricles,** extend outward from the atria (figure 18.3). The lower chambers, the **ventricles,** force the blood out of the heart into arteries. (Note: Veins are blood vessels that carry the blood toward the heart; arteries carry the blood away from the heart chambers.)

The atrium and ventricle on the right side are separated from those on the left by a *septum.* The atrium on each side communicates with its corresponding ventricle through an opening called the **atrioventricular orifice,** which is guarded by an *atrioventricular valve (A-V valve).*

Grooves on the surface of the heart mark the divisions between its chambers, and they also contain major blood vessels that supply the heart tissues. The deepest of these grooves is the **atrioventricular** (coronary) **sulcus,** which encircles the heart between the atrial and ventricular portions. Two **interventricular** (anterior and posterior) **sulci** indicate the location of the septum that separates the right and left ventricles.

Recently it was discovered that muscle cells associated with the atria secrete a peptide hormone called *atrial natriuretic factor* (ANF). This hormone seems to be released when the muscle cells are stretched excessively, as may occur when the volume of blood within the atria is increased. The effect of ANF is to inhibit the release of a substance (renin) from the kidneys (chapter 20) and to inhibit the release of aldosterone from the adrenal cortex (chapter 13). As a result of these actions, the excretion of sodium ions and water from the kidneys is increased, leading to a decrease in blood volume and blood pressure.

CHART 18.1 Wall of the heart

Layer	Composition	Function
Epicardium (visceral pericardium)	Serous membrane of connective tissue covered with epithelium and including blood capillaries, lymph capillaries, and nerve fibers	Forms a protective outer covering
Myocardium	Cardiac muscle tissue separated by connective tissues and including blood capillaries, lymph capillaries, and nerve fibers	Produces muscular contractions that force blood from the heart chambers
Endocardium	Membrane of epithelium and connective tissues, including elastic and collagenous fibers, blood vessels, and specialized muscle fibers	Forms a protective inner lining of the chambers and valves

1. Describe the layers of the heart wall.
2. Name and locate the four chambers of the heart.
3. Name the orifices that occur between the upper and the lower chambers of the heart.
4. Name the structure that separates the right and left sides of the heart.

The right atrium receives blood from two large veins: the *superior vena cava* and the *inferior vena cava.* These veins return blood that is low in oxygen from various body parts. A smaller vein, the *coronary sinus,* also drains blood into the right atrium from the wall of the heart.

The atrioventricular orifice between the right atrium and the right ventricle is guarded by a large **tricuspid valve,** which is composed of three leaflets, or cusps, as its name implies. This valve permits the blood to move from the right atrium into the right ventricle and prevents it from passing in the opposite direction. The cusps fold passively out of the way when the blood pressure is greater on the atrial side, and they close passively when the pressure is greater on the ventricular side (figures 18.6 and 18.7).

Strong, fibrous strings, called *chordae tendineae,* are attached to the cusps on the ventricular side. These strings originate from small mounds of cardiac muscle tissue, the **papillary muscles,** that project inward from the walls of the ventricle. When the tricuspid valve closes, the chordae tendineae and papillary muscles prevent the cusps from swinging back into the atrium.

The right ventricle has a thinner muscular wall than the left ventricle. This right chamber pumps the blood a fairly short distance to the lungs against a relatively low resistance to blood flow. The left ventricle, on the other hand, must force the blood to all the other parts of the body against a much greater resistance to flow.

When the muscular wall of the right ventricle contracts, the blood inside its chamber is put under increasing pressure, and the tricuspid valve closes passively. As a result, the only exit is through the *pulmonary trunk,* which divides to form the left and right *pulmonary arteries.* At the base of this trunk is a **pulmonary valve** (pulmonary semilunar valve), which consists of three cusps. This valve opens as the right ventricle contracts. However, when the ventricular muscles relax the blood begins to back up in the pulmonary trunk. This causes the pulmonary valve to close, thus preventing a return flow into the ventricular chamber.

The left atrium receives the blood from the lungs through four *pulmonary veins*—two from the right lung and two from the left lung. The blood passes from the left atrium into the left ventricle through the atrioventricular orifice, which is guarded by a valve. This valve consists of two leaflets, and it is appropriately named the **bicuspid,** or **mitral** (shaped like a miter, a turbanlike headdress), **valve.** It prevents the blood from flowing back into the left atrium from the ventricle. As with the tricuspid valve, the cusps of the bicuspid valve are prevented from swinging back by the papillary muscles and the chordae tendineae.

When the left ventricle contracts, the bicuspid valve closes passively, and the only exit is through a large artery called the *aorta.* Its branches distribute blood to all parts of the body.

At the base of the aorta is an **aortic valve** (aortic semilunar valve) that consists of three cusps. (See figure 18.8.) It opens and allows blood to leave the left ventricle as it contracts. When the ventricular muscles relax, this valve closes and prevents blood from backing up into the ventricle. Chart 18.2 summarizes the heart valves.

A heart disorder that affects up to 6% of the U.S. population is *mitral valve prolapse.* In this condition, one (or both) of the cusps of the mitral valve is stretched so that it bulges into the left atrium during ventricular contraction. Although in most cases the valve continues to function adequately, sometimes blood regurgitates into the left atrium, causing some degree of disability. The cause of mitral valve prolapse is unknown.

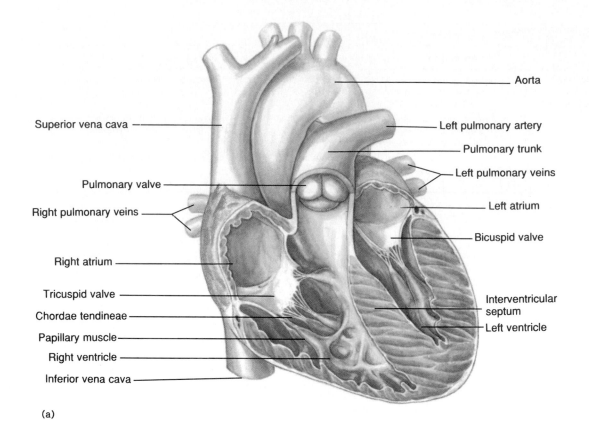

Superior vena cava

Pulmonary valve

Right pulmonary veins

Right atrium

Tricuspid valve

Chordae tendineae

Papillary muscle

Right ventricle

Inferior vena cava

Aorta

Left pulmonary artery

Pulmonary trunk

Left pulmonary veins

Left atrium

Bicuspid valve

Interventricular septum

Left ventricle

(a)

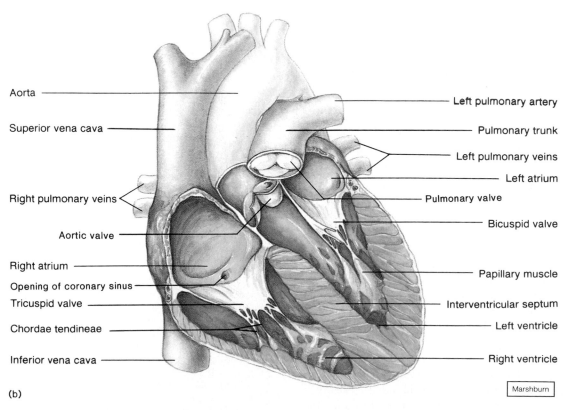

Aorta

Superior vena cava

Right pulmonary veins

Aortic valve

Right atrium

Opening of coronary sinus

Tricuspid valve

Chordae tendineae

Inferior vena cava

Left pulmonary artery

Pulmonary trunk

Left pulmonary veins

Left atrium

Pulmonary valve

Bicuspid valve

Papillary muscle

Interventricular septum

Left ventricle

Right ventricle

Marshburn

(b)

Figure 18.6   Frontal sections of the heart (a) showing the connection between the right ventricle and the pulmonary trunk, and (b) showing the connection between the left ventricle and the aorta.

1. Which blood vessels carry blood into the right atrium?
2. Where does the blood go after it leaves the right ventricle?
3. Which blood vessels carry blood into the left atrium?
4. What prevents blood from flowing back into the ventricles when they are relaxed?

Skeleton of the Heart

At their proximal ends, the pulmonary trunk and aorta are surrounded by rings of dense fibrous connective tissue. These rings are continuous with others that en-circle the atrioventricular orifices. They provide firm attachments for the heart valves and for various muscle fibers. In addition, they prevent the outlets of the atria and ventricles from dilating during myocardial contraction. The fibrous rings, together with other masses of dense fibrous tissue in the upper portion of the septum between the ventricles (interventricular septum), constitute the *skeleton of the heart* (figure 18.9).

Path of Blood through the Heart

Blood that is relatively low in oxygen concentration and relatively high in carbon dioxide concentration enters the right atrium through the venae cavae and the coronary sinus. As the right atrial wall contracts, the blood passes through the right atrioventricular orifice and enters the chamber of the right ventricle (figure 18.10).

When the right ventricular wall contracts, the tricuspid valve closes the right atrioventricular orifice, and the blood moves into the pulmonary trunk and its branches (pulmonary arteries). From these vessels, blood enters the capillaries associated with the alveoli of the lungs. Gas exchanges occur between the blood

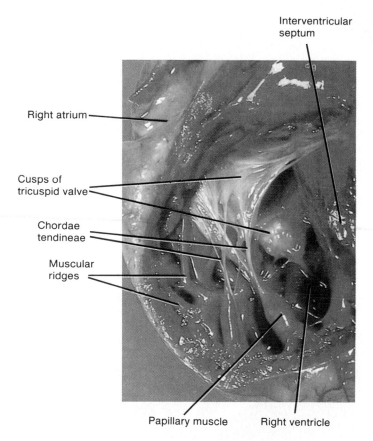

Figure 18.7 Photograph of a human tricuspid valve.

Right atrium

Cusps of tricuspid valve

Chordae tendineae

Muscular ridges

Interventricular septum

Papillary muscle   Right ventricle

Figure 18.8 Photograph of the pulmonary and aortic valves of the heart (superior view). How would you explain the difference in thickness of the blood vessel walls?

CHART 18.2	Valves of the heart				
Valve	Location	Function	Valve	Location	Function
Tricuspid valve	Right atrioventricular orifice	Prevents blood from moving from right ventricle into right atrium during ventricular contraction	Bicuspid (mitral) valve	Left atrioventricular orifice	Prevents blood from moving from left ventricle into left atrium during ventricular contraction
Pulmonary valve	Entrance to pulmonary trunk	Prevents blood from moving from pulmonary trunk into right ventricle during ventricular relaxation	Aortic valve	Entrance to aorta	Prevents blood from moving from aorta into left ventricle during ventricular relaxation

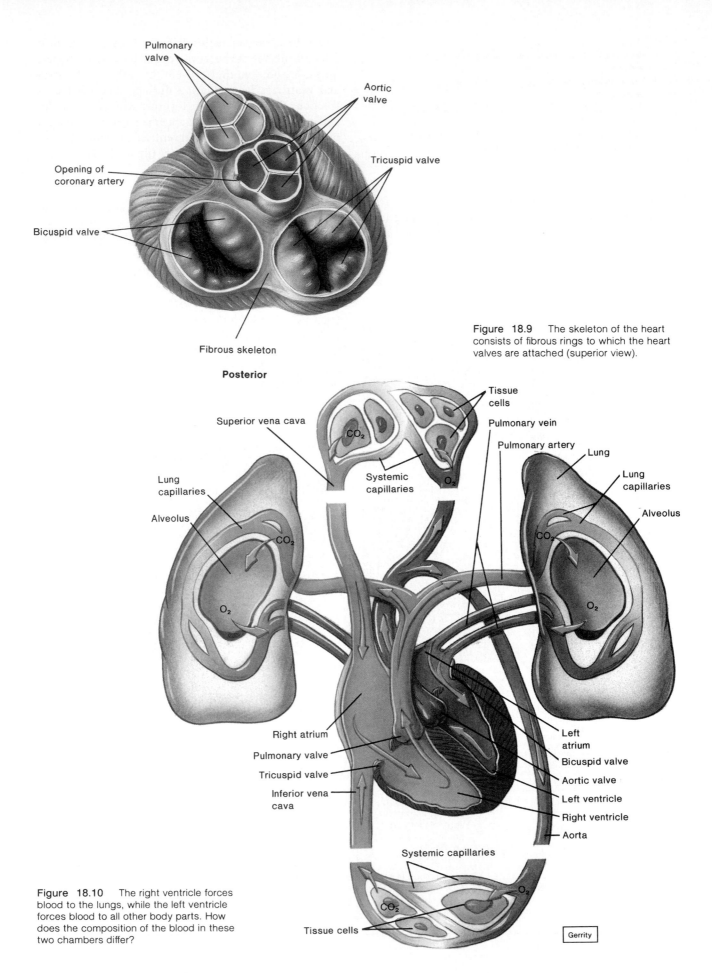

Pulmonary valve

Aortic valve

Opening of coronary artery

Tricuspid valve

Bicuspid valve

Fibrous skeleton

Posterior

Figure 18.9 The skeleton of the heart consists of fibrous rings to which the heart valves are attached (superior view).

Tissue cells

CO_2

Superior vena cava

Pulmonary vein

Pulmonary artery

Lung

Systemic capillaries

Lung capillaries

O_2

Lung capillaries

Alveolus

CO_2

Alveolus

CO_2

O_2

O_2

Right atrium

Left atrium

Pulmonary valve

Bicuspid valve

Tricuspid valve

Aortic valve

Inferior vena cava

Left ventricle

Right ventricle

Aorta

Systemic capillaries

O_2

CO_2

Figure 18.10 The right ventricle forces blood to the lungs, while the left ventricle forces blood to all other body parts. How does the composition of the blood in these two chambers differ?

Tissue cells

Gerrity

658 • Processing and Transporting

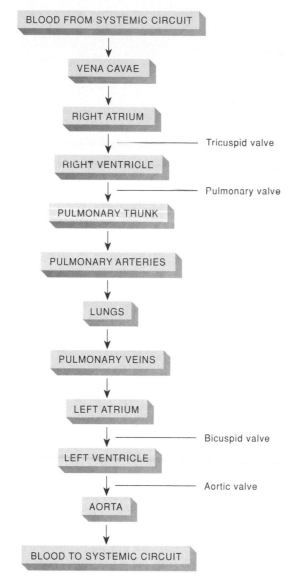

Figure 18.11 Path of blood through the heart and pulmonary circuit.

The flowchart shows:

BLOOD FROM SYSTEMIC CIRCUIT

↓

VENA CAVAE

↓

RIGHT ATRIUM

↓ — Tricuspid valve

RIGHT VENTRICLE

↓ — Pulmonary valve

PULMONARY TRUNK

↓

PULMONARY ARTERIES

↓

LUNGS

↓

PULMONARY VEINS

↓

LEFT ATRIUM

↓ — Bicuspid valve

LEFT VENTRICLE

↓ — Aortic valve

AORTA

↓

BLOOD TO SYSTEMIC CIRCUIT

in the capillaries and the air in the alveoli. The freshly oxygenated blood, which is now relatively low in carbon dioxide, returns to the heart through the pulmonary veins that lead to the left atrium.

The left atrial wall contracts, and the blood moves through the left atrioventricular orifice and into the chamber of the left ventricle. When the left ventricular wall contracts, the bicuspid valve closes the left atrioventricular orifice, and the blood passes into the aorta and its branches. Figure 18.11 summarizes the path taken by the blood as it moves through the heart and the pulmonary circuit.

Blood Supply to the Heart

Blood is supplied to the tissues of the heart by the first two branches of the aorta, called the right and left **coronary arteries.** Their openings lie just beyond the aortic valve (figures 18.8 and 18.12).

One branch of the left coronary artery, the *circumflex artery,* follows the atrioventricular sulcus between the left atrium and the left ventricle. Its branches supply blood to the walls of the left atrium and the left ventricle. Another branch of the left coronary artery, the *anterior interventricular artery* (or *left anterior descending artery*), travels in the anterior interventricular sulcus, and its branches supply the walls of both ventricles.

The right coronary artery passes along the atrioventricular sulcus between the right atrium and the right ventricle. It gives off two major branches—a *posterior interventricular artery,* which travels along the posterior interventricular sulcus and supplies the walls of both ventricles, and a *marginal artery,* which passes along the lower border of the heart. Branches of the marginal artery supply the walls of the right atrium and the right ventricle (figures 18.13 and 18.14).

If a branch of a coronary artery becomes abnormally constricted or obstructed by a thrombus or embolus, the myocardial cells it supplies may experience a blood oxygen deficiency called *ischemia.* As a result of ischemia, the person may experience a painful condition called *angina pectoris.* The discomfort usually occurs during physical activity, when the myocardial cell's need for oxygen exceeds the blood oxygen supply. However, the pain is commonly relieved by rest. Angina pectoris may also be triggered by an emotional disturbance. It may cause a sensation of heavy pressure, tightening, or squeezing in the chest. Although it is usually felt in the region behind the sternum (retrosternally) or in the anterior portion of the upper thorax, the pain may radiate to other parts, including the neck, jaw, throat, arm, shoulder, elbow, back, or upper abdomen. Angina pectoris may be accompanied by profuse perspiration (diaphoresis), difficulty in breathing (dyspnea), nausea, or vomiting.

Sometimes a portion of the heart dies because a blood clot forms and completely obstructs a coronary artery or one of its branches (coronary thrombosis). This condition, a *myocardial infarction,* (more commonly called a heart attack) is one of the leading causes of death.

Because the heart must beat continually to supply blood to the body tissues, the myocardial cells require a constant supply of freshly oxygenated blood. The myocardium contains many capillaries fed by branches

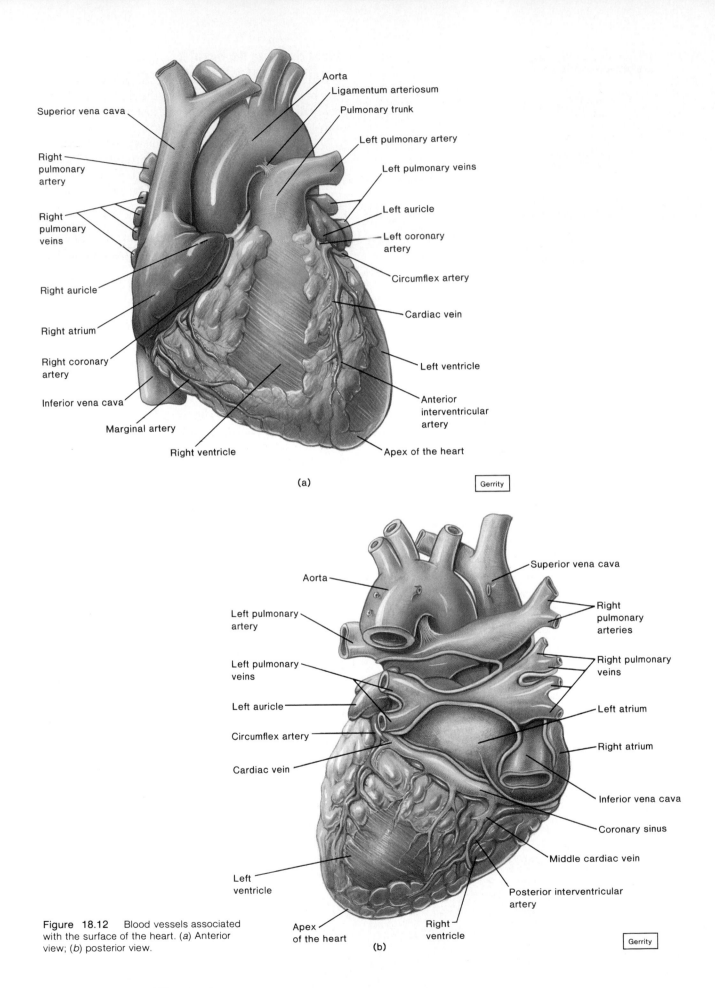

Superior vena cava

Right pulmonary artery

Right pulmonary veins

Right auricle

Right atrium

Right coronary artery

Inferior vena cava

Marginal artery

Right ventricle

Aorta

Ligamentum arteriosum

Pulmonary trunk

Left pulmonary artery

Left pulmonary veins

Left auricle

Left coronary artery

Circumflex artery

Cardiac vein

Left ventricle

Anterior interventricular artery

Apex of the heart

Gerrity

(a)

Aorta

Left pulmonary artery

Left pulmonary veins

Left auricle

Circumflex artery

Cardiac vein

Left ventricle

Apex of the heart

Superior vena cava

Right pulmonary arteries

Right pulmonary veins

Left atrium

Right atrium

Inferior vena cava

Coronary sinus

Middle cardiac vein

Posterior interventricular artery

Right ventricle

Gerrity

(b)

Figure 18.12 Blood vessels associated with the surface of the heart. (*a*) Anterior view; (*b*) posterior view.

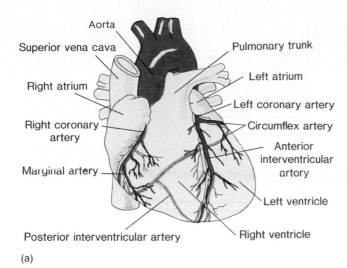

(a)

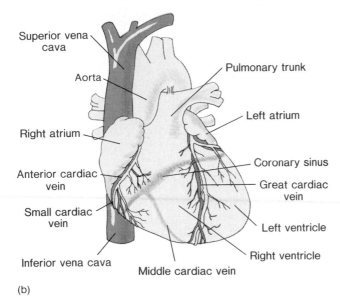

(b)

Figure 18.13 (a) Blood is supplied to heart tissues by branches of the coronary arteries. (b) Blood is drained from heart tissues by branches of the cardiac veins.

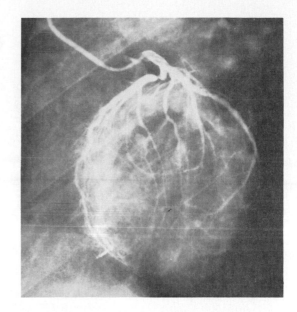

Figure 18.14 An angiogram (X-ray film) of the coronary arteries.

of the coronary arteries. The smaller branches of these arteries usually have interconnections (anastomoses) between vessels that provide alternate pathways for blood.

In most body parts, blood flow in arteries reaches a peak during ventricular contraction. However, blood flow in the vessels of the myocardium is poorest during ventricular contraction. This is because the muscle fibers of the myocardium compress nearby vessels as they contract, and this action interferes with blood flow. Also, the openings into the coronary arteries are closed during ventricular contraction. Conversely, during ventricular relaxation, the myocardial vessels are no longer compressed and the orifices of the coronary arteries are open. Consequently, blood flow into the myocardium increases.

The blood that has passed through the capillaries of the myocardium is drained by branches of the **cardiac veins,** whose paths roughly parallel those of the coronary arteries. As figure 18.13b shows, these veins join the **coronary sinus,** an enlarged vein on the posterior surface of the heart in the atrioventricular sulcus. The coronary sinus empties into the right atrium. Figure 18.15 summarizes the path of the blood that supplies the tissues of the heart.

In the *heart transplantation* procedure, the recipient's failing heart is removed, except for the posterior walls of the right and left atria and their connections to the venae cavae and pulmonary veins. The donor heart is prepared similarly and is attached to the atrial cuffs remaining in the recipient's thorax. Finally, the recipient's aorta and pulmonary arteries are connected to those of the donor heart.

1. What structures make up the skeleton of the heart?
2. Review the path of blood through the heart.
3. What vessels supply blood to the myocardium?
4. How does blood return from the cardiac tissues to the right atrium?

Actions of the Heart

Although the previous discussion described the actions of the heart chambers separately, they do not function independently. Instead, their actions are regulated so that the atrial walls contract while the ventricular walls are relaxed, and ventricular walls contract while the atrial walls are relaxed. Such a series of events constitutes a complete heartbeat or **cardiac cycle.** At the end

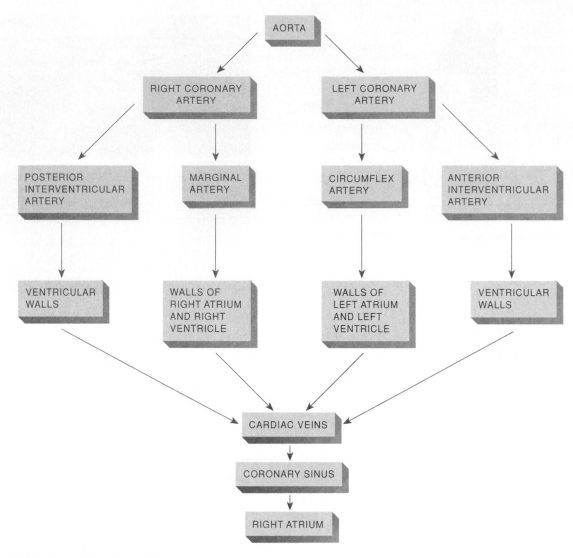

Figure 18.15 Path of blood through the coronary circulation.

Cardiac Cycle

During a cardiac cycle, the pressure within the chambers of the heart rises and falls. For example, when the atria are relaxed, the blood flows into them from the large, attached veins. As these chambers fill, the pressure inside gradually increases. About 70% of the entering blood flows directly into the ventricles through the atrioventricular orifices before the atrial walls contract. Then, during atrial contraction (atrial systole), the atrial pressure rises suddenly, forcing the remaining 30% of the atrial contents into the ventricles. This is followed by atrial relaxation (atrial diastole) (figure 18.16).

As the ventricles contract (ventricular systole), the A-V valves guarding the atrioventricular orifices

of each cycle, the atria and the ventricles remain relaxed for a moment, and then a new cardiac cycle begins.

close passively and begin to bulge back into the atria, causing the atrial pressure to rise sharply. At the same time, the papillary muscles contract, and by pulling on the chordae tendineae, they prevent the cusps of the A-V valves from bulging too far into the atria. The atrial pressure soon falls, however, as the blood flows out of the ventricles into the arteries. During the ventricular contraction, the A-V valves remain closed, and the atrial pressure gradually increases as the atria fill with the blood. When the ventricles relax (ventricular diastole), the A-V valves open passively, the blood flows through them into the ventricles, and the atrial pressure drops to a low point.

Pressure in the ventricles is low while they are filling, but when the atria contract, the ventricular pressure increases slightly. Then, as the ventricles contract, the ventricular pressure rises sharply, and as soon as the pressure exceeds that in the atria, the A-V valves close. The ventricular pressure continues to increase until it exceeds the pressure in the pulmonary trunk

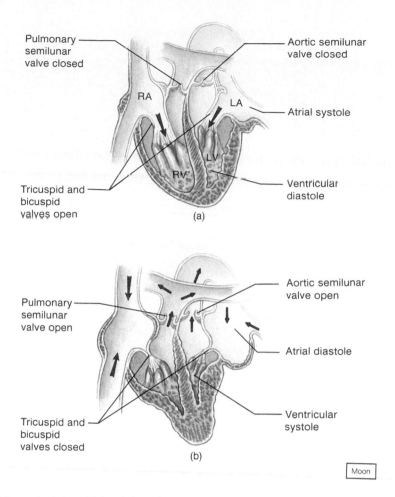

Pulmonary semilunar valve closed

Aortic semilunar valve closed

RA

LA

Atrial systole

Tricuspid and bicuspid valves open

LV

RV

Ventricular diastole

(a)

Pulmonary semilunar valve open

Aortic semilunar valve open

Atrial diastole

Tricuspid and bicuspid valves closed

Ventricular systole

(b)

Moon

Figure 18.16 (a) The atria empty during atrial systole and (b) fill with blood during atrial diastole.

and aorta. Then, the pulmonary and aortic valves open, and blood is ejected from the ventricles into these arteries. When the ventricles are nearly empty, the ventricular pressure begins to drop, and it continues to drop as the ventricles relax. When the ventricular pressure is less than that in the arteries, the pulmonary and aortic valves are closed by arterial blood flowing back toward the ventricles. As soon as the ventricular pressure falls below that of the atria, the A-V valves open, and the ventricles begin to fill once more. The graph in figure 18.17 summarizes some of the changes that occur in the left ventricle during a cardiac cycle.

Heart Sounds

The sounds associated with a heartbeat can be heard with a stethoscope and are described as *lub*-dup sounds. These sounds are due to vibrations in the heart tissues that are produced as the blood flow is suddenly speeded or slowed with the contraction and relaxation of the heart chambers, and with the opening and closing of the valves.

The first part of a heart sound (*lub*) occurs during the ventricular contraction, when the A-V valves are

closing. The second part (dup) occurs during ventricular relaxation, when the pulmonary and aortic valves are closing (figure 18.17).

Sometimes during inspiration, the interval between the closure of the pulmonary and the aortic valves is long enough that a sound related to each of these events can be heard. In this case, the second heart sound is said to be *split*.

Heart sounds are of particular interest because they provide information concerning the condition of the heart valves. For example, an inflammation of the endocardium (endocarditis) may cause changes in the shapes of the valvular cusps (valvular stenosis). Then, when the cusps close the closure may be incomplete, and some blood may leak back through the valve. If this happens, an abnormal sound called a *murmur* may be heard. The seriousness of a murmur depends on the amount of valvular damage. Fortunately for those who have serious problems, it may be possible to repair the damaged valves or to replace them by open heart surgery.

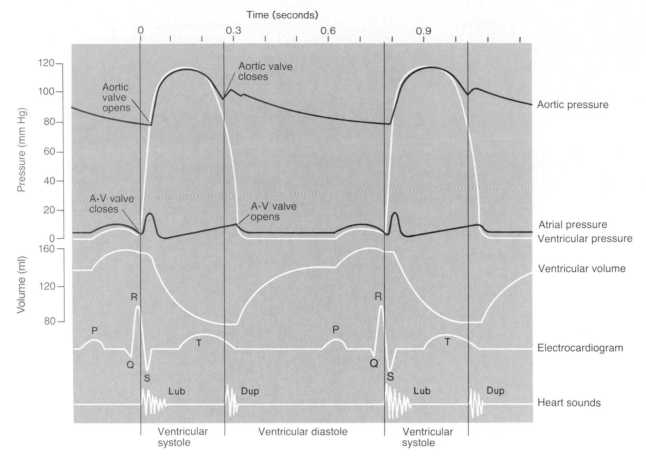

Figure 18.17 A graph of some of the changes that occur in the left ventricle during a cardiac cycle.

Using a stethoscope, it is possible to hear sounds associated with the aortic and pulmonary valves by listening from the second intercostal space on either side of the sternum. The *aortic sound* is heard on the right, and the *pulmonic sound* is heard on the left.

The sound associated with the bicuspid (mitral) valve can be heard from the fifth intercostal space at the nipple line on the left. The sound of the tricuspid valve can be heard at the tip of the sternum (figure 18.18).

Cardiac Muscle Fibers

As mentioned in chapter 9, cardiac muscle fibers function much like those of skeletal muscles. In cardiac muscle, however, the fibers are interconnected in branching networks that spread in all directions through the heart. When any portion of this net is stimulated, an impulse travels to all of its parts, and the whole structure contracts as a unit.

A mass of merging cells that act as a unit is called a **functional syncytium.** There are two such structures

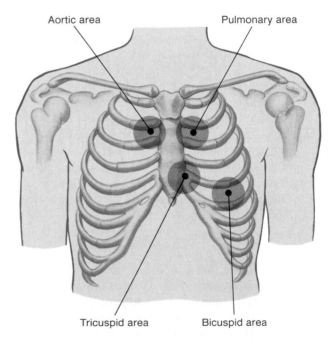

Figure 18.18 Thoracic regions where the sounds of each heart valve are heard most easily.

in the heart—one in the atrial walls and another in the ventricular walls. These masses of cardiac muscle fibers are separated from each other by portions of the heart's fibrous skeleton, except for a small area in the right atrial floor. In this region, the *atrial syncytium* and the *ventricular syncytium* are connected by fibers of the cardiac conduction system.

---

1. Describe the pressure changes that occur in the atria and ventricles during a cardiac cycle.
2. What causes heart sounds?
3. What is meant by a functional syncytium?
4. Where are the functional syncytia of the heart located?

Cardiac Conduction System

Throughout the heart are clumps and strands of specialized cardiac muscle tissue whose fibers contain only a few myofibrils. Instead of contracting, these parts initiate and distribute impulses (cardiac impulses) throughout the myocardium. They comprise the **cardiac conduction system,** which coordinates the events of the cardiac cycle.

A key portion of this conduction system is called the **sinoatrial node** (S-A node). It consists of a small, elongated mass of specialized cardiac muscle tissue just beneath the epicardium. It is located in the posterior wall of the right atrium, below the opening of the superior vena cava, and its fibers are continuous with those of the atrial syncytium.

The membranes of the nodal cells are in contact with one another, and these cells have the ability to excite themselves. That is, without being stimulated by nerve fibers or any other outside agents, the nodal cells initiate impulses that spread into the surrounding myocardium and stimulate the cardiac muscle fibers to contract. Furthermore, this activity is rhythmic. The S-A node initiates one impulse after another, seventy to eighty times a minute. Thus, it is responsible for the rhythmic contractions of the heart and is often called the **pacemaker.**

As a cardiac impulse travels from the S-A node into the atrial syncytium, the right and left atria contract almost simultaneously. Instead of passing directly into the ventricular syncytium, which is separated from the atrial syncytium by the fibrous skeleton of the heart, the cardiac impulse passes along fibers of the conduction system that are continuous with atrial muscle fibers. These conducting fibers lead to a mass of specialized cardiac muscle tissue called the **atrioventricular node** (A-V node). This node is located in the floor of the right atrium near the septum that separates the atria (interatrial septum) and just beneath the endocardium. It

provides the only normal conduction pathway between the atrial and ventricular syncytia.

The fibers that conduct the cardiac impulse into the A-V node (junctional fibers) have very small diameters, and because small fibers conduct impulses slowly, they cause the impulse to be delayed. The impulse is delayed still more as it travels through the A-V node, and this delay allows time for the atria to empty and the ventricles to fill with blood.

Once the cardiac impulse reaches the distal side of the A-V node, it passes into a group of large fibers that make up the **A-V bundle** (atrioventricular bundle or bundle of His), and the impulse moves rapidly through them. The A-V bundle enters the upper part of the interventricular septum, and divides into right and left branches that lie just beneath the endocardium. About halfway down the septum, the branches give rise to enlarged **Purkinje fibers.**

The base of the aorta, which contains the aortic valves, is enlarged and protrudes somewhat into the interatrial septum close to the A-V bundle. Consequently, inflammatory conditions, such as bacterial endocarditis involving the aortic valves (aortic valvulitis), may also affect the A-V bundle.

If a portion of the bundle is damaged, it may no longer conduct impulses normally. As a result, cardiac impulses may reach the two ventricles at different times so that they fail to contract together. This condition is called a *bundle branch block.*

The Purkinje fibers spread from the interventricular septum into the papillary muscles, which project inward from the ventricular walls, and then continue downward to the apex of the heart. There they curve around the tips of the ventricles and pass upward over the lateral walls of these chambers. Along the way, the Purkinje fibers give off many small branches, which become continuous with cardiac muscle fibers. These parts of the conduction system are shown in figure 18.19 and are summarized in figure 18.20.

The muscle fibers in the ventricular walls are arranged in irregular whorls, so that when they are stimulated by the impulses on the Purkinje fibers, the ventricular walls contract with a twisting motion (figure 18.21). This action squeezes or wrings the blood out of the ventricular chambers and forces it into the arteries.

---

1. What kinds of tissues make up the cardiac conduction system?
2. How is a cardiac impulse initiated?
3. How is this impulse transmitted from the right atrium to the other heart chambers?

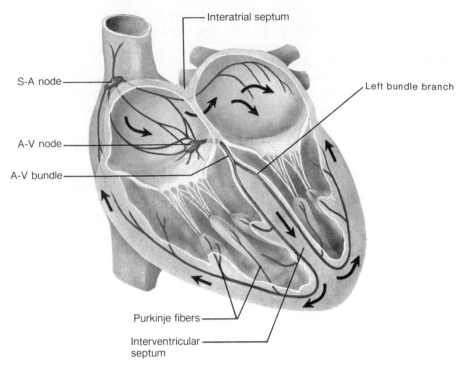

Interatrial septum

S-A node

A-V node

A-V bundle

Left bundle branch

Purkinje fibers

Interventricular septum

Figure 18.19 The cardiac conduction system. What is the function of this system?

Regulation of the Cardiac Cycle

The primary function of the heart is to pump the blood to the body cells, and when the needs of these cells change, the quantity of the blood pumped must change also. For example, during strenuous exercise, the amount of blood required by the skeletal muscles increases greatly, and the rate of the heartbeat increases in response to this need. Since the S-A node normally controls the heart rate, changes in this rate often involve factors that affect the pacemaker. These include the motor impulses carried on the parasympathetic and sympathetic nerve fibers. (See figures 11.39 and 11.40.)

The parasympathetic fibers that supply the heart arise from neurons in the medulla oblongata and make up parts of the *vagus nerves.* Most of these fibers branch to the S-A and A-V nodes. When the nerve impulses reach their endings, these fibers secrete acetylcholine, which causes a decrease in S-A and A-V nodal activity. As a result, the rate of heartbeat decreases.

The vagus nerves seem to carry impulses continually to the S-A and A-V nodes. These impulses impose a braking action on the heart. Consequently, parasympathetic activity can cause the heart rate to change in either direction. An increase in the impulses causes a slowing of the heart, and a decrease in the impulses releases the parasympathetic brake and allows the heartbeat to increase.

Sympathetic fibers reach the heart by means of the *accelerator nerves,* whose branches join the S-A and A-V nodes as well as other areas of the atrial and ventricular myocardium. The endings of these fibers se-

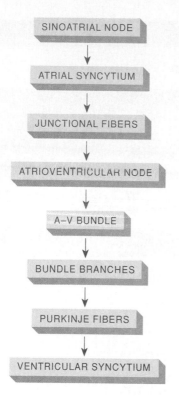

```
SINOATRIAL NODE
        ↓
ATRIAL SYNCYTIUM
        ↓
JUNCTIONAL FIBERS
        ↓
ATRIOVENTRICULAR NODE
        ↓
A–V BUNDLE
        ↓
BUNDLE BRANCHES
        ↓
PURKINJE FIBERS
        ↓
VENTRICULAR SYNCYTIUM
```

Figure 18.20 Components of the cardiac conduction system.

crete norepinephrine in response to nerve impulses, and this substance causes an increase in the rate and the force of myocardial contractions.

A normal balance between the inhibitory effects of the parasympathetic fibers and the excitatory effects of the sympathetic fibers is maintained by the *cardiac control center* of the medulla oblongata. In this region of the brain, masses of neurons function as *cardioinhibitor* and *cardioaccelerator reflex centers*. These centers receive sensory impulses from various parts of the circulatory system and relay motor impulses to the heart in response.

For example, receptors that are sensitive to stretch are located in certain regions of the aorta (aortic sinus and aortic arch) and in the carotid arteries (carotid sinuses). These receptors, called *pressoreceptors* (baroreceptors) can detect changes in blood pressure. Thus, if the pressure rises, the receptors are stretched, and they signal the cardioinhibitor center in the medulla.

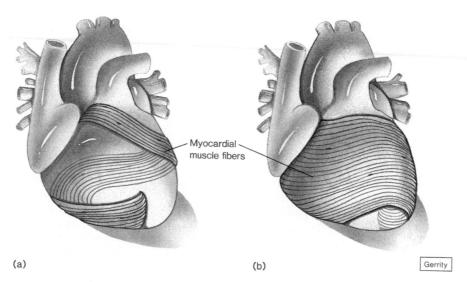

(a) (b) Gerrity

Myocardial muscle fibers

Figure 18.21 The muscle fibers within the ventricular walls are arranged in patterns of whorls. The fibers of groups (a) and (b) surround both ventricles in these anterior views of the heart.

The Electrocardiogram

A recording of the electrical changes that occur in the *myocardium* during a cardiac cycle is called an **electrocardiogram** (ECG). These changes result from the depolarization and repolarization associated with the action potentials occurring in contracting cardiac muscle fibers. Because the body fluids can conduct electrical currents, such changes can be detected on the surface of the body.

To record an ECG, metal electrodes are placed at certain locations on the skin. These electrodes are connected by wires to an instrument (shown in figure 18.22) that responds to very weak electrical changes by causing a pen or stylus to mark on a moving strip of paper. When the instrument is operating, up-and-down movements of the pen correspond to electrical changes occurring within the body as a result of myocardial activity.

Because the paper moves past the pen at a known rate, the distance between pen deflections can be used to measure the time elapsing between various phases of the cardiac cycle.

As figure 18.23 illustrates, the normal ECG pattern includes several deflections, or *waves,* during each cardiac cycle. Between cycles, the muscle fibers remain polarized, and no detectable electrical changes occur. Consequently, the pen does not move and simply marks along the baseline as the paper passes through the instrument. However, when the S-A node triggers a cardiac impulse the atrial fibers are stimulated to depolarize, and an electrical change occurs. As a result, the pen is deflected, and when this electrical change is completed, the pen returns to the base position. This first pen movement produces a *P wave* that is caused by a depolarization of the atrial fibers just before they contract.

When the cardiac impulse reaches the ventricular fibers, they are stimulated to depolarize rapidly. Because the ventricular walls are much more extensive than those of the atria, the amount of electrical change is greater, and the pen is deflected to a greater degree than before. As before, when the electrical change is completed, the pen returns to the baseline, leaving a mark called the *QRS complex,* which usually consists of a *Q wave,* an *R wave,* and an *S wave.* This complex appears just prior to the contraction of the ventricular walls.

Near the end of the ECG pattern, the pen is deflected once again, producing a *T wave.* This wave is caused by the electrical changes occurring as the ventricular muscle fibers become repolarized relatively slowly. The record of the atrial repolarization is missing from the pattern because the atrial fibers repolarize at the same time that the ventricular fibers depolarize. Thus, the recording of the atrial repolarization is obscured by the QRS complex (figures 18.24 and 18.25).

ECG patterns are especially important because they allow a physician to assess the heart's ability to conduct impulses and, therefore, to judge its condition. For example, the time period between the beginning of a P wave and the beginning of a QRS complex (*P-Q interval,* or *P-R interval* if the initial wave is upright) indicates how long it takes for the cardiac impulse to travel from the S-A node through the A-V node. If ischemia or other problems involving the fibers of the A-V conduction pathways are present, this P-Q interval sometimes increases. Similarly, if the AV-bundle is injured, the duration of the QRS complex may increase, because it may take longer for an impulse to spread throughout the ventricular walls (figure 18.26).

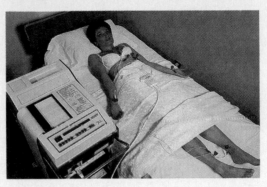

Figure 18.22   This instrument is used to record an ECG.

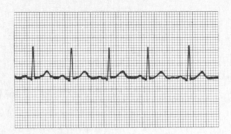

Figure 18.23   A normal ECG.

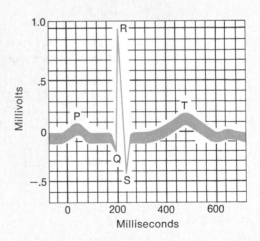

Figure 18.24   In an ECG pattern, the P wave results from depolarization of the atria; the QRS complex results from depolarization of the ventricles; and the T wave results from repolarization of the ventricles.

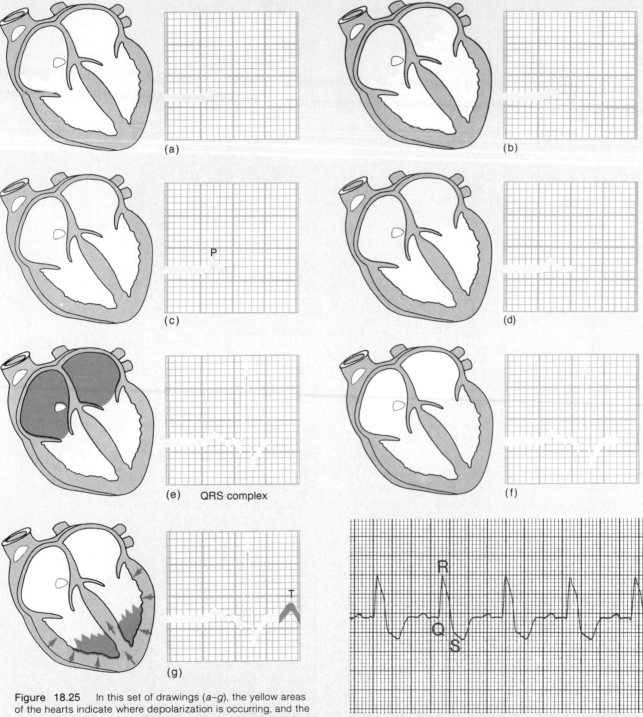

(a)

(b)

(c)

P

(d)

(e)   QRS complex

(f)

(g)

Figure 18.25 In this set of drawings (a–g), the yellow areas of the hearts indicate where depolarization is occurring, and the green areas indicate where tissues are repolarizing; the portion of the ECG pattern produced at each step is shown by the colored line on the graph paper.

T

R

Q
S

Figure 18.26 A prolonged QRS complex may result from damage to the A-V bundle fibers.

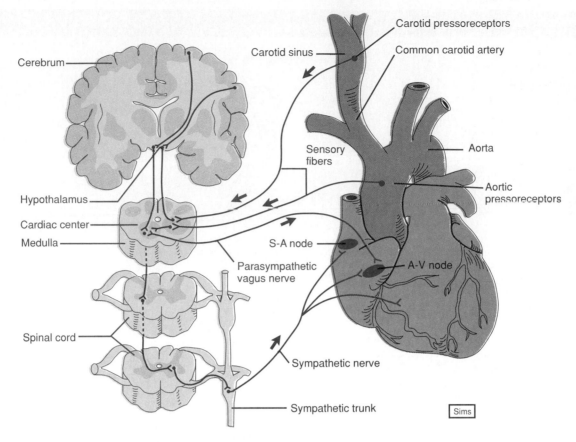

Figure 18.27 The activities of the S-A and A-V nodes can be altered by autonomic nerve impulses.

In response, the medulla sends *parasympathetic* motor impulses to the heart, causing the heart rate and force of contraction to decrease. This action also causes blood pressure to drop toward the normal level (figure 18.27).

Another regulatory reflex involves pressoreceptors in the venae cavae near the entrances to the right atrium. If the blood pressure increases abnormally in these vessels, the receptors signal the cardioaccelerator center, and *sympathetic* impulses flow to the heart. As a result, the heart rate and force of contraction increase, and the venous blood pressure is reduced.

The cardiac control center can also be influenced by impulses from the cerebrum or hypothalamus. Such impulses may cause the heart rate to decrease, as occurs when a person faints following an emotional upset, or they may cause the heart rate to increase during a period of anxiety.

Two other factors that influence the heart rate are temperature change and the presence of various ions. Heart action is increased by a rising body temperature, which accounts for the fact that heart rate usually increases during fever. On the other hand, cardiac activity is decreased by abnormally low body temperature. Consequently, a patient's body temperature is sometimes deliberately lowered (hypothermia) to slow the heart during surgery.

Of the ions that influence heart action, the most important are potassium (K^+) and calcium (Ca^{++}) ions. Although homeostatic mechanisms normally maintain the concentrations of these substances within narrow ranges, these mechanisms sometimes fail, and the consequences can be serious or even fatal.

An excess of *potassium ions* (hyperkalemia) seems to alter the usual polarized state of the cardiac muscle fibers, and the result is a decrease in the rate and force of contractions. In fact, if the potassium ion concentration is very high, the conduction of cardiac impulses may be blocked, and heart action may suddenly stop (cardiac arrest). Conversely, if the potassium concentration drops below normal (hypokalemia), the heart may develop a serious abnormal rhythm (arrhythmia). This condition can also be life threatening.

Excessive *calcium ions* (hypercalcemia) cause increased heart actions, and there is a danger that the heart will undergo a prolonged contraction. Conversely, a low calcium ion concentration (hypocalcemia) depresses heart action. This effect occurs because calcium ions help initiate the muscle contraction mechanism, as described in chapter 9.

Arrhythmias

Although slight variations in heart actions sometimes occur normally, marked changes in the usual rate or rhythm may suggest cardiovascular disease. Such abnormal actions, termed **cardiac arrhythmias,** include the following:

1. **Tachycardia.** Tachycardia is characterized by an abnormally fast heartbeat, usually over one hundred beats per minute. This condition may be caused by such factors as an increase in body temperature, nodal stimulation by sympathetic fibers, the presence of various drugs or hormones, heart disease, excitement, exercise, anemia, or shock. Figure 18.28 shows the ECG of a tachycardic heart.

2. **Bradycardia.** Bradycardia means a slow heart rate, usually one that is less than sixty beats per minute. It may be caused by decreased body temperature, nodal stimulation by parasympathetic impulses, or the presence of certain drugs. It also may occur during sleep. Athletes sometimes have unusually slow heartbeats because their hearts have developed the ability to pump a greater than normal volume of blood with each beat (figure 18.29).

3. **Premature heartbeats.** A premature beat is one that occurs before it is expected in a normal series of cardiac cycles. Such an occurrence is probably caused by cardiac impulses originating from unusual (ectopic) regions of the heart. That is, the impulse originates from a site other than the S-A node. Cardiac impulses may arise from ischemic tissues or from muscle fibers that are irritated by the effects of disease or drugs.

4. **Flutter.** A heart chamber is said to flutter when it is contracting regularly, but at a very rapid rate, such as 250–350 contractions per minute. Although normal hearts may flutter occasionally, this condition is more likely to be due to damage to the myocardium (figure 18.30).

5. **Fibrillation.** Fibrillation is also characterized by rapid heart actions, but, unlike flutter, the accompanying contractions are *uncoordinated* because small regions of the myocardium contract and relax independently of all the other areas. As a result, the myocardium fails to contract as a whole, and the walls of the fibrillating chambers are completely ineffective in pumping blood.

   Although a person may survive atrial fibrillation because blood can continue to move

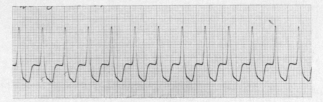

Figure 18.28 Tachycardia is characterized by a rapid heartbeat.

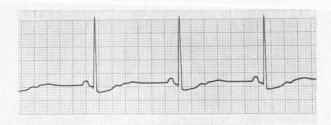

Figure 18.29 Bradycardia is characterized by a slow heartbeat.

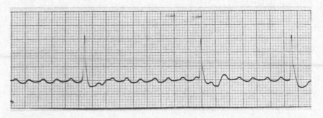

Figure 18.30 Atrial flutter is characterized by an abnormally rapid rate of atrial contraction.

through the heart if the ventricles are functional, ventricular fibrillation is very likely to cause death. In fact, ventricular fibrillation is the most common cause of sudden cardiac death—that is, the unexpected collapse and death of a seemingly healthy person.

The ventricles can be stimulated to fibrillate by a variety of factors, including ischemia associated with obstruction of a coronary artery (coronary occlusion), electric shock, traumatic injury to the heart or chest wall, or toxic effects of various drugs. Once the ventricles are fibrillating, normal cardiac rhythm is not likely to be restored unless the heart can be *defibrillated*. Under such conditions, if circulation is not reestablished within a few minutes, the person may suffer irreparable damage to the cerebral cortex and may die.

Defibrillation is accomplished by exposing a fibrillating myocardium to a strong electric current for a short time. This action causes all of the muscle fibers within the myocardium to become depolarized simultaneously, so that all contractile activity ceases momentarily. It is hoped that, within the next instant, the S-A node will begin to function, and a normal rhythm will be reestablished. Such a restoration of the normal cardiac rhythm is called *cardioversion.*

6. **Arrhythmias due to conduction disorders.** Any interference or block in cardiac impulse conduction may cause arrhythmia. The type of arrhythmia produced, however, varies with the location and extent of the block. Related to this is the fact that certain cardiac tissues other than the S-A node can function as pacemakers.

The S-A node usually initiates seventy to eighty heartbeats per minute. If the S-A node is damaged, impulses originating in the A-V node may travel upward into the atrial myocardium and downward into the ventricular walls, stimulating them to contract. Under the influence of the A-V node acting as a *secondary pacemaker,* the heart may continue to pump blood, but at a rate of forty to sixty beats per minute. Similarly, the Purkinje fibers can initiate cardiac impulses, causing the heart to contract fifteen to forty times per minute.

A patient who has such a disorder of the cardiac conduction system might be treated with an *artificial pacemaker.* This device includes an electrical pulse generator and a lead wire that communicates with a portion of the myocardium. The pulse generator contains a permanent battery that serves as an energy source and a microprocessor that can sense the cardiac rhythm. The microprocessor can also signal the heart to increase or decrease its rate of contraction, as is needed.

An artificial pacemaker can be surgically implanted beneath the patient's skin, and its functions can be adjusted from the outside by means of an external programmer.

1. What nerves supply parasympathetic fibers to the heart? What nerves supply sympathetic fibers?
2. How do parasympathetic and sympathetic impulses help control the heart rate?
3. How do changes in body temperature affect the heart rate?

Blood Vessels

The blood vessels are organs of the cardiovascular system, and they form a closed circuit of tubes that carries blood from the heart to the body cells and back again. These vessels include arteries, arterioles, capillaries, venules, and veins. The arteries and arterioles conduct blood away from the ventricles of the heart and lead to the capillaries. The capillaries function to exchange substances between the blood and the body cells, and the venules and veins return blood from the capillaries to the atria.

Arteries and Arterioles

Arteries are strong, elastic vessels that are adapted for carrying the blood away from the heart under relatively high pressure. These vessels subdivide into progressively thinner tubes and eventually give rise to fine branches called **arterioles.**

The wall of an artery consists of three distinct layers, or *tunics,* shown in figure 18.31. The innermost layer (tunica intima) is composed of a layer of simple squamous epithelium, called *endothelium,* resting on a connective tissue membrane that is rich in elastic and collagenous fibers.

In addition to separating the flowing blood from the blood vessel wall, endothelium helps prevent blood clotting by secreting substances that inhibit platelet aggregation (chapter 17). Endothelium also may play a role in regulating local blood pressure by secreting substances that can cause blood vessel dilation, as well as substances that can cause blood vessel constriction.

The middle layer (tunica media) makes up the bulk of the arterial wall. It includes smooth muscle fibers, which encircle the tube, and a thick layer of elastic connective tissue.

The outer layer (tunica adventitia) is relatively thin and consists chiefly of connective tissue with irregularly arranged elastic and collagenous fibers. This layer attaches the artery to the surrounding tissues. It also contains minute vessels (vasa vasorum) that give rise to capillaries and provide blood to the more external cells of the artery wall.

The endothelial lining of an artery provides a smooth surface that allows blood cells and platelets to flow through without being damaged. The connective tissues give the vessel a tough elasticity that enables it to withstand the force of blood pressure and, at the same time, to stretch and accommodate the sudden increase

Artery

Vein

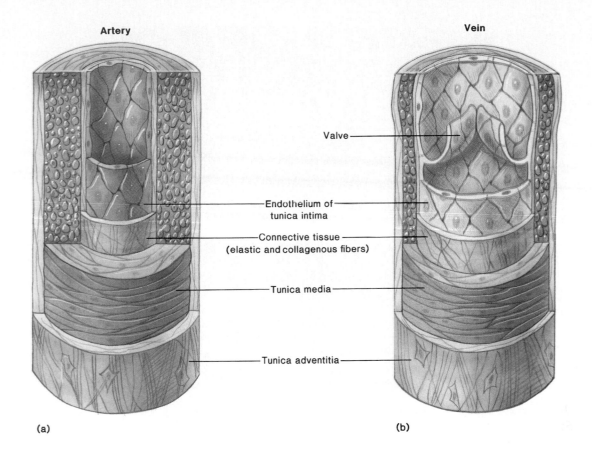

Valve

Endothelium of
tunica intima

Connective tissue
(elastic and collagenous fibers)

Tunica media

Tunica adventitia

(a)

(b)

Figure 18.31 (a) The wall of an artery and (b) the wall of a vein.

in blood volume that accompanies each ventricular contraction.

The smooth muscles in the walls of arteries and arterioles are innervated by the sympathetic branches of the autonomic nervous system. Impulses on these *vasomotor fibers* cause the smooth muscles to contract, thus reducing the diameter of the vessel. This action is called **vasoconstriction.** If such vasomotor impulses are inhibited, the muscle fibers relax and the diameter of the vessel increases. In this case, the vessel is said to undergo **vasodilation.** Changes in the diameters of arteries and arterioles greatly influence the flow and pressure of the blood.

Arterioles, which are microscopic continuations of arteries, give off branches called *metarterioles* that, in turn, join capillaries. Although the walls of the larger arterioles have three layers similar to those of arteries, these walls become thinner and thinner as the arterioles approach the capillaries. The wall of a very small arteriole consists only of an endothelial lining and some smooth muscle fibers, surrounded by a small amount of connective tissue (figures 18.32 and 18.33).

The arteriole and metarteriole walls are adapted for vasoconstriction and vasodilation in that their muscle fibers respond to impulses from the autonomic

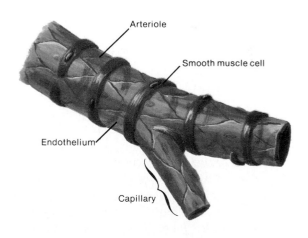

Arteriole

Smooth muscle cell

Endothelium

Capillary

Figure 18.32 Small arterioles have smooth muscle fibers in their walls; capillaries lack these fibers.

nervous system by contracting or relaxing. Thus, these vessels help control the flow of blood into the capillaries.

Sometimes metarterioles are connected directly to venules, and the blood entering them can bypass the

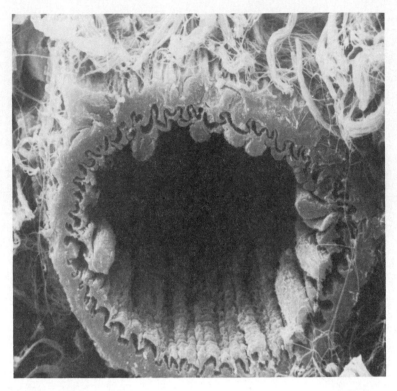

Figure 18.33 Scanning electron micrograph of an arteriole cross section (×3,900). (*Tissues and Organs: A Text-Atlas of Scanning Electron Microscopy,* by R. G. Kessel and R. H. Kardon. © 1979 W.H. Freeman and Company.)

capillaries. These connections between arteriole and venous pathways, shown in figure 18.34, are called *arteriovenous shunts.*

1. Describe the wall of an artery.
2. What is the function of the smooth muscle in the arterial wall?
3. How is the structure of an arteriole different from that of an artery?

Capillaries

Capillaries are the smallest blood vessels. They form the connections between the smallest arterioles and the smallest venules. Capillaries are essentially extensions of the inner linings of these larger vessels, in that their walls consist of endothelium—a single layer of squamous epithelial cells (figure 18.31). These thin walls form the semipermeable membranes through which substances in the blood are exchanged for substances in the tissue fluid surrounding body cells.

Capillary Permeability

The openings or pores in the capillary walls are thin slits occurring where two adjacent endothelial cells overlap. The sizes of these pores, and consequently the permeability of the capillary wall, vary from tissue to tissue. For example, the pores are relatively small in the capillaries of smooth, skeletal, and cardiac muscle, while those in capillaries associated with endocrine glands and the lining of the small intestine are larger. Among the capillaries with the largest openings are those of the liver, spleen, and red bone marrow. The pores in the walls of these vessels commonly allow large protein molecules and even intact cells to pass through as they enter or leave the blood (figure 18.35).

In the brain, the endothelial cells of the capillary walls are more tightly fused by continuous tight junctions than are those in other body regions. Consequently, some substances, such as hormones, amino acids, and certain ions, which readily leave capillaries in other tissues, enter brain tissues only slightly or not at all. Essential nutrients, on the other hand, are carried through the endothelial cells by facilitated diffusion mechanisms (chapter 3), which are specialized to move particular molecules into the brain tissues.

This resistance to the passage of some substances through the capillary walls is called the *blood-brain barrier.* It protects the brain tissues from the extreme changes that sometimes occur in the concentrations of blood components. The blood-brain barrier is also of clinical interest because it prevents certain drugs from entering the brain tissues or cerebrospinal fluid in sufficient amounts to effectively treat some diseases.

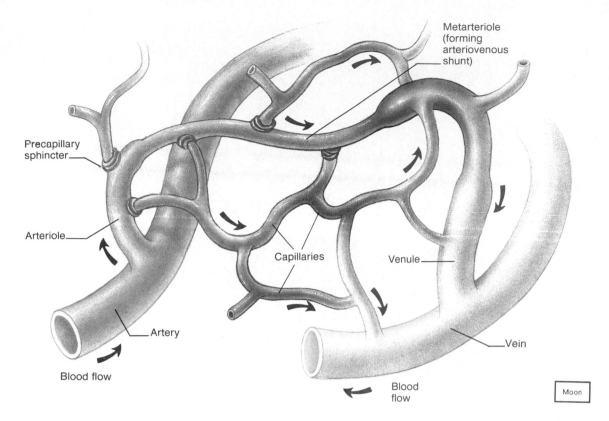

Metarteriole
(forming
arteriovenous
shunt)

Precapillary
sphincter

Arteriole

Capillaries

Venule

Artery

Vein

Blood flow

Blood
flow

Moon

Figure 18.34 Some metarterioles provide arteriovenous shunts by connecting arterioles directly to venules.

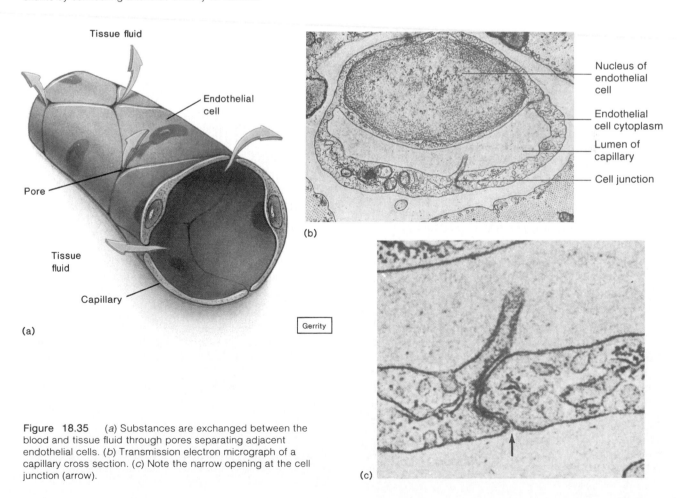

Tissue fluid

Endothelial
cell

Pore

Tissue
fluid

Capillary

(a)

Gerrity

Nucleus of
endothelial
cell

Endothelial
cell cytoplasm

Lumen of
capillary

Cell junction

(b)

(c)

Figure 18.35 (a) Substances are exchanged between the blood and tissue fluid through pores separating adjacent endothelial cells. (b) Transmission electron micrograph of a capillary cross section. (c) Note the narrow opening at the cell junction (arrow).

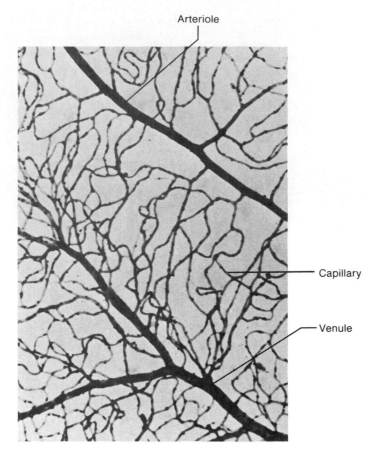

Arteriole

Capillary

Venule

Figure 18.36 Light micrograph of a capillary network.

The capillaries associated with the pituitary gland, pineal gland, and some portions of the hypothalamus are exceptional in that they lack the characteristics of the blood-brain barrier. Consequently, these parts of the brain are able to respond readily to changes in the chemical composition of plasma, as described in chapter 11.

Arrangement of Capillaries

The density of the capillaries within the tissues varies directly with the tissues' rates of metabolism. Thus, muscle and nerve tissues, which use relatively large quantities of oxygen and nutrients, are richly supplied with capillaries; however, cartilaginous tissues, the epidermis, and the cornea, in which metabolic rates are very slow, lack capillaries.

The patterns of capillary arrangement also differ in various body parts. For example, some capillaries pass directly from arterioles to venules, but others lead to highly branched networks (figure 18.36). Such arrangements make it possible for the blood to follow different pathways through a tissue and meet the varying demands of its cells. During periods of exercise, for example, the blood can be directed into the capillary networks of the skeletal muscles, where the cells are experiencing an increasing need for oxygen and nu-

trients. At the same time, the blood can bypass some of the capillary nets in the tissues of the digestive tract, where the demand for blood is less critical. Conversely, when a person is relaxing after a meal, blood can be shunted from the inactive skeletal muscles into the capillary networks of the digestive organs, where it is needed to support the processes of digestion and absorption.

Regulation of Capillary Blood Flow

The distribution of blood in the various capillary pathways is regulated mainly by the smooth muscles that encircle the capillary entrances. As figure 18.34 shows, these muscles form *precapillary sphincters,* which may close a capillary by contracting, or open it by relaxing. How the precapillary sphincters are controlled is not well understood, but they seem to respond to the demands of the cells supplied by their individual capillaries. When the cells have low concentrations of oxygen and nutrients, the sphincter relaxes; and when the cellular needs have been met, the sphincter may contract again.

1. Describe the wall of a capillary.
2. What is the function of a capillary?
3. How is blood flow into capillaries controlled?

Exchanges in the Capillaries

The vital function of exchanging gases, nutrients, and metabolic by-products between the blood and the tissue fluid surrounding body cells occurs in the capillaries. The substances exchanged move through the capillary walls primarily by the processes of diffusion, filtration, and osmosis described in chapter 3. Of these processes, diffusion provides the most important means of transfer.

By diffusion, molecules and ions move from regions where they are more highly concentrated toward regions where they are in lower concentration. Because blood entering certain capillaries carries relatively high concentrations of oxygen and nutrients, these substances diffuse through the capillary walls and enter the tissue fluid. Conversely, the concentrations of carbon dioxide and other wastes are generally greater in such tissues, and the wastes tend to diffuse into the capillary blood.

The paths followed by these substances depend primarily on their solubilities in lipids. Those that are soluble in lipid, such as oxygen, carbon dioxide, and fatty acids, can diffuse through most areas of the cell membranes that make up the capillary wall because the membranes are largely lipid. Lipid-insoluble substances, such as water, sodium ions, and chloride ions, diffuse through pores in the cell membranes and through the slitlike openings between the endothelial cells that form the capillary wall (figure 18.35).

Plasma proteins generally remain in the blood because they are not soluble in the lipid portions of the capillary membranes, and their molecular size is too great to permit diffusion through the membrane pores or slitlike openings between the endothelial cells of most capillaries.

As described in chapter 3, *filtration* involves forcing molecules through a membrane by *hydrostatic pressure.* In the capillaries, the force responsible for filtration is provided by the blood pressure generated by contractions of the ventricular walls.

Blood pressure is also responsible for moving blood through the arteries and arterioles. However, this pressure tends to decrease as the distance from the heart increases, because friction (peripheral resistance) between the blood and the vessel walls slows the flow. For this reason, blood pressure is greater in the arteries than in the arterioles, and greater in the arterioles than in the capillaries. It is similarly greater at the arteriole end of a capillary than at the venule end.

Blood components generally fail to pass through the walls of arteries and arterioles because these walls are too thick. However, when blood reaches the thin-walled capillaries, the hydrostatic pressure of the blood increases the rate at which substances diffuse through the capillary wall. This filtration effect occurs primarily at the arteriole ends of capillaries, while diffusion takes place along their entire lengths.

The plasma proteins, which remain in the capillaries, help make the *osmotic pressure* of the blood greater (hypertonic) than that of the tissue fluid. Although the capillary blood has a greater osmotic attraction for water than does the tissue fluid, this attraction is overcome by the greater force of the blood pressure. As a result, the net movement of water and dissolved substances is outward at the arteriole end of the capillary by filtration.

More specifically, the force, including *hydrostatic pressure,* that tends to move fluid outward at the arteriole end of a capillary is 41.3 millimeters of mercury (mm Hg), while the *osmotic pressure* of the blood, which tends to move fluid inward, is only 28 mm Hg. Consequently, there is a net movement of water and dissolved substances outward. However, the blood pressure decreases as the blood moves through the capillary, and, at the venule end, the outward force equals 21.3 mm Hg, while the osmotic pressure of the blood remains unchanged at 28 mm Hg. Thus, there is a net movement of water and dissolved materials into the venule end of the capillary by osmosis. This process is shown in figure 18.37.

Normally, more fluid leaves the capillaries than returns to them, and the excess is collected and returned to the venous circulation by *lymphatic vessels.* This mechanism is discussed in chapter 19.

Sometimes unusual events cause an increase in the permeability of the capillaries, and an excessive amount of fluid is likely to enter the spaces between the tissue cells (interstitial spaces). This may occur, for instance, following a traumatic injury to the tissues or in response to certain chemicals such as *histamine,* either of which circumstance increases membrane permeability. In any case, so much fluid may leak out of the capillaries that the lymphatic drainage is overwhelmed, and the affected tissues become swollen (edematous) and painful.

If the right ventricle of the heart is failing—that is, if it is unable to pump the blood out of the chamber as rapidly as it enters—other parts of the body may develop edema. This occurs because the blood backs up into the veins, venules, and capillaries, and the blood pressure in these vessels increases. As a result of this increased *back pressure,* the osmotic pressure of the blood in the venule ends of the capillaries is less effective in attracting water from tissue fluid, and the tissues tend to become edematous. This is true particularly of the tissues in the lower extremities if the person is upright, or in the back if the person is supine. In the terminal stages of heart failure, edema may become widespread, and fluid may begin to accumulate in the peritoneal cavity of the abdomen. This condition is called *ascites.*

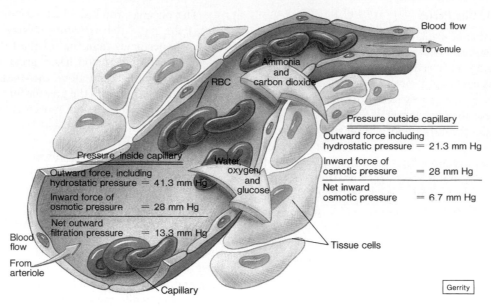

Blood flow
To venule

Ammonia
and
carbon dioxide

RBC

Pressure outside capillary

Outward force including
hydrostatic pressure = 21.3 mm Hg

Inward force of
osmotic pressure = 28 mm Hg

Net inward
osmotic pressure = 6.7 mm Hg

Pressure inside capillary

Outward force, including
hydrostatic pressure = 41.3 mm Hg

Inward force of
osmotic pressure = 28 mm Hg

Net outward
filtration pressure = 13.3 mm Hg

Water,
oxygen,
and
glucose

Blood
flow

From
arteriole

Capillary

Tissue cells

Gerrity

Figure 18.37 Substances leave the capillaries because of a net outward filtration pressure; substances enter the capillaries because of a net inward force of osmotic pressure.

---

1. What forces are responsible for the exchange of substances between the blood and the tissue fluid?
2. Why is the fluid movement out of a capillary greater at its arteriole end than at its venule end?
3. Since more fluid leaves the capillary than returns to it, how is the remainder returned to the vascular system?

Venules and Veins

Venules are the microscopic vessels that continue from the capillaries and merge to form **veins.** The veins, which carry blood back to the atria, follow pathways that roughly parallel those of the arteries.

The walls of veins are similar to those of arteries in that they are composed of three distinct layers. However, the middle layer of the venous wall is poorly developed. Consequently, veins have thinner walls that contain less smooth muscle and less elastic tissue than those of comparable arteries (figures 18.31 and 18.38).

Many veins, particularly those in the arms and legs, contain flaplike *valves,* which project inward from their linings. These valves, shown in figure 18.39, are usually composed of two leaflets that close if the blood begins to back up in a vein. In other words, the valves aid in returning the blood to the heart, because the valves open as long as the flow is toward the heart, but close if it is in the opposite direction.

In addition to providing pathways for the blood returning to the heart, the veins function as *blood reservoirs,* which can be drawn on in times of need. For example, if a hemorrhage accompanied by a drop in arterial blood pressure occurs, the muscular walls of the veins are stimulated reflexly by sympathetic nerve impulses. The resulting venous constrictions help in-

crease the blood pressure. This mechanism ensures a nearly normal blood flow even when as much as 25% of the blood volume has been lost. Figure 18.40 illustrates the relative volumes of blood in the veins and other blood vessels.

The characteristics of the blood vessels are summarized in chart 18.3.

---

1. How does the structure of a vein differ from that of an artery?
2. What are the functions of veins and venules?

Blood Pressure

Blood pressure is the force exerted by the blood against the inner walls of the blood vessels. Although such a force occurs throughout the vascular system, the term *blood pressure* most commonly refers to pressure in arteries supplied by branches of the aorta (systemic arteries).

Arterial Blood Pressure

The arterial blood pressure rises and falls in a pattern corresponding to the phases of the cardiac cycle. That is, when the ventricles contract (ventricular systole), their walls squeeze the blood inside their chambers and force it into the pulmonary trunk and aorta. As a result, the pressures in these arteries increase sharply. The maximum pressure achieved during ventricular contraction is called the **systolic pressure.** When the ventricles relax (ventricular diastole), the arterial pressure drops, and the lowest pressure that remains in the arteries before the next ventricular contraction is termed the **diastolic pressure.**

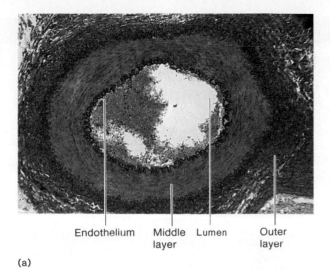

Endothelium | Middle layer | Lumen | Outer layer

(a)

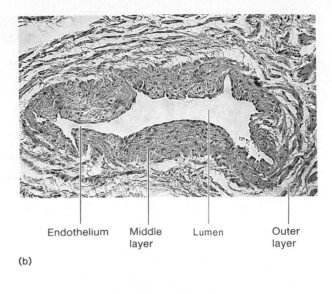

Endothelium | Middle layer | Lumen | Outer layer

(b)

Figure 18.38 Note the structural differences in these cross sections of (a) an artery (×100) and (b) a vein (×160).

(a) (b)

Toward heart

Pedigo

Figure 18.39 (a) Venous valves allow blood to move toward the heart, but (b) prevent blood from moving away from the heart.

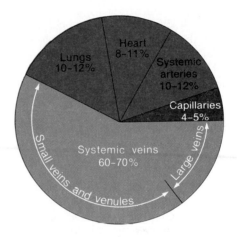

Heart 8–11%
Lungs 10–12%
Systemic arteries 10–12%
Capillaries 4–5%
Systemic veins 60–70%
Small veins and venules
Large veins

Figure 18.40 Most of the blood volume is contained within the veins and venules. (From *Circulation,* by Bjorn Folkow and Eric Neil. Copyright © 1971 by Oxford University Press, Inc. Reprinted by permission.)

CHART 18.3	Characteristics of blood vessels	
Vessel	**Type of wall**	**Function**
Artery	Thick, strong wall with three layers—an endothelial lining, a middle layer of smooth muscle and elastic tissue, and an outer layer of connective tissue	Carries relatively high pressure blood from the heart to arterioles
Arteriole	Thinner wall than an artery, but with three layers; smaller arterioles have an endothelial lining, some smooth muscle tissue, and a small amount of connective tissue	Connects an artery to a capillary; helps control the blood flow into a capillary by undergoing vasoconstriction or vasodilation
Capillary	Single layer of squamous epithelium	Provides a membrane through which nutrients, gases, and wastes are exchanged between the blood and tissue cells; connects an arteriole to a venule
Venule	Thinner wall; less smooth muscle and elastic tissue than in an arteriole	Connects a capillary to a vein
Vein	Thinner wall than an artery, but with similar layers; the middle layer is more poorly developed; some with flaplike valves	Carries relatively low pressure blood from a venule to the heart; valves prevent a backflow of blood; serves as blood reservoir

Blood Vessel Disorders

It is estimated that nearly half of all deaths in the United States are due to the arterial disease called **atherosclerosis.** This condition is characterized by an accumulation of soft masses of fatty materials, particularly cholesterol, on the inside of the arterial walls. Such deposits are called *plaque,* and, as they develop, they tend to protrude into the lumens of the vessels and interfere with blood flow (figure 18.41). Furthermore, plaque often forms a surface that can initiate the formation of a blood clot. As a result, persons with atherosclerosis may develop thrombi or emboli that cause blood deficiency (*ischemia*) or tissue death (*necrosis*) downstream from the obstruction.

The walls of affected arteries also tend to undergo degenerative changes during which they lose their elasticity and become hardened or *sclerotic.* This stage of the disease is called *arteriosclerosis.* When this occurs, there is danger that a sclerotic vessel will rupture under the force of blood pressure.

Although the cause of atherosclerosis is not well understood, the disease is often associated with excessive use of saturated fats and refined carbohydrates in the diet, elevated blood pressure, cigarette smoking, obesity, and lack of physical exercise. Emotional and genetic factors may also increase the susceptibility to atherosclerosis. (See chapter 15.)

Sometimes the wall of an artery is so weakened by the effects of disease that blood pressure causes a region of the artery to become dilated, forming a pulsating sac called an **aneurysm.** Once an aneurysm begins to form, it tends to continue increasing in size. If the resulting sac develops by a longitudinal splitting of the middle layer of the arterial wall, it is called a *dissecting aneurysm.* An aneurysm may cause symptoms by pressing on nearby organs, or it may rupture and produce a great loss of blood.

Although most aneurysms seem to be caused by arteriosclerosis, they may also occur as a consequence of trauma, high blood pressure, infections, or congenital defects in blood vessels. Common sites of aneurysms include the thoracic and abdominal aorta, and an arterial circle at the base of the brain (circle of Willis).

Phlebitis, or inflammation of a vein, is a relatively common disorder, and although it may occur in association

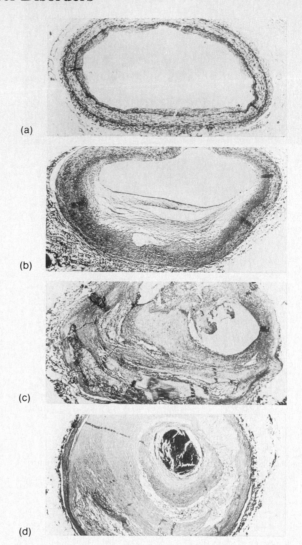

(a)

(b)

(c)

(d)

Figure 18.41 As atherosclerosis develops, masses of fatty materials accumulate beneath the inner linings of certain arteries and arterioles, and protrude into their lumens. (*a*) Normal arteriole; (*b, c,* and *d*) accumulation of plaque on the inner wall of the arteriole.

with an injury or infection or as an aftermath of surgery, it sometimes develops for no apparent reason.

If the inflammation is restricted to a superficial vein, such as the greater or lesser saphenous veins, the blood flow may be rechanneled through other vessels. But if it occurs in a deep vein, such as the tibial, peroneal, popliteal, or femoral veins, the consequences can be quite serious, particularly if the blood within the affected vessel clots and blocks normal circulation. This condition is called *thrombophlebitis*. As a result of thrombophlebitis, there is a risk that a blood clot within a vein will become detached, move along with the venous blood, pass through the heart, and become lodged in the pulmonary arterial system within a lung. Such an obstruction is called a *pulmonary embolism*.

Varicose veins are distinguished by the presence of abnormal and irregular dilations in superficial veins, particularly those in the lower legs. This condition is usually associated with prolonged, increased back pressure within the affected vessels due to the force of gravity, as occurs when a person stands. The problem can also be aggravated when venous blood flow is obstructed by crossing the legs or by sitting in a chair so that its edge presses against the area behind the knee.

Excessive back pressure causes the veins to stretch and their diameters to increase. Because the valves within these vessels do not change size, they soon lose their abilities to block the backward flow of blood, and blood tends to accumulate in the enlarged regions.

Increased venous pressure is also accompanied by rising pressure within the venules and capillaries that supply the veins. Consequently, tissues in affected regions typically become edematous and painful.

Although some people seem to inherit a weakness in their venous valves, *varicose veins* are most common in persons who stand or sit for prolonged periods. Pregnancy and obesity also seem to favor the development of this condition. The discomfort associated with varicose veins can sometimes be relieved by elevating the legs above the level of the heart or by putting on support hosiery before arising in the morning, thus preventing the vessel dilation that occurs upon standing. Occasionally, the intravenous injection of a substance that destroys veins (sclerosing agent) or surgical removal of the affected veins may be necessary.

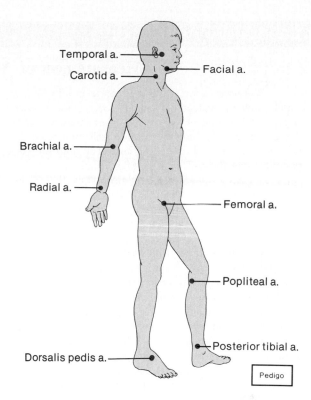

Figure 18.42 Sites where arterial pulse is most easily detected. (*a.* stands for *artery.*)

The surge of blood entering the arterial system during a ventricular contraction causes the elastic walls of the arteries to swell, but the pressure drops almost immediately as the contraction is completed, and the arterial walls recoil. This alternate expanding and recoiling of an arterial wall can be felt as a *pulse* in an artery that runs close to the surface. Figure 18.42 shows several sites where a pulse can be detected. The radial artery, for example, courses near the surface at the wrist and is commonly used to sense a person's radial pulse.

The radial pulse rate is equal to the rate at which the left ventricle is contracting, and for this reason, it can be used to determine the heart rate. A pulse can also reveal something about blood pressure, because an elevated pressure produces a pulse that feels full, while a low pressure produces a pulse that is easily compressed.

1. Distinguish between systolic and diastolic blood pressure.
2. What cardiac event is responsible for the systolic pressure? For the diastolic pressure?
3. What causes a pulse in an artery?

Measurement of Arterial Blood Pressure

Systemic arterial blood pressure usually is measured using an instrument called a *sphygmomanometer* (figure 18.43). This device consists of an inflatable rubber cuff connected by tubing to a compressible bulb and a glass tube containing a column of mercury. The bulb is used to pump air into the cuff, and the pressure produced is indicated by a rise in the mercury column. Thus, the pressure in the cuff can be expressed in millimeters of mercury (mm Hg). A pressure of 100 mm Hg, for example, would be enough to force the mercury column upward for a distance of 100 mm.

To measure arterial blood pressure, the cuff of the sphygmomanometer is usually wrapped around the upper arm so that it surrounds the brachial artery. Air is pumped into the cuff until the cuff pressure exceeds the pressure in that artery. As a result, the vessel is squeezed closed, and its blood flow is stopped. At this moment, if the diaphragm of a stethoscope is placed over the brachial artery at the distal border of the cuff, no sounds can be heard from the vessel because the blood flow has been interrupted. As air is slowly released from the cuff, the air pressure inside it decreases. When the cuff pressure is approximately equal to the systolic blood pressure within the brachial artery, the artery opens enough for a small amount of blood to spurt through. This movement produces a sharp sound (Korotkoff's sound) that can be heard through the stethoscope, and the height of the mercury column when this first tapping sound is heard represents the *arterial systolic pressure* (SP).

As the cuff pressure continues to drop, a series of increasingly louder sounds can be heard. Then, when the cuff pressure is approximately equal to that within the fully opened artery, the sounds become abruptly muffled and disappear. The height of the mercury column when the sounds become abruptly muffled represents the *arterial diastolic pressure* (DP).

The results of a blood pressure measurement are reported as a fraction, such as 120/80. In this notation, the

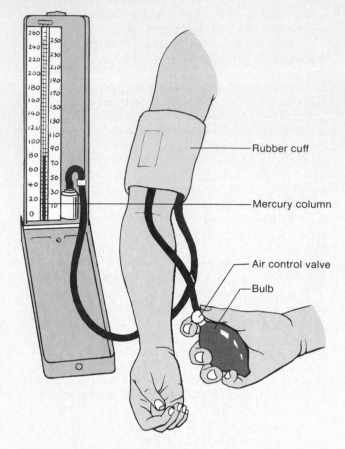

Figure 18.43 A sphygmomanometer is used to measure arterial blood pressure.

Labels: Rubber cuff; Mercury column; Air control valve; Bulb

Factors That Influence Arterial Blood Pressure

Arterial pressure depends on a variety of factors, including heart action, blood volume, resistance to flow, and the viscosity of the blood (figure 18.45).

Heart Action

In addition to producing the blood pressure by forcing the blood into the arteries, the heart action determines how much blood will enter the arterial system with each ventricular contraction, as well as the rate of this fluid output.

The volume of blood discharged from the ventricle with each contraction is called the **stroke volume** and equals about 70 milliliters. The volume discharged from the ventricle per minute is called the **cardiac output.** It is calculated by multiplying the stroke volume by the heart rate in beats per minute. (Cardiac output = Stroke volume × Heart rate.) Thus, if the stroke volume is 70 milliliters and the heart rate is 72 beats per minute, the cardiac output is 5,040 milliliters per minute.

The blood pressure varies with the cardiac output. If either the stroke volume or the heart rate increases, so does the cardiac output, and, as a result, the blood pressure rises. Conversely, if the stroke volume or the heart rate decreases, the cardiac output and blood pressure decrease also.

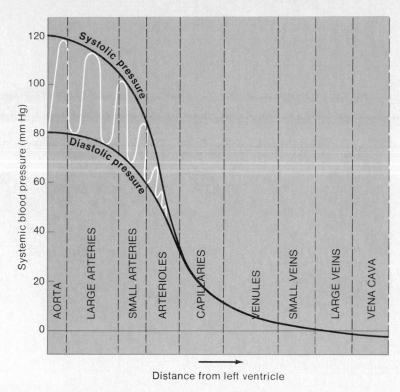

Figure 18.44 Blood pressure decreases as the distance from the left ventricle increases.

upper number indicates the systolic pressure in mm Hg (SP), and the lower number indicates the diastolic pressure in mm Hg (DP). Figure 18.44 shows how these pressures decrease as distance from the left ventricle increases.

The difference between the systolic and diastolic pressures (SP − DP), which is called the *pulse pressure* (PP), is generally about 40 mm Hg.

The average pressure in the arterial system is also of interest because it represents the force that is effective throughout the cardiac cycle for driving blood to the tissues. This force, called the *mean arterial pressure,* is approximated by adding the diastolic pressure and one-third of the pulse pressure (DP+ 1/3PP).

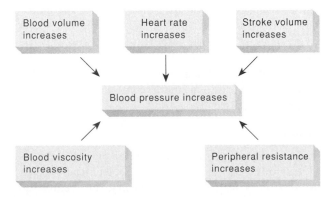

Figure 18.45 Some of the factors that influence arterial blood pressure.

Blood Volume

The **blood volume** is equal to the sum of the blood cell and plasma volumes in the vascular system. Although the blood volume varies somewhat with age, body size, and sex, it usually remains about 5 liters for adults.

The blood pressure is directly proportional to the volume of the blood within the cardiovascular system. Thus, any changes in the blood volume are accompanied by changes in the blood pressure. For example, if blood volume is reduced by a hemorrhage, the blood pressure drops. If the normal blood volume is restored by a blood transfusion, the normal pressure may be reestablished.

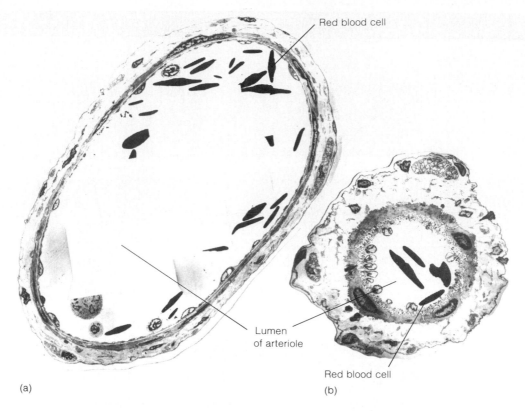

Red blood cell

Lumen
of arteriole

Red blood cell

(a)

(b)

Figure 18.46 (*a*) Relaxation of smooth muscle in the arteriole wall produces dilation, while (*b*) contraction of the smooth muscle causes constriction.

Arterial blood volume, and consequently arterial blood pressure, also vary directly with cardiac output. Thus, if the stroke volume or the heart rate increases, the arterial system may be forced to accept a greater volume of blood than can flow into the peripheral blood vessels. Consequently, the blood volume that the arteries must accommodate is increased even when the ventricles are relaxed. For this reason, an increased cardiac output is reflected in an elevated *diastolic pressure.* On the other hand, an increase in the force of ventricular contraction produces an elevated *systolic pressure.*

> Blood volume can be determined by injecting a known volume of an indicator, such as radioactive iodine, into the blood. After a time that allows for thorough mixing, a blood sample is withdrawn, and the concentration of the indicator is measured. The total blood volume is calculated using the formula: Blood volume = Amount of indicator injected/ Concentration of indicator in blood sample.

Peripheral Resistance

Friction between the blood and the walls of the blood vessels produces a force called **peripheral resistance,** which hinders blood flow. This force must be overcome by blood pressure if the blood is to continue flowing.

Consequently, factors that alter the peripheral resistance cause changes in the blood pressure.

For example, if the smooth muscles in the walls of arterioles contract, the peripheral resistance of these constricted vessels increases. Blood tends to back up into the arteries supplying the arterioles, and the arterial pressure rises. Dilation of the arterioles has the opposite effect—peripheral resistance lessens, and the arterial blood pressure drops in response (figure 18.46).

As mentioned previously, the arterial walls are quite elastic, and when the ventricles discharge a surge of blood, the arteries swell. Almost immediately, the elastic tissues recoil, and the vessel walls press against the blood inside. This action helps force the blood onward against the peripheral resistance caused by the arterioles and capillaries. It also tends to convert the intermittent flow of blood, which is characteristic of the arterial system, into a more continuous movement through the capillaries.

Viscosity

The viscosity of a fluid is a physical property that is related to the ease with which its molecules flow past one another. The greater the viscosity, the greater the resistance to flowing.

The presence of blood cells and plasma proteins increases the viscosity of the blood. Since the greater the blood's resistance to flowing, the greater the force needed to move it through the vascular system, it is not surprising that the blood pressure rises as the blood viscosity increases and drops as the viscosity decreases.

Although the viscosity of blood normally remains relatively stable, any condition that alters the concentrations of blood cells or plasma proteins may cause changes in viscosity. For example, anemia and hemorrhage may be accompanied by a decreasing viscosity and a consequent decrease in the blood pressure. If red blood cells are present in abnormally high numbers (polycythemia), the viscosity increases and an increase in the blood pressure is likely.

1. How is cardiac output calculated?
2. What is the relationship between cardiac output and blood pressure?
3. How does blood volume affect blood pressure?
4. What is the relationship between peripheral resistance and blood pressure? Between viscosity and blood pressure?

Control of Blood Pressure

Two important mechanisms for maintaining normal arterial pressure involve the regulation of the cardiac output and the peripheral resistance.

The *cardiac output* depends on the volume of blood discharged from the left ventricle with each contraction (stroke volume) and the rate of heartbeat. These actions are affected by mechanical, neural, and chemical factors.

For example, the volume of blood entering the ventricle affects the stroke volume. As the blood enters, myocardial fibers in the ventricular wall are mechanically stretched. Within limits, the greater the length of these fibers, the greater the force with which they contract. This relationship between fiber length and force of contraction is called *Starling's law of the heart.* Because of it, the heart can respond to the immediate demands placed on it by the varying quantities of blood that return from the venous system. In other words, the more blood that enters the heart from the veins, the stronger the ventricular contraction, the greater the stroke volume, and the greater the cardiac output (figure 18.47). The less blood that returns from the veins, the weaker the ventricular contraction, the lesser the stroke volume, and the lesser the cardiac output.

This mechanism ensures that the volume of blood discharged from the heart is equal to the volume entering its chambers. Consequently, the volume of blood that enters the right atrium from the venae cavae is normally equal to the volume that leaves the left ventricle and enters the aorta.

The neural regulation of heart rate was described previously in this chapter. To review, pressoreceptors

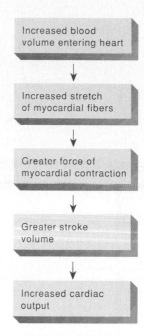

Figure 18.47 Cardiac output is related to the volume of blood entering the heart.

(baroreceptors) located in the walls of the aortic arch and carotid sinuses are sensitive to changes in the blood pressure. If the arterial pressure increases, nerve impulses travel from the receptors to the *cardiac center* of the medulla oblongata. This center relays parasympathetic impulses to the S-A node in the heart, and the rate of heartbeat decreases in response. As a result of this *cardioinhibitor reflex,* the cardiac output is reduced, and the blood pressure decreases toward the normal level. Figure 18.48 summarizes this mechanism.

Conversely, if the arterial blood pressure decreases, the *cardioaccelerator reflex,* involving sympathetic impulses to the S-A node, is initiated, and the heart beats faster. This response increases the cardiac output, and the arterial pressure increases.

As mentioned in chapter 13, *epinephrine* causes the heart rate to increase and consequently alters cardiac output and blood pressure. Other factors that cause an increase in heart rate and an increase in blood pressure include emotional responses, such as fear and anger; physical exercise; and an increase in body temperature.

Peripheral resistance is regulated primarily by changes in the diameters of arterioles. Because blood vessels with smaller diameters offer a greater resistance to blood flow, factors that cause arteriole vasoconstriction bring about an increase in peripheral resistance, and factors causing vasodilation produce a decrease in resistance.

The *vasomotor center* of the medulla oblongata continually sends sympathetic impulses to the smooth

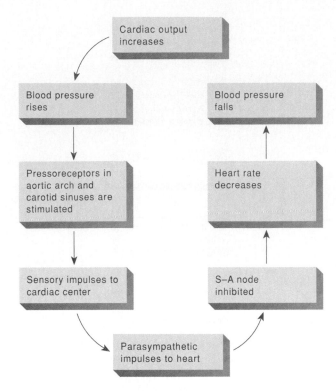

Figure 18.48 This mechanism helps regulate blood pressure by inhibiting the S-A node.

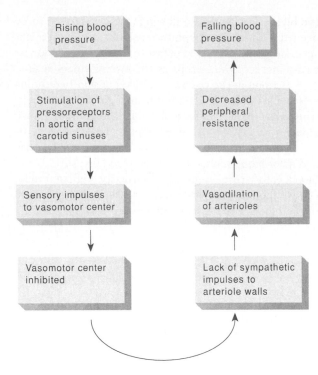

Figure 18.49 This mechanism helps regulate blood pressure by causing arteriole vasodilation.

muscles in the arteriole walls. As a result, these muscles are kept in a state of tonic contraction, which helps maintain the peripheral resistance associated with a normal blood pressure. Because the vasomotor center is responsive to changes in the blood pressure, it can cause an increase in the peripheral resistance by increasing its outflow of sympathetic impulses, or it can cause a decrease in such resistance by decreasing its sympathetic outflow. In the latter case, the vessels undergo vasodilation as the sympathetic stimulation is decreased.

For instance, as figure 18.49 illustrates, whenever the arterial blood pressure suddenly increases, the pressoreceptors in the aortic arch and carotid sinuses signal the vasomotor center, and the sympathetic outflow to the arteriole walls is reduced. The resulting vasodilation causes a decrease in the peripheral resistance, and the blood pressure decreases toward the normal level.

Control of vasoconstriction and vasodilation by the vasomotor center is especially important in the arterioles of the *abdominal viscera* (splanchnic region). These vessels, if fully dilated, could accept nearly all the blood of the body and cause the arterial pressure to approach zero. Thus, control of their diameters is essential in the regulation of normal peripheral resistance.

Chemical substances, including carbon dioxide, oxygen, and hydrogen ions, also influence peripheral resistance by affecting the smooth muscles in the walls of arterioles and metarterioles, and the actions of precapillary sphincters. For example, an increasing P_{CO_2}, a decreasing P_{O_2}, and a decreasing pH cause relaxation of these muscles and a consequent decrease in the blood pressure. In addition, epinephrine and norepinephrine cause vasoconstriction of many vessels, followed by an increase in blood pressure, even though epinephrine causes vasodilation of the vessels within skeletal muscles.

1. What factors affect cardiac output?
2. Define Starling's law of the heart.
3. What is the function of the pressoreceptors in the walls of the aortic arch and carotid sinuses?
4. How does the vasomotor center control the diameter of arterioles?

Venous Blood Flow

Blood pressure decreases as the blood moves through the arterial system and into the capillary networks. In fact, little pressure remains at the venule ends of capillaries (figure 18.44). Therefore, blood flow through the venous system is not the direct result of heart action, but depends on other factors, such as skeletal muscle contraction, breathing movements, and vasoconstriction of veins.

For example, when skeletal muscles contract, they thicken and press on nearby vessels, squeezing the blood

Hypertension

Hypertension, or high blood pressure, is characterized by a persistently elevated arterial pressure, and is one of the more common diseases of the cardiovascular system.

When the cause of high blood pressure is unknown, as is often the case, it may be called *essential* (also primary or idiopathic) *hypertension*. Sometimes the elevated pressure is related to another problem, such as arteriosclerosis or kidney disease. When its cause is known, the condition is called *secondary hypertension*.

Arteriosclerosis, for example, is accompanied by decreasing elasticity of the arterial walls, and is followed by narrowing of the lumens of these vessels. Both of these effects promote an increase in the blood pressure.

Kidney diseases often produce changes that interfere with blood flow to kidney cells. In response, the affected tissues may release an enzyme called **renin** that acts on certain plasma proteins to form a substance called **angiotensin.**

Angiotensin is a powerful vasoconstrictor whose action increases the peripheral resistance in the arterial system, and this causes the arterial pressure to rise. Angiotensin also causes *aldosterone* to be released from the adrenal cortex. This hormone promotes the retention of sodium ions and water by the kidneys, and the resulting increase in blood volume causes an additional increase in blood pressure (figure 18.50).

Normally, this mechanism ensures that a decrease in blood flow to the kidneys is followed by an increase in arterial pressure, which, in turn, results in an increase in blood flow to the kidneys. If the decreased blood flow is the result of a disease, such as atherosclerosis, the mechanism may cause high blood pressure and promote further deterioration of the arterial system.

Although the mechanism is not well understood, high sodium intake seems to cause vasoconstriction in some individuals, thus leading to increased blood pressure. Another factor that seems to promote hypertension is obesity, which is accompanied by an increased volume of body fluids and increased heart action needed to provide additional blood flow. Psychological stress, which activates sympathetic nerve impulses that cause generalized vasoconstriction, may also lead to hypertension.

The consequences of prolonged, uncontrolled hypertension can be very serious. For example, the left ventricle must contract with greater force than usual in order to discharge a normal volume of blood against increased arterial pressure. As a result of the increased work load, the myocardium tends to thicken, and the heart enlarges. At first,

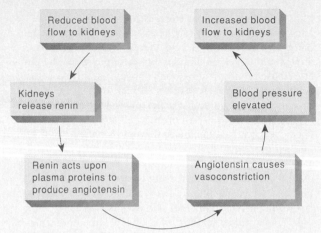

Figure 18.50 This mechanism acts to elevate blood pressure.

such hypertrophy may benefit cardiac function. However, as the cardiac muscle tissue increases, the coronary blood vessels may not increase to the same extent. Consequently, the vessels may become less and less able to supply the myocardium with sufficient blood. This change may be accompanied by degeneration of muscle fibers and a replacement of muscle with fibrous tissue. In time, the enlarged heart is weakened and may fail.

Hypertension also enhances the development of atherosclerosis. As a result, various arteries, including those of the coronary circuit, tend to accumulate plaque that may cause arterial occlusions. The patient then may suffer a *coronary thrombosis* or a *coronary embolism.* Similar changes in the arteries of the brain increase the chances of a *cerebral vascular accident* (CVA), which is due to a cerebral thrombosis, embolism, or hemorrhage, and is more commonly called a stroke.

The treatment of hypertension may include exercising regularly, controlling body weight, reducing stress, and limiting the diet to foods that are low in sodium and high in potassium. It also may involve the use of drugs, such as diuretics and/or inhibitors of sympathetic nerve activity. Diuretics typically increase the urinary excretion of sodium and water, reducing the volume of body fluids. Sympathetic inhibitors block the synthesis of neurotransmitters, such as norepinephrine, or block various receptor sites of effector cells.

---

inside. As mentioned previously, many veins, particularly those in the arms and legs, contain flaplike *semilunar valves* that project inward from their linings. These valves offer little resistance to the blood flowing toward the heart, but they close if blood moves in the opposite direction. Consequently, as *skeletal muscles*

exert pressure on veins with valves, some blood is moved from one valve section to another. This massaging action of contracting skeletal muscles helps push the blood through the venous system toward the heart. (See figure 18.51.)

Cardiovascular Adjustment to Exercise

When a person engages in strenuous exercise, a number of changes occur in the circulatory system that help increase the blood flow, and therefore the oxygen availability, to the skeletal muscles. For example, when the skeletal muscles are resting, only about one-fifth of their capillaries are open. As exercise begins, oxygen concentration in the tissues of active skeletal muscles decreases, and, in response, essentially all of the capillaries in the muscles open. At the same time, blood flow into the tissues of active muscles greatly increases due to vasodilation stimulated by such factors as sympathetic nerve impulses, decreasing oxygen concentration, increasing carbon dioxide concentration, and decreasing pH within the muscle tissues. The blood flow to the myocardium and skin also increases.

In certain other parts of the body, sympathetic nerve impulses cause vasoconstriction, and, as a result, blood flow into the abdominal viscera and inactive skeletal muscles is reduced. The blood flow to the brain and kidneys, however, remains nearly unchanged, so that these vital organs continue to receive oxygen and nutrients at the rate required to support their needs.

When the walls of the veins in visceral organs constrict, blood is moved out of these blood reservoirs, adding to the venous return to the heart. This increased venous return is also aided by the pumping action associated with an increased respiratory rate during exercise.

The heart rate increases in response to decreased parasympathetic stimulation of the S-A and A-V nodes and to sympathetic reflexes triggered by the stimulation of proprioceptors in the skeletal muscles and of stretch receptors in the lungs. As the volume of blood returning to the heart increases, the ventricular walls are stretched, and this stimulates them to contract with greater force. Consequently, with each heartbeat, the stroke volume increases and blood pressure increases.

A number of physiological changes occur in response to physical conditioning that may improve athletic performance. These include increases in heart pumping efficiency, blood volume, blood hemoglobin concentration, and mitochondrial content of muscle fibers. As a result of such changes, oxygen can be delivered to and utilized by muscle tissue more effectively. In the case of an athlete, the structure of the heart typically changes in response to the demands placed upon it. A marathon runner's heart, for example, must provide a sustained increase in cardiac output during competition. As a result of the training for such endurance type exercise, the heart may enlarge as much as 40% or more; its myocardial mass increases; the sizes of its ventricular cavities expand; and the walls of its ventricles thicken. Concurrently, the heart's stroke volume increases, and its rate of beat usually decreases.

Respiratory movements provide another means of moving venous blood. During inspiration, the pressure within the thoracic cavity is reduced as the diaphragm contracts and the rib cage moves upward and outward. At the same time, the pressure within the abdominal cavity is increased as the diaphragm presses downward on the abdominal viscera. Consequently, the blood tends to be squeezed out of the abdominal veins and forced into thoracic veins. Backflow of blood into the legs is prevented by valves in the veins of the legs.

During exercise, these respiratory movements act together with skeletal muscle contractions to increase the return of venous blood to the heart.

Another mechanism that promotes return of venous blood involves constriction of the veins. When venous pressure is low, sympathetic reflexes stimulate smooth muscles in the walls of veins to contract. Such *venoconstriction* increases venous pressure and forces more blood toward the heart.

Also, as mentioned previously, the veins provide a blood reservoir that can adapt its capacity to changes in blood volume (figure 18.40). If some blood is lost and the blood pressure decreases, venoconstriction can help return the blood pressure to normal by forcing blood out of this reservoir.

Central Venous Pressure

Because all the veins, except those of the pulmonary circuit, drain into the right atrium, the pressure within this heart chamber is called *central venous pressure*.

This pressure is of special interest because it affects the pressure within the peripheral veins. For example, if the heart is beating weakly, the central venous pressure increases, and blood tends to back up in the venous network, causing its pressure to increase also. However, if the heart is beating forcefully, the central venous pressure and the pressure within the venous network decrease.

Sometimes, as a result of disease or injury, blood or tissue fluid may accumulate in the pericardial cavity and cause an increase in pressure within this cavity. This condition, called *acute cardiac tamponade,* can be life threatening. As the pressure around the heart increases, it may compress the heart, interfere with the flow of blood into its chambers, and prevent the pumping action of the chambers. An early symptom of acute cardiac tamponade may be increased central venous pressure, accompanied by visible engorgement of the veins in the neck.

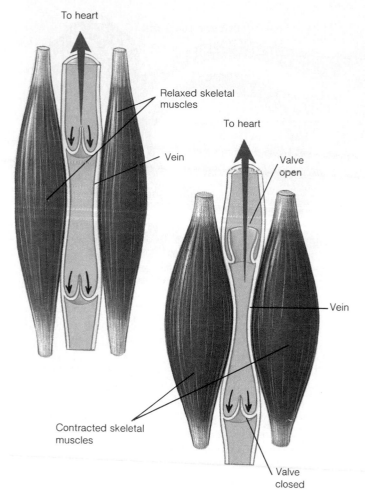

Figure 18.51 The massaging action of skeletal muscles helps move blood through the venous system toward the heart.

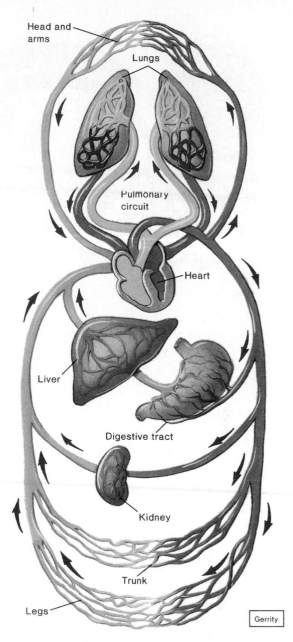

Figure 18.52 The pulmonary circuit consists of the vessels that carry blood between the heart and the lungs; all other vessels are included in the systemic circuit.

Other factors that increase the flow of blood into the right atrium, and thus cause the central venous pressure to become elevated, include an increase in blood volume or widespread venoconstriction.

1. What is the function of the venous valves?
2. How do skeletal muscles affect venous blood flow?
3. How do respiratory movements affect venous blood flow?
4. What factors stimulate venoconstriction?

Paths of Circulation

The blood vessels of the cardiovascular system can be divided into two major pathways—a pulmonary circuit and a systemic circuit. The **pulmonary circuit** consists of those vessels that carry the blood from the heart to the lungs and back to the heart. The **systemic circuit** is responsible for carrying the blood from the heart to all other parts of the body and back again (figure 18.52).

The circulatory pathways described in the following sections are those of an adult. The fetal pathways, which are somewhat different, are described in chapter 23.

Pulmonary Circuit

The blood enters the pulmonary circuit as it leaves the right ventricle through the pulmonary trunk. The pulmonary trunk extends upward and posteriorly from the heart, and about 5 centimeters above its origin, it divides into the right and left pulmonary arteries. These

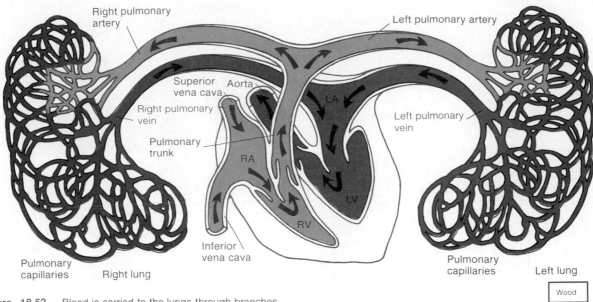

Figure 18.53 Blood is carried to the lungs through branches of the pulmonary arteries, and it returns to the heart through pulmonary veins.

branches penetrate the right and left lungs, respectively. Within the lungs, they divide into *lobar branches* (three on the right side and two on the left) that accompany the main divisions of the bronchi into the lobes of the lungs. After repeated divisions, the lobar branches give rise to arterioles that continue into the capillary networks associated with the walls of the alveoli (figure 18.53).

The blood in the arteries and arterioles of the pulmonary circuit has a relatively low concentration of oxygen and a relatively high concentration of carbon dioxide. As explained in chapter 16, gas exchanges occur between the blood and the air as the blood moves through the *pulmonary capillaries.*

Because the right ventricle contracts with less force than the left ventricle, the arterial pressure in the pulmonary circuit is less than that in the systemic circuit. Consequently, the pulmonary capillary pressure is relatively low.

The force tending to move fluid out of a pulmonary capillary is 23 mm Hg, while the force tending to pull fluid into it is 22 mm Hg; thus, such a capillary has a net filtration pressure of 1 mm Hg. This pressure is responsible for a slight, continuous flow of fluid into the narrow interstitial space between the pulmonary capillary and the alveolus.

The epithelial cells of the alveolar membranes are so tightly joined that ions of sodium, chlorine, and potassium, as well as molecules of glucose and urea, that enter the interstitial space usually fail to enter the alveoli. This helps maintain a relatively high osmotic pressure in the interstitial fluid. Consequently, any water that gets into the alveoli is rapidly moved back into the interstitial space by osmosis. This mechanism

helps keep the alveoli from filling with fluid, and they are said to remain *dry* (figure 18.54).

Fluid in the interstitial space may be drawn back into the pulmonary capillaries by the osmotic pressure of the blood, or it may be returned to the circulation by means of lymphatic vessels as described in chapter 19.

As a result of the gas exchanges occurring between the blood and the alveolar air, the blood entering the venules of the pulmonary circuit is rich in oxygen and low in carbon dioxide. These venules merge to form small veins, and they, in turn, converge to form still larger ones. Four *pulmonary veins,* two from each lung, return blood to the left atrium, and this completes the vascular loop of the pulmonary circuit.

A condition called *pulmonary edema* sometimes accompanies a failing left ventricle or a damaged bicuspid valve. For example, if the left ventricle is weak, it may be unable to move the normal volume of blood into the systemic circuit. As the blood backs up into the pulmonary circuit, the pressure in the pulmonary capillaries increases, and the interstitial spaces may become flooded with fluid. Increasing pressure in the interstitial fluid may cause the alveolar membranes to rupture, and fluid may enter the alveoli more rapidly than it can be removed. As a result, the alveolar surface available for gas exchange is reduced, and the person may suffocate.

Systemic Circuit

The freshly oxygenated blood received by the left atrium is forced into the systemic circuit by the contraction of the left ventricle. This circuit includes the

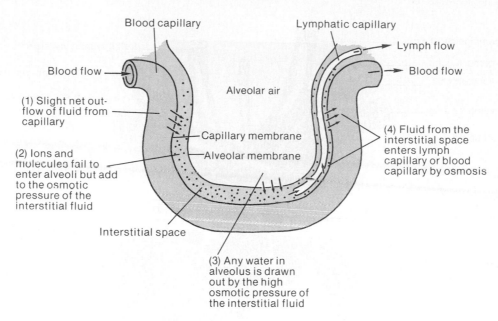

Blood capillary

Lymphatic capillary

Lymph flow

Blood flow

Blood flow

Alveolar air

(1) Slight net out-flow of fluid from capillary

(4) Fluid from the interstitial space enters lymph capillary or blood capillary by osmosis

Capillary membrane

Alveolar membrane

(2) Ions and molecules fail to enter alveoli but add to the osmotic pressure of the interstitial fluid

Interstitial space

(3) Any water in alveolus is drawn out by the high osmotic pressure of the interstitial fluid

Figure 18.54   The alveoli normally remain dry because the cells of the alveolar wall are tightly joined and water is drawn out of the alveoli by the high osmotic pressure of the interstitial fluid.

aorta and its branches that lead to all of the body tissues, as well as the companion system of veins that returns the blood to the right atrium.

1. Distinguish between the pulmonary and systemic circuits of the cardiovascular system.
2. Trace a drop of blood through the pulmonary circuit from the right ventricle.
3. Explain why the alveoli normally remain dry.

Arterial System

The **aorta** is the largest artery in the body. It extends upward from the left ventricle, arches over the heart to the left, and descends just in front of the vertebral column.

Principal Branches of the Aorta

The first portion of the aorta is called the *ascending aorta*. Located at its base are the three cusps of the aortic valve, and opposite each cusp is a swelling in the aortic wall called an **aortic sinus.** The right and left *coronary arteries* spring from two of these sinuses. Blood flow into these arteries is intermittent and is driven by the elastic recoil of the aortic wall following a contraction of the left ventricle.

As mentioned previously, several small structures called **aortic bodies** occur within the epithelial lining of the aortic sinuses. These bodies contain pressoreceptors that control blood pressure, and chemoreceptors that are sensitive to the blood concentrations of oxygen and carbon dioxide.

Three major arteries originate from the *arch of the aorta* (aortic arch). They are the brachiocephalic (innominate) artery, the left common carotid artery, and the left subclavian artery.

The **brachiocephalic** (brak″e-o-se-fal′ik) **artery** supplies blood to the tissues of the arm and head, as its name suggests. It is the first branch from the aortic arch, and rises upward through the mediastinum to a point near the junction of the sternum and the right clavicle. There it divides, giving rise to the right **common carotid** (kah-rot′id) **artery,** which carries blood to the right side of the neck and head, and the right **subclavian** (sub-kla′ve-an) **artery,** which leads into the right arm. Branches of the subclavian artery also supply blood to parts of the shoulder, neck, and head.

The left *common carotid artery* and the left *subclavian artery* are respectively the second and third branches of the aortic arch. They supply blood to regions on the left side of the body corresponding to those supplied by their counterparts on the right. (See figure 18.55 and reference plates 71, 72, and 73.)

Although the upper part of the *descending aorta* is positioned to the left of the midline, it gradually moves medially and finally lies directly in front of the vertebral column at the level of the twelfth thoracic vertebra.

The portion of the descending aorta above the diaphragm is known as the **thoracic** (tho-ras′ik) **aorta,** and it gives off numerous small branches to the thoracic wall and the thoracic visceral organs. These branches, the *bronchial, pericardial,* and *esophageal arteries,* supply blood to the structures for which they

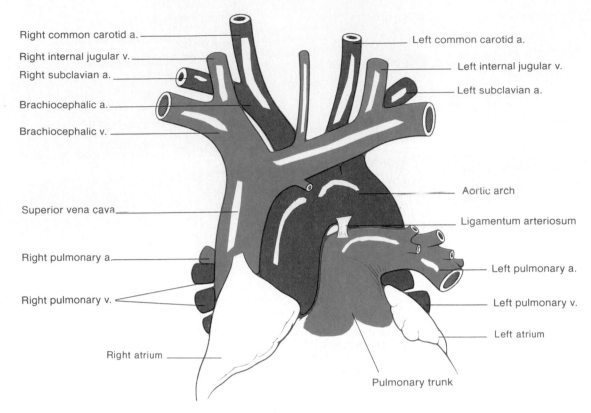

Right common carotid a.

Right internal jugular v.

Right subclavian a.

Brachiocephalic a.

Brachiocephalic v.

Superior vena cava

Right pulmonary a.

Right pulmonary v.

Right atrium

Left common carotid a.

Left internal jugular v.

Left subclavian a.

Aortic arch

Ligamentum arteriosum

Left pulmonary a.

Left pulmonary v.

Left atrium

Pulmonary trunk

Figure 18.55 The major blood vessels associated with the heart.

were named. Other branches become *mediastinal arteries,* supplying various tissues within the mediastinum, and *posterior intercostal arteries* that pass into the thoracic wall.

Below the diaphragm, the descending aorta becomes the **abdominal aorta,** and it gives off branches to the abdominal wall and various abdominal visceral organs. These branches include the following:

1. **Celiac** (se′le-ak) **artery.** This single vessel gives rise to the left *gastric, splenic,* and *hepatic arteries,* which supply upper portions of the digestive tube, the spleen, and the liver, respectively. (Note: The hepatic artery supplies the liver with about one-third of its blood flow, and this blood is oxygen-rich. The remaining two-thirds of the liver's blood flow arrives by means of the portal vein and is oxygen-poor.)

2. **Phrenic** (fren′ik) **arteries.** These paired arteries supply blood to the diaphragm.

3. **Superior mesenteric** (mes″en-ter′ik) **artery.** The superior mesenteric is a large, unpaired artery that branches to many parts of the intestinal tract, including the jejunum, ileum, cecum, ascending colon, and transverse colon.

4. **Suprarenal arteries.** This pair of vessels supplies blood to the adrenal glands.

5. **Renal** (re′nal) **arteries.** The renal arteries pass laterally from the aorta into the kidneys. Each artery then divides into several lobar branches within the kidney tissues.

6. **Gonadal** (go′nad-al) **arteries.** In a female, paired *ovarian arteries* arise from the aorta and pass into the pelvis to supply the ovaries. In a male, *spermatic arteries* originate in similar locations. They course downward and pass through the body wall by way of the *inguinal canal* to supply the testes.

7. **Inferior mesenteric artery.** Branches of this single artery lead to the descending colon, the sigmoid colon, and the rectum.

8. **Lumbar arteries.** Three or four pairs of lumbar arteries arise from the posterior surface of the aorta in the region of the lumbar vertebrae. These arteries supply various muscles of the skin and the posterior abdominal wall.

9. **Middle sacral** (sa′kral) **artery.** This small, single vessel descends medially from the aorta along the anterior surfaces of the lower lumbar vertebrae. It carries blood to the sacrum and coccyx.

The abdominal aorta terminates near the brim of the pelvis, where it divides into right and left *common*

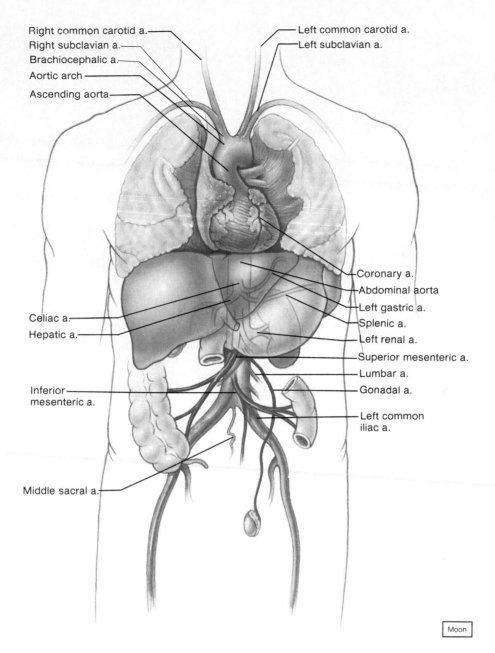

Figure 18.56 The principal branches of the aorta.

iliac arteries. These vessels supply blood to lower regions of the abdominal wall, the pelvic organs, and the lower extremities. (See figures 18.56 and 18.57.)

Chart 18.4 summarizes the main branches of the aorta.

Arteries to the Neck, Head, and Brain

Blood is supplied to parts within the neck, head, and brain through branches of the subclavian and common carotid arteries (figures 18.58 and 18.59). The main divisions of the subclavian artery to these regions are the vertebral, thyrocervical, and costocervical arteries.

The common carotid artery communicates to these parts by means of the internal and external carotid arteries.

The **vertebral** (ver'te-bral) **arteries** arise from the subclavian arteries in the base of the neck near the tips of the lungs. They pass upward through the foramina of the transverse processes of the cervical vertebrae and enter the skull by way of the foramen magnum. Along their paths, these vessels supply blood to vertebrae and to the ligaments and muscles associated with them.

Within the cranial cavity, the vertebral arteries unite to form a single *basilar artery*. This vessel passes along the ventral brain stem and gives rise to branches

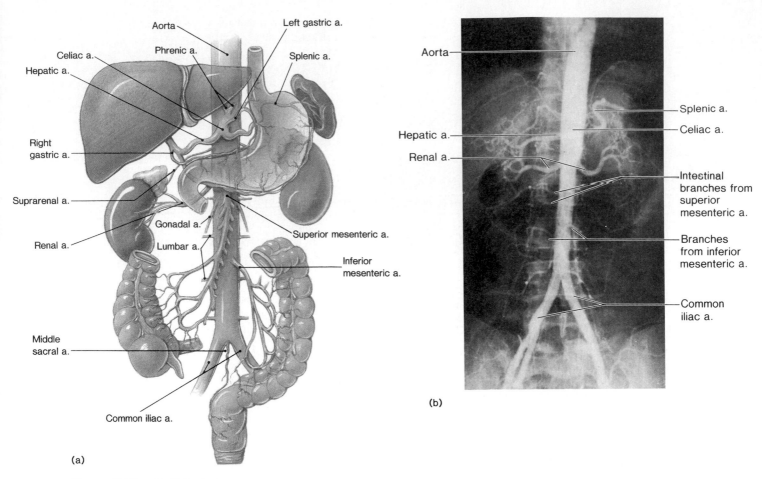

Figure 18.57 (a) Abdominal aorta and its major branches; (b) angiogram (X-ray film) of the abdominal aorta.

Portion of aorta	Major branch	General regions or organs supplied	Portion of aorta	Major branch	General regions or organs supplied
Ascending aorta	Right and left coronary arteries	Heart	Abdominal aorta	Celiac artery	Organs of upper digestive system
Arch of aorta	Brachiocephalic artery	Right arm, right side of head		Phrenic artery	Diaphragm
	Left common carotid artery	Left side of head		Superior mesenteric artery	Portions of small and large intestines
	Left subclavian artery	Left arm		Suprarenal artery	Adrenal gland
Descending aorta				Renal artery	Kidney
Thoracic aorta	Bronchial artery	Bronchi		Gonadal artery	Ovary or testis
	Pericardial artery	Pericardium		Inferior mesenteric artery	Lower portions of large intestine
	Esophageal artery	Esophagus		Lumbar artery	Posterior abdominal wall
	Mediastinal artery	Mediastinum		Middle sacral artery	Sacrum and coccyx
	Posterior intercostal artery	Thoracic wall		Common iliac artery	Lower abdominal wall, pelvic organs, and leg

CHART 18.4 The aorta and its principal branches

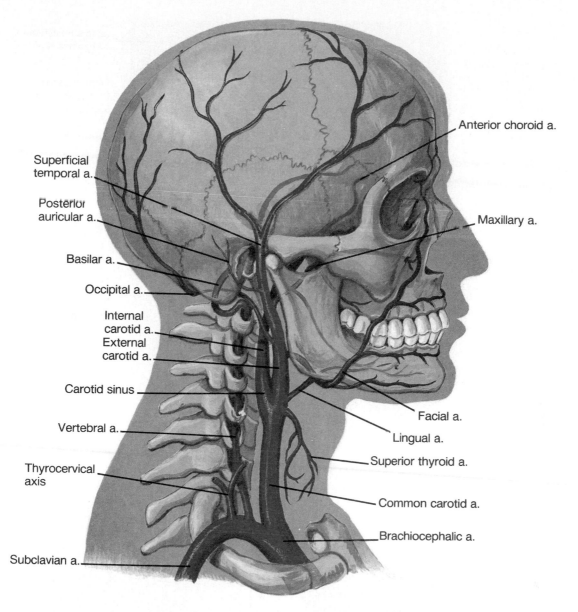

Superficial
temporal a.

Posterior
auricular a.

Basilar a.

Occipital a.

Internal
carotid a.

External
carotid a.

Carotid sinus

Vertebral a.

Thyrocervical
axis

Subclavian a.

Anterior choroid a.

Maxillary a.

Facial a.

Lingual a.

Superior thyroid a.

Common carotid a.

Brachiocephalic a.

Figure 18.58 The main arteries of the head and neck. (Note that the clavicle has been removed.)

leading to the pons, midbrain, and cerebellum. The basilar artery terminates by dividing into two *posterior cerebral arteries* that supply portions of the occipital and temporal lobes of the cerebrum. The posterior cerebral arteries also help form an arterial circle at the base of the brain, the **circle of Willis,** which forms a connection between the vertebral artery and internal carotid artery systems (figure 18.60). The union of these systems provides alternate pathways through which blood can reach brain tissues in the event of an arterial occlusion.

The **thyrocervical** (thi″ro-ser′vĭ-kal) **arteries** are short vessels that give off branches to the thyroid glands, parathyroid glands, larynx, trachea, esophagus, and pharynx, as well as to various muscles in the neck, shoulder, and back.

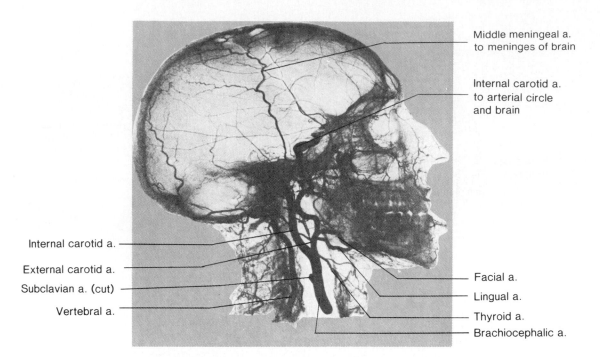

Figure 18.59 An angiogram of the arteries associated with the head.

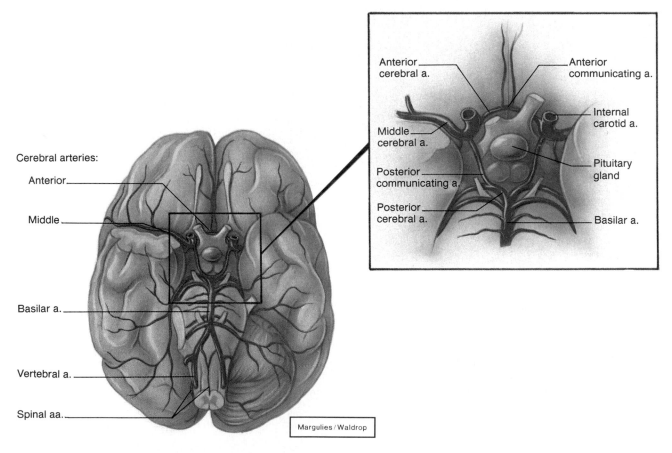

Margulies/Waldrop

Figure 18.60 The circle of Willis is formed by the anterior cerebral arteries, which are connected by the anterior communicating artery, and by the posterior cerebral arteries, which are connected to the internal carotid arteries by the posterior communicating arteries.

The **costocervical** (kos"to-ser'vĭ-kal) **arteries,** which are the third vessels to branch from the subclavians, carry blood to muscles in the neck, back, and thoracic wall.

The left and right *common carotid arteries* ascend deeply within the neck on either side. At the level of the upper laryngeal border, they divide to form the internal and external carotid arteries.

The **external carotid artery** courses upward on the side of the head, giving off branches to various structures in the neck, face, jaw, scalp, and base of the skull. The main vessels that originate from this artery include the following:

1. *Superior thyroid artery* to the hyoid bone, larynx, and thyroid gland.
2. *Lingual artery* to the tongue, muscles of the tongue, and salivary glands beneath the tongue.
3. *Facial artery* to the pharynx, palate, chin, lips, and nose.
4. *Occipital artery* to the scalp on the back of the skull, the meninges, the mastoid process, and various muscles in the neck.
5. *Posterior auricular artery* to the ear and the scalp over the ear.

The external carotid artery terminates by dividing into *maxillary* and *superficial temporal arteries*. The maxillary artery supplies blood to the teeth, gums, jaws, cheek, nasal cavity, eyelids, and meninges. The temporal artery extends to the parotid salivary gland and to various surface regions of the face and scalp.

The **internal carotid artery** follows a deep course upward along the pharynx to the base of the skull. Entering the cranial cavity, it provides the major blood supply to the brain. Its major branches include the following:

1. *Ophthalmic artery* to the eyeball and to various muscles and accessory organs within the orbit.
2. *Posterior communicating artery* that forms part of the circle of Willis.
3. *Anterior choroid artery* to the choroid plexus within the lateral ventricle of the brain and to a variety of nerve structures within the brain.

The internal carotid artery terminates by dividing into *anterior* and *middle cerebral arteries*. The middle cerebral artery passes through the lateral sulcus and supplies the lateral surface of the cerebrum, including the primary motor and sensory areas of the face and arms, the optic radiations, and the speech area. (See chapter 11.) The anterior cerebral artery extends anteriorly between the cerebral hemispheres and supplies the medial surface of the brain.

Near the base of each internal carotid artery is an enlargement called a **carotid sinus.** Like the aortic

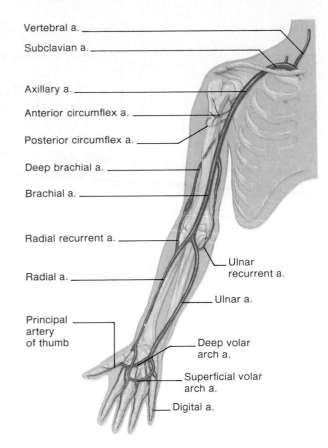

Figure 18.61 The main arteries of the shoulder and arm.

sinuses, these structures contain pressoreceptors that function in the reflex control of the blood pressure. A number of small epithelial masses, called **carotid bodies,** also occur in the wall of the carotid sinus. These bodies are very vascular and contain chemoreceptors that act with the chemoreceptors of the aortic bodies to regulate various circulatory and respiratory actions.

Arteries to the Shoulder and Arm

The subclavian artery, after giving off branches to the neck, continues into the upper arm (figure 18.61). It passes between the clavicle and the first rib, and becomes the axillary artery.

The **axillary artery** supplies branches to structures in the axilla and the chest wall, including the skin of the shoulder, part of the mammary gland, the upper end of the humerus, the shoulder joint, and various muscles in the back, shoulder, and chest. As this vessel leaves the axilla, it becomes the brachial artery.

The **brachial artery** courses along the humerus to the elbow. It gives rise to a *deep brachial artery* that curves posteriorly around the humerus and supplies the triceps muscle. Shorter branches pass into the muscles on the anterior side of the upper arm, while others descend on each side to the elbow and interconnect with

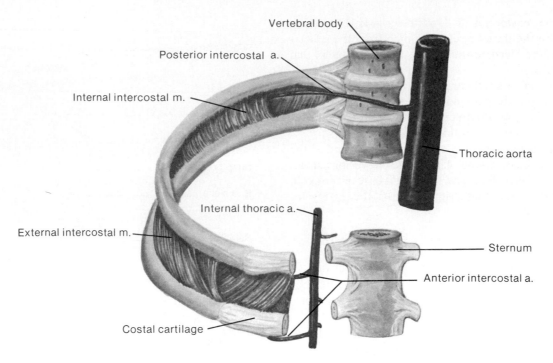

Figure 18.62 Arteries that supply the thoracic wall. (*a.* stands for *artery; m.* for *muscle.*)

arteries in the forearm. The resulting arterial network allows blood to reach the lower arm even if a portion of the distal brachial artery becomes obstructed.

Within the elbow, the brachial artery divides into an ulnar artery and a radial artery. The **ulnar** (ul′nar) **artery** leads downward on the ulnar side of the forearm to the wrist. Some of its branches join the anastomosis around the elbow joint, while others supply blood to flexor and extensor muscles in the lower arm.

The **radial** (ra′de-al) **artery,** a continuation of the brachial artery, travels along the radial side of the forearm to the wrist. As it nears the wrist, it comes close to the surface and provides a convenient vessel for taking the pulse (radial pulse). Various branches of the radial artery join the anastomosis of the elbow and supply the lateral muscles of the forearm.

At the wrist, the branches of the ulnar and radial arteries join to form an interconnecting network of vessels. Arteries arising from this network supply blood to structures in the wrist, hand, and fingers.

Arteries to the Thoracic and Abdominal Walls

Blood reaches the thoracic wall through several vessels, including branches from the subclavian artery and the thoracic aorta (figure 18.62).

The subclavian artery contributes to this supply through a branch called the **internal thoracic artery.** This vessel originates in the base of the neck and passes downward on the pleura and behind the cartilages of the upper six ribs. It gives off two *anterior intercostal*

arteries to each of the upper six intercostal spaces; these two arteries supply the intercostal muscles, other intercostal tissues, and the mammary glands.

The *posterior intercostal arteries* arise from the thoracic aorta and enter the intercostal spaces between the third through the eleventh ribs. These arteries give off branches that supply the intercostal muscles, the vertebrae, the spinal cord, and various deep muscles of the back.

The blood supply to the anterior abdominal wall is provided primarily by branches of the *internal thoracic* and *external iliac arteries.* Structures in the posterior and lateral abdominal wall are supplied by paired vessels originating from the abdominal aorta, including the *phrenic* and *lumbar arteries* mentioned previously.

Arteries to the Pelvis and Leg

The abdominal aorta divides to form the **common iliac arteries** at the level of the pelvic brim, and these vessels provide blood to the pelvic organs, gluteal region, and legs.

Each common iliac artery descends a short distance and divides into an internal (hypogastric) branch and an external branch. The **internal iliac artery** gives off numerous branches to various pelvic muscles and visceral structures, as well as to the gluteal muscles and the external genitalia. Important branches of this vessel are shown in figure 18.63. They include the following:

1. *Iliolumbar artery* to the ilium and muscles of the back.

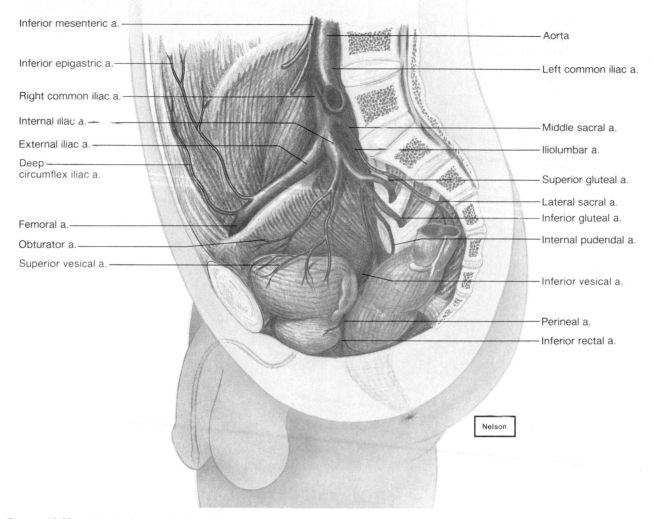

Inferior mesenteric a.

Inferior epigastric a.

Right common iliac a.

Internal iliac a.

External iliac a.

Deep circumflex iliac a.

Femoral a.

Obturator a.

Superior vesical a.

Aorta

Left common iliac a.

Middle sacral a.

Iliolumbar a.

Superior gluteal a.

Lateral sacral a.

Inferior gluteal a.

Internal pudendal a.

Inferior vesical a.

Perineal a.

Inferior rectal a.

Nelson

Figure 18.63 Arteries that supply the pelvic region.

2. *Superior* and *inferior gluteal arteries* to the gluteal muscles, pelvic muscles, and skin of the buttocks.
3. *Internal pudendal artery* to muscles in the distal portion of the alimentary canal, the external genitalia, and the hip joint.
4. *Superior* and *inferior vesical arteries* to the urinary bladder. In males, these vessels also supply the seminal vesicles and the prostate gland.
5. *Middle rectal artery* to the rectum.
6. *Uterine artery* to the uterus and vagina in females.

The **external iliac artery** provides the main blood supply to the legs (figure 18.64). It passes downward along the brim of the pelvis and gives off two large branches—an *inferior epigastric artery* and a *deep circumflex iliac artery*. These vessels supply the muscles and skin in the lower abdominal wall.

Midway between the symphysis pubis and the anterior superior iliac spine of the ilium, the external iliac artery becomes the femoral artery.

The **femoral** (fem'or-al) **artery,** which passes fairly close to the anterior surface of the upper thigh, gives off many branches to muscles and superficial tissues of the thigh. These branches also supply the skin of the groin and the lower abdominal wall. Important subdivisions of the femoral artery include the following:

1. *Superficial circumflex iliac artery* to the lymph nodes and skin of the groin.
2. *Superficial epigastric artery* to the skin of the lower abdominal wall.
3. *Superficial* and *deep external pudendal arteries* to the skin of the lower abdomen and external genitalia.
4. *Profunda femoris artery* (the largest branch of the femoral artery) to the hip joint and various muscles of the thigh.

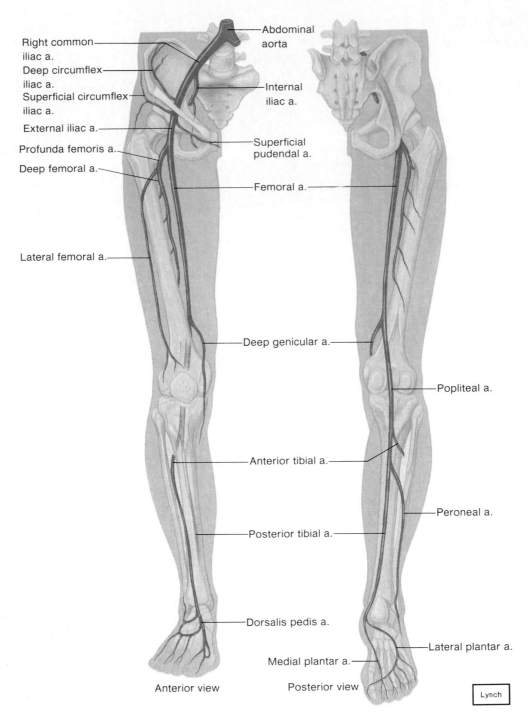

Right common iliac a.

Deep circumflex iliac a.

Superficial circumflex iliac a.

External iliac a.

Profunda femoris a.

Deep femoral a.

Lateral femoral a.

Abdominal aorta

Internal iliac a.

Superficial pudendal a.

Femoral a.

Deep genicular a.

Popliteal a.

Anterior tibial a.

Peroneal a.

Posterior tibial a.

Dorsalis pedis a.

Lateral plantar a.

Medial plantar a.

Anterior view

Posterior view

Lynch

Figure 18.64 Main branches of the external iliac artery.

5. *Deep genicular artery* to distal ends of thigh muscles and to an anastomosis around the knee joint.

As the femoral artery reaches the proximal border of the space behind the knee (popliteal fossa), it becomes the **popliteal** (pop-lit′e-al) **artery.** Branches of this artery supply blood to the knee joint and to certain muscles in the thigh and calf. Also, many of its branches join the anastomosis of the knee and help provide alternate pathways for blood in the case of arterial obstructions. At the lower border of the popliteal fossa, the popliteal artery divides into the anterior and posterior tibial arteries.

The **anterior tibial** (tib′e-al) **artery** passes downward between the tibia and the fibula, giving off branches to the skin and muscles in anterior and lateral regions of the lower leg. It also communicates with the anastomosis of the knee and with a network of arteries around the ankle. This vessel continues into the foot as the *dorsalis pedis artery,* which supplies blood to the foot and toes.

The **posterior tibial artery,** the larger of the two popliteal branches, descends beneath the calf muscles, giving off branches to the skin, muscles, and other tissues of the lower leg along the way. Some of these vessels join the anastomoses of the knee and ankle. As it passes between the medial malleolus and the heel, the posterior tibial artery divides into the *medial and lateral plantar arteries.* Branches from these arteries supply blood to tissues of the heel, foot, and toes.

The largest branch of the posterior tibial artery is the *peroneal artery,* which travels downward along the fibula and contributes to the anastomosis of the ankle.

The major vessels of the arterial system are shown in figure 18.65.

---

1. Describe the structure of the aorta.
2. Name the vessels that arise from the aortic arch.
3. Name the branches of the thoracic and abdominal aorta.
4. Which vessels supply blood to the head? To the arm? To the abdominal wall? To the leg?

A healthy artery has a smooth inner lining, and the blood flow through it is undisturbed. However, as a consequence of atherosclerosis, regions of the inner artery lining become thickened, and this may cause the blood flow to become turbulent and the blood velocity to change. Such an arterial abnormality can sometimes be detected using a noninvasive procedure that combines ultrasonography (chapter 1), which produces an image of the artery, with an instrument (Doppler device) that senses the velocity and turbulence of the blood. This technique is commonly used to assess the condition of the carotid and femoral arteries.

Venous System

Venous circulation is responsible for returning the blood to the heart after exchanges of gases, nutrients, and wastes have occurred between the blood and body cells.

Characteristics of Venous Pathways

The vessels of the venous system begin with the merging of capillaries into venules, venules into small veins, and small veins into larger ones. Unlike the arterial pathways, however, those of the venous system are difficult to follow. This is because the vessels are commonly interconnected in irregular networks, so that many unnamed tributaries may join to form a relatively large vein.

On the other hand, the larger veins typically parallel the courses taken by named arteries, and these veins often have the same names as their companions in the arterial system. Thus, with some exceptions, the name of a major artery also provides the name of the vein next to it. For example, the renal vein parallels the renal artery, the common iliac vein accompanies the common iliac artery, and so forth.

The veins that carry the blood from the lungs and myocardium back to the heart have already been described. The veins from all the other parts of the body converge into two major pathways, which lead to the right atrium. They are the *superior* and *inferior venae cavae.*

Veins from the Head, Neck, and Brain

The blood from the face, scalp, and superficial regions of the neck is drained by the **external jugular** (jug'u-lar) **veins.** These vessels descend on either side of the neck, passing over the sternocleidomastoid muscles and beneath the platysma. They empty into the *right* and *left subclavian veins* in the base of the neck (figure 18.66).

The **internal jugular veins,** which are somewhat larger than the external jugular veins, arise from numerous veins and venous sinuses of the brain and from deep veins in various parts of the face and neck. They pass downward through the neck beside the common carotid arteries and also join the subclavian veins. These unions of the internal jugular and subclavian veins form large **brachiocephalic** (innominate) **veins** on each side. These vessels then merge in the mediastinum and give rise to the **superior vena cava,** which enters the right atrium.

The superior vena cava sometimes becomes compressed by the growth of a lung cancer, the enlargement of a lymph node, or an aortic aneurysm. This condition may interfere with return of blood from the upper body to the heart. It is characterized by pain, shortness of breath, distension of veins draining into the superior vena cava, and swelling of tissues in the face, head, and arms. In severe cases, compression of the superior vena cava may result in such restricted blood flow to the brain that the person's life is threatened.

Veins from the Arm and Shoulder

The arm is drained by a set of deep veins and a set of superficial ones. The deep veins generally parallel the arteries in each region and are given similar names, such as the *radial vein, ulnar vein, brachial vein,* and *axillary vein.* The superficial veins are interconnected in complex networks just beneath the skin. They also

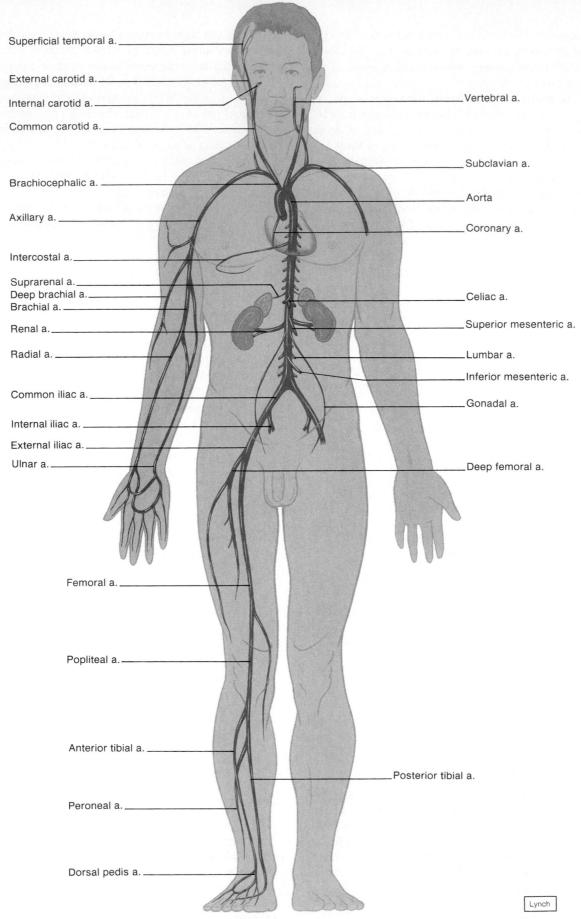

Superficial temporal a.

External carotid a.

Internal carotid a.

Common carotid a.

Vertebral a.

Subclavian a.

Brachiocephalic a.

Aorta

Axillary a.

Coronary a.

Intercostal a.

Suprarenal a.
Deep brachial a.
Brachial a.

Celiac a.

Renal a.

Superior mesenteric a.

Radial a.

Lumbar a.

Inferior mesenteric a.

Common iliac a.

Gonadal a.

Internal iliac a.

External iliac a.

Ulnar a.

Deep femoral a.

Femoral a.

Popliteal a.

Anterior tibial a.

Posterior tibial a.

Peroneal a.

Dorsal pedis a.

Lynch

Figure 18.65   Major vessels of the arterial system.

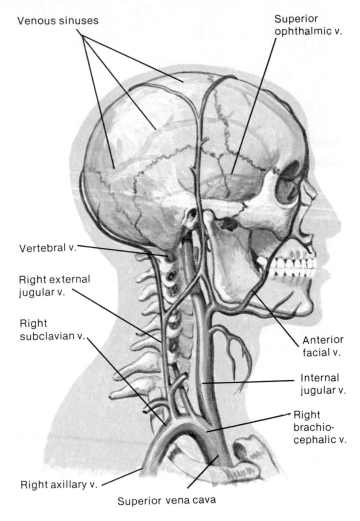

Figure 18.66 The main veins of the head and neck. (Note that the clavicle has been removed.) (*v.* stands for *vein.*)

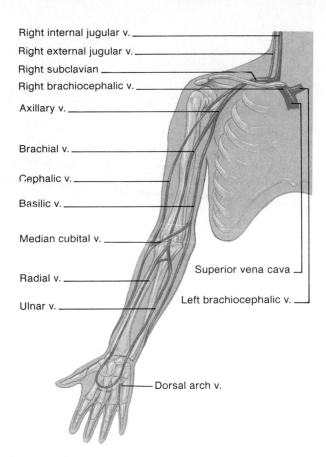

Figure 18.67 The main veins of the arm and shoulder.

communicate with the deep vessels of the arm, providing many alternate pathways through which the blood can leave the tissues (figure 18.67).

The main vessels of the superficial network are the basilic and cephalic veins. They arise from anastomoses in the hand and wrist on the ulnar and radial sides, respectively.

The **basilic vein** passes along the back of the forearm on the ulnar side for a distance and then curves forward to the anterior surface below the elbow. It continues ascending on the medial side until it reaches the middle of the upper arm. There it penetrates the tissues deeply and joins the *brachial vein.* As the basilic and brachial veins merge, they form the *axillary vein.*

The **cephalic vein** courses upward on the lateral side of the arm from the hand to the shoulder. In the shoulder it pierces the tissues and empties into the axillary vein. Beyond the axilla, the axillary vein becomes the *subclavian vein.*

In the bend of the elbow, a *median cubital vein* ascends from the cephalic vein on the lateral side of the

arm to the basilic vein on the medial side. This vein is usually visible and is relatively large. It is often used as a site for *venipuncture,* when it is necessary to remove a sample of blood for examination or to add fluids to the blood.

Veins from the Abdominal and Thoracic Walls

The abdominal and thoracic walls are drained mainly by tributaries of the brachiocephalic and azygos veins. For example, the *brachiocephalic vein* receives blood from the *internal thoracic vein,* which generally drains the tissues supplied by the internal thoracic artery. Some *intercostal veins* also empty into the brachiocephalic vein (figure 18.68).

The **azygos** (az'i-gos) **vein** originates in the dorsal abdominal wall and ascends through the mediastinum on the right side of the vertebral column to join the superior vena cava. It drains most of the muscular tissue in the abdominal and thoracic walls.

Tributaries of the azygos vein include the *posterior intercostal veins* on the right side, which drain the intercostal spaces, and the *superior* and *inferior hemiazygos veins,* which receive blood from the posterior intercostal veins on the left. The right and left

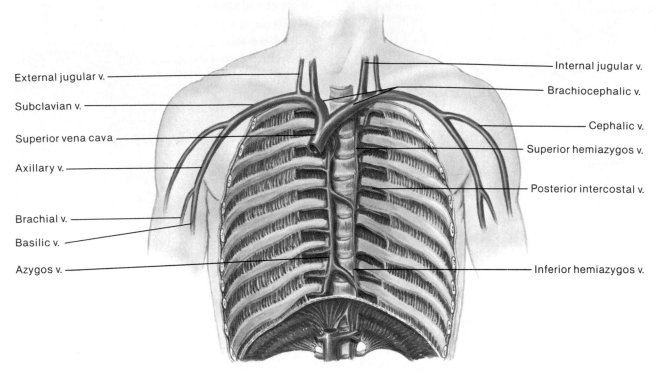

External jugular v.

Subclavian v.

Superior vena cava

Axillary v.

Brachial v.

Basilic v.

Azygos v.

Internal jugular v.

Brachiocephalic v.

Cephalic v.

Superior hemiazygos v.

Posterior intercostal v.

Inferior hemiazygos v.

Figure 18.68 Veins that drain the thoracic wall.

ascending lumbar veins, with tributaries that include vessels from the lumbar and sacral regions, also connect to the azygos system.

Veins from the Abdominal Viscera

Although veins usually carry the blood directly to the atria of the heart, those that drain the abdominal viscera are exceptions (figure 18.69). They originate in the capillary networks of the stomach, intestines, pancreas, and spleen, and carry the blood from these organs through a **portal** (por′tal) **vein** to the liver. There the blood enters capillarylike **hepatic sinusoids** (hĕ-pat′ik si′nŭ-soids). This unique venous pathway is called the **hepatic portal system.** (See figure 18.70.)

The tributaries of the portal vein include the following vessels:

1. Right and left *gastric veins* from the stomach.
2. *Superior mesenteric vein* from the small intestine, ascending colon, and transverse colon.
3. *Splenic vein* from a convergence of several veins draining the spleen, the pancreas, and a portion of the stomach. Its largest tributary, the *inferior mesenteric vein,* brings blood upward from the descending colon, sigmoid colon, and rectum.

About 80% of the blood flowing to the liver in the hepatic portal system comes from the capillaries in the stomach and intestines, and is oxygen-poor, but is rich in nutrients. As discussed in chapter 14, the liver acts on these nutrients in a variety of ways. For example, the liver helps regulate the blood glucose concentration by converting excesses of glucose into glycogen and fat for storage, or by converting glycogen into glucose when the blood glucose concentration drops below normal.

Similarly, the liver helps regulate the blood concentrations of recently absorbed amino acids and lipids by modifying their molecules into forms usable by cells, by oxidizing them, or by changing them into storage forms. The liver also stores certain vitamins and detoxifies harmful substances.

The blood in the portal vein nearly always contains bacteria that have entered through intestinal capillaries. These microorganisms are removed by the phagocytic action of large *Kupffer cells* lining the hepatic sinusoids. Consequently, nearly all of the bacteria have been removed from the portal blood before it leaves the liver.

After passing through the hepatic sinusoids of the liver, the blood in the hepatic portal system is carried through a series of merging vessels into **hepatic veins.** These veins empty into the *inferior vena cava,* and thus the blood is returned to the general circulation.

Other veins also empty into the inferior vena cava as it ascends through the abdomen. They include the *lumbar, gonadal, renal, suprarenal,* and *phrenic veins.* Generally, these vessels drain regions that are supplied by arteries with corresponding names.

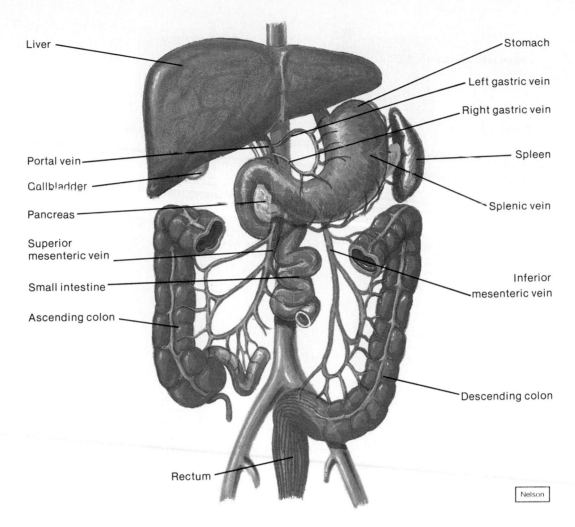

Figure 18.69 Veins that drain the abdominal viscera.

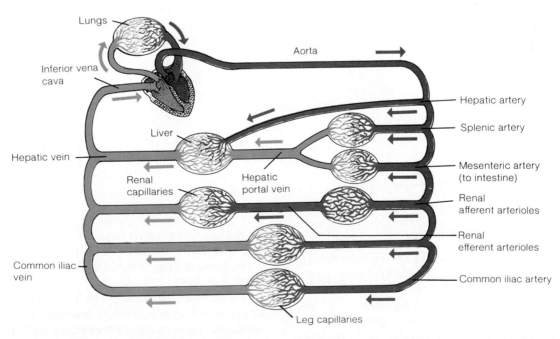

Figure 18.70 In this schematic drawing of the circulatory system, note how the hepatic portal vein drains one set of capillaries and leads to another set.

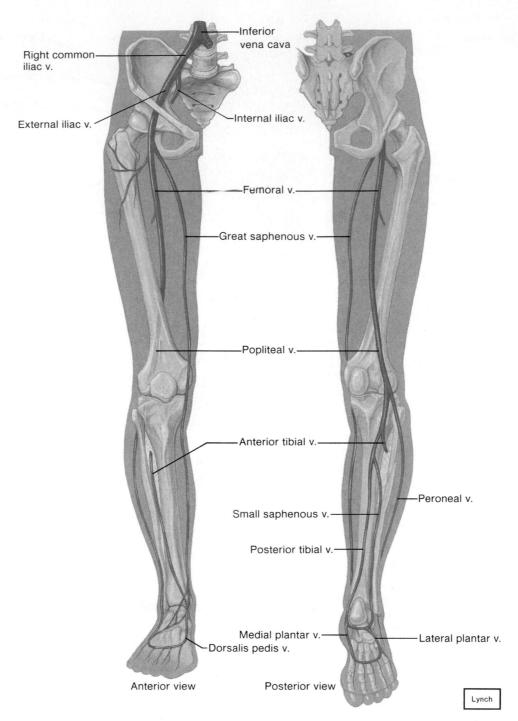

Figure 18.71   The main veins of the leg and pelvis.

Veins from the Leg and Pelvis

As in the arm, veins that drain the blood from the leg can be divided into deep and superficial groups (figure 18.71).

The deep veins of the lower leg, such as the *anterior* and *posterior tibial veins,* have names that correspond with the arteries they accompany. At the level of the knee, these vessels form a single trunk, the **popliteal vein.** This vein continues upward through the thigh as the **femoral vein,** which, in turn, becomes the **external iliac vein.**

The superficial veins of the foot and leg interconnect to form a complex network beneath the skin. These vessels drain into two major trunks: the small and great saphenous veins.

The **small saphenous** (sah-fe′nus) **vein** begins in the lateral portion of the foot and passes upward behind the lateral malleolus. It ascends along the back of the calf, enters the popliteal fossa, and joins the *popliteal vein.*

The **great saphenous vein,** which is the longest vein in the body, originates on the medial side of the foot. It ascends in front of the medial malleolus and extends upward along the medial side of the leg. In the thigh just below the inguinal ligament, it penetrates deeply and joins the femoral vein. Near its termination, the great saphenous vein receives tributaries from a number of vessels that drain the upper thigh, groin, and lower abdominal wall.

In addition to communicating freely with each other, the saphenous veins communicate extensively with the deep veins of the leg. As a result, there are many pathways by which blood can be returned to the heart from the lower extremities.

> Whenever a person stands, the saphenous veins are subjected to increased blood pressure because of gravitational forces. When such pressure is prolonged, these veins are especially prone to developing abnormal dilations and becoming *varicose veins.* The saphenous veins commonly serve as a source of blood vessel material for vascular bypass surgery.

In the pelvic region, the blood is carried away from organs of the reproductive, urinary, and digestive systems by vessels leading to the **internal iliac vein.** This vein is formed by tributaries corresponding to the branches of the internal iliac artery, such as the *gluteal, pudendal, vesical, rectal, uterine,* and *vaginal veins.* Typically, these veins have many interconnections and form complex networks (plexuses) in the regions of the rectum, urinary bladder, and prostate gland (in the male) or uterus and vagina (in the female).

The internal iliac veins originate deep within the pelvis and ascend to the pelvic brim. There they unite with the right and left *external iliac veins* to form the **common iliac veins.** These vessels, in turn, merge to produce the *inferior vena cava* at the level of the fifth lumbar vertebra.

Figure 18.72 shows the major vessels of the venous system.

---

1. Name the veins that return the blood to the right atrium.
2. What major veins drain the blood from the head? From the arm? From the abdominal viscera? From the leg?

Clinical Terms Related to the Cardiovascular System

anastomosis (ah-nas″to-mo′sis) An interconnection between two blood vessels, sometimes produced surgically.

aneurysm (an′u-rizm) A saclike swelling in the wall of a blood vessel, usually an artery.

angiospasm (an′je-o-spazm″) A muscular spasm in the wall of a blood vessel.

arteriography (ar″te-re-og′rah-fe) The injection of radiopaque solution into the vascular system for an X-ray examination of arteries.

asystole (a-sis′to-le) A condition in which the myocardium fails to contract.

cardiac tamponade (kar′de-ak tam″po-nād′) Compression of the heart by an accumulation of fluid within the pericardial cavity.

congestive heart failure (kon-jes′tiv hart fāl′yer) A condition in which the heart is unable to pump an adequate amount of blood to the body cells.

cor pulmonale (kor pul-mo-na′le) A heart-lung disorder characterized by pulmonary hypertension and hypertrophy of the right ventricle.

embolectomy (em″bo-lek′to-me) The removal of an embolus through an incision in a blood vessel.

endarterectomy (en″dar-ter-ek′to-me) The removal of the inner wall of an artery to reduce an arterial occlusion.

palpitation (pal″pi-ta′shun) Awareness of a heartbeat that is unusually rapid, strong, or irregular.

pericardiectomy (per″i-kar″de-ek′to-me) An excision of the pericardium.

phlebitis (fle-bi′tis) An inflammation of a vein, usually in the legs.

phlebotomy (fle-bot′o-me) An incision of a vein for the purpose of withdrawing blood.

sinus rhythm (si′nus rithm) The normal cardiac rhythm regulated by the S-A node.

thrombophlebitis (throm″bo-fle-bi′tis) The formation of a blood clot in a vein in response to inflammation of the venous wall.

valvotomy (val-vot′o-me) An incision of a valve.

venography (ve-nog′rah-fe) An injection of radiopaque solution into the vascular system for X-ray examination of veins.

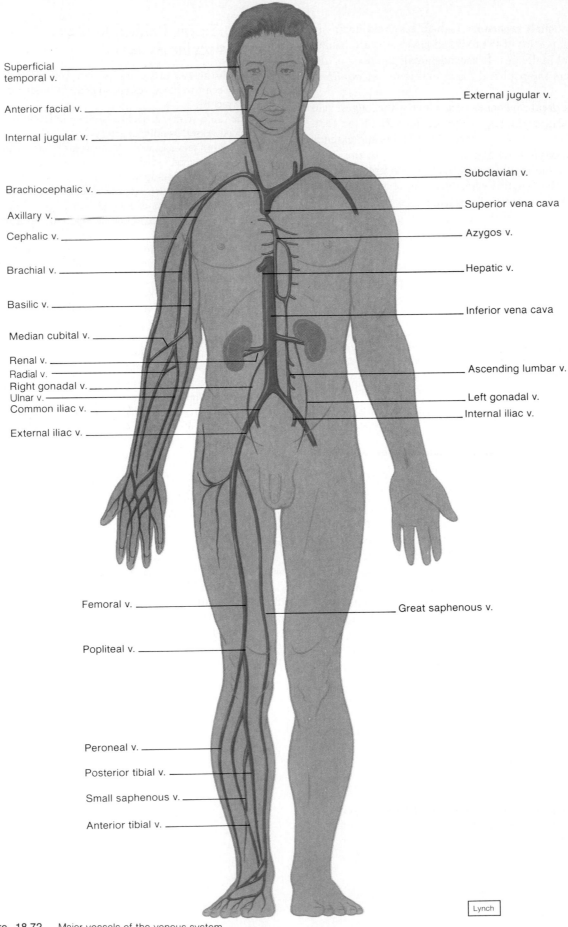

Superficial temporal v.

Anterior facial v.

Internal jugular v.

Brachiocephalic v.

Axillary v.

Cephalic v.

Brachial v.

Basilic v.

Median cubital v.

Renal v.

Radial v.

Right gonadal v.

Ulnar v.

Common iliac v.

External iliac v.

External jugular v.

Subclavian v.

Superior vena cava

Azygos v.

Hepatic v.

Inferior vena cava

Ascending lumbar v.

Left gonadal v.

Internal iliac v.

Femoral v.

Great saphenous v.

Popliteal v.

Peroneal v.

Posterior tibial v.

Small saphenous v.

Anterior tibial v.

Lynch

Figure 18.72   Major vessels of the venous system.

Chapter Summary

Introduction (page 652)

The cardiovascular system is vital for providing oxygen and nutrients to tissues, and for removing wastes.

Structure of the Heart (page 652)

1. Size and location of the heart
   a. The heart is about 14 centimeters long and 9 centimeters wide.
   b. It is located within the mediastinum and rests on the diaphragm.
2. Coverings of the heart
   a. The heart is enclosed in a layered pericardium.
   b. The pericardial cavity is a potential space between the visceral and parietal layers of the pericardium.
3. Wall of the heart
   a. The wall of the heart is composed of three layers.
   b. These layers include an epicardium, a myocardium, and an endocardium.
4. Heart chambers and valves
   a. The heart is divided into four chambers—two atria and two ventricles—that communicate through atrioventricular orifices on each side.
   b. Right chambers and valves
      (1) The right atrium receives blood from the venae cavae and coronary sinus.
      (2) The right atrioventricular orifice is guarded by the tricuspid valve.
      (3) The right ventricle pumps blood into the pulmonary trunk.
      (4) The base of the pulmonary trunk is guarded by a pulmonary valve.
   c. Left chambers and valves
      (1) The left atrium receives blood from the pulmonary veins.
      (2) The left atrioventricular orifice is guarded by the bicuspid valve.
      (3) The left ventricle pumps blood into the aorta.
      (4) The base of the aorta is guarded by an aortic valve.
5. Skeleton of the heart
   a. The skeleton of the heart consists of fibrous rings that enclose the bases of the pulmonary artery, aorta, and atrioventricular orifices.
   b. The fibrous rings provide attachments for valves and muscle fibers, and prevent the orifices from dilating excessively during ventricular contractions.
6. Path of blood through the heart
   a. Blood that is relatively low in oxygen concentration and high in carbon dioxide concentration enters the right side of the heart from the venae cavae and is pumped into the pulmonary circulation.
   b. After the blood is oxygenated in the lungs and some of its carbon dioxide is removed, it returns to the left side of the heart through the pulmonary veins.
   c. From the left ventricle, it moves into the aorta.
7. Blood supply to the heart
   a. Blood is supplied to the myocardium through the coronary arteries.
   b. It is returned to the right atrium through the cardiac veins and coronary sinus.

Actions of the Heart (page 661)

1. Cardiac cycle
   a. The atria contract while the ventricles relax; the ventricles contract while the atria relax.
   b. Pressure within the chambers rises and falls in repeated cycles.
2. Heart sounds
   a. Heart sounds can be described as *lub*-dup.
   b. Heart sounds are due to the vibrations produced by the blood and valve movements.
   c. The first part of the sound occurs as A-V valves are closing, and the second part is associated with the closing of pulmonary and aortic valves.
3. Cardiac muscle fibers
   a. Cardiac muscle fibers are interconnected to form a functional syncytium.
   b. If any part of the syncytium is stimulated, the whole structure contracts as a unit.
   c. Except for a small region in the floor of the right atrium, the atrial syncytium is separated from the ventricular syncytium by the fibrous skeleton.
4. Cardiac conduction system
   a. This system is composed of specialized cardiac muscle tissue and functions to initiate and conduct depolarization waves through the myocardium.
   b. Impulses from the S-A node pass slowly to the A-V node; impulses travel rapidly along the A-V bundle and Purkinje fibers.
   c. Muscle fibers in the ventricular walls are arranged in whorls that squeeze blood out of the ventricles when they contract.
5. Regulation of the cardiac cycle
   a. The heartbeat is affected by physical exercise, body temperature, and concentration of various ions.
   b. The S-A and A-V nodes are innervated by branches of sympathetic and parasympathetic nerve fibers.
      (1) Parasympathetic impulses cause heart action to decrease; sympathetic impulses cause heart action to increase.
      (2) Autonomic impulses are regulated by the cardiac center in the medulla oblongata.

Blood Vessels (page 672)

The blood vessels form a closed circuit of tubes that transport blood between the heart and body cells. The tubes include arteries, arterioles, capillaries, venules, and veins.

1. Arteries and arterioles
   a. The arteries are adapted to carry relatively high pressure blood away from the heart.
   b. The arterioles are branches of arteries.
   c. The walls of arteries and arterioles consist of layers of endothelium, smooth muscle, and connective tissue.
   d. The smooth muscles are innervated by autonomic fibers that can stimulate vasoconstriction or vasodilation.
2. Capillaries
   Capillaries form connections between arterioles and venules. The capillary wall consists of a single layer of cells that forms a semipermeable membrane.
   a. Capillary permeability
      (1) The openings in the capillary walls are thin slits between adjacent endothelial cells.
      (2) The sizes of the openings vary from tissue to tissue.
      (3) Endothelial cells of brain capillaries are tightly fused, forming a blood-brain barrier; substances are moved through this barrier by facilitated diffusion.
   b. Arrangement of capillaries
      Capillary density varies directly with tissue metabolic rates.
   c. Regulation of capillary blood flow
      (1) Capillary blood flow is regulated by precapillary sphincters.
      (2) Precapillary sphincters open when cells are low in oxygen and nutrients, and close when cellular needs are met.
3. Exchanges in capillaries
   a. Gases, nutrients, and metabolic by-products are exchanged between the capillary blood and the tissue fluid.
   b. Diffusion provides the most important means of transport.
   c. Diffusion pathways depend on lipid solubilities.
   d. Plasma proteins generally remain in the blood.
   e. Filtration, which is due to the hydrostatic pressure of blood, causes a net outward movement of fluid at the arterial end of a capillary.
   f. Osmosis causes a net inward movement of fluid at the venule end of a capillary.
   g. Some factors cause fluids to accumulate excessively in the tissues.
4. Venules and veins
   a. Venules continue from capillaries and merge to form veins.
   b. Veins carry blood to the heart.
   c. Venous walls are similar to arterial walls, but are thinner and contain less muscle and elastic tissue.

Blood Pressure (page 678)

Blood pressure is the force exerted by blood against the insides of the blood vessels.

1. Arterial blood pressure
   a. The arterial blood pressure is produced primarily by heart action; it rises and falls with phases of the cardiac cycle.
   b. Systolic pressure occurs when the ventricle contracts; diastolic pressure occurs when the ventricle relaxes.
2. Factors that influence arterial blood pressure
   a. Heart action, blood volume, resistance to flow, and blood viscosity influence arterial blood pressure.
   b. Arterial pressure increases as cardiac output, blood volume, peripheral resistance, or blood viscosity increases.
3. Control of blood pressure
   a. Blood pressure is controlled in part by the mechanisms that regulate cardiac output and peripheral resistance.
   b. Cardiac output depends on the volume of blood discharged from the ventricle with each beat and on the rate of heartbeat.
      (1) The more blood that enters the heart, the stronger the ventricular contraction, the greater the stroke volume, and the greater the cardiac output.
      (2) Heart rate is regulated by the cardiac center of the medulla oblongata.
   c. Regulation of peripheral resistance involves changes in the diameter of arterioles, factors that are controlled by the vasomotor center of the medulla oblongata.
4. Venous blood flow
   a. Venous blood flow is not a direct result of heart action; it depends on skeletal muscle contraction, breathing movements, and venoconstriction.
   b. Many veins contain flaplike valves that prevent blood from backing up.
   c. Venous constriction can increase venous pressure and blood flow.
5. Central venous pressure
   a. Central venous pressure is the pressure in the right atrium.
   b. It is influenced by factors that alter the flow of blood into the right atrium.
   c. It affects pressure within the peripheral veins.

Paths of Circulation (page 689)

1. Pulmonary circuit
   a. The pulmonary circuit is composed of vessels that carry blood from the right ventricle to the lungs, pulmonary capillaries, and vessels that lead back to the left atrium.
   b. Pulmonary capillaries contain lower pressure than those of the systemic circuit.
   c. Tightly joined epithelial cells of alveoli walls prevent most substances from entering the alveoli.
   d. Water is rapidly drawn out of alveoli into the interstitial fluid by osmotic pressure, so alveoli remain dry.
2. Systemic circuit
   a. The systemic circuit is composed of vessels that lead from the heart to the body cells and back to the heart.
   b. It includes the aorta and its branches as well as the system of veins that return blood to the right atrium.

Arterial System (page 691)

1. Principal branches of the aorta
   a. The branches of the ascending aorta include the right and left coronary arteries.
   b. The branches of the aortic arch include the brachiocephalic, left common carotid, and left subclavian arteries.
   c. The branches of the descending aorta include the thoracic and abdominal groups.
   d. The abdominal aorta terminates by dividing into right and left common iliac arteries.
2. Arteries to the neck, head, and brain
   These include branches of the subclavian and common carotid arteries.
3. Arteries to the shoulder and arm
   a. The subclavian artery passes into the upper arm, and in various regions is called the axillary and brachial artery.
   b. Branches of the brachial artery include the ulnar and radial arteries.
4. Arteries to the thoracic and abdominal walls
   a. The thoracic wall is supplied by branches of the subclavian artery and thoracic aorta.
   b. The abdominal wall is supplied by branches of the abdominal aorta and other arteries.
5. Arteries to the pelvis and leg
   The common iliac artery supplies the pelvic organs, gluteal region, and leg.

Venous System (page 701)

1. Characteristics of venous pathways
   a. The veins are responsible for returning blood to the heart.
   b. Larger veins usually parallel the paths of major arteries.
2. Veins from the head, neck, and brain
   a. These regions are drained by the jugular veins.
   b. Jugular veins unite with subclavian veins to form the brachiocephalic veins.

3. Veins from the arm and shoulder
   a. The arm is drained by sets of superficial and deep veins.
   b. The major superficial veins are the basilic and cephalic veins.
   c. The median cubital vein in the bend of the elbow is often used as a site for venipuncture.
4. Veins from the abdominal and thoracic walls
   These walls are drained by tributaries of the brachiocephalic and azygos veins.
5. Veins from the abdominal viscera
   a. The blood from the abdominal viscera generally enters the hepatic portal system and is carried to the liver.
   b. The blood in the portal system is rich in nutrients.
   c. The liver helps regulate the blood concentrations of glucose, amino acids, and lipids.
   d. Phagocytic cells in the liver remove bacteria from the portal blood.
   e. From the liver, the blood is carried by hepatic veins to the inferior vena cava.
6. Veins from the leg and pelvis
   a. These regions are drained by sets of deep and superficial veins.
   b. The deep veins include the tibial veins, and the superficial veins include the saphenous veins.

Clinical Application of Knowledge

1. Based on your understanding of the way capillary blood flow is regulated, do you think it is wiser to rest or to exercise following a heavy meal? Give a reason for your answer.
2. If a patient develops a blood clot in the femoral vein of the left leg, and a portion of the clot breaks loose, where is the blood flow likely to carry the embolus? What symptoms is this condition likely to produce?
3. When a person strains to lift a heavy object, intrathoracic pressure is increased. What do you think will happen to the rate of venous blood returning to the heart during such lifting? Why?
4. Why is a ventricular fibrillation more likely to be life threatening than an atrial fibrillation?
5. Cirrhosis of the liver is a disease commonly associated with alcoholism. In this condition, the blood flow through the hepatic blood vessels is often obstructed. As a result of such obstruction, the blood backs up, and the capillary pressure greatly increases in the organs drained by the hepatic portal system. What effects might this increasing capillary pressure produce, and which organs would be affected by it?
6. If a cardiologist inserted a catheter into a patient's right femoral artery, which arteries would the tube have to pass through in order to reach the entrance of the left coronary artery?

Review Activities

1. Describe the general structure, function, and location of the heart.
2. Describe the pericardium.
3. Compare the layers of the cardiac wall.
4. Identify and describe the locations of the chambers and the valves of the heart.
5. Describe the skeleton of the heart, and explain its function.
6. Trace the path of the blood through the heart.
7. Trace the path of the blood through the coronary circulation.
8. Describe a cardiac cycle.
9. Describe the pressure changes that occur in the atria and ventricles during a cardiac cycle.
10. Explain the origin of heart sounds.
11. Describe the arrangement of the cardiac muscle fibers.
12. Distinguish between the S-A node and the A-V node.
13. Explain how the cardiac conduction system functions in controlling the cardiac cycle.
14. Discuss how the nervous system functions in the regulation of the cardiac cycle.
15. Describe two factors other than the nervous system that affect the cardiac cycle.
16. Distinguish between an artery and an arteriole.
17. Explain how vasoconstriction and vasodilation are controlled.
18. Describe the structure and function of a capillary.
19. Describe the function of the blood-brain barrier.
20. Explain how the blood flow through a capillary is controlled.
21. Explain how diffusion functions in the exchange of substances between blood plasma and tissue fluid.
22. Explain why water and dissolved substances leave the arteriole end of a capillary and enter the venule end.
23. Describe the effect of histamine on a capillary.
24. Distinguish between a venule and a vein.
25. Explain how veins function as blood reservoirs.
26. Distinguish between systolic and diastolic blood pressures.
27. Name several factors that influence the blood pressure, and explain how each produces its effect.
28. Describe how blood pressure is controlled.
29. List the major factors that promote the flow of venous blood.
30. Define *central venous pressure*.
31. Distinguish between the pulmonary and systemic circuits of the cardiovascular system.
32. Trace the path of blood through the pulmonary circuit.
33. Explain why the alveoli normally remain dry.
34. Describe the aorta, and name its principal branches.
35. On a diagram, locate and identify the major arteries that supply the abdominal visceral organs.
36. On a diagram, locate and identify the major arteries that supply parts of the head, neck, and brain.
37. On a diagram, locate and identify the major arteries that supply parts of the shoulder and arm.
38. On a diagram, locate and identify the major arteries that supply parts of the thoracic and abdominal walls.
39. On a diagram, locate and identify the major arteries that supply parts of the pelvis and leg.
40. Describe the relationship between the major venous pathways and the major arterial pathways.
41. On a diagram, locate and identify the major veins that drain parts of the head, neck, and brain.
42. On a diagram, locate and identify the major veins that drain parts of the arm and shoulder.
43. On a diagram, locate and identify the major veins that drain parts of the abdominal and thoracic walls.
44. On a diagram, locate and identify the major veins that drain parts of the abdominal viscera.
45. Review the actions of the liver on nutrients carried in the portal veins.
46. On a diagram, locate and identify the major veins that drain parts of the leg and pelvis.

Coronary Artery Disease

Dave R., a 52-year-old overweight accountant, had been having chest pains occasionally for several months. As a rule, the pain occurred during his usual weekend tennis match, but it was mild, and he attributed it to overeating or indigestion. Also, the discomfort almost always diminished after the game. Recently, however, the pain seemed more severe and was lasting longer. Dave decided to consult with his physician about the problem.

The physician explained that Dave was probably experiencing *angina pectoris,* a symptom of *coronary artery disease* (CAD), and suggested that he undergo an *exercise stress test.* This test required Dave to walk on a treadmill, whose speed and incline were increased while he exercised. During the test, an *electrocardiogram* (ECG) was recorded continuously (figure 18.22), and Dave's blood pressure was monitored. Near the end of the test, when his heart had reached the desired rate, a small quantity of *radioactive thallium-201* was injected into a vein. A *scintillation counter* (figure 2.2) was used to scan Dave's heart to determine if the blood carrying the thallium had been uniformly distributed to the myocardium by branches of his coronary arteries (figure 18.13a).

On the basis of the test results, a cardiologist decided that Dave was developing CAD. It was also determined that he had elevated blood pressure (hypertension), and a blood test revealed excessive blood cholesterol (hypercholesterolemia).

Dave was advised to stop smoking, to reduce his intake of foods high in saturated fats, cholesterol, and sodium, and to exercise regularly, rather than on weekends only. He was given medications to reduce his blood pressure and to relieve the pain of the angina when it occurred. Dave was also cautioned to avoid stressful situations whenever possible, to reduce his present body weight, and to maintain a weight corresponding to the standard for his age and height (chart 15.6).

Six months later, in spite of faithful compliance with medical advice, Dave suffered a heart attack—a sign that the blood flow to some part of his myocardium had been obstructed, producing an oxygen deficiency (ischemia). He was at home at the time of the attack, which began as severe, crushing chest pain accompanied by shortness of breath and sweating. Paramedics were summoned, and they stabilized Dave's condition and arranged to have him transported to a hospital.

At the hospital, a cardiologist concluded from an ECG that Dave's heart attack (acute myocardial infarction) was caused by a blood clot obstructing a coronary artery (occlusive coronary thrombosis). The cardiologist intravenously administered a blood clot dissolving substance (thrombolytic agent), with the hope that it would remove the clot and reopen the vessel. Once again, an ECG was recorded continuously. After some time, the cardiologist concluded from the ECG that the blood vessel remained partially obstructed, and he ordered a *coronary angiogram* (figure 18.14). In this X-ray procedure, which was conducted in the *catheterization laboratory* (cath lab), a thin plastic catheter was passed through a guiding sheath that had been inserted into the femoral artery (figure 18.64) of Dave's right inguinal area. From there, the catheter was pushed into the aorta until it reached the region of the opening to the left coronary artery, and later the region of the opening to the right coronary artery. The progress of the catheter was monitored by means of *X-ray fluoroscopy,* and each time the catheter was in proper position, a radiopaque dye (contrast medium) was released from its distal end into the blood. X-ray images that revealed the path of the dye as it entered a coronary artery and its branches were recorded on videotape and on motion picture film, which were later analyzed "frame by frame." A single severe narrowing was discovered near the origin of Dave's left anterior descending artery. The cardiologist decided to perform *percutaneous transluminal coronary angioplasty* (PTCA) in order to enlarge the opening (lumen) of that vessel.

The PTCA was performed by passing another plastic catheter through the guiding sheath used for the angiogram. This second tube had a tiny deflated balloon at its tip, and when the balloon was located in the region of the arterial narrowing, it was inflated for a short time with relatively high pressure. The inflating balloon compressed the atherosclerotic plaque (atheroma), which was responsible for the obstruction, against the arterial wall. It also stretched the blood vessel wall, thus increasing the diameter of its lumen (recanalization). The blood flow to the myocardial tissue downstream from the obstruction improved immediately.

Dave was told that although the coronary obstruction was lessened for the moment, the disease that produced it was still active and that the obstruction might recur (reocclusion). If it did, he might have to undergo another PTCA or have *coronary bypass graft surgery.* In this latter procedure, a portion of a vein, such as the great saphenous vein in the leg, is removed, and a section of the vein is surgically connected between the aorta and the affected coronary artery at a point beyond the obstruction.

Once again, Dave was advised to avoid saturated fats, cholesterol, and sodium in his diet, to exercise regularly, to give up smoking, to take medication to control hypertension, to reduce stress when possible, and to maintain a desirable body weight, all in the hope of retarding the progress of his coronary artery disease. Dave was also advised to have the progress of his disease monitored by undergoing exercise stress tests periodically.

Review Questions

1. How would you explain to the patient the origin of the pain associated with angina pectoris?
2. Why would stressful situations be likely to aggravate the symptoms of CAD?
3. Why might a patient experience shortness of breath during a heart attack?
4. Why is it important that any obstruction within a coronary artery be reduced as quickly as possible?
5. How would you explain to a patient that PTCA does not cure coronary artery disease?

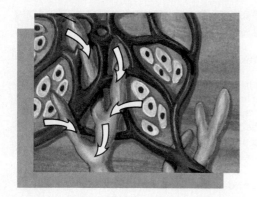

19

Lymphatic System and Immunity

*W*hen substances are exchanged between the blood and the tissue fluid, more fluid normally leaves the blood capillaries than returns to them. If the fluid remaining in the tissue spaces were allowed to accumulate, the hydrostatic pressure in the tissues would increase abnormally. The *lymphatic system* helps prevent such an imbalance by providing pathways through which tissue fluid can be transported as lymph from the tissue spaces to the veins, where it becomes part of the blood.

The lymphatic system also helps defend the tissues against infections by filtering particles from the lymph and by supporting the activities of the lymphocytes, which furnish *immunity,* or resistance, to the effects of specific disease-causing agents ■

After you have studied this chapter, you should be able to:

1. Describe the general functions of the lymphatic system.

2. Describe the location of the major lymphatic pathways.

3. Describe how tissue fluid and lymph are formed and explain the function of lymph.

4. Explain how lymphatic circulation is maintained and describe the consequence of lymphatic obstruction.

5. Describe a lymph node and its major functions.

6. Describe the location of the major chains of lymph nodes.

7. Discuss the functions of the thymus and spleen.

8. Distinguish between specific and nonspecific immunity, and provide examples of each.

9. Explain how two major types of lymphocytes are formed and how they function in immune mechanisms.

10. Name the major types of immunoglobulins and discuss their origins and actions.

11. Distinguish between primary and secondary immune responses.

12. Distinguish between active and passive immunity.

13. Explain how allergic reactions and tissue rejection reactions are related to immune mechanisms.

14. Complete the review activities at the end of this chapter. Note that these items are worded in the form of specific learning objectives. You may want to refer to them before reading the chapter.

---

allergen (al'er-jen)

antibody (an'tĭ-bod''e)

antigen (an'tĭ-jen)

clone (klōn)

complement (kom'plĕ-ment)

hapten (hap'-ten)

immunity (ĭ-mu'nĭ-te)

immunoglobulin (im''u-no-glob'u-lin)

interferon (in''tcr fēr'on)

lymph (limf)

lymphatic pathway (lim-fat'ik path'wa)

lymph node (limf nōd)

lymphocyte (lim'fo-sīt)

macrophage (mak'ro-fāj)

pathogen (path'o-jen)

reticuloendothelial tissue (rĕ-tik''u-lo-en''do-the'le-al tish'u)

spleen (splēn)

thymus (thi'mus)

vaccine (vak'sēn)

---

auto-, self: autoimmune disease—condition in which the immune system attacks the body's own tissues.

-gen, to be produced: allergen—a substance that stimulates an allergic response.

humor-, fluid: humoral immunity—immunity resulting from antibodies in body fluids.

immun-, free: immunity—resistance to (freedom from) a specific disease.

inflamm-, setting on fire: inflammation—a condition characterized by localized redness, heat, swelling, and pain in the tissues.

nod-, knot: nodule—a small mass of lymphocytes surrounded by connective tissue.

patho-, disease: pathogen—a disease-causing agent.

The lymphatic system is closely associated with the cardiovascular system because it includes a network of vessels that assist in circulating the body fluids. These vessels transport excess fluid away from the interstitial spaces that exist between the cells in most tissues and return it to the blood stream. The organs of the lymphatic system also help defend the body against invasion by disease-causing agents (figure 19.1).

Lymphatic Pathways

The lymphatic pathways begin as lymphatic capillaries. These tiny tubes merge to form lymphatic vessels, which, in turn, lead to larger vessels that unite with the veins in the thorax.

Lymphatic Capillaries

Lymphatic capillaries are microscopic, closed-ended tubes. They extend into the interstitial spaces, forming complex networks that parallel the networks of the blood capillaries (figure 19.2). The walls of the lym-phatic capillaries, like those of the blood capillaries, consist of a single layer of squamous epithelial cells. This thin wall makes it possible for tissue fluid (inter-stitial fluid) from the interstitial space to enter the lymphatic capillary. Once the fluid is inside a lymphatic capillary, it is called **lymph.**

The villi of the small intestine contain specialized lymphatic capillaries called *lacteals,* described in chapter 14. These vessels are responsible for trans-porting recently absorbed fats away from the digestive tract.

Lymphatic Vessels

Lymphatic vessels, which are formed by the merging of lymphatic capillaries, have walls similar to those of veins. That is, their walls are composed of three layers: an endothelial lining, a middle layer of smooth muscle and elastic fibers, and an outer layer of connective tissue. Also, like veins, the lymphatic vessels have flap-like *valves,* which help prevent the backflow of lymph. Figure 19.3 shows one of these valves.

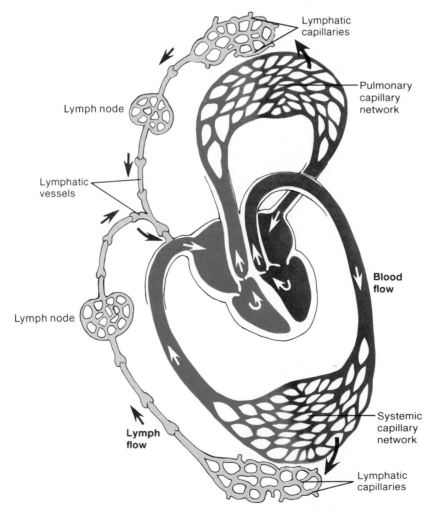

Figure 19.1   Lymphatic vessels transport fluid from interstitial spaces to the blood stream.

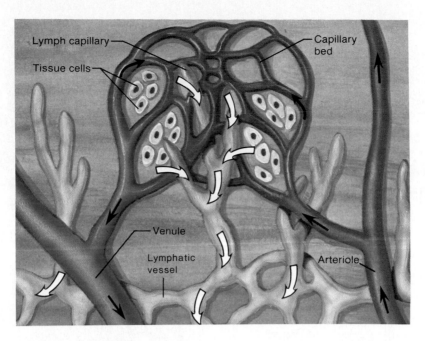

Figure 19.2 Lymph capillaries are microscopic, closed-ended tubes that begin in the interstitial spaces of most tissues.

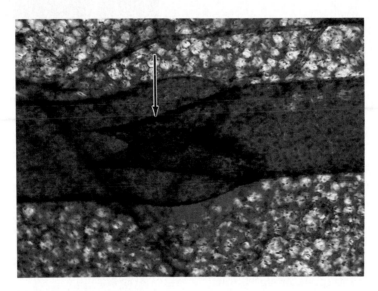

Figure 19.3 A light micrograph of the flaplike valve (arrow) within a lymphatic vessel (×25). What is the function of this valve?

The larger lymphatic vessels lead to specialized organs called **lymph nodes,** and after leaving these structures, the vessels merge to form still larger lymphatic trunks.

Lymphatic Trunks and Collecting Ducts

The **lymphatic trunks,** which drain lymph from relatively large regions of the body, are named for the regions they serve. For example, the *lumbar trunk* drains lymph from the legs, lower abdominal wall, and pelvic organs; the *intestinal trunk* drains organs of the abdominal viscera; the *intercostal* and *bronchomediastinal trunks* drain lymph from portions of the thorax; the *subclavian trunk* drains the arm; and the *jugular trunk* drains portions of the neck and head. These lymphatic trunks then join one of two **collecting ducts**— the thoracic duct or the right lymphatic duct. Figure 19.4 shows the location of the major lymphatic trunks and collecting ducts, and figure 19.5 shows a lymphangiogram, or X-ray film, of some lymphatic vessels and lymph nodes.

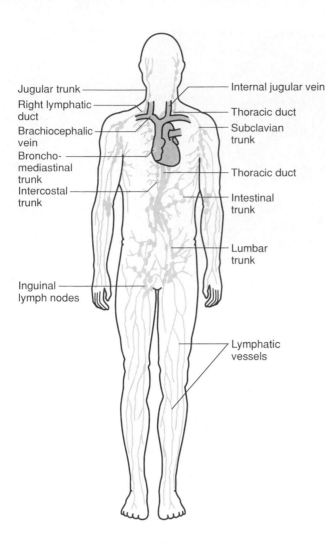

Figure 19.4 Lymphatic vessels merge into larger lymphatic trunks, which in turn drain into collecting ducts.

Jugular trunk
Right lymphatic duct
Brachiocephalic vein
Broncho-mediastinal trunk
Intercostal trunk
Inguinal lymph nodes

Internal jugular vein
Thoracic duct
Subclavian trunk
Thoracic duct
Intestinal trunk
Lumbar trunk
Lymphatic vessels

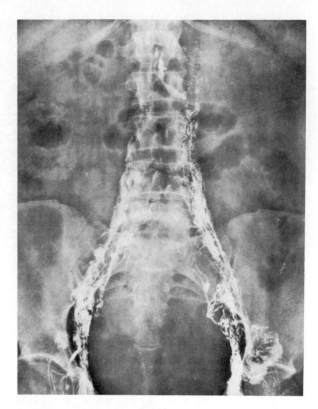

Figure 19.5 A lymphangiogram (X-ray film) of the lymphatic vessels and lymph nodes of the pelvic region.

The **thoracic duct** is the larger and longer of the two collecting ducts. It begins in the abdomen, passes upward through the diaphragm beside the aorta, ascends in front of the vertebral column through the mediastinum, and empties into the left subclavian vein near the junction of the left jugular vein. This duct drains lymph from the intestinal, lumbar, and intercostal trunks, as well as from the left subclavian, left jugular, and left bronchomediastinal trunks.

The **right lymphatic duct** originates in the right thorax at the union of the right jugular, right subclavian, and right bronchomediastinal trunks. It empties into the right subclavian vein near the junction of the right jugular vein.

After leaving the two collecting ducts, lymph enters the venous system and becomes part of the plasma just before the blood returns to the right atrium.

Thus, lymph from the lower body regions, the left arm, and the left side of the head and neck enters the thoracic duct; lymph from the right side of the head and neck, the right arm, and the right thorax enters the right lymphatic duct (figure 19.6). Figure 19.7 summarizes the lymphatic pathway.

The skin is richly supplied with lymphatic capillaries. Consequently, if the skin is broken, or if something is injected into it (such as venom from a stinging insect), foreign substances are likely to enter the lymphatic system relatively rapidly.

1. What is the general function of the lymphatic system?
2. Through what lymphatic vessels would lymph pass in traveling from a leg to the blood stream?

Tissue Fluid and Lymph

Lymph is essentially tissue fluid that has entered a lymphatic capillary. Thus, the formation of lymph is closely associated with the formation of tissue fluid.

Tissue Fluid Formation

As explained in chapter 17, *tissue fluid* originates from blood plasma. Tissue fluid is composed of water and dissolved substances that leave the blood capillaries as a result of diffusion and filtration.

Although tissue fluid contains various nutrients and gases found in plasma, it generally lacks proteins

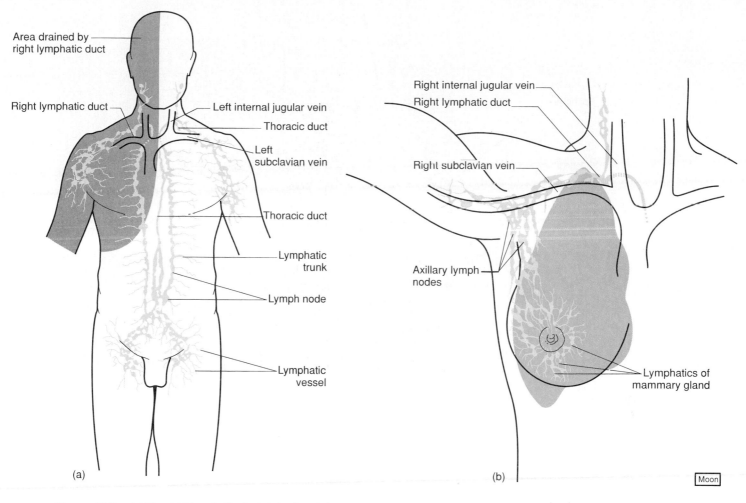

Area drained by right lymphatic duct

Right lymphatic duct

Left internal jugular vein

Thoracic duct

Left subclavian vein

Thoracic duct

Lymphatic trunk

Lymph node

Lymphatic vessel

(a)

Right internal jugular vein

Right lymphatic duct

Right subclavian vein

Axillary lymph nodes

Lymphatics of mammary gland

(b)

Moon

Figure 19.6 (a) The right lymphatic duct drains lymph from the upper right side of the body, while the thoracic duct drains lymph from the rest of the body. (b) Lymph drainage of the right breast.

of large molecular size. Some proteins with smaller molecules do leak out of the blood capillaries and enter the interstitial spaces. Usually, these smaller proteins are not reabsorbed when water and other dissolved substances move back into the venule ends of these capillaries by diffusion and osmosis. As a result, the protein concentration of the tissue fluid tends to rise, causing the *osmotic pressure* of the fluid to rise also.

Lymph Formation

As the osmotic pressure of the tissue fluid rises, this pressure interferes with the osmotic reabsorption of water by the blood capillaries. The volume of fluid in the interstitial spaces then tends to increase, as does the pressure within the spaces.

This increasing interstitial pressure is responsible for forcing some of the tissue fluid into the lymphatic capillaries, where it becomes lymph (figure 19.2).

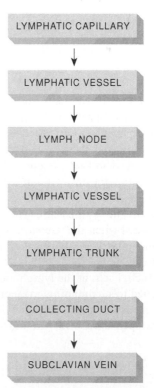

LYMPHATIC CAPILLARY

↓

LYMPHATIC VESSEL

↓

LYMPH NODE

↓

LYMPHATIC VESSEL

↓

LYMPHATIC TRUNK

↓

COLLECTING DUCT

↓

SUBCLAVIAN VEIN

Figure 19.7 The lymphatic pathway.

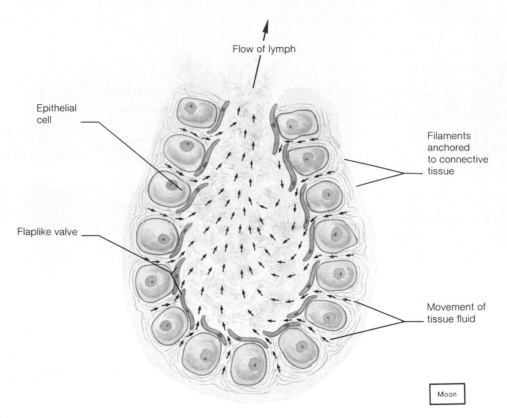

Flow of lymph

Epithelial cell

Filaments anchored to connective tissue

Flaplike valve

Movement of tissue fluid

Moon

Figure 19.8 Tissue fluid enters lymphatic capillaries through flaplike valves between adjacent epithelial cells.

Function of Lymph

Most of the protein molecules that leak out of the blood capillaries are carried away by lymph and are returned to the blood stream. At the same time, lymph transports various foreign particles, such as bacterial cells or viruses that may have entered the tissue fluids, to lymph nodes.

Although these proteins and foreign particles cannot easily enter blood capillaries, the lymphatic capillaries are especially adapted to receive them. Specifically, the epithelial cells that form the walls of these vessels are arranged so that the edge of one cell overlaps the edge of an adjacent cell, but is not attached to it. This arrangement, shown in figure 19.8, creates flaplike valves in the lymphatic capillary wall, which are pushed inward when the pressure is greater on the outside of the capillary, but close when the pressure is greater on the inside.

The epithelial cells of the lymphatic capillary wall are also attached to surrounding connective tissue cells by thin filaments, so that the lumen of a lymphatic capillary remains open even when the outside pressure is increased.

1. How would you explain the relationship between tissue fluid and lymph?
2. How does the presence of protein in tissue fluid affect the formation of lymph?
3. What are the major functions of lymph?

Movement of Lymph

Although the entrance of lymph into the lymphatic capillaries is influenced by the *osmotic pressure* of the tissue fluid, the movement of lymph through the lymphatic vessels is controlled largely by *muscular activity.*

Flow of Lymph

Lymph, like venous blood, is under relatively low hydrostatic pressure and may not flow readily through the lymphatic vessels without the aid of outside forces. These forces include contraction of the skeletal muscles, pressure changes due to the action of the breathing muscles, and contraction of the smooth muscles in the walls of the larger lymphatic vessels.

As the *skeletal muscles* contract, they compress the lymphatic vessels. This squeezing action causes the lymph inside a vessel to move, but because the lymphatic vessels contain valves that prevent backflow, the lymph can only move toward a collecting duct. Similarly, the smooth muscles in the walls of the larger lymphatic vessels may contract and compress the lymph inside. This action also helps force the fluid onward.

The *breathing muscles* aid the circulation of lymph (as they do that of venous blood) by creating a relatively low pressure in the thorax during inhalation. At the same time, the pressure in the abdominal cavity is increased by the contracting diaphragm. Consequently, lymph (and venous blood as well) is squeezed out of the abdominal vessels and forced into the thoracic vessels. Once again, the backflow of lymph (and blood) is prevented by valves within the lymphatic (and blood) vessels.

Since the actions of skeletal and breathing muscles promote the circulation of lymph, it is not surprising that the flow of lymph is greatest during periods of physical exercise. In addition, lymph formation increases still more because of the elevated perfusion of fluid from the blood into the interstitial spaces that occurs during exercise.

Obstruction of Lymph Movement

The continuous movement of fluid from interstitial spaces into blood capillaries and lymphatic capillaries causes the volume of fluid in these spaces to remain stable. However, conditions sometimes occur that interfere with lymph movement, and tissue fluids accumulate in the interstitial spaces, causing *edema.*

For example, lymphatic vessels may be obstructed as a result of surgical procedures in which portions of the lymphatic system are removed. Because the affected pathways can no longer drain lymph from the tissues, proteins tend to accumulate in the interstitial spaces. This causes an increase in the osmotic pressure of the tissue fluid, which, in turn, promotes the accumulation of water within the tissues.

Because lymphatic vessels tend to carry particles away from tissues, these vessels may also transport cancer cells and promote their spread to other sites (metastasis). For this reason, the lymphatic tissues (lymph nodes) in the axillary regions are commonly biopsied during the surgical removal of cancerous breast tissue. When lymphatic tissue is removed during the procedure, lymphatic drainage from the arm and other nearby tissues is sometimes obstructed, and these parts may become edematous following surgery.

1. What factors promote the flow of lymph?
2. What is the consequence of lymphatic obstruction?

Lymph Nodes

The **lymph nodes** (lymph glands) are structures located along the lymphatic pathways. They contain large numbers of *lymphocytes* and *macrophages,* which are vital in the defense against invading microorganisms.

Structure of a Lymph Node

The lymph nodes vary in size and shape; however, they are usually less than 2.5 centimeters in length and are somewhat bean-shaped. A section of a typical lymph node is illustrated in figure 19.9.

The indented region of a bean-shaped node is called the **hilum,** and it is the portion through which the blood vessels and nerves connect with the structure. The lymphatic vessels leading to a node (afferent vessels) enter separately at various points on its convex surface, but the lymphatic vessels leaving the node (efferent vessels) exit from the hilum.

Each lymph node is enclosed by a capsule of white fibrous connective tissue. This tissue also extends into the node and partially subdivides it into compartments, which contain dense masses of lymphocytes and macrophages. These masses, called **nodules,** represent the structural units of the node.

The spaces within the node, called **lymph sinuses,** provide a complex network of chambers and channels through which lymph circulates as it passes through the node. Lymph enters a lymph node through an *afferent lymphatic vessel,* moves slowly through the lymph sinuses, and leaves through an *efferent lymphatic vessel* (figure 19.10).

Sometimes the lymphatic vessels become inflamed due to a bacterial infection. When this happens in the superficial lymphatic vessels, painful reddish streaks may appear beneath the skin—for example, in an arm or a leg. This condition is called *lymphangitis,* and it is usually followed by *lymphadenitis,* an inflammation of the lymph nodes. The affected nodes may become greatly enlarged and quite painful.

Nodules also occur singly or in groups associated with the mucous membranes of the respiratory and digestive tracts. The *tonsils,* described in chapter 14, are composed of partially encapsulated lymph nodules. Also, aggregations of nodules called *Peyer's patches* are scattered throughout the mucosal lining of the ileum of the small intestine.

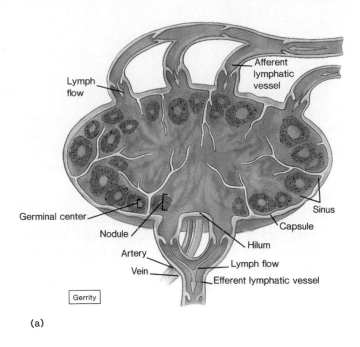

(a)

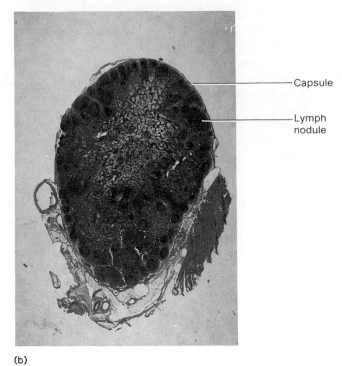

(b)

Figure 19.9 (*a*) A section of a lymph node. What factors promote the flow of lymph through a node? (*b*) Light micrograph of a lymph node (×2.5).

Locations of Lymph Nodes

The lymph nodes generally occur in groups or chains along the paths of the larger lymphatic vessels. Although they are widely distributed throughout the body, lymph nodes are absent in the tissues of the central nervous system.

The major locations of the lymph nodes, shown in figure 19.11, are as follows:

1. **Cervical region.** Nodes in the cervical region occur along the lower border of the mandible, in front of and behind the ears, and deep within the neck along the paths of the larger blood vessels. These nodes are associated with the lymphatic vessels that drain the skin of the scalp and face, as well as the tissues of the nasal cavity and pharynx.
2. **Axillary region.** Nodes in the underarm region receive lymph from vessels that drain the arms, the wall of the thorax, the mammary glands (breasts), and the upper wall of the abdomen.
3. **Inguinal region.** Nodes in the inguinal region receive lymph from the legs, the external genitalia, and the lower abdominal wall.
4. **Pelvic cavity.** Within the pelvic cavity, nodes occur primarily along the paths of the iliac blood vessels. They receive lymph from the lymphatic vessels of the pelvic viscera.

5. **Abdominal cavity.** Nodes within the abdominal cavity occur in chains along the main branches of the mesenteric arteries and the abdominal aorta. These nodes receive lymph from the abdominal viscera.
6. **Thoracic cavity.** Nodes of the thoracic cavity occur within the mediastinum and along the trachea and bronchi. They receive lymph from the thoracic viscera and from the internal wall of the thorax.

The supratrochlear lymph nodes (cubital lymph nodes), which are located superficially on the medial side of the elbow, often become enlarged in children as a result of infections associated with numerous cuts and scrapes on the hands.

Functions of Lymph Nodes

As mentioned previously, the lymph nodes contain large numbers of *lymphocytes*. The lymph nodes, in fact, are centers for lymphocyte production, although such cells are produced in red bone marrow as well. Lymphocytes act against foreign particles, such as bacterial cells and viruses, which are carried to the lymph nodes by the lymphatic vessels.

The lymph nodes also contain *macrophages*, which engulf and destroy foreign substances, damaged

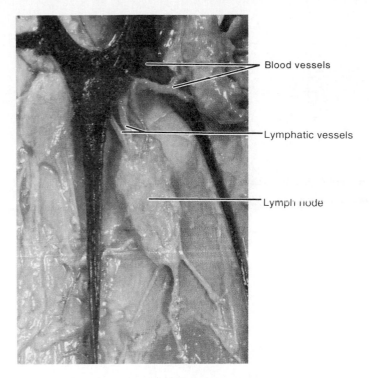

Figure 19.10 Lymph enters and leaves a lymph node through lymphatic vessels.

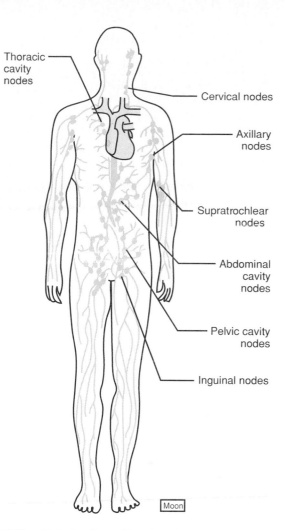

Figure 19.11 Major locations of lymph nodes.

cells, and cellular debris. The functions of the cells associated with the lymph nodes are described in a subsequent section of this chapter.

Thus, the primary functions of lymph nodes are the production of lymphocytes that help defend the body against microorganisms, and the filtration of potentially harmful foreign particles and cellular debris from lymph before it is returned to the blood stream.

1. How would you distinguish between a lymph node and a lymph nodule?
2. In what body regions are lymph nodes most abundant?
3. What are the major functions of lymph nodes?

Thymus and Spleen

Two other lymphatic organs, whose functions are closely related to those of the lymph nodes, are the thymus and the spleen.

Thymus

The **thymus** (thymus gland), shown in figure 19.12, is a soft, bilobed structure whose lobes are surrounded by connective tissue. It is located within the mediastinum, in front of the aortic arch and behind the upper part of the sternum, extending from the root of the neck to the

pericardium. Although the thymus varies in size from person to person, it is usually relatively large during infancy and early childhood. After puberty, it tends to decrease in size, and in an adult, it may be quite small. In an elderly person, the thymus is often largely replaced by adipose and connective tissue.

The thymus is composed of lymphatic tissue, which is subdivided into *lobules* by connective tissues extending inward from its surface (figure 19.13). The lobules contain large numbers of lymphocytes that developed from precursor cells originating in the bone marrow. The majority of these cells (thymocytes) remain inactive; however, some of them mature into a group of cells (T-lymphocytes) that leaves the thymus and functions in providing immunity.

Epithelial cells within the thymus may secrete a protein hormone called *thymosin,* which is thought to stimulate the maturation of the T-lymphocytes after they leave the thymus and migrate to other lymphatic tissues.

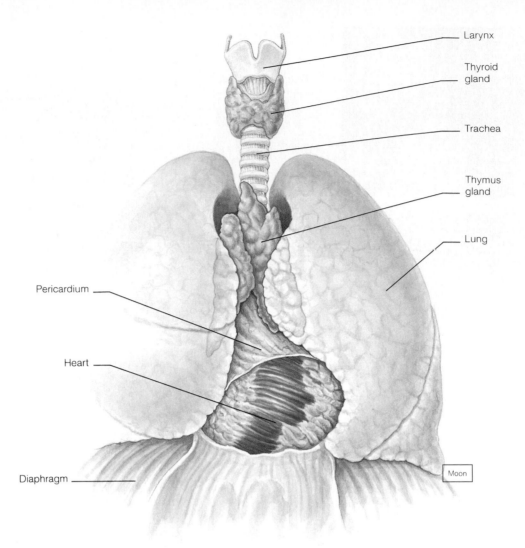

Larynx

Thyroid
gland

Trachea

Thymus
gland

Lung

Pericardium

Heart

Diaphragm

Moon

Figure 19.12 The thymus gland is a bilobed organ located between the lungs and above the heart.

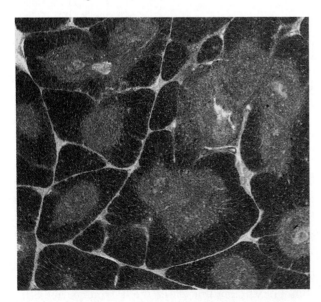

Figure 19.13 A cross section of the thymus gland (×10). Note how the organ is subdivided into lobules.

Spleen

The **spleen** is the largest of the lymphatic organs. It is located in the upper left portion of the abdominal cavity, just beneath the diaphragm and behind the stomach. (See reference plates 4, 5, and 6.)

The spleen resembles a large lymph node in some respects. For example, it is enclosed in connective tissue that extends inward from the surface and partially subdivides the organ into chambers, or *lobules*. It also has a *hilum* on one surface through which blood vessels and nerves enter. However, unlike the sinuses of a lymph node, the spaces (venous sinuses) within the chambers of the spleen are filled with *blood* instead of lymph.

Within the lobules of the spleen, the tissues are called *pulp* and are of two types: white pulp and red pulp. The *white pulp* is distributed throughout the spleen in tiny islands. This tissue is composed of nodules (splenic nodules), which are similar to those found in lymph nodes and contain large numbers of lymphocytes. The *red pulp,* which fills the remaining spaces

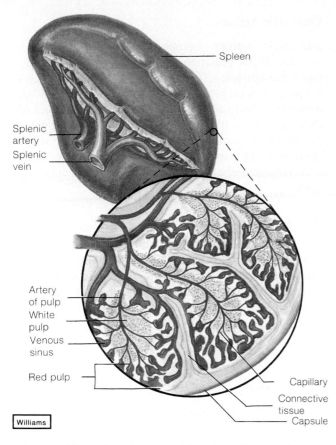

Figure 19.14 The spleen resembles a large lymph node.

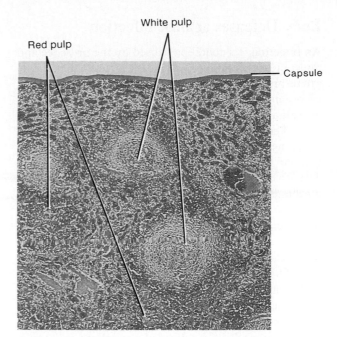

Figure 19.15 Light micrograph of the spleen (×40).

of the lobules, surrounds the venous sinuses. This pulp contains relatively large numbers of red blood cells, which are responsible for its color, along with many lymphocytes and macrophages (figures 19.14 and 19.15).

The blood capillaries within the red pulp are quite permeable. Red blood cells can squeeze through the pores in these capillary walls and enter the venous sinuses. The older, more fragile red blood cells may rupture as they make this passage, and the resulting cellular debris is removed by phagocytic macrophages located within the splenic sinuses.

The macrophages also engulf and destroy foreign particles, such as bacteria, that may be carried in the blood as it flows through the sinuses. Thus, the spleen filters blood much as the lymph nodes filter lymph.

The lymphocytes of the spleen, like those of the thymus, lymph nodes, and nodules, help defend the body against infections.

Chart 19.1 summarizes the characteristics of the major organs of the lymphatic system.

1. Why are the thymus and spleen considered organs of the lymphatic system?
2. What are the major functions of the thymus and the spleen?

CHART 19.1	Major organs of the lymphatic system	
Organ	Location	Function
Lymph nodes	In groups or chains along the paths of larger lymphatic vessels	Center for lymphocyte production; house lymphocytes that act against foreign particles in lymph; house macrophages that engulf and destroy foreign particles and cellular debris carried in lymph
Thymus	Within the mediastinum behind the upper portion of the sternum	Houses lymphocytes; changes undifferentiated lymphocytes into T-lymphocytes
Spleen	In upper left portion of abdominal cavity beneath the diaphragm and behind the stomach	Houses macrophages that filter foreign particles, damaged red blood cells, and cellular debris from the blood; houses lymphocytes

During fetal development, pulp cells of the spleen produce blood cells, much as red bone marrow cells do in later life. As the time of birth approaches, this splenic function ceases. However, in certain diseases, such as *erythroblastosis fetalis*, in which large numbers of red blood cells are destroyed, the splenic pulp cells may resume their hematopoietic activity.

Body Defenses against Infection

An **infection** is a condition caused by the presence and multiplication of a disease-causing agent. Such agents are termed **pathogens,** and they include viruses and microorganisms such as bacteria, fungi, and protozoans, as well as other parasitic forms of life.

The human body is equipped with a variety of defense mechanisms that prevent the entrance of pathogens or destroy them if they enter the tissues. This defensive capacity is called **immunity.** Some immune mechanisms are quite general and provide *nonspecific immunity,* in that they protect against many types of pathogens. These mechanisms include species resistance, mechanical barriers, the actions of enzymes, interferon, inflammation, and phagocytosis.

Other defense mechanisms are very precise in their actions, providing protection against particular disease-causing agents. These mechanisms are responsible for the type of resistance called *specific immunity.* They involve the actions of specialized cells, such as lymphocytes and macrophages, that can recognize the presence of foreign substances in the body and respond against them.

Nonspecific Immunity

Species Resistance

Species resistance refers to the fact that a given kind of organism, or *species* (such as the human species, *Homo sapiens*), develops diseases that are unique to it. At the same time, a species may be resistant to diseases that affect other species, because its tissues somehow fail to provide the temperature or chemical environment that is needed by a particular pathogen. For example, humans are subject to infections by the microorganisms that cause measles, mumps, gonorrhea and syphilis, but other animal species are generally resistant to these diseases. Similarly, humans are resistant to certain forms of malaria and tuberculosis that affect various birds.

Mechanical Barriers

The *skin* and *mucous membranes* lining the tubes of the respiratory, digestive, urinary, and reproductive systems create **mechanical barriers** that prevent the entrance of some infectious agents. As long as these barriers remain unbroken, many pathogens are unable to penetrate them. In addition, the mucus-coated ciliated epithelium, described in chapter 16, that lines the respiratory passages entraps particles and sweeps them out of the airways and into the pharynx, where they are swallowed.

Enzymatic Actions

Enzymatic actions against disease-causing agents are due to the presence of enzymes in various body fluids. Gastric juice, for example, contains the protein-splitting enzyme *pepsin,* and has a low pH due to the presence of hydrochloric acid. The combined effect of these substances is lethal to many pathogens that enter the stomach. Similarly, tears contain the enzyme *lysozyme,* which has an antibacterial action against certain pathogens that may get onto the surfaces of the eyes.

Interferon

Interferon is the name given to a group of hormonelike peptides produced by certain cells, including lymphocytes and fibroblasts, in response to the presence of viruses or the cells of certain tumors. Although the effect of interferon is nonspecific, it inhibits the proliferation of viruses, stimulates phagocytosis, and enhances the activity of certain other cells that help to resist infections and the growth of tumors.

Inflammation

Inflammation is a tissue response that may accompany tissue invasion by pathogens (chapter 5).

The major symptoms of inflammation include localized redness, swelling, heat, and pain. The *redness* is a result of blood vessel dilation and the consequent increase of blood volume within the affected tissues (hyperemia). This effect, coupled with an increase in the permeability of the nearby capillaries, is responsible for tissue *swelling* (edema). The *heat* is due to the presence of blood from the deeper body parts, which is generally warmer than that near the surface, and the *pain* results from the stimulation of nearby pain receptors.

White blood cells tend to accumulate at the sites of inflammation, and some of these cells help control pathogens by *phagocytosis.* (See figures 3.29 and 3.30.) In the case of bacterial infections, the resulting mass of white blood cells, bacterial cells, and damaged tissue cells may form a thick fluid called *pus.*

Body fluids (exudate) also tend to collect in inflamed tissues. These fluids contain *fibrinogen* and other blood factors that promote clotting. As a result of clotting, a network of fibrin threads may develop within the affected region. Later, *fibroblasts* may appear and form fibers around the area until it is enclosed in a sac of fibrous connective tissue. This action inhibits the spread of pathogens and toxic substances to adjacent tissues.

Once an infection has been controlled, phagocytic cells remove dead cells and other debris from the

CHART 19.2	Major actions that may occur during an inflammation response

Blood vessels dilate.

Capillary permeability increases.
Tissues become red, swollen, warm, and painful.

White blood cells invade the region.
Pus may form as white blood cells, bacterial cells, and cellular debris accumulate.

Body fluids seep into the area.
A clot containing threads of fibrin may form.

Fibroblasts appear.
A connective tissue sac may be formed around the injured tissues.

Phagocytes are active.
Bacteria dead cells and other debris are removed.

Cells reproduce.
Newly formed cells replace injured ones.

CHART 19.3	Types of nonspecific immunity

Type	Description
Species resistance	A species or organism is resistant to certain diseases to which other species are susceptible.
Mechanical barriers	Unbroken skin and mucous membranes prevent the entrance of some infectious agents.
Enzymatic actions	Enzymes present in various body fluids act against pathogens.
Interferon	A group of hormonelike peptides (produced by cells in response to the presence of viruses or certain tumors) that interferes with the reproduction of viruses, stimulates phagocytosis, and enhances the activity of cells that resist infections and the growth of tumors.
Inflammation	A tissue response to injury that helps prevent the spread of infectious agents into nearby tissues.
Phagocytosis	Neutrophils and macrophages engulf and destroy foreign particles and cells.

site of inflammation. At the same time, the dead cells are replaced by cellular reproduction.

The process of inflammation is summarized in chart 19.2.

Phagocytosis

As mentioned in chapter 17, the most active phagocytic cells of the blood are *neutrophils* and *monocytes*. These wandering cells can leave the blood stream by squeezing between the cells of the blood vessel walls (diapedesis). They are attracted to sites of inflammation by the chemicals released from injured tissues (chemotaxis). Neutrophils are able to engulf and digest smaller particles, while monocytes can phagocytize somewhat larger ones.

Monocytes give rise to *macrophages* (histiocytes), which become fixed in various tissues and attached to the inner walls of blood and lymphatic vessels. These relatively nonmotile, phagocytic cells, which can divide and produce new macrophages, are found in such organs as the lymph nodes, spleen, liver, and lungs. This diffuse group of phagocytic cells constitutes the **reticuloendothelial tissue** (reticuloendothelial system, or RES).

As a result of reticuloendothelial activities, foreign particles are removed from the lymph as it moves from the interstitial spaces to the blood stream. Any such foreign particles that reach the blood are likely to be removed by the phagocytes located in the vessels and tissues of the spleen, liver, or bone marrow.

Chart 19.3 summarizes the types of nonspecific immunity.

1. What is meant by an infection?
2. Explain six nonspecific defense mechanisms.
3. Define reticuloendothelial tissue.

Specific Immunity

Specific immunity is resistance to particular foreign agents, such as pathogens, or to the toxins they release. It involves a number of *immune mechanisms,* in which certain cells recognize the presence of particular foreign substances and act to eliminate them. The cells that function in immune mechanisms include lymphocytes and macrophages.

Origin of Lymphocytes

During fetal development (before birth), undifferentiated lymphocytes (stem cells) are released from the red bone marrow and carried away by the blood. About half of these cells reach the thymus, where they remain for a time. Within the thymus, these cells (thymocytes) undergo special processing and become differentiated. Thereafter, they are called *T-lymphocytes,* or *T-cells* (thymus-derived lymphocytes). Later, the T-cells are transported away from the thymus by the blood, wherein they comprise 70%–80% of the circulating lymphocytes. They tend to reside in various organs of the lymphatic system and are particularly abundant in the lymph nodes, the thoracic duct, and the white pulp of the spleen.

Lymphocytes that have been released from the red bone marrow and do not reach the thymus are processed in another part of the body (probably the fetal

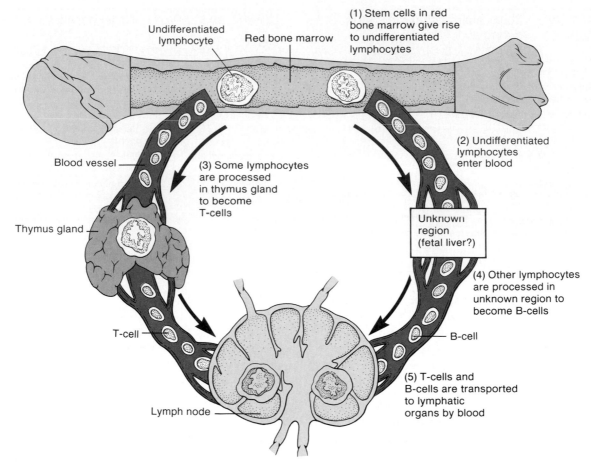

Figure 19.16   Bone marrow releases undifferentiated lymphocytes, which after processing, become T-lymphocytes and B-lymphocytes.

liver), and they differentiate into *B-lymphocytes* or *B-cells* (bone marrow-derived lymphocytes).

B-cells are distributed by the blood and constitute 20%–30% of the circulating lymphocytes. They settle in the lymphatic organs with the T-cells and are abundant in the lymph nodes, spleen, bone marrow, secretory glands, intestinal lining, and reticuloendothelial tissue (figures 19.16 and 19.17).

Antigens

Prior to birth, body cells somehow make an inventory of the proteins and various other large molecules that are present in the body. Subsequently, the substances that are present (self-substances) can be distinguished from foreign substances (nonself-substances). More specifically, T-cells and B-cells develop receptors on their surfaces that allow them to recognize foreign substances. When such substances are recognized, the lymphocytes can produce immune reactions against them.

Foreign substances to which lymphocytes respond are called **antigens.** Such substances include pro-

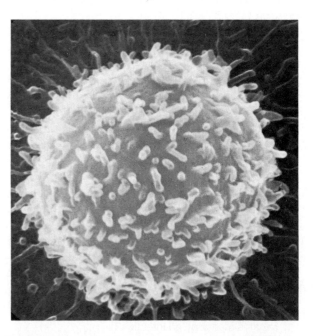

Figure 19.17   Scanning electron micrograph of a human circulating lymphocyte (×36,000).

CHART 19.4	A comparison of T-cells and B-cells	
Characteristic	T-cells	B-cells
Origin of undifferentiated cell	Red bone marrow	Red bone marrow
Site of differentiation	Thymus	Region outside the thymus (probably the fetal liver)
Primary locations	Lymphatic tissues, 70%–80% of the circulating lymphocytes	Lymphatic tissues, 20%–30% of the circulating lymphocytes
Primary functions	Responsible for cell-mediated immunity in which T-cells interact directly with antigen-bearing agents.	Responsible for antibody-mediated immunity in which B-cells interact indirectly with antigen-bearing agents by producing antibodies.

teins, polysaccharides, and some glycolipids. As a group, antigens are large, complex molecules, however, sometimes a smaller molecule that cannot stimulate an immune response by itself combines with a larger one and forms a combination that is antigenic. Such a small molecule is called a **hapten.** When lymphocytes are stimulated by such an antigenic combination, they can react either to the hapten or to the larger molecule of the combination. Substances that serve as haptens occur in various drugs, such as penicillin, in household and industrial chemicals, and in dust particles and products of animal skins (dander).

1. What is meant by specific immunity?
2. How do T-cells and B-cells originate?
3. What is the difference between an antigen and a hapten?

Functions of Lymphocytes

T-cells and B-cells respond to the antigens they recognize in different ways. For example, some T-cells attach themselves to foreign, antigen-bearing cells, such as bacterial cells, and interact with these foreign cells directly—that is, with cell-to-cell contact. This type of response is called **cell-mediated immunity** (CMI).

T-cells (and certain other cells, such as macrophages) also can synthesize and secrete a variety of polypeptides, called *lymphokines* (cytokines). These products, in turn, enhance various cellular responses to antigens. For example, the lymphokines called *interleukin-1* and *interleukin-2* stimulate the synthesis of several lymphokines from other T-cells. In addition, interleukin-1 helps activate resting T-cells, while interleukin-2 causes T-cells to proliferate and activates a certain type of T-cell (cytotoxic T-cells). Other lymphokines stimulate the production of leukocytes in the red bone marrow (colony-stimulating factor), cause growth and maturation of B-cells (interleukin-4), and activate macrophages (gamma-interferon).

T-cells may also secrete toxic substances (lymphotoxins) that are lethal to their antigen-bearing target cells, growth-inhibiting factors that prevent target-cell

growth, or interferon that inhibits the proliferation of viruses and tumor cells.

B-cells, on the other hand, act indirectly against antigens by producing and secreting globular proteins (immunoglobulins) called **antibodies.** The antibodies, in turn, are carried by the body fluids and react in various ways to destroy specific antigens or antigen-bearing particles. This type of response is called **antibody-mediated immunity** (AMI), or humoral immunity. Chart 19.4 compares the characteristics of T-cells and B-cells.

1. Define immunity.
2. Explain how T-cells and B-cells originate.
3. Explain the difference between an antigen and a hapten.
4. What are the functions of a T-cell? Of a B-cell?

Clones of Lymphocytes

Within the populations of T-cells and B-cells of each person are millions of varieties. The members of each variety originate from a single early cell, so that the members are all alike. Also, the members of each variety have a particular type of antigen-receptor on their surface membrane that is capable of responding only to a specific antigen. The term **clone** is used to refer to the members of each variety as a group.

Activation of B-cells

Before a lymphocyte can respond to the presence of an antigen, the lymphocyte must be activated. A B-cell, for example, becomes activated when it encounters an antigen whose molecular shape fits the shape of the B-cell's antigen receptors. In response to the receptor-antigen combination, the B-cell proliferates by mitosis, and its clone is enlarged.

Some of the newly formed members of the activated B-cell's clone become *plasma cells,* and they make use of their DNA information and protein-synthesizing mechanism (chapter 4) to produce antibody molecules. These antibody molecules are similar in structure to the antigen-receptor molecules that were present on the original B-cell's surface membrane. Thus, the antibodies are able to combine with the antigen-bearing agent that has invaded the body and react against it.

It is estimated that an individual's B-cells are able to produce as many as 10 million to one billion different varieties of antibodies, each of which reacts against a specific antigen. This ability provides a defense against a very large number of different disease-causing agents.

1. What is a clone of lymphocytes?
2. How is a B-cell activated?
3. What is the function of a plasma cell?

Antibody Molecules

The antibodies produced and secreted by B-cells are all soluble globular proteins. These proteins are called **immunoglobulins,** and they constitute the *gamma globulin* fraction of the plasma proteins (chapter 17).

Some persons partially or completely lack the ability to form immunoglobulins. As a consequence, they cannot produce normal concentrations of antibodies and are very susceptible to infectious diseases. This condition, which may be due to inheritance or acquired as a result of environmental factors, is termed *agammaglobulinemia.*

Each immunoglobulin molecule is composed of four chains of amino acids that are linked together by pairs of sulfur atoms (disulfide bonds). Two of these chains are identical light chains (L-chains), and two are identical heavy chains (H-chains). The heavy chains contain about twice as many amino acids as the light chains. Each of the five major types of immunoglobulin molecules contains a particular kind of heavy chain.

As with other proteins, the sequence of amino acids within the peptide chains is responsible for the unique, three-dimensional structure of each immunoglobulin molecule. This special structure, in turn, is related to the physiological properties of the molecule. For example, the terminal portions at one end of each of the heavy and light chains contain variable sequences of amino acids (variable regions), and these parts are specialized to react with the shape of a specific antigen molecule. The remaining portions of the chains (constant regions) are responsible for other properties of the immunoglobulin molecule, such as its ability to bond to cellular structures or to combine with chemical substances (figure 19.18).

Types of Immunoglobulins

Of the five major types of immunoglobulins, three constitute the bulk of the circulating antibodies. They are called immunoglobulin G, which accounts for about

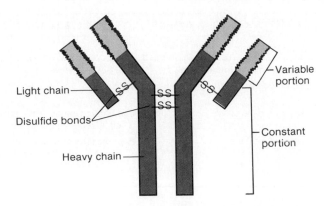

Figure 19.18 An immunoglobulin molecule consists basically of two identical light chains of amino acids and two identical heavy chains of amino acids.

80% of the antibodies; immunoglobulin A, which makes up about 13%; and immunoglobulin M, which is responsible for about 6%. Immunoglobulin D and immunoglobulin E account for the remainder of the antibodies.

Immunoglobulin G (IgG) occurs in plasma and tissue fluids, and is particularly effective against bacterial cells, viruses, and various toxins. IgG also activates a group of enzymes called *complement,* which is described in the following section.

Immunoglobulin A (IgA) is commonly found in the secretions of various exocrine glands, and occurs in milk, tears, nasal fluid, gastric juice, intestinal juice, bile, and urine.

At birth, infants normally possess no serum immunoglobulins that have been synthesized by their own cells. On the other hand, IgG can pass from the mother's blood through the placenta and into the infant's blood. Consequently, after receiving such maternal antibodies, a newborn has some protection against those conditions for which its mother is immune. This protection lasts for several months, during which time the maternal IgG in the infant's blood disappears, and the infant's cells begin synthesizing IgG for themselves.

On the other hand, IgA does not pass through the placenta. However, it can pass from a nursing mother to her infant by means of milk and other secretions from the breasts. These antibodies help protect the infant against various digestive and respiratory disturbances that otherwise might cause serious problems in early infancy.

Immunoglobulin M (IgM) is a type of antibody that develops in the blood plasma, apparently in response to contact with certain antigens in foods or bacteria. The agglutinins anti-A and anti-B, described in chapter 17, are examples of IgM. IgM also activates complement.

Immunoglobulin D (IgD) is found on the surfaces of most B-lymphocytes, especially those of in-

CHART 19.5	Characteristics of major immunoglobulins	
Type	Occurrence	Major function
IgG	Tissue fluid and plasma	Acts against bacterial cells, viruses, and toxins; activates complement
IgA	Secretions of exocrine glands	Acts against bacterial cells and viruses
IgM	Plasma	Reacts with antigens occurring naturally on some red blood cell membranes following certain blood transfusions; activates complement
IgD	Surface of most B-lymphocytes	Poorly understood, but apparently of limited significance
IgE	Secretions of exocrine glands	Promotes allergic reactions

fants. Although little is known about immunoglobulin D, the significance of this antibody is thought to be quite limited.

Immunoglobulin E (IgE) occurs in various exocrine secretions along with IgA. Its best understood action is associated with allergic reactions that are described in a subsequent section of this chapter.

Chart 19.5 summarizes the major immunoglobulins and their functions.

1. What is an immunoglobulin?
2. Describe the structure of an immunoglobulin molecule.
3. Name the five major types of immunoglobulins.

Actions of Antibodies

In general, antibodies react to antigens by attacking them directly, by activating a set of enzymes (complement) that attack the antigens, or by stimulating changes in local areas that help prevent the spread of the antigens.

In the case of direct attacks, antibodies may combine with antigens and cause them to clump together (agglutination) or to form insoluble substances (precipitation). Such actions make it easier for phagocytic cells to engulf the antigen-bearing agents and eliminate them. In other instances, antibodies cover the toxic portions of antigen molecules and neutralize their effects (neutralization), or they cause cell membranes to rupture (lysis), thereby destroying invading cells. However, under normal conditions, direct attack by antibodies is apparently of limited value in protecting the body against infections. Of greater importance is the antibody action of activating *complement*.

Complement is a group of inactive enzymes (complement system) that occurs in plasma and other body fluids. When certain IgG or IgM antibodies combine with antigens, reactive sites on the constant regions of antibody molecules become exposed, and this triggers a series of reactions leading to the activation of the complement enzymes. The activated enzymes produce a variety of effects, including attracting macrophages and neutrophils into the region (chemotaxis), causing antigen-bearing agents to clump together, altering cell membranes in ways that make the cells more susceptible to phagocytosis (opsonization), affecting the membranes of foreign cells so that they rupture (lysis), and altering the molecular structure of viruses, thereby making them harmless. Other enzymes promote the inflammation reaction, which helps prevent the spread of antigens to nearby tissues (figure 19.19).

Several human disorders are associated with hereditary deficiencies of certain components of the complement system. As a result of such deficiencies, the insoluble antibody-antigen complexes tend to accumulate abnormally in the blood. These substances may be deposited in organs, such as the kidneys, and may interfere with their functions. Other disorders related to complement deficiencies include systemic lupus erythematosus (SLE) and juvenile rheumatoid arthritis.

Immunoglobulin E also promotes inflammation reactions, but these reactions tend to be excessive and harmful to tissues in some individuals. This antibody is usually attached to the membranes of widely distributed mast cells. (See chapter 5.) When antigens are present, they combine with the antibodies, and the resulting antigen-antibody complexes stimulate the mast cells to release various substances. These substances, which include histamine, cause the changes associated with inflammation, such as vasodilation and edema.

Chart 19.6 summarizes the actions of antibodies.

1. In what general ways do antibodies act?
2. What is the function of complement?
3. How is complement activated?

Activation of T-cells

Although a B-cell can by itself recognize the antigen for which it is specialized to react, a T-cell requires the presence of another kind of cell, called an *accessory cell* (antigen-presenting cell), before the T-cell can

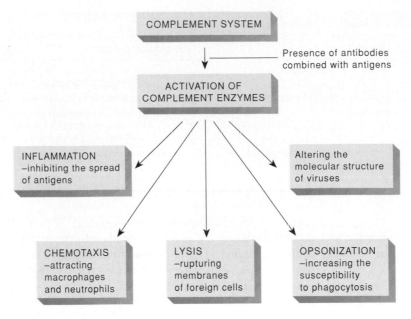

Figure 19.19 Actions of the complement system.

CHART 19.6	Actions of antibodies	
General action	Type of effect	Description
Direct attack		
	Agglutination	Causes antigens to clump together
	Precipitation	Causes antigens to form insoluble substances
	Neutralization	Causes antigens to lose toxic properties
	Lysis	Causes cell membranes to rupture
Activation of complement		
	Chemotaxis	Attracts macrophages and neutrophils into region
	Opsonization	Alters cell membranes so cells are more susceptible to being phagocytized
	Inflammation	Promotes local tissue changes that help prevent the spread of antigens

become activated. Macrophages, B-cells, and several other types of body cells can serve as accessory cells.

For example, when an antigen-bearing agent, such as an invading bacterium, is phagocytized by a macrophage, the agent is digested by the macrophage's lysosomes. Antigens that are released by the lysosomal digestive process are then moved to the macrophage's surface membrane and displayed on the membrane in association with certain protein molecules, called the *major histocompatibility complex* (MHC). A specialized type of T-cell, called a *T-helper cell,* may contact such a displayed antigen. If the antigen fits and combines with the T-helper cell's antigen receptors, the T-helper cell becomes activated.

When an activated T-helper cell encounters a B-cell that has already combined with an identical antigen, the T-helper cell interacts with the B-cell. As a result of this interaction, the T-helper cell releases some lymphokines, such as interleukin-4 and B-cell differentiation factor. These lymphokines, in turn, enhance the B-cell reaction to the antigen by stimulating the B-cell to proliferate, thus enlarging its clone of antibody-producing cells (figures 19.20 and 19.21). The lymphokines may also attract macrophages and leukocytes into inflamed tissues and help retain them there. Chart 19.7 summarizes the steps leading to antibody production as a result of B-cell and T-cell activities.

A second type of T-cell, called a *cytotoxic T-cell,* is specialized to recognize antigens that are produced in the cytoplasm of various body cells. For example, cells of tumors and cells that contain viruses usually produce specific kinds of antigens that are displayed on their surface membranes in association with MHC molecules. If a cytotoxic T-cell encounters an antigen that fits and combines with its antigen receptors, the cytotoxic T-cell becomes activated.

Once activated, a cytotoxic T-cell (killer T-cell) proliferates, enlarging its clone. Cytotoxic T-cells can bind to the surfaces of antigen-bearing cells, and the cytotoxic T-cells then release a protein that causes

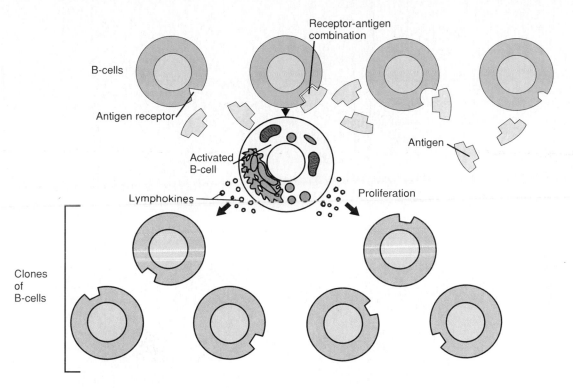

Figure 19.20 When a B-cell encounters an antigen that fits its antigen receptor, it becomes activated and proliferates, thus enlarging its clone.

Labels in figure: B-cells; Antigen receptor; Receptor-antigen combination; Antigen; Activated B-cell; Lymphokines; Proliferation; Clones of B-cells

CHART 19.7	Steps in the production of antibodies

B-cell activity
1. Antigen-bearing agents enter body tissues.
2. B-cell becomes activated when it encounters an antigen that fits its antigen receptors.
3. Activated B-cell proliferates, enlarging its clone.
4. Some of the newly formed B-cells become plasma cells.
5. Plasma cells synthesize and secrete antibodies whose molecular structure is similar to the activated B-cell's antigen receptors.
6. Antibodies combine with antigen-bearing agents, thus helping to destroy them.

T-cell activity
1. Antigen-bearing agents enter body tissues.
2. Accessory cell, such as a macrophage, phagocytizes antigen-bearing agent, and the macrophage's lysosomes digest the agent.
3. Antigens from the digested antigen-bearing agents are displayed on the surface membrane of the accessory cell.
4. T-helper cell becomes activated when it encounters a displayed antigen that fits its antigen receptors.
5. Activated T-helper cell releases lymphokines when it encounters a B-cell that has previously combined with an identical antigen-bearing agent.
6. Lymphokines stimulate the B-cell to proliferate.
7. Some of the newly formed B-cells become antibody-secreting plasma cells.
8. Antibodies combine with antigen-bearing agents, thus helping to destroy them.

porelike openings to appear in the antigen-bearing cell's membranes, thus destroying such cells. In this manner, cytotoxic T-cells, which continually monitor the body cells, can recognize and eliminate certain tumor cells and cells infected with various viruses.

Immune Responses

When B-cells or T-cells become activated after first encountering the antigens for which they are specialized to react, their actions constitute a **primary immune response.** During such a response, the plasma cells that are formed release antibodies into the lymph. The antibodies are transported to the blood and then to all body parts, where they help destroy antigen-bearing agents. The production and release of the antibodies continues for several weeks.

Following a primary immune response, some of the B-cells that were produced during proliferation of the clone remain dormant and serve as *memory cells* (figure 19.22). In this way, if the identical antigen is encountered in the future, the clones of these memory cells increase in size, and they can respond rapidly to the antigen to which they were previously sensitized. Such a subsequent reaction is called a **secondary immune response.**

As a result of a primary immune response, detectable concentrations of antibodies usually appear in

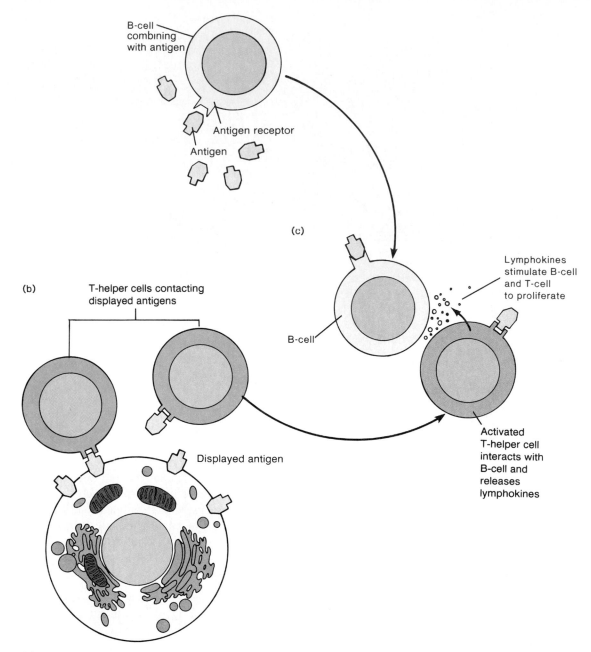

B-cell
combining
with antigen

Antigen receptor

Antigen

(c)

Lymphokines
stimulate B-cell
and T-cell
to proliferate

(b)

T-helper cells contacting
displayed antigens

B-cell

Displayed antigen

Activated
T-helper cell
interacts with
B-cell and
releases
lymphokines

(a)   **Macrophage displaying antigen**

Figure 19.21   (*a*) After digesting antigen-bearing agents, a
macrophage displays antigens on its surface. (*b*) T-helper cells
become activated when they contact displayed antigens that fit
their antigen receptors. (*c*) An activated T-helper cell interacts
with a B-cell that has combined with an identical antigen and
causes the B-cell to proliferate.

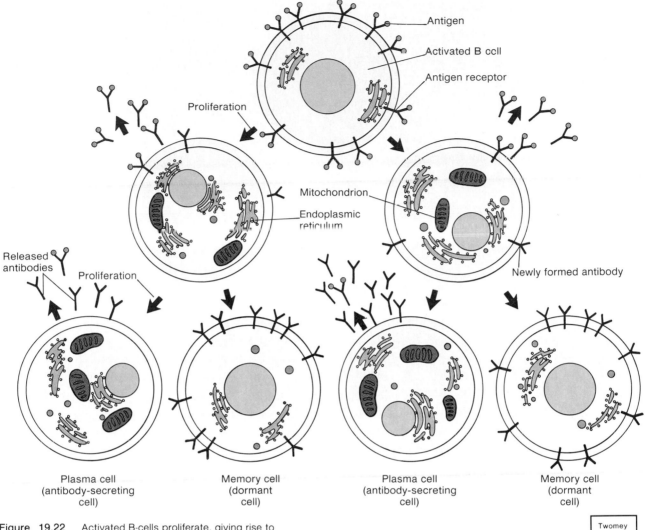

Antigen

Activated B cell

Antigen receptor

Proliferation

Mitochondrion

Endoplasmic reticulum

Released antibodies

Proliferation

Newly formed antibody

Plasma cell
(antibody-secreting
cell)

Memory cell
(dormant
cell)

Plasma cell
(antibody-secreting
cell)

Memory cell
(dormant
cell)

Twomey

Figure 19.22 Activated B-cells proliferate, giving rise to antibody-secreting plasma cells and dormant memory cells.

the body fluids within five to ten days following an exposure to antigens. If the identical antigen is encountered some time later, a secondary immune response may produce additional antibodies within a day or two (figure 19.23). Although such newly formed antibodies may persist in the body for only a few months, or perhaps a few years, the memory cells live much longer. Consequently, the ability to produce a secondary immune response may be long-lasting.

Because of its location, the skin is commonly invaded by antigen-carrying agents, and some of the cells associated with the skin seem to function in the immune system. For example, a relatively large number of T-cells normally infiltrate the skin, and the common epidermal cells (keratinocytes) are known to secrete interleukin-1, which is the lymphokine that helps activate T-cells. In addition, experimental evidence suggests that certain other epidermal cells may present antigens to T-cells, thus acting as accessory cells.

1. How do T-cells become activated?
2. What is the function of lymphokines?
3. How do cytotoxic T-cells destroy antigen-bearing cells?
4. What is the difference between a primary and a secondary immune response?

Types of Specific Immunity

One type of immunity is called *naturally acquired active immunity.* It occurs when a person who has been exposed to a live pathogen develops a disease and becomes resistant to that pathogen as a result of a primary immune response.

Another type of active immunity can be produced in response to a **vaccine.** Such a substance contains an antigen that can stimulate a primary immune response against a particular disease-causing agent, but does not produce the severe symptoms of that disease.

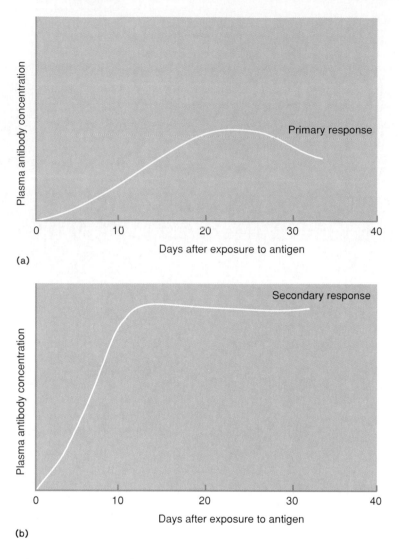

Figure 19.23 (a) A primary immune response produces a lesser concentration of antibodies than does (b) a secondary immune response.

For example, a vaccine might contain bacteria or viruses that have been killed or weakened so that they cannot cause a serious infection, or it may contain a toxin of an infectious organism that has been chemically altered to destroy its toxic effects. In any case, the antigens present in the vaccine still retain the characteristics needed to stimulate a primary immune response. Thus, a person who has been vaccinated is said to develop *artificially acquired active immunity*.

Vaccines are available to stimulate active immunity against a variety of diseases. These diseases include typhoid fever, cholera, whooping cough, diphtheria, tetanus, polio, measles (rubeola), rubella (German measles), mumps, influenza, hepatitis B, and bacterial pneumonia (pneumococcal infection).

Sometimes a person who has been exposed to infection needs protection against a disease-causing microorganism, but lacks the time needed to develop active immunity. In such a case, it may be possible to inject the person with ready-made antibodies. These antibodies may be obtained from *gamma globulin* separated from the blood plasma of persons who have already developed immunity against the particular disease.

A person who receives an injection of gamma globulin is said to have *artificially acquired passive immunity*. This type of immunity is called *passive* because the antibodies involved are not produced by the recipient's cells. Such immunity is relatively short-term, seldom lasting more than a few weeks. Furthermore, because the recipient's lymphocytes have not been activated against the pathogens for which the protection

Acquired Immune Deficiency Syndrome

Acquired immune deficiency syndrome (AIDS) was first recognized in 1981, and its incidence is roughly doubling each year. It is caused by a retrovirus—a virus that contains RNA, instead of DNA, as its genetic material. The virus is called *human immunodeficiency virus* (HIV-1), and it is widespread throughout the world. (A related virus, called HIV-2, which can also cause AIDS, occurs mainly in West Africa.) This virus infects T-helper cells—the cells required to activate B-lymphocytes and induce the production of antibodies—as well as some macrophages, B-cells, endothelial cells, and neuroglial cells.

When the virus infects a T-helper cell, its viral RNA can direct the synthesis of viral DNA (viral genes), which, in turn, may be incorporated into the host cell's DNA. Then the virus may become dormant, failing to produce any noticeable effects for some time. During this latency period, which may last for several years, the infected person may show no symptoms of AIDS, but can transmit viruses to uninfected persons.

If the viral genes in the infected T-helper cell are activated by some unknown factor, the genes induce the production of new AIDS viruses that can leave the host cell, infect other T-helper cells, and kill them. As a consequence of widespread T-helper cell destruction, the person's immune system is greatly impaired, and the group of symptoms (syndrome) that characterize AIDS develops. These symptoms include enlargement of lymph nodes, weight loss, and fever, as well as severe, opportunistic infections caused by a variety of bacteria, protozoa (such as *Pneumocystis carinii*), fungi, and viruses, and the appearance of certain forms of cancer (Kaposi's sarcoma, carcinoma of the skin and rectum, and B-cell lymphoma). Some patients develop serious neurological dysfunctions, including varying degrees of paralysis in the extremities, various sensory deficiencies, loss of deep tendon reflexes, exaggerated sensitivity to touch, and weakness in the urethral and anal sphincter muscles.

AIDS is most easily transmitted by the direct introduction of the HIV virus into the blood, as may occur during certain sexual activities, such as anal intercourse, in which the skin or mucous membranes are damaged. The virus can also be transmitted by contaminated hypodermic needles or by the transfusion of virus-containing blood or blood products and may be transmitted as a result of vaginal or oral sexual activity.

AIDS is apparently not spread by casual contact with AIDS patients. In fact, a study involving several hundred family members with AIDS patients living at home revealed that the AIDS virus had not been transmitted to any of the household members, except by sexual contact.

The persons who comprise the largest groups of AIDS patients are homosexual and bisexual males (about 70% of the AIDS patients in recent years), heterosexual intravenous drug abusers (19%), and heterosexuals (male or female) whose sexual partners are infected (4%). The remaining 7% of AIDS patients do not belong to one of these groups and have acquired the disease from some other source, such as a transfusion of, or skin exposure to, contaminated blood or blood products. The unborn fetuses of infected mothers, for example, acquire the disease in about 50% of the cases. As a result, one of the more rapidly growing groups of AIDS patients is comprised of children born to infected mothers who are intravenous drug users or whose sexual partners are intravenous drug users. In about 3% of AIDS patients, the source of the virus remains undetermined.

Antibodies appear in the blood of persons infected by the AIDS virus, and these antibodies can usually be detected by means of a simple blood test performed between two weeks and three months following exposure to the virus. If antibodies do not develop within several months following exposure to the virus, it is assumed that the person has not become infected.

At present, no vaccine is available to prevent AIDS infections, and there is no cure for the disease. To prevent contact with the AIDS virus (as well as other blood-borne viruses, such as the hepatitis B virus), healthcare workers are advised to avoid puncturing their skin with needles contaminated with blood or blood products. They should also take special precautions to prevent skin and mucous membrane contact with body fluids of all patients, including vaginal secretions, semen, cerebrospinal fluid, synovial fluid, serous fluids from the pleural, peritoneal, or pericardial cavities, and any other body fluids that contain visible blood.

According to the Surgeon General of the U.S. Public Health Service, the most certain way for members of the general population to avoid becoming infected with the AIDS virus (and to control the AIDS epidemic in the United States) is for individuals to avoid promiscuous sexual practices, to maintain mutually faithful monogamous sexual relationships, and to avoid injecting illicit drugs. In the absence of monogamy, the best way for a sexually active person to prevent getting AIDS is by using a latex condom during sexual intercourse, although this method is not 100% effective in preventing infections.

---

was needed, the person continues to be susceptible to those pathogens in the future.

During pregnancy, certain antibodies (IgG) are able to pass from the maternal blood into the fetal blood stream. This transfer is accomplished by receptor-mediated endocytosis (see chapter 3) and involves the presence of receptor sites on cells of the fetal yolk sac. (See chapter 23.) These receptor sites bind to a region that is common to the molecular structure of IgG molecules. Then, after entering the fetal cells, the antibodies are secreted into the fetal blood. As a result, the fetus acquires a limited amount of immunity against

CHART 19.8 Types of immunity

Type	Stimulus	Result
Naturally acquired active immunity	Exposure to live pathogens	Symptoms of a disease and stimulation of an immune response
Artificially acquired active immunity	Exposure to a vaccine containing weakened or dead pathogens	Stimulation of an immune response without the severe symptoms of a disease
Artificially acquired passive immunity	Injection of gamma globulin containing antibodies	Immunity for a short time without stimulating an immune response
Naturally acquired passive immunity	Antibodies passed to fetus from mother with active immunity	Short-term immunity for infant, without stimulating an immune response

the pathogens for which the mother has developed active immunities. In this case, the fetus is said to have *naturally acquired passive immunity,* which may remain effective for six months to a year after birth.

The types of immunity are summarized in chart 19.8.

Competence of the immune system tends to decline with advancing age. Consequently, elderly persons have a higher incidence of tumors and an increased susceptibility to infections, such as influenza, pneumonia, and tuberculosis. This decline in effectiveness is primarily due to a loss of T-cells. B-cell activity, on the other hand, is relatively unchanged with age.

The *tuberculin skin test* is used to detect individuals who have tuberculosis (TB) or who have had it (or a closely related infection) in the past. The test makes use of a tuberculin preparation called *purified protein derivative* (PPD), which is introduced into the superficial layers of the test subject's skin (Mantoux test). If the person's T-cells have been sensitized to the antigens of the mycobacteria that cause tuberculosis, an allergic reaction (positive test result) occurs within forty-eight to seventy-two hours. In a positive reaction, a localized region of the skin and subcutaneous tissue becomes hardened (indurated). The absence of this reaction (negative result) signifies that the person's T-cells have not previously been exposed to the mycobacterial antigens.

Allergic Reactions

Allergic reactions are closely related to immune responses in that both may involve the sensitizing of lymphocytes or the combining of antigens with antibodies. Allergic reactions, however, are likely to be excessive and to cause damage to tissues.

One form of allergic reaction can occur in almost anyone, but another form affects only those people who have inherited from their parents an ability to produce exaggerated immune responses.

A *delayed-reaction allergy* is an allergic response that may occur in anyone. It results from repeated exposure of the skin to certain chemical substances—commonly, household or industrial chemicals or some cosmetics. As a consequence of these repeated contacts, T-cells eventually become activated by the presence of the foreign substance, and a large number of T-cells collect in the skin. Their actions and the actions of macrophages they attract cause the release of various chemical factors, which, in turn, cause eruptions and inflammation of the skin (dermatitis). This reaction is called *delayed* because it usually takes about forty-eight hours to occur.

In other cases, an allergic reaction may occur within minutes after contact with a nonself-substance. Persons with this *immediate-reaction allergy* have an inherited ability to synthesize abnormally large amounts of IgE antibodies in response to the presence of certain antigens. In this instance, the allergic reaction involves the activation of B-cells, and the antigen that triggers the reaction is called an **allergen.**

In an immediate-reaction allergy, the B-cells become activated when the allergen is first encountered, and subsequent exposures trigger allergic reactions. The IgE involved is attached to the membranes of widely distributed *mast cells* and *basophils,* and when an allergen-antibody reaction occurs, these cells release a variety of substances, such as *histamine, prostaglandin D_2,* and *leukotrienes.* These substances, in turn, cause a variety of physiological effects, including the dilation of blood vessels, increased vascular permeability that leads to swollen tissues, contraction of bronchial and intestinal smooth muscles, and increased mucus production. The result is a severe inflammation reaction that is responsible for the symptoms of the allergy, such as hives, hay fever, asthma, eczema, or gastric disturbances (figure 19.24).

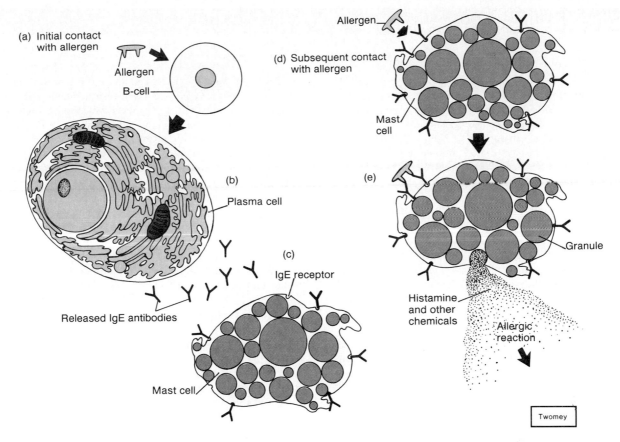

(a) Initial contact with allergen

Allergen

B-cell

(b)

Plasma cell

(c)

IgE receptor

Released IgE antibodies

Mast cell

(d) Subsequent contact with allergen

Allergen

Mast cell

(e)

Granule

Histamine and other chemicals

Allergic reaction

Twomey

Figure 19.24 In immediate-reaction allergy: (a) B-cells are activated when they contact an allergen; (b) an activated B-cell becomes an antibody-secreting plasma cell; (c) the antibodies become attached to the membranes of mast cells; (d) subsequently, when the allergen is encountered, it combines with the antibodies of the mast-cell membranes; and (e) the mast cell releases substances that cause the symptoms of the allergic reaction.

In most individuals, such severe responses to non-self-substances are thought to be inhibited by a special group of T-cells called *suppressor cells.* However, in a person with an immediate-reaction allergy, this control function of the T-cells seems to be less effective. Even so, the suppressor cells may eventually interfere with the production of IgE, and the allergic reaction is terminated.

Transplantation and Tissue Rejection

It is occasionally desirable to transplant some tissue or an organ, such as the skin, kidney, heart, or liver, from one person to another to replace a nonfunctional, damaged, or lost body part. In these cases, there is a danger that the recipient's cells may recognize the donor's tissues as being foreign. This triggers the recipient's immune mechanisms, which may act to destroy the donor tissue. Such a response is called a **tissue rejection reaction.**

Tissue rejection involves the activities of lymphocytes and both cell-mediated and antibody-mediated responses—responses similar to those that occur when

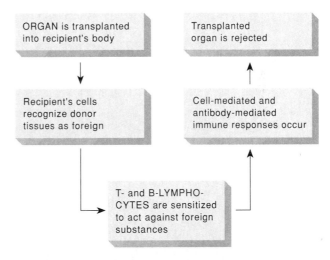

Figure 19.25 An organ transplant may be followed by a tissue rejection reaction.

any nonself-substances are encountered (figure 19.25). The greater the antigenic difference between the proteins of the recipient tissues and the donor tissues, the more rapid and severe the rejection reaction is likely to be. Thus, the reaction can sometimes be minimized by matching the recipient and donor tissues. This means

Autoimmune Diseases

As mentioned previously, immune responses are usually directed toward nonself-substances or foreign molecules, while self-substances are tolerated by the immune mechanism. Occasionally, something happens to change this, and the tolerance to self-substances is lost. As a result, cell-mediated and antibody-mediated immune responses may be directed toward a person's own tissues. Such a condition is called an **autoimmune disease** (autoallergy).

Although their causal mechanism is not well understood, autoimmune diseases are more common in older people, and they may involve viral or bacterial infections. Some investigators believe that the release of abnormally large quantities of antigens may occur when infectious agents cause tissue damage. These agents may also cause body proteins to change into forms that stimulate antibody production. At the same time, the activity of suppressor cells, which are thought to limit this type of reaction, may be repressed.

After such antibodies appear, the concentration of the antibodies sometimes fluctuates greatly. Also, they may persist indefinitely, or they may disappear.

Some autoimmune diseases affect specific organs, while others produce more generalized effects. For example, in the disease *autoimmune thyroiditis,* the effects are directed toward the thyroid gland. In this condition, antibodies are produced by B-cells that respond abnormally to self-proteins associated with thyroid hormones (thyroglobulin) and to substances in thyroid epithelial cell membranes. As a result of the antibody actions, the gland may become inflamed, and much of its tissue may be destroyed.

Systemic lupus erythematosus (SLE), which occurs most commonly in young women, is an example of an autoimmune disease that affects many organs and tissues. In this case, B-cells produce antibodies that react with cell nuclei and various cytoplasmic substances. The resulting antigen-antibody combinations tend to accumulate in membranes of the kidneys, heart, lungs, blood vessels, and joints, creating functional problems in these organs. Also, antibodies that react with red blood cells, platelets, and lymphocytes may be produced. One of the early symptoms of this disease is an inflammation of the skin over the nose and cheeks.

Other autoimmune diseases include myasthenia gravis, thyrotoxicosis (Graves disease), chronic atrophic gastritis, primary adrenal atrophy, juvenile rheumatoid arthritis, insulin-dependent diabetes mellitus, multiple sclerosis, and systemic sclerosis (scleroderma).

---

locating a donor whose tissues are antigenically similar to those of the person needing a transplant—a procedure much like matching the blood of a donor with that of a recipient before giving a blood transfusion.

> Nearly every human cell contains a variety of antigens on the surface of its outer membrane. These antigens, which are inherited, comprise the *major histocompatibility complex* (MHC) that was described previously. The types of these antigens present on the cell membrane vary from person to person. The antigens of the MHC that occur on the white blood cells are called *human leukocyte antigens* (HLA), and they are used to match the tissues of a donor and a recipient before a tissue or organ transplant is performed.
>
> It is primarily the MHC antigens that the host immune system may react against. Thus, as a rule, the more closely the antigens of the donor and recipient match, the greater the chance that a tissue or organ transplant will succeed. (Note: Only identical twins have a perfect match for MHC antigens.)

Another approach to reducing the rejection of transplanted tissue involves the use of *immunosuppressive drugs.* These substances interfere with the recipient's immune mechanisms in various ways. For example, a drug may suppress the formation of antibodies or the production of T-cells, thereby reducing the humoral and cellular responses.

Unfortunately, the use of immunosuppressive drugs leaves the recipient relatively unprotected against infections. Although the drug may prevent a tissue rejection reaction, the recipient may develop a serious infectious disease, such as pneumonia, that is difficult to control and may cause death.

> One of the more commonly used immunosuppressive drugs is *cyclosporine.* This drug, which is a product of a fungus, seems to depress the secretion of lymphokines by T-helper cells and thus interfere with the tissue rejection reaction. At the same time, cyclosporine allows suppressor T-cells and neutrophils to continue functioning so that some portions of the immune system remain intact.

1. Explain the difference between active and passive immunities.
2. In what ways is an allergic reaction related to an immune reaction?
3. In what ways is a tissue rejection reaction related to an immune response?

Clinical Terms Related to the Lymphatic System and Immunity

anaphylaxis (an''ah-fi-lak'sis) Immediate hypersensitivity to the presence of a foreign substance.

asplenia (ah-sple'ne-ah) The absence of a spleen.

autograft (aw'to-graft) The transplantation of tissue from one part of the body to another part of the same body.

histocompatibility (his''to-kom-pat''i-bil'i-te) Compatibility between the tissues of a donor and the tissues of a recipient, based on antigenic similarities.

homograft (ho'mo-graft) The transplantation of tissue from one person to another person.

immunocompetence (im''u-no-kom'pe-tens) The ability to produce an immune response to the presence of antigens.

immunodeficiency (im''u-no-de-fish'en-se) A lack of the ability to produce an immune response.

lymphadenectomy (lim-fad''ĕ-nek'to-me) The surgical removal of lymph nodes.

lymphadenopathy (lim-fad''ĕ-nop'ah-the) Enlargement of the lymph nodes.

lymphadenotomy (lim-fad''ĕ-not'o-me) An incision of a lymph node.

lymphocytopenia (lim''fo-si''to-pe'ne-ah) An abnormally low concentration of lymphocytes in the blood.

lymphocytosis (lim''fo-si''to'sis) An abnormally high concentration of lymphocytes in the blood.

lymphoma (lim-fo'mah) A tumor composed of lymphatic tissue.

lymphosarcoma (lim''fo-sar-ko'mah) A cancer within the lymphatic tissue.

splenectomy (sple-nek'to-me) The surgical removal of the spleen.

splenitis (sple-ni'tis) An inflammation of the spleen.

splenomegaly (sple''no-meg'ah-le) An abnormal enlargement of the spleen.

splenotomy (sple-not'o-me) An incision of the spleen.

thymectomy (thi-mek'to-me) The surgical removal of the thymus.

thymitis (thi-mi'tis) An inflammation of the thymus.

Chapter Summary

Introduction (page 716)

The lymphatic system is closely associated with the cardiovascular system. It transports excess tissue fluid to the blood stream and helps defend the body against disease-causing agents.

Lymphatic Pathways (page 716)

1. Lymphatic capillaries
   a. Lymphatic capillaries are microscopic, closed-ended tubes that extend into interstitial spaces.
   b. They receive lymph through their thin walls.
   c. Lacteals are lymphatic capillaries in the villi of the small intestine.
2. Lymphatic vessels
   a. Lymphatic vessels are formed by the merging of lymphatic capillaries.
   b. They have walls similar to veins and possess valves that prevent backflow of lymph.
   c. Larger lymphatic vessels lead to lymph nodes and then merge into lymphatic trunks.
3. Lymphatic trunks and collecting ducts
   a. Lymphatic trunks drain lymph from relatively large body regions.
   b. Trunks lead to two collecting ducts within the thorax.
   c. Collecting ducts join the subclavian veins.

Tissue Fluid and Lymph (page 718)

1. Tissue fluid formation
   a. Tissue fluid originates from blood plasma and includes water and dissolved substances that have passed through the capillary wall.
   b. It generally lacks proteins of large molecular size, but some smaller protein molecules leak into interstitial spaces.
   c. As the protein concentration of tissue fluid increases, osmotic pressure increases also.
2. Lymph formation
   a. Rising osmotic pressure in tissue fluid interferes with the return of water to the blood capillaries.
   b. Increasing pressure within interstitial spaces forces some tissue fluid into lymphatic capillaries, and this fluid becomes lymph.
3. Function of lymph
   a. Lymph returns protein molecules to the blood stream.
   b. It transports foreign particles to the lymph nodes.

Movement of Lymph (page 720)

1. Flow of lymph
   a. Lymph is under low pressure and may not flow readily without aid from external forces.
   b. Forces that aid in the movement of lymph include squeezing action of skeletal muscles and low pressure in the thorax created by breathing movements.
2. Obstruction of lymph movement
   a. Any condition that interferes with the flow of lymph results in edema.
   b. Obstruction of lymphatic vessels due to surgery also results in edema.

Lymph Nodes (page 721)

1. Structure of a lymph node
   a. Lymph nodes are usually bean-shaped, with blood vessels, nerves, and efferent lymphatic vessels attached to the indented region; afferent lymphatic vessels enter at points on the convex surface.
   b. Lymph nodes are enclosed in connective tissue that extends into the nodes and subdivides them into nodules.
   c. Nodules contain masses of lymphocytes and macrophages, as well as spaces through which lymph flows.
2. Locations of lymph nodes
   a. Lymph nodes generally occur in groups or chains along the paths of larger lymphatic vessels.
   b. They occur primarily in cervical, axillary, and inguinal regions, and within the pelvic, abdominal, and thoracic cavities.
3. Functions of lymph nodes
   a. Lymph nodes are centers for production of lymphocytes that act against foreign particles.
   b. They contain macrophages that remove foreign particles from lymph.

Thymus and Spleen (page 723)

1. Thymus
   a. The thymus is a soft, bilobed organ located within the mediastinum.
   b. It tends to decrease in size after puberty.
   c. It is composed of lymphatic tissue, which is subdivided into lobules.
   d. Lobules contain lymphocytes, most of which are inactive, that develop from precursor cells in bone marrow.
   e. Some lymphocytes leave the thymus and function in providing immunity.
   f. The thymus may secrete a hormone called thymosin, which stimulates lymphocytes that have migrated to other lymphatic tissues.

2. Spleen
   a. The spleen is located in the upper left portion of the abdominal cavity.
   b. It resembles a large lymph node that is encapsulated and subdivided into lobules by connective tissue.
   c. Spaces within lobules are filled with blood.
   d. It contains numerous macrophages and lymphocytes, which filter foreign particles and damaged red blood cells from the blood.

Body Defenses against Infection (page 726)

Infection is caused by the presence and multiplication of pathogens. The body is equipped with specific and nonspecific defenses against infection.

Nonspecific Immunity (page 726)

1. Species resistance
   Each species of organism is resistant to certain diseases that may affect other species, but susceptible to diseases that other species may be able to resist.
2. Mechanical barriers
   a. Mechanical barriers include skin and mucous membranes.
   b. As long as mechanical barriers remain unbroken, they prevent entrance of some pathogens.
3. Enzymatic actions
   a. Enzymes of gastric juice are lethal to some pathogens.
   b. Enzymes in tears have antibacterial actions.
4. Interferon
   a. Interferon is a group of hormonelike peptides produced by certain cells in response to the presence of viruses or tumor cells.
   b. It can interfere with the proliferation of viruses; it stimulates phagocytosis, and enhances the activity of cells that help resist infections and the growth of tumors.
5. Inflammation
   a. Inflammation is a tissue response to damage, injury, or infection.
   b. The response includes localized redness, swelling, heat, and pain.
   c. Chemicals released by damaged tissues attract various white blood cells to the site of inflammation.
   d. Clotting may occur in body fluids that accumulate in affected tissues.
   e. Fibrous connective tissue may form a sac around the injured tissue and thus prevent the spread of pathogens.

6. Phagocytosis
    a. The most active phagocytes in blood are neutrophils and monocytes; monocytes give rise to macrophages, which remain fixed in tissues.
    b. Phagocytic cells associated with the linings of blood vessels in the bone marrow, liver, spleen, and lymph nodes constitute the reticuloendothelial tissue.
    c. Phagocytes remove foreign particles from tissues and body fluids.

Specific Immunity (page 727)

1. Origin of lymphocytes
    a. Lymphocytes originate in red bone marrow and are released into the blood before they become differentiated.
    b. Some reach the thymus where they become T-cells.
    c. Those failing to reach the thymus become B-cells after being processed in some other region of the body.
    d. Both T-cells and B-cells tend to reside in organs of the lymphatic system.
2. Antigens
    a. Before birth, body cells make an inventory of the proteins and other large molecules present in the body.
    b. After the inventory, lymphocytes develop receptors that allow them to differentiate between foreign substances and self-substances.
    c. Antigens are foreign substances that combine with T-cell and B-cell surface receptors and stimulate these cells to cause an immune reaction.
    d. Haptens are small molecules that can combine with larger ones, forming combinations that are antigenic.
3. Functions of lymphocytes
    a. Some T-cells interact with antigen-bearing agents directly, providing cell-mediated immunity.
    b. T-cells secrete lymphokines, such as interleukins, that enhance cellular responses to antigens.
    c. T-cells may also secrete substances that are toxic to their target cells, growth inhibiting factors, or interferon.
    d. B-cells interact with antigen-bearing agents indirectly, providing antibody-mediated immunity.
4. Clones of lymphocytes
    a. Within the populations of T-cells and B-cells there are millions of varieties.
    b. The members of each variety respond only to a specific antigen.
    c. As a group, the members of each variety form a clone.

5. Activation of B-cells
    a. A B-cell is activated when it encounters an antigen that fits the B-cell's antigen receptors.
    b. An activated B-cell proliferates, and its clone is enlarged.
    c. Some activated B-cells become antibody-producing plasma cells.
    d. The antibody molecules react against the antigen-bearing agent that stimulated their production.
    e. An individual's B-cells provide defense against a very large number of disease-causing agents.
6. Antibody molecules
    a. Antibodies are proteins called immunoglobulins.
    b. They constitute the gamma globulin fraction of plasma.
    c. Each immunoglobulin molecule consists of four chains of amino acids linked together.
    d. Variable regions at the ends of these chains are specialized to react with antigens.
7. Types of immunoglobulins
    a. The five major types of immunoglobulins are IgG, IgA, IgM, IgD, and IgE.
    b. IgG, IgA, and IgM make up most of the circulating antibodies.
8. Actions of antibodies
    a. Antibodies attack antigens directly, activate complement, or stimulate local tissue changes that are unfavorable to antigen-bearing agents.
    b. Direct attacks occur by means of agglutination, precipitation, neutralization, or lysis.
    c. Activated enzymes of complement attract phagocytes, alter cells so they become more susceptible to phagocytosis, and promote inflammation.
9. Activation of T-cells
    a. T-cell activation requires the presence of an accessory cell.
    b. When a macrophage acts as an accessory cell, it phagocytizes an antigen-bearing agent, digests the agent, and displays the antigens on its surface membrane in association with certain protein molecules (MHC).
    c. A T-helper cell becomes activated when it encounters displayed antigens for which it is specialized to react.
    d. When an activated T-helper cell contacts a B-cell that has combined with the antigen for which the activated T-helper cell is specialized to react, the T-cell and the B-cell interact.
    e. In response to the T-cell and B-cell interaction, the T-helper cell releases lymphokines that cause the B-cell to proliferate and macrophages to be attracted into the affected tissues.
    f. Cytotoxic T-cells are specialized to react against antigens produced by tumor cells or cells infected with viruses.

10. Immune responses
    a. When B-cells or T-cells first encounter an antigen for which they are specialized to react, the reaction is called a primary immune response.
        (1) During this response, antibodies are produced for several weeks.
        (2) Some B-cells and T-cells remain dormant as memory cells.
    b. A secondary immune response occurs rapidly if the same antigen is encountered later.
11. Types of specific immunity
    a. A person who encounters a pathogen and has a primary immune response develops naturally acquired active immunity.
    b. A person who receives vaccine containing a dead or weakened pathogen develops artificially acquired active immunity.
    c. A person who receives an injection of gamma globulin that contains ready-made antibodies has artificially acquired passive immunity.
    d. When antibodies pass through a placental membrane from a pregnant woman to her fetus, the fetus develops naturally acquired passive immunity.
    e. Active immunity lasts much longer than passive immunity.
12. Allergic reactions
    a. Allergic reactions involve antigens combining with antibodies; such reactions are likely to be excessive or violent, and may cause tissue damage.
    b. Delayed-reaction allergy, which can occur in anyone and cause inflammation of the skin, results from repeated exposure to antigenic substances.
    c. A person with immediate-reaction allergy has an inherited ability to produce an abnormally large amount of IgE.
    d. Allergic reactions may damage mast cells, which, in turn, release histamine and serotonin.
    e. Released chemicals are responsible for the symptoms of the allergic reaction: hives, hay-fever, asthma, eczema, or gastric disturbances.
    f. An allergic reaction is usually terminated by suppressor cells that inhibit the production of IgE.
13. Transplantation and tissue rejection
    a. If tissue is transplanted from one person to another, the recipient's cells may recognize the donor's tissue as foreign and act against it.
    b. Tissue rejection reaction may be reduced by matching the donor and recipient tissues or by using immunosuppressive drugs.
    c. Immunosuppressive drugs interfere with the recipient's immune mechanisms, but cause the recipient to be very susceptible to infection.

Clinical Application of Knowledge

1. Based on your understanding of the functions of lymph nodes, how would you explain the fact that enlarged nodes are often removed for microscopic examination as an aid to diagnosing certain disease conditions?
2. Why is it true that an injection into the skin is, to a large extent, an injection into the lymphatic system?
3. Explain why vaccination provides long-lasting protection against a disease, while gamma globulin provides only short-term protection.
4. When a breast is removed surgically for the treatment of breast cancer, the lymph nodes in the nearby axillary region are sometimes excised also. Why is this procedure likely to be followed by swelling of the arm on the treated side?
5. If an infant was found to be lacking a thymus because of a developmental disorder, what could be predicted about the infant's susceptibility to infections? Why?
6. Although the AIDS virus is not transmitted by casual contact with infected persons, what precautions should healthcare workers take when caring for such patients?

Review Activities

1. Explain how the lymphatic system is related to the cardiovascular system.
2. Trace the general pathway of lymph from the interstitial spaces to the blood stream.
3. Identify and describe the locations of the major lymphatic trunks and collecting ducts.
4. Distinguish between tissue fluid and lymph.
5. Describe the primary functions of lymph.
6. Explain why physical exercise promotes lymphatic circulation.
7. Explain how a lymphatic obstruction leads to edema.
8. Describe the structure of a lymph node, and list its major functions.
9. Locate the major body regions occupied by lymph nodes.
10. Describe the structure and functions of the thymus.
11. Describe the structure and functions of the spleen.
12. Distinguish between specific and nonspecific body defenses against infection.
13. Explain what is meant by *species resistance*.
14. Name two mechanical barriers to infection.
15. Describe how enzymatic actions function as defense mechanisms.
16. Define *interferon,* and explain its action.
17. List the major symptoms of inflammation, and explain why each occurs.

18. Identify the major phagocytic cells in the blood and other tissues.
19. Explain where the reticuloendothelial tissue is located.
20. Review the origin of T-cells and B-cells.
21. Distinguish between an antigen and an antibody.
22. Define *hapten.*
23. Explain what is meant by *cell-mediated immunity.*
24. Define *lymphokine.*
25. Explain what is meant by *antibody-mediated immunity.*
26. Define *clone of lymphocytes*
27. Explain how a B-cell is activated.
28. Explain the function of plasma cells.
29. Describe an immunoglobulin molecule.
30. Distinguish between the variable region and the constant region of an immunoglobulin molecule.
31. List the major types of immunoglobulins, and describe their main functions.
32. Explain four mechanisms by which antibodies may attack antigens directly.
33. Explain the function of complement.
34. Describe how T-cells become activated.
35. Distinguish between a primary and a secondary immune response.
36. Distinguish between active and passive immunity.
37. Define *vaccine.*
38. Explain how a vaccine produces its effect.
39. Describe how a fetus may obtain antibodies from the maternal blood.
40. Explain the relationship between an allergic reaction and an immune response.
41. Distinguish between an antigen and an allergen.
42. List the major events leading to a delayed-reaction allergic response.
43. Describe how an immediate-reaction allergic response may occur.
44. Explain the relationship between a tissue rejection and an immune response.
45. Describe two methods used to reduce the severity of a tissue rejection reaction.

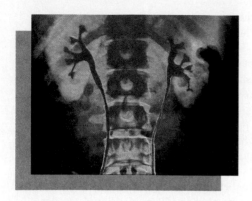

Urinary System

*T*he body cells form a variety of wastes as by-products of metabolic processes, and if these substances accumulate, their effects are likely to be toxic.

Body fluids, such as blood and lymph, carry wastes away from the tissues that produce them. Other parts remove these wastes from the blood and transport them to the outside. The respiratory system, for example, removes carbon dioxide from the blood, while the *urinary system* removes various salts and nitrogenous wastes. In both systems, the wastes are carried to the outside through tubular organs.

The urinary system also helps to maintain the normal concentrations of water and electrolytes within the body fluids, to regulate the pH and volume of body fluids, and to control red blood cell production and blood pressure ■

Chapter Objectives	Key Terms	Aids to Understanding Words

Chapter Objectives

After you have studied this chapter, you should be able to:

1. Name the organs of the urinary system and list their general functions.

2. Describe the locations of the kidneys and the structure of a kidney.

3. List the functions of the kidneys.

4. Trace the pathway of blood through the major vessels within a kidney.

5. Describe a nephron and explain the functions of its major parts.

6. Explain how glomerular filtrate is produced and describe its composition.

7. Explain how various factors affect the rate of glomerular filtration and how this rate is regulated.

8. Discuss the role of tubular reabsorption in urine formation.

9. Explain why the osmotic concentration of the glomerular filtrate changes as it passes through a renal tubule.

10. Describe a countercurrent mechanism and explain how it helps concentrate urine.

11. Define *tubular secretion* and explain its role in urine formation.

12. Describe the structure of the ureters, urinary bladder, and urethra.

13. Discuss the process of micturition and explain how it is controlled.

14. Complete the review activities at the end of this chapter. Note that the items are worded in the form of specific learning objectives. You may want to refer to them before reading the chapter.

Key Terms

afferent arteriole (af'er-ent ar-te're-ōl)

autoregulation (aw''to-reg''u-la'shun)

countercurrent mechanism (kown'ter-kur'ent mek''ah-nizm)

detrusor muscle (de-truz'or mus'l)

efferent arteriole (ef'er-ent ar-te're-ōl)

glomerulus (glo-mer'u-lus)

juxtaglomerular apparatus (juks''tah-glo-mer'u-lar ap''ah-ra'tus)

micturition (mik''tu-rish'un)

nephron (nef'ron)

peritubular capillary (per''i-tū'bu-lar kap' ĭ-ler''e)

renal corpuscle (re'nal kor'pusl)

renal cortex (re'nal kor'teks)

renal medulla (re'nal mĕ-dul'ah)

renal plasma threshold (re'nal plaz'mah thresh'old)

renal tubule (re'nal tu'būl)

retroperitoneal (re''tro-per''ĭ-to-ne'al)

Aids to Understanding Words

calyc-, small cup: major *calyc*es—cuplike subdivisions of the renal pelvis.

cort-, covering: renal *cort*ex—a shell of tissue surrounding the inner region of a kidney.

detrus-, to force away: *detrus*or muscle—a muscle within the bladder wall that causes urine to be expelled.

glom-, little ball: *glom*erulus—a cluster of capillaries within a renal corpuscle.

juxta-, near to: *juxta*medullary nephron—a nephron located near the renal medulla.

mict-, to pass urine: *mict*urition—the process of expelling urine from the bladder.

nephr-, pertaining to the kidney: *nephr*on—the functional unit of a kidney.

papill-, nipple: renal *papill*ae—small elevations that project into a renal calyx.

ren-, kidney: *ren*al cortex—outer region of a kidney.

trigon-, a triangular shape: *trigon*e—a triangular area on the internal floor of the bladder.

The urinary system consists of the following parts: a pair of glandular *kidneys,* which remove substances from the blood, form urine, and help regulate various metabolic processes; a pair of tubular *ureters,* which transport urine away from the kidneys; a saclike *urinary bladder,* which serves as a urine reservoir; and a tubular *urethra,* which conveys urine to the outside of the body. These organs are shown in figures 20.1 and 20.2.

Kidneys

A **kidney** is a reddish brown, bean-shaped organ with a smooth surface. It is about 12 centimeters long, 6 centimeters wide, and 3 centimeters thick in an adult, and it is enclosed in a tough, fibrous capsule (tunic fibrosa).

Location of the Kidneys

The kidneys lie on either side of the vertebral column in a depression high on the posterior wall of the abdominal cavity.

Although the positions of the kidneys may vary slightly with changes in posture and with breathing movements, their upper and lower borders are generally at the levels of the twelfth thoracic and third lumbar vertebrae, respectively. The left kidney is usually about 1.5 to 2 centimeters higher than the right one.

The kidneys are positioned *retroperitoneally,* which means they are behind the parietal peritoneum and against the deep muscles of the back. They are held in position by connective tissue (renal fascia) and masses of adipose tissue (renal fat) that surround them. (See figure 20.3 and reference plates 58, 59.)

Structure of a Kidney

The lateral surface of each kidney is convex, but its medial side is deeply concave. The resulting medial depression leads into a hollow chamber called the **renal sinus.** The entrance to this sinus is termed the *hilum,* and through it pass various blood vessels, nerves, lymphatic vessels, and the ureter (figure 20.1).

The superior end of the ureter is expanded to form a funnel-shaped sac called the **renal pelvis,** which is located inside the renal sinus. The pelvis is subdivided into two or three tubes, called *major calyces* (sing. *calyx*), and they, in turn, are subdivided into several (eight to fourteen) *minor calyces* (figure 20.4).

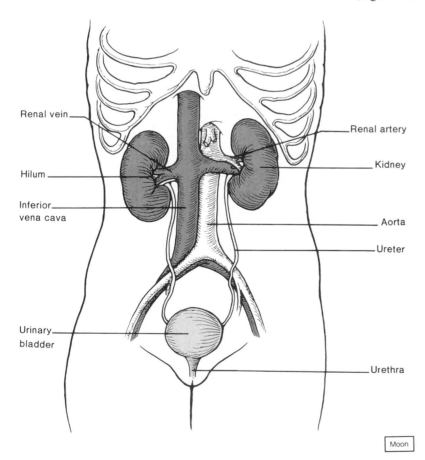

Renal vein

Hilum

Inferior vena cava

Urinary bladder

Renal artery

Kidney

Aorta

Ureter

Urethra

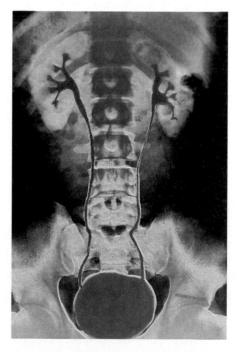

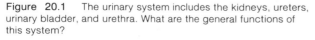

Figure 20.1 The urinary system includes the kidneys, ureters, urinary bladder, and urethra. What are the general functions of this system?

Figure 20.2 What features of the urinary system do you recognize in this falsely colored X-ray film?

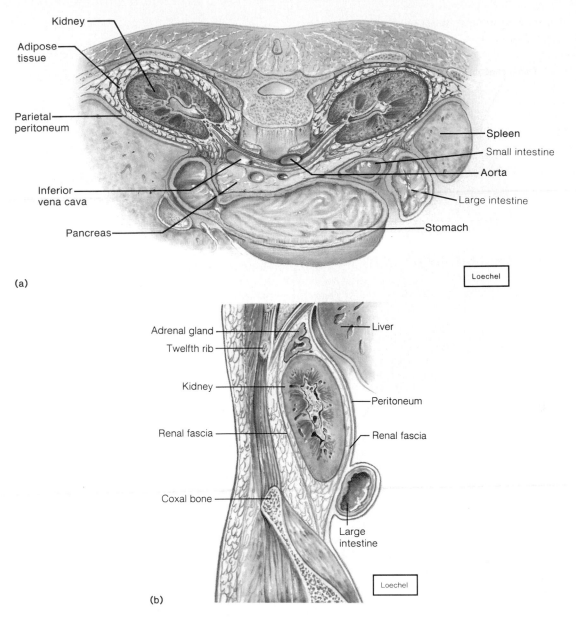

Labels in figure (a):
Kidney
Adipose tissue
Parietal peritoneum
Inferior vena cava
Pancreas
Spleen
Small intestine
Aorta
Large intestine
Stomach
Loechel
(a)

Labels in figure (b):
Adrenal gland
Twelfth rib
Kidney
Renal fascia
Coxal bone
Liver
Peritoneum
Renal fascia
Large intestine
Loechel
(b)

Figure 20.3 (a) Transverse sections of the kidneys, which are located behind the parietal peritoneum, surrounded and supported by adipose and other connective tissue; (b) sagittal section of a kidney.

A series of small elevations project into the renal sinus from its wall. These projections are called *renal papillae,* and each of them is pierced by tiny openings that lead into a minor calyx.

The substance of the kidney includes two distinct regions: an inner medulla and an outer cortex. The **renal medulla** is composed of conical masses of tissue called *renal pyramids,* whose bases are directed toward the convex surface of the kidney, and whose apexes form the renal papillae. The tissue of the medulla appears striated due to the presence of microscopic tubules leading from the cortex to the renal papillae.

The **renal cortex,** which appears somewhat granular, forms a shell around the medulla. Its tissue dips into the medulla between adjacent renal pyramids, forming *renal columns.* The granular appearance of the cortex is due to the random arrangement of tiny tubules associated with the **nephrons,** the functional units of the kidney.

Functions of the Kidneys

The kidneys remove metabolic wastes from the blood and excrete them to the outside. They also carry on a variety of equally important regulatory activities including helping control the rate of red blood cell formation by secreting the hormone *erythropoietin* (see chapter 17), helping regulate the blood pressure by secreting the enzyme *renin* (see chapter 18), and helping

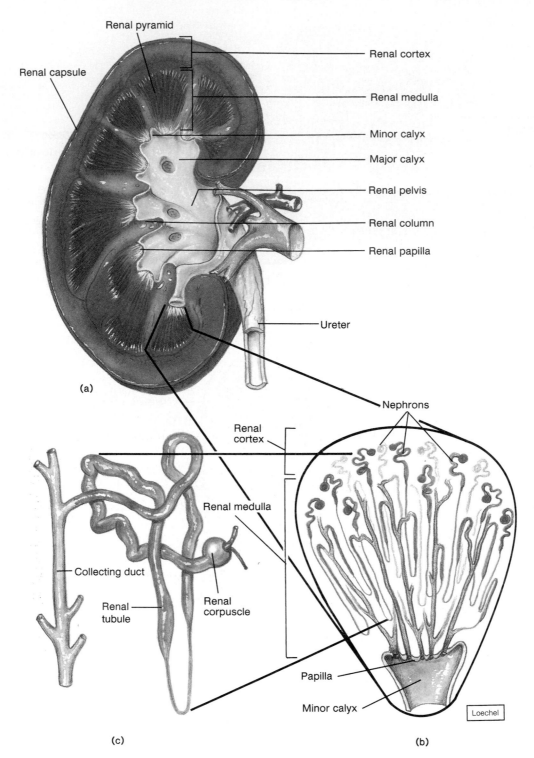

Renal pyramid

Renal capsule

Renal cortex

Renal medulla

Minor calyx

Major calyx

Renal pelvis

Renal column

Renal papilla

Ureter

(a)

Nephrons

Renal cortex

Renal medulla

Collecting duct

Renal tubule

Renal corpuscle

Papilla

Minor calyx

Loechel

(c)

(b)

Figure 20.4 (*a*) Longitudinal section of a kidney; (*b*) a renal pyramid containing nephrons; (*c*) a single nephron.

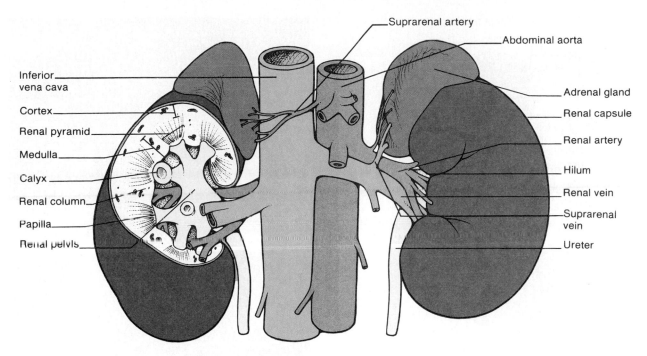

Figure 20.5 Blood vessels associated with the kidneys and adrenal glands.

regulate the absorption of calcium ions by activating *vitamin D*. (See chapter 13.)

The kidneys also help regulate the volume, composition, and pH of body fluids. These functions involve complex mechanisms that lead to the formation of urine. They are discussed in a subsequent section of this chapter, and are explored in still more detail in chapter 21.

1. Where are the kidneys located?
2. Describe the structure of a kidney.
3. Name the functional unit of the kidney.
4. What are the general functions of the kidneys?

Renal Blood Vessels

Blood is supplied to the kidneys by means of the **renal arteries,** which arise from the abdominal aorta (figure 20.5). These arteries transport a relatively large volume of blood; in fact, when a person is at rest, the renal arteries usually carry from 15% to 30% of the total cardiac output into the kidneys.

A renal artery enters a kidney through the hilum and gives off several branches, called the *interlobar arteries,* which pass between the renal pyramids. At the junction between the medulla and the cortex, the interlobar arteries branch to form a series of incomplete arches, the *arciform arteries* (arcuate arteries), which,

in turn, give rise to *interlobular arteries.* The lateral branches of the interlobular arteries, called **afferent arterioles,** lead to the nephrons.

Venous blood is returned through a series of vessels that correspond generally to the arterial pathways. For example, the venous blood passes through interlobular, arciform, interlobar, and renal veins. The **renal vein** then joins the inferior vena cava as it courses through the abdominal cavity. (Branches of the renal arteries and veins are shown in figures 20.6 and 20.7.)

Patients with end-stage renal disease are sometimes treated with a *kidney transplant.* In this surgical procedure, a kidney from a living donor or a cadaver, whose tissues are antigenically similar (histocompatible) to those of the recipient, is placed in the depression on the medial surface of the right or left ilium (iliac fossa). The renal artery and vein of the donor kidney are connected to the recipient's iliac artery and vein, respectively, and the kidney's ureter is attached to the dome of the recipient's urinary bladder.

Nephrons

Structure of a Nephron

A kidney contains about one million nephrons, each consisting of a **renal corpuscle** and a **renal tubule** (figure 20.4).

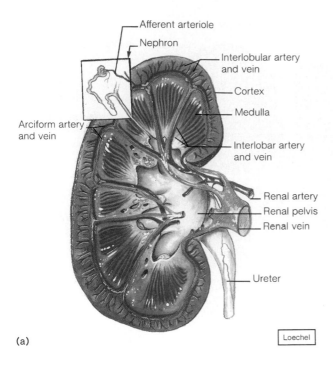

Afferent arteriole

Nephron

Interlobular artery and vein

Cortex

Medulla

Arciform artery and vein

Interlobar artery and vein

Renal artery

Renal pelvis

Renal vein

Ureter

(a)

Loechel

Figure 20.6 (a) Main branches of the renal artery and vein; (b) corrosion cast of the renal arterial system.

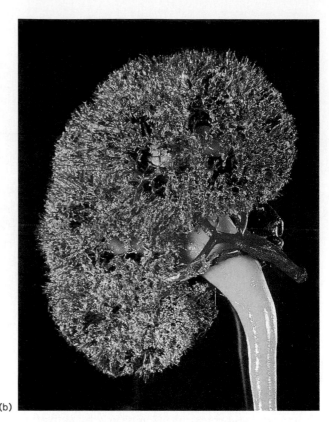

(b)

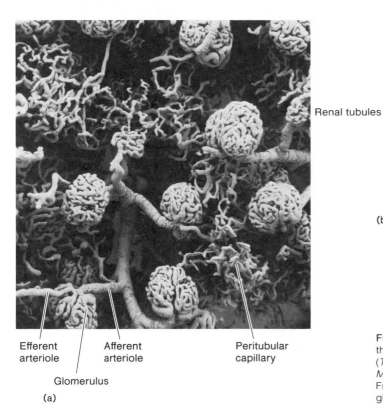

Efferent arteriole

Afferent arteriole

Peritubular capillary

Glomerulus

(a)

Renal tubules

Glomerulus

Glomerular capsule

(b)

Figure 20.7 (a) A scanning electron micrograph of a cast of the renal blood vessels associated with the glomeruli (×260). (*Tissues and Organs: A Text-Atlas of Scanning Electron Microscopy*, by R. G. Kessel and R. H. Kardon. © 1979 W. H. Freeman and Company.); (b) scanning electron micrograph of a glomerular capsule surrounding a glomerulus.

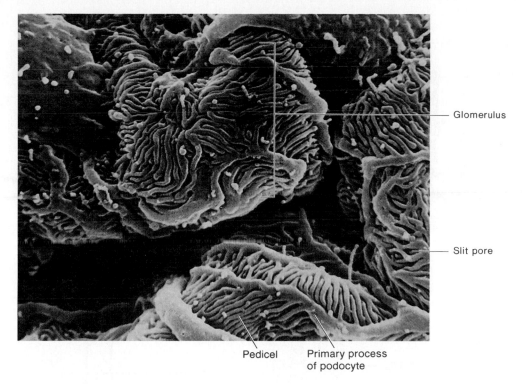

Glomerulus

Slit pore

Pedicel

Primary process
of podocyte

Figure 20.8 Scanning electron micrograph of a glomerulus
(×8,000). Note the slit pores between the pedicels.

A renal corpuscle (Malpighian corpuscle) is composed of a tangled cluster of blood capillaries called a **glomerulus,** which is surrounded by a thin-walled, saclike structure called a **glomerular capsule** (Bowman's capsule).

The glomerular capsule is an expansion at the closed end of a renal tubule. It is composed of two layers of squamous epithelial cells: a visceral layer that closely covers the glomerulus, and an outer parietal layer that is continuous with the visceral layer and with the wall of the renal tubule.

The cells of the parietal layer are typical squamous epithelial cells; however, those of the visceral layer are highly modified epithelial cells called *podocytes*. Each podocyte has several primary processes extending from its cell body, and these processes, in turn, bear numerous secondary processes, or *pedicels*. The pedicels of each cell interdigitate with those of adjacent podocytes, and the clefts between them form a complicated system of *slit pores*. (See figure 20.8.)

The renal tubule leads away from the glomerular capsule and becomes highly coiled. This coiled portion of the tubule is named the *proximal convoluted tubule*.

The proximal convoluted tubule dips toward the renal pelvis to become the *descending limb of the loop of Henle*. The tubule then curves back toward its renal corpuscle and forms the *ascending limb of the loop of Henle*.

The ascending limb returns to the region of the renal corpuscle, where it becomes highly coiled again and is called the *distal convoluted tubule*. This distal portion is shorter than the proximal tubule, and its convolutions are less complex.

Several distal convoluted tubules merge in the renal cortex to form a *collecting duct,* which, in turn, passes into the renal medulla, becoming larger and larger as it is joined by other collecting ducts. The resulting tube (papillary duct) empties into a minor calyx through an opening in a renal papilla. The parts of a nephron are shown in figures 20.9 and 20.10.

Juxtaglomerular Apparatus

Near its beginning, the distal convoluted tubule passes between the afferent and efferent arterioles and contacts them. At the point of contact, the epithelial cells of the distal tubule are quite narrow and densely packed. These cells comprise a structure called the *macula densa*.

Close by, in the walls of the arterioles near their attachments to the glomerulus, are some large, smooth muscle cells. They are called *juxtaglomerular cells,* and together with the cells of the macula densa, they constitute the **juxtaglomerular apparatus** (complex). This structure plays an important role in regulating the flow of blood through various renal vessels (figure 20.11).

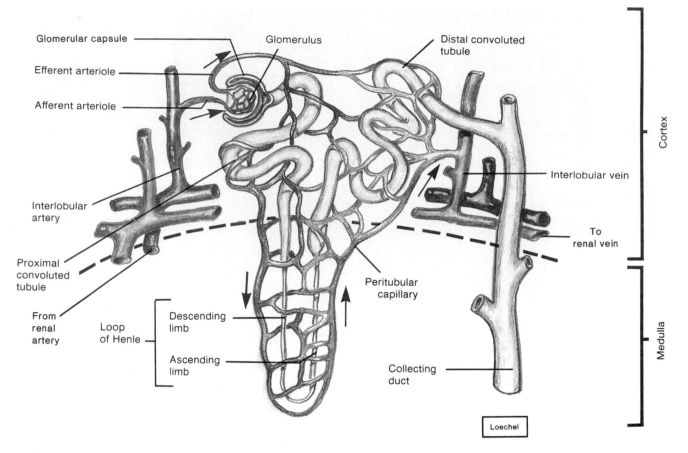

Figure 20.9 Structure of a nephron and the blood vessels associated with it.

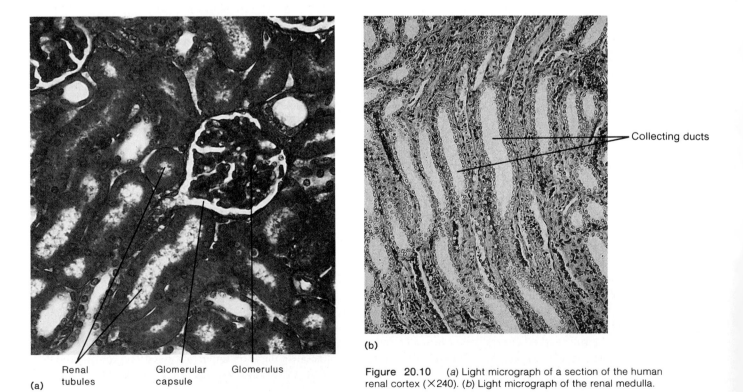

(a)

Renal tubules   Glomerular capsule   Glomerulus

(b)

Collecting ducts

Figure 20.10 (a) Light micrograph of a section of the human renal cortex (×240). (b) Light micrograph of the renal medulla.

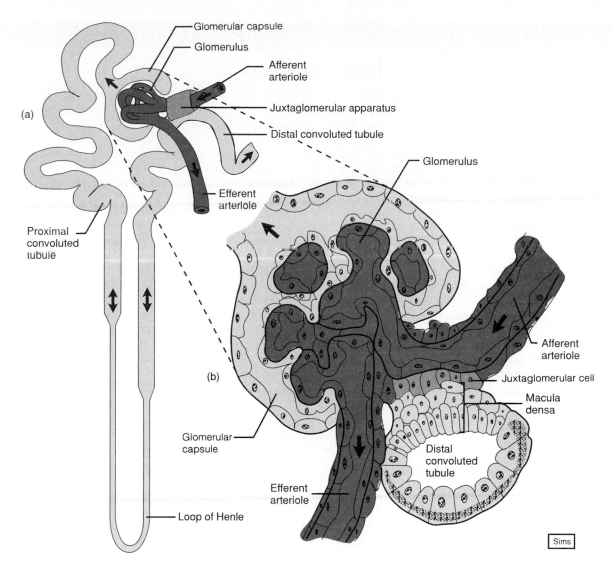

Figure 20.11 (*a*) Location of the juxtaglomerular apparatus. (*b*) Enlargement of a section of the juxtaglomerular apparatus, which consists of the macula densa and the juxtaglomerular cells.

Labels for image:
- Glomerular capsule
- Glomerulus
- Afferent arteriole
- Juxtaglomerular apparatus
- Distal convoluted tubule
- Efferent arteriole
- Proximal convoluted tubule
- Loop of Henle
- (a)
- (b)
- Glomerulus
- Afferent arteriole
- Juxtaglomerular cell
- Macula densa
- Distal convoluted tubule
- Glomerular capsule
- Efferent arteriole
- Sims

Cortical and Juxtamedullary Nephrons

Most nephrons have corpuscles located in the renal cortex near the surface of the kidney. These are called *cortical nephrons,* and they have relatively short loops of Henle that usually do not reach the renal medulla.

Another group, called *juxtamedullary nephrons,* have corpuscles close to the renal medulla, and their loops of Henle extend deep into the medulla. Although they represent only about 20% of the total, these nephrons play an important role in regulating the process of concentrating urine (figure 20.12).

Blood Supply of a Nephron

The cluster of capillaries that forms a glomerulus arises from an **afferent arteriole.** After passing through the capillary of the glomerulus, the blood enters an **efferent arteriole** (rather than a venule), whose diameter is somewhat less than that of the afferent vessel.

Because of its small diameter, the efferent arteriole creates some resistance to blood flow. This causes blood to back up into the glomerulus, producing a relatively high pressure in the glomerular capillary.

The efferent arteriole branches into a complex, freely interconnecting network of capillaries, which surrounds the various portions of the renal tubule. This network is called the **peritubular capillary system,** and the blood it contains is under relatively low pressure (figures 20.9 and 20.12).

Special branches of this system, which receive blood primarily from the efferent arterioles of the juxtamedullary nephrons, form capillary loops called *vasa*

Glomerulonephritis

N*ephritis* refers to an inflammation of the kidney; *glomerulonephritis* refers to an inflammation affecting the glomeruli. This latter condition may occur in acute or chronic forms and can lead to renal failure.

Acute glomerulonephritis (AGN) usually results from an abnormal immune reaction that develops one to three weeks following a certain type of bacterial infection (beta-hemolytic streptococci). As a rule, the infectious condition occurs in some other part of the body and does not affect the kidneys directly. Instead, the presence of bacterial antigens triggers an immune reaction, and antibodies are produced against these antigens. As a consequence of antigen-antibody reactions, insoluble immune complexes are formed (see chapter 19), and they are carried by the blood to the kidneys. The antigen-antibody complexes tend to be deposited in the glomerular capillaries, causing them to be blocked.

At the same time that the immune reaction is taking place, an inflammation reaction is also triggered (see chapters 5 and 19), and large numbers of white blood cells are attracted into the region, blocking the capillaries still more. Those capillaries remaining open may become excessively permeable, and if this happens, plasma proteins and red blood cells may be lost into the urine.

Most glomerulonephritis patients eventually regain normal kidney function; however, in severe cases, renal functions may fail completely, and without treatment, the person is likely to die within a week or so.

Chronic glomerulonephritis is a progressive disease in which increasing numbers of nephrons are slowly damaged until finally the kidneys are unable to perform their usual functions. This condition is usually associated with certain diseases other than streptococcal infections, and once again, it involves the formation of antigen-antibody complexes that precipitate and accumulate in the glomeruli. The resulting inflammation is prolonged, and it is accompanied by tissue changes in which the glomerular membranes are slowly replaced by fibrous tissue. As this happens, the functions of the nephrons are permanently lost, and eventually the kidneys may fail.

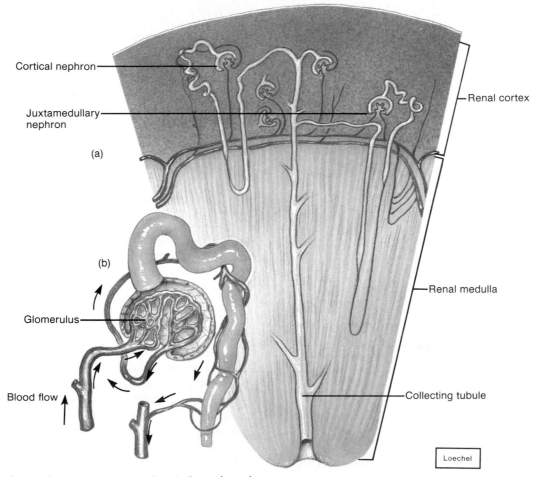

Figure 20.12 (*a*) Cortical nephrons are close to the surface of a kidney; juxtamedullary nephrons are near the renal medulla. (*b*) Enlarged view of a cortical nephron.

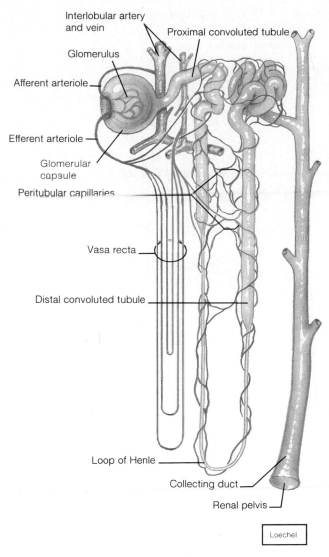

Figure 20.13 The capillary loop of the vasa recta is closely associated with the loop of Henle of a juxtamedullary nephron.

Labels on figure:
- Interlobular artery and vein
- Glomerulus
- Afferent arteriole
- Proximal convoluted tubule
- Efferent arteriole
- Glomerular capsule
- Peritubular capillaries
- Vasa recta
- Distal convoluted tubule
- Loop of Henle
- Collecting duct
- Renal pelvis
- Loechel

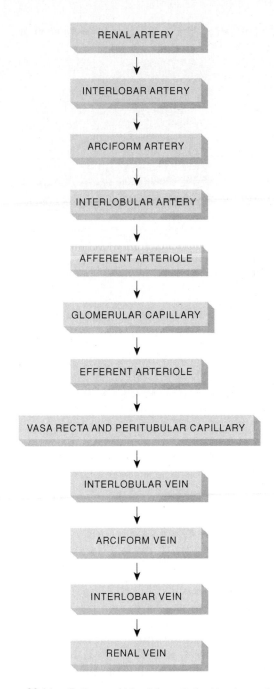

Flowchart:
RENAL ARTERY → INTERLOBAR ARTERY → ARCIFORM ARTERY → INTERLOBULAR ARTERY → AFFERENT ARTERIOLE → GLOMERULAR CAPILLARY → EFFERENT ARTERIOLE → VASA RECTA AND PERITUBULAR CAPILLARY → INTERLOBULAR VEIN → ARCIFORM VEIN → INTERLOBAR VEIN → RENAL VEIN

Figure 20.14 Pathway of blood through the blood vessels of the kidney and nephron.

recta. These loops dip into the renal medulla and are closely associated with the loops of the juxtamedullary nephrons.

After flowing through the vasa recta, the blood is returned to the renal cortex, where it joins blood from other branches of the peritubular capillary system and enters the venous system of the kidney (figure 20.13).

Figure 20.14 summarizes the pathway followed by the blood as it passes through the blood vessels of the kidney and nephron.

1. Describe the system of vessels that supplies blood to the kidney.
2. Name the parts of a nephron.
3. What parts comprise the juxtaglomerular apparatus?
4. Distinguish between a cortical nephron and a juxtamedullary nephron.
5. Describe the blood supply of a nephron.

Urine Formation

The functions of the nephrons include removing waste substances from the blood, and regulating water and electrolyte concentrations within the body fluids. The end product of these functions is **urine,** which is excreted to the outside of the body, containing wastes, excess water, and excess electrolytes.

Urine formation involves the following processes: *filtration* into renal tubules of various substances from

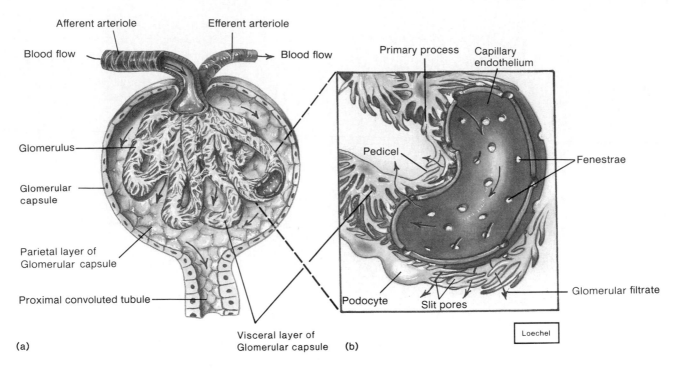

Afferent arteriole   Efferent arteriole

Blood flow

Blood flow

Glomerulus

Glomerular capsule

Parietal layer of Glomerular capsule

Proximal convoluted tubule

Visceral layer of Glomerular capsule

(a)

Primary process   Capillary endothelium

Pedicel

Fenestrae

Glomerular filtrate

Podocyte   Slit pores

Loechel

(b)

Figure 20.15   (*a*) The first step in urine formation is the filtration of substances through the glomerular membrane into the glomerular capsule. (*b*) The glomerular filtrate passes through fenestrae of the capillary endothelium.

the blood plasma within glomerular capillaries; *reabsorption* into the plasma of some of these substances; and *secretion* into the renal tubules of other substances from the plasma within the peritubular capillaries.

Glomerular Filtration

Urine formation begins when water and various dissolved substances are filtered out of the glomerular capillaries and into the glomerular capsules. The filtration of these materials through the capillary walls is much like the filtration that occurs at the arteriole ends of other capillaries throughout the body. The glomerular capillaries, however, are many times more permeable than the capillaries in other tissues, due to the presence of numerous tiny openings (fenestrae) in their walls (figure 20.15).

Filtration Pressure

As in the case of other capillaries, the main force responsible for moving substances through the glomerular capillary wall is the pressure of the blood inside (glomerular hydrostatic pressure). This movement is also influenced by the osmotic pressure of the blood plasma in the glomerulus and by the hydrostatic pressure inside the glomerular capsule. An increase in either of these pressures will oppose movement out of the capillary and thus reduce filtration. The net pressure acting to force substances out of the glomerulus is called **fil-**

tration pressure, and it can be calculated by subtracting the sum of the opposing forces from the hydrostatic pressure of the glomerulus:

$$\text{Filtration pressure} = \text{Glomerular hydrostatic pressure} - \left(\text{Glomerular plasma osmotic pressure} + \text{Capsular hydrostatic pressure} \right)$$

The resulting **glomerular filtrate** is received by the glomerular capsule, and it has about the same composition as the filtrate that becomes tissue fluid elsewhere in the body. That is, glomerular filtrate is largely water and contains essentially the same substances as the blood plasma, except for the larger protein molecules, which the filtrate lacks. More specifically, glomerular filtrate contains water, glucose, amino acids, urea, uric acid, creatine, creatinine, and ions of sodium, chlorine, potassium, calcium, bicarbonate, phosphate, and sulfate. The relative concentrations of some of the substances in the blood plasma, glomerular filtrate, and urine are shown in chart 20.1.

The concentrations of certain components of the blood plasma can be used to evaluate kidney functions. For example, if the kidneys are functioning inadequately, the plasma concentrations of urea (as indicated by a blood urea nitrogen test) and of creatinine may increase as much as tenfold above normal.

CHART 20.1	Relative concentrations of certain substances in plasma, glomerular filtrate, and urine		
	Concentrations (mEq/l)		
Substance	Plasma	Glomerular filtrate	Urine
Sodium (Na⁺)	142	142	128
Potassium (K⁺)	5	5	60
Calcium (Ca⁺²)	4	4	5
Magnesium (Mg⁺²)	3	3	15
Chlorine (Cl⁻)	103	103	134
Bicarbonate (HCO₃⁻)	27	27	14
Sulfate (SO₄⁻²)	1	1	33
Phosphate (PO₄⁰)	2	2	40

	Concentrations (mg/100 ml)		
Substance	Plasma	Glomerular filtrate	Urine
Glucose	100	100	0
Urea	26	26	1,820
Uric acid	4	4	53
Creatinine	1	1	196

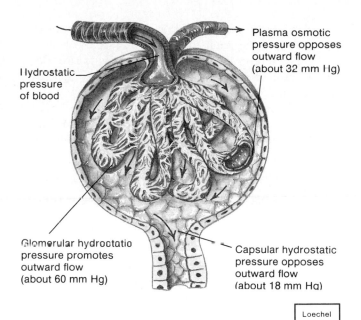

Figure 20.16   The rate of glomerular filtration is affected by the hydrostatic pressure of the plasma, the osmotic pressure of the plasma, and the hydrostatic pressure of the fluid in the glomerular capsule.

Filtration Rate

The rate of glomerular filtration is directly proportional to the filtration pressure. Consequently, the factors that affect the glomerular hydrostatic pressure, glomerular plasma osmotic pressure, or hydrostatic pressure in the glomerular capsule will also affect the rate of filtration (figure 20.16).

For example, since the glomerular capillary is located between two arterioles—the *afferent* and *efferent arterioles*—any change in the diameters of these vessels is likely to cause a change in the glomerular hydrostatic pressure, and will be accompanied by a change in the glomerular filtration rate. The afferent arteriole, through which the blood enters the glomerulus, may constrict as a result of mild stimulation by sympathetic nerve impulses. If this occurs, the blood flow diminishes, the glomerular hydrostatic pressure decreases, and the filtration rate drops. If, on the other hand, the efferent arteriole (through which the blood leaves the glomerulus) becomes constricted, blood backs up into the glomerulus, the glomerular hydrostatic pressure increases, and the filtration rate rises. Converse effects are produced by vasodilation of these vessels.

As mentioned previously, the osmotic pressure of the glomerular plasma is another factor that influences filtration pressure and affects the rate of filtration. In capillaries, the blood hydrostatic pressure, acting to force water and dissolved substances outward, is opposed by the effect of the plasma osmotic pressure that attracts water inward. Actually, as filtration occurs

through the capillary wall, proteins remaining in the plasma cause the osmotic pressure within the glomerular capillary to rise. When this pressure reaches a certain high level, filtration ceases. Conversely, conditions that tend to decrease plasma osmotic pressure, such as a decrease in plasma protein concentration, cause an increase in the filtration rate.

If the arterial blood pressure drops excessively, as may occur during *shock,* the glomerular hydrostatic pressure may decrease below the level required for filtration. At the same time, the epithelial cells of the renal tubules may fail to receive sufficient nutrients to maintain their high rates of metabolism. As a result, cells may die (tubular necrosis), and renal functions may be lost. Changes such as these may result in chronic renal failure.

For these reasons, the rate of blood flow through the glomerulus also affects the filtration rate. More specifically, if the blood is flowing slowly, a larger proportion of the plasma filters out of the glomerulus, and as the plasma osmotic pressure rises, the filtration rate decreases. On the other hand, if the flow is rapid, there is less change in the plasma osmotic pressure, and the filtration rate remains higher.

The hydrostatic pressure in the glomerular capsule is still another factor that may affect the filtration pressure and the rate of filtration. This capsular pressure sometimes changes as a result of an obstruction, such as a stone in a ureter or an enlarged prostate gland pressing on the urethra. If this occurs, fluids tend to back up into the renal tubules and cause the hydrostatic pressure in the glomerular capsules to rise. Because any increase in capsular pressure opposes glomerular filtration, the rate of filtration may decrease significantly.

In an average adult, the glomerular filtration rate for the nephrons of both kidneys is about 125 milliliters per minute, or 180,000 milliliters (180 liters) in twenty-four hours. Assuming that the blood plasma volume is about 3 liters, the production of 180 liters of filtrate in twenty-four hours means that all of the plasma must be filtered through the glomeruli about sixty times each day (figure 20.17).

Since this twenty-four-hour volume is nearly 45 gallons, it is obvious that not all of it is excreted as urine. Instead, most of the fluid that passes through the renal tubules is reabsorbed and reenters the plasma.

The volume of plasma filtered by the kidneys is also related to the amount of *surface area* within the glomerular capillaries. This surface area is estimated to be about 2 square meters—approximately equal to the surface area of an adult's skin.

---

1. What general processes are involved in urine formation?
2. How is filtration pressure calculated?
3. What factors influence the rate of glomerular filtration?

Regulation of Filtration Rate

The regulation of the glomerular filtration rate involves the *juxtaglomerular apparatus,* which was described previously (figure 20.11), and two negative feedback mechanisms. These mechanisms are triggered whenever the filtration rate is decreasing. For example, as the rate decreases, the concentration of chloride ions reaching the *macula densa* in the distal convoluted tubule also decreases. In response, the macula densa signals the smooth muscles in the wall of the afferent arteriole to relax, and the vessel becomes dilated. This action allows more blood to flow into the glomerulus, increasing the glomerular pressure: as a consequence, the filtration rate rises toward its previous level.

At the same time that the macula densa signals the afferent arteriole to dilate, it stimulates the *juxtaglomerular cells* to release *renin.* (See chapter 13.) This enzyme causes a plasma globulin (angiotensinogen) to form a substance called *angiotensin I.* Angiotensin I is converted quickly to *angiotensin II* by an enzyme called *converting enzyme* that is present in the plasma and lungs.

Angiotensin II is a vasoconstrictor, and it stimulates the smooth muscle cells in the wall of the efferent arteriole to contract, constricting the vessel. As a result of this action, blood tends to back up into the

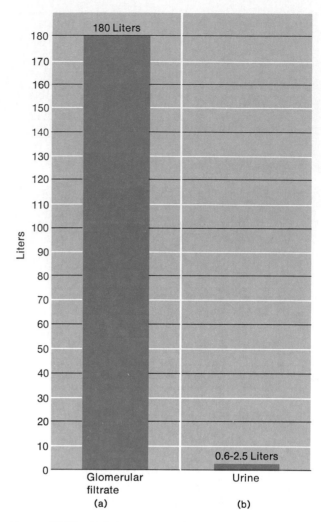

Figure 20.17 Relative amounts of (*a*) glomerular filtrate and (*b*) urine formed in twenty-four hours.

glomerulus, and, as the glomerular hydrostatic pressure increases, the filtration rate also increases (figure 20.18).

Elevated blood pressure (hypertension) is sometimes associated with a renal disorder in which there is an excessive release of renin, followed by an increased formation of the vasoconstrictor angiotensin II. Patients with this form of high blood pressure are often treated with a drug that prevents the formation of angiotensin II by inhibiting the action of the enzyme that converts angiotensin I into angiotensin II. Such a drug is called an angiotensin converting-enzyme inhibitor (ACE inhibitor).

These two mechanisms operate together to ensure a constant blood flow through the glomerulus and a relatively stable glomerular filtration rate, in spite of marked changes occurring in the arterial blood pressure. This phenomenon, by which a mechanism within

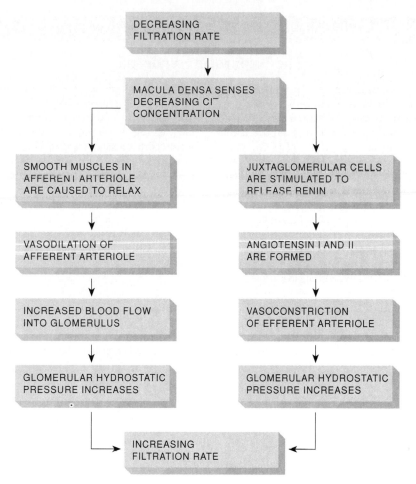

DECREASING
FILTRATION RATE

↓

MACULA DENSA SENSES
DECREASING Cl⁻
CONCENTRATION

SMOOTH MUSCLES IN
AFFERENT ARTERIOLE
ARE CAUSED TO RELAX

↓

VASODILATION OF
AFFERENT ARTERIOLE

↓

INCREASED BLOOD FLOW
INTO GLOMERULUS

↓

GLOMERULAR HYDROSTATIC
PRESSURE INCREASES

JUXTAGLOMERULAR CELLS
ARE STIMULATED TO
RELEASE RENIN

↓

ANGIOTENSIN I AND II
ARE FORMED

↓

VASOCONSTRICTION
OF EFFERENT ARTERIOLE

↓

GLOMERULAR HYDROSTATIC
PRESSURE INCREASES

INCREASING
FILTRATION RATE

Figure 20.18   Two mechanisms involving the macula densa ensure a constant blood flow through the glomerulus.

an organ or tissue maintains a constant blood flow through that part even though the arterial blood pressure is changing, is called **autoregulation.**

Tubular Reabsorption

If the composition of the glomerular filtrate entering the renal tubule is compared with that of the urine leaving the tubule, it is clear that changes occur as the fluid passes through the tubule (chart 20.1). For example, glucose is present in the filtrate, but absent in the urine. Also, urea and uric acid are considerably more concentrated in urine than they are in the glomerular filtrate. Such changes in fluid composition are largely the result of **tubular reabsorption,** a process by which substances are transported out of the glomerular filtrate, through the epithelium of the renal tubule, and into the blood of the peritubular capillary (figure 20.19).

   Because the efferent arteriole is narrower than the peritubular capillary, the blood flowing from the former into the latter is under relatively low pressure. Also, the wall of the peritubular capillary is more permeable

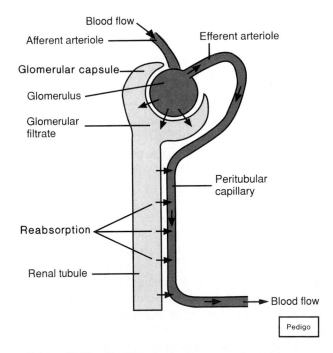

Figure 20.19   Reabsorption is the process by which substances are transported from the glomerular filtrate into the blood of the peritubular capillary. What substances are reabsorbed in this manner?

The Nephrotic Syndrome

The *nephrotic syndrome* is a set of symptoms that often appears in patients with renal diseases. This syndrome is characterized by considerable loss of plasma proteins into the urine (proteinuria), widespread edema, and increased susceptibility to infections.

Plasma proteins are lost into the urine because of increased permeability of the glomerular membranes, which accompanies various renal disorders, such as glomerulonephritis. As a consequence of a decreasing plasma protein concentration (hypoproteinemia), the plasma osmotic pressure falls. This may lead to widespread, severe edema as a large volume of fluid accumulates in the interstitial spaces within the tissues and in various body spaces, such as the abdominal cavity, pleural cavity, pericardial cavity, and joint cavities.

Also, as edema develops, the blood volume decreases and the blood pressure drops. These changes may trigger the release of aldosterone from the adrenal cortex (see chapter 13), which, in turn, stimulates the kidneys to conserve sodium ions and water. This action reduces the urine output and may aggravate the edema.

The nephrotic syndrome sometimes appears in young children who have *lipoid nephrosis*. The cause of this condition is unknown, but it is characterized by changes occurring in the epithelial cells of the glomeruli. As a result of these changes, the cells of the glomerular membranes become enlarged and distorted, allowing proteins to leak through.

---

than that of other capillaries. Both of these factors enhance the rate of fluid reabsorption from the renal tubule.

Although tubular reabsorption occurs throughout the renal tubule, most of it occurs in the proximal convoluted portion. The epithelial cells in this portion have numerous microscopic projections called *microvilli* that form a "brush border" on their free surfaces. (See chapter 14.) These tiny extensions greatly increase the surface area exposed to the glomerular filtrate and enhance the reabsorption process. (See figure 14.42.)

Various segments of the renal tubule are adapted to reabsorb specific substances, using particular modes of transport. Glucose reabsorption, for example, occurs primarily through the walls of the proximal tubule by active transport. Water also is reabsorbed rapidly through the epithelium of the proximal tubule by osmosis; however, portions of the distal tubule are almost impermeable to water. This characteristic of the distal tubule is important in the regulation of urine concentration and volume, as is described in a subsequent section of this chapter.

As described in chapter 3, an active transport mechanism depends on the presence of carrier molecules in a cell membrane. These carriers transport passenger molecules through the membrane, release them, and return to the other side to transport more passenger molecules. Such a mechanism has a *limited transport capacity;* that is, it can only transport a certain number of molecules in a given amount of time because the number of carriers is limited.

Usually all of the glucose in the glomerular filtrate is reabsorbed, because there are enough carrier molecules to transport it. Sometimes, however, the plasma glucose concentration increases, and if it reaches

a critical level, called the *renal plasma threshold,* there will be more glucose molecules in the filtrate than the active transport mechanism can handle. As a result, some glucose will remain in the filtrate and be excreted in the urine.

The appearance of glucose in the urine is called *glucosuria* (or *glycosuria*). This condition may occur following the administration of glucose intravenously or in a patient with insulin-dependent (type I) diabetes mellitus. If the cause is diabetes mellitus, the blood glucose concentration rises because of insufficient insulin from the pancreas.

Amino acids also enter the glomerular filtrate and are reabsorbed in the proximal convoluted tubule, apparently by three different active transport mechanisms. Each mechanism is thought to reabsorb a different group of amino acids, whose members have molecular similarities. As a result of their actions, only a trace of amino acids usually remains in the urine.

Although the glomerular filtrate is nearly free of protein, some *albumin* may be present. These proteins have relatively small molecules, and they are reabsorbed by *pinocytosis* through the brush border of epithelial cells lining the proximal convoluted tubule. Once they are inside an epithelial cell, the proteins are converted to amino acids and moved into the blood of the peritubular capillary.

Other substances reabsorbed by the epithelium of the proximal convoluted tubule include creatine, lactic acid, citric acid, uric acid, ascorbic acid (vitamin C), phosphate ions, sulfate ions, calcium ions, potassium ions, and sodium ions. As a group, these substances are reabsorbed by active transport mechanisms with limited transport capacities. Such a substance usually does not appear in the urine until its concentration in the glomerular filtrate exceeds its particular threshold.

Sodium and Water Reabsorption

Substances that remain in the renal tubule tend to become more and more concentrated as water is reabsorbed from the filtrate. Most water reabsorption occurs *passively* by osmosis in the proximal convoluted tubule and is closely associated with the active reabsorption of sodium ions. In fact, if sodium reabsorption increases, water reabsorption increases; if sodium reabsorption decreases, water reabsorption decreases also.

About 70% of the *sodium ion reabsorption* occurs in the proximal segment of the renal tubule by active transport (sodium pump mechanism). When the positively charged ions (Na^+) are moved through the tubular wall, negatively charged ions, including chloride ions (Cl^-), phosphate ions (PO_4^{-3}), and bicarbonate ions (HCO_3^-), accompany them. This movement of negatively charged ions is due to the electrochemical attraction between particles of opposite electrical charge. It is termed **passive transport** because it does not require a direct expenditure of cellular energy.

As more and more sodium ions are actively transported into the peritubular capillary along with various negatively charged ions, the concentration of solutes within the peritubular blood is increased. Furthermore, since water moves through cell membranes from regions of lesser solute concentration (hypotonic) toward regions of greater solute concentration (hypertonic), water is transported by osmosis from the renal tubule into the peritubular capillary. Because of the movement of solutes and water into the peritubular capillary, the volume of fluid within the renal tubule is greatly reduced (figure 20.20).

1. How is the peritubular capillary adapted for reabsorption?
2. What substances present in the glomerular filtrate are not normally present in the urine?
3. What mechanisms are responsible for reabsorption of solutes from the glomerular filtrate?
4. Define the renal plasma threshold.
5. Describe the role of passive transport in urine formation.

Regulation of Urine Concentration and Volume

Sodium ions continue to be reabsorbed by active transport as the tubular fluid moves through the loop of Henle, the distal convoluted tubule, and the collecting duct. Consequently, almost all the sodium that enters the renal tubule as part of the glomerular filtrate may be reabsorbed before the urine is excreted.

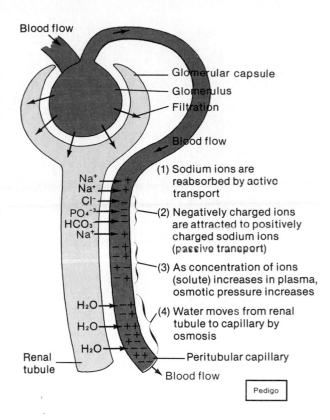

Figure 20.20 Water reabsorption by osmosis occurs in response to the reabsorption of sodium by active transport.

When sodium ions are reabsorbed, water molecules follow the sodium ions and, thus, are reabsorbed passively by osmosis in various segments of the renal tubule. As this occurs, the osmotic concentration of the tubular fluid changes. These changes result from a *countercurrent mechanism* and from the actions of certain hormones.

The **countercurrent mechanism** involves the loops of Henle of the juxtamedullary nephrons. The descending and ascending limbs of these U-shaped structures lie parallel and very close to one another. The mechanism is named for the fact that fluid moving down the descending limb produces a current that is counter to that of the fluid moving up in the ascending limb.

The tubular fluid in the proximal segment is *isotonic* to the plasma of the peritubular capillary blood; however, it becomes *hypertonic* when it moves through the descending limb. Then, as it moves through the ascending limb, the tubular fluid becomes *hypotonic*.

These changes in osmotic concentration are due primarily to activity in the epithelium of the ascending limb. Specifically, the epithelial lining in the upper portion of this limb is thickened (thick segment) and is impermeable to water. However, the epithelium does carry on the active reabsorption of chloride ions, and sodium ions follow them passively. As these sodium and chloride ions (NaCl) accumulate, the interstitial fluid outside the ascending limb becomes *hypertonic,* while

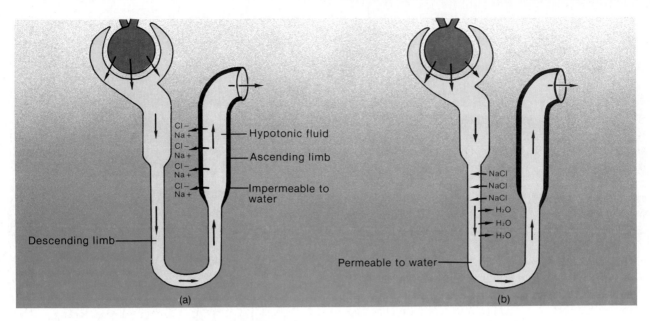

Figure 20.21 (a) Fluid in the ascending limb of the loop of Henle becomes hypotonic as solute is reabsorbed. (b) Fluid in the descending limb becomes hypertonic as it loses water by osmosis and gains solute by diffusion.

the tubular fluid inside becomes *hypotonic* because it is losing its solute.

The epithelium of the descending limb is thin (thin segment), lacks microvilli, and is quite permeable. Because this segment is surrounded by the hypertonic fluid formed by the ascending limb, water tends to leave the descending limb by osmosis. Thus, the contents of the descending limb become more and more concentrated, or *hypertonic,* to the plasma of the peritubular capillary.

Also, since the concentration of NaCl in the medullary interstitial fluid is high, NaCl reenters the descending limb by diffusion. This makes the fluid in the descending limb still more concentrated (hypertonic) (figure 20.21).

Thus, NaCl moves from the descending limb into the ascending limb, is reabsorbed into the medullary interstitial fluid, and diffuses back into the descending limb again. Each time this circuit is completed, the concentration of NaCl increases, or multiplies. For this reason, the mechanism is called a *countercurrent multiplier.*

As a result of this mechanism, the NaCl concentration of the medullary interstitial fluid is greatest near the tip of the loops of Henle and decreases progressively toward the renal cortex. This concentration gradient is important to the process of concentrating urine (figure 20.22), as will be explained in a subsequent section of this chapter.

The maintenance of the NaCl concentration gradient is aided by another countercurrent mechanism operating in the *vasa recta,* which is described in a previous section of this chapter. In this case, blood flows relatively slowly down the descending portion of the vasa recta, and NaCl enters it by diffusion. Then, as the blood moves back up toward the renal cortex, most of the NaCl diffuses from the blood and reenters the medullary interstitial fluid. Consequently, little NaCl is carried away from the renal medulla (figure 20.23).

The tubular fluid reaching the distal convoluted tubule is hypotonic to its surroundings. The cells lining this segment and the collecting duct that follows continue to reabsorb sodium and chloride ions under the influence of aldosterone, which is secreted by the adrenal cortex (chapter 13). However, the lining cells of the distal convoluted tubules are quite impermeable to water. Thus, water tends to accumulate inside the tubule and may be excreted as dilute urine.

As discussed in chapter 13, antidiuretic hormone (ADH) is produced by specialized neurons in the *hypothalamus.* This hormone is released from the posterior lobe of the pituitary gland in response to a decreasing concentration of water in the blood or to decreasing blood volume and pressure. When ADH reaches the kidney, it causes an increase in the permeability of the epithelial cell linings of the distal convoluted tubule and the collecting duct; consequently, water moves rapidly out of these segments by osmosis. Thus, the urine volume is reduced, and the urine becomes more concentrated (hypertonic).

This concentrating of the urine also continues as it passes down the *collecting duct,* because the hyper-

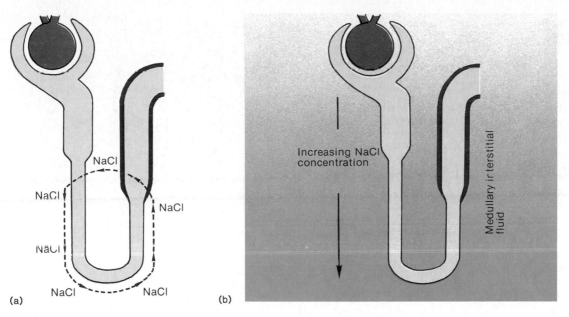

(a)   **(b)**

Figure 20.22 (a) As NaCl completes the countercurrent circuit again and again, it becomes more concentrated. (b) As a result, an NaCl concentration gradient is established in the medullary interstitial fluid.

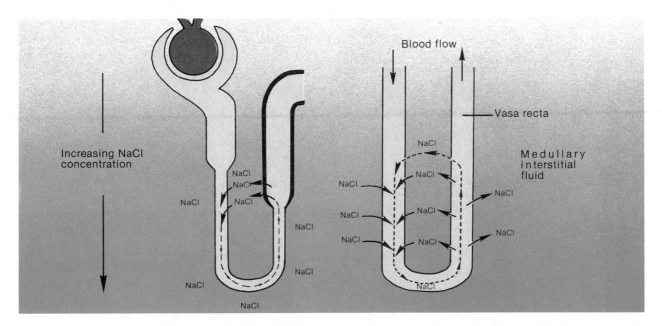

Figure 20.23 A countercurrent mechanism in the vasa recta helps maintain the NaCl concentration gradient in the medullary interstitial fluid.

tonic interstitial fluid of the renal medulla that surrounds the collecting duct draws water from it by osmosis (figure 20.24).

To summarize, ADH stimulates the production of concentrated urine, which contains soluble wastes and other substances in a minimum of water; it also inhibits the loss of body fluids whenever there is a danger of dehydration. If the water concentration of the body fluids is excessive, ADH secretion is decreased. In the absence of ADH, the epithelial linings of the distal segment and the collecting duct become less permeable to water, thus less water is reabsorbed, and the urine tends to be more dilute. Chart 20.2 summarizes the role of ADH in urine production.

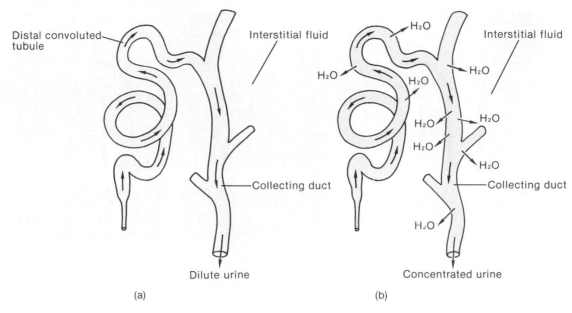

Distal convoluted tubule

Interstitial fluid

Collecting duct

Dilute urine

(a)

Interstitial fluid

H_2O

H_2O

H_2O

H_2O

H_2O

H_2O

H_2O

H_2O

Collecting duct

H_2O

Concentrated urine

(b)

Figure 20.24 (a) The distal convoluted tubule and collecting duct are impermeable to water, so water may be excreted as dilute urine. (b) If ADH is present, however, these segments become permeable, and water is reabsorbed by osmosis into the hypertonic medullary interstitial fluid.

CHART 20.2	Role of ADH in the regulation of urine concentration and volume

1. Concentration of water in the blood decreases.
2. Osmoreceptors in the hypothalamus of the brain are stimulated by an increase in the osmotic pressure of the body fluid.
3. Hypothalamus signals the posterior pituitary gland to release ADH.
4. Blood carries ADH to the kidneys.
5. ADH causes the distal convoluted tubules and collecting ducts to increase water reabsorption by osmosis.
6. Urine becomes more concentrated and the urine volume decreases.

Urea and Uric Acid Excretion

Urea is a by-product of amino acid catabolism. Consequently, the plasma concentration of urea is directly related to the amount of protein in the diet. Urea enters the renal tubule by filtration, and about 50% of it is reabsorbed (passively) by diffusion, but the remainder is excreted in the urine.

Urea is also involved in more complex movements within the kidney, including a countercurrent multiplier mechanism that causes it to be concentrated in the medullary interstitial fluid. As a result, urea becomes more concentrated in the urine than would otherwise be possible.

Uric acid, which results from the metabolism of certain organic bases (purines) in nucleic acids, is reabsorbed by active transport. Although this mechanism seems able to reabsorb all the uric acid normally present in glomerular filtrate, about 10% of the amount filtered is excreted in the urine. This amount is apparently *secreted* into the renal tubule.

> *Gout* is a disorder in which the plasma concentration of uric acid becomes abnormally high. Since uric acid is a relatively insoluble substance, it tends to precipitate when it is present in excess. As a result, crystals of uric acid may be deposited in the joints and other tissues, where they produce inflammation and extreme pain. The joints of the great toes are affected most commonly, but joints in the hands and feet may also be involved.
>
> This condition, which often seems to be inherited, is sometimes treated by administering drugs that inhibit the reabsorption of uric acid and, thus, cause its excretion to increase.

1. Describe a countercurrent mechanism.
2. What role does the hypothalamus play in regulating urine concentration and volume?
3. Explain how urea and uric acid are excreted.

Tubular Secretion

Tubular secretion (tubular excretion) is the process by which certain substances are transported from the plasma of the peritubular capillary into the fluid of the renal tubule. As a result, the amount of a particular substance excreted in the urine may be greater than the amount filtered from the plasma in the glomerulus (figure 20.25).

Some substances are secreted by active transport mechanisms similar to those that function in reabsorption. However, the *secretory mechanisms* transport substances in the opposite direction. For example, certain organic compounds, including penicillin, creatinine, and histamine, are actively secreted into the tubular fluid by the epithelium of the proximal convoluted segment.

Hydrogen ions are also actively secreted. In this case, the proximal segment of the renal tubule is specialized to secrete large quantities of hydrogen ions between the plasma and the tubular fluid, where the hydrogen ion concentrations are similar. The distal segment and collecting duct are specialized to transport small quantities of hydrogen ions between the plasma and the urine, where hydrogen ion concentrations may vary. The secretion of hydrogen ions plays an important role in regulating the pH of body fluids, as is explained in chapter 21.

Although most of the *potassium ions* in the glomerular filtrate are actively reabsorbed in the proximal convoluted tubule, some may be secreted passively in the distal segment and collecting duct. During this process, the active reabsorption of sodium ions out of the tubular fluid produces a negative electrical charge within the tube. Because positively charged potassium ions (K^+) are attracted to regions that are negatively charged, these ions move through the tubular epithelium and enter the tubular fluid. (See figure 20.26.)

To summarize, urine is formed as a result of the following: glomerular filtration of materials from blood plasma; reabsorption of substances, such as glucose, amino acids, proteins, creatine, lactic acid, citric acid, uric acid, ascorbic acid, phosphate ions, sulfate ions, calcium ions, potassium ions, sodium ions, water, and urea; and secretion of substances, such as penicillin, creatinine, histamine, phenobarbital, hydrogen ions, ammonia, and potassium ions (chart 20.3).

1. Define *tubular secretion.*
2. What substances are actively secreted? Passively secreted?
3. How does the reabsorption of sodium affect the secretion of potassium?

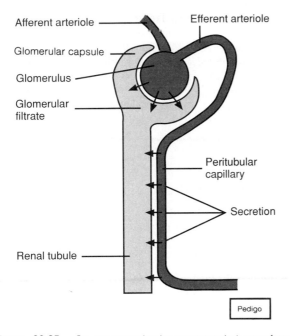

Figure 20.25 Secretory mechanisms move substances from the plasma of the peritubular capillary into the fluid of the renal tubule.

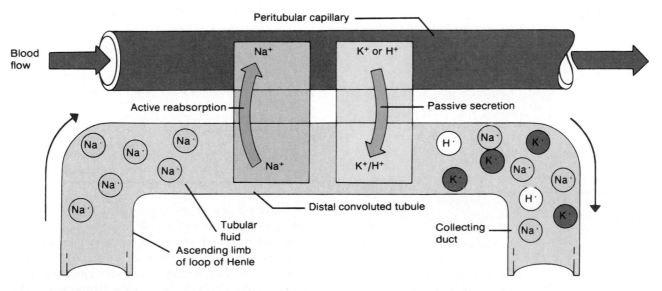

Figure 20.26 Passive secretion of potassium ions (or hydrogen ions) may occur in response to the active reabsorption of sodium ions.

Plasma Clearance

Because the kidneys remove substances from the plasma, the rate at which a particular substance is removed is indicative of kidney efficiency. This rate of removal is called *plasma clearance.*

Various tests of plasma clearance are used to detect glomerular damage or to judge the progress of renal diseases. One such test is called the *inulin clearance test,* and it makes use of *inulin* (not to be confused with insulin), which is a complex polysaccharide that occurs in certain plant roots. During the test, a known amount of inulin is infused into the blood at a constant rate. The inulin passes freely through the glomerular membranes, so that its concentration in the glomerular filtrate equals that of the plasma. In the renal tubule, inulin is neither reabsorbed to any significant degree nor is it secreted. Consequently, the rate at which it appears in the urine can be used to calculate the rate of glomerular filtration or the rate of clearance from the plasma.

Similarly, *creatinine,* a substance that is produced at a constant rate as a result of muscle metabolism, is removed from the blood by the kidneys. Because nearly all of the creatinine filtered by the kidneys normally appears in the urine, a change in the rate of creatinine excretion may reflect a renal disorder. The test used to compare a patient's blood and urine creatinine concentrations is called the *creatinine clearance test.*

Another plasma clearance test involves the use of *para-aminohippuric acid* (PAH). This substance filters freely through the glomerular membranes. However, unlike inulin, any PAH remaining in the plasma after filtration is secreted into the renal tubule by the epithelial cells of the proximal convoluted portions. For this reason, the rate of PAH clearance can be used to calculate the rate of plasma blood flow through the kidneys. Then, if the hematocrit is known (see chapter 17), the rate of total blood flow through the kidneys can be calculated.

CHART 20.3	Functions of nephron parts
Part	**Function**
Renal Corpuscle	
Glomerulus	Filtration of water and dissolved substances from the plasma
Glomerular capsule	Receives the glomerular filtrate
Renal Tubule	
Proximal convoluted tubule	Reabsorption of glucose, amino acids, creatine, lactic acid, citric acid, uric acid, ascorbic acid, phosphate ions, sulfate ions, calcium ions, potassium ions, and sodium ions by active transport
	Reabsorption of proteins by pinocytosis
	Reabsorption of water by osmosis
	Reabsorption of chloride ions and other negatively charged ions by electrochemical attraction
	Active secretion of substances such as penicillin, histamine, creatinine, and hydrogen ions
Descending limb of the loop of Henle	Reabsorption of water by osmosis
Ascending limb of the loop of Henle	Reabsorption of chloride ions by active transport and passive reabsorption of sodium ions
Distal convoluted tubule	Reabsorption of sodium ions by active transport
	Reabsorption of water by osmosis.
	Active secretion of hydrogen ions
	Passive secretion of potassium ions by electrochemical attraction

Composition of Urine

The composition of urine varies considerably from time to time because of variations in dietary intake and physical activity. In addition to containing about 95% water, urine usually contains *urea* from the catabolic metabolism of amino acids, *uric acid* from the metabolism of nucleic acids, and *creatinine* from the metabolism of creatine. It may also contain a trace of *amino acids,* as well as a variety of *electrolytes* whose concentrations tend to vary directly with the amounts included in the diet (chart 20.1). The normal concentrations of various components of urine are listed in Appendix C on page 936.

Abnormal constituents of urine include glucose, proteins, hemoglobin, ketones, and various blood cells. However, the significance of such substances in urine may depend on the amounts present and on other factors. For example, glucose may appear in urine following a large intake of carbohydrates, proteins may appear following vigorous physical exercise, and ketones may appear following a prolonged fast. Also, some pregnant women have glucose in their urine toward the end of pregnancy.

The volume of urine produced by the kidneys usually varies between 0.6 and 2.5 liters per day. The exact volume is influenced by such factors as fluid intake, environmental temperature, relative humidity of the surrounding air, and a person's emotional condition, respiratory rate, and body temperature. An output of 50–60 cubic centimeters of urine per hour is

Figure 20.27 Cross section of a ureter (×160).

considered normal, and an output of less than 30 cubic centimeters per hour may be an indication of kidney failure.

> The kidneys of infants and young children are unable to concentrate urine and conserve water as effectively as those of adults. Consequently, such young persons produce relatively large volumes of urine and tend to lose water rapidly, which may lead to dehydration.

1. List the normal constituents of urine.
2. What is the normal hourly output of urine? The minimal hourly output?

Elimination of Urine

After being formed by nephrons, urine passes from the collecting ducts through openings in the renal papillae and enters the major and minor calyces of the kidney. From there it passes through the renal pelvis and is conveyed by a ureter to the urinary bladder. Urine is excreted to the outside by means of the urethra.

Ureters

Each **ureter** is a tubular organ about 25 centimeters long, which begins as the funnel-shaped renal pelvis. It extends downward behind the parietal peritoneum and parallel to the vertebral column. Within the pelvic cavity, it courses forward and medially to join the urinary bladder from underneath.

The wall of a ureter is composed of three layers. The inner layer, or *mucous coat,* includes several thicknesses of transitional epithelial cells and is continuous with the linings of the renal tubules and the urinary bladder. The middle layer, or *muscular coat,* consists largely of smooth muscle fibers arranged in circular and longitudinal bundles. The outer layer, or *fibrous coat,* is composed of connective tissue (figure 20.27).

> Because the linings of the ureters and the urinary bladder are continuous, infectious agents such as bacteria may ascend from the bladder into the ureters. An inflammation of the bladder, which is called *cystitis,* occurs more commonly in women than in men because the female urethral pathway is shorter. An inflammation of the ureter is called *ureteritis.*

Although the ureter is simply a tube leading from the kidney to the urinary bladder, its muscular wall helps move the urine. Muscular peristaltic waves, originating in the renal pelvis, force the urine along the length of the ureter. These waves are initiated by the presence of urine in the renal pelvis, and their frequency is related to the rate of urine formation. If the rate of urine formation is high, a peristaltic wave may occur every few seconds; if the rate is low, a wave may occur every few minutes.

When such a peristaltic wave reaches the urinary bladder, it causes a jet of urine to spurt into the bladder. The opening through which the urine enters is covered by a flaplike fold of mucous membrane. This fold acts as a valve, allowing urine to enter the bladder from the ureter but preventing it from backing up from the bladder into the ureter.

If a ureter becomes obstructed, as when a small kidney stone (renal calculus) is present in its lumen, strong peristaltic waves are initiated in the proximal portion of the tube. Such waves may help move the stone into the bladder. At the same time, the presence of a stone usually stimulates a sympathetic reflex (ureter-orenal reflex) that results in constriction of the renal arterioles and reduces the production of urine in the kidney on the affected side.

Kidney stones, which are usually composed of calcium oxalate, calcium phosphate, uric acid, or magnesium phosphate, sometimes form in the renal pelvis. If such a stone passes into a ureter, it may produce severe pain. This pain commonly begins in the region of the kidney and tends to radiate into the abdomen, pelvis, and legs. The pain may also be accompanied by nausea and vomiting.

Although about 60% of kidney stone patients pass their stones spontaneously, the others must have the stones removed. In the past, such removal required surgery or tubular instruments that could be passed through the tubes of the urinary tract and used to capture or crush the stones. More recently, kidney stones have often been fragmented by shock waves generated outside the body. The resulting sandlike fragments can then be eliminated with the urine. This procedure, called *extracorporeal shock-wave lithotripsy* (ESWL), involves placing the patient in a stainless steel tub filled with water. The shock waves are produced underwater by a spark-gap electrode, and the waves are focused on the stones by means of a reflector that concentrates the shock-wave energy.

1. Describe the structure of a ureter.
2. How is urine moved from the renal pelvis to the urinary bladder?
3. What prevents urine from backing up from the urinary bladder into the ureters?
4. How does an obstruction in a ureter affect urine production?

Urinary Bladder

The **urinary bladder** is a hollow, distensible, muscular organ. It is located within the pelvic cavity, behind the symphysis pubis, and beneath the parietal peritoneum (figure 20.28 and reference plate 52). In a male, it lies against the rectum posteriorly, and in a female, it contacts the anterior walls of the uterus and vagina.

Although the bladder is somewhat spherical, its shape is altered by the pressures of surrounding organs. When it is empty, the inner wall of the bladder is thrown into many folds, but as it fills with urine, the wall be-

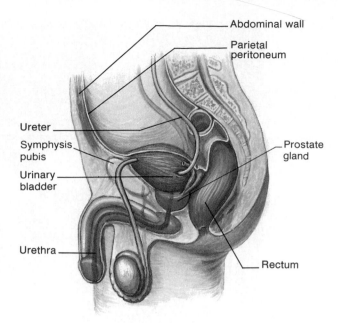

Figure 20.28 The urinary bladder is located within the pelvic cavity and behind the symphysis pubis. In a male, it lies against the rectum.

comes smoother. At the same time, the superior surface of the bladder expands upward into a dome.

When it is greatly distended, the bladder pushes above the pubic crest and into the region between the abdominal wall and the parietal peritoneum. The dome can reach the level of the umbilicus and press against the coils of the small intestine.

The internal floor of the bladder includes a triangular area called the *trigone,* which has an opening at each of its three angles (figure 20.29). Posteriorly, at the base of the trigone, the openings are those of the ureters. Anteriorly, at the apex of the trigone, there is a short, funnel-shaped extension called the *neck* of the bladder. This part contains the opening into the urethra. The trigone generally remains in a fixed position, even though the rest of the bladder changes shape during distension and contraction.

The wall of the urinary bladder consists of four layers. The inner layer, or *mucous coat,* includes several thicknesses of transitional epithelial cells, similar to those lining the ureters and the upper portion of the urethra. (See chapter 5.) The thickness of this tissue changes as the bladder expands and contracts. Thus, during distension, the tissue appears to be only two or three cells thick, but during contraction, it appears to be five or six cells thick. (See figure 5.7.)

The second layer of the wall is the *submucous coat.* It consists of connective tissue and contains many elastic fibers.

The third layer of the bladder wall, the *muscular coat,* is composed primarily of coarse bundles of smooth

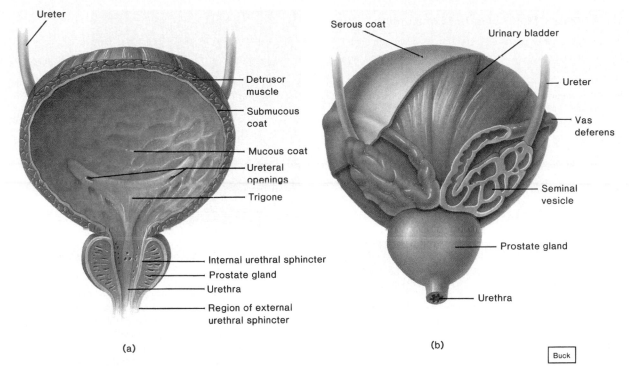

Ureter

Detrusor
muscle

Submucous
coat

Mucous coat

Ureteral
openings

Trigone

Internal urethral sphincter

Prostate gland

Urethra

Region of external
urethral sphincter

(a)

Serous coat

Urinary bladder

Ureter

Vas
deferens

Seminal
vesicle

Prostate gland

Urethra

(b)

Buck

Figure 20.29   A male urinary bladder. (*a*) Frontal section;
(*b*) posterior view.

muscle fibers. These bundles are interlaced in all directions and at all depths, and together they comprise the **detrusor muscle.** The portion of the detrusor muscle that surrounds the neck of the bladder forms an *internal urethral sphincter.* Sustained contraction of this sphincter muscle prevents the bladder from emptying until the pressure within the bladder increases to a certain level. The detrusor muscle is supplied with parasympathetic nerve fibers that function in the micturition reflex.

The outer layer of the wall, the *serous coat,* consists of the parietal peritoneum. This layer occurs only on the upper surface of the bladder. Elsewhere, the outer coat is composed of fibrous connective tissue (figure 20.30).

1. Describe the trigone of the urinary bladder.
2. Describe the structure of the bladder wall.
3. What kind of nerve fibers supply the detrusor muscle?

Micturition

Micturition (urination) is the process by which urine is expelled from the urinary bladder. It involves the contraction of the detrusor muscle, and may be aided by contractions of muscles in the abdominal wall and pelvic floor, and by fixation of the thoracic wall and diaphragm. Micturition also involves the relaxation of the *external urethral sphincter.* This muscle, which is part

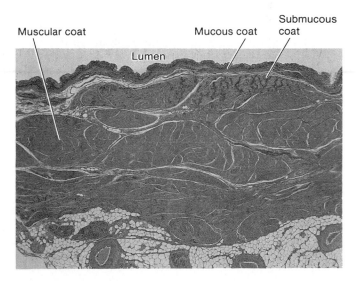

Muscular coat

Mucous coat

Submucous
coat

Lumen

Figure 20.30   Light micrograph of the human urinary bladder wall (×6).

of the urogenital diaphragm (chapter 9), surrounds the urethra about 3 centimeters from the bladder and is composed of voluntary skeletal muscle tissue.

The need to urinate is usually stimulated by distension of the bladder wall as it fills with urine. When the wall expands, stretch receptors are stimulated, and the micturition reflex is triggered.

The *micturition reflex center* is located in the sacral portion of the spinal cord. When the reflex center

CHART 20.4 Major events of micturition

1. Urinary bladder becomes distended as it fills with urine.
2. Stretch receptors in the bladder wall are stimulated, and they signal the micturition center in the sacral spinal cord.
3. Parasympathetic nerve impulses travel to the detrusor muscle, which responds by contracting rhythmically.
4. The need to urinate is sensed as urgent.
5. Urination is prevented by voluntary contraction of the external urethral sphincter and by inhibition of the micturition reflex by impulses from the brain stem and the cerebral cortex.
6. Following the decision to urinate, the external urethral sphincter is relaxed, and the micturition reflex is facilitated by impulses from the pons and the hypothalamus.
7. The detrusor muscle contracts, and urine is expelled through the urethra.
8. Neurons of the micturition reflex center fatigue, the detrusor muscle relaxes, and the bladder begins to fill with urine again.

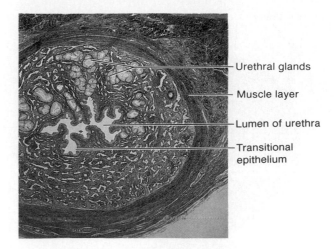

Figure 20.31 Cross section of the urethra (×10).

Labels: Urethral glands; Muscle layer; Lumen of urethra; Transitional epithelium

Damage to the spinal cord above the sacral region may result in the loss of voluntary control of urination. However, if the micturition reflex center and its sensory and motor fibers are uninjured, micturition may continue to occur reflexly. In this case, the bladder collects urine until its walls are stretched enough to trigger a micturition reflex, and the detrusor muscle contracts in response. This condition is called an *automatic bladder*.

is signaled by sensory impulses from the stretch receptors, *parasympathetic* motor impulses travel out to the detrusor muscle, which undergoes rhythmic contractions in response. This action is accompanied by a sensation of urgency.

Although the urinary bladder may hold as much as 600 milliliters of urine, the desire to urinate is usually experienced when it contains about 150 milliliters. Then, as the volume of urine increases to 300 milliliters or more, the sensation of fullness becomes increasingly uncomfortable.

As the bladder fills with urine, and its internal pressure increases, contractions of its wall become more and more powerful. When these contractions become strong enough to force the internal urethral sphincter to open, another reflex begins to operate. This second reflex signals the external urethral sphincter to relax, and the bladder may empty.

However, because the external urethral sphincter is composed of skeletal muscle, it can be consciously controlled. Thus, the sphincter muscle ordinarily remains contracted until a decision is made to urinate. This control is aided by nerve centers in the *brain stem* and *cerebral cortex* that are able to inhibit the micturition reflex. When a person decides to urinate, the external urethral sphincter is allowed to relax, and the micturition reflex is no longer inhibited. Nerve centers within the *pons* and the *hypothalamus* function to make the micturition reflex more effective. Consequently, the detrusor muscle contracts, and urine is excreted to the outside through the urethra. Within a few moments, the neurons of the micturition reflex seem to fatigue, the detrusor muscle relaxes, and the bladder begins to fill with urine again.

The micturition process is outlined in chart 20.4.

Urethra

The **urethra** is a tube that conveys urine from the urinary bladder to the outside of the body. Its wall is lined with mucous membrane and contains a relatively thick layer of smooth muscle tissue, whose fibers are generally directed longitudinally. It also contains numerous mucous glands, called *urethral glands,* which secrete mucus into the urethral canal (figure 20.31).

In a female, the urethra is about 4 centimeters long. It passes forward from the bladder, courses below the symphysis pubis, and empties between the labia minora. Its opening, the *external urethral orifice* (urinary meatus), is located anterior to the vaginal opening and about 2.5 centimeters posterior to the clitoris (figure 20.32).

In a male, the urethra, which functions both as a urinary canal and a passageway for cells and secretions from various reproductive organs, can be divided into three sections: the prostatic urethra, the membranous urethra, and the penile urethra (figure 20.32 and reference plate 60).

The **prostatic urethra** is about 2.5 centimeters long and passes from the urinary bladder through the *prostate gland,* which is located just below the bladder. Ducts from various reproductive structures join the urethra in this region (figure 20.32).

The **membranous urethra** is about 2 centimeters long. It begins just distal to the prostate gland, passes

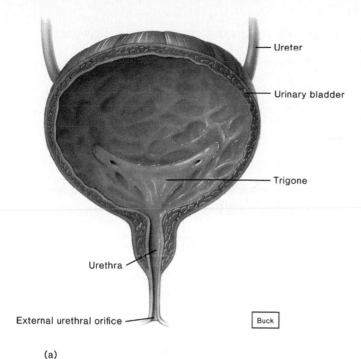

- Ureter
- Urinary bladder
- Trigone
- Urethra
- External urethral orifice

Buck

(a)

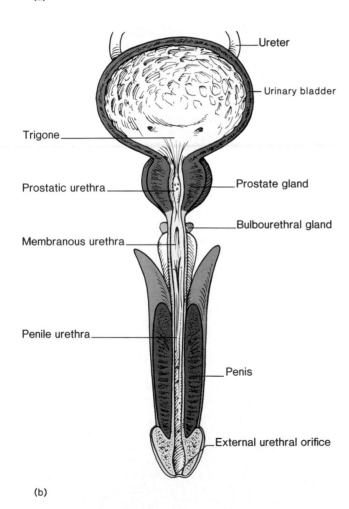

- Ureter
- Urinary bladder
- Prostate gland
- Bulbourethral gland
- Penis
- External urethral orifice

- Trigone
- Prostatic urethra
- Membranous urethra
- Penile urethra

(b)

Figure 20.32 (a) Longitudinal section of the female urinary bladder and urethra; (b) longitudinal section of the male urinary bladder and urethra.

through the urogenital diaphragm, and is surrounded by the fibers of the external urethral sphincter muscle.

The **penile urethra** is about 15 centimeters long and passes through the corpus spongiosum of the penis, where it is surrounded by erectile tissue. This portion of the urethra terminates with the *external urethral orifice* at the tip of the penis.

1. Describe the process of micturition.
2. How is it possible to consciously inhibit the micturition reflex?
3. Describe the structure of the urethra.
4. How does the urethra of a male differ from that of a female?

Clinical Terms Related to the Urinary System

anuria (ah-nu're-ah) An absence of urine due to failure of kidney function or to an obstruction in a urinary pathway.

bacteriuria (bak-te''re-u're-ah) Bacteria in the urine.

cystectomy (sis-tek'to-me) The surgical removal of the urinary bladder.

cystitis (sis-ti'tis) An inflammation of the urinary bladder.

cystoscope (sis'to-skōp) An instrument used for visual examination of the interior of the urinary bladder.

cystotomy (sis-tot'o-me) An incision of the wall of the urinary bladder.

diuresis (di''u-re'sis) An increased production of urine.

diuretic (di''u-ret'ik) A substance that causes an increased production of urine.

dysuria (dis-u're-ah) Painful or difficult urination.

enuresis (en''u-re'sis) Uncontrolled urination.

hematuria (hem''ah-tu're-ah) Blood in the urine.

incontinence (in-kon'ti-nens) An inability to control urination and/or defecation reflexes.

nephrectomy (nĕ-frek'to-me) The surgical removal of a kidney.

nephrolithiasis (nef''ro-lĭ-thi'ah-sis) The presence of a stone (or stones) in the kidney.

nephroptosis (nef''rop-to'sis) A movable or displaced kidney.

oliguria (ol''i-gu're-ah) A scanty output of urine.

polyuria (pol''e-u're-ah) An excessive output of urine.

pyelolithotomy (pi''ĕ-lo-lĭ-thot'o-me) The removal of a stone from the renal pelvis.

pyelonephritis (pi''ĕ-lo-nĕ-fri'tis) An inflammation of the renal pelvis.

pyelotomy (pi''ĕ-lot'o-me) An incision into the renal pelvis.

pyuria (pi-u're-ah) Pus (excessive white blood cells) in the urine.

uremia (u-re'me-ah) Condition in which substances ordinarily excreted in the urine accumulate in the blood.

ureteritis (u-re''ter-i'tis) An inflammation of the ureter.

urethritis (u''re-thri'tis) An inflammation of the urethra.

Chapter Summary

Introduction (page 748)

The urinary system consists of the kidneys, ureters, urinary bladder, and urethra.

Kidneys (page 748)

1. Location of the kidneys
   a. The kidneys are on either side of the vertebral column, high on the posterior wall of the abdominal cavity.
   b. They are positioned behind the parietal peritoneum, and held in place by adipose and connective tissue.
2. Structure of a kidney
   a. A kidney contains a hollow renal sinus.
   b. The ureter expands into the renal pelvis, which, in turn, is divided into major and minor calyces.
   c. Renal papillae project into the renal sinus.
   d. Kidney tissue is divided into a medulla and a cortex.
3. Functions of the kidneys
   a. The kidneys remove metabolic wastes from the blood and excrete them to the outside.
   b. They also help regulate red blood cell production, blood pressure, calcium ion absorption, and the volume, composition, and pH of the blood.
4. Renal blood vessels
   a. Arterial blood flows through the renal artery, interlobar arteries, arciform arteries, interlobular arteries, and afferent arterioles.
   b. Venous blood returns through a series of vessels that correspond to those of the arterial pathways.
5. Nephrons
   a. Structure of a nephron
      (1) A nephron is the functional unit of the kidney.
      (2) It consists of a renal corpuscle and a renal tubule.
         (a) The corpuscle consists of a glomerulus and a glomerular capsule.
         (b) Portions of the renal tubule include the proximal convoluted tubule, the loop of Henle (ascending and descending limbs), the distal convoluted tubule, and the collecting duct.
      (3) The collecting duct empties into the minor calyx of the renal pelvis.
   b. Juxtaglomerular apparatus
      (1) The juxtaglomerular apparatus is located at the point of contact between the distal convoluted tubule and the afferent and efferent arterioles.
      (2) It consists of the macula densa and the juxtaglomerular cells.
   c. Cortical and juxtamedullary nephrons
      (1) Cortical nephrons are the most numerous and have corpuscles near the surface of the kidney.
      (2) Juxtamedullary nephrons have corpuscles near the medulla.
   d. Blood supply of a nephron
      (1) The glomerular capillary receives blood from the afferent arteriole and passes it to the efferent arteriole.
      (2) The efferent arteriole gives rise to the peritubular capillary system, which surrounds the renal tubule.
      (3) Capillary loops, called vasa recta, dip down into the medulla.

Urine Formation (page 757)

Nephrons remove wastes from the blood and regulate water and electrolyte concentrations. Urine is the end product of these functions, which involve filtration, reabsorption, and secretion of substances from renal tubules.

1. Glomerular filtration
   a. Urine formation begins when water and dissolved materials are filtered out of the glomerular capillary.
   b. The glomerular capillaries are much more permeable than the capillaries in other tissues.
2. Filtration pressure
   a. Filtration is due mainly to hydrostatic pressure inside the glomerular capillaries.
   b. The osmotic pressure of the blood plasma and hydrostatic pressure in the glomerular capsule also affect filtration.
   c. Filtration pressure is the net force acting to move material out of the glomerulus and into the glomerular capsule.
   d. The composition of the filtrate is similar to that of tissue fluid.
3. Filtration rate
   a. The rate of filtration varies with the filtration pressure.
   b. Filtration pressure changes with the diameters of the afferent and efferent arterioles.
   c. As the osmotic pressure in the glomerulus increases, filtration decreases.
   d. Filtration rate varies with the rate of blood flow through the glomerulus.
   e. As the hydrostatic pressure in a glomerular capsule increases, the filtration rate decreases.
   f. The kidneys produce about 125 milliliters of glomerular fluid per minute, most of which is reabsorbed.
   g. The volume of filtrate varies with the surface area of the glomerular capillary.
4. Regulation of filtration rate
   a. When the filtration rate decreases, the macula densa causes the afferent arteriole to dilate, increasing blood flow through the glomerulus and increasing the filtration rate.
   b. The macula densa also causes the juxtaglomerular cells to release renin, which triggers a series of changes leading to constriction of the efferent arteriole, increasing the glomerular hydrostatic pressure and the filtration rate.

c. Autoregulation is the ability of an organ or tissue to maintain a constant blood flow when the arterial blood pressure is changing.

5. Tubular reabsorption
  a. Substances are selectively reabsorbed from the glomerular filtrate.
  b. The peritubular capillary is adapted for reabsorption.
    (1) It carries low pressure blood.
    (2) It is very permeable.
  c. Most reabsorption occurs in the proximal tubule, where the epithelial cells possess microvilli.
  d. Various substances are reabsorbed in particular segments of the renal tubule by different modes of transport.
    (1) Glucose and amino acids are reabsorbed by active transport.
    (2) Water is reabsorbed by osmosis.
    (3) Proteins are reabsorbed by pinocytosis.
  e. Active transport mechanisms have limited transport capacities.
  f. If the concentration of a substance in the filtrate exceeds its renal plasma threshold, the excess is excreted in the urine.
  g. Substances that remain in the filtrate are concentrated as water is reabsorbed.
  h. Sodium ions are reabsorbed by active transport.
    (1) As positively charged sodium ions are transported out of the filtrate, negatively charged ions accompany them.
    (2) Water is passively reabsorbed by osmosis as sodium ions are actively reabsorbed.

6. Regulation of urine concentration and volume
  a. Most of the sodium ions are reabsorbed before the urine is excreted.
  b. Sodium ions are concentrated in the renal medulla by the countercurrent mechanism.
    (1) Chloride ions are actively reabsorbed in the ascending limb, and sodium ions follow them passively.
    (2) Tubular fluid in the ascending limb becomes hypotonic as it loses solutes.
    (3) Water leaves the descending limb by osmosis, and NaCl enters this limb by diffusion.
    (4) Tubular fluid in the descending limb becomes hypertonic as it loses water and gains NaCl.
    (5) As NaCl repeats this circuit, its concentration in the medulla increases.
  c. The vasa recta countercurrent mechanism helps maintain the NaCl concentration in the medulla.
  d. The distal tubule and collecting duct are impermeable to water, which therefore tends to be excreted in urine.
  e. ADH from the posterior pituitary gland causes the permeability of the distal tubule and collecting duct to increase, and thus promotes the reabsorption of water.

7. Urea and uric acid excretion
  a. Urea is a by-product of amino acid metabolism.
    (1) It is reabsorbed passively by diffusion.
    (2) About 50% of the urea is excreted in urine.
    (3) A countercurrent mechanism helps in the excretion of urea.
  b. Uric acid results from the metabolism of nucleic acids.
    (1) Most is reabsorbed by active transport.
    (2) Some is secreted into the renal tubule.

8. Tubular secretion
  a. Tubular secretion is the process by which certain substances are transported from the plasma to the tubular fluid.
  b. Some substances are secreted actively.
    (1) These include various organic compounds and hydrogen ions.
    (2) Hydrogen ions are secreted by the proximal and distal segments of the renal tubule.
  c. Potassium ions are secreted passively in the distal segment and collecting duct where they are attracted by the negative charge that develops in the lumen of the tubule.

9. Composition of urine
  a. Urine is about 95% water, and it usually contains urea, uric acid, and creatinine.
  b. It may contain a trace of amino acids and varying amounts of electrolytes, depending upon the dietary intake.
  c. The volume of urine varies with the fluid intake and with certain environmental factors.

Elimination of Urine (page 769)

1. Ureters
  a. The ureter is a tubular organ that extends from each kidney to the urinary bladder.
  b. Its wall has mucous, muscular, and fibrous layers.
  c. Peristaltic waves in the ureter force urine to the bladder.
  d. Obstruction in the ureter stimulates strong peristaltic waves and a reflex that causes the kidney to decrease urine production.

2. Urinary bladder
  a. The urinary bladder is a distensible organ that stores urine and forces it into the urethra.
  b. The openings for the ureters and urethra are located at the three angles of the trigone in the floor of the urinary bladder.
  c. Muscle fibers in the wall form the detrusor muscle.
  d. A portion of the detrusor muscle forms an internal urethral sphincter.

3. Micturition
  a. Micturition is the process by which urine is expelled.
  b. It involves contraction of the detrusor muscle and relaxation of the external urethral sphincter.
  c. Micturition reflex
    (1) Stretch receptors in the bladder wall are stimulated by distension.

(2) The micturition reflex center in the sacral portion of the spinal cord sends parasympathetic motor impulses to the detrusor muscle.

(3) As the bladder fills, its internal pressure increases, and the internal urethral sphincter is forced open.

(4) A second reflex causes the external urethral sphincter to relax, unless its contraction is maintained by voluntary control.

(5) The control of urination is aided by nerve centers in the brain stem and cerebral cortex.

4. Urethra
   a. The urethra conveys urine from the bladder to the outside.
   b. In females, it empties between the labia minora.
   c. In males, it conveys products of reproductive organs as well as urine.
      (1) There are three portions of the male urethra: prostatic, membranous, and penile.
      (2) The urethra empties at the tip of the penis.

Clinical Application of Knowledge

1. If an infant is born with narrowed renal arteries, what effect would this condition have on the volume of urine produced? Explain your answer.

2. How would you explain the observation that people with nephrotic syndrome, in which plasma proteins are lost into the urine, have increased susceptibilities to infections?

3. If a patient who has had major abdominal surgery receives intravenous fluids equal to the volume of blood lost during surgery, would you expect the volume of urine produced to be greater or less than normal? Why?

4. If a physician prescribed oral penicillin therapy for a patient with an infection of the urinary bladder, how would you describe for the patient the route by which the drug would reach the bladder?

5. If the blood pressure of a patient who is in shock as a result of a severe injury decreases excessively, how would you expect the volume of urine produced by the patient's kidneys to change? Why?

6. Inflammation of the urinary bladder is more common in women than in men. How might this observation be related to the anatomy of the female and male urethras?

Review Activities

1. Name the organs of the urinary system, and list their general functions.

2. Describe the external and internal structure of a kidney.

3. List the functions of the kidneys.

4. Name the vessels through which the blood passes as it travels from the renal artery to the renal vein.

5. Distinguish between a renal corpuscle and a renal tubule.

6. Name the parts through which fluid passes as it travels from the glomerulus to the collecting duct.

7. Describe the location and structure of the juxtaglomerular apparatus.

8. Distinguish between cortical and juxtamedullary nephrons.

9. Distinguish among filtration, reabsorption, and secretion as they relate to urine formation.

10. Define *filtration pressure*.

11. Compare the composition of the glomerular filtrate with that of the blood plasma.

12. Explain how the diameters of the afferent and efferent arterioles affect the rate of glomerular filtration.

13. Explain how changes in the osmotic pressure of the blood plasma may affect the rate of glomerular filtration.

14. Explain how the hydrostatic pressure of a glomerular capsule affects the rate of glomerular filtration.

15. Describe two mechanisms by which the juxtaglomerular apparatus helps regulate the filtration rate.

16. Define *autoregulation*.

17. Discuss how tubular reabsorption is a selective process.

18. Explain how the peritubular capillary is adapted for reabsorption.

19. Explain how the epithelial cells of the proximal convoluted tubule are adapted for reabsorption.

20. Explain why active transport mechanisms have limited transport capacities.

21. Define *renal plasma threshold*, and explain its significance in tubular reabsorption.

22. Explain how amino acids and proteins are reabsorbed.

23. Describe the effect of sodium reabsorption on the reabsorption of negatively charged ions.

24. Explain how sodium ion reabsorption affects water reabsorption.

25. Explain how hypotonic tubular fluid is produced in the ascending limb of the loop of Henle.

26. Explain why fluid in the descending loop of Henle is hypertonic.

27. Describe the function of ADH.

28. Explain how urine may become concentrated as it moves through the collecting duct.

29. Compare the processes by which urea and uric acid are reabsorbed.

30. Explain how the renal tubule is adapted to secrete hydrogen ions.

31. Explain how potassium ions may be secreted passively.

32. List the more common substances found in urine and their sources.

Chronic Kidney Failure

Charles B., a 43-year-old muscular dairyman, had been feeling unusually tired for several weeks. Occasionally he had also felt dizzy, and he had been finding it increasingly difficult to sleep at night. More recently he had begun to experience a burning pain in his lower back, just below his rib cage, and he had noticed that his urine was quite dark in color. In addition, his feet and ankles seemed swollen, and his wife told him that his face looked puffy. She encouraged him to consult with their family physician about these symptoms.

In the course of his examination, the physician determined that Charles had elevated blood pressure (hypertension) and that the regions of his kidneys were sensitive to pressure. A urinalysis revealed excessive protein (proteinuria) and the presence of blood (hematuria). Blood tests indicated elevated blood urea nitrogen (BUN), elevated serum creatinine, and decreased serum protein (hypoproteinemia) concentrations.

Charles was told that he probably had *chronic glomerulonephritis,* an inflammation of the capillaries within the glomeruli of the renal nephrons (figure 20.9), and that this was a progressive degenerative disease for which there was no cure. This diagnosis was later confirmed when a tiny sample of kidney tissue was removed (biopsy) and examined microscopically (figure 20.10).

In spite of medical treatment and careful attention to his diet, Charles's condition deteriorated rapidly. When it appeared that most of his kidney function had been lost (end-stage renal disease, or ESRD), he was advised to begin artificial kidney treatments (hemodialysis).

To prepare Charles for hemodialysis, a vascular surgeon created a fistula in his left forearm by surgically connecting an artery to a vein. The greater pressure of the blood in the artery that now flowed directly into the vein caused the vein to enlarge, making it more accessible.

During hemodialysis treatment, a hollow needle was inserted into the vein of the fistula near its arterial connection. This allowed the blood to flow, with the aid of a blood pump, through a tube leading to the blood compartment of a dialysis machine. Within this compartment, the blood passed over a semipermeable membrane. On the opposite side of the membrane, there was a dialysate solution, whose composition was controlled. Negative pressure on the dialysate side of the membrane, created by a vacuum pump, increased the movement of fluid through the membrane. At the same time, waste substances and excessive electrolytes diffused from the blood through the membrane and entered the dialysate solution. The blood was then returned through a tube to the vein of the fistula.

In order to maintain favorable blood concentrations of waste substances, electrolytes, and water, Charles had to undergo hemodialysis three times per week, with each treatment lasting three to four hours. During the treatments, he was given an anticoagulate to prevent blood clotting. He was also provided with an antibiotic drug to control infections and an antihypertensive drug to reduce his blood pressure.

Charles was advised to carefully control his intake of water, sodium, potassium, proteins, and total calories between treatments. It was also suggested that he might be interested in another option for the treatment of ESRD— a kidney transplant—which could free him from the time-consuming dependence on hemodialysis.

Review Questions

1. Why is swelling of tissues commonly associated with kidney failure?
2. What mechanism might cause increased blood pressure to accompany kidney disease?
3. Why are protein and blood likely to appear in the urine of a patient with glomerulonephritis?
4. What factors might be considered when determining the time needed for each hemodialysis treatment?
5. Why is it important for a patient being treated with hemodialysis to control water intake? Protein intake?

---

33. List some of the factors that affect the volume of urine produced each day.
34. Describe the structure and function of a ureter.
35. Explain how the muscular wall of the ureter aids in moving urine.
36. Discuss what happens if a ureter becomes obstructed.
37. Describe the structure and location of the urinary bladder.
38. Define *detrusor muscle.*
39. Distinguish between the internal and external urethral sphincters.
40. Describe the micturition reflex.
41. Explain how the micturition reflex can be voluntarily controlled.
42. Compare the urethra of a female with that of a male.

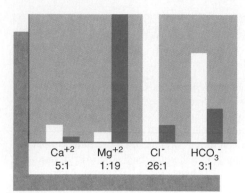

Ca^{+2}
5:1

Mg^{+2}
1:19

Cl$^-$
26:1

HCO$_3^-$
3:1

21

Water, Electrolyte, and Acid-Base Balance

*C*ell functions and, indeed, cell survival depend upon homeostasis—the existence of a stable cellular environment. In such an environment, the body cells are continually supplied with oxygen and nutrients, and the waste products resulting from their metabolic activities are continually carried away. At the same time, the concentrations of water and dissolved electrolytes, and the pH in the cellular fluids remain relatively constant. This condition requires the maintenance of *water, electrolyte,* and *acid-base balance* ■

Chapter Objectives

After you have studied this chapter, you should be able to:

1. Explain what is meant by water and electrolyte balance, and discuss the importance of this balance.

2. Describe how the body fluids are distributed within compartments, how the fluid composition differs between compartments, and how the fluids move from one compartment to another.

3. List the routes by which water enters and leaves the body, and explain how water input and output are regulated.

4. Explain how electrolytes enter and leave the body, and how the input and output of electrolytes are regulated.

5. Explain what is meant by acid-base balance.

6. Describe how hydrogen ion concentrations are expressed.

7. List the major sources of hydrogen ions in the body.

8. Distinguish between strong and weak acids and bases.

9. Explain how changing pH values of the body fluids are minimized by chemical buffer systems, the respiratory center, and the kidneys.

10. Complete the review activities at the end of this chapter. Note that the items are worded in the form of specific learning objectives. You may want to refer to them before reading the chapter.

Key Terms

acid (as'id)

acidosis (as''i-do'sis)

alkalosis (al''kah-lo'sis)

base (bās)

buffer system (buf'er sis'tem)

electrolyte balance (e-lek'tro-līt bal'ans)

extracellular (ek''strah-sel'u-lar)

intracellular (in''trah-sel'u-lar)

osmoreceptor (oz''mo-re-sep'tor)

transcellular (trans-sel'u-lar)

water balance (wot'er bal'ans)

Aids to Understanding Words

de-, separation from: *de*hydration—the removal of water from the cells or body fluids.

edem-, swelling: *edem*a—swelling due to an abnormal accumulation of extracellular fluid.

-emia, a blood condition: hypoprotein*emia*—an abnormally low concentration of blood plasma protein.

extra-, outside: *extra*cellular fluid—the fluid outside of the body cells.

im- (or **in-**), not: *im*balance—a condition in which factors are not in equilibrium.

intra-, within: *intra*cellular fluid—the fluid within the body cells.

neutr-, neither one nor the other: *neutr*al—a solution that is neither acidic nor basic.

-osis, a state of: acid*osis*—condition in which hydrogen ion concentration is abnormally high.

-uria, a urine condition: keto*uria*—presence of ketone bodies in the urine.

779

The term *balance* suggests a state of equilibrium, and in the case of water and electrolytes, it means that the quantities entering the body are equal to the quantities leaving it. Maintaining such a balance requires mechanisms to ensure that lost water and electrolytes will be replaced and that any excesses will be expelled. As a result, the quantities within the body remain relatively stable at all times.

It is important to remember that water balance and electrolyte balance are interdependent because the electrolytes are dissolved in the water of the body fluids. Consequently, anything that alters the concentrations of the electrolytes will necessarily alter the concentration of the water by adding solutes to it or by removing solutes from it. Likewise, anything that changes the concentration of the water will change the concentrations of the electrolytes by making them either more concentrated or more diluted.

Distribution of Body Fluids

Water and electrolytes are not uniformly distributed throughout the tissues. Instead, they occur in regions, or *compartments,* that contain fluids of varying compositions. The movement of water and electrolytes between these compartments is regulated so that their distribution remains relatively stable.

Fluid Compartments

The body of an average adult male is about 63% water by weight. This water (about 40 liters), together with its dissolved electrolytes, is distributed into two major compartments: an intracellular fluid compartment and an extracellular fluid compartment (figure 21.1).

> While the average adult male body is about 63% water by weight, the average adult female body is only about 52% water. This difference is related to the fact that adipose tissue has a relatively low water content and that females generally have more subcutaneous adipose tissue than males. Also, obese people of either sex contain proportionally less body water than do lean people.

The **intracellular fluid compartment** includes all the water and electrolytes that are enclosed by cell membranes. In other words, intracellular fluid is the fluid within the cells, and, in an adult, it represents about 63% by volume of the total body water.

The **extracellular fluid compartment** includes all the fluid outside the cells—within the tissue spaces (interstitial fluid), the blood vessels (plasma), and the lymphatic vessels (lymph). A specialized fraction of the extracellular fluid is separated from other extracellular fluids by various layers of epithelium. It is called **transcellular fluid** and includes *cerebrospinal fluid* of the

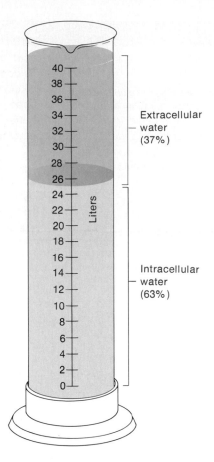

Figure 21.1 Of the 40 liters of water in the body of an average adult male, about 63% is intracellular and 37% is extracellular.

central nervous system, *aqueous and vitreous humors* of the eyes, *synovial fluid* of the joints, *serous fluid* within the various body cavities, and fluid *secretions* of the glands. The fluids of the extracellular compartment constitute about 37% by volume of the total body water (figure 21.2).

Composition of Body Fluids

Extracellular fluids generally have similar compositions. They are characterized by relatively high concentrations of sodium, chloride, and bicarbonate ions, and lesser concentrations of potassium, calcium, magnesium, phosphate, and sulfate ions. The blood plasma fraction of extracellular fluid contains considerably more protein than do either interstitial fluid or lymph.

Intracellular fluid contains relatively high concentrations of potassium, phosphate, and magnesium ions. It includes a somewhat greater concentration of sulfate ions, and lesser concentrations of sodium, chloride, and bicarbonate ions than does extracellular fluid. Intracellular fluid also has a greater concentration of protein than plasma. These relative concentrations are shown in figure 21.3.

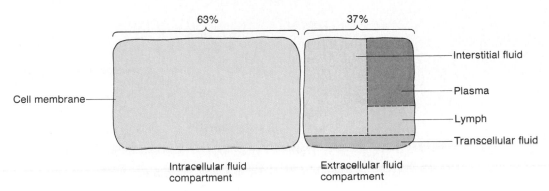

Total body water

63%   37%

Cell membrane

Interstitial fluid

Plasma

Lymph

Transcellular fluid

Intracellular fluid
compartment

Extracellular fluid
compartment

Figure 21.2   Fluid within the intracellular compartment is
separated from fluid in the extracellular compartment by cell
membranes. What membranes separate the various components
of the extracellular fluid compartment?

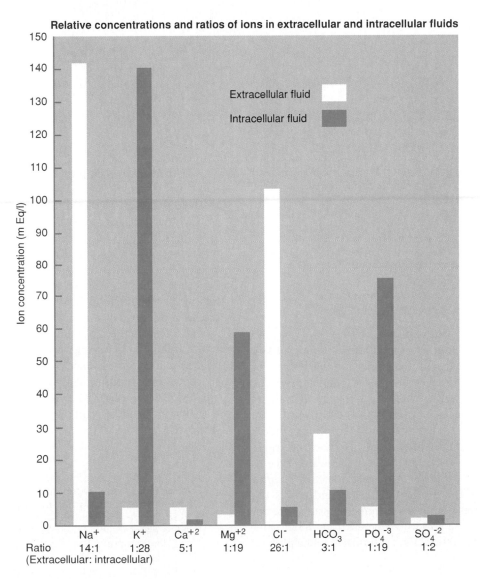

Figure 21.3   Extracellular fluid has relatively high
concentrations of sodium, chloride, and bicarbonate ions.
Intracellular fluid has relatively high concentrations of potassium,
magnesium, and phosphate ions.

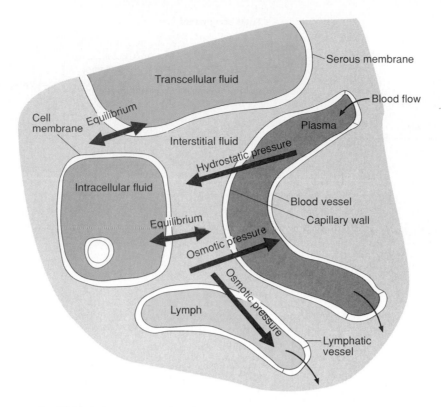

Figure 21.4 Net movements of fluids between compartments result from differences in hydrostatic and osmotic pressures.

1. How are fluid balance and electrolyte balance interdependent?
2. Describe the normal distribution of water within the body.
3. Which electrolytes are in higher concentrations in extracellular fluids? In intracellular fluid?
4. How does the concentration of protein vary in the various body fluids?

Movement of Fluid between Compartments

The movement of water and electrolytes from one fluid compartment to another is regulated largely by two factors: hydrostatic pressure and osmotic pressure. For example, as explained in chapter 18, fluid leaves the plasma at the arteriole ends of capillaries and enters the interstitial spaces because of the net outward force of *hydrostatic pressure* (blood pressure). Fluid returns to the plasma from the interstitial spaces at the venule ends of capillaries because of the net inward force of *osmotic pressure*. Likewise, as mentioned in chapter 19, fluid leaves the interstitial spaces and enters the lymph capillaries due to the osmotic pressure that develops within these spaces. As a result of the circulation of lymph, interstitial fluid is returned to the plasma.

The movement of fluid between the intracellular and extracellular compartments is similarly controlled by such pressures. However, because hydrostatic pressure within the cells and surrounding interstitial fluid is ordinarily equal and remains stable, any net fluid movement that occurs is likely to be the result of changes in osmotic pressure (figure 21.4).

For example, because the sodium ion concentration in the extracellular fluids is especially high, a decrease in this concentration will cause a net movement of water from the extracellular compartment into the intracellular compartment by osmosis. As a consequence, the cells will tend to swell. Conversely, if the concentration of sodium ions in the interstitial fluid increases, the net movement of water will be outward from the intracellular compartment, and the cells will shrink as they lose water.

1. What factors control the movement of water and electrolytes from one fluid compartment to another?
2. How does the sodium ion concentration within body fluids affect the net movement of water between the compartments?

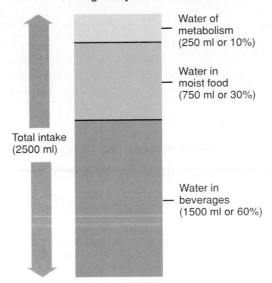

Average daily intake of water

Water of metabolism (250 ml or 10%)

Water in moist food (750 ml or 30%)

Total intake (2500 ml)

Water in beverages (1500 ml or 60%)

Figure 21.5 Major sources of body water.

Water Balance

Water balance exists when the total intake of water is equal to the total loss of water.

Water Intake

Although the volume of water gained each day varies from individual to individual, an average adult living in a moderate environment takes in about 2,500 milliliters.

Of this amount, probably 60% will be obtained from drinking water or beverages, while another 30% will be gained from moist foods. The remaining 10% will be a by-product of the oxidative metabolism of various nutrients, which is called **water of metabolism** (figure 21.5).

Regulation of Water Intake

The primary regulator of water intake is thirst, and although the thirst mechanism is poorly understood, it seems to involve the osmotic pressure of extracellular fluids and a *thirst center* in the hypothalamus of the brain.

As water is lost from the body, the osmotic pressure of the extracellular fluids increases. *Osmoreceptors* in the thirst center are thought to be stimulated by such a change, and as a result, the hypothalamus causes the person to feel thirsty and to seek water. The feeling of thirst is usually accompanied by dryness of the mouth, which is related to the loss of extracellular water and a consequent decreased flow of saliva.

The thirst mechanism is normally triggered whenever the total body water is decreased by 1% or 2%. As a person drinks water in response to thirst, the act of drinking and the resulting distension of the stomach wall seem to trigger nerve impulses that inhibit the thirst mechanism. Thus, the person usually stops drinking long before the swallowed water has been absorbed. This inhibition helps prevent the person from drinking more than is required to replace the quantity lost. Thus, it avoids development of an imbalance of the opposite type. Chart 21.1 summarizes the steps in this mechanism.

1. What is meant by water balance?
2. Where is the thirst center located?
3. What mechanism stimulates fluid intake? What mechanism inhibits it?

Water Output

Water normally enters the body only through the mouth, but it can be lost by a variety of routes. These include obvious losses in urine, feces, and sweat (sensible perspiration), as well as less obvious losses, which occur by diffusion of water through the skin (insensible perspiration) and by the evaporation of water from the lungs during breathing.

CHART 21.1 Regulation of water intake

1. The body loses 1% to 2% of its water.

2. Osmoreceptors in the thirst center are stimulated by an increase in the osmotic pressure of extracellular fluid due to water loss.

3. Activity in the hypothalamus causes the person to feel thirsty and to seek water in response.

4. The act of drinking and the distension of the stomach by water stimulate nerve impulses that inhibit the thirst center.

5. Water is absorbed through the walls of the stomach and small intestine.

6. The osmotic pressure of extracellular fluid decreases.

Average daily output of water

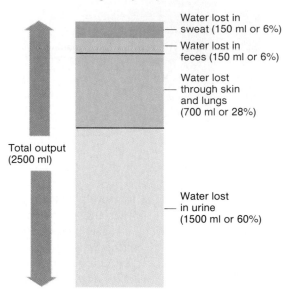

Water lost in sweat (150 ml or 6%)

Water lost in feces (150 ml or 6%)

Water lost through skin and lungs (700 ml or 28%)

Total output (2500 ml)

Water lost in urine (1500 ml or 60%)

Figure 21.6 Routes by which body water is lost. What factors influence water loss by each of these routes?

If an average adult takes in 2,500 milliliters of water each day, then 2,500 milliliters must be eliminated if water balance is to be maintained. Of this volume, perhaps 60% will be lost in urine, 6% in feces, and 6% in sweat. About 28% will be lost by diffusion through the skin and evaporation from the lungs (figure 21.6). These percentages will, of course, vary with such environmental factors as temperature and relative humidity, and with physical exercise.

Water lost by sweating is a necessary part of the body's temperature control mechanism; water lost in feces accompanies the elimination of undigested food materials; and water lost by diffusion and evaporation is largely unavoidable. Therefore, the primary means of regulating water output involves urine production.

Athletes who engage in endurance events may lose a relatively large volume of body water from sweating, particularly under hot and humid conditions. If more than about 2% of the body weight is lost in this manner, severe demands are placed on the person's ability to regulate body temperature. (See chapter 6.) Also, athletic performance is diminished, and the person may experience muscle cramps, nausea, and varying degrees of weakness.

To prevent these effects of dehydration, it is essential that an athlete replace body water as it is lost. This can usually be accomplished by drinking water before and during endurance events. It may be necessary, however, to drink more than is needed to satisfy thirst in order to completely replace the water lost.

CHART 21.2 Events in regulation of water balance

In the case of dehydration	In the case of excessive water intake
1. Extracellular fluid becomes more concentrated.	1. Extracellular fluid becomes osmotically less concentrated.
2. Osmoreceptors in the hypothalamus are stimulated by a change in the osmotic pressure of body fluid.	2. Osmoreceptors in the hypothalamus are stimulated by this change.
3. The hypothalamus signals the posterior pituitary gland to release ADH into the blood.	3. The posterior pituitary gland decreases the release of ADH.
4. Blood carries ADH to the kidneys.	4. Renal tubules decrease water reabsorption.
5. ADH causes the distal convoluted tubules and collecting ducts to increase water reabsorption.	5. Urine output increases, and excess water is excreted.
6. Urine output decreases, and water is conserved.	

Regulation of Water Balance

As discussed in chapter 20, the volume of water excreted in the urine is regulated mainly by activity in the distal convoluted tubules and collecting ducts of the nephrons. The epithelial linings of these segments of the renal tubule remain relatively impermeable to water unless antidiuretic hormone (ADH) is present.

As explained in chapter 13, the release of ADH involves osmoreceptors in the hypothalamus. If the blood plasma becomes more concentrated because of excessive water loss, these osmoreceptors tend to become dehydrated and to shrink. This change triggers impulses that signal the posterior pituitary gland to release ADH. The ADH is carried by the blood to the kidneys where it causes an increase in the permeability of the distal tubule and collecting duct. Consequently, water reabsorption increases, and water is conserved. This action resists further osmotic change in the plasma. In fact, the *osmoreceptor-ADH mechanism* can reduce a normal urine production of 1,500 milliliters per day to about 500 milliliters per day at times when the body is tending to become dehydrated.

On the other hand, if excessive amounts of water are taken into the body, the plasma becomes less concentrated, and the osmoreceptors swell as they receive extra water by osmosis. In this instance, the release of ADH is inhibited, and the distal segment and collecting duct remain impermeable to water. Consequently, less water is reabsorbed and a greater volume of urine is produced. Chart 21.2 summarizes the steps in this mechanism.

Diuretics are substances that promote the production of urine. A number of common substances, such as caffeine in coffee and tea, have diuretic effects, as do a variety of drugs used to reduce the volume of the body fluids.

Diuretics produce their effects in different ways. Some, such as alcohol and various narcotic drugs, promote urine formation by inhibiting the release of ADH. Certain other substances inhibit the reabsorption of sodium ions and other solutes in various portions of the renal tubules. As a consequence, the osmotic pressure of the tubular fluid increases, and the reabsorption of water by osmosis and its return to the blood are reduced.

1. By what routes is water lost from the body?
2. What is the primary regulator of water loss?
3. What types of water loss are unavoidable?
4. What role does the hypothalamus play in the regulation of water balance?

Water Balance Disorders

Among the more common disorders involving an imbalance in the water of body fluids are dehydration, water intoxication, and edema.

Dehydration is a deficiency condition that occurs when the output of water exceeds the intake. This condition may develop following excessive sweating, or as a result of prolonged water deprivation accompanied by continued water output. In either case, as water is lost, the extracellular fluid becomes increasingly more concentrated, and water tends to leave cells by osmosis (figure 21.7). Dehydration may also accompany illnesses in which excessive fluids are lost as a result of prolonged vomiting or diarrhea. During dehydration, the skin and mucous membranes of the mouth feel dry, and body weight is lost. Also a high fever may develop as the body temperature-regulating mechanism becomes less effective due to a lack of water needed for sweating. In severe cases, as waste products accumulate in the extracellular fluid, symptoms of cerebral disturbances, including mental confusion, delirium, and coma, may develop.

The kidneys of infants are less able to conserve water than are those of adults. Consequently, infants are more likely to become dehydrated than adults. Elderly people are also especially susceptible to developing water imbalances because the sensitivity of their thirst mechanisms tends to decrease with age, and physical disabilities may make it difficult for them to obtain adequate fluids.

The treatment for dehydration involves replacing the lost water and electrolytes. It is important to note that if the water alone is replaced, the extracellular fluid will become more dilute than normal, and this may produce a condition called *water intoxication.*

Water intoxication is characterized by the presence of hypotonic extracellular fluid. This condition may occur as a result of administering pure water to a dehydrated person, or it may develop in a person who drinks water for a prolonged period faster than the kidneys can excrete the excess.

In either instance, as the water is absorbed, the extracellular fluid becomes hypotonic to the cells. Then the cells tend to swell as water enters them in abnormal quantities by osmosis (figure 21.8).

The symptoms of water intoxication are related mainly to a decrease in the extracellular sodium ion concentration, and they include painful muscular contractions (heat cramps) and convulsions, confusion, and coma associated with the swelling of brain tissues. Treatment of this condition usually involves restricting water intake and administering hypertonic salt solutions.

Edema is characterized by an abnormal accumulation of extracellular fluid within the interstitial spaces. It can be caused by a variety of factors, including a decrease in the plasma protein concentration (hypoproteinemia), obstructions in lymphatic vessels, increased venous pressure, and increased capillary permeability.

Hypoproteinemia may result from such conditions as liver disease, in which there is a failure to synthesize plasma proteins; kidney disease (glomerulonephritis), in which the glomerular capillaries are damaged, allowing protein to escape into the urine; or starvation, in which the intake of amino acids is insufficient to support the synthesis of plasma proteins.

In each of these instances, the plasma protein concentration is decreased, and this is reflected in decreased plasma osmotic pressure. Consequently, the normal return of tissue fluid to the venule ends of capillaries is reduced, and tissue fluid tends to accumulate in the interstitial spaces.

As discussed in chapter 19, *lymphatic obstructions* may result from various surgical procedures or from certain

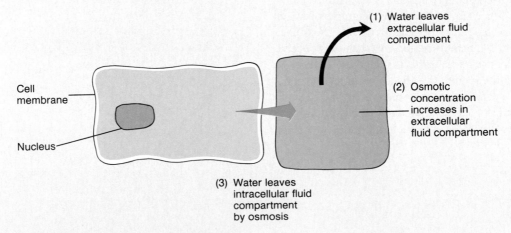

(1) Water leaves extracellular fluid compartment

(2) Osmotic concentration increases in extracellular fluid compartment

Cell membrane

Nucleus

(3) Water leaves intracellular fluid compartment by osmosis

Figure 21.7 If excessive amounts of extracellular fluids are lost, cells become dehydrated by osmosis. What happens to the concentration of electrolytes within the intracellular compartments at the same time?

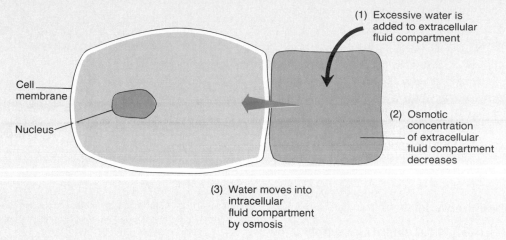

Figure 21.8 If excessive water is added to the extracellular fluid compartment, cells gain water by osmosis.

(1) Excessive water is added to extracellular fluid compartment

Cell membrane

Nucleus

(2) Osmotic concentration of extracellular fluid compartment decreases

(3) Water moves into intracellular fluid compartment by osmosis

CHART 21.3 Factors associated with edema

Factor	Cause	Effect
Low plasma protein concentration	Liver disease and failure to synthesize proteins; kidney disease and loss of proteins in urine; lack of protein in diet due to starvation	Plasma osmotic pressure decreases; less fluid enters venule ends of capillaries by osmosis
Obstruction of lymph vessels	Surgical removal of portions of lymphatic pathways; certain parasitic infections	Back pressure in lymph vessels interferes with movement of fluid from interstitial spaces into lymph capillaries
Increased venous pressure	Venous obstructions or faulty venous valves	Back pressure in veins interferes with return of fluid from interstitial spaces into venule ends of capillaries
Inflammation	Tissue damage	Capillaries become abnormally permeable; fluid leaks from plasma into interstitial spaces

parasitic infections in which the parasites inhabit lymphatic vessels. In such cases, back pressure develops in the lymphatic vessels, interfering with the normal movement of tissue fluid into them. At the same time, proteins that are ordinarily removed by lymphatic circulation tend to accumulate in the interstitial spaces. The presence of these proteins causes the osmotic pressure of the interstitial fluid to rise, and this effect causes still more fluid to be attracted into the interstitial spaces.

If the outflow of blood from the liver into the vena cava is blocked, the venous pressure within the liver and portal blood vessels increases greatly. As a result, fluid with a high concentration of protein tends to be exuded from the surfaces of the liver and intestine into the peritoneal cavity. This causes a rise in the osmotic pressure of the abdominal fluid, which, in turn, attracts more water into the peritoneal cavity by osmosis. This condition, called *ascites,* is characterized by an uncomfortable distension of the abdomen.

Edema may also result from increased capillary permeability accompanying an *inflammation reaction.* As described in chapters 5 and 19, this reaction occurs in response to tissue damage and usually involves the release of chemicals, such as histamine, from damaged cells. Histamine causes an increase in capillary permeability, so that excessive amounts of fluid tend to leak out of the capillary and enter the interstitial spaces.

Chart 21.3 summarizes the factors that result in edema.

Electrolyte Balance

An **electrolyte balance** exists when the quantities of electrolytes (substances that release ions in water) gained by the body are equal to those lost (figure 21.9).

Electrolyte Intake

The electrolytes of greatest importance to cellular functions are those that release the ions of sodium, potassium, calcium, magnesium, chloride, sulfate, phosphate, and bicarbonate. These substances are obtained primarily from foods, but they may also occur in drinking water and other beverages. In addition, some electrolytes occur as by-products of various metabolic reactions.

Regulation of Electrolyte Intake

Ordinarily, a person obtains sufficient electrolytes by responding to hunger and thirst. However, when there is a severe electrolyte deficiency, a person may experience *salt craving,* which is a strong desire to eat salty foods.

Electrolyte Output

Some electrolytes are lost from the body by perspiration. The quantities of electrolytes leaving will vary with the amount of perspiration produced. Greater quantities are lost on warmer days and during times of strenuous exercise. Also, varying amounts of electrolytes are lost in the feces. The greatest electrolyte output, however, occurs as a result of kidney function and urine production.

1. What electrolytes are most important to cellular functions?
2. What mechanisms ordinarily regulate electrolyte intake?
3. By what routes are electrolytes lost from the body?

Regulation of Electrolyte Balance

The concentrations of positively charged ions, such as sodium (Na^+), potassium (K^+), and calcium (Ca^{+2}), are particularly important. For example, certain concentrations of these ions are necessary for the conduction of nerve impulses, the contraction of muscle fibers, and the maintenance of cell membrane permeability. Thus, it is vital that their concentrations be regulated.

Sodium ions account for nearly 90% of the positively charged ions in extracellular fluids. The primary mechanism regulating these ions involves the kidneys and the hormone *aldosterone.* As explained in chapter 13, this hormone is secreted by the adrenal cortex. Its presence causes an increase in sodium ion reabsorption in the collecting ducts of the renal tubules.

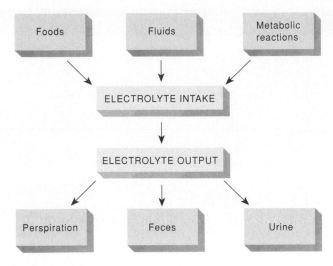

Figure 21.9 Electrolyte balance exists when the intake of electrolytes from all sources equals the output of electrolytes.

> If adrenal cortical hormones are deficient, the person develops *Addison's disease.* In this condition, an aldosterone deficiency causes the normal concentration of extracellular sodium to be depleted within a few days. Without treatment, the person is likely to live only a short time. (See chapter 13.)

Aldosterone also functions in regulating *potassium ions.* In fact, the most important stimulus for aldosterone secretion is a rising potassium ion concentration, which seems to stimulate the cells of the adrenal cortex directly. This hormone enhances the reabsorption of sodium ions and, at the same time, causes the secretion of potassium ions into the renal filtrate.

As discussed in chapter 20, this action occurs because of the negative electrical charge that develops within the collecting duct as sodium ions are actively transported outward. The positively charged potassium ions (K^+) in the extracellular fluid of the kidney are attracted to this negative charge and are passively secreted into the urine. Consequently, as sodium ions are conserved, potassium ions are excreted (figure 21.10).

As mentioned in chapter 13, the concentration of *calcium ions* in extracellular fluids is regulated mainly by the parathyroid glands. Whenever the calcium ion concentration drops below normal, these glands are stimulated directly, and they secrete *parathyroid hormone* in response.

Parathyroid hormone causes increased activity in bone-resorbing cells (osteocytes and osteoclasts). As a result of their actions, the concentrations of calcium and phosphate ions in the extracellular fluids increase.

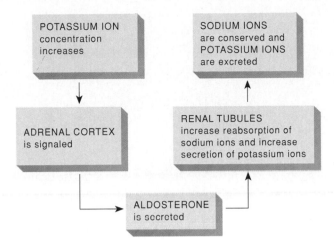

Figure 21.10 If the concentration of potassium ions increases, the kidneys act to conserve sodium ions and excrete potassium ions.

Parathyroid hormone also stimulates the absorption of calcium from the intestine indirectly. (See chapter 13.) Concurrently, it causes the kidneys to conserve calcium ions and increase the excretion of phosphate ions. Thus, the net effect of the hormone is to increase the *calcium ion* concentration of the extracellular fluid, but to maintain a normal *phosphate ion* concentration (figure 21.11).

Abnormal increases in blood calcium (hypercalcemia) sometimes result from hyperparathyroidism, in which excessive secretion of PTH causes increased bone resorption (chapter 13). Hypercalcemia may also be caused by the presence of cancers, particularly those originating in bone marrow, breasts, lungs, or prostate glands. Usually the increase in calcium occurs when ions are released from bone tissue that is being invaded and destroyed by cancerous growths. In other cases, however, the blood calcium concentration increases when cancer cells produce substances that have physiological effects similar to parathyroid hormone. Such hormonelike substances are most commonly associated with lung cancers. This condition may be accompanied by weakness and fatigue, as well as by impaired mental functions, headache, nausea, increased urine volume (polyuria), and increased thirst (polydipsia).

Abnormal decreases in blood calcium (hypocalcemia) may result from a reduced availability of PTH following the surgical removal of the parathyroid glands. It may also occur when vitamin D is deficient, either because of decreased absorption following gastrointestinal surgery or excessive excretion due to kidney disease. Because hypocalcemia is likely to be accompanied by spasms of muscles (tetany), including those within the airways, and by cardiac arrhythmias, it can be life threatening. This condition is usually treated by administering calcium salts and high doses of vitamin D to promote calcium absorption.

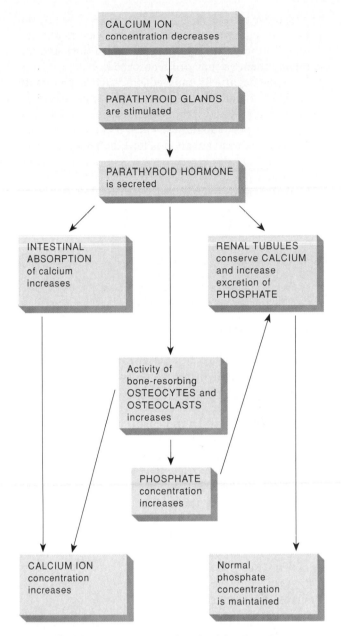

Figure 21.11 If the concentration of calcium ions decreases, the parathyroid glands cause an increase in calcium ion concentration, but maintain a normal concentration of phosphate ions.

The renal tubule also helps regulate calcium ion concentration by conserving calcium through reabsorption when the extracellular concentration is low, and by allowing excess calcium to be excreted when the concentration is high.

Generally, the concentrations of negatively charged ions are controlled secondarily by the regulatory mechanisms that control positively charged ions. For example, chloride ions (Cl^-), the most abundant negatively charged ions in the extracellular fluids, are passively reabsorbed from the renal tubules in response to the active reabsorption of sodium ions. That is, the negatively charged chloride ions are electrically attracted to the positively charged sodium ions and accompany them as they are reabsorbed.

Some negatively charged ions, such as phosphate ions (PO_4^{-3}) and sulfate ions (SO_4^{-2}), also are regulated partially by active transport mechanisms that have limited transport capacities. Thus, if the extracellular phosphate ion concentration is low, the phosphate ions in the renal tubules are conserved. On the other hand, if the renal plasma threshold is exceeded, the excess phosphate will be excreted in the urine.

1. What is the action of aldosterone?
2. How are the concentrations of sodium and potassium ions controlled?
3. How is calcium regulated?
4. What mechanism regulates the concentrations of most negatively charged ions?

Acid-Base Balance

As discussed in chapter 2, electrolytes that ionize in water and release hydrogen ions are called *acids.* Substances that combine with hydrogen ions are called *bases.* The concentrations of acids and bases within the body fluids must be controlled if homeostasis is to be maintained. The regulation of hydrogen ions is very important, because slight changes in hydrogen ion concentrations can alter the rates of enzyme-controlled metabolic reactions, cause shifts in the distribution of other ions, or modify the actions of various hormones. Thus, acid-base balance is primarily concerned with the regulation of hydrogen ion concentrations, which are measured in pH units. (See chapter 2 for a detailed explanation of pH.)

Sources of Hydrogen Ions

Most of the hydrogen ions in the body fluids originate as by-products of metabolic processes, although small quantities may enter in foods.

The major metabolic sources of hydrogen ions include the following:

1. **Aerobic respiration of glucose.** This process results in the production of carbon dioxide and water. Carbon dioxide diffuses out of the cells and reacts with water in the extracellular fluids to form *carbonic acid:*

$$CO_2 + H_2O \longrightarrow H_2CO_3$$

The resulting carbonic acid then ionizes to release hydrogen ions and bicarbonate ions:

$$H_2CO_3 \longrightarrow H^+ + HCO_3^-$$

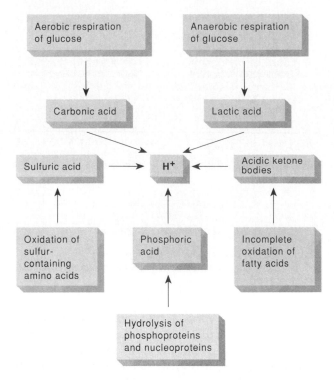

Figure 21.12 Some of the metabolic processes that serve as sources of hydrogen ions.

2. **Anaerobic respiration of glucose.** When glucose is utilized anaerobically, *lactic acid* is produced, and hydrogen ions are added to the body fluids.
3. **Incomplete oxidation of fatty acids.** The incomplete oxidation of fatty acids results in the production of *acidic ketone bodies,* which cause the hydrogen ion concentration to increase.
4. **Oxidation of amino acids containing sulfur.** The oxidation of sulfur-containing amino acids results in the production of *sulfuric acid* (H_2SO_4), which ionizes to release hydrogen ions and sulfate ions.
5. **Breakdown (hydrolysis) of phosphoproteins and nucleoproteins.** Phosphoproteins and nucleoproteins contain phosphorus. Their oxidation results in the production of *phosphoric acid* (H_3PO_4), which ionizes to release hydrogen ions.

The acids produced by various metabolic processes vary in strength, thus their effects on the hydrogen ion concentration of the body fluids vary also (figure 21.12).

1. Explain why the regulation of hydrogen ion concentration is so important.
2. What are the major sources of hydrogen ions in the body?

Sodium and Potassium Imbalances

Extracellular fluids usually have high sodium ion concentrations, and intracellular fluid normally has high potassium ion concentrations. The renal regulation of sodium is closely related to that of potassium because active reabsorption of sodium (under the influence of aldosterone) is accompanied by passive secretion (and excretion) of potassium. Thus, it is not surprising that conditions involving sodium ion imbalance are closely related to those involving potassium ion imbalance.

Such disorders can be summarized as follows:

1. **Low sodium concentration** (hyponatremia). Possible causes of sodium deficiencies include prolonged sweating, vomiting, or diarrhea; renal disease in which sodium is reabsorbed inadequately; adrenal cortex disorders in which aldosterone secretion is insufficient to promote the reabsorption of sodium (Addison's disease); and excessive intake of drinking water.

    Possible effects of *hyponatremia* include the development of extracellular fluid that is hypotonic to cells and promotes the movement of water into the cells by osmosis. This is accompanied by the symptoms of water intoxication described in a previous section of this chapter.

2. **High sodium concentration** (hypernatremia). Possible causes of elevated sodium concentration include excessive water loss by diffusion and evaporation, as may occur during high fever, or increased water loss accompanying diabetes

insipidus, in which ADH secretion is insufficient to maintain water conservation by the renal tubules.

   Possible effects of *hypernatremia* include disturbances of the central nervous system, such as confusion, stupor, and coma.

3. **Low potassium concentration** (hypokalemia). Possible causes of potassium deficiency include excessive release of aldosterone by the adrenal cortex (Cushing's syndrome), which causes increased renal secretion of potassium; use of diuretic drugs that promote the excretion of potassium; kidney disease; and prolonged vomiting or diarrhea.

   Possible effects of *hypokalemia* include muscular weakness or paralysis, respiratory difficulty, and severe cardiac disturbances, such as atrial or ventricular arrhythmias.

4. **High potassium concentration** (hyperkalemia). Possible causes of elevated potassium concentration include renal disease in which potassium excretion is decreased; use of drugs that promote renal conservation of potassium; insufficient secretion of aldosterone by the adrenal cortex (Addison's disease); or a shift of potassium from the intracellular fluid to the extracellular fluid, a change that accompanies an increase in hydrogen ion concentration (acidosis).

   Possible effects of *hyperkalemia* include paralysis of the skeletal muscles and severe cardiac disturbances, such as cardiac arrest.

Strengths of Acids and Bases

Acids that ionize more completely are *strong acids,* and those that ionize less completely are *weak acids.* For example, the hydrochloric acid (HCl) of gastric juice is a strong acid, but the carbonic acid (H_2CO_3) produced when carbon dioxide reacts with water is weak.

The extent to which such acids ionize can be indicated by the lengths of the arrows in chemical equations. That is, for hydrochloric acid, a long arrow indicates a considerable amount of ionization; for carbonic acid, a short arrow indicates a smaller amount of ionization:

$$HCl \rightleftharpoons H^+ + Cl^-$$
$$H_2CO_3 \rightleftharpoons H^+ + HCO_3$$

The arrows pointing to the left indicate that these chemical reactions are *reversible,* and once again, the lengths of the arrows represent the degrees to which the reverse reactions tend to proceed.

Bases are substances that, like hydroxyl ions (OH^-), will combine with hydrogen ions. *Chloride ions* (Cl^-) and *bicarbonate ions* (HCO_3^-) are also bases. Furthermore, since chloride ions combine less readily with hydrogen ions (as indicated by the short arrow pointing to the left in the preceding equation), they are *weak* bases, but the bicarbonate ions, which combine more readily with hydrogen ions, are *strong bases.*

Once again, the extent to which these substances react with hydrogen ions can be indicated by the lengths of the arrows in chemical equations, as follows:

$$Cl^- + H^+ \rightleftharpoons HCl$$
$$HCO_3^- + H^+ \rightleftharpoons H_2CO_3$$

Regulation of Hydrogen Ion Concentration

The concentration of hydrogen ions in the body fluid, as measured by pH, is regulated primarily by acid-base buffer systems, the respiratory center in the brain stem, and the nephrons in the kidneys.

Acid-Base Buffer Systems

Acid-base buffer systems occur in all the body fluids and are usually composed of sets of two or more chemical substances. Such chemicals can combine with acids or bases when they occur in excess. More specifically, the substances of a buffer system can convert strong acids, which tend to release large quantities of hydrogen ions, into weak acids, which release fewer hydrogen ions. Likewise, these buffers can combine with strong bases and change them into weak bases. Such activity helps minimize the pH changes in body fluids.

The three most important acid-base buffer systems in body fluids are these:

1. **Bicarbonate buffer system.** The bicarbonate buffer system, which is present in intracellular and extracellular body fluids, consists of carbonic acid (H_2CO_3) and sodium bicarbonate ($NaHCO_3$). If a strong acid, like hydrochloric acid, is present, it reacts with the sodium bicarbonate. The products of the reaction are carbonic acid, which is a weaker acid, and sodium chloride. Consequently, an increase in the hydrogen ion concentration in the body fluid is minimized:

$$HCl + NaHCO_3 \longrightarrow H_2CO_3 + NaCl$$
(strong acid)   (weak acid)

On the other hand, if a strong base like sodium hydroxide (NaOH) is present, it reacts with the carbonic acid. The products are sodium bicarbonate ($NaHCO_3$), which is a weaker base, and water. Thus, a shift toward a more basic (alkaline) state is minimized:

$$NaOH + H_2CO_3 \longrightarrow NaHCO_3 + H_2O$$
(strong base)   (weak base)

2. **Phosphate buffer system.** The phosphate acid-base buffer system is also present in the intracellular and extracellular body fluids. It is particularly important as a regulator of the hydrogen ion concentrations in the tubular fluid of the nephrons and in urine. This buffer system consists of two phosphate compounds—sodium monohydrogen phosphate (Na_2HPO_4) and sodium dihydrogen phosphate (NaH_2PO_4).

    If a strong acid is present, it reacts with the sodium monohydrogen phosphate to produce a weaker acid (sodium dihydrogen phosphate) and sodium chloride:

$$HCl + Na_2HPO_4 \longrightarrow NaH_2PO_4 + NaCl$$
(strong acid)   (weak acid)

If a strong base is present, it reacts with the sodium dihydrogen phosphate, and again the products are a weak base (sodium monohydrogen phosphate) and water:

$$NaOH + NaH_2PO_4 \longrightarrow Na_2HPO_4 + H_2O$$
(strong base)   (weak base)

3. **Protein buffer system.** The protein acid-base buffer system consists of the plasma proteins, such as albumins, and various proteins within the cells, including the hemoglobin of red blood cells.

    As described in chapter 2, proteins are composed of amino acids bound together in complex chains. Some of these amino acids have freely exposed groups of atoms, called *carboxyl groups*. Under certain conditions, a carboxyl group ($-COOH$) can become ionized, and a hydrogen ion is released:

$$-COOH \longrightarrow -COO^- + H^+$$

Some of the amino acids within a protein molecule also contain freely exposed *amino groups* ($-NH_2$). Under certain conditions, these amino groups can accept hydrogen ions:

$$-NH_2 + H^+ \longrightarrow -NH_3^+$$

Thus, protein molecules can function as acids by releasing hydrogen ions from their carboxyl groups, or as bases by accepting hydrogen ions into their amino groups. This special property allows protein molecules to operate as an acid-base buffer system. In the presence of excess hydrogen ions, the $-COO^-$ portions of the protein molecules accept hydrogen ions and become $-COOH$ groups again. This action decreases the number of free hydrogen ions in body fluid and minimizes the pH change.

In the presence of excess hydroxyl ions (OH^-), the $-NH_3^+$ groups within protein molecules give up hydrogen ions and become $-NH_2$ groups again. These hydrogen ions then combine with the hydroxyl ions to form water molecules. Once again, the pH change is reduced to a minimum, because water is a neutral substance:

$$H^+ + OH^- \longrightarrow H_2O$$

The *hemoglobin* in red blood cells provides an example of a specific protein that functions to buffer hydrogen ions. As explained in chapter 16, carbon dioxide, which is produced by the cellular oxidation of glucose,

CHART 21.4 Chemical acid-base buffer systems

Buffer system	Constituents	Actions
Bicarbonate system	Sodium bicarbonate ($NaHCO_3$)	Converts a strong acid into a weak acid
	Carbonic acid (H_2CO_3)	Converts a strong base into a weak base
Phosphate system	Sodium monohydrogen phosphate (Na_2HPO_4)	Converts a strong acid into a weak acid
	Sodium dihydrogen phosphate (NaH_2PO_4)	Converts a strong base into a weak base
Protein system (and amino acids)	$-COO^-$ group of a molecule	Accepts hydrogen ions in the presence of excess acid
	$-NH_3^+$ group of a molecule	Releases hydrogen ions in the presence of excess base

diffuses through the capillary wall and enters the plasma and red blood cells. The red cells contain an enzyme, *carbonic anhydrase*, that speeds the reaction between carbon dioxide and water, producing carbonic acid:

$$CO_2 + H_2O \longrightarrow H_2CO_3$$

The carbonic acid quickly dissociates, releasing hydrogen ions and bicarbonate ions (HCO_3^-):

$$H_2CO_3 \longrightarrow H^+ + HCO_3^-$$

The hydrogen ions are accepted by the hemoglobin molecules, and thus, the pH change that would otherwise occur is reduced to a minimum.

Individual amino acids in body fluids can also function as acid-base buffers by accepting or giving up hydrogen ions because every amino acid has an amino group ($-NH_2$) and a carboxyl group ($-COOH$) as part of its molecule.

To summarize, acid-base buffer systems take in hydrogen ions when body fluids are becoming more acidic, and give up hydrogen ions when the fluids are becoming more basic (alkaline). The buffer systems accomplish this by converting stronger acids into weaker ones or by converting stronger bases into weaker ones, as summarized in chart 21.4.

Although buffer systems cannot prevent a pH change, they are able to reduce it to a minimum. Fur-thermore, the various acid-base buffer systems in body fluids act together to buffer each other. Consequently, whenever the hydrogen ion concentration begins to change, the chemical balances within all of the buffer systems change too, and the drift in pH is resisted.

Neurons are particularly sensitive to changes in the pH of body fluids. For example, if the interstitial fluid becomes more alkaline than normal (alkalosis), the neurons tend to become more excitable and the person may experience seizures. Conversely, if conditions become more acidic (acidosis), neuron activity tends to be depressed, and the person's level of consciousness may decrease.

1. What is the difference between a strong acid or base and a weak acid or base?
2. How does a chemical buffer system help regulate the pH?
3. List the major buffer systems of the body.

The Respiratory Center

The **respiratory center** in the brain stem helps regulate hydrogen ion concentrations in the body fluids by controlling the rate and depth of breathing. Specifically, if the body cells increase their production of carbon dioxide, as occurs during periods of physical exercise, the production of carbonic acid increases. As the carbonic acid dissociates, the concentration of hydrogen ions increases, and the pH of the fluids tends to drop. (See chapter 16.)

Such an increasing concentration of carbon dioxide, and the consequent increase in hydrogen ion concentration in the blood plasma, stimulate chemosensitive areas within the respiratory center.

In response, the respiratory center causes the depth and rate of breathing to increase, so that a greater amount of carbon dioxide is excreted through the lungs. This loss of carbon dioxide is accompanied by a drop in the hydrogen ion concentration in the body fluids, because the released carbon dioxide comes from carbonic acid (figure 21.13):

$$H_2CO_3 \longrightarrow CO_2 + H_2O$$

Conversely, if the body cells are less active, the concentrations of carbon dioxide and hydrogen ions in the plasma remain relatively low. As a consequence, the breathing rate and depth are decreased. In time, this may result in an accumulation of carbon dioxide in the body fluids, and as the pH drops, the respiratory center is stimulated to increase the rate and depth of breathing once again.

Thus, the activity of the respiratory center is altered in response to shifts in the plasma pH, reducing these shifts to a minimum. Because most of the hy-

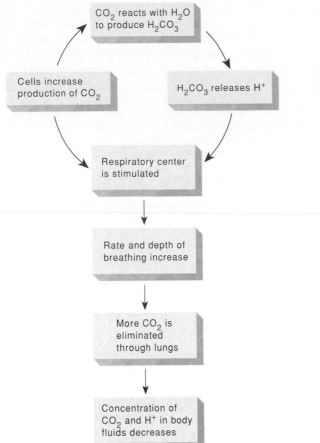

Figure 21.13 An increase in carbon dioxide production is followed by an increase in carbon dioxide elimination.

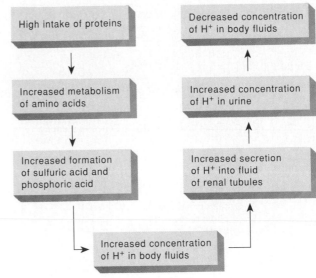

Figure 21.14 If the concentration of hydrogen ions in body fluids increases, the renal tubules increase their secretion of hydrogen ions into the urine.

drogen ions in the body fluids originate from carbonic acid produced when carbon dioxide reacts with water, the respiratory regulation of hydrogen ion concentration is of considerable importance.

The Kidneys

The nephrons of the kidneys help regulate the hydrogen ion concentration of the body fluids by secreting hydrogen ions. As discussed in chapter 20, these ions are secreted into the urine of the renal tubules by the epithelial cells that line the proximal and distal convoluted tubules and the collecting ducts.

This mechanism is important in balancing the quantities of sulfuric acid, phosphoric acid, and various organic acids that appear in body fluids as by-products of metabolic processes.

The metabolism of certain amino acids, for example, results in the formation of sulfuric and phosphoric acids. Consequently, a diet high in proteins (which give rise to amino acids when they are digested) will be accompanied by an excessive production of acids. The kidneys compensate for such gains in acids by altering the rate of hydrogen ion secretion into the urine. Thus, a shift in the pH of body fluids is resisted (figure 21.14).

Once hydrogen ions reach the urine, they are buffered by phosphates that were filtered into the fluid of the renal tubule. Ammonia also aids in this buffering action.

Through deamination of certain amino acids, the cells of the renal tubules produce ammonia (NH_3). It diffuses readily through the cell membranes and enters the urine. Because ammonia is a weak base, it can accept hydrogen ions, and *ammonium ions* (NH_4^+) result:

$$H^+ + NH_3 \longrightarrow NH_4^+$$

Cell membranes are quite impermeable to ammonium ions. Consequently, these ions are trapped in the urine as they form, and they are excreted from the body with the urine.

As a result of a poorly understood mechanism, a prolonged increase in the hydrogen ion concentration of body fluids is accompanied by an increase in the production of ammonia by cells of the renal tubules. This mechanism helps to transport excessive hydrogen ions to the outside and to control the pH of the urine.

The various regulators of hydrogen ion concentration operate at different rates. Acid-base buffers, for example, function rapidly and can convert strong acids or bases into weak acids or bases almost immediately. For this reason, these chemical buffer systems are sometimes called the body's *first line of defense* against shifts in pH.

Physiological buffer systems, such as the respiratory and renal mechanisms, function more slowly, and constitute *secondary defenses*. The respiratory mechanism may require several minutes to begin resisting a change in pH, and the renal mechanism may require one to three days to regulate a changing hydrogen ion concentration (figure 21.15).

1. How does the respiratory system help regulate acid-base balance?
2. How do the kidneys respond to excessive concentrations of hydrogen ions?
3. How do the rates at which chemical and physiological buffer systems act differ?

Clinical Terms Related to Water and Electrolyte Balance

acetonemia (as″ĕ-to-ne′me-ah) The presence of abnormal amounts of acetone in the blood.

acetonuria (as″ĕ-to-nu′re-ah) The presence of abnormal amounts of acetone in the urine.

albuminuria (al-bu″mĭ-nu′re-ah) The presence of albumin in the urine.

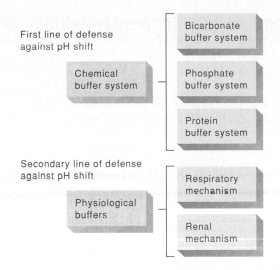

Figure 21.15 Chemical buffers act rapidly, while physiological buffers may require several minutes to begin resisting a change in pH.

anasarca (an″ah-sar′kah) A widespread accumulation of tissue fluid.

antacid (ant-as′id) A substance that neutralizes an acid.

anuria (ah-nu′re-ah) The absence of urine excretion.

azotemia (az″o-te′me-ah) An accumulation of nitrogenous wastes in the blood.

diuresis (di″u-re′sis) An increased production of urine.

glycosuria (gli″ko-su′re-ah) The presence of excessive sugar in the urine.

hyperglycemia (hi″per-gli-se′me-ah) An abnormally high concentration of blood sugar.

hyperkalemia (hi″per-kah-le′me-ah) The presence of excessive potassium in the blood.

hypernatremia (hi″per-na-tre′me-ah) The presence of excessive sodium in the blood.

hyperuricemia (hi″per-u″rĭ-se′me-ah) The presence of excessive uric acid in the blood.

hypoglycemia (hi″po-gli-se′me-ah) An abnormally low concentration of blood sugar.

ketonuria (ke′to-nu′re-ah) The presence of ketone bodies in the urine.

ketosis (ke″to′sis) Acidosis due to the presence of excessive ketone bodies in the body fluids.

proteinuria (pro″te-ĭ-nu′re-ah) The presence of protein in the urine.

uremia (u-re′me-ah) A toxic condition resulting from the presence of excessive amounts of nitrogenous wastes in the blood.

Acid-Base Imbalances

Ordinarily, the hydrogen ion concentration of body fluids is maintained within very narrow pH ranges by the actions of chemical and physiological buffer systems. However, abnormal conditions may disturb the acid-base balance. For example, the pH of arterial blood is normally 7.35–7.45, and if this value drops below 7.35, the person is said to have *acidosis*. If the pH rises above 7.45, the condition is called *alkalosis*. Such shifts in the pH of body fluids may be life threatening, and, in fact, a person usually cannot survive if the pH drops to 6.8 or rises to 8.0 for more than a few hours (figure 21.16).

Acidosis results from an accumulation of acids or a loss of bases, both of which are accompanied by abnormal increases in the hydrogen ion concentrations of body fluids. Conversely, alkalosis results from a loss of acids or an accumulation of bases accompanied by a decrease in hydrogen ion concentrations (figure 21.17).

The two major types of acidosis are *respiratory acidosis* and *metabolic acidosis*. Respiratory acidosis is caused by factors that produce an increase in carbon dioxide, accompanied by an increase in the concentration of carbonic acid, which is the respiratory acid. Metabolic acidosis is due to an abnormal accumulation of any other acids in the body fluids or to a loss of bases.

Similarly, the two major types of alkalosis are *respiratory alkalosis* and *metabolic alkalosis*. Respiratory alkalosis is caused by an excessive loss of carbon dioxide and a consequent loss of carbonic acid. Metabolic alkalosis is due to an excessive loss of hydrogen ions or to the gain of bases.

Since **respiratory acidosis** involves an accumulation of carbon dioxide, it can be caused by factors that hinder pulmonary ventilation (figure 21.18). These include the following:

1. Injury to the respiratory center of the brain stem, followed by decreased rate and depth of breathing.
2. Obstructions in air passages that interfere with the movement of air into the alveoli.
3. Diseases that decrease gas exchanges, such as pneumonia, or those that cause a reduction in the surface area of the respiratory membrane, such as emphysema.

As a result of any of these conditions, the amount of carbonic acid occurring in the body fluids increases, the concentration of hydrogen ions increases, and the pH value drops. This shift in pH may be resisted by the action of chemical buffers, such as hemoglobin. At the same time, the respiratory center may be stimulated by the increasing concentrations of carbon dioxide and hydrogen ions, so that the breathing rate and depth are increased and the carbon dioxide concentration is thereby lowered. Also, the kidneys may begin to secrete increasing quantities of hydrogen ions.

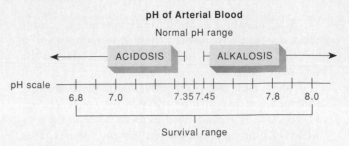

pH of Arterial Blood

Figure 21.16   If the pH of the arterial blood drops to 6.8 or rises to 8.0 for more than a few hours, the person usually cannot survive.

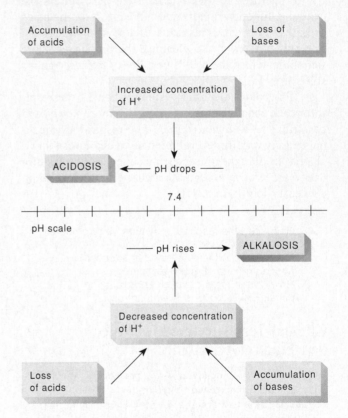

Figure 21.17   Acidosis results from an accumulation of acids or a loss of bases. Alkalosis results from a loss of acids or an accumulation of bases.

Eventually, the pH of the body fluids may return to normal as a result of the actions of these chemical and physiological buffers. When this happens, the acidosis is said to be *compensated*.

The symptoms of respiratory acidosis include depression of activities in the central nervous system, character-

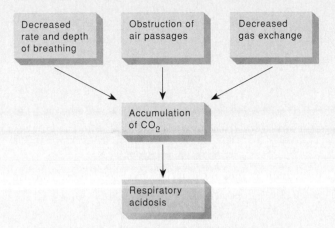

Figure 21.18 Some of the factors that lead to respiratory acidosis.

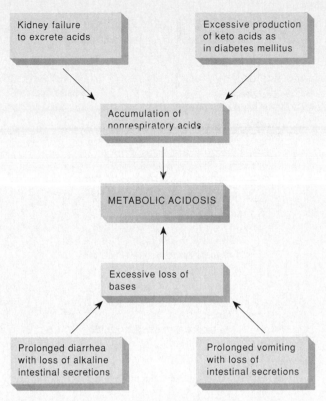

Figure 21.19 Some of the factors that lead to metabolic acidosis.

ized by drowsiness, disorientation, and stupor. These signs are usually accompanied by evidence of respiratory insufficiency, such as labored breathing and cyanosis. In *uncompensated acidosis,* the person may become comatose and die.

Metabolic acidosis involves either an accumulation of nonrespiratory acids or the loss of bases (figure 21.19). Factors that may lead to this condition include the following:

1. Kidney disease in which there is reduced glomerular filtration and a failure to excrete the acids produced by metabolic processes (uremic acidosis).
2. Prolonged vomiting in which the alkaline contents of the upper intestine are lost in addition to the stomach contents. (If only the stomach contents are lost, the result is metabolic alkalosis.)
3. Prolonged diarrhea in which alkaline intestinal secretions are lost excessively (especially in infants).
4. Diabetes mellitus, in which some fatty acids are converted into ketone bodies. These ketone bodies include *acetoacetic acid, beta-hydroxybutyric acid,* and *acetone.* Normally these substances are produced in relatively small quantities, and they are oxidized by cells as energy sources. However, if fats are being utilized at an abnormally high rate, as may occur in diabetes mellitus, ketone bodies may accumulate faster than they can be oxidized. At such times, these compounds may be excreted in the urine (ketonuria); in addition, acetone, which is volatile, may be excreted by the lungs and impart a fruity odor to the breath. More seriously, the accumulation of acetoacetic acid and beta-hydroxybutyric acid may cause a drop in pH (ketonemic acidosis).

These acids also tend to combine with bicarbonate ions in the urine. As a result, bicarbonate ions are excreted excessively, and this loss interferes with the function of the bicarbonate acid-base buffer system.

In each case, the pH tends to shift toward lower values. However, this shift is resisted by the following factors: the actions of chemical buffer systems, which accept excessive hydrogen ions; the respiratory center, which causes the breathing rate and depth to increase; and the kidneys, which secrete increasing quantities of hydrogen ions.

Respiratory alkalosis develops as a result of *hyperventilation,* described in chapter 16. Hyperventilation is accompanied by excessive loss of carbon dioxide and consequent decreases in carbonic acid and hydrogen ion concentrations (figure 21.20).

Hyperventilation may occur during periods of anxiety, although it may also accompany fever or poisoning from salicylates, such as aspirin. At high altitudes, hyperventilation may occur in response to low oxygen pressure. Also, musicians, such as bass tuba players, who must provide a large volume of air when playing sustained passages, sometimes experience the effects of hyperventilation. In each case,

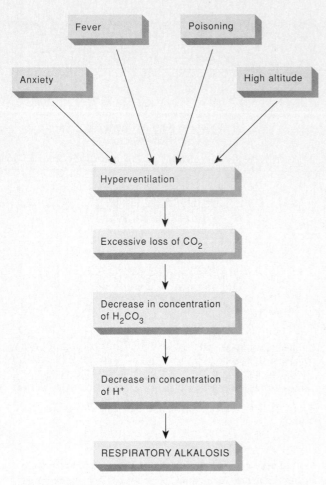

Figure 21.20   Some of the factors that lead to respiratory alkalosis.

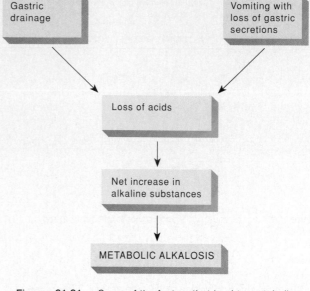

Figure 21.21   Some of the factors that lead to metabolic alkalosis.

carbon dioxide is lost excessively by rapid, deep breathing, and the pH value of the body fluids increases.

This shift in pH is resisted by chemical buffers, such as hemoglobin, that release hydrogen ions. Also, since the concentrations of carbon dioxide and hydrogen ions are lower, the respiratory center is stimulated to a lesser degree. As a result, the tendency to hyperventilate is inhibited, thus reducing further loss of carbon dioxide. At the same time, the kidneys decrease their secretion of hydrogen ions, and the urine becomes alkaline as bases are excreted.

The symptoms of respiratory alkalosis include lightheadedness, agitation, dizziness, and tingling sensations. In severe cases, impulses may be triggered spontaneously on peripheral nerves, and muscles may respond with tetanic contractions (chapter 9).

Metabolic alkalosis results from excessive loss of hydrogen ions or from a gain in bases, both of which are accompanied by a rise in the pH of the blood (alkalemia) (figure 21.21).

This condition may occur following gastric drainage (lavage), prolonged vomiting in which only the stomach contents are lost, or the use of certain diuretic drugs. Because gastric juice is very acidic, its loss leaves the body fluids with a net increase of basic substances and a pH shift toward alkaline values. Metabolic alkalosis may also develop as a result of ingesting an excessive amount of antacids, such as sodium bicarbonate, taken to relieve the symptoms of indigestion. The symptoms of metabolic alkalosis include a decrease in the breathing rate and depth, which, in turn, results in an increased concentration of carbon dioxide in the blood.

Chapter Summary

Introduction (page 780)

The maintenance of water and electrolyte balance requires that the quantities of these substances entering the body equal the quantities leaving it. Altering the water balance necessarily affects the electrolyte balance.

Distribution of Body Fluids (page 780)

1. Fluid compartments
   a. The intracellular fluid compartment includes the fluids and electrolytes enclosed by cell membranes.
   b. The extracellular fluid compartment includes all fluids and electrolytes outside cell membranes.
      (1) Interstitial fluid within tissue spaces
      (2) Plasma within blood
      (3) Lymph within lymphatic vessels
      (4) Transcellular fluid within body cavities
2. Composition of body fluids
   a. Extracellular fluids
      (1) Extracellular fluids are characterized by high concentrations of sodium, chloride, and bicarbonate ions, with lesser amounts of potassium, calcium, magnesium, phosphate, and sulfate ions.
      (2) Plasma contains more protein than do either interstitial fluid or lymph.
   b. Intracellular fluid contains relatively high concentrations of potassium, phosphate, and magnesium ions; it also contains a greater concentration of sulfate ions and lesser concentrations of sodium, chloride, and bicarbonate ions than extracellular fluid contains.
3. Movement of fluid between compartments
   a. Fluid movements are regulated by hydrostatic pressure and osmotic pressure.
      (1) Fluid leaves plasma because of hydrostatic pressure and returns to plasma because of osmotic pressure.
      (2) Fluid enters the lymph vessels because of osmotic pressure.
      (3) Fluid movement in and out of the cells is regulated primarily by osmotic pressure.
   b. Sodium ion concentrations are especially important in the regulation of fluid movements.

Water Balance (page 783)

1. Water intake
   a. The volume of water taken in varies from person to person.
   b. Most water is gained as liquid or in moist foods.
   c. Some water is produced by oxidative metabolism.
2. Regulation of water intake
   a. The thirst mechanism is the primary regulator of water intake.
   b. The act of drinking and the resulting distension of the stomach inhibit the thirst mechanism.

3. Water output
   a. Water is lost in a variety of ways.
      (1) It is excreted in the urine, feces, and sweat.
      (2) Insensible loss occurs through evaporation from the skin and lungs.
   b. The primary regulation of water output involves urine production.
4. Regulation of water balance
   a. Water balance is regulated mainly by the distal convoluted tubules and collecting ducts of the nephrons.
      (1) ADH from the posterior pituitary gland stimulates water reabsorption in these segments.
      (2) The mechanism involving ADH can reduce normal output of 1,500 milliliters to 500 milliliters per day.
   b. If excess water is taken in, the ADH mechanism is inhibited.

Electrolyte Balance (page 788)

1. Electrolyte intake
   a. The most important electrolytes in the body fluids are those that release ions of sodium, potassium, calcium, magnesium, chloride, sulfate, phosphate, and bicarbonate.
   b. These ions are obtained in foods and beverages or as by-products of metabolic processes.
2. Regulation of electrolyte intake
   a. Electrolytes are usually obtained in sufficient quantities in response to hunger and thirst mechanisms.
   b. In a severe electrolyte deficiency, a person may experience a salt craving.
3. Electrolyte output
   a. Electrolytes are lost through perspiration, feces, and urine.
   b. Quantities lost vary with temperature and physical exercise.
   c. The greatest electrolyte loss occurs as a result of kidney functions.
4. Regulation of electrolyte balance
   a. Concentrations of sodium, potassium, and calcium ions in the body fluids are particularly important.
   b. The regulation of sodium ions involves the secretion of aldosterone from the adrenal glands.
   c. The regulation of potassium ions also involves aldosterone.
   d. The regulation of calcium ions involves parathyroid hormone.
   e. In general, negatively charged ions are regulated secondarily by the mechanisms that control positively charged ions.
      (1) Chloride ions are passively reabsorbed in renal tubules as sodium ions are actively reabsorbed.
      (2) Some negatively charged ions, such as phosphate ions, are reabsorbed partially by active transport mechanisms with limited capacities.

Acid-Base Balance (page 790)

Acids are electrolytes that release hydrogen ions.
Bases are substances that combine with hydrogen ions.

1. Sources of hydrogen ions
   a. Aerobic respiration of glucose
      (1) Aerobic respiration of glucose produces carbon dioxide, which reacts with water to form carbonic acid.
      (2) Carbonic acid dissociates to release hydrogen ions and bicarbonate ions.
   b. Anaerobic respiration of glucose gives rise to lactic acid.
   c. Incomplete oxidation of fatty acids gives rise to acidic ketone bodies.
   d. Oxidation of sulfur-containing amino acids gives rise to sulfuric acid.
   e. Hydrolysis of phosphoproteins and nucleoproteins gives rise to phosphoric acid.
2. Strengths of acids and bases
   a. Acids vary in the extent to which they ionize.
      (1) Strong acids, such as hydrochloric acid, ionize more completely.
      (2) Weak acids, such as carbonic acid, ionize less completely.
   b. Bases vary in strength also.
      (1) Strong bases, such as hydroxyl ions, combine readily with hydrogen ions.
      (2) Weak bases, such as chloride ions, combine with hydrogen ions less readily.
3. Regulation of hydrogen ion concentration
   a. Acid-base buffer systems
      (1) Buffer systems are composed of sets of two or more substances.
      (2) They function to convert strong acids into weaker acids, or strong bases into weaker bases.
      (3) They include the bicarbonate buffer system, phosphate buffer system, and protein buffer system.
      (4) Buffer systems minimize pH changes.
   b. Respiratory center
      (1) The respiratory center is located in the brain stem.
      (2) It helps regulate pH by controlling the rate and depth of breathing.
      (3) As carbon dioxide and hydrogen ion concentrations increase, chemosensitive areas associated with the respiratory center are stimulated; the breathing rate and depth increase, and the carbon dioxide concentration decreases.
      (4) If the carbon dioxide and hydrogen ion concentrations are low, the respiratory center inhibits breathing.
   c. Kidneys
      (1) Nephrons help regulate pH by secreting hydrogen ions.
      (2) Hydrogen ions in urine are buffered by phosphates.
      (3) Ammonia produced by renal cells helps transport hydrogen ions to the outside of the body.
   d. Chemical buffers act rapidly; physiological buffers act more slowly.

Clinical Application of Knowledge

1. An elderly, semiconscious patient is tentatively diagnosed as having acidosis. What components of the arterial blood will be most valuable in determining if the acidosis is of respiratory origin?
2. Some time ago, several newborn infants died due to an error in which sodium chloride was substituted for sugar in their formula. What symptoms would this produce? Why do you think infants are more prone to the hazard of excess salt intake than adults?
3. Explain the threat to fluid and electrolyte balance in the following situation: A patient is being nutritionally maintained on concentrated solutions of hydrolyzed protein that are administered through a gastrostomy tube.
4. Describe what might happen to the plasma pH of a patient as a result of:
   a. prolonged diarrhea
   b. suction of the gastric contents
   c. hyperventilation
   d. hypoventilation
5. During radiation therapy, the mucosa of the stomach and intestines may be damaged. Explain the effect that this might have on the patient's electrolyte balance.
6. If the right ventricle of a patient's heart is failing, and the venous pressure increases as a result, what changes might occur in the patient's extracellular fluid compartments?

Review Activities

1. Explain how water balance and electrolyte balance are interdependent.
2. Name the body fluid compartments, and describe their locations.
3. Explain how the fluids within these compartments differ in composition.
4. Describe how fluid movements between the compartments are controlled.
5. Prepare a list of sources of normal water gain and loss to illustrate how the input of water equals the output of water.
6. Define *water of metabolism.*
7. Explain how water intake is regulated.
8. Explain how the nephrons function in the regulation of water output.
9. List the most important electrolytes in the body fluids.
10. Explain how electrolyte intake is regulated.
11. List the routes by which electrolytes leave the body.
12. Explain how the adrenal cortex functions in the regulation of electrolyte output.
13. Describe the role of the parathyroid glands in regulating electrolyte balance.
14. Describe the role of the renal tubule in regulating electrolyte balance.
15. Distinguish between an acid and a base.
16. List five sources of hydrogen ions in the body fluids, and name an acid that originates from each source.
17. Distinguish between a strong acid and a weak acid, and name an example of each.
18. Distinguish between a strong base and a weak base, and name an example of each.
19. Explain how an acid-base buffer system functions.
20. Describe how the bicarbonate buffer system resists changes in pH.
21. Explain why a protein has acidic as well as basic properties.
22. Describe how a protein functions as a buffer system.
23. Describe the function of hemoglobin as a buffer of carbonic acid.
24. Explain how the respiratory center functions in the regulation of the acid-base balance.
25. Explain how the kidneys function in the regulation of the acid-base balance.
26. Describe the role of ammonia in the transport of hydrogen ions to the outside of the body.
27. Distinguish between a chemical buffer system and a physiological buffer system.

Human Life Cycle

T he chapters of unit 5 are concerned with the reproduction, growth, and development of the human organism. They describe how the organs of the male and female reproductive systems function to produce an embryo and how this offspring grows and develops as it passes through the phases of its life cycle before and after birth. The final chapter of this unit explains the genetic determination of individual traits.

The prenatal period of human development begins with the fertilization of an egg cell and ends at birth.

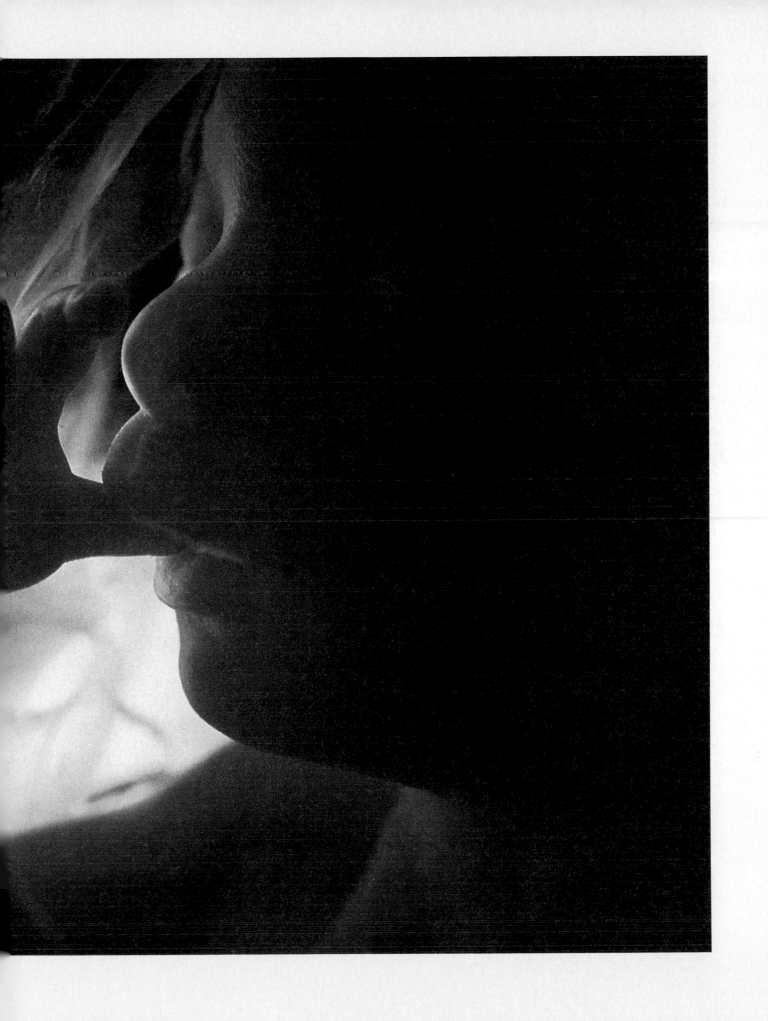

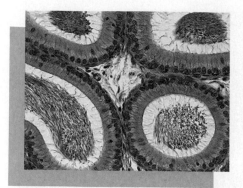

Reproductive Systems

*T*he male and female *reproductive systems* are specialized to produce offspring. These systems are unique in that their functions are not necessary for the survival of each individual. Instead, their functions are vital to the continuation of the human species.

The organs of the male and female reproductive systems are adapted to produce sex cells, to sustain these cells, and to transport them to a location where fertilization may occur. Some of the reproductive organs also secrete hormones that play vital roles in the development and maintenance of sexual characteristics and in the regulation of reproductive physiology. In addition, specialized organs of the female system are adapted to support the life of an offspring that may develop from a fertilized egg and to transport this offspring to the outside of the female's body ■

Chapter Objectives

After you have studied this chapter, you should be able to:

1. State the general functions of the reproductive systems.

2. Name the parts of the male reproductive system and describe the general functions of each part.

3. Trace the path followed by sperm cells from their site of formation to the outside.

4. Describe the structure of the penis and explain how its parts function to produce an erection.

5. Explain how hormones control the activities of the male reproductive organs and the development of male secondary sexual characteristics.

6. State the general functions of the female reproductive system.

7. Name the parts of the female reproductive system and describe the general functions of each part.

8. Describe how hormones control the activities of the female reproductive system and the development of female secondary sexual characteristics.

9. Describe the major events that occur during a menstrual cycle.

10. Describe the hormonal changes that occur in the maternal body during pregnancy.

11. Describe the birth process and explain the role of hormones in this process.

12. List several methods of birth control and describe the relative effectiveness of each method.

13. Complete the review activities at the end of this chapter. Note that the items are worded in the form of specific learning objectives. You may want to refer to them before reading the chapter.

Key Terms

androgen (an'dro-jen)

cleavage (klēv'ij)

contraception (kon''trah-sep'shun)

ejaculation (e-jak''u-la'shun)

emission (e-mish'un)

estrogen (es'tro-jen)

fertilization (fer''ti-li-za'shun)

follicle (fol'i-kl)

gonadotropin (go-nad''o-trōp'in)

implantation (im''plan-ta'shun)

menopause (men'o-pawz)

menstrual cycle (men'stroo-al si'kl)

oogenesis (o''o-jen'ĕ-sis)

orgasm (or'gazm)

ovulation (o''vu-la'shun)

placenta (plah-sen'tah)

pregnancy (preg'nan-se)

progesterone (pro-jes'tĕ-rōn)

puberty (pu'ber-te)

semen (se'men)

spermatogenesis (sper''mah-to-jen'ĕ-sis)

testosterone (tes-tos'te-rōn)

zygote (zi'gōt)

Aids to Understanding Words

andr-, man: androgens—male sex hormones.

contra-, opposed to: contraception—the prevention of fertilization.

crur-, lower part: crura—diverging parts at the base of the penis by which it is attached to the pelvic arch.

ejacul-, to shoot forth: ejaculation—the process by which semen is expelled from the male reproductive tract.

fimb-, a fringe: fimbriae—irregular extensions on the margin of the infundibulum of the uterine tube.

follic-, small bag: follicle—the ovarian structure that contains an egg cell.

genesis, origin: spermatogenesis—process by which sperm cells are formed.

gubern-, to guide: gubernaculum—fibromuscular cord that guides the descent of a testis.

labi-, lip: labia minora—flattened, longitudinal folds that extend along the margins of the vestibule.

mens-, month: menstrual cycle—the monthly female reproductive cycle.

mons, mountain: mons pubis—the rounded elevation of fatty tissue overlying the symphysis pubis in a female.

puber-, adult: puberty—the time when a person becomes able to function reproductively.

The organs of the male and female reproductive systems are concerned with the process of producing offspring, and each organ is adapted to perform specialized tasks. For example, some reproductive organs produce sex cells, and others sustain these cells or transport them from one location to another. Still other organs produce and secrete hormones, which regulate the formation of the sex cells and play roles in the development and maintenance of male and female sexual characteristics.

Organs of the Male Reproductive System

The organs of the male reproductive system are specialized to produce and maintain the male sex cells, or *sperm cells*; to transport these cells, together with various supporting fluids, to the female reproductive tract; and to secrete male sex hormones.

The *primary sex organs* (gonads) of this system are the two *testes* in which the sperm cells (spermatozoa) and the male sex hormones are formed. The other structures of the male reproductive system are termed *accessory sex organs* (secondary sex organs), and they include two groups: the internal reproductive organs and the external reproductive organs (figure 22.1).

Testes

The **testes** are ovoid structures about 5 centimeters in length and 3 centimeters in diameter. Each testis is suspended by a spermatic cord and is contained within the cavity of the saclike *scrotum*. (See figures 22.1 and reference plate 52.)

Descent of the Testes

In a male fetus, the testes originate from masses of tissue located behind the parietal peritoneum, near the developing kidneys. Usually about a month or two before birth, these organs descend to regions in the lower abdominal cavity and pass through the abdominal wall into the scrotum.

The descent of the testes is stimulated by the male sex hormone *testosterone,* which is secreted by the developing testes. The actual movement of these organs seems to be aided by a fibromuscular cord called the **gubernaculum.** This cord is attached to the developing testis and extends into the inguinal region of the abdominal cavity. It passes through the abdominal wall and is fastened to the skin on the outside. As the testis descends, apparently guided by the gubernaculum, it passes through the **inguinal canal** of the abdominal wall and enters the scrotum, where it remains anchored by the gubernaculum. Each testis carries with it a developing *vas deferens* and various blood vessels and nerves.

These structures later form parts of the **spermatic cord** by which the testis is suspended in the scrotum (figure 22.2).

If the testes fail to descend into the scrotum, they will not produce sperm cells. In fact, in this condition, called *cryptorchidism,* the cells that normally produce sperm cells degenerate, and the male is sterile. This failure to produce sperm cells is apparently caused by the unfavorable temperature of the abdominal cavity, which is a few degrees higher than the scrotal temperature.

During the descent of a testis, a pouch of peritoneum, called the *vaginal process,* moves through the inguinal canal and into the scrotum. In about one-quarter of males, this process remains more or less open. Such an opening provides a potential passageway through which a loop of intestine may be forced by excessive abdominal pressure, producing a condition called an *indirect inguinal hernia.* If the protruding intestinal loop is so tightly constricted within the inguinal canal that its blood supply is interrupted, the condition is called a *strangulated hernia.* Without prompt treatment, the strangulated tissues may die.

1. What are the primary sex organs of the male reproductive system?
2. Describe the descent of the testes.
3. What happens if the testes fail to descend into the scrotum?

Structure of the Testes

A testis is enclosed by a tough, white, fibrous capsule called the *tunica albuginea.* Along its posterior border, the connective tissue thickens and extends into the organ, forming a mass called the *mediastinum testis.* From this structure, thin layers of connective tissue, called *septa,* pass into the testis and subdivide it into about 250 *lobules.*

A lobule contains one to four highly coiled, convoluted **seminiferous tubules,** each of which is approximately 70 centimeters long when uncoiled. These tubules course posteriorly and unite to form a complex network of channels called the *rete testis.* The rete testis is located within the mediastinum testis and gives rise to several ducts that join a tube called the *epididymis.* The epididymis, in turn, is coiled on the outer surface of the testis.

The seminiferous tubules are lined with a specialized stratified epithelium, which includes the *spermatogenic cells* that give rise to the sperm cells. Other specialized cells, called **interstitial cells** (cells of Leydig), are located in the spaces between the seminiferous tubules. Interstitial cells produce and secrete the male sex hormones (figures 22.3 and 22.4).

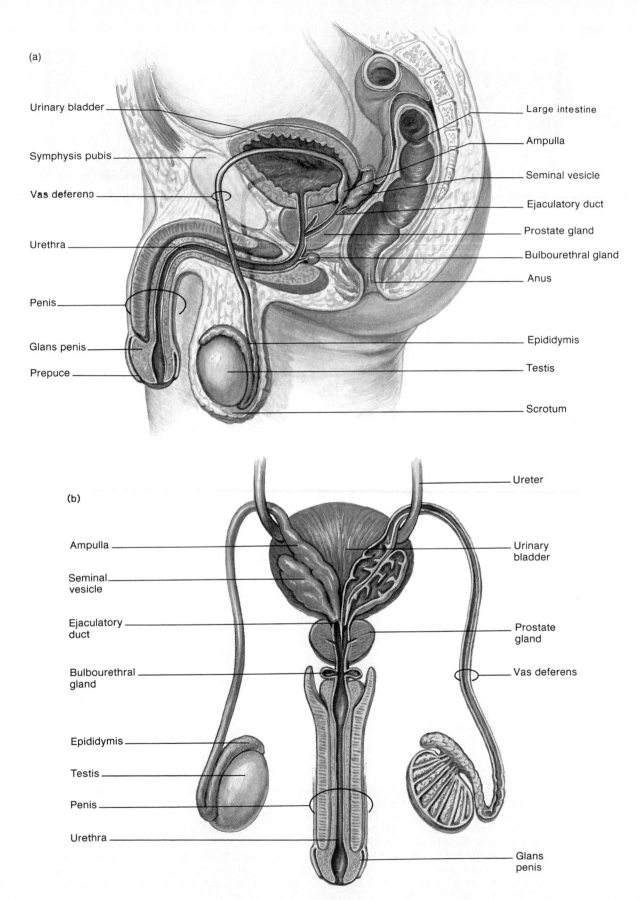

(a)

Urinary bladder

Symphysis pubis

Vas deferens

Urethra

Penis

Glans penis

Prepuce

Large intestine

Ampulla

Seminal vesicle

Ejaculatory duct

Prostate gland

Bulbourethral gland

Anus

Epididymis

Testis

Scrotum

(b)

Ampulla

Seminal vesicle

Ejaculatory duct

Bulbourethral gland

Epididymis

Testis

Penis

Urethra

Ureter

Urinary bladder

Prostate gland

Vas deferens

Glans penis

Figure 22.1 (a) Sagittal view of male reproductive organs; (b) posterior view.

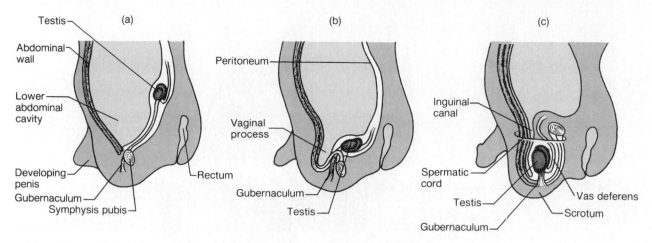

Figure 22.2 During fetal development, each testis descends through an inguinal canal and enters the scrotum (*a–c*). What is the function of the gubernaculum when this occurs?

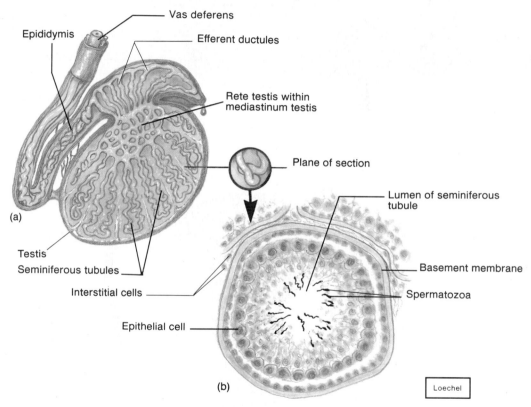

Figure 22.3 (*a*) Sagittal section of a testis; (*b*) cross section of a seminiferous tubule.

The epithelial cells of the seminiferous tubules sometimes give rise to *testicular cancer,* one of the more common types of cancer occurring in young men. In most cases, the first sign of this condition is a painless enlargement of a testis or a scrotal mass that seems to be attached to a testis. When testicular cancer is suspected, a tissue sample is usually removed (biopsied) and examined microscopically. If cancer cells are present, the affected testis is surgically removed (orchiectomy). Depending upon the type of cancerous tissue present and the extent of the disease, a patient may be treated with radiation or with chemotherapy employing one or more drugs. As a result of such treatment, the cure rate for testicular cancer is high.

Figure 22.4 Scanning electron micrograph of cross sections of human seminiferous tubules (about ×100). (*Tissues and Organs: A Text-Atlas of Scanning Electron Microscopy*, by R. G. Kessel and R. H. Kardon. © 1979 W. H. Freeman and Company.)

1. Describe the structure of a testis.
2. Where are the sperm cells produced within the testes?
3. What cells produce male sex hormones?

Formation of Sperm Cells

The epithelium of the seminiferous tubules consists of two types of cells: supporting cells (Sertoli's cells) and spermatogenic cells. The *supporting cells* are tall, columnar cells that extend the full thickness of the epithelium from its base to the lumen of the seminiferous tubule. Numerous thin processes project from these cells, filling the spaces between nearby spermatogenic cells. They support, nourish, and regulate the *spermatogenic cells,* which give rise to sperm cells (spermatozoa).

In a young male, the spermatogenic cells are undifferentiated and are called *spermatogonia.* Each of these cells contains 46 chromosomes in its nucleus, which is the usual number for human cells. (See figure 22.5.)

During early adolescence, certain hormones stimulate the spermatogonia to become active. Some of

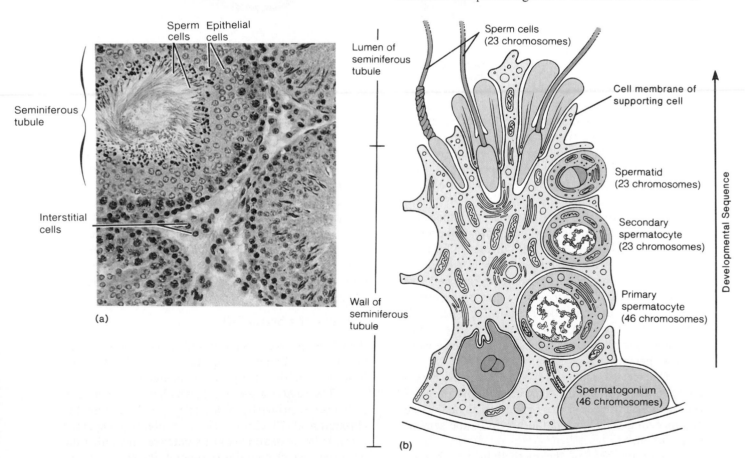

Figure 22.5 (*a*) Light micrograph of seminiferous tubules (×200). (*b*) Spermatogonia give rise to primary spermatocytes by mitosis; the spermatocytes, in turn, give rise to sperm cells by meiosis.

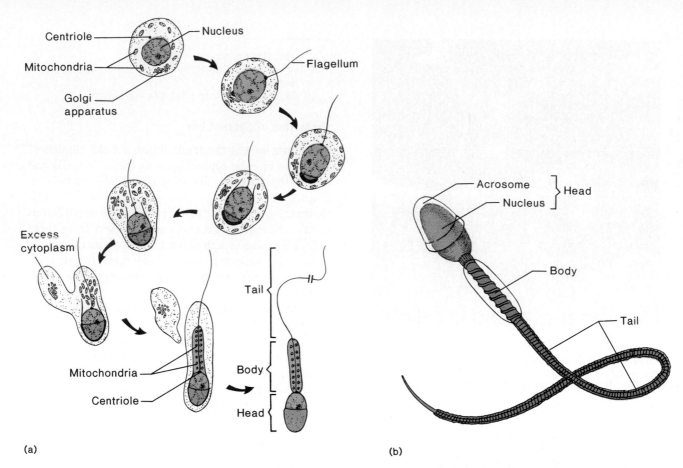

Centriole

Nucleus

Mitochondria

Golgi apparatus

Flagellum

Excess cytoplasm

Tail

Mitochondria

Body

Centriole

Head

(a)

Acrosome } Head
Nucleus

Body

Tail

(b)

Figure 22.6 (a) The head of the sperm develops largely from the nucleus of the formative cell. (b) Parts of a mature sperm cell.

them undergo mitosis, giving rise to new spermatogonia and providing a reserve supply of these undifferentiated cells. Others enlarge and become *primary spermatocytes* that then divide by a special type of cell division called **meiosis,** which is described in detail in chapter 24.

In the course of meiosis, the primary spermatocytes each divide to form two *secondary spermatocytes.* Each of these cells, in turn, divides to form two *spermatids,* which mature into sperm cells. Also during meiosis, the number of chromosomes in each cell is reduced by one-half. Consequently, for each primary spermatocyte that undergoes meiosis, four sperm cells with 23 chromosomes in each of their nuclei are formed. This process by which sperm cells are produced is called **spermatogenesis** (figure 22.5).

The spermatogonia are located near the base of the germinal epithelium. As spermatogenesis occurs, cells in more advanced stages are pushed along the sides of supporting cells toward the lumen of the seminiferous tubule.

Near the base of the epithelium, membranous processes from adjacent supporting cells are fused by specialized junctions (occluding junctions) into complexes that divide the tissue into two layers. The sper-

matogonia are on one side of this barrier, and the cells in more advanced stages are on the other side. This membranous complex seems to help maintain a favorable environment for the development of sperm cells by preventing certain large molecules from moving from the interstitial fluid of the basal epithelium into the region of the differentiating cells.

Spermatogenesis occurs continually throughout the reproductive life of a male. The resulting sperm cells collect in the lumen of each seminiferous tubule. Then they pass through the rete testis to the epididymis, where they remain for a time and mature.

Structure of a Sperm Cell

A mature sperm cell is a tiny, tadpole-shaped structure about 0.06 millimeters long. It consists of a flattened head, a cylindrical body, and an elongated tail.

The *head* of a sperm cell, which is oval in outline, is composed primarily of a nucleus, and contains the chromatin of 23 chromosomes in highly compacted form. It has a small part at its anterior end called the *acrosome,* which contains lysosomal-like enzymes that aid the sperm cell in penetrating an egg cell at the time of fertilization (figure 22.6). This process is described in a subsequent section of this chapter.

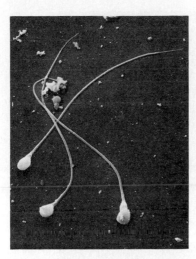

Figure 22.7 Scanning electron micrograph of human sperm cells (×4,000).

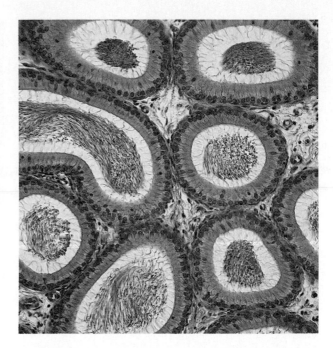

Figure 22.8 Cross section of a human epididymis (×50).

The *body* (midpiece) of the sperm cell contains a central, filamentous core and a large number of mitochondria arranged in a spiral. The *tail* (flagellum) consists of several longitudinal fibrils enclosed in an extension of the cell membrane. The tail also contains a relatively large quantity of ATP, which is thought to provide the energy for the lashing movement that propels the sperm cell through fluid. The scanning electron micrograph in figure 22.7 shows a few mature sperm cells.

1. Explain the function of the supporting cells in the seminiferous tubules.
2. Describe the process of spermatogenesis.
3. Describe the structure of a sperm cell.

Male Internal Accessory Organs

The *internal accessory organs* of the male reproductive system include the epididymides, vasa deferentia, ejaculatory ducts, and urethra, as well as the seminal vesicles, prostate gland, and bulbourethral glands.

Epididymis

Each **epididymis** (pl. *epididymides*) is a tightly coiled, threadlike tube that is about 6 meters long (figures 22.1, 22.8, and reference plate 52). This tube is connected to ducts within a testis. It emerges from the top of the testis, descends along its posterior surface, and then courses upward to become the *vas deferens.*

The inner lining of the epididymis is composed of pseudostratified columnar cells that bear nonmotile cilia. These cells are thought to secrete glycogen and other substances, which help sustain the lives of the stored sperm cells and promote their maturation.

When immature sperm cells reach the epididymis, they are nonmobile. However, as they travel through the epididymis, as a result of rhythmic peristaltic contractions, they undergo *maturation.* Following this aging process, the sperm cells are capable of moving independently and fertilizing egg cells. However, they usually do not engage in swimming motions until after ejaculation, which is described in a subsequent section of this chapter.

Vas Deferens

Each **vas deferens** (pl. *vasa deferentia*), also called ductus deferens, is a muscular tube about 45 centimeters long that is lined with pseudostratified columnar epithelium (figure 22.9). It begins at the lower end of the epididymis and passes upward along the medial side of a testis to become part of the spermatic cord. It passes through the inguinal canal, enters the abdominal cavity outside the parietal peritoneum, and courses over the pelvic brim. From there, it extends backward and medially into the pelvic cavity, where it ends behind the urinary bladder.

Near its termination, the vas deferens becomes dilated into a portion called the *ampulla.* Just outside the prostate gland, the tube becomes slender again and unites with the duct of a seminal vesicle. The fusion of these two ducts forms an **ejaculatory duct,** which passes through the substance of the prostate gland and empties into the urethra through a slitlike opening (figure 22.1).

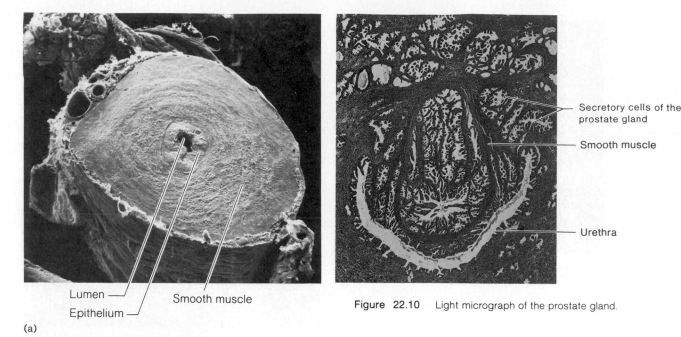

Lumen ——
Epithelium ——
Smooth muscle

(a)

Figure 22.10   Light micrograph of the prostate gland.

Secretory cells of the prostate gland

Smooth muscle

Urethra

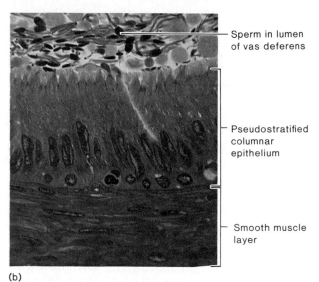

Sperm in lumen of vas deferens

Pseudostratified columnar epithelium

Smooth muscle layer

(b)

Figure 22.9   (a) Scanning electron micrograph of a cross section of the vas deferens (×70). (*Tissues and Organs: A Text-Atlas of Scanning Electron Microscopy,* by R. G. Kessel and R. H. Kardon. © 1979 W. H. Freeman and Company.); (b) light micrograph of the wall of the vas deferens (×250).

Seminal Vesicle

A **seminal vesicle** is a convoluted, saclike structure about 5 centimeters long that is attached to the vas deferens near the base of the urinary bladder (figure 22.1).

The glandular tissue lining the inner wall of the seminal vesicle secretes a slightly alkaline fluid. This fluid is thought to help regulate the pH of the tubular contents as sperm cells are conveyed to the outside. The secretion of the seminal vesicle also contains a variety of nutrients. For example, it is rich in *fructose,* a mono-

saccharide that provides the sperm cells with an energy source. It also contains *prostaglandins,* which may stimulate muscular contractions within the female reproductive organs and thus aid the movement of sperm cells toward the female egg cell.

At the time of ejaculation, the contents of the seminal vesicles are emptied into the ejaculatory ducts, greatly increasing the volume of the fluid that is discharged from the vas deferens.

1. Describe the structure of the epididymis.
2. Trace the path of the vas deferens.
3. What is the function of a seminal vesicle?

Prostate Gland

The **prostate gland** is a chestnut-shaped structure about 4 centimeters across and 3 centimeters thick that surrounds the beginning of the urethra, just below the urinary bladder (figures 22.1 and 22.10). It is enclosed by connective tissue and is composed of many branched tubular glands. These glands are separated by septa of connective tissue and smooth muscle that extend inward from the capsule. Their ducts open into the urethra.

The prostate gland secretes a thin, milky fluid with an alkaline pH. This secretion neutralizes the sperm cell-containing fluid (semen), which is acidic due to an accumulation of metabolic wastes produced by the stored sperm cells. It also enhances the motility of the sperm cells, which remain relatively immobile in the acidic contents of the epididymis. In addition, the prostatic fluid helps neutralize the acidic secretions of the vagina, thus helping to sustain sperm cells that enter the female reproductive tract.

Male Infertility

In a male, *infertility* is the lack of the ability to induce fertilization of an egg cell. This condition can result from a variety of disorders. For example, if during development the testes fail to descend into the scrotum, the higher temperature of the abdominal cavity or inguinal canal prevents the formation of sperm cells by causing the cells in the seminiferous tubules to degenerate. Similarly, certain diseases, such as mumps, may cause an inflammation of the testes (orchitis) and produce infertility by destroying the cells of the seminiferous tubules.

Other males are infertile because of a deficiency of sperm cells in their semen. Normally, a milliliter of this fluid contains about 120 million sperm cells, and although the estimates of the number of cells necessary for fertility vary, 20 million sperm cells per milliliter in a release of 3 to 5 milliliters is often cited as the minimum for fertility.

Even though a single sperm cell is needed to fertilize an egg cell, many sperm must be present at the time. This is because each sperm cell can release enzymes that are stored in its acrosome. A certain concentration of these enzymes is apparently necessary to remove the layers of cells that normally surround an egg cell, thus exposing the egg cell for fertilization.

Still other males seem to be infertile because of the quality of the sperm cells they produce. In such cases, the sperm cells may have poor motility, or they may have abnormal shapes related to the presence of defective genetic material.

The prostate gland releases its secretions into the urethra as a result of smooth muscle contractions in its capsular wall. This release occurs as the contents of the vas deferens and the seminal vesicles are entering from the ejaculatory ducts, and thus the volume of the semen is increased still more.

> Although the prostate gland is relatively small in male children, it begins to grow in early adolescence and reaches its adult size a few years later. As a rule, its size remains unchanged between the ages of twenty and fifty. In older males, however, the prostate gland commonly enlarges. As this happens, the gland may squeeze the urethra and interfere with urination.
>
> The treatment of an abnormally enlarged prostate gland is usually surgical. If the obstruction created by the gland is slight, the procedure may be performed through the urethral canal and is called a *transurethral prostatic resection.*
>
> The prostate gland is a common site of cancer in older males. Such cancers are usually stimulated to grow more rapidly by the male sex hormone, testosterone, and are inhibited by the female sex hormone, estrogen. Consequently, treatment for this type of cancer may involve removing the testes (the main source of testosterone), administering drugs that block the action of testosterone, or administering estrogen. Although such treatment usually does not stop the cancer, it may slow its development.

Bulbourethral Glands

The **bulbourethral glands** (Cowper's glands) are two small structures, each about a centimeter in diameter, that are located below the prostate gland lateral to the membranous urethra and enclosed by fibers of the external urethral sphincter muscle (figure 22.1).

These glands are composed of numerous tubes whose epithelial linings secrete a mucuslike fluid. This fluid is released in response to sexual stimulation and provides some lubrication to the end of the penis in preparation for sexual intercourse (coitus). Most of the lubricating fluid for intercourse, however, is secreted by female reproductive organs.

Semen

The fluid conveyed by the urethra to the outside during ejaculation is called **semen**. It consists of sperm cells from the testes and secretions of the seminal vesicles, prostate gland, and bulbourethral glands. Semen has a slightly alkaline pH (about 7.5) and contains a variety of nutrients. It also contains prostaglandins, which enhance sperm cell survival and movement through the female reproductive tract.

The volume of semen released at one time varies from 2 to 6 milliliters, and the average number of sperm cells present in the fluid is about 120 million per milliliter.

Sperm cells remain immobile while they are in the ducts of the testis and epididymis, but become activated as they are mixed with the secretions of accessory glands. However, the sperm cells remain unable to fertilize an egg cell until they enter the female reproductive tract. The development of this ability is called *capacitation,* and it involves changes that weaken the acrosomal membranes of the sperm cells.

Although sperm cells are able to live for many weeks in the ducts of the male reproductive tract, they tend to survive for only a day or two after being expelled to the outside even when they are maintained at body temperature. On the other hand, sperm cells can be stored and kept viable for years if they are frozen at a temperature below $-100°$ C.

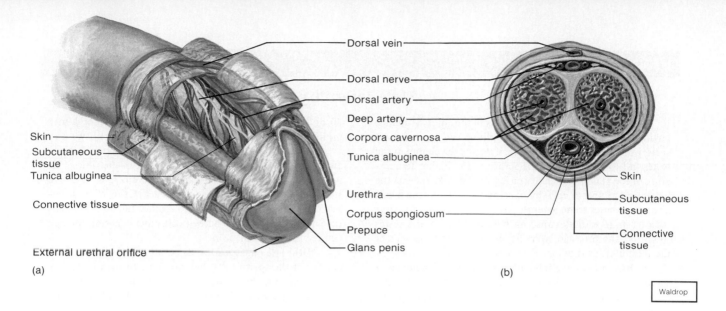

Dorsal vein
Dorsal nerve
Dorsal artery
Deep artery
Corpora cavernosa
Tunica albuginea
Skin
Subcutaneous tissue
Connective tissue
Skin
Subcutaneous tissue
Tunica albuginea
Connective tissue
Urethra
Corpus spongiosum
Prepuce
Glans penis
External urethral orifice

(a)

(b)

Waldrop

Figure 22.11 (*a*) Interior structure of the penis; (*b*) cross section of the penis.

1. Where is the prostate gland located?
2. What is the function of the prostate gland's secretion?
3. What is the function of the bulbourethral glands?
4. What are the characteristics of semen?

Male External Reproductive Organs

The male external reproductive organs are the scrotum, which encloses the testes, and the penis, through which the urethra passes.

Scrotum

The **scrotum** is a pouch of skin and subcutaneous tissue that hangs from the lower abdominal region behind the penis.

Although its subcutaneous tissue lacks fat, the scrotal wall contains a layer of smooth muscle fibers that constitute the *dartos muscle.* When these muscle fibers are contracted, the scrotal skin becomes wrinkled and is held close to the testes; when the fibers are relaxed, the scrotum hangs more loosely.

The scrotum is subdivided into chambers by a medial septum, and each chamber is occupied by a testis. Each chamber also contains a serous membrane, which provides a covering for the front and sides of the testis and the epididymis. This covering helps ensure that the testis will move smoothly within the scrotum (figure 22.1).

Penis

The **penis** is a cylindrical organ that conveys urine and semen through the urethra to the outside. It is also spe-

cialized to become enlarged and stiffened by a process called *erection,* so that it can be inserted into the female vagina during sexual intercourse.

The *body,* or shaft, of the penis is composed of three columns of erectile tissue, which include a pair of dorsally located *corpora cavernosa* and a single *corpus spongiosum* below. The penis is enclosed by skin, a thin layer of subcutaneous tissue, and a layer of connective tissue. In addition, each column is surrounded by a tough capsule of white fibrous connective tissue called a *tunica albuginea* (figure 22.11).

The corpus spongiosum, through which the urethra extends, is enlarged at its distal end to form a sensitive, cone-shaped **glans penis.** This glans covers the ends of the corpora cavernosa and bears the urethral opening—the *external urethral orifice.* The skin of the glans is very thin and hairless. Also, a loose fold of skin called the *prepuce* (foreskin) begins just behind the glans and extends forward to cover it as a sheath. The prepuce is sometimes removed by a surgical procedure called *circumcision.*

At the *root* of the penis, the columns of erectile tissue become separated. The corpora cavernosa diverge laterally in the perineum and are firmly attached to the medial surfaces of the pubic arch by connective tissue. These diverging parts form the *crura* of the penis. The single corpus spongiosum is enlarged between the crura as the *bulb* of the penis, which is attached to membranes of the perineum.

1. Describe the structure of the penis.
2. What is circumcision?
3. How is the penis attached to the perineum?

Erection, Orgasm, and Ejaculation

The masses of erectile tissue within the body of the penis contain networks of vascular spaces (venous sinusoids). These spaces are lined with endothelium and are separated from each other by bands (trabeculae) of smooth muscle and connective tissue.

Ordinarily, the vascular spaces remain small as a result of partial contractions in the smooth muscle fibers that surround them. During sexual stimulation, the smooth muscles become relaxed, and *parasympathetic nerve impulses* pass from the sacral portion of the spinal cord to the arteries leading into the penis, causing them to dilate. At the same time, the increasing pressure of the arterial blood entering the vascular spaces compresses the veins of the penis. As a result, the veins are partially occluded, and the flow of venous blood away from the penis is reduced. Consequently, blood accumulates in the erectile tissues, and the penis swells, elongates, and becomes erect (figure 22.12).

The culmination of sexual stimulation is called **orgasm** and involves a pleasurable feeling of physiological and psychological release. Also, orgasm in the male is accompanied by emission and ejaculation.

Emission is the movement of sperm cells from the testes and secretions from the prostate gland and seminal vesicles into the urethra, where they are mixed to form semen. Emission occurs in response to *sympathetic nerve impulses* traveling from the spinal cord, which stimulate peristaltic contractions in smooth muscles within the walls of the testicular ducts, epididymides, vasa deferentia, and ejaculatory ducts. At the same time, other sympathetic impulses stimulate rhythmic contractions of the seminal vesicles and prostate gland.

As the urethra fills with semen, sensory impulses are stimulated and pass into the sacral portion of the spinal cord. In response, motor impulses are transmitted from the cord to certain skeletal muscles at the base of the erectile columns of the penis, causing them to contract rhythmically. This increases the pressure within the erectile tissues and aids in forcing the semen through the urethra to the outside—a process called **ejaculation.**

The sequence of events during emission and ejaculation is regulated so that the fluid from the bulbourethral glands is expelled first. This is followed by the release of fluid from the prostate gland, the passage of the sperm cells, and finally, the ejection of fluid from the seminal vesicles (figure 22.13).

Immediately after ejaculation, sympathetic impulses cause vasoconstriction of the arteries that supply the erectile tissue, thus reducing the inflow of blood. The smooth muscles within the walls of the vascular spaces partially contract again, and the veins of the penis carry the excess blood out of these spaces. The

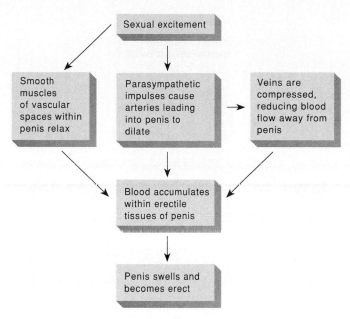

Figure 22.12   Mechanism that causes erection of the penis.

penis gradually returns to its former flaccid condition, and usually another erection and ejaculation cannot be triggered for a period of ten to thirty minutes or longer.

The functions of the male reproductive organs are summarized in chart 22.1.

Spontaneous emissions and ejaculations commonly occur in adolescent males during sleep. These *nocturnal emissions* are caused by changes in hormonal concentrations that accompany adolescent development and sexual maturation.

1. How is blood flow into the erectile tissues of the penis controlled?
2. Distinguish between orgasm, emission, and ejaculation.
3. Review the events associated with emission and ejaculation.

Hormonal Control of Male Reproductive Functions

Male reproductive functions are controlled largely by hormones secreted by the *hypothalamus,* the *anterior pituitary gland,* and the *testes.* These hormones are responsible for the initiation and maintenance of sperm cell production and for the development and maintenance of male sexual characteristics.

Hypothalamic and Pituitary Hormones

Prior to ten years of age, the young male body is reproductively immature. During this period, the body remains childlike, and the spermatogenic cells of the

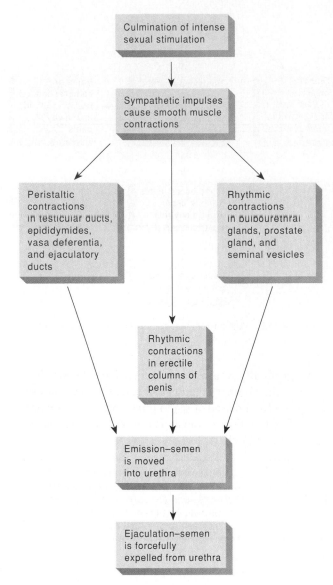

Culmination of intense sexual stimulation

↓

Sympathetic impulses cause smooth muscle contractions

↓

Peristaltic contractions in testicular ducts, epididymides, vasa deferentia, and ejaculatory ducts

Rhythmic contractions in bulbourethral glands, prostate gland, and seminal vesicles

Rhythmic contractions in erectile columns of penis

↓

Emission—semen is moved into urethra

↓

Ejaculation—semen is forcefully expelled from urethra

Figure 22.13 Mechanism that results in emission and ejaculation.

CHART 22.1	Functions of the male reproductive organs
Organ	**Function**
Testis	
Seminiferous tubules	Production of sperm cells
Interstitial cells	Production and secretion of male sex hormones
Epididymis	Storage and maturation of sperm cells; conveys sperm cells to the vas deferens
Vas deferens	Conveys sperm cells to the ejaculatory duct
Seminal vesicle	Secretes an alkaline fluid containing nutrients and prostaglandins; the fluid helps neutralize the acidic semen
Prostate gland	Secretes an alkaline fluid that helps neutralize the acidic semen and enhances the motility of sperm cells
Bulbourethral gland	Secretes a fluid that lubricates the end of the penis
Scrotum	Encloses and protects the testes
Penis	Conveys urine and semen to outside of body; inserted into the vagina during sexual intercourse; the glans penis is richly supplied with sensory nerve endings associated with feelings of pleasure during sexual stimulation

testes remain undifferentiated. Then a series of changes occurs, leading to the development of a reproductively functional adult. Although the mechanism that initiates such changes is not well understood, it involves the hypothalamus.

As explained in chapter 13, the hypothalamus secretes gonadotropin-releasing hormone (GnRH), which enters the blood vessels leading to the anterior pituitary gland. In response, the anterior pituitary gland secretes the **gonadotropins** called *luteinizing hormone* (LH) and *follicle-stimulating hormone* (FSH). LH, which in males is called interstitial cell-stimulating hormone (ICSH), promotes the development of the interstitial cells (cells of Leydig) of the testes, and they, in turn, secrete male sex hormones. FSH causes the supporting cells (Sertoli's cells) of the seminiferous tu-

bules to proliferate, grow, mature, and become responsive to the effects of the male sex hormone *testosterone.* Then, in the presence of FSH and testosterone, these supporting cells stimulate the spermatogenic cells to undergo spermatogenesis, giving rise to sperm cells (figure 22.14). The supporting cells also secrete a hormone called *inhibin,* which inhibits the anterior pituitary gland, and thus prevents the oversecretion of FSH.

Male Sex Hormones

As a group, the male sex hormones are termed **androgens,** and although most of them are produced by the interstitial cells of the testes, small amounts are synthesized in the adrenal cortex (see chapter 13).

The hormone called **testosterone** is the most abundant of the androgens, and when it is secreted, it is transported in the blood loosely attached to plasma proteins. As in the case of other steroid hormones, testosterone affects its target cells by combining with receptor molecules in the nuclei. (See chapter 13.) However, in many target cells, such as those in the prostate gland, seminal vesicles, and male external accessory organs, testosterone is converted to another androgen called **dihydrotestosterone,** which, in turn, stimulates the cells of these organs.

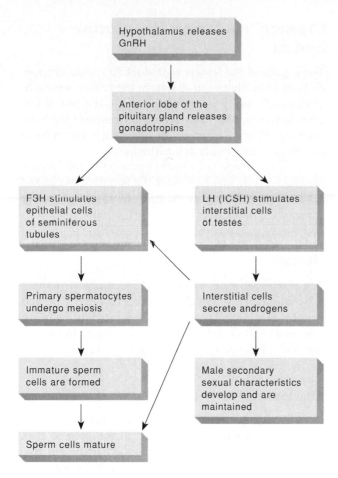

Figure 22.14 Mechanism by which the hypothalamus controls the maturation of sperm cells and the development of male secondary sexual characteristics.

Androgen molecules that fail to become fixed in target cells are usually changed by the liver into forms that can be excreted in bile or urine.

Although the secretion of testosterone begins during fetal development and continues for a few weeks following birth, it nearly ceases during childhood. However, sometime between the ages of thirteen and fifteen, androgen production usually increases rapidly. This phase in development, during which an individual becomes reproductively functional, is called **puberty.** After puberty, testosterone secretion continues throughout the life of a male.

Actions of Testosterone

Testosterone is first produced by cells of the embryonic testes after about eight weeks of development. It stimulates the formation of the male reproductive organs, including the penis, scrotum, prostate gland, seminal vesicles, and various ducts. Still later, it causes the testes to descend into the scrotum, as described previously.

During puberty, testosterone stimulates enlargement of the testes and various accessory organs of the reproductive system, and it causes the development of the male *secondary sexual characteristics.* (Note: The primary male sexual characteristic is the presence of the testes.) These secondary sexual characteristics are special features associated with the adult male body, and they include the following:

1. Increased growth of body hair, particularly on the face, chest, axillary region, and pubic region, but sometimes accompanied by decreased growth of hair on the scalp.
2. Enlargement of the larynx and thickening of the vocal folds, accompanied by the development of a lower pitched voice.
3. Thickening of the skin.
4. Increased muscular growth accompanied by the development of broader shoulders and a relatively narrow waist.
5. Thickening and strengthening of the bones.

Other actions of testosterone include increasing the rate of cellular metabolism and the production of red blood cells, so that the average number of red blood cells in a cubic millimeter of blood is usually greater in males than in females. Testosterone also stimulates sexual activity by influencing certain portions of the brain.

Androgens are steroids that have masculinizing (androgenic) and growth-promoting (anabolic) effects in certain tissues. Synthetic forms of the androgen testosterone are sometimes used to treat disease conditions, including certain forms of anemia, breast cancer, and osteoporosis. These and similar drugs, which are commonly called *anabolic steroids,* have been used by some athletes in the hope of increasing their body mass and enhancing their athletic performances. Most often, the anabolic steroids used by athletes are obtained illicitly, and some of these drugs have no medically approved use in humans.

Although anabolic steroids can increase muscle mass when used together with proper dieting and training exercise, the drugs are associated with a variety of undesirable and dangerous side effects. These effects include adverse changes in liver functions, increased risk of heart and blood vessel diseases, upsets in the normal balance of sex hormones, and severe psychological disorders.

Regulation of Male Sex Hormones

The extent to which the male secondary sexual characteristics develop is directly related to the amount of testosterone secreted by the interstitial cells. This quantity is regulated by a negative feedback system involving the hypothalamus (figure 22.15).

As the concentration of testosterone in the blood increases, the hypothalamus becomes inhibited, and its stimulation of the anterior pituitary gland by GnRH is

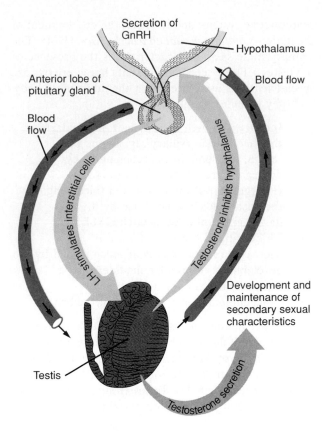

Figure 22.15 A negative feedback mechanism operating between the anterior lobe of the pituitary gland and the testes controls the concentration of testosterone.

Labels in figure: Secretion of GnRH; Hypothalamus; Anterior lobe of pituitary gland; Blood flow; Blood flow; LH stimulates interstitial cells; Testosterone inhibits hypothalamus; Development and maintenance of secondary sexual characteristics; Testis; Testosterone secretion

decreased. As the pituitary's secretion of LH (ICSH) is reduced in response, the amount of testosterone released by the interstitial cells is also reduced.

As the blood concentration of testosterone drops, the hypothalamus becomes less inhibited, and it once again stimulates the anterior pituitary gland to release LH. The increasing secretion of LH causes the interstitial cells to release more testosterone, and its blood concentration increases.

Thus, the concentration of testosterone in the male body is regulated so that it remains relatively constant.

> During the midlife of a male, the target-cell sensitivity to testosterone and the plasma concentration of testosterone usually decline. Consequently, even though sexual activity may continue into old age, males typically experience a decrease in sexual functions as they grow older. This decrease is sometimes called the *male climacteric.*

1. What hormone initiates the changes associated with male sexual maturity?
2. Describe several of the male secondary sexual characteristics.
3. List the functions of testosterone.
4. Explain how the secretion of male sex hormones is regulated.

Organs of the Female Reproductive System

The organs of the female reproductive system are specialized to produce and maintain the female sex cells, or *egg cells;* to transport these cells to the site of fertilization; to provide a favorable environment for a developing offspring; to move the offspring to the outside; and to produce female sex hormones.

The *primary sex organs* (gonads) of this system are the *ovaries,* which produce the female sex cells and sex hormones. The other parts of the system comprise the internal and external *accessory organs.*

Ovaries

The **ovaries** are solid, ovoid structures measuring about 3.5 centimeters in length, 2 centimeters in width, and 1 centimeter in thickness. They are located, one on each side, in a shallow depression (ovarian fossa) in the lateral wall of the pelvic cavity (figure 22.16).

Attachments of the Ovaries

Each ovary is attached to several ligaments that help hold it in position. The largest of these, formed by a fold of peritoneum, is called the *broad ligament.* It is also attached to the uterine tubes and the uterus.

At its upper end, the ovary is held by a small fold of peritoneum, called the *suspensory ligament,* that contains the ovarian blood vessels and nerves. At its lower end, it is attached to the uterus by a rounded, cordlike thickening of the broad ligament called the *ovarian ligament* (figure 22.17).

Descent of the Ovaries

Like the testes in a male fetus, the ovaries in a female fetus originate from masses of tissue behind the parietal peritoneum, near the developing kidneys. During development, these structures descend to locations just below the pelvic brim, where they remain attached to the lateral pelvic wall.

Structure of the Ovaries

The tissues of an ovary can be subdivided into two rather indistinct regions, an inner *medulla* and an outer *cortex.*

The ovarian medulla is largely composed of loose connective tissue, and contains numerous blood vessels, lymphatic vessels, and nerve fibers. The ovarian cortex is composed of more compact tissue, and has a somewhat granular appearance due to the presence of tiny masses of cells called *ovarian follicles.*

The free surface of the ovary is covered by a layer of cuboidal epithelium cells (germinal epithelium). Just

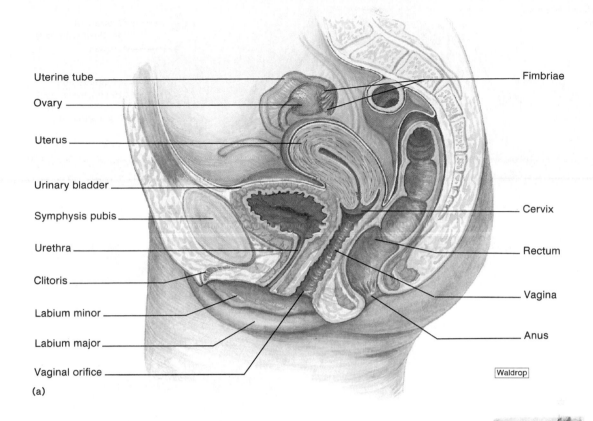

Uterine tube

Ovary

Uterus

Urinary bladder

Symphysis pubis

Urethra

Clitoris

Labium minor

Labium major

Vaginal orifice

Fimbriae

Cervix

Rectum

Vagina

Anus

Waldrop

(a)

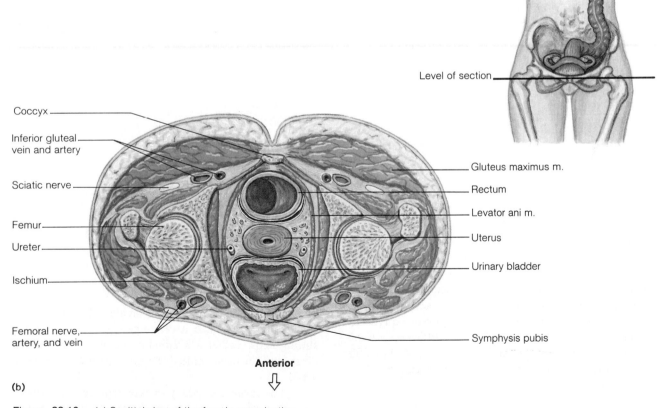

Level of section

Coccyx

Inferior gluteal
vein and artery

Sciatic nerve

Femur

Ureter

Ischium

Femoral nerve,
artery, and vein

Gluteus maximus m.

Rectum

Levator ani m.

Uterus

Urinary bladder

Symphysis pubis

Anterior

(b)

Figure 22.16   (a) Sagittal view of the female reproductive
organs; (b) transverse section of the female pelvic cavity.

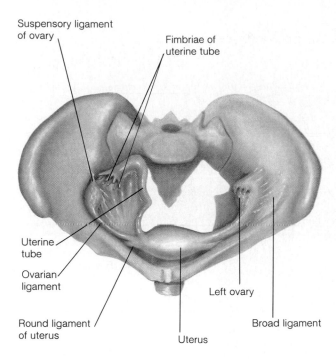

Figure 22.17 The ovaries are located on each side against the lateral walls of the pelvic cavity. (Note: The right uterine tube is retracted to reveal the ovarian ligament.)

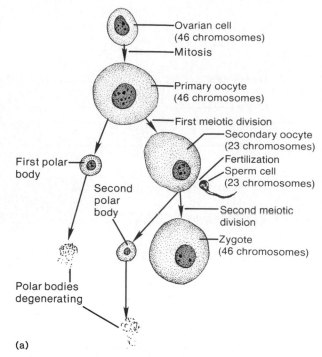

(a)

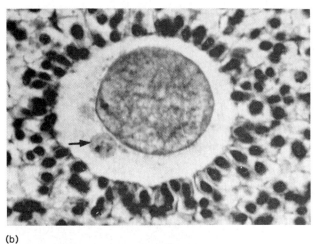

(b)

Figure 22.18 (a) During oogenesis, a single egg cell (secondary oocyte) results from the meiosis of a primary oocyte. If the egg cell is fertilized, it forms a second polar body and becomes a zygote; (b) light micrograph of a secondary oocyte and a polar body (arrow).

beneath this epithelium is a layer of dense connective tissue called the *tunica albuginea.*

1. What are the primary sex organs of the female?
2. Describe the descent of the ovary.
3. Describe the structure of an ovary.

Primordial Follicles

During prenatal development (before birth), small groups of cells in the outer region of the ovarian cortex form several million **primordial follicles.** Each of these structures consists of a single, large cell called a *primary oocyte,* which is closely surrounded by a layer of flattened epithelial cells called *follicular cells.*

Early in development, the primary oocytes begin to undergo *meiosis,* but the process soon halts and is not continued until puberty. Once the primordial follicles have appeared, no new ones are formed. Instead, the number of oocytes in the ovary steadily declines, as many of the oocytes degenerate. Of the several million oocytes formed originally, only a million or so remain at the time of birth, and perhaps 400,000 are present at puberty. Of these, probably fewer than 400 or 500 will be released from the ovary during the reproductive life of a female.

Oogenesis

Beginning at puberty, some of the primary oocytes are stimulated to continue meiosis. As in the case of sperm cells, the resulting cells have one-half as many chro-

mosomes (23) in their nuclei as their parent cells. (See chapter 24.)

When a primary oocyte divides, the division of the cellular cytoplasm is very unequal. One of the resulting cells, called a *secondary oocyte,* is relatively large, and the other, called the *first polar body,* is very small.

The large secondary oocyte represents a future *egg cell* (ovum) in that it can be fertilized by uniting with a sperm cell. If this happens, the oocyte divides unequally to produce a tiny *second polar body* and a relatively large fertilized egg cell, or **zygote.** (See figure 22.18.)

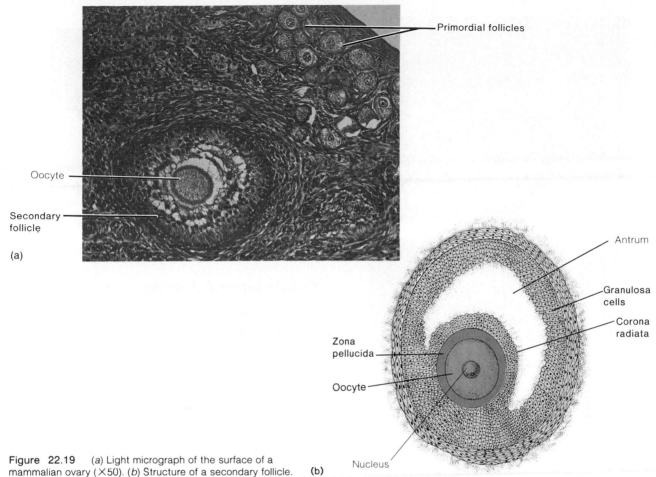

Oocyte

Secondary follicle

(a)

Primordial follicles

Antrum

Granulosa cells

Corona radiata

Zona pellucida

Oocyte

Nucleus

Figure 22.19 (a) Light micrograph of the surface of a mammalian ovary (×50). (b) Structure of a secondary follicle. **(b)**

Thus, the result of this process, which is called **oogenesis,** is one secondary oocyte, or future egg cell, and one polar body. After being fertilized, the secondary oocyte divides to produce a second polar body and a zygote that can give rise to an embryo. The polar bodies have no further function, and they soon degenerate.

1. How does the timing of egg cell production differ from that of sperm cell production?
2. Describe the major events of oogenesis.

Maturation of a Follicle

At *puberty,* the anterior pituitary gland secretes increased amounts of FSH, and the ovaries enlarge in response. At the same time, many of the primordial follicles undergo maturation (figure 22.19).

During this maturation process, the oocyte of a follicle enlarges, and the surrounding follicular cells proliferate by *mitosis,* giving rise to a stratified epithelium composed of cells called *granulosa cells.* A layer of glycoprotein, called the *zona pellucida,* gradually separates the oocyte from the granulosa cells; at this stage, the structure is called a *primary follicle.*

Meanwhile, the ovarian cells outside the follicle become organized into two cellular layers—an *inner vascular layer* (theca interna), composed largely of loose connective tissue and blood vessels, and an *outer fibrous layer* (theca externa), composed of tightly packed connective tissue cells.

The follicular cells continue to proliferate, and when there are six to twelve layers of cells, several irregular, fluid-filled spaces appear among them. These spaces soon join together to form a single cavity (antrum), and the oocyte is pressed to one side of the follicle. At this stage, the follicle has a diameter of about 0.2 millimeters and is called a *secondary follicle.* (See figures 22.19 and 22.20.)

Ten to fourteen days after the process begins, the follicle reaches maturity. The *mature follicle* (preovulatory, or Graafian, follicle) is about 10 millimeters or more across, and its fluid-filled cavity bulges outward on the surface of the ovary, like a blister. The oocyte within the mature follicle is a large, spherical cell, surrounded by a relatively thick, extracellular coat (zona pellucida), to which a mantle of follicular cells (corona radiata) is attached. Processes from these follicular cells extend through the zona pellucida and may supply the oocyte with nutrients.

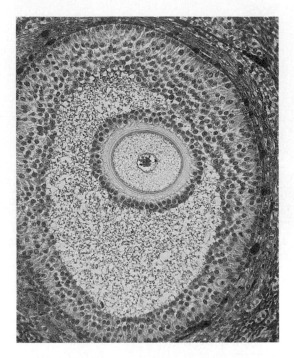

Figure 22.20 What features can you identify in this light micrograph of a maturing follicle (×250)?

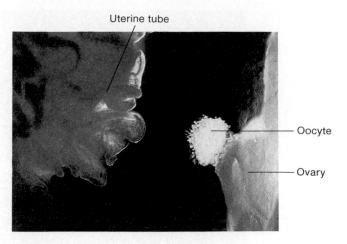

Figure 22.21 Light micrograph of a follicle during ovulation.

Although as many as twenty primary follicles may begin the process of maturation at any one time, one follicle (dominant follicle) usually outgrows the others. Typically, only the dominant follicle reaches full development, and the other follicles degenerate.

Ovulation

As a follicle matures, its primary oocyte undergoes oogenesis, giving rise to a secondary oocyte and a first polar body. These cells are released from the follicle by the process called ovulation.

Ovulation is stimulated by hormones from the anterior pituitary gland, which cause the mature follicle to swell rapidly and its wall to weaken. Eventually the wall ruptures, and the follicular fluid, accompanied by the oocyte, oozes outward from the surface of the ovary. The expulsion of a mammalian oocyte is shown in figure 22.21.

After ovulation, the oocyte and one or two layers of follicular cells surrounding it are usually propelled to the opening of a nearby *uterine tube*. If the oocyte is not fertilized by union with a sperm cell within a relatively short time, it will degenerate. The maturation of a follicle and the release of an oocyte are illustrated in figures 22.22 and 22.23.

1. What changes occur in a follicle and its oocyte during maturation?
2. What causes ovulation?
3. What happens to an oocyte following ovulation?

Female Internal Accessory Organs

The *internal accessory organs* of the female reproductive system include a pair of uterine tubes, a uterus, and a vagina.

Uterine Tubes

The **uterine tubes** (fallopian tubes, or oviducts) are suspended by portions of the broad ligament, and have openings near the ovaries. Each tube, which is about 10 centimeters long and 0.7 centimeters in diameter, passes medially to the uterus, penetrates its wall, and opens into the uterine cavity.

Near each ovary, a uterine tube expands to form a funnel-shaped *infundibulum,* which partially encircles the ovary medially. On its margin, the infundibulum bears a number of irregular, branched extensions called *fimbriae* (figure 22.24). Although the infundibulum generally does not touch the ovary, one of the larger extensions (ovarian fimbria) is connected directly to the ovary.

The wall of a uterine tube consists of an inner mucosal layer, a middle muscular layer, and an outer covering of peritoneum. The mucosal layer is drawn into numerous longitudinal folds and is lined with simple columnar epithelial cells, some of which are *ciliated* (figure 22.25). The epithelium secretes mucus, and the cilia beat toward the uterus. These actions help draw the egg cell and expelled follicular fluid into the infundibulum following ovulation.

Ciliary action also aids the transport of the egg cell down the uterine tube, and peristaltic contractions of the tube's muscular layer help force the egg along.

Uterus

The **uterus** receives the embryo that results from a fertilized egg cell, and sustains its life during development. It is a hollow, muscular organ, shaped somewhat like an inverted pear.

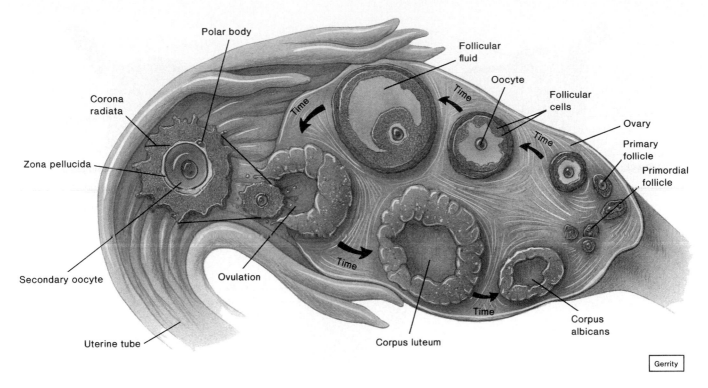

Figure 22.22 As a follicle matures, the egg cell enlarges and becomes surrounded by a mantle of follicular cells and fluid. Eventually, the mature follicle ruptures and the egg cell is released.

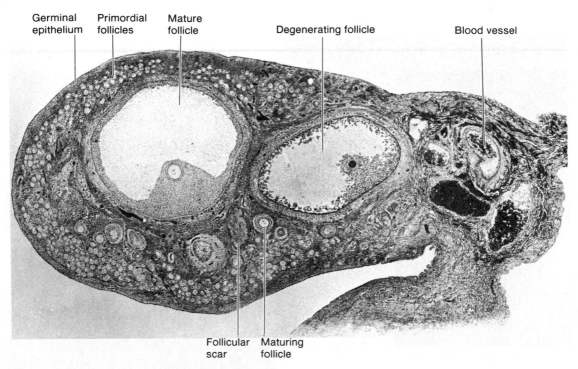

Figure 22.23 Light micrograph of a mammalian (monkey) ovary (×42).

The *broad ligament,* which is also attached to the ovaries and uterine tubes, extends from the lateral walls of the uterus to the pelvic walls and floor, creating a septum across the pelvic cavity (figure 22.24). A flattened band of tissue within the broad ligament, called the *round ligament,* connects the upper end of the uterus to the pelvic wall (figures 22.17 and 22.24).

Although the size of the uterus changes greatly during pregnancy, in its nonpregnant, adult state, it is about 7 centimeters long, 5 centimeters wide (at its

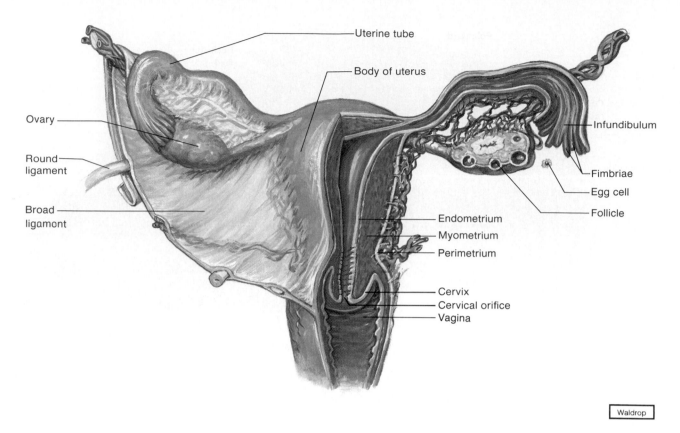

Uterine tube

Body of uterus

Ovary

Round ligament

Broad ligamont

Infundibulum

Fimbriae

Egg cell

Follicle

Endometrium

Myometrium

Perimetrium

Cervix

Cervical orifice

Vagina

Waldrop

Figure 22.24 The funnel-shaped infundibulum of the uterine tube partially encircles the ovary. What factors aid the movement of an egg cell into the infundibulum following ovulation?

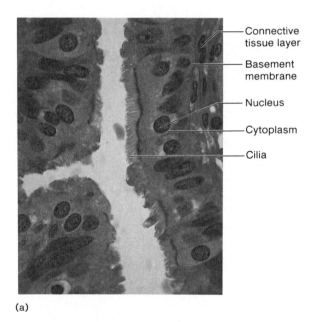

Connective tissue layer

Basement membrane

Nucleus

Cytoplasm

Cilia

(a)

(b)

Figure 22.25 (a) Light micrograph of a uterine tube. (b) Scanning electron micrograph of ciliated cells that line the uterine tube (×650).

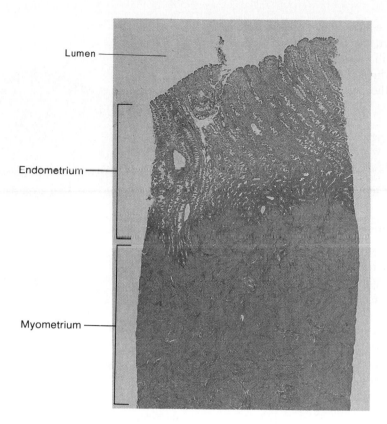

Lumen

Endometrium

Myometrium

Figure 22.26 Light micrograph of the uterine wall (×10).

broadest point), and 2.5 centimeters in diameter. The uterus is located medially within the anterior portion of the pelvic cavity, above the vagina, and is usually bent forward over the urinary bladder.

The upper two-thirds, or *body,* of the uterus has a dome-shaped top, called the *fundus,* and is joined by the uterine tubes, which enter its wall at its broadest part.

The lower one-third of the uterus is called the **cervix.** This tubular part extends downward into the upper portion of the vagina. The cervix surrounds the opening called the *cervical orifice* (ostium uteri), through which the uterus is connected with the vagina.

Cancer developing within the tissues of the uterine cervix can usually be detected by means of a relatively simple and painless procedure called the *Pap* (Papanicolaou) *smear test.* This technique involves scraping off a tiny sample of cervical tissue, smearing the sample on a glass slide, staining it, and examining it for the presence of abnormal cells.

Because this test can reveal certain types of cervical cancers in the early stages of development, when they may be cured completely, the American Cancer Society recommends that women between ages twenty and sixty-five have a Pap test every three years.

The uterine wall is relatively thick and is composed of three layers: the endometrium, the myometrium, and the perimetrium. (See figure 22.26.) The **endometrium** is the inner mucosal layer lining the uterine cavity. This lining is covered with columnar epithelium and contains numerous tubular glands. The **myometrium,** a very thick, muscular layer, consists largely of bundles of smooth muscle fibers arranged in longitudinal, circular, and spiral patterns, and interlaced with connective tissues. During the monthly female reproductive cycles and during pregnancy, the endometrium and myometrium undergo extensive changes. (These changes are described in a subsequent section of this chapter.) The **perimetrium** consists of an outer serosal layer, which covers the body of the uterus and part of the cervix.

Vagina

The **vagina** is a fibromuscular tube, about 9 centimeters in length, extending from the uterus to the outside. It conveys uterine secretions, receives the erect penis during sexual intercourse, and transports the offspring during the birth process.

The vagina extends upward and back into the pelvic cavity. It is posterior to the urinary bladder and urethra, anterior to the rectum, and attached to these parts by connective tissues. The upper one-fourth of the vagina is separated from the rectum by a pouch (rectouterine pouch). The tubular vagina also surrounds the end of the cervix, and the recesses that occur between the vaginal wall and the cervix are termed *fornices* (sing. *fornix*).

> The fornices are clinically important because they are relatively thin-walled and allow the internal abdominal organs to be palpated during a physical examination. Also, the posterior fornix, which is somewhat longer than the others, provides a surgical access to the peritoneal cavity through the vagina.

The *vaginal orifice* is partially closed by a thin membrane of connective tissue and stratified squamous epithelium called the **hymen.** A central opening of varying size allows uterine and vaginal secretions to pass to the outside.

The vaginal wall consists of three layers. The inner *mucosal layer* consists of stratified squamous epithelium and is drawn into numerous longitudinal and transverse ridges (vaginal rugae). This layer is devoid of mucous glands; the mucus found in the lumen of the vagina comes from glands of the cervix.

The middle *muscular layer* consists mainly of smooth muscle fibers arranged in longitudinal and circular patterns. At the lower end of the vagina is a thin band of striated muscle. This band helps close the vaginal opening; however, a voluntary muscle (bulbospongiosus) is primarily responsible for closing this orifice.

The outer *fibrous layer* consists of dense fibrous connective tissue interlaced with elastic fibers, and it attaches the vagina to surrounding organs.

1. How is an egg cell moved along a uterine tube?
2. Describe the structure of the uterus.
3. What is the function of the uterus?
4. Describe the structure of the vagina.

Female External Reproductive Organs

The *external accessory organs* of the female reproductive system include the labia majora, the labia minora, the clitoris, and the vestibular glands. As a group, these structures that surround the openings of the urethra and vagina compose the **vulva.** They are shown in figure 22.27.

Labia Majora

The **labia majora** (sing. *labium majus*) enclose and protect the other external reproductive organs. They correspond to the scrotum of the male, and are composed primarily of rounded folds of adipose tissue and a thin layer of smooth muscle, covered by skin. On the

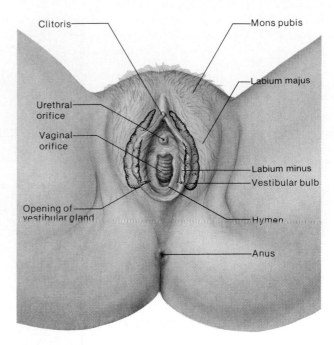

Figure 22.27 Female external reproductive organs and vestibular bulbs.

outside, this skin includes numerous hairs, sweat glands, and sebaceous glands, while on the inside, it is thinner and hairless.

The labia majora lie closely together and are separated longitudinally by a cleft (pudendal cleft), which includes the urethral and vaginal openings. At their anterior ends, the labia merge to form a medial, rounded elevation of adipose tissue called the *mons pubis,* which overlies the symphysis pubis. At their posterior ends, the labia are somewhat tapered, and they merge into the perineum near the anus.

Labia Minora

The **labia minora** (sing. *labium minus*) are flattened longitudinal folds located within the cleft between the labia majora. These folds extend along either side of the vestibule. They are composed of connective tissue that is richly supplied with blood vessels, causing a pinkish appearance. This tissue is covered with stratified squamous epithelium.

Posteriorly, the labia minora merge with the labia majora, while anteriorly, they converge to form a hood-like covering around the clitoris.

Clitoris

The **clitoris** is a small projection at the anterior end of the vulva between the labia minora. Although most of it is embedded in the surrounding tissues, it is usually about 2 centimeters long and 0.5 centimeters in diameter. The clitoris corresponds to the male penis and has a similar structure. More specifically, it is composed of two columns of erectile tissue called *corpora cavernosa*. These columns are separated by a septum and are surrounded by a covering of dense fibrous connective tissue.

At the root of the clitoris, the corpora cavernosa diverge to form *crura,* which, in turn, are attached to the sides of the pubic arch. At its anterior end, a small mass of erectile tissue forms a **glans,** which is richly supplied with sensory nerve fibers.

Vestibule

The **vestibule** of the vulva is the space enclosed by the labia minora. The vagina opens into the posterior portion of the vestibule, and the urethra opens in the midline, just in front of the vagina and about 2.5 centimeters behind the glans of the clitoris.

A pair of **vestibular glands** (Bartholin's glands), which correspond to the male bulbourethral glands, lie on either side of the vaginal opening. Their ducts open into the vestibule near the lateral margins of the vaginal orifice.

Beneath the mucosa of the vestibule on either side is a mass of vascular erectile tissue. These structures, shown in figure 22.27, are called the *vestibular bulbs.* They are separated from each other by the vagina and the urethra, and they extend forward from the level of the vaginal opening to the clitoris.

1. What is the male counterpart of the labia majora? Of the clitoris?
2. What structures are located within the vestibule?

Erection, Lubrication, and Orgasm

Erectile tissues located in the clitoris and around the vaginal entrance respond to sexual stimulation. Following such stimulation, *parasympathetic nerve impulses* pass out from the sacral portion of the spinal cord, causing the arteries associated with the erectile tissues to dilate. As a result, the inflow of blood increases, and the erectile tissues swell. At the same time, the vagina begins to expand and elongate.

If the sexual stimulation is sufficiently intense, parasympathetic impulses stimulate the vestibular glands to secrete mucus into the vestibule. This secretion moistens and lubricates the tissues surrounding the vestibule and the lower end of the vagina, thus facilitating the insertion of the penis into the vagina. Also, since mucus usually continues to be secreted from these glands during sexual intercourse, it helps prevent irritation of tissues that might occur if the vagina remained dry.

The clitoris is abundantly supplied with sensory nerve fibers, which are especially sensitive to local

CHART 22.2	Functions of the female reproductive organs
Organ	Function
Ovary	Produces egg cells and female sex hormones
Uterine tube	Conveys egg cell toward uterus; site of fertilization; conveys developing embryo to uterus
Uterus	Protects and sustains life of embryo during pregnancy
Vagina	Conveys uterine secretions to outside of body; receives erect penis during sexual intercourse; transports fetus during birth process
Labia majora	Enclose and protect other external reproductive organs
Labia minora	Form margins of vestibule; protect openings of vagina and urethra
Clitoris	Glans is richly supplied with sensory nerve endings associated with feeling of pleasure during sexual stimulation
Vestibule	Space between labia minora that includes vaginal and urethral openings
Vestibular glands	Secrete fluid that moistens and lubricates vestibule

stimulation. The culmination of such stimulation is the pleasurable sense of physiological and psychological release called *orgasm.*

Just prior to orgasm, the tissues of the outer third of the vagina become engorged with blood and swell. This action increases the friction on the penis during intercourse. As orgasm occurs, a series of reflexes involving the sacral and lumbar portions of the spinal cord are initiated.

In response to these reflexes, the muscles of the perineum contract rhythmically, and the muscular walls of the uterus and uterine tubes become active. These muscular contractions are thought to aid the transport of sperm cells through the female reproductive tract toward the upper ends of the uterine tubes.

Following orgasm, the flow of blood into the erectile tissues is reduced, and the muscles of the perineum and reproductive tract tend to relax. Consequently, the organs return to a state similar to that prior to sexual stimulation.

The various functions of the female reproductive organs are summarized in chart 22.2.

1. What events result from parasympathetic stimulation of the female reproductive organs?
2. What changes take place in the vagina just prior to and during female orgasm?
3. What is the response of the uterus and the uterine tubes to orgasm?

Hormonal Control of Female Reproductive Functions

Female reproductive functions are controlled largely by hormones secreted by the *hypothalamus,* the *anterior pituitary gland,* and the *ovaries.* These hormones are responsible for the development and maintenance of female secondary sexual characteristics, the maturation of female sex cells, and the changes that occur during the monthly reproductive cycle.

Female Sex Hormones

A female child's body remains reproductively immature until about ten years of age. At that time, the hypothalamus begins to secrete increasing amounts of gonadotropin-releasing hormone (GnRH), which, in turn, stimulates the anterior pituitary gland to release the gonadotropins FSH and LH. These hormones play primary roles in the control of female sex cell maturation and in the production of female sex hormones.

Several different female sex hormones are secreted by various tissues, including the ovaries, the adrenal cortices, and the placenta (during pregnancy). These hormones belong to two major groups, called **estrogen** and **progesterone.**

The primary source of *estrogen* (in a nonpregnant female) is the ovaries, although some estrogen is also synthesized in adipose tissue from adrenal androgens. At puberty, under the influence of the anterior pituitary gland, the ovaries secrete increasing amounts of the hormone. Estrogen stimulates enlargement of various accessory organs, including the vagina, uterus, uterine tubes, ovaries, and the external structures. Estrogen is also responsible for the development and maintenance of female *secondary sexual characteristics.* These are listed in figure 22.28 and include the following:

1. Development of the breasts and the ductile system of the mammary glands within the breasts.
2. Increased deposition of adipose tissue in the subcutaneous layer generally and in the breasts, thighs, and buttocks particularly.
3. Increased vascularization of the skin.

The ovaries are also the primary source of *progesterone* (in a nonpregnant female). This hormone promotes changes that occur in the uterus during the female reproductive cycle. In addition, it affects the mammary glands and helps regulate the secretion of gonadotropins from the anterior pituitary gland.

Certain other changes that occur in females at puberty seem to be related to *androgen* (male sex hormone) concentrations. For example, increased growth of hair in the pubic and axillary regions seems to be due to the presence of androgen secreted by the adrenal

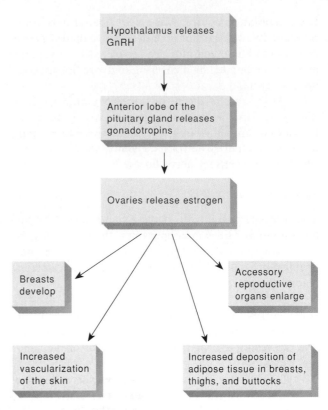

Figure 22.28 Mechanism by which female secondary sexual characteristics are stimulated to develop.

cortices. Conversely, the development of the female skeletal configuration, which includes narrow shoulders and broad hips, seems to be related to a low concentration of androgen.

Female athletes who are trained for endurance events, such as the marathon, typically maintain about 6% body fat. Male endurance athletes usually have about 4% body fat. This difference of 50% in proportion of body fat is related to the actions of sex hormones. The male hormone testosterone tends to promote the deposition of protein throughout the body and especially in skeletal muscles; the female hormone estrogen causes an increased deposition of adipose tissue in the breasts, thighs, buttocks, and the subcutaneous layer of the skin.

1. What factors initiate sexual maturation in a female?
2. Name the two major female sex hormones.
3. What is the function of estrogen?
4. What is the function of androgen in a female?

Female Reproductive Cycle

The female reproductive cycle, or **menstrual cycle,** is characterized by regular, recurring changes in the uterine lining, which culminate in menstrual bleeding (menses). Such cycles usually begin near the thirteenth year of life and continue into middle age, after which the cycles cease.

Women athletes sometimes experience disturbances in their menstrual cycles, ranging from diminished menstrual flow (oligomenorrhea) to the complete stoppage of menses (amenorrhea). The incidence of menstrual disorders generally increases with the intensity and duration of exercise periods, occurring most commonly in athletes who perform the most strenuous activities and who follow the most intense training schedules. This effect seems to be related to a loss of adipose tissue and a consequent decline in estrogen, which is synthesized in these tissues from adrenal androgens.

A female's first menstrual cycle (menarche) is initiated after the ovaries and other organs of the female reproductive control system have become mature and responsive to certain hormones. Then, the hypothalamic secretion of gonadotropin-releasing hormone (GnRH) stimulates the anterior pituitary gland to release threshold levels of FSH (follicle-stimulating hormone) and LH (luteinizing hormone). As its name implies, FSH acts upon the ovary to stimulate the maturation of a *follicle,* and during this development, the granulosa cells of the follicle produce increasing amounts of estrogen and some progesterone. LH also plays a role in the production of estrogen by stimulating certain ovarian cells (theca interna) to secrete the necessary precursor molecules (testosterone) from which estrogen is synthesized.

In a young female, estrogen stimulates the development of various secondary sexual characteristics. The estrogen secreted during subsequent menstrual cycles is responsible for continuing the development of these traits and maintaining them. Chart 22.3 summarizes the hormonal control of female secondary sexual characteristics.

An increasing concentration of estrogen during the first week or so of a menstrual cycle causes changes in the uterine lining, including thickening of the glandular endometrium (proliferative phase). Meanwhile, the developing follicle has completed its maturation, and by the fourteenth day of the cycle, the follicle appears on the surface of the ovary as a blisterlike bulge.

Within the follicle, the granulosa cells, which surround the oocyte and connect it to the inner wall, have become loosened. Also, the follicular fluid has accumulated rapidly.

While the follicle was maturing, the estrogen that it secreted inhibited the release of LH from the anterior pituitary gland, but allowed the LH to be stored in the gland. The estrogen also caused the anterior pituitary cells to become more sensitive to the action of

CHART 22.3 Hormonal control of female secondary sexual characteristics

1. The hypothalamus releases GnRH, which stimulates the anterior pituitary gland.
2. The anterior pituitary gland secretes FSH and LH.
3. FSH stimulates the maturation of a follicle.
4. Granulosa cells of the follicle produce and secrete estrogen; LH stimulates certain cells to secrete estrogen precursor molecules.
5. Estrogen is responsible for the development and maintenance of most of the female secondary sexual characteristics.
6. Concentrations of androgen affect other secondary sexual characteristics, including skeletal growth and growth of hair.
7. Progesterone, secreted by the ovaries, affects cyclical changes in the uterus and mammary glands.

GnRH, which is released from the hypothalamus in rhythmic pulses about ninety minutes apart.

Near the fourteenth day of follicular development, the anterior pituitary cells finally respond to the pulses of GnRH and to the increasing quantity of progesterone released from the follicle. In response, the anterior pituitary cells release the LH that they have stored. The resulting surge in LH concentration, which lasts for about thirty-six hours, causes the bulging follicular wall to weaken and rupture. At the same time, the oocyte and follicular fluid escape from the ovary in the process of *ovulation.*

Following ovulation, the remnants of the follicle and the theca interna within the ovary undergo rapid changes. The space occupied by the follicular fluid fills with blood, which soon clots, and under the influence of LH, the follicular and thecal cells enlarge greatly to form a temporary glandular structure within the ovary, called a **corpus luteum** (figure 22.22).

Although the follicular cells secrete some progesterone during the first part of the menstrual cycle, corpus luteum cells secrete large quantities of progesterone and estrogen during the last half of the cycle. Consequently, as a corpus luteum becomes established, the blood concentration of progesterone increases sharply.

Progesterone acts on the endometrium of the uterus, causing it to become more vascular and glandular. It also stimulates the uterine glands to secrete increasing quantities of glycogen and lipids (secretory phase). As a result, the endometrial tissues of the uterus become filled with fluids containing nutrients and electrolytes, which provide a favorable environment for the development of an embryo.

Estrogen and progesterone inhibit the release of LH from the anterior pituitary gland. The corpus luteum also secretes the hormone *inhibin,* which inhibits the secretion of FSH. Consequently, no other follicles are stimulated to develop during the time that the corpus luteum is active. However, if the oocyte that was released at ovulation is not fertilized by a sperm cell, the corpus luteum begins to degenerate (regress) about the twenty-fourth day of the cycle. Eventually it is replaced by fibrous connective tissue, and the remnant of such a corpus luteum is called a *corpus albicans* (figure 22.22).

When the corpus luteum ceases to function, the concentrations of estrogen and progesterone decline rapidly, and in response, the blood vessels in the endometrium become constricted. This action reduces the supply of oxygen and nutrients to the thickened uterine lining, and these lining tissues (decidua) soon disintegrate and slough off. At the same time, blood escapes from damaged capillaries, creating a flow of blood and cellular debris, which passes through the vagina as the *menstrual flow* (menses). This flow usually begins about the twenty-eighth day of the cycle and continues for three to five days, while the estrogen concentration is relatively low.

The beginning of the menstrual flow marks the end of a menstrual cycle and the beginning of a new cycle. This cycle is diagrammed in figure 22.29 and summarized in chart 22.4.

Because the blood concentrations of estrogen and progesterone are low at the beginning of the menstrual cycle, the hypothalamus and anterior pituitary gland are no longer inhibited. Consequently, the concentrations of FSH and LH soon increase, and a new follicle is stimulated to mature. As this follicle secretes estrogen, the uterine lining undergoes repair, and the endometrium begins to thicken again.

> Some women experience a set of unpleasant changes between the time of ovulation and the beginning of the menstrual flow. This disorder is called *premenstrual syndrome* (PMS), and its symptoms may include increased emotional tension, irritability, headache, fatigue, breast soreness, and swelling of hands, ankles, and abdomen. The cause of PMS is unknown, but it is believed to involve hormonal imbalance.

Menopause

After puberty, menstrual cycles normally continue to occur at more or less regular intervals into the late forties or early fifties, at which time they usually become increasingly irregular. Then, in a few months or years, the cycles cease altogether. This period in life is called **menopause** (female climacteric).

The cause of menopause seems to be an aging of the ovaries. After about thirty-five years of cycling, few primary follicles remain to be stimulated by pituitary gonadotropins. Consequently, the follicles no longer mature, ovulation does not occur, and the blood concentration of estrogen decreases greatly, although many women continue to synthesize some estrogen from adrenal androgens.

As a result of reduced estrogen concentration and lack of progesterone, the female secondary sexual characteristics undergo varying degrees of change. The vagina, uterus, and uterine tubes may decrease in size, as may the external reproductive organs. The pubic and axillary hair may become thinner, and the breasts may regress. Other changes that occur commonly in response to low estrogen concentration include thinning of the epithelial linings associated with urinary and reproductive organs, increased loss of bone matrix (osteoporosis), and thinning of the skin. Because the pituitary secretions of FSH and LH are no longer inhibited, these hormones may be released continuously for some time.

> About 50% of women reach menopause by age fifty, and 85% reach it by age fifty-two. Of these, perhaps 20% have no unusual symptoms—they simply stop menstruating. However, about 50% of menopausal women experience unpleasant vasomotor symptoms, including sensations of heat in the face, neck, and upper body called "hot flashes." Such a sensation may last for thirty seconds to five minutes and may be accompanied by chills and sweating. Menopausal women may also experience varying degrees of headache, backache, and fatigue.
>
> The cause of these vasomotor symptoms is not well understood, but they may involve changes in the rhythmic secretion of GnRH by the hypothalamus in response to declining concentrations of sex hormones.
>
> To reduce the unpleasant side effects of menopause, women are often treated with *hormone replacement therapy* (HRT). Such treatment usually involves administering estrogens combined with progesterone. This therapy also reduces the excessive loss of bone tissue that is common in postmenopausal women.

1. Trace the events of the female menstrual cycle.
2. What effect does progesterone have on the endometrium of the uterus?
3. What causes the menstrual flow?
4. What are some changes that may occur at menopause?

Pregnancy

Pregnancy is the condition characterized by the presence of a developing offspring within the uterus. It results from the union of an egg cell and a sperm cell—an event called **fertilization.**

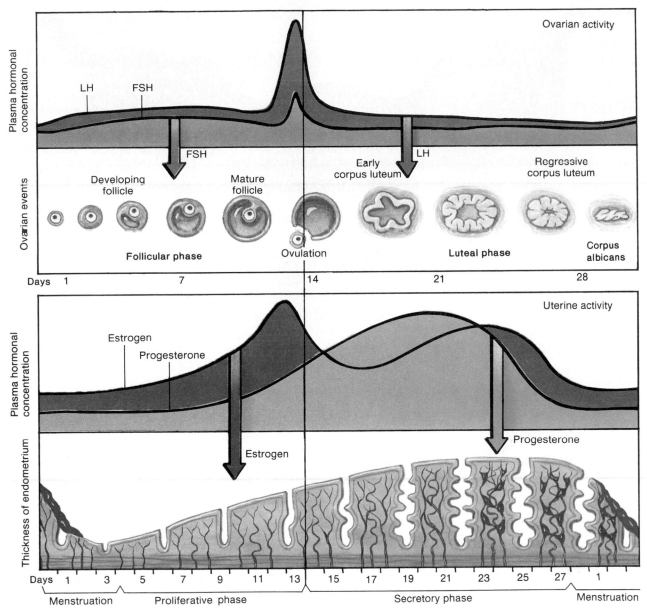

Figure 22.29 Major events in the female ovarian and menstrual cycles.

CHART 22.4 Major events in a menstrual cycle

1. The anterior pituitary gland secretes FSH and LH.

2. FSH stimulates maturation of a follicle.

3. Granulosa cells of the follicle produce and secrete estrogen.

   a. Estrogen maintains secondary sexual traits.

   b. Estrogen causes the uterine lining to thicken.

4. The anterior pituitary gland releases a surge of LH, which stimulates ovulation.

5. Follicular and thecal cells become corpus luteum cells, which secrete estrogen and progesterone.

   a. Estrogen continues to stimulate uterine wall development.

   b. Progesterone stimulates the uterine lining to become more glandular and vascular.

   c. Estrogen and progesterone inhibit the secretion of LH, and inhibin inhibits the secretion of FSH from the anterior pituitary gland.

6. If the egg cell is not fertilized, the corpus luteum degenerates and no longer secretes estrogen and progesterone.

7. As the concentrations of luteal hormones decline, blood vessels in the uterine lining constrict.

8. The uterine lining disintegrates and sloughs off, producing a menstrual flow.

9. The anterior pituitary gland, which is no longer inhibited, again secretes FSH and LH.

10. The menstrual cycle is repeated.

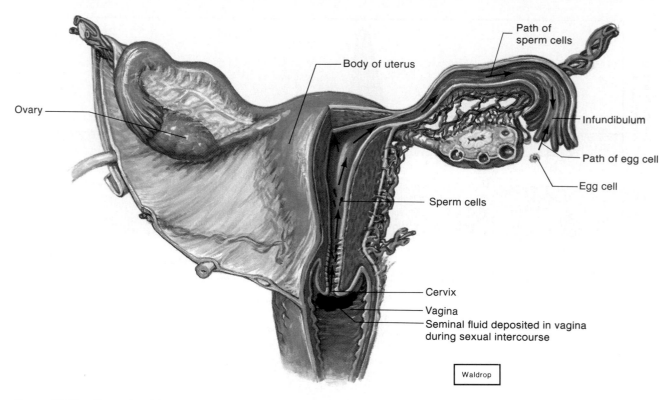

Path of
sperm cells

Body of uterus

Ovary

Infundibulum

Path of egg cell

Egg cell

Sperm cells

Cervix

Vagina

Seminal fluid deposited in vagina
during sexual intercourse

Waldrop

Figure 22.30 The paths of the egg and sperm cells through the female reproductive tract. What factors aid the movements of these cells?

Transport of Sex Cells

Ordinarily, before fertilization can occur, an egg cell (secondary oocyte) must be released by ovulation and enter a uterine tube.

During sexual intercourse, semen containing sperm cells is deposited in the vagina near the cervix. To reach the egg cell, the sperm cells must then move upward through the uterus and uterine tube. This movement is aided by the lashing of each sperm's tail and by muscular contractions within the walls of the uterus and uterine tube, which are stimulated by prostaglandins in the semen. Also, under the influence of high estrogen concentrations during the first part of the menstrual cycle, the uterus and cervix contain a thin, watery secretion that promotes sperm transport and survival. Conversely, during the latter part of the cycle, when the progesterone concentration is relatively high, these parts secrete a viscous fluid that is unfavorable for sperm transport and survival (figure 22.30).

The sperm transport mechanism is relatively inefficient, however, because even though as many as 300 million to 500 million sperm cells may be deposited in the vagina by a single ejaculation, only a few hundred of them ever reach an egg cell.

Sperm cells are thought to reach the upper portions of the uterine tube within an hour following sexual intercourse. Although many sperm cells may reach an egg cell, only one sperm cell actually fertilizes the egg (figure 22.31).

Studies indicate that an egg cell may survive for only twelve to twenty-four hours following ovulation, while sperm cells may live up to seventy-two hours within the female reproductive tract. Consequently, sexual intercourse probably must occur no more than seventy-two hours before ovulation or twenty-four hours following ovulation if fertilization is to take place.

Fertilization

When a sperm cell reaches an egg cell, it invades the follicular cells that adhere to the egg's surface (corona radiata) and binds to the *zona pellucida* that surrounds the egg cell membrane. The acrosome of a sperm cell attached to the zona pellucida releases a trypsin-like proteinase that helps the motile sperm penetrate the zona pellucida (figure 22.32).

The sperm cell's plasma membrane fuses with that of the egg, and the sperm movement ceases. At the same time, the egg cell membrane becomes unresponsive to other sperm cells. This rapid change seems to involve a depolarization of the egg cell membrane. The union of the egg and sperm cell membranes also triggers an action in some lysosomal-like granules (cortical granules) that occur just beneath the egg cell membrane. These granules release enzymes that cause the zona pellucida to harden. This reduces the chance that other sperm cells will reach the egg, and it forms a protective layer around the newly formed fertilized egg cell.

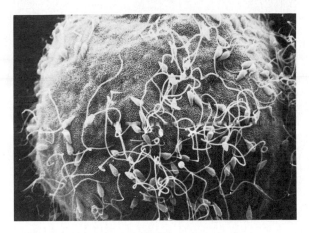

Figure 22.31 Scanning electron micrograph of sperm cells on the surface of an egg cell (×1,200).

After the sperm cell has fused with the egg cell membrane, it passes through the membrane and enters the cytoplasm. During this process, the sperm cell loses its tail, and the nucleus within its head swells. As explained previously, the egg cell (secondary oocyte) then divides unequally to form a relatively large cell and a tiny second polar body, which is later expelled. Next, the nucleus of the egg cell and that of the sperm cell come together in the center of the larger cell. Their nuclear membranes disappear, and their chromosomes combine, thus completing the process of **fertilization.** This process is diagrammed in figure 22.32.

Because the sperm cell and the egg cell each provide 23 chromosomes, the product of fertilization is a cell with 46 chromosomes—the usual number in a human cell. This cell, called a **zygote,** is the first cell of the future offspring.

1. What factors enhance the motility of sperm cells within the female reproductive tract?
2. Where in the female reproductive system does fertilization normally take place?
3. List the events that occur during fertilization.

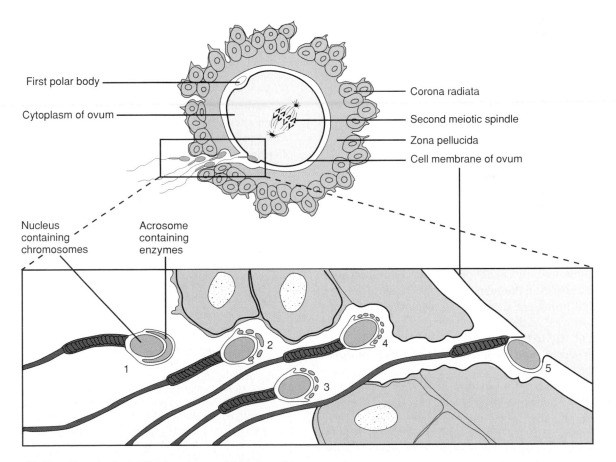

First polar body

Cytoplasm of ovum

Corona radiata

Second meiotic spindle

Zona pellucida

Cell membrane of ovum

Nucleus containing chromosomes

Acrosome containing enzymes

1

2

3

4

5

Figure 22.32 Steps in the fertilization process: (*1*) Sperm cell reaches corona radiata surrounding the egg cell. (*2*) Acrosome of sperm cell releases protein-digesting enzyme. (*3* and *4*) Sperm cell penetrates zona pellucida surrounding egg cell. (*5*) Sperm cell's plasma membrane fuses with egg cell membrane.

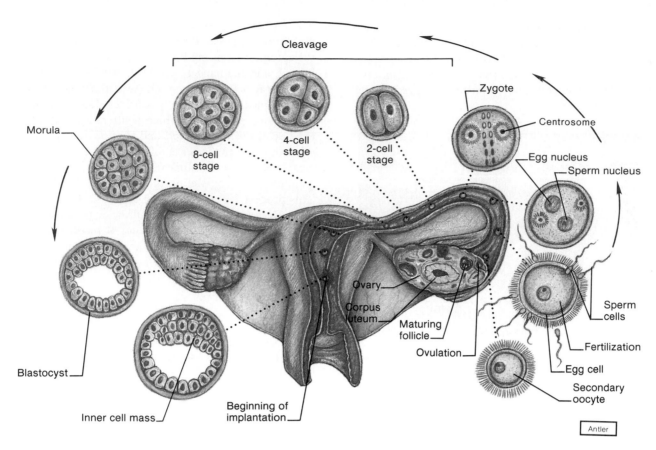

Figure 22.33 Stages in early human development.

Early Embryonic Development

About thirty hours after forming, the zygote undergoes *mitosis,* giving rise to two cells (blastomeres). These cells, in turn, divide into four cells, which divide into eight cells, and so forth. With each subsequent division, the resulting cells are smaller and smaller. This phase of development is termed **cleavage.**

Meanwhile, the tiny mass of cells is moved through the uterine tube to the uterine cavity. This movement is aided by the action of cilia of the tubular epithelium and by weak peristaltic contractions of smooth muscles in the tubular wall. Secretions from the epithelial lining are thought to provide the developing organism with nutrients.

The trip to the uterus takes about three days, and by then, the structure consists of a solid ball (morula) of about sixteen cells.

The structure remains free within the uterine cavity for about three days. During this stage, the zona pellucida of the original egg cell degenerates, and the structure, which now consists of a hollow ball of cells (blastocyst), begins to attach itself to the uterine lining. By the end of the first week of development, it is superficially *implanted* in the endometrium (figure 22.33).

About the time of implantation, certain cells within the blastocyst become organized into a group (inner cell mass) that will give rise to the body of the offspring. This marks the beginning of the *embryonic period* of development. The offspring is termed an **embryo** until the end of the eighth week, after which it is called a **fetus** up to the time of birth.

Eventually, the outer cells of the embryo, together with cells of the maternal endometrium, form a complex vascular structure called the **placenta.** This organ attaches the embryo to the uterine wall and exchanges nutrients, gases, and wastes between the maternal blood and the embryonic blood. The placenta also secretes hormones. The placenta is described in more detail in chapter 23.

Occasionally, developing offspring may become implanted in tissues outside the uterus, such as those of a uterine tube, an ovary, the cervix, or an organ in the abdominal cavity. The result is called an *ectopic pregnancy.* Most commonly, this condition occurs within a uterine tube and is termed a *tubal pregnancy.*

In a tubal pregnancy, the tube usually ruptures as the embryo enlarges. This is accompanied by severe pain and heavy vaginal bleeding. Treatment involves prompt surgical removal of the embryo and repair or removal of the damaged uterine tube.

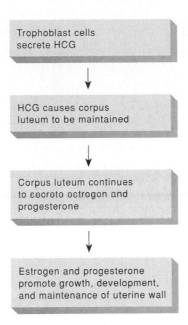

Figure 22.34 Mechanism that prevents the loss of the uterine wall during the early stages of pregnancy.

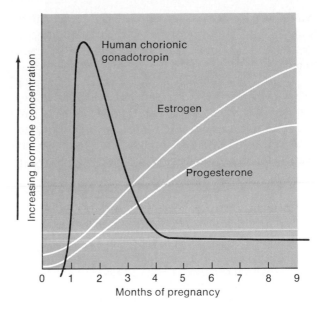

Figure 22.35 Relative concentrations of three hormones in the blood during pregnancy.

1. What is a zygote? A morula?
2. How does an embryo become implanted in the uterine wall?
3. How does a developing offspring obtain nutrients and oxygen?

Hormonal Changes during Pregnancy

During a typical menstrual cycle, the corpus luteum degenerates about two weeks after ovulation. Consequently, estrogen and progesterone concentrations decline rapidly, the uterine lining is no longer maintained, and the endometrium sloughs off as menstrual flow. If this occurs following implantation, the embryo is lost (spontaneously aborted).

The mechanism that normally prevents such a termination of pregnancy involves a hormone called HCG (human chorionic gonadotropin). This hormone is secreted by a layer of embryonic cells (trophoblast) that surrounds the developing embryo and is later involved with the formation of the placenta. (See chapter 23.) HCG has properties similar to those of LH, and it causes the corpus luteum to be maintained and to continue secreting estrogen and progesterone. Thus, the uterine wall continues to grow and develop. At the same time, the release of FSH and LH from the anterior pituitary gland is inhibited, so that normal menstrual cycles cease (figure 22.34).

The secretion of HCG continues at a high level for about two months, then declines to a relatively low level by the end of four months. Although the corpus luteum is maintained throughout the pregnancy, its function as a source of hormones becomes less impor-

tant after the first three months (first trimester). This is due to the fact that the placenta is usually well developed by then, and the placental tissues secrete sufficient estrogen and progesterone (figure 22.35).

HCG secretion by the embryonic tissues begins shortly after fertilization and increases to a peak in about fifty to sixty days. Thereafter, the concentration of HCG drops to a much lower level and remains relatively stable throughout the pregnancy.

Since HCG is excreted in the urine, its presence in urine can be used to detect the presence of an embryo. Such a pregnancy test may indicate positive results as early as eight to ten days after fertilization.

For the remainder of the pregnancy, *placental estrogen* and *placental progesterone* maintain the uterine wall. The placenta also secretes a hormone called **placental lactogen.** This hormone is thought to stimulate breast development and prepare the mammary glands for milk secretion, a function that is aided by placental estrogen and placental progesterone. Placental progesterone and a polypeptide hormone called *relaxin* from the corpus luteum seem to inhibit the smooth muscles in the myometrium so that uterine contractions are suppressed until the birth process begins.

The high concentration of placental estrogen during pregnancy causes an enlargement of the vagina and the external reproductive organs. Also, relaxin causes a relaxation of the ligaments holding the symphysis pubis and sacroiliac joints together. This action, which usually occurs during the last week of pregnancy, allows for a greater movement at these joints and thus aids the passage of the fetus through the birth canal.

Other hormonal changes that occur during pregnancy include increased secretions of aldosterone from the adrenal cortex and of parathyroid hormone from the parathyroid glands. Aldosterone promotes the renal reabsorption of sodium, leading to fluid retention, and parathyroid hormone helps to maintain a high concentration of maternal blood calcium. (See chapter 13.)

Chart 22.5 summarizes the hormonal changes of pregnancy.

---

1. What mechanism is responsible for maintaining the uterine wall during pregnancy?
2. What is the source of HCG during the first few months of pregnancy?
3. What is the source of the hormones that sustain the uterine wall during pregnancy?
4. What other hormonal changes occur during pregnancy?

Other Changes during Pregnancy

A number of other changes occur in a woman's body as a result of the increased demands of a growing fetus. For example, as the fetus increases in size, the uterus enlarges greatly, and instead of being confined to its normal location in the pelvic cavity, it extends upward and may eventually reach the level of the ribs. At the same time, the abdominal organs are displaced upward and compressed against the diaphragm. Also, as the uterus enlarges, it tends to press on the urinary bladder and causes the woman to feel a need to urinate frequently.

As the placenta grows and develops, it requires more blood, and as the fetus enlarges, it needs more oxygen and produces greater amounts of wastes that must be excreted. Consequently, the mother's blood volume, cardiac output, breathing rate, and urine production all tend to increase in response to fetal demands.

The fetal need for increasing amounts of nutrients is reflected in the need for an increased dietary intake by the mother. Her intake must supply adequate vitamins, minerals, and proteins for herself and the fetus. The fetal tissues have a greater capacity to capture available nutrients than do the maternal tissues. Consequently, if the mother's diet is inadequate, her body will usually show symptoms of a deficiency condition before fetal growth is adversely affected.

Birth Process

Pregnancy usually continues for forty weeks (280 days, or ten average menstrual cycles), or about nine calendar months (ten lunar months), if it is measured from

---

CHART 22.5   Hormonal changes during pregnancy

1. Following implantation, the embryonic cells begin to secrete HCG.
2. HCG causes the corpus luteum to be maintained and to continue secreting estrogen and progesterone.
3. As the placenta develops, it secretes large quantities of estrogen and progesterone.
4. Placental estrogen and progesterone act to:
   a. Stimulate the uterine lining to continue development.
   b. Maintain the uterine lining.
   c. Inhibit the secretion of FSH and LH from the anterior pituitary gland.
   d. Stimulate the development of the mammary glands.
   e. Inhibit uterine contractions (progesterone).
   f. Cause the enlargement of the reproductive organs (estrogen).
5. Relaxin from the corpus luteum also inhibits uterine contractions and causes the pelvic ligaments to relax.
6. The placenta secretes placental lactogen that stimulates breast development.
7. Aldosterone from the adrenal cortex promotes reabsorption of sodium.
8. Parathyroid hormone from the parathyroid glands helps maintain a high concentration of maternal blood calcium.

---

the beginning of the last menstrual cycle. The pregnancy terminates with the *birth process* (parturition).

Although the mechanism that initiates birth is not well understood, a variety of factors are involved. For example, progesterone seems to suppress uterine contractions during pregnancy, and as the placenta ages, the concentration of progesterone within the uterus declines. Also, progesterone is known to inhibit the synthesis of a prostaglandin that promotes uterine contractions. Thus, a decreasing progesterone concentration may initiate the birth process.

Also, as discussed in chapter 13, the stretching of the uterine and vaginal tissues late in pregnancy is thought to initiate nerve impulses to the hypothalamus. The hypothalamus, in turn, signals the posterior pituitary gland, which responds by releasing the hormone **oxytocin.**

Oxytocin is a powerful stimulator of uterine contractions, and its effect, combined with the greater excitability of the myometrium due to the decline in progesterone secretion, aids labor, at least in its later stages.

Labor is the term for the process in which muscular contractions force the fetus through the birth canal. Once labor starts, rhythmic contractions that begin at the top of the uterus and travel down its length force the contents of the uterus toward the cervix.

Since the fetus is usually positioned with its head downward, labor contractions force the head against the cervix. This action causes the cervix to stretch,

Female Infertility

It is estimated that about 60% of infertile marriages are the result of female disorders. One of the more common of these disorders is hyposecretion of gonadotropic hormones from the anterior pituitary gland, followed by failure of the female to ovulate (anovulation).

This type of anovulatory cycle can sometimes be detected by testing the female's urine for the presence of *pregnanediol,* a product of progesterone metabolism. Since the concentration of progesterone normally rises following ovulation, no increase in pregnanediol in the urine during the latter part of the menstrual cycle suggests a lack of ovulation.

The treatment of such a disorder may include the administration of the hormone HCG, which is obtained from human placentas. As mentioned, this substance has effects similar to those of LH and can stimulate ovulation. Another ovulatory substance, HMG (human menopausal gonadotropin), which can be obtained from the urine of postmenopausal women, contains a mixture of LH and FSH. HMG also may be used to treat females with gonadotropin deficiencies. However, either HCG or HMG may overstimulate the ovaries and cause many follicles to release egg cells simultaneously, resulting in multiple births later.

Another cause of female infertility is *endometriosis,* in which tissue resembling the inner lining of the uterus (endometrium) is present abnormally in the abdominal cavity. Some researchers believe that small pieces of the endometrium may move up through the uterine tubes during menses and become implanted in the abdominal cavity. In any case, once this tissue is present in the cavity, it undergoes changes similar to those that take place in the uterine lining during the menstrual cycle. However, when the tissue begins to break down at the end of the cycle, it cannot be expelled to the outside. Instead, its products remain in the abdominal cavity where they may irritate its lining (peritoneum) and cause considerable abdominal pain. These products also tend to stimulate the formation of fibrous tissue (fibrosis), which, in turn, may encase the ovary, preventing ovulation mechanically, or may obstruct the uterine tubes.

Still other women become infertile as a result of infections, such as gonorrhea, which may cause the uterine tubes to become inflamed and obstructed, or may stimulate the production of viscous mucus that can plug the cervix and prevent the entrance of sperm cells.

---

which is thought to elicit a reflex that stimulates still stronger labor contractions. Thus, a *positive feedback system* operates in which uterine contractions produce more intense uterine contractions until a maximum effort is achieved (figure 22.36). At the same time, dilation of the cervix reflexly stimulates an increased release of oxytocin from the posterior pituitary gland.

> During childbirth, the tissues between the vulva and anus (perineum) are sometimes torn by the stretching that occurs as the infant passes through the birth canal. For this reason, an incision may be made along the midline of the perineum from the vestibule to within 1.5 centimeters of the anus before the birth is completed. This procedure, called an *episiotomy,* ensures that the perineal tissues are cut cleanly rather than torn.

As labor continues, abdominal wall muscles are stimulated to contract by a positive feedback mechanism, and they also aid in forcing the fetus through the cervix and vagina to the outside.

Chart 22.6 summarizes some of the factors involved in promoting labor, and figure 22.37 illustrates the steps of the birth process.

Following the birth of the fetus (usually within ten to fifteen minutes), the placenta, which remains inside the uterus, separates from the uterine wall and is expelled by uterine contractions through the birth

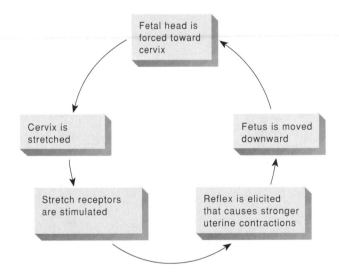

Figure 22.36 The birth process involves this positive feedback mechanism.

canal. This expulsion, termed the *afterbirth,* is accompanied by bleeding, because vascular tissues are damaged in the process. However, the loss of blood is usually minimized by continued contraction of the uterus that compresses the bleeding vessels. This contraction is stimulated by the action of oxytocin.

For several weeks following childbirth, the uterus becomes smaller by a process called *involution.* Also, its endometrium sloughs off and is discharged through the vagina. This is followed by the return of an epithelial lining characteristic of a nonpregnant female.

1. List some of the physiological changes that occur in a woman's body during pregnancy.
2. Describe the role of progesterone in initiating labor.
3. Explain how dilation of the cervix affects labor.
4. How is bleeding controlled naturally after the placenta is expelled?

Mammary Glands

The **mammary glands** are accessory organs of the female reproductive system that are specialized to secrete milk following pregnancy.

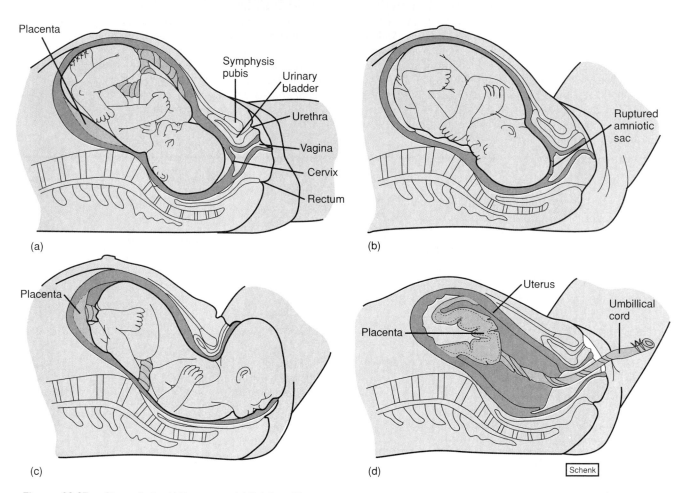

Figure 22.37 Stages in the birth process. (*a*) Fetal position before labor; (*b*) dilation of the cervix; (*c*) expulsion of the fetus; (*d*) expulsion of the placenta.

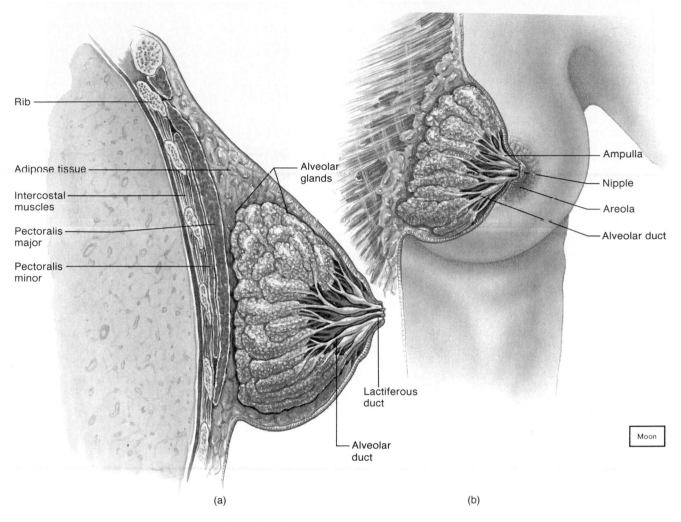

Rib

Adipose tissue

Intercostal
muscles

Pectoralis
major

Pectoralis
minor

Alveolar
glands

Lactiferous
duct

Alveolar
duct

Ampulla

Nipple

Areola

Alveolar duct

Moon

(a)

(b)

Figure 22.38 Structure of the breast. (*a*) Sagittal section;
(*b*) anterior view.

Location of the Glands

The mammary glands are located in the subcutaneous
tissue of the anterior thorax within the hemispherical
elevations called *breasts*. The breasts overlie the *pectoralis major* muscles and extend from the second to the
sixth ribs and from the sternum to the axillae.

A *nipple* is located near the tip of each breast at
about the level of the fourth intercostal space, and it is
surrounded by a circular area of pigmented skin called
the *areola* (figure 22.38).

Structure of the Glands

A mammary gland is composed of fifteen to twenty irregularly shaped lobes. Each lobe contains glands (alveolar glands) and a duct (lactiferous duct) that leads
to the nipple and opens to the outside. The lobes are
separated by dense connective and adipose tissues.
These tissues also support the glands and attach them
to the fascia of the underlying pectoral muscles. Other
connective tissue, which forms dense strands called
suspensory ligaments, extends inward from the dermis
of the breast to the fascia, helping support the weight
of the breast.

Breast cancer, which is one of the more common types of cancer in women, usually begins as a small, painless lump.

Because early diagnosis of such a tumor is of prime importance in successful treatment, the American Cancer Society recommends that women beyond the age of twenty examine their breasts each month, paying particular attention to the upper, outer portions. The examination should be made just after menstruation when the breasts are usually soft, and any lump that is discovered should be checked immediately by a physician.

It is also recommended that after age thirty-five women have their breasts examined at regular intervals with *mammography*—a breast cancer detection technique that makes use of low dosage X ray.

More specifically, the American Cancer Society recommends that women between the ages of thirty-five and forty have a baseline mammogram, that those between the ages of forty and forty-nine have a mammogram every one or two years, and that those over the age of fifty have a mammogram yearly. A breast cancer can be detected on a *mammogram* (an X-ray film of the breast) when the growing tumor is relatively small, perhaps two years before a lump can be felt. (See figure 22.39.)

Whenever a questionable lump is detected by physical examination or mammography, a biopsy should be performed so that the tissue can be observed microscopically. As a rule, such a microscopic examination is necessary before a breast cancer can be diagnosed with certainty.

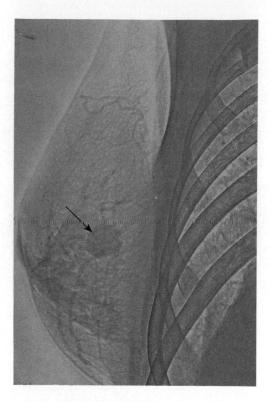

Figure 22.39 Mammogram (X-ray film) of a breast with a tumor (arrow).

Development of the Breasts

The mammary glands of male and female children are similar. As children reach *puberty*, the male glands fail to develop, while ovarian hormones stimulate the female glands to develop. As a result, the alveolar glands and ducts enlarge, and fat is deposited, so that the breasts become surrounded by adipose tissue, except for the region of the areola.

During pregnancy, placental estrogen and placental progesterone stimulate further development of the mammary glands. Estrogen causes the ductile systems to grow and branch, and to have large quantities of fat deposited around them. Progesterone, on the other hand, stimulates the development of the alveolar glands at the ends of the ducts. These changes are also promoted by the presence of placental lactogen.

As a consequence of hormonal activity, the breasts may double in size during pregnancy. At the same time, the adipose tissue of the breasts is largely replaced by glandular tissue. Beginning about the fifth week of pregnancy, the anterior pituitary gland releases increasing amounts of *prolactin*. Thus, the concentration of this hormone is relatively high throughout pregnancy. However, placental progesterone inhibits milk production, and placental lactogen seems to block the action of prolactin. Consequently, even though the

mammary glands become capable of secreting milk, none is produced. The micrographs in figure 22.40 compare the mammary gland tissues of a nonpregnant woman with those of a woman who is producing milk.

Production and Secretion of Milk

Following childbirth and the expulsion of the placenta, the maternal blood concentrations of placental hormones decline rapidly, and the action of prolactin is no longer inhibited.

Prolactin stimulates the mammary glands to secrete large quantities of milk. This hormonal effect does not occur until two or three days following birth, and in the meantime, the glands secrete a thin, watery fluid called *colostrum*. Although colostrum is relatively rich in proteins, its concentrations of carbohydrates and fats are lower than those of milk.

The milk produced under the influence of prolactin does not flow readily through the ductile system of the mammary gland, but must be actively ejected by contraction of specialized *myoepithelial cells* surrounding the alveolar glands. The contraction of these cells and the consequent ejection of milk through the ducts result from a reflex action (figure 22.41).

This reflex is elicited when the breast is suckled or the nipple or areola is otherwise mechanically stimulated. Then, impulses from sensory receptors within the breasts travel to the hypothalamus, which signals

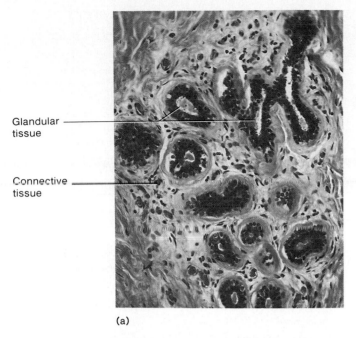

Glandular tissue

Connective tissue

(a)

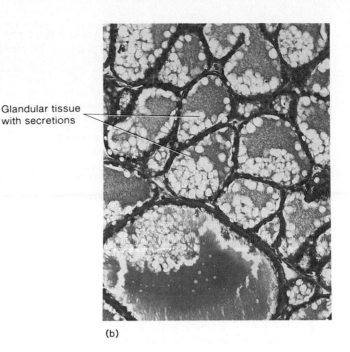

Glandular tissue with secretions

(b)

Figure 22.40 (a) Light micrograph of a mammary gland in a nonpregnant woman (×63); (b) light micrograph of an active (lactating) mammary gland (×63).

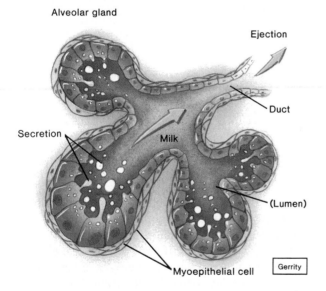

Alveolar gland

Ejection

Duct

Secretion

Milk

(Lumen)

Myoepithelial cell

Gerrity

Figure 22.41 Milk is ejected from an alveolar gland by the action of myoepithelial cells.

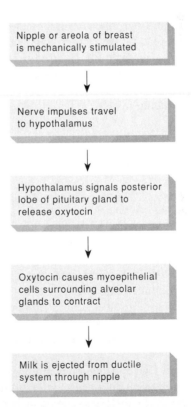

Nipple or areola of breast is mechanically stimulated

↓

Nerve impulses travel to hypothalamus

↓

Hypothalamus signals posterior lobe of pituitary gland to release oxytocin

↓

Oxytocin causes myoepithelial cells surrounding alveolar glands to contract

↓

Milk is ejected from ductile system through nipple

Figure 22.42 Mechanism that causes the ejection of milk from the breasts. What happens to milk production if the milk is not regularly removed from the breast?

the posterior pituitary gland to release oxytocin. The oxytocin reaches the breast by means of the blood and stimulates the myoepithelial cells to contract. Consequently, milk is ejected into a suckling infant's mouth in about thirty seconds (figure 22.42).

Sensory impulses triggered by mechanical stimulation of the nipples also signal the hypothalamus to allow a continued secretion of prolactin. Thus, prolactin is released as long as milk is removed from the breasts. However, if milk is not removed regularly, the

CHART 22.7 Hormonal control of the mammary glands

Before pregnancy (beginning of puberty)	Following childbirth
Ovarian hormones secreted during menstrual cycles stimulate alveolar glands and ducts of mammary glands to develop.	1. Placental hormonal concentrations decline, so that the action of prolactin is no longer inhibited.
During pregnancy	2. The breasts begin producing milk.
1. Estrogen causes the ductile system to grow and branch.	3. Mechanical stimulation of the breasts causes release of oxytocin from the posterior pituitary gland.
2. Progesterone stimulates development of alveolar glands.	4. Oxytocin stimulates ejection of milk from the ducts.
3. Placental lactogen promotes development of the breasts.	5. As long as milk is removed, more prolactin is released; if milk is not removed, milk production ceases.
4. Prolactin is secreted throughout pregnancy, but milk production is inhibited by placental progesterone.	

hypothalamus inhibits the secretion of prolactin, and within about one week, the mammary glands lose their capacity to produce milk.

Although it is possible for a woman to become pregnant during the period that she breast-feeds her child, menstrual cycles seem to be inhibited, at least for a time, while the mammary glands are active. Although the mechanism responsible for this effect is not well understood, it is thought that prolactin may suppress the release of gonadotropins from the anterior pituitary gland. In any event, menstrual cycles may not resume for some time following the birth of an infant that is breast-fed. On the other hand, after several months of breast-feeding, FSH is usually released, and the monthly reproductive cycles are reestablished.

Chart 22.7 summarizes the hormonal effects involved in producing milk.

1. Describe the structure of a mammary gland.
2. How does pregnancy affect the mammary glands?
3. What stimulates the mammary glands to produce milk?
4. What causes milk to flow into the ductile system of a mammary gland?

Birth Control

Birth control is the voluntary regulation of the number of offspring produced and the time they will be conceived. This control involves a method of **contraception** designed to avoid the fertilization of an egg cell following sexual intercourse or to prevent the implantation of an embryo.

A variety of contraceptive methods are commonly practiced, including coitus interruptus, rhythm method, mechanical barriers, chemical barriers, oral contraceptives, intrauterine devices, and surgical methods. The relative effectiveness of these methods is summarized in chart 22.8.

Coitus Interruptus

Coitus interruptus involves withdrawing the penis from the vagina before ejaculation, thus preventing the en-

CHART 22.8 Effectiveness of contraceptive methods (from various studies)

Method	Failure rate*
Abstinence	0
Tubal ligation	Less than 0.5
Vasectomy	Less than 0.5
Combined birth control pills	2.5–3.0
Contraceptive implant	0.5–2.5
IUD	4.5–6.0
Condom	9.5–12.0
Diaphragm (with spermicide)	14.5–19.0
Creams, foams, and jellies	12.0–21.0
Coitus interruptus (withdrawal)	6.5–23.0
Rhythm method (natural family planning)	15.5–24.0
Chance (no protection)	90.0

*Failure rate equals the number of women per hundred who use the method and become pregnant within one year.

trance of sperm cells into the female reproductive tract. This method of contraception often proves unsatisfactory and may result in pregnancy, since some males find it emotionally difficult to withdraw just prior to ejaculation. Also, small quantities of semen containing sperm cells may be expelled from the penis into the vagina before ejaculation occurs.

Rhythm Method

The *rhythm method* (also called timed coitus or natural family planning), like coitus interruptus, uses no artificial devices or chemicals. Instead, it requires abstinence from sexual intercourse a few days before and a few days after ovulation.

Since ovulation theoretically occurs on the fourteenth day of a twenty-eight-day menstrual cycle, it might seem easy to avoid intercourse near ovulation. However, few women have absolutely regular menstrual cycles, and the lengths of the cycles vary from time to time. Furthermore, the variable part of the cycle

(a)

(b)

(c)

(d)

(e)

Figure 22.43 Devices and substances used for birth control. (a) Condom; (b) diaphragm; (c) spermicidal gel; (d) oral contraceptive; (e) IUD.

occurs before ovulation, because regardless of the length of a cycle, the menstrual flow almost always begins thirteen to fifteen days following ovulation. Inasmuch as the length of a cycle cannot be predicted ahead of time, it is almost impossible to predict the time of ovulation accurately. Thus, it is not surprising that the rhythm method results in a relatively high rate of pregnancy.

Another disadvantage of the rhythm method is that it requires adherence to a particular pattern of behavior and restricts spontaneity in sexual activity.

The effectiveness of the rhythm method can sometimes be increased by measuring and recording the woman's body temperature when she awakes each morning for several months. Since the body temperature typically rises about 0.6 degrees Fahrenheit immediately following ovulation, this procedure may allow a woman to more accurately predict the "unsafe times" in her reproductive cycle. On the other hand, many women apparently do not show a change in body temperature at ovulation, and furthermore, the body temperature may vary in response to other factors such as illnesses or emotional upsets.

1. Why is coitus interruptus an unreliable method of contraception?
2. Describe the idea behind the rhythm method of contraception.
3. What factors make the rhythm method less reliable than some other methods of contraception?

Mechanical Barriers

Mechanical barriers are sometimes employed to prevent sperm cells from entering the female reproductive tract during sexual intercourse. One such device used by males is a *condom*. It consists of a thin latex sheath that is placed over the erect penis before intercourse to prevent semen from entering the vagina upon ejaculation (figure 22.43).

The condom provides a relatively inexpensive method of contraception, and it may also help protect the user against contracting venereal diseases. However, men often feel that a condom decreases the sensitivity of the penis during intercourse. Also, its use has the disadvantage of interrupting the sex act.

Another mechanical barrier, used by females, is the *diaphragm*. It is a cup-shaped structure with a flexible ring forming the rim. The diaphragm is inserted into the vagina so that it covers the cervix and thus prevents the entrance of sperm cells into the uterus (figure 22.43).

To be effective, a diaphragm must be fitted for size by a physician, be inserted properly, and be used in conjunction with a chemical spermicide that is applied to the surface adjacent to the cervix and to the rim of the diaphragm. Also, the diaphragm must be left in position for several hours following sexual intercourse.

Although the diaphragm can be inserted into the vagina sometime before sexual contact is to occur so that its use does not interrupt the sex act, some women find the insertion of the diaphragm distasteful or uncomfortable.

Chemical Barriers

Chemical barriers for the purpose of contraception include a variety of creams, foams, and jellies with spermicidal properties. Within the vagina, such chemicals create an environment that is unfavorable for sperm cells (figure 22.43).

Chemical barriers are fairly easy to use, but have a relatively high failure rate when used alone. They are best used along with a diaphragm.

Oral Contraceptives

An *oral contraceptive* is commonly called the pill, because this method employs a series of drug-containing tablets taken by a woman (figure 22.43).

The most common forms of oral contraceptives contain synthetic estrogen-like and progesterone-like substances. When taken daily, these drugs seem to disrupt the normal pattern of gonadotropin secretion and prevent the surge in LH release that triggers ovulation. They also interfere with the build-up of the uterine lining that is necessary for implantation of embryos. (See chapter 23.)

Successful use of oral contraceptives requires motivation and planning. However, if they are used correctly, they prevent pregnancy nearly 100% of the time. However, they commonly cause nausea, retention of body fluids, increased pigmentation of the skin, and breast tenderness. Also, some women, particularly those over thirty-five years of age who smoke, may develop intravascular blood clots. Still others may develop liver disorders or high blood pressure as a result of using these substances.

A large dose of high potency estrogens is sometimes used as a "morning-after pill"—a treatment designed to prevent implantation of an embryo after sexual intercourse. Such a treatment acts by promoting powerful contractions of smooth muscles in the female reproductive tract that may cause a fertilized egg or early embryo to be expelled. However, if an embryo has become implanted already, this treatment may injure the embryo, which then may continue to develop.

Contraceptive Implants

One of the more recently developed forms of contraception is the *contraceptive implant*. Such an implant consists of a set of small progesterone-containing capsules or rods, which are inserted surgically under the skin of a woman's upper arm or scapular region. The progesterone, which is released slowly from the implant, prevents ovulation in much the same way as do oral contraceptives.

A contraceptive implant is effective for a period of up to five years, and its contraceptive action can be reversed by having the device removed.

1. Describe two methods of contraception that use mechanical barriers.
2. How can the effectiveness of chemical contraceptives be increased?
3. What substances are contained in oral contraceptives?
4. How is pregnancy prevented by an oral contraceptive? A contraceptive implant?

Intrauterine Devices

An *intrauterine device,* or *IUD,* is a small, solid object, often with exposed copper parts, that can be placed within the uterine cavity by a physician. Such a device is relatively effective in preventing pregnancy. It is believed that the IUD interferes with the implantation of embryos (interception) within the uterine wall, perhaps by causing inflammatory reactions in the uterine tissues (figure 22.43).

Unfortunately, an IUD may be expelled from the uterus spontaneously or may produce unpleasant or dangerous side effects, such as pain or excessive bleeding. It may also injure the uterus or produce other serious health problems, and should be checked at regular intervals by a physician.

Surgical Methods

Surgical methods of contraception involve sterilization of either the male or female.

In the male, a small section of each vas deferens is removed near the epididymis, and the cut ends of the ducts are tied. This procedure is called a *vasectomy,* and it is a relatively simple operation that produces few side effects, although it may cause some pain for a week or two.

After a vasectomy, sperm cells cannot leave the epididymis, thus they are not included in the semen. However, sperm cells may already be present in portions of the ducts distal to the cuts. Consequently, the sperm count of the semen released by ejaculation may not reach zero for several weeks.

The corresponding procedure in the female is called *tubal ligation.* In this instance, the uterine tubes are cut and tied so that sperm cells cannot reach an egg cell.

Neither a vasectomy nor a tubal ligation produces any changes in the hormonal concentrations or the sexual drives of the individuals involved. These sterilization procedures, shown in figure 22.44, provide the most reliable forms of contraception.

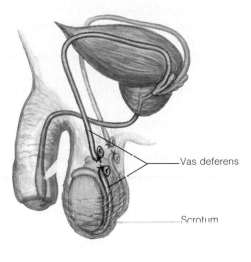

(a)

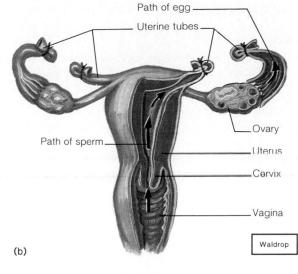

Path of egg
Uterine tubes
Path of sperm
Ovary
Uterus
Cervix
Vagina
Waldrop

(b)

Figure 22.44 (*a*) Vasectomy involves removal of a portion of each vas deferens. (*b*) Tubal ligation involves removal of a portion of each uterine tube.

Although it is sometimes possible to reverse the effects of vasectomy or tubal ligation by surgically reconnecting the severed tubes, such procedures require difficult microsurgery that has a relatively low rate of success.

1. How is an IUD thought to prevent pregnancy?
2. Describe the surgical methods of contraception for a female and for a male.

Clinical Terms Related to the Reproductive Systems

abortion (ah-bor'shun) The spontaneous or deliberate termination of pregnancy; a spontaneous abortion is commonly termed a miscarriage.

amenorrhea (a-men''o-re'ah) The absence of menstrual flow, usually due to a disturbance in hormonal concentrations.

cesarean section (sĕ-sa're-an sek'shun) The delivery of a fetus through an abdominal incision.

conization (ko''nĭ-za'shun) The surgical removal of a cone of tissue from the cervix for examination.

curettage (ku''rĕ-tahzh') A surgical procedure in which the cervix is dilated and the endometrium of the uterus is scraped (commonly called D and C for dilation and curettage).

dysmenorrhea (dis''men-ŏ-re'ah) Painful menstruation.

eclampsia (ĕ-klamp'se-ah) A condition characterized by convulsions and coma that sometimes accompanies toxemia of pregnancy.

endometriosis (en''do-mĕ''tre-o'sis) A condition in which tissue resembling the endometrium of the uterine lining is present and growing in the abdominal cavity.

endometritis (en''do-mĕ-tri'tis) An inflammation of the uterine lining.

epididymitis (ep''ĭ-did''ĭ-mi'tis) An inflammation of the epididymis.

gestation (jes-ta'shun) The entire period of pregnancy.

hematometra (hem''ah-to-me'trah) An accumulation of menstrual blood within the uterine cavity.

hyperemesis gravidarum (hi''per-em'ĕ-sis grav'i-dar-um) Excessive vomiting associated with pregnancy; morning sickness.

hysterectomy (his''tĕ-rek'to-me) The surgical removal of the uterus.

mastitis (mas''ti'tis) An inflammation of a mammary gland.

oophorectomy (o''of-o-rek'to-me) The surgical removal of an ovary.

oophoritis (o''of-o-ri'tis) An inflammation of an ovary.

orchiectomy (or''ke-ĕk'to-me) The surgical removal of a testis.

orchitis (or-ki'tis) An inflammation of a testis.

prostatectomy (pros''tah-tek'to-me) The surgical removal of a portion of or all of the prostate gland.

prostatitis (pros''tah-ti'tis) An inflammation of the prostate gland.

salpingectomy (sal''pin-jek'to-me) The surgical removal of a uterine tube.

toxemia of pregnancy (tok-se'me-ah) A condition characterized by a group of metabolic disorders sometimes occurring during pregnancy.

vaginitis (vaj''ĭ-ni'tis) An inflammation of the vaginal lining.

varicocele (var'ĭ-ko-sēl'') Distension of the veins within the spermatic cord.

Chapter Summary

Introduction (page 806)

Various reproductive organs produce sex cells and sex hormones, help sustain the lives of these cells, or transport them from place to place.

Organs of the Male Reproductive System (page 806)

The male reproductive organs are specialized to produce and maintain sperm cells, transport these cells, and produce male sex hormones. The primary male sex organs are the testes, which produce sperm cells and male sex hormones. Accessory organs include the internal and external reproductive organs.

Testes (page 806)

1. Descent of the testes
   a. Testes originate behind the parietal peritoneum near the level of the developing kidneys.
   b. The gubernaculum guides the descent of the testes into the lower abdominal cavity and through the inguinal canal.
   c. Undescended testes fail to produce sperm cells because of the relatively high abdominal temperature.
2. Structure of the testes
   a. The testes are composed of lobules separated by connective tissue and filled with seminiferous tubules.
   b. The seminiferous tubules unite to form the rete testis that joins the epididymis.
   c. The seminiferous tubules are lined with epithelium, which produces sperm cells.
   d. The interstitial cells that produce male sex hormones occur between the seminiferous tubules.
3. Formation of sperm cells
   a. The epithelium lining the seminiferous tubules includes supporting cells and spermatogenic cells.
      (1) The supporting cells support and nourish the spermatogenic cells.
      (2) The spermatogenic cells give rise to spermatogonia.
   b. Sperm cells are produced from spermatogonia by the process of spermatogenesis.
      (1) The number of chromosomes in sperm cells is reduced by one-half (46 to 23) by meiosis.
      (2) Spermatogenesis produces four sperm cells from each primary spermatocyte.
   c. Membranous processes of adjacent supporting cells form a barrier within the epithelium.
      (1) The barrier separates early and advanced stages of spermatogenesis.
      (2) It helps provide a favorable environment for differentiating cells.

4. Structure of a sperm cell
   a. Sperm head contains a nucleus with 23 chromosomes.
   b. Sperm body contains many mitochondria.
   c. Sperm tail propels the cell.

Male Internal Accessory Organs (page 811)

1. Epididymis
   a. The epididymis is a tightly coiled tube on the outside of the testis that leads into the vas deferens.
   b. It stores and nourishes immature sperm cells and promotes their maturation.
2. Vas deferens
   a. The vas deferens is a muscular tube that forms part of the spermatic cord.
   b. It passes through the inguinal canal, enters the abdominal cavity, courses medially into the pelvic cavity, and ends behind the urinary bladder.
   c. It fuses with the duct from the seminal vesicle to form the ejaculatory duct.
3. Seminal vesicle
   a. The seminal vesicle is a saclike structure attached to the vas deferens.
   b. It secretes an alkaline fluid that contains nutrients and prostaglandins.
   c. This secretion is added to sperm cells during ejaculation.
4. Prostate gland
   a. This gland surrounds the urethra just below the urinary bladder.
   b. It secretes a thin, milky fluid, which neutralizes semen and vaginal secretions, and enhances the motility of sperm cells.
5. Bulbourethral glands
   a. These glands are two small structures beneath the prostate gland.
   b. They secrete a fluid that serves as a lubricant for the penis in preparation for sexual intercourse.
6. Semen
   a. Semen is composed of sperm cells and secretions of the seminal vesicles, prostate gland, and bulbourethral glands.
   b. This fluid is slightly alkaline and contains nutrients and prostaglandins.
   c. Semen activates sperm cells, but these sperm cells are unable to fertilize egg cells until they enter the female reproductive tract.

Male External Reproductive Organs (page 814)

1. Scrotum
   a. The scrotum is a pouch of skin and subcutaneous tissue that encloses the testes.
   b. The dartos muscle in the scrotal wall causes the skin of the scrotum to be held close to the testes or to hang loosely.

2. Penis
   a. The penis conveys urine and semen.
   b. It is specialized to become erect for insertion into the vagina during sexual intercourse.
   c. Its body is composed of three columns of erectile tissue surrounded by connective tissue.
   d. The root of the penis is attached to the pelvic arch and membranes of the perineum.
3. Erection, orgasm, and ejaculation
   a. During erection, the vascular spaces within the erectile tissue become engorged with blood as arteries dilate and veins are compressed.
   b. Orgasm is the culmination of sexual stimulation and is accompanied by emission and ejaculation.
   c. The movement of semen occurs as the result of sympathetic reflexes involving peristaltic contraction of smooth muscles in the walls of tubular organs.
   d. Following ejaculation, the penis becomes flaccid.

Hormonal Control of Male Reproductive Functions (page 815)

1. Hypothalamic and pituitary hormones
   The male body remains reproductively immature until the hypothalamus releases GnRH, which stimulates the anterior pituitary gland to release gonadotropins.
   a. FSH stimulates spermatogenesis.
   b. LH (ICSH) stimulates the interstitial cells to produce male sex hormones.
   c. Inhibin prevents oversecretion of FSH.
2. Male sex hormones
   a. Male sex hormones are called androgens.
   b. Testosterone is the most important androgen.
   c. Testosterone is converted into dihydrotestosterone in some organs.
   d. Androgens that fail to become fixed in tissues are metabolized in the liver and excreted.
   e. Androgen production increases rapidly at puberty.
3. Actions of testosterone
   a. Testosterone stimulates the development of the male reproductive organs and causes the testes to descend.
   b. It is responsible for the development and maintenance of male secondary sexual characteristics.
4. Regulation of male sex hormones
   a. Testosterone concentration is regulated by a negative feedback mechanism.
      (1) As the concentration rises, the hypothalamus is inhibited and the anterior pituitary secretion of gonadotropins is reduced.
      (2) As the concentration falls, the hypothalamus signals the anterior pituitary to secrete gonadotropins.
   b. The concentration of testosterone remains relatively stable from day to day.

Organs of the Female Reproductive System (page 818)

The primary female sex organs are the ovaries, which produce female sex cells and sex hormones. Accessory organs include internal and external organs.

Ovaries (page 818)

1. Attachments of the ovaries
   a. The ovaries are held in position by several ligaments.
   b. These ligaments include broad, suspensory, and ovarian ligaments.
2. Descent of the ovaries
   a. The ovaries descend from behind the parietal peritoneum near the developing kidneys.
   b. They are attached to the pelvic wall just below the pelvic brim.
3. Structure of the ovaries
   a. The ovaries are subdivided into a medulla and a cortex.
   b. The medulla is composed of connective tissue, blood vessels, lymphatic vessels, and nerves.
   c. The cortex contains ovarian follicles and is covered by cuboidal epithelium.
4. Primordial follicles
   a. During development, groups of cells in the ovarian cortex form millions of primordial follicles.
   b. Each primordial follicle contains a primary oocyte and a layer of flattened epithelial cells.
   c. The primary oocyte begins to undergo meiosis, but the process is soon halted and is not continued until puberty.
   d. The number of oocytes steadily declines throughout the life of a female.
5. Oogenesis
   a. Beginning at puberty, some oocytes are stimulated to continue meiosis.
   b. When a primary oocyte undergoes oogenesis, it gives rise to a secondary oocyte in which the original chromosome number is reduced by one-half (from 46 to 23).
   c. A secondary oocyte can be fertilized to produce a zygote.
6. Maturation of a follicle
   a. At puberty, FSH initiates the maturation of follicles.
   b. During maturation, the oocyte enlarges, the follicular cells proliferate, and a fluid-filled cavity appears and produces a secondary follicle.
   c. Ovarian cells surrounding the follicle form two layers.
   d. Usually only one follicle reaches full development.
7. Ovulation
   a. Ovulation is the release of an oocyte from an ovary.
   b. The oocyte is released when its follicle ruptures.
   c. After ovulation, the oocyte is drawn into the opening of the uterine tube.

Female Internal Accessory Organs (page 822)

1. Uterine tubes
   a. These tubes convey egg cells toward the uterus.
   b. The end of each uterine tube is expanded, and its margin bears irregular extensions.
   c. The movement of an egg cell into the opening is aided by ciliated cells that line the tube and by peristaltic contractions in the wall of the tube.
2. Uterus
   a. The uterus receives the embryo and sustains its life during development.
   b. The cervix of the uterus is partially enclosed by the vagina.
   c. The uterine wall includes the endometrium, myometrium, and perimetrium.
3. Vagina
   a. The vagina connects the uterus to the vestibule.
   b. It receives the erect penis, conveys uterine secretions to the outside, and transports the fetus during birth.
   c. The vaginal orifice is partially closed by a thin membrane, the hymen.
   d. Its wall consists of mucosa, muscularis, and outer fibrous coat.

Female External Reproductive Organs (page 826)

1. Labia majora
   a. The labia majora are rounded folds of adipose tissue and skin that enclose and protect the other external reproductive parts.
   b. The upper ends form a rounded, elevation over the symphysis pubis.
2. Labia minora
   a. The labia minora are flattened, longitudinal folds between the labia majora.
   b. They are well supplied with blood vessels.
3. Clitoris
   a. The clitoris is a small projection at the anterior end of the vulva; it corresponds to the male penis.
   b. It is composed of two columns of erectile tissue.
   c. Its root is attached to the sides of the pubic arch.
4. Vestibule
   a. The vestibule is the space between the labia majora that encloses the vaginal and urethral openings.
   b. The vestibular glands secrete mucus into the vestibule during sexual stimulation.
5. Erection, lubrication, and orgasm
   a. During periods of sexual stimulation, the erectile tissues of the clitoris and vestibular bulbs become engorged with blood and swollen.
   b. The vestibular glands secrete mucus into the vestibule and vagina.
   c. During orgasm, the muscles of the perineum, uterine wall, and uterine tubes contract rhythmically.

Hormonal Control of Female Reproductive Functions (page 828)

Hormones from the hypothalamus, anterior pituitary gland, and ovaries play important roles in the control of sex cell maturation, and the development and maintenance of female secondary sexual characteristics.

1. Female sex hormones
   a. A female body remains reproductively immature until about ten years of age when gonadotropin secretion increases.
   b. The most important female sex hormones are estrogen and progesterone.
      (1) Estrogen is responsible for the development and maintenance of most female secondary sexual characteristics.
      (2) Progesterone functions to cause changes in the uterus.
2. Female reproductive cycle
   a. The female reproductive cycle is called a menstrual cycle; the cycle is characterized by regularly recurring changes in the uterine lining culminating in menstrual flow.
   b. A menstrual cycle is initiated by FSH, which stimulates the maturation of a follicle.
   c. Granulosa cells of a maturing follicle secrete estrogen, which is responsible for maintaining the secondary sexual traits and causing the uterine lining to thicken.
   d. Ovulation is triggered when the anterior pituitary gland releases a relatively large amount of LH.
   e. Following ovulation, the follicular cells and thecal cells give rise to the corpus luteum.
      (1) The corpus luteum secretes progesterone, which causes the uterine lining to become more vascular and glandular.
      (2) If an oocyte is not fertilized, the corpus luteum begins to degenerate.
      (3) As the concentrations of estrogen and progesterone decline, the uterine lining disintegrates, causing menstrual flow.
   f. During this cycle, estrogen and progesterone inhibit the release of LH, and inhibin inhibits the release of FSH; as the concentrations of these hormones decline, the anterior pituitary secretes FSH and LH again, stimulating a new menstrual cycle.
3. Menopause
   a. Eventually the ovaries cease responding to FSH, and cycling ceases.
   b. Menopause is characterized by a low estrogen concentration and a continuous secretion of FSH and LH.
   c. The female reproductive organs undergo varying degrees of regressive changes.

Pregnancy (page 830)

1. Transport of sex cells
   a. Movement of the egg cell to the uterine tube is aided by ciliary action.
   b. To move, a sperm cell lashes its tail; its movement within the female body is aided by muscular contractions in the uterus and uterine tube.
2. Fertilization
   a. With the aid of an enzyme, a sperm cell penetrates the zona pellucida.
   b. When a sperm cell reaches an egg cell membrane, the entrance of any other sperm cells is prevented by changes in the egg cell membrane and the zona pellucida.
   c. When the nuclei of a sperm and an egg cell fuse, the process of fertilization is complete.
   d. The product of fertilization is a zygote with 46 chromosomes.
3. Early embryonic development
   a. Cells undergo mitosis, giving rise to smaller and smaller cells.
   b. The developing offspring moves down the uterine tube to the uterus, where it becomes implanted in the endometrium.
   c. The offspring is called an embryo from the second through the eighth week of development; thereafter it is called a fetus.
   d. Eventually the embryonic and maternal cells together form a placenta.
4. Hormonal changes during pregnancy
   a. Embryonic cells produce HCG that causes the corpus luteum to be maintained.
   b. Placental tissue produces high concentrations of estrogen and progesterone.
      (1) Estrogen and progesterone maintain the uterine wall and inhibit the secretion of FSH and LH.
      (2) Progesterone and relaxin inhibit contractions of uterine muscles.
      (3) Estrogen causes enlargement of the vagina.
      (4) Relaxin helps relax the ligaments of the pelvic joints.
   c. The placenta secretes placental lactogen that stimulates the development of the breasts and mammary glands.
   d. During pregnancy, increasing secretion of aldosterone promotes retention of sodium and body fluid, and increasing secretion of parathyroid hormone helps maintain a high concentration of maternal blood calcium.
5. Other changes during pregnancy
   a. The uterus enlarges greatly.
   b. The woman's blood volume, cardiac output, breathing rate, and urine production increase.
   c. The woman's dietary needs increase, but if intake is inadequate, fetal tissues have priority for use of available nutrients.
6. Birth process
   a. Pregnancy usually lasts forty weeks.
   b. During pregnancy, placental progesterone inhibits uterine contractions.
   c. A variety of factors are involved with the birth process.
      (1) A decreasing concentration of progesterone and the release of prostaglandins may initiate the birth process.
      (2) The posterior pituitary gland releases oxytocin.
      (3) Uterine muscles are stimulated to contract, and labor begins.
      (4) A positive feedback mechanism causes stronger contractions and greater release of oxytocin.
   d. Following the birth of the infant, placental tissues are expelled.

Mammary Glands (page 838)

1. Location of the glands
   a. The mammary glands are located in the subcutaneous tissue of the anterior thorax within the breasts.
   b. The breasts extend between the second and sixth ribs, and from sternum to axillae.
2. Structure of the glands
   a. The mammary glands are composed of lobes that contain tubular glands.
   b. The lobes are separated by dense connective and adipose tissues.
   c. The mammary glands are connected to the nipple by ducts.
3. Development of the breasts
   a. Male breasts remain nonfunctional.
   b. Estrogen stimulates female breast development.
      (1) Alveolar glands and ducts enlarge.
      (2) Fat is deposited around and within the breasts.
   c. During pregnancy, the breasts change.
      (1) Estrogen causes the ductile system to grow.
      (2) Progesterone causes development of alveolar glands.
      (3) Prolactin is released during pregnancy, but no milk is produced because progesterone inhibits milk production.
4. Production and secretion of milk
   a. Following childbirth, the concentrations of placental hormones decline.
      (1) The action of prolactin is no longer blocked.
      (2) The mammary glands begin to secrete milk.
   b. Reflex response to mechanical stimulation of the nipple causes the posterior pituitary to release oxytocin, which causes milk to be ejected from the alveolar ducts.
   c. As long as milk is removed from glands, more milk is produced; if milk is not removed, production ceases.
   d. During the period of milk production, the menstrual cycle is partially inhibited.

Birth Control (page 842)

Voluntary regulation of the number of children produced and the time they are conceived is called birth control. This usually involves some method of contraception.

1. Coitus interruptus
   a. Coitus interruptus is withdrawal of the penis from the vagina before ejaculation.
   b. Some semen may be expelled from the penis before ejaculation.
2. Rhythm method
   a. Abstinence from sexual intercourse a few days before and after ovulation is the rhythm method.
   b. It is almost impossible to predict the time of ovulation accurately.
3. Mechanical barriers
   a. The condom is used by males.
   b. The diaphragm is used by females.
4. Chemical barriers
   a. Spermicidal creams, foams, and jellies are chemical barriers to conception.
   b. These provide an unfavorable environment in the vagina for sperm survival.
5. Oral contraceptives
   a. Tablets that contain synthetic estrogen-like and progesterone-like substances are taken by the woman.
   b. They disrupt the normal pattern of gonadotropin secretion, and prevent ovulation and the normal build-up of the uterine lining.
   c. When used correctly, this method is almost 100% effective.
   d. Some women develop undesirable side effects.
6. Contraceptive implant
   a. A contraceptive implant consists of a set of progesterone-containing capsules or rods that are inserted under the skin.
   b. Progesterone released from the implant prevents ovulation.
   c. The implant is effective for years and its action can be reversed by having it removed.
7. Intrauterine devices
   a. An IUD is a solid object inserted in the uterine cavity.
   b. It is thought to prevent pregnancy by interfering with implantation.
   c. It may be expelled spontaneously or produce undesirable side effects.
8. Surgical methods
   a. These involve sterilization procedures.
      (1) Vasectomy is performed in males.
      (2) Tubal ligation is performed in females.
   b. Surgical methods are the most reliable forms of contraception.

Clinical Application of Knowledge

1. What changes, if any, might occur in the secondary sexual characteristics of an adult male following removal of one testis? Following removal of both testes? Following removal of the prostate gland?
2. How would you explain the fact that new mothers sometimes experience cramps in their lower abdomens when they begin to nurse their babies?
3. If a woman who is considering having a tubal ligation asks, "Will the operation cause me to go through my change of life early?" how would you answer?
4. What effect would it have on a woman's menstrual cycles if a single ovary were removed surgically? What effect would it have if both ovaries were removed?
5. Which methods of contraception are theoretically the most effective in preventing unwanted pregnancies? Which methods are theoretically the least effective?
6. As a male reaches adulthood, what will be the consequences if his testes have remained undescended since birth? Why?

Review Activities

1. List the general functions of the male reproductive system.
2. Distinguish between the primary and accessory male reproductive organs.
3. Describe the descent of the testes.
4. Define *cryptorchidism*.
5. Describe the structure of a testis.
6. Explain the function of the supporting cells in the testis.
7. List the major steps in spermatogenesis.
8. Describe a sperm cell.
9. Describe the epididymis, and explain its function.
10. Trace the path of the vas deferens from the epididymis to the ejaculatory duct.
11. On a diagram, locate the seminal vesicles, and describe the composition of their secretion.
12. On a diagram, locate the prostate gland, and describe the composition of its secretion.
13. On a diagram, locate the bulbourethral glands, and explain the function of their secretion.
14. Describe the composition of semen.
15. Define *capacitation*.
16. Describe the structure of the scrotum.
17. Describe the structure of the penis.
18. Explain the mechanism that produces an erection of the penis.
19. Distinguish between emission and ejaculation.

20. Explain the mechanism of ejaculation.
21. Explain the role of GnRH in the control of male reproductive functions.
22. Distinguish between androgen and testosterone.
23. Define *puberty*.
24. Describe the actions of testosterone.
25. List several male secondary sexual characteristics.
26. Explain how the concentration of testosterone is regulated.
27. List the general functions of the female reproductive system.
28. Distinguish between the primary and accessory female reproductive organs.
29. Describe how the ovaries are held in position.
30. Describe the descent of the ovaries.
31. Describe the structure of an ovary.
32. Define *primordial follicle*.
33. List the major steps in oogenesis.
34. Distinguish between a primary and a secondary follicle.
35. Describe a mature follicle.
36. Define *ovulation*.
37. On a diagram, locate the uterine tubes, and explain their function.
38. Describe the structure of the uterus.
39. Describe the structure of the vagina.
40. Distinguish between the labia majora and the labia minora.
41. On a diagram, locate the clitoris, and describe its structure.
42. Define *vestibule*.
43. Describe the process of erection in the female reproductive organs.
44. Define *orgasm*.
45. Explain the role of GnRH in regulating female reproductive functions.
46. List several female secondary sexual characteristics.
47. Define *menstrual cycle*.
48. Explain how a menstrual cycle is initiated.
49. Summarize the major events in a menstrual cycle.
50. Define *menopause*.
51. Describe how male and female sex cells are transported within the female reproductive tract.
52. Describe the process of fertilization.
53. List the major functions of the placenta.
54. Explain the major hormonal changes that occur in the maternal body during pregnancy.
55. Describe the major nonhormonal changes that occur in the maternal body during pregnancy.
56. Describe the role of progesterone in initiating the birth process.
57. Discuss the events that occur during the birth process.
58. Describe the structure of a mammary gland.
59. Explain the roles of prolactin and oxytocin in milk production and secretion.
60. Define *contraception*.
61. List several methods of contraception, and explain how each interferes with normal reproductive functions.

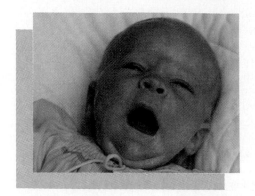

23

Human Growth
and Development

*T*he products of the female and male reproductive systems are egg cells and sperm cells, respectively. When an egg cell and a sperm cell unite, a zygote is formed by the process of fertilization. Such a single-celled zygote is the first cell of an offspring, and it is capable of giving rise to an adult of the subsequent generation. The processes by which this is accomplished are called *growth* and *development* ■

Chapter Objectives

After you have studied this chapter, you should be able to:

1. Distinguish between growth and development.

2. Describe the major events that occur during the period of cleavage.

3. Explain how the primary germ layers originate and list the structures produced by each layer.

4. Describe the formation and function of the placenta.

5. Define *fetus* and describe the major events that occur during the fetal stage of development.

6. Trace the general path of blood through the fetal circulatory system.

7. Describe the major circulatory and physiological adjustments that occur in the newborn.

8. Name the stages of development that occur between the neonatal period and death, and list the general characteristics of each stage.

9. Complete the review activities at the end of this chapter. Note that the items are worded in the form of specific learning objectives. You may want to refer to them before reading the chapter.

Key Terms

amnion (am′ne-on)

chorion (ko′re-on)

cleavage (klēv′ij)

embryo (em′bre-o)

fetus (fe′tus)

germ layer (jerm la′er)

neonatal (ne″o-na′tal)

placenta (plah-sen′tah)

postnatal (pōst-na′tal)

prenatal (pre-na′tal)

senescence (sĕ-nes′ens)

umbilical cord (um-bil′ĭ-kal kord)

zygote (zi′got)

Aids to Understanding Words

allant-, sausage-shaped: *allant*ois—tubelike structure that extends from the yolk sac into the connecting stalk of an embryo.

chorio-, skin: *chorio*n—outermost membrane that surrounds the fetus and other fetal membranes.

cleav-, to divide: *cleav*age—period of development characterized by a division of the zygote into smaller and smaller cells.

lacun-, a pool: *lacun*a—space between the chorionic villi that fills with maternal blood.

morul-, mulberry: *morul*a—embryonic structure consisting of a solid ball of about sixteen cells, thus looking somewhat like a mulberry.

nat-, to be born: pre*nat*al—the period of development before birth.

sen-, old: *sen*escence—process of growing old.

troph-, nourishment: *troph*oblast—cellular layer that surrounds the inner cell mass and helps nourish it.

umbil-, the navel: *umbil*ical cord—structure attached to the fetal navel (umbilicus) that connects the fetus to the placenta.

G rowth refers to an increase in size. In a human, growth usually reflects an increase in cell numbers as a result of *mitosis,* followed by enlargement of the newly formed cells and enlargement of the body.

Development, on the other hand, is the continuous process by which an individual changes from one life phase to another (figure 23.1). These life phases include a **prenatal period,** which begins with the fertilization of an egg cell and ends at birth, and a **postnatal period,** which begins at birth and ends with death.

It is usually not possible to determine the actual time of fertilization because reliable records concerning sexual activities are seldom available. However, the approximate time can be calculated by adding fourteen days to the date of the onset of the last menstruation. This time can then be used to calculate the fertilization age of an embryo. The expected time of birth can be estimated by adding 266 days to the fertilization date. Most fetuses are born within ten to fifteen days of this calculated time.

Prenatal Period

The prenatal period of development usually lasts for forty weeks (ten lunar months) and can be divided into a period of cleavage, an embryonic stage, and a fetal stage.

Period of Cleavage

As described in chapter 22, fertilization normally occurs within a uterine tube. About thirty hours after a *zygote* is formed, it undergoes mitosis, giving rise to two new cells. These cells, in turn, divide to form four cells, which then divide into eight cells, and so forth. With each subsequent division, the resulting cells are smaller and smaller. This distribution of the zygote's contents into smaller and smaller cells is called *cleavage,* and the cells produced in this way are called *blastomeres* (figures 23.2 and 23.3).

The mass of cells that is formed by cleavage is still enclosed in the zona pellucida of the original egg cell, and in about three days it consists of a solid ball of about sixteen cells called a *morula* (figures 23.4 and 23.5).

The morula enters the uterine cavity and remains unattached for about three days. During this time, the zona pellucida degenerates, and the morula develops a fluid-filled central cavity. Once this cavity appears, the morula becomes a hollow ball of cells called a **blastocyst** (figure 23.4).

Within the blastocyst, cells in one region group together to form an *inner cell mass* that eventually gives rise to the **embryo proper**—the body of the developing offspring. The cells forming the wall of the blastocyst make up the *trophoblast,* which forms structures that assist the embryo in its development.

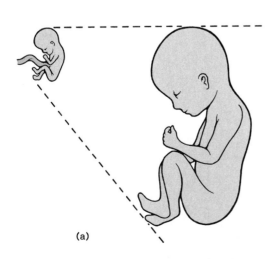

(a)

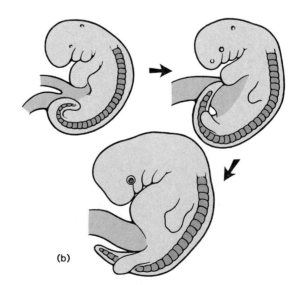

(b)

Figure 23.1 (*a*) Growth involves an increase in size. (*b*) Development refers to the process of changing from one phase of life to another.

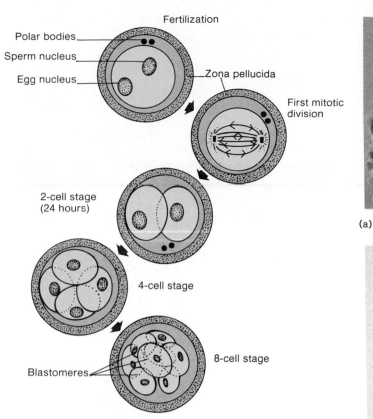

Fertilization

Polar bodies

Sperm nucleus

Egg nucleus

Zona pellucida

First mitotic division

2-cell stage (24 hours)

4-cell stage

8-cell stage

Blastomeres

Figure 23.2 During the period of cleavage, the cells divide by mitosis and become smaller and smaller.

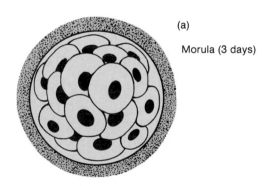

(a)

Morula (3 days)

(b)

Section of blastocyst (6 days)

Inner cell mass

Fluid-filled cavity

Trophoblast

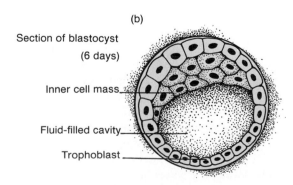

Figure 23.4 (a) A morula consists of a solid ball of cells. (b) A blastocyst has a fluid-filled cavity.

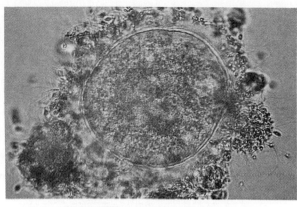

(a)

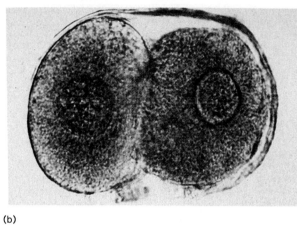

(b)

Figure 23.3 (a) A light micrograph of a human egg cell surrounded by follicular cells and sperm cells; (b) two-cell stage of development (×1,000).

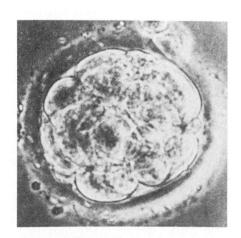

Figure 23.5 Light micrograph of a human morula.

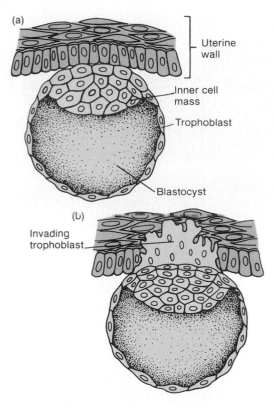

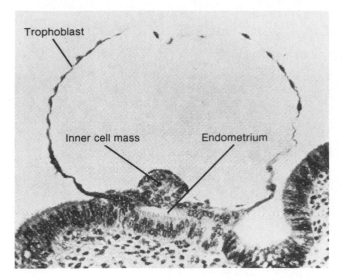

Figure 23.7 Light micrograph of a blastocyst from a monkey in contact with the endothelium of the uterine wall (×300).

Figure 23.6 (a) About the sixth day of development, the blastocyst contacts the uterine wall and (b) begins to become implanted within the wall.

Sometimes two ovarian follicles release egg cells simultaneously, and if both are fertilized, the resulting zygotes can develop into fraternal (dizygotic) twins. Such twins are no more alike genetically than any brothers or sisters from the same parents. In other instances, twins develop from a single fertilized egg (monozygotic twins). This may happen if two inner cell masses form within a blastocyst and each produces an embryo. Twins of this type usually share a single placenta, and they are identical genetically. Thus, they are always the same sex and are very similar in appearance.

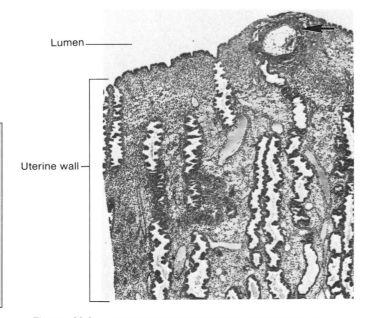

Figure 23.8 Light micrograph of a human embryo (arrow) undergoing implantation in the uterine wall.

About the sixth day, the blastocyst begins to attach itself to the uterine lining (figures 23.6 and 23.7). This attachment is apparently aided by its secretion of proteolytic enzymes that digest a portion of the endometrium. The blastocyst sinks slowly into the resulting depression, becoming completely buried in the uterine lining. At the same time, the uterine lining is stimulated to thicken below the implanting blastocyst, and cells of the trophoblast begin to produce tiny, fingerlike processes (microvilli) that grow into the endometrium. This process of **implantation** begins near the end of the first week and is completed during the second week of development (figures 23.8 and 23.9).

As discussed in chapter 22, the trophoblast of the blastocyst secretes the hormone HCG, which has properties similar to those of LH and causes the corpus luteum to be maintained during the early stages of pregnancy. HCG also is thought to help protect the blastocyst against being rejected as a foreign (nonself) substance by the maternal immunological system (see chapter 19) and to stimulate the synthesis of hormones from the developing placenta.

In Vitro Fertilization

A woman who is infertile because her uterine tubes are blocked may become pregnant by means of *in vitro fertilization*. In one such technique, oocytes are removed from the woman's ovary and mixed with sperm in a laboratory dish to achieve fertilization (*in vitro* means "in glass" and refers to the artificial environment of the fertilization dish). Later, the resulting embryos are transferred into the woman's uterus for development.

The in vitro fertilization procedure usually begins at menses with the administration of HMG (human menopausal gonadotropin) or some other substance that will induce the development of ovarian follicles. The growth of the follicles can be monitored using ultrasonography, and when they have reached a certain diameter, the patient is given HCG (human chorionic gonadotropin) to induce ovulation.

Oocytes released from the ovary are harvested with the aid of a *laparoscope*—an optical instrument used to examine the abdominal interior. The oocytes are incubated at 37° C in a buffered medium with a pH of 7.4, and when they are mature, they are mixed in a laboratory dish with sperm that have been washed to remove various inhibitory factors.

Fertilized eggs are incubated in a special medium. After fifty to sixty hours, when the developing embryos have reached the 8-cell or 16-cell stage, normal embryos are transferred through the woman's cervix and into her uterus with the aid of a specially designed catheter. She is then treated with progesterone to promote a favorable uterine environment for implantation of the embryos. As a result of this procedure, successful implantation occurs in about 20%–30% of the cases.

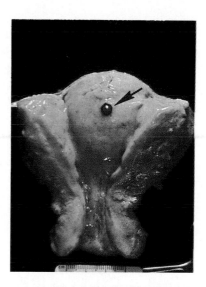

Figure 23.9 An embryo appears as a red blister bulging out from the wall of this dissected uterus.

Usually the blastocyst becomes implanted in the upper posterior wall of the uterus. However, sometimes implantation occurs in the lower portion near the cervix, and as the placenta develops, it may partially or totally cover the opening of the cervix. This condition is called *placenta previa*, and it is likely to produce complications, since any expansion of the cervix may damage placental tissues and cause bleeding.

1. Distinguish between growth and development.
2. What changes characterize the period of cleavage?
3. How does a blastocyst become attached to the endometrium of the uterus?
4. In what ways does the endometrium respond to the activities of the blastocyst?

Embryonic Stage

The **embryonic stage** extends from the second week through the eighth week of development. It is characterized by the formation of the placenta, the development of the main internal organs, and the appearance of the major external body structures.

Early in this stage, the cells of the inner cell mass become organized into a flattened **embryonic disk,** which consists of two distinct layers: an outer *ectoderm* and an inner *endoderm*. A short time later, a third layer of cells, the *mesoderm,* forms between the ectoderm and endoderm. These three layers of cells are called the **primary germ layers,** and they are responsible for forming all the body organs. (See figure 23.10.)

More specifically, *ectodermal cells* give rise to the nervous system, portions of special sensory organs, the epidermis, hair, nails, glands of the skin, and linings of the mouth and anal canal. *Mesodermal cells* form all types of muscle tissue, bone tissue, bone marrow, blood, blood vessels, lymphatic vessels, various connective tissues, internal reproductive organs, kidneys, and the epithelial linings of the body cavities. *Endodermal cells* produce the epithelial linings of the digestive tract, respiratory tract, urinary bladder, and urethra (figure 23.11).

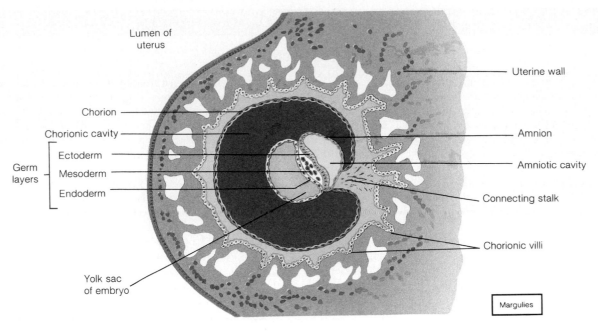

Figure 23.10 Early in the embryonic stage of development, the three primary germ layers are formed.

During the fourth week of development, the flat embryonic disk is transformed into a cylindrical structure, which is attached to the developing placenta by a *connecting stalk* (body stalk) (figures 23.12 and 23.13). By this time, the head and jaws are appearing, the heart is beating and forcing blood through blood vessels, and tiny buds, which will give rise to the arms and legs, are forming (figure 23.14).

During the fifth through the seventh weeks, as shown in figure 23.15, the head grows rapidly and becomes rounded and erect. The face, which is developing the eyes, nose, and mouth, becomes more humanlike. The arms and legs elongate, and fingers and toes appear (figures 23.16 and 23.17).

By the end of the seventh week, all the main internal organs have become established, and as these structures enlarge, they affect the shape of the body. Consequently, the body takes on a humanlike appearance.

Meanwhile, the embryo continues to become implanted within the uterus. As mentioned, early in this process, slender projections grow out from the trophoblast into the surrounding endometrium. These extensions, which are called **chorionic villi,** become branched, and by the end of the fourth week, they are well formed.

While the chorionic villi are developing, the embryonic blood vessels appear within them, and these vessels are continuous with those passing through the connecting stalk to the body of the embryo. At the same time, irregular spaces called **lacunae** are eroded around and between the villi. These spaces become filled with the maternal blood that escapes from eroded endometrial blood vessels.

A thin membrane separates embryonic blood within the capillary of a chorionic villus from maternal blood in a lacuna. This membrane, called the **placental membrane,** is composed of the epithelium of the villus and the epithelium of the capillary (figure 23.18). Through this membrane, exchanges take place between the maternal blood and the embryonic blood. Oxygen and nutrients diffuse from the maternal blood into the embryonic blood, and carbon dioxide and other wastes diffuse from the embryonic blood into the maternal blood. Various substances also move through the placental membrane by active transport and pinocytosis.

Because most drugs are able to pass freely through the placental membrane, substances ingested by the mother may affect the fetus. Thus, fetal drug addiction may occur following the mother's use of various addicting drugs, such as heroin.

Similarly, depressant drugs administered to the mother during labor can produce effects within the fetus and may, for example, depress the activity of its respiratory system.

1. What major events occur during the embryonic stage of development?
2. What tissues and structures develop from ectoderm? From mesoderm? From endoderm?
3. Describe the structure of a chorionic villus.
4. How are substances exchanged between the embryonic blood and the maternal blood?

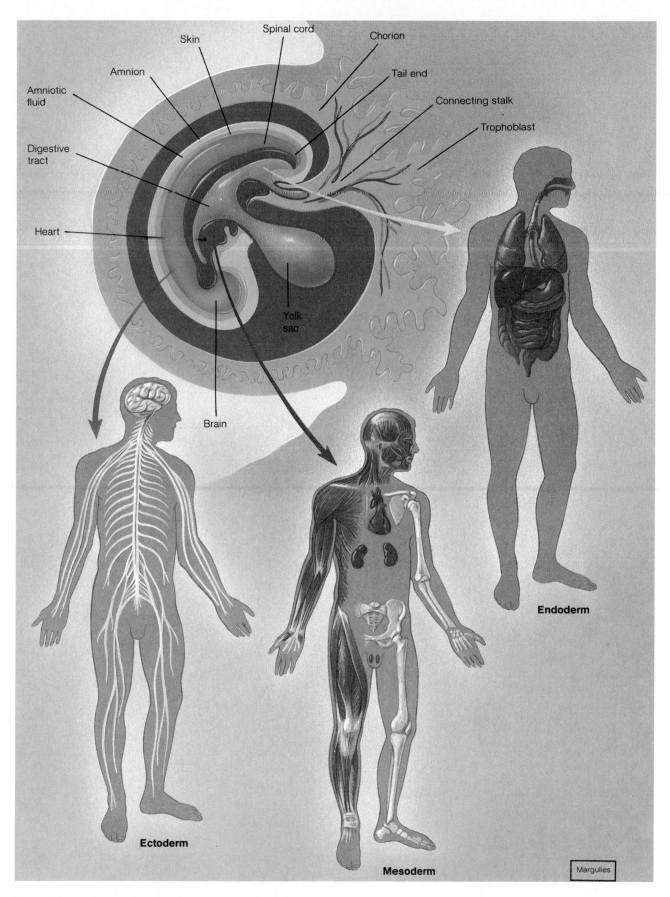

Amniotic fluid

Digestive tract

Heart

Amnion

Skin

Spinal cord

Chorion

Tail end

Connecting stalk

Trophoblast

Yolk sac

Brain

Ectoderm

Mesoderm

Endoderm

Margulies

Figure 23.11 Each of the primary germ layers is responsible for the formation of a particular set of organs.

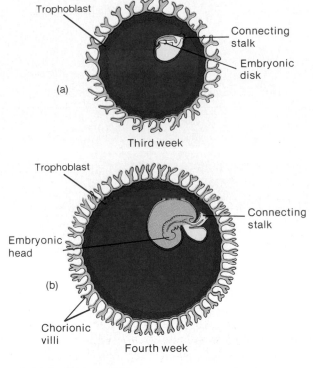

Figure 23.12 During the fourth week, (a) the flat embryonic disk becomes (b) a cylindrical structure.

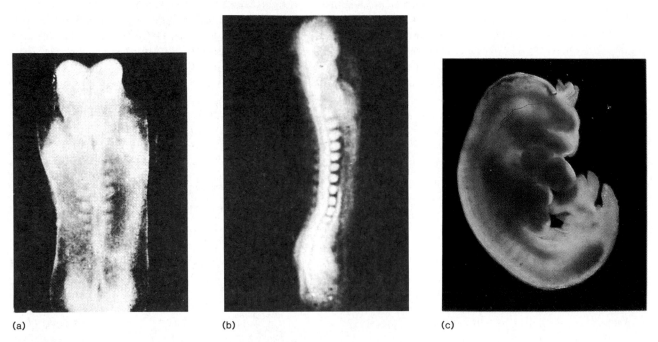

(a)   (b)   (c)

Figure 23.13 (a) A human embryo at three weeks; (b) at 3.5 weeks; (c) at about four weeks.

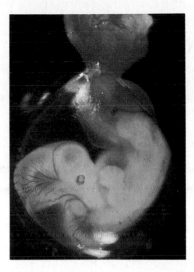

Figure 23.14 Human embryo after about five weeks of development.

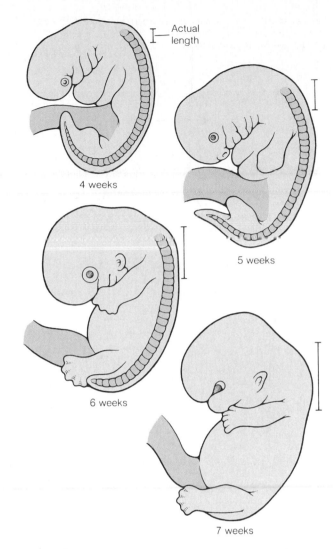

Figure 23.15 In the fifth through the seventh weeks of development, the embryonic body and face develop a humanlike appearance.

Until about the end of the eighth week, the chorionic villi cover the entire surface of the former trophoblast, which is now called the **chorion.** However, as the embryo and the chorion surrounding it continue to enlarge, only those villi that remain in contact with the endometrium endure. The others degenerate, and the portions of the chorion to which they were attached become smooth. Thus, the region of the chorion still in contact with the uterine wall is restricted to a disk-shaped area that becomes the **placenta.**

The embryonic portion of the placenta is composed of the chorion and its villi; the maternal portion is composed of the area of the uterine wall (decidua basalis) to which the villi are attached. When it is fully formed, the placenta appears as a reddish brown disk, about 20 centimeters long and 2.5 centimeters thick. It usually weighs about 0.5 kilogram. Figure 23.19 shows the structure of the placenta.

While the placenta is forming from the chorion, another membrane, called the **amnion,** develops around the embryo. This second membrane begins to appear during the second week. Its margin is attached around the edge of the embryonic disk, and fluid called **amniotic fluid** fills the space between the amnion and the embryonic disk.

As the embryo is transformed into a cylindrical structure, the margins of the amnion are folded around it so that the embryo becomes enclosed by the amnion and surrounded by amniotic fluid. As this process continues, the amnion envelops the tissues on the underside

of the embryo, by which it is attached to the chorion and the developing placenta. In this manner, as figure 23.20 illustrates, the **umbilical cord** is formed.

As the placenta develops, it synthesizes a relatively large quantity of progesterone from cholesterol obtained from the maternal blood. This placental progesterone is then used by cells associated with the developing fetal adrenal glands to synthesize estrogens. The estrogens, in turn, promote changes in the maternal uterus and breasts, and influence the metabolism and development of various fetal organs.

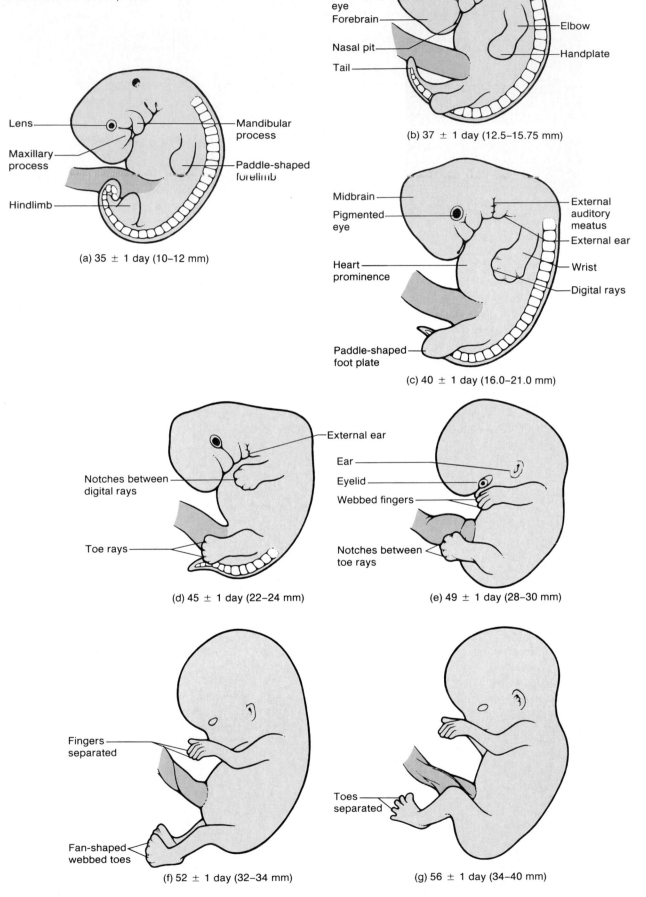

Figure 23.16 (a–c) Changes occurring during the fifth week of development; (d–g) changes occurring during the sixth and seventh weeks of development.

Lens

Maxillary process

Mandibular process

Hindlimb

Paddle-shaped forelimb

(a) 35 ± 1 day (10–12 mm)

Developing eye

Forebrain

Nasal pit

Tail

Developing ear

Elbow

Handplate

(b) 37 ± 1 day (12.5–15.75 mm)

Midbrain

Pigmented eye

Heart prominence

Paddle-shaped foot plate

External auditory meatus

External ear

Wrist

Digital rays

(c) 40 ± 1 day (16.0–21.0 mm)

External ear

Notches between digital rays

Toe rays

(d) 45 ± 1 day (22–24 mm)

Ear

Eyelid

Webbed fingers

Notches between toe rays

(e) 49 ± 1 day (28–30 mm)

Fingers separated

Fan-shaped webbed toes

(f) 52 ± 1 day (32–34 mm)

Toes separated

(g) 56 ± 1 day (34–40 mm)

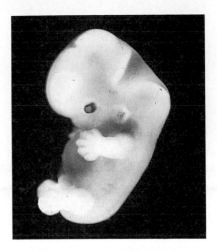

Figure 23.17 Six-week-old human embryo.

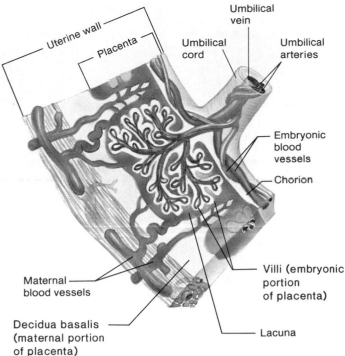

Figure 23.19 The placenta consists of an embryonic portion and a maternal portion.

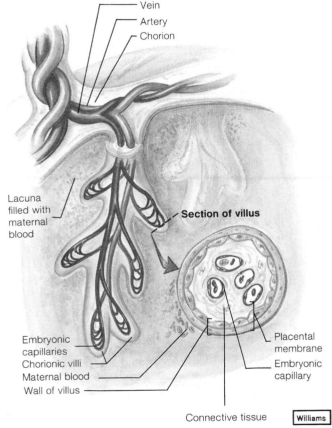

Figure 23.18 As is illustrated in the section of villus (lower part of figure), the placental membrane consists of the epithelial wall of an embryonic capillary and the epithelial wall of a chorionic villus. What is the significance of this membrane?

When the umbilical cord is fully developed, it is about 1 centimeter in diameter and about 55 centimeters in length. It begins at the umbilicus of the embryo and is inserted into the central region of the placenta. The cord contains three blood vessels—two *umbilical arteries* and one *umbilical vein*—through which blood passes between the embryo and the placenta (figure 23.19).

The umbilical cord also suspends the embryo in the *amniotic cavity,* and the amniotic fluid provides a watery environment in which the embryo can grow freely without being compressed by surrounding tissues. The amniotic fluid also protects the embryo from being jarred by the movements of the mother's body.

Eventually, the amniotic cavity becomes so enlarged that the membrane of the amnion contacts the thicker chorion around it, and the two membranes become fused into an *amniochorionic membrane* (figures 23.21 and 23.22).

In addition to the amnion and chorion, two other embryonic membranes appear during development. They are the yolk sac and the allantois.

The **yolk sac** appears during the second week, and it is attached to the underside of the embryonic disk (figures 23.20 and 23.21). It forms blood cells in the early stages of development and gives rise to the cells that later become sex cells. Portions of the yolk sac also enter into the formation of the embryonic digestive tube. Part of this membrane becomes incorporated into the umbilical cord, while the remainder lies in the cavity between the chorion and the amnion near the placenta.

The **allantois** forms during the third week as a tube extending from the early yolk sac into the connecting stalk of the embryo. It, too, forms blood cells and gives rise to the umbilical arteries and vein (figures 23.20 and 23.21).

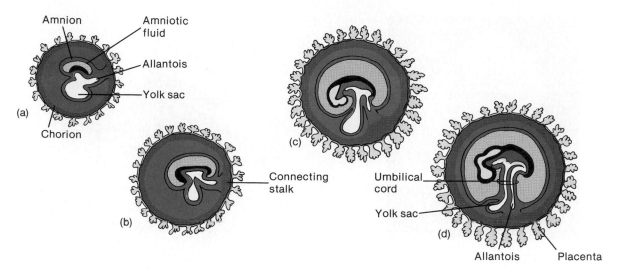

Figure 23.20 (a–c) As the amnion develops, it surrounds the embryo, and (d) the umbilical cord is formed. What structures comprise this cord?

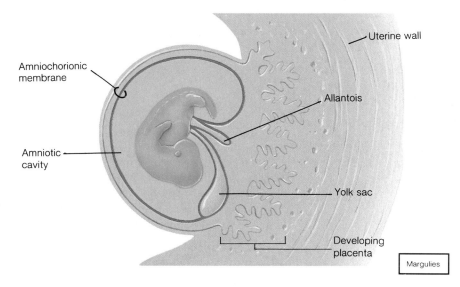

Figure 23.21 As the amniotic cavity enlarges, the amnion contacts the chorion, and the two membranes fuse to form the amniochorionic membrane.

The *embryonic stage* is completed at the end of the eighth week. It is the most critical period of development, for during this time the embryo becomes implanted within the uterine wall and all the essential external and internal body parts are formed. Any disturbances in the developmental processes occurring during the embryonic stage are likely to result in major malformations or malfunctions.

By the beginning of the eighth week, the embryo is usually 30 millimeters long and weighs less than 5 grams. Although its body is quite unfinished, it is recognizable as a human being (figure 23.23).

The eighth week of development occurs two weeks after a pregnant woman has missed her second menstrual period. Since these first few weeks are critical periods in development, it is important that a woman seek health care as soon as she thinks she may be pregnant.

1. Describe the development of the amnion.
2. What blood vessels are found in the umbilical cord?
3. What is the function of amniotic fluid?
4. What is the significance of the yolk sac?

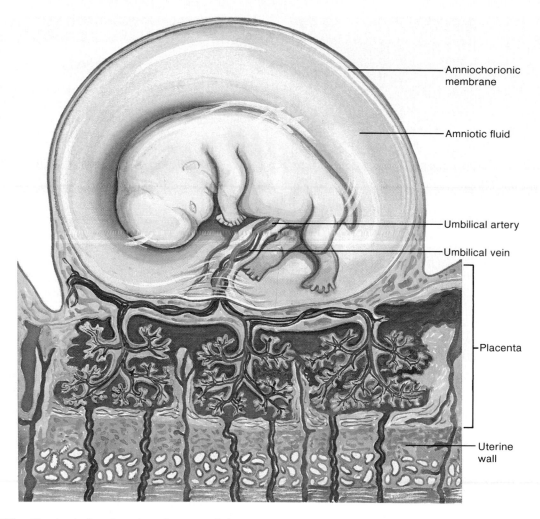

Figure 23.22 The developing placenta as it appears during the seventh week of development.

Labels on figure:
- Amniochorionic membrane
- Amniotic fluid
- Umbilical artery
- Umbilical vein
- Placenta
- Uterine wall

Fetal Stage

The **fetal stage** begins at the end of the eighth week of development and lasts until birth. During this period, the offspring is called a **fetus,** and although the existing body structures continue to grow and mature, only a few new parts appear. The rate of growth is great, however, and the body proportions change considerably. For example, at the beginning of the fetal stage, the head is disproportionately large, and the legs are relatively short (figure 23.24).

During the third lunar month, the growth in body length is accelerated, but the growth of the head slows. (Note: A lunar month equals twenty-eight days.) The arms achieve the relative length they will maintain throughout development, and ossification centers appear in most of the bones. By the twelfth week, the external reproductive organs are distinguishable as male or female. Figure 23.25 illustrates how these external reproductive organs of the male and female differentiate from similar fetal structures.

In the fourth lunar month, the body grows very rapidly and reaches a length of 13 to 17 centimeters. The legs lengthen considerably, and the skeleton continues to ossify.

In the fifth lunar month, the rate of growth decreases somewhat. The legs achieve their final relative proportions, and the skeletal muscles become active so that the mother may feel fetal movements (quickening). Some hair appears on the head, and the skin becomes covered with fine, downy hair (lanugo). The skin is also coated with a cheesy mixture of sebum from the sebaceous glands and dead epidermal cells (vernix caseosa).

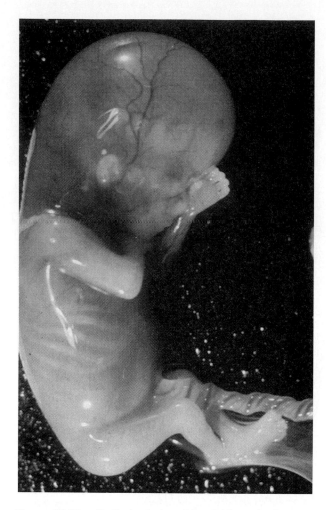

Figure 23.23 By the beginning of the eighth week of development, the embryonic body is recognizable as a human.

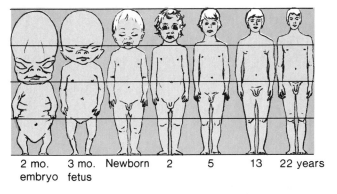

2 mo. 3 mo. Newborn 2 5 13 22 years
embryo fetus

Figure 23.24 During development, the body proportions change considerably. (From L. Patten, *Human Embryology.* Copyright © 1933 McGraw-Hill Publishing Company. Used by permission.)

Teratogens

Factors that can cause congenital malformations by affecting an embryo during its period of rapid growth and development are called *teratogens*. Such agents include a variety of drugs, certain viruses and microorganisms, and high-level ionizing radiation.

The molecules of many drugs are able to pass freely from maternal blood to fetal blood through the placental membrane. Consequently, substances ingested by the mother-to-be may affect her embryo. For example, when a pregnant woman drinks alcohol, her embryo is exposed to an alcohol concentration equal to that of her blood, and the alcohol can produce a wide range of effects on the developing offspring.

More specifically, alcohol is able to produce a set of physical and mental abnormalities called the *fetal alcohol syndrome* (FAS). The symptoms of this syndrome include prenatal and postnatal growth retardation, abnormal facial features, reduced head size, organ malformations, and mental disorders. It is thought to result when a pregnant woman drinks as little as 3 ounces of alcohol per day or engages in a single drinking binge. Consequently, the Surgeon General of the United States has recommended that women who are pregnant, and those attempting to become pregnant, should abstain from drinking alcohol.

The rubella virus that causes German measles also can cause congenital malformations. If a pregnant woman develops a rubella infection during the first four or five weeks of embryonic development, the time when the embryonic eyes, ears, and heart are forming, these organs may be malformed. As a result, the offspring may be blind (due to cataracts), be deaf, and have various heart disorders. Exposure to the rubella virus during later stages of development may result in functional defects in the central nervous system.

If there is any chance that a woman may be pregnant, in addition to avoiding exposure to ionizing radiation and infectious microorganisms, it is prudent for her to avoid taking any drugs—even seemingly harmless ones—without the advice of her physician.

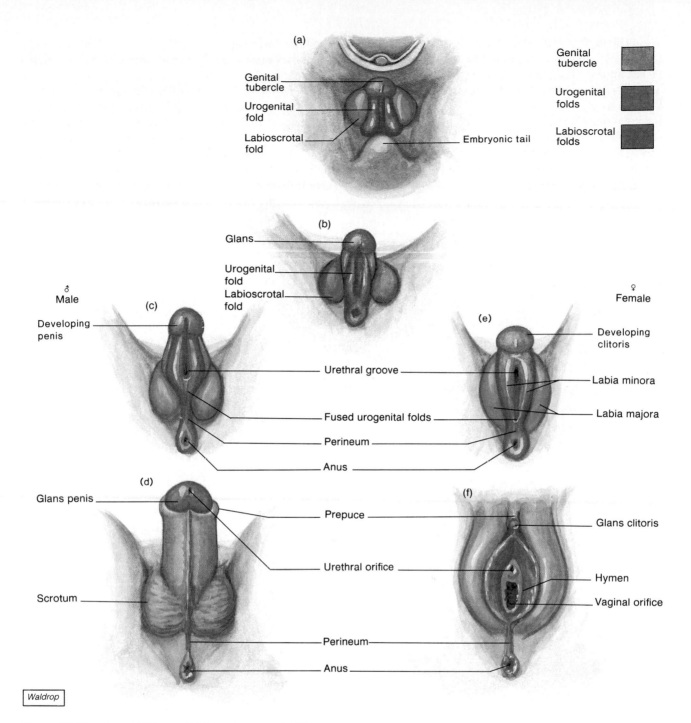

Genital tubercle

Urogenital fold

Labioscrotal fold

Embryonic tail

Genital tubercle

Urogenital folds

Labioscrotal folds

(b)

Glans

Urogenital fold

Labioscrotal fold

♂
Male

(c)

Developing penis

(e)

Developing clitoris

Urethral groove

Labia minora

♀
Female

Fused urogenital folds

Labia majora

Perineum

Anus

(d)

Glans penis

Prepuce

(f)

Glans clitoris

Scrotum

Urethral orifice

Hymen

Vaginal orifice

Perineum

Anus

Waldrop

Figure 23.25 (*a* and *b*) The genital tubercle, urogenital fold, and labioscrotal folds that appear during the fourth week of development may differentiate into (*c* and *d*) male external reproductive organs or (*e* and *f*) female external reproductive organs.

During the sixth lunar month, the body gains a substantial amount of weight. The eyebrows and eyelashes appear. The skin is quite wrinkled and translucent. It is also reddish, due to the presence of dermal blood vessels.

In the seventh lunar month, the skin becomes smoother as fat is deposited in the subcutaneous tissues. The eyelids, which fused together during the third month, reopen. At the end of this month, a fetus is about 37 centimeters in length.

> If a fetus is born prematurely, its chance of surviving increases directly with its age and weight. One factor that affects the chance of survival is the development of the lungs. Thus, fetuses have increased chances of surviving if their lungs are sufficiently developed so that they have the thin respiratory membranes necessary for rapid exchange of oxygen and carbon dioxide, and if the lungs produce enough surfactant to reduce the alveolar surface tension. (See chapter 16.) A fetus of less than 24 weeks or weighing less than 600 grams at birth seldom survives, even when given intensive medical care.

In the eighth lunar month, the fetal skin is still reddish and somewhat wrinkled. The testes of males descend from regions near the developing kidneys, through the inguinal canal, and into the scrotum. (See chapter 22.)

During the ninth lunar month, the fetus reaches a length of about 47 centimeters. The skin is smooth, and the body appears chubby due to an accumulation of subcutaneous fat. The reddishness of the skin fades to pinkish or bluish pink.

At the end of the tenth lunar month, the fetus is said to be *full term*. It is about 50 centimeters long and weighs 2.7 to 3.6 kilograms. The skin has lost its downy hair, but is still coated with sebum and dead epidermal cells. The scalp is usually covered with hair; the fingers and toes have well-developed nails; and the skull bones are largely ossified. As figure 23.26 shows, the fetus is usually positioned upside down with its head toward the cervix (*vertex position*).

Chart 23.1 summarizes the stages of prenatal development.

> The size and shape of the uterus may be used as an index of fetal growth. Thus, failure of the uterus to increase in size progressively during pregnancy suggests retarded fetal development, while a sudden increase in growth may indicate the presence of more than one fetus, the formation of excessive amniotic fluid (hydramnios), or an abnormal growth of chorionic tissue (hydatid mole).

1. What major changes characterize the fetal stage of development?
2. When can the sex of a fetus be determined?
3. How is a fetus usually positioned within the uterus at the end of the tenth lunar month?

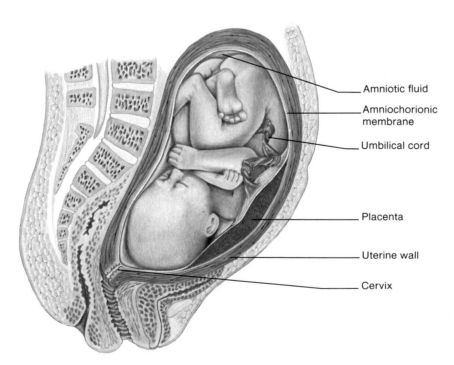

Amniotic fluid

Amniochorionic membrane

Umbilical cord

Placenta

Uterine wall

Cervix

Figure 23.26 A full-term fetus usually becomes positioned with its head near the cervix.

CHART 23.1 Stages of prenatal development

Stage	Time period	Major events
Period of cleavage	First week	Cells undergo mitosis, blastocyst forms; inner cell mass appears; blastocyst becomes implanted in uterine wall
Embryonic stage	Second through eighth week	Inner cell mass becomes embryonic disk; primary germ layers form; embryo proper becomes cylindrical; main internal organs and external body structures appear; placenta and umbilical cord form; embryo proper is suspended in amniotic fluid
Fetal stage	Ninth week to birth	Existing structures continue to grow; ossification centers appear in bones; reproductive organs develop; arms and legs achieve final relative proportions; muscles become active; skin is covered with sebum and dead epidermal cells; head becomes positioned toward cervix

Fetal Blood and Circulation

Throughout the fetal stage of development, the maternal blood supplies the fetus with oxygen and nutrients, and carries away its wastes. These substances diffuse between the maternal and fetal blood through the placental membrane, and they are carried to and from the fetal body by the umbilical blood vessels (figure 23.27). Consequently, the fetal blood and vascular system must be adapted to intrauterine life in special ways. For example, the concentration of oxygen-carrying hemoglobin in the fetal blood is about 50% greater than in the maternal blood. Also, the fetal hemoglobin is chemically slightly different and has a greater attraction for oxygen than maternal hemoglobin. Thus, at a particular oxygen pressure, fetal hemoglobin can carry 20%–30% more oxygen than maternal hemoglobin.

In the fetal circulatory system, the *umbilical vein* transports blood rich in oxygen and nutrients from the placenta to the fetal body. This vein enters the body through the umbilical ring and travels along the anterior abdominal wall to the liver. About half the blood it carries passes into the liver, and the rest enters a vessel called the **ductus venosus**, which bypasses the liver.

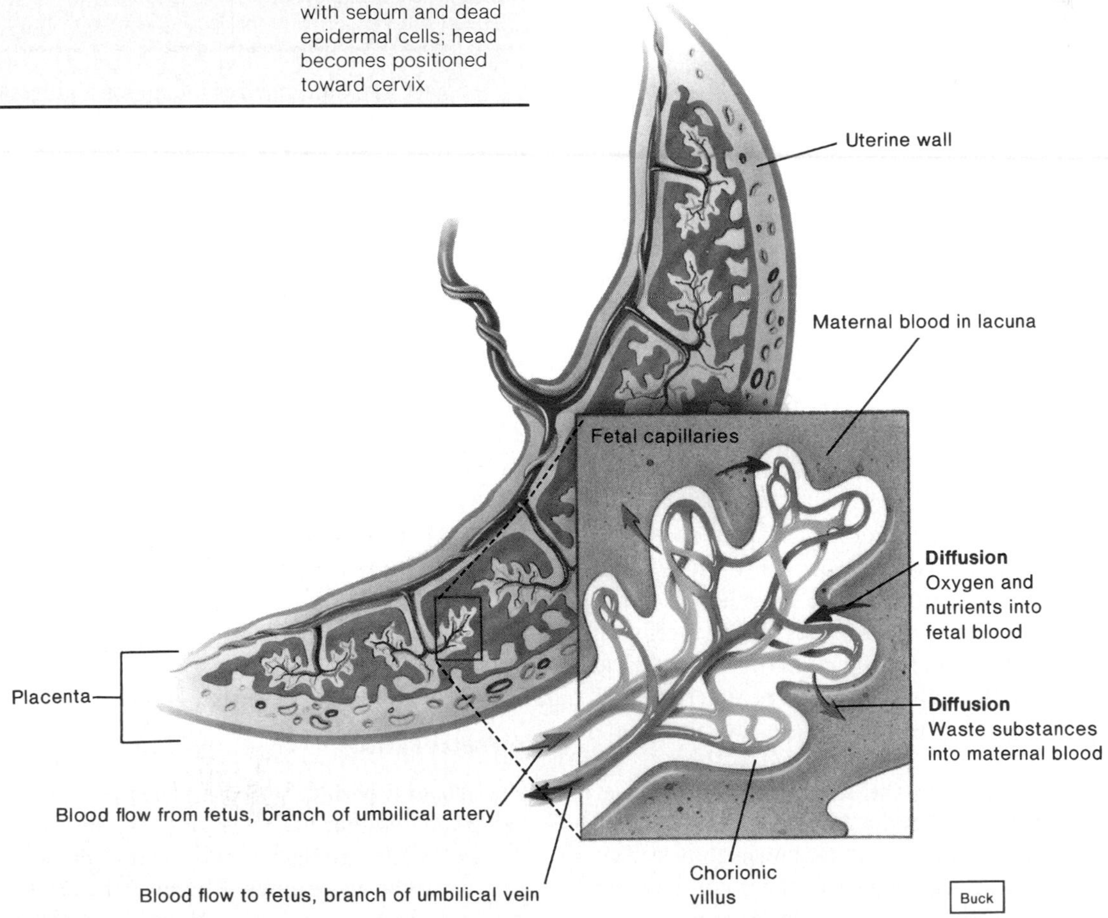

Figure 23.27 Oxygen and nutrients enter the fetal blood from the maternal blood by diffusion. Waste substances enter the maternal blood from the fetal blood by diffusion.

The ductus venosus travels a short distance and joins the inferior vena cava. There, the oxygenated blood from the placenta is mixed with the deoxygenated blood from the lower parts of the fetal body. This mixture continues through the vena cava to the right atrium.

In an adult heart, the blood from the right atrium enters the right ventricle and is pumped through the pulmonary trunk and pulmonary arteries to the lungs. In the fetus, however, the lungs are nonfunctional, and the blood largely bypasses them. More specifically, as the blood from the inferior vena cava enters the fetal right atrium, a large proportion of it is shunted directly into the left atrium through an opening in the atrial septum. This opening is called the **foramen ovale,** and the blood passes through it because the blood pressure in the right atrium is somewhat greater than that in the left atrium. Furthermore, a small valve (septum primum) located on the left side of the atrial septum overlies the foramen ovale and helps prevent blood from moving in the reverse direction.

The rest of the fetal blood entering the right atrium, including a large proportion of the deoxygenated blood entering from the superior vena cava, passes into the right ventricle and out through the pulmonary trunk.

Only a small volume of blood enters the pulmonary circuit, because the lungs are collapsed, and their blood vessels have a high resistance to blood flow. However, enough blood reaches the lung tissues to sustain them.

Most of the blood in the pulmonary trunk bypasses the lungs by entering a fetal vessel called the **ductus arteriosus,** which connects the pulmonary trunk to the descending portion of the aortic arch. As a result of this connection, the blood with a relatively low oxygen concentration, which is returning to the heart through the superior vena cava, bypasses the lungs. At the same time, it is prevented from entering the portion of the aorta that provides branches leading to the heart and brain.

The more highly oxygenated blood that enters the left atrium through the foramen ovale is mixed with a small amount of deoxygenated blood returning from the pulmonary veins. This mixture moves into the left ventricle and is pumped into the aorta. Some of it reaches the myocardium through the coronary arteries, and some reaches the brain tissues through the carotid arteries.

The blood carried by the descending aorta is partially oxygenated and partially deoxygenated. Some of it is carried into the branches of the aorta that lead to various parts of the lower regions of the body. The rest passes into the *umbilical arteries,* which branch from the internal iliac arteries and lead to the placenta. There the blood is reoxygenated. (See figure 23.28.)

CHART 23.2	Fetal circulatory adaptations
Adaptation	**Function**
Fetal blood	Has greater oxygen-carrying capacity than maternal blood
Umbilical vein	Carries oxygenated blood from the placenta to the fetus
Ductus venosus	Conducts about half the blood from the umbilical vein directly to the inferior vena cava, thus bypassing the liver
Foramen ovale	Conveys a large proportion of the blood entering the right atrium from the inferior vena cava, through the atrial septum, and into the left atrium, thus bypassing the lungs
Ductus arteriosus	Conducts some blood from the pulmonary trunk to the aorta, thus bypassing the lungs
Umbilical arteries	Carry the blood from the internal iliac arteries to the placenta

The umbilical cord usually contains two arteries and one vein. In a small percentage of newborns, there is only one umbilical artery. Since this condition is often associated with various other cardiovascular disorders, the vessels within the severed cord are routinely counted following a birth.

Chart 23.2 summarizes the major features of fetal circulation. At the time of birth, important adjustments must occur in the circulatory system when the placenta ceases to function and the newborn begins to breathe.

1. Which umbilical vessel carries oxygen-rich blood to the fetus?
2. What is the function of the ductus venosus?
3. How does fetal circulation allow blood to bypass the lungs?
4. What characteristic of the fetal lungs tends to shunt blood away from them?

Postnatal Period

The postnatal period of development lasts from birth until death. It can be divided into the neonatal period, infancy, childhood, adolescence, adulthood, and senescence.

Neonatal Period

The **neonatal period,** which extends from birth to the end of the first four weeks, begins very abruptly at birth (figure 23.29). At that moment, physiological adjustments must be made quickly, because the newborn must suddenly do for itself those things that the mother's body has been doing for it. Thus, the newborn (neonate) must carry on respiration, obtain nutrients, digest

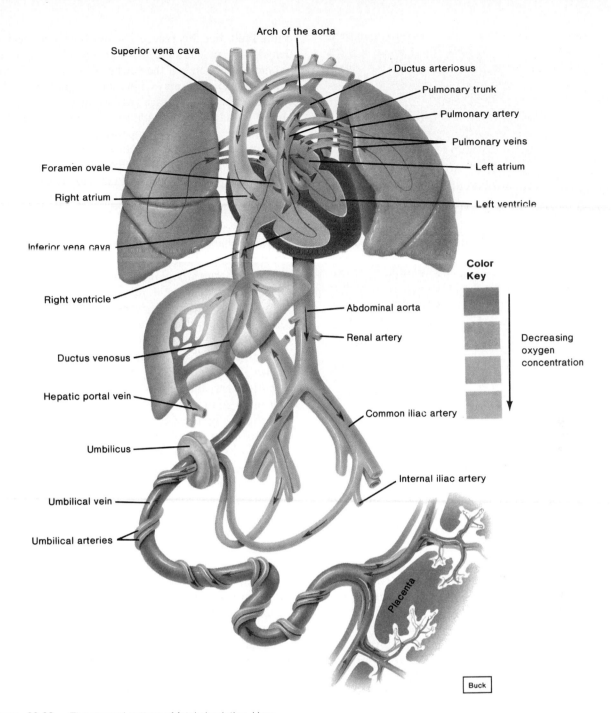

Figure 23.28 The general pattern of fetal circulation. How does this pattern of circulation differ from that of an adult?

nutrients, excrete wastes, regulate body temperature, and so forth. However, its most immediate need is to obtain oxygen and excrete carbon dioxide, so its first breath is critical.

The first breath must be particularly forceful, because the newborn's lungs are collapsed, and the airways are small, offering considerable resistance to air movement. Also, surface tension tends to hold the moist membranes of the lungs together. However, the lungs of a full-term fetus continuously secrete *surfactant* (see chapter 16), which reduces surface tension. Thus, after the first powerful breath begins to expand the lungs, breathing becomes easier.

Mechanical stimulation of a newborn's lungs, caused by the mother's muscular uterine contractions during natural birth, seems to increase the quantity of surfactant secreted. Related to this is the observation that respiratory distress syndrome occurs more commonly in infants whose births do not include labor.

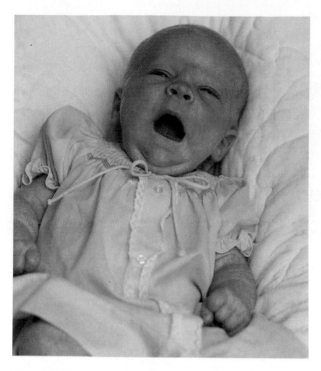

Figure 23.29 The neonatal period extends from birth to the end of the fourth week after birth.

It is not clear whether the newborn's first breath is stimulated by one or several factors. Those that may be involved include an increasing concentration of carbon dioxide, a decreasing pH, low oxygen concentration, a drop in body temperature, and mechanical stimulation that occurs during and after the birth process. Also, in response to the stress experienced by the fetus during the birth process, the fetal blood concentrations of epinephrine and norepinephrine rise significantly (chapter 13). These hormones seem to promote normal breathing by increasing the secretion of surfactant and causing the airways to dilate.

Prior to birth, the fetus depends primarily on glucose and fatty acids obtained from the mother's blood as energy sources. The newborn, on the other hand, is suddenly without an external source of nutrients, and its mother's milk will not be produced for two to three days. Furthermore, the newborn has a relatively high rate of metabolism, and its liver, which is not fully mature, may be unable to supply enough glucose to support its metabolic needs. Consequently, the newborn typically utilizes stored fat as an energy source.

1. Define the neonatal period of development.
2. Why must the first breath of an infant be particularly forceful?
3. What factors seem to stimulate the first breath?
4. What does a newborn use for an energy supply during the first few days after birth?

As a rule, the newborn's kidneys are unable to produce concentrated urine, so they excrete a relatively dilute fluid. For this reason, the newborn may become dehydrated and develop a water and electrolyte imbalance. Also, some of the newborn's homeostatic control mechanisms may function imperfectly. For example, the temperature regulating system may be unable to maintain a constant body temperature. As a consequence, the body temperature may respond to slight stimuli by fluctuating above or below the normal level during the first few days of life.

When the placenta ceases to function and breathing begins, changes also occur in the newborn's circulatory system. For example, following birth, the umbilical vessels constrict. The arteries close first, and if the umbilical cord is not clamped or severed for a minute or so, blood continues to flow from the placenta to the newborn through the umbilical vein, adding to the newborn's blood volume.

The proximal portions of the umbilical arteries persist in the adult as the *superior vesical arteries* that supply blood to the urinary bladder. The more distal portions become solid cords (lateral umbilical ligaments). The umbilical vein becomes the cordlike *ligamentum teres* that extends from the umbilicus to the liver in an adult.

Similarly, the ductus venosus constricts shortly after birth and is represented in the adult as a fibrous cord (ligamentum venosum), which is superficially embedded in the wall of the liver.

The foramen ovale closes as a result of blood pressure changes occurring in the right and left atria as the fetal vessels constrict. More precisely, as blood ceases to flow from the umbilical vein into the inferior vena cava, the blood pressure in the right atrium falls. Also, as the lungs expand with the first breathing movements, the resistance to blood flow through the pulmonary circuit decreases, more blood enters the left atrium through the pulmonary veins, and the blood pressure in the left atrium increases.

As the pressure in the left atrium rises and that in the right atrium falls, the valve (septum primum) on the left side of the atrial septum closes the foramen ovale. In most individuals, this valve gradually fuses with the tissues along the margin of the foramen. In an adult, the site of the previous opening is marked by a depression called the *fossa ovalis*.

The ductus arteriosus, like the other fetal vessels, constricts after birth. This constriction seems to be stimulated by a substance called *bradykinin,* which is released from the lungs during their initial expansion. Bradykinin is thought to act when the oxygen concentration of the aortic blood rises as a result of breathing. After the ductus arteriosus has closed, blood can no longer bypass the lungs by moving from the pulmonary trunk directly into the aorta. In an adult, the ductus arteriosus is represented by a cord called the *ligamentum arteriosum.*

The ductus arteriosus sometimes fails to close. Although the cause of this condition is usually unknown, it often occurs in newborns whose mothers were infected with rubella virus (German measles) during the first three months of embryonic development.

After birth, the metabolic rate and oxygen consumption of the neonatal tissues increase, in large part because of the need to maintain the body temperature. If the ductus arteriosus remains open, the neonate's blood oxygen concentration may remain too low to adequately supply the body tissues, including the myocardium. As a result, the heart may fail, even though the myocardium is normal. This type of heart failure is a relatively common cause of infant death, occurring during the first month.

The changes in the newborn's circulatory system do not occur very rapidly. Although the constriction of the ductus arteriosus may be functionally complete within fifteen minutes, the permanent closure of the foramen ovale may take up to a year. These circulatory changes are illustrated in figure 23.30 and summarized in chart 23.3.

1. How do the kidneys of a newborn differ from those of an adult?
2. What portions of the umbilical arteries continue to function into adulthood?
3. What is the fate of the foramen ovale? Of the ductus arteriosus?
4. When is the closure of the ductus arteriosus functionally complete?

Infancy

The period of development extending from the end of the first four weeks to one year is called **infancy.** During this time, the infant grows rapidly and may triple its birth weight. Its teeth begin to erupt through the gums, and its muscular and nervous systems mature so that coordinated muscular activities become possible. Con-

sequently, the infant is soon able to sit, creep, and stand; to reach for objects and hold them; and to follow objects visually (figure 23.31).

Infancy is also characterized by the beginning of the ability to communicate with others. The infant

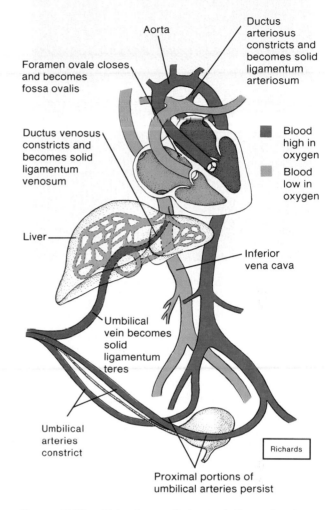

Figure 23.30 Major changes that occur in the newborn's circulatory system.

CHART 23.3	Circulatory adjustments in the newborn	
Structure	Adjustment	In the adult
Umbilical vein	Becomes constricted	Becomes ligamentum teres that extends from the umbilicus to the liver
Ductus venosus	Becomes constricted	Becomes ligamentum venosum that is superficially embedded in the wall of the liver
Foramen ovale	Is closed by valvelike septum primum as blood pressure in right atrium decreases and pressure in left atrium increases	Valve fuses along margin of foramen ovale and is marked by a depression called the fossa ovalis
Ductus arteriosus	Constricts in response to bradykinin released from the expanding lungs	Becomes ligamentum arteriosum that extends from the pulmonary trunk to the aorta
Umbilical arteries	Distal portions become constricted	Distal portions become lateral umbilical ligaments; proximal portions function as superior vesical arteries

Figure 23.31 Infancy extends from the end of the fourth week after birth until one year of age.

Figure 23.32 Childhood begins at the end of the first year and is completed at puberty.

learns to smile, laugh, and respond to some sounds. By the end of the first year, the infant may be able to say two or three words.

Because infancy (as well as childhood) is a period of rapid growth, the infant has particular nutritional needs. For example, in addition to an energy source, the body requires proteins to provide the amino acids necessary to form new tissues; calcium and vitamin D to promote the development and ossification of skeletal structures; iron to support blood cell formation; and vitamin C for the production of structural tissues such as cartilage and bone.

As mentioned previously, fetal hemoglobin is slightly different from adult hemoglobin and has a greater affinity for oxygen. As a rule, however, fetal hemoglobin is not produced after birth, and by the age of four months, most of the circulating hemoglobin is the adult type.

Childhood

The period called **childhood** begins at the end of the first year and is completed at puberty (figure 23.32). During this period, growth continues at a high rate. The primary teeth appear and are subsequently replaced by secondary ones. The child develops a high degree of voluntary muscular control and learns to walk, run, and

climb. Bladder and bowel controls are established. The child learns to communicate effectively by speaking, and later, usually learns to read, write, and reason objectively. At the same time, the child is maturing emotionally.

1. Define infancy.
2. What developmental changes characterize infancy?
3. Define childhood.
4. What developmental changes characterize childhood?

Adolescence

Adolescence is the period of development between puberty and adulthood. As discussed in chapter 22, puberty is a period of anatomical and physiological changes that result in reproductively functional individuals. For the most part, these changes are hormonally controlled, and they include the appearance of secondary sexual characteristics as well as growth spurts in the muscular and skeletal systems.

Females usually experience these changes somewhat earlier than males, so that early in adolescence, females may be taller and stronger than their male peers (figure 23.33). On the other hand, females achieve their full growth at earlier ages, and in late adolescence, the

Figure 23.33 Adolescence is the period between puberty and adulthood.

average male is taller and stronger than the average female. (See figure 23.34.)

The periods of rapid growth in adolescence, which usually begin between the ages of eleven and thirteen in females and between thirteen and fifteen in males, cause increased demands for certain nutrients. In addition to energy sources, foods must provide ample amounts of proteins, vitamins, and minerals to support the growth of new tissues.

Adolescence is also characterized by increasing levels of motor skill, intellectual ability, and emotional maturity.

Adulthood

Adulthood (maturity) extends from adolescence to old age. Since the anatomy and physiology of the adult has been the basis for chapters 1–22, little more needs to be said about this period of development.

Ordinarily, the adult remains relatively unchanged anatomically or physiologically for many years. However, sometime after the age of thirty, depending on hereditary and environmental factors, degenerative changes usually begin to occur (figure 23.35). It has been estimated that the average person in this stage of development loses about 0.8% of his or her functional

capacity each year. For example, skeletal muscles tend to lose strength as more and more connective tissue appears within the muscles; the circulatory system becomes less efficient as the lumens of arterioles and arteries become narrowed due to accumulations of fatty deposits; the skin tends to become loose and wrinkled as elastic fibers in the dermis undergo changes; and the hair on the head turns gray as melanin-producing enzyme systems cease to function.

Also, the capacity of the sex cell-producing tissues decline. Females experience menopause in later adulthood, and males lose some sexual functions because of decreased secretion of male sex hormones.

> Because older people are generally less active, they need fewer calories to maintain desirable body weight, but at the same time, they need essential nutrients. Thus, if older adults eat simply to satisfy hunger, they may develop symptoms of malnutrition resulting either from deficiencies of such nutrients as iron, magnesium, calcium, vitamin A, vitamin D, or vitamin C, or from excesses of sodium, chlorine, or phosphorus.

Senescence

Senescence is the process of growing old, and it culminates in the death of the individual. This process involves a continuation of the degenerative changes that begin during adulthood (figure 23.36). As a result, the body becomes less and less able to cope with the demands placed on it by the individual and by the environment.

The cause of senescence is not well understood. However, the degenerative changes that occur seem to involve a variety of factors. Some are due to the prolonged use of body parts. For example, the cartilages covering the ends of bones at joints may wear away, leaving the joints stiff and painful.

Other degenerative changes are caused by disease processes that interfere with vital functions, such as gas exchanges or blood circulation. Still others seem to be due to cellular changes, including alterations in metabolic rates and in distribution of body fluids. The rate of cell division may decline, and the ability of cells to carry on immune responses may decrease. As a result, the person becomes less able to repair damaged tissue and more susceptible to disease.

Senescence is typically accompanied by decreasing efficiency of the central nervous system. As a result, the person may experience a loss of intellectual functions. Also, the physiological coordinating capacity of the nervous system may decrease, and homeostatic mechanisms may fail to operate effectively.

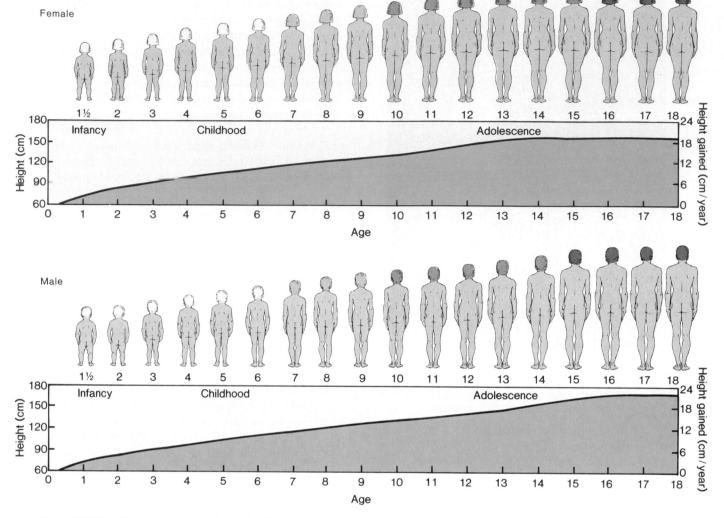

Figure 23.34 Physical changes and growth in height that occur in males and females between infancy and adolescence. (From "Growing Up" by J. M. Tanner. Copyright © 1973 by Scientific American, Inc. All rights reserved.)

Sensory functions usually decrease with age also. Thus, perhaps one-third of older persons have significant hearing losses; difficulty focusing on objects at varying distances; and problems with depth perception. Typically, the senses of taste and smell diminish with age, and many older persons have decreased abilities to perceive environmental temperature accurately. They may also have impaired perception of touch and pressure as well as decreased sensitivity to painful stimuli.

Death usually results, not from these degenerative changes, but from mechanical disturbances in the cardiovascular system, failure of the immune system, or disease processes that affect vital organs.

Throughout the industrialized world, the average life span for women is several years longer than that for men. Although the reason for this difference is not completely understood, males seem to have an increased tendency to develop cardiovascular diseases. Also, at least in their earlier years, males are more likely to die as a result of accident, homicide, suicide, lung cancer, pulmonary disease, and cirrhosis of the liver.

The average *life expectancy* has increased in relatively recent times in developed countries as a result of improved control of infectious diseases. However, the human *life span* has not changed significantly for thousands of years. This life span seems to be fixed between ninety and one hundred years.

Causes of Aging

Aging can be defined as the structural and functional changes that occur normally in a person as time passes, and that gradually produce functional impairments leading to death.

Although aging is considered a natural part of development, there are no generally accepted explanations for it. On the other hand, both cellular and environmental factors seem to contribute to the process.

Some researchers believe that cells are programmed to age, just as they are programmed to undergo other developmental changes. For example, body cells seem able to divide a limited number of times before they lose their capacity to reproduce and finally die. Furthermore, it is known from studies of human cells cultured in the laboratory that the number of times a cell divides is roughly related to the age of its donor. The loss of the body cells' ability to reproduce could contribute to a decline in body functions by preventing the replacement of worn or damaged cells.

Other researchers emphasize the effects of *mutations* as a cause of aging. They theorize that a decline in cellular functions may involve changes in DNA molecules that occur as a result of exposure to certain environmental factors. (See chapter 4.) Then, if the mutations occur faster than the DNA repair system can correct them, molecular errors may accumulate so that, in time, cellular metabolism is disturbed. This type of damage might be particularly serious in tissues, such as skeletal muscle and nerve tissue, whose cells are usually not replaced by cell division, or in cells that provide vital control functions.

Still another possible cause of aging involves the action of highly unstable substances called *free radicals*. These substances are sometimes produced in cells as a result of metabolic processes, and they may be encountered in the environment. In either case, they can damage DNA molecules and can react with other cell structures. For example, they can cause chemical bonds to form abnormally between protein molecules. The formation of such cross-linkages may contribute to the aging changes that occur in collagen. This protein is quite flexible—a quality that is very important for maintaining the normal characteristics of many organs. (See chapter 5.) However, with the passage of time, collagen tends to become stiffer and less flexible, and this aging-related change may be promoted by the action of free radicals.

Other factors that are thought to contribute to the aging process include an accumulation of certain cellular wastes, such as lipofuscin in neurons, and the effects of stress acting over prolonged periods. (See chapter 13.)

Figure 23.35 Adulthood extends from adolescence to old age.

Figure 23.36 The changes that characterize old age begin during adulthood and culminate with death.

CHART 23.4 Stages in postnatal development

Stage	Time period	Major events
Neonatal period	Birth to end of fourth week	Newborn begins to carry on respiration, obtain nutrients, digest nutrients, excrete wastes, regulate body temperature, and make circulatory adjustments
Infancy	End of fourth week to one year	Growth rate is high; teeth begin to erupt; muscular and nervous systems mature so that coordinated activities are possible; communication begins
Childhood	One year to puberty	Growth rate is high; deciduous teeth erupt and are replaced by permanent teeth; high degree of muscular control is achieved; bladder and bowel controls are established; intellectual abilities mature
Adolescence	Puberty to adulthood	Person becomes reproductively functional and emotionally more mature; growth spurts occur in skeletal and muscular systems; high levels of motor skills are developed; intellectual abilities increase
Adulthood	Adolescence to old age	Person remains relatively unchanged anatomically and physiologically; degenerative changes begin to occur
Senescence	Old age to death	Degenerative changes continue; body becomes less and less able to cope with the demands placed upon it; death usually results from mechanical disturbances in the cardiovascular system or from disease processes that affect vital organs

Chart 23.4 summarizes the major phases of postnatal life and their characteristics, while chart 23.5 lists some of the aging-related changes.

1. What changes characterize adolescence?
2. Define adulthood.
3. What changes characterize adulthood?
4. What changes characterize senescence?

Clinical Terms Related to Growth and Development

abruptio placentae (ab-rup′she-o plah-cen′ta) A premature separation of the placenta from the uterine wall.

amniocentesis (am″ne-o-sen-te′sis) A technique in which a sample of amniotic fluid is withdrawn from the amniotic cavity by inserting a hollow needle through the mother's abdominal wall.

dizygotic twins (di″zi-got′ik twinz) Twins resulting from the fertilization of two ova by two sperm cells.

hydatid mole (hi′dah-tid mōl) A type of uterine tumor that originates from placental tissue.

hydramnios (hi-dram′ne-os) The presence of excessive amniotic fluid.

intrauterine transfusion (in″trah-u′ter-in trans-fu′zhun) Transfusion administered by injecting blood into the fetal peritoneal cavity before birth.

lochia (lo′ke-ah) Vaginal discharge following childbirth.

meconium (mĕ-ko′ne-um) The first fecal discharge of a newborn.

monozygotic twins (mon″o-zi-got′ik twinz) Twins resulting from the fertilization of one ovum by one sperm cell.

perinatology (per″i-na-tol′o-je) Branch of medicine concerned with the fetus after twenty-five weeks of development and with the newborn for the first four weeks after birth.

postpartum (pōst-par′tum) Occurring after birth.

teratology (ter″ah-tol′o-je) The study of abnormal development and congenital malformations.

trimester (tri-mes′ter) Each third of the total period of pregnancy.

ultrasonography (ul″trah-son-og′rah-fe) Technique used to visualize the size and position of fetal structures by means of ultrasonic sound waves.

Chapter Summary

Introduction (page 854)

Growth refers to an increase in size; development is the process of changing from one phase of life to another.

Prenatal Period (page 854)

1. Period of cleavage
   a. Fertilization occurs in a uterine tube and results in a zygote.
   b. The zygote undergoes mitosis, and the newly formed cells divide.
   c. Each subsequent division produces smaller and smaller cells.
   d. A solid ball of cells (morula) is formed, and it becomes a hollow ball called a blastocyst.
   e. The inner cell mass that gives rise to the embryo proper forms within the blastocyst.
   f. The blastocyst becomes implanted in the uterine wall.
      (1) Enzymes digest the endometrium around the blastocyst.
      (2) Fingerlike processes from the blastocyst penetrate into the endometrium.
   g. The period of cleavage lasts through the first week of development.
   h. The trophoblast secretes HCG, which helps maintain the corpus luteum, helps protect the blastocyst against being rejected, and stimulates the developing placenta to secrete hormones.

CHART 23.5 Aging-related changes

Organ system	Aging-related changes
Integumentary system	Degenerative loss of collagenous and elastic fibers in dermis; decreased production of pigment in hair follicles; reduced activity of sweat and sebaceous glands
	Skin tends to become thinner, wrinkled, and dry; hair turns gray and then white
Skeletal system	Degenerative loss of bone matrix
	Bones become thinner, less dense, and more likely to fracture; stature may shorten due to compression of intervertebral disks and vertebrae
Muscular system	Loss of skeletal muscle fibers; degenerative changes in neuromuscular junctions
	Loss of muscular strength
Nervous system	Degenerative changes in neurons; loss of dendrites and synaptic connections; accumulation of lipofuscin in neurons; decreases in sensory sensitivities
	Decreasing efficiency in processing and recalling information; decreasing ability to communicate; diminished senses of smell and taste; loss of elasticity of lenses and consequent loss of ability to accommodate for close vision
Endocrine system	Reduced hormonal secretions
	Decreased metabolic rate; reduced ability to cope with stress; reduced ability to maintain homeostasis
Digestive system	Decreased motility in gastrointestinal tract; reduced secretion of digestive juices
	Reduced efficiency of digestion
Respiratory system	Degenerative loss of elastic fibers in lungs; reduced number of alveoli
	Reduced vital capacity; increase in dead air space; reduced ability to clear airways by coughing
Cardiovascular system	Degenerative changes in cardiac muscle; decrease in lumen diameters of arteries and arterioles
	Decreased cardiac output; increased resistance to blood flow; increased blood pressure
Lymphatic system	Decrease in efficiency of immune system
	Increased incidence of infections and neoplastic diseases; increased incidence of autoimmune diseases
Excretory system	Degenerative changes in kidneys; reduction in number of functional nephrons
	Reductions in filtration rate, tubular secretion, and reabsorption
Reproductive systems	
Female	Degenerative changes in ovaries; decrease in secretion of sex hormones
	Menopause; regression of secondary sexual characteristics
Male	Reduced secretion of sex hormones; enlargement of prostate gland; decrease in sexual energy

2. Embryonic stage
   a. The embryonic stage extends from the second through the eighth weeks.
   b. It is characterized by the development of the placenta and the main internal and external body structures.
   c. The cells of the inner cell mass become arranged into primary germ layers.
      (1) Ectoderm gives rise to the nervous system, portions of the skin, the lining of the mouth, and the lining of the anal canal.
      (2) Mesoderm gives rise to muscles, bones, blood vessels, lymphatic vessels, reproductive organs, kidneys, and linings of body cavities.
      (3) Endoderm gives rise to linings of the digestive tract, respiratory tract, urinary bladder, and urethra.
   d. The embryonic disk becomes cylindrical and is attached to the developing placenta by the connecting stalk.
   e. The embryo develops head, face, arms, legs, and mouth, and appears more humanlike.
   f. Chorionic villi develop and are surrounded by spaces filled with maternal blood.
   g. The placental membrane consists of the epithelium of the villi and the epithelium of the capillaries inside the villi.
      (1) Oxygen and nutrients diffuse from the maternal blood through the placental membrane and into the fetal blood.
      (2) Carbon dioxide and other wastes diffuse from the fetal blood through the placental membrane and into the maternal blood.

h. The placenta develops in the disk-shaped area where the chorion remains in contact with the uterine wall.
   (1) The embryonic portion consists of the chorion and its villi.
   (2) The maternal portion consists of the uterine wall to which villi are attached.
i. A fluid-filled amnion develops around the embryo.
j. The umbilical cord is formed as the amnion envelops the tissues attached to the underside of the embryo.
   (1) The umbilical cord includes two arteries and a vein
   (2) It suspends the embryo in the amniotic cavity.
k. The chorion and amnion become fused.
l. The yolk sac forms on the underside of the embryonic disk.
   (1) It gives rise to blood cells and cells that later form sex cells.
   (2) It helps form the digestive tube.
m. The allantois extends from the yolk sac into the connecting stalk.
   (1) It forms blood cells.
   (2) It gives rise to the umbilical vessels.
n. By the beginning of the eighth week, the embryo is recognizable as a human.

3. Fetal stage
a. This stage extends from the end of the eighth week and continues until birth.
b. Existing structures grow and mature; only a few new parts appear.
c. The body enlarges, arms and legs achieve final relative proportions, the skin is covered with sebum and dead epidermal cells, the skeleton continues to ossify, the muscles become active, and fat is deposited in subcutaneous tissue.
d. The fetus is full term at the end of the tenth lunar month.
   (1) It is about 50 centimeters long and weighs 6–8 pounds.
   (2) It is positioned with its head toward the cervix.

4. Fetal blood and circulation
a. Blood is carried between the placenta and the fetus by umbilical vessels.
b. Fetal blood carries a greater concentration of oxygen than does maternal blood.
c. Blood enters the fetus through the umbilical vein and partially bypasses the liver by means of the ductus venosus.
d. Blood enters the right atrium and partially bypasses the lungs by means of the foramen ovale.
e. Blood entering the pulmonary trunk partially bypasses the lungs by means of the ductus arteriosus.
f. Blood enters the umbilical arteries from the internal iliac arteries.

Postnatal Period (page 870)

1. Neonatal period
a. This period extends from birth to the end of the fourth week.
b. The newborn must begin to carry on respiration, obtain nutrients, excrete wastes, and regulate its body temperature.
c. The first breath must be powerful in order to expand the lungs.
   (1) Surfactant reduces surface tension.
   (2) The first breath is stimulated by a variety of factors.
d. The liver is immature and unable to supply sufficient glucose, so the newborn depends primarily on stored fat as an energy source.
e. Immature kidneys cannot concentrate urine very well.
   (1) The newborn may become dehydrated.
   (2) Water and electrolyte imbalances may develop.
f. Homeostatic mechanisms may function imperfectly and body temperature may be unstable.
g. The circulatory system undergoes changes when placental circulation ceases.
   (1) Umbilical vessels constrict.
   (2) The ductus venosus constricts.
   (3) The foramen ovale is closed by a valve as blood pressure in the right atrium falls and pressure in the left atrium rises.
   (4) Bradykinin released from the expanding lungs stimulates constriction of the ductus arteriosus.

2. Infancy
a. Infancy extends from the end of the fourth week to one year of age.
b. Infancy is a period of rapid growth.
   (1) The muscular and nervous systems mature, and coordinated activities become possible.
   (2) Communication begins.
c. Rapid growth depends on an adequate intake of proteins, vitamins, and minerals in addition to energy sources.

3. Childhood
a. Childhood extends from the end of the first year to puberty.
b. It is characterized by rapid growth, development of muscular control, and establishment of bladder and bowel control.

4. Adolescence
a. Adolescence extends from puberty to adulthood.
b. It is characterized by physiological and anatomical changes that result in a reproductively functional individual.
c. Females may be taller and stronger than males in early adolescence, but the situation reverses in late adolescence.
d. Adolescents develop high levels of motor skills, their intellectual abilities increase, and they continue to mature emotionally.

5. Adulthood
   a. Adulthood extends from adolescence to old age.
   b. The adult remains relatively unchanged physiologically and anatomically for many years.
   c. After age thirty, degenerative changes usually begin to occur.
      (1) Skeletal muscles lose strength.
      (2) The circulatory system becomes less efficient.
      (3) The skin loses its elasticity.
      (4) The capacity to produce sex cells declines.
6. Senescence
   a. Senescence is the process of growing old.
   b. Degenerative changes continue, and the body becomes less able to cope with demands placed upon it.
   c. Changes occur because of prolonged use, effects of disease, and cellular alterations.
   d. An aging person usually experiences losses in intellectual functions, sensory functions, and physiological coordinating capacities.
   e. Death usually results from mechanical disturbances in the cardiovascular system or from disease processes that affect vital organs.

Clinical Application of Knowledge

1. How would you explain the observation that twins resulting from a single fertilized egg cell can exchange body parts by tissue or organ transplant procedures without experiencing rejection reactions?
2. One of the more common congenital cardiac disorders is a ventricular septum defect in which an opening remains between the right and left ventricles. What problem would such a defect create as blood moves through the heart?
3. What symptoms may appear in a newborn if its ductus arteriosus fails to close?
4. How would you explain the observation that it is sometimes difficult to determine the racial origin of a newborn by its skin color during the first few days following birth?
5. Why is it important for a middle-aged adult who has neglected physical activity for many years to have a physical examination before beginning an exercise program?
6. If an aged relative came to live with you, what special provisions could you make in your household environment and routines that would demonstrate your understanding of the changes brought on by aging processes?

Review Activities

1. Define *growth* and *development*.
2. Describe the process of cleavage.
3. Distinguish between a blastomere and a blastocyst.
4. Describe the formation of the inner cell mass, and explain its significance.
5. Describe the process of implantation.
6. List three functions of HCG.
7. Explain how the primary germ layers form.
8. List the major body parts derived from ectoderm.
9. List the major body parts derived from mesoderm.
10. List the major body parts derived from endoderm.
11. Describe the formation of the placenta, and explain its functions.
12. Define *placental membrane*.
13. Distinguish between the chorion and the amnion.
14. Explain the function of amniotic fluid.
15. Describe the formation of the umbilical cord.
16. Explain how the yolk sac and the allantois are related, and list the functions of each.
17. Explain why the embryonic period of development is so critical.
18. Define *fetus*.
19. List the major changes that occur during the fetal stage of development.
20. Describe a full-term fetus.
21. Compare the properties of fetal hemoglobin with those of maternal hemoglobin.
22. Explain how the fetal circulatory system is adapted for intrauterine life.
23. Trace the pathway of blood from the placenta to the fetus and back to the placenta.
24. Distinguish between a newborn and an infant.
25. Explain why a newborn's first breath must be particularly forceful.
26. List some of the factors that are thought to act as stimuli for the first breath.
27. Explain why newborns tend to develop water and electrolyte imbalances.
28. Describe the circulatory changes that occur in the newborn.
29. Describe the characteristics of an infant.
30. Distinguish between a child and an adolescent.
31. Define *adulthood*.
32. List some of the degenerative changes that begin during adulthood.
33. Define *senescence*.
34. List some of the factors that seem to promote senescence.

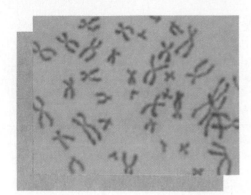

CHAPTER

24

Human Genetics

*T*he complex changes an embryo undergoes as it grows and develops are controlled by mechanisms that ensure that the resulting offspring will resemble its parents. At the same time, these mechanisms guarantee that the offspring will differ from its parents and will become an individual with unique characteristics. The study of the similarities and differences between parents and their offspring, and of the mechanisms responsible, is called *genetics* ∎

After you have studied this chapter, you should be able to:

1. Explain how heredity and environment influence the development of individual characteristics.

2. Distinguish between genes and chromosomes, and explain why genes and chromosomes are paired.

3. Explain why gene expression may differ with different allelic combinations.

4. Describe how environmental factors may influence gene expression.

5. Outline the process of meiosis and explain why cells resulting from this process contain different combinations of genetic information.

6. Explain the patterns by which single traits are transmitted from parents to offspring.

7. Describe how chromosomes control the inheritance of gender and how sex chromosomes act.

8. Explain the pattern by which a trait controlled by an X chromosome-linked gene is inherited.

9. Define *nondisjunction* and explain its significance in the appearance of chromosome disorders.

10. Explain what is meant by genetic counseling.

11. Complete the review activities at the end of this chapter. Note that the items are worded in the form of specific learning objectives. You may want to refer to them before reading the chapter.

allele (ah-lēl')

autosome (aw'to-sōm)

dominant (dom'ĭ-nant)

gene (jēn)

genetics (jĕ-net'iks)

genotype (je'no-tīp)

heredity (hĕ-red'ĭ-te)

heterozygous (het''er-o-zi'gus)

homozygous (ho''mo-zi'gus)

meiosis (mi-o'sis)

multiple allele (mul'tĭ-pl ah-lēl')

mutation (mu-ta'shun)

nondisjunction (non''dis-jungk'shun)

phenotype (fe'no-tīp)

recessive (re-ses'iv)

sex chromosome (seks kro'mo-sōm)

alb-, white: *alb*inism—condition in which normal skin pigments are lacking.

centr-, a point: *centr*omere—region at which paired chromatids are held together.

hem-, blood: *hem*ophilia—condition in which bleeding is prolonged due to a defect in the blood clotting mechanism.

hetero-, different: *hetero*zygous—condition in which the members of a gene pair are different.

homo-, same: *homo*logous chromosomes—a pair of chromosomes that contain similar genetic information.

pheno-, visible: *pheno*type—the physical appearance that results from the way genes are expressed in an individual.

syn-, together: *syn*apsis—process by which homologous chromosomes become tightly intertwined.

tri-, three: *tri*somy—condition in which one kind of chromosome is triply represented.

lthough genetics is a relatively new field of study, knowledge of genetic mechanisms has expanded rapidly in recent times. It is known that individual characteristics result from the interaction of two major factors: heredity and environment.

Heredity is the transmission of genetic information from parents to offspring. This information is passed in the form of DNA molecules contained in the nuclei of egg and sperm cells, and it provides chemical instructions for human development. **Environment,** on the other hand, includes the chemical, physical, and biological factors in the surroundings of an individual that influence his or her characteristics.

Genes and Chromosomes

As explained in chapter 4, genetic information in the form of nucleotide sequences within DNA molecules tells a cell how to construct specific kinds of protein molecules, which, in turn, function as structural materials, enzymes, and other vital substances.

Genes

The nucleotide sequence of a DNA molecule that contains the information for producing one kind of protein molecule is called a **gene.** The genes within the egg and sperm cells that unite to form a zygote instruct the zygote how to synthesize particular proteins. Some of these proteins act as enzymes, which, in turn, promote the metabolic reactions necessary for the growth and development of a unique individual.

It takes one gene to produce a particular kind of enzyme. Therefore, many sets of genes are required for the synthesis of enzymes responsible for an offspring's specific gender, eye color, hair color, skin color, blood type, and so forth.

> The total number of genes in a human set is unknown. However, many kinds of enzymes, structural proteins, and proteins with special functions occur in cells. Since at least one gene is needed for the synthesis of each kind of protein, some investigators estimate that the human set includes as many as 100,000 genes.

Chromosome Numbers

Within a cell, the DNA molecules, and therefore the genes, are found within *chromosomes.* These structures, which are described in chapter 3, consist of proteins (histones) as well as DNA. They appear as rod-shaped bodies in the nucleus when a cell undergoes cell division (figures 3.33 and 24.1).

The number of chromosomes within the cells of organisms varies with species. For the human species (*Homo sapiens*), the number is 46. Thus, a human zygote contains 46 chromosomes, 23 of which were re-

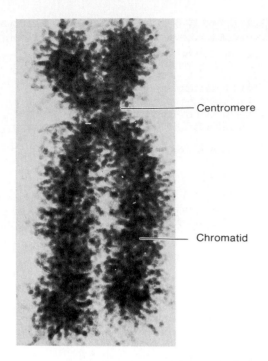

Figure 24.1 A transmission electron micrograph of a human chromosome from the prophase stage of mitosis in which the chromosome is composed of two identical chromatids united by a centromere (×33,824).

ceived from its female parent by means of an egg cell and 23 from its male parent by means of a sperm cell (figure 24.2). Chart 24.1 lists the number of chromosomes present in the body cells of several animal species.

> Modern laboratory techniques make it possible to determine the nucleotide sequence of a gene and to locate its approximate position within a specific chromosome. Consequently, many human genes have been sequenced, and about 2,100 genes have been associated with particular chromosomes. Of these, nearly 450 genes have been related to particular diseases. For example, the gene responsible for the disease called *cystic fibrosis* has been located on human chromosome number 7. The gene that causes the disease *neurofibromatosis* is on chromosome number 17, and the one causing *Huntington's disease* is on chromosome number 4.

Chromosome Pairs

The 23 chromosomes that a zygote receives from its female parent contain a complete set of genetic information for the growth and development of an offspring, as do the 23 chromosomes it receives from its male parent. Those chromosomes from the parents that contain similar genetic material are said to comprise **homologous pairs.** That is, the chromosome of the maternal set containing the information for a particular set of traits and the chromosome of the paternal

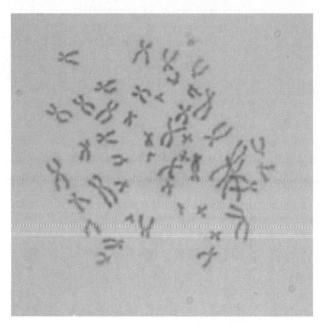

Figure 24.2 Light micrograph of a set of human chromosomes (×1,000).

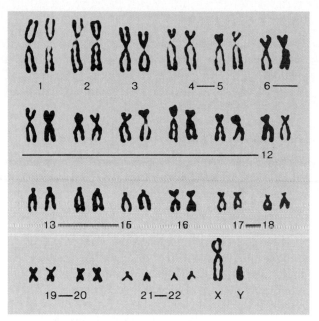

Figure 24.3 A karyotype of human chromosomes.

CHART 24.1	Number of chromosomes in the body cells of various animals	
Animal	Scientific name	Chromosome number
Domestic cat	*Felis domesticus*	38
Cattle	*Bos taurus*	60
Chicken	*Gallus domesticus*	78
Chimpanzee	*Pan troglodytes*	48
Dog	*Canis familiaris*	78
Frog	*Rana esculenta*	26
Human	*Homo sapiens*	46
Mouse	*Mus musculus*	40
Monkey	*Macaca mulatta*	42
Pig	*Sus scrofa*	40
Rat	*Rattus norvegicus*	42
Sheep	*Ovis aries*	54
Toad	*Bufo vulgaris*	22

set containing information for these same traits make up a homologous pair. The chromosomes shown in figure 24.3 are arranged in homologous pairs; such an arrangement of chromosomes is called a **karyotype.**

During interphase, the stage in a cell's life cycle between mitotic divisions (described in chapter 3), the chromosomes and the genes they contain become duplicated. One member of each duplicated chromosome is transmitted to each new cell, so that the new cells receive identical sets of homologous chromosomes and therefore receive identical sets of genes.

Gene Pairs

Since the chromosomes within a cell are paired, the genes within the chromosomes are paired also. Because the members of a gene pair are parts of homologous chromosomes, one member of each pair originates from the female parent and the other originates from the male parent.

Sometimes the members of a gene pair are alike. That is, they contain exactly the same genetic information. For example, both may contain information for the synthesis of a particular enzyme needed for the production of normal skin pigment (melanin). A zygote receiving such a pair of identical genes is called **homozygous** for the genes involved, and the zygote could develop into an individual with normal skin pigmentation.

In other instances, the members of a gene pair are different, because a mutation has occurred in one of them (chapter 4). Such an alternate form of a gene is called an **allele.** For example, one member of a gene pair may contain information for the synthesis of the skin pigment enzyme, while the allele lacks information for enzyme production. When a zygote contains a gene pair whose members differ, it is said to be **heterozygous** for the genes involved.

1. Define *gene.*
2. What is the relationship between genes and chromosomes?
3. What is an allele?
4. What is meant by homozygous? By heterozygous?

Gene Expression

The particular combination of genes present in a zygote and the cells that result when it divides constitutes a **genotype.** The appearance of the individual that develops as a result of the ways the genes are expressed is termed the **phenotype.**

Dominant and Recessive Genes

A zygote that is *homozygous* for normal skin pigmentation will develop into an offspring with normally colored skin. Similarly, a zygote that is *heterozygous* for skin pigmentation—containing one normal gene and one mutant gene—will also develop into an offspring with normal skin pigment because one gene (normal pigmentation) is expressed, but its allele (lack of pigmentation) is not. In such a case, the expressed gene is called **dominant,** and the unexpressed gene is called **recessive.**

For convenience, dominant genes are symbolized with capital letters (A), and recessive genes are symbolized with small letters (a). Thus, the *genotype* of an individual who is homozygous for normal pigmentation due to a dominant gene is symbolized AA, while a heterozygous pair is symbolized Aa. In both instances, the *phenotypes* of the individuals would be normal pigmentation, since a dominant gene is present in each genotype.

A zygote that receives two recessive genes for skin pigmentation has the genotype aa and will develop into an individual who lacks skin pigment. The phenotype resulting from such a genotype is the condition called *albinism* (figure 6.14).

Thus, in the instance of a gene pair involving dominant and recessive genes, three genotypes are possible. They are *homozygous dominant* (AA), *heterozygous* (Aa), and *homozygous recessive* (aa). However, only two phenotypes are possible, since the dominant gene is expressed whenever it is present—in both the homozygous dominant (AA) and the heterozygous (Aa) individuals. The recessive gene is expressed only in the homozygous recessive (aa) condition.

Chart 24.2 lists some of the traits that are determined by dominant and recessive genes.

Incomplete Dominance

Sometimes the members of a gene pair occur in two forms that are expressed differently, and neither is dominant to the other. In other words, when such genes are paired in a heterozygous individual, both genes are partially expressed. Genes of this type are said to blend, or to be *incompletely dominant.*

For example, two forms of a gene needed for the synthesis of hemoglobin are incompletely dominant. A person with the genotype H^1H^1 develops normal he-

CHART 24.2	Traits determined by single pairs of dominant and recessive genes	
Phenotype due to expression of dominant gene	Phenotype due to expression of recessive gene
Full lips	Thin lips
Dark hair	Light hair
Free earlobes	Attached earlobes
Bridge of nose convex	Bridge of nose concave
Dark eyes (brown iris)	Light eyes (blue iris)
Farsightedness	Normal vision
Astigmatism	Normal vision
Extra fingers or toes (polydactyly)	Normal numbers of fingers and toes
Freckles	Lack of freckles
Ability to roll tongue into U-shape	Lack of this ability
Dimples in cheeks	Lack of dimples
Feet with normal arches	Flatfeet
Otosclerosis with hearing loss	Normal hearing

moglobin, while a person with the genotype H^2H^2 develops hemoglobin with abnormal molecules. The amino acid sequences in some of the polypeptide portions (beta chains) of these abnormal molecules differ from the normal sequence. More specifically, within the 146 amino acids in the normal sequence, the abnormal form has a valine in place of a glutamic acid, so that the abnormal form differs from the normal form in a single amino acid. As a consequence, the abnormal molecules tend to form long chains when they are exposed to low oxygen concentrations, and they cause the red blood cells containing them to become distorted or sickle-shaped. Consequently, a person with the genotype H^2H^2 is said to inherit *sickle-cell anemia* (figure 24.4).

A heterozygous individual with the genotype H^1H^2 develops some normal and some abnormal hemoglobin. In this case, the red blood cells may exhibit sickling if oxygen concentrations are relatively low, but usually the red cells remain normal. This condition is called *sickle-cell trait.*

The gene for sickle-cell anemia occurs most commonly in persons of Central African ancestry. About 8% of the United States population with such ancestry possess this gene. As a result, about one child in 170 produced by this group develops sickle-cell anemia. This condition is characterized by symptoms associated with lack of tissue oxygen, which result from the blockage of capillaries by sickled cells. These symptoms include severe joint and abdominal pain, skin ulcers, and chronic kidney disease.

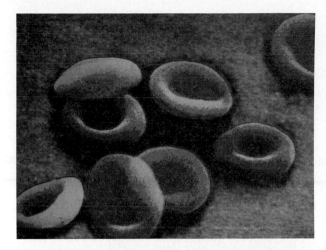

(a)

(b)

Figure 24.4 (a) Falsely colored scanning electron micrograph of normal red blood cells, (b) red blood cells from a person with sickle-cell anemia (×1,650).

These blood types are determined by various combinations of three allelic genes that can be symbolized I^A, I^B, and i^O. In this instance, genes I^A and I^B are equally dominant and are said to be **codominant genes.** They are both dominant to gene i^O. Since each person inherits only two genes for ABO blood type, and there are three forms of the gene, six genotypes are possible. They are: $I^A I^A$, $I^A i^O$, $I^B I^B$, $I^B i^O$, $I^A I^B$, and $i^O i^O$.

If gene I^A is present, agglutinogen A develops on the red blood cell membranes, and if gene I^B is present, agglutinogen B is produced. The presence of gene i^O causes either no agglutinogen to be produced or such a weak agglutinogen that it is usually insignificant. Thus, persons with genotypes $I^A I^A$ or $I^A i^O$ have blood type A. Those with genotypes $I^B I^B$ or $I^B i^O$ have blood type B. Genotype $I^A I^B$ produces type AB blood, while genotype $i^O i^O$ results in type O blood.

Thus, as far as the ABO blood group is concerned, every person belongs to one of six genotypes and has one of four phenotypes:

Genotypes	Phenotypes
$I^A I^A$, $I^A i^O$	Blood type A
$I^B I^B$, $I^B i^O$	Blood type B
$I^A I^B$	Blood type AB
$i^O i^O$	Blood type O

Blood and other tissue types are sometimes used as genetic evidence to prove whether a particular male is or is not the father of a particular child. In such cases, when the child's genotype includes a gene that is not contained in the genotype of either the male or the mother, the male is clearly not the father.

1. Distinguish between genotype and phenotype.
2. What is meant by dominant gene? By recessive gene?
3. Define incomplete dominance.
4. Explain why there are six genotypes but only four phenotypes in the ABO blood group.

Polygenic Inheritance

Although some traits—such as the presence or absence of skin pigment, the production of normal or abnormal hemoglobin, and the development of a certain blood type—seem to be determined by single pairs of genes, other traits are controlled by several pairs. Inheritance of this type is called *polygenic,* and examples of such inheritance include body height and the concentration of skin pigment.

Height seems to be determined by at least four pairs of genes located in different positions within the chromosome set. Each pair causes a small effect in the development of height.

Multiple Alleles

In the previous examples, only two forms of a gene were involved. But in other instances, genes occur in several forms; such a group is called **multiple alleles.**

With multiple alleles, each person carries a single pair of genes, but since the different forms of the gene can occur in various combinations, several genotypes are possible. The inheritance of ABO blood type, for example, involves multiple alleles.

As described in chapter 17, a person's ABO blood type is related to the *agglutinogens* present on the red blood cell membranes. Specifically, two agglutinogens are involved: *agglutinogen A* and *agglutinogen B.* If the red cells have only agglutinogen A, the blood type is A; if only agglutinogen B is present, the blood type is B; if both agglutinogens A and B are present, the blood type is AB; and if neither agglutinogen A nor B is present, the blood type is O.

Figure 24.5 Variations in height are due to different combinations of several pairs of genes.

Using the symbol T for a gene causing tallness and t for a gene causing shortness, the genotype of a very tall person might be $T^1T^1\ T^2T^2\ T^3T^3\ T^4T^4$, and the genotype of a very short person might be $t^1t^1\ t^2t^2\ t^3t^3\ t^4t^4$. Persons with various combinations of these genes, such as $T^1T^1\ T^2t^2\ T^3T^3\ t^4t^4$ or $T^1t^1\ T^2t^2\ t^3t^3\ t^4t^4$, would grow to intermediate heights. In fact, because many different combinations of these genes are possible, people exhibit considerable variation in height (figure 24.5).

Similarly, the amount of melanin produced in the skin seems to be determined by several pairs of genes, and as before, at least four pairs may be involved. Consequently, the genotype of a person with very dark skin might be symbolized $M^1M^1\ M^2M^2\ M^3M^3\ M^4M^4$, and one with very light skin might be $m^1m^1\ m^2m^2\ m^3m^3\ m^4m^4$. People with various combinations of genes for dark and light skin would develop skin of intermediate colors and would exhibit a broad range of variation (figure 24.6).

Environmental Factors

Although the traits of height and concentration of skin pigment are determined by several pairs of genes, the development of these and other hereditary traits is influenced by environmental factors as well. For example, whether or not an individual will achieve the height that is hereditarily possible is affected by such environmental factors as diet and disease. Even if a person has the genotype for extreme tallness, the person may only grow to an intermediate height if the diet is

Figure 24.6 Variations in skin color are due in part to different combinations of several pairs of genes. What other factors influence skin color?

inadequate in essential nutrients or if the body is affected by a disease that interferes with bone growth and development.

Similarly, the amount of pigment deposited in the skin is influenced by environmental factors. If the skin is exposed to ultraviolet light from sunlight or a sunlamp, the skin becomes more heavily pigmented. Conversely, if such exposure is avoided, the concentration of pigment is decreased (figure 24.7).

To summarize, the growth and development of a zygote are controlled by the paired genes it receives from an egg cell and a sperm cell. These genes encode genetic information for the synthesis of specific enzyme molecules, and the enzymes, in turn, promote the metabolic reactions that cause the development of such characteristics as gender, blood type, eye color, hair color, skin color, height, concentration of skin pigment, and so forth.

The way a particular gene is expressed depends on the nature of the gene and the way it is combined with other genes. Gene expression is also influenced by environmental factors that may enhance or inhibit the development of a trait.

> Some genes, called *lethal genes,* cause abnormal development that ultimately results in the death of the individual who inherits them.
>
> For example, if the lethal genes cause abnormal heart development, an embryo might die about the fourth week, when its heart normally begins to function. If the lethal genes cause abnormal development of the kidneys, however, death might not occur until after birth, when the newborn must function without the aid of its mother's organs.

1. How does polygenic inheritance make possible many variations of a given trait?
2. How may environmental factors influence gene expression?

Chromosomes and Sex Cell Production

Although the offspring of a set of parents have many characteristics in common, they also differ because they originate from different egg and sperm cells and develop in different environments. Furthermore, different egg and sperm cells (from the same parents) produce zygotes with different genotypes, because the sex cells of each individual carry unique combinations of genes. This uniqueness results from the process of meiosis.

Meiosis

Meiosis occurs during *spermatogenesis* and *oogenesis,* and as a result of it, the chromosome numbers of the

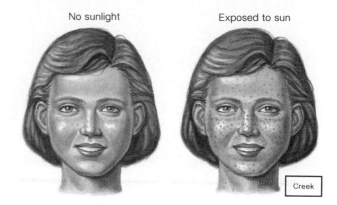

No sunlight   Exposed to sun

Creek

Figure 24.7   The gene responsible for freckles is more fully expressed when the skin is exposed to sunlight.

resulting sex cells are reduced by one-half. (See chapter 22.) This process involves two successive divisions, called the *first* and *second meiotic divisions.* In the first meiotic division, the homologous chromosomes of the parent cell are separated, and the chromosome number is reduced in the newly formed cells. During the second meiotic division, the chromosomes act much as they do in *mitosis,* and the result is the formation of sex cells (figure 24.8).

Like mitosis, described in chapter 3, meiosis is a continuous process without marked interruptions between steps. However, for convenience, the process can be divided into stages as follows:

1. **First meiotic prophase.** This stage (figure 24.9) resembles prophase in mitosis, although it lasts somewhat longer and is more complex. During it, the individual chromosomes appear as thin threads within the nucleus. The threads become shorter and thicker, the nucleoli disappear, the nuclear membrane fades away, and the spindle fibers become organized.

   Throughout this stage, the individual chromosomes, which were duplicated during the interphase prior to the beginning of meiosis, are composed of two identical chromatids. Each pair of chromatids is held together by a central region called the *centromere.*

   As the prophase continues, homologous chromosomes approach each other, lie side by side, and become tightly intertwined. This process of pairing, shown in figure 24.10, is called *synapsis,* and during synapsis the chromatids of the homologous chromosomes contact one another at various points. In fact, the chromatids may break in one or more places and exchange parts, forming chromatids with new combinations of genetic information.

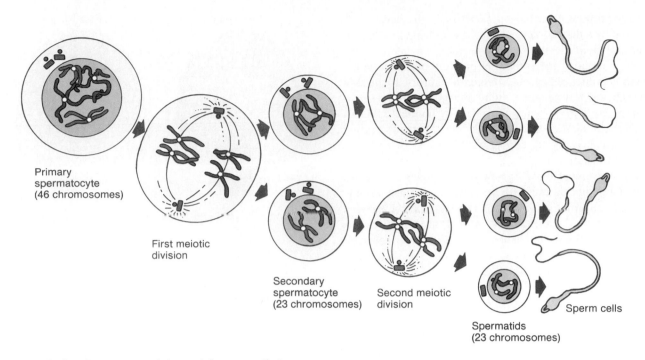

Figure 24.8 Spermatogenesis is a meiotic process that involves two successive divisions.

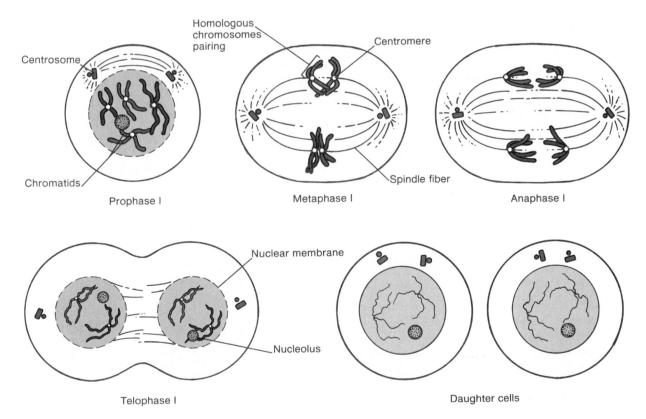

Figure 24.9 Stages in the first meiotic division.

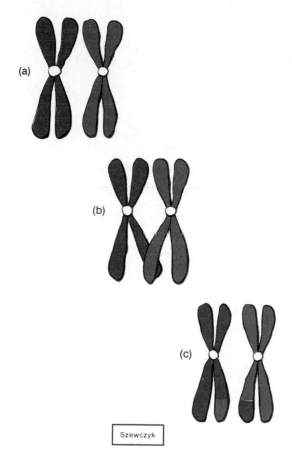

Figure 24.10 (a) Pairing of homologous chromosomes; (b) chromatids crossing over; (c) results of crossing over.

For example, since one chromosome of a homologous pair is of maternal origin and the other is of paternal origin, an exchange of parts between homologous chromatids results in chromatids that contain genetic information from both maternal and paternal sources. This process of exchanging genetic information is called **crossing over.**

2. **First meiotic metaphase.** During the first metaphase, shown in figure 24.9, the synaptic pairs of chromosomes become lined up about midway between the poles of the developing spindle. Each pair consists of two chromosomes and four chromatids. Each member of a chromosome pair becomes attached to a spindle fiber that is associated with one pole of the spindle.

3. **First meiotic anaphase.** In mitosis, the centromeres that hold the chromatids of each chromosome pair together separate during anaphase, and the chromatids are free to move toward opposite ends of the spindle. However, during the first meiotic division, the centromeres do not separate (figure 24.9), and as a result, the homologous chromosomes become separated, and the chromatids of each

chromosome move together to one end of the spindle. Thus, each new cell receives only one member of a homologous pair of chromosomes, and the chromosome number is thereby reduced by one-half.

4. **First meiotic telophase.** Meiotic telophase is similar to mitotic telophase in that the parent cell divides into two new cells (figure 24.9). Also during this phase, nuclear membranes appear around the chromosomes, the nucleoli reappear, and the spindle fibers fade away.

The first meiotic telophase is followed by a short *interphase* and the beginning of the second meiotic division. This division is essentially the same as mitosis, and it can also be divided into *prophase, metaphase, anaphase,* and *telophase.* These stages are shown in figure 24.11.

During the **second meiotic prophase,** the chromosomes reappear. They are still composed of pairs of chromatids, and the chromatids are held together by centromeres. Near the end of this phase, the chromosomes move into positions midway between the poles of the developing spindle.

In the **second meiotic metaphase,** the double-stranded chromosomes become attached to spindle fibers, and during the **second meiotic anaphase,** the centromeres separate so that the chromatids are free to move to opposite poles of the spindle. As a result of the **second meiotic telophase,** each cell produced by the first meiotic division divides to form two cells.

Results of Meiosis

In spermatogenesis, meiosis results in the formation of *four sperm cells,* each of which contains 23 single-stranded chromosomes (chapter 22). These chromosomes contain a complete set of genetic information but only one gene of each type, whereas the original parent cell (spermatogonium) contained a pair of each type of gene. Furthermore, the genetic information varies from sperm cell to sperm cell because of the crossing over that occurred during the first meiotic prophase. For example, as figure 24.12 shows, one sperm cell may contain genes for the development of eye color and hair color from an original maternal chromosome and a gene for blood type from an original paternal chromosome. Another sperm cell may contain a gene for maternal eye color and paternal hair color and blood type, while a third sperm cell contains genes for maternal hair color and blood type and a paternal gene for eye color.

In oogenesis, meiosis results in *one egg cell* (secondary oocyte) and a nonfunctional polar body (chapter 22). If the egg cell is fertilized, a second polar body is produced (figure 24.13). Like sperm cells, egg cells

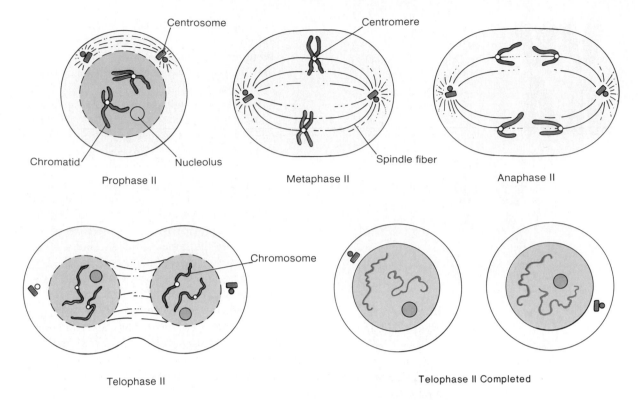

Centrosome

Centromere

Chromatid

Nucleolus

Prophase II

Spindle fiber

Metaphase II

Anaphase II

Chromosome

Telophase II

Telophase II Completed

Figure 24.11 Stages in the second meiotic division.

contain 23 single-stranded chromosomes whose combinations of genetic information differ from egg cell to egg cell because of crossing over.

1. Distinguish between meiosis and mitosis.
2. Describe the major events that occur during meiosis.
3. What is the final result of meiosis?

Inheritance of Single Traits

Genes occur in pairs, but as a result of meiosis, a sex cell contains only one member of each gene pair. Consequently, some sex cells contain the original *maternal gene* of a gene pair, and the other sex cells contain the original *paternal gene* of such a pair.

Dominant and Recessive Inheritance

As mentioned previously, some genes occur in two forms—dominant and recessive. In pigmentation, for example, the dominant gene for the development of normal pigment can be symbolized as A, while its recessive allele can be symbolized as a. Thus, a person who is homozygous dominant for this trait has the genotype AA and will appear normal. A person who is heterozygous has the genotype Aa and also will appear normal, since a dominant gene for pigmentation is present. A person who is homozygous recessive, however, has the genotype aa and will have *albinism*.

Because a person with genotype AA has two identical genes, all the sex cells produced by this individual will receive a dominant gene (A). Likewise, all of the sex cells produced by a person with genotype aa will receive a recessive gene (a). However, a person with the genotype Aa will produce equal numbers of two kinds of sex cells. One kind will have a dominant gene (A), and the other kind will have a recessive gene (a) (figure 24.14).

When information concerning the genotypes of parents is available, it is often possible to predict what types of offspring can be produced as well as the proportions in which they are likely to occur. For example, since a male parent with genotype AA produces only A-bearing sperm cells, and a female parent with genotype aa produces only a-bearing egg cells, all the zygotes resulting from the fusion of such sex cells will have the genotype Aa:

	Male Parent	Female Parent
Genotype:	AA	aa
Phenotype:	Normal pigment	Albinism
Sex cells:	All A	All a

Egg cells

a

Possible zygotes:

Sperm cells

A | Aa

Offspring genotypes: All Aa
Offspring phenotypes: All normal pigment

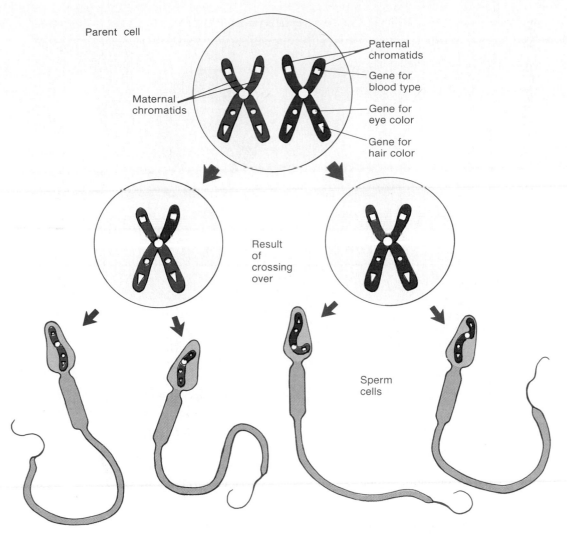

Figure 24.12 As a result of crossing over, the genetic information held within the sperm cells (and the egg cells) varies from cell to cell.

Oogenesis

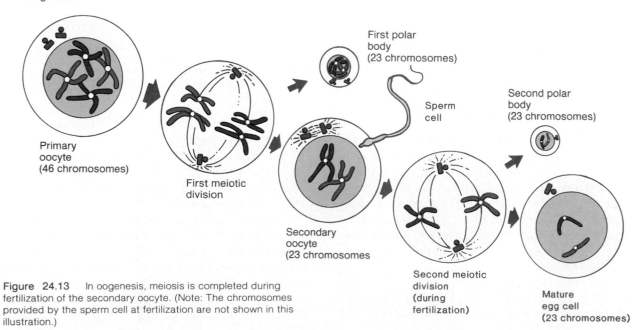

Figure 24.13 In oogenesis, meiosis is completed during fertilization of the secondary oocyte. (Note: The chromosomes provided by the sperm cell at fertilization are not shown in this illustration.)

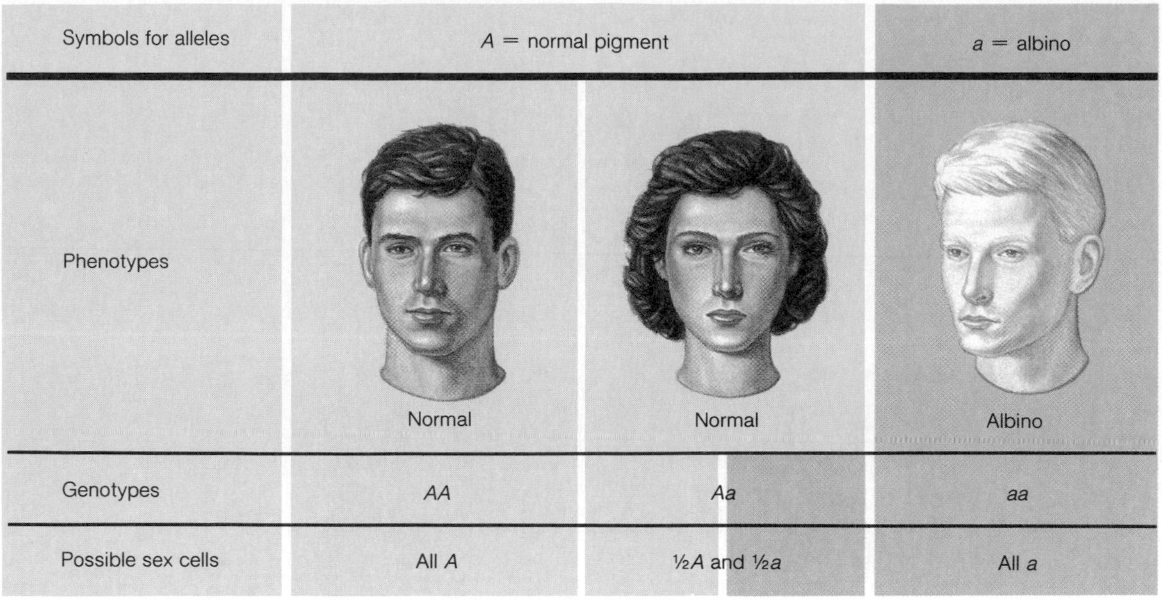

Symbols for alleles	A = normal pigment		a = albino
Phenotypes	Normal	Normal	Albino
Genotypes	*AA*	*Aa*	*aa*
Possible sex cells	All *A*	½*A* and ½*a*	All *a*

Figure 24.14 In the case of alleles *A* and *a* for skin pigmentation, there are two possible phenotypes and three possible genotypes. The sex cells produced by each person reflect the composition of his or her genotype.

On the other hand, if the male parent has the genotype *Aa* and the female parent is *aa*, offspring of two possible genotypes can be produced—*Aa* and *aa:*

	Male Parent	Female Parent
Genotype:	*Aa*	*aa*
Phenotype:	Normal pigment	Albinism
Sex cells:	½ *A* and ½ *a*	All *a*

Egg cells

		a
Possible zygotes:	½ *A*	½ *Aa*
	½ *a*	½ *aa*

Offspring genotypes: ½ *Aa* and ½ *aa*
Offspring phenotypes: ½ normal pigment and ½ albinism

In this case, ½ of the sperm cells carry the gene *A* and ½ carry the gene *a*. Since sperm cells have equal chances of fertilizing an egg cell, it is expected that ½ of the offspring from these parents will have the genotype *Aa* and ½ will have the genotype *aa*. Because the presence of a dominant gene (*A*) produces normal pigment, those offspring with genotype *Aa* will have normal pigmentation, and those with genotype *aa* will have albinism.

A more complex situation arises if both the parents have the genotype *Aa*. In this instance, each parent produces two kinds of sex cells in equal numbers: ½ *A* and ½ *a*—thus the offspring resulting from such parents can have three possible genotypes—*AA, Aa,* or *aa:*

	Male Parent	Female Parent
Genotype:	*Aa*	*Aa*
Phenotype:	Normal pigment	Normal pigment
Sex cells:	½ *A* and ½ *a*	½ *A* and ½ *a*

Egg cells

		½ *A*	½ *a*
Possible zygotes:	½ *A*	¼ *AA*	¼ *Aa*
	½ *a*	¼ *Aa*	¼ *aa*

Offspring genotypes: ¼ *AA,* ½ *Aa,* ¼ *aa*
Offspring phenotypes: ¾ normal pigment and ¼ albinism

In this case, ¼ (25%) of the offspring are expected to have the genotype *AA,* ½ (50%) are expected to be *Aa,* and ¼ (25%) are expected to be *aa.* However, since those genotypes *AA* or *Aa* will appear normal, only ¼ of the offspring from such parents are expected to have albinism (*aa*).

Because chance events, such as the fertilization of a particular kind of egg cell by a particular kind of sperm cell, occur independently, the genotype of one child has no effect upon the genotypes of other children produced by the same set of parents. Thus, each child conceived by parents with genotypes *Aa* and *Aa* has a 25% chance of being *AA* (normal), a 50% chance of being *Aa* (normal, but carrying the albinism gene), and a 25% chance of having *aa* (albinism).

Incompletely Dominant Inheritance

As mentioned previously, two forms of a particular gene function in the production of hemoglobin. One of the alleles (H^1) is responsible for the formation of normal hemoglobin, and the other (H^2) results in abnormal molecules that cause *sickling* of red blood cells. Since genes H^1 and H^2 are *incompletely dominant,* a person with genotype H^1H^1 has normal hemoglobin; a person with genotype H^1H^2 develops sickle-cell trait; and a person with genotype H^2H^2 suffers from sickle-cell anemia.

The kinds of offspring expected from parents with various combinations of incompletely dominant genes can be determined as follows:

1. Parent genotypes H^1H^1 and H^2H^2:

	Male Parent	Female Parent
Genotype:	H^1H^1	H^2H^2
Phenotype:	Normal	Sickle-cell anemia
Sex cells:	All H^1	All H^2

		Egg cells
		H^2
Possible zygotes: Sperm cells	H^1	H^1H^2

Offspring genotypes: All H^1H^2
Offspring phenotypes: All sickle-cell trait

2. Parent genotypes H^1H^2 and H^2H^2:

	Male Parent	Female Parent
Genotype:	H^1H^2	H^2H^2
Phenotype:	Sickle-cell trait	Sickle-cell anemia
Sex cells:	½ H^1 and ½ H^2	All H^2

		Egg cells
		H^2
Possible zygotes: Sperm cells	½ H^1	½ H^1H^2
	½ H^2	½ H^2H^2

Offspring genotypes: ½ H^1H^2 and ½ H^2H^2
Offspring phenotypes: ½ sickle-cell trait and ½ sickle-cell anemia

3. Parent genotypes H^1H^2 and H^1H^2:

	Male Parent	Female Parent
Genotype:	H^1H^2	H^1H^2
Phenotype:	Sickle-cell trait	Sickle-cell trait
Sex cells:	½ H^1 and ½ H^2	½ H^1 and ½ H^2

		Egg cells	
		½ H^1	½ H^2
Possible zygotes: Sperm cells	½ H^1	¼ H^1H^1	¼ H^1H^2
	½ H^2	¼ H^1H^2	¼ H^2H^2

Offspring genotypes: ¼ H^1H^1, ½ H^1H^2, ¼ H^2H^2
Offspring phenotypes: ¼ normal, ½ sickle-cell trait, ¼ sickle-cell anemia

---

1. What offspring genotypes can be produced by parents with the genotypes *AA* and *aa*? By parents with genotypes *AA* and *Aa*?
2. How many offspring genotypes can be produced by parents with genotypes *Aa* and *Aa*? How many offspring phenotypes?
3. If both parents have the sickle-cell trait, what will be the proportions of the genotypes and phenotypes of offspring that can be produced?

Sex Chromosomes

The gender of an individual is inherited in much the same manner as other traits. However, in this instance, whole chromosomes are involved rather than a few paired genes.

Gender Inheritance

Among the 46 chromosomes of a human body cell, there are 22 pairs called **autosomes** (chromosomes other than sex chromosomes) and one pair of **sex chromosomes.** The sex chromosomes are responsible for the development of the contrasting characteristics associated with *maleness* and *femaleness*. In other words, the kinds of sex chromosomes present in a zygote determine whether the embryonic reproductive organs (which are essentially neutral in the early stages of development) will form into male or female structures.

Sex chromosomes are of two types. One type, which is relatively large, is called an **X chromosome,** and the other, which is relatively small, is called a **Y chromosome.** A female karyotype of sex chromosomes consists of two X chromosomes; consequently, all egg cells contain a single X chromosome. The male karyotype consists of an X and a Y chromosome, thus equal numbers of two types of sperm cells are produced. One-half of the sperm cells possess an X chromosome, and

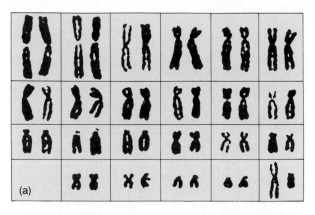

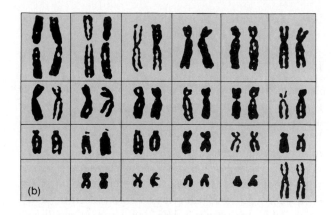

Figure 24.15 (a) A karyotype of human male chromosomes and (b) a karyotype of human female chromosomes arranged in pairs of homologous chromosomes. How are the sets different?

one-half possess a Y chromosome. This difference is shown in figure 24.15.

The gender of an offspring is determined by the type of sperm cell that fertilizes the egg cell. If the sperm involved carries an X chromosome, it will be paired with the X chromosome of the egg cell, and the XX combination will result in the development of a female. On the other hand, if a Y-bearing sperm cell fertilizes the egg, the XY combination results in the development of a male offspring. Since X-bearing and Y-bearing sperm cells occur in equal numbers, it is expected that equal numbers of males and females will result from the chance meeting of egg and sperm cells.

	Male Parent	Female Parent
Sex chromosomes:	XY	XX
Sex cells:	½ X and ½ Y	All X

Egg cells

		X
Sperm cells	½ X	½ XX
	½ Y	½ XY

Possible zygotes:

Offspring gender ratio: ½ males (XY) and ½ females (XX)

However, for reasons that are not well understood, an unusually large proportion of males are born. In fact, in the United States population, the ratio of males to females at birth is 106:100.

Even though greater numbers of males than females are born, the death rate among males is higher. Consequently, the gender ratio between males and females changes as they age. This change in gender ratio is reflected in the following data:

Age	Gender Ratio (males:females)
Birth	106:100
18 years	100:100
50 years	85:100
85 years	50:100
100 years	20:100

Actions of Sex Chromosomes

A zygote with an XY combination of sex chromosomes apparently gives rise to a male offspring because its Y chromosome contains information for the synthesis of a particular kind of protein molecule. This protein, called the *H-Y antigen,* appears on the surfaces of cells produced as the offspring develops. Its presence on the cell membranes of the undifferentiated reproductive organs somehow promotes their development into testes. At the same time, the antigen stimulates cells within the immature testes to proliferate and form spermatogonia. However, the spermatogonia are inhibited by action of the developing supporting cells (Sertoli's cells) of the testes, and sperm cell formation does not occur until puberty. (See chapter 22.)

Once the formation of the testes has been initiated, the function of the Y chromosome in determining the gender of the offspring seems to be completed. Further masculinization of the developing reproductive structures is controlled by hormones secreted by the testes as they form.

A zygote with an XX combination of sex chromosomes gives rise to a female offspring apparently without the aid of a substance corresponding to the H-Y antigen. Instead, in the presence of two X chromosomes, the embryonic reproductive organs enlarge for several weeks and then begin to differentiate into ovaries. Concurrently, several million primordial follicles are formed. (See chapter 22.)

> Sometimes there are areas in chromosomes that are especially susceptible to breaking abnormally. Such areas are called *fragile sites.* Although most fragile sites seem to be unrelated to clinical disorders, an X chromosome with a fragile site called *fragile X* is associated with the most common form of inherited mental retardation (fragile X-linked mental retardation).

X Chromosome-linked Inheritance

In addition to their importance in gender determination, the X chromosomes contain genes responsible for the development of certain traits unrelated to gender. However, since these traits are determined by genes located on the X chromosomes, the patterns by which they are inherited differ in each gender. For example, such traits are more likely to be expressed in males (XY) than in females (XX).

This phenomenon is related to the fact that the X and Y sex chromosomes are quite different in size, and most of the genes on the X chromosome do not have corresponding genes on the relatively small Y chromosome. Consequently, in males, the genes of the single X chromosome are expressed even though they may be recessive, while in females, a trait determined by such a recessive gene is expressed only if it is present on both X chromosomes (homozygous recessive). Traits determined by recessive genes located on X chromosomes are called **X-linked.**

> Some traits are determined by genes located within a portion of the Y chromosome for which there is no homologous region in an X chromosome. These traits appear only in males and are passed from fathers to their sons. This type of genetic transmission is called *holandric inheritance.* An example of such a trait is *hypertrichosis,* which is characterized by excessively hairy ears.

Color blindness is an example of an X-linked trait. In this instance, normal color vision is due to the presence of a dominant gene (C). This gene is needed for the development of functional color receptors (cones) in the retina of the eye. The recessive allele (c) causes the production of defective receptors.

A male will have normal color vision if his X chromosome carries the dominant gene, which can be symbolized X^C. However, if a male receives one recessive gene for color blindness (X^c), he will be color-blind, since there is no corresponding gene on the Y chromosome to overcome its effect.

A female, on the other hand, will have normal color vision if either of her X chromosomes carries the dominant gene (X^C). Thus, both the homozygous dominant (X^CX^C) and the heterozygous female (X^CX^c) will develop normal vision. Only a homozygous recessive female (X^cX^c) will be color-blind.

A color-blind male only occurs when the female parent carries recessive genes for color blindness, since the X chromosome of a male offspring is always received from his mother:

	Normal Male Parent	Color-blind Female Parent
Sex chromosomes:	X^CY	X^cX^c
Sex cells:	½ X^C and ½ Y	All X^c

Egg cells
X^c

Possible zygotes:	Sperm cells X^C	½ X^CX^c
	Y	½ X^CY

Offspring genotypes:
Females— X^CX^c
Males— X^cY

Offspring phenotypes:
Females—normal
Males—color-blind

Although all the daughters in this case will have normal vision, they will carry the gene for color blindness.

In the case of a homozygous dominant female parent (X^CX^C) and a male parent with color blindness (X^cY), all of the children will develop normal vision. However, all the daughters of a color-blind male parent will receive the recessive gene for color blindness, since all daughters receive the X chromosome of the father. These female offspring will be heterozygous (X^CX^c) and will be carriers of the gene for color blindness.

                    Color-blind   Normal
                    Male Parent   Female Parent
Sex chromosomes: X^cY X^cX^c
Sex cells: ½ X^c and ½ Y All X^c

Egg cells
X^c

Possible zygotes:	½ X^c	½ X^cX^c
	½ Y	½ X^cY

(Sperm cells)

Offspring genotypes:
Females—X^cX^c
Males—X^cY

Offspring phenotypes:
Females—normal
Males—normal

If the female parent has normal vision but carries the gene for color blindness (X^CX^c), and the male parent has normal vision (X^CY), all the female offspring will have normal vision, but ½ of them are expected to be carriers of the recessive gene. At the same time, of the male offspring ½ are expected to have normal vision (X^CY) and ½ are expected to be color-blind (X^cY).

                    Normal   Normal
                    Male Parent   Female Parent
Genotypes: X^CY X^CX^c
Sex cells: ½ X^C and ½ Y   ½ X^C and ½ X^c

Egg cells
X^C X^c

		X^C	X^c
Possible zygotes:	½ X^C	¼ X^CX^C	¼ X^CX^c
	½ Y	¼ X^C	¼ X^CY

(Sperm cells)

Offspring genotypes:
Females—½ X^CX^C, ½ X^CX^c
Males—½ X^CY, ½ X^cY

Offspring phenotypes:
Females—normal
Males—½ normal, ½ color-blind

Hemophilia, sometimes called the bleeder's disease, is also due to an X-linked recessive gene. In this instance, a single recessive gene in the male (X^h) or two recessive genes in the female (X^hX^h) cause a defect in the blood clotting mechanism. As a result, prolonged bleeding may accompany an otherwise minor injury.

CHART 24.3	Some traits related to X-linked recessive genes
Trait	**Characteristics**
Brown enamel	Tooth enamel appears brown rather than white
Coloboma iridis	A fissure in the iris of the eye causes the pupil to appear slitlike rather than round
Color blindness, deutan type	Decreased sensitivity to green light; the most common type of color blindness
Color blindness, protan type	Decreased sensitivity to red light
Congenital night blindness	Vision is poor in dim light, but relatively normal in bright light; this condition is inherited, not due to a deficiency of vitamin A
Hemophilia A	Deficiency of blood clotting factor VIII; classic hemophilia
Hemophilia B	Deficiency of blood clotting factor IX; Christmas disease
Microphthalmia	Eyes fail to develop normally and instead remain small and nonfunctional
Optic atrophy	Blindness results from degeneration of the optic nerves

Like color blindness, hemophilia is more common in males. The recessive gene is received by male offspring from their mothers and is passed from affected fathers to their daughters.

Actually, two major forms of hemophilia are caused by recessive genes carried on the X chromosomes. *Hemophilia A* (classic hemophilia) is characterized by the lack of plasma protein called *antihemophilic globulin* (blood clotting factor VIII, or von Willebrand's factor). *Hemophilia B* (Christmas disease) is due to a deficiency of another blood substance called *plasma thromboplastin component* (blood clotting factor IX, or Christmas factor).

Chart 24.3 lists some other traits related to X-linked genes.

An inherited disease called *pseudohemophilia* (von Willebrand's disease) is caused by the presence of a dominant gene on an autosome. In this condition, the platelets are defective, and bleeding is prolonged. However, since the responsible gene is on an autosome, this disease is not X-linked.

1. How is the gender of an offspring determined?
2. How do sex chromosomes promote the differentiation of reproductive organs?
3. What is meant by X-linked inheritance?
4. Why do X-linked traits appear most commonly in males?

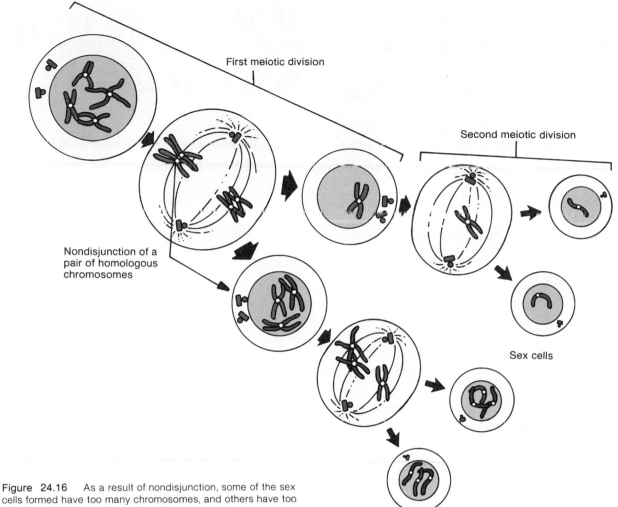

First meiotic division

Second meiotic division

Nondisjunction of a
pair of homologous
chromosomes

Sex cells

Figure 24.16 As a result of nondisjunction, some of the sex cells formed have too many chromosomes, and others have too few.

Pattern baldness is a hereditary trait that is due to the expression of autosomal *sex-influenced genes.* Such genes act differently in males than in females; that is, they act as dominant genes in the hormonal environment of an adult male and as recessive genes in the hormonal environment of an adult female. Consequently, pattern baldness occurs much more frequently in males than in females. Although some females may develop this condition, it usually is less obvious and appears later in life than in males.

Chromosome Disorders

Chromosome disorders are characterized by the presence of abnormal numbers of chromosomes. Since the human karyotype usually contains 46 chromosomes, any other number is considered abnormal.

Cause of Chromosome Disorders

As described previously, homologous chromosomes undergo pairing (synapsis) during meiosis. Later, these homologous pairs separate and move to opposite ends of the spindle. As a result, each newly formed cell receives one chromosome of a homologous pair.

However, if one pair of homologous chromosomes fails to separate during meiosis, both members of that pair will move into one new cell, while the other new cell will receive no chromosome of that type. This failure of homologous chromosomes to separate is called **nondisjunction.** When it occurs, some of the resulting cells have too many chromosomes, while others have too few (figure 24.16).

Sex Chromosome Disorders

If nondisjunction occurs during *oogenesis,* some of the resulting egg cells may receive two X chromosomes (XX), while others receive none (figure 24.17). Both of these types of egg cells can be fertilized by sperm cells, and the zygotes formed can undergo development.

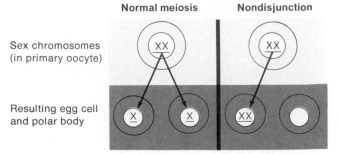

Figure 24.17   An egg cell containing two X chromosomes can result from nondisjunction.

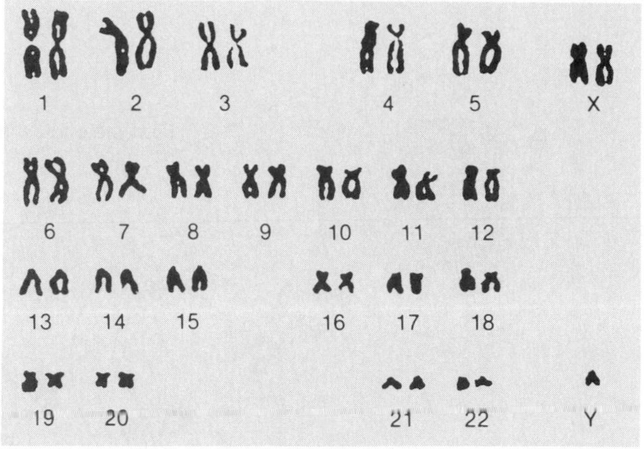

Apparently, only one X chromosome of an XX pair is functional, and in many female cells, the other nonfunctional X chromosome is found lying against the nuclear membrane, where it forms a tiny structure called a *Barr body*.

In some hospitals, cells of the amnion of each newborn are stained and observed for the presence of Barr bodies. In this way, it is possible to determine if there are abnormal numbers of X chromosomes in the cells.

If an XX egg cell is fertilized by a Y-bearing sperm cell, the resulting zygote will have 47 chromosomes and the karyotype XXY (designated: 47, XXY). This abnormal combination of sex chromosomes causes the development of an individual with male sex organs. However, the testes usually fail to produce sperm cells, so that the person is sterile. This condition, called *Klinefelter's syndrome*, is also characterized by tall stature, a somewhat feminine musculature, and partial development of the breasts (figure 24.18).

An XX egg cell fertilized by an X-bearing sperm cell becomes a zygote with the karyotype XXX (47, XXX). The individual who develops from such a zygote is a relatively normal female, and although XXX females are sometimes sterile, others can produce offspring. Some XXX females are mentally retarded.

Females with the karyotypes 48, XXXX and 49, XXXXX are known to exist, and although a consistent pattern of traits is lacking in such individuals, the chances of developing mental retardation seem to increase markedly as the number of X chromosomes in the karyotype increases.

Similarly, males with the karyotypes 48, XXXY, 48, XXYY, 49, XXXXY, and 49, XXXYY are known. Such individuals are phenotypically similar to those with Klinefelter's syndrome. Another abnormal male karyotype involves the presence of two Y chromosomes. Males with this karyotype (47, XYY) are usually over six feet tall and are often mentally retarded.

Figure 24.18   A person with Klinefelter's syndrome has the sex chromosome karyotype 47,XXY. How might such a combination come about?

If an egg cell without an X chromosome (O) is fertilized by an X-bearing sperm cell, the resulting 45, XO zygote develops into an individual with *Turner's syndrome* (figure 24.19). In this instance, the person is female, but usually fails to mature reproductively. The sex organs remain juvenile, the breasts fail to develop, the stature is short, and loose folds of skin appear in the posterior region of the neck.

As they age, older persons seem to develop increasing numbers of body cells with 45 chromosomes. In males, this abnormal number may involve a lack of Y chromosomes, and in females, cells tend to miss an X chromosome. These losses apparently result from abnormal mitotic divisions, but their significance to the aging process is unknown.

Autosomal Abnormalities

Nondisjunction sometimes occurs among homologous pairs of autosomes, and as a result, a cell may receive an extra autosome. For example, if nondisjunction occurs during oogenesis, an egg cell may appear with two chromosomes of the same type. If such an egg cell is fertilized by a normal sperm cell, the zygote will have three chromosomes of one kind—a condition that is called *trisomy*.

In humans, *Down syndrome* results from trisomy in which a zygote receives three autosomes of a particular type (chromosome 21) (figure 24.20). This condition, also called *trisomy-21*, is characterized by a small, round head, mental retardation, short stature, stubby hands and feet, and a protruding tongue. In addition, the reproductive organs usually remain underdeveloped, and the heart may be malformed.

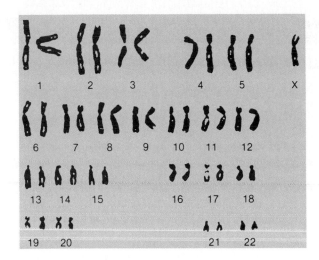

Figure 24.19 What is abnormal in this karyotype from a person with Turner's syndrome?

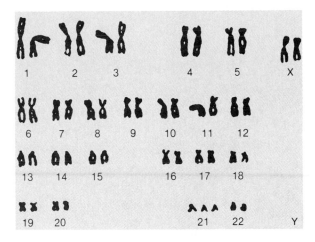

Figure 24.20 A person with Down syndrome has three members of chromosome 21 in the karyotype.

Although the cause of nondisjunction leading to Down syndrome is unknown, this condition is more likely to occur in an offspring of an older mother than in one of a younger mother. In fact, Down syndrome occurs only about once in two thousand live births among women of early childbearing age, but it occurs about once in fifty live births among women over forty years of age.

Other chromosome abnormalities resulting from triply represented autosomes include *Edward's syndrome* (trisomy 18) in which the fetus develops heart and kidney defects, and *Patau's syndrome* (trisomy 13) in which the infant is usually blind and has heart abnormalities. With either abnormality, the offspring seldom lives longer than a few weeks following birth.

1. What is meant by nondisjunction?
2. What are some of the sex chromosome disorders that are due to nondisjunction?
3. Define *trisomy*.
4. What are the consequences of trisomy?

Medical Genetics

Medical genetics is the branch of genetics concerned with the relationship between inheritance and disease.

Genetic Diseases

Genetic diseases comprise a diverse group of disorders including abnormalities of blood cells (such as sickle-cell anemia), defects in blood clotting mechanisms (such as hemophilia), and mental retardation (such as Down syndrome). The causes of such diseases are *mutations*—changes in the genes or chromosomes that can be passed from parent to offspring by means of sex cells. (See chapter 4.)

Genetic diseases can be subdivided into three main groups. Those within the first group produce their effects during embryonic or fetal development, and they are usually caused by chromosomal abnormalities, as when a chromosome is either missing or present in excess. Such diseases commonly result in miscarriages. Those fetuses that survive usually have physical or neurological damage, as in the case of Down syndrome.

Diseases of the second group are usually due to disorders of single genes that control various metabolic processes. These *monogenic diseases* express themselves after birth, when the newborn is forced to rely upon its own life processes. In the disease called *phenylketonuria,* for example, an affected newborn is unable to metabolize the amino acid phenylalanine. Prior to its birth, the fetus was protected by its mother's ability to metabolize this substance, but as a newborn, its developing nervous system may be damaged by the presence of phenylalanine.

The third group of genetic diseases produce their effects during adolescence or adulthood. Although some of these conditions are caused by single-gene disorders, most of them involve interactions between genetic factors and the environment. Diseases of this type include certain forms of hypertension, peptic ulcers, and excessive blood cholesterol (hypercholesterolemia).

Genetic Counseling

As a group, genetic diseases are difficult to treat successfully, for although it may be possible to identify the gene or chromosome involved with a particular disorder, the mechanism by which that genetic factor initiates its effect may not be understood. Consequently, at present, the most effective approach for dealing with genetic diseases is by preventing them.

Detection of Genetic Disorders

Sometimes it is possible to detect individuals with particular genetic disorders through blood tests. *Sickle-cell trait*, for example, can be detected by exposing red blood cells to low oxygen concentrations. For this test, a drop of blood is mixed with a substance that deoxygenates the blood, and if the red cells contain abnormal hemoglobin, they assume a sickle shape.

Blood tests can also reveal carriers (heterozygotes) of the recessive genes involved in the development of the fol-

lowing diseases: *hemophilia,* in which bleeding is prolonged; *Tay-Sach's disease,* which leads to early blindness, deterioration of mental and physical abilities, and death; and *Cooley's anemia* (thalassemia major), which is accompanied by severe defects in the red blood cells and invariably leads to death.

If prospective parents are concerned about the genetic condition of a developing fetus (prenatal diagnosis), information can sometimes be obtained by testing fetal cells.

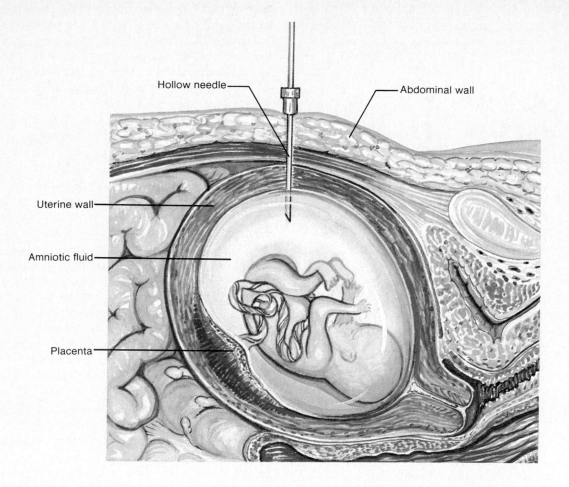

Hollow needle

Abdominal wall

Uterine wall

Amniotic fluid

Placenta

Figure 24.21   Fetal cells can be obtained for examination by means of amniocentesis.

Genetic counseling is a service by which people who have produced children with hereditary diseases, or who are members of families in which such disorders have occurred, can obtain information about the probability of genetic disorders being transmitted to their offspring. Sometimes a genetic counselor can make rather precise predictions. For example, a counselor

could tell parents who both have the sickle-cell trait that their chances of having a child with sickle-cell anemia are one in four, or 25%. However, if one parent is normal for hemoglobin and the other parent has sickle-cell anemia, the chance of their producing a child with sickle-cell anemia is zero, although all of their children will inherit the sickle-cell trait.

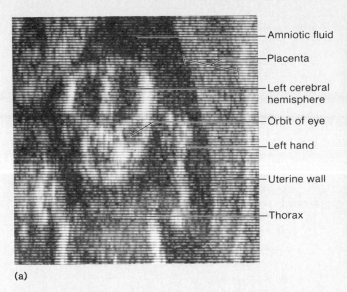

— Amniotic fluid

— Placenta

— Left cerebral hemisphere

— Orbit of eye

— Left hand

— Uterine wall

— Thorax

(a)

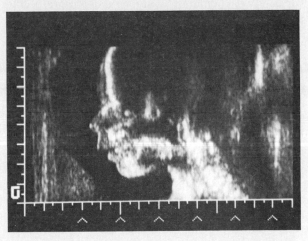

Figure 24.22 A fetus within a uterus as revealed by ultrasonography. (*a*) Frontal view; (*b*) side view.

This procedure, called *amniocentesis,* can be performed in a physician's office. With the aid of ultrasonography (chapter 1), a small quantity of *amniotic fluid* is withdrawn from the amniotic cavity by passing a hollow needle through the maternal abdominal and uterine walls (figure 24.21). When this is done between the fourteenth and sixteenth weeks of development, cells of fetal origin (fibroblasts) are usually present in the fluid, and they can be cultured and tested for chromosomal defects and certain single-gene abnormalities. The amniotic fluid obtained by amniocentesis can also be analyzed for the presence of a specific protein (alpha-fetoprotein, or AFP) that is associated with serious defects in the developing nervous system, such as spina bifida or anencephaly.

Another technique sometimes used to test fetal cells is called *chorionic villi biopsy.* This procedure employs sonography to guide a catheter into a pregnant woman's uterus and uses suction to obtain a sample of the hairlike projections (villi) from the embryonic membrane (chorion) surrounding the young offspring. The cells of the sample can then be analyzed for chromosomal defects and certain biochemical abnormalities. Such a biopsy must be performed

between eight and ten weeks of pregnancy and has the advantage of providing enough tissue to examine immediately without culturing cells.

Fetal cells can also be tested for various hemoglobin disorders, including thalassemia and sickle-cell anemia. Even though fetal fibroblasts obtained by amniocentesis do not synthesize hemoglobin, they contain a complete set of human DNA molecules. These molecules can be removed from the cells, and certain portions of their gene structures can be analyzed. This technique utilizes special enzymes (restriction enzymes) that break the DNA nucleotide sequences within the DNA molecules at particular points, producing precise patterns of fragments. The nature of these fragments reveals the presence of normal or abnormal genes for hemoglobin synthesis.

Ultrasonography is commonly used to examine a fetus for the presence of gross anatomic disorders, such as the absence of the braincase and brain (anencephaly) or certain fetal heart defects (figure 24.22). Similarly, disorders in fetal development may be observed directly. In this procedure, called *fetoscopy,* a tubular fiberoptic device is inserted into the mother's uterus through her cervix or abdominal wall, and the fetus is examined visually.

The accuracy of such a prediction depends on correct diagnosis of the genetic disorder, understanding of the pattern by which the disorder is inherited, and knowledge of the parents' genotypes. Unfortunately, some hereditary traits behave differently in different families. For example, a trait may act as an autosomal recessive in one family and as an

X-linked recessive in another. Consequently, a genetic counselor may have to determine how a particular trait is transmitted within a particular family before making a prediction.

In addition, the counselor must be concerned with the psychological effects such a prediction may have upon the individuals involved, and must help them weigh the probability of a genetic defect occurring against the probability of desirable traits that might appear in their child—traits that could compensate to some degree for the possible defect. At the same time, the counselor must try to remain objective and avoid imposing a particular set of values upon the prospective parents, since the final decision about childbearing is theirs.

1. How are genetic diseases subdivided into groups?
2. What is the purpose of genetic counseling?
3. What factors affect the accuracy of predictions made by genetic counselors?

Hereditary Disorders of Clinical Interest

cri du chat syndrome (kre du shah sin' drōm) Characterized by peculiar mewing (cat cry) vocal sounds, microcephaly, and severe mental retardation; due to a partial deletion in one autosome (probably number 5).

cystic fibrosis (sis'tik fi-bro'sis) Characterized by formation of thick mucus in the pancreas and lungs that interferes with normal digestion and breathing; an expression of autosomal recessive genes.

familial cretinism (fah-mil'e-al kre'ti-nizm) Characterized by lack of thyroid secretions due to a defect in the iodine transport mechanism. Untreated children are dwarfed, sterile, and usually mentally retarded; an expression of autosomal recessive genes.

galactosemia (gah-lak''to-se'me-ah) Characterized by an inability to metabolize galactose, a component of milk sugar. This results in cataract, mental retardation, and damaged liver; an expression of autosomal recessive genes.

gout (gowt) Characterized by an accumulation of uric acid in the blood and tissues due to abnormal metabolism of substances called purines; an expression of an autosomal dominant gene.

hepatolenticular degeneration (Wilson's disease) (hep''ah-to-len-tik'u-lar de-jen''ĕ-ra'shun) Characterized by an increase in the absorption of copper and the accumulation of copper in the brain and liver. This results in degenerative changes in the brain and in cirrhosis of the liver; an expression of autosomal recessive genes.

hereditary hemochromatosis (he-red'ĭ-ter''e he''mo-kro''mah-to'sis) Characterized by an accumulation of iron in the liver, pancreas, and heart. This results in cirrhosis of the liver, diabetes, and heart failure; an expression of a sex-influenced autosomal dominant gene.

Huntington's chorea (hunt'ing-tunz ko-re'ah) Characterized by uncontrolled twitching of voluntary muscles and deterioration of mental capacities. This condition usually does not appear until after maturity, so affected persons often transmit the mutant gene to their children before the symptoms develop; an expression of an autosomal dominant gene.

Marfan's syndrome (mar-fahnz' sin'drom) Characterized by abnormally long extremities, dislocation of the lenses, and congenital defects of the cardiovascular system; an expression of an autosomal dominant gene.

nephrogenic diabetes insipidus (nef''ro-jen'ik di''ah-be'tēz in-sip'ĭ-dus) Characterized by the production of large volumes of very dilute urine and consequent dehydration due to failure of the distal convoluted tubules to reabsorb water in response to ADH; an expression of an X-linked recessive gene.

phenylketonuria (PKU) (fen''il-ke''to-nu're-ah) Characterized by an inability to normally metabolize the amino acid phenylalanine. This results in nerve and brain damage, and is accompanied by mental retardation; an expression of autosomal recessive genes.

pseudohypertrophic muscular dystrophy (su''do-hi''per-trof'ik mus'ku-lar dis'tro-fe) Characterized by progressive atrophy of the muscles, which usually begins during childhood and leads to death in adolescence; an expression of an X-linked recessive gene.

retinitis pigmentosa (ret''ĭ-ni'tis pig''men-to'sa) Characterized by progressive atrophy of the retina, causing blindness; an expression of an X-linked recessive gene.

Tay-Sach's disease (ta saks' di-zēz') Characterized by early blindness, deterioration of mental and physical abilities, and death; an expression of autosomal recessive genes.

thalassemia major (Cooley's anemia) (thal''ah-se'me-ah mā'jĕr) Characterized by very thin and fragile red blood cells, resulting in anemia and a short life expectancy; an expression of autosomal recessive genes.

Chapter Summary

Introduction (page 884)

Individual characteristics result from the interaction of heredity and environment. Heredity is the transmission of genetic information from parents to offspring; environment includes chemical, physical, and biological factors in an individual's surroundings.

Genes and Chromosomes (page 884)

1. Genes
   a. A gene is a portion of a DNA molecule that contains information for producing one kind of protein molecule.
   b. Genes within a zygote instruct it to synthesize particular proteins, which function as enzymes that promote metabolic reactions needed for growth and development.
2. Chromosome numbers
   a. Human body cells normally contain 46 chromosomes.
   b. Twenty-three chromosomes of a zygote are received from the female parent, and 23 are received from the male parent.

3. Chromosome pairs
   a. Chromosomes from male and female parents contain complete sets of genetic information.
   b. Within these chromosomes, those that contain similar genetic information comprise homologous chromosomes.
   c. During mitosis, newly formed cells receive identical pairs of homologous chromosomes.
4. Gene pairs
   a. Because chromosomes are paired, the genes within the chromosomes are paired also.
   b. A zygote that contains a pair of identical genes is homozygous; if the members of a pair are different, the zygote is heterozygous.
   c. An alternate form of a gene is called an allele.

Gene Expression (page 886)

The combination of genes present in an individual's cells constitutes a genotype; the appearance of the individual is called its phenotype.

1. Dominant and recessive genes
   a. In the heterozygous condition, if one gene is expressed and its allele is unexpressed, the expressed gene is called dominant and the allele is called recessive.
   b. Dominant genes are symbolized by capital letters and recessive genes are symbolized by small letters.
   c. Three genotypes are possible in a gene pair involving dominant and recessive genes, but only two phenotypes are possible.
2. Incomplete dominance
   a. When the members of a gene pair are expressed differently and neither is dominant to the other, they are called incompletely dominant.
   b. Incompletely dominant genes result in three possible genotypes and three possible phenotypes.
3. Multiple alleles
   a. When genes occur in several forms, the group of genes is called multiple alleles.
   b. The inheritance of ABO blood types involves multiple alleles.
   c. In ABO blood type inheritance, every person belongs to one of six genotypes and has one of four phenotypes.
4. Polygenic inheritance
   a. Some traits are determined by single gene pairs, while others are determined by several pairs of genes.
   b. Body height and concentration of skin pigment are examples of traits determined by polygenic inheritance.
   c. Polygenic inheritance results in a broad range of phenotypes, so that people vary greatly in their heights and in the concentration of pigment in their skins.

5. Environmental factors
   a. Environmental factors influence the ways genes are expressed.
   b. Body height is influenced by genes and also by such environmental factors as diet and disease.
   c. Concentration of skin pigment is influenced by genes and also by exposure to ultraviolet light.

Chromosomes and Sex Cell Production (page 889)

Sex cells carry unique combinations of genes as a result of meiosis.

1. Meiosis
   a. Meiosis involves two successive divisions.
      (1) In the first meiotic division, homologous chromosomes of the parent cell are separated and the chromosome number is reduced by one-half.
      (2) In the second meiotic division, the chromosomes act as they do in mitosis.
      (3) Meiosis results in the formation of sex cells.
   b. Meiosis is a continuous process without marked interruptions between steps, but for convenience, it can be divided into stages.
      (1) Stages of the first meiotic division include prophase, metaphase, anaphase, and telophase.
      (2) Stages of the second meiotic division also include prophase, metaphase, anaphase, and telophase.
2. Results of meiosis
   a. In spermatogenesis, four sperm cells are produced.
      (1) Each sperm cell contains 23 chromosomes.
      (2) The genetic information within the sperm cells varies from cell to cell as a result of crossing over.
   b. In oogenesis, one egg cell (secondary oocyte) and one nonfunctional polar body are produced.
      (1) Each egg cell contains 23 chromosomes.
      (2) The genetic information within the egg cells varies from cell to cell as a result of crossing over.

Inheritance of Single Traits (page 892)

1. Dominant and recessive inheritance
   a. In pigmentation, the gene for normal pigment is symbolized as A, and its recessive allele is symbolized as a.
      (1) AA is homozygous dominant. Aa is heterozygous, and aa is homozygous recessive.
      (2) AA and Aa individuals appear normal, while aa individuals have albinism.
   b. When the genotypes of parents are known, it is possible to predict the types of offspring they can produce.

2. Incompletely dominant inheritance
   a. Three genotypes and three phenotypes can result from a pair of incompletely dominant genes.
   b. When the genotypes of the parents are known, it is possible to predict the types of offspring they can produce.

Sex Chromosomes (page 895)

1. Gender inheritance
   a. A karyotype of human chromosomes consists of 22 pairs of autosomes and one pair of sex chromosomes.
   b. The sex chromosomes are responsible for the development of characteristics associated with maleness and femaleness.
      (1) A male karyotype of sex chromosomes consists of an X chromosome and a Y chromosome.
      (2) A female karyotype of sex chromosomes consists of two X chromosomes.
   c. The gender of an offspring is determined by the type of sperm cell that fertilizes the egg cell.
2. Actions of sex chromosomes
   a. The Y chromosome of an XY combination causes the synthesis of the H-Y antigen.
   b. The H-Y antigen promotes the differentiation of embryonic reproductive organs into testes and stimulates the formation of spermatogonia.
   c. The immature testes secrete hormones that control further masculinization of the reproductive organs.
   d. Embryonic reproductive organs resulting from an XX-bearing zygote differentiate into ovaries without the aid of a substance corresponding to the H-Y antigen.
3. X chromosome-linked inheritance
   a. Sex chromosomes contain genes responsible for certain traits that are unrelated to gender.
   b. The patterns in which these traits are inherited differ in each gender.
      (1) In males, genes on the single X chromosome are expressed even if they are recessive.
      (2) In females, such recessive genes are expressed only if they are present on both X chromosomes.
   c. Traits determined by recessive genes located on X chromosomes are called X-linked.
      (1) Color blindness and hemophilia are examples of X-linked traits.
      (2) Males receive X-linked traits from their mothers.
      (3) All the daughters of a male parent expressing an X-linked trait will receive the recessive gene for that trait.

Chromosome Disorders (page 899)

Disorders are characterized by the presence of abnormal numbers of chromosomes.

1. Cause of chromosome disorders
   a. If homologous chromosomes fail to separate during meiosis, one new cell will receive two chromosomes of one type while the other new cell will receive no chromosomes of that type.
   b. Such a failure of homologous chromosomes to separate is called nondisjunction.
2. Sex chromosome disorders
   a. Nondisjunction during oogenesis can result in egg cells with two X chromosomes and egg cells with no X chromosomes.
   b. Such abnormal egg cells can be fertilized, and the resulting zygotes may develop into offspring.
      (1) If an XX egg cell is fertilized by a Y-bearing sperm cell, the resulting XXY combination leads to Klinefelter's syndrome.
      (2) If an XX egg cell is fertilized by an X-bearing sperm cell, the resulting XXX combination produces a relatively normal female.
      (3) If an egg cell without an X chromosome is fertilized by an X-bearing sperm cell, the result leads to Turner's syndrome.
3. Autosomal abnormalities
   a. Nondisjunction may occur among autosomes and result in two chromosomes of the same type within an egg cell.
   b. If such an egg cell is fertilized by a normal sperm cell, the resulting zygote will have three chromosomes of one kind, a condition called trisomy.
   c. Down syndrome, Edward's syndrome, and Patau's syndrome are due to triply represented autosomes.

Medical Genetics (page 901)

1. Genetic diseases
   a. Hereditary diseases include a diverse group of disorders.
   b. Such diseases are caused by mutations.
   c. Genetic diseases can be subdivided into the following three groups:
      (1) Diseases that produce their effects during embryonic or fetal development, such as Down syndrome;
      (2) Diseases that express themselves after birth, such as phenylketonuria;
      (3) Diseases that produce their effects in adolescence or adulthood, such as certain forms of hypertension.
2. Genetic counseling
   a. Genetic counseling is a service by which individuals can obtain information about the chances of transmitting genetic disorders to their offspring.

b. Genetic counselors can sometimes precisely predict the probability of a future offspring being affected by an inherited disease.

c. The accuracy of such predictions depends upon the following:
 (1) Correct diagnosis of the genetic disorder;
 (2) Understanding of the pattern by which the disorder is inherited;
 (3) Knowledge of the parents' genotypes.

Clinical Application of Knowledge

1. Using the principles of human genetics, how could you support the stand taken by many civil and religious authorities forbidding marriages between persons who are first or second cousins?

2. If a young couple with the sickle-cell trait who have a child with sickle-cell anemia said to you, "Our first child has sickle-cell anemia, but now we can have three more children before we have another one with this disease," how would you respond?

3. If a 49-year-old woman, who thought she was menopausal and thus failed to seek medical attention when she began missing her menstrual periods, finds that she is five months pregnant, what special kinds of tests are likely to be ordered by her physician? Why?

4. Using a knowledge of genetics and blood testing, why is it more difficult to prove that a particular male is the father of a particular child than to prove that he is not the father?

5. What factors influence the changes in the gender ratio (males:females) as a population ages?

6. Why do you think the procedure of chorionic villi biopsy presents a greater risk to a fetus than does ultrasonography?

Review Activities

1. Identify two major factors that influence the development of individual characteristics.
2. Define *gene*.
3. Discuss the origin of the 46 chromosomes in a human zygote.
4. Define *homologous chromosomes*.
5. Distinguish between homozygous and heterozygous.
6. Distinguish between genotype and phenotype.
7. Explain what is meant by a *dominant gene* and its *recessive allele*.
8. Define *incomplete dominance*.

9. Explain how ABO blood type inheritance is controlled by multiple alleles.
10. Describe how environmental factors may influence the expression of the genes that control pigmentation of the skin.
11. Outline the process of meiosis.
12. Explain the significance of synapsis during meiosis.
13. Describe the genotypes and phenotypes of the offspring expected from the following parents. (*A* represents the dominant gene for normal pigmentation, and *a* represents its recessive allele.)
 a. *AA* and *Aa*
 b. *Aa* and *aa*
 c. *Aa* and *Aa*
14. Describe the genotypes and phenotypes of the offspring expected from the following parents. (H^1 represents the normal gene for hemoglobin production, and H^2 represents the incompletely dominant allele that causes the formation of abnormal hemoglobin.)
 a. H^1H^1 and H^2H^2
 b. H^1H^1 and H^1H^2
 c. H^1H^2 and H^1H^2
15. Distinguish between autosomes and sex chromosomes.
16. Explain how gender is inherited.
17. Explain why one-half of the human zygotes are expected to develop into males and one-half into females.
18. Explain how the presence of a Y chromosome causes embryonic reproductive organs to differentiate into testes.
19. Define *X-linked genes*.
20. Explain why X-linked genes are always expressed more frequently in males than in females.
21. Describe the genotypes and phenotypes of the offspring expected from the following parents:
 a. Color-blind male and normal (homozygous dominant) female.
 b. Normal male and color-blind female.
22. Distinguish between hemophilia A and hemophilia B.
23. Explain how nondisjunction may lead to chromosome disorders.
24. Explain how an individual with a sex chromosome combination of 47, XXY might be produced.
25. Explain how an individual with a sex chromosome karyotype 45, XO might be produced.
26. Define *autosomal trisomy*.
27. List the three main groups of genetic diseases.
28. Describe the services provided by a genetic counselor.

Reference Plates

Human Cadavers

*T*he following set of illustrations includes medial sections, horizontal sections, and regional dissections of human cadavers. These photographs will help you visualize the spatial and proportional relationships between the major anatomic structures of actual specimens. The photographs can also serve as the basis for a review of the information you have gained from your study of the human organism.

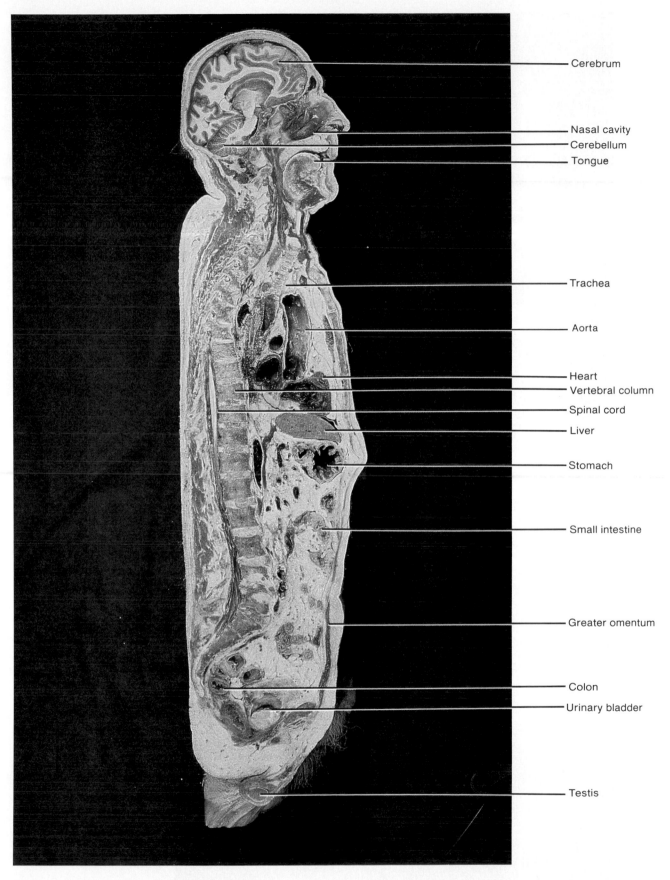

Cerebrum

Nasal cavity
Cerebellum
Tongue

Trachea

Aorta

Heart
Vertebral column
Spinal cord
Liver
Stomach

Small intestine

Greater omentum

Colon
Urinary bladder

Testis

Plate 48 Sagittal section of the head and trunk.

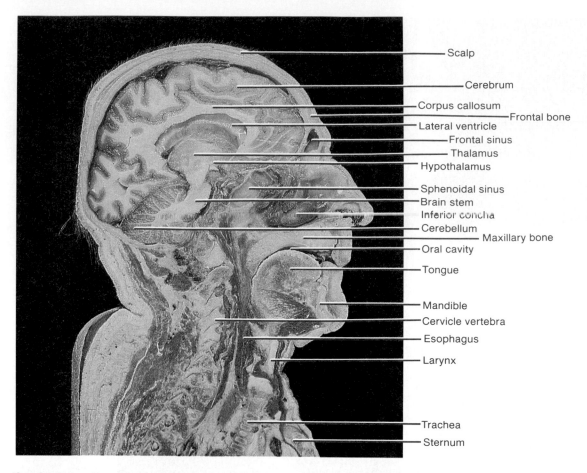

Scalp

Cerebrum

Corpus callosum
Frontal bone
Lateral ventricle
Frontal sinus
Thalamus
Hypothalamus

Sphenoidal sinus
Brain stem
Inferior concha
Cerebellum
Maxillary bone
Oral cavity
Tongue

Mandible
Cervicle vertebra
Esophagus
Larynx

Trachea
Sternum

Plate 49   Sagittal section of the head and neck.

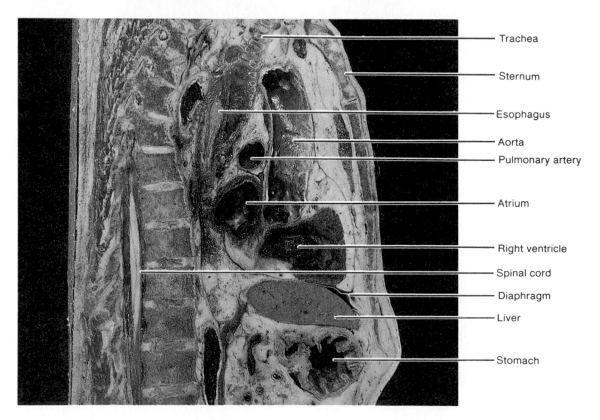

Trachea

Sternum

Esophagus

Aorta
Pulmonary artery

Atrium

Right ventricle
Spinal cord
Diaphragm
Liver

Stomach

Plate 50   Viscera of the thoracic cavity, sagittal section.

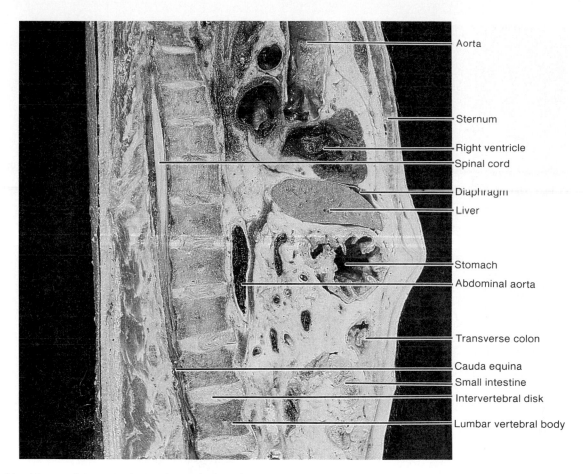

Aorta

Sternum

Right ventricle

Spinal cord

Diaphragm

Liver

Stomach

Abdominal aorta

Transverse colon

Cauda equina

Small intestine

Intervertebral disk

Lumbar vertebral body

Plate 51 Viscera of the abdominal cavity, sagittal section.

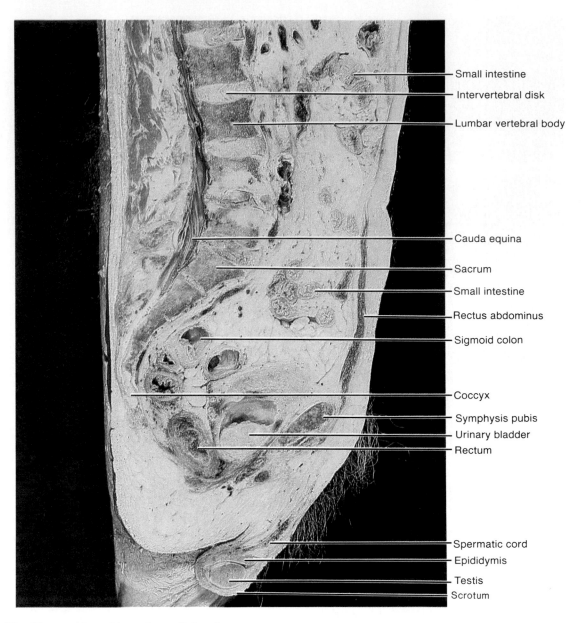

— Small intestine

— Intervertebral disk

— Lumbar vertebral body

— Cauda equina

— Sacrum

— Small intestine

— Rectus abdominus

— Sigmoid colon

— Coccyx

— Symphysis pubis

— Urinary bladder

— Rectum

— Spermatic cord

— Epididymis

— Testis

— Scrotum

Plate 52 Viscera of the pelvic cavity, sagittal section.

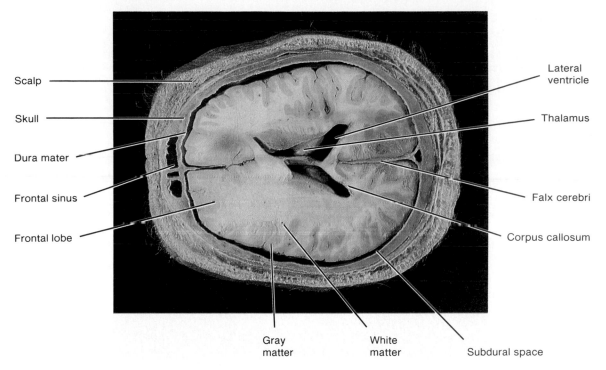

Scalp

Skull

Dura mater

Frontal sinus

Frontal lobe

Lateral
ventricle

Thalamus

Falx cerebri

Corpus callosum

Gray
matter

White
matter

Subdural space

Plate 53 Horizontal section of the head above the eyes,
superior view.

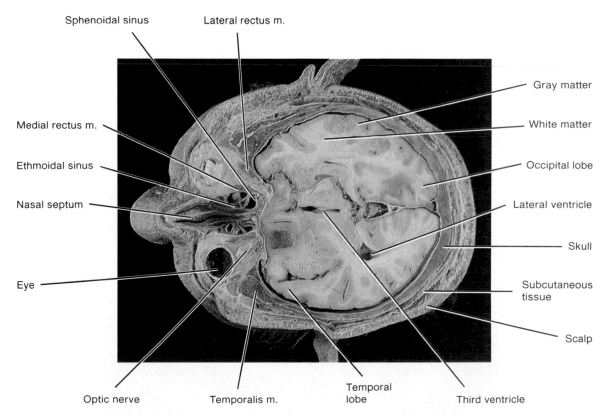

Sphenoidal sinus

Lateral rectus m.

Medial rectus m.

Ethmoidal sinus

Nasal septum

Eye

Optic nerve

Temporalis m.

Temporal
lobe

Third ventricle

Gray matter

White matter

Occipital lobe

Lateral ventricle

Skull

Subcutaneous
tissue

Scalp

Plate 54 Horizontal section of the head at the level of the
eyes, superior view.

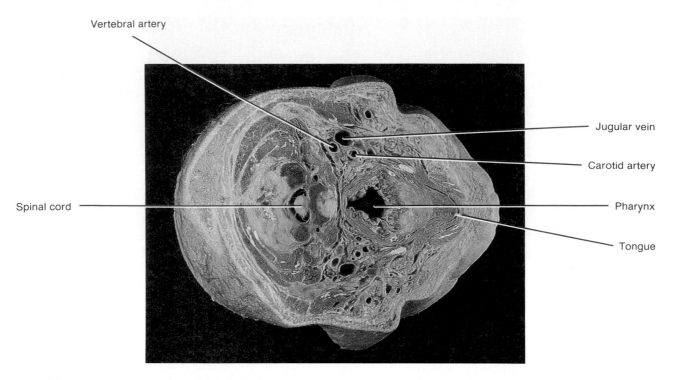

Plate 55   Horizontal section of the neck, inferior view.

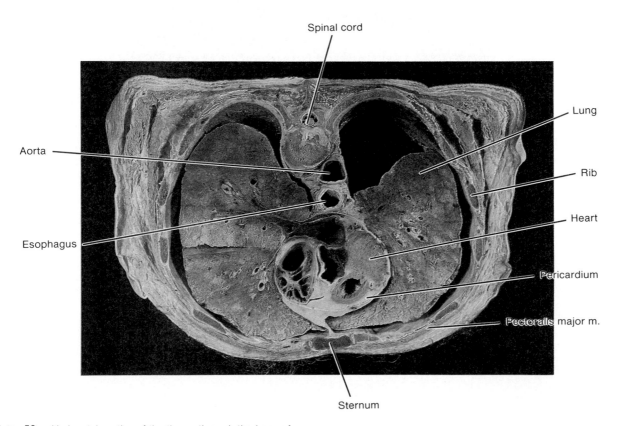

Plate 56   Horizontal section of the thorax through the base of
the heart, inferior view.

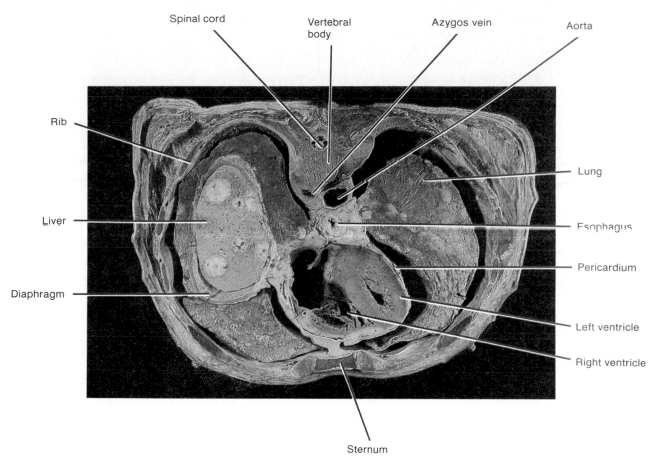

Spinal cord

Vertebral body

Azygos vein

Aorta

Rib

Lung

Liver

Esophagus

Pericardium

Diaphragm

Left ventricle

Right ventricle

Sternum

Plate 57 Horizontal section of the thorax through the heart, inferior view.

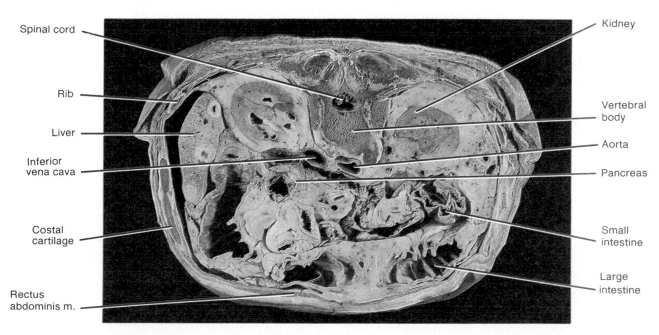

Spinal cord

Kidney

Rib

Vertebral body

Liver

Aorta

Inferior vena cava

Pancreas

Costal cartilage

Small intestine

Rectus abdominis m.

Large intestine

Plate 58 Horizontal section of the abdomen through the kidneys, inferior view.

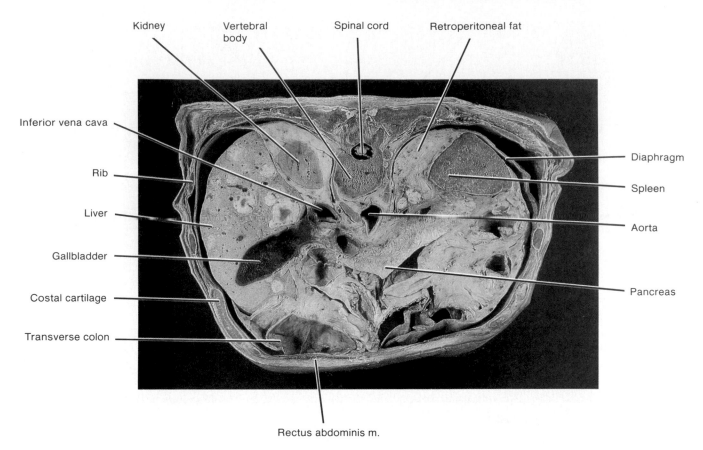

Kidney   Vertebral body   Spinal cord   Retroperitoneal fat

Inferior vena cava

Rib

Liver

Gallbladder

Costal cartilage

Transverse colon

Diaphragm

Spleen

Aorta

Pancreas

Rectus abdominis m.

Plate 59   Horizontal section of the abdomen through the pancreas, inferior view.

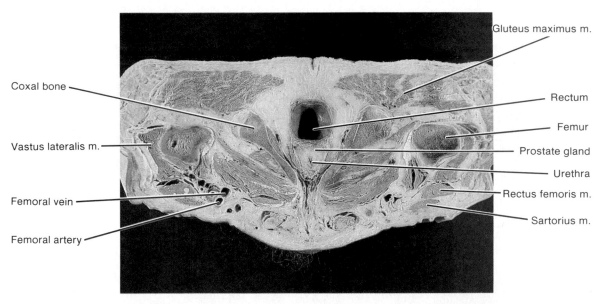

Gluteus maximus m.

Coxal bone

Vastus lateralis m.

Femoral vein

Femoral artery

Rectum

Femur

Prostate gland

Urethra

Rectus femoris m.

Sartorius m.

Plate 60   Horizontal section of the male pelvic cavity, superior view.

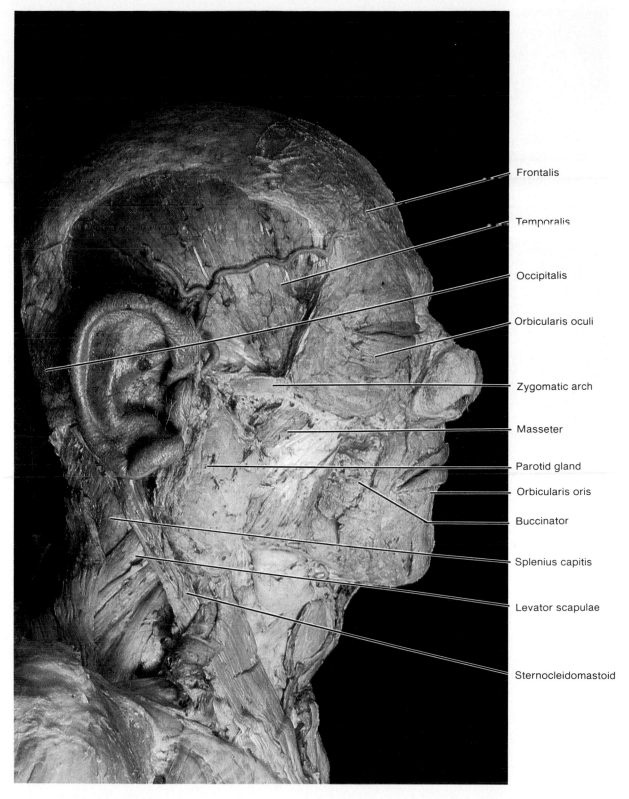

Frontalis

Temporalis

Occipitalis

Orbicularis oculi

Zygomatic arch

Masseter

Parotid gland

Orbicularis oris

Buccinator

Splenius capitis

Levator scapulae

Sternocleidomastoid

Plate 61 Lateral view of the head.

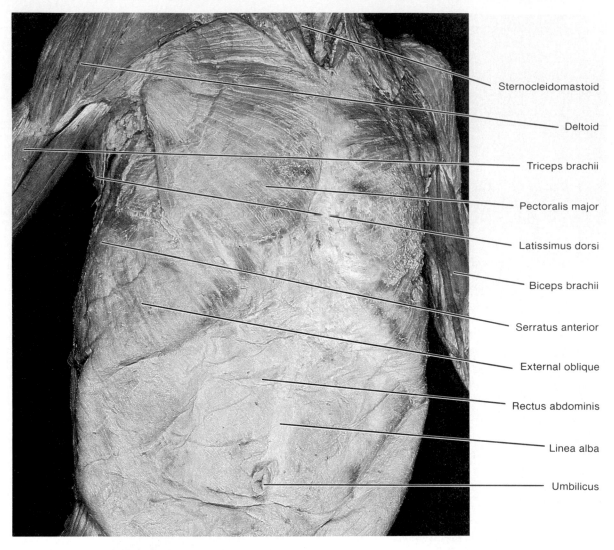

Sternocleidomastoid

Deltoid

Triceps brachii

Pectoralis major

Latissimus dorsi

Biceps brachii

Serratus anterior

External oblique

Rectus abdominis

Linea alba

Umbilicus

Plate 62 Anterior view of the trunk.

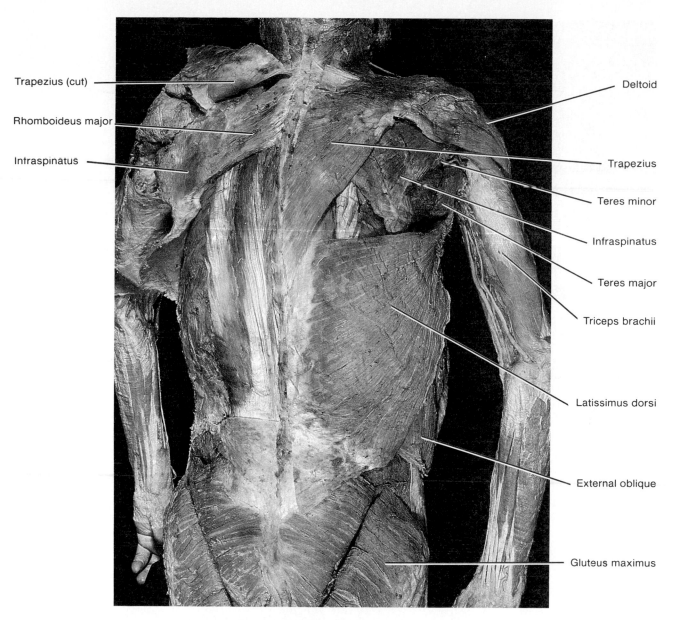

Trapezius (cut)

Rhomboideus major

Intraspinatus

Deltoid

Trapezius

Teres minor

Infraspinatus

Teres major

Triceps brachii

Latissimus dorsi

External oblique

Gluteus maximus

Plate 63 Posterior view of the trunk, with deep thoracic muscles exposed on the left.

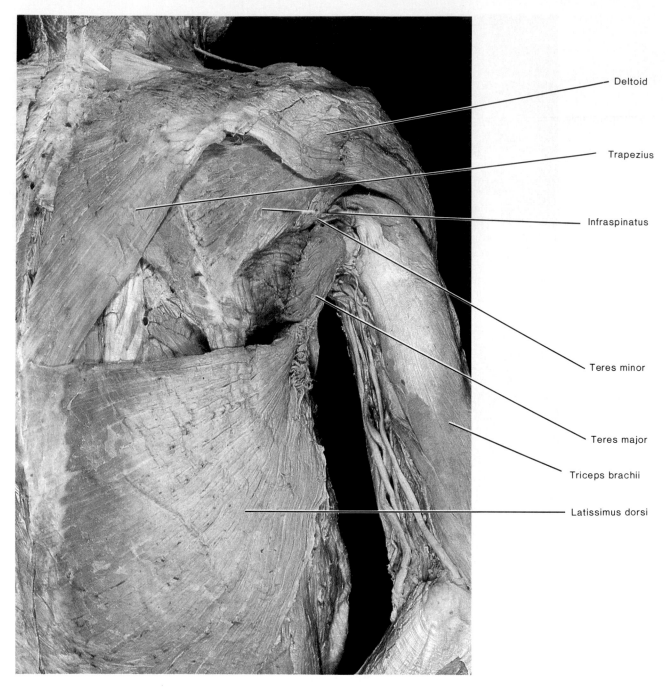

Deltoid

Trapezius

Infraspinatus

Teres minor

Teres major

Triceps brachii

Latissimus dorsi

Plate 64 Posterior view of the right thorax and upper arm.

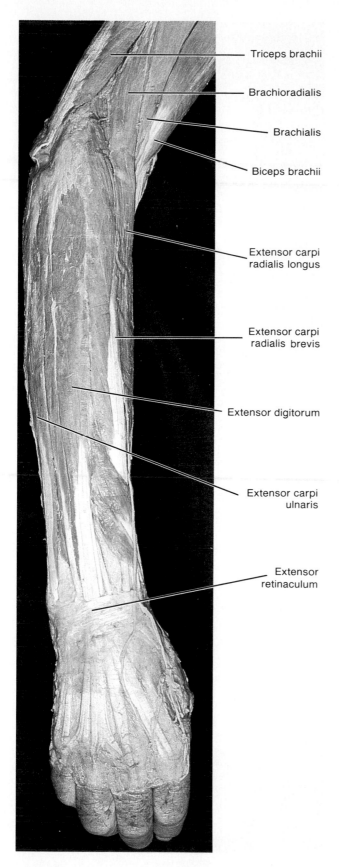

Triceps brachii

Brachioradialis

Brachialis

Biceps brachii

Extensor carpi
radialis longus

Extensor carpi
radialis brevis

Extensor digitorum

Extensor carpi
ulnaris

Extensor
retinaculum

Plate 65 Posterior view of the right forearm and hand.

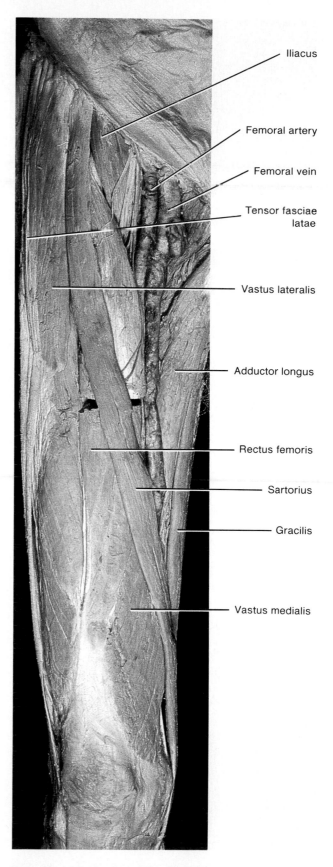

Iliacus

Femoral artery

Femoral vein

Tensor fasciae
latae

Vastus lateralis

Adductor longus

Rectus femoris

Sartorius

Gracilis

Vastus medialis

Plate 66 Anterior view of the right thigh.

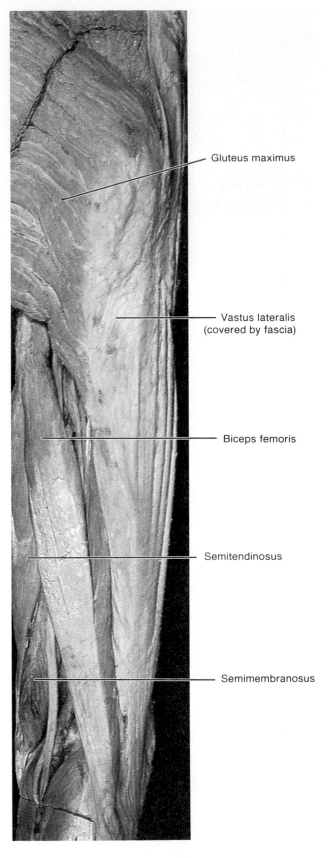

Gluteus maximus

Vastus lateralis
(covered by fascia)

Biceps femoris

Semitendinosus

Semimembranosus

Plate 67 Posterior view of the right thigh.

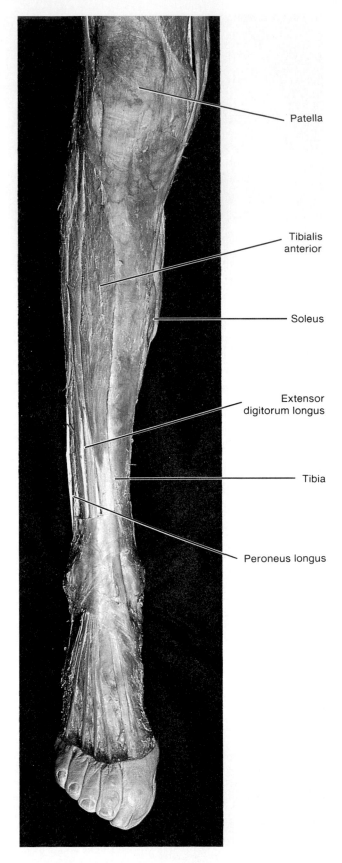

Patella

Tibialis
anterior

Soleus

Extensor
digitorum longus

Tibia

Peroneus longus

Plate 68 Anterior view of the right leg.

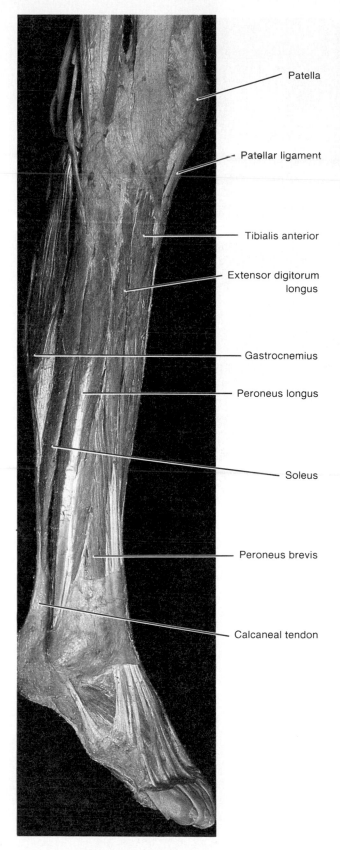

Patella

Patellar ligament

Tibialis anterior

Extensor digitorum
longus

Gastrocnemius

Peroneus longus

Soleus

Peroneus brevis

Calcaneal tendon

Plate 69 Lateral view of the right leg.

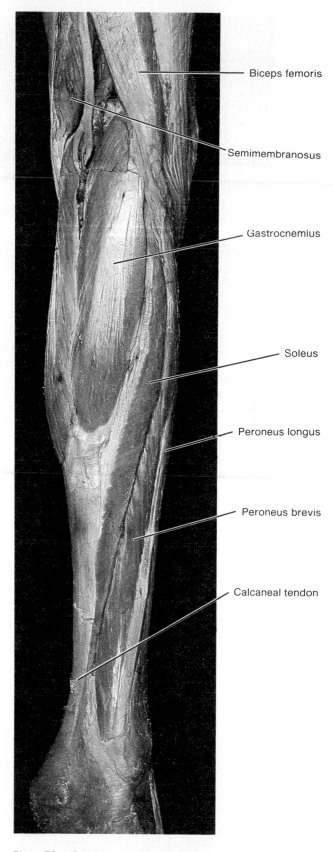

Biceps femoris

Semimembranosus

Gastrocnemius

Soleus

Peroneus longus

Peroneus brevis

Calcaneal tendon

Plate 70 Posterior view of the right leg.

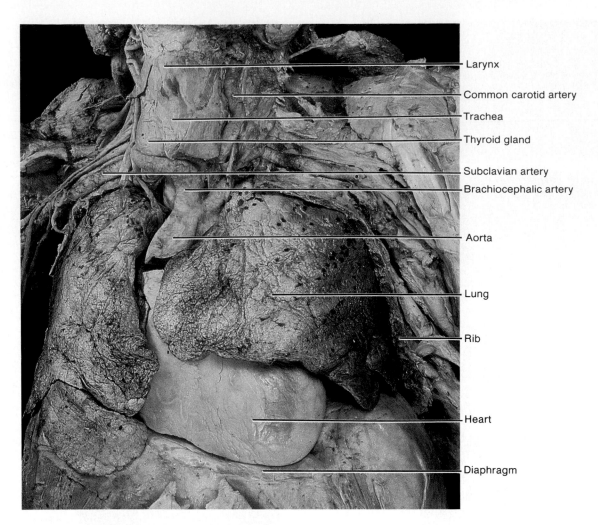

Larynx

Common carotid artery

Trachea

Thyroid gland

Subclavian artery

Brachiocephalic artery

Aorta

Lung

Rib

Heart

Diaphragm

Plate 71 Thoracic viscera, ventral view. (Brachiocephalic vein has been removed to expose the aorta.)

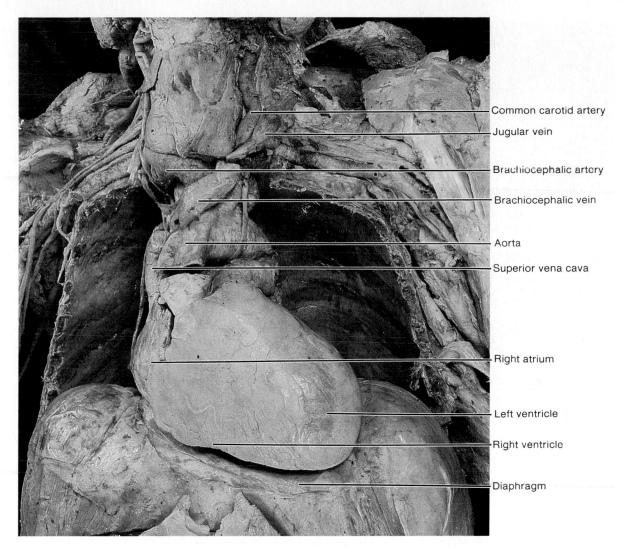

Common carotid artery

Jugular vein

Brachiocephalic artery

Brachiocephalic vein

Aorta

Superior vena cava

Right atrium

Left ventricle

Right ventricle

Diaphragm

Plate 72 Thorax with the lungs removed, ventral view.

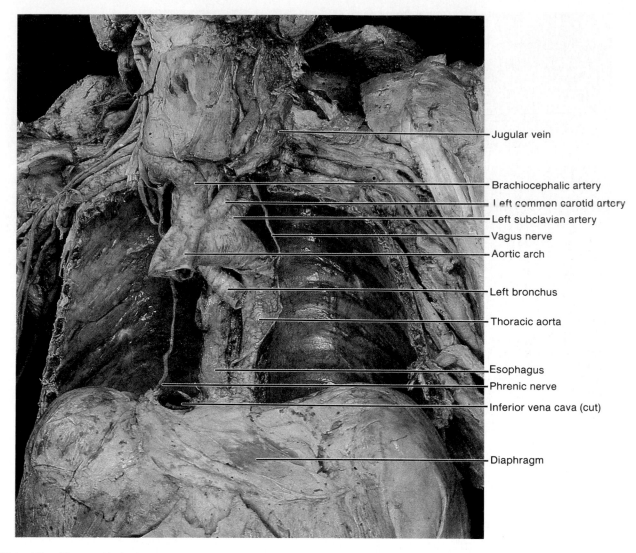

Jugular vein

Brachiocephalic artery

Left common carotid artery

Left subclavian artery

Vagus nerve

Aortic arch

Left bronchus

Thoracic aorta

Esophagus

Phrenic nerve

Inferior vena cava (cut)

Diaphragm

Plate 73 Thorax with the heart and lungs removed, ventral view.

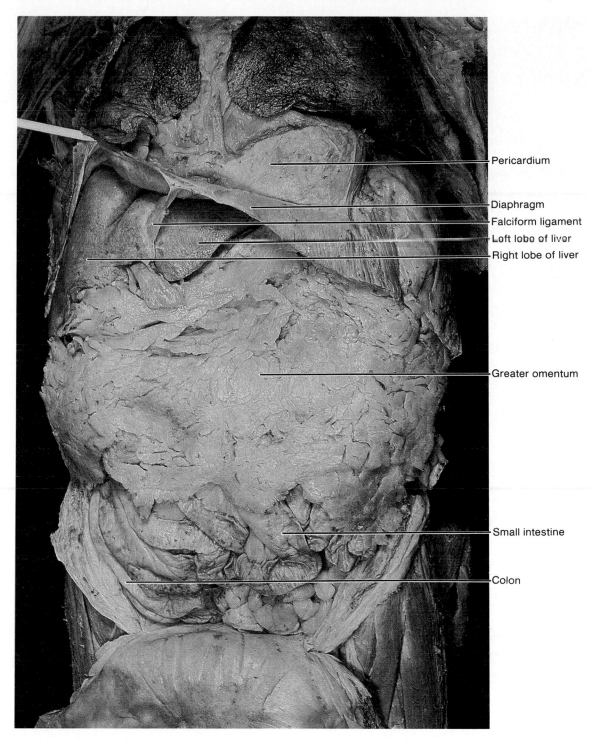

Pericardium

Diaphragm

Falciform ligament

Loft lobe of livor

Right lobe of liver

Greater omentum

Small intestine

Colon

Plate 74 Abdominal viscera, ventral view.

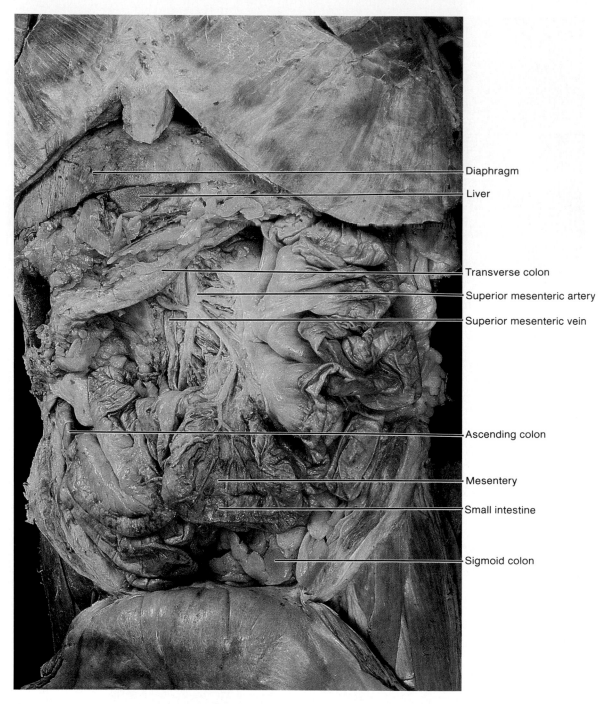

Diaphragm

Liver

Transverse colon

Superior mesenteric artery

Superior mesenteric vein

Ascending colon

Mesentery

Small intestine

Sigmoid colon

Plate 75 Abdominal viscera with the greater omentum removed. (Small intestine has been displaced to the left.)

Appendixes

Appendix A
Periodic Table of Elements

1a	IIa	IIIb	IVb	Vb	VIb	VIIb	VIIIb			IB	IIIB	IIIa	IVa	Va	VIa	VIIa	0
1 H 1.008																	2 He 4.00
3 Li 6.94	4 Be 9.01											5 B 10.81	6 C 12.01	7 N 14.00	8 O 15.99	9 F 18.99	10 Ne 20.18
11 Na 22.99	12 Mg 24.31											13 Al 26.98	14 Si 28.09	15 P 30.97	16 S 32.06	17 Cl 35.45	18 Ar 39.95
19 K 39.10	20 Ca 40.08	21 Sc 44.96	22 Ti 47.90	23 V 50.94	24 Cr 51.99	25 Mn 54.94	26 Fe 55.85	27 Co 58.93	28 Ni 58.71	29 Cu 63.54	30 Zn 65.37	31 Ga 69.72	32 Ge 72.59	33 As 74.92	34 Se 78.96	35 Br 79.91	36 Kr 83.80
37 Rb 85.47	38 Sr 87.62	39 Y 88.91	40 Zr 91.22	41 Nb 92.91	42 Mo. 95.94	43 Tc (99)	44 Ru 101.97	45 Rh 102.91	46 Pd 106.4	47 Ag 107.87	48 Cd 112.40	49 In 114.82	50 Sn 118.69	51 Sb 121.75	52 Te 127.60	53 I 126.90	54 Xe 131.30
55 Cs 132.91	56 Ba 137.34	57–71 see below	72 Hf 178.49	73 Ta 180.95	74 W 183.85	75 Re 186.2	76 Os 190.2	77 Ir 192.2	78 Pt 195.09	79 Au 196.97	80 Hg 200.59	81 Tl 204.37	82 Pb 207.19	83 Bi 208.98	84 Po (210)	85 At (210)	86 Rn (222)
87 Fr (223)	88 Ra (226)	89–103 see below	104 Rf (261)	105 Ha (260)	106 • 263												

*newly produced

57 La 138.91	58 Ce 140.12	59 Pr 140.91	60 Nd 144.24	61 Pm (147)	62 Sm 150.35	63 Eu 151.96	64 Gd 157.25	65 Tb 158.92	66 Dy 162.50	67 Ho 164.93	68 Er 167.26	69 Tm 168.93	70 Yb 173.04	71 Lu 174.97
89 Ac (227)	90 Th 232.04	91 Pa (231)	92 U 238.03	93 Np (237)	94 Pu (242)	95 Am (243)	96 Cm (247)	97 Bk (247)	98 Cf (251)	99 Es (254)	100 Fm (253)	101 Md (256)	102 No (254)	103 Lw (257)

Values in parentheses are approximate.

Key

1	—— Atomic number
H	—— Symbol of element
1.008	—— Atomic weight

Appendix B
Units of Measurement and Their Equivalents

Apothecaries' Weights and Their Metric Equivalents

1 grain (gr) =
0.05 scruple (s)
0.017 dram (dr)
0.002 ounce (oz)
0.0002 pound (lb)
0.065 gram (g)
65. milligrams (mg)

1 scruple (s) =
20. grains (gr)
0.33 dram (dr)
0.042 ounce (oz)
0.004 pound (lb)
1.3 grams (g)
1,300. milligrams (mg)

1 dram (dr) =
60. grains (gr)
3. scruples (s)
0.13 ounce (oz)
0.010 pound (lb)
3.9 grams (g)
3,900. milligrams (mg)

1 ounce (oz) =
480. grains (gr)
24. scruples (s)
8. drams (dr)
0.08 pound (lb)
31.1 grams (g)
31,100. milligrams (mg)

1 pound (lb) =
5,760. grains (gr)
288. scruples (s)
96. drams (dr)
12. ounces (oz)
373. grams (g)
373,000. milligrams (mg)

Apothecaries' Volumes and Their Metric Equivalents

1 minim (min) =
0.017 fluid dram (fl dr)
0.002 fluid ounce (fl oz)
0.0001 pint (pt)
0.06 milliliter (ml)
0.06 cubic centimeter (cc)

1 fluid dram (fl dr) =
60. minims (min)
0.13 fluid ounce (fl oz)
0.008 pint (pt)
3.70 milliliters (ml)
3.70 cubic centimeters (cc)

1 fluid ounce (fl oz) =
480. minims (min)
8. fluid drams (fl dr)
0.06 pint (pt)
29.6 milliliters (ml)
29.6 cubic centimeters (cc)

1 pint (pt) =
7,680. minims (min)
128. fluid drams (fl dr)
16. fluid ounces (fl oz)
473. milliliters (ml)
473. cubic centimeters (cc)

Metric Weights and Their Apothecaries' Equivalents

1 gram (g) =
0.001 kilogram (kg)
1,000. milligrams (mg)
1,000,000. micrograms (μg)
1,000,000,000. nanograms (ng)
1,000,000,000,000. picograms (pg)
15.4 grains (gr)
0.032 ounce (oz)

1 kilogram (kg) =
1,000. grams (g)
1,000,000. milligrams (mg)
1,000,000,000. micrograms (μg)
32. ounces (oz)
2.2 pounds (lb)

1 milligram (mg) =
0.000001 kilogram (kg)
0.001 gram (g)
1,000. micrograms (μg)
0.0154 grains (gr)
0.000032 ounce (oz)

Metric Volumes and Their Apothecaries' Equivalents

1 liter (l) =
1,000. milliliters (ml)
1,000. cubic centimeters (cc)
2.1 pints (pt)
270. fluid drams (fl dr)
34. fluid ounces (fl oz)

1 milliliter (ml) =
0.001 liter (l)
1. cubic centimeter (cc)
16.2 minims (min)
0.27 fluid dram (fl dr)
0.034 fluid ounce (fl oz)

Approximate Equivalents of Household Measures

1 teaspoon (tsp) =
4. milliliters (ml)
4. cubic centimeters (cc)
1. fluid dram (fl dr)

1 tablespoon (tbsp) =
15. milliliters (ml)
15. cubic centimeters (cc)
0.5 fluid ounce (fl oz)
3.7 teaspoons (tsp)

1 cup (c) =
240. milliliters (ml)
240. cubic centimeters (cc)
8. fluid ounces (fl oz)
0.5 pint (pt)
16. tablespoons (tbsp)

1 quart (qt) =
960. milliliters (ml)
960. cubic centimeters (cc)
2. pints (pt)
4. cups (c)
32. fluid ounces (fl oz)

Conversion of Units from One Form to Another

Refer to the preceding equivalency lists when converting one unit to another equivalent unit.

To convert a unit shown in bold type to one of the equivalent units listed immediately below it, multiply the first number (bold type unit) by the appropriate equivalent unit listed below it.

Sample Problems:

1. Convert 320 grains into scruples (1 gr = 0.05 s).

$$320 \text{ gr} \times \frac{0.05 \text{ s}}{1 \text{ gr}} = 16.0 \text{ s}$$

2. Convert 320 grains into drams (1 gr = 0.017 dr).

$$320 \text{ gr} \times \frac{0.017 \text{ dr}}{1 \text{ gr}} = 5.44 \text{ dr}$$

3. Convert 320 grains into grams (1 gr = 0.065 g).

$$320 \text{ gr} \times \frac{0.065 \text{ g}}{1 \text{ gr}} = 20.8 \text{ g}$$

Body Temperatures in °Fahrenheit and °Celsius

°F	°C	°F	°C
95.0	35.0	100.0	37.8
95.2	35.1	100.2	37.9
95.4	35.2	100.4	38.0
95.6	35.3	100.6	38.1
95.8	35.4	100.8	38.2
96.0	35.5	101.0	38.3
96.2	35.7	101.2	38.4
96.4	35.8	101.4	38.6
96.6	35.9	101.6	38.7
96.8	36.0	101.8	38.8
97.0	36.1	102.0	38.9
97.2	36.2	102.2	39.0
97.4	36.3	102.4	39.1
97.6	36.4	102.6	39.2
97.8	36.6	102.8	39.3
98.0	36.7	103.0	39.4
98.2	36.8	103.2	39.6
98.4	36.9	103.4	39.7
98.6	37.0	103.6	39.8
98.8	37.1	103.8	39.9
99.0	37.2	104.0	40.0
99.2	37.3	104.2	40.1
99.4	37.4	104.4	40.2
99.6	37.6	104.6	40.3
99.8	37.7	104.8	40.4
		105.0	40.6

To convert °F to °C

Subtract 32 from °F and multiply by 5/9.

_____ °F − 32 × 5/9 = _____ °C

To convert °C to °F

Multiply °C by 9/5 and add 32.

_____ °C × 9/5 + 32 = _____ °F

Appendix C
Laboratory Tests of Clinical Importance

Common tests performed on blood		
Test	Normal values* (adult)	Clinical significance
Albumin (serum)	3.2–5.5 gm/100 ml	Values increase in multiple myeloma and decrease with proteinuria and as a result of severe burns.
Albumin-globulin ratio, or A/G ratio (serum)	1.5:1 to 2.5:1	Ratio of albumin to globulin is lowered in kidney diseases and malnutrition.
Ammonia	12–55 μ mol/l	Values increase in severe liver disease, pneumonia, shock, and congestive heart failure.
Amylase (serum)	4–25 units/ml	Values increase in acute pancreatitis, intestinal obstructions, and mumps. They decrease in chronic pancreatitis, cirrhosis of the liver, and toxemia of pregnancy.
Bilirubin, total (serum)	0–1.0 mg/100 ml	Values increase in conditions causing red blood cell destruction or biliary obstruction.
Blood urea nitrogen, or BUN (plasma or serum)	8–25 mg/100 ml	Values increase in various kidney disorders and decrease in liver failure and during pregnancy.
Calcium (serum)	8.5–10.5 mg/100 ml	Values increase in hyperparathyroidism, hypervitaminosis D, and respiratory conditions that cause a rise in CO_2 concentration. They decrease in hypoparathyroidism, malnutrition, and severe diarrhea.
Carbon dioxide (serum)	24–30 mEq/l	Values increase in respiratory diseases, intestinal obstruction, and vomiting. They decrease in acidosis, nephritis, and diarrhea.
Chloride (serum)	100–106 mEq/l	Values increase in nephritis, Cushing's syndrome, dehydration, and hyperventilation. They decrease in metabolic acidosis, Addison's disease, diarrhea, and following severe burns.
Cholesterol, total (serum)	120–220 mg/100 ml (below 200 mg/100 ml recommended by the American Heart Association)	Values increase in diabetes mellitus and hypothyroidism. They decrease in pernicious anemia, hyperthyroidism, and acute infections.
Cholesterol, high-density lipoprotein (HDL)	Women: 30–80 mg/100 ml Men: 30–70 mg/100 ml	Values increase in liver disease. Decreased values are associated with an increased risk of atherosclerosis.
Cholesterol, low-density lipoprotein (LDL)	62–185 mg/100 ml	Increased values are associated with an increased risk of atherosclerosis.
Creatine (serum)	0.2–0.8 mg/100 ml	Values increase in muscular dystrophy, nephritis, severe damage to muscle tissue, and during pregnancy.
Creatinine (serum)	0.6–1.5 mg/100 ml	Values increase in various kidney diseases.
Erythrocyte count, or red cell count (whole blood)	Men: 4,600,000–6,200,000/cu mm Women: 4,200,000–5,400,000/cu mm Children: 4,500,000–5,100,000/cu mm (varies with age)	Values increase as a result of severe dehydration or diarrhea, and decrease in anemia, leukemia, and following severe hemorrhage.

*These values may vary with hospital, physician, and type of equipment used to make measurement.

Common tests performed on blood—*continued*

Test	Normal values* (adult)	Clinical significance
Ferritin (serum)	Men: 10–270 μg/100 ml Women: 5–280 μg/100 ml	Values correlate with total body iron store. They decrease with iron deficiency.
Globulin (serum)	2.3–3.5 gm/100 ml	Values increase as a result of chronic infections.
Glucose (plasma)	70–110 mg/100 ml	Values increase in diabetes mellitus, liver diseases, nephritis, hyperthyroidism, and pregnancy. They decrease in hyperinsulinism, hypothyroidism, and Addison's disease.
Hematocrit (whole blood)	Men: 40–54 ml/100 ml Women: 37–47 ml/100 ml Children: 35–49 ml/100 ml (varies with age)	Values increase in polycythemia due to dehydration or shock. They decrease in anemia and following severe hemorrhage.
Hemoglobin (whole blood)	Men: 14–18 gm/100 ml Women: 12–16 gm/100 ml Children: 11.2–16.5 gm/100 ml (varies with age)	Values increase in polycythemia, obstructive pulmonary diseases, congestive heart failure, and at high altitudes. They decrease in anemia, pregnancy, and as a result of severe hemorrhage or excessive fluid intake.
Iron (serum)	50–150 μg/100 ml	Values increase in various anemias and liver disease. They decrease in iron deficiency anemia.
Iron-binding capacity (serum)	250–410 μg/100 ml	Values increase in iron deficiency anemia and pregnancy. They decrease in pernicious anemia, liver disease, and chronic infections.
Lactic acid (whole blood)	0.6–1.8 mEq/l	Values increase with muscular activity and in congestive heart failure, severe hemorrhage, and shock.
Lactic dehydrogenase, or LDH (serum)	45–90 U/l	Values increase in pernicious anemia, myocardial infarction, liver diseases, acute leukemia, and widespread carcinoma.
Lipids, total (serum)	450–850 mg/100 ml	Values increase in hypothyroidism, diabetes mellitus, and nephritis. They decrease in hyperthyroidism.
Magnesium	1.3–2.1 mEq/l	Values increase in renal failure, hypothyroidism, and Addison's disease. They decrease in renal disease, liver disease, and pancreatitis.
Mean corpuscular hemoglobin (MCH)	26–32 pg/RBC	Values increase in macrocytic anemia. They decrease in microcytic anemia.
Mean corpuscular volume (MCV)	86–98 μ mm^3/RBC	Values increase in liver disease and pernicious anemia. They decrease in iron deficiency anemia.
Osmolality	275–295 mosm/kg	Values increase in dehydration, hypercalcemia, and diabetes mellitus. They decrease in hyponatremia, Addison's disease, and water intoxication.
Oxygen saturation (whole blood)	Arterial: 96–100% Venous: 60–85%	Values increase in polycythemia and decrease in anemia and obstructive pulmonary diseases.
pH (whole blood)	7.35–7.45	Values increase due to vomiting, Cushing's syndrome, and hyperventilation. They decrease as a result of hypoventilation, severe diarrhea, Addison's disease, and diabetic acidosis.
Phosphatase, acid (serum)	Women: 0.01–0.56 Sigma U/ml Men: 0.13–0.63 Sigma U/ml	Values increase in cancer of the prostate gland, hyperparathyroidism, certain liver diseases, myocardial infarction, and pulmonary embolism.
Phosphatase, alkaline (serum)	13–39 U/l	Values increase in hyperparathyroidism (and in other conditions that promote resorption of bone), liver diseases, and pregnancy.

*These values may vary with hospital, physician, and type of equipment used to make measurement.

Common tests performed on blood—*continued*

Test	Normal values* (adult)	Clinical significance
Phosphorus (serum)	3.0–4.5 mg/100 ml	Values increase in kidney diseases, hypoparathyroidism, acromegaly, and hypervitaminosis D. They decrease in hyperparathyroidism.
Platelet count (whole blood)	150,000–350,000/cu mm	Values increase in polycythemia and certain anemias. They decrease in acute leukemia and aplastic anemia.
Potassium (serum)	3.5–5.0 mEq/l	Values increase in Addison's disease, hypoventilation, and conditions that cause severe cellular destruction. They decrease in diarrhea, vomiting, diabetic acidosis, and chronic kidney disease.
Protein, total (serum)	6.0–8.4 gm/100 ml	Values increase in severe dehydration and shock. They decrease in severe malnutrition and hemorrhage.
Prothrombin time (serum)	12–14 sec (one stage)	Values increase in certain hemorrhagic diseases, liver disease, vitamin K deficiency, and following the use of various drugs.
Red cell distribution width (RDW)	8.5–11.5 microns	Variation in cell width changes with pernicious anemia.
Sedimentation rate, erythrocyte (whole blood)	Men: 1–13 mm/hr Women: 1–20 mm/hr	Values increase in infectious diseases, menstruation, pregnancy, and as a result of severe tissue damage.
Serum glutamic pyruvic transaminase (SGPT)	Women: 4–17 U/l Men: 6–24 U/l	Values increase in liver disease, pancreatitis, and acute myocardial infarction.
Sodium (serum)	135–145 mEq/l	Values increase in nephritis and severe dehydration. They decrease in Addison's disease, myxedema, kidney disease, and diarrhea.
Thromboplastin time, partial (plasma)	35–45 sec	Values increase in deficiencies of blood factors VIII, IX, and X.
Thyroid-stimulating hormone (TSH)	0.5–5.0 μU/ml	Values increase in hypothyroidism and decrease in hyperthyroidism.
Thyroxine, or T_4 (serum)	4–12 μg/100 ml	Values increase in hyperthyroidism and pregnancy. They decrease in hypothyroidism.
Transaminases, or SGOT (serum)	7–27 units/ml	Values increase in myocardial infarction, liver disease, and diseases of skeletal muscles.
Triglycerides	40–150 mg/100 ml	Values increase in liver disease, nephrotic syndrome, hypothyroidism, and pancreatitis. They decrease in malnutrition and hyperthyroidism.
Triiodothyronine, or T3 (serum)	75–195 ng/100 ml	Values increase in hyperthyroidism and decrease in hypothyroidism.
Uric acid (serum)	Men: 2.5–8.0 mg/100 ml Women: 1.5–6.0 mg/100 ml	Values increase in gout, leukemia, pneumonia, toxemia of pregnancy, and as a result of severe tissue damage.
White blood cell count, differential (whole blood)	Neutrophils 54–62% Eosinophils 1–3% Basophils 0–1% Lymphocytes 25–33% Monocytes 3–7%	Neutrophils increase in bacterial diseases; lymphocytes and monocytes increase in viral diseases; eosinophils increase in collagen diseases, allergies, and in the presence of intestinal parasites.
White blood cell count, total (whole blood)	5,000–10,000/cu mm	Values increase in acute infections, acute leukemia, and following menstruation. They decrease in aplastic anemia and as a result of drug toxicity.

*These values may vary with hospital, physician, and type of equipment used to make measurement.

Common tests performed on urine

Test	Normal values*	Clinical significance
Acetone and acetoacetate	0	Values increase in diabetic acidosis.
Albumin, qualitative	0 to trace	Values increase in kidney disease, hypertension, and heart failure.
Ammonia	20–70 mEq/l	Values increase in diabetes mellitus and liver diseases.
Bacterial count	Under 10,000/ml	Values increase in urinary tract infection.
Bile and bilirubin	0	Values increase in melanoma and biliary tract obstruction.
Calcium	Under 300 mg/24 hr	Values increase in hyperparathyroidism and decrease in hypoparathyroidism
Creatinine (24 hours)	15–25 mg/kg body weight/day	Values increase in infections, and decrease in muscular atrophy, anemia, leukemia, and kidney diseases.
Creatinine clearance (24 hour)	100–140 ml/min	Values increase in renal diseases.
Glucose	0	Values increase in diabetes mellitus and various pituitary gland disorders.
Hemoglobin	0	Blood may occur in urine as a result of extensive burns, crushing injuries, hemolytic anemia, or blood transfusion reactions.
17-hydroxycorticosteroids	3–8 mg/24 hr	Values increase in Cushing's syndrome and decrease in Addison's disease.
Osmolality	850 osmol/kg	Values increase in hepatic cirrhosis, congestive heart failure, and Addison's disease. They decrease in hypokalemia, hypercalcemia, and diabetes insipidus.
pH	4.6–8.0	Values increase in urinary tract infections and chronic renal failure. They decrease in diabetes mellitus, emphysema, and starvation.
Phenylpyruvic acid	0	Values increase in phenylketonuria.
Specific gravity (SG)	1.003–1.035	Values increase in diabetes mellitus, nephrosis, and dehydration. They decrease in diabetes insipidus, glomerulonephritis, and severe renal injury.
Urea	25–35 gm/24 hr	Values increase as a result of excessive protein breakdown. They decrease as a result of impaired renal function.
Urea clearance	Over 40 ml blood cleared of urea/min	Values increase in renal diseases.
Uric acid	0.6–1.0 gm/24 hr as urate	Values increase in gout and decrease in various kidney diseases.
Urobilinogen	0–4 mg/24 hr	Values increase in liver diseases and hemolytic anemia. They decrease in complete biliary obstruction and severe diarrhea.

*These values may vary with hospital, physician, and type of equipment used to make measurements.

Appendix D
Answers to Clinical Application of Knowledge Questions

Chapter 1, page 28

1. In some states, death is defined as "The cessation of the vital functions of the body or the cessation of the heartbeat and respirations." Definitions of death are related to the characteristics of life in that death is characterized by an absence of brainwaves, absence of reflexes, nonresponsiveness and unreceptivity, lack of spontaneous muscular movements, absence of blood pressure, lack of pulse, and the absence of spontaneous respirations.
2. Treatment of: Body temperature—increased fluid intake, hypothermia blanket, antipyretic medications, antibiotics if fever is due to an infection, rest, and maintaining nutritional intake.
   Blood oxygen—oral tracheal suction to keep airway open, administering oxygen. Water content—increased fluid intake, administering intravenous fluid, controlling room temperature and humidity.
3. The individual with the tumor in the dorsal cavity would probably develop symptoms first. Since the dorsal cavity is smaller, the tumor would exert pressure against the brain or spinal cord, causing symptoms of visual disturbances, headaches, behavioral changes, disequilibrium, disturbances in gait, seizures, incoordination, or weakness in the extremities.
4. The organs in the umbilical region that could be the source of pain include the small intestine (obstruction), pancreas (pancreatitis), abdominal aorta (aneurysm), transverse colon (diverticulitis), or a distended urinary bladder.
5. An unconscious person would need: oral tracheal suctioning (to keep the airway open and administer oxygen); nasogastric, gastric, or intravenous feedings; urinary catheter (for urine collection); turning in bed (to prevent decubitis ulcers, or bedsores); elastic stockings (to prevent blood clots and thrombophlebitis); stool softeners or suppositories; range-of-motion exercises to the extremities; and possibly a hypothermia or hyperthermia blanket to maintain body temperature.
6. Ultrasonography would be preferable since it does not use radiation, but uses very high frequency sound waves.

Chapter 2, page 61

1. Examples of acidic substances: vinegar (acetic acid), aspirin (acetylsalicylic acid), rust remover (oxalic acid), toilet bowl cleaner (muriatic acid), and automobile battery fluid (sulfuric acid).
   Examples of basic substances: ammonia, lye, drain and oven cleaner (sodium hydroxide), milk of magnesia (magnesium hydroxide), and antacids (sodium bicarbonate).
   Examples of acidic foods: meat (protein in general), whole wheat (grain in general), eggs, cheese, cranberries, plums, cola drinks, and tomatoes.
   Examples of basic foods: milk (fresh), vegetables, raisins, apricots, and nuts.
2. Examples of sources of saturated fats: butter, shortening, red meats, poultry, egg yolk, cheese, ice cream, whole milk, cream, seafood, crackers made with palm or coconut oil, and lard.
   Examples of sources of unsaturated fats: olives, olive oil, margarine (from corn oil), foods prepared with cottonseed, peanut, soybean, or safflower oil.
3. Radiation therapy uses precisely controlled doses of radioactivity that are targeted to irradiate only the tumor. A patient subjected to CT scanning does not become a radiation source.
4. Although radiation therapy may damage tissues, the benefits derived from medical uses are thought to outweigh the health risks.
5. The body utilizes carbohydrates to provide energy for metabolic reactions. Proteins are important in that they comprise structural components of tissues such as collagen, muscle fibers, cell membranes, elastin, and keratin, and also have functional roles as antibodies, enzymes, and certain hormones.
6. Clinical laboratory tests: Serum sodium, potassium, chloride, carbon dioxide, glucose, BUN (blood urea nitrogen), acid phosphatase, alkaline phosphatase, calcium, cholesterol, triglycerides, iron, total protein, cardiac enzymes (CPK, SGOT, LDH), liver enzymes (SGPT), arterial blood gases, and urine (pH, glucose, ketones, blood, protein).

Chapter 3, page 95

1. a. Osmosis. The hypertonic drug solution would cause water to leave the cells, causing swelling of the tissues, exerting pressure on nerves, and causing pain.
   b. Filtration. A low blood pressure would not result in enough hydrostatic pressure to force water and dissolved particles (glomerular filtrate) from the glomerular capillaries into the kidney tubules.
   c. Diffusion. Urea, an end product of protein metabolism, would need to diffuse from a higher concentration in the patient's blood to a lower concentration in the dialyzing fluid if it is to be removed from the blood.

2. Cell membranes contain phospholipids, into which fat-soluble substances (organic solvents) would easily diffuse.
3. White blood cells normally increase in number during an infection. Exposure to an X ray may destroy these cells, along with their ability to phagocytose bacteria and produce antibodies against bacteria and viruses.
4. The rhythmic beating of cilia normally aids in moving respiratory secretions up into the pharynx where they are swallowed or expectorated. If cilia are immobile, secretions remain in the lower respiratory tract where they may harbor microorganisms, resulting in infections.
5. Explanation to a patient: A phagocytic cell is capable of surrounding and engulfing a bacterium, taking it into the cell itself where chemicals known as enzymes decompose the bacterium. Phagocytic cells work in large numbers to destroy bacteria.
6. a. growth—cells continually undergo mitosis, increasing their numbers; they may also increase in size;
   b. wound healing—cells undergo mitosis only when necessary to replace those that have been injured;
   c. cancer—immature, undifferentiated cells undergo uncontrolled, continued mitosis, often metastasizing, or separating from the original site of the malignancy to distant organs of the body.

Chapter 4, page 123

1. Since enzymes catalyze metabolic reactions at only a specific pH, if the pH of body fluids is altered, enzymes will become denatured and cell metabolism will be altered.
2. The individual's cells utilize glucose as their primary and most efficient source of metabolic energy. With less carbohydrate available to provide glucose, the cells would make increased use of fatty acids, glycerol, and amino acids as sources of energy. The individual's urine would have a low pH (acidic), and would contain ketone bodies.

3. Enzymes and their associated (vitamin) coenzymes are reused in metabolic reactions. Excessive amounts of coenzymes are normally excreted.
4. Mutations (DNA changes) could occur in the cells of the developing embryo, resulting in birth defects or deformities.
5. Oxygen concentration would decrease and carbon dioxide concentration would increase. The result would be a lowering of blood pH, or acidosis.
6. Mutagenic factors would include X ray, radioisotopes, and chemotherapy drugs.

Chapter 5, page 144

1. Cartilage and (dense) fibrous connective tissue have a limited blood supply, limiting the number of phagocytes, nutrients, and red blood cells (which carry oxygen) reaching the site, thus slowing healing.
2. Connective tissues occur throughout the body and are associated with organs of virtually every body system.
3. Swelling—due to increased capillary permeability, which allows fluid to escape into the interstitial spaces;
   redness—due to vasodilation of arterioles and venules supplying the tissue with blood;
   pain—due to swelling and release of chemicals, which respectively compress and irritate nerve endings in the area;
   warmth—due to vasodilation of arterioles and venules, increasing the flow of blood to the tissues.
4. a. respiratory tubes—coughing, sputum production, difficulty breathing (dyspnea), and fever (if respiratory infection is present).
   b. digestive tract—gastric pain, vomiting, diarrhea, and nausea.
5. There would be a decrease or absence of ciliary action and mucus secretion, which would inhibit removal of inhaled dust particles and pathogens through the cough reflex. Inhaled air would not be warmed or humidified before reaching the lungs.

Chapter 6, page 166

1. Dehydration, electrolyte imbalance, inability to regulate body temperature, susceptibility to infection, decreased ability to sense touch, and vitamin D deficiency.
   Environmental modifications could include: aseptic environment, hydration through oral or intravenous routes, controlling room temperature, and balanced diet.
2. Ability to regulate body temperature would be decreased and the infant would be susceptible to hypothermia.
3. The superficial burn would irritate and thus stimulate nerve endings (pain receptors); the deeper burn would destroy them.
4. The subcutaneous injection would be absorbed more rapidly since there are more blood vessels in subcutaneous tissue than in the intradermal area.
5. Evaporation—sponge baths with cool water
   conduction—hypothermia blanket
   convection—lightweight covers and loose garments
   radiation—cool room temperature
6. Exercise and hot weather raise the body temperature, which increases metabolic rate, which, in turn, results in loss of water and electrolytes through sweating. Water and fruit juices replace lost fluids and electrolytes.

Chapter 7, page 226

1. Lead—prohibit sale and import of paint, dishes, and toys containing lead, use only unleaded gasoline, and screen for lead poisoning in schools;
   Radium—periodically check health workers for level of exposure to radium, avoid or minimize direct contact with patients receiving radiation therapy or radioactive implants, regular maintenance of radiology equipment and proper training in its use, monitoring homes for levels of radon gas, which is a disintegration product of radium.
   Strontium—exerting control over atomic testing and nuclear power plant emissions.
2. Children's bones are more flexible and resilient than those of adults; thus they are less susceptible to complete fractures.

3. Difficulties with coordination, balance, and gait.
4. As an individual ages, intervertebral disks become thinner and more rigid, resulting in compression fractures of the anteriorly located vertebral bodies. These events, in turn, cause the vertebral column to shorten and develop an exaggerated thoracic curvature.
5. Rate of development—If the posterior fontanel has closed two months after birth, the sphenoid three months after birth, the mastoid twelve months after birth, and the anterior by eighteen months or two years of age, the infant is probably developing "on time." Delayed closing of these fontanels may indicate slower than normal development.
Intracranial pressure—Depressed or "sunken in" fontanels may indicate decreased intracranial pressure or dehydration. Bulging fontanels may indicate increased intracranial pressure or overhydration.
6. Women have less bone mass than men, and lose bone mass at a faster rate and at an earlier age than men.
Postmenopausal women have decreased blood concentrations of female sex hormones, resulting in further loss of bone mass and mineral content.
The risk of developing osteoporosis may be reduced by: increasing dietary calcium intake or taking a calcium supplement, regular exercise, abstaining from alcohol and smoking, and, in some postmenopausal women, hormone replacement therapy.

Chapter 8, page 266

1. Ligaments are comprised of dense fibrous connective tissue, which has a poor blood supply. Cartilage is a rigid type of connective tissue that lacks a direct blood supply, resulting in slow cellular respiration and slow healing.
2. The shoulder and hip joints are ball-and-socket joints in which the heads of the humerus and femur respectively fit into and are protected by bony sockets. These sockets, in turn, are surrounded by joint capsules, ligaments, and muscles that protect against injuries. The knee

joint is of the condyloid and hinge types, has shallow articular surfaces, and is protected only by ligaments, thus making it more susceptible to injuries.
3. The hip joint could be more satisfactorily replaced. The socket of the hip joint is deeper, its ligaments are larger and stronger, and it is surrounded by muscles. It also does not bear as much weight as the knee joint.
4. There would be a loss of calcium, phosphorus, and matrix from the bones, resulting in bone destruction and brittleness.
5. Keeping the joints mobile prevents stiffness, which leads to immobility. It also prevents contractures, or atrophy and shortening of muscles, which may result in permanent flexion of joints.
6. Ligaments holding the articulating surfaces of the shoulder together are rather weak. Dislocating the shoulder would stretch or strain these ligaments, further weakening them, making future dislocations more likely.

Chapter 9, page 320

1. Warm-ups increase athletic performance by increasing oxygen consumption during exercise (thus decreasing oxygen debt), increasing flexibility of joints, increasing capillary blood flow to muscles, and increasing heart actions.
2. The levator ani and coccygeus muscles.
3. Active range-of-motion exercises of joints and muscles performed by the caregiver (that flex, extend, and rotate muscles and joints); also periodically changing the patient's position in bed.
4. Heat application and its resulting vasodilation result in increased blood flow and oxygen to the site, promoting aerobic metabolism and the removal of lactic acid and other metabolic waste substances from the area.
5. The gluteus medius muscle covers the superior lateral portion of the ilium. Bone markings that are used to establish the location of this muscle are the greater trochanter of the femur, the anterior superior iliac spine, posterior iliac spine, and crest of the ilium.

6. Passively moving or electrically stimulating the injured muscle would help prevent muscle atrophy and contractures.

Chapter 10, page 358

1. Axons of peripheral nerves are able to regenerate due to Schwann cells realigning themselves within the connective tissue endoneurium that forms a tube in which regenerating axonal sprouts are guided back to their original connections with the myoneural junction. Schwann cells also form new myelin around the growing axon. Motor nerve fibers within the spinal cord lack connective tissue sheaths, and the myelin-producing cells fail to produce new myelin following an injury.
2. Low potassium ion concentration (hypokalemia) results in hyperpolarization (increased negativity) of nerve and muscle fiber membranes, which prevents transmission of nerve impulses. Administration of potassium ions causes these membranes to become less negative, reestablishing the resting membrane potential, thus restoring the nerve fibers' original excitability.
3. Myelin sheaths in the central nervous system permit impulse conduction along the axons of nerve fibers. Therefore, motor nerve impulses would not be conducted to myoneural junctions in individuals with multiple sclerosis due to scarring and degeneration of the myelin sheath.
4. Calcium. The developing fetus needs calcium from the mother for development of bones and teeth. This may result in a blood calcium deficiency in the mother, causing sodium channels to remain open and excessive diffusion of sodium ions through nerve fiber membranes, which leads to continuous impulse transmission to skeletal muscles, resulting in tetanic muscle spasms.
5. Neuroglia are more numerous and are capable of mitotic division; mature neurons are incapable of undergoing mitosis.
6. If the biceps and triceps reflexes are present, the injury would be below C7. If the biceps and triceps reflexes are absent, the

injury would be above C5. If the biceps reflex is present and the triceps reflex is absent, the injury would be between C5 and C7.

Chapter 11, page 412

1. The sample should be withdrawn from the subarachnoid space between L3 and L4, or L4 and L5. The patient should be positioned on his side with knees drawn up to the abdomen and his chin flexed down toward his chest (to increase spaces between the vertebrae).
2. Right occipital lobe—loss of left half of visual field in each eye (left homonymous hemianopsis); Right temporal lobe—partial hearing loss in both ears, sensory aphasia if dominant hemisphere.
3. Descending (motor) anterior, lateral, and corticospinal tracts in the spinal cord would be affected on the same side as the injury, causing paralysis. Ascending (sensory) lateral spinothalmic tracts for temperature and pain transmission cross over to the opposite side of the spinal cord before ascending, resulting in loss of these sensations below the level of the injury on that side of the body.
4. Obstruction of tiny blood vessels of the cerebrum could cause a transient ischemic attack (temporary) or a stroke (permanent), resulting in weakness or paralysis of a limb, partial blindness, confusion, difficulty concentrating, memory loss, and decreased cognitive ability.
5. The cause of the CVA will largely determine treatment. If it involved an obstructed blood vessel, carotid endarterectomies and anticoagulant therapy may be used in selected cases.
   If it involved a ruptured blood vessel, treatment may include antihypertensive and anticonvulsant medications; surgical removal of the hematoma or clipping of the vessel may be necessary if the bleeding was due to a ruptured aneurysm.
6. Sweating, dilation of pupils, dry mouth, and increased pulse, blood pressure and respiration.

Chapter 12, page 458

1. The heat receptors in the skin have adapted to increased stimulation (by conducting fewer impulses to the brain).
2. The bullet would probably be causing localized, rather than widespread, stimulation of pain receptors, due to acute (A-delta) fibers being stimulated. Crushing of the skin would stimulate more free nerve endings there and in subcutaneous tissues, resulting in stimulation of both acute and chronic (C) pain fibers
3. Symptoms would include vertigo (dizziness), nausea, vomiting, headache, disequilibrium, tinnitus, and deafness.
4. Since the cells of the retina would become separated from their supply of oxygen and nutrients, there would be progressive loss of vision and eventual blindness.
5. Shorter, horizontally directed auditory tubes would enable microorganisms to enter the middle ear more easily, since the mucous membrane of the pharynx is continuous with that of the middle ear.
6. Explanation to a patient: Nerve impulses transmitting the sensation of pain from the heart follow common nerve pathways to the spinal cord and then to the brain as those originating from the skin of the left arm, shoulder, and the base of the neck.

Chapter 13, page 497

1. The person would require more sugar. Stored glycogen could become depleted through glucagon stimulating the conversion of glycogen into glucose during exercise. Exercise, in turn, lowers blood glucose level by facilitating glucose transport into cells (essentially the same action as insulin) and by decreasing the rate of carbohydrate metabolism.
2. Prolonged cortisol administration may result in: elevated blood glucose level (diabetes mellitus), edema, and hypertension (due to retention of sodium), redistribution of body fat, muscle wasting (due to protein catabolism), osteoporosis (due to loss of calcium from bones), and decreased antibody response (which would impair wound healing and increase susceptibility to infections).

3. Cool environment, well-balanced diet, quiet surroundings, relaxation, and simple diversional activities.
4. Replacement hormones would include: glucocorticoids (cortisol), thyroxine, and sex hormones (estrogens, progesterone, testosterone; FSH, LH, and GH if a younger adult). If hypophysectomy was performed to slow metastatic breast cancer, sex hormones would not be administered.
   Why? Cortisol is necessary for life in order to regulate glucose, protein, and fat metabolism, and decrease inflammation. In adults, thyroxine is needed for regulation of the rate of glucose, protein, and fat metabolism. Sex hormones would maintain secondary sex characteristics and in a younger individual, might restore fertility.
5. Aldosterone secretion would cause the blood concentration of sodium ions to increase and potassium ions to decrease.
6. Growth hormone—Administration of GH would directly stimulate target cells to break down fats, and increase blood sugar levels and protein synthesis. GH would also directly stimulate bone tissue growth and mediate increased cartilage formation. Growth hormone-releasing hormone— Selection of GRH for administration would depend on the child's having a functioning anterior pituitary gland, which, in turn, would secrete GH. GH would then exert its stimulating effects on target cells and tissues.

Chapter 14, page 546

1. Gastric secretions would be greatly reduced. Gastrin, which normally is the stimulus to secretory activity in the stomach, would be decreased; protein digestion, which normally begins in the stomach, would be decreased or absent due to the lack of pepsin and HCl; gastric lipase, which primarily digests butterfat, would be greatly decreased or not secreted at all. The patient would also be prone to pernicious anemia as a result of the removal of parietal cells of the stomach.

Removal of 95% of the stomach would also reduce its storage capacity. The patient would need to eat several small meals per day and drink fluids between, rather than with, meals to avoid hypoglycemia and fluid loss through diarrhea.

2. The cystic duct from the gallbladder empties into the common bile duct, which along with the main pancreatic duct empties into the duodenum at the hepatopancreatic sphincter. Cholecystitis (especially if accompanied by gallstones that block the common bile duct) often causes reflux of bile into the pancreatic duct, which activates pancreatic enzymes, leading to pancreatitis.

3. In moderation, alcohol stimulates gastrin release, which would increase secretion of pepsin and HCl. This, in turn, would stimulate gastric peristalsis and pancreatic enzyme secretion, increasing digestive activity.

4. Prolonged vomiting would result in alkalosis; prolonged diarrhea would result in acidosis.

5. Vomiting and symptoms of dehydration and malnutrition (weight loss, dry mucous membranes, sunken fontanels, and loss of skin elasticity).

6. Wheat and rye flour contain gluten, which when eaten by individuals with this particular malabsorption syndrome, results in loss of cells from villi that are responsible for absorption of nutrients from the small intestine. If gluten continues to be ingested, the condition will persist, resulting in weakness, weight loss, anorexia, and inability to absorb fat-soluble vitamins (A, D, E, and K), resulting in their deficiency.

Chapter 15, page 579

1. Blood sugar (glucose) concentration would remain stable due to glycogenolysis, which converts glycogen into glucose, and gluconeogenesis, in which certain amino acids can be deaminated and converted into glucose.

2. To calculate caloric content of nutrients, multiply grams of carbohydrate by 4.1 (Cal./gm), grams of lipid by 9.5 (Cal./gm), and grams of protein by 4.1 (Cal./gm).

According to 1980 National Research Council figures, carbohydrates should comprise 50% of daily caloric intake, proteins 15–20%, and fats 30–35%.

Daily energy needs (Calories) per day are:

Age	Males	Females
15–18	2,800	2,100
19–22	2,900	2,100
23–50	2,700	2,000

3. Answers will vary.

4. Fish (contains all the essential amino acids) and proteins of plant origin (contain selected amino acids, when eaten in variety provide all essential amino acids) such as rice, beans, peas, nuts, and seeds.

5. Avoid eating egg yolks, liver and other organ meats; eat meat, cheese, butter, and whole milk in moderation; select a diet high in fruits, vegetables, bread, and cereals.

6. The twelve-year-old boy would require more calories, protein, carbohydrates, fats, and B vitamins.

Chapter 16, page 615

1. Air entering a tracheostomy would not be warmed, filtered for removal of microorganisms and dust particles, or humidified by the nasal passages or oropharynx. The patient may experience irritation around and inside the tracheostomy opening (due to dryness), coughing, and respiratory infections.

2. Residual volume would be increased due to overdistension of the lungs and loss of elastic recoil force from destruction of alveolar walls. This would increase pressure in the pleural cavity and inhibit expiration of air from the lungs. Expiratory reserve volume would be decreased due to loss of elastic recoil force of the alveoli. Vital capacity would be decreased due to residual volume taking up more of the thoracic gas volume (total lung capacity). Total lung capacity would remain the same or would be increased due to overdistension of the lungs, increased anterior-posterior diameter of the chest wall, and increased residual volume.

3. Concentration of blood oxygen would increase due to deep and rapid inhaling of air, which is 21% oxygen.

4. The mixture of oxygen and carbon dioxide would be more beneficial. Carbon dioxide would normally stimulate the respiratory center and oxygen would prevent tissue hypoxia.

5. At high altitudes, the partial pressure of oxygen is less, which would stimulate chemoreceptors in the carotid and aortic bodies, resulting in an increased respiratory rate. If the individual is unable to breathe at this increased rate, oxygen will diffuse less rapidly from the alveoli into the blood and the tissues, including the brain, and the person will become hypoxic and lose consciousness.

6. Pursed lip breathing slows the flow rate of exhaled air. This creates a back pressure in the airways, which keeps the smaller airways open, preventing their collapse and obstruction. Thus, the lungs will "empty" more completely.

Chapter 17, page 648

1. The hematocrit would increase. Dehydration results from fluid loss exceeding fluid intake. This loss is replenished by the movement of water (and electrolytes) from plasma into the interstitial spaces and cells. Red blood cells are too large to pass through capillary walls, leaving a proportionately greater concentration of red blood cells in a lesser volume of plasma.

2. The white cell count would decrease. The patient's environment would need to be modified to decrease chances of infection (frequent hand washing, avoiding individuals with colds, administration of antibiotics if a fever is present, and avoiding raw foods). If hospitalized, additional precautions would include care givers wearing a mask, a gown, and gloves, and giving the patient a private room.

3. A deficient dietary intake of iron caused by lack of finances (foods high in iron, such as meats, seafood, and green leafy vegetables, may be expensive), physical limitations making it difficult to shop, lack of motivation to cook for oneself, and loss of teeth.

4. Prothrombin and fibrinogen (and some of the clotting factors) are synthesized in the liver. Bile salts from the liver are also needed in order for vitamin K to be absorbed in the small intestine. A deficiency of any of these will result in excessive bleeding.
5. The white blood cells formed in leukemia are immature and remain undifferentiated, and therefore have little or no ability to phagocytose bacteria or produce antibodies to combat infection.
6. No. The Rh-negative fetus would not have receptors for anti-Rh agglutinins on its red blood cell membranes.

Chapter 18, page 711

1. It would be wiser to rest. With exercise, blood flow to the myocardium, skin, and active muscles increases due to vasodilation (stimulated by sympathetic nerve stimulation, decreased partial pressure of oxygen, and decreasing pH). Conversely, sympathetic stimulation causes vasoconstriction and reduced blood flow to the digestive organs, which would impede digestion and absorption of nutrients if exercise were undertaken at this time, and could also result in cramping.
2. Blood flow would probably carry the embolus to the pulmonary vessels. Symptoms would include dyspnea, chest pain, tachycardia, cough, spitting up blood (hemoptysis), and restlessness (due to cerebral hypoxia).
3. The rate of venous return would decrease. Increased intrathoracic pressure compresses the walls of the large veins in the chest (superior and inferior venae cavae).
4. The atria pump blood to the ventricles. During rest and moderate activity, filling of the ventricles is not dependent on atrial contraction. The ventricles pump arterial blood to the lungs and all body organs and tissues. During ventricular fibrillation, the ventricular myocardium quivers but does not contract as a whole, resulting in cessation of cardiac output.
5. Increased capillary pressure could cause varicosities in the thin-walled esophageal veins (known as esophageal varices),

which may rupture, resulting in massive hemorrhage. It could also cause hemorrhoids, caput medusae (dilated cutaneous veins radiating out from the umbilicus), ascites, peripheral edema, and splenomegaly.
6. Femoral artery, external iliac artery, common iliac artery, aorta (abdominal and thoracic), aortic arch, ascending aorta, and left coronary artery.

Chapter 19, page 744

1. Lymph nodes function to filter out and thus trap microorganisms and cancer cells as they travel between tissues and the blood stream. Lymph node biopsy can therefore be used in diagnosing leukemias, cancers of the lymphatic system, metastatic cancers, and fungal infections (especially of the lungs).
2. The dermis of the skin contains many lymph capillaries.
3. Vaccination (vaccines) contains antigens that stimulate the body to produce antibodies for protection against a particular pathogen in the future. Thus, it provides active immunity. Gamma globulin contains antibodies not produced by the lymphocytes of the individual receiving the injection. Thus, no immunological memory is established as in the case of active immunity.
4. Lymph nodes lie along lymphatic pathways. Thus, excising them interrupts the flow of lymph from tissues distal to the excisional site. The accumulation of tissue fluid and increased osmotic pressure in the fluid (due to the presence of plasma proteins, which lead out of lymph vessels) causes swelling.
5. The infant's susceptibility to infections would increase. The thymus gland is the site of T-cell maturation. T-cell deficiency results in the individual having little or no resistance to intracellular virus, fungal, or bacterial infections.
6. Precautions would include: wearing gloves and a gown when handling items contaminated with blood or other body fluids; wearing gloves, a gown, a mask, and eye protection when performing procedures involving extensive contact with blood or other body fluids; exercising care in handling and prompt disposal

of needles and syringes in puncture-resistant, sealable containers; use of mouthpieces in performing mouth-to-mouth resuscitation; and promptly cleaning up spills of blood and body fluids with bleach or detergent (while wearing gloves).

Chapter 20, page 776

1. The volume of urine would decrease. Narrowed renal arteries would decrease blood flow to the glomeruli, which, in turn, would result in decreased glomerular filtration rate, glomerular filtrate, and urine.
2. Plasma proteins include the gamma globulin fraction, which includes immunoglobulins. Immunoglobulins produce antibodies that provide a defense against infection.
3. The volume of urine produced would probably be greater than normal. Loss of blood would mean loss of plasma proteins, which, in turn, would decrease plasma osmotic pressure. Intravenous fluids would maintain blood volume and blood pressure, but would contain no plasma proteins to oppose hydrostatic pressure, which promotes production of glomerular filtrate and urine.
4. Since penicillin is excreted unchanged and concentrated in the urine, the penicillin will act while the urine is in the urinary bladder. Therefore, the route to the bladder is through the kidneys.
   Absorption into capillaries of the small intestine, venules, superior mesenteric vein, portal vein, liver, hepatic vein, inferior vena cava, heart, aorta, renal arteries, kidneys (glomeruli, glomerular capsule, tubules, collecting ducts, renal pelvis), ureters, urinary bladder.
5. The volume of urine would decrease. (Shock results in sympathetic vasoconstriction of peripheral vessels and decreased glomerular filtration rate. In response to decreased glomerular filtration rate, the renin-angiotensin mechanism is activated, which increases circulating blood volume.)
6. The female urethra is shorter (4 cm long vs. 19 cm in the male) than that of the male. Therefore, microorganisms could ascend into the bladder more readily.

Chapter 21, page 800

1. Partial pressure of carbon dioxide would be increased and partial pressure of oxygen would be decreased. Oxygen saturation of red blood cells would also be decreased.
2. Symptoms would include: dry, sticky mucous membranes; lethargy (with irritability when disturbed), tremors; convulsions; muscle rigidity; and edema. Infants have proportionately more extracellular than intracellular water (hence, they have less reserve to draw on to "dilute" the sodium), their kidneys are immature and less able to conserve water, and they have greater body surface area in proportion to weight, resulting in increased fluid loss through perspiration.
3. Concentrated protein feedings through a gastrostomy tube would pull water osmotically into the intestine from the blood and extracellular fluid, causing diarrhea, dehydration, and increased serum sodium level.
4. Plasma pH would:
   a. decrease
   b. increase
   c. increase
   d. decrease
5. Absorption would be impaired, therefore the patient would be prone to hyponatremia and hypokalemia.
6. Fluid would accumulate in the extracellular fluid compartments, resulting in edema, ascites, and increased venous pressure.

Chapter 22, page 850

1. There would be no significant decrease in secondary sex characteristics following removal of one testis.
   Removal of both testes would result in decreased rate of growth of the beard and body hair due to significant reduction in testosterone production.
   Removal of the prostate gland would have no effect on secondary sex characteristics.
2. The posterior pituitary hormone oxytocin is secreted to stimulate ejection of milk from the alveolar ducts. Oxytocin secretion also stimulates uterine contractions, hence the sensation of cramps.

3. No. Menstruation is dependent on hormone secretion (by the anterior pituitary gland and ovaries). Tubal ligation has no effect on hormone secretion.
4. If a single ovary is removed, there is no change in the menstrual cycles; if both ovaries are removed, the menstrual cycles would cease, due to significant reduction in production of estrogen and progesterone.
5. The most effective methods of contraception are tubal ligation, vasectomy, contraceptive implant, and oral contraceptives. The least effective methods of contraception are coitus interruptus, rhythm method, diaphragms, and jellies.
6. The consequences would be infertility. The higher temperature in the abdominal cavity inhibits sperm formation.

Chapter 23, page 881

1. Identical twins are genetically identical. Therefore, they have identical antigens that would not stimulate cell-mediated and antibody-mediated immune responses.
2. Problems could include mixing of oxygenated and deoxygenated blood in the right and left ventricles, resulting in cyanosis and hypoxia of body organs. It could also result in increased distension and workload of the right ventricle, which would impede circulation of blood to the lungs and result in already-oxygenated blood being recirculated through the lungs.
3. Symptoms could include those of congestive heart failure (cyanosis, fatigue, and peripheral and pulmonary edema).
4. Pigment development and transfer of pigment (melanin) into neighboring cells of the epidermis is incomplete at birth.
5. The physically inactive adult's skeletal muscles will have begun to lose strength; joint mobility and bone density would be decreased; vital capacity would be decreased; and narrowing of arterial lumens could lead to hypertension and less efficient pumping of blood from the heart.

6. Environmental provisions could include: reducing background noise; marking stairways clearly; providing night lights and higher wattage reading lamps; avoiding use of scatter rugs; applying nonslip strips in bathtubs; avoiding electric heaters; providing firm chairs; serving a low-sugar, low-salt, low-fat, high-fiber, high-calcium diet; encouraging fluid (water) intake; and clearly labeling medication containers.

Chapter 24, page 907

1. The marriage of first cousins would increase the probability of the expression of deleterious genes in their offspring. (Recessive alleles for undesirable or even lethal traits would have a far greater chance of becoming homozygous.)
2. There is a one in four (25%) chance of each conception producing a child with sickle-cell anemia.
3. Tests would include amniocentesis and/or chorionic villi biopsy. Advanced maternal age increases the chance of bearing a child afflicted with Down syndrome.
4. Since blood types are inherited, if a supposed father possesses a blood genotype not congruous with that of the child, he cannot be the child's father. Standard blood typing tests are not specific enough to positively identify individuals.
5. Factors influencing the sex ratio include: greater numbers of males are aborted or stillborn (X-linked lethal recessive genes have been implicated here), greater numbers of males also succumb to childhood diseases, are killed in wars, and tend to die from cardiovascular diseases at an earlier age (estrogens tend to protect women from atherosclerotic vascular changes until after menopause).
6. Chorionic villi biopsy is an invasive procedure that carries the risk of bleeding and inducing premature labor.

Glossary

Each word in this glossary is followed by a phonetic guide to pronunciation.

In this guide, any unmarked vowel that ends a syllable or stands alone as a syllable has the long sound. Thus, the word *play* would be spelled *pla.*

Any unmarked vowel that is followed by a consonant has the short sound. The word *tough,* for instance, would be spelled *tuf.*

If a long vowel appears in the middle of a syllable (followed by a consonant), it is marked with the macron (ˉ), the sign for a long vowel. Thus, the word *plate* would be phonetically spelled *plāt.*

Similarly, if a vowel stands alone or ends a syllable, but has the short sound, it is marked with a breve (˘).

a

abdomen (ab-do′men) Portion of the body between the diaphragm and the pelvis.

abduction (ab-duk′shun) Movement of a body part away from the midline.

absorption (ab-sorp′shun) The taking in of substances by cells or membranes.

accessory organs (ak-ses′o-re or′ganz) Organs that supplement the functions of other organs; accessory organs of the digestive and reproductive systems.

accommodation (ah-kom″o-da′shun) Adjustment of the lens for close vision.

acetone (as′e-tōn) One of the ketone bodies produced as a result of the oxidation of fats.

acetylcholine (as″ĕ-til-ko′lēn) Substance secreted at the axon ends of many neurons that transmits a nerve impulse across a synapse.

acetyl coenzyme A (as′ĕ-til ko-en′zīm) An intermediate compound produced during the oxidation of carbohydrates and fats.

acid (as′id) A substance that ionizes in water to release hydrogen ions.

acidosis (as″ĭ-do′sis) Condition in which there is a relative increase in the acid content of body fluids.

ACTH Adrenocorticotropic hormone.

actin (ak′tin) A protein in a muscle fiber that, together with myosin, is responsible for contraction and relaxation.

action potential (ak′shun po-ten′shal) The sequence of electrical changes occurring when a nerve cell membrane is exposed to a stimulus that exceeds its threshold.

activation energy (ak″tĭ-va′shun en′er-je) Energy needed to initiate a chemical reaction.

active site (ak′tiv sīt) Region of an enzyme molecule that combines temporarily with a substrate.

active transport (ak′tiv trans′port) Process that requires an expenditure of energy to move a substance across a cell membrane; usually moved against the concentration gradient.

acupuncture (ak′u-pungk″chūr) Procedure in which needles are inserted into various tissues to control pain sensations.

adaptation (ad″ap-ta′shun) Adjustment to environmental conditions.

adduction (ah-duk′shun) Movement of a body part toward the midline.

adenoids (ad′ĕ-noids) The pharyngeal tonsils located in the nasopharynx.

adenosine diphosphate (ah-den′o-sēn di-fos′fāt) ADP; molecule produced when the terminal phosphate is lost from a molecule of adenosine triphosphate.

adenosine triphosphate (ah-den′o-sēn tri-fos′fāt) An organic molecule that stores energy and releases energy for use in cellular processes.

adenylate cyclase (ah-den′i-lāt si′klās) An enzyme that is activated when certain hormones combine with receptors on cell membranes, causing ATP to become cyclic AMP.

ADH Antidiuretic hormone.

adipose tissue (ad′ĭ-pōs tish′u) Fat-storing tissue.

adolescence (ad″o-les′ens) Period of life between puberty and adulthood.

ADP Adenosine diphosphate.

adrenal cortex (ah-dre′nal kor′teks) The outer portion of the adrenal gland.

adrenal glands (ah-dre′nal glandz) Endocrine glands located on the superior portions of the kidneys.

adrenaline (ah-dren′ah-lin) Epinephrine.

adrenal medulla (ah-dre′nal me-dul′ah) The inner portion of the adrenal gland.

adrenergic fiber (ad″ren-er′jik fi′ber) A nerve fiber that secretes norepinephrine at the terminal end of its axon.

adrenocorticotropic hormone (ah-dre″no-kor″te-ko-trōp′ik hor′mōn) ACTH; hormone secreted by the anterior lobe of the pituitary gland that stimulates activity in the adrenal cortex.

adulthood (ah-dult′hood) Period of life that extends from adolescence to old age.

aerobic respiration (a″er-ōb′ik res″pi-ra′shun) Phase of cellular respiration that requires the presence of oxygen.

afferent arteriole (af′er-ent ar-te′re-ōl) Vessel that supplies blood to the glomerulus of a nephron within the kidney.

agglutination (ah-gloo″ti-na′shun) Clumping together of blood cells in response to a reaction between an agglutinin and an agglutinogen.

agglutinin (ah-gloo′ti-nin) A substance that reacts with an agglutinogen; an antibody.

agglutinogen (ag″loo-tin′o-jen) A substance that stimulates the formation of agglutinins; a foreign substance.

agranulocyte (a-gran′u-lo-sīt) A nongranular leukocyte.

albumin (al-bu′min) A plasma protein that helps regulate the osmotic concentration of the blood.

aldosterone (al-dos′ter-ōn) A hormone, secreted by the adrenal cortex, that functions in regulating sodium and potassium ion concentrations and water balance.

alimentary canal (al″i-men′tar-e kah-nal′) The tubular portion of the digestive tract that leads from the mouth to the anus.

alkaline (al′kah-līn) Pertaining to or having the properties of a base or alkali; basic.

alkaline tide (al′kah-lin tīd) An increase in the blood concentration of bicarbonate ions following a meal.

alkaloid (al′kah-loid) A group of organic substances that are usually bitter in taste and have toxic effects.

alkalosis (al″kah-lo′sis) Condition in which there is a relative increase in the alkaline content of body fluids.

allantois (ah-lan′to-is) A structure that appears during embryonic development and functions in the formation of umbilical blood vessels.

allele (ah-lēl) An alternative form of a gene that corresponds to positions on homologous chromosomes.

allergen (al′er-jen) A foreign substance capable of stimulating an allergic reaction.

all-or-none response (al′or-nun′ re-spons′) Phenomenon in which a muscle fiber contracts completely when it is exposed to a stimulus of threshold strength.

alpha receptor (al′fah re-sep′tor) Receptor on effector cell membrane that combines with epinephrine or norepinephrine.

alpha-tocopherol (al″fah-to-kof′er-ol) Vitamin E.

alveolar ducts (al-ve′o-lar dukts′) Fine tubes that carry air to the air sacs of the lungs.

alveolar pores (al-ve′o-lar pōrz) Minute openings in the walls of air sacs, which permit air to pass from one alveolus to another.

alveolar process (al-ve′o-lar pros′es) Projection on the border of the jaw in which the bony sockets of the teeth are located.

alveolus (al-ve′o-lus) An air sac of a lung; a saclike structure.

amacrine cell (am′ah-krin sel) A retinal neuron whose fibers pass laterally between other retinal cells.

amine (am′in) A type of nitrogen-containing organic compound, including the hormones secreted by the adrenal medulla.

amino acid (ah-me′no as′id) An organic compound of relatively small molecular size that contains an amino group ($-NH_2$) and a carboxyl group ($-COOH$); the structural unit of a protein molecule.

amniocentesis (am″ne-o-sen-te′sis) A procedure in which a sample of amniotic fluid is removed through the abdominal wall of a pregnant woman.

amnion (am′ne-on) An embryonic membrane that encircles a developing fetus and contains amniotic fluid.

amniotic cavity (am″ne-ot′ik kav′i-te) Fluid-filled space enclosed by the amnion.

amniotic fluid (am″ne-ot′ik floo′id) Fluid within the amniotic cavity that surrounds the developing fetus.

ampulla (am-pul′ah) An expansion at the end of each semicircular canal that contains a crista ampullaris.

amylase (am′i-lās) An enzyme that functions to hydrolyze starch.

anabolic metabolism (an″ah-bol′ik mĕ-tab′o-lizm) Metabolic process by which larger molecules are formed from smaller ones; anabolism.

anaerobic respiration (an-a″er-ōb′ik res″pi-ra′shun) Phase of cellular respiration that occurs in the absence of oxygen.

anal canal (a′nal kah-nal′) The last two or three inches of the large intestine that open to the outside as the anus.

anaphase (an′ah-fāz) Stage in mitosis during which duplicate chromosomes move to opposite poles of the cell.

anaplasia (an″ah-pla′ze-ah) A change in which mature cells become more primitive; a failure to differentiate.

anastomosis (ah-nas″to-mo′sis) A union of nerve fibers or blood vessels to form an intercommunicating network.

anatomy (ah-nat′o-me) Branch of science dealing with the form and structure of body parts.

androgen (an′dro-jen) A male sex hormone such as testosterone.

anemia (ah-ne′me-ah) A condition characterized by a deficiency of red blood cells or a deficiency of hemoglobin.

aneurysm (an′u-rizm) A saclike expansion of a blood vessel wall.

angiotensin (an″je-o-ten′sin) A vasoconstricting substance that is produced when blood flow to the kidneys is reduced, causing an increase in the blood pressure.

anoxia (an-ok′se-ah) Condition in which the oxygen concentration of the tissues is abnormally low.

antagonist (an-tag′o-nist) A muscle that acts in opposition to a prime mover.

antebrachium (an″te-bra′ke-um) The forearm.

antecubital (an″te-ku′bi-tal) The region in front of the elbow joint.

anterior (an-te′re-or) Pertaining to the front; the opposite of posterior.

anterior pituitary (an-te′re-or pi-tu′i-tār″e) The front lobe of the pituitary gland.

antibody (an′ti-bod′e) A specific substance produced by cells in response to the presence of an antigen; it reacts with the antigen.

antibody-mediated immunity (an′ti-bod″e me′de-ā-tid ĭ-mu′ni-te) Resistance to disease-causing agents resulting from the production of specific antibodies by B-lymphocytes; humoral immunity.

anticoagulant (an″ti-ko-ag′u-lant) A substance that inhibits the action of the blood clotting mechanism.

anticodon (an″ti-ko′don) A set of three nucleotides of a transfer RNA molecule.

antidiuretic hormone (an″ti-di′u-ret′ik hor′mōn) Hormone released from the posterior lobe of the pituitary gland that enhances the conservation of water by the kidneys; ADH.

antigen (an′ti-jen) A substance that stimulates cells to produce antibodies.

antioxidant (an″ti-ok′si-dant) A substance that inhibits the oxidation of another substance.

antithrombin (an″ti-throm′bin) A substance that inhibits the action of thrombin and thus inhibits the blood clotting mechanism.

anus (a′nus) Inferior outlet of the digestive tube.

aorta (a-or′tah) Major systemic artery that receives blood from the left ventricle.

aortic body (a-or′tik bod′e) A structure associated with the wall of the aorta that contains a group of chemoreceptors.

aortic sinus (a-or′tik si′nus) Swelling in the wall of the aorta that contains pressoreceptors.

aortic valve (a-or′tik valv) Flaplike structures in the wall of the aorta near its origin that prevent blood from returning to the left ventricle of the heart.

apneustic area (ap-nu′stik a′re-ah) A portion of the respiratory control center located in the pons.

apocrine gland (ap′o-krin gland) A type of sweat gland that responds during periods of emotional stress.

aponeurosis (ap″o-nu-ro′sis) A sheetlike tendon by which certain muscles are attached to other parts.

appendicular (ap″en-dik′u-lar) Pertaining to the arms or legs.

appendix (ah-pen′diks) A small, tubular appendage that extends outward from the cecum of the large intestine; vermiform appendix.

aqueous humor (a′kwe-us hu′mor) Watery fluid that fills the anterior and posterior chambers of the eye.

arachnoid granulation (ah-rak′noid gran″u-la′shun) Fingerlike structures that project from the subarachnoid space of the meninges into blood-filled dural sinuses and function in the reabsorption of cerebrospinal fluid.

arachnoid mater (ah-rak'noid ma'ter) Delicate, weblike middle layer of the meninges.

arbor vitae (ar'bor vi'ta) Treelike pattern of white matter seen in a section of cerebellum.

areola (ah-re'o-lah) Pigmented region surrounding the nipple of the mammary gland or breast.

arrector pili muscle (ah-rek'tor pil'i mus'l) Smooth muscle in the skin associated with a hair follicle.

arrhythmia (ah-rith'me-ah) Abnormal heart action characterized by a loss of rhythm.

arterial pathway (ar-te're-al path'wa) Course followed by blood as it travels from the heart to the body cells.

arteriole (ar-te're-ōl) A small branch of an artery that communicates with a capillary network.

arteriosclerosis (ar-te"re-o-sklē-ro'sis) Condition in which the walls of arteries thicken and lose their elasticity; hardening of the arteries.

artery (ar'ter-e) A vessel that transports blood away from the heart.

arthritis (ar-thri'tis) Condition characterized by inflammation of joints.

articular cartilage (ar-tik'u-lar kar'tī-lij) Hyaline cartilage that covers the ends of bones in synovial joints.

articulation (ar-tik"u-la'shun) The joining together of parts at a joint.

ascending colon (ah-send'ing ko'lon) Portion of the large intestine that passes upward on the right side of the abdomen from the cecum to the lower edge of the liver.

ascending tracts (ah-send'ing trakts) Groups of nerve fibers in the spinal cord that transmit sensory impulses upward to the brain.

ascorbic acid (as-kor'bik as'id) One of the water-soluble vitamins; vitamin C.

assimilation (ah-sim"ĭ-la'shun) The action of changing absorbed substances into forms that differ chemically from those entering.

association area (ah-so"se-a'shun a're-ah) Region of the cerebral cortex related to memory, reasoning, judgment, and emotional feelings.

astigmatism (ah-stig'mah-tizm) Visual defect due to errors in refraction caused by abnormal curvatures in the surface of the cornea or lens.

astrocyte (as'tro-sīt) A type of neuroglial cell that functions to connect neurons to blood vessels.

atherosclerosis (ath"er-o-sklē-ro'sis) Condition in which fatty substances accumulate abnormally on the inner linings of arteries.

atmospheric pressure (at"mos-fer'ik presh'ur) Pressure exerted by the weight of the air; about 760 mm of mercury at sea level.

atom (at'om) Smallest particle of an element that has the properties of that element.

atomic number (ah-tom'ik num'ber) Number equal to the number of protons in an atom of an element.

atomic weight (ah-tom'ik wāt) Number approximately equal to the number of protons plus the number of neutrons in an atom of an element.

ATP Adenosine triphosphate.

ATPase Enzyme that causes ATP molecules to release the energy stored in the terminal phosphate bonds.

atrioventricular bundle (a"tre-o-ven-trik'u-lar bun'dl) Group of specialized fibers that conduct impulses from the atrioventricular node to the ventricular muscle of the heart; A-V bundle.

atrioventricular node (a"tre-o-ven-trik'u-lar nōd) Specialized mass of cardiac muscle fibers located in the interatrial septum of the heart; functions in the transmission of the cardiac impulses from the sinoatrial node to the ventricular walls; A-V node.

atrioventricular orifice (a"tre-o-ven-trik'u-lar or'i-fis) Opening between the atrium and the ventricle on one side of the heart.

atrioventricular sulcus (a"tre-o-ven-trik'u-lar sul'kus) A groove on the surface of the heart that marks the division between an atrium and a ventricle.

atrioventricular valve (a"tre-o-ven-trik'u-lar valv) Cardiac valve located between an atrium and a ventricle.

atrium (a'tre-um) A chamber of the heart that receives blood from veins.

atrophy (at'ro-fe) A wasting away or decrease in size of an organ or tissue.

audiometer (aw"de-om'ē-ter) An instrument used to measure the acuity of hearing.

auditory (aw'di-to"re) Pertaining to the ear or to the sense of hearing.

auditory ossicle (aw'di-to"re os'i-kl) A bone of the middle ear.

auditory tube (aw'di-to"re tub) The tube that connects the middle ear cavity to the pharynx; eustachian tube.

auricle (aw'ri-kl) An earlike structure; the portion of the heart that forms the wall of an atrium.

autoimmune disease (aw"to-ĭ-mūn' di-zēz') Disorder characterized by an immune response directed toward a person's own tissues; autoallergy.

autonomic nervous system (aw"to-nom'ik ner'vus sis'tem) Portion of the nervous system that functions to control the actions of the visceral organs and skin.

autoregulation (aw"to-reg"u-la'shun) Phenomenon by which a mechanism within an organ or tissue maintains a constant blood flow in spite of changing arterial blood pressure.

autosome (aw'to-sōm) A chromosome other than a sex chromosome.

A-V bundle (bun'dl) A group of fibers that conduct cardiac impulses from the A-V node to the Purkinje fibers; bundle of His.

A-V node (nōd) Atrioventricular node.

axial skeleton (ak'se-al skel'ē-ton) Portion of the skeleton that supports and protects the organs of the head, neck, and trunk.

axillary (ak'sĭ-ler"e) Pertaining to the armpit.

axon (ak'son) A nerve fiber that conducts a nerve impulse away from a neuron cell body.

axonal transport (ak'so-nal trans'port) The process by which substances are carried from the neuron cell body to the distal end of an axon.

b

basal ganglion (ba'sal gang'gle-on) Mass of gray matter located deep within a cerebral hemisphere of the brain.

basal metabolic rate (ba'sal met"ah-bol'ic rāt) Rate at which metabolic reactions occur when the body is at rest.

base (bās) A substance that ionizes in water to release hydroxyl ions (OH) or other ions that combine with hydrogen ions.

basement membrane (bās'ment mem'brān) A layer of nonliving material that anchors epithelial tissue to underlying connective tissue.

basophil (ba'so-fil) White blood cell characterized by the presence of cytoplasmic granules that become stained by basophilic dye.

beta oxidation (ba'tah ok"si-da'shun) Chemical process by which fatty acids are converted to molecules of acetyl coenzyme A.

beta receptor (ba'tah re-sep'tor) Receptor on effector cell membrane that combines mainly with epinephrine and only slightly with norepinephrine.

bicarbonate ion (bi-kar'bon-āt i'on) HCO_3^-.

bicuspid tooth (bi-kus'pid tōōth) A premolar that is specialized for grinding hard particles of food.

bicuspid valve (bi-kus'pid valv) Heart valve located between the left atrium and the left ventricle; mitral valve.

bile (bīl) Fluid secreted by the liver and stored in the gallbladder.

bilirubin (bil"ĭ-roo'bin) A bile pigment produced as a result of hemoglobin breakdown.

biliverdin (bil"ĭ-ver'din) A bile pigment produced as a result of hemoglobin breakdown.

biochemistry (bi"o-kem'is-tre) Branch of science dealing with the chemistry of living organisms.

biofeedback (bi"o-fēd'bak) Procedure in which electronic equipment is used to help a person learn to consciously control certain visceral responses.

biotin (bi'o-tin) A water-soluble vitamin; a member of the vitamin B complex.

bipolar neuron (bi-po'lar nu'ron) A nerve cell whose cell body has only two processes, one serving as an axon and the other as a dendrite.

blastocyst (blas'to-sist) An early stage of embryonic development that consists of a hollow ball of cells.

B-lymphocyte (lim'fo-sīt) Lymphocyte that reacts against foreign substances in the body by producing and secreting antibodies; B-cell.

BMR Basal metabolic rate.

Bowman's capsule (bo'manz kap'sūl) Proximal portion of a renal tubule that encloses the glomerulus of a nephron; glomerular capsule.

brachial (bra'ke-al) Pertaining to the arm.

bradycardia (brad''e-kar'de-ah) An abnormally slow heart rate or pulse rate.

brain stem (brān stem) Portion of the brain that includes the midbrain, pons, and medulla oblongata.

brain wave (brān wāv) Recording of fluctuating electrical activity occurring in the brain.

Broca's area (bro'kahz a're-ah) Region of the frontal lobe that coordinates complex muscular actions of the mouth, tongue, and larynx, making speech possible.

bronchial tree (brong'ke-al trē) The bronchi and their branches that carry air from the trachea to the alveoli of the lungs.

bronchiole (brong'ke-ōl) A small branch of a bronchus within the lung.

bronchus (brong'kus) A branch of the trachea that leads to a lung.

buccal (buk'al) Pertaining to the mouth and the inner lining of the cheeks.

buffer (buf'er) A substance that can react with a strong acid or base to form a weaker acid or base, and thus resist a change in pH.

bulbourethral glands (bul''bo-u-re'thral glandz) Glands that secrete a viscous fluid into the male urethra at times of sexual excitement; Cowper's glands.

bursa (bur'sah) A saclike, fluid-filled structure, lined with synovial membrane, that occurs near a joint.

bursitis (bur-si'tis) Inflammation of a bursa.

c

calcification (kal''sĭ-fĭ-ka'shun) The process by which salts of calcium are deposited within a tissue.

calcitonin (kal''sĭ-to'nin) Hormone secreted by the thyroid gland that helps regulate the level of blood calcium.

calorie (kal'o-re) A unit used in the measurement of heat energy and the energy values of foods.

calorimeter (kal''o-rim'ĕ-ter) A device used to measure the heat energy content of foods; bomb calorimeter.

canaliculus (kan''ah-lik'u-lus) Microscopic canals that interconnect the lacunae of bone tissue.

cancellous bone (kan'sĕ-lus bōn) Bone tissue with a latticework structure; spongy bone.

capacitation (kah-pas''i-ta'shun) The development by a sperm cell of the ability to fertilize an egg cell.

capillary (kap'ĭ-ler''e) A small blood vessel that connects an arteriole and a venule.

carbaminohemoglobin (kar''bah-me'no-he''mo-glo'bin) Compound formed by the union of carbon dioxide and hemoglobin.

carbohydrate (kar''bo-hi'drāt) An organic compound that contains carbon, hydrogen, and oxygen, with a 2:1 ratio of hydrogen to oxygen atoms.

carbonic anhydrase (kar-bon'ik an-hi'drās) Enzyme that promotes the reaction between carbon dioxide and water to form carbonic acid.

carbon monoxide (kar'bon mon-ok'sīd) A toxic gas that combines readily with hemoglobin to form a relatively stable compound, CO.

carboxypeptidase (kar-bok''se-pep'ti-dās) A protein-splitting enzyme found in pancreatic juice.

cardiac conduction system (kar'de-ak kon-duk'shun sis'tem) System of specialized cardiac muscle fibers that conducts cardiac impulses from the S-A node into the myocardium.

cardiac cycle (kar'de-ak si'kl) A series of myocardial contractions that constitute a complete heartbeat.

cardiac muscle (kar'de-ak mus'el) Specialized type of muscle tissue found only in the heart.

cardiac output (kar'de-ak owt'poot) A quantity calculated by multiplying the stroke volume in milliliters by the heart rate in beats per minute.

cardiac veins (kar'de-ak vāns) Blood vessels that return blood from the venules of the myocardium to the coronary sinus.

carina (kah-ri'nah) A cartilaginous ridge located between the openings of the right and left bronchi.

carotene (kar'o-tēn) A yellow, orange, or reddish pigment that occurs in plants and from which vitamin A can be synthesized.

carotid bodies (kah-rot'id bod'ēz) Masses of chemoreceptors located in the wall of the internal carotid artery near the carotid sinus.

carpals (kar'pals) Bones of the wrist.

carpus (kar'pus) The wrist; the wrist bones as a group.

cartilage (kar'tĭ-lij) Type of connective tissue in which cells are located within lacunae and are separated by a semi-solid matrix.

catabolic metabolism (kat''ah-bol'ik mĕ-tab'o-lism) Metabolic process by which large molecules are broken down into smaller ones; catabolism.

catalase (kat'ah-lās) An enzyme that causes the decomposition of hydrogen peroxide.

catalyst (kat'ah-list) A substance that increases the rate of a chemical reaction, but is not permanently altered by the reaction.

cataract (kat'ah-rakt) Condition characterized by loss of transparency of the lens of the eye.

catecholamine (kat''ĕ-kol'am-in) A type of organic compound that includes epinephrine and norepinephrine.

cauda equina (kaw'da ek-wīn'a) A group of spinal nerves that extends below the distal end of the spinal cord.

cecum (se'kum) A pouchlike portion of the large intestine to which the small intestine is attached.

celiac (se'le-ak) Pertaining to the abdomen.

cell (sel) The structural and functional unit of an organism.

cell body (sel bod'e) Portion of a nerve cell that includes a cytoplasmic mass and a nucleus, and from which the nerve fibers extend.

cell-mediated immunity (sel me'de-ā-tĭd i mu'nĭ-te) Resistance to invasion by foreign cells characterized by direct attack of T-lymphocytes.

cellular respiration (sel'u-lar res''pi-ra'shun) Process by which energy is released from organic compounds within cells.

cellulose (sel'u-lōs) A polysaccharide that is very abundant in plant tissues, but cannot be digested by human enzymes.

cementum (se-men'tum) Bonelike material that fastens the root of a tooth into its bony socket.

central canal (sen'tral kah-nal') Tube within the spinal cord that is continuous with the ventricles of the brain and contains cerebrospinal fluid.

central nervous system (sen'tral ner'vus sis'tem) Portion of the nervous system that consists of the brain and spinal cord; CNS.

centriole (sen'tre-ōl) A cellular organelle that functions in the organization of the spindle during mitosis.

centromere (sen'tro-mēr) Portion of a chromosome to which the spindle fiber attaches during mitosis.

centrosome (sen'tro-sōm) Cellular organelle consisting of two centrioles.

cephalic (sĕ-fal'ik) Pertaining to the head.

cerebellar cortex (ser''ĕ-bel'ar kor'teks) The outer layer of the cerebellum.

cerebellum (ser''ĕ-bel'um) Portion of the brain that coordinates skeletal muscle movement.

cerebral aqueduct (ser'ĕ-bral ak'wĕ-dukt'') Tube that connects the third and fourth ventricles of the brain.

cerebral cortex (ser'ĕ-bral kor'teks) Outer layer of the cerebrum.

cerebral hemisphere (ser'ĕ-bral hem'ĭ-sfēr) One of the large, paired structures that together constitute the cerebrum of the brain.

cerebrospinal fluid (ser''ĕ-bro-spi'nal floo'id) Fluid that occupies the ventricles of the brain, the subarachnoid space of the meninges, and the central canal of the spinal cord.

cerebrovascular accident (ser''e-bro-vas'ku-lar ak'si-dent) A condition in which the blood flow to the brain is suddenly interrupted; a stroke.

cerebrum (ser′ĕ-brum) Portion of the brain that occupies the upper part of the cranial cavity and is concerned with higher mental functions.

cerumen (sĕ-roo′men) Waxlike substance produced by cells that line the canal of the external ear.

cervical (ser′vĭ-kal) Pertaining to the neck or to the cervix of the uterus.

cervix (ser′viks) Narrow, inferior end of the uterus that leads into the vagina.

chemoreceptor (ke″mo-re-sep′tor) A receptor that is stimulated by the presence of certain chemical substances.

chemotaxis (ke″mo-tak′sis) Phenomenon by which leukocytes are attracted by chemicals released from damaged cells.

chief cell (chēf sel) Cell of gastric gland that secretes various digestive enzymes, including pepsinogen.

childhood (chīld′hood) Phase of the human life cycle that begins at the end of the first year and ends at puberty.

chloride shift (klo′rīd shift) Movement of chloride ions from the blood plasma into red blood cells as bicarbonate ions diffuse out of the red blood cells into the plasma.

cholecystokinin (ko″le-sis″to-ki′nin) Hormone secreted by the small intestine that stimulates the release of pancreatic juice from the pancreas and bile from the gallbladder.

cholesterol (ko-les′ter-ol) A lipid produced by body cells that is used in the synthesis of steroid hormones and is excreted into the bile.

cholinergic fiber (ko″lin-er′jik fi′ber) A nerve fiber that secretes acetylcholine at the terminal end of its axon.

cholinesterase (ko″lin-es′ter-ās) An enzyme that causes the decomposition of acetylcholine.

chondrin (kon′drin) A protein that occurs in the intercellular substance of cartilage tissues.

chondrocyte (kon′dro-sīt) A cartilage cell.

chorion (ko′re-on) Embryonic membrane that forms the outermost covering around a developing fetus and contributes to the formation of the placenta.

chorionic villi (ko″re-on′ik vil′i) Projections that extend from the outer surface of the chorion and help attach an embryo to the uterine wall.

choroid coat (ko′roid kōt) The vascular, pigmented middle layer of the wall of the eye.

choroid plexus (ko′roid plek′sus) Mass of specialized capillaries from which cerebrospinal fluid is secreted into a ventricle of the brain.

chromaffin cell (kro-maf′in sel) Cell of the adrenal medulla that produces, stores, and secretes epinephrine and norepinephrine.

chromatid (kro′mah-tid) A member of a duplicate pair of chromosomes.

chromatin (kro′mah-tin) Nuclear material that gives rise to chromosomes during mitosis.

chromosome (kro′mo-sōm) Rodlike structure that appears in the nucleus of a cell during mitosis; contains the genes responsible for heredity.

chylomicron (ki″lo-mi′kron) A microscopic droplet of fat, found in the blood following the digestion of fats.

chyme (kīm) Semifluid mass of food materials that passes from the stomach to the small intestine.

chymotrypsin (ki″mo-trip′sin) A protein-splitting enzyme found in pancreatic juice.

cilia (sil′e-ah) Microscopic, hairlike processes on the exposed surfaces of certain epithelial cells.

ciliary body (sil′e-er″e bod′e) Structure associated with the choroid layer of the eye that secretes aqueous humor and contains the ciliary muscle.

circadian rhythm (ser″kah-de′an rithm) A pattern of repeated behavior associated with the cycles of night and day.

circle of Willis (sir′kl uv wil′is) An arterial ring located on the ventral surface of the brain.

circular muscles (ser′ku-lar mus′lz) Muscles whose fibers are arranged in circular patterns, usually around an opening or in the wall of a tube; sphincter muscles.

circumduction (ser″kum-duk′shun) Movement of a body part, such as a limb, so that the end follows a circular path.

cisternae (sis-ter′ne) Enlarged portions of the sarcoplasmic reticulum near the actin and myosin filaments of a muscle fiber.

citric acid cycle (sit′rik as′id si′kl) A series of chemical reactions by which various molecules are oxidized and energy is released from them; Krebs cycle.

cleavage (klēv′ij) The early successive divisions of embryonic cells into smaller and smaller cells.

clitoris (kli′to-ris) Small erectile organ located in the anterior portion of the female vulva; corresponds to the penis of the male.

clone (klōn) A group of cells that originate from a single cell and are all alike.

CNS Central nervous system.

coagulation (ko-ag″u-la′shun) The clotting of blood.

cocarboxylase (ko″kar-bok′sĭ-lās) A coenzyme that is synthesized from thiamine and acts in the oxidation of carbohydrates.

cochlea (kok′le-ah) Portion of the inner ear that contains the receptors of hearing.

codon (ko′don) A set of three nucleotides of a messenger RNA molecule.

coenzyme (ko-en′zīm) A nonprotein substance that is necessary to complete the structure of an enzyme molecule.

coenzyme A (ko-en′zīm) Acetyl coenzyme A.

cofactor (ko′fak-tor) A nonprotein substance that must be combined with the protein portion of an enzyme before the enzyme can act.

collagen (kol′ah-jen) Protein that occurs in the white fibers of connective tissues and in the matrix of bone.

collateral (ko-lat′er-al) A branch of a nerve fiber or blood vessel.

colon (ko′lon) The large intestine.

color blindness (kul′er blīnd′nes) An inability to distinguish colors normally.

colostrum (ko-los′trum) The first secretion of the mammary glands following the birth of an infant.

common bile duct (kom′mon bīl dukt) Tube that transports bile from the cystic duct to the duodenum.

compartment (com-part′ment) A space occupied by a group of muscles, blood vessels, and nerves that is enclosed by fasciae.

complement (kom′plĕ-ment) A group of enzymes that are activated by the combination of antibody with antigen and enhance the reaction against foreign substances within the body.

complete protein (kom-plēt′ pro′te-in) A protein that contains adequate amounts of the essential amino acids to maintain body tissues and to promote normal growth and development.

compound (kom′pownd) A substance composed of two or more elements joined by chemical bonds.

concussion (kon-kush′un) The loss of consciousness due to a violent blow to the head.

condom (kon′dum) A latex sheath used to cover the penis during sexual intercourse; used as a contraceptive.

conduction (kon-duk′shun) Process by which body heat moves into the molecules of cooler objects in contact with the body surface.

condyle (kon′dīl) A rounded process of a bone, usually at the articular end.

cones (kōns) Color receptors located in the retina of the eye.

congenital (kon-jen′ĭ-tal) Any condition that exists at the time of birth.

conjunctiva (kon″junk-ti′vah) Membranous covering on the anterior surface of the eye.

connective tissue (kō-nek′tiv tish′u) One of the basic types of tissue that includes bone, cartilage, and various fibrous tissues.

contraception (kon″trah-sep′shun) The prevention of fertilization of the egg cell or the development of an embryo.

contralateral (kon″trah-lat′er-al) Positioned on the opposite side of something else.

convection (kon-vek′shun) The transmission of heat from one substance to another through the circulation of heated air particles.

convergence (kon-ver′jens) The coming together of nerve impulses from different parts of the nervous system so that they reach the same neuron.

convolution (kon"vo-lu'shun) An elevation on the surface of a structure caused by an infolding of the structure upon itself.

cornea (kor'ne-ah) Transparent anterior portion of the outer layer of the eye wall.

coronary artery (kor'o-na''re ar'ter-e) An artery that supplies blood to the wall of the heart.

coronary sinus (kor'o-na''re si'nus) A large vessel on the posterior surface of the heart into which the cardiac veins drain.

corpus callosum (kor'pus kah-lo'sum) A mass of white matter within the brain, composed of nerve fibers connecting the right and left cerebral hemispheres.

corpus luteum (kor'pus lu'te-um) Structure that forms from the tissues of a ruptured ovarian follicle and functions to secrete female hormones.

corpus striatum (kor'pus stri-a'tum) Portion of the cerebrum that includes certain basal ganglia.

cortex (kor'teks) Outer layer of an organ such as the adrenal gland, cerebrum, or kidney.

cortical nephron (kor'tĭ-kl nef'ron) A nephron with its corpuscle located in the renal cortex.

cortisol (kor'tĭ-sol) A glucocorticoid secreted by the adrenal cortex.

costal (kos'tal) Pertaining to the ribs.

covalent bond (ko'va-lent bond) Chemical bond formed by the sharing of electrons between atoms.

cranial (kra'ne-al) Pertaining to the cranium.

cranial nerve (kra'ne-al nerv) Nerve that arises from the brain.

creatine phosphate (kre'ah-tin fos'fāt) A substance present in muscle that acts to store energy.

crenation (kre-na'shun) Shrinkage of a cell caused by contact with a hypertonic solution.

crest (krest) A ridgelike projection of a bone.

cretinism (kre'tĭ-nizm) A condition resulting from a lack of thyroid secretion in an infant.

cricoid cartilage (kri'koid kar'tĭ-lij) A ringlike cartilage that forms the lower end of the larynx.

crista ampullaris (kris'tah am-pul-lah'ris) Sensory organ located within a semicircular canal that functions in the sense of dynamic equilibrium.

crossing over (kros'ing o'ver) The exchange of genetic material between homologous chromosomes during meiosis.

crossmatching (kros'mach''ing) A procedure used to determine whether donor and recipient blood samples will agglutinate.

cubital (ku'bi-tal) Pertaining to the forearm.

cuspid (kus'pid) A canine tooth.

cutaneous (ku-ta'ne-us) Pertaining to the skin.

cyanocobalamin (si''ah-no-ko-bal' ah-min) Vitamin B$_{12}$.

cyanosis (si''ah-no'sis) A condition characterized by a bluish coloration of the skin due to a decreased blood oxygen concentration.

cyclic AMP (sik'lik AMP) A substance produced from ATP that causes a variety of changes in cells.

cyclosporine (si''klo-spor'in) An immunosuppressive drug that suppresses the action of T-helper cells.

cystic duct (sis'tik dukt) Tube that connects the gallbladder to the common bile duct.

cytocrine secretion (si'to-krin se-kre'shun) Process by which melanocytes transfer granules of melanin into adjacent epithelial cells.

cytoplasm (si'to-plazm) The contents of a cell surrounding its nucleus.

d

deamination (de-am''ĭ-na'shun) Chemical process by which amino groups (NH$_2$) are removed from amino acid molecules.

deciduous teeth (de-sid'u-us tēth) Teeth that are shed and replaced by permanent teeth; primary teeth.

decomposition (de-kom''po-zish'un) The breakdown of molecules into simpler compounds.

defecation (def''ĕ-ka'shun) The discharge of feces from the rectum through the anus.

dehydration (de''hi-dra'shun) Excessive loss of water.

dehydration synthesis (de''hi-dra'shun sin'thĕ-sis) Anabolic process by which molecules are joined together to form larger molecules.

dendrite (den'drīt) Nerve fiber that transmits impulses toward a neuron cell body.

densitometer (den''si-tom'e-ter) An instrument used to measure the density of bone tissue.

dental caries (den'tal kar'ēz) Process by which teeth become decalcified and decayed.

dentin (den'tin) Bonelike substance that forms the bulk of a tooth.

deoxyhemoglobin (de-ok''si-he''mo-glo'bin) Hemoglobin that lacks oxygen.

depolarization (de-po''lar-ĭ-za'shun) The loss of an electrical charge on the surface of a membrane.

dermatome (der'mah-tōm) An area of the body supplied by sensory nerve fibers associated with a particular dorsal root of a spinal nerve.

dermis (der'mis) The thick layer of the skin beneath the epidermis.

descending colon (de-send'ing ko'lon) Portion of the large intestine that passes downward along the left side of the abdominal cavity to the brim of the pelvis.

descending tracts (de-send'ing trakts) Groups of nerve fibers that carry nerve impulses downward from the brain through the spinal cord.

desmosome (des'mo-sōm) A specialized junction between cells, which serves as a "spot weld."

detrusor muscle (de-trūz'or mus'l) Muscular wall of the urinary bladder.

dextrose (dek'strōs) Glucose.

diabetes insipidus (di''ah-be'tēz in-sip'ĭ-dus) Condition characterized by an abnormally great production of urine due to a deficiency of antidiuretic hormone.

diabetes mellitus (di''ah-be'tēz mel-li'tus) Condition characterized by a high blood glucose level and the appearance of glucose in the urine due to a deficiency of insulin.

dialysis (di-al'ĭ-sis) Process by which smaller molecules are separated from larger ones in a liquid.

diapedesis (di''ah-pĕ-de'sis) Process by which leukocytes squeeze between the cells that make up the walls of blood vessels.

diaphragm (di'ah-fram) A sheetlike structure composed largely of skeletal muscle and connective tissue that separates the thoracic and abdominal cavities; also a caplike device inserted in the vagina to be used as a contraceptive.

diaphysis (di-af'ĭ-sis) The shaft of a long bone.

diastole (di-as'to-le) Phase of the cardiac cycle during which a heart chamber wall is relaxed.

diastolic pressure (di-a-stol'ik presh'ur) Arterial blood pressure during the diastolic phase of the cardiac cycle.

diencephalon (di''en-sef'ah-lon) Portion of the brain in the region of the third ventricle that includes the thalamus and hypothalamus.

differentiation (dif''er-en''she-a'shun) Process by which cells become structurally and functionally specialized during development.

diffusion (dĭ-fu'zhun) Random movement of molecules from a region of higher concentration toward one of lower concentration.

digestion (di-jest'yun) The process by which larger molecules of food substances are broken down into smaller molecules that can be absorbed; hydrolysis.

dihydrotestosterone (di-hi''dro-tes-tos'ter-ōn) Hormone produced from testosterone that stimulates certain cells of the male reproductive system.

dipeptide (di-pep'tīd) A molecule composed of two amino acids bound together.

disaccharide (di-sak'ah-rīd) A sugar produced by the union of two monosaccharide molecules.

distal (dis'tal) Further from the midline or origin; opposite of proximal.

diuretic (di''u-ret'ik) A substance that causes an increased production of urine.

DNA Deoxyribonucleic acid.

dominant gene (dom'ĭ-nant jēn) The gene of a gene pair that is expressed while its allele is not expressed.

dorsal root (dor'sal root) The sensory branch of a spinal nerve by which it joins the spinal cord.

dorsal root ganglion (dor'sal root gang'gle-on) Mass of sensory neuron cell bodies located in the dorsal root of a spinal nerve.

dorsum (dors'um) Pertaining to the back surface of a body part.

ductus arteriosus (duk'tus ar-te''re-o'sus) Blood vessel that connects the pulmonary artery and the aorta in a fetus.

ductus venosus (duk'tus ven-o'sus) Blood vessel that connects the umbilical vein and the inferior vena cava in a fetus.

duodenum (du''o do'num) The first portion of the small intestine that leads from the stomach to the jejunum.

dural sinus (du'ral si'nus) Blood-filled channel formed by the splitting of the dura mater into two layers.

dura mater (du'rah ma'ter) Tough outer layer of the meninges.

dynamic equilibrium (di-nam'ik e''kwi-lib're-um) The maintenance of balance when the head and body are suddenly moved or rotated.

e

eccrine gland (ek'rin gland) Sweat gland that functions in the maintenance of body temperature.

ECG Electrocardiogram; EKG.

ectoderm (ek'to-derm) The outermost layer of the primary germ layers, responsible for forming certain embryonic body parts.

edema (ĕ-de'mah) An excessive accumulation of fluid within the tissue spaces.

effector (ĕ-fek'tor) Organ, such as a muscle or gland, that responds to stimulation.

efferent arteriole (ef'er-ent ar-te're-ōl) Arteriole that conducts blood away from the glomerulus of a nephron.

ejaculation (e-jak''u-la'shun) Discharge of sperm-containing semen from the male urethra.

elastin (e-las'tin) Protein that comprises the yellow, elastic fibers of connective tissue.

electrocardiogram (e-lek''tro-kar'de-o-gram'') A recording of the electrical activity associated with the heartbeat; ECG or EKG.

electrolyte (e-lek'tro-līt) A substance that ionizes in water solution.

electrolyte balance (e-lek'tro-līt bal'ans) Condition that exists when the quantities of electrolytes entering the body equal those leaving it.

electron (e-lek'tron) A small, negatively charged particle that revolves around the nucleus of an atom.

electrovalent bond (e-lek''tro-va'lent bond) Chemical bond formed between two ions as a result of the transfer of electrons.

element (el'ĕ-ment) A basic chemical substance.

embolus (em'bo-lus) A substance, such as a blood clot or bubble of gas, that is carried by the blood and obstructs a blood vessel.

embryo (em'bre-o) An organism in its earliest stages of development.

emission (e-mish'un) The movement of sperm cells from the vas deferens into the ejaculatory duct and urethra.

emphysema (em''fĭ-se'mah) A condition characterized by abnormal enlargement of the air sacs of the lungs.

emulsification (e-mul''sĭ-fĭ'ka'shun) Process by which fat globules are caused to break up into smaller droplets by the action of bile salts.

enamel (e-nam'el) Hard covering on the exposed surface of a tooth.

endocardium (en''do-kar'de-um) Inner lining of the heart chambers.

endochondral bone (en''do-kon'dral bōn) Bone that begins as hyaline cartilage that is subsequently replaced by bone tissue.

endocrine gland (en'do-krin gland) A gland that secretes hormones directly into the blood or body fluids.

endocytosis (en''do-si-to'sis) Physiological process by which substances may move through a cell membrane and involves the formation of tiny, cytoplasmic vacuoles.

endoderm (en'do-derm) The innermost layer of the primary germ layers responsible for forming certain embryonic body parts.

endolymph (en'do-limf) Fluid contained within the membranous labyrinth of the inner ear.

endometrium (en''do-me'tre-um) The inner lining of the uterus.

endomysium (en''do-mis'e-um) The sheath of connective tissue surrounding each skeletal muscle fiber.

endoneurium (en''do-nu're-um) Layer of loose connective tissue that surrounds individual nerve fibers of a nerve.

endoplasmic reticulum (en-do-plaz'mic rĕ-tik'u-lum) Cytoplasmic organelle composed of a system of interconnected membranous tubules and vesicles.

endorphin (en-dor'fin) A neuropeptide that occurs in the pituitary gland and has a pain-suppressing action.

endothelium (en''do-the'le-um) The layer of epithelial cells that forms the inner lining of blood vessels and heart chambers.

energy (en'er-je) An ability to cause something to move and thus to do work.

energy balance (en'er-je bal'ans) Condition that exists when the caloric intake of the body equals its caloric output.

enkephalin (en-kef'ah-lin) A neuropeptide that occurs in the brain and spinal cord; it inhibits pain impulses, thus relieving pain sensations.

enzyme (en'zīm) A protein that is synthesized by a cell and acts as a catalyst in a specific cellular reaction.

eosinophil (e''o-sin'o-fil) White blood cell characterized by the presence of cytoplasmic granules that become stained by acidic dye.

ependyma (ĕ-pen'dĭ-mah) Membrane, composed of neuroglial cells, that lines the ventricles of the brain.

epicardium (ep''ĭ-kar'de-um) The visceral portion of the pericardium located on the surface of the heart.

epicondyle (ep''ĭ-kon'dīl) A projection of a bone located above a condyle.

epidermis (ep''ĭ-der'mis) Outer epithelial layer of the skin.

epididymis (ep''ĭ-did'ĭ-mis) Highly coiled tubule that leads from the seminiferous tubules of the testis to the vas deferens.

epidural space (ep''ĭ-du'ral spās) The space between the dural sheath of the spinal cord and the bone of the vertebral canal.

epigastric region (ep''ĭ-gas'trik re'jun) The upper middle portion of the abdomen.

epiglottis (ep''ĭ-glot'is) Flaplike cartilaginous structure located at the back of the tongue near the entrance to the trachea.

epimysium (ep''ĭ-mis'e-um) The outer sheath of connective tissue surrounding a skeletal muscle.

epinephrine (ep''ĭ-nef'rin) A hormone secreted by the adrenal medulla during times of stress.

epineurium (ep''ĭ-nu're-um) Outermost layer of connective tissue surrounding a nerve.

epiphyseal disk (ep''ĭ-fiz'e-al disk) Cartilaginous layer within the epiphysis of a long bone that functions as a growing region.

epiphysis (ĕ-pif'ĭ-sis) The end of a long bone.

epiploic appendage (ep''i-plo'ik ah-pen'dij) Small collections of fat in the serous layer of the large intestinal wall.

epithelium (ep''ĭ-the'le-um) The type of tissue that covers all free body surfaces.

equilibrium (e''kwi-lib're-um) A state of balance between two opposing forces.

erythroblast (ĕ-rith'ro-blast) An immature red blood cell.

erythroblastosis (ĕ-rith''ro-blas-to'sis) Condition characterized by the presence of erythroblasts in the circulating blood.

erythrocyte (ĕ-rith'ro-sīt) A red blood cell.

erythropoiesis (ĕ-rith''ro-poi-e'sis) Red blood cell formation.

erythropoietin (ĕ-rith''ro-poi'ĕ-tin) Substance released by the kidneys and liver that promotes red blood cell formation.

esophageal hiatus (ĕ-sof''ah-je'al hi-a'tus) Opening in the diaphragm through which the esophagus passes.

esophagus (ĕ-sof'ah-gus) Tubular portion of the digestive tract that leads from the pharynx to the stomach.

essential amino acid (ĕ-sen′shal ah-me′no as′id) Amino acid required for health that cannot be synthesized in adequate amounts by body cells.

essential fatty acid (ĕ-sen′shal fat′e as′id) Fatty acid required for health that cannot be synthesized in adequate amounts by body cells.

estrogen (es′tro-jen) Hormone that stimulates the development of female secondary sexual characteristics.

eustachian tube (u-sta′ke-an tūb) Tube that connects the middle ear to the pharynx; auditory tube.

evaporation (e″vap′o-ra-shun) Process by which a liquid changes into a gas.

eversion (e-ver′zhun) Movement in which the sole of the foot is turned outward.

exchange reaction (eks-chānj′ re-ak′shun) A chemical reaction in which parts of two kinds of molecules trade positions.

excretion (ek-skre′shun) Process by which metabolic wastes are eliminated.

exocrine gland (ek′so-krin gland) A gland that secretes its products into a duct or onto a body surface.

expiration (ek″spi-ra′shun) Process of expelling air from the lungs.

extension (ek-sten′shun) Movement by which the angle between parts at a joint is increased.

extracellular (ek″strah-sel′u-lar) Outside of cells.

extrapyramidal tract (ek″strah-pi-ram′i-dal trakt) Nerve tracts, other than the corticospinal tracts, that transmit impulses from the cerebral cortex into the spinal cord.

extremity (ek-strem′i-te) A limb; an arm or leg.

f

facet (fas′et) A small, flattened surface of a bone.

facilitated diffusion (fah-sil′i-tāt″id di-fu′zhun) Diffusion in which substances are moved through membranes from a region of higher concentration to a region of lower concentration by carrier molecules.

fallopian tube (fah-lo′pe-an tūb) Tube that transports an egg cell from the region of the ovary to the uterus; oviduct or uterine tube.

fascia (fash′e-ah) A sheet of fibrous connective tissue that encloses a muscle.

fasciculus (fah-sik′u-lus) A small bundle of muscle fibers.

fat (fat) Adipose tissue; or an organic substance whose molecules contain glycerol and fatty acids.

fatty acid (fat′e as′id) An organic substance that serves as a building block for a fat molecule.

feces (fe′sēz) Material expelled from the digestive tract during defecation.

ferritin (fer′i-tin) An iron-protein complex in which iron is stored in liver cells.

fertilization (fer″ti-li-za′shun) The union of an egg cell and a sperm cell.

fetoscopy (fe-tos′ko-pe) The direct observation of a fetus using a fiber optic device.

fetus (fe′tus) A human embryo after eight weeks of development.

fibril (fi′bril) A tiny fiber or filament.

fibrillation (fi″bri-la′shun) Uncoordinated contraction of muscle fibers.

fibrin (fi′brin) Insoluble, fibrous protein formed from fibrinogen during blood coagulation.

fibrinogen (fi-brin′o-jen) Plasma protein that is converted into fibrin during blood coagulation.

fibrinolysin (fi″bri-nol′i-sin) A protein-splitting enzyme that can digest the substance of a blood clot.

fibroblast (fi′bro-blast) Cell that functions to produce fibers and other intercellular materials in connective tissues.

filtration (fil-tra′shun) Movement of material through a membrane as a result of hydrostatic pressure.

fissure (fish′ūr) A narrow cleft separating parts, such as the lobes of the cerebrum.

flaccid paralysis (flak′sid pah-ral′i-sis) A condition characterized by total loss of tone in the muscles innervated by damaged nerve fibers.

flagella (flah-jel′ah) Relatively long, motile processes that extend out from the surface of a cell.

flexion (flek′shun) Bending at a joint so that the angle between bones is decreased.

follicle (fol′i-kl) A pouchlike depression or cavity.

follicle-stimulating hormone (fol′i-kl stim′u-la″ting hor′mōn) A substance secreted by the anterior pituitary gland that stimulates the development of an ovarian follicle in a female or the production of sperm cells in a male; FSH.

follicular cells (fo-lik′u-lar selz) Ovarian cells that surround a developing egg cell and secrete female sex hormones.

fontanel (fon″tah-nel′) Membranous region located between certain cranial bones in the skull of a fetus or infant.

foramen (fo-ra′men) An opening, usually in a bone or membrane (pl. *foramina*).

foramen magnum (fo-ra′men mag′num) Opening in the occipital bone of the skull through which the spinal cord passes.

foramen ovale (fo-ra′men o-val′e) Opening in the interatrial septum of the fetal heart.

forebrain (fōr′brān) The anteriormost portion of the developing brain that gives rise to the cerebrum and basal ganglia.

formula (fōr′mu-lah) A group of symbols and numbers used to express the composition of a compound.

fossa (fos′ah) A depression in a bone or other part.

fovea (fo′ve-ah) A tiny pit or depression.

fovea centralis (fo′ve-ah sen-tral′is) Region of the retina, consisting of densely packed cones, that is responsible for the greatest visual acuity.

fracture (frak′chur) A break in a bone.

frenulum (fren′u-lum) A fold of tissue that serves to anchor and limit the movement of a body part.

frontal (frun′tal) Pertaining to the forehead.

FSH Follicle-stimulating hormone.

g

galactose (gah-lak′tōs) A monosaccharide component of the disaccharide lactose.

gallbladder (gawl′blad-er) Saclike organ associated with the liver that stores and concentrates bile.

gamete (gam′ēt) A sex cell; either an egg cell or a sperm cell.

ganglion (gang′gle-on) A mass of neuron cell bodies, usually outside the central nervous system.

gastric gland (gas′trik gland) Gland within the stomach wall that secretes gastric juice.

gastric juice (gas′trik jōos) Secretion of the gastric glands within the stomach.

gastric lipase (gas′trik li′pās) Fat-splitting enzyme that occurs in small quantities in gastric juice.

gastrin (gas′trin) Hormone secreted by the stomach lining that stimulates the secretion of gastric juice.

gene (jēn) Portion of a DNA molecule that contains the information needed to synthesize an enzyme.

genetic code (jĕ-net′ik kōd) System by which information for synthesizing proteins is built into the structure of DNA molecules.

genetics (jĕ-net′iks) The study of the mechanism by which characteristics are passed from parents to offspring.

genotype (je′no-tīp) The combination of genes present within a zygote or within the cells of an individual.

germinal epithelium (jer′mi-nal ep″i-the′le-um) Tissue within an ovary that gives rise to sex cells.

germ layers (jerm la′ers) Layers of cells within an embryo that form the body organs during development: ectoderm, mesoderm, and endoderm.

globin (glo′bin) The protein portion of a hemoglobin molecule.

globulin (glob′u-lin) A type of protein that occurs in blood plasma.

glomerulus (glo-mer′u-lus) A capillary tuft located within the glomerular capsule of a nephron.

glottis (glot′is) Slitlike opening between the true vocal folds or vocal cords.

glucagon (gloo′kah-gon) Hormone secreted by the pancreatic islets of Langerhans that causes the release of glucose from glycogen.

glucocorticoid (gloo″ko-kor′tĭ-koid) Any one of a group of hormones secreted by the adrenal cortex that influence carbohydrate, fat, and protein metabolism.

gluconeogenesis (gloo″ko-ne″o-jen′ĕ-sis) The synthesis of glucose from noncarbohydrate materials, such as amino acid molecules.

glucose (gloo′kōs) A monosaccharide found in the blood that serves as the primary source of cellular energy.

glucosuria (gloo″ko-su′re-ah) The presence of glucose in urine.

gluteal (gloo′te-al) Pertaining to the buttocks.

glycerol (glis′er-ol) An organic compound that serves as a building block for fat molecules.

glycogen (gli′ko-jen) A polysaccharide that functions to store glucose in the liver and muscles.

glycolysis (gli-kol′ĭ-sis) The conversion of glucose to pyruvic acid during cellular respiration.

glycoprotein (gli″ko-pro′te-in) A substance composed of a carbohydrate combined with a protein.

goblet cell (gob′let sel) An epithelial cell that is specialized to secrete mucus.

goiter (goi′ter) A condition characterized by the enlargement of the thyroid gland.

Golgi apparatus (gol′je ap″ah-ra′tus) A cytoplasmic organelle that functions in preparing cellular products for secretion.

Golgi tendon organ (gol′jē ten′dun or′gan) Sensory receptors occurring in tendons close to muscle attachments that are involved in reflexes that help maintain posture.

gomphosis (gom-fo′sis) Type of joint in which a cone-shaped process is fastened in a bony socket.

gonad (go′nad) A sex-cell-producing organ; an ovary or testis.

gonadotropin (go-nad″o-trōp′in) A hormone that stimulates activity in the gonads.

granulocyte (gran′u-lo-sīt) A leukocyte that contains granules in its cytoplasm.

gray matter (grā mat′er) Region of the central nervous system that generally lacks myelin and thus appears gray.

groin (groin) Region of the body between the abdomen and thighs.

growth (grōth) Process by which a structure enlarges.

growth hormone (grōth hōr′mōn) A hormone released by the anterior lobe of the pituitary gland that promotes the growth of the organism; GH or somatotropin.

h

hair follicle (hār fol′ĭ-kl) Tubelike depression in the skin in which a hair develops.

half-life (haf′līf) The time it takes for one-half of the radioactivity of an isotope to be released.

hapten (hap′ten) A small molecule that, in combination with a larger one, forms an antigenic substance and can later stimulate an immune reaction by itself.

haustra (haws′trah) Pouches in the wall of the large intestine.

hematocrit (he-mat′o-krit) The volume percentage of red blood cells within a sample of whole blood.

hematoma (he″mah-to′mah) A mass of coagulated blood within tissues or a body cavity.

hematopoiesis (hem″ah-to-poi-e′sis) The production of blood and blood cells; hemopoiesis.

heme (hem) The iron-containing portion of a hemoglobin molecule.

hemocytoblast (he″mo-si′to-blast) A cell that gives rise to blood cells.

hemoglobin (he″mo-glo′bin) Pigment of red blood cells responsible for the transport of oxygen.

hemolysis (he-mol′ĭ-sis) The rupture of red blood cells accompanied by the release of hemoglobin.

hemopoiesis (he″mo-poi-e′sis) The production of blood and blood cells; hematopoiesis.

hemorrhage (hem′ō-rij) Loss of blood from the circulatory system; bleeding.

hemostasis (he″mo-sta′sis) The stoppage of bleeding.

heparin (hep′ah-rin) A substance that interferes with the formation of a blood clot; an anticoagulant.

hepatic (hĕ-pat′ik) Pertaining to the liver.

hepatic lobule (hĕ-pat′ik lob′ūl) A functional unit of the liver.

hepatic sinusoid (hĕ-pat′ik si′nŭ-soid) Vascular channel within the liver.

heredity (hĕ-red′ĭ-te) The transmission of genetic information from parent to offspring.

heterozygote (het″er-o-zi′gōt) An individual who possesses different alleles in a gene pair.

heterozygous (het″er-o-zi′gus) Pertaining to a heterozygote.

hindbrain (hīnd′brān) Posterior-most portion of the developing brain that gives rise to the cerebellum, pons, and medulla oblongata.

histamine (his′tah-min) A substance released from cells subjected to stressful conditions.

histology (his-tol′o-je) The study of the structure and function of tissues.

homeostasis (ho″me-o-sta′sis) A state of equilibrium in which the internal environment of the body remains relatively constant.

homozygote (ho″mo-zi′gōt) An individual possessing identical genes in a gene pair.

homozygous (ho″mo-zi′gus) Pertaining to a homozygote.

hormone (hor′mōn) A substance secreted by an endocrine gland that is transmitted in the blood or body fluids.

humoral immunity (hu′mor-al ĭ-mu′ni-te) Resistance to the effects of specific disease-causing agents due to the presence of circulating antibodies; antibody-mediated immunity.

hydrolysis (hi-drol′ĭ-sis) The splitting of a molecule into smaller portions by the addition of a water molecule.

hydrostatic pressure (hi″dro-stat′ik presh′ur) Pressure exerted by fluids, such as blood pressure.

hydroxyapatite (hi-drok″se-ap′ah-tīt) A type of crystalline calcium phosphate found in bone matrix.

hydroxyl ion (hi-drok′sil i′on) OH^-.

hymen (hi′men) A membranous fold of tissue that partially covers the vaginal opening.

hyperglycemia (hi″per-gli-se′me-ah) An excessive level of blood glucose.

hyperkalemia (hi″per-kah-le′me-ah) An excessive concentration of blood potassium.

hypernatremia (hi″per-nah-tre′me-ah) An excessive concentration of blood sodium.

hyperparathyroidism (hi″per-par″ah-thi′roi-dizm) An excessive secretion of parathyroid hormone.

hyperplasia (hi″per-pla′ze-ah) An increased production and growth of new cells.

hyperpolarization (hi″per-po″lar-i-za′shun) An increase in the negativity of the resting potential of a cell membrane.

hypertension (hi″per-ten′shun) Excessive blood pressure.

hyperthyroidism (hi″per-thi′roi-dizm) An excessive secretion of thyroid hormones.

hypertonic (hi″per-ton′ik) Condition in which a solution contains a greater concentration of dissolved particles than the solution with which it is compared.

hypertrophy (hi-per′tro-fe) Enlargement of an organ or tissue.

hyperventilation (hi″per-ven″tĭ-la′shun) Breathing that is abnormally deep and prolonged.

hypervitaminosis (hi″per-vi″tah-mĭ-no′sis) Excessive intake of vitamins.

hypochondriac region (hi″po-kon′dre-ak re′jun) The portion of the abdomen on either side of the middle or epigastric region.

hypogastric region (hi″po-gas′trik re′jun) The lower middle portion of the abdomen.

hypoglycemia (hi″po-gli-se′me-ah) Abnormally low concentration of blood glucose.

hypokalemia (hi″po-kah-le′me-ah) A low concentration of blood potassium.

hyponatremia (hi″po-nah-tre′me-ah) A low concentration of blood sodium.

hypoparathyroidism (hi″po-par″ah-thi′roi-dizm) An undersecretion of parathyroid hormone.

hypophysis (hi-pof′i-sis) The pituitary gland.

hypoproteinemia (hi″po-pro″te-ĭ-ne′me-ah) A low concentration of blood proteins.

hypothalamus (hi″po-thal′ah-mus) A portion of the brain located below the thalamus and forming the floor of the third ventricle.

hypothyroidism (hi″po-thi′roi-dizm) A low secretion of thyroid hormones.

hypotonic (hi″po-ton′ik) Condition in which a solution contains a lesser concentration of dissolved particles than the solution to which it is compared.

hypoxia (hi-pok′se-ah) A deficiency of oxygen in the tissues.

i

ileocecal valve (il″e-o-se′kal valv) Sphincter valve located at the distal end of the ileum where it joins the cecum.

ileum (il′e-um) Portion of the small intestine between the jejunum and the cecum.

iliac region (il′e-ak re′jun) Portion of the abdomen on either side of the lower middle, or hypogastric, region.

ilium (il′e-um) One of the bones of a coxal bone or hipbone.

immunity (i-mu′ni-te) Resistance to the effects of specific disease-causing agents.

immunoglobulin (im″u-no-glob′u-lin) Globular plasma proteins that function as antibodies of immunity.

immunosuppressive drugs (im″u-no-su-pres′iv drugz) Substances that inhibit the formation of antibodies.

implantation (im″plan-ta′shun) The embedding of an embryo in the lining of the uterus.

impulse (im′puls) A wave of depolarization conducted along a nerve fiber or muscle fiber.

incisor (in-si′zor) One of the front teeth that is adapted for cutting food.

inclusion (in-kloo′zhun) A mass of lifeless chemical substance within the cytoplasm of a cell.

incomplete protein (in″kom-plēt′ pro′tēn) A protein that lacks adequate amounts of essential amino acids.

infancy (in′fan-se) Period of life from the end of the first four weeks to one year of age.

inferior (in-fēr′e-or) Situated below something else; pertaining to the lower surface of a part.

inflammation (in″flah-ma′shun) A tissue response to stress that is characterized by dilation of blood vessels and an accumulation of fluid in the affected region.

infrared ray (in″frah-red′ ra) A form of radiation energy, with wavelengths longer than visible light, by which heat moves from warmer surfaces to cooler surroundings.

infundibulum (in″fun-dib′u-lum) The stalk by which the pituitary gland is attached to the base of the brain.

ingestion (in-jes′chun) The taking of food or liquid into the body by way of the mouth.

inguinal (ing′gwi-nal) Pertaining to the groin region.

inguinal canal (ing′gwi-nal kah-nal′) Passage in the lower abdominal wall through which a testis descends into the scrotum.

inhibin (in′hib′in) A hormone secreted by cells of the testes and ovaries that inhibits the secretion of FSH from the anterior pituitary gland.

inorganic (in″or-gan′ik) Pertaining to chemical substances that lack carbon and hydrogen.

insertion (in-ser′shun) The end of a muscle that is attached to a movable part.

inspiration (in″spi-ra′shun) Act of breathing in; inhalation.

insula (in′su lah) A cerebral lobe located deep within the lateral sulcus.

insulin (in′su-lin) A hormone secreted by the pancreatic islets of Langerhans that functions in the control of carbohydrate metabolism.

integumentary (in-teg-u-men′tar-e) Pertaining to the skin and its accessory organs.

intercalated disk (in-ter″kah-lāt′ed disk) Membranous boundary between adjacent cardiac muscle cells.

intercellular (in″ter-sel′u-lar) Between cells.

intercellular fluid (in″ter-sel′u-lar floo′id) Tissue fluid located between cells other than blood cells.

interferon (in″ter-fēr′on) A substance produced by cells that inhibits the multiplication of viruses and growth of tumors.

interleukin (in″ter-lu′kin) A substance released by phagocytic cells that causes changes in the set point of the body's thermostat.

interneuron (in″ter-nu′ron) A neuron located between a sensory neuron and a motor neuron; intercalated; internuncial, or association neuron.

interphase (in′ter-fāz) Period between two cell divisions when a cell is carrying on its normal functions.

interstitial cell (in″ter-stish′al sel) A hormone-secreting cell located between the seminiferous tubules of the testis.

interstitial fluid (in″ter-stish′al floo′id) Same as intercellular fluid.

intervertebral disk (in″ter-ver′tĕ-bral disk) A layer of fibrocartilage located between the bodies of adjacent vertebrae.

intestinal gland (in-tes′ti-nal gland) Tubular gland located at the base of a villus within the intestinal wall.

intestinal juice (in-tes′ti-nal jōōs) The secretion of the intestinal glands.

intracellular (in″trah-sel′u-lar) Within cells.

intracellular fluid (in″trah-sel′u-lar floo′id) Fluid within cells.

intracellular junction (in″trah-sel′u-lar jungk′shun) A connection between the membranes of adjacent cells.

intramembranous bone (in″trah-mem′brah-nus bōn) Bone that forms from membranelike layers of primitive connective tissue.

intrathecal injection (in″trah-the′kal injek′shun) The infusion of a substance directly into the cerebrospinal fluid.

intrauterine device (in″trah-u′ter-in de-vīs′) A solid object placed in the uterine cavity for purposes of contraception; IUD.

intrinsic factor (in-trin′sik fak′tor) A substance produced by the gastric glands that promotes the absorption of vitamin B_{12}.

inversion (in-ver′zhun) Movement in which the sole of the foot is turned inward.

involuntary (in-vol′un-tār″e) Not consciously controlled; functions automatically.

iodopsin (i″o-dop′sin) A light-sensitive pigment within the cones of the retina.

ion (i′on) An atom or a group of atoms with an electrical charge.

ionization (i″on i za′shun) Chemical process by which substances dissociate into ions.

ipsilateral (ip″si-lat′er-al) Positioned on the same side as something else.

iris (i′ris) Colored muscular portion of the eye that surrounds the pupil and regulates its size.

irritability (ir″i-tah-bil′i-te) The ability of an organism to react to changes taking place in its environment.

ischemia (is-ke′me-ah) A deficiency of blood in a body part.

isometric contraction (i″so-met′rik kon-trak′shun) Muscular contraction in which the muscle fails to shorten.

isotonic contraction (i″so-ton′ik kon-trak′shun) Muscular contraction in which the muscle shortens.

isotonic solution (i″so-ton′ik so-lu′shun) A solution that has the same concentration of dissolved particles as the solution with which it is compared.

isotope (i′so-tōp) An atom that has the same number of protons as other atoms of an element but has a different number of neutrons in its nucleus.

IUD An intrauterine device.

j

jejunum (jĕ-joo′num) Portion of the small intestine located between the duodenum and the ileum.

joint (joint) The union of two or more bones; an articulation.

juxtaglomerular apparatus (juks″tah-glo-mer′u-lār ap″ah-ra′tus) Structure located in the walls of arterioles near the glomerulus that plays an important role in regulating renal blood flow.

juxtamedullary nephron (juks″tah-med′u-lār-e nef′ron) A nephron with its corpuscle located near the renal medulla.

k

karotype (kar′eotip) A set of chromosomes arranged in homologous pairs.

keratin (ker′ah-tin) Protein present in the epidermis, hair, and nails.

keratinization (ker′ah-tin″i-za′shun) The process by which cells form fibrils of keratin and become hardened.

kernicterus (ker-nik′ter-us) Condition in which bilirubin precipitates in the brain tissues of an infant affected with erythroblastosis fetalis.

ketogenesis (ke″to-jen′ĕ-sis) The formation of ketone bodies.

ketone body (ke′tōn bod′e) Type of compound produced during fat catabolism, including acetone, acetoacetic acid, and betahydroxybutyric acid.

ketosis (ke″to′sis) A condition in which the concentration of ketone bodies in body fluids is abnormally increased.

kilocalorie (kil′o-kal″o-re) One thousand calories.

kilogram (kil′o-gram) A unit of weight equivalent to 1,000 grams.

Krebs cycle (krebz si′kl) The citric acid cycle.

Kupffer cell (koop′fer sel) Large, fixed phagocyte in the liver that removes bacterial cells from the blood.

kyphosis (ki-fo′sis) An abnormally increased convex curvature in the thoracic portion of the vertebral column.

l

labor (la′bor) The process of childbirth.

labyrinth (lab′ĭ-rinth) The system of interconnecting tubes within the inner ear, which includes the cochlea, vestibule, and semicircular canals.

lacrimal gland (lak′rĭ-mal gland) Tear-secreting gland.

lactase (lak′tās) Enzyme that converts lactose into glucose and galactose.

lactate (lak′tāt) Lactic acid.

lactation (lak-ta′shun) The production of milk by the mammary glands.

lacteal (lak′te-al) A lymphatic capillary associated with a villus of the small intestine.

lactic acid (lak′tik as′id) An organic substance formed from pyruvic acid during anaerobic respiration.

lactose (lak′tōs) A disaccharide that occurs in milk; milk sugar.

lacuna (lah-ku′nah) A hollow cavity.

lamella (lah-mel′ah) A layer of matrix in bone tissue.

laryngopharynx (lah-ring″go-far′ingks) The lower portion of the pharynx near the opening to the larynx.

larynx (lar′ingks) Structure located between the pharynx and trachea that houses the vocal cords.

latent period (la′tent pe′re-od) Time lapse between the application of a stimulus and the beginning of a response in a muscle fiber.

lateral (lat′er-al) Pertaining to the side.

leukocyte (lu′ko-sīt) A white blood cell.

leukocytosis (lu″ko-si-to′sis) An abnormally large increase in the number of white blood cells.

leukopenia (lu″ko-pe′ne-ah) An abnormally low number of leukocytes in the blood.

lever (lev′er) A simple mechanical device consisting of a rod, fulcrum, weight, and a source of energy that is applied to some point on the rod.

ligament (lig′ah-ment) A cord or sheet of connective tissue by which two or more bones are bound together at a joint.

limbic system (lim′bik sis′tem) A group of interconnected structures within the brain that function to produce various emotional feelings.

linea alba (lin′e-ah al′bah) A narrow band of tendinous connective tissue located in the midline of the anterior abdominal wall.

lingual (ling′gwal) Pertaining to the tongue.

lipase (li′pās) A fat-digesting enzyme.

lipid (lip′id) A fat, oil, or fatlike compound that usually has fatty acids in its molecular structure.

lipoprotein (lip″o-pro′te-in) A complex of lipid and protein.

lordosis (lor-do′sis) An abnormally increased concave curvature in the lumbar portion of the vertebral column.

lumbar (lum′bar) Pertaining to the region of the loins.

lumen (lu′men) Space within a tubular structure such as a blood vessel or intestine.

luteinizing hormone (lu′te-in-īz″ing hor′mōn) A hormone secreted by the anterior pituitary gland that controls the formation of corpus luteum in females and the secretion of testosterone in males; LH (ICSH in males).

lymph (limf) Fluid transported by the lymphatic vessels.

lymph node (limf nōd) A mass of lymphoid tissue located along the course of a lymphatic vessel.

lymphocyte (lim′fo-sīt) A type of white blood cell that functions to provide immunity.

lymphokine (lim′fo-kīn) A substance secreted by T-cells that enhances cellular responses to antigens.

lysosome (li′so-sōm) Cytoplasmic organelle that contains digestive enzymes.

m

macrocyte (mak′ro-sīt) A large red blood cell.

macromineral (mak′ro-min″er-al) An inorganic substance that is necessary for metabolism and is one of a group that accounts for 75% of the mineral elements within the body; major mineral.

macrophage (mak′ro-fāj) A large phagocytic cell.

macroscopic (mak″ro-skop′ik) Large enough to be seen with the unaided eye.

macula (mak′u-lah) A group of hair cells and supporting cells associated with an organ of static equilibrium.

macula lutea (mak′u-lah lu′te-ah) A yellowish depression in the retina of the eye that is associated with acute vision.

malabsorption (mal″ab-sorp′shun) The failure to absorb the nutrients following digestion of food substances.

malignant (mah-lig′nant) The power to threaten life; cancerous.

malnutrition (mal″nu-trish′un) A condition resulting from an improper diet.

maltase (mawl′tās) An enzyme that converts maltose into glucose.

maltose (mawl′tōs) A disaccharide composed of two glucose molecules.

mammary (mam′ar-e) Pertaining to the breast.

marrow (mar′o) Connective tissue that occupies the spaces within bones.

mast cell (mast sel) A cell to which antibodies, formed in response to allergens, become attached.

mastication (mas″tĭ-ka′shun) Chewing movements.

matrix (ma′triks) The intercellular substance of connective tissue.

matter (mat′er) Anything that has weight and occupies space.

meatus (me-a′tus) A passageway or channel, or the external opening of such a passageway.

mechanoreceptor (mek″ah-no-re-sep′tor) A sensory receptor that is sensitive to mechanical stimulation such as changes in pressure or tension.

medial (me′de-al) Toward or near the midline.

mediastinum (me″de-ah-sti′num) Tissues and organs of the thoracic cavity that form a septum between the lungs.

medulla (mĕ-dul′ah) The inner portion of an organ.

medulla oblongata (mĕ-dul′ah ob″long-gah′tah) Portion of the brain stem located between the pons and the spinal cord.

medullary cavity (med′u-lār″e kav′ĭ-te) Cavity within the diaphysis of a long bone occupied by marrow.

megakaryocyte (meg″ah-kar′e-o-sīt) A large bone marrow cell that functions to produce blood platelets.

meiosis (mi-o′sis) Process of cell division by which egg and sperm cells are formed.

melanin (mel′ah-nin) Dark pigment normally found in skin and hair.

melanocyte (mel′ah-no-sīt″) Melanin-producing cell.

melatonin (mel″ah-to′nin) A hormone thought to be secreted by the pineal gland.

memory cell (mem′o-re sel) B-lymphocyte or T-lymphocyte produced in response to a primary immune response that remains dormant and can respond rapidly if the same antigen is encountered in the future.

menarche (mĕ-nar′ke) The first menstrual period.

meninges (mĕ-nin′jēz) A group of three membranes that covers the brain and spinal cord (sing. *meninx*).

menisci (men-is′si) Pieces of fibrocartilage that separate the articulating surfaces of bones in the knee.

menopause (men'o-pawz) Termination of the menstrual cycles.

menstrual cycle (men'stroo-al si'kl) The female reproductive cycle that is characterized by regularly reoccurring changes in the uterine lining.

menstruation (men"stroo-a'shun) Loss of blood and tissue from the uterine lining at the end of a female reproductive cycle.

mesentery (mes'en-ter"e) A fold of peritoneal membrane that attaches an abdominal organ to the abdominal wall.

mesoderm (mez'o-derm) The middle layer of the primary germ layers, responsible for forming certain embryonic body parts.

messenger RNA (mes'in-jer RNA) Molecule of RNA that transmits information for protein synthesis from the nucleus of a cell to the cytoplasm.

metabolic rate (met"ah-bol'ic rāt) The rate at which chemical changes occur within the body.

metabolism (mě-tab'o-lizm) All of the chemical changes that occur within cells considered together.

metacarpals (met"ah-kar'pals) Bones of the hand between the wrist and finger bones.

metaphase (met'ah-fāz) Stage in mitosis when chromosomes become aligned in the middle of the spindle.

metastasis (mě-tas'tah-sis) The spread of disease from one body region to another; a characteristic of cancer.

metatarsals (met"ah-tar'sals) Bones of the foot between the ankle and toe bones.

microfilament (mi"kro-fil'ah-ment) Tiny rod of protein that occurs in cytoplasm and functions in causing various cellular movements.

microglia (mi-krog'le-ah) A type of neuroglial cell that helps support neurons and acts to carry on phagocytosis.

micromineral (mi'kro-min"er-al) An essential mineral substance that is present in a minute amount within the body; trace element.

microscopic (mi"kro-skop'ik) Too small to be seen with the unaided eye.

microtubule (mi"kro-tu'būl) A minute, hollow rod found in the cytoplasm of cells.

microvilli (mi"kro-vil'i) Tiny, cylindrical processes that extend outward from some epithelial cell membranes and increase the membrane surface area.

micturition (mik"tu-rish'un) Urination.

midbrain (mid'brān) A small region of the brain stem located between the diencephalon and the pons.

mineralocorticoid (min"er-al-o-kor'tǐ-koid) Any one of a group of hormones secreted by the adrenal cortex that influences the concentrations of electrolytes in body fluids.

mitochondrion (mi"to-kon'dre-on) Cytoplasmic organelle that contains enzymes responsible for aerobic respiration (pl. *mitochondria*).

mitosis (mi-to'sis) Process by which body cells divide to form two identical cells.

mitral valve (mi'tral valv) Heart valve located between the left atrium and the left ventricle; bicuspid valve.

mixed nerve (mikst nerv) Nerve that includes both sensory and motor nerve fibers.

molar (mo'lar) A rear tooth with a somewhat flattened surface adapted for grinding food.

molding (mold'ing) The process by which the shape of the fetal skull changes slightly during birth.

molecule (mol'ě-kūl) A particle composed of two or more atoms joined together.

monoamine inhibitor (mon"o-am'in in-hib'i-tor) A substance that inhibits the action of the enzyme monoamine oxidase.

monocyte (mon'o-sīt) A type of white blood cell that functions as a phagocyte.

monosaccharide (mon"o-sak'ah-rīd) A simple sugar, such as glucose or fructose, that represents the structural unit of a carbohydrate.

morula (mor'u-lah) An early stage in embryonic development; a solid ball of cells.

motor area (mo'tor a're-ah) A region of the brain from which impulses to muscles or glands originate.

motor end plate (mo'tor end plāt) Specialized portion of a muscle fiber membrane at a neuromuscular junction.

motor nerve (mo'tor nerv) A nerve that consists of motor nerve fibers.

motor neuron (mo'tor nu'ron) A neuron that transmits impulses from the central nervous system to an effector.

motor unit (mo'tor unit) A motor neuron and the muscle fibers associated with it.

mucosa (mu-ko'sah) The membrane that lines tubes and body cavities that open to the outside of the body; mucous membrane.

mucous cell (mu'kus sel) Glandular cell that secretes mucus.

mucous membrane (mu'kus mem'brān) Mucosa.

mucus (mu'kus) Fluid secretion of the mucous cells.

multiple alleles (mul'tǐ-pl ah-lēls') Condition in which a gene occurs in several forms.

multiple motor unit summation (mul'tǐ-pl mo'tor u'nit sum-mā'shun) A sustained muscle contraction of increasing strength resulting from responses by many motor units.

multipolar neuron (mul"tǐ-po'lar nu'ron) Nerve cell that has many processes arising from its cell body.

muscle spindle (mus'el spin'dul) Modified skeletal muscle fiber that can respond to changes in muscle length.

mutagenic (mu'tah-jen'ik) Pertaining to a factor that can cause mutations.

mutation (mu-ta'shun) A change in the genetic information of a chromosome.

myelin (mi'ě-lin) Fatty material that forms a sheathlike covering around some nerve fibers.

myocardium (mi"o-kar'de-um) Muscle tissue of the heart.

myofibril (mi"o-fi'bril) Contractile fibers found within muscle cells.

myoglobin (mi"o-glo'bin) A pigmented compound found in muscle tissue that acts to store oxygen.

myogram (mi'o-gram) A recording of a muscular contraction.

myometrium (mi"o-me'tre-um) The layer of smooth muscle tissue within the uterine wall.

myoneural junction (mi"o-nu'ral jungk'shun) Site of union between a motor neuron axon and a muscle fiber.

myopia (mi-o'pe-ah) Nearsightedness.

myosin (mi'o-sin) A protein that, together with actin, is responsible for muscular contraction and relaxation.

myxedema (mik"sě-de'mah) Condition resulting from a deficiency of thyroid hormones in an adult.

n

nasal cavity (na'zal kav'ǐ-te) Space within the nose.

nasal concha (na'zal kong'kah) Shell-like bone extending outward from the wall of the nasal cavity; a turbinate bone.

nasal septum (na'zal sep'tum) A wall of bone and cartilage that separates the nasal cavity into two portions.

nasopharynx (na"zo-far'ingks) Portion of the pharynx associated with the nasal cavity.

negative feedback (neg'ah-tiv fēd'bak) A mechanism that is activated by an imbalance and acts to correct it.

neonatal (ne"o-na'tal) Pertaining to the period of life from birth to the end of four weeks.

nephron (nef'ron) The functional unit of a kidney, consisting of a renal corpuscle and a renal tubule.

nerve (nerv) A bundle of nerve fibers.

neurilemma (nu"ri-lem'ah) Sheath on the outside of some nerve fibers due to the presence of Schwann cells.

neurofibrils (nu"ro-fi'brils) Fine cytoplasmic threads that extend from the cell body into the processes of neurons.

neuroglia (nu-rog'le-ah) The supporting tissue within the brain and spinal cord, composed of neuroglial cells.

neuromodulator (nu"ro-mod'u-lā-tor) A substance that alters a neuron's response to a neurotransmitter.

neuromuscular junction (nu"ro-mus'ku-lar jungk'shun) Myoneural junction.

neuron (nu'ron) A nerve cell that consists of a cell body and its processes.

neuropeptide (nu"ro-pep'tīd) A peptide that occurs in the brain and seems to function as a neurotransmitter or neuromodulator.

neurotransmitter (nu″ro-trans-mit′er) Chemical substance secreted by the terminal end of an axon that stimulates a muscle fiber contraction or an impulse in another neuron.

neutral (nu′tral) Neither acid nor alkaline.

neutron (nu′tron) An electrically neutral particle found in an atomic nucleus.

neutrophil (nu′tro-fil) A type of phagocytic leukocyte.

niacin (ni′ah-sin) A vitamin of the B-complex group; nicotinic acid.

niacinamide (ni″ah-sin′ah-mīd) The physiologically active form of niacin.

nicotinic acid (nik″o-tin′ik as′id) Niacin.

Nissl bodies (nis′l bod′ēz) Membranous sacs that occur within the cytoplasm of nerve cells and have ribosomes attached to their surfaces.

nitrogen balance (ni′tro-jen bal′ans) Condition in which the quantity of nitrogen ingested equals the quantity excreted.

nondisjunction (non″dis-jungk′shun) The failure of a pair of chromosomes to separate during meiosis.

nonelectrolyte (non″e-lek′tro-līt) A substance that does not dissociate into ions when it is dissolved.

nonprotein nitrogenous substance (non-pro′te-in ni-troj′ě-nus sub′stans) A substance, such as urea or uric acid, that contains nitrogen but is not a protein.

norepinephrine (nor″ep-i-nef′rin) A neurotransmitter substance released from the axon ends of some nerve fibers.

nuclease (nu′kle-ās) An enzyme that causes nucleic acids to decompose.

nucleic acid (nu-kle′ik as′id) A substance composed of nucleotides bonded together; RNA or DNA.

nucleolus (nu-kle′o-lus) A small structure that occurs within the nucleus of a cell and contains RNA and proteins.

nucleoplasm (nu′kle-o-plazm″) The contents of the nucleus of a cell.

nucleosome (nu′kle-o-sōm) A beadlike particle within a chromatin fiber composed of DNA and protein.

nucleotide (nu′kle-o-tīd″) A component of a nucleic acid molecule, consisting of a sugar, a nitrogenous base, and a phosphate group.

nucleus (nu′kle-us) A cellular organelle that is enclosed by a double-layered, porous membrane and contains DNA; the dense core of an atom that is composed of protons and neutrons.

nutrient (nu′tre-ent) A chemical substance that must be supplied to the body from its environment.

nutrition (nu-trish′un) The study of the sources, actions, and interactions of nutrients.

o

obesity (o-bēs′ĭ-te) An excessive accumulation of adipose tissue; usually the condition of exceeding the desirable weight by more than 20%.

occipital (ok-sip′ĭ-tal) Pertaining to the lower, back portion of the head.

olfactory (ol-fak′to-re) Pertaining to the sense of smell.

olfactory nerves (ol-fak′to-re nervz) The first pair of cranial nerves, which conduct impulses associated with the sense of smell.

oligodendrocyte (ol″ĭ-go-den′dro-sīt) A type of neuroglial cell that functions to connect neurons to blood vessels and to form myelin.

oocyte (o′o-sīt) An immature egg cell.

oogenesis (o″o-jen′ě-sis) The process by which an egg cell forms from an oocyte.

ophthalmic (of-thal′mik) Pertaining to the eye.

optic (op′tik) Pertaining to the eye.

optic chiasma (op′tik ki-az′mah) X-shaped structure on the underside of the brain formed by a partial crossing over of fibers in the optic nerves.

optic disk (op′tik disk) Region in the retina of the eye where nerve fibers leave to become part of the optic nerve.

oral (o′ral) Pertaining to the mouth.

organ (or′gan) A structure consisting of a group of tissues that performs a specialized function.

organelle (or″gah-nel′) A living part of a cell that performs a specialized function.

organic (or-gan′ik) Pertaining to carbon-containing substances.

organism (or′gah-nizm) An individual living thing.

orifice (or′ĭ-fis) An opening.

origin (or′ĭ-jin) End of a muscle that is attached to a relatively immovable part.

oropharynx (o″ro-far′ingks) Portion of the pharynx in the posterior part of the oral cavity.

osmoreceptor (oz″mo-re-sep′tor) Receptor that is sensitive to changes in the osmotic pressure of body fluids.

osmosis (oz-mo′sis) Diffusion of water through a selectively permeable membrane due to the existence of a concentration gradient.

osmotic (oz-mot′ik) Pertaining to osmosis.

osmotic pressure (oz-mot′ik presh′ur) The amount of pressure needed to stop osmosis; the potential pressure of a solution due to the presence of nondiffusible solute particles in the solution.

osseous tissue (os′e-us tish′u) Bone tissue.

ossification (os″ĭ-fi-ka′shun) The formation of bone tissue.

osteoblast (os′te-o-blast″) A bone-forming cell.

osteoclast (os′te-o-klast″) A cell that causes the erosion of bone.

osteocyte (os′te-o-sīt) A mature bone cell.

osteon (os′te-on) A cylinder-shaped unit containing bone cells that surround an osteonic canal; Haversian system.

osteonic canal (os′te-o-nik ka-nal′) A tiny channel in bone tissue that contains a blood vessel; Haversian canal.

otolith (o′to-lith) A small particle of calcium carbonate associated with the receptors of equilibrium.

otosclerosis (o″to-sklě-ro′sis) Abnormal formation of spongy bone within the ear that may interfere with the transmission of sound vibrations to hearing receptors.

oval window (o′val win′do) Opening between the stapes and the inner ear.

ovarian (o-va′re-an) Pertaining to the ovary.

ovary (o′var e) The primary reproductive organ of a female; an egg-cell-producing organ.

oviduct (o′vĭ-dukt) A tube that leads from the ovary to the uterus; uterine tube or fallopian tube.

ovulation (o″vu-la′shun) The release of an egg cell from a mature ovarian follicle.

ovum (o′vum) A mature egg cell.

oxidase (ok′sĭ-dās) An enzyme that promotes oxidation.

oxidation (ok″sĭ-da′shun) Process by which oxygen is combined with a chemical substance; the removal of hydrogen or the loss of electrons; the opposite of reduction.

oxygen debt (ok′sĭ-jen det) The amount of oxygen that must be supplied following physical exercise to convert accumulated lactic acid to glucose.

oxyhemoglobin (ok″sĭ-he″mo-glo′bin) Compound formed when oxygen combines with hemoglobin.

oxytocin (ok″sĭ-to′sin) Hormone released by the posterior lobe of the pituitary gland that causes contraction of smooth muscles in the uterus and mammary glands.

p

pacemaker (pās′māk-er) Mass of specialized cardiac muscle tissue that controls the rhythm of the heartbeat; the sinoatrial node.

pain receptor (pān re″sep′tor) Sensory nerve ending associated with the feeling of pain.

palate (pal′at) The roof of the mouth.

palatine (pal′ah-tīn) Pertaining to the palate.

palmar (pahl′mar) Pertaining to the palm of the hand.

pancreas (pan′kre-as) Glandular organ in the abdominal cavity that secretes hormones and digestive enzymes.

pancreatic (pan″kre-at′ik) Pertaining to the pancreas.

pantothenic acid (pan″to-the′nik as′id) A vitamin of the B-complex group.

papilla (pah-pil′ah) Tiny, nipplelike projection.

papillary muscle (pap′ĭ-ler″e mus′el) Muscle that extends inward from the ventricular walls of the heart and to which the chordae tendineae are attached.

paracrine (par'ah-krin) A type of endocrine secretion in which the hormone has its effect locally in nearby cells.

paradoxical sleep (par''ah-dok'se-kal slēp) Sleep in which some areas of the brain are active, producing dreams and rapid eye movements.

paralysis (pah-ral'i-sis) Loss of ability to control voluntary muscular movements, usually due to a disorder of the nervous system.

parasympathetic division (par''ah-sim''pah-thet'ik di-vizh'un) Portion of the autonomic nervous system that arises from the brain and sacral region of the spinal cord.

parathormone (par''ah-thōr'mōn) Hormone secreted by the parathyroid glands that helps regulate the level of blood calcium and phosphate ions; parathyroid hormone or PTH.

parathyroid glands (par''ah-thi'roid glandz) Small endocrine glands that are embedded in the posterior portion of the thyroid gland.

paravertebral ganglia (par''ah-ver'tē-bral gang'gle-ah) Sympathetic ganglia that form chains along the sides of the vertebral column.

parietal (pah-ri'ē-tal) Pertaining to the wall of an organ or cavity.

parietal cell (pah-ri'ē-tal sel) Cell of a gastric gland that secretes hydrochloric acid and intrinsic factor.

parietal pleura (pah-ri'ē-tal ploo'rah) Membrane that lines the inner wall of the thoracic cavity.

parotid glands (pah-rot'id glandz) Large salivary glands located on the sides of the face just in front and below the ears.

partial pressure (par'shal presh'ur) The pressure produced by one gas in a mixture of gases.

parturition (par''tu-rish'un) The process of childbirth.

pathogen (path'o-jen) Any disease-causing agent.

pathology (pah-thol'o-je) The study of disease.

pectoral (pek'tor-al) Pertaining to the chest.

pectoral girdle (pek'tor-al ger'dl) Portion of the skeleton that provides support and attachment for the arms.

pelvic (pel'vik) Pertaining to the pelvis.

pelvic girdle (pel'vik ger'dl) Portion of the skeleton to which the legs are attached.

pelvis (pel'vis) Bony ring formed by the sacrum and coxal bones.

penis (pe'nis) External reproductive organ of the male through which the urethra passes.

pepsin (pep'sin) Protein-splitting enzyme secreted by the gastric glands of the stomach.

pepsinogen (pep-sin'o-jen) Inactive form of pepsin.

peptidase (pep'ti-dās) An enzyme that causes the breakdown of polypeptides.

peptide (pep'tīd) Compound composed of two or more amino acid molecules joined together.

peptide bond (pep'tīd bond) Bond that forms between the carboxyl group of one amino acid and the amino group of another.

perception (persep'shun) Mental awareness of sensory stimulation; sensation.

pericardial (per''ī-kar'de-al) Pertaining to the pericardium.

pericardium (per''ī-kar'de-um) Serous membrane that surrounds the heart.

perichondrium (per''ī-kon'dre-um) Layer of fibrous connective tissue that encloses cartilaginous structures.

perilymph (per'ī-limf) Fluid contained in the space between the membranous and osseous labyrinths of the inner ear.

perimetrium (per-i-me'tre-um) The outer serosal layer of the uterine wall.

perimysium (per''ī-mis'e-um) Sheath of connective tissue that encloses a bundle of striated muscle fibers.

perineal (per''ī-ne'al) Pertaining to the perineum.

perineum (per''ī-ne'um) Body region between the scrotum or urethral opening and the anus.

perineurium (per''ī-nu're-um) Layer of connective tissue that encloses a bundle of nerve fibers within a nerve.

periodontal ligament (per''e-o-don'tal lig'ah-ment) Fibrous membrane that surrounds a tooth and attaches it to the bone of the jaw.

periosteum (per''e-os'te-um) Covering of fibrous connective tissue on the surface of a bone.

peripheral (pē-rif'er-al) Pertaining to parts located near the surface or toward the outside.

peripheral nervous system (pē-rif'er-al ner'vus sis'tem) The portions of the nervous system outside the central nervous system.

peripheral protein (pe-rif'er-al pro'te-in) A globular protein associated with the inner surface of the cell membrane.

peripheral resistance (pē-rif'er-al re-zis'tans) Resistance to blood flow due to friction between the blood and the walls of the blood vessels.

peristalsis (per''ī-stal'sis) Rhythmic waves of muscular contraction that occur in the walls of various tubular organs.

peritoneal (per''ī-to-ne'al) Pertaining to the peritoneum.

peritoneal cavity (per''ī-to-ne'al kav'ī-te) The potential space between the parietal and visceral peritoneal membranes.

peritoneum (per''ī-to-ne'um) A serous membrane that lines the abdominal cavity and encloses the abdominal viscera.

peritubular capillary (per''ī-tu'bu-lar kap'ī-ler''e) Capillary that surrounds a renal tubule and functions in reabsorption and secretion during urine formation.

permeable (per'me-ah-bl) Open to passage or penetration.

peroxisome (pē-roks'ī-sōm) Membranous cytoplasmic vesicle that contains enzymes responsible for the production and decomposition of hydrogen peroxide.

pH The negative logarithm of the hydrogen ion concentration used to indicate the acid or alkaline condition of a solution.

phagocytosis (fag''o-si-to'sis) Process by which a cell engulfs and digests solid substances.

phalanx (fa'langks) A bone of a finger or toe (pl. *phalanges*).

pharynx (far'ingks) Portion of the digestive tube between the mouth and the esophagus.

phenotype (fe'no-tīp) The appearance of an individual due to the action of a particular set of genes.

phospholipid (fos''fo-lip'id) A lipid that contains two fatty acid molecules and a phosphate group combined with a glycerol molecule.

photoreceptor (fo''to-re-sep'tor) A nerve ending that is sensitive to light energy.

physiology (fiz''e-ol'o-je) The branch of science dealing with the study of body functions.

pia mater (pi'ah ma'ter) Inner layer of meninges that encloses the brain and spinal cord.

pineal gland (pin'e-al gland) A small structure located in the central part of the brain.

pinocytosis (pin''o-si-to'sis) Process by which a cell engulfs droplets of fluid from its surroundings.

pituitary gland (pi-tu'ī-tār''e gland) Endocrine gland that is attached to the base of the brain and consists of anterior and posterior lobes; the hypophysis.

placenta (plah-sen'tah) Structure by which an unborn child is attached to its mother's uterine wall and through which it is nourished.

plantar (plan'tar) Pertaining to the sole of the foot.

plasma (plaz'mah) Fluid portion of circulating blood.

plasma cell (plaz'mah sel) Antibody-producing cell that is formed as a result of the proliferation of sensitized B-lymphocytes.

plasma protein (plaz'mah pro'te-in) Any of several proteins normally found dissolved in blood plasma.

platelet (plāt'let) Cytoplasmic fragment formed in the bone marrow that functions in blood coagulation.

pleural (ploo'ral) Pertaining to the pleura or membranes investing the lungs.

pleural cavity (ploo'ral kav'ī-te) Potential space between the pleural membranes.

pleural membranes (ploo'ral mem'brānz) Serous membranes that enclose the lungs.

plexus (plek'sus) A network of interlaced nerves or blood vessels.

pneumotaxic area (nu''mo-tax'ik a're-ah) A portion of the respiratory control center located in the pons of the brain.

polar body (po'lar bod'e) Small, nonfunctional cell produced as a result of meiosis during egg cell formation.

polarization (po''lar-i-za'shun) The development of an electrical charge on the surface of a cell membrane due to an unequal distribution of ions on either side of the membrane.

polycythemia (pol''e-si-the'me-ah) An excessive concentration of red blood cells.

polymorphonuclear leukocyte (pol''e-mor''fo-nu'kle-ar lu'ko-sīt) A leukocyte or white blood cell with an irregularly lobed nucleus.

polypeptide (pol''e-pep'tīd) A compound formed by the union of many amino acid molecules.

polysaccharide (pol''e-sak'ah-rīd) A carbohydrate composed of many monosaccharide molecules joined together.

pons (ponz) A portion of the brain stem above the medulla oblongata and below the midbrain.

popliteal (pop''li-te'al) Pertaining to the region behind the knee.

positive feedback (poz'ĭ-tiv fēd'bak) Process by which changes cause more changes of a similar type, producing unstable conditions; vicious cycle.

posterior (pos-tēr'e-or) Toward the back; opposite of anterior.

postganglionic fiber (pōst''gang-gle-on'ik fi'ber) Autonomic nerve fiber located on the distal side of a ganglion.

postnatal (pōst-na'tal) After birth.

precursor (pre-ker'sor) Substance from which another substance is formed.

preganglionic fiber (pre''gang-gle-on'ik fi'ber) Autonomic nerve fiber located on the proximal side of a ganglion.

pregnancy (preg'nan-se) The condition in which a female has a developing offspring in her uterus.

prenatal (pre-na'tal) Before birth.

presbycusis (pres''bĭ-ku'sis) Loss of hearing that accompanies old age.

presbyopia (pres''be-o'pe-ah) Condition in which the eye loses its ability to accommodate due to loss of elasticity in the lens; farsightedness of age.

pressoreceptor (pres''o-re-sep'tor) A receptor that is sensitive to changes in pressure.

primary reproductive organs (pri'ma-re re''pro-duk'tiv or'ganz) Sex-cell-producing parts; testes in males and ovaries in females.

prime mover (prīm mōōv'er) Muscle that is mainly responsible for a particular body movement.

profibrinolysin (pro''fi-bri-no-li'sin) The inactive form of fibrinolysin.

progeny (proj'e-ne) The product of the reproductive process; offspring.

progesterone (pro-jes'tĕ-rōn) A female hormone secreted by the corpus luteum of the ovary and by the placenta.

projection (pro-jek'shun) Process by which the brain causes a sensation to seem to come from the region of the body being stimulated.

prolactin (pro-lak'tin) Hormone secreted by the anterior pituitary gland that stimulates the production of milk from the mammary glands; PRL.

pronation (pro-na'shun) Movement in which the palm of the hand is moved downward or backward.

prophase (pro'fāz) Stage of mitosis during which chromosomes become visible.

proprioceptor (pro''pre-o-sep'tor) A sensory nerve ending that is sensitive to changes in tension of a muscle or tendon.

prostacyclin (pros''tah-si'klin) A substance released from endothelial cells that inhibits platelet adherence.

prostaglandins (pros''tah-glan'dins) A group of compounds that have powerful, hormonelike effects.

prostate gland (pros'tāt gland) Gland surrounding the male urethra below the urinary bladder that adds its secretion to semen during ejaculation.

protein (pro'tēn) Nitrogen-containing organic compound composed of amino acid molecules joined together.

prothrombin (pro-throm'bin) Plasma protein that functions in the formation of blood clots.

proton (pro'ton) A positively charged particle found in an atomic nucleus.

protraction (pro-trak'shun) A forward movement of a body part.

proximal (prok'sĭ-mal) Closer to the midline or origin; opposite of distal.

puberty (pu'ber-te) Stage of development in which the reproductive organs become functional.

pulmonary (pul'mo-ner''e) Pertaining to the lungs.

pulmonary circuit (pul'mo-ner''e ser'kit) System of blood vessels that carries blood between the heart and the lungs.

pulse (puls) The surge of blood felt through the walls of arteries due to the contraction of the ventricles of the heart.

pupil (pu'pil) Opening in the iris through which light enters the eye.

Purkinje fibers (pur-kin'je fi'berz) Specialized muscle fibers that conduct the cardiac impulse from the A-V bundle into the ventricular walls.

pyloric sphincter muscle (pi-lor'ik sfingk'ter mus'l) Sphincter muscle located between the stomach and the duodenum; pylorus.

pyramidal cell (pĭ-ram'ĭ-dal sel) A large, pyramid-shaped neuron found within the cerebral cortex.

pyridoxine (pir''ĭ-dok'sēn) A vitamin of the B-complex group; vitamin B_6.

pyrogen (pi'ro-jen) A substance that causes an increase in body temperature.

pyruvic acid (pi-roo'vik as'id) An intermediate product of carbohydrate oxidation.

radiation (ra''de-a'shun) A form of energy that includes visible light, ultraviolet light, and X rays; the means by which body heat is lost in the form of infrared rays.

radioactive (ra''deoak'tiv) A property of an element by which its atoms decompose by releasing energy or pieces of themselves.

rate-limiting enzyme (rāt lim'ī-ting en'zīm) An enzyme, usually present in limited amount, that controls the rate of a metabolic pathway by regulating one of its steps.

receptor (re'sep'tor) Part located at the distal end of a sensory dendrite that is sensitive to stimulation.

recessive gene (re-ses'iv jēn) An allele of a gene pair that is not expressed while the other member of the pair is expressed.

recruitment (re-krōōt'ment) Increase in number of motor units activated as intensity of stimulation increases.

rectum (rek'tum) The terminal end of the digestive tube between the sigmoid colon and the anus.

red marrow (red mar'o) Blood-cell-forming tissue located in spaces within bones.

red muscle (red mus'el) Slow-contracting postural muscles that contain an abundance of myoglobin.

referred pain (re-ferd' pān) Pain that feels as if it is originating from a part other than the site being stimulated.

reflex (re'fleks) A rapid, automatic response to a stimulus.

reflex arc (re'fleks ark) A nerve pathway, consisting of a sensory neuron, interneuron, and motor neuron, that forms the structural and functional bases for a reflex.

refraction (re-frak'shun) A bending of light as it passes from one medium into another medium with a different density.

refractory period (re-frak'to-re pe're-od) Time period following stimulation during which a neuron or muscle fiber will not respond to a stimulus.

relaxin (re-lak'sin) Hormone from corpus luteum that inhibits uterine contractions during pregnancy.

releasing hormone (re-le'sing hor'mōn) A substance secreted by the hypothalamus whose target cells are in the anterior pituitary gland.

renal (re'nal) Pertaining to the kidney.

renal corpuscle (re'nal kor'pusl) Part of a nephron that consists of a glomerulus and a glomerular capsule; Malpighian corpuscle.

renal cortex (re'nal kor'teks) The outer portion of a kidney.

renal medulla (re'nal mĕ-dul'ah) The inner portion of a kidney.

renal pelvis (re'nal pel'vis) The hollow cavity within a kidney.

renal tubule (re'nal tu'būl) Portion of a nephron that extends from the renal corpuscle to the collecting duct.

renin (re'nin) Enzyme released from the kidneys that triggers a mechanism leading to a rise in blood pressure.

reproduction (re''pro-duk'shun) The process by which an offspring is formed.

resorption (re-sorp'shun) The process by which something is lost as a result of physiological activity.

respiration (res''pi-ra'shun) Cellular process by which energy is released from nutrients; breathing.

respiratory center (re-spi'rah-to''re sen'ter) Portion of the brain stem that controls the depth and rate of breathing.

respiratory membrane (re-spi'rah-to''re mem'brān) Membrane composed of a capillary wall and an alveolar wall through which gases are exchanged between the blood and the air.

response (re-spons') The action resulting from a stimulus.

resting potential (res'ting po-ten'shal) The difference in electrical charge between the inside and outside of an undisturbed nerve cell membrane.

reticular formation (re-tik'u-lar fōr-ma' shun) A complex network of nerve fibers within the brain stem that functions in arousing the cerebrum.

reticulocyte (re-tik'u-lo-sīt) A young red blood cell that has a network of fibrils in its cytoplasm.

reticuloendothelial tissue (re-tik''u-lo-en''do-the'le-al tish'u) Tissue composed of widely scattered phagocytic cells.

retina (ret'ĭ-nah) Inner layer of the eye wall that contains the visual receptors.

retinal (ret'ĭ-nal) A form of vitamin A; retinene.

retinene (ret'ĭ-nēn) Substance used in the production of rhodopsin.

retraction (re-trak'shun) Movement of a part toward the back.

retroperitoneal (ret''ro-per''ĭ-to-ne'al) Located behind the peritoneum.

rhodopsin (ro-dop'sin) Light-sensitive substance that occurs in the rods of the retina; visual purple.

rhythmicity area (rith-mis'ĭ-te a're-ah) A portion of the respiratory control center located in the medulla.

riboflavin (ri''bo-fla'vin) A vitamin of the B-complex group; vitamin B_2.

ribonucleic acid (ri''bo-nu-kle'ik as'id) A nucleic acid that contains ribose sugar; RNA.

ribose (ri'bōs) A five-carbon sugar found in RNA molecules.

ribosome (ri'bo-sōm) Cytoplasmic organelle that consists largely of RNA and functions in the synthesis of proteins.

RNA Ribonucleic acid.

rod (rod) A type of light receptor that is responsible for colorless vision.

rotation (ro-ta'shun) Movement by which a body part is turned on its longitudinal axis.

rouleaux (roo-lo') Groups of red blood cells stuck together like stacks of coins.

round window (rownd win'do) A membrane-covered opening between the inner ear and the middle ear.

rugae (roo'je) Thick folds in the inner wall of the stomach that disappear when the stomach is distended.

S

saccule (sak'ūl) A saclike cavity that makes up part of the membranous labyrinth of the inner ear.

sagittal (saj'ĭ-tal) A plane or section that divides a structure into right and left portions.

salivary gland (sal'ĭ-ver-e gland) A gland, associated with the mouth, that secretes saliva.

salt (sawlt) A compound produced by a reaction between an acid and a base.

saltatory conduction (sal'tah-tor-e kon-duk'shun) A type of nerve impulse conduction in which the impulse seems to jump from one node to the next.

S-A node (nōd) Sinoatrial node.

sarcolemma (sar''ko-lem'ah) The cell membrane of a muscle fiber.

sarcomere (sar'ko-mēr) The structural and functional unit of a myofibril.

sarcoplasm (sar'ko-plazm) The cytoplasm within a muscle fiber.

sarcoplasmic reticulum (sar''ko-plaz'mik re-tik'u-lum) Membranous network of channels and tubules within a muscle fiber, corresponding to the endoplasmic reticulum of other cells.

saturated fatty acid (sat'u-rat''ed fat'e as'id) Fatty acid molecule that lacks double bonds between the atoms of its carbon chain.

Schwann cell (shwahn sel) A type of neuroglial cell that surrounds a fiber of a peripheral nerve, forming the neurilemmal sheath and myelin.

sclera (skle'rah) White fibrous outer layer of the eyeball.

scoliosis (sko''le-o'sis) Abnormal lateral curvature of the vertebral column.

scrotum (skro'tum) A pouch of skin that encloses the testes.

sebaceous gland (se-ba'shus gland) Gland of the skin that secretes sebum.

sebum (se'bum) Oily secretion of the sebaceous glands.

secretin (se-kre'tin) Hormone secreted from the small intestine that stimulates the release of pancreatic juice from the pancreas.

semen (se'men) Fluid discharged from the male reproductive tract at ejaculation that contains sperm cells and the secretions of various glands.

semicircular canal (sem''ĭ-ser'ku-lar kah-nal') Tubular structure within the inner ear that contains the receptors responsible for the sense of dynamic equilibrium.

seminiferous tubule (sem''ĭ-nif'er-us tu'būl) Tubule within the testes in which sperm cells are formed.

semipermeable (sem''ĭ-per'me-ah-bl) Condition in which a membrane is permeable to some molecules and not to others; selectively permeable.

senescence (se-nes'ens) The process of growing old.

sensation (sen-sa'shun) A feeling resulting from the interpretation of sensory nerve impulses by the brain.

sensory area (sen'so-re a're-ah) A portion of the cerebral cortex that receives and interprets sensory nerve impulses.

sensory nerve (sen'so-re nerv) A nerve composed of sensory nerve fibers.

sensory neuron (sen'so-re nu'ron) A neuron that transmits an impulse from a receptor to the central nervous system.

serotonin (se''ro-to'nin) A vasoconstricting substance that is released by blood platelets when blood vessels are broken and thus helps control bleeding.

serous cell (se'rus sel) A glandular cell that secretes a watery fluid with a high enzyme content.

serous fluid (se'rus floo'id) The secretion of a serous membrane.

serous membrane (ser'us mem'brān) Membrane that lines a cavity without an opening to the outside of the body.

serum (se'rum) The fluid portion of coagulated blood.

sesamoid bone (ses'ah-moid bōn) A round bone that may occur in tendons adjacent to joints.

sex chromosome (seks kro'mo-sōm) A chromosome responsible for the development of characteristics associated with maleness or femaleness; an X or Y chromosome.

sigmoid colon (sig'moid ko'lon) S-shaped portion of the large intestine between the descending colon and the rectum.

simple sugar (sim'pl shoog'ar) A monosaccharide.

sinoatrial node (si''no-a'tre-al nōd) Group of specialized tissue in the wall of the right atrium that initiates cardiac cycles; the pacemaker; S-A node.

sinus (si'nus) A cavity or hollow space in a bone or other body part.

skeletal muscle (skel'e-tal mus'l) Type of muscle tissue found in muscles attached to skeletal parts.

smooth muscle (smooth mus'el) Type of muscle tissue found in the walls of hollow visceral organs; visceral muscle.

sodium pump (so'de-um pump) The active transport mechanism that functions to concentrate sodium ions on the outside of a cell membrane.

solute (sol'ūt) The substance that is dissolved in a solution.

solvent (sol'vent) The liquid portion of a solution in which a solute is dissolved.

somatic cell (so-mat'ik sel) Any cell of the body other than the sex cells.

somatomedin (so''mah-to-me'din) A substance released from the liver in response to growth hormone that promotes the growth of cartilage.

somatostatin (so-mat'o-sta''tin) A hormone secreted by the islet of Langerhans cells that inhibits the release of growth hormone.

somatotropin (so''mah-to-tro'pin) Growth hormone.

spastic paralysis (spas'tik pah-ral'ĭ-sis) A form of paralysis characterized by an increase in muscular tone without atrophy of the muscles involved.

special sense (spesh'al sens) Sense that involves receptors associated with specialized sensory organs, such as the eyes and ears.

spermatid (sper'mah-tid) An intermediate stage in the formation of sperm cells.

spermatocyte (sper-mat'o-sīt) An early stage in the formation of sperm cells.

spermatogenesis (sper''mah-to-jen'ĕ-sis) The production of sperm cells.

spermatogonium (sper''mah-to-go'ne-um) Undifferentiated spermatogenic cell found in the germinal epithelium of a seminiferous tubule.

spermatozoa (sper''mah-to-zo'ah) Male reproductive cells; sperm cells.

sphincter (sfingk'ter) A circular muscle that functions to close an opening or the lumen of a tubular structure.

sphygmomanometer (sfig'mo-mah-nom'ĕ-ter) Instrument used for measuring blood pressure.

spinal (spi'nal) Pertaining to the spinal cord or to the vertebral canal.

spinal cord (spi'nal kord) Portion of the central nervous system extending downward from the brain stem through the vertebral canal.

spinal nerve (spi'nal nerv) Nerve that arises from the spinal cord.

spleen (splēn) A large, glandular organ located in the upper left region of the abdomen.

spongy bone (spunj'e bōn) Bone that consists of bars and plates separated by irregular spaces; cancellous bone.

squamous (skwa'mus) Flat or platelike.

starch (starch) A polysaccharide that is common in foods of plant origin.

static equilibrium (stat'ik e''kwi-lib're-um) The maintenance of balance when the head and body are motionless.

stereocilia (ste''re-o-sil'e-ah) The hairlike processes of the hair cells within the organ of Corti.

sterility (stĕ-ril'ĭ-te) An inability to produce offspring.

steroid (ste'roid) An organic substance whose molecules include complex rings of carbon and hydrogen atoms.

stimulus (stim'u-lus) A change in the environmental conditions that is followed by a response by an organism or cell.

stomach (stum'ak) Digestive organ located between the esophagus and the small intestine.

strabismus (strah-biz'mus) A condition characterized by lack of visual coordination; crossed eyes.

stratified (strat'ĭ-fīd) Arranged in layers.

stratum corneum (stra'tum kor'ne-um) Outer horny layer of the epidermis.

stratum germinativum (stra'tum jer'mĭ-na''ti-vum) The deepest layer of the epidermis in which the cells undergo mitosis; stratum basale.

stress (stres) Condition produced by factors causing potentially life threatening changes in the body's internal environment.

stressor (stres'or) A factor capable of stimulating a stress response.

stroke volume (strōk vol'ūm) The amount of blood discharged from the ventricle with each heartbeat.

structural formula (struk'cher-al fōr'mu-lah) A representation of the way atoms are united within a molecule, using the symbols for each element present and lines to indicate chemical bonds.

subarachnoid space (sub''ah-rak'noid spās) The space within the meninges between the arachnoid mater and the pia mater.

subcutaneous (sub''ku-ta'ne-us) Beneath the skin.

sublingual (sub-ling'gwal) Beneath the tongue.

submaxillary (sub-mak'sĭ-ler''e) Below the maxilla.

submucosa (sub''mu-ko'sah) Layer of connective tissue that underlies a mucous membrane.

substrate (sub'strāt) The substance upon which an enzyme acts.

sucrase (su'krās) Digestive enzyme that acts upon sucrose.

sucrose (soo'krōs) A disaccharide; table sugar.

sulcus (sul'kus) A shallow groove, such as that between adjacent convolutions on the surface of the brain.

summation (sum-ma'shun) Phenomena in which the amount of change in membrane potential is directly related to the intensity of stimulation.

superficial (soo''per-fish'al) Near the surface.

superior (su-pe're-or) Pertaining to a structure that is higher than another structure.

supination (soo''pĭ-na'shun) Rotation of the forearm so that the palm faces upward when the arm is outstretched.

suppressor cell (sŭ-pres'or sel) A special type of T-lymphocyte that functions to interfere with the production of antibodies responsible for allergic reactions.

surface tension (ser'fas ten'shun) Force that tends to hold moist membranes together due to an attraction that water molecules have for one another.

surfactant (ser-fak'tant) Substance produced by the lungs that reduces the surface tension within the alveoli.

suture (soo'cher) An immovable joint, such as that between adjacent flat bones of the skull.

sympathetic nervous system (sim''pah-thet'ik ner'vus sis'tem) Portion of the autonomic nervous system that arises from the thoracic and lumbar regions of the spinal cord.

symphysis (sim'fĭ-sis) A slightly movable joint between bones separated by a pad of fibrocartilage.

synapse (sin'aps) The junction between the axon end of one neuron and the dendrite or cell body of another neuron.

synaptic knob (si-nap'tik nob) Tiny enlargement at the end of an axon that secretes a neurotransmitter substance.

synchondrosis (sin''kon-dro'sis) Type of joint in which bones are united by bands of hyaline cartilage.

syncytium (sin-sish'e-um) A mass of merging cells.

syndesmosis (sin''des-mo'sis) Type of joint in which the bones are united by relatively long fibers of connective tissue.

syndrome (sin'drōm) A group of symptoms that together characterize a disease condition.

synergist (sin'er-jist) A muscle that assists the action of a prime mover.

synovial fluid (sĭ-no've-al floo'id) Fluid secreted by the synovial membrane.

synovial joint (sĭ-no've-al joint) A freely movable joint.

synovial membrane (si-no've-al mem'brān) Membrane that forms the inner lining of the capsule of a freely movable joint.

synthesis (sin'thĕ-sis) The process by which substances are united to form more complex substances.

system (sis'tem) A group of organs that act together to carry on a specialized function.

systemic circuit (sis-tem'ik ser'kit) The vessels that conduct blood between the heart and all body tissues except the lungs.

systole (sis'to-le) Phase of the cardiac cycle during which a heart chamber wall is contracted.

systolic pressure (sis-tol'ik presh'ur) Arterial blood pressure during the systolic phase of the cardiac cycle.

t

tachycardia (tak''e-kar'de-ah) An abnormally rapid heartbeat.

target tissue (tar'get tish'u) Specific tissue on which a hormone acts.

tarsus (tar'sus) The bones that form the ankle.

taste bud (tāst bud) Organ containing the receptors associated with the sense of taste.

telophase (tel'o-fāz) Stage in mitosis during which newly formed cells become separate structures.

tendon (ten'don) A cordlike or bandlike mass of white fibrous connective tissue that connects a muscle to a bone.

testis (tes'tis) Primary reproductive organ of a male; a sperm-cell-producing organ.

testosterone (tes-tos'tĕ-rōn) Male sex hormone secreted by the interstitial cells of the testes.

tetanus (tet'ah-nus) A continuous, forceful muscular contraction.

thalamus (thal'ah-mus) A mass of gray matter located at the base of the cerebrum in the wall of the third ventricle.

thermoreceptor (ther''mo-re-sep'tor) A sensory receptor that is sensitive to changes in temperature; a heat receptor.

thiamine (thi'ah-min) Vitamin B₁.

thoracic (tho-ras'ik) Pertaining to the chest.

threshold potential (thresh'old po-ten'shal) Level of potential at which an action potential or nerve impulse is produced.

threshold stimulus (thresh'old stim'u-lus) The level of stimulation that must be exceeded to elicit a nerve impulse or a muscle contraction.

thrombin (throm'bin) Blood clotting enzyme that causes fibrinogen to become fibrin.

thrombocyte (throm'bo-sīt) A blood platelet.

thrombocytopenia (throm''bo-si''to-pe'ne-ah) A low number of platelets in the circulating blood.

thrombus (throm'bus) A blood clot in a blood vessel that remains at its site of formation.

thymosin (thi'mo-sin) A group of peptides secreted by the thymus gland that affects the production of certain types of white blood cells.

thymus (thi'mus) A glandular organ located in the mediastinum behind the sternum and between the lungs.

thyroglobulin (thi''ro-glob'u-lin) Substance secreted by cells of the thyroid gland that serves to store thyroid hormones.

thyroid gland (thi'roid gland) Endocrine gland located just below the larynx and in front of the trachea that secretes thyroid hormones.

thyrotropin (thi''ro-trōp'in) Hormone secreted by the anterior pituitary gland that stimulates the thyroid gland to secrete hormones; TSH.

thyroxine (thi-rok'sin) A hormone secreted by the thyroid gland.

tidal volume (tīd'al vol'ūm) Amount of air that enters the lungs during a normal, quiet inspiration.

tissue (tish'u) A group of similar cells that performs a specialized function.

T-lymphocyte (lim'fo-sīt) Lymphocytes that interact directly with antigen-bearing particles and are responsible for cell-mediated immunity; T-cell.

trabecula (trah-bek'u-lah) Branching bony plate that separates irregular spaces within spongy bone.

trachea (tra'ke-ah) Tubular organ that leads from the larynx to the bronchi.

transcellular fluid (trans''sel'u-lar floo'id) A portion of the extracellular fluid, including the fluid within special body cavities.

transferrin (trans-fer'rin) Blood plasma protein that functions to transport iron.

transfer RNA (trans'fer RNA) Molecule of RNA that carries an amino acid to a ribosome in the process of protein synthesis.

transverse colon (trans-vers' ko'lon) Portion of the large intestine that extends across the abdomen from right to left below the stomach.

transverse tubule (trans-vers' tu'būl) Membranous channel that extends inward from a muscle fiber membrane and passes through the fiber.

tricuspid valve (tri-kus'pid valv) Heart valve located between the right atrium and the right ventricle.

triglyceride (tri-glis'er-īd) A lipid composed of three fatty acids combined with a glycerol molecule.

triiodothyronine (tri''i-o''do-thi'ro-nēn) One of the thyroid hormones.

trisomy (tri'so-me) Condition in which a cell contains three chromosomes of a particular type instead of two.

trochanter (tro-kan'ter) A broad process on a bone.

trochlea (trok'le-ah) A pulley-shaped structure.

trophoblast (trof'o-blast) The outer cells of a blastocyst that help form the placenta and other embryonic membranes.

tropic hormone (trōp'ik hor'mōn) A hormone that has an endocrine gland as its target tissue.

tropomyosin (tro''po-mi'o-sin) Protein that functions in blocking muscle contraction until calcium ions are present.

troponin (tro'po-nin) Protein that functions with tropomyosin to block muscle contraction until calcium ions are present.

trypsin (trip'sin) An enzyme in pancreatic juice that acts to break down protein molecules.

trypsinogen (trip-sin'o-jen) Substance secreted by pancreatic cells that becomes trypsin as a result of enzymatic action.

tubercle (tu'ber-kl) A small, rounded process on a bone.

tuberosity (tu''bĕ-ros'ĭ-te) An elevation or protuberance on a bone.

twitch (twich) A brief muscular contraction followed by relaxation.

tympanic membrane (tim-pan'ik mem'brān) A thin membrane that covers the auditory canal and separates the external ear from the middle ear; the eardrum.

u

umbilical cord (um-bil'ĭ-kal kord) Cordlike structure that connects the fetus to the placenta.

umbilical region (um-bil'ĭ-kal re'jun) The central portion of the abdomen.

umbilicus (um-bil'ĭ-kus) Region to which the umbilical cord was attached; the navel.

unipolar neuron (u''ni-po'lar nu'ron) A neuron that has a single nerve fiber extending from its cell body.

unsaturated fatty acid (un-sat'u-rāt''ed fat'e as'id) Fatty acid molecule that has one or more double bonds between the atoms of its carbon chain.

urea (u-re'ah) A nonprotein nitrogenous substance produced as a result of protein metabolism.

ureter (u-re'ter) A muscular tube that carries urine from the kidney to the urinary bladder.

urethra (u-re'thrah) Tube leading from the urinary bladder to the outside of the body.

uterine (u'ter-in) Pertaining to the uterus.

uterine tube (u'ter-in tūb) Tube that extends from the uterus on each side toward an ovary and functions to transport sex cells; fallopian tube or oviduct.

uterus (u'ter-us) Hollow muscular organ located within the female pelvic in which a fetus develops.

utricle (u'tri-kl) An enlarged portion of the membranous labyrinth of the inner ear.

uvula (u'vu-lah) A fleshy portion of the soft palate that hangs down above the root of the tongue.

v

vaccine (vak'sēn) A substance that contains antigens and is used to stimulate the production of antibodies.

vacuole (vak'u-ōl) A space or cavity within the cytoplasm of a cell.

vagina (vah-ji'nah) Tubular organ that leads from the uterus to the vestibule of the female reproductive tract.

Valsalva's maneuver (val-sal'vahz mah-noo'ver) Increasing the intrathoracic pressure by forcing air from the lungs against a closed glottis.

varicose veins (var'i-kos vānz) Abnormally swollen and enlarged veins, especially in the legs.

vasa recta (va'sah rek'tah) A branch of the peritubular capillary that receives blood from the efferent arterioles of juxtamedullary nephrons.

vascular (vas'ku-lar) Pertaining to blood vessels.

vas deferens (vas def'er-ens) Tube that leads from the epididymis to the urethra of the male reproductive tract (pl. *vasa deferentia*).

vasoconstriction (vas''o-kon-strik'shun) A decrease in the diameter of a blood vessel.

vasodilation (vas''o-di-la'shun) An increase in the diameter of a blood vessel.

vasopressin (vas''o-pres'in) Antidiuretic hormone.

vein (vān) A vessel that carries blood toward the heart.

vena cava (vēn'ah kāv'ah) One of two large veins that convey deoxygenated blood to the right atrium of the heart.

ventral root (ven'tral rōōt) Motor branch of a spinal nerve by which it is attached to the spinal cord.

ventricle (ven'tri-kl) A cavity, such as those of the brain that are filled with cerebrospinal fluid, or those of the heart that contain blood.

venule (ven'ūl) A vessel that carries blood from capillaries to a vein.

vermiform appendix (ver'mi-form ah-pen'diks) Appendix.

vesicle (ves'ĭ-kal) Membranous, cytoplasmic sac formed by an action of the cell membrane.

villus (vil'us) Tiny, fingerlike projection that extends outward from the inner lining of the small intestine.

visceral (vis'er-al) Pertaining to the contents of a body cavity.

visceral peritoneum (vis'er-al per''ĭ-to-ne'um) Membrane that covers the surfaces of organs within the abdominal cavity.

visceral pleura (vis'er-al ploo'rah) Membrane that covers the surfaces of the lungs.

viscosity (vis-kos'ĭ-te) The tendency for a fluid to resist flowing due to the internal friction of its molecules.

vital capacity (vi'tal kah-pas'ĭ-te) The maximum amount of air a person can exhale after taking the deepest breath possible.

vitamin (vi'tah-min) An organic substance other than a carbohydrate, lipid, or protein that is needed for normal metabolism but cannot be synthesized in adequate amounts by the body.

vitreous body (vit'reus bod'e) The collagenous fibers and fluid that occupy the posterior cavity of the eye.

vitreous humor (vit're-us hu'mor) The substance that occupies the space between the lens and the retina of the eye.

vocal cords (vo'kal kordz) Folds of tissue within the larynx that produce vocal sounds when they vibrate.

Volkmann's canal (fōlk'mahnz kah-nal') A transverse channel that interconnects Haversian canals within compact bone.

volt (vōlt) Unit used to measure differences in electrical potential.

voluntary (vol'un-tār''e) Capable of being consciously controlled.

vulva (vul'vah) The external reproductive parts of the female that surround the opening of the vagina.

W

water balance (wot'er bal'ans) Condition in which the quantity of water entering the body is equal to the quantity leaving it.

water intoxication (wot'er in-tok''sĭ-ka'shun) Condition in which the extracellular body fluids become abnormally diluted due to the presence of excessive water.

water of metabolism (wot'er uv mĕ-tab'o-lizm) Water produced as a by-product of metabolic processes.

wave summation (wāv sum-ma'shun) A sustained muscle contraction that occurs when a series of twitches fuse together; summation of twitches.

white muscle (whīt mus'el) Fast-contracting skeletal muscle.

X

X-linked trait (eks-linkt' trāt) Trait determined by a recessive gene located on an X chromosome.

Y

yellow marrow (yol'o mar'o) Fat storage tissue found in the cavities within certain bones.

Z

zonular fiber (zon'u-lar fi'ber) A delicate fiber within the eye that extends from the ciliary process to the lens capsule.

zygote (zi'gōt) Cell produced by the fusion of an egg and sperm; a fertilized egg cell.

zymogen granule (zi-mo'jen gran'ūl) A cellular structure that stores inactive forms of protein-splitting enzymes in a pancreatic cell.

Credits

LINE ART

Chapter 1
Figs. 1.6 and 1.18: From Kent M. Van De Graaff, *Human Anatomy*, 3d ed. Copyright © 1992 Wm. C. Brown Communications, Inc., Dubuque, Iowa. All Rights Reserved. Reprinted by permission. **Fig. 1.7:** From Kent M. Van De Graaff and Stuart Ira Fox, *Concepts of Human Anatomy and Physiology*. Copyright © 1986 Wm. C. Brown Communications, Inc., Dubuque, Iowa. All Rights Reserved. Reprinted by permission. **1.16b:** From Kent M. Van De Graaff and Stuart Ira Fox, *Concepts of Human Anatomy and Physiology*, 3d ed. Copyright © 1992 Wm. C. Brown Communications, Inc., Dubuque, Iowa. All Rights Reserved. Reprinted by permission.

Chapter 2
Fig. 2.3b: From Kent M. Van De Graaff and Stuart Ira Fox, *Concepts of Human Anatomy and Physiology*, 2d ed. Copyright © 1989 Wm. C. Brown Communications, Inc., Dubuque, Iowa. All Rights Reserved. Reprinted by permission. **2.27a:** From Stuart Ira Fox, *Human Physiology*, 3d ed. Copyright © 1990 Wm. C. Brown Communications, Inc., Dubuque, Iowa. All Rights Reserved. Reprinted by permission.

Chapter 3
Fig. 3.22: From Kent M. Van De Graaff and Stuart Ira Fox, *Concepts of Human Anatomy and Physiology*, 3d ed. Copyright © 1992 Wm. C. Brown Communications, Inc., Dubuque, Iowa. All Rights Reserved. Reprinted by permission. **3.27:** From Stuart Ira Fox, *Human Physiology*, 3d ed. Copyright © 1990 Wm. C. Brown Communications, Inc., Dubuque, Iowa. All Rights Reserved. Reprinted by permission.

Chapter 4
Fig. 4.27: From Stuart Ira Fox, *Human Physiology*, 3d ed. Copyright © 1990 Wm. C. Brown Communications, Inc., Dubuque, Iowa. All Rights Reserved. Reprinted by permission.

Chapter 5
Fig. 5.8: From Kent M. Van De Graaff, *Human Anatomy*, 3d ed. Copyright © 1992 Wm. C. Brown Communications, Inc., Dubuque, Iowa. All Rights Reserved. Reprinted by permission.

5.9: From Kent M. Van De Graaff, *Human Anatomy*, 2d ed. Copyright © 1988 Wm. C. Brown Communications, Inc., Dubuque, Iowa. All Rights Reserved. Reprinted by permission.

Chapter 6
Fig. 6.16: From Kent M. Van De Graaff and Stuart Ira Fox, *Concepts of Human Anatomy and Physiology*, 3d ed. Copyright © 1992 Wm. C. Brown Communications, Inc., Dubuque, Iowa. All Rights Reserved. Reprinted by permission.

Chapter 7
Figs. 7.18, 7.20, 7.22, 7.23, 7.28, 7.31: From Kent M. Van De Graaff, *Human Anatomy*, 3d ed. Copyright © 1992 Wm. C. Brown Communications, Inc., Dubuque, Iowa. All Rights Reserved. Reprinted by permission.

Chapter 8
Figs. 8.8, 8.9a, c–f: From Kent M. Van De Graaff, *Human Anatomy*, 3d ed. Copyright © 1992 Wm. C. Brown Communications, Inc., Dubuque, Iowa. All Rights Reserved. Reprinted by permission.

Chapter 9
Figs. 9.20 and 9.21: From Kent M. Van De Graaff and Stuart Ira Fox, *Concepts of Human Anatomy and Physiology*, 3d ed. Copyright © 1992 Wm. C. Brown Communications, Inc., Dubuque, Iowa. All Rights Reserved. Reprinted by permission. **Figs. 9.27, 9.32e, 9.37:** From Kent M. Van De Graaff and Stuart Ira Fox, *Concepts of Human Anatomy and Physiology*, 2d ed. Copyright © 1989 Wm. C. Brown Communications, Inc., Dubuque, Iowa. All Rights Reserved. Reprinted by permission.

Chapter 10
Figs. 10.1 and 10.25: From Kent M. Van De Graaff and Stuart Ira Fox, *Concepts of Human Anatomy and Physiology*, 2d ed. Copyright © 1989 Wm. C. Brown Communications, Inc., Dubuque, Iowa. All Rights Reserved. Reprinted by permission. **10.4b:** From Stuart Ira Fox, *Human Physiology*, 3d ed. Copyright © 1990 Wm. C. Brown Communications, Inc., Dubuque, Iowa. All Rights Reserved. Reprinted by permission. **10.21:** From Kent M. Van De Graaff and Stuart Ira Fox, *Concepts of Human Anatomy and Physiology*, 3d ed. Copyright © 1992 Wm. C. Brown

Communications, Inc., Dubuque, Iowa. All Rights Reserved. Reprinted by permission.

Chapter 11
Figs. 11.2, 11.4, 11.27, 11.28, 11.31, 11.32, 11.38: From Kent M. Van De Graaff, *Human Anatomy*, 3d ed. Copyright © 1992 Wm. C. Brown Communications, Inc., Dubuque, Iowa. All Rights Reserved. Reprinted by permission. **Figs. 11.14 and 11.36:** From Kent M. Van De Graaff and Stuart Ira Fox, *Concepts of Human Anatomy and Physiology*, 2d ed. Copyright © 1989 Wm. C. Brown Communications, Inc., Dubuque, Iowa. All Rights Reserved. Reprinted by permission. **Fig. 11.29:** From Kent M. Van De Graaff, *Human Anatomy*, 2d ed. Copyright © 1988 Wm. C. Brown Communications, Inc., Dubuque, Iowa. All Rights Reserved. Reprinted by permission.

Chapter 12
Figs. 12.12, 12.15a, 12.31, 12.32, 12.36b: From Kent M. Van De Graaff and Stuart Ira Fox, *Concepts of Human Anatomy and Physiology*, 3d ed. Copyright © 1992 Wm. C. Brown Communications, Inc., Dubuque, Iowa. All Rights Reserved. Reprinted by permission. **Fig. 12.15b:** From Stuart Ira Fox, *Human Physiology*, 4th ed. Copyright © 1993 Wm. C. Brown Communications, Inc., Dubuque, Iowa. All Rights Reserved. Reprinted by permission. **Figs. 12.17 and 12.18:** From Stuart Ira Fox, *Human Physiology*, 3d ed. Copyright © 1990 Wm. C. Brown Communications, Inc., Dubuque, Iowa. All Rights Reserved. Reprinted by permission.

Chapter 13
Fig. 13.13: From Kent M. Van De Graaff and Stuart Ira Fox, *Concepts of Human Anatomy and Physiology*, 3d ed. Copyright © 1992 Wm. C. Brown Communications, Inc., Dubuque, Iowa. All Rights Reserved. Reprinted by permission. **13.33:** From Kent M. Van De Graaff and Stuart Ira Fox, *Concepts of Human Anatomy and Physiology*, 2d ed. Copyright © 1989 Wm. C. Brown Communications, Inc., Dubuque, Iowa. All Rights Reserved. Reprinted by permission.

Chapter 14
Figs. 14.1, 14.10, 14.26, 14.31: From Kent M. Van De Graaff, *Human Anatomy*, 3d ed. Copyright © 1992 Wm. C. Brown Communications, Inc., Dubuque, Iowa. All Rights Reserved. Reprinted by permission. **Figs. 14.3, 14.33, 14.48, 14.50:** From Kent M. Van De Graaff and Stuart Ira Fox, *Concepts of Human Anatomy and Physiology*, 3d ed. Copyright © 1992 Wm. C. Brown Communications, Inc., Dubuque, Iowa. All Rights Reserved. Reprinted by permission.

Chapter 16
Figs. 16.1, 16.12, 16.22, 16.31: From Kent M. Van De Graaff, *Human Anatomy*, 3d ed. Copyright © 1992 Wm. C. Brown Communications, Inc., Dubuque, Iowa. All Rights Reserved. Reprinted by permission. **16.11:** From Kent M. Van De Graaff and Stuart Ira Fox, *Concepts of Human Anatomy and Physiology*, 2d ed. Copyright © 1989 Wm. C. Brown Communications, Inc., Dubuque, Iowa. All Rights Reserved. Reprinted by permission.

Chapter 18
Figs. 18.5, 18.12a, 18.13, 18.31: From Kent M. Van De Graaff, *Human Anatomy*, 3d ed. Copyright © 1992 Wm. C. Brown Communications, Inc., Dubuque, Iowa. All Rights Reserved. Reprinted by permission. **Figs. 18.34 and 18.60:** From Kent M. Van De Graaff and Stuart Ira Fox, *Concepts of Human Anatomy and Physiology*, 3d ed. Copyright © 1992 Wm. C. Brown Communications, Inc., Dubuque, Iowa. All Rights Reserved. Reprinted by permission. **Figs. 18.43 and 18.70:** From Stuart Ira Fox, *Human Physiology*, 3d ed. Copyright © 1990 Wm. C. Brown Communications, Inc., Dubuque, Iowa. All Rights Reserved. Reprinted by permission. **18.55:** From Kent M. Van De Graaff, *Human Anatomy*, 2d ed. Copyright © 1988 Wm. C. Brown Communications, Inc., Dubuque, Iowa. All Rights Reserved. Reprinted by permission. **Fig. 18.63:** From Kent M. Van De Graaff and Stuart Ira Fox, *Concepts of Human Anatomy and Physiology*. Copyright © 1986 Wm. C. Brown Communications, Inc., Dubuque, Iowa. All Rights Reserved. Reprinted by permission.

Chapter 19

Fig. 19.2: From Kent M. Van De Graaff, *Human Anatomy*, 3d ed. Copyright © 1992 Wm. C. Brown Communications, Inc., Dubuque, Iowa. All Rights Reserved. Reprinted by permission.

Chapter 20

Fig. 20.5: From Kent M. Van De Graaff, *Human Anatomy*, 2d ed. Copyright © 1988 Wm. C. Brown Communications, Inc., Dubuque, Iowa. All Rights Reserved. Reprinted by permission. **20.11:** From Kent M. Van De Graaff and Stuart Ira Fox, *Concepts of Human Anatomy and Physiology*, 3d ed. Copyright © 1992 Wm. C. Brown Communications, Inc., Dubuque, Iowa. All Rights Reserved. Reprinted by permission. **20.26:** From Stuart Ira Fox, *Human Physiology*, 3d ed. Copyright © 1990 Wm. C. Brown Communications, Inc., Dubuque, Iowa. All Rights Reserved. Reprinted by permission. **20.32b:** From Kent M. Van De Graaff, *Human Anatomy*, 3d ed. Copyright © 1992 Wm. C. Brown Communications, Inc., Dubuque, Iowa. All Rights Reserved. Reprinted by permission.

Chapter 22

Fig. 22.3: From Kent M. Van De Graaff, *Human Anatomy*, 2d ed. Copyright © 1988 Wm. C. Brown Communications, Inc., Dubuque, Iowa. All Rights Reserved. Reprinted by permission. **Figs. 22.5b, 22.11, 22.24, 22.30, 22.37, and 22.38:** From Kent M. Van De Graaff and Stuart Ira Fox, *Concepts of Human Anatomy and Physiology*, 3d ed. Copyright © 1992 Wm. C. Brown Communications, Inc., Dubuque, Iowa. All Rights Reserved. Reprinted by permission. **Fig. 22.32:** From Kent M. Van De Graaff, *Human Anatomy*, 3d ed. Copyright © 1992 Wm. C. Brown Communications, Inc., Dubuque, Iowa. All Rights Reserved. Reprinted by permission.

Chapter 23

Figs. 23.16 and 23.25: From Kent M. Van De Graaff, *Human Anatomy*, 3d ed. Copyright © 1992 Wm. C. Brown Communications, Inc., Dubuque, Iowa. All Rights Reserved. Reprinted by permission. **23.22:** From Stuart Ira Fox, *Human Physiology*, 3d ed. Copyright © 1990 Wm. C. Brown Communications, Inc., Dubuque, Iowa. All Rights Reserved. Reprinted by permission.

Chapter 24

Fig. 24.21: From Stuart Ira Fox, *Human Physiology*, 3d ed. Copyright © 1990 Wm. C. Brown Communications, Inc., Dubuque, Iowa. All Rights Reserved. Reprinted by permission.

PHOTOGRAPHS

Plates 37–47: © Dr. Sheril D. Burton, **48–75:** © Wm. C. Brown Publishers/Karl Rubin Photographer.

Chapter/Unit Openers

Unit Opener 1: © Peter Arnold; **2:** © Comstock; **3:** © Peter Arnold; **4:** © Peter Arnold; **5:** © Lennart Nilsson.

Chapter 1

Fig. 1.1: © From the works of Andreas Vesalius of Brussels by J. B. de C. M. Saunders and Charles P. O'Malley, Dover Publications, Inc., NY 1973; **1.16A–C:** © Photo Researchers, Inc. **1.21:** © Alexander Tsiaras/Photo Researchers, Inc.; **1.22:** © Lutheran Hospital/Peter Arnold, Inc.; **1.23:** © CWRI/Science Library.

Chapter 2

Fig. 2.2: © Mark Antman/Image Works, Inc.; **2.3A:** © SIU 1988/Peter Arnold, Inc.; **2.16:** © Courtesy of Ealing Corp., South Natick; **2.18:** Courtesy of Ealing Corp., South Natick; **2.31A&B:** (A) © Science Photo Library/Photo Researchers, Inc. (B) © CNRI/Science Photo Library/Photo Researchers, Inc.; **2.32:** © Monte Bucksbaum/University of California.

Chapter 3

Fig. 3.4A: © Gordon Leedale/Biophoto Assoc.; **3.7:** © W. Orme Rod/Visuals Unlimited; **3.8A:** © Martin M. Rotker/Taurus Photos, Inc.; **3.8B:** © Keith R. Porter/University of Colorado; **3.8C:** © Dr. Anderjs Liepins/Science Photo Source/Photo Researchers, Inc.; **3.9A:** © Fawcett/Photo Researchers, **3.10A:** © Gordon Leedale/Biophoto Assoc.; **3.12A:** © Keith Porter/Science Source/Photo Researchers, Inc.; **3.13:** © K. G. Murti/Visuals Unlimited; **3.14A:** © Gordon Leedale/Biophoto Assoc.; **3.15A:** © Reproduced with permission of the American Lung Assoc.; **3.16:** © Manfred Kage/Peter Arnold, Inc.; **3.17:** © Keith Porter; **3.18B:** © Dr. Joseph Grall. Reproduced from *Journal of Cell Biology*. Copyright permission of Rockefeller University Press; **3.19A:** © Stephen L. Wolfe; **3.24A–C:** © S. J. Singer; **3.34B–3.38B:** © Edwin Reschke; **3.39A–C:** © From *Scanning Electron Microscopy in Biology* by R. G. Kessel and C. Y. Shih Springer Verlag; **3.40B:** © Edwin Reschke.

Chapter 4

Fig. 4.19B: © Courtesy of Ealing Corp., South Natick.

Chapter 5

Figs. 5.1B, 5.2B: © Edwin Reschke; **5.3B:** © Manfred Kage/Peter Arnold, Inc.; **5.4:** © Fawcett/Hirokawa/Heuser/Photo Researching; **5.5B:** © Edwin Reschke; **5.6B:** © Fred Hossler/Visuals Unlimited; **5.7A:** © Edwin Reschke; **5.10:** © Edwin Reschke/Peter Arnold, Inc.; **5.11:** © David M. Phillips/Visuals Unlimited; **5.12:** © Jean-Paul Revel; **5.13:** © Veronica Burmeister/Visuals Unlimited; **5.14A&B:** (A) © Courtesy of Robert Trelstad (B) © From T. Tsuji, R. M. Lavker and A. M. Kligman, *J. Microscope* 115:165–173, 1978; **5.15B:** © Edwin Reschke; **5.16B:** © Edwin Reschke; **5.17B:** © Edwin Reschke; **5.18:** © Alfred Owczarzak/Taurus Photos, Inc.; **5.19:** © Alfred Owczarzak/Taurus Photos, Inc.; **5.20B, 5.21.B:** © Edwin Reschke; **5.22B:** © John Cunningham/Visuals Unlimited; **5.23B:** © Victor B. Eichler; **5.24B, 5.25B, 5.26B:** © Edwin Reschke; **5.27B, 5.28B:** © Manfred Kage/Peter Arnold, Inc.

Chapter 6

Figs. 6.1, 6.3B: © Edwin Reschke; **6.4:** © Victor Eichler/BioArt; **6.5A:** © M. Schlwia/Visuals Unlimited; **6.6:** © Photo Researchers; **6.7B:** © Victor Eichler; **6.8:** © CNRI/Photo Researchers, Inc.; **6.9:** © Per H. Kjeldson/Visuals Unlimited; **6.12:** © John D. Cunningham/Visuals Unlimited; **6.14:** © Nancy M. Hamilton/Photo Researchers, Inc.

Chapter 7

Fig. 7.3A: © Lester Bergman & Associates; **7.4:** © R. G. Kessel and R. H. Kardon, W. H. Freeman; **7.6:** © Biophoto Associates/Photo Researchers, Inc.; **7.7:** © Biophoto Associates/Photo Researchers, Inc.; **7.8B:** © Victor B. Eichler/Bio-Art; **7.9:** © BioPhoto Associates/Photo Researchers, Inc.; **7.11:** © James Shaffer; **7.34A&B:** © Jan Halaska/Photo Researchers; **7.39:** © Courtesy of Utah Valley Hospital, Department of Radiology; **7.43:** © Victor Eichler/Bio-Art; **7.46:** © Courtesy of Eastman Kodak; **7.49, 7.53, 7.56, 7.59, 7.64A:** © Martin M. Rotker/Taurus Photos, Inc.; **7.64B:** © Ted Conde.

Chapter 8

Fig. 8.13B: © Paul Reiman; **8.15B:** © Paul Reiman; **8.18B:** © Paul Reiman; **8.20B:** © Paul Reiman.

Chapter 9

Fig. 9.3: © From *Tissues and Organs: A Text-Atlas of Scanning Electron Microscopy*, R. G. Kessel and R. K. Kardon, 1979 W. H. Freeman; **9.5:** © Courtesy of H. E. Huxley, Brandeis University; **9.7B:** © Victor Eichler/Bio-Art; **9.18:** © Courtesy of Dr. Paul Heidger, University of Iowa; **9.PRTA-C:** © John C. Mese.

Chapter 10

Fig. 10.2: © Ed Reschke; **10.5A:** © H. Webster and John Hubbard, *The Vertebrate Peripheral Nervous System* (1974) Plenum Press, 1974; **10.5B:** © Biophoto Associates/Photo Researchers; **10.7:** © From *Tissues and Organs: A Text-Atlas of Scanning Electron Microscopy*, R. G. Kessel and R. H. Kardon, 1979 W. H. Freeman; **10.20:** © Don Fawcett/ Photo Researchers, Inc.

Chapter 11

Fig. 11.4B: © Per H. Kjeldson, University of Michigan at Ann Arbor; **11.18A&B:** © Courtesy of Utah Valley Hospital; **11.22:** © Courtesy of Igatu Shoin, LTD.; **11.25:** © From *Tissues and Organs: A Text-Atlas of Scanning Electron Microscopy*, R. G. Kessel and R. H. Kardon, 1979 W. H. Freeman.

Chapter 12

Fig. 12.2A,B: © Ed Reschke; **12.7:** © Dwight Kuhn; **12.9:** © Victor Eichler/Bio-Art; **12.16A&B:** © (A) John D. Cunningham/Visuals Unlimited © (B) Fred Hossler/Visuals Unlimited; **12.21:** © Dean E. Hillman; **12.29:** © Science Photo Library/Photo Researchers; **12.35:** © Per H. Kjeldson, University of Michigan at Ann Arbor; **12.36A:** © Carroll W. Weiss/Camera M.D. Studios; **12.42C:** © Frank S. Werblin.

Chapter 13

Fig. 13.12: © John D. Cunningham/Visuals Unlimited; **13.14:** © Department of Illustrations, Washington University School of Medicine and American Journal of Medicine 20 (Jan. '56) pg. 133; **13.18:** © Fred Hossler/Visuals Unlimited; **13.20:** © F. A. Davis Co.; **13.21:** © Courtesy of F. A. Davis Co., Philadelphia and Dr. R. H. Kampmeier; **13.22:** © Lester V. Bergman and Associates; **13.24:** © BioPhoto Associates/Photo Researchers, Inc.; **13.28A:** © John D. Cunningham/Visuals Unlimited; **13.28B:** © R. C. Calentine/Visuals Unlimited; **13.34:** © Ed Reschke.

Chapter 14

Fig. 14.9B: © BioPhoto Associates/Science Source/Photo Researchers, Inc.; **14.12A:** © Bruce Iverson; **14.12B:** © BioPhoto Associates/Science Source/Photo Researchers, Inc.; **14.12C:** © Bruce Iverson; **14.16:** © Edwin Reschke; **14.19:** © Courtesy of Utah Valley Hospital, Department of Radiology; **14.21:** © Edwin Reschke; **14.32:** © Carroll H. Weiss/M.D. Camera Studios; **14.34:** © Armed Forces Institute of Pathology; **14.41:** © Manfred Kage/Peter Arnold, Inc.; **14.42B:** © K. R. Porter; **14.49:** © James Shaffer; **14.51:** © Ed Reschke/Peter Arnold, Inc.; **14.52:** © Edwin Reschke.

Chapter 15

Fig. 15.6: © UNICEF; **15.8A&B:** © (A) James Shaffer © (B) Bruce Curtis/Peter Arnold, Inc.; **15.10:** © World Health Organization.

Chapter 16

Figs. 16.4A, 16.4B: © Courtesy of Eastman Kodak; **16.7C:** © CNRI/Phototake, Inc.; **16.10:** © Edwin Reschke; **16.13:** © John Watney/Photo Library; **16.15:** © Dwight Kuhn; **16.17:** © From *Tissues and Organs: A Text-Atlas of Scanning Electron Microscopy* by R. G. Kessel and R. H. Kardon, 1979 W. H. Freeman; **16.18:** © American Lung Association; **16.26:** © Edward Letteau/Photo Researchers, Inc.; **16.27A&B:** © Victor Eichler; **16.33:** © Reproduced with permission of the American Lung Assoc.; **16.35:** © Murray, John F. *The Normal Lungs* 2/E, W. B. Saunders Co. 1986.

Chapter 17

Fig. 17.5: © Edwin Reschke; **17.3B:** © Bill Longcore/Science Source/Photo Researchers, Inc.; **17.9A:** © Martin M. Rotker/Science Photo Library/Taurus Photos, Inc.; **17.9B:** Dr. G. F. Leedale/Biophoto Associates/Photo Researchers, Inc; **17.10:** © Warren Rosenberg, Iona College/BPS; **17.11:** © Ed Reschke/Peter Arnold, Inc.; **17.12:** © Alfred Owzarzak/Taurus Photos, Inc.; **17.13, 17.14:** © Edwin Reschke; **17.17A&B:** © Victor Eichler/Bio-Art; **17.21:** © Secchi-Lecaque/Roussel Photo Researchers; **17.23A:** © Carroll M. Weiss/ M.D. Camera Studios.

Chapter 18

Fig. 18.2: © Courtesy of Igaku-Shoin, LTD.; **18.7, 18.8:** © Courtesy of Igaku-Shoin, LTD.; **18.14:** © Courtesy of Eastman Kodak; **18.22:** © Bob Coyle; **18.33:** *Tissues and Organs: A Text-Atlas of Scanning Electron Microscopy*, by R. G. Kessel and R. H. Kardon. Copyright 1979 W. H. Freeman and Company; **18.35B&C:** © D. W. Fawcett; **18.36:** © T. Kuwabara from Bloom, W., and D. W. Fawcett, *A Textbook of Histology* 10/E. 1975 W. B. Saunders; **18.38:** © (A) Edwin Reschke © (B) Victor Eichler/Bio-Art; **18.41A–D:** © American Heart Association, Dallas, TX; **18.46A&B:** © Courtesy Dr. Patricia Phelps; **18.57B:** © Courtesy Utah Valley Hospital; **18.59:** © From Practische OntlecdKundi from J. Dankmeyer, H. G. Lambers, and J. M. F. Landsmees, Bohn, Scheltema & Holkema.

Chapter 19

Fig. 19.3: © Ed Reschke; **19.5:** © Courtesy of Eastman Kodak; **19.9B:** © John Watney/Photo Researchers, Inc.; **19.10:** © Igaku-Shoin, LTD; **19.13:** © John D. Cunningham; **19.15:** © BioPhoto Associates/Photo Researchers, Inc.; **19.17:** © Courtesy of Memorial Sloan-Kettering Cancer Center.

Chapter 20

Fig. 20.2: © CNRI/Science Photo Library/Photo Researchers, Inc.; **20.6B:** © Lester V. Bergman & Associates; **20.7A:** © *Tissues and Organs: A Text-Atlas of Scanning Electron Microscopy*, by R. G. Kessel and R. H. Kardon. Copyright 1979 W. H. Freeman and Company; **20.7B:** © R. B. Wilson; **20.8:** © David M.

Index